中国金属学会　编

第十四届
中国钢铁年会论文集
（摘要）

Proceedings of the 14th CSM Steel Congress

北　京
冶金工业出版社
2023

内 容 简 介

本论文集共收录第十四届中国钢铁年会 1054 篇特邀报告和论文摘要、552 篇论文全文，共 813 万字。全书内容涉及采矿、选矿、炼焦化学、炼铁、炼钢、连铸、电冶金与废钢铁、轧制与热处理、表面与涂镀、金属材料深加工、粉末冶金、钢铁材料数字化、汽车用钢、特殊钢、高温合金、低合金钢、电工钢、非晶合金、冶金能源、冶金环保与资源利用、冶金固废资源综合利用、冶金设备与工程技术、冶金自动化与智能化、冶金物流、冶金流程工程学等方面，全面反映了近两年来我国及世界钢铁行业科研、生产、管理等方面的最新成果，是一本内容全面、新颖，具有较高学术水平的专业论文集。本书可供钢铁行业的科研人员、工程技术人员、管理人员、高校师生等学习参考。

本论文集以纸质图书和电子版方式出版。纸质图书为收录论文的摘要集（含特邀报告），电子版为收录论文的全文。

图书在版编目(CIP)数据

第十四届中国钢铁年会论文集：摘要／中国金属学会编．—北京：冶金工业出版社，2023.10
ISBN 978-7-5024-9456-8

Ⅰ.①第… Ⅱ.①中… Ⅲ.①钢铁工业—学术会议—文集 Ⅳ.①TF4-53

中国国家版本馆 CIP 数据核字(2023)第 182534 号

第十四届中国钢铁年会论文集（摘要）

出版发行	冶金工业出版社	电　话	（010）64027926
地　址	北京市东城区嵩祝院北巷 39 号	邮　编	100009
网　址	www.mip1953.com	电子信箱	service@mip1953.com

责任编辑　李培禄　美术编辑　彭子赫　版式设计　郑小利
责任校对　王永欣　责任印制　禹　蕊

北京虎彩文化传播有限公司印刷
2023 年 10 月第 1 版，2023 年 10 月第 1 次印刷
880mm×1230mm　1/16；58.25 印张；2 彩页；1800 千字；860 页
定价 350.00 元（全套）

投稿电话　（010）64027932　投稿信箱　tougao@cnmip.com.cn
营销中心电话　（010）64044283
冶金工业出版社天猫旗舰店　yjgycbs.tmall.com
(本书如有印装质量问题，本社营销中心负责退换)

第十四届中国钢铁年会组委会

年 会 主 席　张晓刚

执 行 主 席　田志凌

委　　　　员（以姓氏笔画为序）

　　　　　　于 勇　王新江　左 良　曲 阳

　　　　　　杨仁树　沈 彬　张少明　张志祥

　　　　　　赵民革　胡望明　侯 军　唐立新

　　　　　　戴志浩

年 会 秘 书 长　王新江

年 会 秘 书 处　中国金属学会学术工作部

《第十四届中国钢铁年会论文集》编委会

主　任　田志凌

副主任　王新江

编　委　（以姓氏笔画为序）

丁　波　王　华　龙红明　朱晓华　刘　旭　刘国栋

米振莉　孙彦广　苏　航　杜　涛　李　晶　杨　荃

杨　勇　杨志勇　杨利彬　吴树建　沙孝春　宋卫东

张　继　张　琪　张建良　张峻伟　陈　刚　陈　杰

尚成嘉　郦秀萍　贺可太　梅富强　董　瀚　谢少荣

戴竟舸

秘书组　（以姓氏笔画为序）

王　杰　王　敏　王卫卫　王宏涛　王俊莉　史成斌

包　娜　刘　洋　刘　倩　刘征建　刘洪萍　阮竹恩

孙　健　孙文强　孙建伦　孙继洋　苏　岚　李秀程

李雪鸣　张　辉　张　鹏　陈　兵　陈佳男　罗光敏

侯冬梅　姚同路　贺东风　董树勇　谢锦丽　雷少武

满廷慧　谯朝晖

前　言

第十四届中国钢铁年会于 2023 年 10 月 25—26 日在重庆召开。2023 年是全面贯彻落实党的二十大精神的开局之年，也是实施"十四五"规划承上启下的关键之年。我国正向着全面建成社会主义现代化强国的第二个百年奋斗目标迈进，作为国民经济支撑产业和制造业强国建设的基础行业，近两年来，我国钢铁工业取得了辉煌的成就和举世瞩目的国际影响，钢铁科技发展也收获了累累硕果。碳达峰与碳中和目标的提出，关乎经济社会发展方式的转变，也给我国钢铁工业带来了巨大的挑战和机遇，在解决绿色低碳对行业发展制约的同时，将会大力推动钢铁行业的高质量发展。本届年会以"**绿色低碳，高质量发展**"为主题，围绕钢铁生产全流程的基础理论、工艺技术、产品设计、制造和应用技术，研讨钢铁领域科技创新的方向和路径，促进钢铁行业向高端化、绿色化、智能化、低碳发展，加快钢铁强国建设，把中华民族伟大复兴的历史伟业推向前进。

本届年会的征文工作得到全国冶金与材料领域的专家、学者和科技人员的积极支持，共收到 1007 篇投稿论文和 161 篇特邀报告摘要。经专家评审并根据作者意愿，年会论文摘要集收录 1054 篇，全文集收录 552 篇。内容涉及采矿、选矿、炼焦化学、炼铁、炼钢、连铸、电冶金与废钢铁、轧制与热处理、表面与涂镀、金属材料深加工、粉末冶金、钢铁材料数字化、汽车用钢、特殊钢、高温合金、低合金钢、电工钢、非晶合金、冶金能源、冶金环保与资源利用、冶金固废资源综合利用、冶金设备与工程技术、冶金自动化与智能化、冶金物流、冶金流程工程学等方面，在此表示感谢！

本届年会论文全文以电子版方式出版，特邀报告和录用论文的摘要以纸质版方式出版。

由于论文集编辑、出版时间较紧，疏漏与错误之处，恳请读者批评指正。

中国金属学会
2023 年 10 月

目　录

大会特邀报告

The Impact of Decarbonisation Efforts on the Global Steel Industry ·················· Edwin BASSON　　I-1

日本制铁为实现碳中和所作的努力 ···················· 福田 和久（FUKUDA KAZUHISA）　　I-1

钢铁工业碳中和的思考与探索 ··· 毛新平　　I-1

POSCO's Green & Digital Challenges for Carbon Neutral ····················· Kisoo KIM　　I-2

第十三届冶金青年科技奖获奖人特邀报告

优特钢长材转型的系统性研究 ·· 马志勇，杨勇强　　II-1
Systematic Research on the Transformation of Long Product and Special Steel
·· MA Zhiyong, YANG Yongqiang

高强高韧钛合金研发及钢钛联合技术趋势 ·············· 马英杰，黄森森，雷家峰，杨　锐　　II-1
The Development of High Strength and Toughness Titanium Alloy and Technical Tendency of Titanium-steel Uniting ································ MA Yingjie, HUANG Sensen, LEI Jiafeng, YANG Rui

双相不锈钢冲击韧性研究进展及工艺技术实践
········ 丰　涵，宋志刚，何建国，吴晓涵，王宝顺，吴明华，李国平，周灿栋，高　虹　　II-2
Research Progress and Technology Practice of Impact Toughness of Duplex Stainless Steel
····················· FENG Han, SONG Zhigang, HE Jianguo, WU Xiaohan, WANG Baoshun,
　　　　　　　　　　　　WU Minghua, LI Guoping, ZHOU Candong, GAO Hong

高温合金疲劳-蠕变-环境交互作用下裂纹扩展机制的研究
·· 王民庆，杜金辉，邓　群，田志凌，朱　静　　II-4
The Mechanism of Crack Propagation for Superalloy under Fatigue-Creep-Environment Condition
································ WANG Minqing, DU Jinhui, DENG Qun, TIAN Zhiling, ZHU Jing

连铸坯表面温度在线测量与智能控制技术 ······· 龙木军，陈登福，张晏铭，黄云伟，张思远，谭　锴　　II-5
On-line Measurement and Intelligent Control Technology of Continuous Casting Slab Surface Temperature
·········· LONG Mujun, CHEN Dengfu, ZHANG Yanming, HUANG Yunwei, ZHANG Siyuan, TAN Kai

HB600级耐磨钢组织和硬度均匀性对耐磨性能的影响 ························· 闫强军，邓想涛　　II-6
Effect of Microstructure and Hardness Uniformity on the Wear Resistance Performance of HB600 Steel
··· YAN Qiangjun, DENG Xiangtao

钢铁行业（含氯类）新污染物生成与控制技术研究及CO_2捕集矿化技术进展
········ 苏　伟，邢　奕，田京雷，侯长江，王嘉庆，孙宇佳，孙方舟，崔永康，郭文韬　　II-7
Research on the Generation and Control Technology of New Pollutants in the Iron and Steel Industry and the Progress of CO_2 Capture and Mineralization Technology
····················· SU Wei, XING Yi, TIAN Jinglei, HOU Changjiang, WANG Jiaqing, SUN Yujia,
　　　　　　　　　　　　SUN Fangzhou, CUI Yongkang, GUO Wentao

攀西钛资源开发与利用研究进展 ………………………………………… 李开华，李 亮，朱福兴 II-8
Research Progress on the Development and Utilization of Titanium Resources in Panxi
………………………………………………………………… LI Kaihua, LI Liang, ZHU Fuxing

高精度冷连轧数字孪生与信息物理系统（CPS）关键技术研发及应用 ………………… 李 旭 II-9
Development and Application of Key Technologies for Digital Twin and Cyber-Physical Systems (CPS) for High-precision Cold Tandem Rolling ……………………………………………………………… LI Xu

基于无废城市建设的长流程钢厂固废管理模式与技术路径 ………………………………… 李红红 II-10
Solid Waste Management Mode and Technical Path of Long-process Steel Plant Based on the Concept of Zero-waste Construction …………………………………………………………………… LI Honghong

无底层绿色环保涂层高磁感取向硅钢的开发与推广应用研究 ………… 杨佳欣，申明辉，宋 刚 II-11
Development and Application Promotion of Glassless High Permeability Grain-oriented Silicon Steel with Green and Environmentally Friendly Coating ……………… YANG Jiaxin, SHEN Minghui, SONG Gang

Ni 对低合金钢海洋大气腐蚀产物氧化还原行为影响 ……………………… 杨建炜，刘智勇 II-11
Effect of Ni on Oxidation and Reduction Behavior of Corrosion Product Formed on Low Alloy Steel in Marine Atmosphere ……………………………………………………… YANG Jianwei, LIU Zhiyong

Cr 和 RE 元素对 HRB400 螺纹钢耐蚀性能的影响规律研究
………………………………………… 陈永峰，陈天奇，刘 涛，杨伟勇，刘 超 II-12
Study on the Effect Pattern of Cr and RE Elements on the Corrosion Resistance of HRB400 Rebars
………………………………… CHEN Yongfeng, CHEN Tianqi, LIU Tao, YANG Weiyong, LIU Chao

鞍钢钢铁产品数字化研发实践与展望 ……………………………………… 林 利，王军生 II-14
Anshan Iron and Steel Products Digital R&D Practice and Prospects ………… LIN Li, WANG Junsheng

钢液初始凝固坯壳捕渣机理及夹渣缺陷控制研究 ………… 钟小灿，周乐君，罗 豪，王万林 II-14
Mechanism and Controlling of the Slag Capture in the Initial Solidification Shell during the Continuous Casting Process ……………………………… ZHONG Xiaocan, ZHOU Lejun, LUO Hao, WANG Wanlin

分层供热低碳富氢烧结技术的研发与应用 ………… 周浩宇，叶恒棣，张俊杰，甘 敏 II-16
Research and Application of Stratified Heat Supply for Homogeneous and Low Carbon Sintering
………………………………………… ZHOU Haoyu, YE Hengdi, ZHANG Junjie, GAN Min

金属材料的微生物腐蚀机理与防治 ……………………………………………………… 徐大可 II-17
Microbiologically Influenced Corrosion of Metals: Mechanism and Mitigation ……………… XU Dake

镍铁渣增值利用研究进展 …………………………………………………………………… 彭志伟 II-18
Research Advances in Value-added Utilization of Ferronickel Slag ……………………… PENG Zhiwei

抽锭电渣重熔关键技术、装备的研发及推广应用 ……… 臧喜民，姜周华，刘福斌，耿 鑫，李万明 II-18
Research and Application on Technology and Equipment of Electroslag Remelting Withdrawal Process
………………………………… ZANG Ximin, JIANG Zhouhua, LIU Fubin, GENG Xin, LI Wanming

地下铁矿山高阶段嗣后胶结充填理论与应用研究 ………………… 谭玉叶，宋卫东，张兴才 II-19
Application and Theory Study on the High-stage Subsequent Cemented Filling in Underground Iron Mine
………………………………………………………… TAN Yuye, SONG Weidong, ZHANG Xingcai

分会场特邀报告

金属矿山智能化膏体充填开采——以谦比希铜矿为例 …………………………………… 吴爱祥 III-1
适应时代发展，建设六维矿山 ………………………………………………………………… 连民杰 III-1

智能矿山建设的探索与实践	武拴军	III-1
实施双碳规划引领，加快绿色转型发展——武钢资源可持续发展路径探索	鲁炳强	III-2
我国金属矿智能化开采现状与发展	王李管	III-2
难选铁矿石选矿技术发展及展望	陈雯	III-2
磁选设备的研发及应用现状	王俊莲，常自勇，王晓莉，王化军	III-3
铬铁矿石选矿技术现状及趋势	孙炳泉	III-4
提高旋流器分级精度的应用研究进展	刘培坤	III-4
新型磁铁矿精选设备——永磁组合箱式磁选机	袁致涛，孟庆有，周明亮	III-4
瓷球磨矿在铁矿山中的工业应用	吴彩斌	III-5
铁矿石磨矿-分级数学模型研究与实践	崔宝玉，赵强	III-5
战略金属硫化矿海水浮选基础及应用	李育彪	III-6
重载高效浓密机在选矿厂的应用	曾建红	III-6
低成本高炉用焦与精细化配煤新技术体系	程欢，肖璐颖，廖飞，胡文佳	III-7
高强度高反应性焦炭的制备技术	徐国忠，钟祥云，王明登，封一飞，白滨，李超，刘洋（女），刘洋（男），张雅茹，白金锋	III-8
智慧炼焦解决方案的研究与应用	王慧璐	III-8
超高温连铸坯热送与直接轧制关键技术	朱苗勇，蔡兆镇，罗森，祭程，王卫领	III-9
Key Technology for Hot Delivery and Direct Rolling of Ultra-high Temperature Continuous Casting Slab	ZHU Miaoyong, CAI Zhaozhen, LUO Sen, JI Cheng, WANG Weiling	
特大断面圆坯连铸技术发展与产品质量控制	张家泉，董娟，刘谦	III-10
Super Large Section Round Bloom Casting and the As-cast Product Quality Control	ZHANG Jiaquan, DONG Juan, LIU Qian	
低碳背景下PMO技术的创新发展与应用	翟启杰 等	III-11
Innovative Development and Application of PMO Technology in the Theme of Low Carbon	ZHAI Qijie et al	
钢液脱硫机理的实验、理论和模拟仿真研究	张立峰	III-12
连铸保护渣关键问题及研究进展	文光华	III-13
Key Problems and Research Progress in Mold Powders for Continuous Casting	WEN Guanghua	
鞍钢创新质量管理理念与实践	孙群	III-14
Innovative Quality Management Concept and Practice of Angang	SUN Qun	
游离晶对铸锭组织形成的影响	韩青有	III-14
Floating Grains and Their Influence on the Solidification Structure of a Billet	HAN Qingyou	
等轴晶薄带连铸与短流程电工钢	周成，周游，玄东坡，姜天亮，范文浩，毛焱，松希郎卡智美	III-15
Strip Casting with Equiaxed Grains and Electrical Steels by the Short Processes	ZHOU Cheng, ZHOU You, XUAN Dongpo, JIANG Tianliang, FAN Wenhao, MAO Yan, SONGXI Langkazhimei	
我国电炉短流程炼钢发展研究	王滨，余璐	III-17
Research on the Development of EAF Steelmaking Process in China	WANG Bin, YU Lu	
我国电炉短流程炼钢资源现状与发展	杨勇，姚同路	III-19

Current Situation and Development of EAF Short Process Steelmaking Resources in China
······YANG Yong, YAO Tonglu

宝钢电炉绿色低碳冶金探索与进展······李治平，徐迎铁，孟庆玉 III-21

超高功率柔性 IGBT 直流电源和智能化电弧炉炼钢技术······张豫川，吴学涛，杨宁川 III-21

Ultra-high Power Flexible IGBT DC Power Supply and Intelligent Electric Arc Furnace Steelmaking Technology
······ZHANG Yuchuan, WU Xuetao, YANG Ningchuan

低碳电炉炼钢工艺典型特征和实践路径······王中丙 III-22

Typical Characteristics and Practical Path of Low Carbon Electric Arc Furnace Steelmaking
······WANG Zhongbing

特冶技术进步推进镍基合金发展及未来展望······杨玉军 III-23

Advancement of Special Metallurgical Technology Promotes the Development and Future Prospects of Nickel-based Alloys······YANG Yujun

特殊钢冶炼过程 D/DS 夹杂物控制技术探讨······姜周华，董君伟，田家龙 III-24

Discussion on Control Technology of D/DS Inclusions in Special Steel Making Process
······JIANG Zhouhua, DONG Junwei, TIAN Jialong

连铸大压下高质量特厚板生产及铁素体轧制技术最新进展······康永林 III-25

板形控制技术的最新研究进展······何安瑞 III-25

首钢高硅硅钢连轧的工艺构想与实践······马家骥 III-25

现代冷轧板带绿色制造与智能制造······王业科，陶涛，牛强，刘显军 III-26

Green Manufacturing and Intelligent Manufacturing of Modern Cold-rolled Strip
······WANG Yeke, TAO Tao, NIU Qiang, LIU Xianjun

长材生产线技术革新与进步······杨勇强 III-27

无加热直接轧制关键技术与工业化······许宏安 III-27

热轧钢筋低成本绿色生产技术······王卫卫 III-27

钢管技术和产品发展······成海涛 III-28

宝钢低碳绿色冷轧涂镀钢板工艺技术······任玉苓 III-28

大厚度高韧性海洋工程用钢工艺研究及工业实践······赵坦 III-28

中薄板坯高等级管线钢工艺研究及工业化实践······林利 III-29

高速棒材生产线智能化创新建设及其关键技术
······徐言东，王晓晨，程知松，余伟，张勇军，刘洋，白金龙 III-29

Innovation and Key Technologies of Intelligent High Speed Bar Production Line Construction
······XU Yandong, WANG Xiaochen, CHENG Zhisong, YU Wei, ZHANG Yongjun, LIU Yang, BAI Jinlong

唐钢特殊镀层产品特性简介······李建英 III-30

宝钢建筑用涂镀产品未来发展······任玉苓 III-30

QH 涂镀板环保长效保护涂层······张俊 III-31

卷钢涂层辐射固化技术现状及展望······马源 III-31

结构钢大气环境腐蚀与试验评价技术······苏艳 III-32

中国金属制品行业现状及绿色低碳技术进展······毛海波，谷宇，米振莉，汪凯 III-32

Present Situation and Green Low-Carbon Technology's Progress in the Chinese Steel Wire Products Industry
······MAO Haibo, GU Yu, MI Zhenli, WANG Kai

B750HL 波形梁护栏辊弯成形策略研究······邢刚，韩飞，孙玮隆 III-33

Research on Roll Forming Strategy of B750HL Corrugated Beam Guardrail
······XING Gang, HAN Fei, SUN Weilong

架空输电线路耐候钢杆塔关键技术及工程应用······黄 耀 III-34
Key Technologies and Engineering Applications of Weathering Resistant Steel Poles and Towers for Overhead Transmission Lines ······HUANG Yao

长寿命、轻量化盘扣式脚手架开发······方 幸，米振莉，赵爱民，熊靖芸，王贺伟，华 维 III-35
Development of Long-life, Lightweight Turnbuckle Scaffold
······FANG Xing, MI Zhenli, ZHAO Aimin, XIONG Jingyun, WANG Hewei, HUA Wei

超高强钢汽车 A 柱上边梁的热气胀成形制造实践
······阮尚文，张 猛，成 刚，吴彦欣，米振莉，程鹏志 III-36

定制辊压成型技术与中国制造业的高质量发展······晏培杰 III-36

C276 哈氏合金的高温热变形行为研究······谢广明，王丰睿，孟 晨，王玉前 III-37
Study on Hot Deformation Behavior of C276 Hastelloy at High Temperature
······XIE Guangming, WANG Fengrui, MENG Chen, WANG Yuqian

Cost-effective Fabrication of Advanced Heat Sink Materials from Powder ······YANG Fei III-38

316L 不锈钢选区激光熔化增材制备过程的数值仿真研究
······安希忠，咬登治，王 炬，吴 琼，吴宇航 III-38

激光增材制造高 Co-Ni 二次硬化超高强度钢的组织设计和强韧性平衡策略······冉先喆 III-39

复杂氧化物颗粒强化 ODS 钢的增材制造成形及强韧化机制研究
······刘 彬，曹远奎，张宇阳，张瑞谦，刘 咏 III-39

激光增材制造飞机起落架用超高强不锈钢缺陷、组织与性能研究······刘壮壮 III-40

实现向铝合金中引入形变孪晶和 9R 相：合金化致层错能降低及塑性变形的晶粒尺寸效应
······周登山，张景藩，庞学永，张博文，李 跃，孙彬涵，VALIEV Ruslan Z.，张德良 III-40

增材制造可降解锌镁合金······李亚庚 III-41

增材制造高品质 H13 模具钢的数值模拟、组织与力学性能研究
······杨 鑫，张兆洋，徐圣航，刘石球，黄明浩，丁 超，汤慧萍 III-41
Numerical Simulation, Microstructure, and Mechanical Properties of H13 Steel Fabricated by Additive Manufacturing
······YANG Xin, ZHANG Zhaoyang, XU Shenghang, LIU Shiqiu, HUANG Minghao, DING Chao, TANG Huiping

冶金低热值煤气高效超临界发电技术及应用······熊敬超 III-42
The Development and Application of High-efficiency Supercritical Power Generation Technology for Metallurgy Low Calorific Value Gas ······XIONG Jingchao

金属材料数字化研发平台架构与应用······苏 航，刘和平，侯雅青，杨 丽，孙 旭，贺笃鹏 III-43

AIM：实验室小样本数据与工业大数据的融合分析······王晨充，李虎威，徐 伟 III-43

冷轧带钢力学性能"数据钢卷"的实践······唐成龙 III-44

金属材料高通量增材制备技术研究
······侯雅青，苏 航，应 华，张 浩，李发发，何亚洲，宋有朋，王炫东，张荣宝 III-44

长输管道环焊缝超声相控阵缺陷数据分析技术研究······李学达，韩 彬，牛盛源，肖文涛，高振华 III-45

基于机器学习的轴承钢碳化物自动识别与评价······张莎莎，朱晓林，于帅鹏，束长青，姚正军 III-46

集成计算与金属结构材料设计 ………………………………………………………… 鲁晓刚 III-46

鞍钢钢铁产品数字化研发实践与展望 ……………………………………… 林 利，王军生 III-47
Anshan Iron and Steel Products Digital R&D Practice and Prospects ………… LIN Li, WANG Junsheng

钢铁材料基因工程数据库与大数据应用 …………………… 姜 雪，张 雷，何 杰，宿彦京 III-48

新能源汽车对汽车用钢发展的机遇与挑战 ……………………………………………… 王 利 III-48
Opportunities and Challenges of New Energy Vehicles for the Development of Automotive Steels
.... WANG Li

高性能汽车钢材料多层级复杂组织调控 …………… 金学军，李 伟，许元涛，刘仕龙 III-49
Design of Hierarchical Complex Structure for High-performance Automotive Steel Materials
.... JIN Xuejun, LI Wei, XU Yuantao, LIU Shilong

增强成形性高强汽车钢的组织性能研究 ………… 赵征志，褚晓红，孙 航，杨玉环，唐 荻 III-50
Study on Microstructure and Properties of High Strength Automotive Steel with Improved Formability
.... ZHAO Zhengzhi, CHU Xiaohong, SUN Hang, YANG Yuhuan, TANG Di

高成形性汽车用超高强钢开发及应用 ……………………………… 刘华赛，韩 赟，阳 锋 III-52
Recent Progress on Development and Application of Ultra-high Strength Steel with High Formability
.... LIU Huasai, HAN Yun, YANG Feng

热轧和冷轧中锰汽车钢的超塑性变形行为与机制研究
.... 曹文全，王 昌，满廷慧，王存宇，董 瀚，翁宇庆 III-54
Superplasticity Behavior and Mechanism of the Hot/Cold Rolled Medium-Mn Autosteel
.... CAO Wenquan, WANG Chang, MAN Tinghui, WANG Cunyu, DONG Han, WENG Yuqing

先进高强汽车钢板的集成计算材料工程研究
.... 何燕霖，郑伟森，张 宇，林 利，徐 鑫，刘仁东，李 麟 III-55

人工智能驱动与计算模拟辅助的承温1200℃单晶高温合金智能设计
.... 杨 帆，蔺思源，赵文月，茹 毅，裴延玲，李树索，宫声凯 III-55
AI-driven and Computational Simulation Assisted Intelligent Design of Single-crystal Superalloys for 1200℃
   Ultra-high Temperature Application
.... YANG Fan, LIN Siyuan, ZHAO Wenyue, RU Yi, PEI Yanling, LI Shusuo, GONG Shengkai

高温合金定向凝固晶体生长与CET转变 …………………… 李应举，邱 义，冯小辉，杨院生 III-56
Directional Solidification Crystal Growth and CET Transformation of Superalloys
.... LI Yingju, QIU Yi, FENG Xiaohui, YANG Yuansheng

特种冶炼行业信息化发展思路 …………………………………… 朱勤天，陈正阳，毕中南 III-57
Thoughts on the Informationization Development of China's Special Smelting Industry
.... ZHU Qintian, CHEN Zhengyang, BI Zhongnan

基于机器学习的高性能铝合金成分设计 …………………… 杨 健，张 佼，戴永兵，隽永飞 III-58
Composition Design of High-performance Aluminum Alloys Based on Machine Learning
.... YANG Jian, ZHANG Jiao, DAI Yongbing, JUAN Yongfei

数值模拟在高温合金构件生产和制造上的应用 …………………………………………… 孙志民 III-59
Application of Numerical Simulation in Super-alloy Products' Design and Manufacture ……… SUN Zhimin

数智化在宝武特冶高温合金锻造生产中的应用与探索 ………………………… 曹金华，侯智鹏 III-60
Application and Exploration on the Digital Intelligence Technology in the Superalloys Forging Production of
   Baowu Special Metallurgy ……………………………………………… CAO Jinhua, HOU Zhipeng

先进材料数据库及应用技术 ………………………………………… 张洪梅，程兴旺，赵平洛 III-61
ZHANG Hongmei, CHENG Xingwang, ZHAO Pingluo

机器学习辅助高熵合金成分设计 ………………………………………………………… 薛德祯 III-62
Machine Learning-assisted Composition Design of High-entropy Alloys ……………… XUE Dezhen

440MPa级高性能船体钢技术研究进展 ………………………………… 罗小兵，师仲然，柴希阳 III-62
Research of 440 MPa High-performance Hull Steel ………… LUO Xiaobing, SHI Zhongran, CHAI Xiyang

输氢管材与氢气相容性研究 ……………………………………… 李　拔，汪　兵，贾书君，刘清友 III-63
Study on Compatibility of Hydrogen Transport Pipe and Hydrogen
……………………………………………………… LI Ba, WANG Bing, JIA Shujun, LIU Qingyou

免喷丸热成形钢组织性能调控及抗氧化性能 …………………………… 郑昊青，常智渊，苏冠侨 III-64
Development and Application of Antioxidant and Shot Peening Free Hot-formed Steel
……………………………………………………… ZHENG Haoqing, CHANG Zhiyuan, SU Guanqiao

超低Nb技术及其在结构板材及型材上的应用 ………………………………………………… 王厚昕 III-65
Study on Ultra Low Nb and Its Application in Commodity Products with YS 355MPa of Both Flat and H Beams
………………………………………………………………………………………… WANG Houxin

极薄取向硅钢制造与应用 ………………………………………………………………… 刘宝志 III-66
Manufacture and Application of Ultra-thin GO Steel ……………………………………… LIU Baozhi

双碳时代中国硅钢产业发展机遇与挑战 …………………………………………………… 吴树建 III-67
Opportunities and Challenges for the Development of China's Silicon Steel Industry in the Dual Carbon Era
………………………………………………………………………………………… WU Shujian

无底层绿色环保涂层高磁感取向硅钢的开发与推广应用研究 ………… 杨佳欣，申明辉，宋　刚 III-67
Development and Application Promotion of Glassless High Permeability Grain-oriented Silicon Steel with
Green and Environmentally Friendly Coating ……… YANG Jiaxin, SHEN Minghui, SONG Gang

高饱和磁化强度Fe-Si-B-C系非晶纳米晶合金组织结构与性能调控
………………………………………………………… 惠希东，吕　旷，李育洛，王伟民 III-68

超稳定的Ce基金属玻璃 …………………………………………………………………… 张　博 III-69

模压电感用软磁复合材料研发 …………………………………………… 王　丽，王明旭，史贵丙 III-69

Rejuvenation to Relaxation Transition and Liquid Memory Effect in Metallic Glasses with Different Energy States
………………………………………………… WANG Lingling, WANG Zheng, HU Lina III-70

基于熔体性质非晶合金微观结构-性能一体化调控 …………………………………………… 张海峰 III-70

人工智能辅助非晶合金结构与性能研究 ……………………………………………………… 管鹏飞 III-70

弛豫非晶合金的再韧化 ……………………………………………………………………… 李　毅 III-71

高性能软磁非晶合金材料的开发及应用 ……………… 孙保安，周　靖，李雪松，白海洋，汪卫华 III-71
Research on High-performance Amorphous Magnetic Alloys
………………………… SUN Baoan, ZHOU Jing, LI Xuesong, BAI Haiyang, WANG Weihua

退火和低温处理态软磁Fe-Co-B系金属玻璃的结构调制与性能表征
………………………………………………………… 王　艳，肖宗奇，刘国强，翟思成 III-72

Ni-Nb二元合金玻璃形成能力和结晶行为差异的结构起源研究 ………………… 卢文飞，沈　军 III-73

一种具有位错-析出相骨架的超强度和延展性的高熵合金 ……………… 贾延东，王　刚，穆永坤 III-73
A High-entropy Alloy with Dislocation-Precipitate Skeleton for Ultrastrength and Ductility
……………………………………………………… JIA Yandong, WANG Gang, MU Yongkun

含钴铁基非晶纳米晶合金新成分开发探索 ············· 董帮少，邢彦兴，崔宏祥，周少雄 III-74

金属玻璃薄膜稳定性和不均匀性及性能调控 ·································· 罗 强 III-75

激光 3D 打印非晶合金复合材料的相变增韧行为研究 ····················· 张鹏程，柳 林 III-75

低碳绿色钢铁冶金关键技术研发与应用 ··· 叶恒棣 III-76
Development and Application of Low-carbon Green Iron and Steel Metallurgy Key Technology
·· YE Hengdi

钢渣在建材领域资源化应用关键问题分析 ··· 田志红 III-77
Analysis of Key Problems in Resource Utilization of Steel Slag in Building Materials Field
·· TIAN Zhihong

基于 5G 技术的钢铁企业多维度生态环保管控系统开发与应用 ········ 姜德旺，张亚夫，朱宴恒 III-77
Development and Application of Multi-dimensional Ecological Environmental Control System for Iron and
    Steel Enterprises Based on 5G Technology ··········· JIANG Dewang, ZHANG Yafu, ZHU Yanheng

钢铁行业高质量绿色低碳发展路径探讨 ··· 陈 洁 III-79
Research and Analysis of High-quality and Green-low-carbon Development Path of Steel Industry
·· CHEN Jie

鞍钢节能环保技术开发与实践 ·················· 马光宇，蔡秋野，王向锋，李卫东，陈 鹏 III-80
Development and Practice of Energy Conservation and Environmental Protection Technologies in Ansteel
·· MA Guangyu, CAI Qiuye, WANG Xiangfeng, LI Weidong, CHEN Peng

钢铁行业超低排放有组织监测评估技术与实践 ··· 刘通浩 III-81
Technology and Practice of Organized Monitoring and Evaluation of Ultra-low Emission in Steel Industry
·· LIU Tonghao

钢铁企业水系统现状及节水降碳优化策略分析 ··· 余云飞 III-82
Current Situation of Water System in Iron and Steel Enterprises and Analysis of Optimal Strategy of Saving
    Water and Reducing Carbon ··· YU Yunfei

钢铁行业超低排放与除尘系统节能技术发展现状 ·················· 王 珲，张 璞，杨雅娟 III-83
Development Status of Ultra-low Emission and Energy-saving Technology of Dust Removal System in the
    Steel Industry ··· WANG Hui, ZHANG Pu, YANG Yajuan

熔融钢渣高效负能处理技术与装备 ··· 王宇鹏 III-83
Technology and Equipment for High Efficient Negative Energy Treatment of Molten Steel Slag
·· WANG Yupeng

全流程钢厂水系统智慧管控与零排放关键技术开发应用 ······························· 桂其林 III-84
Development and Application of Key Technologies for Intelligent Control and Zero Discharge of Steel Mill
    Water System in the Whole Process ··· GUI Qilin

钢铁冶金大综固废综合利用与 CO_2 减排 ··· 郭占成 III-85
Comprehensive Utilization of Solid Waste and CO_2 Emission Reduction in Iron and Steel Metallurgy
·· GUO Zhancheng

含锌粉尘球团还原行为研究 ······················· 朱德庆，石书冰，潘 建，郭正启，杨聪聪 III-85
Study on the Reduction Behavior of Zinc-bearing Dust Pellets
································ ZHU Deqing, SHI Shubing, PAN Jian, GUO Zhengqi, YANG Congcong

基于无废城市建设的长流程钢厂固废管理模式与技术路径 ····························· 李红红 III-86
Solid Waste Management Mode and Technical Path of Long-process Steel Plant Based on the Concept of
    Zero-waste Construction ·· LI Honghong

大型波纹形耐压壳筒节结构优化与轧制理论研究 ……………… 彭 艳，孙建亮，张 旭，吴润泽，单存尧 III-87
Theoretical Research on the Structural Optimization Design and Rolling of Corrugated Pressure-resistant Shell
　Heavy Cylinder ……………………… PENG Yan, SUN Jianliang, ZHANG Xu, WU Runze, SHAN Cunyao
中厚板试样取样生产智能控制系统研究 …………………………………………………………… 辛小臣 III-88
Research on Intelligent Control System for Sampling Production of Medium and Thick Plate Samples
　………………………………………………………………………………………………… XIN Xiaochen
中冶赛迪轧钢数字化解决方案及实践 …………………………………………………… 李 华，陶 涛 III-89
带式输送机的智能化断带保护技术的研究与应用 ………… 徐言东，白金龙，郭 强，张勇军，白英超 III-90
　………………………………… XU Yandong, BAI Jinlong, GUO Qiang, ZHANG Yongjun, BAI Yingchao
首钢智能制造技术研究与探索 ……………………………………………………………………… 于 孟 III-91
Research and Exploration of Intelligent Manufacturing Technology in Shougang ………………… YU Meng
钢铁企业全要素资源优化与智慧运营核心关键技术创新 ………………………………………… 汝金同 III-92
钢厂板坯库调度问题的研究进展评述 …………………………………… 郑 忠，胡 赟，高小强 III-94
基于工业互联网平台的钢铁行业 AI 应用 …………………………………………………………… 李 胜 III-94
中厚板板型检测系统开发与应用 …………………………………………………………………… 田 勇 III-95
数字驱动的宽厚板自动转钢系统研发与应用 ……………………………………………………… 何纯玉 III-95
中冶赛迪智能制造解决方案探索与实践 …………………………………………………………… 童 心 III-96
中冶赛迪炼钢智能制造探索与实践 ………………………………………………………………… 王智君 III-96
数字能源技术助力钢企创新发展的探索与实践 …………………………………………………… 吴 杉 III-96
板坯库智能作业管理系统设计的实现与应用 ……………………………………………………… 张 鑫 III-97
数字能源精益管控一体化 …………………………………………………………………………… 张冰峰 III-98
国际物流与物流国际化 ……………………………………………………………………………… 樊俊花 III-98
International Logistics and Logistics Internationalization ……………………………………… FAN Junhua
冶金产品智能化储运技术研究 ……………………………………………………………………… 刘武胜 III-99
Research on Intelligent Storage and Transportation Technology for Metallurgical Products ……LIU Wusheng
数字化时代钢铁行业供应链金融创新发展与风险管理 …………………………………………… 卢 强 III-99
Innovative Development and Risk Management of Supply Chain Finance in the Steel Industry in the Digital Era
　……………………………………………………………………………………………………… LU Qiang
网络货运平台赋能大宗商品供应链数字化转型升级的创新实践 ………………………………… 王后建 III-100
Innovative Practice of Empowering the Digital Transformation and Upgrading of Bulk Commodity Supply Chains
　through Online Freight Platforms …………………………………………………………… WANG Houjian
现代"紧凑型"钢铁制造流程的极致设计-冶金流程学理论在邯钢项目上的应用实践
　……………………………………………………… 李 铁，王新东，刘亚峰，付 康，李传民 III-101
The Ultimate Design of Modern "Compact" Steel Manufacturing Processes-Application Practice of Metallurgical
　Process Theory in Handan Steel Project
　…………………………………… LI Tie, WANG Xindong, LIU Yafeng, FU Kang, LI Chuanmin
钢铁冶金铁前工艺流程优化研究与实践 …………………………………… 王兆才，叶恒棣，魏进超 III-102
Research and Practice on Process Optimization of Pre-ironmaking in Ferrous Metallurgy
　………………………………………………………………… WANG Zhaocai, YE Hengdi, WEI Jinchao

1 矿业工程

1.1 采 矿

李楼铁矿采场充填密闭工艺 ……付宇峰 1-1
Filling and Sealing Technology of Stope in Lilou Iron Mine ……FU Yufeng

李楼铁矿大结构采场充填采矿方法工业试验 ……宋 亮 1-1
Industrial Test of Filling Mining Method in Large Structure Stope of Lilou Iron Mine …… SONG Liang

浅谈矿山地下水防治技术现状及发展趋势 ……王 波 1-2
Brief Discussion on the Current Situation and Development Trend of Mine Groundwater Prevention and Control Technology …… WANG Bo

预注浆工艺在松散破碎带巷道施工中的应用研究 ……李伟群，陈利泰 1-3
Research of Advance Grouting Technology in Loose Broken Zone in Mine Excavation
…… LI Weiqun, CHEN Litai

基于膨胀支柱技术的孤立采场安全高效采矿方法 ……李元辉，熊志朋，赵福权 1-3
Safe and Efficient Mining Method in Isolated Stope Based on Expandable Pillar Technology
…… LI Yuanhui, XIONG Zhipeng, ZHAO Fuquan

浅谈数码电子雷管在袁家村铁矿的应用 ……贾传鹏 1-4
The Application of Digital Electronic Detonator in Yuanjiacun Iron Mine …… JIA Chuanpeng

矿山溜井堵塞率预测的 GA-BP 神经网络法 ……朱俊阁，于宏宇，王浩东，马强英，路增祥 1-5
GA-BP Neural Network Method for Predicting the Blockage Rate of Mine Orepass
…… ZHU Junge, YU Hongyu, WANG Haodong, MA Qiangying, LU Zengxiang

地下金属矿山大块产生原因以及处理措施 ……徐 涛，郭武英 1-6
The Causes of the Occurrence of Large Blocks in Underground Metal Mines and the Treatment Measures
…… XU Tao, GUO Wuying

浅谈井下金属矿山发生悬顶的原因以及处理措施 ……徐 涛，郭武英 1-7
The Causes of Overhanging in Underground Metal Mines and the Treatment Measures are Discussed
…… XU Tao, GUO Wuying

排土场排水系统实践探索与思考 ……刘 景，徐飞飞 1-7

某矿切割井一次爆破成井研究及应用 ……夏文浩，宋卫东，张少鹏 1-8
Research and Application of Blasting to Form a 10.5m Cutting Raise by One Step in a Mine

1.2 选 矿

蒙库铁矿含铁围岩再选资源化利用工业实践 ……胡家仓，王国平 1-9

NaCl 浓度对油酸钠体系的泡沫稳定性影响研究 ……智 慧，董振海，王 欢，满晓霏 1-9
Study on the Effect of NaCl Concentration on the Stability of Foam of Sodium Oleate System
…… ZHI Hui, DONG Zhenhai, WANG Huan, MAN Xiaofei

深槽型大倾角带式输送机铁矿石输送试验研究 ……万正道，张玉达，赵 斌 1-10
Experimental Study on Iron Ore Conveying with Deep-slot and Large Inclined Angle Belt Conveyor
…… WAN Zhengdao, ZHANG Yuda, ZHAO Bin

基于离散元仿真的圆锥破碎机腔型优化 ……………… 韩睿智，宋仁伯，王永金，刘丹丹，母 镕 1-11
铁精矿中硫浮选脱除与综合利用试验研究 …………………………………… 刘兴华，董英择 1-12
Experimental Study on Flotation Removal and Comprehensive Utilization of Sulfur in Iron Concentrate
　　　　　　　　　　　　　　　　　　　　　　　………… LIU Xinghua, DONG Yingze
某高磷鲕状赤铁矿闪速磁化焙烧技术开发研究
　　　　　　　　　　…………… 李家林，陈 雯，刘 旭，刘小银，陆晓苏，彭泽友 1-12
磁化焙烧工艺选别细粒难选赤铁矿尾矿的试验研究 ………………………… 杨晓峰，董振海 1-13
湖北某铁矿工艺矿物学及选矿试验研究 ……………………… 李育彪，王忠红，李万青 1-14
白云鄂博低品位铌矿石选矿富集技术研究 ……………… 王 祥，王 伟，周喻林，陈 雯 1-15
Research Progress on Beneficiation Technology of Low-grade Niobium Ore in Baiyun Ebo
　　　　　　　　　　　…………… WANG Xiang, WANG Wei, ZHOU Yulin, CHEN Wen
高硫高锌磁铁矿深度除杂制备超纯铁精矿的工业实践 ……… 廖振鸿，陈 雯，王秋林，刘 旭 1-15
硫酸烧渣悬浮磁化焙烧提铁技术研究 ………………………… 李文博，王 硕，唐志东 1-16
不同高分子絮凝剂对东鞍山细粒赤铁矿聚团-磁选的影响 ……… 李文博，张 爽，程绍凯，周立波 1-17
氟碳铈矿高梯度磁吸附行为研究 ………………………………… 李文博，孙俊燕，张小龙 1-18
独居石在还原焙烧中浮选性质变化研究 ……………………… 李文博，瞿 瑞，赵 熙 1-18
辽西低品位钒钛磁铁矿综合利用 ……………… 刘金生，刑振兴，黄 壮，程功金，丁学勇，薛向欣 1-19
Comprehensive Utilization of Low-grade Vanadium-titanium Magnetite in Western Liaoning Province
　　…… LIU Jinsheng, XING Zhenxing, HUANG Zhuang, CHENG Gongjin, DING Xueyong, XUE Xiangxin
某超大型磁赤混合铁矿高效 SAB 流程选择及实践 ………………………… 张耀斌，牛建昆 1-20
Selection and Practice of High Efficiency SAB Process for a Superlarge Magnetic Hematite Mixed Iron Mine
　　　　　　　　　　　　　　　　　　　　　　　………… ZHANG Yaobin, NIU Jiankun
攀西某钒钛磁铁矿入库尾矿工艺矿物学及有价资源回收探索试验研究
　　　　　　　　　　　　　…………… 蔡先炎，李 硕，陈 勇，陈福林，黄 延 1-21
Exploring Experimental Study on Mineralogy and Valuable Resource Recovery of Inbound Tailings of a Vanadium
　　Titanium Magnetite in Panxi …………… CAI Xianyan, LI Shuo, CHEN Yong, CHEN Fulin, HUANG Yan
攀西地区钒钛磁铁矿碎磨技术发展现状
　　　　　　　…………… 李 硕，蔡先炎，黄 延，王志杰，陈福林，吴 宁，王志林，杨道广 1-22
Development Status of Crushing and Grinding Technology of Vanadium Titano-Magnetite in Panxi Area
　　　　　　　　………… LI Shuo, CAI Xianyan, HUANG Yan, WANG Zhijie, CHEN Fulin,
　　　　　　　　　　　　　WU Ning, WANG Zhilin, YANG Daoguang

2 焦化及节能环保

干熄焦年修复产调控技术优化研究与应用 ……………… 朱庆庙，李 勇，侯士彬，赵振兴，赵 锋 2-1
Research and Application on Optimization of Annual Restoration and Production Control Technology of CDQ
　　　　　　　　　　………… ZHU Qingmiao, LI Yong, HOU Shibin, ZHAO Zhenxing, ZHAO Feng
工业废弃树脂配煤炼焦技术应用研究 ……………………… 朱庆庙，未福宇，侯士彬，赵振兴，赵 锋 2-1
Application of Coking Technology with Industrial Waste Resin Blending Coal
　　　　　　　　　　………… ZHU Qingmiao, WEI Fuyu, HOU Shibin, ZHAO Zhenxing, ZHAO Feng
真空碳酸钾脱硫及 TOP 湿法制酸生产问题的查找与处理 ………… 杨天宇，张 欣，韩修尚，徐利民 2-2

浅谈焦油蒸馏精细化操作	孔德刚，王英达，孙喜民	2-3
焦化项目中备煤工段除尘系统的设计计算	邹　阳，杨雪莲，杨　仙，李梦珂，李传东	2-3
Design Calculations of Dust Removal System in Coking Project		
	ZOU Yang, YANG Xuelian, YANG Xian, LI Mengke, LI Chuandong	
焦化废水石油类测定中不同吸附方法的比较	刘晓桃，宋文杰，尚志春，杜　彬，陈　伟，张　宇	2-4
Comparison of Different Adsorption Methods in Determination of Petroleum in the Coking Wastewater		
	LIU Xiaotao, SONG Wenjie, SHAGN Zhichun, DU Bin, CHEN Wei, ZHANG Yu	
焦炉煤气制 LNG 联产液氨工艺脱碳增产技术的研究	杨　鹏，栾树龙	2-4
Research on Decarbonization and Production Increase Technology of Coke Oven Gas to LNG Combined with Liquid Ammonia Production Process	YANG Peng, LUAN Shulong	
关于干熄焦生产运行的几点新建议和新思考	侯士彬，李秋镝，李　勇，马俊尧，高　微	2-6
Several New Suggestions and New Suggestions about the CDQ Circulating		
	HOU Shibin, LI Qiudi, LI Yong, MA Junyao, GAO Wei	
关于五炼焦作业区与西部炼焦作业区焦炉煤气热值存在差异的原因分析		
	侯士彬，李秋镝，李　勇，马俊尧，高　微	2-6
焦炉煤气干法精脱硫工艺的优化设计与应用	李　伟，王博林，韩渝京，李　鹏	2-7
Optimization Design and Application of Dry Fine Desulfurization Process for Coke Oven Gas		
	LI Wei, WANG Bolin, HAN Yujing, LI Peng	
循环氨水余热用于工艺制冷的实践	王思维，邱全山，周　诚，张晓萍，李　杰，胡　俊	2-7
Practice of Circulating Ammonia Waste Heat used in Process Refrigeration		
	WANG Siwei, QIU Quanshan, ZHOU Cheng, ZHANG Xiaoping, LI Jie, HU Jun	
硫铵母液焦油资源化利用技术的研发	朱晓渊，颜长青，施　亮，王传荟，潘红良	2-8
Research and Development of Technology for Resource Utilization of Tar from Ammonium Sulfate Liquor		
	ZHU Xiaoyuan, YAN Changqing, SHI Liang, WANG Chuanhui, PAN Hongliang	
二氧化铈/氮掺杂氮化碳光催化剂可见光下深度处理焦化废水研究		
	安　宁，胡绍伟，陈　鹏，王　永，王　飞，刘　芳	2-9
Cerium Dioxide/Nitrogen Doped Graphite Carbon Nitride for Photocatalytic Advanced Treatment of Coking Wastewater under Visible Light		
	AN Ning, HU Shaowei, CHEN Peng, WANG Yong, WANG Fei, LIU Fang	
煤岩及焦炭光学组织评价焦炭宏观性能	张文成	2-10
Evaluation of Coke Macroscopic Properties by Optical Organization of Coke and Coal Rank		
	ZHANG Wencheng	
高炉内焦炭降解行为及质量评价方法研究	陈静波，张生富，姜　涛	2-11
7m 焦炉电机车及 APS 控制系统优化	乔文超，张允东，孔　弢，赵　锋，陈　鹏	2-11
Optimization of Seven-meter Coke Oven Electric Motor and APS Control System		
	QIAO Wenchao, ZHANG Yundong, KONG Tao, ZHAO Feng, CHEN Peng	
7.63m 焦炉超超低排放的实践	殷喜和	2-12
The Practice of 7.63m Coke Oven with Ultra-low Emission	YIN Xihe	
焦炉废气快速分析在焦化中的应用	邱　程	2-13
浅谈焦炉停产事故状态下的工艺调整	邱　程，郭　飞	2-13
基于焦炉测温及加热控制系统的探究与成果	邱　程，田龙振	2-13
青钢 7m 焦炉煤气量与变频器电流数据拟合分析	邱　程，郭　飞，何　波	2-14

Fitting Analysis of Gas Volume and Converter Current Data of Qinggang's 7m Coke Oven

························ QIU Cheng, GUO Fei, HE Bo

青特钢焦化低水分熄焦生产实践 ················· 邱 程，郭 飞，王永亮 2-14

Low Moisture Coke Quenching in Qinggang Coking Plant Transformation and Production Practice

························ QIU Cheng, GUO Fei, WANG Yong liang

焦炉智能巡检机器人应用实践 ················· 邱 程，郭 飞 2-15

配合煤堆密度在提高焦炉装煤量方面的应用 ················· 邱 程 2-15

两项煤气净化工艺改进方案 ················· 杨 爽 2-15

基于煤岩学方法分析焦炭"夹生"问题 ················· 王 越，庞克亮，吴昊天，谷致远 2-16

Analysis of "Incompletely Mature" Coke Based on Petrographic Method

························ WANG Yue, PANG Keliang, WU Haotian, GU Zhiyuan

改质沥青的生产工艺及发展潜能概述 ················· 徐志强，张立伟，闫双双，王 斌，王晓楠 2-16

Overview of the Production Process and Development Prospect of Modified Pitch

························ XU Zhiqiang, ZHANG Liwei, YAN Shuangshuang, WANG Bin, WANG Xiaonan

干燥煤结焦过程传热传质行为的数值模拟研究 ················· 仇 灏，刘和平 2-17

武钢 7.63m 焦炉蓄热室积灰分析与判断 ················· 陈细涛，严铁军，张 军，常红兵，陈 鹏 2-18

Analysis of Ash in Regenerator Chamber of WISCO 7.63m Coke Oven

························ CHEN Xitao, YAN Tiejun, ZHANG Jun, CHANG Hongbing, CHEN Peng

焦化厂原料煤全自动检化验系统设计 ················· 贺世泽，吴 波 2-19

Design of Automatic Test System for Raw Coal in Coking Plant ················· HE Shize, WU Bo

焦炉操作工艺对焦炭质量影响的分析 ················· 陈 鹏，严铁军，常红兵，陈细涛，张 军 2-19

Analysis of Influence of Coke Oven Operation on Coke Quality

························ CHEN Peng, YAN Tiejun, CHANG Hongbing, CHEN Xitao, ZHANG Jun

二氧化钛光催化剂改性研究 ················· 刘子娟，吴伟伟，李 达，贾楠楠，刘庆佩，王晓楠 2-20

Study on Modification of Titanium Dioxide Photocatalyst

························ LIU Zijuan, WU Weiwei, LI Da, JIA Nannan, LIU Qingpei, WANG Xiaonan

高灰中硫瘦煤资源开发与利用 ················· 贺 佳，史永林，李昊堃，王钟议 2-21

Development and Utilization of High Ash and Medium Sulphur Lean Coal

························ HE Jia, SHI Yonglin, LI Haokun, WANG Zhongyi

焦化废水深度处理技术的创新与实践 ················· 李 斌，潘东山，王云峰 2-21

Innovation and Practice of Advanced Treatment Technology for Coking Wastewater

························ LI Bin, PAN Dongshan, WANG Yunfeng

联用工艺在煤化工综合废水处理中的应用 ················· 李 斌，潘东山，王云峰 2-22

Application of Combined Process in Comprehensive Wastewater Treatment of Coal Chemical Industry

························ LI Bin, PAN Dongshan, WANG Yunfeng

焦炉机车除尘新技术的应用与创新 ················· 郭有林，成雪松，郭永贵 2-23

Application and Innovation of New Dust Removal Technology for Coke Oven Locamotive

························ GUO Youlin, CHENG Xuesong, GUO Yonggui

氨水中焦油渣回收利用工艺优化 ················· 刘卫杰 2-23

Optimization of the Recovery and Utilization Process of Tar Residue in Ammonia Water ················· LIU Weijie

焦化厂循环氨水的余热利用 ················· 张 磊，付冠华 2-24

190t/h 干熄焦余热锅炉过热器故障分析 ················· 曾石川 2-24

The Failure Analysis of CDQ Boiler Superheater ································· ZENG Shichuan

西门子 OUC 通讯方式在焦化环保除尘中的应用研究 ······················· 杨 光，向 勇　2-25
Research on the Application of Siemens OUC Communication Method in Dust Removal of Coking Environmental
　　Protection ·· YANG Guang, XIANG Yong

基氏流动度曲线特征在炼焦煤混煤判定中的应用 ············· 宋子逵，项 茹，刘 睿，丁 海　2-26
Application of the Characteristics of Gieseler Fluidity Curve in the Determination of Coking Coal Blending
　　··· SONG Zikui, XIANG Ru, LIU Rui, DING Hai

3 炼铁与原料

3.1 烧结与球团

烧结烟气中 NO_x 的产生机理及排放规律探研 ············· 桂林峰，苏亚刚，李 彬，黄永安　3-1
Research on the Generation Mechanism and Emission Law of NO_x in Sintering Flue Gas
　　··· GUI Linfeng, SU Yagang, LI Bin, HUANG Yong'an

带式焙烧机球团智能管控系统的研究与实现 ········· 刘富春，韩基祥，任 伟，李 丹，周检平　3-1
Research and Implementation of Intelligent Control System for Pellet of Belt Baking Machine
　　··· LIU Fuchun, HAN Jixiang, REN Wei, LI Dan, ZHOU Jianping

提高混匀矿稳定率的实践 ····························· 黎兴文，陈 伟，李加龙，张锡宝，赵 彧　3-2
Practice on Improving the Stability Rate of Mixed Ore
　　··· LI Xingwen, CHEN Wei, LI Jialong, ZHANG Xibao, ZHAO Yu

昆钢 B 型料场扬尘综合治理应用探析 ············· 陶传昌，王志富，侯林坤，苏 宝，谢志刚　3-3
Exploring the Application of Comprehensive Dust Control in Kunming Steel's B-type Material Yard
　　··· TAO Chuanchang, WANG Zhifu, HOU Linkun, SU Bao, XIE Zhigang

提高烧结料层上部成品率试验研究 ············· 武 轶，张晓萍，李帮平，袁 鹏，熊德怀，朱贺民　3-3

邯钢 360m² 烧结机降低煤气消耗生产实践 ··· 王岳飞，王永林　3-4
Production Practice of Reducing Gas Consumption in 360m² Sintering Machine of Handan I&S Co.
　　··· WANG Yuefei, WANG Yonglin

鞍钢鲅鱼圈球团回转窑结圈研究及控制实践 ········· 任 伟，国泉峰，王 亮，张 伟，韩子文　3-4
The Ring Study on Angang Bayuquan Rotary Pelletizing Kiln and Its Control
　　··· REN Wei, GUO Quanfeng, WANG Liang, ZHANG Wei, HAN Ziwen

烧结智能优化配矿的发展浅析 ········· 刘 杰，王耀祖，马贤国，陈付振，刘征建，周明顺，张建良　3-5
Analysis on the Development of Sintering Intelligent Optimal Ore Blending ·········· LIU Jie, WANG Yaozu,
　　MA Xianguo, CHEN Fuzhen, LIU Zhengjian, ZHOU Mingshun, ZHANG Jianliang

碱度对高铬型钒钛磁铁矿熔剂型球团中温还原性能与反应动力学机理的影响
　　··· 陈泊键，姜 涛，温 婧，朱凤湘，胡 鹏　3-6
Reducibility Optimization and Reaction Mechanism of High-Chromium Vanadium-Titanium Magnetite Flux Pellets
　　··· CHEN Bojian, JIANG Tao, WEN Jing, ZHU Fengxiang, HU Peng

塞拉利昂铁矿粉的烧结性能评价研究 ············· 段立祥，高洪庄，张大伟，刘 杰，张明洲　3-7
Research of Sierra Leone Ore's Sintering Performance
　　··· DUAN Lixiang, GAO Hongzhuang, ZHANG Dawei, LIU Jie, ZHANG Mingzhou

澳洲某赤褐铁矿磨矿特性及其氧化球团制备行为研究
 ·· 杨聪聪，曲士娟，潘 建，朱德庆，郭正启，李思唯 3-8
改善赤褐铁矿球团生产过程和成品质量的方法
 ····················· 刘胜歌，张俊杰，张巧玉，贾来辉，裴元东，刘 桐，宋亚龙 3-9
高比例赤褐铁矿碱性球团实践应用与研究
 ············ 刘胜歌，张俊杰，张巧玉，贾来辉，裴元东，顾建苛，刘 桐，刘建波 3-10
带式球团配加含铁含碳除尘灰生产实践 ················· 胡守景，王业飞，秦立浩 3-11
Production Practice of Adding Iron and Carbon Containing Dust Removal Ash to Belt Pellets
 ·· HU Shoujing, WANG Yefei, QIN Lihao
高压辊磨预处理国内铁精矿强化球团性能研究 ····························· 高茂林 3-11
Improvement the Pellet Performances of Domestic Iron Concentrate Pretreated by High Pressure Grinding Roller
 ·· GAO Maolin
红磁混合精矿生产低碱度球团生产实践 ··· 祁 立 3-12
海砂型钒钛磁铁矿颗粒表面性能与球团制备研究 ········ 邢振兴，刘金生，薛向欣，程功金，杨 合 3-12
Study on Particle Surface Properties and Pellet Preparation of Sea Sand Type Vanadium Titanomagnetite
 ······················· XING Zhenxing, LIU Jinsheng, XUE Xiangxin, CHENG Gongjin, YANG He
循环风流动状态对球团热工制度的影响 ··· 张 晨 3-13
The Influence of the Circulating Airflow State on the Thermal System of the Pelletizing Process
 ··· ZHANG Chen
基于神经网络和NSGA-Ⅱ的烧结参数多目标优化 ····························· 臧疆文 3-14
Multi-objective Optimization of Sintering Parameters Based on Neural Network and NSGA-Ⅱ
 ··· ZANG Jiangwen
某进口赤褐铁矿磨矿特性及其氧化球团制备行为研究
 ·· 杨聪聪，曲士娟，潘 建，朱德庆，郭正启，李思唯 3-15
MgO对CO分阶段还原赤铁矿过程膨胀性的影响 ············ 张 芳，彭 军，常宏涛，王永斌 3-16
The Effect of MgO on the Swelling Property of Hematite Reduced by CO Step-by-step
 ··· ZHANG Fang, PENG Jun, CHANG Hongtao, WANG Yongbin

3.2 高炉炼铁

高炉开停炉技术创新实践 ··· 陈生利 3-18
Technological Innovation Practice of Blast Furnace Opening and Stopping ················ CHEN Shengli
高炉铁水收得率攻关实践 ··························· 陈生利，彭 厅，余 骏 3-18
Research Practice of Recovery Rate of Molten Iron in Blast Furnace
 ··· CHEN Shengli, PENG Ting, YU Jun
韶钢7号高炉铁口喷溅治理操作实践 ·································· 匡洪锋 3-19
Operation Practice of Iron Spitter Control for No. 7 Blast Furnace at Shaoguan Iron and Steel Group Co., Ltd.
 ·· KUANG Hongfeng
提高铁水硅偏差命中率控制措施 ·· 余 骏，陈生利 3-20
Measures to Improve the Error Hit Ratio of Molten Iron and Silicon ············ YU Jun, CHEN Shengli
梅钢 4070m^3 高炉炉墙结厚分析与炉况恢复 ················ 张光强，韩宏松，占世梅 3-20
昆钢 2500m^3 高炉提高利用系数生产实践 ······················ 麻德铭，李 淼，李晓东 3-21

Production Practice of Improving Utilization Factor of Kunming Steel's 2500m³ Blast Furnace
······ MA Deming, LI Miao, LI Xiaodong

红钢 3 号高炉安全快速停炉实践 ······ 张　波，李晓东　3-21
Safe and Quick Shutdown Practice of No.3 BF at Honggang ······ ZHANG Bo, LI Xiaodong

配加智利粗粉生产实践 ······ 张　月，伊凤永　3-22
Smelting Practice of Adding Chilean Coarse Powder to Vanadium-titanium Ore Blast Furnace

ECIA 环保料场创新设计 ······ 艾　宇，陈尚伦，王沛庆，刘占稳　3-22
Innovative Design of Eco-friendly Stockyard ······ AI Yu, CHEN Shanglun, WANG Peiqing, LIU Zhanwen

高块矿比冶炼在兴澄高炉的生产实践 ······ 张宏星，陈龙智，史志苗　3-23
Production Practice of High Lump Ore Ratio Smelting in Xingcheng Blast Furnace
······ ZHANG Hongxing, CHEN Longzhi, SHI Zhimiao

昆钢 1 号高炉精准快速停炉实践 ······ 李晓东，张　波　3-24
Practice of Accurate and Rapid Shutdown of Kunming Iron and Steel No.1 Blast Furnace
······ LI Xiaodong, ZHANG Bo

基于高炉透气指数的焦炭质量评价与控制方法 ······ 刘起航　3-24

昆钢新区 1 号高炉降低燃料比实践 ······ 段贵军，李　淼，李晓东　3-25
Practice of Reducing Fuel Ratio for No.1 Blast Furnace in Kunming Iron and Steel New Area
······ DUAN Guijun, LI Miao, LI Xiaodong

高炉炼铁煤粉制备系统节能探讨 ······ 汤楚雄　3-26
Discussion on Energy Saving of Pulverized Coal Preparation System for Blast Furnace Ironmaking
······ TANG Chuxiong

高炉喷吹煤结构优化方法研究与应用 ······ 宋灿阳，朱浩民，张晓萍，李　杰，王思维　3-26
Research and Application of Optimization Method of Blast Furnace Coal Injection Structure
······ SONG Canyang, ZHU Haomin, ZHANG Xiaoping, LI Jie, WANG Siwei

宝武多基地高炉使用钢铁循环材料生产实践 ······ 王　臣，朱仁良，朱　彤　3-27

城市碳氢固废高效粒化及粒化产物用于高炉喷吹研究
······ 王　广，刘思贤，聂志睿，乔宇峰，王静松，薛庆国　3-28

鞍钢 11 号高炉风口破损分析 ······ 周正义，张立朝，姜庆喜，唐安平，潘世奇　3-29
Analysis on Damage of No.11 Blast Furnace Tuyere in Ansteel
······ ZHOU Zhengyi, ZHANG Lichao, JIANG Qingxi, TANG Anping, PAN Shiqi

鞍钢 11 号高炉长期闷炉后快速开炉达产实践 ······ 周正义，张立朝，姜庆喜，唐安平，潘世奇　3-29
Rapid Blowing-in Production Practice after Long-term Smouldering in No. 11 Blast Furnace of Ansteel
······ ZHOU Zhengyi, ZHANG Lichao, JIANG Qingxi, TANG Anping, PAN Shiqi

基于风口图像的高炉状态智能诊断技术及应用 ······ 李　强，王子佳，邹宗树　3-30

高炉定量化综合操作技术集成创新（上篇） ······ 林安川，邱贵宝，刘晓兰，蒋玉波　3-31
Integrated Innovation of Quantitative Integrated Operation Technology of Blast Furnace（Ⅰ）
······ LIN Anchuan, QIU Guibao, LIU Xiaolan, JIANG Yubo

长流程下全氧富氢碳循环低碳炼铁新工艺的可行性分析 ······ 李海峰，陈靖然，王小艾　3-32

高炉冶炼喷吹天然气经济性探讨 ······ 魏钦帅，罗　凯，潘洪福　3-33
Discussion on the Economy of Natural Gas Injection in Blast Furnace
······ WEI Qinshuai, LUO Kai, PAN Hongfu

越南和发榕橘二期绿色智能环保原料场工艺设计及特点
······ 王亚伟，陈尚伦，艾　宇，程　浩，刘　怡，孙贵舟，刘占稳　3-34

Process Design and Characteristics of Green Intelligent & Environmental Protection Raw Material Yard of Vietnam's HoaPhat Dung Quat Steel JSC of Phase Ⅱ
………… WANG Yawei, CHEN Shanglun, AI Yu, CHENG Hao, LIU Yi, SUN Guizhou, LIU Zhanwen

蓄铁式主沟结构参数优化模拟研究……………………………………………………董晓森，郑　魁　3-34
Simulation Study on Optimization of Structural Parameters of the Main Channel of Iron Storage Type
……………………………………………………………………………………… DONG Xiaosen, ZHENG Kui

沙钢 2500m³ 高炉用激光料面探测仪研究料面下降速度分布曲线
………………………………………………谭天雷，黄　军，杜友尧，高征铠，高　泰　3-35
Study on the Distribution Curve of Burden Surface Profile Falling Speed Using the Blast Furnace Laser Burden Surface Detector for a 2500m³ Blast Furnace at Shagang
……………………………… TAN Tianlei, HUANG Jun, DU Yourao, GAO Zhengkai, GAO Tai

高球团矿配比条件下炉喉矿焦比变化规律及其优化
………………………………王　斌，王　炜，郑　魁，陈世锦，陈绪亨，赵立辉　3-36
The Variation Law of Furnace Throat Ore/Coke under High Pellet Charging Ratio and Its Optimization
……………… WANG Bin, WANG Wei, ZHENG Kui, CHEN Shijin, CHEN Xuheng, ZHAO Lihui

高炉铜冷却壁渣皮微观结构及导热性能分析……………陈世锦，王　炜，陈绪亨，王　斌，宋大为　3-37
Microstructure and Thermal Conductivity Analysis of Blast Furnace Copper Stave Slag Crust
……………………………… CHEN Shijin, WANG Wei, CHEN Xuheng, WANG Bin, SONG Dawei

高炉定量化综合操作技术集成创新（下篇）………………………………林安川，刘晓兰，蒋玉波　3-38
Integrated Innovation of Quantitative Integrated Operation Technology of Blast Furnace（Ⅱ）
………………………………………………………… LIN Anchuan, LIU Xiaolan, JIANG Yubo

金属化炉料模拟高炉条件下的反应行为………………………闫瑞军，储满生，柳政根，刘培军　3-39
Reaction Behaviors of Metallized Burden under Simulated Blast Furnace Conditions
………………………………………… YAN Ruijun, CHU Mansheng, LIU Zhenggen, LIU Peijun

武钢有限 8 号高炉更换冷却壁快速恢复实践………………………………………林　巍，陈畏林　3-40
Fast Recovery Practice of Cooling Stave Replacement in No.8 Blast Furnace……… LIN Wei, CHEN Weilin

富氢低碳冶炼工况下高炉热量分配模型…………王　刚，许　俊，贺　坤，邹忠平，赵运建，牛　群　3-41
Heat Distribution Model under Hydrogen-rich Low-carbon Conditions in Blast Furnace
………………… WANG Gang, XU Jun, HE Kun, ZOU Zhongping, ZHAO Yunjian, NIU Qun

首钢通钢 2 号高炉开炉快速达产实践………………………谢绍飞，杨晓辉，刘文明，于海新　3-41
Practice of Rapid Production of Shougang Tonggang No. 2 Blast Furnace by Blowing in
………………………………………… XIE Shaofei, YANG Xiaohui, LIU Wenming, YU Haixin

武钢有限 6 高炉加废钢的生产实践……………………………………………余珊珊，李勇波　3-42
Production Practice of Adding Scrap Steel to No.6 Blast Furnace in Wisco……… YU Shanshan, LI Yongbo

首钢通钢高炉减少风口小套破损技术研究………………………………谢绍飞，郑文书，王德军　3-43
Reducing the Damage of Tuyere Small Sleeves in Shougang Tonggang Blast Furnace Technical Study
……………………………………………………… XIE Shaofei, ZHENG Wenshu, WANG Dejun

南钢铁区质量一体化智能管控模型的应用与实践………………………方　磊，诸　刚，刘永辉　3-43
Application and Practice of Quality Integration Intelligent Management Model in Nangang Iron District
………………………………………………………………… FANG Lei, ZHU Gang, LIU Yonghui

焦炭在 H_2O+CO_2 气氛中气孔结构的溶损演化行为 ………孙　章，韩嘉伟，窦明辉，吕叶飞，付佳旭　3-44

高炉主沟温度运行状态监控及分析…………………………谢　皓，颜秋余，雷　磊，李自帅，姚元辉　3-45

煤粉在小套内运动轨迹的模拟研究 ································· 孙俊杰　3-46
Simulation Study of Pulverized Coal Movement Trajectory in the Tuyere ············· SUN Junjie
武钢 4 号高炉降低燃料比操作实践 ························ 金　锁，郑剑洪　3-46
Operation Practice of Fuel Ratio Reduction in No. 4 Blast Furnace of WISCO
·· JIN Suo, ZHENG Jianhong
高炉风口回旋区传热/传质行为及动力学机理的 DEM-CFD 研究 ······· 李　萌，安希忠　3-47
基于高炉风口物料特征及炉缸温度测量判断炉缸活跃性
······················ 王元生，张庆喜，郑华伟，鲍俊芳，陈令坤，任玉明　3-48
Judging the Activity of Hearth Based on the Characteristics of Tuyere Material in Blast Furnace and
　Temperature Measurement of Furnace Hearth ················ WANG Yuansheng, ZHANG Qingxi,
　　　　　　　　　　　　　　ZHENG Huawei, BAO Junfang, CHEN Lingkun, REN Yuming
高炉回旋区内煤粉燃烧行为的三维数值模拟研究 ············ 吴　丽，温良英，张生富　3-49
玉钢高炉提高钒钛铁水质量实践 ································· 杨　敬　3-49
Yugang Blast Furnace to Improve the Quality of Vanadium Iron Quality Practice ········· YANG Jing
玉钢炼铁厂钒钛矿冶炼炉前技术进步 ································· 杨　敬　3-50
Technical Progress in Front of Vanadium Titanium Ore Smelting Furnace of Yugang Iron Making Plant
·· YANG Jing
小颗粒烧结矿在无钟高炉中的优化利用模拟研究
······················ 曾　旺，滕　飞，刘　放，倪　俊，周　恒，吴胜利，寇明银　3-51
面向闭环的高炉加减风智能调控系统开发与应用 ······· 闫炳基，国宏伟，牟原金，梁合兰　3-52
基于智能感知技术的高炉炉前可视化平台开发与搭建 ······· 国宏伟，陈令坤，梁合兰，闫炳基，许　浩　3-53
钢铁企业原料场绿色高效储运技术的研究 ································· 毕　琳　3-53
矿焦耦合冶金性能：1 试验及评价方法 ········ 汪　琦，宋阳升，李廷乐，李哲熙，EVANS Tim　3-54
Ore and Coke Coupling Metallurgical Properties: Part 1 Test and Evaluation Method
························· WANG Qi, SONG Yangsheng, LI Tingle, LI Zhexi, EVANS Tim
昆钢 2 号高炉铁口喷溅治理实践 ························ 陈芳林，麻德铭　3-55
Practice of Treating Iron Spitter at the Iron Mouth of Kunming Steel's 2# Blast Furnace
·· CHEN Fanglin, MA Deming
大型高炉低碳冶炼策略之提高顶压 ············ 储　健，徐　云，陈庆林，张　南，代　兵　3-56
Increased Top Pressure as a Low-Carbon Smelting Strategy for Large Blast Furnaces
························· CHU Jian, XU Yun, CHEN Qinglin, ZHANG Nan, DAI Bing
昆钢 1 号高炉开炉及达产生产实践 ························ 卢郑汀，胡玉清，王　楠　3-57
Opening and Production Practice of 1 # Blast Furnace in Kunming Steel
·· LU Zhengting, HU Yuqing, WANG Nan
沙钢高炉经济高效护炉技术研究与应用 ············ 雷　鸣，杜　屏，魏红超，焦克新，张建良　3-57
Research and Application of Economical and Efficient Blast Furnace Maintenance Technology of Shasteel
················ LEI Ming, DU Ping, WEI Hongchao, JIAO Kexin, ZHANG Jianliang
H_2 对钒钛球团软熔滴落行为的影响研究
······················ 马凯辉，方云鹏，刘凌岭，唐文博，张立利，蒋　胜，陈　茅　3-58
泰钢高炉长寿生产实践与技术探讨 ············ 薛玉卿，陈培敦，李　胜，王学军，孙其德，王　涛　3-59

Discussion on Practice and Technology of Blast Furnace Longevity Production in Taishan Steel
·················· XUE Yuqing, CHEN Peidun, LI Sheng, WANG Xuejun, SUN Qide, WANG Tao

泰钢一号高炉长期焖炉快速恢复实践 ·················· 薛玉卿，李　胜，孙其德，王　涛，韩传宝　3-60

The Practice of Fast Recovery of Long-term Stewed Furnace in No. 1 Blast Furnace of Taishan Steel
·················· XUE Yuqing, LI Sheng, SUN Qide, WANG Tao, HAN Chuanbao

MgO 对包钢高炉渣系流变特性及熔化性的影响
·················· 冯　聪，王雅军，李玉柱，白晓光，郑占斌，于恒亮　3-61

Effects of MgO on Viscous and Melting Behaviors of Blast Furnace Slag in Baotou Steel
·················· FENG Cong, WANG Yajun, LI Yuzhu, BAI Xiaoguang, ZHENG Zhanbin, YU Hengliang

高炉布料落点精准预测系统的应用实践 ·················· 李　胜，薛玉卿，杨立春　3-61

Application Practice of Accurate Prediction System of Blast-furnace Distribution Drop Point
·················· LI Sheng, XUE Yuqing, YANG Lichun

高温对焦炭宏观性能和微观结构的影响 ·················· 付晓微，何志军，高立华，杨立春　3-62

3.3　非高炉炼铁

HIsmelt 渣铁间硅分配比共存理论预报模型研究 ·················· 李　林，张福明，郭汉杰　3-63

Study on IMCT Prediction Model of Silicon Distribution Ratio for Slag-metal of HIsmelt
·················· LI Lin, ZHANG Fuming, GUO Hanjie

HIsmelt 渣铁间磷分配比共存理论预报模型研究 ·················· 李　林，张福明，郭汉杰，林　松，徐海如　3-64

Study on IMCT Prediction Model of Phosphorus Distribution Ratio for Slag-metal of HIsmelt
·················· LI Lin, ZHANG Fuming, GUO Hanjie, LIN Song, XU Hairu

氨气还原不同种类铁矿石的低温冶金性能研究 ·················· 刘玥君，李先春，李　丽，林建廷　3-64

氨气直接还原铁矿石的冶金性能研究 ·················· 刘玥君，李先春，李　丽，林建廷　3-65

绿电低温电解铁 ·················· 杨海涛，王　鑫，谭钰桦，程佳鑫，朱庆山　3-66

基于 HIsmelt 工艺 $CaO-SiO_2-MgO-Al_2O_3-TiO_2-FetO$ 渣系脱硫能力的研究
·················· 刘晏廷，刘　然，高艳甲，王小艾　3-67

八钢冶金煤气脱碳系统运行总结 ·················· 田　果，王　鑫，朱利刚，刘西坤　3-68

Operation Analysis of Metallurgical Gas Decarbonization System of Bayi Iron and Steel Co.
·················· TIAN Guo, WANG Xin, ZHU Ligang, LIU Xikun

基于焦炉煤气的直接还原技术污染物协同治理生产实践 ·················· 李晓兵　3-69

Production Practice of Collaborative Treatment of Pollutants Based on Direct Reduction Technology of Coke Oven Gas
·················· LI Xiaobing

$NaHCO_3$ 对铌精矿微波碳热还原的影响研究 ·················· 文　明，刘雅琦　3-69

4　炼钢与连铸

4.1　炼　钢

基于沸腾出钢工艺的 SPHC 钢渣洗脱磷研究 ·················· 李　超，黄　岩，尚德义，冉茂铎　4-1

Study on Slag-washing Phosphorus Elution of SPHC Steel Based on Boiling Steel Tapping Process
.... LI Chao, HUANG Yan, SHANG Deyi, RAN Maoduo

电感耦合等离子体发射光谱法测定钛铁中微量元素含量 ·················· 任玲玲，谭胜楠，戚振南   4-2
Determination of Silicon, Manganese, Phosphorus, Aluminum, Chromium, Zirconium in Ferrotitanium by
    Inductively Coupled Plasma Atomic Emission Spectrometry ···· REN Lingling, TAN Shengnan, QI Zhennan

真空脱气炉关键设备管理对钢水质量提升浅谈 ·················· 翟昱明，王玉翠，苏小宇   4-2
Study on Maintenance and Management of VD Vacuum Degasification Furnace for Steelmaking
.... ZHAI Yuming, WANG Yucui, SU Xiaoyu

重庆钢铁无取向硅钢 RH 精炼温度控制实践 ···· 文　敏，陈露涛，刘向东，姚前光，张　俊，吴　伟   4-3

洁净钢生产现状及钢包引流新工艺研究
    ······· 李德军，杨　光，金　喆，李博洋，张维维，黄玉平，栗　红，赵成林   4-3
Clean Steel Production Status and New Process of Ladle-tapping
.... LI Dejun, YANG Guang, JIN Zhe, LI Boyang, ZHANG Weiwei,
                                                         HUANG Yuping, LI Hong, ZHAO Chenglin

转炉高废钢比冶炼的关键单元技术进展 ········· 杨　光，李德军，李博洋，赵成林，金　喆，朱晓雷   4-4
Development of Key Unit Technology with High Scrap Ratio in Converter Steelmaking
.... YANG Guang, LI Dejun, LI Boyang, ZHAO Chenglin, JIN Zhe, ZHU Xiaolei

管线钢精炼渣物相及岩相组成对炉渣性能的影响研究 ···· 吕春风，滕行泽，尚德礼，廖相巍，赵成林   4-5
Study on the Influence of Physical and Petrographic Composition of Slag on Slag Properties in Pipeline
    Steel Refining ············· LV Chunfeng, TENG Xingze, SHANG Deli, LIAO Xiangwei, ZHAO Chenglin

转炉吹炼过程锰元素反应规律的研究 ············ 赵成林，王　鹏，黄玉平，齐志宇，张维维，王一名   4-6
Study of Manganese Reaction Rule at Different Blowing Stages in Converter
············ ZHAO Chenglin, WANG Peng, HUANG Yuping, QI Zhiyu, ZHANG Weiwei, WANG Yiming

基于出钢冶金与渣冶金的低成本精炼工艺 ························ 汪洪峰，李月林，贾军艳   4-6
Low Cost Refining Process Based on Tapping and Slag Metallurgy
.... WANG Hongfeng, LI Yuelin, JIA Junyan

炼钢厂降低铁合金消耗的工艺措施 ··· 李　京   4-7
Technological Measures for Reducing Ferroalloy Consumption in Steel Making Plant ················LI Jing

富氢熔态还原炼钢工艺路径下钢水渗氢及脱氢研究 ················ 李明明，杨　宇，邵　磊，邹宗树   4-8

高废钢比对转炉炼钢工艺影响及其对策 ···································· 李伟东，王金辉，何海龙   4-8
Effect of High Scrap Ratio on Smelting Process by Converter and Relevant Countermeasures
.... LI Weidong, WANG Jinhui, HE Hailong

基于高清彩色 CCD 相机的 RH 炉内监控与测温系统 ···· 李保华，周东东，徐　科，雷雨田，化越飞   4-9
Furnace Monitoring and Thermometry System of RH Refining Based on High-definition Color CCD Camera
.... LI Baohua, ZHOU Dongdong, XU Ke, LEI Yutian, HUA Yuefei

基于共存理论的精炼熔渣脱硫热力学模型研究 ······························ 苏建铭，孙　群，温荣宇   4-10
Thermodynamic Model for Desulfurization of Refining Slag Based on Coexistence Theory
.... SU Jianming, SUN Qun, WEN Rongyu

稀土 Ce 对 NM450 钢中夹杂物的影响 ················· 郭树豪，刘金瑞，高　军，樊立峰，崔　衡   4-10

转炉底吹 O_2-CO_2-石灰粉工艺的应用实践研究 ···· 王春阳，董　凯，朱　荣，赵长亮，关顺宽，丁国慧   4-11
The Application Practice of Bottom-blown O_2-CO_2-CaO Process in Steelmaking
······· WANG Chunyang, DONG Kai, ZHU Rong, ZHAO Changliang, GUAN Shunkuan, DING Guohui

冷硬产品夹杂缺陷分析及质量控制 ·························· 辛　鑫，付　薇，高洪涛，黄秋菊   4-12

Analysis and Quality Control of Inclusion Defects in Cold Hard Products
································· XIN Xin, FU Wei, GAO Hongtao, HUANG Qiuju

浅析我国钢包用耐火材料的品种及应用 ··· 李仕龙　4-13

60t 脱磷站-AOD 炉-LF-CCM 生产 1Cr13 钢低成本生产实践
································· 冯文甫，张育明，范建英，郭志彬，阎丽珍　4-13

Low Cost Production Practice of 1Cr13 Steel Produced by 60t DEP-AOD-LF-CCM
································· FENG Wenfu, ZAHNG Yuming, FAN Jianying, GUO Zhibin, YAN Lizhen

稀土—碳材质浸入式水口连铸稀土钢的应用实践 ··················· 谌智勇，王小军，任　强，智建国　4-14

Application Practice of Rare-earth Carbon Immersion Continuous Casting of Rare-earth Steel
································· CHEN Zhiyong, WANG Xiaojun, REN Qiang, ZHI Jianguo

稀土处理对不同钢种冲击韧性的影响 ··· 隋亚飞，赵　如，罗　钢，梁　亮　4-15

Effect of Rare Earth Treatment on Impact Energy of Different Steel Grades
································· SUI Yafei, ZHAO Ru, LUO Gang, LIANG Liang

转炉熔池火点区传热行为模拟研究 ··· 蒋　睿，孙建坤，刘　青　4-15

Simulation Study on Heat Transfer Behavior in Hot Spot Zone of Converter Molten Bath
································· JIANG Rui, SUN Jiankun, LIU Qing

120t 转炉高效冶炼提升生产实践 ··················· 罗焕松，周　甫，曾维侃，吴振坤，刘春辉　4-16

Production Practice of Efficient Smelting and Upgrading of 120 tons Converter
································· LUO Huansong, ZHOU Fu, ZENG Weikan, WU Zhenkun, LIU Chunhui

钢包透气砖防钢液渗漏的数值模拟 ··· 娄文涛，张加毓，朱苗勇　4-17

稀土镁处理顺序和钙预处理对钢中稀土夹杂物特征的影响
································· 任泽斌，陈　凯，李　艳，唐　鑫，郑立春，姜周华　4-18

Effect of Rare Earth Magnesium Treatment Sequence and Calcium Pretreatment on the Characteristics of
Rare Earth Inclusions in Steel
································· REN Zebin, CHEN Kai, LI Yan, TANG Xin, ZHENG Lichun, JIANG Zhouhua

不同赋存状态下稀土对钢力学性能的影响机理研究 ··················· 刘香军，杨昌桥，杨吉春，任慧平　4-19

转炉炉膛结构对熔池流动特性影响研究 ··················· 刘福海，朱　荣，董　凯，魏光升，程　钰　4-21

板坯中间包用耐火材料长寿技术 ··················· 董战春，朱静然，白晓卫，王　兴，刘根胜，章荣会　4-21

Long-life Technology of Refractory Materials for Slab Tundish
································· DONG Zhanchun, ZHU Jingran, BAI Xiaowei, WANG Xing, LIU Gensheng, ZHANG Ronghui

RH 单工位双钢包车工艺布置的工程实践 ··· 丁永锷　4-22

The Layout Solution of Single Station RH with Double Ladle Car in Real Project ················ DING Devin

含 TiO_2 精炼渣对 MgO 耐火材料的侵蚀行为研究 ··················· 郝广御，邓志银，魏春新，朱苗勇　4-23

钙处理工艺对车轴钢中夹杂物的影响 ··················· 郭俊波，沈　昶，杨克枝，陆　强　4-24

Effect of Calcium Treatment on Non-metallic Inclusions in Railway Axle Steel
································· GUO Junbo, SHEN Chang, YANG Kezhi, LU Qiang

浅谈 RH 液压系统冲洗方案 ··· 雷丛卉，宁　博　4-24

Discussion on the Flushing Plan of RH Hydraulic System ················ LEI Conghui, NING Bo

邢钢 50t 转炉改善挡渣效果实践 ··················· 韩清连，张育明，霍志斌，和红杰，李智刚，曹红波　4-25

Practice of Improving Slag Blocking Effect in 50t Converter of Xingtai Iron and Steel Co., Ltd.
································· HAN Qinglian, ZHANG Yuming, HUO Zhibin, HE Hongjie, LI Zhigang, CAO Hongbo

RH-BPI 钢液流动行为数值分析 ··· 刘　壮，娄文涛，朱苗勇　4-26

超低碳钢夹杂物的影响因素分析与改进 ··· 舒宏富，赵傲南　4-27
Analysis and Improvement of Influence Factors of Inclusion in Ultra Low Carbon Steel
··· SHU Hongfu, ZHAO Aonan

210t 转炉强供氧冶炼工艺技术研究与应用 ··· 王鹏飞　4-27
Research and Application of Intense Oxygen Supply Smelting Technology in 210 tons Converter
·· WANG Pengfei

二次燃烧氧枪下顶吹转炉气相区燃烧特征数值研究 ··················· 董澎源，郑淑国，朱苗勇　4-28

水平入料不锈钢-碳钢固液复合工艺技术研发 ··············· 吴恩旭，刘益民，赵金龙，纪　红　4-29
Solid-Liquid Composite Process Technological Development of Stainless Steel Clad Carbon Steel
··· WU Enxu, LIU Yimin, ZHAO Jinlong, JI Hong

新型狭缝-弥散透气砖组合高效底吹工艺 ··························· 李向龙，王德永，屈天鹏，陈　刚　4-30
High-efficiency Bottom Blowing Technology Through Slot-porous Matched Dual Tuyeres
··· LI Xianglong, WANG Deyong, QU Tianpeng, CHEN Gang

外部加入纳米 TiO_2 对 HRB400 夹杂物性质的影响研究 ········· 程　翔，李　慧，赵煜昊，孔　辉　4-31
The Effect of TiO_2 Nanopowder Adding on Inclusion Properties of HRB400
··· CHENG Xiang, LI Hui, ZHAO Yuhao, KONG Hui

转炉冶炼过程增硫现象研究与控制 ··· 赵明哲　4-31
Research and Control of Sulfur Increasing Phenomenon in Converter Smelting Process ····· ZHAO Mingzhe

钢中硫化物形貌控制 ··· 张立峰，周秋月，高小勇　4-32

钢液精炼过程钢中夹杂物成分和数量演变 ·· 王举金，张立峰　4-33

钢中非金属夹杂物变形的新认识 ·· 杨　文，张立峰　4-34

不锈钢精炼过程中渣-钢-夹杂物反应机理研究 ·· 任　英，张立峰　4-35

稀土对铝脱氧钙处理低合金耐磨钢中非金属夹杂物的影响 ·········· 郑海翔，任　强，张立峰　4-37
Effect of Rare Earth on Non-metallic Inclusions in Al Deoxidized Ca Treated Low Alloy Wear Resistant Steel
··· ZHENG Haixiang, REN Qiang, ZHANG Lifeng

第一性原理计算钢中夹杂物腐蚀性能的研究 ··· 张　静，张立峰　4-37

含钇不锈钢中 TiN 在 Y_2O_3 表面析出机理研究 ···················· 徐晓峥，张　静，张立峰　4-38

基于神经网络技术的精炼过程中钢液终点硫含量预测 ···························· 王伟健，张立峰　4-39

钢中常见元素固溶对钢的力学性能影响研究 ······························ 孙　莹，张　静，张立峰　4-40
Effect of Common Elements in Steel Dissolved in Steel on Mechanical Properties
··· SUN Ying, ZHANG Jing, ZHANG Lifeng

钛锆复合处理钢中夹杂物诱导针状铁素形核研究 ···················· 姚　浩，刘承军，张立峰　4-41

高纯铁素体不锈钢 TiN 夹杂物控制研究 ························· 周全磊，任　英，杨　文，张立峰　4-42

钢液中夹杂物界面现象研究 ·· 段豪剑，张立峰　4-44

Nb-Ti 元素对铁素体不锈钢中 Fe 基合金性能影响研究 ··················· 张世锟，张　静，张立峰　4-45
Effect of the Nb-Ti Elements on the Properties of Fe-based Alloys in Ferritic Stainless Steel
··· ZHANG Shikun, ZHANG Jing, ZHANG Lifeng

稀土钆及热处理对 C70S6 钢种拉伸性能的影响 ··························· 卫　红，任　强，张立峰　4-46
Effect of Rare Earth Gd and Heat Treatment on Tensile Properties of C70S6 Steel
··· WEI Hong, REN Qiang, ZHANG Lifeng

钛处理对易切削钢中硫化物的影响 ······································ 焦亚兴，高小勇，任　强，张立峰　4-47

Effect of Titanium Treatment on Sulfides in Free Cutting Steel
·················· JIAO Yaxing, GAO Xiaoyong, REN Qiang, ZHANG Lifeng

界面反应对硅硫钢与氧化钙接触角的影响·················· 马天浩，任　强，张立峰　4-47

Effect of Interfacial Reaction on the Contact Angle Between Si-S Steel and Calcium Oxide
·················· MA Tianhao, REN Qiang, ZHANG Lifeng

U71Mn 重轨钢中氧化物夹杂物对 MnS 评级结果的影响·················· 高梓淇，任　强，张立峰　4-48

Effect of Non-metallic Inclusions in U71Mn Heavy Rail Steel on MnS Rating Results
·················· GAO Ziqi, REN Qiang, ZHANG Lifeng

304 不锈钢中 MnO·Cr_2O_3-MnS 夹杂物耐点蚀性能研究·················· 刘　震，任　强，张立峰　4-49

Study on Pitting Resistance of MnO·Cr_2O_3-MnS Inclusions in 304 Stainless Steel
·················· LIU Zhen, REN Qiang, ZHANG Lifeng

304 不锈钢中 Mn-Cr-O 夹杂物的形成机制·················· 齐江涛，王志军，李院高，伊　璞　4-50

The Formation Mechanism of Mn-Cr-O Inclusions in 304 Stainless Steel
·················· QI Jiangtao, WANG ZhiJun, LI Yuangao, YI Pu

炉壁氧枪垂直角对电弧炉炼钢影响的模拟研究·················· 曾召鹏，王　耀，杜习乾，马建超　4-51

Simulation Research on Effect of Vertical Angle of Wall Oxygen Lance on EAF Steelmaking
·················· ZENG Zhaopeng, WANG Yao, DU Xiqian, MA Jianchao

富锰渣的副产品高效利用的试验研究·················· 吴　伟，杨　勇，赵　斌，赵　博，梁　强，赵进宣　4-52

Study on Efficient Use of By-products for Rich Manganese Slag
·················· WU Wei, YANG Yong, ZHAO Bin, ZHAO Bo, LIANG Qiang, ZHAO Jinxuan

双渣法降低铁钢比的实践探索·················· 王少波，鹿　焱，刘　震，谷茂强，朱苗勇　4-53

Exploration of Reducing Iron-to-Steel Ratio Through Double-Slag Method
·················· WANG Shaobo, LU Yan, LIU Zhen, GU Maoqiang, ZHU Miaoyong

100t 转炉低耗高效长材炼钢技术生产实践·················· 张朝发，李双武，张　垚，李　硕，张宝景　4-53

Production Practice of Steelmaking Technology with Low Consumption and High Efficiency for Long Product in 100 ton Converter ·················· ZHANG Chaofa, LI Shuangwu, ZHANG Yao, LI Shuo, ZHANG Baojing

转炉炼钢渣中微观渣相对低磷钢冶炼的影响·················· 吕延春，李海波，刘　洋，王海宝，孔祥涛　4-54

冶金模型定义的数字化炼钢连铸工艺·················· 李立勋，田　伟，李　杨，刘善喜，孙　博，徐　政　4-55

Digital Steelmaking and Continuous Casting Process Defined by Metallurgical Model
·················· LI Lixun, TIAN Wei, LI Yang, LIU Shanxi, SUN Bo, XU Zheng

炼钢炉渣总量对稀土收得率影响·················· 张　胤，张　顺，习望才，张怀军，曹瑞峰，韩春鹏　4-56

Effect of Total Amount of Steelmaking Slag on Yield of Rare Earth
·················· ZHANG Yin, ZHANG Shun, DIAO Wangcai, ZHANG Huaijun, CAO Ruifeng, HAN Chunpeng

基于时空分辨成像光谱分析的炼钢供氧火点区在线监测技术与研究·················· 孟令辉，董　凯　4-57

Research and Development of Oxygen Injection Fire Zone in Steelmaking Online Monitoring Technology Based on the Spatio-temporal Resolution Imaging and Spectroscopy Analysis
·················· MENG Linghui, DONG Kai

Mg 脱氧的氧化物冶金技术·················· 杨　健，李婷婷，徐龙云，潘晓倩，张银辉　4-57

Oxide Metallurgy Technology with Mg Deoxidization
·················· YANG Jian, LI Tingting, XU Longyun, PAN Xiaoqian, ZHANG Yinhui

降低 120t 转炉炉渣铁含量的研究·················· 乔西亚，韩　啸，何志军，杨　鑫　4-58

高品质特殊钢冶炼恒稳定工艺技术研究与应用·················· 杨　平，王海达，陈　列，李　亮，胡金海　4-59

Research and Application of Constant Stability Production Technology for High Carbon Alloy Steel Smelting
　　…………………………………………… YANG Ping, WANG Haida, CHEN Lie, LI Liang, HU Jinhai
高碳合金钢矩形坯均质化技术研究与应用 ……… 王海达，陈　列，张亚楠，杨　平，范世强，杨　铭　　4-60
Research and Application of Homogenization Technology for High Carbon Alloy Steel Rectangular Billet
　　………………………… WANG Haida, CHEN Lie, ZHANG Ya'nan, YANG Ping, FAN Shiqiang, YANG Ming
钢中夹杂物稳定区域图的绘制方法、典型案例与重要问题 ………… 王宏坡，康　健，王　雨　　4-61
基于 Python 的转炉合金化控制系统开发
　　……………………………………… 胡倩倩，董晓雪，朱海琳，施树蓉，韩　啸，张友建，何志军　　4-62
精炼渣氧化性对 Y15 硫易切削钢中夹杂物的影响
　　…………………………………………… 王桂林，刘丽霞，彭　军，张　芳，刘　爽，安胜利　　4-62
稀土 Ce 对 20CrMnTi 钢中 TiN 夹杂析出影响热力学计算及实验研究
　　……………………………………………………… 彭　军，王　健，张　芳，刘　爽，安胜利　　4-63
Thermodynamic Calculation and Experimental Study on the Effects of Rare Earth Ce on TiN Inclusion
　　Precipitation in 20CrMnTi Steel ……… PENG Jun, WANG Jian, ZHANG Fang, LIU Shuang, AN Shengli
Si 添加对高 Mn-N 型 TWIP 不锈钢力学性能的影响 ………………………………………… 曾泽瑶　　4-64
Effect of Si Addition on Mechanical Properties of High Mn-N Type TWIP Stainless Steel ……ZENG Zeyao
水分子在稀土元素掺杂铁表面的吸附研究 ……………………………………………………… 雷　超　　4-65
A Study about the Adsorption of H_2O on the Surface of Fe with Rare Earth Doping ………… LEI Chao

4.2　连　铸

基于铸坯质量提升关键设备精度校准浅谈 ……………………………… 翟昱明，王玉翠，何文浩　　4-65
Discussion on Accuracy Maintenance of Continuous Caster …… ZHAI Yuming, WANG Yucui, HE Wenhao
方坯定重剪切工艺优化与应用 …………………………………… 胡铁军，贺保堂，李光辉，李子辉　　4-66
中厚船板拉伸试样断口分层缺陷分析 ………… 康　伟，田永久，赵启斌，苏小利，方恩俊，廖相巍　　4-66
Analysis on Fracture Delamination Tensile Test of Medium Thick Ship Plate Steel
　　………… KANG Wei, TIAN Yongjiu, ZHAO Qibin, SU Xiaoli, FANG Enjun, LIAO Xiangwei
连铸结晶器智能预报系统开发 ……… 张维维，王丽娟，赵成林，陈　东，栗　红，黄玉平，李德军　　4-67
Development of Intelligent Prediction System for Continuous Casting Mold
　　…………………………… ZHANG Weiwei, WANG Lijuan, ZHAO Chenglin, CHEN Dong,
　　　　　　　　　　　　　　　　　　　　　　　　　　LI Hong, HUANG Yuping, LI Dejun
150 方弹簧钢连铸坯负偏析的研究与控制 ……………………… 孙俊喜，靳国兵，陈　涛，李　杰　　4-68
Research and Control of Negative Segregation in 150 Square Spring Steel Continuous Casting Billets
　　………………………………………………………… SUN Junxi, JIN Guobing, CHEN Tao, LI Jie
C-Mn-Nb-Cr 管线钢高温力学性能研究 ………………… 张英慧，黄明浩，孔祥磊，王　杨，刘　干　　4-69
Study on Mechanical Properties of C-Mn-Nb-Cr Pipeline Steel at High Temperatures
　　………………………… ZHANG Yinghui, HUANG Minghao, KONG Xianglei, WANG Yang, LIU Gan
RAMON 智能出坯系统 ………………………………………………… 谢卫东，王　彬，蒋　琼　　4-70
RAMON Intelligent Casting System …………………… XIE Weidong, WANG Bin, JIANG Qiong
基于小波分析的结晶器液面波动特征解析与控制 ……………………… 崔　衡，王振东，高　煜　　4-70
湘钢特厚板坯连铸技术研究进展 ……… 徐龙云，蒙　刚，师　文，马仲新，邢银超，张明志，杨建华　　4-71
鞍钢高品质汽车用钢板坯新缺陷的研究 ………… 金宏斌，陈　宇，刘佳铭，宋　宇，高洪涛，杜　林　　4-72

Research of New Defects of Ansteel High Quality Automobile Steel Slab
　　……………………… JIN Hongbin, CHEN Yu, LIU Jiaming, SONG Yu, GAO Hongtao, DU Lin
减少冷轧低碳钢轧材表面夹杂缺陷的工业实践……………………杜 林，孙 群，苏建铭，王一名　4-73
Industrial Practice of Reducing Shell Defects on Surface of Cold Rolled Low Carbon Steel
　　……………………………………………… DU Lin, SUN Qun, SU Jianming, WANG Yiming
轮辐钢中心裂纹原因分析及工艺改进…………………………宋 宇，杜 林，金宏斌，陈柏宇　4-73
Cause Analysis of Central Cracks in Wheel Arm Steel and Craft Improvement
　　………………………………………………… SONG Yu, DU Lin, JIN Hongbin, CHEN Baiyu
连铸中间包冲击区不同控流装置的对比研究…………………杨成虎，张江山，刘 青，秦邦明　4-74
基于机器视觉的异形坯表面缺陷在线检测系统
　　……………………………………秦 观，杜昕旻，李保华，张学民，赵立峰，徐 科　4-75
Online Surface Defect Detection System of Hot Profiled Billets Based on Machine Vision
　　……………………………… QIN Guan, DU Xintao, LI Baohua, ZHANG Xuemin, ZHAO Lifeng, XU Ke
中间包内衬冲蚀及夹杂物生成和演化的研究…………王 强，谭 憧，刘 畅，贺 铸，李光强　4-76
Numerical Simulation on Refractory Flow-Induced Erosion and Inclusion Formation in Continuous Casting
　　Tundish…………………… WANG Qiang, TAN Chong, LIU Chang, HE Zhu, LI Guangqiang
板坯连铸结晶器内气泡捕获夹杂物的数值模拟
　　………………………… 刘 畅，张立峰，DAS Lipas，THOMAS Brian G.，王 强，李光强　4-77
Numerical Simulation on Bubbles Capturing Inclusions in Slab Continuous Casting Mold
　　…………… LIU Chang, ZHANG Lifeng, DAS Lipsa, THOMAS Brian G., WANG Qiang, LI Guangqiang
中间包开浇阶段湍流抑制器应力变化行为的数值模拟研究
　　………………………………………… 林 鹏，金 焱，甘菲芳，刘子钰，李晓婷，吴健舟　4-78
Numerical Simulation Study on Stress Change Behavior of the Turbulent Inhibitor during the Opening Stage
　　………………………… LIN Peng, JIN Yan, GAN Feifang, LIU Ziyu, LI Xiaoting, WU Jianzhou
连铸工艺机理数字化及其发展趋势…………………………………………刘和平，仇 灏，杨 丽　4-78
关于低频电磁场下低反应性连铸保护渣吸收 Al_2O_3 夹杂的研究…………白 博，王 雨，丁昭霖　4-79
Study of the Absorption of Al_2O_3 Inclusion by Low Reactivity Mold Flux under Low Frequency
　　Electromagnetic Field……………………………………… BAI Bo, WANG Yu, DING Zhaolin
20CrMnTi 连铸凝固条件下的 TiN 析出行为研究…………………王卫领，商庭瑞，罗 森，朱苗勇　4-80
重型异型坯铸机漏钢原因与解决对策……………李静文，王金坤，吴耀光，刘建坤，朱卫群，张学森　4-81
Measures and Practices of Breakout Prevention of Heavy Beam Blank
　　……………… LI Jingwen, WANG Jinkun, WU Yaoguang, LIU Jiankun, ZHU Weiqun, ZHANG Xuesen
连铸非稳态过程板坯高洁净度综合控制技术……刘 洋，朱志远，谢翠红，王卫华，季晨曦，赵 晶　4-82
板坯连铸结晶器铜板热/力学行为数值模拟研究………………………姜稀银，蔡兆镇，朱苗勇　4-84
Numerical Simulation of Thermal-mechanical Behavior in Slab Continuous Casting Mold Copper Plate
　　………………………………………………… JIANG Jiyin, CAI Zhaozhen, ZHU Miaoyong
中间包功能稀土材料的开发………………………谌智勇，任 强，王小军，智建国，任 磊　4-85
Development of Rare Earth Oxides in Tundish Functional Material
　　………………………………… CHEN Zhiyong, REN Qiang, WANG Xiaojun, ZHI Jianguo, REN Lei
低碳钢板坯连铸结晶器液位瞬时异常波动时频特征解析
　　………………………………………… 孟晓亮，罗 森，周业连，王卫领，朱苗勇　4-86
板坯连铸结晶器电磁搅拌工艺参数优化研究…………卢海彪，雷作胜，钟云波，程常桂，任维丽　4-87

湍流抑制器对两流中间包流场影响的模拟研究
……………………………………刘 崇，刘 纲，张彩东，丁志军，李 杰，田志强　4-87
Simulation Study on the Effect of Turbulence Suppressor on Flow Field in Two-strand Tundish
……………………LIU Chong, LIU Gang, ZHANG Caidong, DING Zhijun, LI Jie, TIAN Zhiqiang

连铸大方坯点状偏析的轻压下调控机制………王海杰，王 璞，汤群伟，陈 刚，李伟涛，张家泉　4-88

重轨钢氢含量控制研究……………………………韩元庭，潘艳华，徐国涛，白 静，刘 鹏，王彦林　4-89
Study on Hydrogen Content Control of Heavy Rail Steel
………………………HAN Yuanting, PAN Yanhua, XU Guotao, BAI Jing, LIU Peng, WANG Yanlin

小方坯电缆钢浸入式水口侵蚀机理分析及改善措施………刘 鹏，杨文清，韩元庭，张昌宁，白 静　4-90
Analysis and Countermeasures on Corrosion Mechanism of SEN in Cable Steel of Small Section Billet
………………………LIU Peng, YANG Wenqing, HAN Yuanting, ZHANG Changning, BAI Jing

H13多合金钢凝固热力学实验和模拟研究………………罗腾飞，王卫领，商庭瑞，罗 森，朱苗勇　4-90

结晶器微区电磁搅拌对大方坯内部质量的影响……………………………………佘勇军，牛 帅　4-91
Effect of Mold Modulated Wave Stirring on Internal Quality of Bloom ………… SHE Yongjun, NIU Shuai

基于大区域碳浓度分布预测的连铸坯低倍组织精细化评级系统
……………………………………侯自兵，郭坤辉，易 凯，彭治强，郭中傲，文光华　4-92
A Fine Grading System of Macrostructure of Continuous Casting Billet Based on Prediction of Carbon
　Concentration Distribution in Large Area
………………… HOU Zibing, GUO Kunhui, YI Kai, PENG Zhiqiang, GUO Zhongao, WEN Guanghua

原位观察下等温冷却对包晶钢凝固行为的影响
……………………………………王伟安，李小明，杨永坤，朱佳雨，王 阳，王建立　4-93
Effect of Isothermal Cooling on Solidification Behavior of Peritectic Steel under In-situ Observation
………… WANG Weian, LI Xiaoming, YANG Yongkun, ZHU Jiayu, WANG Yang, WANG Jianli

钢的热裂机理及扩展条件研究………………………………仲红刚，林增煌，赵 宇，李天宇，翟启杰　4-94

结晶器液压伺服调宽系统可靠性验证………………………雷丛卉，宁 博，刘 洪，徐学华，樊小强　4-94
Reliability Verification of Hydraulic Servo Width Adjustment System for Crystallizers
………………………………LEI Conghui, NING Bo, LIU Hong, XU Xuehua, FAN Xiaoqiang

水口结构改善大圆坯结晶器电磁偏心搅拌的数值模拟
……………………………………朱佳雨，杨永坤，王建立，王伟安，王 阳，李小明　4-95
Numerical Simulation of Nozzle Structure to Improve Eccentric Mold Electromagnetic Stirring in a Round
　Bloom ………… ZHU Jiayu, YANG Yongkun, WANG Jianli, WANG Weian, WANG Yang, LI Xiaoming

基于棒材直接轧制工艺的铸坯温度精准控制研究
………………方 明，张朝晖，习晓峰，郭红民，梁少鹏，薛 魁，王贺龙，吕 明　4-96
Research on Precise Control of Billet Temperature Based on Direct Rolling Technology of Bars
………………………………FANG Ming, ZHANG Zhaohui, XI Xiaofeng, GUO Hongmin,
　　　　　　　　　　　　　　　　　　　　LIANG Shaopeng, XUE Kui, WANG Helong, LV Ming

316L奥氏体不锈钢方坯残留铁素体特征研究……王 洋，张政睿，陈 超，刘 涛，闫琦吉，林万明　4-97
Characteristics of Residual Ferrite in 316L Austenitic Stainless Steel Billet
………………………WANG Yang, ZHANG Zhengrui, CHEN Chao, LIU Tao, YAN Qiji, LIN Wanming

异型坯单水口浇铸结晶器流场温度场特征及优化……………………彭治强，侯自兵，唐 萍，李 昊　4-98

基于铜浴法模拟料层厚度对保护渣熔化行为的影响…………陈富杭，文光华，唐 萍，张 义　4-99

连铸保护渣中炭黑在浆料中的分散行为研究……许彦可，唐 萍，郭 俊，文光华，李 昊，黄晓燕　4-100

钢液中氢含量对连铸结晶器传热影响的机理研究
················郭 俊，唐 萍，李 昊，黄晓燕，付 濯，文光华 4-101
镀锡板 T5 钢结晶器液面波动控制实践 ············ 肖同达，毛会营，张洪雷 4-102
Control Practice of Mould Liquid Level Fluctuation of T5 Tinplate
················XIAO Tongda, MAO Huiying, ZHANG Honglei
电流相位分布对克莱姆式电磁搅拌器搅拌效率的影响············孟纯涛，张 静，张立峰 4-102
连铸坯全凝固长度内夹杂物碰撞聚合与上浮去除的数值模拟研究············李琪蓝，王举金，张立峰 4-103
连铸坯全断面夹杂物成分分布耦合模型 ············ 张月鑫，王举金，张立峰 4-105
钢包底部改造对浇注末期下渣的影响研究 ············ 吕彬玉，张立峰 4-106
板坯结晶器开浇过程多相流动及凝固的数值模拟研究 ············ 李丰辰，陈 威，张立峰 4-107
板坯结晶器电磁搅拌对卷渣的影响 ············ 郑 甫，陈 威，张立峰 4-108
电磁制动技术研究进展与应用 ············ 肖玉宝，秦 哲，金百刚，郭庆涛，张凯伦，潘 栋 4-109
Research Progress and Application of the Electromagnetic Brake Technique
················XIAO Yubao, QIN Zhe, JIN Baigang, GUO Qingtao, ZHANG Kailun, PAN Dong
超大型异形坯连铸结晶器内凝固过程仿真············刘 艳，雷少武，李 伟，耿明山，王 颖 4-110
Simulation of Solidification in Mold for Large Beam Blank's Continuous Casting Process
················LIU Yan, LEI Shaowu, LI Wei, GENG Mingshan, WANG Ying
钢铁冶金过程的水跃行为研究············杨 泽，许浩峰，贾志扬，杨文城，兰文泽，王彦强，陈 进 4-111
Hydraulic Jump Behavior in Iron and Steel Metallurgical Process
······ YANG Ze, XU Haofeng, JIA Zhiyang, YANG Wencheng, LAN Wenze, WANG Yanqiang, CHEN Jin
脉冲电流对 321 不锈钢中 TiN 析出的影响
················梁尚东，魏光旭，肖紫轩，杨成成，朱正海，黄冬建，常立忠 4-112
Effect of Pulsed Current on TiN Precipitation in 321 Stainless Steel
················LIANG Shangdong, WEI Guangxu, XIAO Zixuan, YANG Chengcheng,
ZHU Zhenghai, HUANG Dongjian, CHANG Lizhong
五机五流非对称连铸中间包流场优化及特性研究及优化设计
················王海达，陈 列，成国光，杨 平，范世强 4-112
Optimization of Flow Field and Characteristics of Five-machine and Five-strand Unsymmetrical Continuous
Casting Tundish ············ WANG Haida, CHEN Lie, CHENG Guoguang, YANG Ping, FAN Shiqiang
高锰奥氏体低温钢凝固特性与微观裂纹敏感性 ············ 何 杨，徐 浩，刘建华 4-113
重轨钢连铸大方坯 V 型偏析探究 ············ 李红光，徐明丽 4-114
Study on the V-shape Segregation of Rail Steel Bloom ············ LI Hongguang, XU Mingli

5 电冶金与废钢铁

5.1 电 冶 金

电渣重熔精炼过程数值模拟研究进展············王 强，刘 畅，王 芳，贺 铸，李光强，李宝宽 5-1
Development of Numerical Simulation in Electroslag Metallurgy
················WANG Qiang, LIU Chang, WANG Fang, HE Zhu, LI Guangqiang, LI Baokuan
碳粉喷吹过程气-固混合喷吹动力学研究············刘 崇，刘 纲，王 强，李 杰，张彩东，丁志军 5-2

Study on the Dynamic Characteristics of Gas-solid Mixed Injection in Carbon Powder Blowing Process
······ LIU Chong, LIU Gang, WANG Qiang, LI Jie, ZHANG Caidong, DING Zhijun

Ce-Mg 协同处理对 M50 航空轴承钢中夹杂物演变的影响
······ 王礼超，田家龙，蒋成钢，任 吉，姜周华 5-3
Effect of Ce-Mg Synergistic Treatment on the Inclusion Evolution in M50 Aerospace Bearing Steel
······ WANG Lichao, TIAN Jialong, JIANG Chenggang, REN Ji, JIANG Zhouhua

D2 冷作模具钢中的碳化物形成机理及控制 ······ 李万明，刘峻辰，臧喜民，姜周华 5-3
Formation Mechanism and Control of Carbides in D2 Cold Work Die Steel
······ LI Wanming, LIU Junchen, ZANG Ximin, JIANG Zhouhua

新型 15Cr-30Ni 气阀合金电渣锭均质化控制 ······ 史成斌，张 淮，王世舟，李 晶 5-4

电炉回收废弃镁碳耐火材料用于造泡沫渣研究 ······ 宋生强 5-5
Recycling of Spent MgO-C Refractories as Slag Foaming Additives in EAF ······ SONG Shengqiang

Ce-Mg 处理对 H13 钢洁净度、组织和性能的影响 ······ 任 吉，田家龙，王礼超，蒋成钢，姜周华 5-6
Effect of Ce-Mg Treatment on Cleanliness, Microstructure and Properties of H13 Steel
······ REN Ji, TIAN Jialong, WANG Lichao, JIANG Chenggang, JIANG Zhouhua

基于随机森林的电弧炉终点参数预报研究
······ 陆泓彬，朱红春，姜周华，吴泽文，艾翔宇，杨 策，冯 浩，张树才 5-6
Random Forest-based Prediction of Electric Arc Furnace Endpoint Parameters
······ LU Hongbin, ZHU Hongchun, JIANG Zhouhua, WU Zewen,
AI Xiangyu, YANG Ce, FENG Hao, ZHANG Shucai

加压电渣重熔高氮不锈钢洁净度控制研究 ······ 杨守星，李花兵，冯 浩，姜周华，陈 敏 5-7
Cleanliness Control of High Nitrogen Stainless Steels during Pressurized Electroslag Remelting
······ YANG Shouxing, LI Huabing, FENG Hao, JIANG Zhouhua, CHEN Min

CERI S-Arc™ 复合型废钢预热电炉技术的可行性研究 ······ 张宇航，王 佳，王少臣，李佳辉 5-8

面向"碳中和"的中国电炉流程发展战略研究 ······ 上官方钦，崔志峰，周继程，倪 冰，李 涛 5-9

全废钢电炉-ESP 产线绿色高效冶炼关键技术研究实践 ······ 魏光升，李 挺，朱 荣，廖交志，李 欣 5-10
Research and Application on Green and Efficient Smelting Technology for EAF-ESP Production Process with 100% Scrap Steel ······ WEI Guangsheng, LI Ting, ZHU Rong, LIAO Jiaozhi, LI Xin

全废钢电弧炉尾渣回用造渣工艺研究 ······ 王 岩，李 晶，李 波，刘 燊 5-11

5.2 废钢铁预处理及应用

100t 全废钢 Consteel 电弧炉冶炼能量平衡与能效分析
······ 李龙飞，胡砚斌，张曦东，徐立军，孟华栋，王 明 5-12
Energy Balance and Energy Efficiency Analysis of 100t Consteel Electric Arc Furnace with All Scrap
······ LI Longfei, HU Yanbin, ZHANG Xidong, XU Lijun, MENG Huadong, WANG Ming

面向废钢高效高质利用的智能验质技术研究 ······ 刘丽冉，杨冬靓 5-12
Research on Intelligent Quality Inspection Technology for High-efficiency and High-quality Utilisation of Steel Scrap ······ LIU Liran, YANG Dongliang

铁碳熔池中废钢熔化行为的数值模拟研究 ······ 刘孟珂，马国军，徐 菊，张 翔，郑顶立 5-13
Numerical Simulation of Melting Behavior of Steel Scrap in Iron-carbon Bath
······ LIU Mengke, MA Guojun, XU Ju, ZHANG Xiang, ZHENG Dingli

废钢预热在提高转炉废钢比中的应用与发展 ······ 胡砚斌，黄 芳，侯中晓，何 赛，贾志立 5-15

Application and Development of Scrap Preheating Technology in Improving Scrap Ratio of Converter
　　……………………………………………… HU Yanbin, HUANG Fang, HOU Zhongxiao, HE Sai, JIA Zhili
废钢智能判级系统发展现状及未来趋势 ……………………………………………………… 赵东伟　5-15
Development Status and Future Trend of Scrap AI Grading System ……………… ZHAO Dongwei
基于机器视觉的废钢智能验质方法研究 ………………………………… 徐文广，肖鹏程，王杏娟　5-16
镀锌废钢锌铁分离机理研究 ………… 朱正海，孙前进，骆小刚，李　勇，邹海历，殷志宏，周　俐　5-17

6　轧制与热处理

6.1　板　材

固溶处理工艺对奥氏体不锈钢 S31603 组织性能的影响
　　………………………………………… 邢梦楠，胡昕明，姚　震，欧阳鑫，刘晨希，王　储　6-1
Effect of Solution Treatment on Microstructure and Properties of Austenitic Stainless Steel S31603
　　………………… XING Mengnan, HU Xinming, YAO Zhen, OUYANG Xin, LIU Chenxi, WANG Chu
提高重卷线薄规格运行速度和产能 …………………………………………………… 黄海生　6-2
热处理工艺对 690MPa 级容器钢组织转变的影响 ………… 刘晨希，胡昕明，欧阳鑫，邢梦楠，王　储　6-2
Effect of Heat Treatment Process on Microstructure Transformation of 690MPa Vessel Steel
　　………………… LIU Chenxi, HU Xinming, OUYANG Xin, XING Mengnan, WANG Chu
宝钢 5m 厚板轧机展宽 PVPC 设定模型优化 …………………………………… 张敏文　6-3
Design of Broadsizing PVPC Model Defect in 5m Heavy Plate Mill Baosteel ……… ZHANG Minwen
S6 轧机工作辊常见失效分析与再生技术研究 ……………………… 张　青，侯兴慧，罗　昌　6-3
Research on Common Failure Analysis and Regeneration Technology of S6 Rolling Mill Work Rolls
　　……………………………………………… ZHANG Qing, HOU Xinghui, LUO Chang
降低高强度热轧酸洗板材降级品率 ………… 李子文，杜宏松，卢　振，刘云飞，李艳杰，姬明利　6-4
Reduce the Degradation Rate of High Strength Hot Rolled Pickling Sheet
　　……………………… LI Ziwen, DU Hongsong, LU Zhen, LIU Yunfei, LI Yanjie, JI Mingli
镀锌极限规格产品生产稳定性的研究与改进 ……………………… 张振峰，谢苏源，刘海超　6-5
The Research and Improvement about the Production Stability of Galvanized Limit Specification Products
　　……………………………………… ZHANG Zhenfeng, XIE Suyuan, LIU Haichao
针对无头轧制产线的轧制模型开发和辊形配置研究
　　………………………………… 周冠禹，何安瑞，郭蓝田，李朝阳，李宏军，郭智勇　6-5
冷轧板边部表面色差原因分析及改进措施 ……… 何士国，李江委，王　存，王　杰，刘旺臣，宋运涛　6-6
Cause Analysis and Improvement Measures of Edge Surface Color Difference of Cold-rolled Sheet
　　………………… HE Shiguo, LI Jiangwei, WANG Cun, WANG Jie, LIU Wangchen, SONG Yuntao
550MPa 级热轧高强集装箱板形翘曲的控制研究
　　………………………………… 李江委，王　刚，王　杰，任俊威，王　存，宋运涛　6-7
Cause Analysis and Control Measures of Flatness Warping of 550MPa Grade High Strength Container Steel
　　………………… LI Jiangwei, WANG Gang, WANG Jie, REN Junwei, WANG Cun, SONG Yuntao
冷轧某产品的缺陷发生率差异研究 ……………………………………… 谷开峰，郑　涛　6-7
Analysis of Difference between Defect Occurrence Rate of a Cold-rolled Product
　　……………………………………………………………… GU Kaifeng, ZHENG Tao

浅谈砷元素对酸洗板表面的影响及改善措施 …………………………………………………… 陈　锦　6-8

冷轧机组入口卸套筒功能改进 ……………………… 姜　雨，段晓溪，张旭亮，吴建学，祁　静　6-8

热镀锌光整机高压清洗系统改造与应用 ………… 姜　雨，陈　瑶，王　良，闫彤彤，李文波，张森建　6-9

基于全面自动控制提升的连续退火模型功能优化研究

……………………………………………… 宋利伟，孙荣生，蔡顺达，刘军友，洪天生，姜丽丽　6-9

Research on Function Optimization of Continuous Annealing Model Based on Full Automatic Control Lifting

…………… SONG Liwei, SUN Rongsheng, CAI Shunda, LIU Junyou, HONG Tiansheng, JIANG Lili

低合金钢带钢辊弯开裂原因分析与热轧工艺控制 ………………………… 江　傲，程志诚，余　伟　6-10

中厚板产品钢坯处理技术的研究与应用实践 …………………… 张　健，黄　健，孙殿东，陈军平　6-11

Research and Application of Steel Slab Processing Technology for Medium and Heavy Plate Products

………………………………………… ZHANG Jian, HUANG Jian, SUN Diandong, CHEN Junping

宽薄规格控轧控冷钢板板形缺陷成因分析 ……… 张建平，韩　旭，韩千鹏，庞宗旭，黄　健，李新玲　6-12

Analysis of Shape Defects of Wide and Thin Gauge Controlled Rolling and Controlled Cooling Steel Plate

…………… ZHANG Jianping, HAN Xu, HAN Qianpeng, PANG Zongxu, HUANG Jian, LI Xinling

临界退火温度对中锰钢奥氏体稳定性和力学性能的影响 …………………………… 张哲睿，任　毅　6-13

Effect of Critical Annealing Temperature on Austenite Stability and Mechanical Properties of Medium

Manganese Steel ………………………………………………………………… ZHANG Zherui, REN Yi

酸洗段圆盘剪剪切质量的控制 ………………… 毛玉川，孙荣生，刘英明，张一凡，蔡顺达，王金星　6-14

The Control of the Cutting Quality of Rotary Shear on the Pickling Section

………… MAO Yuchuan, SUN Rongsheng, LIU Yingming, ZHANG Yifan, CAI Shunda, WANG Jinxing

不锈钢复合板 316L+Q500qE 的组织与性能的研究

……………………………………… 李中平，周文浩，史术华，王智聪，罗　登，熊祥江，范　明　6-15

Microstructure and Properties of 316L + Q500qE Stainless Steel Clad Plate Researching

 ·· LI Zhongping, ZHOU Wenhao, SHI Shuhua, WANG Zhicong, LUO Deng, XIONG Xiangjiang, FAN Ming

退火温度对 1500MPa 冷轧 Q&P 钢组织性能的影响

……………………………………………… 蔡顺达，孙荣生，宋利伟，刘军友，阮国庆，钟莉莉　6-16

Effects of Annealing Temperature on Microstructure and Property of Cold-rolled Q&P Steel with Tensile

Strength of 1500MPa

…………… CAI Shunda, SUN Rongsheng, SONG Liwei, LIU Junyou, RUAN Guoqing, ZHONG Lili

回火工艺对低合金高强耐磨钢组织与性能的影响 ………… 王万慧，夏强强，韩会全，辜蕾钢，刘显军　6-16

Effect of Tempering Process on Microstructure and Properties of Low-alloy High-strength Wear-resistant Steel

……………………………… WANG Wanhui, XIA Qiangqiang, HAN Huiquan, GU Leigang, LIU Xianjun

家电彩涂板表面麻点缺陷产生原因分析及预防措施 …… 汤晓东，谢义康，朱彰荣，吴银义，郑　磊　6-17

Analysis of the Causes and Preventive Measures for the Surface Pitting Defects of Household Appliances

………………………… TANG Xiaodong, XIE Yikang, ZHU Zhangrong, WU Yinyi, ZHENG Lei

可实现轴向变张力轧制的张力辊系统 ……………… 马晓宝，彭永恒，李　鹏，王　涛，陈　鹏　6-18

1420UCM 五机架连轧机振动抑制措施研究 ………… 黄　勇，龚　艺，张东方，胡岸松，陈立钢　6-19

Analyzing Reason and Controlling Measure of the Vibration in 1420 UCM Cold Continuous Rolling Mill

………………………………… HUANG Yong, GONG Yi, ZHANG Dongfang, HU Ansong, CHEN Ligang

减少热轧带钢头尾不规则形状延伸技术研究与实践 ……………………… 曾龙华，赵金凯，王　波　6-19

Research and Practice on Reducing Irregular Shape Extension of Hot Rolled Strip

………………………………………………………… ZENG Longhua, ZHAO Jinkai, WANG Bo

热轧钢卷端部缺陷自动识别技术研究与实践 ………………………………………… 曾龙华，高佳瑞　6-20
Research and Practice of Automatic Defect Identification Technology of Hot Rolled Steel Coil End
………………………………………………………………………………… ZENG Longhua, GAO Jiarui
退火工艺对高强 DR 材组织性能的影响 ……………………… 孙超凡，方　圆，孙　晴，王雅晴　6-21
回火温度对 500MPa 级高强钢屈强比和内应力的影响 ………………… 武凤娟，曲锦波，杨　浩　6-21
影响冷轧高强钢板形质量的性能因素 …………………………… 陈　飞，张　栋，孙翼洲，李高峰　6-22
硅对超高强轻质钢热变形行为影响的研究 ……………… 张振山，宋成浩，吴文源，王皓亮，孙振忠　6-23
热轧高强钢平整过程中残余应力的分析及应用 … 刘天武，张彩东，孙　力，潘　进，张志强，丁文红　6-23
Analysis and Application of Residual Stress on Hot Rolled High Strength Steel during Temper Passing Process
……………………… LIU Tianwu, ZHANG Caidong, SUN Li, PAN Jin, ZHANG Zhiqiang, DING Wenhong
高强度宽幅中厚钢板残余应力的低热-磁-振复合时效消减调控方法与机理研究
………………………………………………………… 宋和川，张清东，周晓敏，张勃洋　6-24
传统热连轧变厚轧制功能开发与应用 ……………………………………… 单旭沂，张　勇，刘　苗　6-25
基于三维有限元的 DP590 冷连轧中间辊窜辊板形控制分析
……………………… 刘　需，李连杰，刘天武，马　成，杨　婷，赵轶哲，刘宏强，姜正义　6-25
高性能耐候桥梁钢 Q500qNH 的开发 …… 彭宁琦，高　擎，史术华，范　明，熊祥江，李中平，陈奇明　6-26
Development of Q500qNH High-performance Atmospheric Corrosion Resisting Steel for Bridge
… PENG Ningqi, GAO Qing, SHI Shuhua, FAN Ming, XIONG Xiangjiang, LI Zhongping, CHEN Qiming
960MPa 级工程机械用钢的连续冷却转变规律 … 惠亚军，田志红，肖宝亮，邱　宇，冯　超，黄天华　6-27
Continuous Cooling Transformation Law of 960MPa Grade Construction Machinery Steel
……………………… HUI Yajun, TIAN Zhihong, XIAO Baoliang, QIU Yu, FENG Chao, HUANG Tianhua
热轧板带钢轧制残余应力的研究 ………………………… 王卫卫，胡霄雨，肖金福，张宏亮，丛菁华　6-28
Research on Residual Stress in Hot Rolled Plate and Strip
………………………………… WANG Weiwei, HU Xiaoyu, XIAO Jinfu, ZHANG Hongliang, CONG Jinghua
浅谈轧制油对不锈钢表面质量的影响 ……………………………………… 吴　雪，刘金元，陈　普　6-28
Discussion on the Effect of Rolling Oil on the Surface Quality of Stainless Steel
………………………………………………………………………… WU Xue, LIU Jinyuan, CHEN Pu
基于理论预测模型的轧机单调自动控制系统开发及其不锈钢轧制应用实践
……………………………………… 李慧峰，王中杰，李建民，王　忠，杨永杰，郭新峰　6-29
提高 410S 不锈钢 2B 产品光泽度工艺研究与实践 ………………… 刘金元，张爱建，陈　普　6-29
Research on the Process of Improving the Gloss of 410S Stainless Steel 2B Products
………………………………………………………………… LIU Jinyuan, ZHANG Aijian, CHEN Pu
10Cr17 不锈钢冷轧轧制力优化设计 …………………………………… 张爱建，刘金元，陈　普　6-30
Optimized Design of 10Cr17 Stainless Steel Cold Rolling Force … ZHANG Aijian, LIU Jinyuan, CHEN Pu
回火工艺对热处理 X80 管线钢组织性能的影响
………………………………… 刘文月，李天怡，安　涛，臧　岩，王传军，王超逸　6-31
Influence of Tempering Process on Microstructure and Mechanical Properties of Heat-treated X80 Pipeline Steel
………………………… LIU Wenyue, LI Tianyi, AN Tao, ZANG Yan, WANG Chuanjun, WANG Chaoyi
热处理工艺对吉帕级无缝管用钢组织及性能的影响
………………………………… 安　涛，刘文月，李天怡，臧　岩，李江文，薛　峰　6-31
Effect of Heat Treatment on Microstructure and Mechanical Property of GPa Seamless Pipe Steel
………………………………… AN Tao, LIU Wenyue, LI Tianyi, ZANG Yan, LI Jiangwen, XUE Feng

6.2 长材和钢管

棒材生产线定尺剪切系统的提速改造 ……………………………………………… 吴光行，庞博文　6-32
Speed-up Refurbishment of Fixed-length Shearing System in Bar Production Line
……………………………………………………………………… WU Guangxing, PANG Bowen

高碳钢车轮踏面近表层异常组织控制
…………………………… 国新春，邓荣杰，刘　智，黄孝卿，翟　龙，陶盈龙，张意哲　6-33
Study on Heat Treatment Process of Heavy-duty Wheel
… GUO Xinchun, DENG Rongjie, LIU Zhi, HUANG Xiaoqing, ZHAI Long, TAO Yinglong, ZHANG Yizhe

慢应变速率对110S油井管拉伸性能和断口形貌的影响
…………………… 钟　彬，陈义庆，艾芳芳，高　鹏，李　琳，张圣洁，伞宏宇，苏显栋，沙楷智　6-34
Effect of Slow Strain Rate on Tensile Properties and Fracture Morphology of 110S Oil Well Pipe
………………………………………………… ZHONG Bin, CHEN Yiqing, AI Fangfang, GAO Peng, LI Lin,
　　　　　　　　　　　　　　　　　　　ZHANG Shengjie, SAN Hongyu, SU Xiandong, SHA Kaizhi

智能高速棒材生产线建设的创新与关键技术
………………………… 徐言东，王晓晨，程知松，余　伟，张勇军，刘　洋，白金龙　6-35
Innovation and Key Technologies of Intelligent High Speed Bar Production Line Construction
…… XU Yandong, WANG Xiaochen, CHENG Zhisong, YU Wei, ZHANG Yongjun, LIU Yang, BAI Jinlong

B500B英标钢筋疲劳断裂失效分析 ……………………… 高　敏，贾元海，刘可可，刘效云　6-36
Fatigue Fracture Failure Analysis of B500B British Standard Steel Bars
……………………………………………………… GAO Min, JIA Yuanhai, LIU Keke, LIU Xiaoyun

铁路车轮踏面C型剥离缺陷分析 ………………庞晋龙，李　翔，马俊松，李自钢，陈　刚，徐金辉　6-36
Defects Analysis of C-type Shelling of Railway Wheel Tread
……………………………………… PANG Jinlong, LI Xiang, MA Junsong, LI Zigang, CHEN Gang, XU Jinhui

一种低碳当量易焊接热轧钢轨的开发 ………………冯立果，崔　恺，李钧正，杨正宗，郭朝军　6-37
Development of a Low Carbon Equivalent Hot-rolling Rail for Easy Welding
………………………………………… FENG Liguo, CUI Kai, LI Junzheng, YANG Zhengzong, GUO Chaojun

大口径UNS N08810无缝管冷扩内壁横裂原因分析及优化
…………………………………………… 黎　毅，郭盈春，杨　帆，刘　宇，郭建邦，康喜唐　6-38
Analysis and Optimization of Transverse Cracks on the Inner Wall of Large Diameter UNS N08810 Seamless
　　Pipe during Cold Expansion
……………………………………… LI Yi, GUO Yingchun, YANG Fan, LIU Yu, GUO Jianbang, KANG Xitang

起重机钢轨QU80开发实践 ………………………………………… 余选才，郑赟峰，董茂松　6-38
Development Practice of QU80 Crane Rail ……………… YU Xuancai, ZHENG Yunfeng, DONG Maosong

X80管线钢高温热形变行为及析出研究
………………………… 武晓龙，王娇娇，周玉青，高云哲，弓俊杰，梁爱国，赵林林　6-39
Study on High Temperature Thermal Deformation Behavior and Precipitation of X80 Pipeline Steel
………………………………………………… WU Xiaolong, WANG Jiaojiao, ZHOU Yuqing, GAO Yunzhe,
　　　　　　　　　　　　　　　　　　　　　　　　GONG Junjie, LIANG Aiguo, ZHAO Linlin

合金成分对10.9级高强螺栓钢抗延迟断裂性能影响的研究 ………任学堂，肖丙政，王青峰，张欢欢　6-40
Study on the Effect of Alloy Composition on the Delayed Fracture Resistance of 10.9 Grade High Strength
　　Bolt Steel …………………………… REN Xuetang, XIAO Bingzheng, WANG Qingfeng, ZHANG Huanhuan

高温超导感应加热技术在热挤压钢管生产中的应用前景 ……………………………… 张国栋，商庆华　6-41

PRS 轧机应用特点分析 ………………………… 孔令波，马洪磊，王云涛，薛正国，郑佳星，张　旭　6-41
Application Characteristics Analysis of PRS Rolling Mill
………………… KONG Lingbo, MA Honglei, WANG Yuntao, XUE Zhengguo, ZHENG Jiaxing, Zhang Xu

攀钢高强耐磨过共析钢轨的研制及应用 ……………………… 韩振宇，邹　明，刘丰收，祭　程，王文健　6-42
The Development and Application of Pangang High-Strength & High Wear-Resistance Hypereutectoid Rail
…………………………… HAN Zhenyu, ZOU Ming, LIU Fengshou, JI Cheng, WANG Wenjian

棒材精轧前后控冷工艺设备研究与改进 …………………………………… 戴川涵，陈育生，黄建辉　6-43
Research and Improvement of Controlled Cooling Process Equipment Before and after Bar Finishing Rolling
………………………………… DAI Chuanhan, CHEN Yusheng, HUANG Jianhui

钢结构建筑用高强度厚规格耐候抗震热轧 H 型钢研制
…………………………………… 卜向东，涛　雅，郭利宏，王永明，卢雄慧，宋振东　6-44
Development of High Strength Thick Gauge Weather-resistant and Anti-seismic Hot-rolled H-Beam
……………… BU Xiangdong, TAO Ya, GUO Lihong, WANG Yongming, LU Xionghui, SONG Zhendong

60～50kg/m U75V 热轧异型钢轨跟端热处理工艺研究 ……………… 张凤明，梁正伟，薛虎东　6-45
Study on Heat Treatment Process of 60~50kg/m U75V Hot Rolled Special-shaped Rail Heel End
………………………………………… ZHANG Fengming, LIANG Zhengwei, XUE Hudong

江西台鑫钢铁有限公司双高速棒材轧钢生产线新工艺设备技术应用 ………………………… 陈立胜　6-46
Application of New Technology and Equipment in Double High Speed Bar Production Line of Jiangxi Taixin
Steel Co., Ltd. ……………………………………………………………………………… CHEN Lisheng

稀土轨研究发展与现状 ……………………… 薛虎东，赵桂英，王慧军，郑　瑞，刘　阳　6-46
Development and Current Situation of Rare Earth Orbital Research
………………………… XUE Hudong, ZHAO Guiying, WANG Huijun, ZHENG Rui, LIU Yang

美标、欧标与铁标钢轨标准差异分析 ………………………… 边　影，薛虎东，王嘉伟，彭　飞　6-47
Analysis of American Standard, European Standard and Iron Standard Rail Standard Difference
………………………………… BIAN Ying, XUE Hudong, WANG Jiawei, PENG Fei

轨梁厂二号线码垛台架运输链测距方法改进 ……………………………………………… 刘晓宇　6-48
万能轧机装配精度对钢轨万能轧法的影响 ……………………… 郭大林，陈　琳，段永强，王　琨　6-48
国内棒材高速上钢系统的技术特点及发展趋势 ……………… 张俊义，方针正，董红卫，何巍巍　6-48
Technical Overview and Development Trend of Domestic High-speed Bar System
………………………… ZHANG Junyi, FANG Zhenzheng, DONG Hongwei, HE Weiwei

高速棒材连续工作制飞剪的力学性能和参数分析 ……… 张俊义，方针正，惠秦川，董红卫，何巍巍　6-49
The Mechanical Performance Analysis of Continuous Working Flying Shear in High-speed Bar Line
………………… ZHANG Junyi, FANG Zhenzheng, HUI Qinchuan, DONG Hongwei, HE Weiwei

高速线材升级改造探索与实践 ………………………………………………… 荣　崎，唐心语　6-50
Exploration and Practice of Upgrading and Transforming High Speed Wire Rod …… RONG Qi, TANG Xinyu

矿用锚索用高强度 Φ14mm 82B 盘条网状渗碳体成因分析
……………………………… 陈　列，朱子轩，郭　鑫，陶立志，潘晓坤，张朝磊　6-50
Cause Analysis of Mesh Cementite of High Strength Φ14 mm 82B Wire Rod for Mine Anchor Cable
………………… CHEN Lie, ZHU Zixuan, GUO Xin, TAO Lizhi, PAN Xiaokun, ZHANG Chaolei

热轧钢筋分段气雾冷却工艺及控制模型研究 ……………………… 赵　舸，王卫卫，丁哲毅，潘富强，
　　范仲继，宋　为，严雪蕾，苏进彬，唐建生，周道树，李丰德，林华春，魏　勇，黄标彩　6-51

Research on the Staged Spray Evaporative Cooling Process and Control Model for Hot-rolled Rebar
 ········· ZHAO Ge, WANG Weiwei, DING Zheyi, PAN Fuqiang, FAN Zhongji, SONG Wei, YAN Xuelei, SU Jinbin, TANG Jiansheng, ZHOU Daoshu, LI Fengde, LIN Huachun, WEI Yong, HUANG Biaocai

热处理工艺对超深井用非标油套管组织性能的影响········臧　岩，李天怡，刘文月，安　涛，李江文　6-52
Effect of Heat Treatment Technology on Microstructure and Properties of Non-standard Oil Casing Used in Ultra-deep Wells ················ ZANG Yan, LI Tianyi, LIU Wenyue, AN Tao, LI Jiangwen

2000MPa级以上超高强钢增塑/韧新机制及其应用················李云杰　6-53

7　表面与涂镀

工业化生产中热镀锌板表面小渣点缺陷原因研究·········宋青松，马幸江，黄　宾，陈　彪，段晓溪　7-1
Study on the Causes of Small Slag Defects on the Surface of Hot Galvanized Sheet in Industrial Production
 ················ SONG Qingsong, MA Xingjiang, HUANG Bin, CHEN Biao, DUAN Xiaoxi

高耐蚀锌铝合金镀层Al含量选择研究·················金永清，黄　建，吴林倬　7-1
Study on the Selection of Al Content in High Corrosion Resistance Zinc Aluminum Alloy Coatings
 ················ JIN Yongqing, HUANG Jian, WU Linzhuo

浅析传统镀锌机组集约化智能化的优化设计·········董　斌，杨柏松，李元亭，叶　波，向双发　7-2
Preliminary Analysis of Intelligent System Design for a Traditional Galvanizing Line
 ················ DONG Bin, YANG Baisong, LI Yuanting, YE Bo, XIANG Shuangfa

连续锌铝镁热涂镀机组针对不同锌铝镁配方的设计优化·················杨柏松，环素学　7-3
Strip Coating Line Optimal Design for Different ZAM Coating Contents ····· YANG Baisong, HUAN Suxue

高强厚规格热基镀锌板生产工艺研究·················贾海超　7-3
Study on Production Technology of High Strength and Thick Specification Hot Base Galvanized Sheet
 ················ JIA Haichao

镀锌板上制备氮化钛镀层实验研究·········徐闻慧，杨洪刚，金　勇，吕家舜，徐承明，王永明　7-4
Experomental Study on Preparation of TiN Coating on Galvanized Sheet
 ················ XU Wenhui, YANG Honggang, JIN Yong, LV Jiashun, XU Chengming, WANG Yongming

表面粗糙度对全无铬钝化热镀锌板耐腐蚀性能的影响
 ········· 黎　敏，张　晨，王长成，鲁洋泽，任新意，邵　蓉，刘永壮　7-5
Effect of Surface Roughness on Corrosion Resistance of Chromium-free Passivated Hot Dip Galvanized Sheet
 ··· LI Min, ZHANG Chen, WANG Changcheng, LU Yangze, REN Xinyi, SHAO Rong, LIU Yongzhuang

通过数值模拟方法研究带钢入锌锅速度对产渣量的影响·················刘　尧，赵敬华，吴广新　7-6
一种锆化条纹缺陷及其改进·················陈　光　7-7
冷轧汽车板缺陷在锆化前处理工艺中的传递性及其对表面耐蚀性的影响·················方百友　7-8
The Transferability of Surface Defects and Its Influence on the Corrosion Resistance of Zirconizing Pretreatment for Cold Automotive Sheets················ FANG Baiyou

光整液残留对无铬耐指纹钝化板耐蚀性影响·········邵　蓉，黎　敏，刘永壮，曹建平，李学涛　7-9
Effect of Finishing Fluid Residue on Corrosion Resistance of Chromium-free Fingerprint Passivation Film
 ················ SHAO Rong, LI Min, LIU Yongzhuang, CAO Jianping, LI Xuetao

锌铝镁镀层电化学腐蚀行为及其耐蚀机理研究·················刘　洁，阎元媛，戴竞舸　7-9
Study of Electrochemical Corrosion Behavior and Corrosion Resistant Mechanism of Zn-6Al-3Mg Coating
 ················ LIU Jie, YAN Yuanyuan, DAI Jingge

乳化液斑对冷轧带钢表面质量的影响及控制措施

·················关洪星，陈　平，梁　直，罗　军，瞿作为，唐　华　7-10

Effect of Emulsion Spot on Surfacein Quality of Cold Rolled Strip and Its Purification

·················GUAN Hongxing, CHEN Ping, LIANG Zhi, LUO Jun, QU Zuowei, TANG Hua

连续退火露点对高强钢表面可镀性和镀层附着性的影响·················董志磊，杨向鹏，阎元媛　7-11

Effect of Continuous Annealing Dew Point on the Surface Patability and Coating Adhesion of High-strength Steel

·················DONG Zhilei, YANG Xiangpeng, YAN Yuanyuan

55%Al-Zn-1.6%Si-1.2%Mg 合金镀层的弯曲裂纹分析与改善·················李　超，王　滕，刘茂林，杨　平　7-12

Analysis and Improvement of Bending Cracks of 55%Al-Zn-1.6%Si-1.2%Mg Alloy Coating

·················LI Chao, WANG Teng, LIU Maolin, YANG Ping

预氧化技术在镀锌高强钢表面质量控制中的应用

·················孙卫华，谢英秀，陈庆军，查　凯，宇　凡，刘万春　7-12

厚规格高附着量锌铝镁镀层板用气刀结构与工艺参数的研究·················许秀飞，许　正，张　瑶，廖嘉玮　7-13

锌铝镁产品表面锈蚀缺陷原因分析与控制·················李志庆，刘茂林，陈德春，章一樊，钱　鑫，阮康康　7-14

Analysis and Control of Corrosion Defects on Surface of Zinc-aluminum-magnesium Products

·················LI Zhiqing, LIU Maolin, CHEN Dechun, ZHANG Yifan, QIAN Xin, RUAN Kangkang

涂覆钢板产品研发现状与进展·················曹宏玮，孙　力，张　鹏，杨士弘，邹炎斌，赵秀娟　7-14

Research and Development Direction of Coated Steel Plate Technology

·················CAO Hongwei, SUN Li, ZHANG Peng, YANG Shihong, ZOU Yanbin, ZHAO Xiujuan

电磁驱动下锌锅表面锌液流动行为及其扒渣效果分析····雷作胜，卢海彪，罗贤文，钟云波，任维丽　7-15

无铬钝化镀锌板粉末喷涂脱漆的原因分析及应对措施····徐卫国，董立华，刘　闯，宇　凡，吕明鹏　7-16

稀土元素对锌铝镁镀层组织及耐腐蚀性能的影响

·················宋　帅，弓俊杰，张　青，薛仁杰，王立辉，陈屹松　7-17

Effect of Rare Earth Elements on the Microstructure and Corrosion Resistance of Zinc-Aluminum-Magnesium Coating·················SONG Shuai, GONG Junjie, ZHANG Qing, XUE Renjie, WANG Lihui, CHEN Yisong

冷轧 IF 带钢表面清洁度的控制实践·················薛仁杰，李岚涛，杨士弘　7-17

The Practice of IF Strip Surface Cleanliness Control in Cold Rolling Line

·················XUE Renjie, LI Lantao, YANG Shihong

时效对镀锡板钝化膜结构影响的研究·················张　鹏，李　虎，孙　力，史文礼，张　诚，曹宏玮　7-18

Study on the Effect of Aging on the Structure of Passivation Film of Tinplate

·················ZHANG Peng, LI Hu, SUN Li, SHI Wenli, ZHANG Cheng, CAO Hongwei

合金化镀锌板表面白点缺陷原因分析·················王　川，刘李斌，李　勇，史　昌，黄学启　7-19

Causes of White-spot Defects on the Surface of Galvannealed Sheets

·················WANG Chuan, LIU Libin, Li Yong, SHI Chang, HUANG Xueqi

铸管外喷工艺的应用·················姚震宇，张海艳　7-20

Application of Casting Pipe External Spray Process·················YAO Zhenyu, ZHANG Haiyan

镀锡板表面特性评价技术与应用·················李建中，李　雪，刘包发，王悦鼎　7-20

镀锌模拟实验机锌灰收集过滤装置设计及应用

·················田绍鹏，崔忠信，冯小雷，王学慧，石玉龙，苏立冬　7-21

Design and Application of Zinc Ash Collecting and Filtering Device for Galvanizing Simulator

·················TIAN Shaopeng, CUI Zhongxin, FENG Xiaolei, WANG Xuehui, SHI Yulong, SU Lidong

8 金属材料深加工

韩国光伏发电装置钢梁基础模块化工艺 ……………………………………………… 罗 晔	8-1
Modular Process for Steel Beam Foundations for Photovoltaic Power Installations in Korea ……… LUO Ye	
铁锰因瓦合金的强化研究 ……………………… 刘津伊，严 玲，刘文月，齐祥羽，张 鹏，李广龙	8-1
Study on Strengthening of Fe-Mn Invar Alloy	
…………………… LIU Jinyi, YAN Ling, LIU Wenyue, QI Xiangyu, ZHANG Peng, LI Guanglong	
粗轧微张力控制方法与实践 ………………………………………………………… 刘 伟，闫 鹏	8-2
Optimization of Micro Tension Controlling in Roughing Mill Area ……………… LIU Wei, YAN Peng	
低应力热轧高强钢薄板矫正策略 ……………………………………………… 高智平，张鹏武	8-3
热连轧卷取机镰刀弯防夹死功能开发及应用 ……………………………… 梅荣利，袁 金，李君波	8-3
酸洗钢辊系铁皮缺陷分析及改进 ………………………… 周 坤，梅荣利，袁 金，徐 浩，李君波	8-3
Research and Improve of Roll System Algam Defects on Pickling Steel	
…………………………………… ZHOU Kun, MEI Rongli, YUAN Jin, XU Hao, LI Junbo	
微合金化高强钢热轧卷尾部性能研究 ………… 王 成，童善康，陈志辉，袁 清，甘晓龙，丁 茹	8-4
Investigating the Properties of Coil Tail in Microalloyed Hot-rolled Strip	
…………………… WANG Cheng, TONG Shankang, CHEN Zhihui, YUAN Qing, GAN Xiaolong, DING Ru	
主提升钢丝绳断丝失效分析 ……………………………………………………………… 赵宪海	8-5
Failure Analysis on Wire Breaking in Wire Rope for Mine Hoisting Purposes ……………… ZHAO Xianhai	
一种矿用钢丝绳插接新技术 ……………… 赵宪海，王靖涵，王国辉，王洪亮，刘思洋，李广宇	8-5
The New Technology Relates to Inserting the Steel Wire Rope	
………… ZHAO Xianhai, WANG Jinghan, WANG Guohui, WANG Hongliang, LIU Siyang, LI Guangyu	
耐热套管特殊螺纹接头 BG-TH 开发理论与试验研究 ……………………… 詹先觉，董晓明，高 展	8-6
Research on Thermal Well Premium Casing Connection BG-TH	
………………………………………………… ZHAN Xianjue, DONG Xiaoming, GAO Zhan	
DIL 805 热膨胀仪在先进高强钢中的典型应用	
……………………………………… 李春诚，郭晓静，杨 波，海 超，李科龙，姜乐朋	8-7
Typical Application of the DIL 805 Thermal Dilatometer in Advanced High Strength Steel	
……………………… LI Chuncheng, GUO Xiaojing, YANG Bo, HAI Chao, LI Kelong, JIANG Lepeng	
Cr 对 22MnB5 热成形钢高温氧化行为的影响 ………… 朱 蓉，张 淇，米振莉，吴彦欣，江海涛	8-8
Effects of Cr on High Temperature Oxidation Behavior of 22MnB5 Hot Stamping Steel	
……………………………… ZHU Rong, ZHANG Qi, MI Zhenli, WU Yanxin, JIANG Haitao	
低密度中锰钢脆性断裂研究 ……………… 左汪楠，杨永刚，宿泽轩，袁长辉，吴彦欣，米振莉	8-8
Study on Brittle Fracture of Low Density Medium Manganese Steel	
…………… ZUO Wangnan, YANG Yonggang, SU Zexuan, YUAN Changhui, WU Yanxin, MI Zhenli	
V 对 Q355 钢 CCT 曲线的影响 ………………… 刘欣悦，方 幸，左汪楠，吴彦欣，米振莉，潘 欣	8-9
The Influence of V on the CCT Curve of Q355 Steel	
………………………… LIU Xinyue, FANG Xing, ZUO Wangnan, WU Yanxin, MI Zhenli, PAN Xin	
连铸连轧工艺下 SPHC 热变形行为模拟 ………… 常 江，米振莉，苏 岚，杨永刚，张茂才，刘心爽	8-10
Simulation of Hot Deformation Behavior of SPHC during Continuous Casting and Rolling	
……………… CHANG Jiang, MI Zhenli, SU Lan, YANG Yonggang, ZHANG Maocai, LIU Xinshuang	

超临界二氧化碳输送用 HFW 焊管性能研究……黄晓辉，张锦刚，毛浓召，韦 奉，王博玉，赵红波　8-11

冷轧薄板激光切割翘曲变形分析研究…………曹 政，冉茂宇，吕 冬，芦延鹏，徐 鑫，林森木　8-12
Analysis and Research on Warpage Deformation of Cold Rolled Plate Laser Cutting
　　……………………………………CAO Zheng, RAN Maoyu, LV Dong, LU Yanpeng, XU Xin, LIN Senmu

网格应变分析在冲压成形质量检测中的应用……………赵轶哲，孙 力，韩世绪，杨 婷，吕 浩　8-12
A Application of Grid Strain Analysis in the Quality Analysis of Stamping
　　………………………………………… ZHAO Yizhe, SUN Li, HAN Shixu, YANG Ting, LV Hao

热卷 Q355B 低合金钢表面裂纹研究与应用……赵启帆，向浪涛，王 灿，戴 林，何 璋，陈启发　8-13
Research and Application of Surface Crack in Hot Coil Q355B Low Alloy Steel
　　………………………………… ZHAO Qifan, XIANG Langtao, WANG Can, DAI Lin, HE Zhang, CHEN Qifa

双相钢铁素体相变动力学模型及预测
　　……………………………… 雷明钢，常 江，李嘉旋，何 方，李守华，杨永刚，米振莉　8-14
Construction of a Kinetic Model of Ferrite Phase Transition in Dual Phase Steel
　　………… LEI Minggang, CHANG Jiang, LI Jiaxuan, HE Fang, LI Shouhua, YANG Yonggang, MI Zhenli

Fe-Mn 系列汽车用钢拉深成型性能分析…………………………代永娟，武祥祥，陈佳佳，周 涛　8-15

35MnB 钢高强韧耐磨件的热处理技术
　　……………… 罗 扬，夏占雪，李福勇，段路昭，宋 月，白丽娟，赵 博，冯立果　8-16
Heat Treatment Technology for 35MnB Steel Parts with High Strength, Toughness and Wear Resistance
　　… LUO Yang, XIA Zhanxue, LI Fuyong, DUAN Luzhao, SONG Yue, BAI Lijuan, ZHAO Bo, FENG Liguo

高强度桥梁缆索用盘条的开发…………………………………………王 雷，李月云，张 宇　8-17

压边力及拉延筋对 DC03 板材拉深影响的有限元模拟研究……………陈 健，方进秀，钱健清　8-18
Finite Element Simulation Study on the Effects of Blank Holder Force and Drawing Ribs on the Deep
　　Drawing of DC03 Plate………………………………………CHEN Jian, FANG Jinxiu, QIAN Jianqing

钢铁工艺知识图谱的构建与应用研究………米振莉，袁长辉，杨永刚，宋 勇，倪道军，常 江　8-19
Research on the Construction and Application of Knowledge Graph for Iron and Steel Processes
　　………… MI Zhenli, YUAN Changhui, YANG Yonggang, SONG Yong, NI Daojun, CHANG Jiang

稀土对 Mn13 组织及性能影响…………………………石 帅，赵燕青，高云哲，弓俊杰　8-20
Effect of Mixed Rare Earth La、Ce on the Microstructure and Inclusion Modification of Cast High Manganese
　　Steel………………………………………… SHI Shuai, ZHAO Yanqing, GAO Yunzhe, GONG Junjie

金属管材热气胀成形技术发展及应用
　　……………………… 张 猛，阮尚文，成 刚，程鹏志，吴彦欣，张 淇，米振莉　8-20

定制辊压成型技术与中国制造业的高质量发展…………………………………………晏培杰　8-21

高强度链条用 20Mn2 拉力不足原因分析……………王冬晨，李永超，樊亚鹏，李亚硕，韩光洋　8-22
Analysis of Causes of Insufficient Tensile Force of 20Mn2 for High Strength Chain
　　………………………………… WANG Dongchen, LI Yongchao, FAN Yapeng, LI Yashuo, HAN Guangyang

酸洗冷连轧机组薄规格启动断带问题研究……………………刘前亮，李玉冰，张 倩，杨雪松　8-23
Research on Start-up Strip Break Issues in the Continuous Pickling Line-Tandem Cold Rolling Mill for
　　Thin Specifications……………………………LIU Qianliang, LI Yubing, ZHANG Qian, YANG Xuesong

轨梁厂 CCS 工作辊道减速机失效分析及结构优化……………贾 欣，冯指名，刘志强，张 骞，刘剑锋　8-23
Failure Analysis and Structure Optimization of Rail and Beam Factory CCS Work Roll Table Reducer
　　………JIA Xin, FENG Zhiming, LIU Zhiqiang, ZHANG Qian, LIU Jianfeng

辊弯成型过渡区长度计算方法分析…………………………刘 鹏，李宏图，王海隆，龚学堂　8-24

Analysis of the Calculation Method for the Length of the Transition Zone in Roll Forming
.... LIU Peng, LI Hongtu, WANG Hailong, GONG Xuetang

9 粉末冶金

基于选区激光熔化的镍基单晶高温合金修复研究………张 鹏，沈 超，张 宇，章 林，曲选辉	9-1
注射成形高强钛合金研究……………………………………魏敬浩，陈 刚，秦明礼，曲选辉	9-1
增材制造 NiTi-Nb 医用多孔支架孔径策略设计及性能研究	
………………………………………………刘国浩，王 瑞，谢锦丽，毕中南，孙广宝	9-2
"核壳"结构特种功能粉体及其潜在工程应用………………………………杨亚锋，李少夫	9-3
基于粉末 3D 打印原理的材料高通量制备实验方法与应用………………………张百成	9-3
紧耦合气雾化技术制备铁基非晶粉末的数值模拟与试验研究	
……………………………………刘佳奇，王 璞，董延楠，赵 欢，庞 靖，张家泉	9-3
铜基 CBN 超硬复合材料界面反应行为研究………………………………范永刚，王 聪	9-4
耐高温金属多孔材料的高温防护策略………………张惠斌，万磊磊，郭 菲，马骏梁	9-5
微纳钨极的近终形制备研究……………………………吴昊阳，王 杰，董宏月，秦明礼，曲选辉	9-6
晶界自净化钨镁合金的制备与性能研究………………………王一甲，张大越，李彬周	9-6
Preparation of Grain Boundary Strengthened W-Mg Alloy Via Powder Metallurgy	
………………………………………………………WANG Yijia, ZHANG Dayue, LI Binzhou	
激光熔覆铁基涂层强化热轧导板技术应用………………………王一甲，李彬周，张大越	9-7
不同粒度 Cu 粉的力学性能研究………………………………徐 强，刘一波，杨志威	9-8
PVP 含量对纯钛注射成型的影响……………张韦晨，李 璐，李传勇，Muhammad Dilawer Hayat	9-9

10 先进钢铁材料及应用

10.1 钢铁材料数字化

管线钢中夹杂物追踪及成因分析………………王复越，任 毅，田永久，崔福祥，赵钰琛，张万顺	10-1
Inclusion Tracking and Cause Analysis of Inclusions in Pipeline Steel	
………………WANG Fuyue, REN Yi, TIAN Yongjiu, CUI Fuxiang, ZHAO Yuchen, ZHANG Wanshun	
统计分析技术在 1580 热轧线生产质量改进中的应用	
…………………………………………刘旺臣，王 存，王 杰，李江委，任俊威，刘 磊	10-2
Application of Statistical Analysis Technology in 1580 Hot Rolling Line	
………………………LIU Wangchen, WANG Cun, WANG Jie, LI Jiangwei, REN Junwei, LIU Lei	
点缺陷在 Cr_2O_3/Fe_2O_3 界面氢损伤的作用………………………………………………米志杉	10-2
高硅高铝耐磨钢在中应变速率下的本构模型……杨哲懿，高 磊，王尊呈，王 帅，刘 威，董 洋	10-3
The Mid-range Strain Rate of Constitutive Relationship about Wear-resistant Steel with High Silicon and High Aluminum…… YANG Zheyi, GAO Lei, WANG Zuncheng, WANG Shuai, LIU Wei, DONG Yang	
ICME 框架下汽车用高强钢的设计研发………………郑伟森，赵 宁，何燕霖，李 麟，鲁晓刚	10-4
钒微合金化 TRIP 钢的高通量计算设计………………梅 宇，吴 通，郑伟森，何燕霖，鲁晓刚，李 麟	10-4

高炉数字孪生驾驶舱与操作评价平台的建设 ········· 孟昕阳，徐　震，王宝海，郝云东，孙敏敏，庞克亮　10-5
Construction of Digital Twin Cockpit and Operation Evaluation Platform for Blast Furnace
　　············ MENG Xinyang, XU Zhen, WANG Baohai, HAO Yundong, Sun Minmin, PANG Keliang
结合物理冶金原理和迁移学习的耐热钢蠕变寿命预测及设计 ················· 魏晓蓼，王晨充，徐　伟　10-6
助力热轧棒线材工程精准精细化——中冶南方数字预测系统
　　··· 李杨齐，骆艳萍，李沐泽，柯衡珍，闵　威　10-6
材料和工艺设计多尺度仿真方法及应用
　　································· 杨　丽，苏　航，刘和平，孙　旭，米志杉，程　挺，李　通　10-7
低压脉冲渗碳组织及变形的预测 ···················· 谢东兴，贺笃鹏，韩　瑞，王泽辉，秦湘阁　10-8
InterMat：一种基于区块链的材料数据共享基础设施 ············ 王畅畅，苏　航，段琳娜，李　灏　10-8
InterMat: A Blockchain Based Material Data Sharing Infrastructure
　　·· WANG Changchang, SU Hang, DUAN Linna, LI Hao
质量分级方法在食品接触用不锈钢使用安全性评价方面的应用 ···· 王世宏，苏　航，李　灏，段琳娜　10-9
基于深度学习和专家知识的材料图谱识别 ·············· 王炫东，苏　航，李　南，孟惠民　10-11
Cr5型模具钢水-空交替控冷数值模拟 ············ 李　栋，梁敬斌，尤晓东，贺笃鹏，谢志彬，邵青立　10-11
基于高通量实验的金属材料腐蚀行为研究 ··············· 宋有朋，颜鲁春，庞晓露，宿彦京，乔利杰，高克玮　10-12
高分辨率数字钢卷及应用 ·· 刘安平，彭燕华　10-13
High Resolution Digital Steel Coil and Its Application ·············· LIU Anping, PENG Yanhua
Al对Fe-Cr-Al合金变形行为影响机制的数值模拟研究 ······ 王　鹏，乔英杰，戚　文，都时禹，李学达　10-14
钢铁材料数字化研发应用实践 ······································· 梁　坤，王　卓　10-14
南钢宽厚板全轧程数值模拟系统开发 ······································· 王凌宇，邱保文　10-15
Development of Numerical Simulation System for Wide and Thick Plate Rolling in Nangang
　　··· WANG Lingyu, QIU Baowen
基于相场-神经网络的镍基高温合金析出相演化行为研究
　　·· 郑国才，秦海龙，毕中南，施荣沛，李东风　10-16
表征深冲钢织构演变特征规律的分析测试方法 ············ 张　娜，赵美英，史文义，李智丽　10-17
Analysis and Test Method for Characterizing of Texture Evolution on Deep Drawing Sheet Steel
　　································· ZHANG Na, ZHAO Meiying, SHI Wenyi, LI Zhili
耐热钢中Laves相抗蠕变机制的分子动力学模拟 ············ 王传军，孙　旭，刘文月，刘　鑫，秦　哲　10-17

10.2　汽　车　用　钢

铝硅镀层厚度对热成形钢极限冷弯性能影响 ····· 徐德超，张博明，张士杰，黄　俊，滕华湘，韩　赟　10-18
Influence of Aluminum-silicon Coating Thickness on the Cold Bending Performance of Hot-formed Steel
　　··············· XU Dechao, ZHANG Boming, ZHANG Shijie, HUANG Jun, TENG Huaxiang, HAN Yun
钼对无碳贝氏体非调质钢动态连续冷却转变的影响 ········ 陈　曦，王福明，代文彬，祁永峰，陈学刚　10-19
Effect of Mo Content on Dynamic Continuous Cooling Transformation of Carbide-free Bainitic Non-quenched
　　and Tempered Steel ············ CHEN Xi, WANG Fuming, DAI Wenbin, QI Yongfeng, CHEN Xuegang
低成本高表面质量600MPa级热轧双相钢研制开发
　　································· 董　毅，时晓光，刘仁东，孙成钱，韩楚菲，王俊雄　10-20
Evelopment of 600MPa Grade Hot-rolled Dual-phase Steel with Low Cost and High Surface Quality
　　············ DONG Yi, SHI Xiaoguang, LIU Rendong, SUN Chengqian, HAN Chufei, WANG Junxiong

不同应变速率下超高强热成形钢的动态变形行为 …………………… 马为涛，王鑫玮，陈星翰，宋仁伯 10-21
基于脉冲处理的中锰钢高温力学性能研究 …………………………… 赵文皓，宋仁伯，王永金，赵　帅 10-22
退火工艺对高成形钢组织和力学性能的影响 …… 张　峰，郭金宇，孟静竹，王科强，靳友鹏，张瑞坤 10-23
Effect of Continuous Annealing Process on Microstructure and Mechanical Properties of High Forming Steel
 ………… ZHANG Feng, GUO Jinyu, MENG Jingzhu, WANG Keqiang, JIN Youpeng, ZHANG Ruikun
热处理工艺对中锰钢组织性能的影响 ………… 敬树坤，宋仁伯，霍巍丰，赵　帅，王鑫玮，陈星翰 10-24
超高强度热成形钢的强化机制及氢致延迟开裂行为 …………………………………… 陈伟健，章顺虎 10-25
高成形性 980MPa 级复相钢组织性能研究 …… 谢春乾，刘华赛，韩　赟，刘李斌，王　川，李明远 10-25
Research on the Microstructure and Mechanical Properties of 980MPa Complex Phase Steel with High
 Formability …………… XIE Chunqian, LIU Huasai, HAN Yun, LIU Libin, WANG Chuan, LI Mingyuan
汽车用镀锌热成形钢胶接性能的研究 ………………… 陈屹松，李志昂，王立辉，张馨月，刘丽君 10-26
Study on the Bonding Performance of Galvanized Hot Formed Steel for Automobile
 ……………………………… CHEN Yisong, LI Zhi'ang, WANG Lihui, ZHANG Xinyue, LIU Lijun
30MnCrB5 连续冷却组织转变研究 ………… 张　青，薛仁杰，宋　帅，王立辉，陈屹松，陈　晨 10-27
Study on Transformation of 30MnCrB5 after Continuous Cooling
 ………… ZHANG Qing, XUE Renjie, SONG Shuai, WANG Lihui, CHEN Yisong, CHEN Chen
2000MPa 级超高强热成形钢退火-淬火-回火工艺及组织性能研究
 ……………………………………………… 王鑫玮，宋仁伯，陈星翰，霍巍丰，赵　帅 10-28
Nb 含量对 Fe-4Mn-2Al-0.2C 钢组织性能及屈服行为的影响
 ………………… 霍巍丰，宋仁伯，王永金，赵　帅，苏盛睿，张应超，王鑫玮 10-28
温轧-热处理耦合工艺对热轧 3Mn 钢力学性能的影响研究
 …………………………… 赵　帅，宋仁伯，张应超，霍巍丰，苏盛睿，王永金 10-29
中锰钢中 VC 析出物对奥氏体逆转变行为的影响——纳米尺度原位观察
 ……………………… 张应超，宋仁伯，赵　帅，苏盛睿，霍巍丰，王海波，王永金 10-30
热轧中锰钢对超快加热和低温回火的力学性能敏感性
 ……………………… 苏盛睿，宋仁伯，全书仪，张应超，霍巍丰，赵　帅，王永金 10-30
淬火温度对 Fe-0.2C-2.1Si-2.5Mn 淬火配分钢组织性能的影响 ………… 高鹏飞，刘旭明，赵征志 10-31
The Effect of Quenching Temperature on the Microstructure and Properties of Fe-0.2C-2.1Si-2.5Mn Quenched
 and Partitioning Steel ……………………………… GAO Pengfei, LIU Xuming, ZHAO Zhengzhi
含 V 奥氏体低密度钢退火时的组织性能演变 ………………… 谢志奇，惠卫军，张永健，赵晓丽 10-32
IF 钢成形后波纹度增量的研究 ………………………… 张露星，张　军，葛　浩，施刘健，崔　磊 10-33
Study on Waviness Increment of IF Steel after Forming
 …………………… ZHANG Luxing, ZHANG Jun, GE Hao, SHI Liujian, CUI Lei
低密度高强塑 Fe-Mn-Al-Cr 系奥氏体不锈钢的研究进展 ………………… 周彦君，肖　蕾，邓想涛 10-33
微合金化 35MnB5 热成形钢在不同热处理状态下的氢脆敏感性研究 ……… 林　超，郭晓菲，韦习成 10-34
Effect of Heat Treatment on Hydrogen Embrittlement Sensitivity of 35MnB5 Hot Press Forming Steel
 ………………………………………………… LIN Chao, GUO Xiaofei, WEI Xicheng

10.3　特　殊　钢

碲极微合金化对非调质钢组织与性能影响 ……………………… 徐翔宇，刘年富，王梓菲，付建勋 10-35
生活垃圾焚烧炉用耐热钢炉排铸件对比分析及优化设计 ………… 王春奕，欧红燕，张全新，胡　毅 10-36

Comparative Analysis and Optimum Design of Heat-resistant Cast Steel for MSW Incinerator Grate
·················· WANG Chunyi, OU Hongyan, ZHANG Quanxin, HU Yi

激光切割对弹簧钢 50CrV4 冷成型影响的研究
·················· 杨 玉，王英海，许 成，王润琦，张吉富，任俊威 10-36
Effect of Laser Cutting on Cold Forming of 50CrV4 Steel
·················· YANG Yu, WANG Yinghai, XU Cheng, WANG Runqi, ZHANG Jifu, REN Junwei

热轧卷取温度对弹簧钢氧化铁皮结构及酸洗特性影响研究·················· 李 雯，裴新华 10-37
Effect of Rolling Temperature on the Iron Oxide Scale Characteristics and Pickling Characteristics of the Spring Steel·················· LI Wen, PEI Xinhua

轴承钢热轧盘条表面缺陷的研究·················· 巩延杰，张育明，范振霞，董 庆，赵昊乾 10-38
Study on Surface Defects of Bearing Steel Hot-rolled Rod
·················· GONG Yanjie, ZHANG Yuming, FAN Zhenxia, DONG Qing, ZHAO Haoqian

N 合金化对辙叉用高锰奥氏体钢组织和性能的影响·················· 齐向阳，陈 晨 10-38

S30408 奥氏体不锈钢不同轧制与固溶工艺条件下的组织演变与再结晶行为
·················· 黄 健，庞宗旭，管吉春，范刘群，张建平，孙殿东，王 勇 10-39
Microstructure Evolution and Recrystallization Behavior of S30408 Austenitic Stainless Steel under Different Rolling and Solution Treatment Conditions ·················· HUANG Jian, PANG Zongxu, GUAN Jichun, FAN Liuqun, ZHANG Jianping, SUN Diandong, WANG Yong

消应力处理对双相不锈钢 S32205 组织及力学性能的影响
·················· 张瀚宁，王 勇，李黎明，孙殿东，颜秉宇 10-40
Effect of Stabilization Treatment on Microstructure and Mechanical Properties of S32205 Duplex Stainless Steel
·················· ZHANG Hanning, WANG Yong, LI Liming, SUN Diandong, YAN Bingyu

汽车悬架弹簧钢表面脱碳及对疲劳性能的影响·················· 王程明，孙晓冉，赵 楠，赵中昱，海 岩 10-41
Surface Decarbonization of Spring Steel for Automobile Suspension and Influence on Fatigue Properties
·················· WANG Chengming, SUN Xiaoran, ZHAO Nan, ZHAO Zhongyu, HAI Yan

开坯加热工艺对硅脱氧弹簧钢盘条夹杂物尺寸控制影响·················· 孟耀青，李建立 10-42

高均质特殊钢大构件的微观机制与工程应用·················· 曹艳飞，李殿中，刘宏伟，傅排先 10-43

GCr15 轴承钢钢坯加热开裂原因分折及防止措施·················· 完颜卫国，张树山，许 兴，王福礼，李 勇 10-43

Fe-Ni-Cr 系中熵合金拉拔过程中性能及微观组织演变
·················· 孙智妍，任 帅，赵英利，张志旺，吴迎飞，孙彩凤，张中武 10-44
Mechanical Properties and Microstructure Evolution of Fe-Ni-Cr System Medium Entropy Alloy during Drawing Process ·················· SUN Zhiyan, REN Shuai, ZHAO Yingli, ZHANG Zhiwang, WU Yingfei, SUN Caifeng, ZHANG Zhongwu

热轧不锈钢板氧化层贫铬层的微观结构特征·················· 岳莹莹，李晓亮，刘承军 10-45

异质超细晶/细晶奥氏体不锈钢的显微组织和力学性能·················· 全书仪，宋仁伯，苏盛睿，王永金，王开坤 10-46

镍对 Cr-Mo-V 系热作模具钢淬透性及碳化物析出的影响
·················· 岳建博，田家龙，龚 伟，廖 俊，于 洋，姜周华 10-46
Effect of Ni on Hardenability and Carbide Precipitation in Cr-Mo-V Hot-work Die Steel
·················· YUE Jianbo, TIAN Jialong, GONG Wei, LIAO Jun, YU Yang, JIANG Zhouhua

高温快速回火对中碳超细贝氏体钢轨钢组织、性能和残余奥氏体稳定性的影响
·················· 王建军，李宏光，王庆超，贾德诚，杨志南，张福成 10-47

预冷变形对 GCr15Si1Mo 轴承钢相变、微结构和强韧性的影响机制
······ 贾德诚，张春生，董润洲，王贺霖，杨志南，张福成 10-49

紧固件用钢的高性能化 ······ 陆恒昌，胡 杰，杨现亮，胡春东，史 文，董 瀚 10-49
On High Performance Steels for Fasteners
······ LU Hengchang, HU Jie, YANG Xianliang, HU Chundong, SHI Wen, DONG Han

加压熔炼条件下钙处理对高铝钢微观组织的影响 ······ 李文宇，李 阳，杨 皓，夏 铃，陈浩嘉 10-50
Effect of Calcium Treatment on Microstructure of High-alumina Steel under Pressurized Melting Conditions
······ LI Wenyu, LI Yang, YANG Hao, XIA Ling, CHEN Haojia

海洋工程用高耐蚀钢筋开发 ······ 陈焕德，周 云，杨晓伟，张 宇 10-51

新型低密度高强耐热不锈钢的设计研究 ······ 陈 卉，江志华，倪志铭，金建军 10-52
Research of Low-density High-strength Heat Resistant Stainless Steels
······ CHEN Hui, JIANG Zhihua, NI Zhiming, JIN Jianjun

减振降噪耐蚀阻尼钢的开发 ······ 李江文，侯华兴，李大航，孙美慧，郭呈宇，张 弛 10-52
Development of Vibration and Noise Reduction Corrosion-resistant Damping Steel
······ LI Jiangwen, HOU Huaxing, LI Dahang, SUN Meihui, GUO Chengyu, ZHANG Chi

Φ800mm 大断面连铸圆坯 42CrMo4 凝固传热过程模拟
······ 任春节，冷永磊，尹修刚，任立坤，邢 伟，张英男，刘福斌，姜周华 10-53
Simulation of Solidification and Heat Transfer Process of Φ800mm Large Cross Section Continuous Casting Round Billet 42CrMo4 ······ REN Chunjie, LENG Yonglei, YIN Xiugang,
······ REN Likun, XING Wei, ZHANG Yingnan, LIU Fubin, JIANG Zhouhua

高温扩散时间对 GCr15SiMn 轴承钢碳化物带状的影响 ······ 王 强，陈 列，谢奎龙，夏伟鹤 10-54
Effect of High Temperature Diffusion Time on the Banded Structure of GCr15SiMn Bearing Steel
······ WANG Qiang, CHEN Lie, XIE Kuilong, XIA Weihe

感应加热下非均匀温度场对高锰钢硬化机理的探索 ······ 房启文，闫学峰，杨志南，张福成 10-55

10.4 高温合金及信息化技术

基于反向传播神经网络的高温合金蠕变曲线预测 ······ 何金珊，马博浩，王西涛 10-56

MES 系统在鞍钢热轧 EPS 生产线的应用 ······ 高 松，车志良，黄玉彬，张 喆，孙上海 10-56
The Application of MES in the EPS Production Line of Ansteel Hot Strip Mill
······ GAO Song, CHE Zhiliang, HUANG Yubin, ZHANG Zhe, SUN Shanghai

基于神经网络的烧结工艺设计系统 ······ 陈 浩 10-57
Sintering Process Design Syetem Based on Neural Network ······ CHEN Hao

镍钴基变形高温合金电子束焊接接头的微观组织演变及力学性能的高通量表征
······ 周海晶，谢锦丽，蔡世平，董金鑫，于鸿垚，毕中南 10-58

P92 钢高温蠕变损伤模拟试验研究 ······ 宋宏峰 10-59
The Study on the High Temperature Creep Damage of the P92 Steel ······ SONG Hongfeng

GH4169 合金涡轮盘残余应力演化的全过程模拟仿真 ······ 宋润华，李东风，秦海龙，毕中南，张 继 10-59

选区激光熔化 TiC 增强 CM247LC 高温合金基复合材料的组织与性能
······ 王 瑞，刘国浩，谢锦丽，毕中南 10-60

机器学习和三目标优化算法辅助调控高温合金 γ/γ′ 微观组织
······ 刘 沛，黄海友，文 成，Turab Lookman，宿彦京，秦海龙，谢锦丽 10-61

基于数据与计算智能的高温合金数字化研发平台 ················· 梁　坤，王　卓　10-62
GH4151 合金高温流变行为的研究 ················ 黄科杰，黄　彬，马丹蕊，刘慧鑫，张麦仓　10-62
恢复热处理对单晶高温合金组织演变的影响 ···· 谭科杰，王新广，孟　杰，梁静静，周亦胄，孙晓峰　10-63
磷和硼对优质 GH4738 合金偏析特征的影响 ············ 马亚芬，马丹蕊，刘慧鑫，黄　彬，张麦仓　10-64
抽拉速率及铸件结构对单晶高温合金雀斑形成的影响 ···· 王志成，李嘉荣，刘世忠，王效光，杨万鹏　10-65
基于理论模型和高通量计算的镍基高温合金设计与优化 ············ 程懿歆，徐　斌，谢锦丽　10-66
基于机器学习的镍基高温合金热变形行为建模与分析 ············ 徐　斌，尹海清，谢锦丽　10-67
选区激光熔化高温合金构件变形和内部残余应力的数值模拟 ····················· 孙志民　10-68
γ′相与孪晶交互作用对新型 Ni-Co 基高温合金的拉伸性能影响规律
　　　　　　　　　　　　　　　　　　············ 段继萱，安　腾，谷　雨，于鸿垚，杜金辉，毕中南　10-68
航空航天用高温钛合金及钛基（Ti-Al）材料研究概况及应用
　　　　　　　　　　　　　　　　　　············ 张天馨，岳　颗，于继洋，黄召阁，滕艾均　10-69
选区激光熔化制备宏观金属构件应力演化研究 ············ 孙志民，孔豪豪，秦海龙，谢锦丽，毕中南　10-70
VIM+VAR 双联工艺熔炼 GH2901 合金的控制实践 ············ 柳金瑞，马志伟，苏爱民　10-71
Control Practice of Smelting GH2901 Alloy by VIM+VAR Duplex Process
　　　　　　　　　　　　　　　　　　················· LIU Jinrui, MA Zhiwei, SU Aimin
基于孪晶强化机理的镍钴基高温合金晶体塑性有限元模型 ············ 邓锐杰，甘　斌，李东风　10-71
机器学习辅助增材制造合金内部冶金缺陷调控研究 ············ 贵云玮，付华栋　10-72
Machine Learning-assisted Regulation of Internal Metallurgical Defects in Additively Manufactured Alloys
　　　　　　　　　　　　　　　　　　················· GUI Yunwei, FU Huadong
新型高强殷钢材料的创新研发及应用 ············ 孙中华　10-73
Innovative Research and Application of New-typed High-strength Invar Alloy ·············· SUN Zhonghua

10.5　低合金钢

Ce 对高品质风电用钢组织性能的影响 ················· 宋成浩，张振山，于　浩　10-74
NM450 耐磨钢 CMT 焊接接头微观组织与力学性能
　　　　　　　　　　　　　　　　　　············ 张志强，张露云，王佳骥，胡奉雅，苗　隽，魏　勇　10-75
Microstructure and Properties of CMT Welding Joint of NM450 Wear-resistant Steel
　　　　　　　　　　　············ ZHANG Zhiqiang, ZHANG Luyun, WANG Jiaji, HU Fengya, MIAO Jun, WEI Yong
1400MPa 级在线淬火型热轧超高强钢的研究及应用 ······ 胡学文，彭　欢，王承剑，王海波，石东亚　10-76
Research and Application on 1400MPa Gade Hot-rolled Utra-high Strength Seel Poduced by On-line Qenched
　　　　　　　　　　　············ HU Xuewen, PENG Huan, WANG Chengjian, WANG Haibo, SHI Dongya
Q345qENH 耐候桥梁钢在模拟不同服役环境下的腐蚀行为
　　　　　　　　············ 高　鹏，陈义庆，李　琳，钟　彬，艾芳芳，苏显栋，伞宏宇，沙楷智，张圣洁　10-77
Corrosion Behavior of Q345qENH Weathering Bridge Steel in Different Simulated Service Environments
　　　　　　　　　　　············ GAO Peng, CHEN Yiqing, LI Lin, ZHONG Bin, AI Fangfang,
　　　　　　　　　　　　　　　　SU Xiandong, SAN Hongyu, SHA Kaizhi, ZHANG Shengjie
海洋环境下 Q690 高强钢对接焊缝疲劳性能试验研究 ··················· 魏欢欢　10-78
易焊接煤矿液压支架用钢的研发及推广应用 ············ 衣海龙，刘少华，郑中洪，黄　重，陈尹泽　10-79
回火过程精细组织调控对 P690QL2 钢低温韧性的影响
　　　　　　　　　　　············ 杜　林，王堤鹤，庞启航，张宏亮，王亮亮，朱莹光，李相禹，侯家平　10-80

Effect of Fine Microstructure Control on Low Temperature Toughness of P690QL2 Steel During Tempering
　　Process ·· DU Lin, WANG Dihe, PANG Qihang, ZHANG Hongliang,
　　　　　　　　　　　　　　　　　　　WANG Liangliang, ZHU Yingguang, LI Xiangyu, HOU Jiaping

低合金高强度极地船舶用钢低温断裂韧性及海冰磨损性能·········王超逸，严　玲，常雪婷，安文瑞　10-81
Low Temperature Mechanics and Sea-ice Wear Properties of Low Alloy High Strength Polar Ship Steel
　　··· WANG Chaoyi, YAN Ling, CHANG Xueting, AN Wenrui

电感耦合等离子体发射光谱法测定水电钢中酸溶硼元素含量
　　······································谭胜楠，葛晶晶，任玲玲，戚振南，禹青霄　10-82
Determination of Acid Solvable Boron in Hydroelectric Steel by Inductively Coupled Plasma Emission
　　Spectrometry ·············· TAN Shengnan, Ge Jingjing, REN Lingling, Qi Zhennan, YU Qingxiao

组织对船板钢在模拟热带海洋大气中的耐蚀性影响················高　峰，李　健，周乃鹏，罗小兵　10-83
Effect of Microstructure on Corrosion Resistance of Hull Steel in Simulated Tropical Marine Atmosphere
　　·· GAO Feng, LI Jian, ZHOU Naipeng, LUO Xiaobing

390MPa级抗碰撞船体钢动态再结晶行为研究············陈作宁，师仲然，胡　骞，展之德，罗小兵　10-83
Study on Dynamic Recrystallization Behavior of High Toughness Steel
　　································ CHEN Zuoning, SHI Zhongran, HU Qian, ZHAN Zhide, LUO Xiaobing

20CrMoH齿轮钢棒材的全尺寸成分偏析研究
　　·······································任　帅，孙智妍，年保国，段路昭，罗　扬，冯立果　10-84
Study on Full Size Composition Segregation of 20CrMoH Gear Steel Bar
　　··················· REN Shuai, SUN Zhiyan, NIAN Baoguo, DUAN Luzhao, LUO Yang, FENG Liguo

U71MnH钢轨成分、组织及力学性能优化研究
　　·· 于海鑫，金纪勇，廖德勇，陈　昕，王　冬，张　瑜　10-85
Optimization of Composition, Microstructure and Mechanical Properties of U71MnH Rail
　　·· YU Haixin, JIN Jiyong, LIAO Deyong, CHEN Xin, WANG Dong, ZHANG Yu

冷却方式对耐海水腐蚀钢组织与腐蚀性能的影响··········解德刚，赵　波，吴　红，王善宝，袁　琴　10-86
The Effect of Cooling Methods on the Structure and Corrosion Performance of Seawater Corrosion Resistant Steel
　　·· XIE Degang, ZHAO Bo, WU Hong, WANG Shanbao, YUAN Qin

铝含量对铁素体-贝氏体热轧双相钢的组织性能及磨损行为的影响
　　··陈昊天，宋仁伯，任书濠，王永金　10-87

1000MPa级高强度水电用钢板Q890SF的研制与开发·············潘中德，刘心阳，武会宾，王思聪　10-87
Research and Development of 1000MPa High Strength Steel Plate Q890SF for Hydropower
　　································ PAN Zhongde, LIU Xinyang, WU Huibin, WANG Sicong

高钢级输氢管线钢L360MH的开发及应用 ······李少坡，张　镇，张彩霞，王志勇，马长文，白学军　10-88
Development and Application of High Grade Hydrogen Pipeline Steel L360MH at Shougang Steel
　　··············· LI Shaopo, ZHANG Zhen, ZHANG Caixia, WANG Zhiyong, MA Changwen, BAI Xuejun

690MPa级超低碳贝氏体钢控轧控冷工艺与组织性能研究
　　···梅　涛，何有洪，柴希阳，罗小兵，刘　静，杨才福　10-89

电磁搅拌对ML40Cr合金冷镦钢盘条锭型偏析的影响
　　···马立国，郭大勇，王秉喜，高　航，张　博，潘　阳　10-90
Effects of Electromagnetic Stirring on ML40Cr Cold Heading Steel Ingot Segregation
　　···············MA Liguo, GUO Dayong, WANG Bingxi, GAO Hang, ZHANG Bo, PAN Yang

含 V 低碳超细贝氏体钢在中温回火过程中的微观结构演变及强韧化机制
.... 梁转琴，王建军，李宏光，张向泽，刘长波，孙东云，杨志南，张福成　10-91
板坯表面热喷涂对 45Mn 钢冷弯性能的影响................ 杨　玉，王英海，许　成，王　刚，张吉富　10-92
Effect of Thermal Spraying on Slab Surface on Cold Bending Properties of 45Mn Steel
.... YANG Yu, WANG Yinghai, XU Cheng, WANG Gang, ZHANG Jifu
先进核电设备关键用钢的研制和应用................ 刘慧斌，侯　洪，张汉谦　10-93
含氮微合金钢中氮化物的生成行为研究
.... 杨　睿，李　阳，姜周华，孙　萌，马　帅，马彦硕，李天赐　10-93
Study on Nitride Formation Behavior in Nitrogen-containing Microalloyed Steel
.... YANG Rui, LI Yang, JIANG Zhouhua, SUN Meng, MA Shuai, MA Yanshuo, LI Tianci
热轧工艺对 X80 变形奥氏体和落锤性能的影响
.... 熊雪刚，张开华，陈　述，田光林，任守斌，龚　慧　10-94
Effects of Hot Rolling Process on Deformed Austenite and DWTT Properties of X80
.... XIONG Xuegang, ZHANG Kaihua, CHEN Shu, TIAN Guanglin, REN Shoubin, GONG Hui
稀土元素 Ce 对大型矿用车高强钢板 Q620E 的影响研究................ 卢晓禹，袁晓鸣，黄　利，王少炳　10-95
Effect of Rare Earth Element Ce on High Strength Steel Plate Q620E of Large Mining Vehicle
.... LU Xiaoyu, YUAN Xiaoming, HUANG Li, WANG Shaobing
耐候无缝钢管 Q355 的开发与应用................ 姚晓乐，张学颖，米永峰，姜海龙，孙文秀，何建中　10-96
Development and Application of Weathering Seamless Steel Tube Q355
.... YAO Xiaole, ZHANG Xueying, MI Yongfeng, JIANG Hailong, SUN Wenxiu, HE Jianzhong
Nb、V 对热煨弯管用 X80 管线钢组织性能的影响研究
.... 李天怡，刘文月，安　涛，臧　岩，刘津伊，薛　峰　10-97
Influence of Nb、V Microalloying on Microstructure and Mechanical Properties of X80 Grade Hot Bent
　　Pipeline Steel................ LI Tianyi, LIU Wenyue, AN Tao, ZANG Yan, LIU Jinyi, XUE Feng
V、Nb 加入量对微合金化钢高温热塑性的影响................ 连小金，陈雪慧，罗小兵，杨才福　10-98
Influence of V、Nb Addition on High Temperature Thermoplasticity of Microalloyed Steel
.... LIAN Xiaojin, CHEN Xuehui, LUO Xiaobing, YANG Caifu
大型桥梁用高品质钢材的研发与应用................ 麻　晗，杨　浩，王　雷，陈焕德　10-99
Nb 含量和变形量对高强度调质钢板淬火再加热奥氏体晶粒尺寸及其分布的影响
.... 邹　扬，张苏渊，秦丽晔　10-100

10.6　电　工　钢

不同 Si、Al 含量对无取向硅钢 50W800 的夹杂物、微观组织及电磁性能的影响研究
.... 高兴健，王思琴　10-101
Study on the Influence of Different Si and Al Contents on Inclusions, Microstructure and Electromagnetic
　　Properties of 50W800 Non-oriented Silicon Steel................ GAO Xingjian, WANG Siqin
取向硅钢环保型表面除鳞技术的生产应用................ 夏强强，刘玉堂，何国芳，周建平　10-102
Production Application of Environmentally Friendly Descaling Technology for Oriented Silicon Steel
.... XIA Qiangqiang, LIU Yutang, HE Guofang, ZHOU Jianping
退火工艺对 27AHSW450 冷轧无取向硅钢组织性能的影响
.... 饶彦俊，宋仁伯，赵知洋，张应超，马为涛　10-103

新一代高技术冷连轧机电工钢同板差板形控制研究进展与创新发展趋势
··曹建国，宋纯宁，孙　磊，赵秋芳，夏文辉，孙双涛 10-104
Research Progress and Innovative Development Trends of Transverse Thickness Difference for Shape Control of Electric Steel in the New-generation High-tech Tandem Cold Rolling Mills
················ CAO Jianguo, SONG Chunning, SUN Lei, ZHAO Qiufang, XIA Wenhui, SUN Shuangtao

无取向硅钢热轧常化工艺研究 ·················张　健，张智义，罗　理，胡万卿，姜福建，张仁波 10-105
Reserch on Hot Rolling Normalization Process of Nonoriented
················ ZHANG Jian, ZHANG Zhiyi, LUO Li, HU Wanqing, JIANG Fujian, ZHANG Renbo

高牌号无取向电工钢 50W270 工艺优化研究····刘文鹏，李亚东，李　浩，陈春梅，孙　超，王祥辉 10-106
The Study on Process Optimization of High Grade Non-oriented Electrical Steel 50W270
················ LIU Wenpeng, LI Yadong, LI Hao, CHEN Chunmei, SUN Chao, WANG Xianghui

薄规格取向硅钢脱碳退火工艺研究················王艺橦，刘旭明，郭　函，耿志宇，薛　峰 10-107

薄规格无取向电工钢冷轧工艺优化··············耿志宇，李亚东，孙　超，刘旭明，董林硕，郭　函 10-108
Optimization of Cold Rolling Process of Thin Gauge Non-oriented Electrical Steel
················ GENG Zhiyu, LI Yadong, SUN Chao, LIU Xuming, Dong Linshuo, GUO Han

取向硅钢硅酸镁层质量的影响因素及其控制研究
··张　刚，胡　煜，赵天亮，胡丞杨，宗　俊，吴开明 10-109
Study on the Influence Factors and Control of Quality of Mg_2SiO_4 Layer on the Surface of Grain-oriented Electrical Steel ······ ZHANG Gang, HU Yu, ZHAO Tianliang, HU Chengyang, ZONG Jun, WU Kaiming

30W1500 无取向硅钢磁性能优化研究············郭飞虎，刘　磊，乔家龙，廖德桥，付　兵，仇圣桃 10-110
Optimization of Magnetic Properties of 30W1500 Non Oriented Silicon Steel
················ GUO Feihu, LIU Lei, QIAO Jialong, LIAO Deqiao, FU Bing, QIU Shengtao

硅钢连续退火炉炭套结瘤机理及防控措施·························何明生，张　敬，龚学成，丁　勇 10-110

电工钢板形控制与优化实践···张　庭，钱震茂，傅潇然 10-111
Control and Optimization of Electrical Steel Plate Shape ········ ZHANG Ting, QIAN Zhenmao, FU Xiaoran

不同加工方式对新能源驱动电机用无取向电工钢磁性能的影响
··吴圣杰，岳重祥，钱红伟，麻　晗 10-112
Effect of Processing Methods on the Magnetic Properties of Non-oriented Electrical Steel for Electrical Vehicle
··· WU Shengjie, YUE Chongxiang, QIAN Hongwei, MA Han

10.7 非晶合金

Ga 对 $Fe_{83.3}Si_2B_{10}P_4Cu_{0.7}$ 纳米晶合金结构及软磁性能的影响···韩成府，李福山 10-113
$FeCrNi_2Nb_{0.1}$ 中熵合金的微观结构调控及其对力学性能的影响···李福山，刘孝威 10-114
金属玻璃力学谱的考察以及粘弹性模型的构造···王　昊，张广强，董帮少 10-114
Investigation of the Mechanical Spectrum of Metallic Glass and the Construction of a Viscoelastic Model
··· WANG Hao, ZHANG Guangqiang, DONG Bangshao

FeSiBC 混合 FeSiCr 或 CIP 复合磁粉芯软磁性能改善及机理分析
··王　璞，朱争取，庞　靖，张家泉 10-115
Si/B 比对高 B_s-FeSiBNbCu 纳米晶合金组织结构及磁性能的影响··············王超人，李艳辉，张　伟 10-116
元素添加对 Finemet 系纳米晶合金高频软磁性能影响研究··················郭树深，李艳辉，张　伟 10-116
难熔金属添加对 FeCoNiSiB 高熵块体非晶合金的形成和性能影响研究···············李艳辉，张　伟 10-117

Cr 添加对 FeBCu 纳米晶合金组织结构及磁性能的影响 ············ 郭明月，薛钧升，李艳辉，张　伟 10-118

高 B_s 纳米晶合金非晶前驱体结构调控及其晶化组织和磁性能研究 ······················· 张　伟 10-118

P 与 Hf 元素置换对 $Fe_{84}Nb_7B_9$ 纳米晶合金组织结构及软磁性能的影响
　·· 李特尼格尔，薛钧升，李艳辉，张　伟 10-119

Fe-Si-B-Cu-Al 系纳米晶合金组织结构与软磁性能的研究 ············ 薛钧升，李艳辉，张　伟 10-119

非晶合金的流动模型和临界阈值准则 ······························ 乔珺威，张　浩，王　重 10-120

成型压力对 FeSiBCuNb 纳米晶磁粉芯电磁性能的影响 ············ 史贵丙，王　丽，王明旭 10-121

中柱可替换型一体成型电感的仿真性能分析 ············ 史贵丙，苗泓毅，王　丽，王明旭 10-122

玻璃态物质弛豫基本单元的探测 ·································· 宋丽建，霍军涛，王军强 10-123

11　节能与低碳技术

大型制氧机精氩塔氩固化在线处理方法的研究 ············ 董昕宏，舒　畅，王政林　11-1

Research on Online Processing Method for Argon Solidification in the Precision Argon Tower of Large Oxygen Generators ·································· DONG Xinhong, SHU Chang, WANG Zhenglin

关于提升炼钢余热蒸汽过热度增加发电量的措施 ············ 郑振海，薛晓金，张明扬，张元华　11-1

Measures for Increasing the Superheat of Steam from Waste Heat of Steelmaking to Increase Power Generation
　·················· ZHENG Zhenhai, XUE Xiaojin, ZHANG Mingyang, ZHANG Yuanhua

包钢厂内鱼雷罐车加盖装置的研究与应用 ············ 王少龙，刘　强，刘俊杰　11-2

Research and Application of the Covered Technology of Torpedo Cars in Baotou Steel Factory
　·· WANG Shaolong, LIU Qiang, LIU Junjie

高炉煤气干煤气直送管网技术研究 ············ 吕　勇，贾冬颖，胥中平　11-3

Research on the Technology of Dry Gas Direct Delivery Pipeline Network for Blast Furnace Gas
　·· LV Yong, JIA Dongying, XU Zhongping

减温减压器（RTP）减温水系统改造 ············ 张腾飞，王　晨　11-4

Retrofit of Temperature and Pressure Reducing Device (RTP) Desuperheating Water System
　·· ZHANG Tengfei, WANG Chen

基于煤气调配途径的轧钢加热炉烟气污染物达标排放对策研究
　·· 周劲军，刘自民，唐嘉瑞，周　栩　11-5

Research on Emission Countermeasures of Flue Gas Pollutants in Steel Rolling Heating Furnace Based on Gas Allocation Method ·············· ZHOU Jinjun, LIU Zimin, TANG Jiarui, ZHOU Xu

基于低碳冶金的碳捕集工艺 ············ 潘　阳，马方曙，陈　誉，陈露露，马志强　11-5

Comparison of Carbon Capture Process Configurations for Low Carbon Metallurgy
　·················· PAN Yang, MA Fangshu, CHEN Yu, CHEN Lulu, MA Zhiqiang

煤气加压机的振动分析和处理 ············ 高子丰，郑　晨，李星星，王龙锋　11-6

Vibration Analysis and Treatment of Gas Compressor
　·· GAO Zifeng, ZHENG Chen, LI Xingxing, WANG Longfeng

连续退火机组平整液循环利用技术开发与实践 ············ 何建锋，潘勋平，胡　斌　11-7

Research and Application of Recycling Technology for Cold-rolling Temper Lubricant
　·· HE Jianfeng, PAN Xunping, HU Bin

钢铁流程能-碳解耦与碳排放解析 ············ 蒋滨繁，熊　瑞，程春云，田卓航，夏德宏　11-7

鞍钢 2、3 高炉水冲渣节能减排生产实践 ………………………………………… 王佳熙，孙成国　11-9
Energy Saving and Emission Reduction Production Practice of Water Flushing Slag for Angang 2 and 3 Blast Furnaces ………………………………………………………… WANG Jiaxi, SUN Chengguo
冲渣外排水回收利用 ………………… 王　维，滕福亮，刘　壮，罗　涵，刘百志，王希坤，栗　艳　11-9
高炉区域工业新水回收利用实践 …… 王　维，王丽芳，罗　涵，王希坤，刘百志，周　鑫，刘鹏羽　11-10
高炉煤气醇胺法 CO_2 吸收塔设计 ………………………………… 王博林，李　伟，韩渝京，李　鹏　11-10
Design of CO_2 Absorption Tower with the Method of MEA for Blast Furnace Gas
　………………………………………………………… WANG Bolin, LI Wei, HAN Yujing, LI Peng
钢铁企业余热余能回收利用措施研究 ……………………………………………………… 范金龙　11-11
Research on Measures for the Recovery and Utilization of Residual Heat and Energy in Iron and Steel Enterprises
　……………………………………………………………………………………………… FAN Jinlong
钢铁行业中能源管理系统应用分析 ………………………………………………………… 范金龙　11-11
Application Analysis of Energy Management System in the Steel Industry ……………… FAN Jinlong
高温高压煤气发电的应用 …………………………………………………………… 姚震宇，张海艳　11-12
Application of High Temperature and High Pressure Gas Power Generation …… YAO Zhenyu, ZHANG Haiyan
氨气还原铁氧化物的反应特性及动力学研究 ……………………… 刘玥君，李先春，李　丽，林建廷　11-13
简述钢铁企业能源管理及智能化的发展现状与展望
　…………………………………………… 陈　琛，倪书权，徐继法，邱明英，李加旺，史　光　11-14
低碳发展背景下炼铁工艺革新的新思考 …………………………… 金永龙，孙宇佳，郝良元，田京雷　11-15
New Thinking on Ironmaking Process Innovation under the Background of Low Carbon Development
　…………………………………………… JIN Yonglong, SUN Yujia, HAO Liangyuan, TIAN Jinglei
基于生命周期理论的汽车钢碳足迹评价 …………………………………………… 房晓晴，孙文强　11-16
大型钢铁企业焦炉煤气柜全离线情况下的煤气平衡实践 ………………………… 吕　军，黄　腾　11-16
Gas Balance Practice of Large Iron and Steel Enterprise under Coke Oven Gas Tanks are All Offline
　……………………………………………………………………………… LV Jun, HUANG Teng
转底炉系统风机配置与节能优化分析 ……………………………………………………… 朱红兵　11-17
面向转炉炼钢的能源管控系统设计与实现 ……………… 李霜青，赵　菲，谷　牧，王　宏，郑　磊　11-18
Design and Implementation of Energy Control System for Converter Steelmaking
　………………………………………… LI Shuangqing, ZHAO Fei, GU Mu, WANG Hong, ZHENG Lei
自备电厂煤粉锅炉掺烧生物质试验 …………… 郁　雷，邱全山，刘自民，曹曲泉，顾　骁，赵　华　11-19
Experiment on Blending Biomass in Pulverized Coal Boiler of Self-provided Power Plant
　………………………………… YU Lei, QIU Quanshan, LIU Zimin, CAO Ququan, GU Xiao, ZHAO Hua
复杂工况下钢包全程保温技术装备研发与应用 …………………… 赵金龙，杨轶龙，吴恩旭，张洪亮　11-20
Research and Development and Application of Ladle Insulation Technology and Equipment under Complex Working Conditions …………………… ZHAO Jinlong, YANG Yilong, WU Enxu, ZHANG Hongliang
朝阳钢铁智慧水务管理探索与研究 …………… 黄永梁，潘大龙，王念军，鲁　璐，刘　洋，王艳霞　11-21
Experience Summary of Smart Water Management of Chaoyang Iron and Steel
　………………………… HUANG Yongliang, PAN Dalong, WANG Nianjun, LU Lu, LIU Yang, WANG Yanxia
钢铁企业高炉氧气消耗预测模型开发及应用 …… 刘　凯，刘磊刚，宋岩峰，覃天强，董惠文，王亚腾　11-21
Development and Application of Oxygen Consumption Prediction Model for Blast Furnace in Iron and Steel Enterprises …… LIU Kai, LIU Leigang, SONG Yanfeng, QIN Tianqiang, DONG Huiwen, WANG Yateng
绿色低碳，高质量发展-磁悬浮风机在焦化生化站应用 ………………………… 木塔力甫·玉苏甫，韩　亮　11-22

浅谈 BTG 亚临界发电机组煤气-烟气换热器应用问题解析 ·················· 郝 鑫，王 浩　11-23
Analysis of Application Problems of Gas Flue-Gas Heat Exchanger in BTG Subcritical Power Generation Units
　·· HAO Xin, WANG Hao
焦炉煤气能源绿色转换技术开发与应用 ·· 刁呈振　11-24
Development and Application of Green Conversion Technology for Coke Oven Gas Energy
　·· DIAO Chengzhen
芜湖新兴烧结工序能效提升生产实践 ··········· 魏瑞瑞，楚天福，段龙龙，于恩斌，朱玉停，武 彪　11-24
Wuhu Xinxing Sintering Process Energy Efficiency Improvement Production Practice
　·· WEI Ruirui, CHU Tianfu, DUAN Longlong, YU Enbin, ZHU Yuting, WU Biao
转炉余热蒸汽蓄热系统的技术改造实践 ···················· 程 鑫，饶江平，朱善合，王 军，汪文婷　11-26
Utilization of Technical Transformation on Storage System for Saturated Steam in Steel-making
　··· CHENG Xin, RAO Jiangping, ZHU Shanhe, WANG Jun, WANG Wenting
供热电站蒸汽管网优化应用分析 ·· 操加元，程 海，严 亮，朱善合　11-26
Analysis on Optimum Usage of Steam Pipeline in Heat Power Station
　··· CAO Jiayuan, CHENG Hai, YAN Liang, ZHU Shanhe
转炉汽化冷却蒸汽平衡与高效利用技术研究 ···· 饶江平，朱善合，付刚强，王 军，彭有杰，程 鑫　11-27
Research on Steam Balance of Vaporization Cooling and Efficient Utilization in Steel-making Converter
　················· RAO Jiangping, ZHU Shanhe, FU Gangqiang, WANG Jun, PENG Youjie, CHENG Xin
多变内外部环境下的能源高效利用，降低外购能源费用 ·· 王 剑　11-28
微波场下 H_2 和 CO 还原磁铁矿的微观结构比较 ··········· 洪陆阔，周美洁，艾立群，孙彩娇，佟 帅　11-28
Comparison of Microstructures of Magnetite Reduced by H_2 and CO under Microwave Field
　·· HONG Lukuo, ZHOU Meijie, AI Liqun, SUN Caijiao, TONG Shuai
浅谈氢冶金工程低压供配电系统 ·· 王 伟　11-29

12　冶金环保与资源利用

12.1　冶金环保

复合菌剂增效处理焦化废水的研究与应用 ······· 王 永，王 飞，刘 芳，安 宁，李函霏，张天赋　12-1
The Research and Application on Coking Waste-water Treatment Through Combined Bacteria
　····································· WANG Yong, WANG Fei, LIU Fang, AN Ning, LI Hanfei, ZHANG Tianfu
钢企高炉煤气前端脱硫与煤气用户末端脱硫的技术经济比较 ··············· 任海霞，王志荃，文浩锦　12-1
Technological and Economic Comparison between Front Fine Desulphurization of Blast Furnace Gas and
　Terminal Desulphurization of Gas User ················ REN Haixia, WANG Zhiquan, WEN Haojin
焦炉煤气精脱硫工艺分析及设计优化 ·· 邓万里，杨 静，李 霁　12-2
Process Analysis and Design Points of Fine Desulfurization of Coke Oven Gas
　··· DENG Wanli, YANG Jing, LI Ji
高炉煤气精脱硫系统工艺研究 ·· 胡 伟，董映红，耿云梅，郭利朋　12-3
Research on the Process of Blast Furnace Gas Fine Desulfurization System
　·· HU Wei, DONG Yinghong, GENG Yunmei, GUO Lipeng
一条线材深加工企业工业污水处理线的设计与实践
　·· 赵宪海，金晓谦，祝 贺，唐 盈，范思豪，谈志昊　12-4

Design and Practice of an Industrial Sewage Treatment Line for Wire Rod Deep Processing Enterprises
　　　　　　　　　　　ZHAO Xianhai, JIN Xiaoqian, ZHU He, TANG Ying, FAN Sihao, TAN Zhihao

高线精轧机辊箱稀油污染源的检测方案………………………………卢永清，吴春东，张建业　12-4

酸再生系统常见故障及改进措施………………………………………………………党文文　12-5
Common Faults and Improvement Measures of Acid Regeneration System …………… DANG Wenwen

循环冷却水应急除氰可行性研究…………………………………………………………蒋青利　12-5

聚合硅酸铝钙处理焦化废水中氟化物的实验研究………王　飞，王　永，刘　芳，安　宁，李函霏　12-6
Experimental Study on the Treatment of the Wastewater Containing Fluoride by Using
　　Poly-silicate-aluminum-calcium ……………… WANG Fei, WANG Yong, LIU Fang, AN Ning, LI Hanfei

烧结烟气错流式活性焦脱硫脱硝技术特征与工程实践
　　……………………………………李　林，辻孝典，初　霖，李　健，张　涛，尚久石，李彦哲　12-6
Technical Characteristics and Engineering Practice of Cross-flow Activated Coke Desulfurization and
　　Denitrification for Sintering Flue Gas
　　……………………LI Lin, TSUJI Takanori, CHU Lin, LI Jian, ZHANG Tao, SHANG Jiushi, LI Yanzhe

烧结烟气错流式活性焦干法净化工艺的构建
　　……………………………………李　林，初　霖，辻孝典，李　健，张　涛，尚久石，李彦哲　12-7
Construction of Activated Coke Desulfurization and Denitrification Process for Sintering Flue Gas
　　……………………LI Lin, CHU Lin, TSUJI Takanori, LI Jian, ZHANG Tao, SHANG Jiushi, LI Yanzhe

探地雷达技术在地下供水管道漏损检测中的应用研究
　　………………………………………………陈　鹏，安　宁，胡绍伟，李函霏，王　飞，刘　芳　12-8
Application Research of Ground Penetrating Radar Technology in Leakage Detection of Underground Water
　　Pipeline ………………………………… CHEN Peng, AN Ning, HU Shaowei, LI Hanfei, WANG Fei, LIU Fang

外旋式大型旋流沉淀池在热轧水处理中的改进探讨………赵治国，李　湧，杨富刚，蔡　毅，谭　焕　12-9
Discussion on the Improvement of External Spinning Extra-large Cyclone Well in Hot Rolled Water
　　…………………………………………………… ZHAO Zhiguo, LI Yong, YANG Fugang, CAI Yi, TAN Huan

基于智慧取水的钢厂纯水制备系统废水减排改造实践………………丁宗琪，王慧军，曹小兵，李　璐　12-9
Renovation Practice of Water Extraction System in Steel Plants Based on Intelligent Water Extraction
　　…………………………………………………… DING Zongqi, WANG Huijun, CAO Xiaobing, LI Lu

转炉烟气余热、化学能回收及超低排放技术综述与展望
　　…………………………………………陈　琛，倪书权，徐继法，李加旺，郝景章，徐　蕾，史　光　12-10

钢铁行业CO_2资源化利用技术应用现状及发展趋势
　　………………………………………………王雪琦，王改荣，李鹏阳，张彩东，李兰杰，田志强　12-10
Application Status and Development Trend of CO_2 Resource Utilization Technology in the Steel Industry
　　………………WANG Xueqi, WANG Gairong, LI Pengyang, ZHANG Caidong, LI Lanjie, TIAN Zhiqiang

八钢烧结机头烟气脱硫脱硝超低排放改造浅析………………………………………………王　宁　12-11
Analysis on Ultra-low Emission Transformation of Flue Gas Desulfurization and Denitration in the Sintering
　　Machine Head of Bagang ……………………………………………………………………… WANG Ning

混凝-浸没式超滤短流程工艺处理钢铁综合废水的实验研究………杨建峡，余云飞，陈思雨，邱利祥　12-12
Experimental Study on the Treatment of the Iron and Steel Plant Wastewater by Using Coagulation-Immersion
　　Ultrafiltration Technology ……………………… YANG Jianxia, YU Yunfei, CHEN Siyu, QIU Lixiang

水质悬浮物测定影响因素分析及流程改进……………………………………………………付志军　12-13

Analysis and Improvement of Influencing Factors for Determination of Suspended Solids in Water Quality
······ FU Zhijun
环保抑尘处理技术在钢铁生产全流程的应用和发展······徐言东，韩 爽，白金龙，王占坡，张瑞新 12-13
活性焦脱硫脱硝生产实践及优化建议······邵久刚，毛 瑞，苏 航，李 涛 12-14
Practical Application of Flue Gas Desulfurization and Denitrification by Activated Coke and Optimal Suggestions
······ SHAO Jiugang, MAO Rui, SU Hang, LI Tao
烧结烟气挥发性有机物及氮氧化物协同减排研究······吕大友，申明锐，王毅璠，龙红明 12-15
八钢炼铁厂翻车机负压收尘技术运用实践······王 宁 12-16
Application Practice of Negative Pressure Dust Collection Technology for Dumper at Bayi Iron and Steel Ironmaking Plant ······ WANG Ning
283.15K 和 353.15K 下 Na_2SO_4-$Na_2S_2O_3$-$NaSCN$-H_2O 四元相平衡及其在焦化脱硫废液中的应用
······阳露波，王建山，张 衡，张小龙，于云涛 12-17
Quaternary Phase Equilibria of Na_2SO_4-$Na_2S_2O_3$-$NaSCN$-H_2O at 283.15K and 353.15K and Its Application in Coking Desulfurization Waste Water
······ YANG Lubo, WANG Jianshan, ZHANG Heng, ZHANG Xiaolong, YU Yuntao

12.2 冶金固废资源综合利用

火电厂掺烧固废热力计算前提条件确定应注意的问题分析与解决建议······张鲁峻，贺 宇，徐新文 12-18
钢铁行业历史固废堆场系统性环境深度治理研究与实践······刘剑平，王如意，高 亮 12-18
Research and Practice to Depth Treatment the Solid Wastes Landfill in Steel Industry
······ LIU Jianping, WANG Ruyi, GAO Liang
球磨粒子钢热压块出钢水率检测方法研究······窦 猛，戴 伟，聂文金，皋 萍，王晓东 12-19
Study on the Method of Measuring the Rate of Molten Steel Output of Hot Briquetting Iron of Ball Mill Particle Steel ······ DOU Meng, DAI Wei, NIE Wenjin, GAO Ping, WANG Xiaodong
炼钢中低价物料使用与替代的生产实践······李 超，尚德义，何 冲，冉茂铎 12-20
Production Practice of the Economic Materials Using in Steelmaking
······ LI Chao, SHANG Deyi, HE Chong, RAN Maoduo
宝钢二次资源消纳实践······吴旺平，王跃飞，鲁 健 12-21
Application of Water-saving Utilization for Raw Material Yard in Baosteel
······ WU Wangping, WANG Yuefei, LU Jian
冷轧磁过滤油泥处理技术实验与思考······徐鹏飞，杨大正，耿继双，王 飞，吴文浩 12-21
Experiment and Thought on Magnetic Filtered Oily Cold-rolling Mill Sludge Treatment Technology
······ XU Pengfei, YANG Dazheng, GENG Jishuang, WANG Fei, WU Wenhao
电炉灰循环减量及资源化利用技术探索与实践
······饶 磊，刘自民，桂满城，张耀辉，马孟臣，刘英才 12-22
Exploration and Practice of Electric Furnace Ash Recycling Reduction and Resource Utilization Technology
······ RAO Lei, LIU Zimin, GUI Mancheng, ZHANG Yaohui, MA Mengchen, LIU Yingcai
烧结-高炉法协同处置多源铬渣和有机固废新技术······涂义康，张元波，苏子键，姜 涛 12-23
Collaborative Disposal of Chromium Slag and Organic Solid Waste Via a Novel Technology of Composite Agglomeration Process (CAP) ······ TU Yikang, ZHANG Yuanbo, SU Zijian, JIANG Tao
电炉粉尘资源化转底炉工艺试验研究······李 林，野田悦郎，徐 萌，中山俊孝，辻孝典，李彦哲 12-24

Process Test Research of Rotary Hearth Furnace for Resource Utilization of Electric Arc Furnace Dust
　……………… LI Lin, NODA Etsuro, XU Meng, NAKAYAMA Toshitaka, TSUJI Takanori, LI Yanzhe
高效处置钢铁企业含锌固废技术实践……………………………孙宇佳，金永龙，田京雷，王　倩　12-25
Practice on Efficient Disposal of Zinc Containing Solid Waste from Iron and Steel Enterprises
　………………………………………………… SUN Yujia, JIN Yonglong, TIAN Jinglei, WANG Qian
含铁锌固废资源化再生关键工艺技术研究………………………………陈方元，李菊艳，徐永斌　12-25
Research on Key Process Technologies for Resource Recycling of Solid Waste Containing Iron and Zinc
　……………………………………………………………… CHEN Fangyuan, LI Juyan, XU Yongbin
矿渣-钢尾渣基充填材料配比优化研究………………代梦博，顾宝澍，刘　斌，孙业长，春铁军　12-26
Optimization of the Ratio of Steel Slag Tailings-GGBS Based Backfill Material
　…………………………………… DAI Mengbo, GU Baoshu, LIU Bin, SUN Yechang, CHUN Tiejun
从高炉瓦斯灰中综合回收碳和铁试验研究……………………………………………… 刘兴华　12-27
Experimental Study on Comprehensive Recovery of Carbon and Iron from Blast Furnace Gas Ash
　…………………………………………………………………………………………… LIU Xinghua
钢渣粗、细集料制备沥青混合料及施工特性研究综述
　…………………………………………………………… 王子鹏，马宏伟，孙宇佳，周永祥　12-28
Summary of Research on the Preparation and Construction Characteristics of Asphalt Mixture with Steel Slag
　Coarse and Fine Aggregates ……………… WANG Zipeng, Ma Hongwei, SUN Yujia, ZHOU Yongxiang
热镀锌渣原位回收技术研究……………初仁生，李　研，王　哲，李翔宇，李战军，郭占成　12-28
钢渣磨料制备与研磨机理仿真研究……………裴晶晶，邢宏伟，张玉柱，吴金虎，林文龙，霍文青　12-29
梯级分离钢铁粉尘中 Zn-Pb-In 的工艺与展望……………张良进，张玉柱，龙　跃，杜培培，王贵华　12-30
Process and Prospect of Cascade Swparating Zn-Pb-In in Steel Dust Based
　……………………………… ZHANG Liangjin, ZHANG Yuzhu, LONG Yue, DU Peipei, WANG Guihua
Cr_2O_3 对含铬熔分钛渣黏流特性的影响……………………………曾睿琪，王　楠，李　尉　12-31
Influence of Cr_2O_3 on the Viscous Flow Behavior of Cr-Containing Molten Titanium Slag
　………………………………………………………………………… ZENG Ruiqi, WANG Nan, LI Wei
脱磷渣和转炉渣协同处理与磷的浸出回收……………………………………于耀辉，杜传明　12-32
120t 转炉钢渣热熔压块使用效果研究…………李廷刚，赵和明，甄新刚，张志强，何世长，范祚驿　12-33
The Using Effect Analysis of 120t LD Using Steel Slag Melt Pressure Block
　…………… LI Tinggang, ZHAO Heming, ZHEN Xingang, ZHANG Zhiqiang, HE Shichang, FAN Zuoyi
烟气脱硫制酸废液返回烧结处理对活性炭性能的影响
　………………………李俊杰，陶家杰，齐鹏宇，丁　龙，钱立新，春铁军，龙红明　12-34
Effect of Return Iron Ore Sintering of Acid Waste Water from Flue Gas Desulfurization on Properties of Activated
　Carbon ……LI Junjie, TAO Jiajie, QI Pengyu, DING Long, QIAN Lixin, CHUN Tiejun, LONG Hongming
赤泥微波碳热还原熔分提铁工艺技术研究……施树蓉，胡倩倩，丛云伶，张军红，高立华，何志军　12-35
Research on Microwave Carbon-Thermal Reduction Smelting and Iron Extraction from Red Mud.
　……………… SHI Shurong, HU Qianqian, CONG Yunling, ZHANG Junhong, GAO Lihua, HE Zhijun

13　冶金设备与工程技术

X 射线荧光光谱法分析热压铁块中主次成分……………禹青霄，杨晓倩，孙少海，杨慧贤，牛威斌　13-1

Determination of Major and Minor Components of Hot Briquette Iron by X-ray Fluorescence Spectrometry
……………………………YU Qingxiao, YANG Xiaoqian, SUN Shaohai, YANG Huixian, NIU Weibin
氢化物发生-电感耦合等离子体原子发射光谱法测定钼铁中砷锡锑铋……………………………葛晶晶 13-1
Determination of Arsenic, Tin, Antimony, Bismuth in Molybdenum Iron by Hydride Generation-inductively
　Coupled Plasma Atomic Emission Spectrometry ……………………………………GE Jingjing
Python 与设备点检管理相结合的实践与运用…………………………………………周子珍，茹剑波 13-2
The Practice and Application of Python and Equipment Inspection Management
　…………………………………………………………………………… ZHOU Zizhen, RU Jianbo
浅析大型液压传动系统温度高的问题………………………………………………………梁　婷 13-3
炼钢连铸新型高温复合辊断裂失效机理分析………………………………………………廖礼宝 13-3
Mechanism Analysis on fracture-failure of High Temperature and New Type Composite Roller in
　Continuous-casting Segment ……………………………………………………………LIAO Libao
冶金旋流井中水流作用下抓渣斗的受力分析与防缠绕防护罩装置
　……………………………崔忠信，孙立红，冯小雷，田绍鹏，张大伟，田　兴 13-4
Force Analysis about Water Flow in a Metallurgical Cyclone Well on the Grab and Its Protective Device
　……………CUI Zhongxin, SUN Lihong, FENG Xiaolei, TIAN Shaopeng, ZHANG Dawei, TIAN Xing
全站仪在激光焊机小车行走直线度检测中的应用
　………………… 赵新宝，高　超，兰晓栋，玄利剑，唐晓宇，孟召锦，刘　昊，司　泽 13-5
The Application of Total Station in Detecting the Straightness of Laser Welding Machine Trolley
　………………………………………………………… ZHAO Xinbao, GAO Chao, LAN Xiaodong,
　　　　　　　　　　　　　　　　　XUAN Lijian, TANG Xiaoyu, MENG Zhaojin, LIU Hao, SI Ze
结合视觉与图谱分析的带钢表面质量判定研究
　……………………………吴昆鹏，邓能辉，郭亚男，郭　爽，石　杰，杨朝霖 13-6
Research on Surface Quality Evaluation of Strip Steel by Combining Visual Inspection and Graph Analysis
　　WU Kunpeng, DENG Nenghui, GUO Ya'nan, GUO Shuang, SHI Jie, YANG Chaolin
高压自密封阀门四开环改进……………………………………赵　宇，贾树勋，侯　新，李雪松 13-7
全氢罩式炉加热罩的使用与维修…………………………………………………………吴立军 13-7
The Use and Maintenance of Heating Hood of Full Hydrogen Bell Furnace ………………WU Lijun
基于三维建模分析的热轧飞剪前侧导板改造
　………………… 王骏超，吴长杰，张会明，屈二龙，醋亚辉，尹贵祖，郭维进，东占萃 13-8
Transformation of Entry Guide of Corp-shear on HSM Line Based on 3D Modeling Analysis
　………………… WANG Junchao, WU Changjie, ZHANG Huiming, QU Erlong,
　　　　　　　　　　　　　　　　　　　LI Yahui, YIN Guizu, GUO Weijin, DONG Zhancui
精轧机增加小立辊的设计与实施…………………吴长杰，张会明，东占萃，郭维进，张志桥，王艺霖 13-9
Design and Application of Adding Edge Roll on Finish Mill
　　　WU Changjie, ZHANG Huiming, DONG Zhancui, GUO Weijin, ZHANG Zhiqiao, WANG Yilin
热轧钢板高速飞剪过程的有限元模拟………………………………………李　轲，陈　敏，高爱民 13-9
Finite Element Simulation of Hot Rolled Steel Plate Flying Shear Process at High Speed
　……………………………………………………………………………… LI Ke, CHEN Min, GAO Aimin
轧机油膜轴承润滑系统稳定性控制……………………………………武明明，王金星，宋　波 13-10
冷轧 20 辊轧机张力闭环控制技术研究 …………………秦大伟，张　栋，孙瑞琪，张　岩，刘宝权 13-11

Research on Tension Closed-loop Control Technology of 20-high Cold Rolling Mill
·················· QIN Dawei, ZHANG Dong, SUN Ruiqi, ZHANG Yan, LIU Baoquan

冷连轧带钢变规格控制优化·················潘建华，王弢，柳军 13-11
Optimization of Dynamic Gauge Changing Control for Tandem Cold Mill
·················· PAN Jianhua, WANG Tao, LIU Jun

冷轧厂退火炉辐射管烧嘴失效案例分析·················姚舜 13-12
Failure Case Analysisof Radiant Tube Burner of Annealing Furnace in Cold Rolling Mill ········ YAO Shun

650mm棒材轧机2号立式机架地脚螺栓断裂原因分析与修复措施·················刘宏，戴江波 13-13

连续酸洗机组入口段带钢跑偏分析及解决方法·················李建文，余卫军，魏静 13-13
Analysis and Control Method of Strip Deviation in the Entrance Section of Normalized Pickling Unit
·················· LI Jianwen, YU Weijun, WEI Jing

多区段微尺度辊形电磁调控技术理论研究
·················杨庭松，袁铁衡，雍晟一，孙文权，荆丰伟，杜凤山，许志强，何安瑞 13-14

轧机齿轮箱输入轴轴承外圈偏心套压溃分析·················李建文，熊勇 13-15
Crushing Analysis of Eccentric Outer Ring of Bearing Input Shaft of Rolling Mill Gear Box
·················· LI Jianwen, XIONG Yong

转炉支撑机构现状研究与趋势分析·················余杨，王庆，曾鸣 13-15
Current Research and Trend Analysis of Converter Support Mechanism
·················· YU Yang, WANG Qing, ZENG Ming

转炉运行品质影响因素分析和研究·················余杨，王庆，曾鸣 13-16
Analysis and Research on Influencing Factors of Running Quality of Converter
·················· YU Yang, WANG Qing, ZENG Ming

滚筒飞剪传动计算研究·················王蕾，常铁柱，张海东，韩文，黄晓慧 13-17
Study on Calculation of Roller Flying Shear Transmission
·················· WANG Lei, CHANG Tiezhu, ZHANG Haidong, HAN Wen, HUANG Xiaohui

"一罐到底"铁水罐罐盖开闭盖装置的研制·················项克舜 13-18
The Development of "One Can to the End" Iron Water Can Lid Opening and Closing Device
·················· XIANG Keshun

宝钢冷轧激光焊机机械精度智能检测装备的研发·················周为民，屈军杰，赵春涛，潘红良 13-18
R&D of Intelligent Inspection Equipment for Welder
·················· ZHOU Weimin, QU Junjie, ZHAO Chuntao, PAN Hongliang

中频感应熔炼炉升级改造研究·················徐凤娟，王东海，耿军，赵福星，杨晓勇，王国良 13-19
Research on Upgrading and Renovation of Medium Frequency Induction Melting Furnace
·················· XU Fengjuan, WANG Donghai, GENG Jun, ZHAO Fuxing, YANG Xiaoyong, WANG Guoliang

绿色数智化流体系统推进酸轧联合机组工艺技术变革再造
·················毛召芝，沈志前，张海东，常铁柱，贾大朋，马丽坤，黄晓慧 13-20
Green Digital Intelligent Fluid System Promotes Process Technology Transformation and Reconstruction of PL-TCM ·················· MAO Zhaozhi, SHEN Zhiqian, ZHANG Haidong, CHANG Tiezhu, JIA Dapeng, MA Likun, HUANG Xiaohui

电工钢常化酸洗机组设备组成及特点·················马丽坤，谢天华，耿庆斌，张华 13-21
Equipment Composition and Characteristic of the Normalizing and Pickling Line for Electrical Steel
·················· MA Likun, XIE Tianhua, GENG Qingbin, ZHANG Hua

薄带铸轧线分卷飞剪机的设计与应用·················常铁柱，王蕾，张海东，黄晓慧 13-21

A Type of Flying Shear for Ultra-thin Strip Casting Rolling Line
······CHANG Tiezhu, WANG Lei, ZHANG Haidong, HUANG Xiaohui

160T 铁水车轮对烧轴故障分析及对策 ······ 易文祥 13-22

精品轨梁轧制工艺、装备与控制关键技术研发及应用
······陈莹卷，周 民，樊泽兴，郭显平，谭成楠，马斯江 13-22

无取向硅钢中超低硫检出下限的延伸及其准确性研究
······易 赛，王 鹏，齐 郁，叶 静，张兆雄，李三红 13-23

Study on the Extension and Accuracy of the Detection Lower Limit for Ultra Low Sulfur in Non-oriented Silicon Steel ······ YI Sai, WANG Peng, QI Yu, YE Jing, ZHANG Zhaoxiong, LI Sanhong

智能型多功能油液监测分离的控制技术 ······沈 强，任佩剑，付乾坤，董耀元，曲晓东，王 健 13-24

Control Technology of Intelligent Multifunctional Oil-water Monitoring and Separation
······SHEN Qiang, REN Peijian, FU Qiankun, DONG Yaoyuan, QU Xiaodong, WANG Jian

低碳背景下废钢铁及 LIBS 技术在废钢判级的研究进展 ······ 刘艳丽，刘 洁，安治国 13-25

Research Progress of Scrap Steel and LIBS Technology in Scrap Grading under Low Carbon Background
······ LIU Yanli, LIU Jie, AN Zhiguo

新型低重量容积比钢包设计与应用 ······ 杨 阳，徐 勇 13-26

Design and Application of the Navel Low Weight-Volume Ratio Ladle
······ YANG Yang, XU Yong

中厚板电磁超声探伤设备的开发与应用 ······ 徐 科，田士辉，姜家宏，师英杰 13-26

接触式板形仪应用技术综述和发展趋势浅析 ······ 俞鸿毅，王 劲，石 燕 13-27

Overview of the Application Technology and Analysis of Development Trends of Contact Shape Meter
······ YU Hongyi, WANG Jin, SHI Yan

不锈钢复合板在热风炉炉壳工程施工中的应用 ······ 何泽凤，李成杰 13-28

Application of Stainless Steel Composite Plate in the Construction of the Hot Blast Furnace Shell
······ HE Zefeng, LI Chengjie

浅析多探头中厚板自动探伤仪精度评价方法 ······ 王立坚，黄 毅，闫智平 13-28

Analyses on Evaluation Method of More Probe Automatic Ultrasonic Testing Equipment Accuracy for Heavy Plate ······ WANG Lijian, HUANG Yi, YAN Zhiping

大型转炉耳轴轴承损坏原因分析及措施制定 ······ 白 华，李 伟，向忠辉，章勤奋 13-29

Analysis of the Causes of Damage to the Ear Shaft Bearing of a Large Converter and the Formulation of Measures ······ BAI Hua, LI Wei, XIANG Zhonghui, ZHANG Qinfen

S214 通廊结构问题分析与加固方案 ······ 任 强，廖生伟，谭 磊 13-30

Structural Problem Analysis and Reinforcement Scheme of the S214 Corridor
······ REN Qiang, LIAO Shengwei, TAN Lei

浅谈高炉炉役后期炉身故障的处理方法 ······ 陈建平，席强华，刘 嘉，陈长鑫 13-30

Talking about the Handling Methods of Late-stage Furnace Body Faults in Blast Furnaces
······ CHEN Jianping, XI Qianghua, LIU Jia, CHEN Changxin

某钢铁基地 10kV 供电系统继电保护定值整定优化 ······ 后程瑞，许子义，成奕佳，李金涛，赵志芸 13-31

Optimization of Relay Protection Setting for 10kV Power Supply System in a Certain Steel Base
······ HOU Chengrui, XU Ziyi, CHENG Yijia, LI Jintao, ZHAO Zhiyun

考虑温度载荷的桥式起重机疲劳失效原因分析 ······ 章勤奋，师 骅，孙海峰，邓 杨 13-32

Analysis of Fatigue Failure Causes of Bridge Crane in Considering with Temperature Load
　　……………………………………………………… ZHANG Qinfen, SHI Hua, SUN Haifeng, DENG Yang
钛、铝和不锈钢连续轧制复合机组装备及工艺的研究
　　………… 龚　辉，刘德华，牛　艳，付建辉，胡志忠，杨可托，李文武，段　粉，苗锁周　13-32
板带热轧组合立辊的应用 ……………………………………………… 马忠辉，刘　娜，李艳红　13-33
Application of Combined Vertical Roller for Hot Rolling of Plate and Strip
　　…………………………………………………………………… MA Zhonghui, LIU Na, LI Yanhong
主从控制在冶金装备中的典型应用及失控分析 ………… 程　曦，刘　浩，钟　实，丁继亚，黄亚军　13-33
Typical Application and Uncontrolled Analysis of Master Slave Control in Metallurgical Equipment
　　………………………………………… CHENG Xi, LIU Hao, ZHONG Shi, DING Jiya, HUANG Yajun
基于时频包络谱峭度的轴承故障冲击特征提取方法 ……………………… 刘　晗，程本俊，邬忠尧　13-34
Extraction Method of Bearing Fault Impact Characteristics Based on Time-frequency Envelope Spectral
　　Steepness ………………………………………………… LIU Han, CHENG Benjun, WU Zhongyao
智能巡检机器人在 220kV 变电站系统中研究与应用 ……………………………………… 苟俊涛　13-35
Research and Application of Intelligent Inspection Robot in 220kV Substation System ………… GOU Juntao
旋风除尘卸灰控制系统改进设计 …………………………………………………………… 付守壮　13-35
Improvement Design of Cyclone Dust Removal and Ash Discharge Control System ……… FU Shouzhuang
相控阵探伤技术在车轮样轮校准中的问题分析及解决方法
　　…………………………………………………… 程松林，程旭辉，陶国强，刘海波，张盛华　13-36
冷轧电工钢薄板月牙剪切过程仿真分析研究 ……………… 陈　兵，王　洋，徐厚均，何俊烨　13-37
Finite Element Simulation Analysis of Shear Fracture Process of Electrical Steel Thin Plate Crescent
　　………………………………………………………… CHEN Bing, WANG Yang, XU Houjun, HE Junye
锌合金中稀土元素测定方法研究 ……………… 张　杰，田秀梅，王　伟，李　颖，王一凌　13-37
Study on Determination Method of Rare Earth Elements in Zinc Alloy
　　………………………………………… ZHANG Jie, TIAN Xiumei, WANG Wei, Li Ying, WANG Yiling
桥式起重机典型故障的研究与解决方案 …………………………………………………… 苏二龙　13-38
Research and Solutions for Typical Faults of Overhead Traveling Cranes ……………………… SU Erlong
智能化新能源钢卷车在现代钢厂冷轧车间的应用 …………… 韦富强，孟祥军，张　建，李永卿　13-39
Application of Intelligent New Energy Coil Car in Cold Rolling Workshop of Modern Steel Plant
　　……………………………………………… WEI Fuqiang, MENG Xiangjun, ZHANG Jian, LI Yongqing
UCMW 轧机万向十字接轴常见损坏现象及应对措施 ……………………………… 王承刚，玄利剑　13-40
Common Damage Phenomena and Countermeasures of Universal Cross Joint in UCMW Rolling Mill
　　………………………………………………………………………… WANG Chenggang, XUAN Lijian
TMEIC 激光焊机光束焦点位置对焊缝质量影响分析 ……… 李冠良，唐晓宇，玄利剑，张　静，辛广荣　13-40
Study on the Influence of Beam Focus Position on Weld Quality in TMEIC Laser Welder
　　………………………… LI Guanliang, TANG Xiaoyu, XUAN Lijian, ZHANG Jing, XIN Guangrong
直接还原铁热装热送新装备技术开发 ……………… 李少英，潘宏涛，王少臣，李佳辉，习小军　13-41
轨梁厂 BD 轧辊的堆焊再制造技术 ………………………… 魏占山，孙贻宝，王金生，张金龙　13-42
Remanufacturing Technology of Breaking-down Roll Surface Welding in Rail and Beam Factory
　　……………………………………………… WEI Zhanshan, SUN Yibao, WANG Jinsheng, ZHANG Jinlong
无人自动抓渣天车关键技术研究及应用 …………… 刘海峰，薛　鑫，张　灿，石玉龙，琚小然　13-43

Study and Practice of Key Unmanned Slag Grabbing Crown Technologies
　　　…………………………… LIU Haifeng, XUE Xin, ZHANG Can, SHI Yulong, JU Xiaoran

热轧板带工艺建模及软件开发 ………………………… 王　涛，刘元铭，黄庆学　13-43
Modeling and Software Development for Hot-rolled Production of Plate and Strip
　　　………………………………………………… WANG Tao, LIU Yuanming, HUANG Qingxue

14　冶金自动化与智能化

钢铁企业电力负荷预测方法综述 …………………………………………… 张效华　14-1
A Review of Load Forecasting in Iron and Steel Plants ……………………… ZHANG Xiaohua

一种工业机器人视觉定位技术的原理及应用 ……………………………… 陈　贝　14-1
The Principle and Application of Visual Positioning Technology for Industrial Robot ……… CHEN Bei

连铸大包自动浇钢技术开发与应用 ………………………………… 唐伟新，陈　贝　14-2
Development and Application of Automatic Casting Technology for Continuous Casting Ladle
　　　………………………………………………………………… TANG Weixin, CHEN Bei

基于5G-TSN架构的云化PLC时延保障方法研究 ……… 张　岩，王军生，孙　雷，孙瑞琪，秦大伟　14-3
Research on Delay Guarantee Mechanism of Cloud Programmable Logic Controller Based on 5G-TSN
　　Architecture ………………… ZHANG Yan, WANG Junsheng, SUN Lei, SUN Ruiqi, QIN Dawei

炼钢厂行车实时调度系统研究及应用 ……… 徐林伟，卢　义，张　飞，徐超琼，丁昭祥　14-4
Research and Application of Real-time Operation Scheduling System in Steelworks
　　　………………………… XU Linwei, LU Yi, ZHANG Fei, XU Chaoqiong, DING Zhaoxiang

基于大数据分析的热镀锌机组关键设备故障诊断技术 ……… 王　鲁，吴　军，顾成勇，王学敏　14-4
Fault Diagnosis Technology for Critical Equipment of Hot Dip Galvanizing Unit Based on Big Data Analysis
　　　…………………………………… WANG Lu, WU Jun, GU Chengyong, WANG Xuemin

焦化厂能源管理中心方案研究及应用 ……… 孙　玲，刘亚玲，高凤君，李明杰，樊艳伟　14-5
Research and Application of Energy Management Center Solution in Coking Plant
　　　……………………………… SUN Ling, LIU Yaling, GAO Fengjun, LI Mingjie, FAN Yanwei

武钢CSP转炉在低铁钢比下氮的控制 ……………… 徐培春，邱　晨，陈尚波，叶　飞　14-6

煤气透平机组智能运维解决方案关键技术研究 ……… 蔡正国，李征磊，陈红琳，章天一　14-6
Research on Key Technology of Smart Operation & Maintenance Solution for Gas Turbine Units
　　　………………………… CAI Zhengguo, LI Zhenglei, CHEN Honglin, ZHANG Tianyi

大型钢铁企业生产系统存储双活的构建与实践 ……… 樊贵先，程　志，彭　超，彭　震，胡　炜　14-7
Construction and Practice of Production System Storage in Large Steel Enterprises
　　　…………………………… FAN Guixian, CHENG Zhi, PENG Chao, PENG Zhen, HU Wei

双交叉限幅控制在冷轧连续退火炉的应用 ……… 兰晓栋，杨红伟，周惠娟，李文静，李　锐，王承刚　14-8
The Application of Double Cross Control in Continuous Annealing Furnace
　　　……………… LAN Xiaodong, YANG Hongwei, ZHOU Huijuan, LI Wenjing, LI Rui, WANG Chenggang

宝山基地4号连铸L2过程机改造实践 ……… 阎建兵，黄晔华，陈建尧，陆　华，朱海健　14-8
Renovation Practice of 4＃CC L2 System in Baosteel
　　　………………………… YAN Jianbing, HUANG Yehua, CHEN Jianyao, LU Hua, ZHU Haijian

"手"+"眼"结合的新一代智能化装备 ……… 邓能辉，王孝敏，吴昆鹏，崔广礼　14-9

冶金机器人的可预测碰撞处理 …………………………………………… 孙　睿，窦　刚　14-10

鞍钢 1450 连退线 2 号卷取机带尾定位的控制过程 ························ 张国利，林 彬，郑大勇　14-11

面向智能制造的知识图谱驱动设备故障诊断方法研究 ····· 刘佳伟，王军生，金 鹏，蔡 畅，孙瑞琪　14-11
Research on Knowledge Graph-driven Equipment Fault Diagnosis Method for Intelligent Manufacturing
·················· LIU Jiawei, WANG Junsheng, JIN Peng, CAI Chang, SUN Ruiqi

高可用性缓存管理系统的构建 ··············· 车 兰，赵 勐，刘 星，吴长河，孙上海，金树来　14-12
Construction for High Availability Cache Management System
············ CHE Lan, ZHAO Meng, LIU Xing, WU Changhe, SUN Shanghai, JIN Shulai

鞍钢销售物流管理系统的设计与实现 ·· 韩奇颖　14-13
Design and Implementation of Ansteel Sales Logistics Management System ················ HAN Qiying

基于 BP 神经网络的钢包渣眼演化行为的预测·· 刘晓航，王胩杰，刘 畅，贺 铸，李光强，王 强　14-14
Prediction of Evolution Behavior of Slag Eye in Steel Ladle Based on BP Neural Network
·············· LIU Xiaohang, WANG Xijie, LIU Chang, HE Zhu, LI Guangqiang, WANG Qiang

面向冶金领域的巡检无人机轨迹跟踪控制系统研究 ···················· 高 艺，刘景亚，吴曼玲　14-15
Research on Trajectory Tracking Control of Quad-rotor for Metallurgical Industry
··· GAO Yi, LIU Jingya, WU Manling

鞍钢股份炼钢总厂智能铁水倒罐间场景开发与应用 ···················· 吴丙恒，吴世龙，颜世蛟　14-15
Development and Application of Intelligent Molten Iron Pouring Tank Scene of AnSteel General Plant
··· WU Bingheng, WU Shilong, YAN Shijiao

鞍钢热轧厂数字化车间建设探索与实践 ························ 董 广，郑英杰，张 平，乔树峰　14-16
Exploration and Practice on the Construction of Digital Workshop in Angang Hot Rolling Mill
························· DONG Guang, ZHENG Yingjie, ZHANG Ping, QIAO Shufeng

面向质量精细管控的型材数字化工厂关键技术与实施方案
·· 董 广，于天琦，王久朋，卜俊男，于 萍　14-17
Key Technologies and Implementation Solution of Digitalized Factory in Profile Steel Enterprises for Quality
Lean Control ············ DONG Guang, YU Tianqi, WANG Jiupeng, BU Junnan, YU Ping

红色工业的探索与发现：图像识别技术在冶金领域的应用与前景 ·································· 张瀚文　14-18

冶金企业人工智能的发展与应用综述 ······················· 陈 琛，倪书权，郝景章，李加旺，史 光　14-19

"工业 4.0" 助推中国钢铁行业迈进数字化智能化新时代 ····························· 佐祥均，代 虹　14-19
Industry 4.0 Boosts the Steel Industry into a New Era of Digitalization and Smart Technology
··· ZUO Xiangjun, DAI Hong

高线卸卷实现自动控制，避免擦划伤问题 ················ 杜鹏阳，黄宝新，郭思聪，王 新，梁 琪　14-20
Automatic Control of High-line Unloading Based on Avoiding Scratch Problems
······················ DU Pengyang, HUANG Baoxin, GUO Sicong, WANG Xin, LIANG Qi

基于 5G 和云化 PLC 的工业智能泵站系统设计
·· 秦大伟，王军生，张 岩，孙瑞琪，刘佳伟，吴晓宁　14-20
Design of Industrial Intelligent Pumping Station System Based on 5G and Cloud PLC
············ QIN Dawei, WANG Junsheng, ZHANG Yan, SUN Ruiqi, LIU Jiawei, WU Xiaoning

宝钢湛钢炼铁厂智慧制造探索与实践 ··· 韩仁义，梁利生　14-21
Exploration and Practice of Intelligent Manufacturing in Iron Making Plant of Baosteel Zhanjiang Iron and
Steel ··· HAN Renyi, LIANG Lisheng

高分辨率数字钢卷快速搜索系统及应用 ··· 刘安平，彭燕华　14-22
High Resolution Digital Steel Coil Fast Search System and Its Application ······· LIU Anping, PENG Yanhua

武钢 CSP 连铸二级过程控制系统 ……………………………………………… 徐　重，尹海峰，马小雨　14-23
CSP Continuous Casting Process Control System of Wisco ………… XU Zhong, YIN Haifeng, MA Xiaoyu
电液直驱结晶器振动优化控制研究与实践 ………… 蔡春扬，方学红，王永猛，刘　玉，龙　灏　14-23
Research and Practice on the Optimal Control about Mould Oscillation by Electro-hydraulic Direct Drive
　　……………………………… CAI Chunyang, FANG Xuehong, WANG Yongmeng, LIU Yu, LONG Hao
无人驾驶混铁车制动系统设计研究 ………………………… 王　康，张文龙，于　磊，葛　光　14-24
Research on Braking System of SmartTPC ……………… WANG Kang, ZHANG Wenlong, YU Lei, GE Guang
无人框架车动力系统设计研究 ……………………………… 王　康，张文龙，于　磊，张子才　14-25
Research on Power System Design of Automated Guided Vehicle
　　………………………………………………… WANG Kang, ZHANG Wenlong, YU Lei, ZHANG Zicai
利用 RFC 接口技术实现 SAP 与轻迈系统数据同步 ……………………………………… 王国靖　14-25
Utilizing RFC Interface Technology to Achieve Data Synchronization between SAP and Qingmai Systems
　　………………………………………………………………………………………… WANG Guojing
钢铁业数字化智能化技术创新实践与思考 ………………………………………… 张　伟，刘　晔　14-26
PDA 高速数据采集系统通讯协议开发 ……………………………………… 刘安平，彭燕华　14-27
Development of Communication Protocol for PDA High Speed Data Acquisition System
　　…………………………………………………………………………… LIU Anping, PENG Yanhua
智能制造系统在某双高棒车间的开发和应用 ……………………………… 赵　铭，张芙蓉　14-27
Development and Application of the Intelligent Manufacturing System in a Double High-speed Bar Line
　　………………………………………………………………………… ZHAO Ming, ZHANG Furong
浅谈水处理分厂 UPS 的演变 ……………………………………………………………… 李　璐　14-28
二十辊轧机自动板形控制系统研究与优化 ………………………… 罗劲松，钱　华，郑　涛　14-28
Research and Optimization of Automatic Flatness Control System for 20 High Mill
　　………………………………………………………………… LUO Jinsong, QIAN Hua, ZHENG Tao
煤焦检测智能分析技术应用与系统开发 ………………………………………………… 欧连发　14-29
基于主变励磁涌流引发电动风机误跳机的分析 ………………………………… 陶瑞基，阮文茜　14-29
Analysis of Misoperation of Electric Fan Caused by Inrush Current of Main Transformer
　　…………………………………………………………………………… TAO Ruiji, RUAN Wenxi
基于精炼炉变压器绕组匝间短路故障的分析 …………………………………… 陶瑞基，王志康　14-30
Analysis of Short Circuit Fault between Turns of Transformer Winding Based on Refining Furnace
　　………………………………………………………………………… TAO Ruiji, WANG Zhikang
智慧高炉集约化管控大数据应用平台研究与开发 ………… 费　静，车玉满，郭天永，姜　喆，孙　波　14-31
Research and Development of Big Data Application Platform for Intelligent Blast Furnace Intensive
　　Management and Control ……………… FEI Jing, CHE Yuman, GUO Tianyong, JIANG Zhe, SUN Bo
转炉全自动出渣系统的开发与应用 ……………………………… 秦佳星，江腾飞，陶　金　14-32
集中空压机站远控滞后现象的分析与优化 ………… 薛冬晨，鲁绍军，李　杨，刘　强，牛勋铎　14-32
Analysis and Optimization of Remote Control Lag in Centralized Air Compressor Station
　　……………………………… XUE Dongchen, LU Shaojun, LI Yang, LIU Qiang, NIU Xunduo
PROFIBUS-DP 网络质量检测技术在钢厂的运用 ………… 丁继亚，钟　实，程　曦，黄亚军，刘　浩　14-33
数字化应变分析技术在汽车零部件冲压成形性能评价中的应用
　　…………………………………………………… 郝志强，刘文博，徐　鑫，梁　笑，李春林　14-33

Application of Digital Strain Analysis Technology in Evaluation of Stamping Forming Performance of Automotive Parts ································ HAO Zhiqiang, LIU Wenbo, XU Xin, LIANG Xiao, LI Chunlin

智慧矿山生产管控平台 ··· 郭武英，霍成全，徐　涛　14-34
Smart Mine Production Control Platform ··································· GUO Wuying, HUO Chengquan, XU Tao

冷轧清洗段刷辊调节系统设计与优化 ·· 黄亚军　14-35
Design and Optimization of Brush Roll Adjustment System for Cold Rolling Cleaning Section
·· HUANG Yajun

高炉鼓风机和TRT集控改造运用实践 ············ 杨　奇，尹丽琼，李　祥，陈红云，陈树海，侯化军　14-35

冶金行业安全生产防控技术及应用 ··· 葛雨田，王　敏，邢立东　14-36

基于数据治理的高炉运行动态评价系统开发与应用
································ 白　雪，陈英博，高大鹏，李洪亮，吕秀丽，薛芷蘅　14-36

"安全生产+经济运行"双核驱动下的钢铁企业能源智能化管控系统开发与应用
·· 白　雪，王念军，周文旭，高大鹏　14-37

厚板4300mm轧机机架辊控制技术 ·· 张　丽，赵中杨，齐　鑫　14-37
Control Technology of Stand Rolls for Heavy Plate Mill 4300mm
·· ZHANG Li, ZHAO Zhongyang, QI Xin

高炉热风炉在线能效分析系统的建立及应用 ············ 孙守斌，刘常鹏，白　雪，鲁　璐，王东山　14-38
Establishment and Application of Online Energy Efficiency Analysis System for Hot Blast Stove
································ SUN Shoubin, LIU Changpeng, BAI Xue, LU Lu, WANG Dongshan

基于物联网技术的井下铲运机遥控改装及应用 ··· 霍成全　14-39
Remote Control Modification and Application of Underground Scraper Based on IoT Technology
·· HUO Chengquan

探索无底柱分段崩落采矿中深孔智能凿岩作业 ···························· 霍成全，王双虎，袁明超　14-39
Exploration of Intelligent Rock Drilling Operations in Segmental Collapse Mining of Bottomless Pillars
································ HUO Chengquan, WANG Shuanghu, YUAN Mingchao

高压设备在线监测技术应用及探索 ····································· 钟　实，周文豪，张丁鹏，洪轶雄　14-40
Application and Exploration of Online Monitoring Technology for High Voltage Equipment
································ ZHONG Shi, ZHOU Wenhao, ZHANG Dingpeng, HONG Yixiong

递归滤波算法在低温卷取控制方面的应用 ········ 杨哲懿，高　磊，王尊呈，王　帅，于生龙，杨　博　14-41

基于GDIA的钢铁连铸安全生产管控系统信息需求数据模型 ········ 张　充，巴烈电，张　伟，赵挺生　14-42
Information Requirement Data Model of Steel Continuous Casting Safety Management and Control System Based on GDIA ················ ZHANG Chong, BA Liedian, ZHANG Wei, ZHAO Tingsheng

激光气体分析仪在转炉煤气回收中的应用 ·· 张英利　14-43
Application of Laser Gas Analyzer in Gas Recovery of Converter ····························· ZHANG Yingli

低压电气控制整合与提升 ··· 李晟京　14-43
Integration and Improvement of Low-Voltage Electrical Control ································ LI Shengjing

轧钢加热炉基础自动化系统的优化与应用 ··· 孔海超　14-44
Optimization and Application of Basic Automation System for Steel Rolling Heating Furnace
·· KONG Haichao

还原竖炉冷却系统复杂控制技术设计与实现 ··· 郝广春　14-45
Design and Implementation of Complex Control Technologies for Cooling System in DR Shaft
·· HAO Guangchun

脱硝氨逃逸在线监测分析系统优化 ··· 贾旭鉴 14-45
Optimization of Online Monitoring and Analysis System for Denitration Ammonia Escape ········ JIA Xujian
基于YOLOv8改进的钢水罐精准位置信息识别模型 ······ 廖文彬，巴烈电，张　伟，赵挺生，张　充 14-46
Improved Semantic Segmentation Model for Ladle Based on YOLOv8
　　··················· LIAO Wenbin, BA Liedian, ZHANG Wei, ZHAO Tingsheng, ZHANG Chong
洗精煤采制样设备的建设研究 ······················· 宫文浩，张　澎，宋立伟，韩　博，付兴海 14-47
Research on the Construction of Sampling Equipment for Clean Coal Mining
　　······················· GONG Wenhao, ZHANG Peng, SONG Liwei, HAN Bo, FU Xinghai
行车无人化技术在热轧库区的应用 ·· 徐明红 14-48
Application of Unmanned Automatic Crane System in Hot Rolling Yards ················ XU Minghong
论信息化系统提升安全管理水平的探索与实践 ··························· 刘晓宇，王　帅，张　靓 14-48
生产协同智能管控中心 ··· 郭武英，徐　涛 14-49
Production Collaboration Intelligent Control Center ······························ GUO Wuying, XU Tao
5G工控网络下云化PLC技术的探索与实践 ··························· 孙瑞琪，王军生，张　岩，秦大伟 14-49
Exploration and Practice of Cloud-based PLC Technology in 5G Industrial Control Network
　　··· SUN Ruiqi, WANG Junsheng, ZHANG Yan, QIN Dawei
简述钢铁工业智能制造技术发展应用 ·· 闫子政，翟玉龙 14-50
Brief Overview of the Development and Application of Intelligent Manufacturing Technology in the Steel
　　Industry ·· YAN Zizheng, ZHAI Yulong

15　冶 金 物 流

公路高架交通灯在中南股份铁路道口的应用 ·· 刘　鹏 15-1
Application of Highway Elevated Traffic Lights at Zhongnan Joint-stock Railway Crossing ········ LIU Peng
物联网和大数据技术在物流系统中的应用 ·· 任立群 15-1
钢铁供应链数字货运平台管理技术的研究 ······················ 张丽莉，刘振清，吴　浩，梁　君，姜　欢 15-2
FAD型铁路货运站场机车自动驾驶系统研究与应用 ······ 郭　兵，田力男，王　晨，李生意，潘广海 15-2
鞍钢鲅鱼圈铁路智能化无人货场研究 ·························· 郭　兵，杨星亮，王　晨，李生意 15-2
鞍钢股份鲅鱼圈分公司新型铁水运输模式探索与分析 ············· 王　晨，郭　兵，田力男，孙东生 15-3
超偏载检测装置在鞍钢鲅鱼圈的研究与运用 ·················· 郭　兵，张晓军，李　丹，张思波 15-3
基于一贯制管理的鞍钢质量管理系统构建 ·· 王　静，闵承鑫 15-4
The Construction of Ansteel Quality Management System Based on the Consistent Management
　　·· WANG Jing, MIN Chengxin
高炉铁水敞口加盖保温技术研究与运行实践
　　··· 徐大勇，马光宇，刘常鹏，袁　玲，李卫东，孙守斌 15-4
Thermal Insulation Technology Research and Operation Practice of Capping for Open Molten Iron Tank of
　　Blast Furnace ······ XU Dayong, MA Guangyu, LIU Changpeng, YUAN Ling, LI Weidong, SUN Shoubin
鞍钢股份智慧供应链应用探索 ·· 侯海云，荣树强，易　茗 15-5
Exploration of Application of Intelligent Supply Chains of Angang Steel Company Limited
　　·· HOU Haiyun, RONG Shuqiang, YI Ming
低铁耗生产模式下铁水运输作业分析 ·· 王　玮，程献权 15-6

ESG 助推中国钢铁企业低碳转型与高质量发展 ………………………………………… 段希雅　15-6
ESG Promotes Low-carbon Transformation and High-quality Development of Chinese Steel Enterprises
　　………………………………………………………………………………………… DUAN Xiya
浅谈人工智能在供应链管理中的应用研究 …………………………………………… 王　浩　15-7
Discussion on the Application of Artificial Intelligence in Supply Chain Management ………… WANG Hao
冶金工矿企业铁路小半径曲线轨道结构创新设计 …………………………………… 江宏法　15-8
Innovative Design on Small-Radius Curved Railway Tracks Used by Metallurgical and Mining Enterprises
　　…………………………………………………………………………………… JIANG Hongfa
车辆全流程管控系统的研究与实现 ……………………………………………… 王　宏，钱　松　15-8
Research and Implementation of the Whole Process of Vehicle Control System
　　……………………………………………………………………………… WANG Hong, QIAN Song
铁路车辆运行中常见故障及应急处理措施 …………………………………………… 李世锋　15-9
Common Failures and Emergency Measures in the Operation of Railway Rolling Stock ………… LI Shifeng
冶金企业内燃机车空调改造的探索与实践 ……………………… 乔玉栋，范秀川，白　帅，罗　健　15-10
Exploration and Practice of Air Conditioning Transformation of Internal Combustion Locomotive in
　　Metallurgical Enterprises ……………………… QIAO Yudong, FAN Xiuchuan, BAI Shuai, LUO Jian
冶金铁路小半径曲线钢轨磨耗的原因及防治 ………………………………… 王建栋，闫　炜　15-11
Causes and Prevention of Wear of Small Radius Curved Rail in Metallurgical Railways

16　冶金流程工程学

SPHC 钢直上浇注的生产实践 ……………………………… 李　超，尹宏军，李文博，冉茂铎　16-1
Production Practice of the Straight Casting in SPHC Steel
　　………………………………………………… LI Chao, YIN Hongjun, LI Wenbo, RAN Maoduo
基于冶金流程工程学的连铸-热轧界面动态设计 ……………………………… 张　琼，耿明山　16-1
Dynamic Design of Continuous Casting-hot Rolling Based on Metallurgical Process Engineering
　　………………………………………………………………………… ZHANG Qiong, GENG Mingshan
板坯赋号规则的优化实践 ……………………………………… 董金刚，陈廷星，谢璞石　16-2
典型钢铁制造流程极致能效与极限碳排放分析 ………………………… 张　琦，籍杨梅，王鑫鑫　16-3
CBAM 对中国钢铁行业的影响和应对策略 ……………… 李　涛，上官方钦，郧秀萍，周继程，倪　冰　16-4
Impact of CBAM on China's Steel Industry and Strategies
　　………………………… LI Tao, SHANGGUAN Fangqin, LI Xiuping, ZHOU Jicheng, NI Bing
订单快速响应的炼钢-连铸-热轧一体化生产计划方法及系统 ……… 郑　忠，王永周，马仕成，高小强　16-5
基于耗散结构优化的钢铁制造流程设计研究 ………………………………………… 张福明　16-6
Research and Design on Steel Manufacturing Process Based on Disatiative Structure Optimization
　　……………………………………………………………………………………… ZHANG Fuming
热轧-冷轧界面装备技术的创新与应用 ………… 韦富强，张　建，郑江涛，孟祥军，王　超，杨建立　16-7
Innovation and Application of Interface Equipment Technology between Hot Rolling and Cold Rolling
　　………… WEI Fuqiang, ZHANG Jian, ZHENG Jiangtao, MENG Xiangjun, WANG Chao, YANG Jianli
基于深度学习的铸坯低倍缺陷智能检测模型及系统
　　………………………… 杨灏瀛，魏　星，汪红兵，赵文慈，闫　岩，温　宁，李　航　16-8

基于钢水温度预定和预测相结合的 LF 炉供电推荐模型 ……………………………… 李晓龙，贺东风　16-8

数据与机理联合驱动的转炉终点碳含量预测模型
　………………………………… 夏一杰，汪红兵，魏　星，石雪卓，温　宁，解　欢　16-9

"碳中和"战略下钢铁制造流程的低碳发展与技术创新 ……………………… 张福明，刘清梅　16-9
Low-carbon Progress and Technological Innovation of Iron and Steel Manufacturing Process under the Strategy
　of "Carbon Neutrality" ……………………………………… ZHANG Fuming, LIU Qingmei

基于"炉-机对应"的炼钢-连铸生产调度问题灰狼优化模型
　………………………………………… 陈　博，邵　鑫，刘　青，王忠刚，高志滨　16-11

钢铁工业极限能耗分析与能效提升措施探讨 …………… 上官方钦，周继程，倪　冰，崔志峰　16-12

炼钢车间天车调度管理系统设计、开发与应用
　……………………………… 冯　凯，徐安军，贺东风，潘　俊，王宇轩，王　磊　16-13

基于 FlexSim 对炼铁-炼钢界面调度规则的仿真优化 …… 王　磊，徐安军，王宇轩，刘　旋，解　欢　16-13

钢铁制造流程能量流网络的有向网分析及建模 ……………… 杜学强，郦秀萍，孙彦广，刘　骁　16-14
Analysis and Modeling of Energy Flow Network for Steel Manufacturing Process Based on Directed Network
　………………………………………… DU Xueqiang, LI Xiuping, SUN Yanguang, LIU Xiao

转炉出钢过程流体流动特性的研究 …………………………………… 袁　飞，刘　旋，徐安军　16-15

钢铁制造流程系统实时生产柔性和稳定性度量方法 …… 高小强，白云航，匡静静，丁卓非，郑　忠　16-17

唐钢新区项目设计特点及运行实绩 ………………………… 李建生，郦秀萍，马新光，韩伟刚　16-17
Design Characteristics and Operational Performance of Tanggang New Area Project
　………………………………………… LI Jiansheng, LI Xiuping, MA Xinguang, HAN Weigang

17 其 他

X 射线荧光光谱法测定预熔渣中 CaF_2 含量 …………………… 牛威斌，禹青霄，葛晶晶　17-1
Determination of CaF_2 Content in Premelted Slag by X-ray Fluorescence Spectrometry
　………………………………………… NIU Weibin, YU Qingxiao, GE Jingjing

新型管线测绘技术在冶金行业中的应用与探索 ………… 刘　光，徐　明，李晓龙，赵逸凡，银振华　17-1
Application and Exploration of New Pipeline Surveying Technology in the Metallurgical Industry
　………………………………………… LIU Guang, XU Ming, LI Xiaolong, ZHAO Yifan, YIN Zhenhua

针状焦预处理工艺对比分析 …………………………… 刘海丰，何　莹，张大奎，穆春丰，刘庆佩　17-2
Comparison and Analysis of Needle Coke Pretreatment Process
　………………………………………… LIU Haifeng, HE Ying, ZHANG Dakui, MU Chunfeng, LIU Qingpei

标准曲线法和标准加入法测定萤石中氟化钙 …… 王泽庭，张　健，王炳琨，康宝军，张国龙，单桂艳　17-3
Determination of Calcium Fluoride in Fluorite by Standard Curve Method and Standard Addition Method
　…… WANG Zeting, ZHANG Jian, WANG Bingkun, KANG Baojun, ZHANG Guolong, SHAN Guiyan

X 射线能量色散检测技术在钼铁合金分析中的研究与应用
　………………………………………… 康宝军，何文英，吴楠楠，唐艳秀，王炳琨，周芊蔚　17-3
Research and Application of X-ray Energy Dispersion Detection Technology in Ferromolybdenum Alloy
　Analysis …… KANG Baojun, HE Wenying, WU Nannan, TANG Yanxiu, WANG Bingkun, ZHOU Qianwei

常规仪表故障与维护技术探讨 ………………………………………………………… 武弘韬　17-4
Discussion on Conventional Instrument Failure and Maintenance Technology ……………… WU Hongtao

联系测量技术在陈台沟铁矿超深井及附属工程施工中的研究与应用 ………………… 刘　骥　17-5
The Research and Application of Contact Survey Technology in the Construction of Super-deep
　　Well and Affiliated Engineering of Chentaigou Iron Mine ……………………… LIU Ji
尾矿库挡砂堤稳定性研究 ………………………………………………………… 王胜建　17-5
Study on the Stability of Tailings Dam …………………………………… WANG Shengjian
高频燃烧红外吸收法测定冶金用二氧化钛中碳 ………… 杨晓倩，郭圣洁，谭胜楠，牛威斌　17-6
Determination of Carbon in Titanium Dioxide for Metallurgy by High Frequency Combustion Infrared
　　Absorption Method …………… YANG Xiaoqian, GUO Shengjie, TAN Shengnan, NIU Weibin
ICP-AES 法测定镀锡板镀层中的铅 ………… 戚振南，任玲玲，杨慧贤，谭胜楠，郭圣洁，葛晶晶　17-7
Determination of Plumbum in Tinplate Layer by ICP-AES
　　……………… QI Zhennan, REN Lingling, YANG Huixian, TAN Shengnan, GUO Shengjie, GE Jingjing
钢水成分查询系统的程序设计与实现 ……………………………………………… 易晓珊　17-7
集中计量管理系统研究与应用 ………………………………………………… 高启胜，翟英杰　17-8
Research and Application of Centralized Measurement Management System …… GAO Qisheng, ZHAI Yingjie
利用不确定度评价熔融炉性能对结果的影响 …… 王丽晖，李　奎，孙　伟，高　巍，郝延楠，潘　忠　17-8
The Uncertainty Was Used to Evaluate the Influence of Melting Furnace Performance on the Results
　　……………………………… WANG Lihui, LI Kui, SUN Wei, GAO Wei, HAO Yannan, PAN Zhong
储氢容器用钢设计及其氢敏感性机理研究 ………………… 范秀如，米志杉，苏　航，杨　丽　17-9
液压系统功率损耗原因及对策分析 …………………… 雷丛卉，宁　博，陈国防，郭　佳，彭立广　17-10
Analysis on the Causes and Countermeasures of Power Loss in Hydraulic Systems
　　………………………………… LEI Conghui, NING Bo, CHEN Guofang, GUO Jia, PENG Liguang
基于高频数据可视化分析的设计与实现 ……………………… 李保琴，张　强，俞烨波　17-10
预处理工艺对 H13 钢碳化物析出行为的影响 ………… 樊明强，赵英利，张利江，赵峥嵘，王建强　17-11
Effect of Pretreatment Technology on Carbide Precipitation Behavior of H13 Steel
　　……………… FAN Mingqiang, ZHAO Yingli, ZHANG Lijiang, ZHAO Zhengrong, WANG Jianqiang
轻量化低应力机车车轮设计及验证 ……………………… 黄孝卿，刘　智，陈　刚，国新春　17-12
Structure Design and Verification of Lightweight and Low Stress Locomotive Wheels
　　………………………………………… HUANG Xiaoqing, LIU Zhi, CHEN Gang, GUO Xinchun
高氮钒铁中钒元素化学分析方法的研究与讨论 ……………………………………… 贺红侠　17-12
Study and Discussion on Chemical Analysis Methods of Vanadium in High Nitrogen Ferric Vanadium
　　……………………………………………………………………………………… HE Hongxia
基于 OpenCV 的石墨球化率测定 ……………………………………… 郑秋薇，邢佳文　17-13
Determination of Nodularity of Graphite Based on OpenCV ………… ZHENG Qiuwei, XING Jiawen
S 钢铁集团加工配送中心加工产线配置策略研究 ………………… 王　兴，陈利成，刘昊林　17-14
Research on the Configuration Strategy of Processing Production Line of S Steel Group's
　　Processing and Distribution Center ……………………… WANG Xing, CHEN Licheng, LIU Haolin
钢铁企业系统诊断与竞争力分析 ……………………… 陈利成，陈琼妮，吴　波，王　兴　17-14
System Diagnosis and Competitiveness Analysis of Steel Enterprises
　　……………………………………… CHEN Licheng, CHEN Qiongni, WU Bo, WANG Xing
Nb 元素对高强度管线钢抗 H_2S 性能的影响研究 ………… 樊艳秋，马长文，李少坡，丁文华，李战军　17-15

大会特邀报告

- ★ 大会特邀报告
  - 第十三届冶金青年科技奖获奖人特邀报告
  - 分会场特邀报告
  - 矿业工程
  - 焦化及节能环保
  - 炼铁与原料
  - 炼钢与连铸
  - 电冶金与废钢铁
  - 轧制与热处理
  - 表面与涂镀
  - 金属材料深加工
  - 粉末冶金
  - 先进钢铁材料及应用
  - 节能与低碳技术
  - 冶金环保与资源利用
  - 冶金设备与工程技术
  - 冶金自动化与智能化
  - 冶金物流
  - 冶金流程工程学
  - 其他

The Impact of Decarbonisation Efforts on the Global Steel Industry

Edwin BASSON

(World Steel Association)

Abstract: The global steel industry has started the process of decarbonisation. This will be a slow process, requiring much investment in equipment, manpower and research and development in years to come.

This is a global challenge that will require global solutions, but evidence around the world indicates that the steel industry has embraced this challenge and is working towards solutions. The solutions will differ between regions, depending on domestic energy systems and input materials availability. Support from the energy and finance industries will be important for long term success in decarbonising the steel industry. The future composition of the steel industry will be very different from the present, with different production technology routes, different energy and input requirements.

The steel industry is fortunate to have already a good idea of the different technology routes that could be used in the decarbonisation challenge. Confidence exists that the industry will meet this challenge successfully in years to come.

日本制铁为实现碳中和所作的努力

福田 和久（FUKUDA KAZUHISA）

（日本制铁株式会社）

摘　要： 将"2050 碳中和愿景"当作今后经营计划的重要组成部分之一，积极推动技术开发及实际操作。本报告除了概述这一愿景以及公司二氧化碳减排方案以外，还将介绍在钢铁生产过程中如何实现脱碳化和 CCUS 方面所作的努力。最后介绍日本制铁公司碳中和进展情况，今后的计划以及目前所面临的课题。

钢铁工业碳中和的思考与探索

毛新平

（北京科技大学，北京　100083）

摘　要： 碳中和是国家的发展战略。钢铁是国民经济的重要基础产业，钢铁工业快速发展有力支撑了国民经济、社会发展和国防建设。钢铁工业碳排放高，在所有制造业中位居首位，碳中和背景下面临巨大挑战。与此同时，以汽车行业为代表的下游行业对生态化的钢铁材料也提出了迫切需求。在此背景下，《钢铁行业碳中和愿景和低碳技术路线图》于 2022 年 8 月正式对外发布，明确提出我国钢铁工业"双碳"六大技术路径。北京科技大学聚焦钢铁工业碳中和的重大需求，成立碳中和研究院，并在富氢/全氢冶金冶炼技术、转炉炼钢减碳技术、近零碳排电炉炼钢工艺技术、再生钢铁的高质化利用、近终形制造高性能钢铁材料等代表性低碳技术开展了大量研究工作，取得了突破性进展。

POSCO's Green & Digital Challenges for Carbon Neutral

Kisoo KIM

(Steel Manufacturing Process Research Center, POSCO Technical Research Laboratories)

Abstract: POSCO's technology development strategy to address climate change includes the development of HyREX, digital transformation, and green products.

The HyREX process, based on the commercially operated FINEX technology for the past 20 years, is the core technology for producing hydrogen-reduced iron to achieve carbon neutrality by 2050. This process utilizes 100% hydrogen to produce direct reduced iron (DRI) and molten iron through electrical smelting furnaces. HyREX is gaining attention as an economically viable hydrogen reduction technology due to the utilization of abundant low-grade iron ores.

To achieve a commercially feasible hydrogen ironmaking process, global technological collaboration is considered a key success factor. As a result, POSCO has hosted the Hydrogen Iron & Steel Making (HyIS) international forum in Seoul (2021) and Stockholm (2022) to promote international cooperation in hydrogen reduction ironmaking.

As the steel industry will require a large amount of hydrogen in the future, POSCO is also focusing on the development of technologies for hydrogen production, storage, transportation, and supply, in order to achieve a hydrogen circular economy. Furthermore, various business opportunities arise from the establishment of green hydrogen distribution and related infrastructure, as well as participation in green hydrogen projects.

POSCO has actively been developing smart factory technologies to enhance productivity and efficiency. One notable example is the implementation of AI-based systems and sensors in the Blast Furnace, which has resulted in significant cost savings and increased production capacity. As a result of these achievements, POSCO was selected as a global lighthouse factory by the World Economic Forum in 2019.

POSCO is also expanding the certification of green products and continuously making efforts to reduce CO_2 emissions. "Greenate" is POSCO's 2050 carbon-neutral master brand, encompassing the three major eco-friendly product brands of 'e-Autopos', 'INNOVILT', and 'Greenable', which contribute to low-carbon mobility, eco-friendly architecture, and the resource circulation economy.

POSCO recognizes the need to adapt to changing market demands and embrace a collaborative approach to R&D to ensure sustainable growth. POSCO will do its best to enhance cooperation among the steel mills of the three Northeast Asian countries to achive the Carbon Neutral 2050.

第十三届冶金青年科技奖获奖人特邀报告

　　大会特邀报告
★　第十三届冶金青年科技奖获奖人特邀报告
　　分会场特邀报告
　　矿业工程
　　焦化及节能环保
　　炼铁与原料
　　炼钢与连铸
　　电冶金与废钢铁
　　轧制与热处理
　　表面与涂镀
　　金属材料深加工
　　粉末冶金
　　先进钢铁材料及应用
　　节能与低碳技术
　　冶金环保与资源利用
　　冶金设备与工程技术
　　冶金自动化与智能化
　　冶金物流
　　冶金流程工程学
　　其他

优特钢长材转型的系统性研究

马志勇，杨勇强

（中冶京诚工程技术有限公司，北京 100176）

摘 要：针对近期长材市场疲软、未来钢产量需求呈下降趋势的外部环境，各钢企纷纷寻求转型和产品结构调整，本文针对细分市场的容量、产品定位、各类产品的品质要求等方面，提出差异化、专业化、精品化的转型发展理念，避免恶性竞争，以利于于钢铁企业高质量发展。

关键词：长材；优特钢；产品结构调整；差异化；专业化；精品化

Systematic Research on the Transformation of Long Product and Special Steel

MA Zhiyong, YANG Yongqiang

(CERI, Beijing 100176, China)

Abstract: In response to the recent sluggishness in the long product market and the downward trend in future steel production demand, various steel enterprises are actively seeking transformation and adjusting their product structures. This article proposes a transformation and development concept focused on differentiation, specialization, and refinement in terms of capacity, product positioning, and quality requirements for various products within specific market segments. This approach aims to avoid cutthroat competition and facilitate the high-quality development of steel companies.

Key words: long product; special steel; product structure adjustment; differentiation; specialization; refinement

高强高韧钛合金研发及钢钛联合技术趋势

马英杰，黄森森，雷家峰，杨 锐

（中国科学院金属研究所，辽宁沈阳 110016）

摘 要：钛合金因其具有强度高、密度低、耐腐蚀等优点，是航空、海洋等重大工程领域的关键结构材料。应用状态下结构钛合金主体相通常是 HCP-α 相，由于 HCP 结构先天塑性变形特征的限制，对于结构钛合金，通过成分、显微结构优化设计来提升材料的塑性、韧性是力学性能优化的主要目标。在合金成分设计方面，通过调控 Al 元素含量，可以促进 HCP-α 相柱面、基面滑移同时开动，提高 HCP 结构的协同变形能力。同时 Al 含量调控还应关注 Ti_3Al 金属间化合物的析出、长大及其对塑性、韧性的影响。由于 α+β 两相钛合金热处理过程中将发生合金元素的扩散再分配，进而影响合金后续不同结构单元的显微组织及力学性能，因此合金微区元素浓度调控是力学性能精确设计的基础。在显微结构设计方面，发展了具有复合型片层结构的钛合金工程化制备技术，进一步提升了合金的韧性。综合以上钛合金成分、显微结构设计技术，团队发展并建立了基于微区结构调控的 α+β 钛合金的工程化设计方法，研制了一系列高强高韧钛合金，应用于我国大飞机、"奋斗者"号载人潜水器、深海科学实验站等关键承载部件。钢钛联合是实现钛合金低成本、大规格制造的关键技术途径，但仍有大量的技术性工作需要逐步开展。

关键词：α+β 钛合金；高强高韧；微区结构单元

The Development of High Strength and Toughness Titanium Alloy and Technical Tendency of Titanium-steel Uniting

MA Yingjie, HUANG Sensen, LEI Jiafeng, YANG Rui

(Institute of Metal Research, Chinese Academy of Sciences, Shenyang 110016, China)

Abstract: Titanium alloys, with high strength, low density and excellent corrosion resistance, have been widely used as the key-structural material in aero and marine industry. For the final applied state, the structural titanium alloys mainly contain the HCP-α phase which shows inherent deficiency for plastic deformation. So, to improve the plasticity and toughness is the main purpose for the research and development of structural titanium alloys by optimizing the component and microstructure. In respect to the alloying strategy, it shows that the addition of aluminum will promote the synergetic activation of basal and prismatic slip, resulting in the improvement of compatible deformation of HCP structure. Meanwhile, the massive formation of Ti_3Al intermetallics with excessive addition of aluminum will obviously decrease the plasticity and toughness of titanium alloys. During the thermal treatments of α+β titanium alloys, alloying element partitioning effect takes place accompanying with the α↔β transformation, which will further affect the microstructure characteristics (phase constitution, microstructure size), plastic deformation and the final mechanical properties of the alloy. So the alloying approach related to element partitioning effect in microzone is necessary for the improvement of mechanical properties. For microstructural optimization, our team has developed the multi-lamellar structure for the engineering application, which further improves the toughness and plasticity of titanium alloy. Based on the above comprehensive alloying and microstructural strategies, our team has developed a serial of high strength and toughness titanium alloys, which have been used to manufacture the key-structural components of aircraft, *Fendouzhe* (the first full ocean depth manned submersible of China) and deep-sea scientific experiment station. Titanium-steel uniting is the fundamental technical process for the low-cost and large dimensional titanium alloy, however, there are many technical works to be solved for the real titanium-steel uniting.

Key words: α+β titanium alloy; high strength and toughness; microzone

双相不锈钢冲击韧性研究进展及工艺技术实践

丰 涵[1]，宋志刚[1]，何建国[1]，吴晓涵[1]，王宝顺[2]，
吴明华[3]，李国平[4]，周灿栋[5]，高 虹[6]

（1. 钢铁研究总院有限公司，北京 100081；2. 浙江久立特材科技股份有限公司，浙江湖州 313028；3. 永兴特种材料科技股份有限公司，浙江湖州 313005；4. 山西太钢不锈钢股份有限公司，山西太原 030003；5. 宝武特种冶金有限公司，上海 201900；
6. 江苏武进不锈股份有限公司，江苏常州 213017）

摘 要：双相不锈钢兼具奥氏体不锈钢的优异耐点蚀性能与铁素体不锈钢的较高强度和耐氯化物应力腐蚀性能，在石油石化、海洋工程、核电、船舶等领域有广泛应用。近年来双相不锈钢产品进一步向低温环境拓展应用，但国内双相不锈钢产品普遍存在冲击韧性低的问题。目前不锈钢断裂机理研究多针对奥氏体钢或铁素体钢，需要完善和发

展多相组织耦合下的双相不锈钢低温冲击韧性研究。

笔者及其团队近年来以 UNS S32750 双相不锈钢为载体，系统研究了基体两相比例（Phase ratio）、相形态（Phase morphology）、晶粒长径比（Aspect ratio）、第二相析出等材料学因素对双相不锈钢低温断裂行为的影响，分析了作用机制。主要研究结果有：（1）固溶态 S32750 双相不锈钢在室温下具有高的冲击韧性，过高的铁素体相比例和粗化的铁素体晶粒损害试验钢冲击韧性：α 相含量从 50%增加到 60%，对应室温冲击功值从 236J 单调递减至 144J，主要体现在裂纹扩展功的下降。（2）γ 相形态由层片状向球状转变、γ 相大长径比晶粒比例从 55%降至 15%，可有效提高试验钢冲击韧性、降低韧脆转变温度，并改善冲击韧性各向异性，提高横向冲击功。（3）试验钢在 900℃时具有非常快的第二相析出速度，并倾向于在高界面能的 α/γ 相界和三叉晶界处发生粗化；1%含量的 σ 相即可严重恶化试验钢的低温冲击韧性，韧脆转变温度上升可达 65℃；Cr_2N 相对试验钢冲击韧性的恶化作用弱于 σ 相。（4）试验钢在 20～-100℃ 范围内的冲击功值随测试温度降低呈 S 形下降，两者符合以下 Boltzmann 关系：$KV=211.1+(-186.8)/(1+e\wedge((T-(-39.9))/11.4))$。（5）测试温度从 US 区降至 DBT 和 LS 区，试验钢冲击功能量构成呈"高裂纹扩展功+高裂纹萌生功" → "中低裂纹扩展功+高裂纹萌生功" → "低裂纹扩展功+低裂纹萌生功"变化。（6）在 DBT 温度区冲断后，试验钢出现同时穿越铁素体和奥氏体的解理断面，铁素体相多呈河流状解理形貌，奥氏体相呈"解理小平面+起源于 γ 晶界处和 α/γ 相界处的撕裂棱"的准解理形貌。

基于以上研究结果，结合我国双相不锈钢生产工艺路线，开发了"双相不锈钢两相双细化工艺技术"、"双相不锈钢中有害相的消除方法"、"控相形态热加工工艺方法"等系列自主知识产权技术，在久立特材、永兴特材、太钢不锈、宝武特冶、武进不锈等重要棒、管、板生产企业得到应用，成功开发出高冲击韧性的 UNS S31803、S32205、S32750、S32760 产品，工业生产的 S32750 双相不锈钢管材-46℃低温冲击韧性提升 60%以上，冲击功值稳定达到 100J 以上。结合冲击韧性提升技术的发展，引导双相不锈钢材料组织调控手段从"相比例"向"相比例+相形态+晶粒尺寸"转变，推动了高质量双相不锈钢产品的国产化和竞争力提升，助力我国高等级双相不锈钢无缝管市占率从不足 10%提高到 75%以上，产生了显著的经济效益和社会价值。

关键词：双相不锈钢；冲击韧性；相比例；相形态；工艺技术

Research Progress and Technology Practice of Impact Toughness of Duplex Stainless Steel

FENG Han[1], SONG Zhigang[1], HE Jianguo[1], WU Xiaohan[1], WANG Baoshun[2], WU Minghua[3], LI Guoping[4], ZHOU Candong[5], GAO Hong[6]

(1. Central Iron & Steel Research Institute Co., Ltd., Beijing 100081, China;
2. Zhejiang Jiuli Hi-Tech Metals Co., Ltd., Huzhou 313028, China; 3. Yongxing Special Stainless-Steel Co., Ltd., Huzhou 313005, China; 4. Taiyuan Iron and Steel (Group) Co., Ltd., Taiyuan 030003, China;
5. Baowu Special Metallurgy Co., Ltd., Shanghai 201900, China;
6. Jiangsu Wujin Stainless Steel Pipe Group Co., Ltd., Changzhou 213017, China)

Abstract: Duplex stainless steel (DSS) proves superior integration of the excellent pitting corrosion of austenitic stainless steel resistance and high strength, chloride stress corrosion resistance of ferritic stainless steel. DSS have been widely used in petroleum and petrochemical, marine engineering, nuclear power, shipbuilding and other fields. In recent years, the application of DSS has been further expanded to low temperature environment, but the domestic DSS generally has the problem of low impact toughness. The research on the fracture mechanism of stainless steel at present is mostly aimed at austenitic steel or ferritic steel. It is necessary to improve and develop the study of low temperature impact toughness on DSS under the coupling of multiphase structure.

Taking UNS S32750 DSS as carrier, the author and his team systematically studied the effects of material factors including phase ratio, phase morphology, aspect ratio and second phase precipitation on the low temperature fracture behavior of DSS, following analyzed the mechanism of action in recent years. The main research results are listed below: (1) The solid solution S32750 DSS shows high impact toughness at room temperature. Exorbitant ferrite (α) ratio and coarsened α grains tend to damage the impact toughness. As the content of α increases from 50% to 60%, the corresponding impact energy at room temperature falls monotonously from 236J to 144J, which mainly reflects in the decrease of crack propagation energy. (2) The morphology of austenite (γ) transforms from lamellar to globular with the length-diameter ratio of grain descended from 55% to 15%, the impact toughness effectively improved, the ductile-brittle transition temperature (DBTT) reduced, the anisotropy of impact toughness optimized and the transverse impact energy increased. (3) The second phases in test steel perform a rapid precipitation rate at 900°C and tend to coarsen at the α/γ phase boundaries or the triangular grain boundaries with high interface energy. 1% σ phase can deteriorate the low temperature impact toughness seriously, which rising DBTT up to 65°C. The worsen effect of Cr_2N on impact behavior is comparatively weaker than that of the σ phase. (4) The impact energy of DSS decreases in an "S" shaped with the temperature range from 20°C to -100°C, which are in accordance with the Boltzmann relationship below: $KV=211.1+(-186.8)/(1+e^{((T-(-39.9))/11.4)})$. (5) With test temperature decreasing from the US zone to the DBT and the LS zones, the variation process of impact energy composition is as follows: High crack propagation energy+ High crack initiation energy→ Medium-low crack propagation energy+ High crack initiation energy→ Low crack propagation energy+ Low crack initiation energy. (6) After broken at the DBT zone, cleavage section appears to pass through α and γ simultaneously. α mostly presents a river-like cleavage morphology. γ shows a quasi-cleavage morphology of "Tiny cleavage plane+ Tearing edge originated from γ grain boundaries and α/γ boundaries".

Based on the research results above and combined with the production process route of DSS in China, the following independent intellectual property rights technologies have been developed: the double refinement process technology of two phases in DSS, the elimination method of detrimental phase in DSS, the thermal processing method of phase-morphology controlled, which have been applied in important rods, tubes and plates production enterprises such as Zhejiang Jiuli Hi-Tech Metals Co., Ltd., Yongxing Special Stainless-Steel Co., Ltd., Taiyuan Iron and Steel (Group) Co. Ltd., Baowu Special Metallurgy Co., Ltd. and Jiangsu Wujin Stainless Steel Pipe Group Co., Ltd.. UNS S31803, S32205, S32750 and S32760 products with high impact behavior have been successfully exploited. The impact toughness of industrial S32750 DSS pipe at –46°C rises by more than 60% with the impact energy more than 100J steadily. Combined with the development of impact toughness improvement technology, the means of microstructure control of DSS are guided to change from "Phase ratio" to "Phase ratio+ Phase morphology+ Grain size", which promotes the localization and competitiveness of high-quality DSS products and helps the market share of high-grade DSS seamless tubes in China increase from less than 10% to more than 75%, resulting in significant economic and social value.

Key words: duplex stainless steel; impact toughness; phase ratio; phase morphology; process technology

高温合金疲劳-蠕变-环境交互作用下裂纹扩展机制的研究

王民庆[1,2]，杜金辉[1,2]，邓　群[1,2]，田志凌[3]，朱　静[4]

（1. 北京钢研高纳科技股份有限公司，北京　100081；2. 高温合金新材料北京市重点实验室，北京　100081；3. 钢铁研究总院，北京　100081；4. 清华大学材料学院，北京　100084）

摘　要： 疲劳-蠕变-环境交互作用下涡轮盘材料裂纹扩展速率以及寿命预测模型的研究一直是高温合金领域的热点和难点。本文以新型航空发动机用涡轮盘材料为研究对象，系统开展了其在疲劳-蠕变-环境交互作用下裂纹扩展机制的研究。结果证实应力促进晶界氧化和氧致动态断裂是引起疲劳-蠕变-环境交互作用下高温裂纹扩展的主要机制。在此基础上，系统研究了不同析出相条件下疲劳裂纹扩展速率的演化规律与机制；并就微量元素对合金疲劳裂纹扩展速率的影响进行了探索研究，发现适量磷对高温疲劳裂纹扩展速率有抑制作用并揭示了其反常作用机制。

关键词：疲劳裂纹扩展速率；氧分压；晶界析出相；磷

The Mechanism of Crack Propagation for Superalloy under Fatigue-Creep-Environment Condition

WANG Minqing[1,2], DU Jinhui[1,2], DENG Qun[1,2], TIAN Zhiling[3], ZHU Jing[4]

(1. Gaona Aero Material Co., Ltd., Beijing 100081, China; 2. Beijing Key Laboratory of Advanced High Temperature Materials, Beijing 100081, China; 3. Central Iron and Steel Research Institute, Beijing 100081, China; 4. School of Materials Science and Engineering, Tsinghua University, Beijing 100084, China)

Abstract: In present paper, a new turbine disc material using in aero-engine, was selected as the research subject and the crack propagation mechanisms under interaction of fatigue-creep-environment were evaluated. The conclusion can be obtained that stress accelerated grain boundary oxidation and dynamic embrittlement by oxygen are the superior mechanisms for the crack propagation under fatigue-creep-environment interaction. Based above research, the study between the precipitation and dwell fatigue crack growth rate were carried out. Meanwhile, the effect of trace elements on the fatigue crack growth rate of the alloy was also investigated. It is found that proper amount of phosphorus can inhibit the fatigue crack growth rate at high temperature, and its mechanism is revealed through research.

Key words: fatigue crack growth rate; oxyen partial pressur; grain boundary precipitate; phosphorus

连铸坯表面温度在线测量与智能控制技术

龙木军，陈登福，张晏铭，黄云伟，张思远，谭　锴

（重庆大学材料科学与工程学院，重庆　400044）

摘　要：钢的连铸过程中，铸坯表面温度是重要的控制参数，其准确性直接影响二冷动态控制的效果，与钢铁连铸生产顺行和铸坯质量息息相关。基于参与性介质的辐射能传递理论，建立了参与性介质环境下的辐射测温数学模型，模拟连铸二冷区环境构建了辐射测量高温物体表面温度的实验装置，从理论计算、实验研究和工厂试验三个方面研究了水蒸气、水雾、水膜和氧化铁皮等对铸坯表面温度辐射测量准确性的影响，总结了参与性介质对单、双色测温结果的影响规律，研发了连铸在线测温装备及系统。综合考虑两种铸坯温度在线获取方法的优缺点，建立了实测温度与传热模型计算温度的融合算法，在此基础上，设计开发了基于在线测温与传热模型融合的连铸二冷智能工艺模型，并在连铸实际生产中进行了应用。结果表明：连铸在线测温系统能够长期、稳定、准确在线测量铸坯表面温度；连铸二冷智能工艺模型互补了实测温度反馈控制和传热模型计算控制的优缺点，模型控制准确、稳定，提升了控制系统的可靠性和智能性。

关键词：连铸坯表面温度；辐射测温；凝固传热模型；温度融合；二冷智能控制

On-line Measurement and Intelligent Control Technology of Continuous Casting Slab Surface Temperature

LONG Mujun, CHEN Dengfu, ZHANG Yanming,
HUANG Yunwei, ZHANG Siyuan, TAN Kai

(School of Materials Science and Engineering, Chongqing University, Chongqing 400044, China)

Abstract: In the process of steel continuous casting, the surface temperature of the slab is an important control parameter, and its accuracy directly affects the effect of secondary cooling dynamic control, which is closely related to the production of steel continuous casting and the quality of the slab. Based on the radiation energy transfer theory of participatory medium, a mathematical model of radiation temperature measurement in participatory medium environment was established. An experimental device for measuring the surface temperature of high-temperature objects by radiation was constructed by simulating the secondary cooling zone environment of continuous casting. The effects of water vapor, water mist, water film and oxide scale on the accuracy of radiation measurement of slab surface temperature were studied from three aspects: theoretical calculation, experimental research and factory test. The influence of participatory medium on the results of single and double color temperature measurement was summarized, and the on-line temperature measurement equipment and system for continuous casting were developed. Considering the advantages and disadvantages of the two methods of on-line acquisition of slab temperature, a fusion algorithm of measured temperature and calculated temperature of heat transfer model is established. On this basis, an intelligent process model of continuous casting secondary cooling based on the fusion of on-line temperature measurement and heat transfer model is designed and developed, and applied in the actual production of continuous casting. The results show that the continuous casting on-line temperature measurement system can measure the surface temperature of the slab on-line for a long time, stably and accurately. The intelligent process model of continuous casting secondary cooling complements the advantages and disadvantages of the measured temperature feedback control and the heat transfer model calculation control. The model control is accurate and stable, and the reliability and intelligence of the control system are improved.

Key words: surface temperature of continuous casting slab; radiative temperature measurement; heat transfer and solidification model; temperature fusion; intelligent control of secondary cooling

HB600级耐磨钢组织和硬度均匀性对耐磨性能的影响

闫强军[1]，邓想涛[2]

（1. 南京钢铁股份有限公司，江苏南京 210035；2. 东北大学，辽宁沈阳 110819）

摘　要： HB600耐磨钢的组织为高碳马氏体，钢板厚度截面的硬度在593~665 HV之间，使其具备良好的耐磨粒磨损性能。将国内某企业生产的30mm厚HB600耐磨钢板（N）与瑞典SSAB同级别同厚度Hardox600钢板（H）对比，N钢板厚度截面硬度均值（628HV）略高于H钢板（606HV），但前者厚度方向组织与硬度的均匀性相对较差。虽然N钢板的松散磨粒磨损性能为H钢板的1.13倍，但后者平行样间磨损失重量的波动较小、磨损面的粗糙度较低。

关键词： 微观组织；硬度；均匀性；耐磨性能

Effect of Microstructure and Hardness Uniformity on the Wear Resistance Performance of HB600 Steel

YAN Qiangjun[1], DENG Xiangtao[2]

(1. Nanjing Iron and Steel Co., Ltd., Nanjing 210035, China; 2. State Key Laboratory of Rolling and Automation, Northeastern University, Shenyang 110819, China)

Abstract: The results show that the microstructure of HB600 wear-resistant steel is high carbon martensite, and the average hardness of steel plate is 593~665HV, which is with excellent wear resistance performance. Comparing HB600 wear-resistant steel plate (N) of a domestic company with the same grade Hardox600 steel plate (H) of the Swedish SSAB company, the cross-sectional hardness of N plate (628HV) is slightly higher than that of H plate (606HV), but the microstructure and hardness uniformity of N plate is less than that of H plate. Although the abrasive wear resistance performance of N plate is 1.13 times that of H plate, the error between the parallel samples of H plate is smaller and the roughness of wear surface is lower compared with N plate.

Key words: microstructure; hardness; uniformity; wear resistance

钢铁行业（含氯类）新污染物生成与控制技术研究及 CO_2 捕集矿化技术进展

苏 伟[1]，邢 奕[1]，田京雷[2]，侯长江[2]，王嘉庆[1]，
孙宇佳[2]，孙方舟[2]，崔永康[1]，郭文韬[1]

（1. 北京科技大学，北京 100083；2. 河钢集团，河北石家庄 056013）

摘 要： 减污降碳协同治理，是实现钢铁行业绿色低碳高质量发展的重要途径。减污方面，国务院办公厅印发新污染物治理行动方案，新污染物治理也将在我国逐步建立长效机制。降碳方面，钢铁行业CCUS是是钢铁等难减排行业低碳转型的可行技术选择以及实现碳中和目标的重要技术保障。本报告从解读行动方案"筛、评、控"角度出发，进而分析钢铁行业多工序的二英类持久性有机污染物、HCl等典型含氯新污染物的生成机制，并探讨其控制技术，分享减排技术应用效果。同时分析钢铁行业 CO_2 捕集矿化技术进展，探讨烟气直接 CO_2 捕集矿化工艺的可行性，并分析不同影响因素对矿化的影响。报告为钢铁行业新污染物的识别与控制、CO_2 的低成本捕集与利用技术的开发提供了参考。

关键词： 新污染物；CO_2 捕集与矿化；持久性有机污染物；HCl

Research on the Generation and Control Technology of New Pollutants in the Iron and Steel Industry and the Progress of CO_2 Capture and Mineralization Technology

SU Wei[1], XING Yi[1], TIAN Jinglei[2], HOU Changjiang[2], WANG Jiaqing[1],
SUN Yujia[2], SUN Fangzhou[2], CUI Yongkang[1], GUO Wentao[1]

(1. University of Science and Technology Beijing, Beijing 100083, China;
2. HBIS Group Co., Ltd., Shijiazhuang 056013, China)

Abstract: Collaborative management of pollution reduction and carbon reduction is an important way to realize the green, low-carbon and high-quality development of the iron and steel industry. In terms of pollution reduction, the General Office of the State Council has issued a new pollutant management action program, and the management of new pollutants will gradually establish a long-term mechanism in China. In terms of carbon reduction, CCUS in the iron and steel industry is a feasible technical option for the low-carbon transformation of iron and steel and other difficult-to-emit industries, as well as an important technological guarantee for realizing the goal of carbon neutrality. This report starts from the perspective of interpreting the "screening, evaluation and control" of the action program, and then analyzes the generation mechanism of dioxin POPs, HCl and other typical chlorine-containing new pollutants in the iron and steel industry in multiple processes, discusses the control technologies, and shares the effect of emission reduction technology application. It also analyzes the progress of CO_2 capture mineralization technology in the iron and steel industry, discusses the feasibility of direct CO_2 capture mineralization process in the flue gas, and analyzes the impact of different influencing factors on mineralization. The report provides references for the identification and control of new pollutants in the iron and steel industry and the development of low-cost CO_2 capture and utilization technologies.

Key words: new pollutants; CO_2 capture and mineralization; POPs; HCl

攀西钛资源开发与利用研究进展

李开华，李 亮，朱福兴

（攀钢集团研究院有限公司，四川攀枝花 617000）

摘 要：介绍了攀西钛资源的概况及特点、冶金分离技术流程，重点阐明了基于攀西钛资源的高炉渣提钛技术和大型熔盐氯化技术的研究进展，分析了航空用高品质海绵钛制备技术的难点和取得成效。结果表明：采用高炉渣提钛技术和大型熔盐氯化技术是实现攀西钛资源高效利用的两条关键可行的技术路线，制备的高品质海绵钛可以实现其在航空等高端领域的应用。

关键词：攀西钛资源；高炉渣提钛；熔盐氯化；海绵钛

Research Progress on the Development and Utilization of Titanium Resources in Panxi

LI Kaihua, LI Liang, ZHU Fuxing

(Pangang Group Research Institute Co., Ltd., Panzhihua 617000, China)

Abstract: The general situation and characteristics of titanium resources in Panxi and the technological process of metallurgical separation are introduced. The research progress of titanium extraction technology from blast furnace slag and large-scale molten salt chlorination technology based on Panxi titanium resources is emphasized. The difficulties and achievements of preparation technology of high quality sponge titanium for aviation are analyzed. The results indicate that the technology of titanium extraction from blast-furnace slag and chlorination of large-scale molten salt are two key and feasible technical routes to realize the efficient utilization of titanium resources in Panxi, and the high-quality titanium sponge prepared can be applied in high-end fields such as aviation.

Key words: Panxi titanium resources; titanium extraction from blast-furnace slag; chlorination in molten salt; titanium sponge

高精度冷连轧数字孪生与信息物理系统（CPS）关键技术研发及应用

李 旭

（东北大学，辽宁沈阳 110819）

摘 要：冷连轧机组作为生产高强钢、硅钢和极薄精密带钢等高端冷轧产品的主力机组，其控制系统是最复杂的工业控制系统之一。冷连轧生产控制精度要求高、动态响应时间短、轧制速度快及工艺参数多，目前仅有德国、日本掌握全套技术。我国冷连轧控制系统长期依赖引进，由于其存在封闭性和功能局限性问题，严重制约了产品的自主研发及拓展。同时，引进系统也存在静态模型精度低、系统间缺少协调控制及模型匹配性差等关键共性问题，导致板形、板厚质量超差及断面形状难以精准控制。本文融合冷轧机理与大数据/机器学习等数据科学，针对冷轧非稳态过程板形、厚度等质量控制的关键难题，开发了数据驱动融合工艺机理的边缘过程数字孪生模型，并以其为核心形成完整的高速过程信息物理系统，大大提高了产品质量及生产效率。相关成果成功应用于唐钢、山东冠州等多条冷连轧产线，典型规格带钢的板形精度控制在 3IU 以内、板厚精度控制在±2μm 以内、超差长度控制在 5m 以内、实现了 0.17mm 厚度 T5 镀锡基板 1450m/min 的高速稳定轧制。

关键词：冷连轧机；板形控制；板厚控制；数字孪生；信息物理系统

Development and Application of Key Technologies for Digital Twin and Cyber-Physical Systems (CPS) for High-precision Cold Tandem Rolling

LI Xu

(Northeastern University, Shenyang 110819, China)

Abstract: Tandem cold rolling is the major production for the high-end cold-rolled product, such as high strength steel, silicon steel and ultra-thin precision strip steel, and its control system is one of the most complex industrial control systems. Since tandem cold rolling has high control precision, short dynamic response time, fast rolling speed and numerous process parameters, only Germany and Japan possess a full set of technology. Tandem cold rolling in our country has been dependent on the imported control system for a long time. As the closure and functional limitations of imported system, the independent research and development of products are seriously restricted. At the same time, the imported system also has some key common problems, such as low static model accuracy, lack of coordination control between systems and poor model matching, resulting in the out-of-tolerance flatness and thickness and the cross-section profile controlling difficultly. In this paper, the mechanism of cold rolling is combined with information science, such as big data and machine learning. An edge process digital twin model coupling data driven with technological principles is developed for the key quality control problems such as the flatness and thickness during unsteady cold rolling process. Furthermore, based on the proposed model, a complete high-speed process cyber-physical system is established and greatly improves product quality and production efficiency. Relevant results have been successfully applied to many cold rolling lines such as Tangsteel and Shandong Guanzhou. The flatness accuracy of typical strip steel is within 3IU, the thickness accuracy of that is within

±2μm and the out-of-tolerance length of that is within 5m. The high-speed and steady rolling of tinned mill black plates (TMBP) with the grade of T5 and the thickness of 0.17mm at 1450m/min is achieved.

Key words: cold rolling mill; plate shape control; plate thickness control; digital twin; cyber-physical systems

基于无废城市建设的长流程钢厂固废管理模式与技术路径

李红红[1,2]

(1. 宝钢湛江钢铁有限公司能源环保部，广东湛江 524072；
2. 宝山钢铁股份有限公司能源环保部，上海 200941)

摘　要： 钢厂工业固废的环境管理是推进"无废城市"建设的重要环节，实现全量资源化利用是钢厂发展循环经济的目标。传统方式是委托社会企业进行综合利用或处置，存在环境风险大、物流量大、监管难等问题；利用"无废"理念，开发了长流程钢厂"源头减量－厂内加工分选－冶金炉窑协同利用－智慧管控"固废管理的创新性技术体系，建设了厂内固废加工、分选、仓储中心，开展了基于钢铁窑炉特性和固废属性协同处理处置的技术路径研究，建立了支撑工业固废全过程管理的系统、数据、技术、产线体系，减少了源头产生量，加大了厂内循环利用量，实现了"固废不出厂"；对防范环境风险、减少固废车辆物流量、实现高水平生态环境保护、缓解城市固废处理压力、优化产城融合和助力无废城市建设具有重要意义。

关键词： 无废城市；长流程钢厂；固废管理；技术路径；协同处理

Solid Waste Management Mode and Technical Path of Long-process Steel Plant Based on the Concept of Zero-waste Construction

LI Honghong[1,2]

(1. Department of Energy and Environment, Baosteel Zhanjiang Iron & Steel Co., Ltd., Zhanjiang 524072, China; 2. Department of Energy and Environment, Baoshan Iron & Steel Co., Ltd., Shanghai 200941, China)

Abstract: The environmental management of industrial solid waste in steel plants is an important part in promoting the construction of zero waste city, and achieving full resource utilization is the goal of steel plants in developing a circular economy. The traditional method is to entrust social enterprises for comprehensive utilization or disposal, which has problems such as high environmental risks, large logistics volume, and difficult supervision. Using the "no waste" concept, an innovative technical system of solid waste management of "source reduction - in-plant processing and sorting - collaborative utilization using metallurgical furnaces-Intelligent management and control system of solid waste management" has been developed in long-process steel plants. The solid waste processing, sorting, and warehousing center in the plant has been built, the technical paths research based on the characteristics of steel kilns and properties of solid waste has been carried out, and the system, data, technology, and production line system supporting the whole process management of industrial solid waste has been established, which has reduced the amount of source generation, increased the amount of recycling in the plant, and realized "No solid waste leave the plant". It is of great significance to prevent environmental risks, reduce the logistics volume of solid waste, achieve a high level of ecological environmental protection, relieve the pressure of urban solid waste treatment, optimize the integration of industry and city, and help build a waste-free city.

Key words: zero-waste city; long-process steel plant; solid waste management; technical path; collaborative utilization

无底层绿色环保涂层高磁感取向硅钢的开发与推广应用研究

杨佳欣，申明辉，宋　刚

（宝钢股份中央研究院武钢有限技术中心，湖北武汉　430080）

摘　要：本研究开展新型大型电机用无底层绿色环保涂层高磁感取向硅钢的研发，并首次成功实现大批量生产及应用。报告详解了无底层绿色环保涂层高磁感取向硅钢的开发与产品质量认证、批量应用过程，以及特殊元素对取向硅钢性能、底层的影响机理，并展望了未来技术发展前景。

关键词：高磁感取向硅钢；底层；绿色环保涂层

Development and Application Promotion of Glassless High Permeability Grain-oriented Silicon Steel with Green and Environmentally Friendly Coating

YANG Jiaxin, SHEN Minghui, SONG Gang

(R&D Center of Wuhan Iron and Steel Company, Research Institute,
Baoshan Iron and Steel Co., Ltd., Wuhan 430080, China)

Abstract: This study completed the research and development of new type glassless high permeability grain-oriented silicon steel with green and environmentally friendly coating for large motor, first mass production and application was successfully implemented. Report introduces the research and development, certification of product quality, batch application process of glassless high permeability grain-oriented silicon steel with green and environmentally friendly coating. In addition, the effects of special element on magnetic properties and glass film, future technological development prospects are introduced.

Key words: high permeability grain-oriented silicon steel; glass film; green and environmentally friendly coating

Ni 对低合金钢海洋大气腐蚀产物氧化还原行为影响

杨建炜[1]，刘智勇[2]

（1. 首钢技术研究院，北京　100043；2. 北京科技大学，北京　100083）

摘　要：针对海洋大气环境用低合金钢的腐蚀机理问题，以及高 Ni 耐蚀钢中 Ni 元素的耐蚀原理开展研究工作。探讨腐蚀过程锈层中各相产物的转变规律以及 Ni 在其中的作用。研究采用薄液膜腐蚀试验装置进行在线电化学测试，结合 XRD 实现电化学测试前后产物的原位观测。同时，为便于研究腐蚀产物转变规律和 Ni 的作用，利用真空箱实现试样在有氧和无氧两种环境中的测试。研究结果表明，Ni 在腐蚀过程中富集于内锈层，使含 Ni 钢的阳极极化电流密度明显低于碳钢。在有氧和无氧两种测试环境中均出现了这一现象，特别是在有氧环境 Ni 抑制腐蚀产物氧化的作用更显著。XRD 分析发现，有氧环境中腐蚀产物极化后形成非晶态的 ferrihydrite，无氧环境极化则形成中间产物绿锈，Ni 可以显著抑制两种氧化产物的转变，特别是抑制非晶态 ferrihydrite 形成更明显。综上所述，Ni 在锈层中富集可明显抑制腐蚀产物的氧化转变，特别是针对有氧环境，延缓产物在腐蚀过程中的转变，进而提高耐蚀性。

关键词：低合金钢；Ni 元素；海洋大气

Effect of Ni on Oxidation and Reduction Behavior of Corrosion Product Formed on Low Alloy Steel in Marine Atmosphere

YANG Jianwei[1], LIU Zhiyong[2]

(1. Shougang Research Institute of Technology, Beijing 100043, China;
2. University of Science and Technology Beijing, Beijing 100083, China)

Abstract: The corrosion mechanism of low-alloy steels used in marine atmospheric environment and the corrosion resistance mechanism of Ni in high-Ni corrosion-resistant steels are studied. The transformation behavior of each phase corrosion product during the corrosion process and the role of Ni in the corrosion products were discussed. In this study, a thin electrolyte film corrosion test apparatus was used to carry out online electrochemical testing, and XRD was combined to realize the in-situ observation of products before and after electrochemical testing. At the same time, in order to facilitate the study of the transformation behavior of corrosion products and the role of Ni, the vacuum box is used to realize the test of samples in both aerobic and oxygen-free environments. The results show that Ni is enriched in the inner rust layer during the corrosion process, so that the anodic polarization current density of Ni-containing steel is significantly lower than that of carbon steel. This phenomenon occurred in both aerobic and anaerobic test environments, especially in aerobic environments, where the effect of Ni in inhibiting the oxidation of corrosion products was more significant. XRD analysis found that the polarization of corrosion products in aerobic environment formed amorphous ferrihydrite, and the polarization of anaerobic environment formed intermediate product green rust, and Ni could significantly inhibit the transition of the two oxidation products, especially the formation of amorphous ferrihydrite. In summary, the enrichment of Ni in the rust layer can significantly inhibit the oxidative transformation of corrosion products, especially for aerobic environments, and delay it. The transformation of the product during the corrosion process, which in turn improves the corrosion resistance.

Key words: low alloy steel; element of Ni; marine atmosphere

Cr 和 RE 元素对 HRB400 螺纹钢耐蚀性能的影响规律研究

陈永峰[1]，陈天奇[2]，刘涛[1,3]，杨伟勇[4]，刘超[2]

(1. 新兴铸管股份有限公司，河北武安 056300；2. 北京科技大学新材料技术研究院国家材料腐蚀与防护科学数据中心，北京 100083；3. 北京科技大学冶金生态与工程学院，北京 100083；4. 芜湖新兴铸管有限责任公司钢铁研究院，安徽芜湖 241000)

摘　要：随着我国经济建设的高质量发展，我国重大装备正在迅速向深海、盐湖、高原、盐碱地、极热或极寒等复杂的环境扩展。此类环境下的重点工程对耐蚀钢的需求得到了进一步扩大。钢筋作为工程建设的关键材料，常会因为环境腐蚀而造成严重的后果。据工程院相关腐蚀调查表明，我国每年钢筋腐蚀损失在5千亿元以上，因此设法提高钢筋的服役安全性，发展高品质耐蚀钢筋已经是势在必行。

本工作以HRB400钢筋为基材，采用Cr和RE等元素对HRB400钢筋进行了合金化调控，结果电化学测试和周浸试验系统地研究了含Cr和RE的HRB400钢筋的耐蚀性。结果表明，随Cr含量的提高，点蚀电位增大，维钝电流密度降低，极化电阻R_p增大，耐点蚀性能显著提高。Cr通过稳定钝化膜提高了钢筋的耐蚀性，不参与蚀坑的形成，而点蚀是由$(RE)_2O_2S$夹杂物的溶解引起的，导致蚀坑底部局部酸化，降低了钝化膜的稳定性。在长期腐蚀过程中，Cr和RE均降低了钢筋的腐蚀速率，这与Cr和RE促进锈层中α-FeOOH的形成有关。总体而言，低合金钢筋的耐蚀性较HRB40提升了2~4倍。此外，腐蚀大数据在线监测表明在海洋环境下低合金耐蚀钢筋较普通HRB400的腐蚀速率降低50%以上。

关键词：HRB400；微合金化；电化学测试；耐蚀钢筋

Study on the Effect Pattern of Cr and RE Elements on the Corrosion Resistance of HRB400 Rebars

CHEN Yongfeng[1], CHEN Tianqi[2], LIU Tao[1,3], YANG Weiyong[4], LIU Chao[2]

(1. Xinxing Ductile Iron Pipes Co., Ltd., Wuan 056300, China; 2. National Materials Corrosion and Protection Data Center, Institute of Advanced Materials & Technology, University of Science and Technology Beijing, Beijing 100083, China; 3. School of Metallurgical and Ecological Engineering, University of Science and Technology Beijing, Beijing 100083, China; 4. Iron and Steel Research Institute, Wuhu Xinxing Ductile Pipes Co., Ltd., Wuhu 241000, China)

Abstract: With the high-quality development of China's economic construction, China's major equipment is rapidly expanding to the deep sea, salt lake, plateau, saline soil, extreme heat or cold and other complex environments. The demand for corrosion-resistant steel for key projects in such environments has been further expanded. Reinforcing steel as a key material for engineering construction, often due to environmental corrosion and cause serious consequences. According to the corrosion survey of the Academy of Engineering, the annual corrosion loss of steel rebar in China is more than 500 billion, so it is imperative to develop high-quality corrosion-resistant steel rebar to improve the service safety of steel rebar.

In this work, 12 kinds of corrosion-resistant steel bars with different microalloying types and contents were designed and developed by using HRB400 steel bars as base material and alloying with elements such as Cr and RE. As a result, electrochemical tests and weekly immersion tests systematically investigated the corrosion resistance of HRB400 steel bars containing Cr and RE. The results show that with the increase of Cr content, the pitting potential increases, the dimensional passivation current density decreases, the polarization resistance Rp increases, and the resistance to pitting corrosion is significantly improved. Cr improves the corrosion resistance of the reinforcement bar by stabilizing the passivation film, and does not take part in the formation of the etch pits, whereas pitting corrosion is induced by the dissolution of $(RE)_2O_2S$ inclusions, which leads to the localized acidification at the bottom of etch pits, and reduces the stability of the passivation film. During long-term corrosion, both Cr and RE reduced the corrosion rate of the bars, which was related to the fact that Cr and RE promoted the formation of α-FeOOH in the rust layer. Overall, the corrosion resistance of low-alloy steel bars was improved by 2-4 times compared with HRB40. In addition, the online monitoring of corrosion big data shows that the corrosion rate of low-alloy corrosion-resistant rebar is reduced by more than 50% compared with that of ordinary HRB400 in the marine environment.

Key words: HRB400; microalloying; electrochemical test; corrosion-resistant steel bar

鞍钢钢铁产品数字化研发实践与展望

林 利，王军生

（鞍钢集团钢铁研究院，辽宁鞍山 114009）

摘 要："材料基因组计划"通过跨尺度高通量计算结合数据库和人工智能技术的快速推动新材料研发，实现研发周期缩短一半、研发成本降低一半"的战略目标。数字化研发是一种手段、工具、平台，利用这一手段实现产品（新产品、现有产品）、工艺、应用服役评价的探索，最终实现成本降低，效率及产品价值提升。目前鞍钢确认了利用数字化研发实现产品开发探索、工艺数字孪生、应用及服役数字化评价的数字化研发目标，并在趋势探索、产品开发、工艺研究、应用性能、服役性能、方案制定、检验表征等维度进行了深入的探索和实践，以达到数字化研发实现产业价值的深度挖掘的目的。

关键词：材料；数字化

Anshan Iron and Steel Products Digital R&D Practice and Prospects

LIN Li, WANG Junsheng

(Iron and Steel Research Institute of Ansteel Group, Anshan 114009, China)

Abstract: The "Materials Genome Project" quickly promotes the research and development of new materials through cross-scale high-throughput computing combined with database and artificial intelligence technology, and realizes the strategic goal of shortening the research and development cycle and reducing research and development costs by half. Digital research and development is a means, tool and platform, use this method to realize the exploration of products (new products, existing products), processes, application and service evaluation, and finally realize cost reduction, efficiency and product value improvement. At present, Ansteel has confirmed the use of digital R&D to realize product development exploration and process digital twins, application and service digital evaluation of digital R & D goals, and in-depth exploration and practice in the dimensions of trend exploration, product development, process research, application performance, service performance, program formulation, inspection and characterization, in order to achieve digital R & D to realize industrial value for the purpose of deep excavation.

Key words: materials; digitization

钢液初始凝固坯壳捕渣机理及夹渣缺陷控制研究

钟小灿[1,2]，周乐君[1,2]，罗 豪[1,2]，王万林[1,2]

（1. 中南大学冶金与环境学院，湖南长沙 410083；
2. 中南大学清洁冶金国际联合研究中心，湖南长沙 410083）

摘 要：无间隙原子钢具有低的屈服点和屈强比、高伸长率、高的塑性应变比、高的加工硬化指数等深冲性能，并且具有无时效性等特点，使得其广泛应用于汽车、家电等制造过程，尤其是汽车外板、内板等。保护渣作为连铸结晶器的功能材料，覆盖钢液表面的保护渣更是起着防止钢液二次氧化、绝热保温、吸收夹杂物、控制传热及润滑铸坯等极为关键的作用。连铸中坯壳凝固前沿是否捕捉保护渣与渣钢界面特性息息相关，为了平衡保护渣捕获夹杂物和铸坯卷渣的冲突需求，依靠改变界面性能影响较大的表面活性剂组分（Na_2O、B_2O_3等）的保护渣设计思路，借助卧滴法技术来探究保护渣成分的改变对渣-钢界面特性的影响。表面张力是保护渣非常重要的物化性质之一，但想要准确测量高温时表面张力比较困难。本研究以 IF 钢及其连铸保护渣为研究对象，首先，研究了 IF 钢粗糙度对界面接触角的影响；之后，基于热力学数据库与 Butler 方程，结合 Matlab 对保护渣的表面张力进行了模型计算；最后，对不同 Na_2O、B_2O_3 含量的 IF 钢连铸保护渣与高铝高强钢之间的界面现象、熔化行为、熔体内氧的状态以及渣钢界面层进行了研究。本研究的主要结论如下：（1）保护渣的润湿性能随着高铝钢基底粗糙度的升高而增强。随着温度的升高，渣钢间的界面接触角不断减小，模型计算结果表明 Na_2O 和 B_2O_3 都能减小表面张力；（2）保护渣中 Na_2O 的增加，使得保护渣的接触角及界面张力减小，同时熔化温度显著降低，界面层厚度增加，并且熔体中的 $O^{2-}+O^-$ 含量上升而 O^0 含量降低，保护渣的硅酸盐熔体结构解聚，从而简化熔体结构；（3）保护渣中 B_2O_3 的增加，使得保护渣的接触角及界面张力减小，同时熔化温度显著降低，界面层厚度增加，B_2O_3 的加入使得部分三维 $[BO_4]^{5-}$ 结构转变为二维 $[BO_3]^{3-}$，熔体结构趋于简单。综上，本研究创新性地从界面角度出发，研究了不同粗糙度的固态钢基底和液态保护渣之间润湿性，对界面性能影响较大的表面活性组分 Na_2O 和 B_2O_3 对钢-渣界面润湿性的影响，以此来加深对坯壳凝固前沿与保护渣之间的界面行为的理解。本研究还基于修正的 Butler 方程，结合热力学数据库，通过模型计算七元及八元系等复杂体系保护渣的表面张力，这对优化现有保护渣或开发新型保护渣时奠定了理论基础。研究结果可以用来优化现有或开发新型保护渣，进而提高 IF 钢铸坯表面质量及连铸生产效率。

关键词：连铸；结晶器；初始凝固坯壳；捕渣机理；夹渣缺陷

Mechanism and Controlling of the Slag Capture in the Initial Solidification Shell during the Continuous Casting Process

ZHONG Xiaocan[1,2], ZHOU Lejun[1,2], LUO Hao[1,2], WANG Wanlin[1,2]

(1. School of Metallurgy and Environment, Central South University, Changsha 410083, China;
2. National Center for International Research of Clean Metallurgy,
Central South University, Changsha 410083, China)

Abstract: Interstitial-free steel has been widely used in fields, such as automobiles, electrical appliances, etc. due to the properties of low yield point and yield ratio, high elongation, high plastic strain ratio, high work hardening. The interface behavior in continuous casting mold affects many processes such as inclusion absorption, slag entrapment and nozzle erosion, which directly affects the production efficiency and quality of casting product. As a functional material of the continuous casting process, the mold flux covers the surface of the molten steel and plays a crucial role in preventing secondary oxidation, thermal insulation, absorbing inclusions, controlling heat transfer and lubricating the shell. Whether the solidification front of the shell in mold can capture the mold flux or not, is mainly determined by the interface properties between the slag and steel. In order to balance the conflicting needs of avoiding the slag capture and inclusions absorption, the surface-active components (Na_2O, B_2O_3, etc.) were focused on the newly designed mold flux. Therefore, this study firstly investigated the effect of roughness of steel substrate on contact angle between mold flux and substrate using sessile drop method. Then, surface tension of the mold flux was calculated based on Butler equation and thermodynamic database. At last, interfacial tension, melting behavior, oxygen state in the melt, and slag steel interface layer were also studied. The main results are listed as following: (1) The wetting ability of the mold flux enhances with the increase of the roughness of steel substrate, and the contact angle of mold flux/steel decreases continuously with temperature increasing. In addition, the model calculation shows that the surface tension of mold flux reduces when both Na_2O and B_2O_3 increase. (2) The increase of Na_2O content in the mold flux leads to a decrease in contact angle and interfacial tension, meanwhile the melting temperature significantly decreases and the thickness of the interface layer increases. XPS analysis indicates that the silicate melt structure of the mold flux depolymerizes, as the $O^{2-}+O^-$ content in the melt increases and the O^0 content decreases. (3) The addition of B_2O_3 in the mold flux leads to a decrease in the contact angle and interfacial tension of mold flux, meanwhile significantly reduces the melting temperature and increases the thickness of the interface layer. The melt structure also tends to be simpler with the addition of B_2O_3, as it can transform some three-dimensional $[BO_4]^{5-}$ structures into two-dimensional $[BO_3]^{3-}$ structures. The results obtained here can be used to guide the optimization or development of new mold flux for IF steel, and improve the surface quality and production efficiency of continuous casting process.

Key words: continuous casting; mold; initial solidified shell; mechanism of slag entrapment; slag defect

分层供热低碳富氢烧结技术的研发与应用

周浩宇[1]，叶恒棣[1]，张俊杰[3]，甘　敏[2]

（1. 中冶长天国际工程有限责任公司，湖南长沙　410205；2. 中南大学，湖南长沙　410083；
3. 中天钢铁，江苏常州　213011）

摘　要：针对当前钢铁冶金中烧结矿低碳生产面临的四大难题，通过研究烧结料层热量产生、消耗、传递、蓄积的规律及影响因素，揭示了能源相态、温度区间及变化梯度、微观气氛与烧结成矿产质量的关系，建立了烧结料层均热数学模型及求解分层补热的数学算法，开发了基于基准配碳加分层补热实现均热低碳富氢烧结的方法，发明了烧结过程能源配加时、空、质、量多维度偶合工艺体系及配送装置系统。以中天550m²烧结生产线为对象开展了小试试验与工业化试验。试验表明：应用分层供热烧结方法后，烧结速度、烧结成品率、转鼓强度、利用系数等质量指标较基准工况均略微上升，小试试验配碳量从4.0%降至3.25%，减碳率18.75%，工业化试验配碳量从52.86kgce/t降至45.89kgce/t，减碳率13.5%。

关键词：钢铁冶金；烧结；低碳生产；分层供热

Research and Application of Stratified Heat Supply for Homogeneous and Low Carbon Sintering

ZHOU Haoyu[1], YE Hengdi[1], ZHANG Junjie[3], GAN Min[2]

(1. Zhongye Changtian International Engineering Co., Ltd., Changsha 410205, China;
2. School of Minerals Processing & Bioengineering, Central South University, Changsha 410083, China;
3. Zhongtian Steel, Changzhou 213011, China)

Abstract: Be directed against the 4 major problems of low carbon sintering, the heat production, heat transfer and heat accumulation in the thick-layer sintering bed were analyzed. A mathematical model of calculating the heat transformation in the ore sintering layer was derived to study the low carbon sintering. Aiming at a homogeneous and low carbon sintering, a stratified heat supply method was proposed. The method gives the heat supply criterion and the related key technical parameters. Taking a sintering machine of 550m² in Zhongtian steel company as the study object, we explored the ultimate coke reduction by sinter pot tests. The test results showed that with the stratified heat supply method, the coke consumption reduced to 3.25% from 4.0% (decreased by 18.5%) in laboratory test and reduced to 45.89kgce/t from 52.86kgce/t in industrial test, while other quality index of sintering speed, yield, tumbler index and productivity received slight improvement.

Key words: ferrous metallurgy; sinter; low carbon production; stratified heat supply

金属材料的微生物腐蚀机理与防治

徐大可

（东北大学材料科学与工程学院，辽宁沈阳　110819）

摘　要：微生物腐蚀（Microbiologically influenced corrosion, MIC）是当今材料领域亟需解决的一个难题。全球每年因微生物腐蚀造成的损失约占腐蚀总损失 20%，微生物腐蚀在美国造成 600 亿美元/年的经济损失，在我国造成的经济损失也超过 4000 亿元。微生物腐蚀是材料、腐蚀、微生物多学科交叉的研究方向，由于缺乏对 MIC 机理的深入认知，人们甚至认为 MIC 是腐蚀领域中的一个"谜"。报告人以微生物腐蚀的机理研究（基因水平）为主导，立足于"应用基础研究指导解决实际工程问题"这一思路，围绕"海洋微生物腐蚀机理及其防治、耐海洋微生物腐蚀材料的设计与开发"这一主题，开展了一系列交叉、深入和创新的研究工作：

（1）微生物腐蚀机理：提出了基于"生物能量学"和"细胞外电子传递"的微生物腐蚀新机理，认为微生物作为主体通过细胞外电子传递参与腐蚀反应过程并从本质上改变了腐蚀发生机制这一全新理论，实现了海洋环境下机理指导的微生物腐蚀检测，为耐微生物腐蚀材料的设计提供了新思路。

（2）微生物腐蚀的防治：上述机理研究明确了黏附在材料表面的细菌生物被膜是导致微生物腐蚀的主要原因，为微生物腐蚀防治提供了明确的靶点。因此申请人采用攻（抗菌材料+ D-氨基酸杀菌剂增效剂+一剂多效的绿色杀菌剂）守（活体生物被膜防腐蚀材料）兼备的思路构建了全方位多维度的海洋微生物腐蚀防治体系。

Microbiologically Influenced Corrosion of Metals: Mechanism and Mitigation

XU Dake

(School of Materials Science and Engineering, Northeastern University, Shenyang 110819, China)

Abstract: Microbiologically influenced corrosion (MIC) is recognized as a knotty problem in the material field nowadays. The economic cost caused by MIC is account for 20% of total corrosion in the world each year. In the United States, MIC leads to $60 billion economic cost, while the cost exceeds to 4 trillion Chinese Yuan in China. MIC is an interdisciplinary research field related to material science, corrosion, and microbiology, but it is still a "Mystery" in the eyes of researchers because of the poor understanding of MIC mechanism. The speaker carried out a series of cross-cutting, in-depth, and innovative research work around the theme of "Marine MIC mechanism and its mitigation, and the design and development of materials with significant resistance performance to marine MIC based on the idea of "Applying fundamental research to solve engineering problems in the realistic world". The detailed work shows below:

(1) The mechanism of MIC: the speaker proposed a new theory of MIC mechanism based on the "Bioenergetics" and "Extracellular electron transfer". It is believed that microbes as the main body participate in the corrosion reaction process through extracellular electron transfer and change the corrosion mechanism essentially, realizing the detection of MIC in the marine environment depending on the guidance from mechanism research and providing a new way for the design of materials with significant resistance to MIC.

(2) The mitigation of MIC: the biofilm adhesion on the surface of materials is the main cause of MIC based on the above-mentioned mechanism, providing a clear target for the mitigation of MIC. The speaker established a comprehensive and multi-dimensional marine MIC mitigation system based on the ideas of attacking (antibacterial materials + D-amino acid

镍铁渣增值利用研究进展

彭志伟

（中南大学，湖南长沙 410083）

摘 要：在剖析镍铁渣主要物理化学特性和利用现状的基础上，提出了基于镁硅协同转化的镍铁渣直接材料化增值利用创新思路，开发出冷态镍铁渣制备镁橄榄石型复相定形耐火材料和 $MgO-Al_2O_3-SiO_2$ 不定形耐火浇注料技术，以及熔态镍铁渣原位改质制备顽火辉石为主晶相，尖晶石$[Mg(Fe,Al,Cr)_2O_4]$为强化晶相的微晶玻璃技术，获得的系列产品性能优于同类商业制品，同时镍铁渣利用率显著提升，其使用配比由传统利用技术最高约 55%提升至 75%~85%，为有效化解镍铁渣存量和增量问题提供了整体方案。

关键词：镍铁渣；增值利用；材料化；耐火材料；微晶玻璃

Research Advances in Value-added Utilization of Ferronickel Slag

PENG Zhiwei

(Central South University, Changsha 410083, China)

Abstract: Based on analysis of the main physicochemical properties and current utilization status of ferronickel slag, an innovative idea of direct materialization-oriented value-added utilization of the slag via synergistic transformation of magnesium and silicon components was proposed. The technologies of preparing forsterite-type biphasic shaped refractory materials and $MgO-Al_2O_3-SiO_2$ unshaped refractory castables from cold ferronickel slag and producing glass-ceramics with enstatite as the main crystalline phase and spinel $[Mg(Fe,Al,Cr)_2O_4]$ as the strengthening crystalline phase by in-situ reforming of molten ferronickel slag were developed. The performance of the obtained products was superior to that of similar commercial products. Meanwhile, the utilization percentage of ferronickel slag increased significantly, with the addition from the maximum of about 55% in traditional utilization technologies to 75%–85%, providing an overall solution for effectively resolving the problems of stockpiling and increment of ferronickel slag.

Key words: ferronickel slag; value-added utilization; materialization; refractory materials; glass-ceramics

抽锭电渣重熔关键技术、装备的研发及推广应用

臧喜民[1]，姜周华[2]，刘福斌[2]，耿 鑫[2]，李万明[3]

（1. 沈阳工业大学，辽宁沈阳 110870；2. 东北大学，辽宁沈阳 110819；
2. 辽宁科技大学，辽宁鞍山 114051）

摘 要：为解决传统电渣重熔"效率低、成本高"的生产难题，基于钢的液-固转变行为、渣的高温力学性能及传氧等物化性能研究，采用双极串联T型结晶器开发出抽锭电渣重熔工艺及装备。在此基础上，开发出了单电极、双

电极、结晶器导电、双电源控制等多种抽锭电渣重熔工艺及装备。通过该工艺技术生产出 300mm×340mm×6000mm、ϕ600mm×6000mm、400mm×2000mm×4000mm、ϕ650/ϕ450mm×6000mm 等多种规格的矩形、圆形、板坯实心钢锭和空心锭。生产出的轴承钢、模具钢、齿条钢等材质的产品的冶金质量与传统电渣重熔钢锭一致，细长型电渣锭可以直接进入精轧，省去锻造开坯工序，生产效率提高近 30%、产品综合生产成本降低约 15%。

关键词：抽锭电渣重熔；渣系；特殊钢

Research and Application on Technology and Equipment of Electroslag Remelting Withdrawal Process

ZANG Ximin[1], JIANG Zhouhua[2], LIU Fubin[2], GENG Xin[2], LI Wanming[3]

(1. Shenyang University of Technology, Shenyang 110870, China; 2. Northeastern University, Shenyang 110819, China; 3. University of Science and Technology Liaoning, Anshan 114051, China)

Abstract: An innovative electroslag remelting withdrawal process and its equipment featured in two-series-electrode and T-type mould was developed to solve the production problem of low efficiency and high cost in traditional electroslag remelting based on the research of liquid-solid transformation behavior of steel, high temperature mechanical properties and oxygen transfer properties of slag. On this basis, a variety of electroslag remelting processes and equipment such as single electrode, two-series-electrode, current-conductive mould and double power-supply devices control have been developed. A range of product specifications such as 300mm×340mm×6000mm, ϕ600mm×6000mm, 400mm×2000mm×4000mm, ϕ650/ϕ450mm×6000mm and other specifications of rectangular, circular, slab solid ingot and hollow ingot were produced through the process technology. The metallurgical quality of produced bearing steel, die steel, rack steel and other materials is consistent with the traditional electroslag remelting ingot. The slender electroslag ingots can be directly rolled, which can eliminate the forging process, increase the production efficiency by nearly 30%, and reduce the comprehensive production cost of the product by about 15%.

Key words: electroslag remelting withdrawal process; slag system; special steel

地下铁矿山高阶段嗣后胶结充填理论与应用研究

谭玉叶[1]，宋卫东[1]，张兴才[2]

（1. 北京科技大学土木与资源工程学院，北京 100083；
2. 武钢资源集团大冶铁矿有限公司，湖北黄石 435006）

摘　要：阶段嗣后充填采矿法具有安全性好、生产能力大等显著优点，成为地下金属矿山充填法开采的首选。然而，逐段分层充填所带来的充填体强度劣化、超细全尾砂沉降效率及制浆质量、成本居高不下等问题成为制约其在地下金属矿山应用的瓶颈问题。针对"分层胶结充填体的本构模型与损伤演化机理"及"复杂条件下充填体与围岩力学行为及相互作用机理"等科学问题，以理论创新为突破口，开展了"超细全尾砂絮凝沉降细观机理、高阶段嗣后分层胶结充填体损伤演化机制、多场耦合条件下围岩-充填体细宏观力学特性"三方面研究，并基于理论创新进行"无动力深锥浓密及充填系统、高阶段嗣后分层胶结充填体结构及强度设计方法、高阶段嗣后高效开采工艺及地压控制"技术集成，形成了"地下铁矿山高阶段嗣后胶结充填开采工艺体系"。研究成果应用以来取得了良好的技术指标，胶凝材料使用总比例降低 22%，阶段高度加高 50%，矿房矿柱尺寸加宽 33%，全员采矿劳动生产率提高 35%，充填成本降低 18.9 元/t，实现了细尾砂全利用及铁矿资源的绿色、安全、高效开采，推动了矿山行业高阶段嗣后充填开采理论及技术研究的进步。研究成果促使大冶铁矿走上安全生产、低成本运营及绿色开发之路，获批"国家示范绿色铁矿山"；并使得 4A 级国家矿山公园及世界文化遗产得到有效保护；促进了资源开发、环境保护、经济效益和社区建设的绿色可持续协调发展，为产业转型和升级创造了条件；丰富了矿山的生态文明、物质文明及精神文明内涵，使百年老矿焕发出新活力。

关键词：地下铁矿山；高阶段嗣后充填开采；充填体强度；超细全尾砂

Application and Theory Study on the High-stage Subsequent Cemented Filling in Underground Iron Mine

TAN Yuye[1], SONG Weidong[1], ZHANG Xingcai[2]

(1. College of Civil Engineering and Resource Engineering, University of Science and Technology Beijing, Beijing 100083, China; 2. Wuhan Iron and Steel Resources Group Daye Iron Mine Co., Ltd., Huangshi 435006, China)

Abstract: The stage subsequent filling mining method has the advantages of good safety and large production capacity, and has become the first choice for filling mining in underground metal mines. However, the problems of strength degradation of filling body, sedimentation efficiency of ultra-fine tailings, high pulping quality and cost caused by step-by-step layered filling have become the bottleneck problems restricting its application in underground metal mines. Aiming at the scientific problems such as' constitutive model and damage evolution mechanism of layered cemented backfill 'and' mechanical behavior and interaction mechanism between backfill and surrounding rock under complex conditions'. Taking the theoretical innovation as the breakthrough point, three aspects of research were carried out, including the microscopic mechanism of flocculation settlement of ultra-fine full tailings, the damage evolution mechanism of high-stage subsequent layered cemented filling body, and the fine macroscopic mechanical properties of surrounding rock-filling body under multi-field coupling conditions. Based on the theoretical innovation, the technology integration of "non-powered deep cone thickening and filling system, high-stage subsequent layered cemented filling body structure and strength design method, high-stage subsequent efficient mining technology and ground pressure control" was carried out, and the "high-stage subsequent cemented filling mining technology system for underground iron mines" was formed. Since the application of the research results, good technical indicators have been achieved. The total proportion of cementitious materials has been reduced by 22%, the stage height has been increased by 50%, the size of room and pillar has been widened by 33%, the labor productivity of full mining has been increased by 35%, and the filling cost has been reduced by 18.9 yuan/t. The full utilization of fine tailings and the green, safe and efficient mining of iron ore resources have been realized, which has promoted the progress of the theory and technology research of high-stage subsequent filling mining in the mining industry. The research results have prompted Daye Iron Mine to embark on the road of safe production, low-cost operation and green development, and have been approved as a' national demonstration green iron mine'; and make the 4A level national mine park and the world cultural heritage are effectively protected; it promotes the green and sustainable coordinated development of resource development, environmental protection, economic benefits and community construction, and creates conditions for industrial transformation and upgrading. It enriches the connotation of ecological civilization, material civilization and spiritual civilization of the mine, and makes the century-old mine glow with new vitality.

Key words: underground iron mine; high stage subsequent filling mining; filling body strength; ultra-fine full tailings

分会场特邀报告

大会特邀报告
第十三届冶金青年科技奖获奖人特邀报告
★ 分会场特邀报告
矿业工程
焦化及节能环保
炼铁与原料
炼钢与连铸
电冶金与废钢铁
轧制与热处理
表面与涂镀
金属材料深加工
粉末冶金
先进钢铁材料及应用
节能与低碳技术
冶金环保与资源利用
冶金设备与工程技术
冶金自动化与智能化
冶金物流
冶金流程工程学
其他

金属矿山智能化膏体充填开采
——以谦比希铜矿为例

吴爱祥

（北京科技大学，北京 100083）

摘 要：近 30 年来，中国膏体充填经历金川二矿-铜绿山-会泽发展三步曲，目前在黑色、有色、黄金、煤炭、非金属等系统迅速发展，并从国内走向国外，在赞比亚、塔吉克斯坦、印度尼西亚等国家推广应用。2013 年，谦比希铜矿西矿体建成了我国在海外建设并运营的第一套膏体充填系统。2020 年，谦比希铜矿东南矿体建成了智能化膏体充填系统。以智能化膏体充填采矿法为基础，辅以智能化采矿设备、高效提升与辅助系统、OPTIMINE 生产管理系统和全矿融合控制系统，谦比希铜矿实现了大盘区无轨智能化安全高效开采。

适应时代发展，建设六维矿山

连民杰

（中钢矿业开发有限公司，北京 100089）

摘 要：中国式现代化的提出为我国矿产资源开发利用带来了新的机遇和挑战，建设智慧矿山已经成为业内的趋势和共识，但智慧矿山建设仅是矿山生产经营管理的一部分，而不是全部。未来中国矿山企业应适应时代发展，紧跟全球科技革命和产业变革浪潮，吸收先进技术红利，顺应安全、低碳、环保新要求，不断革新开采模式，持续创新矿山经营理念，综合考虑各矿山的安全生产，资源综合利用，经济和社会效益，环境保护和周边社区的和谐，更要考虑社会对矿山的认同，对矿工的认同和尊重以及矿山和矿工的社会地位提高等各个方面。因此，中国式现代化矿山建设要考虑六个维度，即本质安全、环境和谐、资源节约、智慧运营、效益可观、社会认同。

智能矿山建设的探索与实践

武拴军

（金川集团股份有限公司龙首矿，甘肃金昌 737100）

摘 要：随着科学技术的进步以及国家对智能矿山建设的大力支持，更多的技术应用到矿山行业当中，但同时也伴随产生了很多限制因素，例如网络建设难度大、数据整合较为困难等问题。龙首矿在建设智能矿山的过程中，以"一个数字化生产运营管控中心、一个视频监控执法中心、六个专业化集控室，不断提升采掘自动化水平"为总体目标，克服诸多难点及困境，逐步完善基础网络建设及融合，同时结合自身生产需求探索实施了一批以 5G+无人驾驶电机车为代表的智能化建设项目，打破信息壁垒，有序推进"少人化""无人化"的智能矿山建设。

实施双碳规划引领，加快绿色转型发展
——武钢资源可持续发展路径探索

鲁炳强

（武钢资源集团有限公司，湖北武汉　430014）

摘　要：本报告介绍了国内外矿山企业碳减排情况、中国宝武双碳目标、武钢资源基本情况及绿色发展需求，分析了武钢资源下属子公司采矿、选矿、球团生产等工序环节的碳排放现状，从结构节能、生产工艺节能、管理降碳等方面提出了关键降碳路径和双碳规划，对远期碳中和技术进行分析和展望，对碳中和路线进行了总体安排，从而确保矿山企业全面实现双碳目标，最终实现可持续发展。

我国金属矿智能化开采现状与发展

王李管

（中南大学，湖南长沙　410083）

摘　要：矿山智能化是实现矿山企业生产过程的安全、高效、低成本的有效途径，也是矿业发展的必然趋势。目前我国智能矿山建设已经初具规模，但是由于受到矿床赋存条件、技术和装备水平的限制，矿山智能化建设难度较大，建设水平有待提升。从未来采矿趋势发展来看，生产技术数字化、生产装备与系统智能化、安全与生产管理信息化将是金属矿智能化开采持续推进的主要方向，金属矿的开采将通过装备的智能化、系统的智能化和自动化、整个企业的智能化这三个层级逐步完成智能化的建设。

难选铁矿石选矿技术发展及展望

陈雯

（长沙矿冶研究院有限责任公司，湖南长沙　410012）

摘　要：2022年我国钢材消费量9.2亿吨，预计2023年需求量9.1亿吨，钢材的巨大消费带动我国对铁矿石的巨大需求。但我国铁矿石对外依存度过高，进口量大，且进口矿运输通道单一、供应链脆弱，安全保障问题突出。同时我国铁矿资源丰富，但优质铁矿资源匮乏、铁矿物嵌布粒度细、矿物组成复杂导致铁矿石利用率低，开发国产铁矿资源对保障国家资源安全具有重要战略意义。

余永富院士"提铁降硅"学术思想引领了铁矿选矿技术的创新浪潮，主要进展聚焦在量的保障、质的提升和有价组分综合利用三方面。质的提升方面，以"铁、硅、铝"三元素综合评价铁精矿质量，通过流程、药剂及装备等的创新解决了铁精矿脱硅（铝）问题，增强了国产铁精矿的市场竞争力。高质量发展战略对铁精矿中硫、磷、钾、钠、锌及有色金属硫化物等杂质的深度脱除提出了新挑战。载体矿物多矿相同步活化浮选技术、选择性失活抑制剂研发以及界面精准调控技术解决了铁精矿深度除杂的世界性选矿难题。针对高硫高锌铁矿，工业上首次获得了TFe 70.8%、S 0.08%、Zn 0.018%、SiO_2 0.5%的高纯铁精矿，盘活首钢秘鲁铁矿1.3亿吨高硫高锌难选磁铁矿，技术水平世界领先。高效碎磨短流程、粗细粒梯级分选技术解决了铁矿物与含铁硅酸盐选择性解离、易泥化难题，药剂、工艺和装备协同创新，使赤铁矿降含铁硅酸盐技术达到世界领先水平。在综合利用方面，借鉴成矿地球化学原理采用冶金手段对白云鄂博铌矿综合利用进行了研究，找到了铌矿相定向转化的热力学窗口，提出了调控富铌矿相品位和粒度的方法，为铁基伴生战略性关键金属资源的综合利用提供了新思路。在量的保障方面，选择性磨矿技术解决了微细粒矿物选择性解离难题，矿泥分散-选择性絮凝解决了极细粒混合铁矿选矿技术瓶颈，提高了我国铁矿可利用资源量。焙烧是菱铁矿利用的唯一手段、10μm以下赤铁矿及褐铁矿利用的主要手段，对比黄梅闪速磁化焙烧与酒钢悬浮焙烧主要选矿指标与能耗指标，闪速磁化焙烧产业化指标好、能耗低、投资低，开辟了复杂难选铁矿资源开发利用新途径。

未来需秉承"能收早收，能丢早丢，能物理分选不浮选，能浮选不焙烧"的原则，将现代智慧手段与选择性理念深度融合，通过精准识别选择性抛尾、精准分析选择性解离、低碳环保高选择性药剂研发和精准调控选择性还原，实现复杂难选铁矿资源的绿色高效开发。

关键词：难选铁矿；发展历程；发展方向

参 考 文 献

[1] 陈雯, 谷广辉, 刘旭, 等. 秘铁高纯铁精矿选矿技术及伴生铜铅锌综合利用[J]. 中国冶金, 2022, 32(3): 128.
[2] 陈雯, 余永富, 冯志力, 等. 60万吨/年难选菱(褐)铁矿闪速磁化焙烧成套技术与装备[J]. 金属矿山, 2017(3): 54-58.
[3] 朱庆山, 李洪钟. 难选铁矿流态化磁化焙烧研究进展与发展前景[J]. 化工学报, 2014, 65(7): 2438-2441.

磁选设备的研发及应用现状

王俊莲[1,2]，常自勇[1,2]，王晓莉[1,2]，王化军[1,3]

（1. 北京科技大学土木与资源工程学院，北京 100083；2. 金属矿山高效开采与安全教育部重点实验室，北京 100083；3. 北京君致清科技有限公司，北京 102208）

摘　要：磁选技术广泛应用于铁矿、钛铁矿、锰矿等磁性矿物选别及非金属矿除杂纯化。随着矿石资源的不断消耗，品位低、嵌布粒度细的难选矿越来越多，研发高性能磁选机是磁选设备发展的重要方向。同时，随着选矿厂规模的不断扩大、对节能降耗要求的不断提高及大数据、云技术的快速发展，磁选设备不断向大型化、节能化和智能化发展。本文总结了近十年来我国干式磁选机、湿式磁选机、联合力场磁选机及超导磁选机的发展概况，并介绍了各种磁选机的应用性能。

关键词：干式磁选机；湿式磁选机；联合力场磁选机；超导磁选机

铬铁矿石选矿技术现状及趋势

孙炳泉

（中钢集团马鞍山矿山研究总院股份有限公司，安徽马鞍山 243000）

摘　要：世界铬矿资源很丰富，而我国却十分匮乏，未来全球铬矿的生产与消费之间将维持供求平衡。近10年的国内外相关研究文献表明，铬铁矿石选矿基础研究得以加强，选矿试验及生产实践中所采用的工艺仍以分级-重选为主，其次是重、磁选联合工艺；开始重视各种强磁选技术、块矿辐射分选技术及难选矿石选冶联合工艺的研究或应用；并对低品位铬铁矿石的利用及尾矿、含铬冶金渣的综合利用开展了大量的试验研究工作。中高品位铬铁矿石选矿工艺流程的高效简化，细粒或超细粒铬铁矿回收技术及设备的研发与应用，低品位铬铁矿石选择性解离技术及重、磁、浮选联合工艺流程的开发与应用，强化尾矿及冶金渣有价成分的回收及其大宗资源化利用，高质量铬铁精矿及特种高值铬铁矿砂加工技术开发等，将是铬矿选矿的主要发展趋势。

提高旋流器分级精度的应用研究进展

刘培坤

（山东科技大学，山东青岛 266590）

摘　要：旋流器对多组分矿物的精细分级、精确分选是学者和工程师们一直关注的热点。本文通过综述螺旋分级机、高频细筛、流化分级机及旋流器等分级设备在选矿领域的应用现状及其优缺点。针对传统旋流器易于出现的反富集问题，进行了旋流器与高频筛组合分级的试验，达到了提高综合分级效率的目的；并围绕旋流场内等沉速度带来的分级精度和效率低的瓶颈问题，通过理论分析、数值模拟与试验相结合，探讨了精细分级、精确分选关键技术可行性。试验表明：设计的带有冲洗水结构的组合曲线型锥体可以有效缓解反富集，基于流场调控研究提出的的新型蜗壳式进料体和翼翅形溢流管结构可以有效降低溢流跑粗；针对现场实际运行过程中存在的物料性质、工艺参数等变化大等实际问题，开发了在线调节溢流管和沉砂嘴的方法，提高了旋流器的操作弹性，为选矿厂分级过程智能化调控提供了可能。最后，围绕绿色、高效选矿生产需求，展望了精细分级装备的发展方向。

关键词：精细分级；旋流器；曲线锥体；智能调控

新型磁铁矿精选设备——永磁组合箱式磁选机

袁致涛，孟庆有，周明亮

（东北大学资源与土木工程学院，辽宁沈阳 110819）

摘　要：针对微细粒级磁铁矿剩磁高、矫顽力大、团聚现象严重等问题，研制出了一种交叉脉动磁场和强剪切流场协同的永磁组合箱式磁选设备。永磁组合箱式磁选机的磁系由阵列镶嵌在不导磁极板上的永磁体组成，分选区域位于两极板的挤压区域，磁块规则排布在极板上，纵向磁块磁极交替排列，横向磁极同极排列，在传动系统的带动下，磁系在分选箱两侧摆动，产生往复脉动磁场。磁性颗粒在磁场作用下易发生磁团聚，微细粒磁铁矿组成的磁团、磁链克服重力并在左右两侧强弱交变的磁场作用下呈折线轨迹下降。此过程中倾斜运动的磁团、磁链与竖直向上的上升水流呈一定夹角，使得磁链及磁团在较强磁场强度下也能被打散，反复的"打散-磁团聚-打散"可消除夹杂的脉石和弱磁性连生体，有助于提高铁精矿品质。采用永磁组合箱式磁选机对贾家堡铁矿进行了试验研究，针对二段磨矿粒度–0.074mm 占 90%，TFe 品位为 53.93%的给矿，可获得精矿 TFe 品位 63.08%、回收率 93.16%的分选指标；针对三段磨矿粒度–0.020mm 占 75%，TFe 品位 62.83%的给矿，可获得精矿 TFe 品位 67.81%、回收率 96.11%的良好指标。现场试验表明，永磁组合箱式磁选机分选微细粒磁铁矿效果显著，优于磁选柱及现场指标。

瓷球磨矿在铁矿山中的工业应用

吴彩斌

（江西理工大学，江西赣州　341000）

摘　要：钢球作为磨矿介质应用已有 100 多年历史，磨矿过程长期存在的能耗高、钢耗高、磨矿产品粒度分布特性差等系列问题，无法得到有效解决。瓷球作为新一代非铁介质，具有高硬度、高韧性、高耐磨等优点，能磨矿、能磨细、能节能，且磨矿产品粒度分布均匀，在二段磨和细磨领域完全可以作为研磨介质取代钢球。课题组在世界上首次提出"无钢球磨矿工艺"，形成瓷球磨矿原理，并在二段磨、再磨段践行瓷球磨矿，使磨矿过程电能下降 40%左右，磨矿介质成本节约 33%左右，过粉碎降低 10%以上，节能、降耗、增效显著，可为选矿厂创造显著的经济、社会和环境效益。

瓷球磨矿研究成果已经在龙桥矿业，宝武罗河矿业、南山矿业、张庄矿业，马坑矿业等铁矿山得到工业化应用。

铁矿石磨矿-分级数学模型研究与实践

崔宝玉，赵　强

（东北大学，辽宁沈阳　110819）

摘　要：磨矿-分级作业是选矿生产流程中最关键的环节，在选矿厂的基建投资和生产费用中占有很高的比例。磨矿-分级作业是整个选矿厂的"瓶颈"，决定选矿生产的处理能力，磨矿-分级产品粒度特性、单体解离度等对后续选别作业的指标乃至整个选矿厂的经济技术指标有重要影响。

铁矿石磨矿-分级作业受矿石性质、矿浆性质、现场工艺参数等多种因素的影响，整个过程复杂多变，当前生产中主要依靠经验相关性来实现该过程的设计、操作和控制。团队将实验室的相关研究与工程实践相结合，概括了一些常用模型的特点，并根据实际数据评估了这些模型的应用精度和广度。结果表明，这些模型在应用过程中均具有各自的优缺点，通常在某一特定工况下具有较高精度，而对其他工况适用性较差。对矿石性质考虑不充分是影响这些模型精度的主要原因。

团队结合现场磨矿-分级工艺，从铁矿石性质出发，深入分析了入磨物料和球磨机排料的粒度特性、力学特性、可磨度等矿石特征，并结合实验室实验、数值模拟等方法，提出了铁矿石磨矿-分级控制模型优化方案，通过将矿石性质参数写入到控制模型中，显著提高了新模型预测结果精度。本研究可为铁矿石磨矿-分级工业实践中数学模型的优化和应用提供借鉴。

战略金属硫化矿海水浮选基础及应用

李育彪

（武汉理工大学，湖北武汉 430070）

摘　要：随着淡水资源的逐渐匮乏和工业废水排放标准的逐渐提高，越来越多选矿厂不得不使用含大量离子的循环水、苦咸水甚至海水进行选矿作业，使用海水代替淡水进行矿物浮选分离具有重要应用前景。但浮选过程中，部分矿物溶解导致浮选矿浆中离子溶度增加，且海水中存在大量"难免离子"，在碱性浮选条件下，会产生亲水性胶体沉淀吸附在矿物表面或黏附在气泡表面，严重恶化浮选分离效果。

　　铜、钼属于战略性金属，是国民经济与科技发展必须的大宗紧缺性基础原材料，对其高效利用属于国家重大战略需求。铜钼硫化矿是铜和钼的主要来源，以辉钼矿和黄铜矿为主，主要通过浮选分离富集而获得。由于黄铜矿和辉钼矿都有较好的天然可浮性，其浮选分离属于世界性难题。因此，研究铜钼硫化矿高效浮选分离方法，对我国社会经济发展具有重要现实意义。传统的铜钼硫化矿浮选分离技术主要是铜钼混浮，再抑铜浮钼。但该工艺需要添加大量抑制剂，流程复杂、生产成本高，且会产生大量有害废水。此外，浮选前的细磨作业流程产生大量具有各向异性的黄铜矿和辉钼矿晶面，这些晶面的氧化速率及与浮选药剂的作用机理并不相同，导致不同的浮选效果。

　　综上所述，基于黄铜矿和辉钼矿的晶面各向异性，开展海水对战略金属铜钼硫化矿浮选分离的基础研究，有利于揭示海水体系中铜钼硫化矿的浮选分离机理、拓展铜钼硫化矿浮选分离新路径，对节约淡水资源、保护矿山环境、提高资源利用效率等都具有重要现实意义，所得到的研究成果将会在海水、苦咸水和选矿回水浮选分离硫化矿的基础理论及应用方面产生积极意义。

　　本研究提出了黄铜矿新鲜解理面原子重组及非均相氧化机理、建立了基于同步辐射技术的黄铜矿原位微纳米高精度检测方法、基于密度泛函理论计算从原子/分子水平揭示了黄铜矿典型晶面与 H_2O 和 O_2 的原位反应机理；揭示了矿浆中常见"难免离子"对黄铜矿浮选的影响机理，深化了分散剂对黄铜矿浮选的调控机制；量化了矿浆中"难免离子"抑制辉钼矿浮选的强弱关系，揭示了分散剂对亲水性沉淀颗粒的调控机制，创新性地提出了辉钼矿浮选抑制受亲水性沉淀物结晶度控制的新思路；揭示了黄铜矿与辉钼矿海水氧化预处理浮选分离机理，建立了基于氧化预处理的铜钼浮选分离方法。本研究成果的进一步实施促进了铜钼硫化矿典型晶面性质、氧化机理、海水浮选分离等领域的理论与实践，推动了硫化矿海水氧化预处理研究领域的发展，为清洁、高效的战略金属铜钼硫化矿海水分离提供理论依据。

重载高效浓密机在选矿厂的应用

曾建红

（长沙矿冶研究院有限责任公司，湖南长沙 410012）

摘　要：长沙矿冶研究院有限责任公司在传统高效浓缩理论的基础上，通过技术创新开发了单位面积处理能力 1000kg/(m²·h)的新型高效重载浓密机。采用该浓密机处理某钛铁矿选矿尾矿，在给矿干矿量 900～1200t/h、给矿浓度 20%～25%条件下，获得了底流浓度 45%～50%、溢流固含量小于 200mg/L 的优良技术指标。项目的成功实施解决了选矿厂场地受限、斜板浓密箱堵塞压板跑浑等系列问题，提高了生产作业率，改善了选别指标，降低了运行成本，取得了显著的经济效益。

低成本高炉用焦与精细化配煤新技术体系

程　欢，肖璐颖，廖　飞，胡文佳

（华北理工大学，河北唐山　063210）

摘　要：炼焦煤的性质及冶金焦炭质量的准确评价是直接影响配煤炼焦与高炉冶炼的关键因素[1]。现有的煤、焦质量评价方法对炼焦煤热解成焦机制和焦炭在高炉中溶损反应动力学模拟存在不足，导致炼焦煤资源利用不合理、配煤炼焦成本过高、高炉运行不稳定等问题较为突出[2, 3]。为突破煤、焦质量精准评价共性问题，在深入研究炼焦煤热解成焦机制、焦炭在高炉中溶损反应动力学等理论基础上，创新性的提出了炼焦煤热塑耦联性和焦炭综合热性能两种分别关于煤、焦质量评价的新方法，并基于这两种方法，开发了低成本高炉用焦与精细化配煤新技术体系。

提出的两种全新方法简介如下：（**1**）**炼焦煤热塑耦联性评价新方法**[4]。研发了一套集单向传热、热塑态透气性、黏稠性、膨胀性同步原位捕捉为一体的炼焦煤热塑耦联性综合测试系统，创新性的提出了热塑区挥发分析出受阻指数 R_f、最大膨胀压力 p_{max} 等多维表征技术指标，充分挖掘了部分传统意义上属于弱黏煤的炼焦潜能，提高了其在炼焦配煤中的配入比例，有效扩大了炼焦煤资源；（**2**）**焦炭综合热性能评价新方法**[5]。研发了集多温度制度测控温、大型支撑式连续热重分析于一体的焦炭溶损反应行为测试装置，创新性的提出了多温度制度定溶损 25%指标体系，先后评价了一批冶金焦的综合热性能，在高炉用焦结构优化、不同冶金焦配吃等方面开展了大量工作，显著降低了多座高炉的用焦成本。

提出的两种新方法有效提升了我国煤、焦质量评价技术的发展与完善，形成了完整的低成本高炉用焦与精细化配煤新技术体系，已在多家钢铁-焦化联合企业实施应用，充分210优化了区域煤、焦资源，大幅降低了配煤成本及冶金焦炭生产成本，高炉冶炼取得了良好效果，创造了显著的经济效益和社会效益。为钢铁及煤焦化工业实现碳达峰、碳中和提供了重要的理论与技术支撑。

参 考 文 献

[1] Díez M A, Alvarez R, Barriocanal C. International Journal of Coal Geology, 2002, 50(1): 389-412.
[2] Lee S, Yu J, Mahoney M, et al. Fuel Processing Technology, 2019, 188: 51-59.
[3] Zhang W, Shi T, Zhang Q, et al. Fuel, 2019, 251: 218-223.
[4] Wang Q, Cheng H, Zhao X, et al. Energy & Fuels, 2018, 32(7): 7438-7443.
[5] Cheng H, Liang Y, Guo R, et al. Fuel, 2021, 283: 118936.

高强度高反应性焦炭的制备技术

徐国忠[1]，钟祥云[1]，王明登[2]，封一飞[2]，白 滨[2]，李 超[1]，
刘 洋（女）[1]，刘 洋（男）[1]，张雅茹[1]，白金锋[1]

（1. 辽宁科技大学化工学院，辽宁省煤化工工程中心，辽宁省煤化工专业技术创新中心，
辽宁鞍山 114051；2. 中冶焦耐（大连）工程技术有限公司，辽宁大连 116085）

摘 要：在国家"双碳"战略发展背景下，钢铁工业生产过程的减碳要求迫在眉睫。基于减碳目标，现有传统高炉炼铁工艺碳排放量占全国碳排放总量15%左右，在诸多制造业门类中碳排放量最高。因此，基于氢气还原的富氢高炉炼铁是减碳的重要技术路线之一。大量模拟和实践研究表明，在富氢高炉中，由于氢气参与还原反应并吸收系统热量，导致炉内温度有所降低，进而影响铁矿石还原速率。为改善富氢高炉炼铁生产能力，需要提高炉内焦炭的反应性，保证焦炭应有的还原速率。同时，氢气还原铁矿石生成水蒸气，其与焦炭发生的溶损反应会破坏焦炭微观结构，导致焦炭宏观冷态强度和热态性能变差；为此，针对富氢高炉冶炼过程，开发高反应性高强焦炭以保证其在高炉中的骨架作用至关重要。

首先，向炼焦配合煤中添加廉价和来源广泛的 CaO 和 MgO，研究 CaO 和 MgO 对焦炭反应性和反应后强度的影响。结果表明，在添加 CaO 或 MgO 的质量分数小于 0.8%时，随着 CaO 或 MgO 添加数量的提高焦炭的反应性提高较为显著，而反应后强度降低幅度更加显著；当 CaO 或 MgO 添加量为 0.4%时，焦炭的反应性提高 5 个百分点，而焦炭的反应后强度却降低近 15 个百分点。显然，只依靠添加 CaO 或 MgO 无法获得高强度高反应性焦炭。为此，我们向配合煤中添加高温沥青，以期减缓反应后强度的降低。结果显示，当沥青添加为 1%时焦炭的反应性降低 2%左右，而反应后强度提高 5%左右。当沥青添加量继续增加，由于过量气孔的生成导致反应性提高，而反应后强度继续降低。显然，在 CaO 或 MgO 的基础上，再添加沥青仍然不能够获得高强度高反应焦炭。

基于上述研究结果及焦炭在高炉中的作用需求，构建了具有二元"核壳"结构的焦炭的模型，即壳焦具有高反应性和低反应强度，以适应富氢高炉热储备区温度降低过程铁还原的要求，壳焦呈现低反应性和高反应后强度，来满足焦炭在高炉内的"骨架"作用。经过 40kg 焦炉优化发现，与某钢铁公司焦化厂配合煤的 40kg 焦炉炼制焦炭的性能相比，在原常规配煤基础上，焦炭反应性 CRI 可以提高 3%~5%，而反应后强度 CSR 提高 2%~3%。该结果表明基于"核壳"二元结构焦炭模型进行配煤炼焦可以获得高强度高反应性焦炭。

关键词：氢冶金；焦炭；碳素溶解反应；反应后强度

智慧炼焦解决方案的研究与应用

王慧璐

（中冶焦耐自动化有限公司，辽宁大连 116085）

摘 要：智慧炼焦是以炼焦工艺原理为理论基础，应用工业大数据、人工智能、物联网、云计算、5G 等新一代信息技术，使各生产装置具有自感知、自决策和自执行等先进功能，并以一种高度柔性与集成的方式实施生产过程的各项活动，在提高劳动生产率的基础上，实现能源消耗、环境指标、生产安全、资源优化配置等更加可控和可持续向好，助力企业高质量发展。本文将重点介绍，中冶焦耐关于智慧炼焦解决方案的多项研究成果。

超高温连铸坯热送与直接轧制关键技术

朱苗勇，蔡兆镇，罗　森，祭　程，王卫领

（东北大学，辽宁沈阳　110819）

摘　要：超低排放、绿色低碳是钢铁行业发展的主题，发展以高拉速连铸为核心内涵的"连铸坯超高温出坯—均热—直接轧制"新工艺流程，是实现钢铁材料制造高效、绿色发展的重要途径。与传统冷装轧制工艺相比，高温铸坯直接轧制工艺可降低能耗 65%、缩短生产周期 90%。实现超高温连铸坯直接轧制流程的前提基础是高拉速连铸制备高温无缺陷坯，但目前我国常规板坯拉速普遍低于 1.8m/min，尚不能满足此新工艺流程的要求，而在实际生产中，提高连铸拉速，将面临大型夹杂、铸坯裂纹、结晶器漏钢等安全质量挑战。此外，传统生产流程中，辊道输送或下线的铸坯，往往是通过加热炉长时间加热来衔接铸-轧界面，以满足铸坯开轧温度与合金固溶的要求，铸坯组织历经了 γ→α→γ 相变；而对于高温铸坯辊道补热方式，补热时间短，铸坯全程为单相奥氏体，组织无明显改变，而且辊道补热后的高温铸坯呈"内热外冷"的逆向温度场特征，铸坯轧制过程形变与组织变化与传统方式有很大的差别。因此，要实现超高温连铸坯热送与直接轧制，需阐明高通量连铸结晶器内复杂界面均匀传热机理与裂纹控制技术原理、液面波动机理及稳定化控制机制、铸轧一体化界面衔接调控机制、"内热外冷"逆向温度场连铸坯轧制微观组织演变规律及其控制机理等关键科学技术问题，奠定解决超高温连铸坯直接轧制所面临的安全、质量、衔接等难题的理论基础。本文将结合团队的研究，阐述分析此新工艺所涉及关键理论与技术，为推动发展铸轧一体化提供指导，促进我国钢铁材料的绿色、高效、高质量制造。

关键词：超高温连铸坯；热送；直接轧制；关键技术

Key Technology for Hot Delivery and Direct Rolling of Ultra-high Temperature Continuous Casting Slab

ZHU Miaoyong, CAI Zhaozhen, LUO Sen, JI Cheng, WANG Weiling

(Northeastern University, Shenyang 110819, China)

Abstract: Ultra-low emission, green and low-carbon is the theme of the development of the steel industry, the development of high speed continuous casting as the key of the new process, namely ultra-high temperature continuous casting slab production, concurrent heating and direct rolling, is an important way to achieve efficient and green development of steel manufacturing. Compared with the traditional cold rolling process, the high temperature casting slab direct rolling process can reduce the energy consumption by 65% and shorten the production cycle by 90%. The premise of direct rolling process with ultra-high temperature continuous casting slab is to prepare high temperature defect free slab with high casting speed. However, at present, the casting speed of conventional slab in China is generally lower than 1.8m/min, which cannot meet the requirements of this new process. In actual production, improving the continuous casting speed will face safety and quality challenges such as large size inclusions, slab cracks and mold leakage. In addition, in the traditional production process, the casting slab delivered by the roller or off the line is often heated by the heating furnace for a long time to connect the cast-rolling interface to meet the requirements of the rolling temperature and the solid solution of the alloys, and the structure of the slab has undergone a γ→α→γ phase transition. As for the concurrent heating method for the high temperature casting slab with roller table, the time is short, the structure of slab is single-phase austenite and has no obvious change in the whole process, and the temperature is higher inside and lower surface after the concurrent heating in roller table, which makes the deformation and structure of the slab during rolling process quite different from the traditional method. Therefore, in order to achieve continuous casting slab hot delivery and direct rolling, it is necessary to clarify the key scientific and technical issues such as uniform heat transfer mechanism, crack control technology principle, liquid level fluctuation mechanism and stabilization control mechanism in high speed continuous casting mold, control mechanism of integrated casting and rolling interface connection, and the evolution law and control mechanism of microstructure of rolling slab with "internal hot and external cold" reverse temperature field. It lays a theoretical foundation for solving the problems of safety, quality and connection in direct rolling of ultra-high temperature continuous casting slab. In this paper, the key theories and technologies involved in this process will be analyzed in combination with the research of the team, to provide guidance for promoting the development of cast-rolling integration, and to promote the green, efficient and high-quality manufacturing of steel in China.

Key words: ultra-high temperature continuous casting slab; hot delivery; direct rolling; key technology

特大断面圆坯连铸技术发展与产品质量控制

张家泉[1]，董 娟[2]，刘 谦[2]

（1. 北京科技大学，北京 100083；2. 江阴兴澄特种钢铁有限公司，江苏江阴 214429）

摘 要： 介绍近年特大断面圆坯连铸技术快速发展与生产应用的行业背景、连铸装备与工艺技术特点以及大兆瓦级风电机组用钢的种类与质量控制要求。指出，加快 8MW 以上陆地风电机组和 13MW 以上海上风电机组建设是国家电力装备绿色低碳创新发展的战略要求，采用特大断面连铸圆坯取代传统钢锭锻造工艺提供风电装备用大单重特殊钢坯料也是当下低碳冶金发展背景下的大势所趋；然而，直径 800mm 以上特大断面圆坯连铸在铸机装备和铸坯质量控制上存在一系列技术挑战。基于浇铸工艺特点和生产实践，讨论了连铸圆坯与模铸钢锭凝固过程的差异、内部质量特征与控制难点以及特大圆坯连铸对工艺装备的要求。提出，应针对风电机组中相关特殊钢产品制备工艺特点与质量控制要求，合理制定连铸工艺、通过调控凝固过程实现特定用途风电用圆坯铸态质量的个性化控制。最后，介绍了兴澄特钢大直径连铸圆坯生产实践与产业化推广进展。

关键词： 特大断面圆坯连铸；特殊钢；质量与控制

Super Large Section Round Bloom Casting and the As-cast Product Quality Control

ZHANG Jiaquan[1], DONG Juan[2], LIU Qian[2]

(1. University of Science and Technology Beijing, Beijing 100083, China;
2. Jiangyin Xingcheng Special Steel Co., Ltd., Jiangyin 214429, China)

Abstract: The current situation of development and application of super large section round bloom casting technology has been illustrated, together with the introduction to its industry background, the characteristics of the continuous casting machine and the demand of users for megawatt wind turbine construction. It is pointed out that the present construction on a large scale of wind turbines system over 8MW inland and/or over 13MW offshore is the national strategic demand for power industry innovative development through green electricity, and the supply of the heavy special steel produced by the unprecedented super large section round bloom casting, instead of traditional large-sized ingot casting, is also the urgent need of green metallurgy development presently. However, there exists lots of challenges for the commission of both the super large casting machine and its bloom casting quality control. Based on the understanding to the casting processes along with production facts, the solidification difference between ingot casting and continuous casting has been discussed, and the tough issues for the bloom inner quality control are revealed accordingly which can only be overcome by an upgrading or innovative continuous casting practices. It is pointed out additionally that the bloom solidification control should made to meet an individualized as-cast quality standard with consideration of the given final hot work process of the turbine steel products. Finally, the production and application of the super large section round bloom casting in Xingcheng Special Steel has been presented.

Key words: super large section round bloom casting; special steels; quality & control

低碳背景下 PMO 技术的创新发展与应用

翟启杰 等

（上海大学先进凝固技术中心，上海 200444）

摘 要： 脉冲磁致振荡（简称 PMO）技术是由上海大学先进凝固技术中心（简称 CAST）原创的凝固组织细化和均质化技术，目前已在多家冶金企业应用。工业应用表明，该技术不仅可以显著细化铸坯凝固组织，降低成分偏析，改善铸坯低倍缺陷，而且可以提高铸坯拉速，放宽工艺参数窗口。报告简要回顾了 PMO 技术的原理，介绍了 PMO 技术与电磁搅拌组合调控在特殊钢矩形坯和圆坯中的应用，分析了其节能效果。在此基础上，报告介绍了 CAST 团队将该技术应用于模铸和高速钢连铸的生产实践。同时探讨了该技术与末端压下均质化技术组合的可能性。

关键词： 连铸；模铸；均质化

Innovative Development and Application of PMO Technology in the Theme of Low Carbon

ZHAI Qijie et al

(Center for Advanced Solidification Technology, Shanghai University, Shanghai 200444, China)

Abstract: Pulse Magneto-Oscillation (PMO) technology is a technology to achieve solidification structure refinement and homogenization, which is originally developed by the team of Center for Advanced Solidification Technology (CAST) affiliated with Shanghai University. Recently, PMO has been applied in multiple metallurgical enterprises in industry scale. Results from industrial applications have shown that this technology can not only significantly refine the solidification structure of the billet, reduce component segregation and improve solidification defects of the billet, but also allow the higher casting speed and broaden the process parameter window. This report will briefly review the state of the art of PMO technology principle, introduce the application of PMO combined with electromagnetic stirring in rectangular and circular special steel billets, and analyze its energy-saving effect. Furthermore, the production practice of CAST team applying this technology to mold casting and high-speed steel continuous casting will also be reported. The possibility of combining this technology with conduction technology is also discussed.

Key words: continuous casting; mold casting; homogenization

钢液脱硫机理的实验、理论和模拟仿真研究

张立峰

（北方工业大学机械与材料工程学院，北京 100144）

摘 要：本文通过实验室试验研究了脱硫剂成分对钢液脱硫效率的影响，定量研究脱硫剂粒子在钢液脱硫过程中成分的变化规律，结果如图1所示。脱硫剂粒子与钢液反应4min后，脱硫剂粒子表面形成了一层CaS外层，说明脱硫剂粒子与钢液迅速发生了脱硫反应。通过添加不同成分和不同尺寸的脱硫剂粒子，确定最有利于钢液脱硫的脱硫剂粒子成分，实现脱硫剂粒子的有效成分设计。

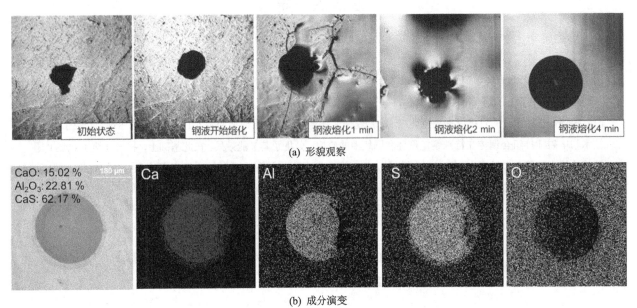

(a) 形貌观察

(b) 成分演变

图1 脱硫剂粒子脱硫过程原位观察

本文以KR铁水预处理和RH精炼为研究对象，分别建立了KR铁水预处理和RH精炼过程钢液脱硫的三维数值模型，采用标准k-ε双方程湍流模型、钢液-空气VOF多相流模型和DPM气泡运动模型，进一步耦合脱硫剂粒子的运动和脱硫动力学，明确了脱硫剂粒子和硫含量在钢液中的三维空间分布规律及其随时间的演变，研究了脱硫剂粒子特征和实际冶炼参数对钢液脱硫效率的影响规律。

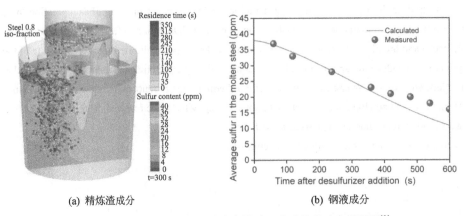

(a) 精炼渣成分   (b) 钢液成分

图2 精炼渣成分对不锈钢洁净度影响工业试验和动力学预测[1]

参 考 文 献

[1] Peng Kaiyu, Sun Yu, Peng Xiaoxuan, et al. Numerical simulation on the desulfurization of the molten steel during RH vacuum refining process by CaO powder injection[J]. Metallurgical and Materials Transactions B, 2023, 54(1): 438-449.

连铸保护渣关键问题及研究进展

文光华

（重庆大学，重庆 400044）

摘 要：在"减碳"关键技术方面，除了关注氢冶金外，如何发挥以高拉速和无缺陷铸坯为标志的连铸高效化作用也应不容忽视。目前，振动结晶器连铸的实际拉坯速度仅为上世纪五十年代创造的纪录（14.5m/min）的 10%～40%。在正常浇铸条件下，结晶器摩擦力和钢坯壳强度不是高拉速的限制因素，而是结晶器内钢液初始凝固不规则行为诱发了偶然性的漏钢（粘结报警）和铸坯缺陷。这种凝固不规则行为除了受结晶器内过大的液面波动影响外，通常归结于结晶器与钢坯壳之间保护渣不规则流入。后者主要受控于保护渣在结晶器内熔化速率及保护渣喂入工艺。为此，课题组基于润湿性和铜浴法，开发了结晶器内熔渣池深度和空心颗粒保护渣熔化速率测试装置，解决了长期困扰人们如何测准熔渣池深度和原位模拟保护渣熔化速率问题。在此基础上，发现保护渣熔化速率与其料层厚度关系呈"M"曲线特征，并解释了间歇式保护渣喂入工艺对熔渣池深度的影响规律。

关键词：连铸；非规则凝固；保护渣；熔化速率；"M"曲线

Key Problems and Research Progress in Mold Powders for Continuous Casting

WEN Guanghua

(Chongqing University, Chongqing 400044, China)

Abstract: In terms of the key technology of "carbon reduction", in addition to paying attention to hydrogen metallurgy, how to play the role of continuous casting efficiency marked by high casting speed and defect-free strands should not be ignored. At present, the actual casting speed of vibrating mold continuous casting is only 10%~40% of the record (14.5m/min) set in the fifties of the last century. Under normal casting conditions, the mold friction and shell strength are not the limiting factors of high casting speed, but the irregular initial solidification of the molten steel in the mold induces accidental breakout (sticker alarm) and strand defects. In addition to being affected by excessive liquid level fluctuations in the mold, this irregular solidification behavior is usually due to the irregular inflow of molten slag between the mold and the shell. The latter is mainly controlled by the feeding process and the melting rate of the mold powders in the mold. To this end, based on the wettability and copper bath method, the research group developed two test devices for the depth of the molten slag pool in the mold and the melting rate of hollow particle powders, which solved the problem of how to measure the depth of molten slag pool and simulate the melting rate of the mold powders in situ for a long time. On this basis, it is found that the relationship between the melting rate of mold powders and the thickness of powder layer exhibits the characteristics of "M" curve, and the influence of intermittent powder addition on the depth of the molten slag pool is explained.

Key words: continuous casting; irregular solidification; mold powder; melting rate; "M" curve

鞍钢创新质量管理理念与实践

孙 群

(鞍钢股份有限公司炼钢总厂，辽宁鞍山 114021)

摘 要：提高产品质量是提高企业核心竞争力，提高企业品牌影响力以及降低企业运营成本、提高企业利润的关键，是制造业所追寻的永恒目标。针对连铸坯生产工艺流程中质量管控存在的问题，提出了"设计引领，流程先行，界面优化，细节管控"的总体质量管理理念，系统介绍了该理念在鞍钢股份炼钢总厂4号线与5号线质量管理过程中的应用情况与实践经验。实践结果表明：鞍钢创新质量管理理念在产线的应用使产线生产效率显著提高，产线品种钢质量指标显著提升，IF钢缺陷率指数和无取向硅钢缺陷率指数均大幅降低。

关键词：质量管理理念；冶金流程工程；界面优化设计

Innovative Quality Management Concept and Practice of Angang

SUN Qun

(General Steelmaking Plant of Ansteel, Anshan 114021, China)

Abstract: Improving product quality is the key to improving the core competitiveness, expanding the brand influence, reducing the operating costs, and expanding the product profits of an enterprise. It is also an eternal goal pursued by the manufacturing industry. In response to the problems of quality control in the production process of continuous casting billets, an overall quality management philosophy of "design leading, process first, interface optimization, and detail control" is proposed. The application and practical experience of this concept in the quality management process of the 4 # and 5 # production lines of Steelmaking Plant are systematically introduced. The practical results show that the application of Angang innovative quality management philosophy in the production line has significantly improved the production efficiency, and the quality indicators of key steel grades in the production line have been remarkable improved. The defect rate index of IF steel has been reduced from 0.8 to 0.3, and the defect rate index of non-oriented silicon steel has been reduced from 0.4 to 0.2.

Key words: quality management philosophy; metallurgical process engineering; interface optimization design

游离晶对铸锭组织形成的影响

韩青有

(东南大学机械系，江苏南京 211189)

摘 要：浇注过程中金属液的过热迅速消失。铸模中金属的温度往往低于近液相线温度。在这样的条件下，凝固过程中的金属液流动和温度场的波动，促使枝晶发生破碎、游离和增殖，产生大量的游离晶，从而显著影响铸锭凝固组织的形成。本文讨论游离晶的形成、其生存条件以及对其相邻晶粒或晶核的影响。通过解剖大型铝合金半连铸铸锭，揭示游离晶在铸锭中的形成、分布及相关机理。作者认为，游离晶可凝固壳顶端形成，可在液相线温度以上短期存活，通过液体流动散布在下游的铸锭中。而在游离晶密度较高的熔池中，枝晶相难以形核，即使形成晶核也会由于粗化过程而难以存活。为此，铸锭凝固时的晶粒尺寸，与游离晶的生长历程相关。控制游离晶的产生应该是控制铸锭组织的一个重要手段。

关键词：半连铸；晶粒生长；对流；凝固理论

Floating Grains and Their Influence on the Solidification Structure of a Billet

HAN Qingyou

(Southeast University, Nanjing 211189, China)

Abstract: Superheat of liquid metal dissipates rapidly during pouring. Temperatures in the liquid metal in the mold are usually below a temperature close to the liquidus of the alloy. Under such conditions, convections and temperature variations during metal solidification promotes fragmentation, floating, and multiplication of dendrites, producing a large amount of floating grains or fragments. These floating grains are bound to affect the formation of solidification structure of the billet. This article discusses the origin of the floating grains, their survival and their influence on the nucleation and growth of neighboring grains. A large-sized DC ingot of an aluminum alloy were sectioned and examined to reveal the floating grains, their distribution in the ingot and associated mechanisms. It is believed that the floating grains are formed at the top of the solid shell, can survive briefly at temperatures above the liquidus, and are transported by the convention to the downstream regions in the ingot. In regions where the density of floating grains is high, nucleation of new grains becomes difficult. Even though new grains can be nucleated, they cannot survive long due to Oswald ripening phenomenon. As a result, grain size in the billet is dependent to the origin and the survival history of the floating grains. Controlling the formation of floating grains ought to be an important way of controlling the final solidification structure of the billet.

Key words: DC casting; grain formation; convection; solidification theory

等轴晶薄带连铸与短流程电工钢

周 成，周 游，玄东坡，姜天亮，范文浩，毛 焱，松希郎卡智美

（北京科技大学材料科学与工程学院，北京 100083）

摘 要：冶炼之后，热轧薄板的生产包括连铸和热轧两大工序，前者制坯，后者通过一定的压缩比来改善铸态组织，以适应不同产品性能的要求，并保证尺寸精度。要求强韧性的产品，希望热轧板为细小的等轴晶组织；用于冷轧基板时，对于要求深冲性能的产品，希望细小的等轴晶组织，对于电工钢，希望较粗大的等轴晶组织。事实上，改善组织的工作从连铸工序就开始了，例如采用电磁搅拌来扩大等轴晶区，从而抑制中心偏析，或提高成品板的成形性能。

薄板生产短流程化，是通过板坯的近终形化和缩减热轧变形量来实现的，因此，薄板短流程生产技术研究的关键课题是：（1）近终形连铸技术；（2）近终形连铸与后续热轧工序的合理匹配。近终形连铸的排热条件更有利于柱状晶的生长，理想的简化热轧工序应消除凝固组织的影响，接近或达到传统热轧板的水平。

最初的薄板坯连铸连轧侧重于投资成本最小化，板坯厚度较薄，大幅减少热轧变形量，实现了简单商品级产品的经济性生产。为了提高产能、产品质量和产品范围，目前 ESP 生产线的板坯厚度增加了约一倍，实现了薄板坯连铸→粗轧→感应补热→精轧的连续生产。

双辊薄带连铸带坯的尺寸近终形程度达到了热轧板水平，也正因为如此，薄带连铸带坯的在线热轧变形量窗口小，不利于消除柱状晶组织对成品薄板组织与性能的影响。单带法的发明者认为，为了消除铸态组织的影响，需要60%~80%的热轧压下率，生产2~4mm厚的热轧板需要10~15mm厚的薄带坯，采用单带法可以实现这一目标。目前单带法薄带坯中试线已生产出高锰钢带坯，经再加热后轧制成卷。

可见，双辊薄带连铸的发展方向应该是：（1）带坯组织的等轴晶化，即组织近终形；（2）增加带坯厚度，提高产量，扩大产品范围。

我们在理论和实验两方面的研究结果表明，采用动量布流的侧注式双辊薄带连铸，可以制备具有等轴晶组织的毫米级连铸薄带坯，形成等轴晶组织的原因在于动量布流造成的熔池涡旋，等轴晶尺寸取决于设备和工艺参数。

以电工钢为研究对象，初步探索了采用等轴晶薄带连铸的短流程生产技术。冷轧电工钢成品规格较薄，连铸薄带坯可以满足其轧制变形量的要求。薄带坯酸洗后直接冷轧，可制备3.0% Si（质量分数，下同）取向与无取向硅钢；薄带坯酸洗后直接温轧，可制备6.5% Si 高硅钢。采用短流程工艺制备的0.35mm厚3.0% Si无取向硅钢，磁性能达到35W400要求；0.23mm厚3.0% Si取向硅钢 B_8=1.92T，$P_{17/50}$=1.31W/kg；0.1mm厚6.5% Si高硅钢，磁性能与JFE相应产品典型值相当，0.08mm厚6.5% Si高硅钢磁性能更优。

关键词：短流程；薄带连铸；等轴晶；电工钢

Strip Casting with Equiaxed Grains and Electrical Steels by the Short Processes

ZHOU Cheng, ZHOU You, XUAN Dongpo, JIANG Tianliang,
FAN Wenhao, MAO Yan, SONGXI Langkazhimei

(School of Materials Science and Engineering, University of
Science and Technology Beijing, Beijing 100083, China)

Abstract: After steelmaking, the production of hot rolled thin plates includes two major processes: continuous casting and hot rolling. The former one produces slabs, while the latter one improves the as-cast microstructure through a certain compression ratio to meet the requirements of different product performance and ensure dimensional accuracy. Products that require high strength and high toughness are expected to have a fine equiaxed grain microstructure in the hot rolled plate. When hot rolled plates are used for cold rolled substrates, for products that require deep drawing performance, it is desired to have a fine equiaxed grain microstructure. As for producing electrical steels, coarse equiaxed grains are desired in the hot rolled plate. In fact, the work of improving the microstructure has started from the continuous casting process, such as using electromagnetic stirring to expand the equiaxed grain zone, thereby suppressing center segregation, or improving the formability of the finished plates.

The short process of thin plate production is achieved through the near net shape of the slab and the reduction of hot rolling deformation. Therefore, the key issues in the research of short process production technology for thin plates are: (1) near net shape casting technology, (2) reasonable matching between near net shape continuous casting and subsequent hot rolling processes. The cooling condition of near net shape casting is conducive to the growth of columnar grains. The ideal simplified process should eliminate the influence of solidification microstructure and approach or reach the level of traditional hot rolled plates.

Initially, thin slab casting and rolling process focused on the minimization of investment costs. The slab thickness was relatively thin, significantly reducing hot rolling deformation, and achieving economic production of commodity grade products. In order to improve production capacity, product quality and product range, the slab thickness of the ESP line has been approximately doubled, and the continuous production of thin slab casting → rough rolling → induction heating → finishing rolling has been realized.

The size of the as-cast strip of twin-roll strip casting has reached the level of the hot rolled plate. Therefore, the online hot rolling deformation window of the as-cast strip is small, which is not conducive to eliminating the influence of columnar grains on the microstructure and properties of the finished plate. The inventors of the belt casting technology believe that a hot rolling reduction rate of 60%-80% is required from thin strips to hot rolled plates, and producing 2-4mm thick hot rolled plates requires 10-15mm thick thin strips. The belt casting technology can achieve this goal. At present, the belt casting pilot line has produced high manganese steel strips, and then rolled into coils after reheated.

It can be seen that the development direction of twin-roll strip casting should be: (1) equiaxed grains of the strip, that is, near net microstructure, and (2) increase strip thickness, increase production, and expand product range.

We adopted the top-side pouring twin-roll casting method and verified theoretically and experimentally that this method can prepare millimeter level thin strip billets with equiaxed grains. The reason for the formation of equiaxed grains is the vortex of the molten pool caused by momentum distribution. The size of the equiaxed grains depends on the equipment and process parameters.

We took electrical steels as the research object, a short process technology using twin-roll casting with equiaxed grains was explored. The cold rolled electrical steel products are relatively thin, and strips of twin-roll casting can meet the requirements for rolling deformation. After acid pickling, the thin strips can be directly cold rolled to prepare 3.0wt Si% grain-oriented silicon steel and non-oriented silicon steel. 6.5wt% Si high silicon steel can be prepared by direct warm rolling after acid pickling of the thin strips. 0.35mm thick 3.0wt% Si non-oriented silicon steel was prepared using the short process technology with magnetic properties meeting the requirements of 35W400. The magnetic properties of 0.23mm thick 3.0wt% Si grain-oriented silicon steel are B_8=1.92T, $P_{17/50}$=1.31W/kg. The magnetic properties of 0.1mm thick 6.5wt% Si high silicon steel are comparable to typical JFE products, while 0.08mm thick 6.5wt% Si high silicon steel has better magnetic properties.

Key words: short process; twin-roll strip casting; equiaxed grain; electrical steels

我国电炉短流程炼钢发展研究

王 滨，余 璐

（中国钢铁工业协会规划发展部，北京 100711）

摘 要： 为掌握我国电炉短流程炼钢发展现状，中国钢铁工业协会组织行业研究力量开展了专题研究，从规模、装备、技术、环保、节能、碳排放、资源、竞争力等方面进行了评估，展望了发展趋势、方向和路径。本报告发现，在碳减排逐渐成为世界钢铁发展主要驱动因素的背景下，"绿色钢铁革命"正在将世界钢铁技术创新和产业变革的演进趋势推向新的高度。世界主要产钢国家都十分重视发展电炉短流程。近几年，我国电炉炼钢得到较快发展，实现了快速高效低成本生产，普钢平均冶炼周期由 50min 降至 35min，吨钢平均电极消耗由 2kg 降至 1kg 以下，金属收得率由 89%提高到 92%~95%，部分技术指标已进入国际先进行列。截至 2022 年底，我国已建成电炉冶炼产能约 1.93 亿吨，未来 3~5 年还将建设 5000 万吨以上。

当前，行业、企业和专家等对长流程炼钢和短流程炼钢使用废钢是否同样降碳，认识不同、角度不同、立场不同，所持观点还有一定的分歧。综合各方意见，本报告认为长、短流程使用废钢都有利于降低吨钢碳排放强度。长流程的废钢比对不同钢种而言，控制在 10%~20%较为合理，若继续提高废钢比，虽然会降低碳排放强度，但由于长流程碳排放主要在炼铁系统，长流程碳排放强度和碳排放总量总体大于电炉短流程。因此，鼓励引导废钢资源尽可能流向全废钢电炉短流程，逐步调整全行业的铁素资源结构、产品结构和流程结构，有利于中国钢铁工业在国际低碳竞争中赢得主动，有利于实现我国钢铁工业碳中和的总体战略。

本报告对我国未来粗钢产量进行预测，中线情形下，到 2025 年、2030 年和 2035 年我国粗钢产量分别为 9.5 亿吨、9 亿吨和 8.5 亿吨。据此预测，2035 年我国社会钢铁蓄积量将达到 178 亿吨，废钢资源量接近 4 亿吨，电炉钢产量比例达到 30%。2040 年，我国废钢资源量达到阶段性峰值后将有所回落。2050 年，我国电炉钢产量比例达到40%并将继续保持。届时，我国钢铁工业将实现流程结构的切换，短流程电炉冶炼产品基本实现全覆盖。

总的来看，发展短流程炼钢是实现低碳绿色发展的必然趋势，钢铁行业在实现"双碳"目标过程中，高炉－转炉和电炉在上下半场的角色和降碳贡献不同，上半场碳排放总量降低重点要依靠长流程，下半场碳中和这场大剧还得靠短流程来收场。无论是上半场还是下半场，废钢资源始终承担着钢铁行业碳减排的重要任务。鼓励引导废钢资源流向全废钢电炉短流程，逐步调整全行业的铁素资源结构、产品结构和流程结构，有利于实现我国钢铁工业"双碳"总体战略。

关键词：电炉炼钢；转炉炼钢；碳达峰；碳中和

Research on the Development of EAF Steelmaking Process in China

WANG Bin, YU Lu

(Planning and Development Department China Iron and Steel Association, Beijing 100711, China)

Abstract: In order to grasp the development status of EAF steelmaking process in China, China Iron and Steel Association(CISA) organized research team to carry out special research, to evaluate the scale, equipment, technology, environmental protection, energy conservation, carbon emissions, resources, competitiveness, and other aspects, and looked forward to the development trend, direction and path. This report finds that under the background of carbon emission reduction gradually becoming the main driving factor of the world's steel industry development, "The Green Steel Revolution" is pushing the evolution trend of the world's steel technology innovation and industrial transformation to a new height. The world's major steel producing countries attach great importance to the development of EAF steelmaking process. In recent years, China's EAF steelmaking process has achieved rapid development, realizing rapid, efficient and low cost production. The average steelmaking period of common carbon steel has been reduced from 50 minutes to 35 minutes, the average electrode consumption per ton of steel has been reduced from 2kg to less than 1kg, the metal recovery rate has increased from 89% to 92%~95%, and some technical indicators have entered the international advanced ranks. By the end of 2022, China has built the EAF Steelmaking capacity of about 193 million tons. In addition，more than 50 million tons will be built in the next three to five years.

At present, because of the different understandings, perspectives, perspectives and positions, there are differences among industries, enterprises, and experts regarding whether scrap steel is more conducive to carbon reduction through BF-BOF steelmaking process or EAF steelmaking process. According to all opinions, this report believes that both the BF-BOF and EAF steelmaking processes are conducive to reducing the carbon emission intensity per ton of steel. For different type of steel，the ratio of scrap steel in the BF-BOF steelmaking process is reasonable to 10%~20%. If the ratio continues to be increased, although the carbon emission intensity will be reduced, but because the carbon emission of the BF-BOF steelmaking process is mainly in the ironmaking system, the carbon emission intensity and total amoun are generally greater than that of the EAF steelmaking processes. Therefore, encouraging and guiding scrap steel resources to used into the EAF steelmaking processes as far as possible, and gradually adjusting the iron resource structure, product structure and process structure of the whole steel industry, is conducive to China's steel industry to win the initiative in the international low-carbon competition, and is conducive to realizing the overall strategy of China's carbon neutrality.

This report forecasts China's future crude steel output. In the case of the middle situation, China's crude steel output in 2025, 2030 and 2035 will be 950 million tons, 900 million tons and 850 million tons respectively. Accordingly, it is predicted that in 2035, the China's social steel storage volume will reach 17.8 billion tons, the scrap steel resource quantity are close to 400 million tons, and the proportion of steel output by EAF steelmaking processes will reach 30%. In 2040, China's scrap steel resource quantity will fall back after reaching the phased peak. In 2050, the proportion of steel output by EAF steelmaking processes in China will be reached 40% and will continue to maintain. At that time, China's steel industry will realize the switch of process structure, the products from EAF steelmaking processes basically achieve full coverage.

Overall, the development of EAF steelmaking process is the inevitable trend of low carbon green development, iron and steel industry in the process of achieving "double carbon" goal. During the process of realizing the goal of carbon peaking and carbon neutrality, BF-BOF and EAF steelmaking process have different carbon reduction contributions. The first half stage focus depends on BF-BOF steelmaking process to reduce total carbon emissions, while the goal of carbon neutrality in the second half stage still needs to be achieved through the EAF steelmaking process. Whether in the first half stage or the second ones, scrap steel resources always undertake the important task of carbon emission reduction in the steel industry. Encourage and guide the scrap steel resources to flow to the EAF steelmaking process, and gradually adjust the iron resource structure, product structure and process structure of the whole steel industry, which is conducive to the realization of the overall strategy of carbon peaking and carbon neutrality of China's steel industry.

Key words: EAF steelmaking process; bf-bof steelmaking process; carbon peaking; carbon neut

我国电炉短流程炼钢资源现状与发展

杨 勇,姚同路

(钢铁研究总院冶金工艺研究所,北京 100081)

摘 要:为探究我国电炉短流程炼钢及资源,本课题组对我国的钢铁工业生产、电炉短流程炼钢、废钢铁、直接还原铁等进行了系统研究,对其现状与发展做了深入分析。世界粗钢产量自 2001 年以来稳步增长,由 8.43 亿吨增长至 2021 年的 19.51 亿吨,增长率 131.4%,与之相对应的是中国粗钢产量,2001 年 1.52 亿吨,2021 年 10.33 亿吨,占世界粗钢产量的 53%。据预测,2030 年以后钢铁蓄积量将会达到 140 亿吨。世界电炉钢比例自 2013 年以来变化不大,2021 年平均比例约 29%,若不含中国,世界其他国家电炉钢比例约为 50%。近年来由于国家有序发展电炉短流程炼钢,特别是取缔"地条钢"以后,电炉钢比例开始回升,2021 年达 10.6%。随着"双碳目标"的不断临近及我国政策的压紧,根据中国废钢铁应用协会和工信部《关于推动钢铁工业高质量发展的指导意见(征求意见稿)》公布的数据显示,到 2025 年,电炉钢产量占粗钢总产量的比例提升至 15%以上,废钢比达到 30%。

未来 10~30 年,我国钢铁工业将逐步进入高废钢比、高炉-转炉长流程、电炉短流程、低碳冶金、氢冶金等多种工艺技术并存的绿色钢铁之路。众多学者对废钢的战略价值做了深入分析,认为废钢作为钢铁循环利用的优良再生资源,是唯一可替代铁矿石用于炼钢的重要原料,也是电炉短流程炼钢的最主要资源。2020 年 9 月,中国政府向世界宣布了 2030 年前实现碳达峰、2060 年前实现碳中和的"双碳"目标。要实现"双碳"目标,我国钢铁行业必须在流程结构上提高废钢-电炉短流程比重,但目前面临的问题是,我国废钢铁资源仍显不足,电炉短流程炼钢成本处于劣势。近年来中国废钢消耗量逐年上升。2017 年废钢消耗量 1.48 亿吨,2021 年已达到 2.26 亿吨,废钢比 21.9%,但仍远低于全球 37%的水平。在 2.26 亿吨的废钢消耗量中,转炉消耗约占 2/3,电炉消耗约占 1/3,说明存在两种流程争夺废钢的情形,当前废钢量明显无法满足大废钢比炼钢需求。据废钢协会预测,2030 年中国废钢资源量为 3.6 亿吨,2035 年为 3.8 亿吨,2040 年达到阶段性峰值水平 4.1 亿吨,2060 年回落至 4 亿吨。

对世界直接还原铁发展状况分析可知,印度、伊朗、俄罗斯等是主要的生产国,目前气基法生产直接还原铁约占 75%。由于我国缺乏天然气资源,在过去几十年里对直接还原法的探索集中在煤基法,生产规模小,生产效率低,直接还原铁产量少。若能研究用焦炉煤气、氢气等生产直接还原铁,开发废钢+直接还原铁+电炉工艺,将会是我国电炉短流程炼钢流程的重大突破。对比世界钢铁工业发展规律可知,未来若干年,随着废钢、直接还原铁产出量越来越多,电炉短流程炼钢资源供不应求的局面会逐步打破,废钢价格逐步走低,废钢和直接还原铁比例将越来越高,电炉短流程炼钢将会得到大力发展。

关键词:废钢资源;电炉短流程;直接还原铁;废钢比

Current Situation and Development of EAF Short Process Steelmaking Resources in China

YANG Yong, YAO Tonglu

(Metallurgical Technology Institute, Central Iron and Steel Research Institute, Beijing 100081, China)

Abstract: In order to explore China's EAF short process steelmaking and resources, research group conducted a systematic study on China's steel industry production, EAF short process steelmaking, scrap, direct reduced iron (DRI), etc., and conducted in-depth analysis of its current situation and development. The world's crude steel production has steadily increased since 2001, from 843 million tons to 1.951 billion tons in 2021, with a growth rate of 131.4%. Correspondingly, China's crude steel production reached 152 million tons in 2001 and 1.033 billion tons in 2021, accounting for 53% of the world's crude steel production. It is predicted that the steel storage volume will reach 14 billion tons after 2030. The proportion of EAF steel in the world has not changed much since 2013, with an average proportion of about 29% in 2021. If excluding China, the proportion of EAF steel in the world is about 50%. In recent years, due to the orderly development of EAF short process steelmaking by the country, especially after the ban on "substandard steel", the proportion of EAF steel has started to rise, reaching 10.6% in 2021. With the approaching of the "dual carbon target" and the tightening of policies in China, according to the data released by China Association of Metalscrap Utilization and Ministry of Industry and Information Technology of the People's Republic of China in the "Guiding Opinions on Promoting High Quality Development of the Steel Industry (Draft for Soliciting Opinions)", by 2025, the proportion of EAF steel production to the total crude steel production will increase to over 15%, and the scrap ratio will reach 30%.

In the next 10 to 30 years, China's steel industry will gradually enter a green steel path where multiple processes and technologies coexist, such as high scrap-ratio, BF-BOF long process, EAF short process, low-carbon metallurgy, hydrogen metallurgy, etc. In recent years, the strategic value of scrap has been deeply analyzed by many researchers, believing that scrap is an excellent renewable resource for steel recycling, which is the only important raw material that can replace iron ore for steelmaking, meanwhile is also the most important resource for EAF short process steelmaking. In September 2020, the Chinese Government announced to the world the "dual carbon" goals of achieving carbon peak by 2030 and carbon neutrality by 2060. For achieving the "dual carbon" goal, China's steel industry must increase the proportion of scrap to EAF short process in the process structure. However, the current problem is that China's scrap resources are still insufficient, and the cost of EAF short process steelmaking is at a disadvantage. For years, the consumption of scrap in China has been increasing year by year. In 2017, the consumption of scrap reached 148 million tons, reaching 226 million tons in 2021, with a scrap ratio of 21.9%, but still far below the global average level of 37%. Among the 226 million tons of scrap consumption, BOF consumption accounts for about 2/3 and EAF consumption accounts for about 1/3, indicating the existence of two processes competing for scrap. The current scrap volume is clearly unable to meet the demand for high scrap ratio steelmaking. According to the prediction of CAMU, China's scrap resources will be 360 million tons in 2030, 380 million tons in 2035, reach a phased peak level of 410 million tons in 2040, and fall back to 400 million tons in 2060.

Based on the analysis of the development status of DRI in the world, it can be seen that India, Iran, Russia, and others are the main producing countries. Currently, the production of DRI using gas-based methods accounts for about 75%. Due to the lack of natural gas resources in China, the exploration of direct reduction method in the past few decades has focused on coal-based method, which has a small production scale, low production efficiency, and low DRI production. If we can study the production of DRI using coke oven gas, hydrogen gas, etc., and develop the process of (Scrap+DRI)-EAF, it will be a significant breakthrough in the EAF short process steelmaking process in China. Comparing the development trend of the world's steel industry, it can be seen that in the coming years, as the output of scrap and DRI increases, the situation of insufficient supply of EAF short process steelmaking resources will gradually be broken, the price of scrap will gradually decrease, the scrap and DRI ratio will become higher, and the EAF short process steelmaking will be vigorously developed.

Key words: scrap resource; EAF short process; DRI; scrap ratio

宝钢电炉绿色低碳冶金探索与进展

李治平[1]，徐迎铁[2]，孟庆玉[1]

（1. 宝山钢铁股份有限公司钢管条钢事业部，上海　200941；
2. 宝山钢铁股份有限公司中央研究院长材所，上海　200941）

摘　要：随着国家"碳达峰、碳中和"发展目标的提出，国家鼓励具备低碳冶金特性的电炉炼钢科学有序发展，对宝钢来说，未来用户端对低碳产品需求日益增长、中国未来废钢资源逐步释放以及未来中国绿电电力供应充分等有利条件支撑了宝钢电炉绿色低碳冶金技术发展。宝钢目前推进的电炉低碳冶金的工程包括：（1）宝钢股份宝山基地150t低碳高效电炉项目；（2）宝钢沙特方合作海尔港基地电炉（以竖炉产生的DRI为主原料）生产厚板的S项目；（3）宝钢股份东山基地湛江零碳排放电炉冶炼板材工程。宝钢未来绿色低碳冶金技术开发紧紧围绕"绿色、低碳、智能、高效"为主题开展一系列技术开发工作，宝钢上述工程项目在炉型选择上充分考虑到二恶英排放等绿色环保以及高效化生产等因素。宝钢电炉低碳冶金开发的技术包括：（1）电炉低碳原料使用技术；（2）电炉高效供电关键技术；（3）电炉余热回收技术；（4）电炉低碳冶金低氮钢关键技术；（5）CO_2-石灰复合喷吹脱氮、脱磷技术；（6）铸坯氢-氧切割关键技术；（7）电炉冶炼生物碳应用技术。如上一系列技术开发目前已经取得重大进展：（1）小型电炉实现40%~60%直接还原铁加废钢的稳定冶炼，实现出钢氮含量22ppm，满足低氮钢要求；（2）实现低碳原料冶炼过程化学能稳定供应，确保了正常吹氧钢水不过氧化，满足了电炉高效生产适应连铸节奏要求；（3）开发的高效供电技术实现电炉降低电耗超30kWh/t，实现电炉高效低耗生产；（4）明确了电炉冶炼炉渣侵蚀耐材机理，为未来还原铁应用后控制炉衬侵蚀提供理论支撑。同时，宝钢电炉低碳冶金也致力于用户端服役低碳排放长材产品研发，包括高强高韧减重产品开发、冷锻产品研发、减免退火产品研发、金刚砂线服务光伏产品研发等。展望未来，宝钢电炉绿色低碳冶金关键技术成功应用后，实现产品成分满足钢种要求以及电炉冶炼周期控制满足高效连铸节奏要求，预计电炉冶炼铸坯平均碳排放低于$200kgCO_2/t$，接近碳中和水平，部分出口产品可以达到低碳排放标准。

关键词：绿色低碳冶金；电弧炉炼钢；碳排放；耐材侵蚀

超高功率柔性IGBT直流电源和智能化电弧炉炼钢技术

张豫川[1]，吴学涛[1]，杨宁川[2]

（1. 中冶赛迪工程技术股份有限公司，重庆　401122；
2. 中冶赛迪技术研究中心有限公司，重庆　401122）

摘　要：本报告研究的内容以进一步降低电炉的电耗和冶炼周期，提高无人化操作水平，降低工人劳动强度为目标，着力于开发新一代高效、低耗、绿色环保大型电弧炉核心供电和智能化装备技术。通过理论分析+实验测试+工程实践来厘清冲击负荷对电源装置稳定性的原理和动态交互匹配的方法，攻克了多电极分控调节、直流电源输出功率实时交互协同调控的瓶颈，开发了新型超高功率柔性IGBT直流电源；以智能感知、图像处理、大数据分析等智能化装备技术为基础，开发了电弧炉炼钢过程废钢连续加料、废钢预热温度、测温取样、合金加料、电极调节、氧枪、出钢自动控制技术，结合电炉外围条件的精细化管控，构建电炉冶炼模型，实现整个电炉炼钢备料、废钢加料、冶炼过程和出钢的智能化操作。通过多个工程项目的验证，大幅提升了智能化控制水平，各项技术指标达到国际先进水平，有力推动了我国短流程炼钢装备和技术的进步。

关键词：冶金；电弧炉；柔性 IGBT 电源；智能化

Ultra-high Power Flexible IGBT DC Power Supply and Intelligent Electric Arc Furnace Steelmaking Technology

ZHANG Yuchuan[1], WU Xuetao[1], YANG Ningchuan[2]

(1. CISDI Engineering Co., Ltd., Chongqing 401122, China;
2. CISDI Research & Development Co., Ltd., Chongqing 401122, China)

Abstract: The research in this report aims to further reduce power consumption and Tap To Tap time of EAF, improve unmanned operation levels and decrease labor intensity. It focuses on the development of a new generation of high efficient, low-consumption, environmentally friendly large-scale electric arc furnace core power supply and intelligent equipment technology. The principles and methods of stabilizing the power supply device under impact load and achieving dynamic interaction matching are clarified through theoretical analysis, experimental testing, and engineering practice. The bottleneck of multi-electrode decentralized control and real-time interactive coordinated control of DC power output power is overcome, and a new type of ultra-high-power flexible IGBT DC power supply is developed. Based on intelligent equipment technologies such as intelligent sensing, image processing, and big data analysis, Intelligent technologies for continuous charging of scrap in the electric arc furnace steelmaking process, preheating temperature of scrap, temperature measurement and sampling, alloy addition, electrode regulation, oxygen lance, and automatic tapping control are developed. Combined with fine control of peripheral conditions of the electric arc furnace, an electric arc furnace smelting model is established to achieve intelligent operation throughout the entire steelmaking process, including raw material preparation, charging of scrap, smelting process, and tapping. Through validation in multiple engineering projects, the level of intelligent control is significantly improved, and various technical indicators have reached international advanced levels, effectively promoting the progress of EAF steelmaking technology in China.

Key words: metallurgy; electric arc furnace; flexible IGBT power supply; intelligence

低碳电炉炼钢工艺典型特征和实践路径

王中丙

（原广州珠江钢铁有限公司，广东广州 510000）

摘 要：大量生产数据表明，低碳电炉炼钢工艺具有低配碳、低渣量和低冶强的特征，特征值分别为 0.2%~0.3%C、100~120kg/t 和 25~30Nm³/t O_2，其实践路径有废钢配料优化与预热技术、加入 HBI/DRI 冶炼技术、泡沫渣控制技术、TPC 终点控制技术、出钢脱氧技术和钢中氮含量控制技术。本文对各项技术要点进行简述。研究表明，电炉"三低"特征不仅使残余元素含量可控，而且显著降低生产成本，起决定性作用的因素是平熔池冶炼。研究指出：低配碳是减少电炉碳排放的有效办法，但不可能做到"无碳"炼钢，目标是使碳最小化或合理化。

关键词：电弧炉炼钢；低碳；泡沫渣；低配碳；平熔池

Typical Characteristics and Practical Path of Low Carbon Electric Arc Furnace Steelmaking

WANG Zhongbing

(Formerly Guangzhou Zhujiang Iron and Steel Co., Ltd., Guangzhou 510000, China)

Abstract: A large amount of production data shows that low carbon electric arc furnace (EAF) steelmaking is characterized by low carbon addition, low slag content and low metallurgical strength. The characteristic values are 0.2%-0.3%C, 100-120kg/t and 25-30Nm^3/t O$_2$, respectively. The practical paths include scrap charging optimization and preheating technology, adding HBI/DRI smelting technology, slag foaming control technology, TPC endpoint control technology, deoxidation technology of tapping and nitrogen content control technology. The key points of various technologies are summarized in this study. The result shows that the "three-low" characteristics of EAF not only make the contents of the residual elements controllable, but also reduce the production cost significantly, and the decisive factor is flat molten pool smelting. The result illustrates that low carbon addition is an effective way to reduce carbon emission from electric arc furnace steelmaking, but it is not possible to achieve "carbon-free" steelmaking, the goal is to minimize or rationalize the carbon content.

Key words: electric arc furnace steelmaking; low carbon; slag foaming; low carbon addition; flat molten pool

特冶技术进步推进镍基合金发展及未来展望

杨玉军

（抚顺特殊钢股份有限公司，辽宁抚顺 113000）

摘 要： 镍基合金生产通常采用真空感应熔炼、电渣重熔、真空电弧重熔等特种冶炼工艺生产。对特冶技术的认知及控制，直接影响产品的冶金质量。本文以真空电弧重熔常见的冶金缺陷为例，分别对年轮状缺陷、两种类型黑斑、三种类型白斑的形成原因进行分析。分别阐述了形成相应缺陷的原材料、重熔电极质量、电磁环境等影响因素、缺陷来源以及控制理念以及对产品性能的影响。并根据镍基合金未来的发展需求，针对后续特冶技术的发展提出相应的设想和诉求。

关键词： 镍基合金；电渣重熔；冶金缺陷；特冶技术；未来发展

Advancement of Special Metallurgical Technology Promotes the Development and Future Prospects of Nickel-based Alloys

YANG Yujun

(Fushun Special Steel Co., Ltd., Fushun 113000, China)

Abstract: Nickel-based alloys are usually produced by special smelting processes such as vacuum induction melting, electroslag remelting, and vacuum arc remelting. The metallurgical quality of products directly affected by the cognition and control of special metallurgy technology. In this study, taking the common metallurgical defects of vacuum arc remelting as an example, the causes of annual ring defects, two types of black spots and three types of white spots are analyzed. The source of defects, the influence factors such as raw materials, quality of remelting electrode and electromagnetic environment, and the control concept as well as the influence on product performance are expounded respectively. And according to the future development needs of nickel-based alloy, the corresponding ideas and demands for the development of subsequent special metallurgy technology are put forward.

Key words: nickel-based alloys; electroslag remelting; metallurgical defects; special metallurgy technology; future development

特殊钢冶炼过程 D/DS 夹杂物控制技术探讨

姜周华，董君伟，田家龙

（东北大学，辽宁沈阳　110819）

摘　要：非金属夹杂物，尤其大颗粒夹杂物是造成特殊钢材料疲劳破坏的主要原因之一，控制夹杂物的数量、类型、尺寸和形状是炼钢过程重要的任务。本文讨论了特殊钢中非金属夹杂物与钢材疲劳寿命的关系，分析了钢中非金属夹杂物的主要来源，发现 D/DS 夹杂物去除和控制是难点。钢水中 Ca 含量增加显著促进铝酸钙类 D/DS 夹杂物的生成，其中精炼渣中 CaO 被钢液中 Al 还原或真空下被碳还原生成 Ca 进入钢液是钢水增钙的重要机制之一。从原材料中控制钙带入、适当降低精炼渣碱度、减少钢渣混合等措施有利于减少 D/DS 夹杂物的生成，控制目标是钢水中 Ca 含量小于 3ppm。电渣重熔过程对去除夹杂物非常有效，但以 CaF_2 为基的熔渣导致渣中 Ca 离子活度高，导致钢中 D/DS 夹杂物含量增加。通过渣系优化可以显著减少电渣重熔钢锭中 D/DS 夹杂物含量。钢中大型夹杂物的形成机理是复杂的，需要根据实际生产情况综合判断和采取措施。

关键词：特殊钢；D/DS 夹杂物；Ca 含量；精炼渣

Discussion on Control Technology of D/DS Inclusions in Special Steel Making Process

JIANG Zhouhua, DONG Junwei, TIAN Jialong

(Northeastern University, Shenyang 110819, China)

Abstract: Non-metallic inclusions, especially large-particle inclusions, are one of the main causes of fatigue failure of special steel materials. Controlling the quantity, type, size and shape of inclusions is an important task in steelmaking process. This paper discusses the relationship between non-metallic inclusions and fatigue life of special steels. The main sources of non-metallic inclusions in steel are analyzed, and it is found that the removal and control of D/DS inclusions are difficult. The increase of Ca content in molten steel significantly promotes the formation of calcium aluminate D/DS inclusions. The reduction of CaO in refining slag by Al in molten steel or the reduction of Ca by carbon under vacuum into molten steel is one of the important mechanisms of calcium increase in molten steel. Controlling calcium intake from raw materials, appropriately reducing the basicity of refining slag and reducing mixing of steel slag are beneficial to reducing the formation of D/DS inclusions, and the control goal is that the Ca content in molten steel is less than 3ppm. The electroslag remelting process is very effective for removing inclusions, but the CaF_2-based slag leads to high Ca ion activity in the slag, resulting in an increase in the content of D/DS inclusions in the steel. The content of D/DS inclusions in electroslag remelting ingot can be significantly reduced by slag system optimization. The formation mechanism of large

inclusions in steel is complicated, so it is necessary to judge and take measures according to the actual production situation.
Key words: special steel; D/DS inclusion; Ca content; refining slag

连铸大压下高质量特厚板生产及铁素体轧制技术最新进展

康永林

（北京科技大学，北京 100031）

摘 要： 针对近期长材市场疲软、未来钢产量需求呈下降趋势的外部环境，各钢企纷纷寻求转型和产品结构调整，本文针对细分市场的容量、产品定位、各类产品的品质要求等方面，提出差异化、专业化、精品化的转型发展理念，避免恶性竞争，以利于钢铁企业高质量发展。（1）通过自主开发连铸大压下关键装备及工艺控制、板坯凝固前沿W形及二冷水优化控制、低压缩比特厚板轧制等技术，突破凝固末端重压下概念，对连铸凝固末端及全凝固后均实施大压下，显著改善了铸坯内部质量，结合轧制工艺优化，形成连铸大压下装备工艺与特厚板轧制成套技术，实现以连铸大压下生产高质量铸坯为基础，以 1.5～2.53 极低压缩比、低碳、低成本生产 120～252mm 高质量特厚板。（2）通过解决低/微碳钢铸坯低温加热（出炉温度 1050℃）、粗轧工艺控制、中间坯冷却控制、铁素体轧制生产效率、精轧工艺优化、带钢表面与边部质量及组织性能控制、冷轧退火工艺优化等一系列技术难题，实现低/微碳钢铁素体轧制批量生产，在大幅度节能降碳、提升板带质量性能取得显著成效。

关键词： 连铸大压下生产；铁素体轧制；节能降碳；提升板带质量性能

板形控制技术的最新研究进展

何安瑞

（北京科技大学，北京 100031）

摘 要： 针对无头轧制、薄带铸轧、高强钢及宽薄规格等板形控制的重大需求，通过快速仿真模型、辊形技术及窜辊策略、板形在线诊断系统等创新，并进行工业应用，在板形质量提升、延长轧制公里数等方面取得良好效果。

关键词： 板形控制；快速仿真模型；辊形技术及窜辊策略；板形在线诊断系统

首钢高硅硅钢连轧的工艺构想与实践

马家骥

（北京首钢股份有限公司，北京 100043）

摘　要：(1) 高硅硅钢连轧的技术逻辑；(2) 小辊径六辊六机架轧机的参数及其配置；(3) 高硅硅钢连轧的实践成果。

关键词：高硅硅钢；小辊径；六辊六机架轧机

现代冷轧板带绿色制造与智能制造

王业科[1]，陶　涛[2]，牛　强[1]，刘显军[1]

(1. 中冶赛迪工程技术股份有限公司，重庆　401122；
2. 中冶赛迪信息技术（重庆）有限公司，重庆　401122)

摘　要：绿色制造是一个综合考虑资源、能源消耗和环境影响的现代制造模式，而智能制造则是融入钢铁制造和运营决策过程中，做到"精准、高效、优质、低耗、安全、环保"，全面提升发展水平，实现钢铁行业高质量发展。本文系统综述了绿色制造中的超高强钢系列、高耐蚀性系列、高功能性系列等绿色产品、应用领域以及对绿色发展的贡献，重点推介了比较有代表性的冷轧绿色工艺及装备技术、节能降耗技术、资源综合利用技术、环境保护技术以及智能制造技术。冷轧智能制造技术是冷轧未来的发展方向，是冷轧绿色制造所追求的目标，智能作业与智能管控是智能制造的两个层级，工业互联网平台则是其运行的纽带。

关键词：冷轧；板带；绿色制造；智能制造

Green Manufacturing and Intelligent Manufacturing of Modern Cold-rolled Strip

WANG Yeke[1], TAO Tao[2], NIU Qiang[1], LIU Xianjun[1]

(1. MCC CISDI Engineering Co., Ltd., Chongqing 401122, China;
2. MCC CISDI Information Technology (Chongqing) Co., Ltd., Chongqing 401122, China)

Abstract: Green manufacturing is a modern manufacturing mode that comprehensively considers resources, energy consumption and environmental impact, while intelligent manufacturing is integrated into the decision-making process of steel manufacturing and operation, so as to achieve "precision, efficiency, high quality, low consumption, safety and environmental protection", comprehensively improve the level of development, and achieve high-quality development of the steel industry. This paper systematically reviews the green products, application fields and contributions to green development such as ultra-high strength steel series, high corrosion resistance series and high functional series in green manufacturing, and focuses on the representative cold rolling green process and equipment technology, energy saving and consumption reduction technology, comprehensive resource utilization technology, environmental protection technology and intelligent manufacturing technology. Cold rolling intelligent manufacturing technology is the future development direction of cold rolling, is the goal pursued by cold rolling green manufacturing, intelligent operation and intelligent control are the two levels of intelligent manufacturing, and the industrial Internet platform is the link of its operation.

Key words: cold rolling; steel plate & strip; green manufacturing; smart manufacturing

长材生产线技术革新与进步

杨勇强

（中冶京诚工程技术有限公司，北京 100176）

摘 要：针对近几年来长材生产线的新技术研发和应用情况，重点介绍高棒单一孔型、螺纹钢控轧控冷技术、柔性化轧制、物料逐支跟踪、在线精整和热处理等先进技术，以及高速飞剪、减定径轧机、轧钢智能化等关键设备。

关键词：长材；新技术研发；轧钢智能化

无加热直接轧制关键技术与工业化

许宏安

（陕西钢铁集团有限公司，陕西西安 710000）

摘 要：双碳背景下，直接轧制工艺因没有加热炉工序而减少了轧制工艺70%的能耗，做到了"七个零和一个一"，可以创造直接轧制模块车间而更适合于自动化智能化的炼轧一体化控制，其优势愈来愈明显，亦是钢铁企业深挖降本增效潜力发力点之一。直接轧制工艺广泛应用的瓶颈仍然集中在：有效利用连铸坯心表温差，采取"细化铸坯芯部晶粒"提升钢材性能，达到节约微合金资源目的，同时消除连铸坯头尾温差保证钢材质量所矛盾。

就为什么要无加热无补热直接轧制，重点列出无加热无补热直接轧制在理论上技术上工艺上装备上四个层次存在的30个关键问题是什么，并对直接轧制关键的问题怎么解决逐一说明，对前沿问题展开汇报。最后在工业化应用十年来总结的基础上探讨下一步的研究与发展。

关键词：直接轧制工艺；关键问题

热轧钢筋低成本绿色生产技术

王卫卫

（钢铁研究总院有限公司，北京 100081）

摘 要：针对目前钢铁行业，尤其是棒线材企业亟需降本增效的现状，通过研发系列棒线材的低成本减量化生产技术，为棒线材生产企业低碳、绿色化发展提供技术参考和实践经验。

关键词：低成本；减量化；低碳；绿色化发展

钢管技术和产品发展

成海涛

(中国金属学会轧钢分会,北京 100081)

摘　要:本文分析了无缝钢管行业取得的成就和存在的问题,对行业健康可持续发展提出了建议。

关键词:无缝钢管;可持续发展

宝钢低碳绿色冷轧涂镀钢板工艺技术

任玉苓

(宝钢股份冷轧厂,上海 201999)

摘　要:冷轧涂镀钢板在汽车、家电、建筑、机械加工等行业有广泛的应用,低碳绿色涂镀产品的应用对社会贡献重大。本文主要介绍了宝钢冷轧热镀高铝锌铝镁机组热处理节能技术、彩涂机组烘烤节能技术、脱脂和平整等工序降低水消耗技术;介绍了通过新产品开发和结构优化提升涂镀产品使用寿命的全生命周期降碳技术;介绍了无铬无重金属绿色产品在高耐蚀涂镀层钢板领域的应用技术。采用节能环保技术生产涂镀钢板促进了冷轧低碳绿色生产工艺的进步。

关键词:热镀高铝锌铝镁;烘烤节能;全生命周期;降碳技术;无铬无重金属绿色产品;低碳绿色生产

大厚度高韧性海洋工程用钢工艺研究及工业实践

赵　坦

(鞍钢集团钢铁研究院/海洋装备用金属材料及其应用国家重点实验室,辽宁鞍山 114033)

摘　要:采用连铸坯生产大厚度厚海工钢存在压缩比不足,轧后加速冷却不均匀等问题,采用传统 TMCP 及低温轧制受到大厚度坯料、轧机轧制力以及扭矩极限等因素制约,随着我国海洋装备逐渐向深远海拓展,大厚度、耐低温、高均质性海工钢的开发面临严峻考验。本报告从大厚度海工钢强韧化机理、断裂失效行为出发,围绕420kg级大厚度高强韧海工钢开发,重点探讨了传统 TMCP 工艺的不足,阐述了高强韧性特厚海工钢原始晶粒、相变组织、晶体取向与断裂行为的影响规律,总结了钢中脆性裂纹影响因素和控制手段,结合鞍钢低压缩比条件大厚度海工钢研发工作,介绍了新型特厚海工钢轧制技术及工业实践应用情况。

关键词:大厚度;海工钢;强韧化机理

中薄板坯高等级管线钢工艺研究及工业化实践

林 利

（鞍钢集团钢铁研究院，辽宁鞍山 114033）

摘 要：为满足国家"北气南下"和"沿海登陆"能源战略，开发出高强高韧 L450M-L625M 卷板系列产品。主要通过低碳高锰 Nb-Cr 系为基础，按需添加 Mo、Ni 和 Cu 元素来优化合金设计；通过奥氏体再结晶组织细化以及未再结晶变形和超快速冷却+层流特殊冷却工艺的合理调控；ASP 流程 170mm 厚连铸坯纯净钢稳定批量冶炼连铸控制技术；板卷头部 650℃以下卷取的高温让头技术；低温强力卷取等组织调控、关键制备工艺和品质保障全流程控制技术，突破了压缩比大于 10 的技术瓶颈，攻克了厚壁管线钢低温 DWTT 韧性控制、纯净钢冶炼及 10 炉连浇、塔型硬弯控制等多项技术难题，形成鞍钢 ASP 特色生产 L450M-L625M 卷板强韧性匹配控制和品质保证工艺等多项创新技术。

关键词：高强高韧；冷却工艺；冶炼连铸控制技术

高速棒材生产线智能化创新建设及其关键技术

徐言东[1,2]，王晓晨[1,2]，程知松[1,2]，余 伟[1,2]，
张勇军[1,2]，刘 洋[1,2]，白金龙[1,2]

（1. 北京科技大学国家板带生产先进装备工程技术研究中心，北京 100083；
2. 北京科技大学高效轧制与智能制造国家工程研究中心，北京 100083）

摘 要：近年来传统棒材生产线在市场的压力下，大多进行了升级改造，正在由局部设备的智能化、少人化向着整体智能化高速棒材生产线迈进。发现其一般都是按照生产过程智能控制、生产管理智能控制的"双智控"架构来进行建设的。北科工研、中冶京诚、中钢设备等设计供货单位通过对长治钢厂高线改高棒、福建三钢一棒改造、济源钢厂二棒改造、珠海粤裕丰棒材成品库、昆钢 4580 棒线材、建龙西钢双高棒等项目进行了智能化创新建设，都取得了较好的效果。这些信息化和智能化手段的应用可促进高速棒材生产线升级，增强市场竞争力。

关键词：高速棒材；智能化；双智控；工业机器人；智能装备

Innovation and Key Technologies of Intelligent High Speed Bar Production Line Construction

XU Yandong[1,2], WANG Xiaochen[1,2], CHENG Zhisong[1,2], YU Wei[1,2],
ZHANG Yongjun[1,2], LIU Yang[1,2], BAI Jinlong[1,2]

(1. National Engineering Technology Research Center of Flat Rolling Equipment, University of Science and Technology Beijing, Beijing 100083, China; 2. National Engineering Research Center for Advanced Rolling and Intelligent Manufacturing, University of Science and Technology Beijing, Beijing 100083, China)

Abstract: In recent years, under the pressure of the market, most of the traditional bar production lines have been upgraded and transformed, and are stepping from the intellectualization and Flexible Manpower Line construction of local equipment to the intellectualization of high-speed bar production line as a whole.It is found that they are generally constructed according to the "Dual Intelligence Control" architecture of intelligent control of production process and production management.Design and supply units such as Institute of Engineering Technology of USTB, Capital Engineering & Research Incorporation, and SINOSTEEL MECC have carried out intelligent innovation construction through such projects as Changzhi Steel Plant's high speed bar production line transformation, Fujian Sangang's No.1 bar production line transformation, Jiyuan Steel Plant's No.2 bar production line transformation, Zhuhai Guangdong Yufen Bar Product Library, Kungang 4580 bar production line and Jianlongxi Steel's double high speed bar production lines, all of which have achieved good results. The application of these informational and intelligent methods can promote the upgrade of high-speed bar production line and enhance the market competitiveness.

Key words: high-speed bar; intelligence; dual intelligence control; industrial robot; intelligent equipment

唐钢特殊镀层产品特性简介

李建英

（河北唐钢冷轧事业部研发中心，河北唐山　063000）

摘　要： 本文简单介绍唐钢冷轧产线总体情况，目前在役特殊镀层的种类。铝硅镀层产品特性介绍，首先是铝硅镀层组织结构特点；其次是铝硅镀层产品不同温度区间下耐高温性以及耐蚀性特点；最后是在不同镀层厚度在热成形工艺加热条件下镀层微观组织结构的变化关系。锌铝镁镀层产品特性，主要介绍低铝低镁系锌铝镁产品组织结构特点、锌铝镁镀层产品在耐蚀性、耐摩擦性、涂装焊接性等方面的优势及典型用途。

宝钢建筑用涂镀产品未来发展

任玉苓

（宝山钢铁股份有限公司冷轧厂，上海　200941）

摘　要： 本文从全品种涂镀产品供应商、全流程智能环保低碳、全行业解决方案三个方面对宝钢建筑用涂镀产品的现状和未来发展进行了介绍。

宝钢股份在涂镀钢板方面具有较大的装备和品种优势，近年开发了锌铝镁系列镀层及其彩涂钢板，不同的锌铝镁镀层在硬度、折弯性能、耐蚀性、耐酸碱性能等有差异，在此基础上开发的彩涂钢板耐蚀性能也存在切口、划线扩蚀速度、平板起泡等级方面的不同趋势。不同产品适应不同的环境和需求是主要发展方向，未来宝钢也将在高强超宽涂镀产品方面进一步拓展。

近年宝钢股份建筑用高耐蚀涂镀钢板生产机组在智能化生产、智慧化管理方面有了一定的开拓，特别实现了在线涂层厚度、在线色差等检测等新技术应用，并率先实现了建筑涂镀产品的全面无铬化，未来将在低碳产品方面继续引领国内建筑围护结构用涂镀钢板的发展。

中国是全球行业门类最全的国家之一，宝钢为近年快速发展的医药、物流、畜牧等行业提供了建筑围护结构用涂镀钢板解决方案。随着国家各项鼓励政策的推出，屋面光伏发展成为新的亮点，宝钢结合光伏行业和钢结构行业需求和发展特点，提供专用宽幅长寿涂镀钢板解决方案。

宝钢建筑用涂镀产品也将继续助力各行各业的发展！

QH 涂镀板环保长效保护涂层

张 俊

（奎克好富顿，上海 201700）

摘 要：随着汽车，家电，建筑等行业的不断发展，对于带钢的综合性能提出了越来越高的要求，单一的镀层金属已经无法满足日益提高的性能需求。多种金属混合镀层在外观，耐腐蚀性，可焊性等方面具有明显的优势，逐渐在市场上取代了单一金属镀层的份额。

传统的六价铬钝化，磷化等表面处理在生产，运输，施工以及终端应用中容易对环境产生不可逆转的破坏，越来越受到世界范围的限制。同时，标准的表面处理工艺通常性能单一，面对复杂的应用环境也面临着无法满足终端用户的困境。

奎克好富顿公司对市场上的多种金属镀层产品特性进行了详细的分析，基于腐蚀化学理论建立了一种长效耐腐蚀保护的模型，并基于该模型设计了两种不同的耐腐蚀涂层。该涂层具备超长耐腐蚀性，对于室外等应用场合提供了一个环保解决方案。同时，该涂层也具备良好的综合性能，可以满足不同终端客户的需求，为多金属镀层产品的推广应用提供了一个优秀的选择。

卷钢涂层辐射固化技术现状及展望

马 源

（宝山钢铁股份有限公司，上海 201900）

摘 要：卷钢预涂层产品由于充分融合了钢材的强度特性、镀层的耐蚀特性以及涂层的耐久和装饰等特性，被广泛应用于建筑、轻工、家电等各个领域。目前国内外主流钢厂普遍采用加热固化的方式，将溶剂型涂料或水性处理剂在带钢表面烘烤成膜的方式进行卷钢涂层产品的生产，这种生产方式存在碳排放量大、能耗高、VOC、速度受限、占地空间大等不足，从"低碳""节能""环保""高效"的发展战略考虑，有碍于卷钢涂层生产的可持续发展。

辐射固化是一项用电来产生能量（紫外线 UV 或电子束 EB），实现无溶剂型涂料在常温下瞬间干燥的一种冷固化技术，由于满足"4E"原则被誉为涂层生产的绿色技术，在木材、纸张、薄膜、光纤等诸多应用领域都显示了自身的优越性，并得到了迅猛发展，但在卷钢涂层领域的应用尚处于起步阶段。

为进一步促进辐射固化技术在卷钢涂层领域的进一步发展，本文从辐射固化的定义、分类、机理、特点等方面对该技术进行了系统介绍，梳理了国际各大主流钢厂对卷钢涂层辐射固化技术的研发和应用现状，并结合当下的"双碳"背景，从低碳、极致成本、差异化特性等角度展望了辐射固化技术在钢铁行业卷钢涂层领域的应用前景，对做什么、怎么做以及工艺过程和产品性能角度的关注点提出了建设性意见。

结构钢大气环境腐蚀与试验评价技术

苏 艳

（中国兵器装备集团第五九研究所有限公司，重庆 401329）

摘 要：结构钢的大气腐蚀是其安全可靠使用重点关注的问题，不同环境、不同材料成份、组织结构所产生的腐蚀行为与机理不尽相同。结构钢环境腐蚀的多样性、复杂性和特殊性是其固有性质在一定外界环境条件下综合反映的结果。

影响结构钢大气腐蚀的因素包括外因和内因两方面。外因是指外界环境因素，影响结构钢腐蚀速率的主要环境因素为润湿时间、二氧化硫和氯化物水平。内因是指结构钢本身的化学成份和组织结构。采用自然环境大气暴露试验，并结合金相分析、断口分析，对比研究了45A、300M 两种结构钢在湿热海洋、亚湿热工业、干热沙漠、寒冷低温环境中的腐蚀行为及其对力学性能的影响规律。金相分析表明，45A 调质态和 300M 微观组织均以回火索氏体为主，含少量铁素体。由于两种结构钢在显微组织、晶粒度、夹杂物分布等微观尺度存在差异，从而在腐蚀萌生和发展上表现出明显不同。海洋大气环境暴露1年，45A 调质态的平均腐蚀速率高达 531μm/a，亚湿热工业环境中约为 28μm/a，仅为前者的 5.3%，而干热沙漠、寒冷低温环境中的平均腐蚀速率与海洋大气环境相差 2 个数量级。300M 也表现出相似的环境敏感性，海洋环境暴露 1 年的腐蚀速率达到 116μm/a，约是亚湿热工业环境的 6 倍，干热沙漠、寒冷低温环境的 15～29 倍。结构钢腐蚀速率随时间的发展逐步下降，最后趋于稳定，腐蚀动力学符合幂函数规律 $D=At^n$。需要注意的是，结构钢腐蚀后表面凹凸不平，局部会产生应力集中效应，对其塑性指标及疲劳性能影响较大。

针对当前实验室中性盐雾、周期浸润等试验考核的局限性，提出了多因素组合/综合模拟加速试验评价新方法，采用图表对比法和加速转换因子法，评估了多因素加速试验的模拟性和加速性，可推广应用于结构钢等金属材料环境适应性快速评价。

关键词：结构钢；大气环境；腐蚀行为；力学性能；试验评价

中国金属制品行业现状及绿色低碳技术进展

毛海波[1]，谷 宇[2]，米振莉[2]，汪 凯[1]

（1. 中钢集团郑州金属制品研究院股份有限公司，河南郑州 450001；
2. 北京科技大学工程技术研究院，北京 100083）

摘 要：简要介绍了中国金属制品的分类及应用领域，并结合其进出口情况、产品品种占比、企业分布情况和产能集中态势分析了该行业的特点。同时，从高强及超高强、耐疲劳等高性能的金属制品产品的研发应用、金属制品加工过程及装备的改进、表面处理技术及装备的改善以及自动化智能化装备的应用等方面总结了我国金属制品行业在绿色低碳生产方面做出的工作，分析了该行业的发展现状。最后，从能源、装备、产品、上下游四个方面提出了未来金属制品绿色低碳技术的发展方向。

关键词：金属制品；绿色低碳；发展现状；发展方向

Present Situation and Green Low-Carbon Technology's Progress in the Chinese Steel Wire Products Industry

MAO Haibo[1], GU Yu[2], MI Zhenli[2], WANG Kai[1]

(1. Sinosteel Zhengzhou Research Institute of Steel Wire Products Co., Ltd., Zhengzhou 450001, China;
2. Institute of Engineering Technology, University of Science and Technology Beijing, Beijing 100083, China)

Abstract: It briefly introduces the classification and application fields of Chinese steel wire products, and analyzes the characteristics of the industry with its import and export situation, the proportion of product varieties, the distribution of enterprises, and the concentration of production capacity. At the same time, it summarizes the work done by the Chinese steel wire products industry in green and low-carbon production and analyzes the development status of the industry in terms of the research and development and application of high-strength and ultra-high-strength, fatigue-resistant, and other high-performance metal products, the improvement of the metal products processing process and equipment, the improvement of the surface treatment technology and equipment, as well as the application of automation and intelligent equipment. Finally, the future development direction of green and low-carbon technology for the steel wire products industry is proposed from four aspects: energy, equipment, products, upstream, and downstream.

Key words: steel wire products; green and low-carbon; development status; development direction

B750HL 波形梁护栏辊弯成形策略研究

邢 刚，韩 飞，孙玮隆

（北方工业大学机械与材料工程学院，北京 100144）

摘 要： 本文以 SB 级高强钢轻量化三波梁护栏作为主要研究对象，采用理论分析、有限元仿真，对辊弯成形过程进行了研究。为了在实际生产中获得满足成形精度的辊弯成形产品，基于已有的辊弯成形机组设计了相应的辊弯成形工艺。利用 Abaqus 有限元仿真软件对高强钢波形梁护栏的成形方案进行有限元模拟，结果表明下山法能够有效抑制板材在成形过程中的缺陷。

关键词： 辊弯成形；高强钢；波形梁护栏

Research on Roll Forming Strategy of B750HL Corrugated Beam Guardrail

XING Gang, HAN Fei, SUN Weilong

(College of Mechanical and Materials Engineering, North China University of Technology, Beijing 100144, China)

Abstract: This paper focuses on the research of lightweight three-wave guardrails made of SB grade high-strength steel, investigating the roll forming process through theoretical analysis and finite element simulations. To achieve the desired forming precision in practical production, a corresponding roll forming process was designed based on existing roll forming equipment. Utilizing Abaqus finite element simulation software, the forming scheme for high-strength steel wave beam guardrails was analyzed, demonstrating that the downhill method effectively suppresses defects during the forming process of the sheet material.

Key words: roll forming; high strength steel; corrugated beam guardrail

架空输电线路耐候钢杆塔关键技术及工程应用

黄 耀

（中国电力科学研究院有限公司，北京 100192）

摘 要： 耐候钢是一种耐大气腐蚀钢，通过添加耐蚀元素，使其在金属基体表面形成稳定锈层，实现钢材自身防腐。报告从耐候钢种研发、成套设计技术、关键加工技术和腐蚀评价体系4个方面系统阐述了耐候钢杆塔关键技术，同时实现了工程应用，并对工程应用的现在进行了追踪与评价。报告的项目成果提高了我国输电线路防腐技术水平和节能环保性，促进了架空输电线路产业结构的转型升级，具有显著的经济、社会效益和推广应用前景。

关键词： 耐候钢；杆塔设计；腐蚀评价；全寿命周期

Key Technologies and Engineering Applications of Weathering Resistant Steel Poles and Towers for Overhead Transmission Lines

HUANG Yao

(China Electric Power Research Institute, Beijing 100192, China)

Abstract: Weathering steel is a type of atmospheric corrosion resistant steel that forms a stable rust layer on the surface of the metal substrate by adding corrosion resistant elements, achieving self-corrosion protection of the steel. The report systematically elaborates on the key technologies of weathering steel poles and towers from four aspects: research and development of weathering steel grades, complete design technology, key processing technology, and corrosion evaluation system. At the same time, engineering applications have been achieved, and the current status of engineering applications has been tracked and evaluated. The project achievements reported have improved the level of anti-corrosion technology and energy conservation and environmental protection of transmission lines in China, promoted the transformation and upgrading of the industrial structure of overhead transmission lines, and have significant economic and social benefits and prospects for promotion and application.

Key words: weathering resistant steel; tower design; corrosion evaluation; full life cycle

长寿命、轻量化盘扣式脚手架开发

方幸[1]，米振莉[1]，赵爱民[2]，熊靖芸[3]，王贺伟[4]，华维[4]

(1. 北京科技大学高效轧制与智能制造国家工程研究中心，北京 100083；
2. 北京科技大学钢铁共性技术协同创新中心，北京 100083；3. 北京机科国创轻量化科学研究院有限公司，北京 100083；4. 磐石建龙钢铁有限公司技术中心，吉林磐石 132300)

摘 要：盘扣式脚手架因其安全可靠、结构稳固以及造型美观等优点，被广泛用于建筑施工中。为了使盘扣式脚手架在使用中更方便且使用时间更长久，对盘扣式脚手架进行长寿命、轻量化研究。本文通过材料性能提升、生产工艺优化、材料结构优化三个方面对盘扣式脚手架的生产进行优化，提出了一套盘扣式脚手架长寿命、轻量化的生产设计方案。结果表明，优化后的专用钢带屈服强度比 Q355 提升约 100MPa，耐蚀性提升 24%。明确了在(300±20)A，(30±2)V，(38±2)rpm 转速条件下，可以获得良好的焊接质量。同时，专用带钢制成的立杆环焊缝强度相比 Q355 进一步提升了 20%左右。通过数值模拟的方式优化了锁头浇注工艺，将双浇道工艺改为单浇道工艺，报废率从 19%降为 3%。通过数值模拟模拟脚手架受力情况，进行结构轻量化研究，制备了新型圆盘与锁头，其在原基础上可减重 7%~15%。
关键词：长寿命；轻量化；盘扣式脚手架

Development of Long-life, Lightweight Turnbuckle Scaffold

FANG Xing[1], MI Zhenli[1], ZHAO Aimin[2], XIONG Jingyun[3], WANG Hewei[4], HUA Wei[4]

(1. National Engineering Research Center for Advanced Rolling and Intelligent Manufacturing, University of Science and Technology Beijing, Beijing 100083, China; 2. Collaborative Innovation Center of Steel Technology, University of Science and Technology Beijing, Beijing 100083, China; 3. Beijing National Innovation Institute of Lightweight Ltd., Beijing 100083, China; 4. Technology Centre of Panshi Jianlong Steel Co., Panshi 132300, China)

Abstract: Turnbuckle scaffold is widely used in building construction because of its safety, reliability, structural stability and aesthetic appearance. In order to make the use of turnbuckle scaffold more convenient and longer, long-life and lightweight turnbuckle scaffold has been studied. In this paper, the production of turnbuckle scaffold is optimised through three aspects: material performance improvement, production process optimisation and material structure optimisation, and a long-life, lightweight production design for turnbuckle scaffold is proposed. The results show that the optimised special steel strip has a yield strength increase of approximately 100 MPa over Q355 and a 24% increase in corrosion resistance. It is clear that good weld quality can be achieved at (300±20)A, (30±2)V and (38±2)rpm speed conditions. At the same time, the strength of the ring weld of the riser made of the special strip is further improved by about 20% compared to Q355. The lock casting process was optimised by means of numerical simulations and the scrap rate was reduced from 19 % to 3 % by replacing the double-pouring process with a single-pouring process. Numerical simulations were carried out to simulate the stresses on the scaffold, and a new type of disc and lock was prepared, which can reduce the weight by 7%~15% from the original base.
Key words: long life; lightweight; turnbuckle scaffold

超高强钢汽车 A 柱上边梁的热气胀成形制造实践

阮尚文[1,3]，张　猛[1,2]，成　刚[1]，吴彦欣[3]，米振莉[3]，程鹏志[1]

（1. 航宇智造（北京）工程技术有限公司，北京　100080；2. 北京航空航天大学能源动力与工程学院，北京　100191；3. 北京科技大学工程技术研究院，北京　100083）

摘　要：轻量化技术是实现节能减排的重要手段，由于管梁结构在比刚度上的先天优势，采取超高强度的材料制成管梁件代替冲压件是实现汽车轻量化和保证汽车安全性的重要途径[1,2]。受制于设备，传统液压成形技术所能制备管件的最高强度很难超过 980MPa，且其室温成形后回弹严重，需要多次修模[3]。管件热气胀技术是一种新兴的在高温状态下借助气体迫使管坯发生快速变形，并在保压状态下进行模内冷却淬火，具有超高强度和尺寸精度的零件制备技术[4]。尽管欧美热气胀技术发展十分迅猛，然而该技术在国内的应用还鲜有报道。

本文主要介绍了航宇智造对某型号汽车的 A 柱上边梁进行全流程国产化热气胀成形试制的实践。该热气胀零件的设计轴线截面最大变化率小于 10%，最小圆角为 9mm，壁厚为 1.6mm。管坯原材料为表面涂有表面铝硅保护涂层的超高强钢，其在常规热冲压处理后的名义抗拉强度不小于 1300MPa，名义断后延伸率不小于 5%。在航宇智造公司自研的 1000t 管材热气胀产线上进行 930~950℃加热保温，并在最大 50MPa 的成形压力下完成零件气胀。通过蓝光成形精度扫描测试表明，其最大尺寸偏差小于 0.5mm；成形后零件进行选区切割，并对材料进行力学性能测试结果显示，三个不同区域的材料抗拉强度的最小值为 1365.9MPa，最大为 1501.7MPa，平均值为 1456MPa。

该实践及其后续验证结果表明，热气胀技术能有效的满足强度和精度要求的车用 A 柱上边梁等关键安全部件的制造要求。相较于传统方法，热气胀技术在相同的材料用量下同时实现了超过了 40%的强度提升和 20%的减重效果，有效实现了该型号车辆的轻量化目标。

参 考 文 献

[1] Trân R, Reblitz J, Haase R, et al. Hydroforming of High-Strength Aluminum Tubes with Thermo-Mechanical Manufacturing Processes[J]. Engineering Proceedings, 2022, 26(1): 3.
[2] 林荣会，宋晓飞. 汽车轻量化研究进展[J]. 青岛理工大学学报, 2018, 39(6): 97-103.
[3] Hartl C. Research and advances in fundamentals and industrial applications of hydroforming[J]. Journal of Materials Processing Technology, 2005, 167(2): 383-392.
[4] Martin S, Wolf S, Martin U, et al. Deformation Mechanisms in Austenitic TRIP/TWIP Steel as a Function of Temperature[J]. Metallurgical and Materials Transactions A, 2014. DOI:10.1007/s11661-014-2684-4.

定制辊压成型技术与中国制造业的高质量发展

晏培杰[1,2]

（1. 南京工业大学先进轻质高性能材料研究中心，江苏南京　210037；
2. 苏州亿创特智能制造有限公司，江苏昆山　215300）

摘　要：定制辊压成型技术是钢铁材料深加工技术的重要分支，如今正面临着制造业发展带来的严峻挑战。定制辊压成型技术可以实现轻量化、节能减排与智能制造等先进制造理念，满足制造业转型升级的迫切需求。本文对比了定制辊压成型技术在国内外的发展现状，阐述其对我国制造业发展的重大意义，同时针对国内定制辊压成型技术发展过程的主要难点，提出了相应的解决方案和实施路径。最后对定制辊压成型技术在重点行业的应用状况以及市场规模做出展望。

国内外发展现状对比。（1）辊压型材的设计和应用技术。欧洲国家对辊压型材的设计和应用已经形成体系，一方面，产品设计工程师将形式各异的型材作为常规结构件进行设计；另一方面，型材供应商对型材的性能及应用工况了解深入，能够指导用户使用。（2）辊压为平台的多工艺集成技术。辊压技术作为一种平台技术，可以与各种不同工艺相集成，以满足产品需要。（3）先进的模具设计技术。（4）轻质材料辊压成型技术。

对我国制造业发展的意义。（1）引领国内辊压行业技术升级。（2）轻质材料辊压成型技术的系统开发。定制辊压成型技术可以根据不同轻质材料的成型性能，给出辊压技术的解决方案，形成量产能力，推动轻质材料的应用和发展。（3）先进型材对智能制造的推动。（4）钢材-定制辊压-产品供应链上下游关系中，定制辊压成型技术起到从材料到产品的桥梁作用，它识别产品的需求，选择超高强钢等合适材料，经过辊压成型技术，满足产品的结构功能一体化，应用于下游产品。

定制辊压成型技术的解决方案。面对制造业升级所带来的诸多新需求，定制辊压型材通过材料优化设计和结构整合优化两种手段改良辊压型材产品。（1）材料优化设计。（2）结构优化设计。（3）产品模块化。本文围绕定制辊压成型技术发展的背景、定义、方案和市场进行了系统综述。定制辊压成型技术未来发展的关键点在于：（1）辊压成型创新技术的研发。（2）定制辊压成型技术的社会重视。（3）产业链上下游的共同价值链形成。也将定制辊压成型技术的典型应用进行了阐述，比如客车车身骨架航空领域，汽车领域、建筑领域、物流领域。

关键词：辊压；材料优化设计

C276 哈氏合金的高温热变形行为研究

谢广明，王丰睿，孟 晨，王玉前

（东北大学轧制技术及连轧自动化国家重点实验室，辽宁沈阳 110819）

摘 要：C276 哈氏合金具有优异的高温强度、抗蠕变和耐蚀性，被广泛应用于石油、化工和核工业等领域中。然而，由于含有大量 Mo 和 W 等难熔元素，以及 Cr、Ni 和 Co 等合金元素，C276 哈氏合金的热加工和再结晶温度均高于常用的镍基合金和奥氏体不锈钢，严重影响了 C276 哈氏合金的热加工性能。为了确定 C-276 哈氏合金的热加工工艺窗口，利用 MMS-200 热模拟试验机对 C276 哈氏合金在 1000~1250℃ 及 0.01~10s^{-1} 的变形条件下进行了单道次热变形实验。结果表明，C276 合金的应力-应变曲线表现出应力随变形温度的升高而降低，随应变速率的增大而增大。结合应力-应变曲线及变形后微观组织结果分析，得到了 C276 哈氏合金最佳热加工窗口应为：变形温度 1150~1200℃ 和应变速率 0.1~1s^{-1}。此外，建立了 C276 合金的热变形本构方程 $\dot{\varepsilon}=3.6836\times10^{19}[\sinh(0.0031108\sigma_p)]\times10^{6.002}\exp(-529/RT)$，计算出其表观激活能 Q 为 529kJ/mol。

关键词：C276 哈氏合金；热变形；微观组织；本构方程

Study on Hot Deformation Behavior of C276 Hastelloy at High Temperature

XIE Guangming, WANG Fengrui, MENG Chen, WANG Yuqian

(State Key Laboratory of Rolling Technology and Continuous Rolling Automation, Northeastern University, Shenyang 110819, China)

Abstract: C276 Hastelloy alloy is widely used in petroleum, chemical, and nuclear industries due to its excellent high-temperature strength, creep resistance, and corrosion resistance. However, the presence of refractory elements such as Mo and W, as well as alloying elements such as Cr, Ni, and Co, in C276 Hastelloy alloy results in higher hot working and recrystallization temperatures compared to commonly used nickel-based alloys and austenitic stainless steels, significantly affecting the hot working performance of C276 Hastelloy alloy. To determine the hot processing window of C276 Hastelloy alloy, single-pass hot deformation experiments were conducted on the C276 Hastelloy alloy using an MMS-200 thermal simulation tester under deformation conditions of 1000-1250℃ and strain rates of 0.01-10 s^{-1}. The results indicate that the stress-strain curve of the C276 alloy shows a decrease in stress with an increase in deformation temperature and an increase with a rising strain rate. Based on the analysis of the stress-strain curve and the microstructure post-deformation, the optimal hot working window for C276 Hastelloy alloy was determined to be a deformation temperature of 1150-1200℃ and a strain rate of 0.1-1s^{-1}. Additionally, a constitutive equation for the hot deformation of C276 alloy was established as $\varepsilon=3.6836\cdot10^{19}[\sinh(0.0031108\sigma_p)]\cdot10^{6.002}\exp(-529/RT)$, where the apparent activation energy Q was calculated to be 529kJ/mol.

Key words: C276 Hastelloy alloy; thermal deformation; microstructure; constitutive equations

Cost-effective Fabrication of Advanced Heat Sink Materials from Powder

YANG Fei

(University of Waikato, Hamilton, 3240, New Zealand)

Abstract: The technical capabilities of modern electronic devices increasingly demand higher frequencies and more power. However, the power and longevity of electronic chips used in high-power electronics are strongly dependent on thermophysical properties of the heat sink, which requires high thermal conductivity and similar thermal expansion to that of chip materials. Materials that have both high thermal conductivity and similar thermal expansion rates to chip materials are therefore in high demand for use as heat sinks in these applications. In this talk, I will present and discuss a possible route to cost-effective fabrication of copper/diamond composites with acceptable thermal conductivity, meeting the application requirement of future high-power electronic devices.

316L 不锈钢选区激光熔化增材制备过程的数值仿真研究

安希忠，咬登治，王 炬，吴 琼，吴宇航

（东北大学冶金学院颗粒技术研究所，辽宁沈阳 110819）

摘 要： 选区激光熔化（SLM）增材制造技术由于所具有的独特优势使其在多个工业领域具有重要的应用。然而，SLM 成形过程中易产生孔隙及粗糙表面等缺陷，其所蕴含的物理机制及动力学很难通过实验来解决，以至于稳定控制材料成形过程及精确优化工艺难度很大。针对这些问题，本工作以 316L 不锈钢（简称 316L）的增材制造为研究对象，通过构建整个 SLM 过程的多尺度数值仿真模型，分别模拟了 316L 粉末的铺粉过程及随后的打印过程。

系统分析了各操作参数对粉床及打印区结构及性能的影响规律，并揭示了其所对应的复杂物理机理。所得结果不仅为降低打印过程中的缺陷及改善产品质量打下坚实的理论基础，同时也为实际3D打印过程中工艺参数的选择及工艺优化提供有效参考。

关键词：316L不锈钢；选区激光熔化增材制造；多尺度数值仿真；铺粉及打印；结构表征及机理分析

激光增材制造高Co-Ni二次硬化超高强度钢的组织设计和强韧性平衡策略

冉先喆[1,2]

（1. 北京航空航天大学大型金属构件增材制造国家工程实验室，北京 100191；
2. 北京航空航天大学宁波创新技术研究院，浙江宁波 315800）

摘 要：在2000MPa级的超高强度钢中，高Co-Ni二次硬化超高强度钢具有最优异的强度和韧性匹配，并主要应用于航空航天和兵器工业先进装备主承力构件制造。由于增材制造过程中极端非平衡冶金移动小熔池和复杂热循环作用，激光增材制造高Co-Ni二次硬化超高强度钢常具有独特的非平衡凝固组织和循环固态相变行为。通过研究成形和后续热处理工艺相关组织性能，获得了合金钢工艺-组织-性能间的关联规律，揭示了不同尺度多层级组织强化机制和薄膜状逆转变奥氏体等多途径增韧机制等科学规律；提出了提高增材制造合金钢强韧性的组织设计方法，并通过后续多步骤热处理工艺获得了成分均匀、细小、等轴晶且略微过时效的回火马氏体组织，实现了目标强度范围内高强度和高韧性平衡。

关键词：激光增材制造；高Co-Ni二次硬化超高强度钢；组织设计；强韧性平衡

复杂氧化物颗粒强化ODS钢的增材制造成形及强韧化机制研究

刘 彬[1]，曹远奎[1]，张宇阳[1]，张瑞谦[2]，刘 咏[1]

（1. 中南大学粉末冶金研究院，湖南长沙 410083；2. 中国核动力研究院，四川成都 528208）

摘 要：纳米氧化物弥散强化(ODS)钢由于其特殊的抗辐照性、高强度和在极端温度和应力条件下的抗氧化/腐蚀性能，是第四代核反应堆燃料包壳的主要候选材料。增材制造可缩短材料制备周期，实现具有复杂结构部件的直接成形，是制造复杂结构ODS钢构件的理想方法。目前文献报道增材制造ODS钢主要存在纳米氧化物颗粒尺寸大，数密度低等缺点，导致力学性能远低于传统粉末冶金方法制备的ODS钢。本研究通过合金设计，在ODS钢中同时引入Zr/Ti活性元素，结合优化的机械合金化与SLM增材制造工艺，在SLM过程中形成了尺寸为5~20nm，分布密度为$3.8\times10^{22}/m^3$的Y-Zr-Ti-O复合纳米颗粒，同时还促进了等轴+柱状异质结构组织的形成，实现了增材制造ODS钢力学性能的显著提高，室温抗拉强度达到1126MPa，800℃下抗拉强度为180MPa，同时具有11%的拉伸塑性。

关键词：氧化物弥散强化；钢；辐照行为；力学性能

激光增材制造飞机起落架用超高强不锈钢缺陷、组织与性能研究

刘壮壮

(北京科技大学新材料技术研究院，北京　100083)

摘　要：新一代大飞机对起落架力学性能、耐腐蚀性能以及结构轻量化提出了更高要求。Custom465(C465)沉淀强化马氏体不锈钢具有优异的强度、韧性和良好的耐腐蚀性能，成为新一代飞机起落架用重要材料。前期研究发现，采用激光粉末床熔融方法成形 C465 钢时容易产生孔隙、裂纹等缺陷，难以满足起落架使用需求。为此，本文分析了 L-PBF 成形 C465 钢过程中裂纹形成机理，通过优化成分开发了适用于增材制造的无裂纹 C465 钢，并研究了激光扫描工艺参数对无裂纹 C465 钢孔隙率和力学性能的影响，制定了适合无裂纹 C465 钢的热处理工艺，使其力学性能超过了锻造 C465 钢的水平，满足飞机起落架对 C465 钢组织和性能的要求。

实现向铝合金中引入形变孪晶和 9R 相：合金化致层错能降低及塑性变形的晶粒尺寸效应

周登山[1,2]，张景藩[1]，庞学永[1]，张博文[1]，李　跃[3]，
孙彬涵[4]，VALIEV Ruslan Z.[5,6]，张德良[1,2]

(1. 东北大学材料科学与工程学院，辽宁沈阳　110819；2. 东北大学轧制技术连轧自动化国家
重点实验室，辽宁沈阳　110819；3. 马普钢铁研究所，德国杜塞尔多夫　40237；4. 华东理工
大学教育部压力系统与安全重点实验室，上海　200237；5. 乌法州航空技术大学，
俄罗斯乌法　450008；6. 圣彼得堡国立大学，俄罗斯圣彼得堡　198504)

摘　要：因面心立方结构纯铝具有极高的层错能值（166mJ/m^2），且可溶合金化元素对其层错能大小的影响甚微，依据传统认知，基本无法利用常规变形方法向粗晶纯铝中引入形变孪晶和 9R 相（一种由高密度层错呈周期排列的结构）。为能在 Al 合金材料中成功引进形变孪晶和 9R 相，我们利用机械合金化技术创制出纳米晶过饱和 AlMgY 固溶合金。显微表征证实了该合金同时含有形变孪晶和 9R 相。通过分析纳米晶 AlMg 合金、三维原子探针和第一性原理计算结果，我们发现 Y 合金化致层错能陡降是纳米晶 AlMgY 中形成形变孪晶的主要原因。另外，我们还发现 9R 相的形成涉及两种机制，即（1）非共格孪晶界的分解和（2）晶界发射肖克莱不全位错。本研究结果将为高性能铝合金及其复合材料的显微结构设计提供一种新思路、新视角。

关键词：机械合金化；纳米晶铝合金；层错能；形变孪晶；9R 相

增材制造可降解锌镁合金

李亚庚

（北京科技大学，材料科学与工程学院，北京 100083）

摘 要：选区激光熔化技术（SLM）在开发定制骨植入物方面表现出巨大的优势。然而目前 SLM 可降解锌合金的研究刚刚起步，需探究其工艺-组织-性能关系，并对其工艺参数与综合性能进行优化与调控。本文首先开展了锌镁合金的 SLM 工艺参数优化，之后对 SLM 锌镁合金的微观组织进行了表征，并开展了力学实验、体外降解实验及生物相容性实验。与 SLM 纯锌进行比较，研究了工艺参数及构建取向对 SLM 锌镁合金微观组织与综合性能的影响。

关键词：增材制造；锌镁合金；显微组织；综合性能

增材制造高品质 H13 模具钢的数值模拟、组织与力学性能研究

杨 鑫[1]，张兆洋[1,2]，徐圣航[1]，刘石球[1]，黄明浩[1]，丁 超[1]，汤慧萍[1]

（1. 先进材料增材制造创新研究中心，浙大城市学院，浙江杭州 310015；
2. 材料科学与工程学院，西安理工大学，陕西西安 710048）

摘 要：围绕高端模具钢对结构功能一体化的需求，开展增材制造高品质 H13 模具钢的数值模拟、组织与力学性能研究。建立 SEBM 成形 H13 钢热履历研究的数值模型，并通过实验验证；随着粉末再利用次数的增加，粉末氧和氮含量分别从 149 和 237ppm 增加到 250 和 394ppm；SEBM 成形的 H13 钢显微组织主要由贝/马氏体复相组织及少量分布在晶界处的残余奥氏体和少量碳化物组成。力学性能随沉积高度变化，顶部抗拉强度为(1338.1 ± 30.8)MPa，延伸率为 2.02%；底部抗拉强度为(1172.4 ± 44.3)MPa，延伸率为 2.60%。研究成果为增材制造技术在模具制造行业的应用提供必要的基础支撑。

关键词：H13 钢；增材制造；微观组织；力学性能

Numerical Simulation, Microstructure, and Mechanical Properties of H13 Steel Fabricated by Additive Manufacturing

YANG Xin[1], ZHANG Zhaoyang[1,2], XU Shenghang[1], LIU Shiqiu[1], HUANG Minghao[1], DING Chao[1], TANG Huiping[1]

(1. Advanced Materials Additive Manufacturing Innovation Research Center, Hangzhou City University, Hangzhou 310015, China; 2. College of Materials Science and Engineering, Xi'an University of Technology, Xi'an 710048, China)

Abstract: A study was conducted to investigate the numerical simulation, microstructure, and mechanical properties of H13 steel fabricated by additive manufacturing, in response to the demand for structural-functional integration in high-end mold steels. A numerical model was established to analyze the thermal history of H13 steel during selective electron beam melting (SEBM), and experimental validation was performed. With an increasing number of powder reuses, the oxygen and nitrogen content in the powder increased from 149ppm and 237ppm to 250ppm and 394ppm, respectively. The microstructure of the H13 steel fabricated via SEBM primarily consisted of a bainite/martensite duplex phase structure, with residual austenite and a small amount of carbides distributed along the grain boundaries. The mechanical properties varied with the deposition height. The top portion exhibited a tensile strength of (1338.1±30.8)MPa and an elongation of 2.02%, while the bottom portion had a tensile strength of (1172.4±44.3)MPa and an elongation of 2.60%. The research findings provide essential foundational support for the application of additive manufacturing in the mold manufacturing industry.

Key words: H13 steel; additive manufacturing; microstructure; mechanical property

冶金低热值煤气高效超临界发电技术及应用

熊敬超

(中冶南方都市环保工程技术股份有限公司，湖北武汉　430205)

摘　要： 本报告讨论了冶金低热值煤气高效超临界发电技术的开发及工程化应用。针对现有低热值煤气锅炉发电技术效率偏低的问题，通过模拟及试验等方法，对超临界煤气发电热力系统、关键装备、智能控制等关键技术等方面进行了研究，首创小型化超临界一次再热煤气发电技术，发电效率达到 44%~45%，相比上一代亚临界煤气发电技术效率提升 8%~10%。本技术已在在宝武鄂钢、鞍钢集团等众多大型钢铁企业推广应用，总装机容量近 2000MW，相比超高压/亚临界煤气锅炉发电技术，项目已装机机组年可节约标煤约 100 万吨，当量减少 CO_2 排放约 200 万吨，SO_2 排放约 1.5 万吨，NO_x 排放约 0.3 万吨。

关键词： 冶金；煤气；超临界

The Development and Application of High-efficiency Supercritical Power Generation Technology for Metallurgy Low Calorific Value Gas

XIONG Jingchao

(Wuhan City Environment Protection Engineering Co., Ltd., Wuhan 430205, China)

Abstract: This work discusses the development and application of high-efficiency supercritical power generation technology for metallurgy low calorific value gas. To improve the generating efficiency of low-calorific value gas boiler power generation technology, the thermal system, key equipment and intelligent control of supercritical gas power generation were studied by simulation and practical tests. As a result, the miniaturized supercritical primary reheat gas power generation technology is first promoted and its power generating efficiency reaches 44%~45%, which increases by 8%~10% compared with the previous generation of subcritical gas power generation technology. The high-efficiency supercritical gas power generation tsechnology has been applied in a number of large steel enterprises such as China Baowu Steel Group Coporation Limited and Angang Steel Group Limited, with a total installed capacity of nearly 2000MW. Compared with ultra-high pressure/subcritical gas boiler power generation technology, the installed supercritical gas power generation units can save about 1 million tons of standard coal per year, reduce CO_2 emissions by about 2million tons per year, SO_2 emissions by about 15000 tons per year, and NO_x emissions by about 3000 tons per year.

Key words: metallurgy; coal gas; supercritical

金属材料数字化研发平台架构与应用

苏 航，刘和平，侯雅青，杨 丽，孙 旭，贺笃鹏

（中国钢研科技集团有限公司数字化研发中心，北京 100081）

摘 要： 材料基因工程 MGE 的工业应用有赖于场景化的企业级、行业级服务平台。基于公网云计算、材料 APP、材料区块链、激光原位合金化等原创性技术，建立了金属材料数字化研发及数据资源共享平台，可实现材料设计、工艺模拟、服役仿真、分布式数据发现、高通量制备等专业级材料数字化研发场景。

通过图形/指令一体化材料多尺度云计算、"云堡垒"智能路由和多操作系统许可证调度技术等的突破，首次建成了面向金属材料的全图形公网云计算平台。支持 40 多种材料、工艺专用软件的远程、排队、移动计算，可以同时支持 40 个大型远程计算任务、500 个排队计算和数据查询任务，支持国产工业软件从许可到服务的模式转变。

建立了材料 APP 集群与云原生系统，支持多种材料 APP 的云端生成和计算，可共享、修改、移植、保护版权。平台已开发众筹 700 多个材料 APP 应用，场景包罗万象，覆盖各种计算尺度和工艺流程。

开发了基于区块链的科研检测数据共享管理系统，满足了材料数据确权管理及安全溯源需求，可有效实现不同团队、部门之间的数据共享共治，促进企业级、行业级科研检测数据矿山的形成，实现 FAIR 原则下的跨团队、跨部门数据可发现、可访问、可操作、可重用。

建立了行业首个块体样品的高通量增材制备系统，突破高速打印条件下元素粉末扩散、均质化难题，可实现一次 200 种不同成分材料宏观试样制备。

金属材料数字化研发平台已被成功应用于高压储氢罐用钢、汽车用 QP 钢、模具钢等材料设计和工艺优化中，大幅提升了产品研发效率，推动了数字化转型，并基于此建立了以悬赏征算、数字化人才认证为代表的行业级数字化研发生态。

关键词： 材料基因组；材料数字化研发；云计算；区块链；APP；高通量制备

AIM：实验室小样本数据与工业大数据的融合分析

王晨充，李虎威，徐 伟

（东北大学轧制技术及连轧自动化国家重点实验室，辽宁沈阳 110819）

摘 要： 随着大数据时代的来临，机器学习（ML）广泛应用于大数据分析，包括钢铁工业领域的性能预报[1-2]。目前对于工业数据的分析，已经积累了大量人工智能分析方法。然而受一种回归策略可扩展性的限制，所建立的工业数据分析系统难以获得多个不同类型钢种的通用性能预测模型，模型扩展能力的提升受到限制[3]。同时实验室小样本数据中形成的机理信息也难以对工业大数据分析提供指导。为了解决该问题，本研究将 ML 分类和回归模型与物理冶金（PM）变量相结合，开发出一种新型工业大数据分析系统。首先从工业生产线获取多个不同类型钢种的生产数据，通过数据预处理获得高质量的可用数据。然后基于 K-最近邻（KNN）算法对七个不同类型的钢种实现数据分类。并根据分类结果，在每个类别中选择不同的机器学习算法建立屈服强度预测模型，以最大限度地提高每个

类别的预测精度。同时考虑到 PM 变量在提高模型精度中的作用,通过引入强相关性的 PM 参数(AC1 温度、AC3 温度和流动应力)以指导 ML 过程的进一步优化。通过对不同回归策略的合理组合,目前的结果清楚地表明,基于数据分类建立的性能预测模型,其可扩展性得到了显著提高。在数据分类中,根据 12 维的化学成分参数,将包含七种不同类型钢种的数据集划分为五个具有不同特征的子数据集,从而降低原始数据库的复杂性。同时与直接使用原始数据集建立的预测模型相比,所提出的工业分析系统具有更高的预测精度和灵活性。对于数据充足且分类准确的钢种,预测误差能控制在较小范围内,预测有效率达到 90%以上。对于其他钢种,预测结果的准确率得到明显提高,避免了数据较少的钢种对整体预测结果造成的干扰。本研究为实验室原型材料向工业化产品的快速转化(AIM)提供了可行方向。

参考文献

[1] Dima A, Bhaskarla S, Becker C, et al. Informatics Infrastructure for the Materials Genome Initiative[J]. JOM-Journal of the Minerals, Metals and Materials Society, 2016, 68(8): 2053-2064.
[2] Olson G B. Designing a New Material World [J]. Science, 2000, 288(5468): 993-998.
[3] Datta S, Pettersson F, Ganguly S, et al. Designing High Strength Multi-phase Steel for Improved Strength-Ductility Balance Using Neural Networks and Multi-objective Genetic Algorithms[J]. ISIJ International, 2007, 47(8): 1195-1203.

冷轧带钢力学性能"数据钢卷"的实践

唐成龙

(宝钢中央研究院数智技术创新中心,上海 201999)

摘　要:目前,钢铁材料力学性能的获取主要依赖破坏实验,但这种做法存在时滞性、数据缺乏代表性、资源耗费大等弊端。随着钢铁行业智能化、自动化程度越来越高,对快速获取钢铁材料力学性能提出挑战。基于铁磁性材料磁畴运动的电磁无损检测技术可镶嵌在产线中,通过获取表征钢铁材料磁畴结构及运动的电磁信号,智能表征钢铁材料的力学性能。宝钢中央研究院数智技术创新中心,联合兰州大学、北京科技大学、南京航空航天大学三所高校,构建前言研究团队,开展电磁无损检测机理研究,致力于电磁无损检测中科学问题(外磁场激励下,电磁特征—磁畴动力学—特征微观组织结构内在物理关系,及建立基于电磁信号的钢铁材料特征微观组织结构、力学性能表征模型)和技术应用问题(以典型钢种为例,实现电磁无损检测技术表征钢铁材料特征微观组织结构、力学性能的应用)。目前,已在宝钢汽车板机组进行实践,智能高效地获取每卷冷轧带钢高密度、高精度力学性能,为实现"冷轧带钢性能数字钢卷"的构建提供可靠数据。

金属材料高通量增材制备技术研究

侯雅青[1],苏　航[1],应　华[2],张　浩[1],李发发[1],何亚洲[1],
宋有朋[1],王炫东[1],张荣宝[3]

(1. 中国钢研科技集团有限公司数字化研发中心,北京　100081;2. 杭州德迪智能科技有限公司,
浙江杭州　310000;3. 北京钢研新材科技有限公司,北京　100081)

摘　要：材料高通量实验领域重要发展方向之一是块体样品的高通量制备技术，以实现对宏观力学性能的直接表征。将金属3D打印技术、高通量制备技术两者有机融合应用于先进结构材料体系的制备开发，实现材料四面体数据的高通量获取，不仅可为材料技术自身的发展提供一项高效方法，也为增材制造技术提供了一个全新的应用领域，科学意义和应用价值巨大。本团队提出LPBF原位合金化技术，即通过将多种单质元素或母合金粉末按照预设的合金成分配比混合，在激光选区熔化成型的同时进行原位冶金，制备出所需成分的复杂结构零件。在此技术基础上研制了一套金属材料高通量制备试验装置，该设备具备4路送粉、动态量化混粉、4通道铺粉模块，可同时实现4种以上粉末、不少于200种材料力学性能样件的一次性制备，已应用于耐蚀合金，高温合金和高熵合金等增材制造专用材料的成分设计和工艺优化。

LPBF原位合金化技术的应用不仅为3D打印专用材料开发提供一种高通量解决方案，还可以应用于多材料打印，在合适的位置打印合适的材料，从而在3D打印结构设计的自由度基础上增加材料选择的自由度。然而与传统LPBF技术相比，该技术工艺的设置不仅要实现样品致密化，还应实现成分的均匀化，更为严苛的工艺窗口限制了该技术的应用。本文基于离散元、有限元和CALPHAD方法设计并构建了一套微观尺度的集成计算模型，以FeCoCrNi中熵合金为材料研究载体，探索异质混合粉末床与激光交互作用过程发生的高速循环熔化过程与微区合金化机理，通过定义粉末床填充率、成分偏差和成分均匀性等参量，分析获得激光原位合金化成型FeCoCrNi系中熵合金的均匀化致密化工艺制定准则。通过与Fe、Co、Cr、Ni元素粉末的原位合金化实验结果对比，验证了该集成计算模型的准确性。该模型的构建基于一套通用的计算方法，可为后续其它金属材料的LPBF原位合金化工艺适用性研究提供理论基础。

关键词：激光原位合金化；增材制造；高通量制备；中熵合金；集成计算；数字化研发

参 考 文 献

[1] Hou Yaqing, Su Hang, et al. An integrated simulation model towards laser powder bed fusion in-situ alloying technology[J]. Materials and Design, 2023.
[2] Hou Yaqing, Su Hang, et al. Fabricating Homogeneous FeCoCrNi High-Entropy Alloys via SLM In Situ Alloying[J]. Metals, 2021.
[3] 侯雅青, 苏航, 等. 金属材料多尺度高通量制备研究进展[J]. 材料导报, 2022.
[4] 侯雅青, 苏航, 等. 一种基于异质粉末SLM打印工艺的金属材料高通量制备方法[P]. 中国: ZL201910562259.8.
[5] 侯雅青, 苏航, 等. 一种高熵合金的高通量增材制备方法[P]. 中国: ZL202111458576.9.

长输管道环焊缝超声相控阵缺陷数据分析技术研究

李学达，韩　彬，牛盛源，肖文涛，高振华

（中国石油大学（华东）材料科学与工程学院，山东青岛　266580）

摘　要：长输管道环焊缝质量的检测多采用无损检测的方式实现[1]，其中超声相控阵无损检测（Phased Array Ultrasonic Testing，PAUT）以其检出率高、精度高成为了环焊缝主要的无损检测方式。深度学习由于高效、智能的优势常与PAUT结合进行缺陷识别[2]。但由于缺陷类型复杂、类型不均衡的小样本性导致了环焊缝PAUT缺陷评判尺度不一、分类不精确等问题，制约了环焊缝PAUT缺陷智能识别的发展[3]。

针对上述问题，首先，使用数学形态学提取环焊缝PAUT缺陷图像的视觉、几何以及语义三种特征。随后，根据数学特征学，将原有的PAUT小样本缺陷更新为PAUT广义小样本缺陷，并将其细化为仅视觉特征缺失的少样本缺陷，视觉、几何特征缺失的小样本缺陷，视觉、几何、语义特征均缺失的零样本缺陷[4]。在此基础上，提出针对不同广义小样本缺陷的深度学习方法。针对少样本缺陷，提出了搭载空洞卷积神经的元学习方法以改善其收敛性较差的问题；针对小样本缺陷，提出了搭载深度最邻近网络的度量学习方法以改善其依赖人工标定的问题；针对零样本缺陷，提出搭载语义反馈模块的属性学习方法以改善其域偏移问题。总体而言，本文构建了以元学习为主体框架，内嵌针对度量学习以及属性学习两个模块的广义小样本缺陷识别方法，将 PAUT 缺陷图像识别的回召率（Recall）及交并（Intersection over Union, IoU）比提升至 91%，识别准确率（Precision）提升至 90.7%。该方法细化了 PAUT 广义小样本缺陷分类，构建了 PAUT 广义小样本缺陷统一识别方法，提高了 PAUT 广义小样本缺陷识别率，推动了环焊缝 PAUT 缺陷智能识别的发展。

参 考 文 献

[1] GAUTHIER B, PAINCHAUD-APRIL G, LE DUFF A, et al. Ultrasonic multi-view data merging using the vector coherence factor[J]. NDT & E International, 2023, 136: 112.
[2] DATTA D, HOSSEINZADEH A Z, CUI R, et al. Railroad Sleeper Condition Monitoring Using Non-Contact in Motion Ultrasonic Ranging and Machine Learning-Based Image Processing[J]. Sensors (Basel), 2023, 23(6): 22-48.
[3] DING L, WAN H P, LU Q B, et al. Using deep learning to identify the depth of metal surface defects with narrowband SAW signals[J]. Optics and Laser Technology, 2023, 157: 1129.
[4] LAN Y C, NATARAJAN J, YANG C H. Study on Porosity and Material Parameters of Hot Moulded Resin Matrix Composites using Nondestructive Laser Ultrasound Technique[J]. Experimental Techniques, 2023, 12: 25-69.

基于机器学习的轴承钢碳化物自动识别与评价

张莎莎，朱晓林，于帅鹏，束长青，姚正军

（南京航空航天大学材料科学与技术学院，江苏南京 211106）

摘　要：轴承钢碳化物的检测与控制对轴承钢的质量评估具有重要意义。目前，轴承钢碳化物带状的严重程度主要由专业检测人员按照标准图谱进行比较法评级来评价。这种方法不仅检测效率低、且评级结果容易受检验人员的主观因素影响。基于此，我们设计了轴承钢碳化物的自动识别与评价体系，这是一种基于机器学习的语义分割和多类别评级模型，其包括碳化物识别和评价模块，对于第一个模块，采用基于卷积神经网络的 DeepLabv3+ 和 U-Net++ 语义分割模型对轴承钢光学显微图像进行训练识别，训练后的模型能够自动识别光学显微图像中的带状碳化物，并利用 OTSU 算法对碳化物颗粒进行阈值分割分析其带宽、碳化物尺寸、碳化物的分布，从而获得分类和判断各种级别轴承钢碳化物的特征信息。对于第二个模块，首先基于轴承钢碳化物的特征信息建立数据集，然后通过比较多种机器学习模型，获得了用于轴承钢碳化物级别评价的最佳机器学习模型。结果表明，轴承钢碳化物自动识别模块碳化物带状平均像素精度可达 97%，评价模块碳化物评级准确率可达 92%（以行业内±0.5 级算为准确的话，准确率达 95% 以上）。相较于传统人工比较法评级，该评价体系在评级效率与准确性方面更有优势，为基于机器学习的钢铁材料复杂组织高效精准识别和评价提供了理论指导和示范。

关键词：轴承钢；碳化物；机器学习；自动识别

集成计算与金属结构材料设计

鲁晓刚

（上海大学材料学院，上海 200444）

摘　要： 自 20 世纪中叶开始，计算热力学（即 CALPHAD 方法）从计算不同成分、温度、压力等条件下材料的热力学性质和相图为核心，逐步向相变动力学领域扩展，建立起集成了计算热力学理论的以扩散相变模拟和形核析出模拟为特色的计算动力学方法，并与相场模拟日渐紧密地结合。同时，依据实验数据、第一性原理计算、经验半经验理论，运用 CALPHAD 优化评估技术，建立适用于多元多相材料的热力学、动力学、热物理性质等集成式材料设计基础参数数据库。本报告将介绍本团队十多年深入研究形成的金属结构材料（包括钢铁、Ni 基高温合金等）集成计算方法，即以计算热力学为桥梁，将电子、原子和分子等微观尺度和宏观层次的计算模拟联系起来，进一步结合大数据解析和机器学习等人工智能方法，构建跨尺度集成计算+数据平台，建立多元多相金属材料的成分/工艺-组织结构-性能之间的定量关系。从而模拟金属结构材料从设计、生产制备到服役的全流程，高效预测其组织结构演变和最终性能，实现材料组织和性能的精确调控。

关键词： CALPHAD；集成计算；金属结构材料

鞍钢钢铁产品数字化研发实践与展望

林　利，王军生

（鞍钢集团钢铁研究院，辽宁鞍山　114009）

摘　要： "材料基因组计划"通过跨尺度高通量计算结合数据库和人工智能技术的快速推动新材料研发，实现研发周期缩短一半、研发成本降低一半的战略目标。数字化研发是一种手段、工具、平台，利用这一手段实现产品（新产品、现有产品）、工艺、应用服役评价的探索，最终实现成本降低，效率及产品价值提升。目前鞍钢确认了利用数字化研发实现产品开发探索、工艺数字孪生、应用及服役数字化评价的数字化研发目标，并在趋势探索、产品开发、工艺研究、应用性能、服役性能、方案制定、检验表征等维度进行了深入的探索和实践，以达到数字化研发实现产业价值的深度挖掘的目的。

关键词： 材料；数字化

Anshan Iron and Steel Products Digital R&D Practice and Prospects

LIN Li, WANG Junsheng

(Iron and Steel Research Institute of Ansteel Group, Anshan 114009, China)

Abstract: The "Materials Genome Project" quickly promotes the research and development of new materials through cross-scale high-throughput computing combined with database and artificial intelligence technology, and realizes the strategic goal of shortening the research and development cycle and reducing research and development costs by half. Digital research and development is a means, tool and platform, use this method to realize the exploration of products (new products, existing products), processes, application and service evaluation, and finally realize cost reduction, efficiency and product value improvement. At present, Ansteel has confirmed the use of digital R&D to realize product development exploration and process digital twins, application and service digital evaluation of digital R & D goals, and in-depth exploration and practice in the dimensions of trend exploration, product development, process research, application performance, service performance, program formulation, inspection and characterization, in order to achieve digital R & D to realize industrial value for the purpose of deep excavation.

Key words: materials; digitization

钢铁材料基因工程数据库与大数据应用

姜雪[1,3]，张雷[2,3]，何杰[3,4]，宿彦京[2,3]

（1. 北京科技大学钢铁共性技术协同创新中心，北京 100083；2. 北京科技大学新材料技术研究院，北京 100083；3. 北京材料基因工程高精尖创新中心，北京 100083；4. 北京科技大学计算机与通信工程学院，北京 100083）

摘 要：材料基因工程是统筹考虑新材料研发-生产-应用全过程的高维影响因素、综合利用高效计算设计、自主/智能实验、大数据和人工智能等技术，提升新材料研发速率，降低研发成本，提高工程化应用水平的前沿共性技术。近年来，随着材料基因工程研究与应用的深入，大数据和人工智能已经成为材料基因工程的核心技术，数据驱动的新材料研发快速发展成为新的研发范式，极大地提升了前沿新材料和颠覆性新材料的研发效率。

材料基因工程数据库以新材料的设计研发为目标，面向新材料发现、设计筛选、工艺优化到服役评价等全流程数据资源，提供采集、存储、发现、访问、集成和分析服务，支撑新材料的理性设计和人工智能技术的应用。材料基因工程数据库具有支持材料计算设计、实验研究、生产应用和科技文献等多源异构数据的采集、定义、存储、融合、查询和互操作的功能，充分考虑以用户为中心的数据模式自主定义和灵活扩展能力，以满足复杂异构材料数据的托管需求，具有兼顾面向人工智能应用的数据融合和服务集成能力，通过数据库与人工智能算法的无缝衔接和交互，实现在线分析与新材料的智能研发。

本文围绕钢铁材料，提出了基于自然语言处理的钢铁材料科技文献数据挖掘技术，实现钢铁牌号、成分、工艺路线与力学性能的高效自动抽取，形成面向人工智能应用的高质量钢铁数据库；研发出基于无模式存储的材料动态容器数据库技术，实现钢铁材料计算设计、实验研究、生产应用和科技文献等多源异构数据复杂异构数据个性化描述、便捷快速采集、标准化存储；构建基于云技术、大数据和机器学习的钢铁材料数字化协作平台，实现数据在线融合、算法在线开发、模型在线交付、软件在线集成、多方协作开发的数据-模型一体化平台，并在钢铁材料研发和生产中应用。

参考文献

[1] Wang W, Jiang X, Tian S, et al. Automated pipeline for superalloy data by text mining[J]. NPJ Computational Materials, 2022, 8(1): 9.
[2] Yan R, Jiang X, Wang W, et al. Materials information extraction via automatically generated corpus[J]. Scientific Data, 2022, 9(1): 401.
[3] Wang W, Jiang X, Tian S, et al. Alloy synthesis and processing by semi-supervised text mining[J]. NPJ Computational Materials, 2023.

新能源汽车对汽车用钢发展的机遇与挑战

王利[1,2]

（1. 宝钢中央研究院，上海 201900；
2. 汽车用钢开发与应用技术国家重点实验室，上海 201900）

摘 要：以电动汽车为主的新能源汽车已成为汽车工业的发展方向，在中国这一趋势更加明显，渗透率已近30%，电动汽车也是汽车工业应对双碳战略的具体技术路线，由于驱动能源的变化，带来了汽车用材料的显著变化，同时相关竞争材料的发展也对汽车钢的发展带来了挑战，如铝合金一体压铸技术等。汽车用钢一直是钢铁领域技术发展最活跃的领域，先进高强度钢仍然是发展方向，本文重点介绍了汽车的这种发展趋势对汽车用钢发展带来的机遇和挑战，并探讨了相应的应对措施。

关键词：新能源汽车；汽车用钢；低碳排放钢；先进高强度钢

Opportunities and Challenges of New Energy Vehicles for the Development of Automotive Steels

WANG Li[1,2]

(1. Research Institute, Baosteel Iron & Steel Co., Ltd., Shanghai, 201900, China; 2. State Key Laboratory of Development and Application Technology of Automotive Steels, Shanghai 201900, China)

Abstract: New energy vehicles, mainly electric vehicles, have become the development direction of the automotive industry. In China, this trend is more obvious, with a penetration rate of nearly 30%. Electric vehicles are also a specific technical route for the automotive industry to respond to the dual carbon strategy. Due to changes in driving energy, significant changes have been made in automotive materials, and the development of related competitive materials has also brought challenges to the development of automotive steel, such as aluminum alloy integrated die-casting technology. Automotive steel has always been the most active field of technological development in the steel industry, and advanced high strength steels (AHSS) are still the development direction. This article focuses on the opportunities and challenges that this development trend of automobiles brings to the development of automotive steel, and explores corresponding countermeasures.

Key words: new energy vehicles; automotive steels; low carbon emission steels; AHSS

高性能汽车钢材料多层级复杂组织调控

金学军[1,2]，李 伟[1,2]，许元涛[1,2]，刘仕龙[1,2]

（1. 上海交通大学材料科学与工程学院，上海 200240；
2. 上海市激光制造与上海市激光制造与材料改性重点实验室，上海 200240）

摘 要：第三代先进高强钢（AHSS），由于兼顾成本效益和重量轻的优点，已成为汽车车身的重要候选材料。对于先进高强钢，难以在屈服强度提高的同时仍保持塑性不降低，"强塑性倒置"一直都是一个无法回避的问题。在追求高强塑性能时，引入的晶体缺陷（位错、晶界、相界等）易诱导氢富集，氢脆敏感性或氢致延迟断裂敏感性也会随强度升高而升高，即"强度与氢脆敏感性"矛盾。"强塑性倒置"与"氢脆"问题已成为制约高强塑钢铁材料广泛应用的一个潜在威胁。在 3.3Ni 低碳中锰钢中，设计了一种新的析出诱导逆相变配分（Tempering-Partitioning, TP）工艺，利用晶内富 Mn fcc M12C 亚稳碳化物相转变诱导形成梯度 Mn 成分、弥散分布和细小尺寸的逆变奥氏体，这种梯度 Mn 成分奥氏体遗传了析出相特性，能够同时兼顾"硬"的颗粒属性与"软"的转变相属性，实现 Orowan 强化与多层级 TRIP 效应的耦合，在高应变速率条件下获得优异的力学性能（屈服强度：1350MPa，总伸长率：30%），成功解决了中锰高强钢中存在的"强塑性倒置"与高应变速率条件下加工硬化能力严重弱化的难题；同时，由于奥氏体具有优异的储氢能力，铁素体晶内弥散分布的纳米奥氏体作为强的氢陷阱位置，可以显著降低位错、晶界、相界等缺陷位置的氢浓度，抑制氢致脆性断裂。利用亚稳碳化物相转变诱导形成了遗传析出相特性的多功能纳米奥氏体（多种强塑化机制、氢陷阱），开发出高强塑、耐高速碰撞、抗氢脆的汽车钢。

关键词：高强钢；强塑性；氢脆；亚稳碳化物；相转变；多功能纳米奥氏体

Design of Hierarchical Complex Structure for High-performance Automotive Steel Materials

JIN Xuejun[1,2], LI Wei[1,2], XU Yuantao[1,2], LIU Shilong[1,2]

(1. School of Materials Science and Engineering, Shanghai Jiaotong University, Shanghai 200240, China;
2. Shanghai Key Laboratory of Materials Laser Processing and Modification,
Shanghai Jiaotong University, Shanghai 200240, China)

Abstract: The third generation advanced high-strength steel (AHSS) has become an important candidate material for automotive bodies due to its advantages of low cost and lightweight. For advanced high-strength steel, it is difficult to maintain plasticity while increasing yield strength, and "strength-plasticity trade-off" has always been an unavoidable problem. When pursuing high strength-plasticity properties, the introduced crystal defects (dislocations, grain boundaries, phase boundaries, etc.) are prone to induce hydrogen enrichment, and the sensitivity to hydrogen embrittlement or hydrogen induced delayed fracture also increases with the increase of strength, which is the contradiction between "strength and hydrogen embrittlement sensitivity". The issues of " strength-plasticity trade-off " and "hydrogen embrittlement" have become a potential threat that restricts the widespread application of high strength-plasticity steel materials. A new precipitation induced reverse phase transformation (TP) process was designed in 3.3Ni low-carbon medium manganese steel, which utilizes the phase transformation of Mn rich fcc $M_{12}C$ metastable carbides in the grain to induce the formation of gradient Mn composition, dispersion distribution, and fine size reversed austenite. This gradient Mn composition austenite inherits the characteristics of precipitates, which can simultaneously combine the "hard" particle properties and the "soft" transformation phase properties, by coupling Orowan strengthening with multi-layer TRIP effect, excellent mechanical properties (yield strength: 1350MPa, total elongation: 30%) were achieved under high strain rate conditions, successfully solving the problem of " strength-plasticity trade-off " and severe weakening of work hardening ability under high strain rate conditions in medium manganese high-strength steel; At the same time, due to the excellent hydrogen storage capacity of austenite, the dispersed nano austenite within the ferrite grain serves as a strong hydrogen trapping sites, which can significantly reduce the hydrogen concentration at defect positions such as dislocations, grain boundaries, and phase boundaries, and inhibit hydrogen induced brittle fracture. Utilizing the transformation of metastable carbides to induce the formation of multifunctional nano austenite with genetic precipitation phase characteristics (multiple strength-plasticity mechanisms, hydrogen traps), automotive steel with high strength-plasticity, resistance to high-speed collision, and hydrogen embrittlement has been developed.

Key words: high strength steel; strength-plasticity; hydrogen embrittlement; metastable carbides; phase transformation; multifunctional nano austenite

增强成形性高强汽车钢的组织性能研究

赵征志，褚晓红，孙 航，杨玉环，唐 荻

（北京科技大学钢铁共性技术协同创新中心，北京 100083）

摘　要： 随着汽车轻量化技术发展的需求，汽车结构设计的日益复杂，对材料的成形性能要求越来越高，即期望材料拉延、扩孔、翻边、弯曲等成形安全裕度越来越大，以满足一些复杂零件的成形要求和提高零件的成品率。因此高强、高成形性汽车用钢的开发也就成为钢铁研究人员的重要任务，也是未来超高强汽车用钢的发展趋势。

通过引入大量的残余奥氏体发挥 TRIP(transformation induced plasticity, TRIP)效应，采用多相、亚稳和多尺度组织提高强塑积，比如 TRIP 钢、Q-P(quenching and partitioning, Q-P)钢和中锰钢，具有极高的强塑积优势，但局部成形性较低，不利于复杂的零件设计，故精准调控残余奥氏体的含量及其稳定性会对提高成形性具有重要意义。继而，增强成形性高强钢被提出，包括 DH 钢（dual phase steels with improved formability, DH 钢）和 CH 钢（complex phase steels with improved formability, CH 钢）。相较于传统的 DP(dual phase, DP)钢和 CP(complex phase steel, CP)钢，DH 和 CH 钢组织均匀细小，并引入少量残余奥氏体，成形性能显著提高。DH 钢和 CH 钢的化学成分在 DP 和 CP 钢的基础上略有调整，提高 Si+Al 含量增加残余奥氏体含量，添加 Nb、Ti 等微合金元素引入第二相发挥析出强化和细晶强化的作用，不会因为化学成分的变化而对其生产工艺进行大幅度的调整，进而提高产品的制造成本。可见，增强成形性高强钢具有较大的研究价值和广阔的应用市场。

通过优化成分、合理调节退火温度和时效温度得到了晶粒细小、含有少量残余奥氏体、性能优异的增强成形性高强钢，并研究了退火工艺对组织性能的影响和残余奥氏体在增强成形性高强钢变形过程中发挥的增强增塑作用。试验钢显微组织为铁素体、贝氏体、马氏体和残余奥氏体，并析出了大量尺寸小于30nm、呈弥散分布的（Nb, Ti）C。研究结果表明，当时效温度为 400℃时，随着退火温度从 860℃升至 920℃，块状马氏体含量减少、尺寸减小，贝氏体、残余奥氏体含量增加，屈服强度和抗拉强度呈降低趋势，延伸率呈先升高再降低趋势；当退火温度为860℃时，随着过时效温度从 260℃升至 460℃，贝氏体和残余奥氏体的含量先增大后减少，马氏体含量则呈相反趋势，导致屈服强度、抗拉强度先降低后升高，延伸率先升高再降低。在增强成形性高强钢中，大部分的残余奥氏体呈块状位于晶界处，少部分的残余奥氏体呈薄膜状分布在贝氏体/马氏体板条间，残余奥氏体含量在 3%~9%之间，尺寸在 0.1~1μm 之间，变形过程中起到了有效的 TRIP 效应。在一定范围内，残余奥氏体含量及其含碳量越高，对塑性的提升作用越大。

关键词： 增强成形性高强钢；残余奥氏体；退火温度；过时效温度；组织性能

Study on Microstructure and Properties of High Strength Automotive Steel with Improved Formability

ZHAO Zhengzhi, CHU Xiaohong, SUN Hang, YANG Yuhuan, TANG Di

(Collaborative Innovation Center of Steel Technology,
University of Science and Technology Beijing, Beijing 100083, China)

Abstract: With the development of automobile lightweight technology, the structure design of automobiles is becoming more and more complex, and the requirements for the formability of materials are getting higher and higher. That is, it is expected that the forming safety margins of drawing, hole expansion, flanging, and bending become larger. To meet the formability requirements of some complex parts and improve the yield of parts, the development of high-strength and high-formability automotive steel has become an important task for steel researchers, and it is also the development trend of ultra-high-strength automotive steel in the future.

By introducing a large amount of retained austenite to exert the TRIP (transformation induced plasticity, TRIP) effect, using multiphase, metastable, and multiscale structures to improve the product of strength and plasticity, such as TRIP steel, Q-P (quenching and partitioning, Q-P) steel and medium manganese steel. It has the advantage of extremely high product of strength and plasticity, but its local formability is lower, which is not conducive to the design of complex parts. Therefore, it is of great significance to accurately control the content and stability of retained austenite to improve formability. Then, high strength steel with improved formability was proposed, including DH steel (dual phase steels with improved formability,

DH steel) and CH steel (complex phase steels with improved formability, CH steels). Compared with the traditional DP (dual phase, DP) steel and CP (complex phase steel, CP) steel, the DH and CH steels have uniform and finer microstructures and a small amount of retained austenite, so the formability is significantly improved. The chemical composition of DH steel and CH steel is slightly adjusted on the basis of DP and CP steel. Increasing the content of Si+Al to increase the content of retained austenite, adding microalloying elements such as Nb and Ti to introduce the second phase to exert precipitation strengthening and grain strengthening, which will not make substantial adjustments to the production process due to changes in the chemical composition, thereby increasing the manufacturing cost of the product. It can be seen that high strength steel with improved formability has great research value and broad application market.

By optimizing the composition, adjusting the annealing temperature and over-aging temperature reasonably, high strength steel with improved formability with fine grains, a small amount of retained austenite, and excellent performance was obtained. At the same time, the strengthening and plasticization mechanism of retained austenite during the deformation were analyzed. The microstructure of the test steel is ferrite, bainite, martensite, and retained austenite, and a large amount of (Nb, Ti)C with a size of less than 30 nm and dispersed distribution was precipitated. The research results show that when the over-aging temperature was 400℃, as the annealing temperature increased from 860℃ to 920℃, the content of massive martensite decreased, the size decreased, the content of bainite and retained austenite increased, and the yield strength and tensile strength showed a decreasing trend, and the elongation first increased and then decreased; when the annealing temperature was 860℃, as the over-aging temperature increased from 260℃ to 460℃, the content of bainite and retained austenite increased first and then decreased, while the content of martensite showed the opposite trend, resulting in the first decrease and then increase of yield strength and tensile strength, and the first increase and then decrease of elongation. In high strength steel with improved formability, most of the retained austenite is located in the grain boundary in blocky, and a small part of the retained austenite is distributed between bainite/martensite laths in filmy. The retained austenite content is 3%～9%, and its size is 0.1~1μm. It has an effective TRIP effect during the deformation process. Within a certain range, the higher content and carbon content of the retained austenite, the greater the plasticization effect.

Key words: high strength steel with improved formability; retained austenite; annealing temperature; over-aging temperature; microstructure and properties

高成形性汽车用超高强钢开发及应用

刘华赛[1,2]，韩赟[1,2]，阳锋[1,2]

（1. 首钢集团有限公司技术研究院，北京 100043；
2. 绿色可循环钢铁流程北京市重点实验室，北京 100043）

摘 要：在国家双碳政策的要求下，汽车制造业节能减排不断推进，对于轻量化汽车用钢的要求不断提高。从钢厂的角度来讲主要需要满足三个平衡：高强度和成型性能的平衡，整体轻量化和安全性的平衡以及高性能和低成本的平衡。目前，主流车型的高强度钢占车身材料的比例已普遍达到60%以上，其中780MPa及以上超高强钢最高比例接近30%，先进高强钢尤其是以相变强化为主的高成形性超高强钢在汽车制造业的应用也愈加被重视。目前以淬火配分钢（QP）钢、具有相变诱导塑性的贝氏体铁素体钢（TBF）钢、增强成形性双相钢（DH）钢为代表的高成形性超高强钢已经在各大主机厂得到商业应用，也成为国内外上各大钢铁供应商竞相开发的材料，本文将针对上述三种已经开发应用的材料开展讨论。

2003年，美国Speer教授根据TRIP效应机理结合钢中碳扩散行为，提出了Q&P热处理工艺。该工艺通过控制碳元素在马氏体和奥氏体之间的扩散，得到了一种具有TRIP效应、高强度与高塑性配合的马氏体钢。QP钢从问世以来就收到广泛关注，各大高校和科研院所开展了广泛的研究工作，取得了丰硕的成果，并且在2012年宝钢开发的 QP 钢得到了商业化应用，目前 QP 钢已经在包括 B 柱、门槛、前纵梁在内的多个零件上得到广泛的应用。

TBF 钢以贝氏体或贝氏体型铁素体作为基体，引入残余奥氏体，同时兼顾高强度和高延伸率。2013 年，雷诺日产集团首次在英菲尼迪 Q50 上使用了神户制钢的 TBF 钢，用于制作 A 柱和 B 柱加强件和上边梁。目前许多原本需用热成形实现的零件，如 B 柱下部，A 柱下部，前纵梁，横梁等零部件等均可以通过 TBF 钢冷成形实现，在相同的强度级别可以降低生产成本以及碳排放，无论在合资品牌还是自主品牌上均具有良好的应用前景。

DH 钢是在传统的 DP 钢的基础上发展起来的，主要是在铁素体和马氏体的组织中引入一定量的残余奥氏体，利用残余奥氏体的 TRIP 效应来改善材料成形能力。2014 年德国汽车工业协会提出了这种材料的使用需求，随后国内外各家钢厂相继开发出了 DH 钢系列产品，例如奥钢联已实现 590-1180MPa 级别 DH 钢的全覆盖，蒂森克虏伯、SSAB 和安赛乐米塔尔等均已开发出 590-980MPa 级别 DH 钢。首钢在国内率先开展了 DH 钢的研发工作，是国内第一家实现 DH 钢产业化应用的钢企，强度级别覆盖 590~1180MPa 级别，产品的断后伸长率较同强度级别双相钢提高 35%以上，目前已批量用于国内十多家主机厂。

关键词：高成形性；超高强钢；开发；应用

Recent Progress on Development and Application of Ultra-high Strength Steel with High Formability

LIU Huasai[1,2], HAN Yun[1,2], YANG Feng[1,2]

(1. Research Institute of Technology, Shougang Group Co., Ltd., Beijing 100043, China; 2. Beijing Key Laboratory of Green Recyclable Process for Iron & Steel Production Technology, Beijing 100043, China)

Abstract: Under the requirements of the national dual carbon policy, energy conservation and emission reduction in the automotive industry are continuously promoted, and the requirements for lightweight automotive steel are constantly increasing. From the perspective of steel suppliers, there are three main balances that need to be achieved: the balance between high strength and high formability, the balance between overall lightweight and vehicle safety, and the balance between high performance and low cost. At present, the proportion of high-strength steel in vehicle has generally reached over 60%, and ultra-high strength steel reached 30%. The application of advanced high-strength steel, especially high formability ultra-high strength steel is also increasingly valued in the automotive industry. At present, high formability ultra-high strength steel, represented by TRIP aided bainitic ferrite steel (TBF), quenching and partitioning steel (QP) steel, and dual phase steel with high formability (DH) steel, has been commercially applied in automobile industry and has become a important material developed by major steel suppliers both domestically and internationally.

In 2003, Professor Speer from the United States proposed the Q&P heat treatment process based on the TRIP effect and carbon diffusion behavior in steel. This process obtains a TRIP aided martensitic steel with high strength and high plasticity by controlling the diffusion of carbon between martensite and austenite. The QP steel developed by Baosteel in 2012 has been commercialized for application. Currently, QP steel has been widely used in multiple parts, including B-pillars, door sills, and front longitudinal beams.

TBF steel uses bainite or bainitic ferrite as the matrix, with introducing residual austenite, while also considering high strength and high elongation. In 2013, Renault Nissan Group first used TBF made by Kobe on the Infiniti Q50 for the production of A-pillar and B-pillar reinforcements and upper edge beams. At present, many parts that originally required hot forming, such as the lower part of the B-pillar, lower part of the A-pillar, front longitudinal beam, crossbeam, etc., can be achieved through cold formed TBF steel, which can reduce costs and carbon emissions at the same strength level. It has good application prospects in both joint venture brands and domestic brands.

DH steel is developed on the basis of traditional DP steel, mainly by introducing a certain amount of residual austenite into the microstructure of ferrite and martensite, and utilizing the TRIP effect of residual austenite to improve the material formability. In 2014, the Verband der Automobilindustrie (VDA) proposed a demand for use, and then steel suppliers developed a series of DH steel products. For example, Voestalpine has achieved full coverage of 590-1180MPa grade DH steel, while ThyssenKrupp, SSAB, and ArcelorMittal have developed 590-980 MPa grade DH steel. Shougang is the first steel enterprise in China to carry out the research and development of DH steel, and is the first to achieve the industrial application of DH steel in China. The strength level covers the range of 590-1180MPa, and the elongation after fracture of the product is more than 35% higher than that of dual phase steel at the same strength level. Currently, it has been used in batches in more than ten main OEMs in China.

Key words: high formability; ultra-high strength steel; development; application

热轧和冷轧中锰汽车钢的超塑性变形行为与机制研究

曹文全[1]，王　昌[1]，满廷慧[2]，王存宇[1]，董　翰[2]，翁宇庆[1]

(1. 钢铁研究总院有限公司，北京　100081；2. 上海大学材料学院，上海　200444)

摘　要： 针对热轧和冷轧中锰汽车钢高温超塑性变形能力探索，通过高温拉伸试验和微观组织结构分析，开展了原始组织和超塑性变形工艺对中锰汽车钢超塑性变形行为的研究，发现在近 750℃拉伸温度及 0.001/s 应变速率下，热轧中锰钢具有≤400%的超塑性变形能力，冷轧中锰钢具有≥1000%的超塑性变形能力。通过对超塑性变形过程中微观组织结构演化研究，发现了热轧中锰钢中马氏体板条等轴化、冷轧中锰钢超细化以及晶粒尺寸随形变量和形变温度提升而粗化的现象。中锰汽车钢的超塑性主要归因于奥氏体逆相变和组织等轴化与超细化，有望为新能源汽车车体整体成型提供颠覆性材料和成型技术。

关键词： 超塑性材料；中锰汽车钢；微观组织结构；奥氏体逆相变；新能源汽车；颠覆性成型技术

Superplasticity Behavior and Mechanism of the Hot/Cold Rolled Medium-Mn Autosteel

CAO Wenquan[1], WANG Chang[1], MAN Tinghui[2],
WANG Cunyu[1], DONG Han[2], WENG Yuqing[1]

(1. Iron & Steel Research Institute, Beijing 100081, China;
2. Shanghai University, Shanghai 200444, China)

Abstract: The high temperature deformation behaviors of both hot rolled medium-Mn auto-steel and the cold rolled medium-Mn auto-steel were examined by tensile test and their deformation microstructure were examined through the TEM and EBSD. It is found that a maximum tensile elongation is about 400% for of the hot rolled auto-steel, but 1000% for the cold rolled auto-steel at deformation temperature of 750℃ and strain rate of 0.001/s, indicating a capacity of superplasticity for both steels Based on the microstructure characterization after superplastic deformation, it is found that the microstructure evolves from martensitic lath to equiaxed grains for hot rolled steel, while from severe deformed matrix to equiaxed grains for the cold rolled steel. The grain size of the medium-Mn steel increased with the increasing of the deformation at high temperature. At last, the superplasticity of the medium-Mn steel is attributed to the austenite reverted transformation and boundary sliding at high temperature, which would provide a disruptive technology for the car body production of the new energy vehicles.

Key words: superplastic materials; medium-mn autosteel; microstructure; austenite reverted transformation; new energy vehicle; disruptive forming technology

先进高强汽车钢板的集成计算材料工程研究

何燕霖[1]，郑伟森[1]，张　宇[2]，林　利[3]，徐　鑫[3]，刘仁东[3]，李　麟[1]

（1. 上海大学材料科学与工程学院，上海　200444；2. 鞍钢北京研究院，北京　102211；
3. 鞍钢股份有限公司，辽宁鞍山　114021）

摘　要：针对先进高强汽车钢板开展了集成计算材料工程（ICME）研究，在把握典型钢种化学成分、热处理工艺、显微组织、力学性能相互影响机制的基础上，优选了影响实验钢力学性能的关键特征参数，采用相图热力学和扩散动力学计算，建立了成分和工艺筛选的高通量集成计算程序；根据马氏体相变热力学，基于 CALPHAD 理论框架，结合 Redlich-Kister 关系式，建立非化学驱动力的热力学模型，并借助相场计算实现了对马氏体转变行为的模拟；将自建数据库以及计算程序搭载至商用软件平台上，构建了优化成分和工艺的 ICME 研究方法，实现了吉帕级高性能原型钢的短周期研发。

关键词：先进高强汽车钢板；集成计算材料工程；马氏体相变；成分和工艺

人工智能驱动与计算模拟辅助的承温 1200℃单晶高温合金智能设计

杨　帆，蔺思源，赵文月，茹　毅，裴延玲，李树索，宫声凯

（北京航空航天大学，北京　100191）

摘　要：为满足先进高推重比航空发动机日益提高的涡轮前温度，快速研发高承温、低密度、低成本的新型单晶高温合金已成为亟待突破的关键课题。本研究建立了一个人工智能预测驱动、计算模拟辅助与试验验证相结合的合金智能设计新框架，成功设计了一种超高温条件下高蠕变抗力的新型单晶高温合金，并对单晶合金的超高温强韧化机理提出了新的见解。通过将 Transformer 引入的元素交互信息表示学习作为先验物理约束，所构建的深度学习模型（Superalloy Transformer-based network for Creep, SaTNC）实现了单晶高温合金宽温域条件下蠕变断裂寿命的准确预测，并表现出具有竞争力的泛化能力，可精准捕捉单晶高温合金蠕变性能对成分的高敏感性。同时，SaTNC 模型所应用的自注意力机制使其具有良好的可解释性，可获取关键合金化元素间的复杂交互作用及其对单晶高温合金蠕变性能的协同影响等材料学见解。针对1200℃高承温能力与低 Re 含量的设计目标，利用 SaTNC 模型所设计的新型单晶合金在 1200℃/80MPa 条件下的蠕变断裂寿命达 170h；由模型指导开发的基于 Re-Mo 协同强化和 Mo-Ta 作用平衡的成分优化策略，被证明实现了 γ′相热强性与 γ/γ′相界面稳定性的同步提升，有效调控了超高温条件下 γ′-γ′ 双相组织的蠕变形筏行为。本研究通过开展第一性原理计算与动力学蒙特卡罗模拟，进一步揭示了深度学习模型所挖掘的合金化元素协同作用机制，明晰了上述单晶合金强韧化新策略的内在微观机理。

关键词：单晶高温合金；智能设计；深度学习；计算材料学

AI-driven and Computational Simulation Assisted Intelligent Design of Single-crystal Superalloys for 1200℃ Ultra-high Temperature Application

YANG Fan, LIN Siyuan, ZHAO Wenyue, RU Yi,
PEI Yanling, LI Shusuo, GONG Shengkai

(Beihang University, Beijing 100191, China)

Abstract: To satisfy the increasing inlet temperature of advanced aero-engine with high thrust-weight ratio, the rapid development of new single-crystal (SX) superalloys applied at ultra-high temperature with low density and low cost is a key issue required to be broken through urgently. This study proposes a novel framework for intelligent alloy design driven by artificial intelligence predictions and combined with computational simulation as well as experimental validation. A new SX superalloy with high creep resistance at ultra-high temperature is successfully designed. New insights into the ultra-high-temperature strengthening mechanisms are also uncovered. With the representation learning of elemental interaction information introduced by Transformer as the core prior physical constraint, the constructed deep learning model (Superalloy Transformer-based network for Creep - SaTNC) achieves accurate prediction of creep rupture life for SX superalloys in a wide temperature range. The SaTNC model exhibits competitive generalization ability and its high capability to capture the sensitivity of creep rupture life on alloy composition has been verified. Moreover, the self-attention mechanism applied in the SaTNC model provides interpretable information, which enables material insights including the interactions among alloying elements and their synergistic effects on the creep rupture life of superalloy to be extracted. Aiming to achieve sufficient creep resistance at 1200℃ with low Re addition, a new SX superalloy is designed by utilizing the SaTNC model and its creep rupture life at 1200℃/80MPa has tested to approach 170h. The compositional optimization strategy developed by the model guidance of Re-Mo synergistic strengthening and Mo-Ta trade-off is proved to achieve the simultaneous enhancements of γ' thermal strength and γ/γ' interfacial stability, which effectively regulates the rafting behavior of SX superalloy at ultra-high temperature. By conducting first-principles calculations and kinetic Monte Carlo simulations, this study further reveals the synergistic mechanism of alloying elements explored by the deep learning model and clarifies the underlying micro-mechanism of the above new strategy for strengthening SX superalloys.

Key words: single-crystal superalloys; intelligent design; deep learning; computational materials science

高温合金定向凝固晶体生长与 CET 转变

李应举，邱　义，冯小辉，杨院生

(中国科学院金属研究所，辽宁沈阳　110016)

摘　要：控制枝晶生长是影响金属材料及铸件性能的重要途径。枝晶以自由生长方式生长，通过控制 CET 转变获得的细晶均质铸锭或铸件在工业领域有重要用途，而枝晶以强制生长方式生长，通过定向生长的单晶或定向合金在航空航天等领域有重要用途。本文采用数值模拟和实验相结合的方法，分别在镍基高温合金增材制造和定向凝固过程中通过抑制枝晶生长获得微细定向柱晶和等轴晶组织。在增材制造过程中，采用计算流体力学与相场法耦合的方法，建立了跨尺度计算框架，模拟计算了 SLM 过程中熔体流动、熔池内组织演变，揭示了枝晶竞争生长机制，发现在激光作用模式为传导模式下可以获得具有单一取向的微细定向柱晶组织。而在高温合金定向凝固过程中，建立了脉冲磁场作用下的能量、动量和质量传递有限元模型，计算获得了存在固/液界面条件下的脉冲磁场作用时凝固过程中的电磁力、熔体流动和焦耳热的分布规律，发现最大电磁力和最大流速均存在于液相界面附近区域这一特点，推导出磁热协同控制凝固作用下 CET 转变判据,计算证明最大电磁对流作用下枝晶可发生断裂,提出了磁热协同控制凝固作用下晶核增殖模型,获得了磁热协同凝固晶粒细化的优化工艺参数。

关键词:枝晶生长;CET 转变;增材制造;数值模拟;晶粒细化

Directional Solidification Crystal Growth and CET Transformation of Superalloys

LI Yingju, QIU Yi, FENG Xiaohui, YANG Yuansheng

(Institute of Metal Research, Chinese Academy of Sciences, Shenyang 110016, China)

Abstract: The control of dendrite growth is an important way to influence the properties of metallic materials and castings. When dendrites growing in a free-growth mode, fine grained ingots or castings obtained by controlling the CET transition have important applications in industry, while dendrites growing in a forced-growth mode, single crystals or directional alloys obtained by directional growth have important applications in aerospace and other fields.

In this paper, numerical simulation and experiment methods are combined to obtain microfine directional columnar and equiaxed grains by inhibiting dendritic growth during additive manufacturing and directional solidification of Ni-based superalloy respectively. In the additive manufacturing process, a cross-scale computational framework was established by coupling computational fluid dynamics with the phase field method to simulate the melt flow and the microstructure evolution in the melt pool during the SLM process, which reveals the competitive growth mechanism of dendrites. It is found that the microfine directional columnar grains with a single orientation can be obtained when the laser interaction mode is the conduction mode. In the process of directional solidification, a finite element model of energy, momentum and mass transfer under the action of pulsed magnetic field was established, and the distribution laws of electromagnetic force, melt flow and Joule heat in the solidification process under the action of pulsed magnetic field in the condition of solid/liquid interface were obtained by calculations. The characteristics of the maximum electromagnetic force and the maximum flow velocity were found to exist in the area near the interface of the liquid phase. The CET transition criterion under magneto-thermal synergistic control of solidification is deduced. The calculation proves that the dendrites can be fractured under the maximum electromagnetic convection, and a model of nucleus proliferation under magneto-thermal synergistic control of solidification is proposed. The optimized process parameters for the grain refinement of magneto-thermal synergistic solidification are obtained.

Key words: dendritic growth; CET transition; additive manufacturing; numerical simulation; grain refinement

特种冶炼行业信息化发展思路

朱勤天,陈正阳,毕中南

(北京钢研高纳科技股份有限公司,北京 100081)

摘 要:特种冶炼是高温合金生产的第一道工序,决定了高温合金的成分、纯净度、冶金缺陷及初始元素偏析等关键特征参数。随着高温合金后端制备技术(锻造、热处理等)相对成熟,更多冶金质量稳定性、成本精益化、使用可靠性等关键问题指向前端冶炼工艺。针对以上问题,国外冶金厂基于 40 年特种冶炼数据库,通过信息化、智能化等技术手段,开发出了高效、稳定的专家系统,实现了特种冶炼工艺的在线精益化控制,高温合金质量稳定性及性价比显著提高。而国内目前的解决措施仍以经验试错为主,导致我国特种冶炼技术与国外的差距日趋增大。国外先进经验表明,信息化技术对实现特种冶炼过程的准确控制、工艺优化以及高温合金冶金质量提升具有重要促进作用。随着我国信息化技术的快速发展,利用工业大数据、人工智能等新一代信息化技术,实现特种冶炼技术的迭代升级,是推动我国高温合金材料赶超国际先进水平的新机遇。本文分析了信息化技术在钢铁及其他领域的应用及发展趋势,提出了国内特种冶金企业利用信息化技术解决现存问题、实现创新发展的基本思路,可为实现特种冶炼行业的信息化、智能化转型提供借鉴。

关键词:特种冶炼;高温合金;信息化技术

Thoughts on the Informationization Development of China's Special Smelting Industry

ZHU Qintian, CHEN Zhengyang, BI Zhongnan

(Gaona Aero Material Co., Ltd., Beijing 100081, China)

Abstract: Special smelting is the first process of superalloy production, which determines the composition, purity, metallurgical defects and initial element segregation of the superalloy. With the relative maturity of the high-temperature alloy back-end preparation technology (forging, heat treatment, etc.), more key issues such as metallurgical quality stability, cost lean, and use reliability point to the front-end smelting process. In response to the above problems, foreign metallurgical plants based on 40 years of special smelting database, through information, intelligent and other technical means, developed an efficient and stable expert system, to achieve online lean control of special smelting process, superalloy quality stability and cost-effective significantly improved. The current domestic solution measures are still based on experience trial and error, resulting in the gap between China's special smelting technology and foreign countries is increasing. Foreign advanced experience shows that information technology plays an important role in the realization of accurate control of special smelting process, process optimization and quality improvement of superalloy metallurgy. With the rapid development of China's information technology, the use of industrial big data, artificial intelligence and other new generation of information technology to achieve the iterative upgrading of special smelting technology is a new opportunity to promote China's high-temperature alloy materials to catch up with the international advanced level. This paper analyzes the application and development trend of information technology in iron and steel and other fields, and puts forward the basic ideas of domestic special metallurgy enterprises using information technology to solve existing problems and achieve innovative development, which can provide reference for realizing the information and intelligent transformation of special smelting industry.

Key words: special smelting; superalloy; information technology

基于机器学习的高性能铝合金成分设计

杨 健,张 佼,戴永兵,隽永飞

(上海市先进高温材料及其精密成型重点实验室,上海交通大学,上海 200240)

摘 要:作为航空工业中最具代表性的铝合金系列,Al-Zn-Mg-Cu 合金具有比强度高、延展性好、抗腐蚀等优点。面对现代航空制造中对构件的轻量化、高规格、高可靠性等要求,Al-Zn-Mg-Cu 合金的性能也亟需进一步优化。面对巨大的成分空间,如何依据性能需求快速筛选最优成分已成为材料工作者亟待解决的问题。传统实验试错耗时费力,近年来数据驱动的机器学习方法凭借其建模灵活、预测准确和泛化能力强等优势,为科研人员带来新的研究视角。然而不同的工业应用场景对材料的性能需求侧重不一,常规的机器学习建模方法难以根据不同的目标性能对合金成分与热处理参数(合称制备策略)进行准确设计。此外,目前在应用机器学习建立成分与性能之间的映射关系时,大多忽略了对已知合金知识的学习和利用,并且建模时涉及的元素局限于数据集中已有的种类。针对上述问题,我们对 Al-Zn-Mg-Cu 合金开展了面向目标性能的制备策略设计、基于知识的强度优化设计以及基于新型强化元素筛选的多性能寻优设计等方面的研究。开发了集成机器学习、从头算分子动力学与第一性原理等方法的合金设计系统，通过该系统设计出了新型高性能 Al-Zn-Mg-Cu 合金，合金抗拉强度达 816MPa，屈服强度达 779MPa，伸长率达 9.1%。新型合金设计系统在新材料研发方面展现出巨大潜力。

关键词：高强 Al-Zn-Mg-Cu 合金；机器学习；合金设计；强化元素筛选；从头算分子动力学；第一性原理

Composition Design of High-performance Aluminum Alloys Based on Machine Learning

YANG Jian, ZHANG Jiao, DAI Yongbing, JUAN Yongfei

(Shanghai Key Laboratory of Advanced High Temperature Materials and Precision Molding, Shanghai Jiaotong University, Shanghai 200240, China)

Abstract: As the most representative aluminum alloy series in the aviation industry, Al-Zn-Mg-Cu alloy has the advantages of high specific strength, good ductility, and corrosion resistance. Faced with the requirements for lightweight, high-standard, and high-reliability components in modern aviation manufacturing, the properties of Al-Zn-Mg-Cu alloys also need to be further optimized. Facing the huge ingredient space, how to quickly screen the optimal ingredients based on performance requirements has become an urgent problem for materials workers. Traditional experiments are time-consuming and laborious by trial and error. In recent years, data-driven machine learning methods have brought new research perspectives to researchers by virtue of their advantages such as flexible modeling, accurate prediction and strong generalization ability. However, different industrial application scenarios have different performance requirements for materials. Conventional machine learning modeling methods are difficult to accurately design alloy composition and heat treatment parameters (collectively referred to as preparation strategies) according to different target properties. In addition, when applying machine learning to establish the mapping relationship between composition and properties, the learning and utilization of known alloy knowledge is mostly ignored, and the elements involved in modeling are limited to the existing types in the data set. In addition, when applying machine learning to establish the mapping relationship between composition and properties, the learning and utilization of known alloy knowledge is mostly ignored, and the elements involved in modeling are limited to the existing types in the data set. In response to the above problems, we have carried out research on the preparation strategy design for the target performance, the strength optimization design based on knowledge, and the multi-performance optimization design based on the screening of new strengthening elements for Al-Zn-Mg-Cu alloy. Developed an alloy design system that integrates machine learning, ab initio molecular dynamics, and first principles. Through this system, a new type of high-performance Al-Zn-Mg-Cu alloy was designed. The tensile strength of the alloy reached 816MPa, and the yield strength reaches 779MPa, the elongation reaches 9.1%. New alloy design systems show great potential in the development of new materials.

Key words: high-strength Al-Zn-Mg-Cu alloy; machine learning; alloy design; strengthening element screening; ab initio molecular dynamics; first-principles

数值模拟在高温合金构件生产和制造上的应用

孙志民

（北京钢研高纳科技股份有限公司，北京 100081）

摘　要：随着计算机技术的发展，数值模拟在各领域的应用越来越广泛。本文以案例分享的方式，介绍了数值模拟在高温合金构件设计、制造领域的应用。案例1是高温合金锻件热处理后的分区控冷模拟，通过流体-热-结构耦合方法预测了锻件的温度场变化进而得到其内部的残余应力变化。通过优化风场获得了不同的锻件冷却速度从而获得不同的残余应力分布。案例2是弹簧的压缩与松弛模拟，通过结构模拟虚拟实现了弹簧的压缩以及松弛过程，从而获得不同结构参数弹簧的承载能力以及松弛特性，为弹簧的结构优化提供依据。案例3是高温合金构件的增材过程模拟，通过热-结构耦合的方法获得了构件的应力和变形，为增材制造的工艺优化、支撑优化提供指导。这些案例从不同侧面反映了数值模拟对高温合金构件设计和制造的提升作用。

关键词：高温合金；数值模拟；工艺优化；结构优化

Application of Numerical Simulation in Super-alloy Products' Design and Manufacture

SUN Zhimin

(Gaona Aero Material Co., Ltd., Beijing 100081, China)

Abstract: With the development of computer technology, numerical simulation is more and more widely used in many fields. In this paper, numerical simulation in super-alloy products' design and manufacture is introduced by case studies. In Case 1, forged pieces' divisional cooling after heat treatment is simulated by coupled method of fluid-thermal-structure. As a result, temperature distribution in the part is achieved. Furthermore, the residual stress in the part is available with the part cooling. Different cooling rate and different residual stress distribution are achieved by optimizing the air velocity and its distribution. In Case 2, spring's compressing and stress relaxation are simulated by FEM (finite element Modeling). As a result, the relationship between loading and geometry parameters and springs' stress relaxation property are available, which will be used to optimize springs' geometry parameters. In Case 3, the distortion and residual stress of super-ally parts in additive manufacture processing is simulated by coupled thermal-structure method, which will helpful to optimize the processing parameters and supports. These cases show how numerical simulation improves super-alloy products design and manufacture.

Key words: super-alloy; numerical simulation; processing optimization; structure optimization

数智化在宝武特冶高温合金锻造生产中的应用与探索

曹金华，侯智鹏

(宝武特种冶金有限公司，上海　201901)

摘　要：随着目前航空、航天及舰船等领域用高温合金需求量的增长和技术要求的提升，高温合金产品一致性、均匀性及稳定性有待进一步提高。报告围绕高温合金的发展应用、热加工生产过程中存在的问题以及数智化在高温合金热加工生产中的应用三个方面展开，重点介绍了宝武特种冶金有限公司近年来在应用数智化技术提升高温合金产品质量稳定性方面所做的探索，希望数智化的推广能对高温合金产品的发展和质量稳定起到促进作用。

关键词：高温合金；锻造；数智化

Application and Exploration on the Digital Intelligence Technology in the Superalloys Forging Production of Baowu Special Metallurgy

CAO Jinhua, HOU Zhipeng

(Baowu Special Metallurgy Co., Ltd., Shanghai 201901, China)

Abstract: With the increasing demand and technical requirements for high-temperature alloys in areas such as aviation, aerospace, and ships, further improvements are needed in the consistency, uniformity, and stability of high-temperature alloy products. This report focuses on three aspects: the development and application of high-temperature alloys, the problems encountered in the hot working production process, and the application of digital intelligence in the hot working production of high-temperature alloys. Emphasis is placed on introducing the recent explorations carried out by Baowu Special Metallurgy Co., Ltd. in applying digital intelligence technology to improve the quality stability of high-temperature alloy products. It is hoped that the promotion of digital intelligence can promote the development and quality stability of high-temperature alloy products.

Key words: superalloy; forging; digital intelligence

先进材料数据库及应用技术

张洪梅，程兴旺，赵平洛

（北京理工大学材料学院，北京 100081）

摘 要：围绕我国先进材料体系结构，开展了材料数据规范研究、数据库架构设计等工作，突破了材料多源异构数据采集汇交技术，构建了我国先进材料性能数据字典，编制了系列材料数据规范，开发了先进材料数据库，具备材料数据采集、存储管理和应用功能，为我国先进材料科学发展和设计选材提供了统一的材料数据平台。并示范应用于新材料设计，实现基于启发式算法对材料高维特征空间进行探索，完成了面向目标性能需求的新型高强度钛合金设计。

关键词：材料数据库；数据规范；数据字典；新材料设计

ZHANG Hongmei, CHENG Xingwang, ZHAO Pingluo

(School of Materials Science and Engineering, Beijing Institute of Technology, Beijing 100081, China)

Abstract: Around of the advanced material system in China, research on material data standard and database design have been carried out. The technology of multi-source heterogeneous data collection and exchange of materials were broken through. China's advanced material performance data dictionary was constructed. And a series of material data standard drafts were compiled. An advanced material database which has the functions of material data collection, storage management, and application, were developed. It provides a unified material data platform for the development of advanced materials science and design material selection in China. And it is demonstrated as an application in new material design, the exploration of high-dimensional feature space of materials based on heuristic algorithms was realized, and the design of a new high-strength titanium alloy for target performance requirements was completed.

Key words: materials database; data standard; data dictionary; new material design

机器学习辅助高熵合金成分设计

薛德祯

(西安交通大学金属材料强度国家重点实验室,陕西西安 710049)

摘　要：高熵合金的设计通常涉及多种元素的巨大组合空间。在这样的空间中进行有效搜索通过传统的试错法、经验法难以实现。本研究将机器学习替代模型与实验设计算法相结合,通过与实验反馈迭代,高效开发具有目标性能的高熵合金。进一步提出了生成和选择新材料特征的策略,并采用该策略来提高对高熵合金的相结构的预测精度。利用这些开发的材料信息学工具包,本研究通过五次实验迭代中发现了强度为 2.8~3.0GPa 的新型高熵合金。研究优化的合金 AlVCrCoNiMo 在室温、800 ℃和 900 ℃下的压缩比屈服强度分别为 395MPa、143MPa 和 104MPa/(cm^3·g)。本研究为有效设计新合金成分提供了一种新的替代方法。

关键词：机器学习；主动学习；高熵合金

Machine Learning-assisted Composition Design of High-entropy Alloys

XUE Dezhen

(State Key Laboratory for Mechanical Behavior of Materials, Xi'an Jiaotong University, Xi'an 710049, China)

Abstract: The design of high entropy alloys often involves a huge combinatorial space spanning compositions that include elements across the periodic table. An efficient search in such a space is usually beyond conventional knowledge-guided trial and error approaches. We formulate a materials design strategy combining a machine learning surrogate model with experimental design algorithms to search for high entropy alloys with targeted properties. A strategy to generate and select new materials features was also proposed and employed to improve the phase prediction accuracy for high entropy alloys. With those developed materials informatics toolkits, we synthesized new high entropy alloys (HEAs) with strengths 2.8~3.0GPa within five experimental iterations. The alloy AlVCrCoNiMo was found to have a compressive specific yield strengths of 395MPa, 143MPa and 104MPa/(cm^3·g) at room temperature, 800℃ and 900℃, respectively. The present study paves an alternative way to efficiently design composition of new alloys.

Key words: machine learning; active learning; high entropy alloys

440MPa 级高性能船体钢技术研究进展

罗小兵，师仲然，柴希阳

(钢铁研究总院有限公司,北京 100081)

摘　要：440MPa 级高强钢是我国民用船舶、海洋工程、军工舰船领域用量最大的钢级,随着我国船舶装备大型化、低成本化、极端环境服役化的发展,钢的易焊接性和耐低温性能要求越来越高,本文系统研究了 Nb-V 复合型、超低碳+Cu 硬化型两种体系下材料的合金成分、轧制工艺等对材料性能的影响，研究结果表明，（1）Nb-V 复合型体系下，再加热温度、精轧开轧温度、终轧温度等轧制工艺参数对铁素体晶粒的细化及低温韧性存在显著影响。通过优化轧制工艺参数，可以实现试验钢在–80℃下保持良好的低温韧性。（2）超低碳+Cu 粒子硬化型体系下，试验钢经过控轧后微观组织为铁素体+少珠光体+弥散分布的 Cu 粒子。在成分设计和组织控制上，利用超低碳后形成的少珠光体组织有助于提高低温韧性；利用 Cu 粒子在轧制过程中的析出提高屈服强度，弥补超低碳后强度的不足，从而实现易焊接性和高强韧性的匹配。

关键词：船体钢；440MPa；易焊接；Cu 析出

Research of 440 MPa High-performance Hull Steel

LUO Xiaobing, SHI Zhongran, CHAI Xiyang

(Central Iron and Steel Research Institute Co., Ltd., Beijing 100081, China)

Abstract: 440MPa high-strength steel is the largest steel grade used in the fields of civil ships, ocean engineering, and military ships in China. With the development of large-scale, low-cost, and extreme environmental service of China's shipbuilding equipment, the requirements for easy weldability and low temperature resistance of steel are becoming higher and higher. This article systematically studies the alloy composition and rolling process of materials under Nb-V composite type and ultra-low carbon+Cu hardening type systems, which affect the material properties. The research results show that: (1) Under the Nb-V composite system, the rolling process parameters such as reheating temperature, finishing rolling opening temperature, and final rolling temperature have a significant impact on the refinement of ferrite grains and low-temperature toughness. By optimizing the rolling process parameters, it is possible to achieve good low-temperature toughness of the test steel at –80℃. (2) In the ultra-low carbon+Cu particle hardening system, the microstructure of the test steel after controlled rolling is ferrite+less pearlite+dispersed Cu particles. In terms of composition design and organizational control, utilizing the low pearlite structure formed after ultra-low carbon can help improve low-temperature toughness; Utilizing the precipitation of Cu particles during the rolling process to improve yield strength and compensate for the lack of strength after ultra-low carbon, thus achieving a match between weldability and high-strength toughness.

Key words: ship hull steel; 440MPa; easy welding; Cu precipitates

输氢管材与氢气相容性研究

李 拔，汪 兵，贾书君，刘清友

（钢铁研究总院有限公司，北京 100081）

摘 要：氢能是同时满足资源、环境和可持续发展要求的二次能源，氢气输送是氢能利用（制氢、储氢、输氢、用氢）中的重要一环，而采用管线输送是最经济和高效的输送方式。输氢管道研发是通过科技创新实现国家"双碳"战略目标的典型范例。鉴于氢气长距离、大规模安全输送的迫切需求，亟需开展纯氢与天然气掺氢长输管道管材相容性研究。本文主要针对高钢级、低氢压掺氢管道和中低钢级、高压纯氢管道，分别介绍了模拟煤制气输送 X80 含氢管道安全性评价和纯氢输送管道抗氢脆性能影响因素的部分研究工作，并对输氢管线钢未来的研究方向进行了展望。

关键词：氢气输送；纯氢；掺氢；管线钢；相容性

Study on Compatibility of Hydrogen Transport Pipe and Hydrogen

LI Ba, WANG Bing, JIA Shujun, LIU Qingyou

(Central Iron and Steel Research Institute Co., Ltd., Beijing 100081, China)

Abstract: Hydrogen energy is a secondary energy source that meets the requirements of resources, the environment and sustainable development. Hydrogen transport is an essential part of hydrogen energy utilization, such as hydrogen production, hydrogen storage, hydrogen transport and hydrogen use, and pipeline transportation is the most economical and efficient transportation method. The research and development of hydrogen transport pipeline is a typical example of realizing the national "dual carbon" strategic goal through scientific and technological innovation. With the urgent need for safe long-distance and large-scale transport of hydrogen, there is a pressing need to investigate the compatibility of pure hydrogen and natural gas/hydrogen mixtures long-distance pipelines. In this paper, we focus on hydrogen-doped pipes with high steel grade and low hydrogen pressure, and pure hydrogen pipes with intermediate or low steel grade and high pressure. Partly research works on the safety evaluation of X80 hydrogen-containing pipelines for simulated coal-to-gas transmission and on the factors affecting the hydrogen brittle resistance of pure hydrogen pipelines are introduced, respectively. Then, future research directions for hydrogen pipe steel are also in prospect.

Key words: hydrogen transmission; pure hydrogen; hydrogen doping; pipeline steel; compatibility

免喷丸热成形钢组织性能调控及抗氧化性能

郑昊青，常智渊，苏冠侨

（攀钢集团攀枝花钢铁研究院有限公司，四川攀枝花 617000）

摘　要： 传统的热成形钢在模具淬火后需进行抛丸处理或酸洗处理，以去除表面氧化铁皮，增加了生产成本。近年来，国内外对抗氧化热成形钢开展了大量研究，以达到免酸洗、免抛丸、抗氧化性能良好的使用需求。研究发现，Si、Cr、Al等元素对于提高热成形钢的抗高温氧化性能有利。另外，有研究表明在钢中添加微量稀土元素（如Ce、Y和La等）能够降低钢的高温氧化速度，改善钢的抗氧化性能，稀土元素能够降低氧化速率，并且表现出优良的附着性能，这有利于减少氧化膜/基体处的缺陷。本研究在22MnB5基础上进行微合金化成分设计，使热成形零部件表面氧化层致密、且和基体紧密结合，达到热成形过程中无氧化皮脱落、热成形后无需抛丸的目的。解决国外热成形关键材料的限制问题，扩大热成形技术在汽车行业的应用。

关键词： 热成形钢；免喷丸；抗氧化性能

Development and Application of Antioxidant and Shot Peening Free Hot-formed Steel

ZHENG Haoqing, CHANG Zhiyuan, SU Guanqiao

(Pangang Group Research Institute Co., Ltd., Panzhihua 617000, China)

Abstract: Traditional hot-formed steel requires shot blasting or acid pickling treatment after mold quenching to remove surface oxide scales, which increases production costs. In recent years, much research has been conducted domestically and internationally on oxidation-resistant hot-formed steel to meet the usage requirements of no acid washing, no shot blasting, and good oxidation resistance. Research has found that elements such as Si, Cr, and Al are beneficial for improving the high-temperature oxidation resistance of hot-formed steel. In addition, studies have shown that adding trace amounts of rare earth elements (such as Ce, Y, and La) to steel can reduce the high-temperature oxidation rate of steel and improve its oxidation resistance. Rare earth elements can reduce the oxidation rate and exhibit excellent adhesion properties, which is beneficial for reducing defects in the oxide film/matrix. This study conducts a microalloying composition design based on 22MnB5, making the surface oxide layer of hot-formed parts dense and tightly bonded to the substrate, achieving the goal of no oxide skin detachment during the hot-forming process and no need for shot blasting after hot forming, addressing the limitations of critical materials for hot forming abroad and expanding the application of hot forming technology in the automotive industry.

Key words: hot-formed steel; no shot peening; oxidation resistance

超低 Nb 技术及其在结构板材及型材上的应用

王厚昕

（中信微合金化技术中心，北京　100076）

摘　要： 铌是一个独特的元素，具有鲜明的细化晶粒作用，已被应用到绝大多数的钢中实现高强韧性及使用性能的改善。要使铌在钢中发挥良好的细晶作用，通常认为需要把添加量提高到 0.02%以上且随着强度尤其韧性要求的提高而不断增加，比如像 X80 管线钢中的 Nb 就往往会达到 0.08%左右。但最新的研究证明，当钢中的 Nb 含量添加到 0.005%就能显著抑制再结晶行为，添加至 0.01%左右便可以使屈服强度 355MPa 级别的低合金结构钢获得良好而稳定的力学性能，冲击韧性可以同时满足 B/C/D 甚至 E 级的要求，形成了超低 Nb 技术，拓宽了微合金化 Nb 元素的应用范围。与此同时，由于具有较高的性能富裕量，根据钢材种类及规格可降低钢中的 Mn 含量至 0.85%~1.25%，从而从源头上减少钢的中心偏析，改善钢的内部质量，并获得综合降本效益。到目前，超低 Nb 技术已经成功被广泛应用到屈服强度 355MPa 级别的结构板材及型材生产，成为这一类量大面广钢材高质量发展的解决方案。本文将系统阐述超低 Nb 技术及其在板材与型材中的典型应用。

关键词： 超低 Nb 技术；低 Mn 微 Nb；中心偏析；高质量发展；Q355

Study on Ultra Low Nb and Its Application in Commodity Products with YS 355MPa of Both Flat and H Beams

WANG Houxin

(CITIC-CBMM Microalloying Center, Beijing 100076, China)

Abstract: Niobium is an unique element with distinct refining grain effect, which has been applied to most of the steel to achieve high strength toughness and performance improvement. In order to make Nb play a good role of grain refinement in steel, it is usually believed that the addition amount needs to be increased to more than 0.02% and continues to increase with the improvement of strength, especially toughness requirements, such as Nb in X80 pipeline steel, which often reaches about 0.08%. However, the latest research proves that when the Nb content in the steel is added upto 0.005%, the recrystallization behavior can be significantly retarded, and the addition to about 0.01% can make the low-alloy structural steel with yield strength of 355MPa obtain good and stable mechanical properties, and the impact toughness can simultaneously meet the requirements of B/C/D or even E class, forming an ultra-low Nb technology. The application range of microalloying Nb element is expanded indeed. At the same time, due to the high margin of mechanical properties, the Mn content in the steel can be reduced down to 0.85%~1.25% according to the type and specification of the steel, so as to reduce the central segregation of the steel from the source, improve the internal quality of the steel, and obtain comprehensive cost reduction benefits. So far, ultra-low Nb technology has been successfully widely used in the production of structural plates and shapes with yield strength of 355MPa, and has become a solution for the high-quality development of this kind of large-scale steel. This paper will systematically describe the ultra low Nb technology and its typical application in plate and shapes.

Key words: ultra low Nb; low Mn micro Nb; centerline segregation; high quality development; Q355

极薄取向硅钢制造与应用

刘宝志

（内蒙古科技大学，内蒙古包头 014017）

摘　要： 极薄取向硅钢是支撑我国现代化建设的重要的软磁材料，在国防军工、电力电子工业、医疗、特高压输变电工程等领域被广泛应用，是国家重要的战略资源，属国家紧缺类产品。主要用于制造高频变压器、饱和电抗器、大功率磁放大器、脉冲变压器、互感器等。其关键生产技术一直受国外封锁，属于"卡脖子"产品，该产品是国家及相关行业发展的迫切需求。报告主要介绍极薄取向硅钢的制造技术及应用。

关键词： 取向硅钢；极薄；软磁材料

Manufacture and Application of Ultra-thin GO Steel

LIU Baozhi

(Inner Mongolia University of Science and Technology, Baotou 014017, China)

Abstract: Ultra thin oriented electrical steel is an important soft magnetic material that supports China's modernization construction. It is widely used in fields such as national defense and military industry, power electronics industry, medical treatment, and ultra-high voltage transmission and transformation engineering. It is an important strategic resource of the country and a scarce product. Mainly used for manufacturing high-frequency transformers, saturated reactors, high-power magnetic amplifiers, pulse transformers, transformers, etc. Its key production technology has always been blocked by foreign countries and belongs to the category of "bottleneck" products. This product is an urgent demand for the development of the country and related industries. The report mainly introduces the manufacturing technology and application of ultra-thin oriented silicon steel.

Key words: oriented silicon steel; extremely thin; soft magnetic material

双碳时代中国硅钢产业发展机遇与挑战

吴树建

(宝山钢铁股份有限公司，上海 201999)

摘　要：作为关系节能减排的重要功能材料，硅钢影响着发电、输变电及用电的效率。近年来，随着我国硅钢制造技术的持续进步，不断取代进口，在多个应用场景中促进我国多个工业领域摆脱卡脖子。在双碳的时代背景下，硅钢产业迎来新的机遇与挑战，本报告重点介绍当前我国硅钢产业运行及市场需求情况。

关键词：硅钢；产业运行；市场需求

Opportunities and Challenges for the Development of China's Silicon Steel Industry in the Dual Carbon Era

WU Shujian

(Baoshan Iron and Steel Co., Ltd., Shanghai 201999, China)

Abstract: As an important functional material related to energy conservation and emission reduction, silicon steel affects the efficiency of power generation, transmission, and consumption. In recent years, with the continuous progress of China's silicon steel manufacturing technology, it has continuously replaced imports and promoted multiple industrial fields in China to break free from bottlenecks in multiple application scenarios. In the context of the dual carbon era, the silicon steel industry is facing new opportunities and challenges. This report focuses on introducing the current operation and market demand of China's silicon steel industry.

Key words: silicon steel; industrial operation; market demand

无底层绿色环保涂层高磁感取向硅钢的开发与推广应用研究

杨佳欣，申明辉，宋　刚

(宝钢股份中央研究院武钢有限技术中心，湖北武汉 430080)

摘　要：本研究开展新型大型电机用无底层绿色环保涂层高磁感取向硅钢的研发，并首次成功实现大批量生产及应用。报告详解了无底层绿色环保涂层高磁感取向硅钢的开发与产品质量认证、批量应用过程，以及特殊元素对取向硅钢性能、底层的影响机理，并展望了未来技术发展前景。

关键词：高磁感取向硅钢；底层；绿色环保涂层

Development and Application Promotion of Glassless High Permeability Grain-oriented Silicon Steel with Green and Environmentally Friendly Coating

YANG Jiaxin, SHEN Minghui, SONG Gang

(R&D Center of Wuhan Iron and Steel Company, Research Institute, Baoshan Iron and Steel Co., Ltd., Wuhan 430080, China)

Abstract: This study completed the research and development of new type glassless high permeability grain-oriented silicon steel with green and environmentally friendly coating for large motor, first mass production and application was successfully implemented. Report introduces the research and development, certification of product quality, batch application process of glassless high permeability grain-oriented silicon steel with green and environmentally friendly coating. In addition, the effects of special element on magnetic properties and glass film, future technological development prospects are introduced.

Key words: high permeability grain-oriented silicon steel; glass film; green and environmentally friendly coating

高饱和磁化强度 Fe-Si-B-C 系非晶纳米晶合金组织结构与性能调控

惠希东[1]，吕旷[1]，李育洛[1]，王伟民[2]

（1. 北京科技大学新金属材料国家重点实验室，北京　100083；
2. 山东大学材料科学与工程学院，山东济南　250100）

摘　要： Fe 基非晶及纳米晶软磁合金作为新一代的绿色节能材料，由于优良的软磁特性在电力电子、汽车、光伏和半导体等领域获得了重要应用。但现有 Fe 基非晶及纳米晶体系中的饱和磁感应强度（B_s）与硅钢相比还存在很大差距，难以满足高功率密度和器件小型化要求。为提高饱和磁化强度，往往需要高的 Fe 含量，这导致了合金系的非晶形成能力（GFA）不足，纳米晶化后的材料脆性严重等问题，对非晶/纳米晶合金的制备工艺提出了更加严苛的要求。如何通过组织结构和性能调控研制出具有高 GFA、良好的工艺和软磁性能的 Fe 基非晶纳米晶合金已成为当前材料科学领域的重要课题。本工作利用第一性原理分子动力学计算模拟了 Fe-Si-B-C 系合金液态到非晶态转变过程中局域原子堆垛结构的演变以及液态合金的动力学性质。基于非晶合金局域有序结构和动力学特征，研究了 C 元素的添加对 FeSi-B-C 系列 GFA、热稳定性和软磁性能的影响。针对高 Fe 含量的 Fe-Si-B-C 体系，通过加入 P 和 Zr 元素，同时调控各类金属元素 Si、B、C 和 P 含量的方法优化了 Fe-Si-B-C-P-Zr 体系的 GFA 及软磁性能。研究了类金属元素 C、Si、B、P 与金属元素 Cu、V 对 Fe-Si-B-C-P-Cu 系非晶纳米晶的 GFA、晶化行为及软磁性能的影响规律。通过对合金系组织结构和软磁性能的优化与调控，成功研制出了具有高 B_s 和低 H_c 的铁基非晶纳米晶合金。

超稳定的 Ce 基金属玻璃

张 博

（松山湖材料实验，广东东莞 523808）

摘 要：作为热力学亚稳态材料，金属玻璃的稳定性一直是其基本性质之一。和氧化物玻璃以及聚合物玻璃相比，金属玻璃的稳定性被认为是最差的。最近松山湖材料实验室和中科院物理所联合团队研究了 Ce70Al10Cu20 金属玻璃在 $0.85T_g$ 下长达 17.7 年超长室温老化样品的稳定性问题，如此接近 T_g 点的高温长时间老化实验在金属玻璃以往的研究中还从未有过。实验发现超长时间高温老化后的 Ce-基非晶合金样品依旧保持着完美的非晶态，表现出极强的抗晶化能力，打破了传统对金属玻璃稳定性差的认识。老化后 Ce 基金属玻璃的 T_g 点升高了 27K，虚化温度 T_f 下降了 46K，这说明其动力性稳定性和热力学稳定性同时得到显著的提升，长时间老化使其进入超稳态。17.7 年室温老化后的超稳 Ce 金属玻璃的热力学稳定性优于大多数通过气相沉积制备的超稳玻璃薄膜，甚至能够与上亿年老化后的琥珀相媲美。分析发现，Ce 基超稳金属玻璃的获得与它的过冷液体的强脆性特性密切相关，Ce 基金属玻璃的过冷液体脆性值在 20~30 之间，是一种典型的强液体玻璃体系。强液体特性一方面使得 Ce 基金属玻璃在老化过程中能够持续快速地通过临近小能谷间的跃迁最终到达较低的能量状态。另一方面，强的液体特性是 Ce 基金属玻璃拥有极低的形核率的主要原因，从而表现出强的抗晶化能力。

模压电感用软磁复合材料研发

王 丽[1,2]，王明旭[1,2]，史贵丙[1,2]

（1. 山东大学机电与信息工程学院，山东威海 264209；
2. 山东大学威海工业技术研究院，山东威海 264209）

摘 要：新一代信息通讯技术革新升级以及第三代半导体材料的应用推动着电子元器件朝向高频化、小型化、集成化发展；其中，软磁材料是制造高频功率电源、功率电感、高效电机等器件的关键。本报告以"材料-器件-产品"为主线对软磁材料发展及应用的研究现况进行论述。首先简要介绍当下人们对于软磁材料需求的背景和现状，并基于软磁材料的发展历史，阐述每种软磁材料的特性，着重分析作为新一代软磁材料的非晶/纳米晶软磁材料所面临的机遇和挑战；然后分别对各种磁芯和磁粉芯的制备工艺、软磁性能和适用频段等进行了对比分析；最后着重介绍我们在成分设计、磁粉和磁粉芯制备、器件结构和电路仿真等方面的研究成果，并对新一代高性能磁芯的研发和应用前景提出展望。

关键词：软磁材料；磁粉芯；磁导率；饱和磁感应强度

Rejuvenation to Relaxation Transition and Liquid Memory Effect in Metallic Glasses with Different Energy States

WANG Lingling, WANG Zheng, HU Lina

(Key Laboratory for Liquid-Solid Structural Evolution and Processing of Materials, Ministry of Education, Shandong University, Jinan 250061, China)

Abstract: Rejuvenation of metallic glasses (MGs) under cryogenic thermal cycling (CTC) has been intensively studied due to its great theoretical and practical value. However, the relationship between energy states of MGs and rejuvenation capacity is still unclear. Here, a systematic study is performed on how the initial state of MGs affects the rejuvenation effect, and the difference in the initial state of MGs is mainly depicted by the fictive temperature T_f, as well as mechanical properties. A "rejuvenation to relaxation" transition upon T_f increasing is found among samples, and the critical energy state ($T_{f,c} \sim 1.1 T_g$) at which the transition occurs are further determined. The transition towards relaxation above $T_{f,c}$ is attributed to the first CTC intervention of the local structures that are smaller than structural units involved in slow β-relaxations. Besides, the CTC-induced recovery of MGs with a low T_f towards high temperature supercooled liquids has been found, and this behavior relates to the memory effect of MGs upon cooling process. When the liquid memory of MGs is awakened by CTC, the recovery to higher energy state (rejuvenation) occurs. This work supplements rejuvenation mechanism of MGs with different initial states, and provides a possibility to predict the evolution direction (rejuvenation or relaxation) of MGs by their energy states (or T_f).

Key words: metallic glasses; rejuvenation; relaxation; liquid memory effect

基于熔体性质非晶合金微观结构–性能一体化调控

张海峰

（东北大学，辽宁沈阳 114009）

摘 要：报告工作从合金熔体性质出发开展了非晶合金成形、组织结构与性能一体化调控研究。揭示了非晶合金熔体结构演化规律及对性能的影响机制，合适的熔体处理工艺提高了合金熔体的稳定性，提高了非晶的形成能力；适量氧的引入促进了 Ti 基非晶合金熔体中二十面体团簇结构形成，在变形过程中，团簇结构和剪切带的交互作用提高了非晶合金的塑性变形能力。氧元素在 Ti 基非晶复合材料中主要偏聚于内生β-Ti(Zr)枝晶相中，促进诱发形变相变行为，既提高了非晶复合材料强度，又提高其塑性变形能力；开发了非晶合金材料薄带连铸成形技术，实现了厚度大于 100μm 非晶及其复合材料的薄带制备。

关键词：非晶材料；熔体性质；结构–性能–成形一体化调控

人工智能辅助非晶合金结构与性能研究

管鹏飞[1,2]

（1. 北京计算科学研究中心，北京 100193；2. 杭州电子科技大学，浙江杭州 310000）

摘 要：非晶合金结构的无序特征与非均匀性的耦合给实验研究其原子尺度的特性及其物性关联带来了巨大挑战，目前的实验研究手段仍然受限于时空分辨率的不足，很难捕捉到其原子尺度特征及对外场响应，基于计算模拟的理论研究为克服这一挑战提供了有效途径。但由于多元素协同精确描述的挑战、计算方法和计算能力的限制，模型体系和真实的非晶合金材料之间还存在着难以逾越的鸿沟。近年来我们充分利用和融合多学科的知识与研究进展，探索和发展了有效的原子精度研究模式，并成功应用于认识复杂非晶合金体系的结构及其性能关联，为无序合金理论框架的建立和材料性能的智能设计提供了可能途径。

关键词：非晶合金；材料设计；计算模拟；人工智能

弛豫非晶合金的再韧化

李 毅

（中国科学院金属研究所，辽宁沈阳 110016）

摘 要：非晶合金是由合金熔体快速冷却而得到，处于一种热力学非平衡态。高能态的非晶合金自发地向低能态转变，这一过程称为弛豫。非晶合金的高温退火，重新配置原子结构向低能量状态转变，引起其自由体积的湮灭，使原子堆积更加致密且比重增加，进而导致其塑性和韧性的急剧降低，使其发生脆化。如何解决非晶合金由弛豫导致的脆性问题，是非晶合金作为结构材料突破应用瓶颈的关键问题之一。我们探索了一种新的无损伤高效快速的年轻化方法，即将弛豫后的非晶合金快速加热到一定的温度进行短暂的加热。处理后的非晶合金不但没有发生弛豫反而实现了年轻化。我们的研究结果表明，利用非晶合金的记忆效应能够简单高效地调控非晶合金的性能和结构，为解决非晶合金的脆性问题提供一种新思路，将拥有更广阔的应用前景。

关键词：非晶合金；年轻化；弛豫；再韧化

高性能软磁非晶合金材料的开发及应用

孙保安[1,2]，周 靖[2]，李雪松[1]，白海洋[1]，汪卫华[1]

（1. 中国科学院物理研究所，北京 100190；2. 松山湖材料实验室，广东东莞 523106）

摘 要：宽禁带半导体的广泛应用要求与芯片匹配的磁性电子元器件向高频、高效和高功率化发展，这对软磁材料的性能提出了更高的要求，如更高的饱和磁感应强度、更低的矫顽力以及高频磁导率等；然而，这些性能在软磁材料里通常相互矛盾，存在互斥关系。对软磁非晶合金来说，这些性能的矛盾和合金玻璃形成能力以及随后的纳米晶化热处理控制难易密切相关。本报告中，我们主要介绍通过调控非晶合金的纳米晶结构以及晶化机制设计开发高性能软磁非晶合金材料的两个工作。首先，基于一种介于传统非晶和纳米晶之间的过渡态结构设计，成功开发出了一种新型高性能软磁非晶合金材料，表现出超高的 B_s（高达 1.94T）和低至 4.3A/m 的 H_c，突破了铁基非晶纳米晶合金体系中 B_s 和 H_c 之间的互斥关系；其次，通过对纳米晶结构和成分分布的精确表征，在铁基非晶合金中发现了一种新的以 Fe-Co 磁性团簇为形核核心的纳米晶化机制，该机制与传统的以 Cu 为核心的纳米晶化机制完全不同。基于该晶化机制，开发出了在高频下具有超高磁导率（36000@100kHz，比 Finemet 高出 44%）的新型软磁合金材料并揭示了其独特的磁畴运动机制。该材料在工业条件下已成功实现百公斤中试并具有较低的成本。

基于该材料，目前已开发出了可用于新能源汽车车载电源、光伏逆变器的高性能共模电感磁芯产品，并成功在企业获得应用。

关键词：非晶合金；纳米晶；软磁性能；磁导率

Research on High-performance Amorphous Magnetic Alloys

SUN Baoan[1,2], ZHOU Jing[2], LI Xuesong[1], BAI Haiyang[1], WANG Weihua[1]

(1. Institute of Physics, Chinese Academy of Sciences, Beijing 100190, China;
2. Songshan Lake Materials Laboratory, Dongguan 523106, China)

Abstract: The widespread application of wide bandgap semiconductors requires the development of magnetic electronic components matched with chips towards high-frequency, high-efficiency, and high-power, which puts forward higher requirements for the performance of soft magnetic materials, such as higher saturation magnetic induction intensity, lower coercivity, and high-frequency magnetic permeability; However, these properties are often contradictory and mutually exclusive in soft magnetic materials. For soft magnetic amorphous alloys, these performance contradictions are closely related to the glass forming ability of the alloy and the difficulty in controlling subsequent nanocrystalline heat treatment. In this report, we mainly introduce two works on developing high-performance soft magnetic amorphous alloy materials by regulating the nanocrystalline structure and crystallization mechanism of amorphous alloys. Firstly, based on a transition state structure design between traditional amorphous and nanocrystalline materials, a new high-performance soft magnetic amorphous alloy material has been successfully developed, exhibiting ultra-high B_s (up to 1.94T) and H_c as low as 4.3A/m, breaking through the mutual exclusion relationship between B_s and H_c in iron based amorphous nanocrystalline alloy systems; Secondly, by accurately characterizing the nanocrystalline structure and composition distribution, a new nanocrystalline mechanism with Fe-Co magnetic clusters as the nucleation core was discovered in iron-based amorphous alloys, which is completely different from the traditional nanocrystalline mechanism with Cu as the core. Based on this crystallization mechanism, an ultra-high magnetic permeability has been developed at high frequencies (36000@100kHz, 44% higher than Finemet, and reveals its unique magnetic domain motion mechanism. This material has successfully achieved a 100 kilogram pilot test under industrial conditions and has a lower cost. Based on this material, high-performance common mode inductance magnetic core products have been developed for use in new energy vehicle on-board power supplies and photovoltaic inverters, and have been successfully applied in enterprises.

Key words: amorphous alloys; nanocrystalline alloys; soft magnetic properties; permeability

退火和低温处理态软磁 Fe-Co-B 系金属玻璃的结构调制与性能表征

王 艳，肖宗奇，刘国强，翟思成

（济南大学，山东济南 250022）

摘 要：为满足复杂服役条件需求，对软磁金属玻璃综合性能的开发正受到越来越广泛的关注。通过退火和低温处理的方式可显著影响 Fe-Co-B 系金属玻璃的软磁行为。与淬态样品相比较，浅低温处理对降低金属玻璃矫顽力的效果最优异，晶化退火有助于较大程度的提升饱和磁化强度。通过适当低温和退火处理的样品，其达到饱和磁化状态速率和显微硬度显著改善。在模拟海水腐蚀的电化学实验中，淬态薄带表面虽可形成较致密钝化层，但其与非晶内表面存在空隙，不利于持续保护薄带表面，而浅低温处理能明显提升综合耐蚀性。通过估算平均最近邻原子间距，发现 Fe-Co-B 系金属玻璃的矫顽力和其有较为可靠的线型关系，并建立相关性数学模型，进行较为深入的机理解释。

关键词：金属玻璃；低温处理；晶化；软磁性能；耐蚀性；显微硬度

Ni-Nb 二元合金玻璃形成能力和结晶行为差异的结构起源研究

卢文飞，沈 军

（福建理工大学，福建福州 350118）

摘 要：在不同非晶合金体系中，玻璃形成能力受限是一个普遍存在的问题。本文系统研究了近共晶成分的 Ni-Nb 二元合金的玻璃形成能力，旨在探索不同成分 Ni-Nb 二元非晶合金之间玻璃形成能力差异的内在起源。通过结合 k-近邻机器学习方法研究了类液体原子结构与抗晶化能力的相关性。结果表明，$Ni_{61}Nb_{39}$、$Ni_{61.5}Nb_{38.5}$ 和 $Ni_{62}Nb_{38}$ 非晶合金的临界直径达到 2mm。在具有良好玻璃形成能力的 Ni-Nb 非晶合金中观察到了非晶-非晶转变。这些合金表现出较低的热力学驱动力和相应较高的脆性系数值。此外，二十面体团簇的含量和类晶体团簇的概率密度与 Ni-Nb 非晶合金的玻璃形成能力和抗结晶性相关。类液体区域中类晶体团簇的大量存在使合金更容易结晶。与此同时，二十面体团簇对类晶体区域产生"钉扎效应"，有效抑制结晶。以上结果为我们理解 Ni-Nb 二元非晶合金的玻璃形成能力的本质提供了线索。

关键词：Ni-Nb 非晶合金；玻璃形成能力；结晶行为；机器学习

一种具有位错-析出相骨架的超强度和延展性的高熵合金

贾延东，王 刚，穆永坤

（上海大学材料科学与工程学院材料研究所，上海 200444）

摘 要：通过引入位错和沉淀相，已证明是改善金属材料力学性能和打破强度-延展性折中的有效方法。然而，在金属材料中很难获得合适的位错和沉淀相的组合，即高密度位错和高体积分数的沉淀相的共存。本研究中利用 3D 打印技术成功实现了在高熵合金中获得高密度位错结构和高体积分数韧性纳米沉淀相的结合。这种 3D 打印的高熵合金具有新颖的位错-沉淀物骨架（DPS）结构和包裹在 DPS 中的高密度韧性纳米沉淀相，其具有约 1.8GPa 的超高拉伸强度和最大延展率约 16%。超高强度主要来自位错-沉淀协同强化，而大延展性主要源于多重堆垛层错（SF）结构的演变。DPS 不仅可以减缓应变过程中的位错运动，而不完全阻碍其运动，更重要的是，DPS 在变形过程中仍具有良好的结构稳定性，避免了边界处的应力集中导致的合金的过早失效。DPS 的形成促进了金属基 3D 打印技术在高性能材料制备中的发展，并为进一步提高高熵合金性能提供了有效途径。

关键词：位错-沉淀骨架；高熵合金；3D 打印；多重层错结构

A High-entropy Alloy with Dislocation-Precipitate Skeleton for Ultrastrength and Ductility

JIA Yandong, WANG Gang, MU Yongkun

(Institute of Materials, Shanghai University, Shanghai 200444, China)

Abstract: The introduction of dislocations and precipitates has proven to be the effective methods to improve the mechanical properties of metallic materials and break strength-ductility trade-off. However, it is difficult to obtain a suitable combination of both strategies in the metal materials, that is, the coexistence of high-density dislocations and high-volume-fraction precipitates. Here, utilizing a three-dimensional (3D) printing technique, we have successfully achieved a combination of high-density dislocation structures and high-volume-fraction ductile nano-precipitates in a high-entropy alloy (HEA). This 3D-printed HEA, with a novelty dislocation-precipitate skeleton (DPS) architecture and high-density ductile nano-precipitations wrapped in the DPS, has an ultra-high tensile strength of ~ 1.8GPa together with the maximum elongation of ~ 16%. The ultra-high strength mainly comes from dislocation-precipitation synergistic strengthening, while the large ductility mainly originates from an evolution of multiple stacking fault (SF) structures. The DPS can not only slow down the dislocation movement during the strain process without completely hindering its motion, but more importantly, the DPS still has good structural stability during the deformation, which avoids any premature failure due to stress concentrations at the boundary. The DPS formation promotes the development of the metal-based 3D printing technique in the preparation of the high-performance materials, and it can provide an efficient pathway for further enhancement of the high-entropy alloy properties.

Key words: dislocation-precipitate skeleton; high-entropy alloy; 3D-printed; multiple stacking fault structures

含钴铁基非晶纳米晶合金新成分开发探索

董帮少[1]，邢彦兴[1]，崔宏祥[2]，周少雄[1]

（1. 江苏集萃安泰创明先进能源材料研究院有限公司，江苏常州 213000；
2. 山东大学（威海）机电与信息工程学院，山东威海 264209）

摘 要： 铁基软磁合金具有高磁导率、低矫顽力和低损耗等众多优点。但其低于硅钢的饱和磁感应强度（B_s）限制了其进一步的应用。提升铁基软磁合金 B_s 的研究大多以提高 Fe 元素的含量或以 Co 元素替代的方式进行，但已有报道结果表明 B_s 提升不够显著。该工作以已实现商业化应用的 FeSiBC 和 FeSiBCuNb 合金为基础，改变了添加 Co 元素的成分设计方式，将成分设计为$(Fe_{82}Si_4B_{13}C_1)_{1-x/100}Co_x$、$Fe_{82-x}Co_xSi_4B_{13}C_1$ 和$(Fe_{74}Si_{15}B_7Cu_1Nb_3)_{1-x/100}Co_x$。采用 XRD、L-TEM、Magneto-Kerr、DSC、B-H 仪、第一性原理分子动力学等手段分析了材料的结构、热性能和软磁性能。获得的主要的研究结果如下：

（1）$(Fe_{82}Si_4B_{13}C_1)_{1-x/100}Co_x$ ($x = 0, 5, 10$ 和 15)非晶合金的 B_s 可从 1.64 T 提高到 1.78 T。特别地，Co 含量（原子数分数）仅为 5%时，B_s 达 1.73T，且 H_c 为 8.9A/m。随着 Co 含量增加，合金的矫顽力增加、磁导率下降。$(Fe_{82}Si_4B_{13}C_1)_{0.90}Co_{10}$ 合金在不含 Cu 等典型形核元素的情况下，经适当热处理后可获得结构可控的非晶纳米晶双相材料。340℃退火后保持非晶结构，表现出典型的软磁特征；380℃退火后为非晶纳米晶双相结构，平均晶粒尺寸为 54nm，晶化相体积分数为 20.2%，表现出低导磁（μ_i=240）和恒导磁特征。第一性原理分子动力学模拟对 Fe(Co)SiBC 熔体的团簇结构研究表明，随着 Co 含量的增加，合金总的双体相关函数第一近邻峰强度增加，熔体中原子团簇的数量增加。

（2）对于 $Fe_{82-x}Co_xSi_4B_{13}C_1$($x$ = 0, 5, 10, 15, 20, 25, 30)合金中，淬态 XRD 和 DSC 的结果表明 Co 的掺杂对合金的 GFA 和对 T_{x1} 没有明显影响，但是会提高居里温度 T_c 和 T_{x2}，使原本的单放热峰趋势变为双放热峰，即提高热稳定性；对条带样品进行普通退火，通过对比发现，380℃-10min 的条件下，样品具有最佳的软磁性能。其中，B_s 随 Co 含量的升高呈现先增加后降低的趋势，H_c 随 Co 含量增加而升高，在 Co 含量（原子数分数）为 15%时，B_s 达到最大，为 1.73 T；淬态非晶条带的磁畴为锯齿状，杂生树枝状畴。380℃退火后，对于无 Co 及 15% Co 合金条带磁畴方向为横向，且磁畴变得更加均匀，15% Co 合金条带的磁畴明显宽于无 Co 合金；对于 30% Co 合金条带，磁畴变得更加杂乱；磁畴变化很好地解释了软磁性能地变化。

（3）$(Fe_{74}Si_{15}B_7Cu_1Nb_3)_{1-x/100}Co_x$($x$ = 0, 2, 4, 6, 8, 10)合金系的淬态带材在 $x \geq 6$ 时发生表面晶化，在 500~560℃退火 60min 时，可以获得稳定的软磁性能，表现出宽的退火工艺窗口。合金的 B_s 可从 1.29T 提升至 1.40T。x = 4 时，合金退火后的 B_s 为 1.35T、H_c 为 2.5A/m，其动态损耗与基础合金相当，但磁感工作点有所提升。结构分析表明，随 Co 含量增加，带材退火后的平均晶粒尺寸均有增大趋势，但仍处于 13~16 nm 的范围。

关键词：非晶纳米晶；软磁性能；含钴；FeSiBC；FeSiBCuNb

金属玻璃薄膜稳定性和不均匀性及性能调控

罗 强

（东南大学材料科学与工程学院，江苏南京 211189）

摘 要：金属玻璃类液体结构特征赋予其优异的物理、力学和化学性能，然而，其复杂的无序结构及由亚稳态本质带来的老化/稳定性问题成为金属玻璃材料科学和应用发展的瓶颈。最近，我们采用磁控溅射技术通过调控衬底温度制备出超稳 Cu-Zr-Al 金属玻璃薄膜，其晶化温度比室温衬底的普通金属玻璃提高 120K，同时硬度提高 30%左右。当衬底温度升到 $0.85T_g$ 时（T_g 为普通快冷玻璃的转变温度），金属玻璃稳定性大幅降低。结合球差校正高角环形暗场扫描透射电子显微镜和纳米压痕刚度成像技术展开系统研究，发现随着衬底温度变化，合金从普通金属玻璃到超稳定金属玻璃的转变本质是特征稳定性团簇发生的逾渗转变。进一步，我们探究了冷热循环对锆基金属玻璃薄膜结构/力学不均匀性、纳米尺度蠕变行为及光学性能的影响。研究结果表明随着循环次数的增加，力学不均匀性呈现先增加后减小的趋势。基于 Maxwell-Voigt 模型分析，冷热循环激活较多软区中弛豫时间长的大尺寸缺陷，对硬区中的缺陷影响较小。此外，15 次循环样品 Cu 及 Ni 元素富集，从而提高光学折射率。本研究为理解金属玻璃薄膜力学/结构不均匀性、稳定性和性能调控影响提供了新思路，有助于设计高性能纳米结构金属玻璃薄膜。

关键词：金属玻璃；不均匀性；稳定性；冷热循环

激光 3D 打印非晶合金复合材料的相变增韧行为研究

张鹏程，柳 林

（华中科技大学材料科学与工程学院、材料成形与模具技术国家重点实验室，湖北武汉 430074）

摘　要：选区激光熔化(SLM)3D 打印技术为制备复杂形状的块体非晶合金提供了一种全新的方法，但在 3D 打印过程中，复杂的热历史会在热影响区中引发晶化，从而导致 3D 打印非晶合金塑韧性差。为解决这一问题，在本工作中，我们选择了可在热影响区中析出具有"相变诱导塑性"效应的 B2-ZrCu 相的 ZrCu 基非晶合金体系进行 3D 打印实验。借助 XRD、DSC、SEM、TEM 和力学性能试验机系统研究了 3D 打印样品的结构和性能。首先，我们研究了 3D 打印二元 $Zr_{50}Cu_{50}$ 非晶复合材料的性能。研究发现，该非晶复合材料在变形过程中表现出明显的加工硬化行为，断裂强度达 1841MPa，塑性应变达 3.17%。然而由于该二元体系非晶形成能力有限，SLM 成形样品中的析出相除 B2-ZrCu 相外，还有部分脆性的金属间化合物，因此塑性提升效果有限。为了进一步提升性能，我们在 $Zr_{50}Cu_{50}$ 体系中添加了非晶相和 B2 相稳定化元素 Al 和 Co，选择 $Zr_{47.5}Cu_{45.5}Al_5Co_2$ 非晶复合材料体系进行实验，制备得到含单一 B2 析出相的非晶复合材料。该体系在变形过程中同样呈现明显的加工硬化行为，且断裂强度达 1851MPa，塑性应变为 4.65%，断裂韧性值为 53.9MPa·$m^{1/2}$。压缩塑性的提升是由热影响区中 B2-ZrCu 相的"TRIP"效应以及热影响区对剪切带扩展的阻碍作用共同导致的。断裂韧性的提升主要取决于缺口处的 B2-ZrCu 相在力的作用下发生马氏体相变吸收能量，从而抑制裂纹的萌生。此工作为激光 3D 打印高强高韧非晶合金及其构件提供了一个新思路。

关键词：增材制造；非晶合金

低碳绿色钢铁冶金关键技术研发与应用

叶恒棣

（中冶长天国际工程有限责任公司，湖南长沙　410205）

摘　要：钢铁工业是能源消耗和污染物排放的重点领域，随着国家、行业节能环保以及"碳达峰、碳中和"相关政策的出台，节能环保相关指标已成为决定企业生存的关键因素。近年来，鞍钢在节能环保领域开展了大量工作，并取得了显著效果。本文首先介绍了鞍钢节能环保工作现状；其次，以高效换热器、纯氧钢包烘烤器等典型技术为例，概述了鞍钢节能降碳技术的开发与应用情况；第三，以废水、固体废弃物处理为例，叙述了鞍钢环保技术的研发与实践情况；最后，综合钢铁行业节能环保技术的发展趋势，对鞍钢的节能环保技术进行展望。

关键词：钢铁工业；节能低碳；环保

Development and Application of Low-carbon Green Iron and Steel Metallurgy Key Technology

YE Hengdi

(MCC Changtian International Engineering Co., Ltd., Changsha 410205, China)

Abstract: The iron and steel industry is the key area of energy consumption and pollutant emission. With the introduction of national and industrial energy conservation and environmental protection and the relevant policies of "carbon peak, carbon neutrality", the relevant indicators of energy conservation and environmental protection have become the key factors determining the survival of enterprises. In recent years, Angang has carried out a lot of work in the field of energy conservation and environmental protection, and achieved remarkable results. This paper first introduces the current situation of energy conservation and environmental protection in Angang. Secondly, the development and application of energy saving and carbon reduction technology in Angang are summarized by taking the typical technology of high efficiency heat exchanger and pure oxygen ladle roaster as examples. Third, taking wastewater and solid waste treatment as an example, the research and development and practice of environmental protection technology in Angang are described. Finally, based on the development trend of energy-saving and environmental protection technology in iron and steel industry, the outlook of Angang's energy-saving and environmental protection technology is carried out.

Key words: steel industry; energy saving and low carbon; environmental protection

钢渣在建材领域资源化应用关键问题分析

田志红

（首钢技术研究院，北京 100043）

摘 要：随着双碳工作的推进，钢渣作为钢铁联合企业的固废，其资源化价值越来越明显，钢渣的大宗利用显得越来越重要了。磁选后钢尾渣可作为烧结熔剂用料、铺路，磨细粉后作为掺合料或者制成胶凝材料直接用于制作混凝土、干混砂浆等建筑用材料。本文从国家政策角度、标准和钢渣物理特性角度出发，围绕上述钢渣大宗利用方向进行了详细分析，给出了钢渣在 Cl^-、Cr^{6+}、游离 CaO、MgO、碱金属、成分波动等方面的要求及控制方向。

关键词：钢渣；大宗利用；资源化应用；建材

Analysis of Key Problems in Resource Utilization of Steel Slag in Building Materials Field

TIAN Zhihong

(Shougang Institute of Technology, Beijing 100043, China)

Abstract: With the advance of dual-carbon work, the resource value of steel slag as solid waste in iron and steel complex enterprises becomes more and more obvious, and the bulk utilization of steel slag becomes more and more important. After magnetic separation, the steel tail slag can be used as the material for sintering flux, paving, grinding powder as an admixture or as a cementing material directly used to make concrete, dry mixed mortar and other building materials. From the Angle of national policy, standard and physical characteristics of steel slag, this paper analyzes in detail the direction of bulk utilization of steel slag, and gives the requirements and control direction of steel slag in Cl^-, Cr^{6+}, free CaO, MgO, alkali metals, composition fluctuation and so on.

Key words: steel slag; bulk utilization; resource application; cementitious material

基于 5G 技术的钢铁企业多维度生态环保管控系统开发与应用

姜德旺[1]，张亚夫[2]，朱宴恒[3]

（1. 北京京诚嘉宇环境科技有限公司，北京 100053；2. 中冶京诚工程技术有限公司，北京 100176；3. 中冶京诚数字科技（北京）有限公司，北京 100176）

摘　要：生态环保管理智能化是实现钢铁企业绿色高质量发展的重要路径之一。21 世纪以来，随着大数据、物联网、5G 技术、人工智能等新一代信息技术的不断发展，为实现钢铁企业生态环境管理智能化提供了可能。目前，部分钢铁企业生态环保管理理念较为落后，管理水平不高；自动化水平较低，现有环境管理系统架构无法有效解决的钢铁企业环保管理监测点位众多、系统网络复杂、数据安全性差、对接数据繁杂等问题。中冶京诚立足于在钢铁领域智能化和绿色化的设计和工程基础，基于多年环评、环保管家经验和先进的钢铁企业环保管理理念，开发出了基于 5G 技术的千万吨级钢铁企业多维度生态环保管控系统。

该系统搭建了"三废+综合环保管理"一体化全要素体系架构，开发"大屏端综合展示+PC 端专项管理+移动端实时监管"立体化管控模式，实现了 4 大项、50 个单元和 3 个不同层级的管控。

该系统提出了基于大数据分析的多目标协同降低环保治理成本控制模型，首次通过多系统大数据分析，达到环保效益和环保成本协同控制最优点。通过系统非同步运行报警及闭环管理，可为千万吨级企业全年节省环保运行费用上千万元。

该系统实现工业互联网平台体系架构与 5G LAN 技术、信创技术栈在千万吨级钢铁企业环保管理系统的融合，开发了生态环保管控平台等 29 项软件，在保证企业数据安全及产品自主可控的基础上，极大简化了系统整体网络架构，降低了维护难度，节约维护成本约 20%，监测数据平均上传带宽提升 10 倍，生产、治理、监测设施数据采集达到百万级/s，满足千万吨级钢铁企业的环保管理需要。

该系统采用 LightGBM 机器集成学习算法、基于视频 AI 智能分析和水质监测结果，开发 PM2.5 浓度预测模型和雨水异常排放识别模型，提高企业环保应急响应能力。

关键词：5G 技术；钢铁企业；生态环保管控系统

Development and Application of Multi-dimensional Ecological Environmental Control System for Iron and Steel Enterprises Based on 5G Technology

JIANG Dewang[1], ZHANG Yafu[2], ZHU Yanheng[3]

(1. Beijing Jingcheng Jiayu Environmental Technology Co., Ltd., Beijing 100053, China;
2. MCC Jingcheng Engineering Technology Co., Ltd., Beijing 100176, China;
3. MCC Jingcheng Digital Technology (Beijing) Co., Ltd., Beijing 100176, China)

Abstract: Intelligent ecological and environmental management is one of the important ways to achieve green and high-quality development of iron and steel enterprises. Since the 21st century, with the continuous development of new generation information technologies such as big data, Internet of Things, 5G technology and artificial intelligence, it has provided the possibility for realizing the intelligent ecological environment management of iron and steel enterprises. At present, some iron and steel enterprises ecological environmental management concept is relatively backward, management level is not high; The level of automation is low, and the existing environmental management system architecture can not effectively solve the problems of numerous environmental management monitoring points, complex system network, poor data security, and complex docking data of iron and steel enterprises. Based on the intelligent and green design and engineering foundation in the steel field, MCC Jingcheng has developed a multi-dimensional ecological and environmental protection management system for tens of millions of tons of steel enterprises based on 5G technology based on years of environmental assessment, environmental management experience and advanced environmental management concepts for steel enterprises.

The system has built an integrated all-factor system architecture of "three wastes + comprehensive environmental management", developed a three-dimensional control mode of "comprehensive display on the large screen + special management on the PC + real-time supervision on the mobile end", and achieved control of 4 major items, 50 units and 3 different levels.

The system proposes a multi-objective collaborative cost control model of environmental protection management based on big data analysis, and for the first time through multi-system big data analysis, the best advantages of environmental protection benefits and environmental protection costs collaborative control are achieved. Through the system asynchronous operation alarm and closed-loop management, tens of millions of tons of enterprises can save tens of millions of yuan of environmental protection operation costs throughout the year.

The system realizes the integration of industrial Internet platform architecture with 5G LAN technology and Xinchuang technology stack in the environmental management system of tens of millions of tons of iron and steel enterprises, and develops 29 pieces of software such as ecological environmental control platform. On the basis of ensuring the security of enterprise data and the autonomy and controllability of products, it greatly simplifies the overall network architecture of the system and reduces the difficulty of maintenance. The maintenance cost is saved by about 20%, the average upload bandwidth of monitoring data is increased by 10 times, and the data collection of production, governance and monitoring facilities reaches millions/s, meeting the environmental management needs of tens of millions of tons of steel enterprises.

The system uses LightGBM machine integrated learning algorithm, intelligent analysis based on video AI and water quality monitoring results to develop PM2.5 concentration prediction models and abnormal rainwater discharge identification models to improve enterprises' environmental emergency response capabilities

Key words: 5G technology; iron and steel enterprises; ecological and environmental control system

钢铁行业高质量绿色低碳发展路径探讨

陈 洁

（中冶检测认证有限公司，北京 100088）

摘 要：钢铁工业是国民经济的重要基础产业，同时我国钢铁工业在全球占据重要地位。我国是全球最大的钢铁生产国，具有能源资源禀赋、产量大、企业数量多、碳排放机理复杂基本特征。随着我国双碳目标的提出，钢铁行业高质量绿色低碳发展已成为必由之路。

近年来，随着国家政策对绿色低碳发展政策的推进，钢铁行业采取积极措施进行超低排放改造、淘汰过剩产能、极致提升能效、降本优化减碳，但钢铁碳排放量仍占全国15%吨左右，而且受到市场经济和国内外需求形势变化的影响，钢铁行业下游需求不及预期，钢企利润持续承压，这对我国钢铁行业的环保、减排、绿色低碳转型提出了挑战。钢铁行业作为"碳达峰""碳中和"目标实现的重要领域和责任主体行业，首当其冲面临的挑战之一就是国家对双碳目标的部署，倒逼钢铁行业经济发展方式转型；我国钢铁生产以高炉-转炉长流程为主，电炉短流程炼钢比例仅10%左右，欧美电炉钢比例达40%~60%，废钢资源有限，限制短流程炼钢工艺发展缓慢；钢铁行业作为高耗能高排放行业，能源结构高碳化，煤、焦炭占能源投入近90%，要大力改进能源消费结构；同时钢铁行业低碳绿色发展顶层设计也已明确，对企业提出了新的竞争要求，企业需努力打造低碳绿色化的核心竞争力。除了我国内部环境对钢铁行业提出的挑战，外部环境的碳关税等也对钢铁行业提出新的挑战。目前，欧盟已经完成立法，正式实施碳边境调节机制，这将进一步增加我国钢铁行业绿色低碳转型的成本和难度。

本报告从双碳背景下钢铁行业面临的挑战、钢铁行业绿色低碳发展现状以及钢铁产品碳评价认证意义的现状等方面，总结分析钢铁行业高质量绿色低碳发展路径，为促进钢铁行业低碳转型起到积极促进与引导作用。

关键词：钢铁行业；碳关税；绿色低碳转型；碳评价认证

Research and Analysis of High-quality and Green-low-carbon Development Path of Steel Industry

CHEN Jie

(Inspection and Certification Co., Ltd., MCC, Beijing 100088, China)

Abstract: The iron and steel industry is an important basic industry of the national economy. Iron and steel industry of China occupies an important position in the world. As the largest steel producer in the world's, we have the basic characteristics of energy and resources endowment, large output, large number of enterprises and complex carbon emission mechanism. With the proposal of China's dual-carbon goal, high-quality and green-low-carbon development of the steel industry has become the only way.

In recent years, with the promotion of the national policy on green and low-carbon development, the steel industry has taken active measures to carry out ultra-low emission transformation, eliminate excess capacity, improve energy efficiency and optimize carbon reduction, but the carbon emissions of steel still account for about 15% of the country. Moreover affected by the market economy and domestic and foreign demand changes, the downstream demand of the steel industry is less than expected. The profits of steel enterprises continue to be under pressure, which challenges the environmental protection, emission reduction, green and low-carbon transformation of China's steel industry. As an important field and responsible industry for the realization of "carbon peak" and "carbon neutral" goals, one of the challenges facing the steel industry is the deployment of the dual carbon goal forcing the transformation of the economic development mode of the steel industry. China's iron and steel production is mainly based on the long process of blast-converter, the proportion of short process of electric furnace steelmaking is only about 10%. The proportion of European and American electric furnace steel is 40%-60%. Limited scrap steel resources limiting the development of short process steelmaking. The steel industry as a high energy consumption and high emission industry has a high carbonization energy structure.Because of coal and coke accounted for nearly 90% of energy input. So we should improve the energy consumption structure vigorously. At the same time, the top-level design of low-carbon green development in the steel industry has also been clear which puts forward new competition requirements for enterprises. Enterprises need to create the core competitiveness of low-carbon green. In addition to the challenges posed by China's internal environment to the steel industry, carbon tariffs in the external environment also pose new challenges to the steel industry. At present, the EU has completed legislation to formally implement the carbon border adjustment mechanism which will further increase the cost and difficulty of the green low-carbon transformation of China's steel industry.

This report summarizes and analyzes the high-quality and green-low-carbon development path of the steel industry from the aspects of the challenges faced by the steel industry under the dual-carbon background and the status quo of green-low-carbon development of the steel industry and the significance of carbon evaluation and certification of steel products. This report plays a positive role in promoting and guiding the low-carbon transformation of the steel industry.

Key words: steel industry; carbon tariff; green-low-carbon transformation; carbon evaluation certification

鞍钢节能环保技术开发与实践

马光宇，蔡秋野，王向锋，李卫东，陈 鹏

(鞍钢集团钢铁研究院，辽宁鞍山 114009)

摘 要： 钢铁工业是能源消耗和污染物排放的重点领域，随着国家、行业节能环保以及"碳达峰、碳中和"相关政策的出台,节能环保相关指标已成为决定企业生存的关键因素。近年来,鞍钢在节能环保领域开展了大量工作,并取得了显著效果。本文首先介绍了鞍钢节能环保工作现状;其次,以高效换热器、纯氧钢包烘烤器等典型技术为例,概述了鞍钢节能降碳技术的开发与应用情况;第三,以废水、固体废弃物处理为例,叙述了鞍钢环保技术的研发与实践情况;最后,综合钢铁行业节能环保技术的发展趋势,对鞍钢的节能环保技术进行展望。

关键词:钢铁工业;节能低碳;环保

Development and Practice of Energy Conservation and Environmental Protection Technologies in Ansteel

MA Guangyu, CAI Qiuye, WANG Xiangfeng, LI Weidong, CHEN Peng

(Ansteel Iron & Steel Research Institutes, Anshan 114009, China)

Abstract: The energy consumption and pollutant emissions of the steel industry have received much attention. With the introduction of national and industry policies related to energy conservation and environmental protection, as well as the "carbon peaking and carbon neutrality", if the enterprises fail to meet the requirements on time, they will have to stop the steel production. In recent years, ANSTEEL has carried out lots of work on energy conservation and environmental protection and achieved remarkable results. Firstly, the current status of energy conservation and environmental protection work in ANSTEEL is introduced. Secondly, the development and application of energy conservation and carbon reduction technologies including high efficiency heat exchanger and ladle baking with pure oxygen in ANSTEEL is summarized. Thirdly, the development and application of environmental protection technologies such as waste water and solid waste treatment is introduced. Finally, according to the development trends of energy conservation and environmental protection technologies in iron and steel industry, the paper gives expectation that ANSTEEL should take further actions to push forward progress and achievement of advanced technologies.

Key words: iron and steel industry; energy conservation and low carbon; environmental protection

钢铁行业超低排放有组织监测评估技术与实践

刘通浩

(中国环境监测总站,北京 100012)

摘 要:报告就钢铁行业超低排放的意义和进展、钢铁行业有组织排放监测评估基本条件及技术要点、钢铁行业有组织排放监测评估报告审核技术要点及钢铁企业自行监测日常管理常见问题与解决方案展开介绍。

关键词:钢铁行业;有组织排放;监测评估

Technology and Practice of Organized Monitoring and Evaluation of Ultra-low Emission in Steel Industry

LIU Tonghao

(China Environmental Monitoring Station, Beijing 100012, China)

Abstract: The report introduces the significance and progress of ultra-low emissions in the steel industry, the basic conditions and technical points of organized emission monitoring and evaluation in the steel industry, the technical points of organized emission monitoring and evaluation report review in the steel industry, and the common problems and solutions of daily management of self-monitoring in steel enterprises.

Key words: the steel industry; organized discharge; monitoring and assessing

钢铁企业水系统现状及节水降碳优化策略分析

余云飞

（中冶赛迪工程技术股份有限公司，重庆 401120）

摘 要：钢铁行业是国民经济的重要支柱产业，在经济发展中具有重要作用。我国是世界最大的钢铁生产国，拥有高产量钢材的同时也面临高耗能、高耗水、高碳排的挑战。为满足国家推动全社会节水等国家战略性层面的节能减排需求，钢铁行业水系统的管理优化和技术升级势在必行。目前国内钢铁企业近年来的吨钢新水耗量及吨钢废水排放量指标的下降趋势正在逐步放缓，已经基本达到极限，难以进一步提升，钢铁企业节水正在进入瓶颈期。通过非典型水源（如海水、雨水等）的充分利用实现"开源"，通过对废水进行深度处理乃至零排放极致回收水资源实现"节流"，并通过智能化手段提升管理水平保证达到最优的"开源节流"效果是未来企业发展的必由之路。

关键词：冶金环保；钢铁企业水系统；节水减排；智能化管控

Current Situation of Water System in Iron and Steel Enterprises and Analysis of Optimal Strategy of Saving Water and Reducing Carbon

YU Yunfei

(Ccid Engineering Technology Co., Ltd., Chongqing 401120, China)

Abstract: The iron and steel industry is an important pillar industry of the national economy and plays an important role in economic development. China is the world's largest steel producer, with a high output of steel, but also faced with high energy consumption, high water consumption, high carbon emission challenges. In order to meet the national strategic level of energy conservation and emission reduction needs such as promoting water saving in the whole society, it is imperative to optimize the management and technological upgrading of water systems in the steel industry. At present, the downward trend of the new water consumption per ton of steel and the discharge of wastewater per ton of steel in domestic iron and steel enterprises in recent years is gradually slowing down, has basically reached the limit, and it is difficult to further improve, and water saving in iron and steel enterprises is entering the bottleneck period. Through the full utilization of atypical water sources (such as seawater, rainwater, etc.) to achieve "open source", through the in-depth treatment of wastewater and even zero discharge of extreme recycling water resources to achieve "throttling", and through intelligent means to improve management level to ensure the optimal "open source throttling" effect is the only way for future enterprise development.

Key words: metallurgical environmental protection; steel enterprise water system; water saving and emission reduction; intelligent management and control

钢铁行业超低排放与除尘系统节能技术发展现状

王珲，张璞，杨雅娟

（中冶建筑研究总院有限公司，北京 100088）

摘　要：系统介绍了钢铁行业超低排放改造相关政策、实施效果与技术应用现状。重点针对除尘系统，分析了节能潜力与技术发展方向，为钢铁行业超低排放技术迭代升级与减污降碳协同增效的落实提供支撑。

关键词：超低排放；节能；技术现状

Development Status of Ultra-low Emission and Energy-saving Technology of Dust Removal System in the Steel Industry

WANG Hui, ZHANG Pu, YANG Yajuan

(Central Research Institute of Building and Construction Co., Ltd., MCC Group, Beijing 100088, China)

Abstract: The relevant policies, implementation effect and technology application status of ultra-low emission transformation in the steel industry are systematically introduced. Focusing on the dust removal system, the energy saving potential and technology development direction are analyzed, which provides support for the implementation of the iterative upgrading of ultra-low emission technologies and the synergistic efficiency of both air pollution and carbon reductions in the steel industry.

Key words: ultra-low emission; energy saving; technology application status

熔融钢渣高效负能处理技术与装备

王宇鹏

（中冶节能环保有限责任公司，北京 100088）

摘　要：浅谈钢铁企业熔融钢渣处理方面最新技术发展方向，介绍中冶环保主导研发的新一代熔融钢渣高效负能处理技术与系统装备。

关键词：熔融钢渣；负能处理技术；环保

Technology and Equipment for High Efficient Negative Energy Treatment of Molten Steel Slag

WANG Yupeng

(System Research Lnstitute of Eco-Etch & Eco-El, Ltd., Mcc Group, Beijing 100088, China)

Abstract: The development direction of the latest technology of molten steel slag treatment in iron and steel enterprises is introduced. The new generation of efficient negative energy treatment technology and system equipment of molten steel slag led by MCC Environmental Protection is introduced.

Key words: molten steel slag; negative energy processing technology; environmental protection

全流程钢厂水系统智慧管控与零排放关键技术开发应用

桂其林

（宝钢湛江钢铁有限公司，广东湛江 524072）

摘 要： 湛江钢铁是宝武集团最新的千万吨级长流程钢铁生产基地，位于中国大陆最南端广东省湛江市东海岛，海岛淡水资源缺乏且周边养殖资源丰富，为深入贯彻习近平生态文明思想，践行宝武集团"三治四化"的工作要求，湛江钢铁立足于长流程钢厂水系统现状，从取水、废水和排水各环节进行系统性分析，开发了多种非常规水源可持续利用技术、"分类处理-柔性平衡-梯级利用"废水回用技术、外排水资源化利用技术和水系统智慧集中管控技术等4项核心技术，形成6项自主创新和5项集成创新关键技术，共输出专利34件和企业技术秘密6项，建立了研发应用国内钢铁行业首创的水系统全流程智慧管控平台，成为国内首家实现全流程废水零排放的千万吨级钢厂。

关键词： 冶金环保；长流程钢厂水系统；非常规水源可持续利用；全流程智慧管控平台

Development and Application of Key Technologies for Intelligent Control and Zero Discharge of Steel Mill Water System in the Whole Process

GUI Qilin

(Baosteel Zhanjiang Iron & Steel Co., Ltd., Zhanjiang 524072, China)

Abstract: Systematic analysis of water intake, waste water and drainage has been carried out, and four core technologies, including a variety of unconventional water source sustainable utilization technologies, wastewater reuse technology of "classification treatment-flexible balance - cascade utilization", external drainage resource utilization technology and water system intelligent centralized management and control technology, have been developed, forming six independent innovation and five integrated innovation key technologies. A total of 34 patents and 6 enterprise technical secrets have been exported, and a full-process intelligent control platform for water system has been established for the research and development and application of the first in the domestic steel industry, becoming the first ten-million-ton steel mill in China to achieve zero discharge of wastewater from the full process.

Key words: metallurgical environmental protection; long process steel water system; sustainable use of unconventional water sources; whole process intelligent control platform

钢铁冶金大综固废综合利用与 CO_2 减排

郭占成

（北京科技大学钢铁冶金新技术国家重点实验室，北京 100083）

摘 要：粉尘、钢渣、脱硫灰、赤泥是大宗冶金固废，目前粉尘的利用存在效率低、能耗高的问题，而钢渣、脱硫灰、赤泥尚未形成规模化利用技术。本文介绍钢铁冶金粉尘湿法、火法相结合的低能耗、有价组分全利用的技术思路；钢渣替代石灰石的烟气脱硫脱硝；脱硫灰与转炉粉尘为原料制备炼钢助熔剂或烧结矿粘结剂铁酸钙及硫酸的技术；铝行业赤泥用于转炉炼钢脱磷的技术。这些技术不仅是固废资源化利用，也与 CO_2 减排密切相关，特别是钢渣替代石灰石的烟气脱硫，如果全国钢渣替代石灰石及于烟气脱硫，其产生量基本与需求量相当，不仅解决了钢渣的出路问题，而且可减排 CO_2 约 6000 万吨。

关键词：冶金粉尘；钢渣；脱硫灰；赤泥；循环利用；CO_2 减排

Comprehensive Utilization of Solid Waste and CO_2 Emission Reduction in Iron and Steel Metallurgy

GUO Zhancheng

(State Key Lab. of Advanced Metallurgy, University of Science and Technology Beijing,
Beijing 100083, China)

Abstract: Dust, steel slag, desulphurized ash and red mud are solid wastes of steelmaking. At present, the utilization of dust has the problems of low efficiency and high energy consumption, but the large-scale utilization technology of steel slag, desulphurized ash and red mud has not yet been formed. This paper introduces the technical idea of low energy consumption and full utilization of valuable components of iron and steel metallurgical dust by combining wet process and pyro process, steel slag replaces limestone for flue gas desulfurization and denitrification, technology for preparing steel flux or sinter binder calcium ferrite and sulfuric acid from desulfurized ash and LD dust, technology of dephosphorization by red mud used in converter steelmaking. These technologies are not only the utilization of solid waste resources, but also closely related to CO_2 emission reduction, especially the replacement of limestone by steel slag for flue gas desulfurization, if the national steel slag replaces limestone and flue gas desulfurization, its production is basically equal to the demand, not only to solve the problem of steel slag outlet, but also to reduce CO_2 by about 60 million tons.

Key words: metallurgical dusts; steel slag; desulfurized ash; red mud; recycle; CO_2 emission reduction

含锌粉尘球团还原行为研究

朱德庆[1,2]，石书冰[1,3]，潘 建[1,2]，郭正启[1,2]，杨聪聪[1,2]

（1. 中南大学资源加工与生物工程学院，湖南长沙 410083；2. 中南大学低碳与氢冶金研究中心，湖南长沙 410083；3. 中钢集团洛阳耐火材料研究院有限公司，河南洛阳 471039）

摘　要： 钢厂含锌粉尘由于锌、铅等重金属含量较高而被列为危险废弃物，其产出量约占粗钢产量的 3%~4%。复杂的化学成份使含锌粉尘难以回收利用，致使其在各钢厂大量堆存，给钢厂带来巨大压力。本文以钢厂三类含锌粉尘为研究对象，系统查明了其物化特性和工艺矿物学特征；分析了含锌粉尘直接还原与碳的活化耦合反应热力学规律；揭示了含锌粉尘中铁氧化物还原和锌、铅、钾、钠还原脱除动力学行为；优化了含锌粉尘含碳球团煤基直接还原工艺，阐明了其直接还原过程的物相转变和微观结构演变规律；由此开发了含锌粉尘金属化球团与活性炭协同制备新工艺，同时制备出比表面积 300.65m^2/g、碘值 655.23mg/g、耐压强度 25.5daN/个、耐磨强度 96.2%的活性炭和抗压强度 905N/个、铁品位 73.59%、铁金属化率 97.03%及锌、铅、钾、钠含量分别为 0.015%、0.008%、0.017%、0.039%的金属化球团。

关键词： 含锌粉尘；含碳球团；金属化球团；活性炭；还原行为

Study on the Reduction Behavior of Zinc-bearing Dust Pellets

ZHU Deqing[1,2], SHI Shubing[1,3], PAN Jian[1,2], GUO Zhengqi[1,2], YANG Congcong[1,2]

(1. School of Resource Processing and Bioengineering, Central South University, Changsha 410083, China;
2. Low Carbon and Hydrogen Metallurgy Research Center of Central South University, Changsha 410083, China; 3. Sinosteel Luoyang Institute of Refractories Research Co., Ltd., Luoyang 471039, China)

Abstract: The zinc-bearing dusts from steel plants which accounted for 3%~4% of crude steel output were classified as hazardous waste and they are hard to be recycled due to their complex composition and high contents of heavy metals such as zinc and lead, as a result, large amounts of zinc-bearing dusts are stockpiled in steel plants and put huge pressure on them. Three kinds of zinc-bearing dusts from steel plant were used as raw material in this paper. The research was carried out to identify mineralogy and physicochemical characteristics of the zinc-bearing dusts. The thermodynamic analysis of coupling process of direct reduction of carbon-containing zinc-bearing dust pellets and activation of carbon was conducted. The kinetic behaviors of reduction of iron oxides and removal of zinc, lead, potassium and sodium from the pellets were revealed. The process of coal-based direct reduction of carbon-containing zinc-bearing dust pellets was optimized. The phases and microstructure transformation of the pellets were clarified. On this basis, a process of synergetic preparation of zinc-bearing dust metallic pellets and activated carbon was developed. The activated carbon with surface area of 300.65m^2/g, iodine absorption value of 655.23mg/g, compressive strength of 255 Newton per granule, abrasive resistance of 96.2% and the metallized pellets with compressive strength of 905 Newton per pellet, iron grade of 73.59%, iron metallization degree of 97.03% and low content of hazardous metals (0.015% Zn, 0.008% Pb, 0.017% K and 0.039% Na) could be simultaneously prepared under the optimal conditions.

Key words: zinc-bearing dusts; carbon-containing pellets; metallic pellets; activated carbon; reduction behavior

基于无废城市建设的长流程钢厂固废管理模式与技术路径

李红红[1,2]

（1. 宝钢湛江钢铁有限公司能源环保部，广东湛江　524072；
2. 宝山钢铁股份有限公司能源环保部，上海　200941）

摘　要： 钢厂工业固废的环境管理是推进"无废城市"建设的重要环节，实现全量资源化利用是钢厂发展循环经济的目标。传统方式是委托社会企业进行综合利用或处置，存在环境风险大、物流量大、监管难等问题；利用"无废"理念，开发了长流程钢厂"源头减量－厂内加工分选－冶金炉窑协同利用－智慧管控"固废管理的创新性技术体系，建设了厂内固废加工、分选、仓储中心，开展了基于钢铁窑炉特性和固废属性协同处理处置的技术路径研究，建立了支撑工业固废全过程管理的系统、数据、技术、产线体系，减少了源头产生量，加大了厂内循环利用量，实现了"固废不出厂"；对防范环境风险、减少固废车辆物流量、实现高水平生态环境保护、缓解城市固废处理压力、优化产城融合和助力无废城市建设具有重要意义。

关键词：无废城市；长流程钢厂；固废管理；技术路径；协同处理

Solid Waste Management Mode and Technical Path of Long-process Steel Plant Based on the Concept of Zero-waste Construction

LI Honghong[1,2]

(1. Department of Energy and Environment, Baosteel Zhanjiang Iron & Steel Co., Ltd., Zhanjiang 524072, China; 2. Department of Energy and Environment, Baoshan Iron & Steel Co., Ltd., Shanghai 200941, China)

Abstract: The environmental management of industrial solid waste in steel plants is an important part in promoting the construction of zero waste city, and achieving full resource utilization is the goal of steel plants in developing a circular economy. The traditional method is to entrust social enterprises for comprehensive utilization or disposal, which has problems such as high environmental risks, large logistics volume, and difficult supervision. Using the "no waste" concept, an innovative technical system of solid waste management of "source reduction - in-plant processing and sorting - collaborative utilization using metallurgical furnaces - Intelligent management and control system of solid waste management" has been developed in long-process steel plants. The solid waste processing, sorting, and warehousing center in the plant has been built, the technical paths research based on the characteristics of steel kilns and properties of solid waste has been carried out, and the system, data, technology, and production line system supporting the whole process management of industrial solid waste has been established, which has reduced the amount of source generation, increased the amount of recycling in the plant, and realized "No solid waste leave the plant". It is of great significance to prevent environmental risks, reduce the logistics volume of solid waste, achieve a high level of ecological environmental protection, relieve the pressure of urban solid waste treatment, optimize the integration of industry and city, and help build a waste-free city.

Key words: zero-waste city; long-process steel plant; solid waste management; technical path; collaborative utilization

大型波纹形耐压壳筒节结构优化与轧制理论研究

彭　艳，孙建亮，张　旭，吴润泽，单存尧

（燕山大学国家冷轧板带装备及工艺工程技术研究中心，河北秦皇岛　066004）

摘　要：大型深潜器和深海空间站等大型深海装备最核心的基础零件是圆柱形耐压壳筒节。耐压壳的抗压性能在很大程度上决定了深潜器最大下潜深度。为实现耐压壳筒节的高性能，高效率，低成本制造要求，提出波纹筒节轧制工艺来制造具有波纹结构的耐压壳筒节。通过屈曲理论对波纹形耐压壳可以承受的最大工作压力进行强度计算。考虑到可能存在的材料缺陷和结构几何缺陷，利用有限元法对整个耐压壳体进行应力和屈曲分析。最终获得合理的波

纹形耐压壳筒节几何尺寸参数。提出波纹筒节轧制工艺制造波纹形耐压壳。基于体积不变原则获得筒节在轧制过程中的形状变化规律，并对轧制稳定成形条件进行数学建模，确定了关键工艺参数的范围。通过有限元仿真和试验进行了相应验证。屈曲理论和有限元结果表明，增大壁厚或波纹高度都可以显著提高波纹壳的抗压性能。波纹形耐压壳的平衡路径均能保持平顺连续状态，表现出一种稳定的屈曲行为。波纹形耐压壳筒节的变形规律与理论分析结果非常吻合。整个轧制过程稳定，主要分为夹持，局部成形和全局成形三个阶段。内波纹形耐压壳的抗压性能有明显提高。最终成形的内波纹筒节波纹形貌良好，整体圆度没有较大损失，满足预期要求。证明了波纹筒节轧制工艺的合理性和可行性。

关键词：波纹筒节轧制；屈曲理论；有限元法；抗压性能；变形机理

Theoretical Research on the Structural Optimization Design and Rolling of Corrugated Pressure-resistant Shell Heavy Cylinder

PENG Yan, SUN Jianliang, ZHANG Xu, WU Runze, SHAN Cunyao

(Yanshan University National Engineering Research Center for Equipment and Technology of Cold Strip Rolling, Qinhuangdao 066004, China)

Abstract: The core basic part of large deep-sea equipment such as large deep submarines and deep-sea space stations is the cylindrical pressure-resistant shell heavy cylinder. The compressive performance of the pressure-resistant shell largely determines the maximum dive depth of the deep. In order to realize the high performance, high efficiency and low cost manufacturing requirements of pressure-resistant shell heavy cylinder, the corrugated heavy cylinder rolling process is proposed to manufacture pressure-resistant shell heavy cylinder with corrugated structure. The strength calculation of the maximum working pressure that the corrugated pressure-resistant shell can withstand is carried out by means of the buckling theory. Considering the possible material defects and structural geometric defects, the finite element method is used to analyze the stress and buckling of the whole pressure-resistant shell. Reasonable geometrical dimensions of the corrugated pressure-resistant shell heavy cylinder are finally obtained. The corrugated heavy cylinder rolling process is proposed to manufacture the corrugated pressure-resistant shell. Based on the principle of volume constant to obtain the shape variation rule of the heavy cylinder in the rolling process, and mathematical modeling of the rolling process stable forming conditions, to determine the range of key process parameters. Corresponding verification is carried out through finite element simulation and test. Buckling theory and finite element results show that increasing either the wall thickness or the corrugation height can significantly improve the compressive performance of the corrugated shell. The equilibrium paths of the corrugated pressure-resistant shell can keep smooth and continuous state, showing a stable buckling behavior. The deformation law of the corrugated pressure-resistant shell heavy cylinder matches very well with the theoretical analysis results. The whole rolling process is stable and mainly divided into three stages: clamping, local forming and global forming. The compressive performance of the inner corrugated pressure-resistant shell is significantly improved. The final inner corrugated heavy cylinder has good corrugation morphology and no large loss of overall roundness, which meets the expected requirements. It proves the rationality and feasibility of the corrugated heavy cylinder rolling process.

Key words: corrugated heavy cylinder rolling process; buckling theory; finite element method; compressive performance; deformation mechanism

中厚板试样取样生产智能控制系统研究

辛小臣

(北京中冶设备研究设计总院有限公司，北京 100029)

摘　要：本研究针对某大型钢铁企业现有中厚板试样取样生产工艺存在的自动化水平低，钢板描号、试样的剪切、排除卡钢等操作均需要人工进行，存在误操作风险及安全隐患；钢板运输过程无监控，易使信息流紊乱、试样剪切区物流缓冲容量小，在剪切及废料收集等环节采用钢板坠落的方式，导致厂区工作噪声大，有害员工健康等众多问题，研究开发中厚板试样取样无人化生产智能控制系统。该系统由生产调度控制系统、桁架控制系统、堆垛机控制系统、试样剪控制系统、机械臂控制系统、打印贴标控制系统、AGV 小车控制系统以及中间库 WMS 系统等子系统组成，实现了中厚板试样取样生产的无人化、智能化作业。

关键词：试样取样；智能控制

Research on Intelligent Control System for Sampling Production of Medium and Thick Plate Samples

XIN Xiaochen

(Beijing Metallurgical Equipment Research Design Institute Co., Ltd., Beijing 100029, China)

Abstract: This study focuses on the low level of automation in the sampling production process of medium and thick plate samples in a large steel enterprise. Manual operations such as steel plate numbering, sample cutting, and removal of stuck steel are required, which poses a risk of misoperation and safety hazards; The process of steel plate transportation is not monitored, which can easily cause information flow disorder and small logistics buffer capacity in the sample cutting area. The use of steel plate falling in the cutting and waste collection processes leads to many problems such as high working noise in the factory area and harmful to employee health. Therefore, an unmanned production intelligent control system for sampling medium and thick plate samples has been developed. This system is composed of subsystems such as production scheduling control system, truss control system, stacker control system, sample shear control system, mechanical arm control system, printing and labeling control system, AGV car control system, and intermediate warehouse WMS system, achieving unmanned and intelligent operation of medium and thick plate sample sampling production.

Key words: sample sampling; intelligent control

参 考 文 献

[1] 丁立言. 仓存自动化[M]. 北京: 清华大学出版社, 2002.
[2] 秦明森. 实用物流技术[M]. 北京: 中国物质出版社, 2001.
[3] 刘琪. 自动化立体库出入库任务调度和货位分配研究[J]. 中国科技信息, 2009(3): 122-123.

中冶赛迪轧钢数字化解决方案及实践

李　华，陶　涛

（中冶赛迪信息技术（重庆）有限公司，重庆　400000）

摘　要：针对轧钢行业当前高质量发展瓶颈，梳理当前各类轧钢产线存在的生产痛点、管理效率以及客户满意度等存在问题，研究如何利用数字化技术结合轧钢厂现状，实现轧钢的高质量、高效率、低成本的运行，进一步提高市场竞争力。中冶赛迪有多个不同工艺轧钢产线的实践案例，其中既有轧钢单工序数字化也有基于统一平台的多工序协同的轧钢数字工厂，中冶赛迪采用基于统一的工业互联网平台底座，以数字质量和数字设备为基础，面向业务建设 8 大主题模块，支持业务人员和管理人员的高质量决策。打造数字钢卷系统，实现钢卷的工艺、设备、能源、成本等多维度数字孪生,以数字质量和数字设备实现质量全过程管控和产线生产稳定性,以优化排程和智能仓储技术提高钢轧界面协同效率,以数字精益系统面向轧钢厂 8 大业务和产线精益运营需求,支持技术和管理高效率运用数字化技术实现高效/精益管控,进一步降本增效,提升产线竞争力。此外基于知识图谱和规则引擎将产线多年的管理和技术经验转化可量化的决策规则,实现只是数字化。此外为支持产线操控智能化和少人化,引入图像 AI 技术覆盖传统监控和检测难以实现的死角,结合自动控制优化实现少人化作业。项目的成功实践充分证明了数字化是解决界面协同、效率提升、质量稳定、能效提升的必要手段,是智能化发展的必经之路。但如何实现数字化则需要有对应的数字化的系统架构和顶层设计,同时需要有效结合工厂的业务,采用数据驱动的各类分析模型,面向岗位提供定制化的智能应用,从而实现高效率、高效益的轧钢运营。

关键词:轧钢;数字化架构;数字轧材;数字质量

带式输送机的智能化断带保护技术的研究与应用

徐言东[1],白金龙[1],郭 强[1],张勇军[1],白英超[2]

(1. 北京科技大学国家板带生产先进装备工程技术研究中心,北京 100083;
2. 沈阳通用电子技术有限公司,辽宁沈阳 110141)

摘 要:目的:由于工业用输送带的长期运转等因素,散料上运胶带输送机断带事故时有发生。断带监测保护系统主要适用于监控到断带事故发生时,对下滑断带的抓捕以防止其造成重大经济损失、人身伤亡事故。本文通过对既有的有代表性的断带监测保护技术的优缺点进行解析,分析了断带监测保护技术所需要聚焦的关键需求与痛点问题,引出了适用于信息化、智能化工厂建设的断带监测保护系统的发展趋势。

方法:通过对带式输送机断带动态过程的分析,确定了带式输送机断带事故预防与处理的三个关键点。其一是在断带处理维度,通过针对性设计保护装置尽可能在输送带承力限度内实现对断带尽可能小的制动距离与尽可能短的制动时间;其二是在事故监测维度,通过设计生产监测与事故监测一体化系统,及时把握输送带的安全态势与进行故障判断,对输送带的异常状态进行常时监测与实时预警;其三是在故障预防维度,通过设计智能化的生产管理系统与设备控制手段,在带式输送机的运用层面进行优化通过降低设备损耗的手段对带式输送机的故障进行预防。

结果:针对断带处理维度,本文提出了一种硬件设计立足于断带运动特征的 PLC 控制型断带捕捉装置;针对事故监测维度,本文设计专用传感器并结合 AI 视频分析技术,提出了一种多传感器融合的输送带状态智能监控系统;针对故障预防维度,本文提出一种监测带式输送机运行数据并实行多维度统计的生产管控平台。以上内容的可行性均有相关专利与临近领域内的研究成果作为支持。

结论:带式输送机具有覆盖生产区域大、事故后果严重的特点,对传统的断带抓捕与故障排除手段均有相当大的挑战,因此,本文通过将断带保护问题分解为事故的处理、监测、预防三个维度,从局部到整体地对智能化断带保护技术进行分层次的剖析并针对各个维度的关键需求设计解决方案,以提高带式输送机区域安全系数,使带式输送机能够更加稳定而高效运行,便于对接厂内的生产管理系统为其提供数据支持,为提高整体生产效率打下坚实基础。

关键词:带式输送机;智能化;断带保护

XU Yandong[1], BAI Jinlong[1], GUO Qiang[1], ZHANG Yongjun[1], BAI Yingchao[2]

(1. National Engineering Technology Research Center of Flat Rolling Equipment,
University of Science and Technology Beijing, Beijing 100083, China;
2. Technology Department, Shenyang General Electronics Technology Co., Ltd., Shenyang 110141, China)

Abstract: Objectives: Due to factors such as the long-term operation of industrial conveyor belts, accidents of belt breakage in bulk material upward conveyor belts often occur. The broken belt monitoring and protection system is mainly suitable for monitoring the catching of sliding broken belts when a broken belt accident occurs to prevent it from causing significant economic losses and personal injury accidents. This article analyzes the advantages and disadvantages of existing representative broken belt monitoring and protection technologies, analyzes the key requirements and pain points that need to be focused on in broken belt monitoring and protection technology, and introduces the development trend of broken belt monitoring and protection systems suitable for informational and intelligence factory construction.

Methods: By analyzing the dynamic process of the belt breakage of belt conveyor, three key points for preventing and handling belt breakage accidents were identified. The first is to design protective devices in a targeted manner in the dimension of handling broken belts, in order to achieve the shortest braking distance and shortest braking time within the limit of the conveyor belt; The second is in the dimension of accident monitoring, by designing an integrated system of production monitoring and accident monitoring, timely grasp the safety situation of the conveyor belt and make accident judgments, and conduct regular monitoring and real-time warning of abnormal conditions of the conveyor belt; The third is to design intelligent production management systems and equipment control methods in the dimension of accident prevention, optimize the application of belt conveyors, and prevent belt conveyor accidents by reducing equipment losses.

Results: For the dimension of broken belt processing, this paper proposes a PLC controlled broken belt capture device with hardware design based on the motion characteristics of broken belts; For the dimension of accident monitoring, this article designs dedicated sensors and combines AI video analysis technology to propose a multi-sensor fusion intelligent monitoring system for conveyor belt status; For the dimension of accident prevention, this article proposes a production control platform that monitors the operation data of belt conveyors and implements multi-dimensional statistics. The feasibility of the above content is supported by relevant patents and research results in adjacent fields.

Conclusions: Belt conveyors have the characteristics of covering a large production area and severe accident consequences, which pose significant challenges to traditional methods of catching and troubleshooting broken belts. Therefore, this article decomposes the problem of broken belt protection into three dimensions: accident handling, monitoring, and prevention. It analyzes the intelligent broken belt protection technology from local to overall levels and designs solutions for key requirements in each dimension, To improve the safety factor of the belt conveyor area, enable the belt conveyor to operate more stably and efficiently, facilitate the connection with the production management system in the factory to provide data support, and lay a solid foundation for improving overall production efficiency.

Key words: belt conveyor; intelligence; belt breakage protection

首钢智能制造技术研究与探索

于 孟

（首钢集团有限公司技术研究院，北京 100043）

摘 要：智能制造已成为促进钢铁制造数字化转型、推动企业实现质量和效率提升的重要手段，当前国内外钢铁企业都在积极开展智能制造技术的研究与落地。近几年，首钢在探索实践过程中，围绕生产过程控制、智能装备开发和智能工厂建设等方面取得了一定的进展。在生产过程控制中，首钢自主开发的自动出钢系统，攻克了钢包车定位、炉口溢渣检测、钢包净空检测、工艺曲线设计以及炉车协同控制等难题，已在迁钢、京唐多个转炉开展应用；在世界首条具有灵活生产模式的第三代薄板坯连铸连轧 MCCR 产线上，自主开发的无头轧制板形模型先进控制技术，实现全品种覆盖，无头凸度控制精度达到 98.5%及以上，处于国际领先水平；自主开发了冷连轧机全机架秒流量厚度控制技术，通过系统自主集成与装备升级，实现了酸轧线厚度高精度稳定控制。在智能检测装备研发中，自主开发的钢卷轮廓质量检测设备及配套控制技术成功应用于热轧-冷轧各产线，实现了全流程隆起缺陷的高精度、定量化检测;开发的带钢翘曲在线检测设备,实现了在镀锡线、热轧平整线的应用,并制定冶金行业标准一项。在智能工厂建设方面,建立了多场景业务应用的面向热轧全工序的车间级工业互联网平台,通过对多产线多工序数据的集成化利用解决了数据孤岛问题,大幅降低了企业成本、提升了生产效率及产品质量;建立了基于多工序协同控制的标准化板形评价体系,开发了热轧板廓和冷轧板形一体化在线质量分析系统,实现了上/下游工序板形质量的协同控制。

关键词: 钢铁;智能制造;过程控制;智能检测装备;智能工厂

Research and Exploration of Intelligent Manufacturing Technology in Shougang

YU Meng

(Shougang Research Institute of Technology, Beijing 100043, China)

Abstract: Intelligent manufacturing has become an important means to promote the digital transformation of steel manufacturing and promote the improvement of quality and efficiency of enterprises. At present, domestic and foreign steel enterprises are actively carrying out the research of intelligent manufacturing technology. In recent years, Shougang has made certain progress in the field of production process control, intelligent equipment development and intelligent factory construction. In the production process control, the automatic tapping technology for converter by Shougang has overcome the difficulties of ladle car positioning, furnace mouth overflow detection, ladle clearance detection, process curve design and furnace car collaborative control, and has been applied in many converters in Shougang Qiangang and Shougang Jingtang. In the Multi-Mode Continuous Casting and Rolling plant of Shougang Jingtang, advanced technology of plate shape model and controlling in endless rolling developed independently and achieved full steel grades coverage. The endless crown control accuracy was above 98.5%, in the international leading level.The second flow thickness control technology of the whole stand of continuous cold rolling mill was developed independently, and the high precision and stable control of the thickness of acid rolling line was realized through the independent integration of the system and equipment upgrading. In the intelligent equipment development, the self-developed steel coil contour quality inspection equipment and control system have been successfully applied to the production line from hot rolling to cold rolling, realizing high-precision and quantitative detection of bulge defects in the whole process. The online testing equipment for strip warping has been developed, which has been applied in tinning line and hot rolling line, and has established a metallurgical industry standard. In the intelligent factory construction, a shop-level industrial Internet platform for the whole process of hot rolling has been established for multi-scenario business applications, and the problem of data isolation has been solved through the integrated utilization of multi-process data of prolific lines, which has greatly reduced enterprise costs and improved production efficiency and product quality. A standardized plate shape evaluation system based on multi-process collaborative control was established, and an integrated online quality analysis system for hot rolled plate profile and cold rolled plate profile was developed to realize collaborative control of plate shape quality in upper and lower processes.

Key words: iron and steel; intelligent manufacturing; process control; intelligent detection equipment; intelligent factory

钢铁企业全要素资源优化与智慧运营核心关键技术创新

汝金同

(南京钢铁股份有限公司,江苏南京 210000)

摘　要： 在数字中国建设的背景下，钢铁行业面临的产业结构升级、绿色低碳、产业链竞争力较弱等多重压力，数字化转型已成为钢铁行业实现高质量发展的必由之路。近年来，南钢聚焦行业共性需求，围绕精益生产、敏捷运营和协同制造等领域，探索新一代数字技术在千万吨级钢铁企业应用的无人区，实现全要素资源优化与智慧运营核心关键技术创新。基础支撑方面，通过面向钢铁企业全价值链的数据空间设计与数据治理技术、企业生产经营全域数据资源集成开发与应用技术、基于实时交互和预演机制的生产经营全业务数字孪生构建技术，筑牢技术底座；精益生产方面，通过基于物流能流协同优化的多制造单元一体化集群管控技术、生产全过程复杂状态动态感知与快速响应机制、柔性制造下全流程生产绩效闭环管控与工艺推优，实现生产过程的成本与质量最优；敏捷运营方面，通过采购风险识别与黑色市场精准预测与动态管控技术、审计/风控精准预测与动态管控技术、以客户为中心的企业研发及生产资源优化重组技术，全面实现企业上下游及内部的风险精准预测与预控；协同制造方面，通过基于区块链的招采和质量的产业链可信运行技术、C端生态建设与产业互联平台、基于产品全生命周期的研发与产供销协同管控，实现产业链上下游互动与全流程价值链智能决策。

关键词： 钢铁行业；技术底座；精益生产；敏捷运营；协同制造

Abstract: In the context of the construction of Digital China, the steel industry faces multiple challenges, including the upgrading of industrial structure, green and low-carbon transition, and weak competitiveness of the full industrial chain. Digital transformation has become the necessary pathway for the steel industry to achieve high-quality development. In recent years, NISCO (Nanjing Iron &Steel Co., Ltd) has focused on industry-specific needs and explored uncharted boundary in the application of next-generation digital technologies of the mega-ton steel enterprises, particularly in areas such as lean production, agile operations, and collaborative manufacturing. This exploration aims to achieve core innovations in diverse resource optimization and smart operations.

In terms of support, a solid technological foundation is being built through the design and governance of data spaces across the entire value chain of steel enterprises, the integration and development of enterprise-wise data resources for production and operations, and the construction of a digital twin for all businesses based on real-time interaction and predictive mechanisms.

In the realm of lean production, cost optimization and quality improvement are achieved by implementing technologies such as integrated cluster control for multiple manufacturing units based on logistics and energy flow synergy, dynamic sensing and rapid response mechanisms for complex states throughout the production process, and closed-loop control of production performance and process optimization under flexible manufacturing, ensuring the optimal balance between cost and quality in the production process.

As for agile operations, precise risk identification and dynamic management techniques for procurement risks and accurate forecast of the non-ferrous market, as well as auditing/risk control of precise forecast and dynamic management techniques, are being deployed. Additionally, customer-centric enterprise research and development and production resource optimization and restructuring technologies are being utilized to comprehensively achieve precise prediction and pre-control of risks both upstream and downstream within the enterprise and internal.

In terms of collaborative manufacturing, industry-chain trust operation technology for procurement and quality based on blockchain, Customer-end ecosystem development and industry interconnectivity platform, and collaborative control throughout the product lifecycle for research, development and production-supply-sale, enable interactive operations between upstream and downstream partners in the industrial chain and intelligent decision-making across the entire value chain.

钢厂板坯库调度问题的研究进展评述

郑 忠[1]，胡 赟[1]，高小强[2]

（1. 重庆大学材料科学与工程学院，重庆 400045；
2. 重庆大学经济与工商管理学院，重庆 400044）

摘 要：在全球信息化浪潮下，低碳绿色智能制造已成为世界工业强国促进经济发展的国家战略[1]。随着钢铁企业并购重组、市场需求复杂多变、节能环保形势严峻，以生产运营管控为核心的智能决策优化成为企业高效低成本运行的关键[2]。钢铁企业以连铸为核心进行生产，板坯库是生产流程中最重要的仓库之一，其主要作用既是连铸生产的板坯存储区，又是热轧生产的原料准备区，对维持连铸工序与热轧工序的不同工艺流程的准连续式生产、协调工序之间的平衡关系起着重要的缓冲作用。板坯库的调度优化对于促进铸轧工序的高效衔接、提高板坯的热送热装率、降低加热过程能耗有重要作用，可实现缩短生产周期、提升产品质量、降低库存成本等目标。

对于不同的钢厂，板坯库的物理布局、容量大小、作业跨的数量以及区域设置均有不同。为了响应多品种、小批量、多规格、高质量的客户需求，板坯库堆放的板坯种类多、数量大、中间处理环节多、流转节奏快，因此板坯库调度问题引起了钢铁企业和学者们的高度关注。现有板坯库调度问题的研究主要包括板坯存储位置调度（板坯入库决策、板坯出库决策、板坯库内倒垛）、板坯库天车调度、物流运行优化、库区管理优化以及板坯库调度系统设计等。研究大多是针对板坯库垛位管理和天车调度等问题，以库区倒垛率最小、天车运行距离最短或多天车运行均衡等为目标，在设备、板坯堆垛和天车运行等约束条件下，建立数学规划模型，应用运筹学的智能算法、概率或模糊优化算法，以及融合了调度规则的各种智能进化求解算法等进行模型求解；对于规模小的确定场景问题，这些算法能够得出较好的调度方案。但是现实工业场景，通常需要计算的板坯数量多、规模大，且扰动频发，场景不确定性变化频繁。因此，现有研究得到的模型算法大多不具备场景适应性，调度方案柔性差，难以应对现实调度过程中的变化；难以在线应用并指导生产。

板坯库的智能调度对钢铁铸轧界面生产的紧凑连续、动态有序运行至关重要[3]，因此需要构建"产存取出"的供需平衡模式下的板坯库智能调度模型及系统。通过板坯库的调度排程，利用数字化与智能化技术，提升板坯库容量变化与连铸、热轧工序计划的协调能力，实现板坯库仓储变化的预测功能；针对板坯库的动态调度，可利用工业互联网、数字孪生技术、调度规则库、流程仿真等技术，建立基于信息物理融合的板坯库调度系统，感知库存、生产、设备等运行状态，板坯的入、出及库内倒垛等信息，适时进行调度优化，构建多时空尺度的板坯库"人-信息-物流"调度系统，实现板坯库的动态调度，提升板坯库的物流管理水平，驱动板坯的高效低碳节能运行管控优化。

参 考 文 献

[1] 孙燕妮, 白晓军. 《中国制造 2025》——中国特色的强国战略[J]. 智能制造, 2020, 10: 43-45.
[2] 郑忠, 张开天, 高小强. 智能钢厂生产运营决策优化的人-信息-物理系统[J]. 中国科学: 技术科学, 2022, 52(2): 2.
[3] 周继程, 上官方钦, 丁毅, 等. 钢铁制造流程"界面"技术与界面能量损失分析[J]. 钢铁, 2020, 55(12): 8.

基于工业互联网平台的钢铁行业 AI 应用

李 胜

（中冶京诚数字科技（北京）有限公司，北京 100176）

摘　要：随着工业互联网在钢铁行业的应用，企业里每天生成的海量数据无法人工分析，数据无法完全转化数据价值。传统的检测手段已经不能满足钢铁行业精细化管理要求；环境恶劣的部分岗位，还存在人工记录数据、目视判断和手工操作等现象；AI 开发门槛高，钢铁行业人才储备不足，无法独立开发 AI 模型。建立低代码 AI 开发平台，降低用户 AI 模型开发门槛，构建基于工业互联网平台的 AI 模型开发，满足企业模型训练、模型发布、模型管理的应用体系势在必行。

AI 平台是 AI 算法与钢铁行业生产应用场景相结合，平台预制了图像分割、物体检测、分类、回归等各种计算模版，提高钢铁类 AI 模型研发速度。当前，中冶京诚数科公司已经完成钢铁全流程的 60+AI 场景的应用，覆盖原料场到轧钢的全工序，涉及物流、质量预测、质量分析等方向。实现原料、铁、钢、铸坯、轧材的智能和准确跟踪；AI 模型与机理模型相结合，提高高炉、转炉、精炼、连铸模型的准确度；AI 模型与视频监控相结合，实现厂区人员安全行为智能检测，实现污染物排放的全天监控，实现皮带类设备安全运行隐患检测等等应用场景。

通过 AI 模型应用，实现传统的炼钢厂区传统的车辆、铁、钢包、铸坯、轧材的跟踪管理实现无人化。实现铁钢、钢轧界面信息的准确衔接。实现特钢领域产品按支跟踪管理的目的，提升了产品质量管理的精细度。

中厚板板型检测系统开发与应用

田　勇

（东北大学，辽宁沈阳　110819）

摘　要：随着市场对钢板产品质量的要求提高，平直度检测技术也随之有了更高的实时性要求。目前国内钢铁企业绝大部分采用传统的离线人工检测方法，效率低且无法实现在线工艺指导；而在线板形检测主要靠人工目测，对操作经验要求高，且与系统无法形成连锁控制。针对上述关键问题，为了缩短板形检测周期、降低经验控制门槛，东北大学自主研发了基于机器视觉的板形在线检测技术，并在湘钢、鞍钢和八钢等中厚板企业得到有效应用，实现在线快速的板形检测，平直度检测精度达到±1mm/m，与矫直或冷却工艺形成板形闭环控制。

数字驱动的宽厚板自动转钢系统研发与应用

何纯玉

（东北大学，辽宁沈阳　110819）

摘　要：转钢控制是宽厚板轧制过程的必备环节，传统控制方式需要人工进行频繁干预，限制了轧制自动化程度的提升。本研究以最小转钢时间为约束条件，对转钢全过程采用步序快速交接、推床预摆和高速自动转钢相结合的方法开发自动转钢系统，此系统基于改进的图像处理算法实时检测钢坯转钢过程的位置与角度信息，实现了对于复杂生产环境下检测数据的平滑处理；通过智能模型学习，建立了自动转钢数字化模型与优化算法，实现了对于转钢速度与转钢精度的自主调优，同时通过高速数据通讯技术无缝衔接转钢系统与基础自动化系统之间的指令交接。数字化转钢系统的现场应用替代了人工转钢控制，使生产工艺过程具备足够高的重现性和精确性，此研究对于提高宽厚板轧线的自动化程度和智能化升级具有普适性意义。

中冶赛迪智能制造解决方案探索与实践

童 心

(中冶赛迪,重庆 401122)

摘 要:本报告分享中冶赛迪信息基于工业互联网平台,构建以三个一体化智能应用集群为代表的钢铁全流程智能工厂典型案例。在示范案例中,我们通过全新的数字化架构,完成了全工序覆盖、全业务支撑、全链路集成,通过数据服务支撑了钢铁企业生产可视化、操作标(精)准化、管理精细化、生产均衡化、生产准时化、改善数据化及协同价值创造等业务场景,让数字化全面融入钢铁企业精益运管体系,以数据驱动业务执行改善。我们深刻认识到,只有将企业日渐丰富的数据投入再生产,结合数字化业务设计,驱动业务管理、组织能力及技术支持持续演进,才能有效实现数据资产化,实现从"管住"到"管好"的管理目标。利用数字化赋能,纵向下沉拉通、横向覆盖联动,整体协同是业务寻优创效的蓝海。全面数字业务化的实践已经开启,下一阶段,我们将持续深化践行数字业务化建设,面向全局,以业务价值流为主线强化智能应用协同耦合;服务岗位,以数岗匹配为路径支撑数据服务业务创效;深化智能,基于工艺机理与数据规律指导业务实践创优;联动决策,联动智能工厂与经营决策助力企业精益敏捷;以人为本,生态化持续演进,构建与业务同步演进的智能工厂。

中冶赛迪炼钢智能制造探索与实践

王智君

(中冶赛迪,重庆 401122)

摘 要:本报告拟介绍基于工业互联网的炼钢一体化管控体系,它是聚焦炼钢生产业务场景,围绕质量、成本、效率构建的炼钢一体化管控体系。主要包含三方面内容:一是以模型化、智能化技术赋能生产作业标准化建设;二是以全面数字化支撑铁钢、钢轧界面生产协同;三是以生产过程精益为目标构建数据分析应用。从而,实现生产透明化、操作标准化、管理精细化、流程精益化和岗位高效化。目前,该体系已在永锋钢铁、镔鑫钢铁、湖南钢铁、云南德胜等多个钢铁企业应用和实施,支撑其在质量稳定、成本控制、生产效率等环节的关键指标稳步提升。

数字能源技术助力钢企创新发展的探索与实践

吴 杉

(中冶赛迪,重庆 401122)

摘 要:钢铁工业是典型的资源、能源密集型行业,具有工艺复杂、多工序流程耦合关联等典型特征,复杂的生产工况要求能源管控系统具备与生产节奏高度匹配、能源与生产大数据联合决策等能力解决能源多源并入、转换和消纳带来的能源流的变化和系统扰动等难题。但当前能源生产操作及管理决策多依赖于岗位人员的知识储备和认知水平，严重制约了能源的集约化、精细化及高效化提升。通过全流程数字能源等技术，推进钢铁工业能源生产集成和高效管控成为钢铁行业绿色、低碳发展转型的一项重要课题。

（1）提出了融合数据驱动模型与机理模型的能量流数字化建模方法，以充分表征生产全流程能源产生、储配、转换及消耗过程及其静态与动态特征，支撑全流程能源一体化智能管控系统的构建。

（2）研发了基于计量校验与仿真、管网运行仿真、能源产耗预测、设备及系统能效水效多维度诊断分析等关键技术的多能源优化调度技术，实现对全厂能源介质发生-输配-消耗的全流程、全要素实时跟踪监视，提高了钢铁企业多能源介质调度决策的可操作性和可执行性，为钢铁企业能源系统的安全生产保供、经济低碳运行提供了技术支撑。

（3）实践了"多能源优化调度技术＋能源集控+精益能效管理"的能源一体化智能闭环管控技术，以安全稳定生产保供为前提，以最佳能效为目标，将优化调度、操作建议直达集控操控岗位，实现岗位数字化、大幅提升劳效、降低生产运营成本。

板坯库智能作业管理系统设计的实现与应用

张 鑫

（北京金自天正智能控制股份有限公司，北京　100005）

摘　要：板坯即是炼钢工序的成品，又是轧钢工序的原料，板坯库的作业是钢轧生产作业过程中的重要环节，是钢轧一体化的重要组成部分。板坯在连铸出坯切割后还需进行喷号、称重、质量判定、环冷、保温、修磨、入库、出库、倒库、移库、倒运等繁杂的作业处理，而这些作业很多都是通过人工实现，这导致在铸机多产量大的钢厂，作业人员及库管人员的工作量巨大，人工跟踪会造成板坯的丢坯、错判、倒运延迟等生产问题。这些问题不仅影响轧钢的产量与热送率，还导致生产成本增加，甚至发生质量事故等问题。因此板坯能否实现自动跟踪管理是钢轧生产作业过程中的重要环节，也是炼轧工艺界面能否顺利衔接的关键。

目前各钢铁企业根据钢轧界面的作业需要，也做了大量的工作，构建了一些管理系统，对板坯进行了综合的跟踪管理，但多数不理想，钢轧作业过程无法统一实时协同管理，导致热送率下降、成本上升，大量产能无法有效释放等问题，因此能实现钢轧一体化统一管理的板坯库智能作业管理系统是十分需要的。

板坯库智能作业管理系统是一个从连铸切割下线开始到轧钢上料入炉的全过程自动化作业跟踪管理系统，其中包括了板坯的切割下线喷号管理、称重作业管理、天车作业管理、环冷保温修磨作业管理、热送辊道上下料作业管理、出入库存管理及作业过程的优化排程等全过程的智能化跟踪管理。

该系统通过机器视觉技术实现了板坯在轨道位置与上下料动作的自动跟踪；通过对天车的自动化改造实现了天车的跟踪与实际的物料位置精准定位，通过 2D/3D 可视化技术实现了库存作业情况的数字化仿真与显示，通过构建策略库以及时间片轮转调度算法，对板坯库各阶段作业进行实时转运排程，动态调度天车进行分拣，另外系统的IT 架构采用了微服务架构，通过面向服务的设计思想，构建了随需应变的、多层分布式集成运行平台。

通过该系统的建设解决了板坯作业过程的全自动跟踪与管理，优化了钢轧界面的物料流转效率，提升了板坯库的管理水平，实现了释放闲置产能、提高热送效率、降低成本，并降低了现场工人的劳动强度。

关键词：板坯库；微服务；3D 库存；智能库存；钢轧一体化；作业管理系统；机器视觉；板坯跟踪

数字能源精益管控一体化

张冰峰

（上海金自天正信息技术有限公司，上海　201999）

摘　要：结合殷瑞钰院士冶金流程工程学"三流一态"理念，经过四十余个钢铁工业企业的工程实践，构建了覆盖钢铁流程工业超低排放及能源动力全介质、全流程的数字能源工业互联网平台AriEPlat，实现钢铁企业能源智能化调度、精益化分析及无人化操作一体化，帮助钢铁企业提升操作效率、减少排放并无线逼近极致能效。通过平台自主开发的智能算法、调度规则优化及流计算引擎，自动识别生产计划、生产工况及环保排放外部约束，实现能源供给及消耗调度策略的智能化推荐，调度指令的自动下达，实现能源本质智能的一体化管控，提升能源利用效率。

国际物流与物流国际化

樊俊花

（北华航天工业学院，河北廊坊　065000）

摘　要：国际物流是加速我国制造业深度向国际化方向发展的助推力。通过优化物流运作，可以提升产品的流通效率和制造业的国际竞争力，进一步推动全球经济的发展。重点分享国际物流的含义，国际物流和国内物流的区别，国际物流的国际性、复杂性、风险和不确定性等特点及《国际货物运输合同公约》《海牙-维也纳国际货物运输合同公约》和《国际海上货物运输公约》等相关的法律法规。

关键词：国际物流；物流国际化

International Logistics and Logistics Internationalization

FAN Junhua

(North China Institute of Aerospace Engineering, Langfang 065000, China)

Abstract: International logistics is a driving force for accelerating the development of China's manufacturing industry towards internationalization. By optimizing logistics operations, the efficiency of product circulation and the international competitiveness of the manufacturing industry can be improved, further promoting the development of the global economy. Focus on sharing the meaning of international logistics, the differences between international logistics and domestic logistics, the international nature, complexity, risks, and uncertainties of international logistics, as well as relevant laws and regulations such as the Convention on Contracts for the International Carriage of Goods, the Hague Vienna Convention on Contracts for the International Carriage of Goods, and the Convention on the International Carriage of Goods by Sea.

Key words: international logistics; internationalization of logistics

冶金产品智能化储运技术研究

刘武胜

（北京起重机研究院有限公司，北京　100000）

摘　要：随着大数据时代的到来，云计算、大数据、智能管控技术在冶金物流行业应用越来越多，在钢铁企业降本增效中收到良好成效。本文介绍钢卷、钢板、型材、线材等冶金产品立体化智能仓储的可行性及储运技术。

关键词：冶金物流；智能仓储；立体化仓储

Research on Intelligent Storage and Transportation Technology for Metallurgical Products

LIU Wusheng

(Beijing Crane Research Institute Co., Ltd., Beijing 100000, China)

Abstract: With the advent of the big data era, cloud computing, big data, and intelligent control technologies are increasingly being applied in the metallurgical logistics industry, achieving good results in reducing costs and increasing efficiency in steel enterprises. This article introduces the feasibility and storage and transportation technology of three-dimensional intelligent storage for metallurgical products such as steel coils, steel plates, profiles, and wires.

Key words: metallurgical logistics; intelligent warehousing; stereoscopic warehousing

数字化时代钢铁行业供应链金融创新发展与风险管理

卢　强

（北京工商大学，北京　100000）

摘　要：作为服务实体经济的重要手段，供应链金融近年来得到了迅猛发展。数字化时代，在钢铁行业转型升级诉求日益凸显的市场环境下，钢铁行业如何构建钢铁供应链数智化、生态化的发展模式，在此基础上通过供应链金融创新赋能相关企业健康发展是一个需要关注的问题。本报告在介绍钢铁行业供应链金融发展现状、运作模式创新的基础上，并结合企业案例揭示和分析如何有效进行风险管理。钢铁行业应抓住机遇，立足企业在供应链中的核心位置运用科技力量做好风险管控，积极推动自身产业全链合理配置资源、提高效能，增强供应链抗风险能力，助力企业在激烈的市场竞争中持续健康发展。

关键词：冶金物流；供应链金融；风险管理

Innovative Development and Risk Management of Supply Chain Finance in the Steel Industry in the Digital Era

LU Qiang

(Beijing Technology and Business University, Beijing 100000, China)

Abstract: As an important means of serving the real economy, supply chain finance has developed rapidly in recent years. In the digital era, in the increasingly prominent demand for transformation and upgrading of the steel industry in the market environment, how to build an intelligent and ecological development model for the steel supply chain in the steel industry, and on this basis, empowering relevant enterprises to develop healthily through supply chain financial innovation is a problem that needs to be paid attention to. On the basis of introducing the current development status and innovative operational models of supply chain finance in the steel industry, this report combines enterprise cases to reveal and analyze how to effectively carry out risk management. The steel industry should seize the opportunity, base itself on the core position of enterprises in the supply chain, use technological strength to do well in risk control, actively promote the rational allocation of resources and improve efficiency throughout the entire industry chain, enhance the supply chain's risk resistance ability, and assist enterprises in sustained and healthy development in fierce market competition.

Key words: metallurgical logistics; supply chain finance; risk management

网络货运平台赋能大宗商品供应链数字化转型升级的创新实践

王后建

（全盛物流科技（山东）有限公司，山东临沂 276000）

摘　要：大宗商品物流总量和运输规模巨大，2022 年我国大宗物流费用规模为 3.4 万亿元，但目前大宗商品供应链存在信息化程度低、上下游协同效率低、订单履约性差、运输过程无监管、资金资源短缺等突出问题。全盛网络货运平台立足年产能规模 1400 万吨的山东临沂临港高端不锈钢与先进特钢产业基地，按照《网络平台道路货物运输经营服务指南》要求，实现信息发布、线上交易、全程监控、金融支付、咨询投诉、统计查询、数据调取、在线评价八大功能，接入交通运输部国家网络货运信息监测平台，接受实时监管，确保业务合规，平台年运输量 350 万吨、年交易额 1.3 亿元，为服务企业降低整体物流费用 5%~10%，提升钢铁、焦炭等行业作业效率 15%，形成了赋能大宗商品供应链数字化转型升级的成熟经验，典型做法获批国家发改委物流业制造业深度融合创新发展案例、工信部国家中小企业公共服务示范平台、山东省重点工业互联网平台、山东省大数据创新应用场景等荣誉并取得显著经济社会效益。

关键词：大宗商品物流；大宗商品供应链；网络货运平台

Innovative Practice of Empowering the Digital Transformation and Upgrading of Bulk Commodity Supply Chains through Online Freight Platforms

WANG Houjian

(Quansheng Logistics Technology (Shandong) Co., Ltd., Linyi 276000, China)

Abstract: The total volume and transportation scale of bulk commodity logistics are enormous. In 2022, the cost of bulk logistics in China was 3.4 trillion yuan. However, there are currently prominent problems in the supply chain of bulk commodities, such as low informatization level, low efficiency of upstream and downstream collaboration, poor order fulfillment, unregulated transportation process, and shortage of financial resources. The Quansheng Network Freight Platform is based on the Shandong Linyi Lingang high-end stainless steel and advanced special steel industry base with an annual production capacity of 14 million tons. In accordance with the requirements of the "Network Platform Road Freight Transport Operation Service Guide", it achieves eight functions: information release, online transactions, full process monitoring, financial payment, consultation and complaint, statistical query, data retrieval, and online evaluation. It is connected to the National Network Freight Information Monitoring Platform of the Ministry of Transport, Accepting real-time supervision to ensure business compliance, the platform has an annual transportation volume of 3.5 million tons and an annual transaction volume of 130 million yuan, reducing overall logistics costs by 5% -10% for service enterprises, improving operational efficiency in industries such as steel and coke by 15%, and forming mature experience in empowering the digital transformation and upgrading of bulk commodity supply chains. Typical practices have been approved by the National Development and Reform Commission for the deep integration and innovative development of the logistics and manufacturing industries Honors such as the National Public Service Demonstration Platform for Small and Medium Sized Enterprises of the Ministry of Industry and Information Technology, Shandong Provincial Key Industrial Internet Platform, and Shandong Provincial Big Data Innovation Application Scenarios have achieved significant economic and social benefits.

Key words: bulk commodity logistics; commodity supply chain; network freight platform

现代"紧凑型"钢铁制造流程的极致设计-冶金流程学理论在邯钢项目上的应用实践

李 铁[1]，王新东[2]，刘亚峰[1]，付 康[1]，李传民[1]

（1. 中冶京诚工程技术有限公司，北京 100176；2. 河北钢铁集团有限公司，河北唐山 114009）

摘 要： 钢铁产业在国民经济中扮演着重要角色，传统的工程设计理论在推动钢铁产业的高质量可持续发展方面面临着挑战，冶金流程学是引导钢铁产业在新时期实现高质量发展的重要理论依据。本文是冶金流程学理论在钢铁联合企业设计的应用实践，一方面旨在通过深入研究制造物理系统的运行规则、优化物理系统的参数、采用绿色"界面技术"、确定物理系统的稳态优化目标来构建静态系统，实现物质流、能量流和信息流的动态-有序、协同-连续、稳定-高效运行，另一方面通过数字系统构建物理系统的数字孪生，载入运行规则、仿真模型、目标函数等，实现钢厂"以虚映实""以虚预实""以虚控实"保障层流运行的状态。据此，本设计打造了邯钢新区项目"紧凑型"钢铁制造流程，给出了冶金流程学指导下的极致设计对生产运行、效率、节能、减污降碳等方面的影响，为企业实现

"绿色化、智能化、品牌化"的世界一流工厂奠定了"实"和"虚"的基础。

关键词：冶金流程工程学；紧凑型；极致设计

The Ultimate Design of Modern "Compact" Steel Manufacturing Processes-Application Practice of Metallurgical Process Theory in Handan Steel Project

LI Tie[1], WANG Xindong[2], LIU Yafeng[1], FU Kang[1], LI Chuanmin[1]

(1. CERI, Beijing 100176, China; 2. HBIS, Tangshan 114009, China)

Abstract: The steel industry, as a foundational industry in the national economy, plays an important role. The traditional engineering design theories face challenges in promoting high-quality and sustainable development in the steel industry. Metallurgical process theory provides an important theoretical basis for guiding the steel industry towards high-quality development in the new era. This study is the application practice of metallurgical process theory in the integrated steel plant engineering. On one hand, it is to construct a static system and achieve dynamic-orderly, coordinated-continuous, and stable-efficient operation of material flow, energy flow, and information flow by thoroughly studying the operational rules of the manufacturing physical system, optimizing its parameters, adopting green "interface technology," and determining the steady-state optimization goals of the physical system. On the other hand, it is to realize stable laminar flow operation of the steel plant by the concepts of "virtual-to-real," "virtual-to-predict-real," and "virtual-to-control-real." through the construction of a digital twin of the physical system by digital system, and loading the operational rules, simulation models, objective functions, etc. Through the aforementioned approaches, this design creates a "compact" steel manufacturing process for the Handan Steel Project. It provides insights into the impact of ultimate design guided by metallurgical process theory on production operations, efficiency, energy conservation, pollution reduction, carbon reduction, and other aspects. This lays the foundation, both in the real and virtual domains, for the realization of a world-class factory that is "green, intelligent, and branded."

Key words: metallurgy process engineering; compact; ultimate design

钢铁冶金铁前工艺流程优化研究与实践

王兆才[1,2]，叶恒棣[1,2]，魏进超[1,2]

（1. 中冶长天国际工程有限责任公司，湖南长沙 410205；
2. 国家烧结球团装备系统工程技术研究中心，湖南长沙 410205）

摘 要：铁前原料制备工序是钢铁冶炼过程中高能耗、高污染的集中环节，是钢铁行业节能减排的重点和难点。本文围绕能源结构优化、质能循环利用、碳污协同治理、跨领域资源协同利用等方面开展烧结球团工序流程优化研究，重点介绍了烧结固体燃料精准制备技术、分层供热低碳富氢烧结技术、烧结与冷却烟气协同循环利用技术、竖-环双级冷却及直联炉罩式余热锅炉技术、烟气多污染物耦合脱除技术、球团嵌入式 SCR-SNCR 高温脱硝技术、烧结工序协同处置冶金与市政固废技术、高配比硫酸渣氧化球团生产技术等技术原理与工程实践效果。通过一系列流程优化技术的综合应用，实现了烧结工序能耗<42kgce/t，吨烧结矿外排废气量减少 50%，烟气多污染物超低排放，固废安全处置并资源化利用。

关键词：烧结；球团；流程优化；节能减排

Research and Practice on Process Optimization of Pre-ironmaking in Ferrous Metallurgy

WANG Zhaocai[1,2], YE Hengdi[1,2], WEI Jinchao[1,2]

(1. Zhongye Changtian International Engineering Co., Ltd., Changsha 410205, China; 2. National Engineering Research Center of Sintering and Pelletizing Equipment System, Changsha 410205, China)

Abstract: The preparation process of raw materials for ironmaking is the centralized link of high energy consumption and high pollution in the ferrous metallurgy, which is the key and difficult point of energy conservation and emission reduction in the iron and steel industry. This work concerned research on process optimization of sintering and pelletizing which focuses on optimization of energy structure, recycling of mass energy, collaborative treatment of carbon pollution, and collaborative utilization of cross-domain resources. The technical principle and effect in engineering practice of several technologies, including precise preparation of sintered solid fuel, multi-layer heat supplying and low carbon H_2-rich sintering, collaborative recycling for flue gas of sintering and cooling process, vertical-circulating double stage cooling and direct hearth type waste heat boiler, coupling removal of multi-pollutant in flue gas, embedded SCR-SNCR denitration of pellet process, collaborative treatment of metallurgical and municipal solid waste in sintering process and oxidized pellet making with high content of sulfuric acid slag. Through the comprehensive application of series process optimization technologies, the energy consumption of sintering<42kgce/t the unit mount of sintering exhaust gas reduced by 50%, ultra-low emission of multi-pollutant in flue gas and safe disposition and resourceful utilization for solid waste were realized.

Key words: sintering; pelletizing; process optimization; energy conservation and emission reduction

1 矿业工程

大会特邀报告
第十三届冶金青年科技奖获奖人特邀报告
分会场特邀报告
★ 矿业工程
焦化及节能环保
炼铁与原料
炼钢与连铸
电冶金与废钢铁
轧制与热处理
表面与涂镀
金属材料深加工
粉末冶金
先进钢铁材料及应用
节能与低碳技术
冶金环保与资源利用
冶金设备与工程技术
冶金自动化与智能化
冶金物流
冶金流程工程学
其他

1.1 采　　矿

李楼铁矿采场充填密闭工艺

付宇峰

（五矿矿业（安徽）工程设计有限公司采矿二室，安徽合肥　230041）

摘　要：本文以安徽开发矿业有限公司李楼铁矿为例，介绍了一种适用于阶段空场嗣后充填采矿方法的采场充填密闭工艺。通过结合该矿山实际情况，从密闭墙位置的选择、密闭墙材料的选择以及密闭墙力学分析计算来进行出密闭墙结构形式和方案的设计；从实际生产中获得的经验阐述了密闭墙的施工程序和施工要求，并且对施工作业管理应注意的细节做出了强调提示。

关键词：充填采矿法；采空区治理；钢结构；采场充填密闭墙

Filling and Sealing Technology of Stope in Lilou Iron Mine

FU Yufeng

(The Second Mining Department of Minmetals Mining (Anhui) Engineering Design Co., Ltd., Hefei 230041, China)

Abstract: Taking Lilou Iron Mine of Anhui Development Mining Co., Ltd as an example, this paper introduces a stope filling sealing technology which is suitable for stage open stope and subsequent filling mining method. Combined with the actual situation of the mine, the structural form and scheme of the closed wall are designed from the selection of the location of the closed wall, the selection of the material of the closed wall and the mechanical analysis and calculation of the closed wall. based on the experience gained from the actual production, this paper expounds the construction procedures and requirements of the closed wall, and emphasizes the details that should be paid attention to in the construction operation management.

Key words: filling mining method; goaf treatment; steel structure; stope filling closed wall

李楼铁矿大结构采场充填采矿方法工业试验

宋　亮

（五矿矿业控股有限公司，安徽合肥　230088）

摘　要：李楼铁矿采用阶段空场嗣后充填采矿法采矿，采场一般垂直矿体走向布置，分矿房和矿柱分两步骤开采，采场结构参数：采场宽度20m，阶段高度100m，长度一般控制在100m以内。垂直水平每25m设1个凿岩水平，采用分段凿岩爆破和松动出矿，阶段电铲集中大量出矿，此前没有如此大结构的充填采矿先例，因此该采矿方法试验能否成功，对实现矿山稳定持续发展具有重要意义。

关键词：采矿方法；采场参数；回采技术指标

Industrial Test of Filling Mining Method in Large Structure Stope of Lilou Iron Mine

SONG Liang

(Minmetals Mining Holdings Co., Ltd., Hefei 230088, China)

Abstract: The Lilou Iron Mine adopts the stage open pit followed by filling mining method for mining. The mining area is generally arranged perpendicular to the direction of the ore body, and is divided into two steps of mining by the ore room and pillars. The structural parameters of the mining area are: the width of the mining area is 20m, the stage height is 100m, and the length is generally controlled within 100m. One rock drilling level is set every 25 meters vertically, and segmented rock drilling, blasting, and loose mining are used. The stage electric shovel concentrates a large amount of ore, and there has been no precedent for filling mining with such a large structure before. Therefore, the success of this mining method test is of great significance for achieving stable and sustainable development of the mine.

Key words: mining methods; stope parameters; mining technical indicators

浅谈矿山地下水防治技术现状及发展趋势

王 波

（鞍山五矿陈台沟矿业有限公司，辽宁鞍山 114000）

摘 要：矿山地下水防治技术是指利用各种技术手段和方法，从防范和控制矿井水灾的角度出发，综合运用水文地质、岩石力学、水力学等学科知识，对矿山水文地质条件、矿井开采工程特点、矿井井下水体特征等方面进行系统研究，制定出符合实际的防治水方案。本文将探讨当前矿山防治水技术的现状以及未来的发展趋势。

关键词：矿山地下水防治技术

Brief Discussion on the Current Situation and Development Trend of Mine Groundwater Prevention and Control Technology

WANG Bo

(Anshan Minmetals Chentaigou Mining Co., Ltd., Anshan 114000, China)

Abstract: Mine groundwater control technology refers to the use of various technical means and methods to prevent and control mine flooding from the perspective of the comprehensive use of hydrogeology, rock mechanics, hydraulics and other disciplines of knowledge, mine hydrogeological conditions, mine mining engineering characteristics, mine underground water characteristics and other aspects of the systematic study, to develop a water control program to meet the actual. This paper will discuss the current situation of water control technology in mines and the future development trend.

Key words: mine groundwater control technology

预注浆工艺在松散破碎带巷道施工中的应用研究

李伟群[1]，陈利泰[2]

(1. 鞍山五矿陈台沟矿业有限公司，辽宁鞍山 114000；
2. 中国华冶科工集团有限公司辽宁矿业分公司，辽宁鞍山 114000)

摘 要：矿山、隧道等工程在施工过程中经常遇到穿越破碎带的情况，在实践中会采取不同的支护方式，其中超前管棚支护是应用比较多的工艺，施工技术也相对成熟。在松散破碎体或破碎带中掘进施工时往往伴随出水、塌落等情况，施工难度比较大，尤其是上山掘进施工时危险性更高，通过采取超前预注浆封闭加固，钢支架+超前管棚施工的工艺小进尺前进，能够比较有效的穿越松散破碎带，完成施工任务。

关键词：松散破碎体；上山掘进；超前预注浆；钢支架；超前管棚

Research of Advance Grouting Technology in Loose Broken Zone in Mine Excavation

LI Weiqun[1], CHEN Litai[2]

(1. Anshan Wukuang Chentaigou Mining Co., Ltd., Anshan 114000, China;
2. Liaoning Mining Branch Company of China Huaye Group Company Limited, Anshan 114000, China)

Abstract: During the construction of mines and tunnels, there are often cases of crossing the broken zone, and different supporting methods are adopted in practice, construction technology is also relatively mature. In the loose broken body or broken belt, the excavation is often accompanied by water, collapse and so on, the construction is more difficult, especially the raise excavation is more dangerous, with the technology of advance grouting and steel support + lead pipe construction advances with small footage, can effectively cross the loose broken zone to complete the construction task.

Key words: loose broken zone; raise excavation; advance grouting; steel support; lead pipe

基于膨胀支柱技术的孤立采场安全高效采矿方法

李元辉[1]，熊志朋[1]，赵福权[2]

(1. 东北大学深部金属矿山安全开采教育部重点实验室，辽宁沈阳 110819；
2. 中国冶金地质总局青岛地质勘查院，山东青岛 266109)

摘 要：针对缓倾斜薄-中厚矿体空场法开采后期形成孤立采场，存在采场地压大、矿石损失率高等突出问题，研发了一种混凝土底座+膨胀支柱组成的大型人工矿柱及其构筑方法。以丹银金矿孤立采场开采为工程背景，采用大型人工矿柱与锚网联合预控顶支护顶板，提出了矿体由上至下分层安全高效采矿方法。现场应用表明：在孤立采场首层开采过程中，顶板荷载逐渐向大型人工矿柱转移，使其承载力逐渐增加，且开采区域与人工矿柱距离越近，支柱承载力增幅越大；下部矿体分层开采过程中，人工矿柱承载力缓慢增加并最终趋于稳定，膨胀矿柱最大承载力为 2091kN，采场顶板和人工矿柱整体稳定性较好。与传统的原岩矿柱支护方案相比，新型采矿方法采出矿石量 30456t，矿石资源回收率提高了一倍以上，经济效益显著，实现了大型孤立采场的安全高效经济开采。

关键词：孤立采场；大型人工矿柱；膨胀支柱；高效采矿方法；地压控制

Safe and Efficient Mining Method in Isolated Stope Based on Expandable Pillar Technology

LI Yuanhui[1], XIONG Zhipeng[1], ZHAO Fuquan[2]

(1. Key Laboratory of Ministry of Education on Safe Mining of Deep Metal Mines, Northeastern University, Shenyang 110819, China; 2. Qingdao Geological Exploration Institute of Metallurgical Geology of China, Qingdao 266109, China)

Abstract: In view of the isolated stope formed in the later stage of open-stoping mining of gently inclined thin-medium thick orebody, there are outstanding problems such as high ground pressure and ore loss rate. In this paper, a large artificial pillar composed of concrete base and expandable pillars and its construction method are developed. Taking the isolated stope mining in Danyin Gold Mine as the engineering background, a safe and efficient mining method of orebody slicing from top to bottom is put forward by adopting large artificial pillar and bolt-mesh combined pre-control roof. The on-site application results indicate that overburden gradually shifts to the large artificial pillars during the first layer mining process, causing their bearing capacity increase gradually. Besides, the closer the distance between mining area and artificial pillar, the greater the increase of pillar bearing capacity. During the slicing mining of lower orebody, the bearing capacity of artificial pillar increases slowly and eventually stabilizes, and the maximum bearing capacity of expandable pillar is 2091 kN. The stability of stope roof and artificial pillars is better. Compared with the traditional original rock pillar support scheme, the new mining method has achieved 30456 tons of ore in the stope, and the recovery rate of ore resources has been increased by more than one time. As a result, the economic benefits are remarkable, thus realizing the safe, efficient and economic mining in large isolated stope.

Key words: isolated stope; large artificial pillar; expandable pillar; efficient mining method; ground pressure control

浅谈数码电子雷管在袁家村铁矿的应用

贾传鹏

（太钢袁家村铁矿，山西吕梁　033500）

摘　要：本文介绍了数码电子雷管在袁家村铁矿的应用。通过对比非电导爆管雷管，数码电子雷管具有明显优势。利用数码电子雷管精度高延期时间可设的特点，在采场不规则区域、黄土边坡不稳定区域、采空区处理、深孔台阶毫秒级逐孔微差起爆等领域开展数码电子雷管的应用实践，展现了数码电子雷管在工程爆破中的应用前景和优势。

关键词：数码电子雷管；爆破；延期时间；金属露天矿

The Application of Digital Electronic Detonator in Yuanjiacun Iron Mine

JIA Chuanpeng

(Tisco Yuanjiacun Iron Mine, Lvliang 033500, China)

Abstract: This paper introduces the application of digital electronic detonator in Yuanjiacun Iron Mine. Digital electronic detonators have obvious advantages over non-conductive detonators. The application practice of digital electronic detonator is carried out in irregular area of stope, unstable area of loess slope, goaf treatment, hole by hole micro-difference initiation of deep hole step, etc., which shows the application prospect and advantage of digital electronic detonator in engineering blasting.

Key words: digital electronic detonator; explosion; delay time; metal open-cast mine

矿山溜井堵塞率预测的 GA-BP 神经网络法

朱俊阁，于宏宇，王浩东，马强英，路增祥

（辽宁科技大学矿业工程学院，辽宁鞍山　114051）

摘　要：预测矿山溜井的堵塞情况并采取措施进行预防，可以有效降低溜井堵塞处理时存在的安全风险，保证溜井运输系统的正常运转。针对单一BP神经网络存在局部极值等缺点，将遗传算法与BP算法相结合，构建了基于GA-BP神经网络的溜井堵塞率预测模型，以溜井放矿相似模拟试验的实验结果作为样本数据，对其进行训练，预测了粉矿含量、贮矿高度、贮矿时间以及含水率4个影响因素下的溜井堵塞率。研究结果表明：（1）GA-BP神经网络模型对溜井堵塞率的预测结果较为准确，与实验结果的相对误差在1%以内。（2）GA-BP神经网络模型对溜井堵塞率的预测的拟合优度值 R^2 达到0.9993，而平均绝对误差为0.34%，平均绝对百分比误差为0.54%，均方误差根为0.35%，拟合程度好，稳定性高。（3）GA-BP神经网络，解决了单一的BP神经网络在权值和阈值上的不足，可实现对溜井堵塞率的有效与精确预测。研究结果对矿山溜井堵塞的预防及安全生产具有一定的参考价值。

关键词：溜井堵塞率；预测模型；GA-BP神经网络；BP神经网络

GA-BP Neural Network Method for Predicting the Blockage Rate of Mine Orepass

ZHU Junge, YU Hongyu, WANG Haodong, MA Qiangying, LU Zengxiang

(School of Mining Engineering, University of Science and Technology Liaoning, Anshan 114051, China)

Abstract: Predicting the blockage situation of mine orepass and taking preventive measures , it can effectively reduce the safety risks during the handling of orepass blockage and ensure the normal operation of the orepass transportation system. To address the shortcomings of a single BP neural network, such as the existence of local extremes, the genetic algorithm is combined with the BP algorithm to build a prediction model of orepass blockage rate based on GA-BP neural network, it is trained using experimental results from similar simulation experiments of orepass as sample data, and the blockage rate of the orepass was predicted under four influencing factors: powder ore content, ore storage height, ore storage time, and water content. The research results show that: (1) The prediction results of the GA-BP neural network model for the blockage rate of the orepass are relatively accurate, with a relative error of less than 1% compared to the experimental results. (2)The goodness-of-fit value R^2 of GA-BP neural network model on the prediction of the blockage rate of the orepass reaches 0.9993, and while the average absolute error was 0.34%, the average absolute percentage error was 0.54%, and the root mean square error was 0.35%, the fitting degree was good and the stability was high. (3) GA-BP neural network, which solves the shortcomings of single BP neural network in terms of weights and thresholds, it can achieve effective and accurate prediction of orepass blockage rate. The research results have certain reference value for the prevention of mine orepass blockage and safety production.

Key words: orepass blockage rate; prediction model; GA-BP neural network; BP neural network

地下金属矿山大块产生原因以及处理措施

徐 涛，郭武英

（富蕴蒙库铁矿有限责任公司，新疆富蕴 836100）

摘 要： 随着矿山开采的深入，地下金属矿山因大块导致的问题愈发突出。大块破碎不仅增加了开采难度，同时也对下一步的开采造成了威胁。蒙库铁矿采用无底柱分段崩落法回采矿石，在蒙库铁矿井下矿开采过程中，因大块率过高严重影响采矿效率和采矿生产，比如因大块集中倒入溜井，最终造成溜井堵塞，处理时间较长，再比如大块处理不及时，堆放在巷道里面，开采过程中不能平稳有效地退采，最终造成矿石贫化率高、出矿率不够、溜井的备用面缺少等各种异常，严重影响生产。为有效解决蒙库铁矿井下矿大块率高的问题，保证安全正常的采矿生产，结合实际情况，现场跟踪落实各道工序，从不同角度分析了大块产生的原因，结合中深孔的设计参数、施工质量、装药质量、炮孔填塞以及制定的工艺指标要求、制度等方向着重讨论了导致地下金属矿山大块产生的原因以及探讨了有效的理论和技术手段对大块现象的处理措施，为井下矿减少大块产生的治理措施提供有效建议和措施。

关键词： 蒙库铁矿；井下矿；大块；无底柱分段崩落法

The Causes of the Occurrence of Large Blocks in Underground Metal Mines and the Treatment Measures

XU Tao, GUO Wuying

(Fuyun Mengku Iron Mine Co., Ltd., Fuyun 836100, China)

Abstract: With the development of mining, the problem of underground metal mines caused by bulk is becoming more and more serious. Fragmentation not only makes mining more difficult, but also poses a threat to the next stage of mining. In the process of underground mining in Mengku iron mine, the mining efficiency and production are seriously affected because of the high lump rate, for example, because the lump is concentrated into the chute, in the end, the chute is blocked, and the processing time is long. For example, large blocks are not processed in time and are piled up in the roadway, and the extraction can not be smoothly and effectively withdrawn during the mining process, finally, the ore dilution rate is high, the ore yield is not enough, the pass of the lack of spare surface and other anomalies, seriously affecting production. In order to effectively solve the problem of high lump rate in Mengku iron mine and ensure safe and regular mining production, combined with the actual situation, the causes of lump production are analyzed from different angles, in the light of the design parameters, construction quality, charge quality, hole filling and the technical index requirements and system of the middle and deep holes, this paper emphatically discusses the reasons leading to the occurrence of bulks in the underground metal mines, and probes into the effective treatment measures for the bulky phenomenon by theoretical and technical means, the effective suggestions and measures are provided for the control measures of reducing block production in underground mine.

Key words: Mengku iron mine; underground mine; block; sublevel caving method without sill pillar

浅谈井下金属矿山发生悬顶的原因以及处理措施

徐 涛，郭武英

（富蕴蒙库铁矿有限责任公司，新疆富蕴 836100）

摘 要： 富蕴蒙库铁矿采用无底柱分段崩落法进行采矿，受采矿方法的影响，爆破后出现悬顶是该采矿方法存在的一大难题，富蕴蒙库铁矿根据矿体的赋存条件及产状特点，结合现场实际情况，分析无底柱分段崩落法发生悬顶的原因以及处理措施，为今后处理悬顶提供参考和借鉴，保障矿山安全生产。

关键词： 金属矿山；悬顶；无底柱分段崩落法

The Causes of Overhanging in Underground Metal Mines and the Treatment Measures are Discussed

XU Tao, GUO Wuying

(Fuyun Mengku Iron Mine Co., Ltd., Fuyun 836100, China)

Abstract: Fuyun Mengku Iron Mine adopts the sublevel caving method without bottom pillars for mining. Due to the influence of the mining method, the occurrence of suspended roof after blasting is a major problem in this mining method. Based on the occurrence conditions and characteristics of the ore body, combined with the actual situation on site, Fuyun Mengku Iron Mine analyzes the reasons for the occurrence of suspended roof in the sublevel caving method without bottom pillars and the treatment measures, providing reference and reference for future treatment of suspended roof, and ensuring safe production of the mine.

Key words: metal mine; suspended roof; non-pillar sublevel caving method

排土场排水系统实践探索与思考

刘 景，徐飞飞

（太钢袁家村铁矿，山西吕梁 033599）

摘 要： 袁家村铁矿 2012 年建成投产，现全年剥离废石和表土达到 9000 多万吨，排土场分上、下盘两个排土场，排土场自然沟谷较多，地形起伏较大，场地范围内最大高差超过 300m，上盘排土场汇水面积 4.05km²，下盘排土场汇水面积 10.4km²。水是影响排土场稳定的重要因素之一，地下水或雨水渗透到排土场内部介质或土场地基，经过不断浸泡和渗透，土场介质和地基土达到饱和后，极易导致排土场滑塌或泥石流。近几年，袁家村铁矿在实践中不断探索，在排土场的上游修筑截洪沟，在下游修筑排水通道，将大气降水或地下水疏通到排土场外缘或下游，减少了排土场的内部及地基的渗水，有效地促进了排土场的稳定。

关键词：土场；排水系统；探索

Abstract: The Yuanjiacun Iron Mine was completed and put into operation in 2012. Currently, more than 90 million tons of waste rock and topsoil have been stripped throughout the year. The waste dump is divided into two areas: the upper and lower walls. The waste dump has many natural valleys and undulating terrain, with a maximum height difference of over 300m within the site area. The catchment area of the upper wall waste dump is 4.05km^2, The catchment area of the lower wall waste dump is 10.4km^2. Water is one of the important factors affecting the stability of the waste dump. Groundwater or rainwater infiltrates into the internal medium or foundation of the waste dump, and after continuous immersion and infiltration, the soil medium and foundation soil reach saturation, which is easily leading to the collapse or debris flow of the waste dump. In recent years, Yuanjiacun Iron Mine has been continuously exploring in practice, building flood interception ditches upstream of the waste dump and drainage channels downstream to divert atmospheric precipitation or groundwater to the outer edge or downstream of the waste dump, reducing water seepage inside the waste dump and the foundation, effectively promoting the stability of the waste dump.

Key words: soil field; drainage system; explore

某矿切割井一次爆破成井研究及应用

夏文浩[1]，宋卫东[1]，张少鹏[1,2]

（1. 北京科技大学土木与资源工程学院，北京 100083；
2. 山东黄金矿业（莱州）有限公司三山岛金矿，山东莱州 261400）

摘 要：针对某矿山普通法成井工作环境恶劣、安全性差、效率低等问题，本文基于中深孔直孔掏槽爆破破岩机理，提出了一种在深部高应力和复杂破碎地质条件下的切割井一次爆破成井方案，通过现场工业试验成功形成了断面3m×3m、高度10.5m的切割井，针对天井"沙漏形"断面的形成原因进行了分析并提出了针对性的改进措施，爆破后的切割井基本满足要求，说明该一次爆破成井技术方案在深部高应力及复杂破碎地质条件下可以有效提高施工效率，保障矿山安全高效生产。

关键词：切割井；爆破成井；直孔掏槽；爆破参数

Research and Application of Blasting to Form a 10.5m Cutting Raise by One Step in a Mine

Abstract: In response to the problems of poor working environment, poor safety, and low efficiency in a certain mine's ordinary method raise completion, based on the rock breaking mechanism of medium deep hole parallel hole cutting blasting and combined with the high stress and complex geological conditions in the deep part of the mining area, a one-step raise blasting plan was designed. Through on-site industrial tests, a cutting raise with a size of 3m×3m and a height of 10.5m was successfully formed. The reasons for the formation of the "hourglass shaped" section of the raise were analyzed and targeted improvement measures were proposed, The cutting raise after blasting basically meets the requirements, indicating that the one-step blasting completion technology can effectively improve construction efficiency and ensure safe

1.2 选 矿

蒙库铁矿含铁围岩再选资源化利用工业实践

胡家仓，王国平

（富蕴蒙库铁矿有限责任公司，新疆富蕴　836100）

摘　要：近年来，随着选矿技术的不断发展，同时铁矿石市场价格逐年呈现上涨的趋势，蒙库铁矿处于露天转地下的关键时期，通过选矿试验及技术交流，综合考虑，启动了蒙库铁矿 9 号矿含铁围岩回收利用项目，对低品位围岩进行了合理的入选品位的选矿工艺研究。根据第三方试验情况，采用了原矿（含铁围岩）—粗碎（200mm）干式磁选—细碎（8mm）干式磁选，作为蒙库铁矿含铁围岩再选的工艺流程方案，同时还确定了成品矿的品位。既可以最大程度保证蒙库铁矿石稳定的供应，又可以最大限度地提高资源利用率，同时实现了客观的经济效益。

关键词：含铁围岩；磁性铁；综合利用

NaCl 浓度对油酸钠体系的泡沫稳定性影响研究

智　慧[1,2]，董振海[1,2]，王　欢[1,2]，满晓霏[1,2]

（1. 鞍钢集团北京研究院有限公司，北京　102200；
2. 鞍钢集团矿业有限公司，辽宁鞍山　114001）

摘　要：泡沫稳定性对矿物浮选过程意义重大，而无机盐离子对泡沫稳定性具有重要的影响。为了研究 NaCl 浓度对油酸钠体系泡沫稳定性的影响，本文在中性条件下以 5×10^{-4} mol/L 油酸钠为起泡剂，利用动态泡沫分析仪分析测量了 NaCl 浓度分别为 0、0.01mol/L、0.1mol/L 时，对泡沫稳定性的影响。结果表明：NaCl 浓度为 0.01mol/L 时，初始产生的气泡小且均匀，具有稳泡作用；NaCl 浓度为 0.1mol/L 时，溶液产生气泡较大且均匀，气泡兼并变大至破灭速度较快且速率均匀；泡沫体积稳定性与 NaCl 浓度之间相关，根据 NaCl 浓度排序其稳定性大小为 0.01mol/L>0.1mol/L≈0mol/L。

关键词：浮选；油酸钠；氯化钠；气泡兼并；泡沫稳定性

Study on the Effect of NaCl Concentration on the Stability of Foam of Sodium Oleate System

ZHI Hui[1,2], DONG Zhenhai[1,2], WANG Huan[1,2], MAN Xiaofei[1,2]

(1. Ansteel Beijing Research Institute Co., Ltd., Beijing 102200, China;
2. Ansteel Group Mining Co., Ltd., Anshan 114001, China)

Abstract: The stability of foam plays an important role in the mineral flotation process. In this paper, the influence of NaCl concentration of 0, 0.01mol/L and 0.1mol/L on the stability of foam was analyzed and measured by dynamic foam analyzer in 5×10^{-4}mol/L neutral sodium oleate solution as frother. The results indicate that: When the NaCl concentration is 0.01mol/L, the initial generated bubbles are small and uniform, and achieve a stabilizing effect. The bubbles generated by a solution with a NaCl concentration of 0.1mol/L are large and uniform, and the bubbles merge and become larger, leading to a faster and more stable burst rate. The volume stability of foam is related to the NaCl concentration. According to the NaCl concentration, the stability is 0.01mol/L>0.1mol/L≈0mol/L.

Key words: flotation; sodium oleate; sodium chloride; bubble merger; foam stability

深槽型大倾角带式输送机铁矿石输送试验研究

万正道,张玉达,赵 斌

(五矿矿业控股有限公司设备动力部,安徽合肥 230000)

摘 要: 介绍了不同槽角、倾角、带速等参数下的大倾角带式输送机铁矿石输送试验,进行了试验设计,介绍了试验过程及结果,测试了中碎后铁矿石的粒级、堆积角,依据试验结果分析18°倾角的运输性,针对供料断流阶段块状物料跳料、断流问题提出防范及缓解措施。

关键词: 深槽型大倾角带式输送机;地下冶金矿山;最大允许倾角;试验;挡料装置

Experimental Study on Iron Ore Conveying with Deep-slot and Large Inclined Angle Belt Conveyor

WAN Zhengdao, ZHANG Yuda, ZHAO Bin

(Minmetals Mining Holding Co., Ltd., Hefei 230000, China)

Abstract: This article introduces the iron ore transportation experiments of a large angle belt conveyor under different parameters such as slot angle, inclination angle, and belt speed. The experimental design is carried out, and the experimental process and results are introduced. The particle size and stacking angle of the iron ore after medium crushing are tested. Based on the experimental results, the transportation performance of an 18 ° inclination angle is analyzed. Preventive and mitigation measures are proposed for the problem of block material jumping and flow interruption during the feeding cut-off stage.

Key words: deep-slot and large inclined angle belt conveyor; underground metallurgical mines; maximum inclination angle; test; device for blocking the sliding of materials

基于离散元仿真的圆锥破碎机腔型优化

韩睿智，宋仁伯，王永金，刘丹丹，母 镕

（北京科技大学材料科学与工程学院，北京 100083）

摘 要：圆锥破碎机是矿石破碎的关键设备，主要用于中细碎。虽然现有的国产圆锥破碎机能够满足基本的生产需求，但与国际先进的破碎机相比，在破碎性能和衬板损耗上仍有一定差距。相关研究表明，根据衬板的耐磨性、物料破碎力以及衬板磨损模型进行破碎腔优化设计可有效降低衬板损耗[1,2]；采用离散元法结合计算机技术对物料的破碎过程进行动态模拟仿真，能够准确、直观地研究破碎机行程、空隙率、啮角和转速等因素对物料破碎效果的影响[3,4]。结合离散元仿真结果对圆锥破碎机进行腔型优化，减少了对物理原型和现场的要求，对圆锥破碎机的设计、测试和优化具有指导意义，为新型圆锥破碎机的研制提供理论基础。

本研究以东鞍山烧结厂的 H8800 型圆锥破碎机为研究对象，通过离散元分析软件 EDEM 建立了腔型参数调整前后的离散元仿真模型，模拟矿石从落料、破碎到排出的全过程，获得不易测量的矿石尺寸和行为等信息，并确定矿石对破碎机衬板的影响，从而探究圆锥破碎机的产品粒度和功耗与腔型参数之间的关系。本研究进行了 0-1.2、9-1.2、18-1.2、20-1.3、20-1.6、20-1.9（高度差-进动角）的仿真实验，用动锥衬板转矩和小颗粒占比分别表征破碎机的功耗和产品粒度，仿真结果见图 1。研究结果发现，当进动角为 1.2°，动定锥顶面距离为 18mm、9mm、0 时，小颗粒占比分别为 19.72%、26.62%、31.59%，动锥衬板转矩分别为 61kN·m、115kN·m、195kN·m；当动定锥顶面距离为 20mm，进动角为 1.9°、1.6°、1.3°时，小颗粒占比分别为 17.90%、18.06%、22.78%，动锥衬板转矩分别为 38kN·m、37kN·m、26kN·m。研究结果表明，动定锥顶面距离减小能提高收集区小颗粒的占比，有效提高破碎产品质量，但会使破碎机功耗显著增加；进动角增大会使粒度分布不均，导致产品质量下降，并且增加破碎机的功耗。

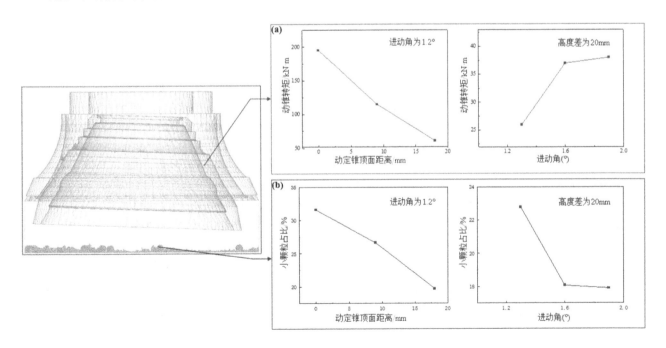

图 1 仿真结果分析

(a) 不同腔型参数下破碎机的功耗；(b) 不同腔型参数下破碎机的产品粒度

参 考 文 献

[1] 高强, 张建华. 矿业工程, 2009, 26(10): 72-75.
[2] 龚姚腾, 张文明, 罗小燕. 北京科技大学学报, 2008, 30(11): 1301-1306.
[3] Johannes Quist, Carl Magnus Evertsson. Minerals Engineering, 2016, 85.
[4] 李东, 樊冰露, 赵倩, 等. 现代矿业, 2017, 33(8): 190-192.

铁精矿中硫浮选脱除与综合利用试验研究

刘兴华，董英择

（长沙矿冶研究院有限责任公司，湖南长沙　410012）

摘　要：含硫矿物作为铁精矿中的有害杂质，在冶炼过程中会生成 SO_2 气体污染环境且严重影响钢铁的热塑性，必需予以脱除。而硫化铁矿物又是用于制造硫酸、提取硫的主要工业原料，具有重要的综合利用价值。本研究以 TFe 品位 64.09%、S 含量 1.65% 的高硫铁精矿作为试验对象，含硫矿物主要为磁黄铁矿和黄铁矿，经过一粗两扫五精浮选闭路流程试验，可获得 TFe 品位 65.01%、S 含量 0.18%、TFe 回收率 97.28% 的低硫铁精矿和 S 品位 40.32%、S 回收率 88.92% 的硫精矿。有效实现了铁精矿脱硫和硫资源的综合利用。

关键词：磁铁矿；磁黄铁矿；黄铁矿；浮选脱硫；综合利用

Experimental Study on Flotation Removal and Comprehensive Utilization of Sulfur in Iron Concentrate

LIU Xinghua, DONG Yingze

(Changsha Research Institute of Mining and Metallurgy Co., Ltd., Changsha 410012, China)

Abstract: As harmful impurities in iron concentrate, sulfur-containing minerals will generate SO_2 gas in the smelting process, which will pollute the environment and seriously affect the thermoplasticity of iron and steel, and must be removed. The iron sulfide mineral is the main industrial raw material for producing sulfuric acid and extracting sulfur, which has important comprehensive utilization value. In this study, high-sulfur iron concentrate with TFe grade of 64.09% and S content of 1.65% was taken as the test object. The sulfur minerals are mainly pyrrhotite and pyrite. After a closed-circuit flotation process test of one roughing, two scavengings and five cleanings, low sulfur iron concentrate with TFe grade 65.01%, S content 0.18% and TFe recovery 97.28% and sulfur concentrate with S grade 40.32% and S recovery 88.92% can be obtained. The desulfurization of iron concentrate and the comprehensive utilization of sulfur resources are realized effectively.

Key words: magnetite; pyrrhotite; pyrite; flotation desulfurization; comprehensive utilization

某高磷鲕状赤铁矿闪速磁化焙烧技术开发研究

李家林，陈雯，刘旭，刘小银，陆晓苏，彭泽友

（长沙矿冶研究院有限责任公司，湖南长沙　410012）

摘　要：我国鲕状赤铁矿资源丰富，保有储量高达 74.45 亿吨；其中宁乡式高磷鲕状赤铁矿探明资源储量约为 37.2 亿吨，由于铁矿物嵌布粒度极细、鲕核包裹严重且伴生大量黏土和含磷矿物，导致单体解离难、杂质含量高及铁磷分离困难等。采用重选、强磁-反浮选或絮凝脱泥-反浮选等常规物理工艺分选，存在铁精矿品位低、磷含量高等问题，磁化焙烧是鲕状赤铁矿大规模工业应用的较优方法。闪速磁化焙烧是针对低品位复杂难选弱磁性含铁资源矿相转化发明的原创性技术，成套技术和装备已于 2016 年成功实现了产业化工业应用，为难选弱磁性铁矿资源大规模绿色、经济、高效开发提供了新途径。

针对某宁乡式高磷鲕状赤铁矿进行了闪速磁化焙烧技术开发研究。工艺矿物学研究表明：铁矿物基本以鲕状结构形式存在；鲕粒内部环带状构造极为发育，由致密核心及环带层层包裹而成，环带数多者可达 10 环以上；环带组成矿物为极微细的赤铁矿和绿泥石、磷灰石、石英等脉石矿物，厚度一般在 5~30μm 范围变化。技术开发研究表明：粒度 0~0.15mm 的高磷鲕状赤铁矿，在焙烧温度 800℃、操作气速 0.30m/s、焙烧时间 3min、CO 体积浓度 5%、添加剂用量 7.5%条件下通过闪速磁化焙烧工艺处理，弱磁性铁矿物磁化率为 91.48%，磁化效果良好。焙烧矿采用阶段磨矿-弱磁-反浮选-中矿化学分选联合工艺处理，可获得综合铁精矿产率 56.53%、TFe 品位 59.63%、P 含量 0.20%、铁回收率 77.18%的选别指标。技术经济分析估算闪速磁化焙烧单位铁精矿制造成本约 450 元/吨，在当前铁矿石市场行情下，闪速磁化焙烧具有盈利能力。技术开发研究成果为高磷鲕状赤铁矿的大规模高效开发提供了新方案。

关键词：鲕状赤铁矿；闪速磁化焙烧；磁化率；降磷

参 考 文 献

[1] 陈雯, 余永富, 冯志力, 等. 60 万吨/年难选菱(褐)铁矿闪速磁化焙烧成套技术与装备[J]. 金属矿山, 2017(3): 54-58.
[2] 张汉泉, 汪凤玲, 李浩. 鲕状赤铁矿磁化焙烧-磁选-反浮选降磷试验[J]. 武汉工程大学学报, 2011, 33(3): 29-31.
[3] 张茹, 李艳军, 刘杰, 等. 东鞍山某鲕状赤铁矿石悬浮焙烧试验[J]. 金属矿山, 2015(12): 55-57.
[4] 王成行, 童雄, 孙吉鹏. 某鲕状赤铁矿磁化焙烧-磁选试验研究[J]. 金属矿山, 2009(5): 57-59.

磁化焙烧工艺选别细粒难选赤铁矿尾矿的试验研究

杨晓峰，董振海

（鞍钢集团北京研究院有限公司，北京　102299）

摘　要：对铁尾矿中有价成分最大限度的回收利用，可实现降低尾矿品位、减少金属流失、提高经济效益的目的。采用物理方法回收细粒难选赤铁矿尾矿，金属回收率不高，为实现铁尾矿资源的高效利用，近年来国内相关研究单位围绕难选矿和赤铁矿尾矿等进行了大量的研究工作，其中磁化焙烧工艺为最有效的技术[1-3]。采用预富集-焙烧工艺对细粒难选赤铁矿尾矿进行选别试验，试验主要包括三部分内容：尾矿的预富集试验、预富集精矿磁化焙烧试验、焙烧产品再磨再选试验。

试验矿样中主要的铁矿物为赤（褐）铁矿，铁分布率为 70.15%；其次为磁性铁，铁分布率 13.58%；碳酸铁铁分布率 13.52%；硅酸铁和硫化铁含量较低，铁分布率分别为 1.97%和 0.76%。故试验矿样中主要回收对象为赤（褐）铁矿、磁铁矿、菱铁矿。

采用"一段粗选、扫选立环磁选预先抛尾-磨矿-弱磁-强磁再选"工艺进行预富集试验，试验矿样在在给矿全铁品位 15.68%的条件下，预富集试验得到了预富集精矿全铁品位 33.19%、精矿产率 28.56%、尾矿品位 8.68%、金属回收率 60.45%的技术指标。

预富集精矿磁化焙烧试验程中考察了焙烧温度、CO 用量、N_2 用量、总气量及给矿量对悬浮焙烧的影响。确定对预富集精矿在焙烧温度为550℃、CO用量为4.0m³/h、N₂用量为2.0m³/h、给矿量为111.70kg/h条件下进行焙烧试验，与焙烧前的预富集精矿相对比，磁性铁分布率由13.99%增大至88.16%，铁矿石磁化焙烧过程中强磁性矿物转化率为87.23%，表明焙烧过程弱磁性铁矿物向强磁性铁矿物转化良好。

获得的焙烧产品经单一磁选工艺选别，在磨矿细度达到-0.025mm含量97%时，可获得全铁品位63.04%，作业回收率80.04%，总铁回收率48.40%，的技术指标；经磁浮联合分选，在磨粒度为和-0.025mm80%时，获得了精矿全铁品位65.10%，作业回收率72.16%，总铁回收率43.62%的技术指标。虽然焙烧产品采用单一磁选工艺与采用磁浮联合工艺相比，精矿产率、金属回收率等选别指标均较好，但其精矿品位低，特别是其要求的磨矿粒度为-0.025mm97%将使磨矿成本大幅度增加。

磁化焙烧工艺选别细粒难选赤铁矿尾矿的试验研究为我国细粒难选赤铁矿尾矿的高效利用提供了新途径，其经济、环境及社会效益较大。

参 考 文 献

[1] 韩跃新, 孙永升, 李艳军, 等. 我国铁矿选矿技术最新进展[J]. 金属矿山, 2015(2): 1-11.
[2] 韩跃新, 高鹏, 李艳军, 等. 我国铁矿资源"劣质能用、优质优用"发展战略研究[J]. 金属矿山, 2016(12): 2-6.
[3] 袁帅, 韩跃新, 高鹏, 等. 难选铁矿石悬浮磁化焙烧技术研究现状及进展[J]. 金属矿山, 2016(12): 9-12.

湖北某铁矿工艺矿物学及选矿试验研究

李育彪，王忠红，李万青

（武汉理工大学资源与环境工程学院，湖北武汉 430070）

摘 要：湖北某铁矿含铁丰富，并含有Cu、S等有价元素。为研究矿石的工艺矿物学特性并制定合理的选矿方案，实现其综合利用。采用化学分析、XRF、XRD、显微镜、电子探针等方法开展了矿石物质组成、矿石结构构造、主要矿物嵌布特征、有用矿物单体解离度等工艺矿物学研究。结果表明：

（1）矿石中全铁含量为28.56%，矿石中铁以磁铁矿、赤铁矿、钛铁矿等氧化铁，黄铁矿、白铁矿、磁黄铁矿等硫化铁及钙铁榴石等含铁硅酸岩形式存在，其中磁铁矿为主要铁矿物。铜以黄铜矿单一形式存在，偶见斑铜矿零星分布。硫主要以硬石膏、黄铁矿、白铁矿和磁黄铁矿形式存在，以黄铁矿为主。铅主要以方铅矿为主，锌主要以闪锌矿为主，铅和锌含量较低。少量稀有金属钛主要是以钛铁矿、锐钛矿、钛榴石形式存在而微量镍和钴均以硫化矿形式赋存于黄铁矿中。脉石矿物主要以石膏等硫酸盐矿物，石榴石等硅酸盐矿物和方解石等碳酸盐矿物为主，此外还有长石、绿泥石、石英、透闪石等矿物存在。金属矿物多呈稠密浸染状-局部聚集呈团块状构造。

（2）矿石中磁铁矿与其他金属矿物嵌布关系密切，需要通过细磨能获得较好的解离度。依据工艺矿物学研究结果，推荐矿石采用"干式磁选抛尾—磨矿—弱磁选铁—浮选铜硫"的选矿工艺流程。通过干式磁选抛尾流程，获得了全铁品位为28.56%的样品；进一步进行"磨矿—弱磁磁选"工艺，并优化磨矿时间、磁选磁场强度及磁选段数方式，最终在磨矿细度60.7%-0.074mm、1800GS的磁场强度及3段磁选管磁选的条件下，获得了TFe品位66.94%，产率67.4%的铁精矿。磁选尾矿中Cu品位为0.012%、S品位为0.33%。

（3）进一步对磁选尾矿进行"铜硫混合浮选—铜硫分离浮选"工艺，通过药剂用量条件及扩大浮选实验，最终获得了铜精矿产率为0.22%，其中铜品位为23.02%；硫精矿产率为9.60%，其中硫品位为53.42%。

白云鄂博低品位铌矿石选矿富集技术研究

王祥，王伟，周喻林，陈雯

（长沙矿冶研究院有限责任公司，湖南长沙 410012）

摘 要：我国白云鄂博地区铌资源储量丰富，但品位低、粒度细、共伴生关系复杂，选矿难度极大，至今未实现工业开发利用。针对区内低品位铌矿石，采用磁选-重选-浮选联合富集工艺，开发了微细粒铌矿重选回收、精准分级预处理、选择性浮选分离等关键技术，可以获得品位 Nb_2O_5 22.37%，回收率 33.56%的铌精矿，铌富集比达 349 倍，且流程中中矿具备较大回收潜力，铌回收率有望进一步提高，可为铌矿的工业开发提供技术支撑。

关键词：铌；精准分级；微细粒重选；选择性浮选；综合利用

Research Progress on Beneficiation Technology of Low-grade Niobium Ore in Baiyun Ebo

WANG Xiang, WANG Wei, ZHOU Yulin, CHEN Wen

(Changsha Research Institute of Mining and Metallurgy Co., Ltd., Changsha 410012, China)

Abstract: The Baiyun Ebo area in China has abundant reserves of niobium resources, but with low grade, fine particle size, complex co associated relationships, and great difficulty in mineral processing, industrial development and utilization have not yet been achieved. For low-grade niobium ore in the area, a combined enrichment process of magnetic separation gravity separation flotation has been adopted, and key technologies such as fine-grained niobium ore gravity separation recovery, precise classification pretreatment, flotation selective separation, etc. have been developed. niobium concentrate with a grade of Nb_2O_5 22.37% and a recovery rate 33.56% can be obtained. The niobium enrichment ratio exceeds 349 times, and the intermediate ore in the process has great recovery potential. The niobium recovery rate is expected to be further improved, It can provide technical support for the industrial development of niobium ore.

Key words: niobium; accurate grading; fine particle gravity separation; selective flotation; comprehensive utilization

高硫高锌磁铁矿深度除杂制备超纯铁精矿的工业实践

廖振鸿，陈雯，王秋林，刘旭

（长沙矿冶研究院有限责任公司，湖南长沙 410012）

摘 要：随着钢铁行业的绿色发展以及国家对环保的愈加重视，制备高品位、低杂质的铁精矿是铁矿选矿的发展趋势。在余永富院士"提铁降硅（杂）"学术思想的指导下，国内选矿行业先后进行了技术改造和工业实践，以提高铁品位，减少 Si、Al、S、P 等杂质，获得了良好的经济收益及社会效益。某公司为了生产满足高端制造业原料要求的高纯铁精矿，需将其所属矿山的高锌（0.8%~1.72%）、高硫（3%~4%）难选磁铁矿石生产的铁精矿中锌杂质含量降低至 0.02%以下（远低于国家标准 0.1%），硫杂质含量降低到 0.2%以下。

该矿石中锌、硫杂质深度脱除的难点在于：（1）载锌矿物多达十多种，锌元素分布广，且大多呈微细粒分布，部分与磁铁矿紧密连生；（2）微细粒磁黄铁矿、闪锌矿、铁闪锌矿表面极易氧化、泥化，极难浮选高效脱除。通过对磁黄铁矿、黄铁矿、闪锌矿、铁闪锌矿的浮选电化学及溶液化学研究，发现极细粒锌、硫载体矿物难以从铁精矿中深度脱除的主要原因是微细粒磁黄铁矿、闪锌矿、铁闪锌矿表面易氧化生成亲水性的氧硫化物，矿物表面易被亲水性的金属羟基化合物罩盖，捕收剂难以与矿物表面的活性金属质点作用，极难浮选。提高闪锌矿、磁黄铁矿的氧化电位与矿物"表面清洗"相协同的活化浮选可大幅提高微细粒极难选磁黄铁矿和闪锌矿的可浮性。以硫酸作活化剂，耗量高，腐蚀设备，污染环境，安全风险大。

针对上述技术瓶颈难题，基于化学清洗与定向氧化耦合原理，重构矿物表面，发明了多活化功能耦合协同的锌硫矿物广谱活化剂 CYA-29，该活化剂具有三方面的特征：（1）引入螯合官能团，将罩盖在磁黄铁矿和闪锌矿表面的亲水物质 $Fe(OH)_3$、$Zn(OH)_2$、$Ca(OH)_2$ 等螯合成可溶性配合物，化学清洗剥离出矿物新鲜表面，便于捕收剂吸附；（2）提高磁黄铁矿、闪锌矿的氧化电位，诱导磁黄铁矿表面氧化生成疏水性的 S_o 或 S_2^0，提高可浮性；（3）提高磁黄铁矿、闪锌矿表面定位离子浓度，压缩或破坏矿物表面水化层，提升捕收剂的作用效能。活化剂 CYA-29 以提高矿物氧化电位、压缩固液界面水化层及化学螯合等协同作用的方式实现对铁精矿中磁黄铁矿、闪锌矿等杂质的高效脱除。所得浮选铁精矿铁品位 71.72%~71.95%、锌含量 0.016%~0.020%、铅含量 0.014%~0.019%、硫含量 0.028%~0.030%。

上述研究突破了高硫、高锌难选磁铁矿石深度脱除锌、硫杂质的技术瓶颈，建成了年产 1000 万吨高纯铁精矿的绿色、高效的大型现代化选矿厂，工业生产所得高纯铁精矿 TFe 70.8%、Zn 0.018%、S 0.08%、SiO_2 0.5%，明显优于铁精矿 C67 一级标准（Fe≮67%，S 0.1%~0.2%，Zn 0.1%~0.2%）。

参 考 文 献

[1] 余永富. 从炼铁工业发展现状讨论我国铁矿山选矿发展方向及值得注意的若干问题[C]//中国科学技术协会，四川省人民政府. 加入 WTO 和中国科技与可持续发展——挑战与机遇、责任和对策(下册). 中国科学技术出版社, 2002: 1.
[2] 韦锦华. "提铁降杂"选矿工艺技术研究与生产实践[J]. 现代矿业, 2009, 25(1): 8-12.
[3] 李文风, 陈雯. 高效铁矿脱硫药剂的研制和应用[J]. 现代矿业, 2010, 26(7): 112-113.
[4] 陈雯. 浮选分离某磁铁矿和富含磁黄铁矿的试验研究[J]. 金属矿山, 2003(5): 33-35.

硫酸烧渣悬浮磁化焙烧提铁技术研究

李文博[1,2,3]，王 硕[1,2]，唐志东[1,2]

（1. 东北大学资源与土木工程学院，辽宁沈阳 110819；
2. 难采选铁矿资源高效开发利用技术国家地方联合工程研究中心，辽宁沈阳 110819；
3. 东北大学轧制技术及连轧自动化国家重点实验室，辽宁沈阳 110819）

摘 要：硫酸烧渣是黄铁矿制备硫酸过程排出的一种粉状废渣，粒度普遍较细[1]。我国硫酸烧渣产量巨大，为了减小硫酸烧渣对环境的危害以及对企业经济效益的负担，为了高效资源化利用硫酸烧渣此类二次资源中的铁[2]，采用悬浮磁化焙烧[3-4]—磁选的方法处理硫酸烧渣，探明了其中各个试验的最佳条件，确定了最优的选别指标。

结果表明：在焙烧温度 480℃、还原气浓度 30%（$CO:H_2=1:3$）、焙烧时间 20min、磁场强度 1200Oe、磨矿细度–0.038mm 含量 80%条件下可以获得精矿铁品位 63.24%、铁回收率 95.05%的磁选精矿。采用 X 射线衍射仪（XRD）、振动样品磁强计（VSM）、X 射线光电子能谱（XPS）、扫描电子显微镜及全自动比表面及孔隙度分析仪 BET 进行了过程机理的分析，分析发现：悬浮磁化焙烧过程发生了赤铁矿向磁铁矿的转变，焙烧产品磁性得到提升，有助于裂缝与孔隙的产生，有助于比表面积、总孔容、平均孔径的增大。本文通过研究悬浮磁化焙烧过程对硫酸烧渣的选别指标与过程机理的影响，对硫酸烧渣选铁具有一定的参考借鉴作用。

参 考 文 献

[1] 王全亮, 周虎强, 代奕华. 广西某硫酸烧渣脱硫选矿工艺研究[J]. 矿冶工程, 2008(5): 44-46, 50.
[2] 李先祥, 张宗华, 张桂芳, 等. 硫酸烧渣综合利用磁选试验研究[J]. 中国矿业, 2005(9): 70-72.
[3] 孔德翠, 刘杰, 张淑敏, 等. 某铁矿石悬浮磁化焙烧—磁选实验研究[J]. 矿产综合利用, 2022(5): 130-134, 146.
[4] 韩跃新, 张强, 孙永升, 等. 难选铁矿石矿相转化清洁高效利用技术新进展[J]. 钢铁研究学报, 2022, 34(12): 1303-1313.

不同高分子絮凝剂对东鞍山细粒赤铁矿聚团-磁选的影响

李文博[1,2,3]，张　爽[1,2]，程绍凯[1,2]，周立波[1,2]

（1. 东北大学资源与土木工程学院，辽宁沈阳　110819；
2. 难采选铁矿资源高效开发利用技术国家地方联合工程研究中心，辽宁沈阳　110819；
3. 东北大学轧制技术及连轧自动化国家重点实验室，辽宁沈阳　110819）

摘　要：我国铁矿资源在大规模开发利用中富矿资源日趋减少，而类似于东鞍山铁矿石等铁品位低、有用矿物嵌布粒度细、易泥化的难选铁矿石显著增多，在连续磨矿工艺中，矿浆中含有大量微细粒铁矿物，因此急需改善细粒级铁矿物强磁回收的效果[1,2]。

相关研究表明，高分子絮凝剂对改善细粒级铁矿物强磁回收的效果具有明显作用[3,4]，因此，以东鞍山细粒赤铁矿为研究对象，探究了直链、支链和交联结构的玉米淀粉、预糊化木薯淀粉、DL-1（醚化改性木薯淀粉）和DL-2（酯化改性木薯淀粉）、改性聚丙烯酰胺（HPAM、PAAS）和海藻酸钠等不同高分子絮凝剂在不同用量下对细粒赤铁矿聚团-磁选的影响，并通过偏光显微镜和红外光谱等方法检测了药剂与赤铁矿和石英团聚前后团聚体的形貌特征和高分子药剂的作用机理，进而分析高分子药剂的选择性和团聚效果。

结果表明：在分级粒度为–43μm 的条件下，高分子药剂的加入能实现微细粒赤铁矿的选择性团聚，采用较优药剂及适宜用量预处理微细粒赤铁矿，磁选精矿铁品位会降低 0.14~0.73 个百分点，但铁回收率提高了 2.13~3.94 个百分点，选矿效率提高了 2.77~3.46 个百分点。对比分析可知，交联淀粉的絮凝效果优于直链和支链淀粉，淀粉通过改性能强化其选择性团聚，改性聚丙烯酰胺和海藻酸钠也对微细粒赤铁矿具有一定的絮凝能力。通过微观形貌和 FTIR 光谱分析高分子药剂与矿物作用前后状态，发现药剂主要通过氢键和化学吸附作用在赤铁矿表面，并实现赤铁矿的选择性团聚，而对石英不发生作用；而由于赤铁矿被絮凝团聚后呈不规则絮状，在较大的聚团中会夹杂少量石英，导致磁选精矿铁品位存在些许降低。

参 考 文 献

[1] SU X G, ZHOU L B, LI W B, et al. Study on selective flocculation-high intensity magnetic separation technology of Donganshan micro-fine iron ore[J]. Conservation and utilization of Mineral Resources, 2021, 41(2): 129-136.
[2] YANG G, ZHOU L B, LI W B, et al. Study on new technology of strengthening fine iron mineral recovery in Donganshan Iron Mine[J]. Metal Mine, 2021, 541(7): 102-109.
[3] LI Y R, YE G H, ZHU B L, et al. Iron recovery from Yunnan Shangchang iron tailings by selective dispersion flocculation-magnetic separation process[J]. Conservation and utilization of Mineral Resources, 2018, 214(2): 63-68.
[4] WANG X, YI D. Experimental study on selective flocculation of oolitic hematite[J]. Multipurpose Utilization of mineral Resources, 2018, 213(5): 79-83.

氟碳铈矿高梯度磁吸附行为研究

李文博[1,2,3]，孙俊燕[1,2]，张小龙[1,2,3]

(1. 东北大学资源与土木工程学院，辽宁沈阳　110819；
2. 难采选铁矿资源高效开发利用技术国家地方联合工程研究中心，辽宁沈阳　110819；
3. 东北大学轧制技术及连轧自动化国家重点实验室，辽宁沈阳　110819)

摘　要：我国稀土资源储量大、类型多、品种全，但有效利用率低导致资源浪费严重。氟碳铈矿是顺磁性矿物，可采用强磁选进行选别，但常规强磁选设备产生的磁感应强度不足以有效回收氟碳铈矿。而高梯度磁选机可提供足够高的磁感应强度，实现了微弱磁性矿物的有效分选[1,2,3]。

相关研究表明，聚磁介质的材质、形状、排列组合形式、结构参数等因素均影响分选效果[4,5]。因此，以四川牦牛坪的氟碳铈矿为研究对象，考查了介质的排列方式、凸极间距、棒间隙等介质结构参数对不同粒级氟碳铈矿磁吸附行为的影响，通过振动样品磁强计探究了颗粒粒度对矿物磁性的影响，并分析了磁性颗粒在分选空间内的受力情况，从而以理论研究为基础进行聚磁介质结构参数的优化，最大限度地提高细粒级弱磁性矿物的回收率。

研究结果表明：在任意凸极间距下，介质呈交错排列时对磁性颗粒的捕集效果明显优于平行排列，介质棒间隙为2mm、3mm时的捕集效果明显优于棒间隙为4mm、5mm；在任意棒间隙下，凸极介质的捕集作用效果优于圆棒介质，尤其在凸极间距为1mm时介质对磁性颗粒捕集效果最佳。通过高梯度磁选过程的理论研究，发现不同粒级的矿物颗粒磁性存在较小的差异，矿物磁性受颗粒粒度的影响较小。通过分析氟碳铈矿在分选空间内的受力情况，发现介质对细粒级矿物颗粒的捕集效果较差；同一矿物粒度下，适当增加磁场强度可以提高有用矿物的回收率；贫连生体所受磁力较小不易被介质棒捕集。

参 考 文 献

[1] Jordens A, Marion C, Kuzmina O, et al. Surface chemistry considerations in the flotation of bastnasite[J]. Minerals Engineering, 2014, 66-68: 119-129.
[2] Xu J Y, Xiong D H, Song S X, et al. Superconducting pulsating high gradient magnetic separation for fine weakly magnetic ores: Cases of kaolin and chalcopyrite[J]. Results in Physics, 2018, 10: 837-840.
[3] Okada S, Mishima F, Akiyama Y, et al. Fundamental study on recovery of resources by magnetic separation using superconducting bulk magnet[J]. Physica C-Superconductivity and Its Applications, 2011, 471(21-22): 1520-1524.
[4] Li W B, Zhou L B, Han Y X, et al. Numerical simulation and experimental verification for magnetic field analysis of thread magnetic matrix in high gradient magnetic separation[J]. Powder Technology, 2019, 355: 300-308.
[5] Zheng X Y, Wang Y H, Lu D F. Effect of matrix shape on the capture of fine weakly magnetic minerals in high-gradient magnetic separation[J]. Ieee Transactions on Magnetics, 2016, 52(9).

独居石在还原焙烧中浮选性质变化研究

李文博[1,2,3]，瞿　瑞[1,2]，赵　熙[1,2]

(1. 东北大学资源与土木工程学院，辽宁沈阳　110819；
2. 难采选铁矿资源高效开发利用技术国家地方联合工程研究中心，辽宁沈阳　110819；
3. 东北大学轧制技术及连轧自动化国家重点实验室，辽宁沈阳　110819)

摘　要：独居石是主要的稀土矿物之一，浮选法是目前从伴生稀土资源中回收独居石的主要方法。近年来，流态化磁化焙烧技术的发展与成功应用使得难选铁矿的回收效果大大改善[1,2]，研究人员采用流态化磁化焙烧技术对白云鄂博难选铁矿以及选铁尾矿展开了一些研究，其中就包括从白云鄂博选铁尾矿中回收独居石。以往研究多针对于选铁完成后尾矿中赋存的独居石表面性质和可浮性的差异，鲜有研究讨论选铁过程中独居石在还原焙烧中表面性质和可浮性的变化。

为探究独居石在经过还原焙烧后表面性质及可浮性的变化，本文采用流态化焙烧以及单矿物浮选进行试验。首先采用 $H_2:CO:N_2=1:3:16$ 的还原气氛在 400～600℃的温度区间内进行还原焙烧，然后使用水杨羟肟酸（SHA）为捕收剂进行单矿物浮选试验。利用 XRD、XPS、接触角分析、红外光谱分析、药剂吸附量分析、扫描电镜对试验结果进行分析。

试验结果表明：还原焙烧对独居石的浮选性质未见不利影响。经过不同时间和温度的还原焙烧处理后的独居石焙烧产物，浮选回收率整体略高于未经焙烧的独居石。独居石表面铁氧化物的还原反应略微改变了部分独居石颗粒表面化学环境，对独居石的浮选回收产生有利影响。

参 考 文 献

[1] Yuan S, Zhou W, Han Y, et al. Individual enrichment of manganese and iron from complex refractory ferromanganese ore by suspension magnetization roasting and magnetic separation[J]. Powder Technology, 2020, 373(prepublish).

[2] Zhou Y, Yang H, Xue X, et al. Separation and recovery of iron and rare earth from bayan obo tailings by magnetizing roasting and (NH4)2SO4 activation roasting[J]. Metals, 2017, 7(6).

辽西低品位钒钛磁铁矿综合利用

刘金生[1,2]，刑振兴[1,2]，黄　壮[1,2]，程功金[1,2]，丁学勇[2]，薛向欣[2]

（1. 东北大学冶金学院，辽宁沈阳　110819；
2. 辽宁省冶金资源循环科学重点实验室，辽宁沈阳　110819）

摘　要：针对辽西低品位钒钛磁铁矿有价组分品位低、矿相组成复杂、嵌布粒度较细等特点，本文提出了磁选预处理工艺，并对预处理铁精矿进行钛铁分离和选铁尾矿的钛铁矿富集分离研究，实现该低品位钒钛磁铁矿的综合利用。结果表明，还原-磁选能进一步将铁精矿中的钛铁分离，提高品位，最佳还原时间为 30min，最佳还原温度为 1125℃，获得了铁品位为 72.86%、回收率为 89.34%的磁性铁精矿。重选-浮选和磁选-浮选实验能回收选铁尾矿中的钛铁矿，重选和强磁处理抛除了大部分的脉石，减少了入浮选矿样的处理量，降低浮选药剂用量，两种工艺流程分别得到 TiO_2 品位 47.74%、回收率 51.26%和 TiO_2 品位为 47.31%、回收率为 57.71%的一级精品钛精矿产品，实现了对钛铁矿的分离与精选。

关键词：低品位；钒钛磁铁矿；还原-磁选；重选-浮选；磁选-浮选

Comprehensive Utilization of Low-grade Vanadium-titanium Magnetite in Western Liaoning Province

LIU Jinsheng[1,2], XING Zhenxing[1,2], HUANG Zhuang[1,2],
CHENG Gongjin[1,2], DING Xueyong[2], XUE Xiangxin[2]

(1. School of Metallurgy, Northeastern University, Shenyang 110819, China; 2. Key Laboratory of Metallurgical Resources Recycling Science of Liaoning Province, Shenyang 110819, China)

Abstract: According to the characteristics of low-grade vanadium-titanium magnetite of Liaoxi with low grade of valuable components, complex composition of ore phase, and fine embedded particle size, pretreatment was proposed in this paper, and titanium-iron separation of pretreated iron concentrates and ilmenite-enriched separation of iron beneficiation tailings were investigated to realize the comprehensive utilization of this low-grade vanadium-titanium magnetite. The results showed that reduction-magnetic separation could further separate the titanium and iron in the iron ore concentrate and improve the grade, the optimal reduction time was 30min and the optimal reduction temperature was 1125℃, and a magnetic iron concentrate with an TFe grade of 72.86% and a recovery rate of 89.34% was obtained. Gravity-flotation and magnetic-flotation separation experiments can recover ilmenite in the iron ore tailing, gravity and strong magnetic separation treatment removed most of the pulsates, which decreased the processing capacity of the ore samples into the flotation and the amount of flotation chemicals. Finally, two process flowsheets were obtained a first-class fine ilmenite concentrate with TiO_2 grade of 47.74%, recovery of 51.26% and TiO_2 grade of 47.31%, recovery of 57.71%, respectively. The separation and selection of ilmenite was realized.

Key words: low-grade; vanadium-titanium magnetite; reduction-magnetic separation; gravity-flotation separation; magnetic-flotation separation

某超大型磁赤混合铁矿高效 SAB 流程选择及实践

张耀斌，牛建昆

（太钢集团岚县矿业有限公司，山西吕梁　033599）

摘　要： 太钢袁家村铁矿属典型微细粒嵌布鞍山式贫红（磁）复杂难选矿石，针对该铁矿矿石类型多、矿石结构复杂、嵌布粒度细等问题，通过国内外多家研究单位进行磨选等工艺方面的试验研究工作，最终流程确定为粗破碎—半自磨—两段球磨—弱磁—强磁—再磨—反浮选流程，在磨矿流程上，舍弃了工艺流程复杂，生产环节多，粉尘污染大，需要生产工人多的传统破碎流程，选择高效 SAB 流程，解决了袁家村铁矿这一嵌布粒度微细，复杂难选的矿石大规模工业应用难题，成为超大型磁赤混合铁矿高效 SAB 流程实践的典范。

关键词： 超大型铁矿；SAB 流程；半自磨；袁家村铁矿；微细粒；复杂难选矿石

Selection and Practice of High Efficiency SAB Process for a Superlarge Magnetic Hematite Mixed Iron Mine

ZHANG Yaobin, NIU Jiankun

(TISCO Lanxian Mining Co., Ltd., Lvliang 033599, China)

Abstract: TISCO Yuanjiacun Iron Mine belongs to a typical fine-grained embedded Anshan style lean red (magnetic) complex and difficult to select ore. In response to the problems of multiple types of ore, complex ore structure, and fine embedded particle size, multiple research units at home and abroad have conducted experimental research on grinding and selection processes. The final process is determined as coarse crushing, semi automatic grinding, two-stage ball milling, weak magnetic, strong magnetic, re grinding, and reverse flotation. In the grinding process, Abandoning the traditional crushing process with complex process flow, multiple production links, high dust pollution, and the need for more production workers, we chose the efficient SAB process to solve the large-scale industrial application problem of Yuanjiacun Iron Mine, a complex and difficult to select ore with fine embedded particle size. Becoming a model for the efficient SAB process practice of ultra large magnetic hematite mixed iron ore.

Key words: ultra large iron ore; SAB process; semi self grinding; yuanjiacun iron mine; fine particles; complex and difficult to select ore

攀西某钒钛磁铁矿入库尾矿工艺矿物学及有价资源回收探索试验研究

蔡先炎[1]，李 硕[1]，陈 勇[2]，陈福林[1]，黄 延[1]

(1. 钒钛资源综合利用国家重点实验室 攀钢集团攀枝花钢铁研究院有限公司，四川攀枝花 617000；2. 攀钢集团新白马矿业有限责任公司，四川攀枝花 617000)

摘 要： 攀西某钒钛磁铁矿入库尾矿含有未回收的有价资源，其工艺矿物学特征情况不明，因此对其开展了化学多元素、矿物组成及含量、化学物相、粒度筛分等工艺矿物学研究，研究结果表明：入库尾矿中主要的有价元素为 TiO_2、Fe、S，含量分别为 3.31%、11.50%、0.478%。尾矿中主要金属矿物为钛铁矿，含量为 4.12%，是 TiO_2 的主要载体矿物，还含有少量的钛磁铁矿和磁黄铁矿。尾矿粒度较细，粗、细粒级中 TiO_2、S 的品位和目的矿物单体解离度差异明显，粗粒级金属矿物与脉石矿物普遍连生，细粒级多为单体。摇床重选分粒级预富集钛铁矿可获得产率 7.07%，TiO_2 品位为 21.12%，回收率为 45.37%的钛粗精矿，预富集效果明显。本研究为该钒钛磁铁矿入库尾矿的合理开发利用提供理论依据。

关键词： 钒钛磁铁矿；入库尾矿；工艺矿物学；预富集

Exploring Experimental Study on Mineralogy and Valuable Resource Recovery of Inbound Tailings of a Vanadium Titanium Magnetite in Panxi

CAI Xianyan[1], LI Shuo[1], CHEN Yong[2], CHEN Fulin[1], HUANG Yan[1]

(1. State Key Laboratory of Vanadium and Titanium Resources Comprehensive Utilization, Panzhihua Iron & Steel Research Institute Co., Ltd. of Pangang Group, Panzhihua 617000, China; 2. New Bai Ma Mining Co., Ltd. of Pangang Group, Panzhihua 617000, China)

Abstract: The inbound tailings of a vanadium titanium magnetite in Panxi contains unrecycled valuable resources, and it's process mineralogy characteristics is unknown. Therefore, the process mineralogy were conducted such as chemical multi-element, mineral composition and content, chemical phase, particle size screening, etc. The research results show that the main valuable elements are TiO_2, Fe, and S in the tailings and the content are 3.31%, 11.50%, and 0.478%, respectively. Iilmenite is main metal mineral in the tailings with a content of 4.12%, which is major carrier mineral of TiO_2, and also contains a small amount of titanium magnetite and pyrrhotite; The particle size of tailings is finer, and the grade of TiO_2, S and the liberation degree of target mineral are obvious difference between coarse and fine particle sizes, and metal minerals are generally associated with gangue in coarse grain size, while are mostly individual in finer grain size. A titanium concentrate can be obtained with a yield of 7.07%, the grade of TiO_2 21.12%, and a recovery rate of 45.37% by using shaking table gravity separation to pre enrich ilmenite, and the effect of preenrichment is evident. The research offer theoretical foundation for the rational development and utilization of inbound tailings of a vanadium titanium magnetite.

Key words: vanadium titanium magnetite; inbound tailings; process mineralogy; preenrichment

攀西地区钒钛磁铁矿碎磨技术发展现状

李硕，蔡先炎，黄延，王志杰，陈福林，
吴宁，王志林，杨道广

（钒钛资源综合利用国家重点实验室 攀钢集团攀枝花钢铁研究院有限公司，
四川攀枝花 617000）

摘要：我国攀西地区的钒钛磁铁矿资源十分丰富，钒钛磁铁矿的主要元素组成为铁、钛、钒、氧，还伴生有大量的铬、钴、硫、镍、锰、镓、钪以及铂族等元素。笔者概述了攀西地区钒钛磁铁矿碎磨技术的发展现状，并介绍了近年来新型碎磨设备在攀西地区钒钛磁铁矿中的研究及应用现状。长期以来，攀西地区钒钛磁铁矿碎磨技术的发展较为缓慢，而碎磨新工艺新设备的工业应用实例也较少。本文不仅分析了钒钛磁铁矿碎磨技术发展缓慢的原因，还指出了效率高、性能优、易于工业应用的新型碎磨工艺与设备的成功研发，是实现攀西地区钒钛磁铁矿选矿技术取得突破性进展的关键所在。

关键词：攀西地区钒钛磁铁矿；碎磨；高压辊磨；微波助磨

Development Status of Crushing and Grinding Technology of Vanadium Titano-Magnetite in Panxi Area

LI Shuo, CAI Xianyan, HUANG Yan, WANG Zhijie, CHEN Fulin,
WU Ning, WANG Zhilin, YANG Daoguang

(State Key Laboratory of Vanadium and Titanium Resources Comprehensive Utilization, Panzhihua Iron & Steel Research Institute Co., Ltd. of Pangang Group, Panzhihua 617000, China)

Abstract: China's Panxi area has abundant resources of vanadium titano-magnetite. The main elements of vanadium titano-magnetite are Fe, Ti, V and O, which also accompanied by a large amount of Cr, Co, S, Ni, Mn, Ga,Sc and antimony. The development status of grinding technology of vanadium titano-magnetite in Panxi area was summarized, and the research and application status of new crushing and grinding equipment in vanadium titano-magnetite in Panxi area in recent years were introduced. For a long time, the grinding technology of vanadium titano-magnetite in Panxi area has developed slowly, and there are few new grinding technology and equipments were applied in industrial. The reasons for the slowly development of grinding technology of vanadium titano-magnetite were analyzed in this paper. Also, the successful development of new grinding technology and equipments with high efficiency, excellent performance and easy industrial application, which is the key to achieve breakthrough progress in vanadium titano-magnetite beneficiation technology in Panxi area was pointed out.

Key words: vanadium titano-magnetite in Panxi area; crushing and grinding; high-pressure roller grinding; microwave grinding

2 焦化及节能环保

大会特邀报告
第十三届冶金青年科技奖获奖人特邀报告
分会场特邀报告
矿业工程
★ 焦化及节能环保
炼铁与原料
炼钢与连铸
电冶金与废钢铁
轧制与热处理
表面与涂镀
金属材料深加工
粉末冶金
先进钢铁材料及应用
节能与低碳技术
冶金环保与资源利用
冶金设备与工程技术
冶金自动化与智能化
冶金物流
冶金流程工程学
其他

干熄焦年修复产调控技术优化研究与应用

朱庆庙[1,2]，李　勇[3]，侯士彬[3]，赵振兴[1,2]，赵　锋[3]

（1. 海洋装备用金属材料及其应用国家重点实验室，辽宁鞍山　114009；2. 鞍钢集团钢铁研究院，辽宁鞍山　114009；3. 鞍钢股份有限公司炼焦总厂，辽宁鞍山　114021）

摘　要：开展干熄焦年修停工及开工控制技术研究，优化停工及开工期间干熄炉内焦炭料位控制。应用红焦热气烘炉方式替代温风干燥烘炉，避免开工期间红焦直接接触冷态砌体，有利于保护干熄炉砌体温度急剧变化，降低烘炉时间，提升干熄焦生产效率。

关键词：干熄焦；年修；复产；调控；料位；应用

Research and Application on Optimization of Annual Restoration and Production Control Technology of CDQ

ZHU Qingmiao[1,2], LI Yong[3], HOU Shibin[3],
ZHAO Zhenxing[1,2], ZHAO Feng[3]

(1. Ansteel Iron & Steel Research Institutes, Anshan 114009, China; 2. State Key Labratory of Metal Material for Marine Equipment and Application, Anshan 114009, China; 3. General Coking Plant of Angang Steel Co., Ltd., Anshan 114021, China)

Abstract: To carry out research on annual maintenance shutdown and start-up control technology of CDQ, and optimize coke material level control in dry quenching furnace during shutdown and start-up. The application of red coke hot air drying oven instead of warm air drying oven can avoid the red coke directly contacting cold masonry during start-up period, which is beneficial to protect the sudden change of masonry temperature of dry quenching furnace, reduce oven time and improve the production efficiency of CDQ.

工业废弃树脂配煤炼焦技术应用研究

朱庆庙[1,2]，末福宇[3]，侯士彬[3]，赵振兴[1,2]，赵　锋[3]

（1. 海洋装备用金属材料及其应用国家重点实验室，辽宁鞍山　114009；
2. 鞍钢集团钢铁研究院，辽宁鞍山　114009；
3. 鞍钢股份有限公司炼焦总厂，辽宁鞍山　114021）

摘　要：水处理产生的固废离子交换树脂是常见的工业固体危险废物之一，难以处理的问题一直困扰研究者们。在分析固废离子交换树脂技术指标的基础上，本文通过固废树脂配煤炼焦试验，检验焦炭质量变化，探讨固废离子交换树脂再利用的技术。结果表明，固废离子交换树脂中会对焦炭质量起到劣化作用；使用固废离子交换树脂回配炼焦，回配比例为 0.1%时，焦炭质量下滑不明显，但回配比例超过 0.1%，对焦炭质量影响加大，焦炭质量明显下降。通过控制固废离子交换树脂的配入量，可以达到稳定焦炭质量，解决固废离子交换树脂不易处理的问题。

关键词：固废离子交换树脂；配煤炼焦；焦炭冷态强度；焦炭热态强度

Application of Coking Technology with Industrial Waste Resin Blending Coal

ZHU Qingmiao[1,2], WEI Fuyu[3], HOU Shibin[3],
ZHAO Zhenxing[1,2], ZHAO Feng[3]

(1. Ansteel Iron & Steel Research Institutes, Anshan 114009, China; 2. State Key Labratory of Metal Material for Marine Equipment and Application, Anshan 114009, China; 3. General Coking Plant of Angang Steel Co., Ltd., Anshan 114021, China)

Abstract: Solid waste Ion-exchange resin from water treatment is one of the most common industrial solid hazardous wastes. Based on the analysis of the technical indexes of the solid waste Ion-exchange resin, this paper examines the coke quality changes through the coking test of the coal blending with the solid waste resin, and probes into the technology of the reuse of the solid waste Ion-exchange resin. The results showed that the quality of coke was deteriorated in the solid waste Ion-exchange resin, and the quality of coke did not decline significantly when the Ion-exchange resin was recombined with 0.1% coke, but the ratio of remix is more than 0.1%, the influence on coke quality is bigger, the coke quality is obviously reduced. By controlling the amount of solid waste Ion-exchange resin, the quality of coke can be stabilized and the problem of difficult disposal of solid waste Ion-exchange resin can be solved.

真空碳酸钾脱硫及 TOP 湿法制酸生产问题的查找与处理

杨天宇，张　欣，韩修尚，徐利民

（鞍钢化学科技有限公司，辽宁鞍山　114021）

摘　要：真空碳酸钾脱硫装置在工艺和设备上都存在一定的问题，例如：脱硫碱洗段由蒸氨废水和浓碱经过混合器后进脱硫碱洗段脱除硫化氢。在生产过程中，由于管道刮碱垢导致至碱洗段废水流量逐渐降低，脱硫碱洗段气液比降低，导致在线硫化氢频繁超标；脱硫真空泵换热器管束由于长时间运行，管束堵塞严重导致循环量下降，降低真空泵真空度；酸气冷却器由于夏季用户低温水使用较多，不能满足低温水使用需求，导致酸气冷却器温度增加，真空泵工作液温度高；诸如以上问题均需要采取针对性措施加以改进和优化，从而提高脱硫效果。本文以鞍钢化学科技有限公司二回收作业区焦炉煤气脱硫净化技术为例，探究真空碳酸钾脱硫工艺的优化方法。

关键词：真空碳酸钾脱硫工艺；优化研究

浅谈焦油蒸馏精细化操作

孔德刚，王英达，孙喜民

（鞍钢化学科技有限公司，辽宁鞍山 114021）

摘 要：鞍钢化学科技焦油蒸馏工艺是 2010 年建成投产，采用常减压结合、直接采出 N93 工业萘、后加碱技术生产高质量沥青产品的生产工艺，焦油蒸馏主要有五个单元，脱水单元、沥青分离单元、精馏单元。2013 年 2 月投产，在 10 年的产生过程中经过多次停产改造，提高生产能力。而单元馏分采出及温度控制还有待优化例如：重油采出位置分别在沥青分离塔塔顶的侧线采出和精馏单元的蒽塔塔底采出，在沥青分离塔采出的目的是为了降低成本，避免重油在精馏单元重复加热，重复采出。原设计煤焦油处理量 $34m^3/h$，沥青分离单元的重油采出量为 $1.5\sim1.6m^3/h$，精馏单元的处理量 $15\sim16m^3/h$。但是随着焦炉大型化改质后，焦油中的沥青含量发生了变化，由原来的 55%降到 45%~48%。为了满负荷达到处理焦油 $34m^3/h$，精馏单元的处理量在 $18.5\sim19.5m^3/h$，相当于负荷提到了 130%，由于塔的设计处理能力为 110%，本文将按各单元操作简明扼要阐述在生产过程中发现的问题和解决方法。

关键词：煤焦油；节能

焦化项目中备煤工段除尘系统的设计计算

邹 阳，杨雪莲，杨 仙，李梦珂，李传东

（陕西省冶金设计研究院有限公司通用室，陕西西安 710032）

摘 要：除尘系统的设计计算是焦化厂除尘系统设计过程中的重要一环，主要包括除尘系统风量计算、末端除尘罩设计计算、除尘风管水力平衡计算三大主要计算内容。本文系统性地总结了规范等参考资料中所涉及的除尘系统风量的计算方法及计算公式，较为详细地介绍了除尘系统末端除尘罩的设计计算，考虑到含尘管道输送介质的特殊性，本文对除尘系统的阻力损失计算以及风机选型也做了详细介绍。本文所述除尘系统的设计计算方法和步骤，可对实际工程中除尘系统的设计提供参考，也可用于判断同类项目除尘系统风量设计及设备选择的合理性。

关键词：风量计算；诱导空气量；受料仓；落料点；物料输送

Design Calculations of Dust Removal System in Coking Project

ZOU Yang, YANG Xuelian, YANG Xian, LI Mengke, LI Chuandong

(Shaanxi Metallurgical Design & Research Institute Co., Ltd., Xi'an 710032, China)

Abstract: The design and calculation of the dust removal system is an important part of the design process of the dust removal system of the coking plant, it mainly includes three main calculation contents: dust removal system air volume calculation, end dust removal hood design calculation, and dust removal duct hydraulic balance calculation. This paper systematically summarizes the calculation methods and calculation formulas of the air volume of the dust removal system involved in the reference materials such as specifications, the design calculation of the dust collection cover at the end of the

焦化废水石油类测定中不同吸附方法的比较

刘晓桃，宋文杰，尚志春，杜 彬，陈 伟，张 宇

（鞍钢集团朝阳钢铁有限公司，辽宁朝阳 122000）

摘 要：以焦化废水石油类测定为例，对比了《红外分光光度法测定石油类和动植物油类》（HJ 637—2018）推荐的振荡吸附和吸附柱两种吸附方法的测定结果，进行了差异及产生原因的分析和探讨，以便找到最适合应用于焦化废水石油类测定的方法。在此基础上，对新标准中一些步骤的可操作性及局限性进行了分析，并总结归纳了影响检验精密度的主要因素及控制措施。

关键词：焦化废水；石油类；红外分光光度法；精密度

Comparison of Different Adsorption Methods in Determination of Petroleum in the Coking Wastewater

LIU Xiaotao, SONG Wenjie, SHAGN Zhichun,
DU Bin, CHEN Wei, ZHANG Yu

(Chaoyang Iron & Steel Co., Ltd. of Ansteel Group Corporation, Chaoyang 122000, China)

Abstract: Taking the determination of petroleum in coking wastewater as an example, a comparison was made between the results of the two adsorption methods recommended in "Infrared Spectroscopy for the Determination of Petroleum and Animal and Plant Oils" (HJ 637-2018), namely oscillating adsorption and adsorption column. The differences and causes were analyzed and discussed in order to find the most suitable method for the determination of petroleum in coking wastewater. On this basis, the operability and limitations of some steps in the new standard were analyzed, and the main factors that affect the precision of inspection and control measures were summarized and summarized.

Key words: coking wastewater; petroleum; infrared spectrophotometry; precision

焦炉煤气制 LNG 联产液氨工艺脱碳增产技术的研究

杨 鹏，栾树龙

（山东钢铁集团日照有限公司，山东日照 276800）

摘　要：焦炉煤气是炼焦的副产品，其主要成分为 H_2、CH_4、CO 和 CO_2，是一种良好的化工原料和高热值燃料。随着环境保护政策愈发严格，各级政府及监管部门相继出台了政策及调控措施，加大了对焦化、化工和钢铁等高污染、高耗能行业的监控。在此背景下，建设焦炉煤气制 LNG 联产合成氨项目，是实现焦炉煤气的综合利用和下游产品的多元化，从根本上治理环境、解决焦化企业困境的极好选择。焦炉煤气制 LNG 联产合成氨工艺中，甲烷化工艺利用煤气中过量的氢气，将 CO 和 CO_2 转化为甲烷，由化学反应方程式可知，1mol CO_2 需消耗 4mol 氢气，只能产生 1mol CH_4，而 4mol 氢气在合成氨工序中可生产 8/3mol NH_3。以每小时处理焦炉煤气量 50000Nm^3 计算，焦炉煤气中的 CO_2 每降低 1%，CH_4 产量降低 357kg，而 NH_3 产量可增加 1004kg，由此产生的经济效益相当可观。因此焦炉煤气脱碳工艺也逐渐进入国内各大 LNG 生产企业的研究范围。本文从脱碳工艺概况，适合焦炉煤气脱碳的工艺方法和应用情况，以及脱碳增产工艺的应用前景进行探讨研究，为相关项目提供参考。

关键词：焦炉煤气；液化天然气；合成氨；甲烷化；MDEA 工艺；低温甲醇洗

Research on Decarbonization and Production Increase Technology of Coke Oven Gas to LNG Combined with Liquid Ammonia Production Process

YANG Peng, LUAN Shulong

(Shandong Iron and Steel Group Rizhao Co., Ltd., Coking Plant, Rizhao 276800, China)

Abstract: Coke oven gas is a byproduct of coking, mainly composed of H_2, CH_4, CO, and CO_2. It is a good chemical raw material and high calorific value fuel. With the increasingly strict environmental protection policies, governments and regulatory departments at all levels have successively introduced policies and regulatory measures, increasing monitoring of high pollution and energy consuming industries such as coking, chemical, and steel. In this context, the construction of a coke oven gas to LNG co production synthetic ammonia project is an excellent choice to achieve the comprehensive utilization of coke oven gas and the diversification of downstream products, fundamentally govern the environment, and solve the difficulties of coking enterprises. In the process of producing LNG from coke oven gas and synthesizing ammonia, the methanation process utilizes excess hydrogen gas to convert CO and CO_2 into methane. According to the chemical reaction equation, one mole of CO_2 requires 4 moles of hydrogen gas to produce only 1 mole of CH_4, while 4 moles of hydrogen gas can produce 8/3 moles of NH_3 in the ammonia synthesis process. Process 50000 Nm^3 of coke oven gas per hour calculation shows that for every 1% decrease in CO_2 in coke oven gas, the CH_4 production decreases by 357kg, while the NH_3 production can increase by 1004kg, resulting in considerable economic benefits. Therefore, the decarbonization process of coke oven gas has gradually entered the research scope of major LNG production enterprises in China. This article explores and studies the decarbonization process overview, suitable process methods and applications for coke oven gas decarbonization, as well as the application prospects of decarbonization and production increase processes, providing reference for related projects.

Key words: coke oven gas; liquified natural gas; synthetic ammonia; methanation; MDEA process; low temperature methanol washing

关于干熄焦生产运行的几点新建议和新思考

侯士彬[1]，李秋镝[3]，李 勇[1]，马俊尧[2]，高 微[1]

（1. 鞍钢炼焦总厂，辽宁鞍山 114031；2. 鞍山钢铁集团有限公司，辽宁鞍山 114031；
3. 鞍山冶金集团工业有限公司鞍山化工分公司，辽宁鞍山 114031）

摘 要：通过对干熄焦系统的运行总结，提出了调节干熄炉炉内料位的新建议，分析了炉墙破损的原因并且提出了干熄焦除尘系统维护的新方法。

关键词：干熄焦；新建议；新思考

Several New Suggestions and New Suggestions about the CDQ Circulating

HOU Shibin[1], LI Qiudi[3], LI Yong[1], MA Junyao[2], GAO Wei[1]

Abstract: According to the Summary of the CDQ Circulating, put forward some new suggestions about coke anticipate of CDQ, analyze the Reason of CDQ chamber breakage of CDQ, put forward some new method about the CDQ emission control system.

Key words: CDQ; new suggestions; new suggestions

关于五炼焦作业区与西部炼焦作业区焦炉煤气热值存在差异的原因分析

侯士彬[1]，李秋镝[3]，李 勇[1]，马俊尧[2]，高 微[1]

（1. 鞍钢炼焦总厂，辽宁鞍山 114031；2. 鞍山钢铁集团有限公司，辽宁鞍山 114031；
3. 鞍山冶金集团工业有限公司鞍山化工分公司，辽宁鞍山 114031）

摘 要：通过对五炼焦作业区与西部炼焦作业区焦炉煤气热值存在差异的原因分析，得出导致五炼焦焦炉煤气热值略低于西部炼焦的根本原因是两个系统所用配合煤煤种差别较大，煤质综合变质程度不同所致。

关键词：焦炉煤气；热值；原因分析

焦炉煤气干法精脱硫工艺的优化设计与应用

李 伟，王博林，韩渝京，李 鹏

（北京首钢国际工程技术有限公司能源环境分公司，北京 100043）

摘 要：焦炉煤气精脱硫是国家环保体系中钢铁企业超低排放要求的重要组成部分，在分析焦炉煤气干法精脱硫存在问题的基础上，提出了优化设计要点，解决了传统干法精脱硫原料气中焦油、水分过多，装卸脱硫剂时填料损失等问题，提出了干法脱硫系统中，脱硫剂遇水时在线处理装置的设计思路，保留并升级了干法精脱硫串、并联操作工艺。达到了焦炉煤气精脱硫的技术目标。

关键词：焦炉煤气；精脱硫；工艺优化；应用

Optimization Design and Application of Dry Fine Desulfurization Process for Coke Oven Gas

LI Wei, WANG Bolin, HAN Yujing, LI Peng

(Beijing Shougang International Engineering Co., Ltd., Beijing 100043, China)

Abstract: Coke oven gas desulfurization is an important part of the steel industry's ultra-low emission requirements in the national environmental protection system. Based on an analysis of the problems with dry desulfurization of coke oven gas, this paper proposes optimization design points to solve problems such as excessive tar and moisture in the raw gas of traditional dry desulfurization, as well as filling loss during loading and unloading of desulfurizing agents. The paper also proposes a design concept for the online treatment device when the desulfurizing agent encounters water in the dry desulfurization system, and retains and upgrades the series and parallel operation processes of dry desulfurization. The technical goal of coke oven gas desulfurization has been achieved.

Key words: coke oven gas; fine desulfurization; optimization design; application

循环氨水余热用于工艺制冷的实践

王思维[1]，邱全山[1]，周 诚[2]，张晓萍[1]，李 杰[1]，胡 俊[1]

（1. 马钢技术中心，安徽马鞍山 243000；2. 马钢煤焦化公司，安徽马鞍山 243000）

摘 要：焦化企业是能源消耗大户，降低产品能耗对企业的可持续发展有重要作用，利用循环氨水作为热源的溴化锂吸收式制冷系统近几年被很多焦化企业所采用。某钢铁公司的7.63m超大型焦炉系统运用该技术在国内外尚属首次，该系统改造后不仅能很好地满足生产工艺的冷水要求，而且降低能耗和设备运行成本以及减少多种污染物的排放，为焦化企业开辟新的节能减排之路。

关键词：循环氨水；吸收式制冷；节能；减排

Practice of Circulating Ammonia Waste Heat used in Process Refrigeration

WANG Siwei[1], QIU Quanshan[1], ZHOU Cheng[2],
ZHANG Xiaoping[1], LI Jie[1], HU Jun[1]

(1. Tech Center of Masteel, Maanshan 243000, China; 2. Coking Plant of Masteel,
Maanshan 243000, China)

Abstract: Coking enterprises are large energy consumers, reducing product energy consumption plays an important role in the sustainable development of enterprises. Lithium bromide absorption refrigeration system using recycled ammonia as a heat source has been adopted by many coking enterprises in recent years. The 7.63m super large coke oven system of a steel company uses this technology for the first time at home and abroad. After the system transformation, it can not only meet the cold water requirements of the production process, but also reduce energy consumption and equipment operating costs and the emission of various pollutants, opening up a new way for coke enterprises to save energy and reduce emission.

Key words: circulating ammonia; absorption refrigeration; energy saving; emission reduction

硫铵母液焦油资源化利用技术的研发

朱晓渊[1]，颜长青[1]，施 亮[1]，王传荟[2]，潘红良[2]

（1. 宝武碳业科技股份有限公司，上海 201999；2. 上海务宝机电科技有限公司，上海 201999）

摘 要： 在硫酸吸收荒煤气中氨气的工艺中，荒煤气中的残余焦油会一并吸收到硫酸洗涤液中，并在硫铵溶液循环罐中积聚，对于这部分焦油，传统的方法是采用人工进行打捞，打捞出的焦油送往配煤车间进行配煤，此工艺不仅不环保，劳动工作量大，而且还存在资源的浪费。本文通过特制的装备，实现硫铵溶液中焦油的在线分离，对于分离出的焦油采用装置中的剩余氨水进行分散，分散后的焦油送至焦油氨水分离器，形成焦油产品，实现硫铵母液焦油的资源化利用。

关键词： 硫铵母液；煤焦油；分离；分散；资源利用

Research and Development of Technology for Resource Utilization of Tar from Ammonium Sulfate Liquor

ZHU Xiaoyuan[1], YAN Changqing[1], SHI Liang[1],
WANG Chuanhui[2], PAN Hongliang[2]

(1. Baowu Carbon Industry Technology Co., Ltd., Shanghai 201999, China; 2. Shanghai Wubao
Mechatronics Technology Co., Ltd., Shanghai 201999, China)

Abstract: In the process of absorbing ammonia gas from raw gas with sulfuric acid, the residual tar in raw gas is absorbed into the sulfuric acid washing solution and accumulated in the ammonium sulfate solution circulation tank. The traditional method for this part of tar is to manually salvage it, and the recovered tar is sent to the coal blending workshop for coal blending. This process is not only environmentally friendly and labor-intensive, but also wastes resources. This article uses specially designed equipment to achieve online separation of tar in ammonium sulfate solution. The separated tar is dispersed using the remaining ammonia water in the device, and the dispersed tar is sent to the tar ammonia water separator to form a tar product, achieving the resource utilization of the ammonium sulfate mother liquor tar.

Key words: ammonium sulfate mother liquor; coal tar; separate; disperse; resource utilization

二氧化铈/氮掺杂氮化碳光催化剂可见光下深度处理焦化废水研究

安　宁[1,2]，胡绍伟[1,2]，陈　鹏[1,2]，王　永[1,2]，王　飞[1,2]，刘　芳[1,2]

（1. 海洋装备用金属材料及其应用国家重点实验室，辽宁鞍山　114009；
2. 鞍钢集团钢铁研究院环境与资源研究所，辽宁鞍山　114009）

摘　要： 焦化废水中毒性物质超过500种，是公认的难处理废水。国内多数焦化企业的生产废水采用生化—物化法处理，最终出水水质指标难达到《炼焦化学工业污染物排放标准》（GB16171—2012）。因此，对焦化废水进行深度处理回用成为行业热点。作为最受青睐的光催化材料，石墨相氮化碳（$g\text{-}C_3N_4$）被广泛用于去除水中有机污染物。本文采用二氧化铈（CeO_2）负载及氮掺杂来优化氮化碳，成功制备具有良好吸光性能及光生载流子分离效率的$CeO_2/N\text{-}g\text{-}C_3N_4$（CeCNN）复合光催化剂。CeCNN2展现了极佳的光催化性能，120min内对焦化废水生化出水COD、UV_{254}的降解效率分别达到了49.79%、69.37%，远高于$g\text{-}C_3N_4$及氮掺杂$g\text{-}C_3N_4$。本研究表明氮掺杂及二氧化铈负载可有效拓宽氮$g\text{-}C_3N_4$光吸收范围，促进光生载流子分离效率，对构建高性能$g\text{-}C_3N_4$基光催化剂具有借鉴意义。

关键词： 焦化废水；氮化碳；二氧化铈；深度处理；光催化机理

Cerium Dioxide/Nitrogen Doped Graphite Carbon Nitride for Photocatalytic Advanced Treatment of Coking Wastewater under Visible Light

AN Ning[1,2], HU Shaowei[1,2], CHEN Peng[1,2], WANG Yong[1,2],
WANG Fei[1,2], LIU Fang[1,2]

(1. State Key Laboratory of Metal Material for Marine Equipment and Application, Anshan 114009, China;
2. Iron & Steel Research Institute of Ansteel Group, Institute of Environment and
Resources, Anshan 114009, China)

Abstract: The coking wastewater contains plenty of toxic substances over 500 species, and it is recognized as the most intractable treated kinds of wastewater. Currently, the production wastewater of most domestic coking enterprises is treated by biochemical-physicochemical method, and the final effluent water quality is still difficult to meet the "Discharge Standard for Coking Chemical Industry Pollutants" (GB16171—2012). Thus, the deep treatment and reuse of coking wastewater has become a hot topic in the industry. In order to make up for the shortage of $g\text{-}C_3N_4$ monomer, CeO_2/N-doped $g\text{-}C_3N_4$ (CeCNN) composite photocatalysts co-modified with nitrogen doping and CeO_2 loading were designed and synthesized with the idea of expanding visible light absorption and promoting photogenerated carrier separation. CeCNN2 exhibits excellent photodegradation performance, the removal rate of COD and UV_{254} in the coking wastewater after biological treatment reached 49.79% and 69.37% within 120 min, respectively, which is much higher than that of g-C_3N_4 and N-doped g-C_3N_4; This work proves that N-doping and CeO_2 loading can effectively broaden the photoresponse range of g-C_3N_4, facilitate the separation of photogenerated electron-hole pairs, and provide a reference for the construction of g-C_3N_4–based photocatalyst with high-efficiency photodegradation activity.

Key words: coking wastewater; g-C_3N_4; CeO_2; advanced treatment; photocatalytic mechanism

煤岩及焦炭光学组织评价焦炭宏观性能

张文成

（宝钢研究院梅钢技术中心，江苏南京 210039）

摘 要：焦炭光学组织结构是焦炭内在质量指标之一，对炼焦煤选择及焦炭性能表征具有重要意义。本文研究了20个单种煤炼焦试验得到单种焦炭，利用光学显微镜进行焦炭光学组织结构分析，并探讨了与煤质指标及焦炭强度指标的关系。研究表明粗粒镶嵌结构及各项异性指数 OTI 与单种煤的镜质组反射率 R 及挥发分显著相关；中粗粒镶嵌结构与焦炭的转鼓强度 DI 相关性显著；焦炭各项异性指数 OTI 与焦炭反应性 CRI 的相关性最显著；中粗粒镶嵌结构与焦炭反应后强度 CSR 的相关性最显著。利用煤的变质程度结合焦炭光学组织可以实现焦炭冷强度和热性能的预测，对于揭示炼焦煤的内在质量差异及焦炭性能评价提供了技术手段。

关键词：焦炭；光学组织结构；煤岩；反应后强度

Evaluation of Coke Macroscopic Properties by Optical Organization of Coke and Coal Rank

ZHANG Wencheng

(Meishan Technology Center, Baosteel Research Institute, Nanjing 210039, China)

Abstract: Coke optical organization structure is one of the intrinsic quality indicators of coke, which is important for coking coal selection and coke performance characterization. In this paper, we analyzed 20 single coke optical structure by optical microscopy, and discussed the relationship with coal quality index and coke strength index. The study showed that the coarse-grained mosaic structure and anisotropy index OTI were significantly correlated with the reflectance R and volatile fraction of the specular group of the single coal. The medium-coarse-grained mosaic structure was significantly correlated with the drum strength DI of the coke. The correlation between each anisotropy index OTI of the coke and the coke reactivity CRI was the most significant. The correlation between the medium-coarse-grained mosaic structure and the post-reaction strength CSR of the coke was the most significant. The prediction of cold strength and thermal properties of coke can be achieved by using the degree of coal metamorphism combined with the optical organization of coke, which provides a technical means to reveal the intrinsic quality differences of coking coal and the evaluation of coke performance.

Key words: coke; texture; coal rank; coke strength after reaction

高炉内焦炭降解行为及质量评价方法研究

陈静波，张生富，姜 涛

（重庆大学材料科学与工程学院，重庆 400044）

摘 要：焦炭是高炉冶炼中至关重要的原燃料，承担着热源、还原剂、骨架及渗碳剂等重要职责。焦炭质量的好坏，对高炉顺行以及冶炼经济性有着重要的影响。在"碳达峰、碳中和"背景下，高炉喷煤、富氧鼓风等低碳冶炼技术的应用以及高炉大型化的推广对焦炭质量提出了更加严格的要求。在焦炭炉内负荷不断增加、优质炼焦煤资源却日益稀缺的压力下，提供有效的焦炭评价指标保证焦炭质量稳定可靠十分重要。然而，现有焦炭质量评价方法严重简化高炉内气氛、温度等反应条件，忽略压力及碱金属催化对焦炭的影响，不符合焦炭在高炉内实际运行情况，导致评价指标与实际应用脱节。因此，亟待深入研究焦炭受到高炉不同部位劣化影响后的降解行为并提出更符合高炉实际运行的评价指标。

本文采用自主研发的高温变压、变气氛焦炭热性能测试装置模拟了高炉块状带、软熔滴落带内动态变化的温度、气氛及压力，采用层层递进的方法定量化研究了高炉不同部位内热应力、碳素溶损及碱金属催化对顶装焦、捣固焦的性能及结构的影响程度，获得了焦炭在高炉内的递变行为规律及造成不同焦炭指标劣化的主要影响因素。结果表明，焦炭失碳主要由碱金属催化下的碳素溶损反应导致。大颗粒焦炭降解主要由热应力及碱金属催化导致，而焦炭细粉生成主要由热应力导致。尽管捣固焦的反应性更低，反应后强度更高，但由于其较弱的抗热应力及抗碱金属能力，在高炉内实际失碳高于顶装焦。导致顶装焦、捣固焦劣化的主要原因不同，顶装焦抗热应力及碱金属催化能力高于捣固焦，捣固焦抗碳素溶损能力高于顶装焦。在此基础上，本文提出了新的焦炭质量评价方法，为高炉焦炭的合理选用提供了理论指导及数据支持。

关键词：焦炭；劣化行为；性能评价；定量分析

7m 焦炉电机车及 APS 控制系统优化

乔文超，张允东，孔 弢，赵 锋，陈 鹏

（鞍钢股份炼焦总厂，辽宁鞍山 114000）

摘 要：鞍钢股份炼焦总厂 JNX3-70-1 型焦炉是国家高技术研究发展示范工程，五期焦炉于 2014 年 10 月开始投产，年产焦炭约 150 万吨。与之配套使用的干熄焦于 2015 年 10 月开始投产。随着超大容积环保型顶装焦炉的开发应用，相应的智能化设备也得以广泛使用，电机车和 APS 就是其中的一部分，电机车设有走行、旋转焦罐，APS 夹紧，提升机提升等控制装置。车内设有操作平台。平台上的两个中控电脑可对电机车进行走行、焦罐旋转接焦、焦罐提升、水熄焦打水等操作。针对操作人员素质和水平的不同，对电机车和 APS 程序及外部电路加以优化改进，不仅对机电设备起到保护作用，还大大降低了设备事故发生率。

关键词：电机车；旋转焦罐；APS；优化

Optimization of Seven-meter Coke Oven Electric Motor and APS Control System

QIAO Wenchao, ZHANG Yundong, KONG Tao, ZHAO Feng, CHEN Peng

(Angang Coking Plant, Anshan 114000, China)

Abstract: The JNX 3-70-1 coke oven of Angang Steel Corporation is a national high-tech research and development demonstration project. The fifth phase coke oven began to be put into operation in October 2014, with an annual output of about 1.5 million tons of coke. The CDcoke was put into production in October 2015. With the development and application of super-large volume environment-friendly top-mounted coke oven, the corresponding intelligent equipment is widely used. Electric locomotive and APS are part of them. The electric motor is equipped with walking, rotary coke tank, APS clamp, lifting machine and other control devices. The car is equipped with an operating platform. The two central control computers on the platform can walk the electric locomotive, rotate the coke tank, lift the coke tank, extinguish the water and draw water. According to the different quality and level of operators, the electric motor and APS program and external circuit are optimized and improved, which not only protects the mechanical and electrical equipment, but also greatly reduces the equipment accident rate.

Key words: electric locomotive; rotary coke tank; APS; optimization

7.63m 焦炉超超低排放的实践

殷喜和

(山西太钢不锈钢股份有限公司焦化厂，山西太原　030003)

摘　要： 本文围绕满足焦炉炼焦加热废气污染物超超低排放要求，从焦炉热工工艺源头入手，分析主要污染物生成机理，详细介绍了温度制度、压力制度的工艺优化措施，取得了明显效果。

关键词： 加热废气；超超低排放；热工工艺优化；温度制度；压力制度

The Practice of 7.63m Coke Oven with Ultra-low Emission

YIN Xihe

(Coking Plant of Shanxi Taiyuan Iron and Steel Co.,Ltd., Taiyuan 030003, China)

Abstract: This paper focuses on meeting the ultra - low emission requirements of waste gas pollutants from coke oven heating , starting from the source of coke oven thermal process , analyzes the generation mechanism of main pollutants , introduces the process optimization measures of temperature system and pressure system in detail , and has achieved obvious results.

Key words: heating exhaust gas; ultra-low emission; thermal process optimization; temperature system; pressure system

焦炉废气快速分析在焦化中的应用

邱 程

（中信泰富特钢青岛特钢，山东青岛　266500）

摘　要：焦炉是结构最复杂的工业窑炉，因受到砌体耐材工艺要求（烘炉投产后炉温维持在 1100℃以上，温度过低容易发生不可逆晶型转化，造成砌体损坏），焦炉自生产开始后就不再进行停产检修，但随着炉龄的增长，焦炉内部出现砖缝串漏，耐火砖坍塌的情况，若仅凭肉眼很难以从焦炉开口处观察到焦炉内部受损情况，所以需要一种能够通过对焦炉废气进行分析进而推断出焦炉内部损伤的具体位置的方法。

浅谈焦炉停产事故状态下的工艺调整

邱　程，郭　飞

（青岛特殊钢铁有限公司焦化厂，山东青岛　266500）

摘　要：介绍了焦炉在停产事故状态下，以集气管，装煤孔，调火，湿熄焦，装煤出焦复产等工艺调整。根据停产时间对各项工艺参数和操作按照焖炉要点展开，确保安全开工。

关键词：集气管；装煤孔；调火；湿熄焦；装煤出焦复产

Abstract: This paper introduces the process adjustment of coke oven in the state of shutdown accident, such as gas collector, coal loading hole, fire regulation, wet coke quenching, coal loading and coke discharge, and resumption of production. According to the shutdown time, carry out various process parameters and operations according to the key points of stewing furnace to ensure safe start-up.

Key words: gas collector; coal loading hole; adjust the fire; wet coke quenching; reproduction of coal loading and coke discharging

基于焦炉测温及加热控制系统的探究与成果

邱　程，田龙振

（青岛特殊钢铁有限公司焦化厂，山东青岛　266500）

摘　要：焦炉作为最复杂的冶金工业窑炉，其燃烧系统结构非常复杂，同时炉温受到装煤推焦、煤气热值等因素的影响，焦炉温度变化规律较难归纳，造成人工测调焦炉炉温的困难。

其次为提高焦炉四大车的利用率，焦炉生产串序由 2-1 小串序优化为 2-1 大串序，焦炉炉温的波动峰谷差变大但规律性较强。为适应当前焦炉生产，需要焦炉自动测温及炉温自动控制系统实现连续测温，且能够根据温度波动进行调节的系统。

K安指标是焦炉持续稳定生产的关键指标,是决定焦炭产量和质量的主要因素之一,所以需要焦炉技术人员不断追求K安指标的最大化。

青钢7m焦炉煤气量与变频器电流数据拟合分析

邱 程,郭 飞,何 波

(青岛特殊钢铁有限公司,山东青岛 266500)

摘 要:文章针对7m大型焦炉正常运行状态下,鼓风机突然跳停的问题,通过对2018年至2021年鼓风机运行状态和焦炭产量、煤气发生量之间的关系进行分析,在确保鼓风机正常运转的情况下,达到最多的焦炭产量。

关键词:焦炉;鼓风机;跳停;煤气发生量;焦炭产量

Fitting Analysis of Gas Volume and Converter Current Data of Qinggang's 7m Coke Oven

QIU Cheng, GUO Fei, HE Bo

(Qingdao Special Steel Co., Ltd., Qingdao 266500, China)

Abstract: This article focuses on the problem of sudden tripping of the blower during the normal operation of a 7-meter large coke oven. By analyzing the relationship between the blower operation status, coke production, and gas generation from 2018 to 2021, the maximum coke production can be achieved while ensuring the normal operation of the blower.

Key words: coke oven; blower; trip stop; gas generation; coke output

青特钢焦化低水分熄焦生产实践

邱 程,郭 飞,王永亮

(青特钢焦化厂,山东青岛 266500)

摘 要:青特钢1、2号焦炉水熄焦系统使用的是从辽阳三盈坤鼎机械制造有限公司引进的低水分熄焦工艺。该工艺具有熄焦过程用水少、焦炭含水量低及水分波动小等特点;从其设计特点和生产实践方面对此项新技术进行了分析和探讨。

关键词:焦炉;低水熄焦工艺;含水量;分析

Low Moisture Coke Quenching in Qinggang Coking Plant Transformation and Production Practice

QIU Cheng, GUO Fei, WANG Yongliang

(Qinggang Coking Plant, Qingdao 266500, China)

Abstract: The coke quenching system of No. 1 and No. 2 coke ovens of Qinggang uses the low moisture coke quenching process introduced from Liaoyang Sanying KunDing Machinery Manufacturing Co., Ltd. The process has the characteristics of less water consumption, low coke water content and small water fluctuation in the process of coke quenching; This new technology is analyzed and discussed from its design characteristics and production practice.

Key words: coke oven; low water quenching process; water content; analysis

焦炉智能巡检机器人应用实践

邱 程，郭 飞

（青岛特殊钢铁有限公司焦化厂，山东青岛 266500）

摘 要：炼焦行业存在安全风险极高、工艺流程复杂、生产环境高危、原料产品多样等特点；焦炉在生产过程中，由于物料、设备等多方面因素影响，焦炉内部易积聚有害气体，如 CO、H_2S 等，若不能及时排放则会引发爆炸事故。为了保障焦炉的安全生产，便需要定期进行巡检巡查。随着 5G、大数据、物联网、人工智能等高新技术的不断发展和应用，炼焦行业的巡检工作已经由传统人工巡查模式逐步向"无人化"巡检模式发展。

关键词：焦炉；无人化；巡检

配合煤堆密度在提高焦炉装煤量方面的应用

邱 程

（青岛特殊钢铁有限公司焦化厂，山东青岛 266500）

摘 要：炼焦作业区自 3 月 15 日经生产核算焦炭产量较上月减少 100t/日，备煤作业区筒仓配煤皮带秤集合量显示的数据也较上月减少，在当日生产炉数不变的情况下，焦炉煤气的产出同样减少。

两项煤气净化工艺改进方案

杨 爽

（中钢设备有限公司，北京 100080）

摘　要：在焦化厂煤气净化车间，根据用户不同，煤气净化程度的不同，而设计不同的煤气净化工艺流程。经过近四十年的发展，现有的煤气净化流程国内均有设计实例。随着环保等法规的要求越来越严格，国内的煤气净化流程已经有趋同的趋势，更加注重环保和节能。近期，针对现有煤气净化工艺的不足，我们提出了如下几项改进方案，希望与焦化专家共同研讨：新型两塔脱硫工艺、蒸氨余热回收工艺。

基于煤岩学方法分析焦炭"夹生"问题

王　越，庞克亮，吴昊天，谷致远

（鞍钢集团北京研究院有限公司，北京　102200）

摘　要：综合利用镜质体反射率、显微组分定量、焦炭光学组织以及气孔结构分析等煤岩学分析手段，研究某批焦炭中出现"夹生炭"的问题。结果发现："夹生炭"的气孔特征及结构参数与正常焦炭相似，但是灰分和挥发分产率略高，焦炭光学组织为光学各向同性，主要以独立分布或者呈细粒状被各向异性组织包裹。"夹生炭"在配合煤中 6～13mm 粒级富集低变质炼焦煤中壳质组所致，而并非焦炭未成熟。低变质炼焦煤中壳质组的碳化行为与镜质组差异较大，反应活性高于镜质组，软化熔融区间差异较大；壳质组富集形成"夹生炭"，对焦炭强度造成不利影响。

关键词：炼焦配煤；壳质组；煤岩分析；焦炭光学组织；焦炭气孔结构

Analysis of "Incompletely Mature" Coke Based on Petrographic Method

WANG Yue, PANG Keliang, WU Haotian, GU Zhiyuan

(Ansteel Beijing Research Institute Co., Ltd., Beijing 102200, China)

Abstract: Petrographic analysis methods, such as vitrinite reflectance, maceral determination, optical texture and pore structure analysis of coke, were used to study "incompletely mature" coke. The results have shown that the parameters of pore structure of "incompletely mature" coke were similar to those of normal coke, but the contents of ash and volatile matter were slightly higher. The optical texture of the "incompletely mature" coke was isotropic, mainly distributed as independent particles or fine dispersed in anisotropic structures. The "incompletely mature" coke were resulted by the high content of liptinite in low rank coking coal enriched in 6-13mm fraction of the blended coal, rather than the immature coke. The carbonization behavior of liptinite in low rank coking coal was quite different from that of vitrinite. The reaction activity of liptinite was higher than that of vitrinite, and the softening and melting behavior was quite different. The "incompletely mature" coke caused by the enrichment of liptinite in low rank coking coal, was harmful to the strength of coke.

Key words: coal blending for coking; liptinite; petrographic analysis; optical texture of coke; coke pore structure

改质沥青的生产工艺及发展潜能概述

徐志强，张立伟，闫双双，王　斌，王晓楠

（鞍钢化学科技有限公司，辽宁鞍山　114000）

摘　要： 煤沥青是煤焦油经蒸馏加工后的重质残余物，是一种极其复杂的混合物，主要组分是稠环芳烃，其收率占煤焦油的55%以上。改质沥青作为煤沥青高温聚合后的产物，在工业上主要作为粘结剂和浸渍剂，用于生产高功率石墨电极和电解铝用预焙阳极，改质沥青的质量在很大程度上影响着我国高品质炭素材料的研发生产，改质沥青的性能得到质的提升，才能助推我国炭素行业实现高质量发展。

关键词： 改质沥青；生产工艺；应用前景

Overview of the Production Process and Development Prospect of Modified Pitch

XU Zhiqiang, ZHANG Liwei, YAN Shuangshuang, WANG Bin, WANG Xiaonan

(Angang Chemical Technology Co., Ltd., Anshan 114000, China)

Abstract: Coal tar pitch is the heavy residue of coal tar after distillation and processing. It is an extremely complex mixture. The main components are condensed aromatic hydrocarbons, and its yield accounts for more than 55% of coal tar. As the product of high temperature polymerization of coal tar pitch, modified pitch is mainly used as a binder and impregnant in industry for the production of high-power graphite electrodes and pre-baked anodes for electrolytic aluminum. The quality of the modified pitch greatly affects the research and development of high-quality carbon materials in China. Only by qualitatively improving the performance of modified pitch can the carbon industry achieve high-quality development.

Key words: modified pitch; production technology; application prospect

干燥煤结焦过程传热传质行为的数值模拟研究

仇　灏，刘和平

（中国钢研科技集团，北京　100081）

摘　要： 我国是世界焦炭生产大国，但焦化行业的特点是高能耗、高污染、高资源消耗，开发绿色炼焦工艺是炼焦行业技术进步的紧迫任务，符合"碳达峰和碳中和"的发展趋势。干燥煤炼焦工艺是在传统湿煤炼焦工艺基础上，将配合煤中水分含量降低至4%以下的新一代炼焦预处理工艺，该工艺从理论上来说，可以有效降低炼焦能耗、缩短炼焦周期并减少剩余氨水[1-3]。但关于干燥煤炼焦工艺降低炼焦能耗，缩短炼焦周期的传热传质机理还不清晰，且国内的干燥煤炼焦工艺仍处于实验研究的阶段，研究周期长、成本高。通过使用数值模拟仿真方法对干燥煤结焦过程中的传热传质行为进行研究，有助于增进对炭化室内煤床成焦过程的认识，并为优化炼焦工艺、降低炼焦耗能提供参考。

本研究建立了三维炭化室内配合煤结焦过程数学模型，通过数值模拟对干燥煤炼焦过程中煤床升温、水分蒸发冷凝及挥发分析出过程等传热传质行为进行了仿真预测并与传统湿煤炼焦工艺进行了对比，从而为干燥煤炼焦工艺的开发及应用提供理论参考。

在SG60型焦炉炭化室计算域中分别对干基堆密度为832kg/m³、水分含量为2.2%的干燥煤和干基堆密度为732kg/m³、水分含量为10.5%的湿煤18.5h结焦过程进行仿真。研究结果表明：干燥煤炼焦工艺对煤床中心升温过程的影响要高于对煤床边缘，且在相同计算条件下，相同炼焦周期下，干燥煤炼焦的产量达到湿煤炼焦的1.13倍。虽然干燥煤炼焦过程中水分蒸发速率要慢于湿煤炼焦，但干燥煤炼焦工艺能够缩短炼焦过程中煤床整体干燥所需的时间，并提高挥发分析出速率，理论上可以提高配合煤的结焦速度。炼焦过程中，煤床中里行气和外行气的析出范围是实时变化的，随着炼焦的进行，里行气析出量逐渐减少，炼焦后期产生的荒煤气均为外行气。在相同的加热条件下计算得到的湿煤炼焦相当耗热量为 2326.61 kJ/kg 干煤，干燥煤炼焦相当耗热量为 2234.96 kJ/kg 干煤，湿煤炼焦工艺的平均炼焦耗热量是干燥煤炼焦工艺 1.041 倍，且干燥煤炼焦工艺能够将"火落时间"缩短 0.3~0.5h，可以说应用干燥煤炼焦工艺能够起到较好的绿色化炼焦效果。

参 考 文 献

[1] 鲍俊芳, 薛改凤, 詹立志, 等. 燃料与化工, 2015, 46(3): 1-2.
[2] 张雪红, 石隆基, 薛改凤, 等. 煤化工, 2020, 48(2): 57-59.
[3] 张雪红, 薛改凤, 詹立志, 等. 燃料与化工, 2015, 46(4): 8-10.

武钢 7.63m 焦炉蓄热室积灰分析与判断

陈细涛[1], 严铁军[2], 张 军[2], 常红兵[1], 陈 鹏[1]

（1. 宝钢股份中央研究院武钢有限技术中心，湖北武汉　430080；
2. 武汉平煤武钢联合焦化有限公司，湖北武汉　430082）

摘　要：因检测到武钢焦化 7.63m 焦炉炉头温度偏低，后检查发现蓄热室内积灰严重，对积灰样品进行取样并开展工业分析、化学成分和光学电子显微分析，发现其主要为游离碳单质。根据焦炉炉体结构和工艺流程，推测产生积灰的物质来源为含碳煤气的不充分燃烧室，提出可能的来源方式，再此基础上进行针对性处理，积灰程度明显减轻，炉头温度得到显著提高。

关键词：7.63m 焦炉；蓄热室；积灰；成分分析

Analysis of Ash in Regenerator Chamber of WISCO 7.63m Coke Oven

CHEN Xitao[1], YAN Tiejun[2], ZHANG Jun[2], CHANG Hongbing[1], CHEN Peng[1]

(1. R&D Center of Wuhan Iron & Steel Co., Ltd., Baosteel Central Research Institute, Wuhan 430080, China; 2. Wuhan Pingmei WISCO United Coking Chemical Co., Ltd., Wuhan 430082, China)

Abstract: Due to the detection of the low end flue temperature of WISCO 7.63m Coke Oven, it was found that there was a lot of ash accumulating in the regenerator chamber. Through Sampling and performing industrial analysis, chemical composition and SEM analysis of the ash samples, it was found that the ash were mainly composed of free carbon elements. Based on the structure and process of the coke oven, it was speculated that the source of ash was the insufficient combustion of gas containing carbon. Possible sources of ash were proposed, and targeted treatment was carried out on this foundation. The ash in the regenerator chamber was reduced significantly, and the end flue temperature was increased notably.

Key words: 7.63m coke oven; regenerator chamber; ash; component analysis

焦化厂原料煤全自动检化验系统设计

贺世泽[1]，吴 波[2]

(1. 宝钢工程技术集团有限公司，上海 200000；
2. 山西太钢不锈钢股份有限公司焦化厂，山西太原 030003)

摘 要：针对焦化厂原料煤特性和质量管理特点，设计建设全自动的采样、制样、化验系统装置，解决了质量难把控、劳动强度大、廉洁风险大等难题。在太钢建设使用三年来，运行稳定，可在行业推广。

关键词：焦化厂原料煤；智能；采、制、化

Design of Automatic Test System for Raw Coal in Coking Plant

HE Shize[1], WU Bo[2]

(1. Baosteel Engineering & Technology Group Co., Ltd., Shanghai 200000, China;
2. Coking Plant of Shanxi Taiyuan Stainless Steel Co., Ltd., Taiyuan 030003, China)

Abstract: According to the characteristics of property and quality management of raw coal in coking plant, the design and construction of automatic sampling, sample preparation and testing system devices have solved the problems of difficult quality control, high labor intensity and high risk of integrity. In the three years of use in TISCO, the operation is stable and can be promoted in the industry.

Key words: raw coal in coking plant; intelligent; sampling, sample preparation and testing

焦炉操作工艺对焦炭质量影响的分析

陈 鹏[1]，严铁军[2]，常红兵[1]，陈细涛[1]，张 军[2]

(1. 宝钢股份中央研究院（武钢有限技术中心）炼铁所，湖北武汉 430081；
2. 武汉钢铁有限公司，湖北武汉 430082)

摘 要：通过 Kendall τ 的非参数性分析法和长达一年的数据对焦炉操作工艺对影响焦炭质量的因素进行数据分析，从数据驱动得出验证了影响焦炭质量的主要因素是入炉煤煤质、焦炭质量影响最为显著的焦炉操作是熄焦方式的改变，在上述条件稳定的情况下，提高焦炉安定性 $K_{安}$，稳定炉温，提高计划操作系数 K_1，适当增加结焦时间均有能有效提升焦炭质量。为评价智慧化发展过程中合理预测评价焦炭质量提供一条新的思路。

关键词：焦炭质量；焦炉操作；数据分析

Analysis of Influence of Coke Oven Operation on Coke Quality

CHEN Peng[1], YAN Tiejun[2], CHANG Hongbing[1], CHEN Xitao[1], ZHANG Jun[2]

(1. Institue of Ironmaking Technology, R&D Center of Wuhan Iron & Steel Co., Ltd., Baosteel Central Research Institute, Wuhan 430081, China; 2. Wuhan Iron & Steel Co., Ltd., Wuhan 430082, China)

Abstract: This paper through the non-parametric analysis method of Kendall τ which used the one-year data, the factors affecting the coke oven operation process are analyzed. Driven by the data, it is verified that the main factors affecting the quality of coke are the quality of coal in the furnace and the most significant influence on the quality of coke is the change of the coke quenching mode. Under the above stable conditions, the coke quality can be effectively improved by improving the stability of coke oven K, stabilizing the furnace temperature, increasing the planned operation coefficient K_1 and appropriately increasing the coking time. It provides a new way to rationally predict and evaluate coke quality in the process of intelligent development.

Key words: coke quality; coke oven operation; data analysis

二氧化钛光催化剂改性研究

刘子娟，吴伟伟，李 达，贾楠楠，刘庆佩，王晓楠

（鞍钢化学科技有限公司，辽宁鞍山 114000）

摘 要：二氧化钛光催化技术已经发展成为新型、高效环境污染治理技术，在废水处理方面得到广泛关注。TiO_2作为光催化技术的核心，具有效率高、稳定性好、价格低廉、无污染等优点，能降解难降解的有机污染物。但二氧化钛较大的禁带宽度及较高的光生电子-空穴复合几率限制其光催化效率。要解决这些难题需要对TiO_2纳米材料进行改性以提高光催化活性，可以利用离子掺杂、半导体复合、贵金属沉积、负载改性等方法抑制光生电子-空穴对的复合，提高可见光利用率。TiO_2进行改性以及制备可重复使用的纳米TiO_2是当前光催化研究领域的两个热点，讨论改性TiO_2的研究成果及其光催化技术在废水污染治理方面的研究方向。

关键词：光催化；二氧化钛；改性；有机废水

Study on Modification of Titanium Dioxide Photocatalyst

LIU Zijuan, WU Weiwei, LI Da, JIA Nannan, LIU Qingpei, WANG Xiaonan

(Ansteel Chemical Technology Co., Ltd., Anshan 114000, China)

Abstract: Titanium dioxide photocatalysis technology has developed into a new and efficient environmental pollution control technology, which has received widespread attention in wastewater treatment. As the core of photocatalytic technology, TiO_2 has the advantages of high efficiency, good stability, low price, and no pollution, and can degrade difficult to degrade organic pollutants. However, the large bandgap width and high probability of photo generated electron hole recombination of titanium dioxide limit its photocatalytic efficiency. To address these challenges, TiO_2 nanomaterials need to be modified to enhance photocatalytic activity. Methods such as ion doping, semiconductor recombination, precious metal deposition, and load modification can be used to suppress the recombination of photo generated electron hole pairs and improve visible light utilization. The modification of TiO_2 and the preparation of reusable nano TiO_2 are currently two hot topics in the field of photocatalysis research. The research achievements of modified TiO_2 and the research direction of photocatalytic technology in wastewater pollution treatment are discussed.

Key words: photocatalysis; titanium dioxide; modification; organic wastewater

高灰中硫瘦煤资源开发与利用

贺 佳，史永林，李昊堃，王钟议

（山西太钢不锈钢股份有限公司技术中心，山西太原 030003）

摘 要： 为降低炼焦配煤成本，扩大炼焦用煤资源，太钢开展高灰中硫瘦煤配煤试验研究及生产实践，利用300kg试验焦炉进行配煤炼焦实验，在7.63m焦炉进行高灰中硫瘦煤配比为22%和24%的工业试生产。试验结果表明：使用3号SM和4号SM高灰中硫瘦煤替代SM后，焦炭冷热强度均有所提高，炼焦配煤成本降低了14余元/吨以上，对焦化企业稳质降本，提高竞争力具有一定的借鉴价值。

关键词： 高灰中硫瘦煤；300kg试验焦炉；7.63 m焦炉；生产实践

Development and Utilization of High Ash and Medium Sulphur Lean Coal

HE Jia, SHI Yonglin, LI Haokun, WANG Zhongyi

(The Technology Centre of Shanxi Taigang Stainless Steel Co., Ltd., Taiyuan 030003, China)

Abstract: To expand the coking coal resources, reduce the cost of coal blending, TISCO carried out the research and production practice of blending high ash and medium sulphur lean coal. The coking experiment of blending coal was carried out by 300kg test coke oven, and in 7.63m coke oven, the high ash and medium sulphur lean coal ratio of 22% and 24% was industrialized respectively. The results indicated: mixing 3#SM and 4#SM high ash and medium sulphur lean coal,the cold and thermal strength of coke was slightly improved. The cost of coal blending was reduced by more than 14.0 yuan/ton, which had certain reference value for coking enterprises to stabilize quality, reduce cost and improve competitiveness.

Key words: high ash and medium sulphur lean coal; 300kg test coke oven; 7.63m coke oven; productive practice

焦化废水深度处理技术的创新与实践

李 斌，潘东山，王云峰

（河钢集团宣钢公司能源公司，河北宣化 075100）

摘 要： 河钢集团宣钢公司焦化废水经过重力除油、气浮、A/O、低温超导进行处理后，废水指标达到GB 16171—2012《炼焦化学工业污染物排放标准》中的间接排放要求，随着国家环境保护政策的不断加强和环境保护标准的不断提高，要求出水满足 GB 18918—2002《城镇污水处理厂污染物排放标准》中一级 A 排放标准的要求，因此必须对现有焦化废水进行深度处理，公司通过对现有焦化废水深度处理工艺进行考察对比，最终采用多介质过滤+臭氧催化氧化+AO-MBR 联用工艺对焦化废水进行深度处理，出水指标优于城镇污水一级 A 的排放标准，为同行业焦化废水深度处理提供了参考。

关键词：焦化废水；深度处理；AO-MBR；臭氧催化氧化；多介质过滤

Innovation and Practice of Advanced Treatment Technology for Coking Wastewater

LI Bin, PAN Dongshan, WANG Yunfeng

(Xuansteel Compang of HBIS Group, Xuanhua 075100, China)

Abstract: After the coking wastewater of Xuansteel compang of HBIS company is treated by gravity oil removal, air floatation, A/O and cryogenic superconductivity, the waste water index meets the indirect discharge requirement of GB 16171—2012"Coking chemical industry pollutant discharge standard". With the continuous strengthening of national environmental protection policy and the continuous improvement of environmental protection standards, the effluent is required to meet the requirements of the Class A discharge standard in the GB 18918—2002"Discharge standard for urban Sewage Treatment pollutants", through the investigation and comparison of the existing coking wastewater advanced treatment process, the company finally adopted multi-media filtration + ozone catalytic oxidation + AO-MBR combined process for the advanced treatment of coking wastewater, the effluent index is better than the discharge standard of Grade A of municipal wastewater, which provides a reference for the advanced treatment of coking wastewater in the same industry.

Key words: coking wastewater; advavcedtreatmen; AO-MBR; ozone catalytic oxidation; multi-media filtration

联用工艺在煤化工综合废水处理中的应用

李 斌，潘东山，王云峰

（河钢集团宣钢公司能源公司，河北宣化 075100）

摘 要：河钢集团宣钢公司随着焦炉煤气深度处理（脱硫、脱苯、脱氨）项目投产，产生了高氨氮的煤化工综合废水，为了保证产生的废水处理后满足 GB18918-2002《城镇污水处理厂污染物排放标准》中一级 A 排放标准的要求，通过采用 SDN+AO-MBR+臭氧氧化联用工艺进行处理，出水指标优于城镇污水一级 A 的排放标准，为煤化工综合废水处理提供了参考。

关键词：煤化工；废水；SDN；AO-MBR；臭氧氧化

Application of Combined Process in Comprehensive Wastewater Treatment of Coal Chemical Industry

LI Bin, PAN Dongshan, WANG Yunfeng

(Xuansteel Company of HBIS Group, Xuanhua 075100, China)

Abstract: The commissioning of the coke oven gas advanced treatment (desulphurization, benzene removal, ammonia removal) project, Xuansteel compang of HBIS has produced coal chemical comprehensive wastewater with high ammonia nitrogen content, in order to ensure that the wastewater generated after treatment meets the requirements of the Class A emission standard in GB 18918—2002 "Urban Sewage Treatment pollutant emission standard", the combined process of SDN + AO-MBR + ozone oxidation was adopted to treat the wastewater, the effluent index is better than the discharge standard of the first class A of urban sewage, which provides a reference for the comprehensive wastewater treatment of coal chemical industry.

Key words: coal chemical industry; waste water; SDN; AO-MBR; ozone catalysis

焦炉机车除尘新技术的应用与创新

郭有林，成雪松，郭永贵

（河钢集团宣钢公司，河北宣化 075100）

摘　要： 针对焦炉机车除尘系统进行了创新和新技术的应用。焦炉机侧除尘系统采用了 U 型管水封除尘技术，通过采用车载 U 型管与水封槽集尘管连接，实现移动吸尘输尘，配套地面脉冲袋式除尘器实现烟气净化；同时对在线除尘其他工艺设备进行创新，对拦焦除尘、除尘预喷涂设备、检修盖密封技术进行了创新，降低了除尘阻力，延续了设备使用寿命；在改善除尘效果和防治焦炉烟尘等方面取得了很大进步。

关键词： 机侧除尘；蝶阀；预喷涂；密封

Application and Innovation of New Dust Removal Technology for Coke Oven Locamotive

GUO Youlin, CHENG Xuesong, GUO Yonggui

(Xuansteel Compang of HBIS Group, Xuanhua 075100, China)

Abstract: In order to improve the dust removal effect of coke oven locomotive, the dust removal system on the side of coke oven machine adopts the U-tube water seal tank dust removal technology. By connecting the vehicle-mounted U-tube with the wanter seal tank dust collection tube, the mobile dust collection and dust transport are realized, and the ground pulse bag filter is matched to realize the flue gas purification. At the same time, other process equipment of on-line dust removal is innovated, and the sealing technology of coke blocking and dust removal, dust removal pre-spraying equipment and maintenance cover is innovated, which reduces the removal resistance and prolongs the service life of the equipment. Great progress has been made in improving dust removal effect and preventing coke oven dust.

Key words: side dust removal; butterfly valve; pre-spraying; sealing

氨水中焦油渣回收利用工艺优化

刘卫杰

（河钢集团宣钢公司，河北宣化 075100）

摘 要：针对目前焦化厂氨水中焦油渣含油、含水偏高的问题，由传统的静置分离工艺转变为设备分离的新工艺。由于旧工艺主要采用的是静置分离，分离效果受生产影响较大，改为新工艺后，在降低氨水中焦油渣含油、含水量的问题的同时，提高了焦油和氨水的产量。另外，相较旧工艺在环保方面、经济效益方面均有较大的提升，实现了焦油渣的环保无害化处理。

关键词：氨水；焦油渣；分离；回收

Optimization of the Recovery and Utilization Process of Tar Residue in Ammonia Water

LIU Weijie

(Xuansteel Compang of HBIS, Xuanhua 075100, China)

Abstract: In response to the current problem of high oil and water content in tar residue in coking plant ammonia water, the traditional static separation process has been transformed into a new process for equipment separation. Due to the old process mainly using static separation, the separation effect is greatly affected by production. After changing to the new process, while reducing the oil and water content of tar residue in ammonia water, the production of tar and ammonia water has been increased. In addition, compared to the old process, there has been a significant improvement in environmental protection and economic benefits, achieving environmentally friendly and harmless treatment of tar residue.

Key words: ammonia water; tar residue; separation; recovery

焦化厂循环氨水的余热利用

张 磊，付冠华

（神华巴彦淖尔能源有限责任公司，内蒙古巴彦淖尔 015000）

摘 要：溴化锂吸收式制冷机组在工厂中的应用十分广泛。本文介绍了一种以循环氨水为驱动热源的溴化锂吸收式制冷机组的原理，阐述了它在焦化厂制冷系统中的应用，比较了它与传统的以蒸汽为热源的制冷机组的节能环保情况。

关键词：溴化锂；循环氨水；焦化厂；节能环保

190t/h 干熄焦余热锅炉过热器故障分析

曾石川

（武汉钢铁有限公司设备管理部，湖北武汉 430080）

摘 要：干熄焦余热锅炉是干熄焦工艺的重要设备，其主要功能是将干熄焦系统循环气体（一般为氮气）余热转化为水的内能，从而达到降低干熄炉焦炭温度，并生产蒸汽的效果。干熄焦系统循环气体含有一定量的焦粉颗粒，运行中冲刷锅炉过热器管造成过热器管壁减薄至泄漏，分析泄露故障原因采取有效的应对措施能有效保障干熄焦生产稳定。

关键词： 干熄焦工艺；高温过热器；故障分析

The Failure Analysis of CDQ Boiler Superheater

ZENG Shichuan

(Wuhan Iron and Steel Co., Ltd., Wuhan 430080, China)

Abstract: The coke dry quenching waste heat boiler is an important equipment in the coke dry quenching process. Its main function is to convert the waste heat of the circulating gas (usually nitrogen) in the dry quenching system into internal energy of water, thereby reducing the coke temperature of the dry quenching furnace and producing steam. The circulating gas of the dry quenching system contains a certain amount of coke powder particles, which wash out the boiler superheater tube during operation, causing the wall of the superheater tube to become thinner and leak. Analyzing the cause of the leakage fault and taking effective measures can effectively ensure the stability of dry quenching production.

Key words: coke dry quenching process; high temperature superheater; failure analysis

西门子 OUC 通讯方式在焦化环保除尘中的应用研究

杨 光，向 勇

（宝钢股份武钢有限公司，湖北武汉 430080）

摘 要： 焦化过程中产生的废气含有大量的有害气体和颗粒物，容易造成环境污染和人体健康伤害。因此，实现高效可靠的除尘系统对于焦化行业至关重要。本研究旨在探究西门子 OUC 通讯方式，解决焦化环保除尘系统点多面广、品牌不一的不利因素，实现各环保设备控制系统以及环保设备与生产设备的互联互通。与传统通讯方式进行相比，OUC 通讯方式给焦化生产带来了效率提升、实时监测和远程控制优势以及系统稳定性的提高。首先，我们介绍了焦化环保除尘的现状和挑战，强调了控制系统的智能化和通讯技术的关键作用；然后详细分析 OUC 通讯方式的原理和特点，最后通过 OUC 通讯方式实现除尘系统的智能化和远程监控，为焦化企业降低了环保成本，提高了生产效率，推动了焦化行业的可持续发展。

关键词： 焦化；除尘；OUC 通讯；智能化；西门子

Research on the Application of Siemens OUC Communication Method in Dust Removal of Coking Environmental Protection

YANG Guang, XIANG Yong

(Baosteel Group Corporation, Wuhan Iron and Steel Corporation Limited, Wuhan 430080, China)

Abstract: The gas generated during the coking process contains a large amount of harmful gases and particulate matter, which can easily cause environmental pollution and harm to human health. Therefore, achieving an efficient and reliable dust removal system is crucial for the coking industry. This study aims to explore the Siemens OUC communication method, in order to address the disadvantages of multiple points and diverse brands in the coking environmental protection dust removal systems. Compared to traditional communication methods, the OUC communication method has brought efficiency improvements, real-time monitoring and remote control advantages, as well as the enhancement of system stability. Firstly, we introduce the current situation and challenges of coking environmental protection dust removal, emphasizing the key role of intelligent control systems and communication technology. Then, we provide a detailed analysis of the principles and characteristics of the OUC communication method. Finally, through the implementation of the OUC communication method, dust removal systems are made intelligent and remotely monitored, which reduces environmental protection costs for coking enterprises, enhances production efficiency, and promotes the sustainable development of the coking industry.

Key words: coking; dust; OUC communication; intelligent; Siemens

基氏流动度曲线特征在炼焦煤混煤判定中的应用

宋子逵[1]，项 茹[1]，刘 睿[2]，丁 海[2]

（1. 宝钢股份中央研究院（青山）炼铁技术研究所，湖北武汉 430080；
2. 武钢有限公司，湖北武汉 430080）

摘 要： 基氏流动度是评价炼焦煤质量的重要指标之一，目前使用较多的包括软化温度、固化温度、固软温度区间、最大流动度温度以及最大流动度，但曲线特征在煤质评价中应用较少。本文通过分析不同种类炼焦煤基氏流动度曲线特征，引入相邻流动度比值的分布峰度，对曲线振动幅度、频度和曲线弧度进行数字化表征，认为其在炼焦煤混煤判定中具有一定作用，有助于综合评价炼焦煤质量。

关键词： 炼焦煤；基氏流动度；峰度；镜质组反射率

Application of the Characteristics of Gieseler Fluidity Curve in the Determination of Coking Coal Blending

SONG Zikui[1], XIANG Ru[1], LIU Rui[2], DING Hai[2]

(1.Institue of Ironmaking Technology, R&D Center of Wuhan Iron & Steel Co., Ltd., Baosteel Central Research Institute, Wuhan 430080, China;
2. Wuhan Iron & Steel Co., Ltd., Wuhan 430080, China)

Abstract: The Gieseler fluidity is one of the important indexes for evaluating the quality of coking coal.Currently, the commonly used indexes include softening temperature, solid-soft temperature range, max fluidity temperature, and max fluidity. However,the curve characteristics are less applied in coal quality evaluation. This paper analyzes the characteristics of Gieseler fluidity curve of different kinds of coking coal, introduces the distribution kurtosis of adjacent fluidity ratios, and digitally characterizes the vibration amplitude, frequency and radian of curve. It is believed that it plays a certain role in the determination of coking coal blending and helps to comprehensively evaluate the quality of coking coal.

Key words: coking coal; gieseler fluidity; kurtosis; vitrinite reflectance

3 炼铁与原料

大会特邀报告
第十三届冶金青年科技奖获奖人特邀报告
分会场特邀报告
矿业工程
焦化及节能环保
★ 炼铁与原料
炼钢与连铸
电冶金与废钢铁
轧制与热处理
表面与涂镀
金属材料深加工
粉末冶金
先进钢铁材料及应用
节能与低碳技术
冶金环保与资源利用
冶金设备与工程技术
冶金自动化与智能化
冶金物流
冶金流程工程学
其他

3.1 烧结与球团

烧结烟气中 NO$_x$ 的产生机理及排放规律探研

桂林峰，苏亚刚，李　彬，黄永安

（武钢集团昆明钢铁股份有限公司炼铁厂，云南安宁　650309）

摘　要： 烧结烟气中 NO$_x$ 的主要来源为固体燃料燃烧过程中产生的挥发性氮及焦炭氮，昆钢新区 300m^2 烧结机生产实践及工业试验表明：使用挥发分与氮含量较低的固体燃料、控制烧结终点适当前移、提高烧结料层厚度均能有效降低烧结烟气中 NO$_x$ 的排放浓度。

关键词： 烧结；烟气；NO$_x$

Research on the Generation Mechanism and Emission Law of NO$_x$ in Sintering Flue Gas

GUI Linfeng, SU Yagang, LI Bin, HUANG Yong'an

(Wuhan Iron and Steel Group Kunming Iron and Steel Co., Ltd., Ironmaking Plant, Anning 650309, China)

Abstract: The main sources of NO$_x$ in sintering flue gas are volatile nitrogen and coke nitrogen generated during the solid fuel combustion process. The production practice and industrial tests of a 300 square meter sintering machine in Kunming Iron and Steel New Area have shown that using solid fuels with low volatile and nitrogen content, controlling the appropriate forward movement of sintering endpoint, and increasing the thickness of sintering material layer can effectively reduce the emission concentration of NO$_x$ in sintering flue gas.

Key words: sintering; smoke; NO$_x$

带式焙烧机球团智能管控系统的研究与实现

刘富春，韩基祥，任　伟，李　丹，周检平

（中钢设备有限公司矿物加工部，北京　100080）

摘　要： 带式焙烧机球团智能管控系统以工艺技术为主线，涵盖自动化、计量、能源、产销等多源数据，贯穿生产全过程管理业务，基于大数据、工业算法、数字孪生等技术，实现所有的"设备"都通过平台互联，所有的"数据"都通过平台处理，所有的"决策"都通过平台生成，所有的"流程"都通过平台实现，所有的"资源"都通过平台分享。通过带式焙烧机生产过程大数据分析和智能管控,强化带式焙烧机球团生产过程稳定，生产流程高效，提高球团产品质量、产量、设备运行率，有效降低能源消耗和人员成本，降低对操作人员全天候连续监控和调整劳动强度，为带式焙烧机球团生产少人化、智能化提供技术方法和系统支撑。本系统共规划设计基于云-边-端的工业大数据采集与异构数据处理系统、智能焙烧模型系统、球团智能管理系统、球团智能决策系统四部分。包含工艺、设备、管理、数学、计算机等多专业协同配合的系统研发。

关键词：球团；带式焙烧机；工业大数据；焙烧模型；智能管控

Research and Implementation of Intelligent Control System for Pellet of Belt Baking Machine

LIU Fuchun, HAN Jixiang, REN Wei, LI Dan, ZHOU Jianping

(Mineral Processing Engineering Department of SINOSTEEL MECC Co., Ltd., Beijing 100080, China)

Abstract: The intelligent control system for belt roaster pellets is based on process technology, covering multiple sources of data such as automation, measurement, energy, production and sales, and runs through the entire production process management business. It is based on technologies such as big data, industrial algorithms, and digital twins. Through big data analysis and intelligent control of the production process of belt roaster, we aim to strengthen the stability and efficiency of the production process of belt roaster pellets, improve the quality, output, and equipment operation rate of pellets, effectively reduce energy consumption and personnel costs, reduce the continuous monitoring and adjustment of labor intensity for operators 24/7, and provide technical methods and system support for the production of belt roaster pellets with fewer people and intelligence.

Key words: balls; belt baking machine; industrial big data; roasting model; intelligent control

提高混匀矿稳定率的实践

黎兴文，陈 伟，李加龙，张锡宝，赵 彧

（武钢集团昆明钢铁股份有限公司炼铁厂，云南安宁 650300）

摘 要：混匀矿作为烧结生产的主要原料，对炼铁和烧结的生产稳定至关重要，本文从料场工艺、混匀矿的重要性、影响混匀矿稳定率的因素、提高混匀矿稳定率的措施等方面对提高混匀矿质量的工艺措施进行了阐述，本文内容为提高混匀矿质量提供了参考依据。

关键词：混匀矿；稳定率；配料；造堆；工艺；质量

Practice on Improving the Stability Rate of Mixed Ore

LI Xingwen, CHEN Wei, LI Jialong, ZHANG Xibao, ZHAO Yu

(Wuhan Iron and Steel Group Kunming Iron and Steel Co., Ltd., Ironmaking Plant, Anning 650300, China)

Abstract: Mixed ore, as the main raw material for sintering production, is crucial for the stability of ironmaking and sintering production. This article elaborates on the process measures to improve the quality of mixed ore from the aspects of material yard technology, the importance of mixed ore, factors affecting the stability rate of mixed ore, and measures to improve the stability rate of mixed ore. The content of this article provides a reference basis for improving the quality of mixed ore.

Key words: mixed ore; stability rate; ingredients; stacking; workmanship; quality

昆钢 B 型料场扬尘综合治理应用探析

陶传昌，王志富，侯林坤，苏　宝，谢志刚

（武钢集团昆明钢铁股份有限公司炼铁厂，云南安宁　650300）

摘　要： 物料堆场的扬尘治理一直是大型工业企业的化环保管控的难点问题，科学高效的管控措施有利于节约生产资源、降低生产成本、改善企业环境。本文通过对料场扬尘治理设施工作机理的分析研究，结合昆钢炼铁厂 B 型料场设备改造及防风抑尘网应用生产经验，旨在通过理论研究及应用实践，对大型物料堆场的扬尘管控治理提供指导意义。

关键词： 物料堆场；扬尘治理；防风抑尘网；环境改善

Exploring the Application of Comprehensive Dust Control in Kunming Steel's B-type Material Yard

TAO Chuanchang, WANG Zhifu, HOU Linkun, SU Bao, XIE Zhigang

(Wuhan Iron and Steel Group Kunming Iron and Steel Co., Ltd., Ironmaking Plant, Anning 650300, China)

Abstract: The dust control of material storage yards has always been a difficult issue in the chemical and environmental control of large industrial enterprises. Scientific and efficient control measures are conducive to saving production resources, reducing production costs, and improving the enterprise environment. This article analyzes and studies the working mechanism of dust control facilities in the material yard, combined with the experience of equipment transformation and wind and dust suppression network application in the B-type material yard of Kunming Iron and Steel Iron and Steel Plant. The aim is to provide guidance for the dust control and management of large material yards through theoretical research and application practice.

Key words: material storage yard; dust control; wind and dust suppression net; environmental improvement

提高烧结料层上部成品率试验研究

武　轶，张晓萍，李帮平，袁　鹏，熊德怀，朱贺民

（马鞍山钢铁股份有限公司技术中心，安徽马鞍山　243000）

摘　要： 本文针对带式烧结机抽风烧结的自动蓄热作用导致料层垂直方向上热量分布严重不均问题，通过烧结杯分层试验探究了各层烧结矿性能差异机理，研究了上部料层强化烧结技术。结果表明：烧结上部料层压实程度低于中下层，且处于"骤热骤冷、高温区保持时间短"的状态，以至于该层返粉率明显高于中下层；通过上部料层单独配

料、布料的方法，从烧结矿碱度、配碳、配矿、混匀制粒、抽风负压及协同耦合等多方面优化，可显著提高上层烧结矿以及整体烧结矿的成品率。采用多项协同耦合技术强化后，上层烧结矿成品率提高 5.31%，增长了 9.19 个百分点，整体烧结矿成品率提高了 3.54%，增长了 5.25 个百分点。

关键词：烧结；返粉；成品率；分层

邯钢 360m² 烧结机降低煤气消耗生产实践

王岳飞，王永林

（河钢集团邯钢公司邯宝炼铁厂，河北邯郸 056100）

摘 要：本文介绍了邯钢邯宝炼铁厂 360m² 烧结机降低点火煤气消耗的生产实践活动，针对烧结机一直煤气单耗高与行业先进水平；通过点火炉微负压点火改造，煤气管道安装自动稳压装置，以及通过改进程序按空煤比实现自动调整等多项技术的应用，实现了焦炉煤气单耗的降低，焦炉煤气单耗从 6.52m³/t 降低到 3.28m³/t 水平，取得了显著的降耗减排效果，为钢铁企业铁前降耗提供借鉴作用。

关键词：烧结机；气料分离；煤气消耗；自动控制

Production Practice of Reducing Gas Consumption in 360m² Sintering Machine of HANDAN I&S Co.

WANG Yuefei, WANG Yonglin

(The Hanbao Steel Factory of Handan Iron and Steel, Handan 056100, China)

Abstract: This paper introduces the practice of reducing ignition gas consumption of 360m² sintering machine in Hanbao iron Smelting plant, aiming at the high single gas consumption of sintering machine and the advanced level of industry. The single consumption of coke oven gas has been reduced from 6.52m³/t to 3.28m³/t through the application of a number of technologies, such as the micro-negative pressure ignition transformation of ignition furnace, the installation of automatic pressure regulator in gas pipeline, and the automatic adjustment according to the ratio of empty coal through the improvement of the program. The single consumption of coke oven gas has been reduced from 6.52m³/t to 3.28m³/t, and remarkable consumption and emission reduction effect has been achieved. It provides reference for iron and steel enterprises to reduce consumption before iron.

Key words: sintering machine; separation of gas and material; coal gas depletion; automatic control

鞍钢鲅鱼圈球团回转窑结圈研究及控制实践

任 伟[1,2]，国泉峰[3]，王 亮[1,2]，张 伟[1,2]，韩子文[1,2]

（1. 海洋装备用金属材料及其应用国家重点实验室，辽宁鞍山 114009；
2. 鞍钢集团钢铁研究院，辽宁鞍山 114009；
3. 鞍钢股份鲅鱼圈分公司炼铁部，辽宁营口 114021）

摘 要：为解决鞍钢鲅鱼圈 200 万吨球团回转窑结圈问题，从生球性能、预热球强度、热工制度和结圈料分析 4 方面进行了分析和研究，认为导致以煤气为主要热源的鲅鱼圈回转窑结圈主要原因是预热球强度低，并因此从原料准备、热工制度等方面进行了优化和改进，正常工作时长从原来的 35 天延长到 40 天左右，停机检修次数也由 2017 年的 10 次降低到 8~9 次。

关键词：回转窑；结圈；预热；工作时长；检修时长

The Ring Study on Angang Bayuquan Rotary Pelletizing Kiln and Its Control

REN Wei[1,2], GUO Quanfeng[3], WANG Liang[1,2], ZHANG Wei[1,2], HAN Ziwen[1,2]

(1. State Key Laboratory of Metal Material for Marine Equipment and Application, Anshan 114009, China; 2. Iron and Steel Research Institute of Angang Group, Anshan 114009, China; 3. Angang Bayuquan Branch Company, Iron Making Plant, Yingkou 114021, China)

Abstract: In order to solve the kiln ring problem for Angang Bayuquan 200mt rotary pelletizing kiln, the basic analysis and study has been carried out including green ball, preheating ball, chemical composition of ring and firing program, and it was concluded that the low preheating ball strength which made samll carcks into kiln from grate was the main reason for ring problem, finally the relative measurement was conducted the likes of raw material preparation and firing program, and the working time increased from 35 days to 40days, the downtime was lower from 10 to 8-9 times per year.

Key words: rotary pelletizing kiln; ring; preheating; working times; downtime

烧结智能优化配矿的发展浅析

刘 杰[1,2]，王耀祖[3]，马贤国[2]，陈付振[2]，

刘征建[3]，周明顺[2]，张建良[3]

（1. 海洋装备用金属材料及其应用国家重点实验室，辽宁鞍山 114009；

2. 鞍钢股份有限公司，辽宁鞍山 114009；

3. 北京科技大学冶金与生态工程学院，北京 100083）

摘 要：概述了烧结智能优化配矿及其研究现状，归纳了国内外炼铁工作者的研究工作，分析了现有的烧结智能配矿的方法和技术，对烧结智能优化配矿的进一步发展和完善进行了展望。指出以吨铁最小能耗和最低成本为优化目标，建立烧结-高炉一体化智能配矿将是烧结优化配矿的发展方向之一。且应充分利用人工智能算法、机器学习结合烧结和高炉工艺的机理和本质，开发一体化配矿系统。

关键词：优化配矿；人工智能算法；模型求解；一体化配矿；烧结

Analysis on the Development of Sintering Intelligent Optimal Ore Blending

LIU Jie[1,2], WANG Yaozu[3], MA Xianguo[2], CHEN Fuzhen[2],
LIU Zhengjian[3], ZHOU Mingshun[2], ZHANG Jianliang[3]

(1. State Key Laboratory of Metal Material for Marine Equipment and Application, Anshan 114009, China;
2. Angang Steel Company Limited, Anshan 114009, China;
3. School of Metallurgical and Ecological Engineering, Beijing 100083, China)

Abstract: The current status of sintering intelligent optimal ore allocation and its research is outlined, the research work of domestic and foreign ironmongers is summarized, the existing methods and technologies of sintering intelligent ore allocation are analyzed, and the further development and improvement of sintering intelligent optimal ore allocation is prospected. It is pointed out that with the minimum energy consumption and lowest cost per ton of iron as the optimization target, the establishment of integrated sinter-blast furnace intelligent ore allocation will be one of the development directions of optimal sintering ore allocation. And the integrated ore distribution system should be developed by making full use of artificial intelligence algorithms and machine learning combined with the mechanism and nature of sintering and blast furnace processes.

Key words: optimal ore allocation; artificial intelligence algorithms; machine learning; integrated ore allocation; sintering

碱度对高铬型钒钛磁铁矿熔剂型球团中温还原性能与反应动力学机理的影响

陈泊键[1]，姜 涛[1]，温 婧[1]，朱凤湘[2]，胡 鹏[2]

（1. 东北大学冶金学院，辽宁沈阳 110819；
2. 攀枝花钢铁研究院矿冶研究所，四川攀枝花 617099）

摘 要： 提高球团矿的入炉比例是高炉高效冶炼高铬型钒钛磁铁矿（HVTM）的有效手段之一。为保证炉渣碱度，对于高铬型钒钛磁铁矿熔剂型球团的研究不可或缺。本文研究了 HVTM 熔剂型球团的中温还原行为，分析了碱度对还原反应机理的影响，同时在 HVTM 球团中配入 30%的普通铁矿以提高其还原表现。结果表明，在 900℃下 30% $CO+70\% N_2$ 的还原性气氛中，碱度由 0.2 提高到 1.8 时，100%HVTM 熔剂型球团的还原度由 60.3%提高到 85.3%，配加 30%普通铁矿的 HVTM 球团（70%HVTM 球团）的还原度由 63.1%提高至 89.5%。随着碱度的提高，100%HVTM 球团的还原反应由随机成核随后生长机理和二维扩散机理共同控制向着随机成核随后生长转变。100%HVTM 熔剂型球团中，碱度不宜超过 1.0。70%HVTM 球团表现出更好的还原性，碱度超过 1.6 时可以满足高炉入炉标准。本研究为实现高炉利用 HVTM 熔剂型球团增添技术储备。

关键词： 高铬型钒钛磁铁矿；熔剂型球团；中温还原性；还原动力学；碱度

Reducibility Optimization and Reaction Mechanism of High-Chromium Vanadium–Titanium Magnetite Flux Pellets

CHEN Bojian[1], JIANG Tao[1], WEN Jing[1], ZHU Fengxiang[2], HU Peng[2]

(1. Northeastern University, School of Metallurgy, Shenyang 110819, China; 2. State Key Laboratory of Vanadium and Titanium Resources Comprehensive Utilization, Panzhihua 617099, China)

Abstract: High-chromium vanadium-titanium magnetite (HVTM) is an exceptional iron source that has not been mined on a large scale. Since using flux pellets facilitates the reduction in emissions and consumption a reasonable way to utilize HVTM is by smelting it into flux pellets in a blast furnace. This study elucidates the reducibility and reaction mechanism of HVTM flux pellets. The influence of basicity on the reduction reaction mechanism was also analyzed, and 30% of conventional iron concentrate was added to optimize the reduction performance of pure HVTM flux pellets. The pellets were treated at 900 ℃ for 180 min in a reducing atmosphere of 30%CO+70%N_2. When the basicity (CaO/SiO_2) increased from 0.2 to 1.8, the reduction degree of pure HVTM pellets (100% HVTM pellets) increased from 60.3% to 85.3%. Meanwhile, the reduction degree of HVTM pellets comprising 30% conventional iron concentrate (70% HVTM pellets) increased from 63.1% to 89.5%. Moreover, 70% HVTM pellets exhibited better reducibility and could meet the blast furnace standards when the basicity exceeded 1.6. The reaction mechanism of 100% HVTM pellets was first controlled by the random nucleation and subsequent growth model as well as the two-dimensional diffusion model. However, after increasing the basicity, the reaction mechanism was only characterized by the random nucleation and subsequent growth model.

Key words: high chromium vanadium-titanium magnetite; flux pellets; reducibility; reduction mechanism; basicity

塞拉利昂铁矿粉的烧结性能评价研究

段立祥，高洪庄，张大伟，刘 杰，张明洲

（鞍钢股份有限公司烧结厂，辽宁鞍山 114021）

摘 要： 为了考查塞拉利昂铁矿粉在鞍钢烧结应用的可行性，在实验室对塞拉利昂矿粉进行了 X 射线衍射和扫描电镜分析，并对其烧结性能、矿相结构等进行了分析和评价。研究结果表明：塞拉利昂铁矿粉硅低铝高，杂质含量少，矿物组成以赤铁矿主，针铁矿，三水铝矿和石英为主；在只配加塞拉利昂粉矿时，随配比提高，利用系数大幅增加，燃耗下降，垂直燃烧速度加快，在配比 20%时成品率和返矿率最优，但转鼓强度下降明显；在 PB 粉和塞拉利昂粉矿混用时，转鼓强度在塞拉利昂粉矿超过 5%后，急剧下降，其他指标以 5%PB+15%塞粉方案最优。随着塞拉利昂粉矿配比的增加，烧结矿微观构成中铁酸钙的数量和形状出现明显变化，裂纹和孔洞大幅增加，赤铁矿多呈粒状分布在烧结矿中，烧结矿的低温还原粉化指标逐渐变差，还原性能稍有下降。

关键词： 塞拉利昂矿；烧结性能；矿相；冶金性能

Research of Sierra Leone Ore's Sintering Performance

DUAN Lixiang, GAO Hongzhuang, ZHANG Dawei, LIU Jie, ZHANG Mingzhou

(Iron and Steel Research Institute of Angang Group, Ansteel Company Limited, Anshan 114021, China)

Abstract: In order to study the feasibility that the Sierra Leone ore was applied to sintering in Angang, the characteristic of Sierra Leone ore was analyzed by the X-ray diffraction and the scanning electron microscope. The sintering performance and the mineralogical structure also were estimated in laboratory. The results show that the Sierra Leone ore has low silicon oxide content and high aluminium oxide content, also low impurity content. The Sierra Leone's mineral composition is mainly composed of hematite, goethite, gibbsite and quartz. When improve the ratio of Sierra Leone ore in sintering that only use it, the productivity is increased largely, the fuel consumption is decreased, the vertical burning speed is accelerated. The yield and return fine ratio have the best indexes when the Sierra Leone ore is 20%, but the sinter tumbler index decreases obviously. The sinter tumbler index decreases sharply after the Sierra Leone ore beyond 5% with the combination of PB ore and Sierra Leone ore. And the others indexes become better when the scheme is 5%PB ore adding 15% Sierra Leone ore. The quantity and shape change obviously that the crack and pores increase largely. And the hematite distributes in sinter with graininess. Also, the low temperature reduction disintegration becomes worse and the reducibility at 900℃ decreases lightly.

Key words: the Sierra Leone ore; sintering performance; mineral structure; metallurgical property

澳洲某赤褐铁矿磨矿特性及其氧化球团制备行为研究

杨聪聪，曲士娟，潘　建，朱德庆，郭正启，李思唯

（中南大学资源加工与生物工程学院，低碳与氢冶金研究中心，湖南长沙　410083）

摘　要： 在当前"双碳"背景下，大力发展球团矿生产、提高高炉入炉比例是完善高炉长流程，支撑钢铁工业绿色低碳转型的有效手段之一。然而，如何扩大球团原料、降低球团生产成本是钢铁企业当前面临的重要课题之一，也直接影响着球团工艺的发展。赤褐铁矿粉通常用作烧结原料，相比细粒铁精矿具有一定价格优势，对于国内球团厂扩大原料来源、指导现场生产和降低球团生产成本均具有重要的现实意义。

本文以澳洲某赤褐铁矿(AO-M)为主要原料，其铁品位和 SiO_2 含量分别为 60.31%和 4.45%，烧损较高(6.08%)，且 Al_2O_3 含量达到了 2.21%，可能对湿式磨矿、生球制备和球团焙烧过程和产品质量产生不利的影响，因而较系统地研究了其湿式磨矿-沉降-过滤特性、原料预处理和球团制备特性，确定了不同 AO-M 配比条件下适宜的焙烧条件，并考察了对球团冶金性能和矿相特征的影响，为明确适宜的 AO-M 配比奠定了重要基础。

AO-M 的 Bond 磨矿功指数为 12.959kW·h/t，与巴西铁矿粉（BO-C、BO-S）、澳洲铁矿粉(AO-P)和非洲铁矿粉（FO-M、FO-S）的磨矿功指数相近，均属易磨矿石；如图 1(I)、(II)可知，单矿矿浆静态沉降及过滤特性试验研究表明，AO-M 相比于其他铁矿的沉降和过滤性能均最差，即使过滤时间超过 4min，滤饼水分仍＞14%，这意味着现场配加 AO-M，将带来矿浆沉降慢、过滤和干燥脱水困难的问题，在实际生产中可通过载体强化沉降和过滤过程。

当固定混合精矿比表面积为 1500cm²/g 左右，AO-M 配比为 0~40%时，在粘结剂用量 0.9%、适宜造球水分(7.9%~8.6%)、造球时间 10~12min 的条件下，均可制备出落下强度≥20 次/(0.5m)、抗压强度≥20N/P、爆裂温度≥330℃ 的合格生球；但提高 AO-M 配比趋于增加生球适宜水分，降低生球爆裂温度，对此应适当调整生球干燥制度，避免因生球热稳定性变差而导致球团开裂、粉化和成品率下降等不利问题。

AO-M 在配矿条件下（磁铁精矿比例 30%~40%）的球团焙烧小型及扩大型试验结果表明：添加 AO-M 需相应提高适宜的预热和焙烧温度，如图1(III)、(IV)可知，AO-M 比例在 20%以内时，适宜的焙烧温度为 1160~1190℃；当 AO-M 比例进一步增加到 30%~40%时，焙烧温度应相应提高到 1220℃以上。通过图 1(V)可知，这可能是因为配加 AO-M 后因结晶水脱除而导致焙烧球团孔隙率及孔径大小有所增加，从而降低了机械性能。同时，配加 20%~40%AO-M 并未对成品球团的冶金性能造成明显的影响，而是均具有良好的冶金性能，RI≥60%，$RDI_{+3.15}$≥97%，RSI≤20%。综合来看，AO-M 在 40%以内均可制备出优质高炉用球团；对于采用链算机-回转窑工艺的球团厂，AO-M 的配比控制在 20%以内为宜。

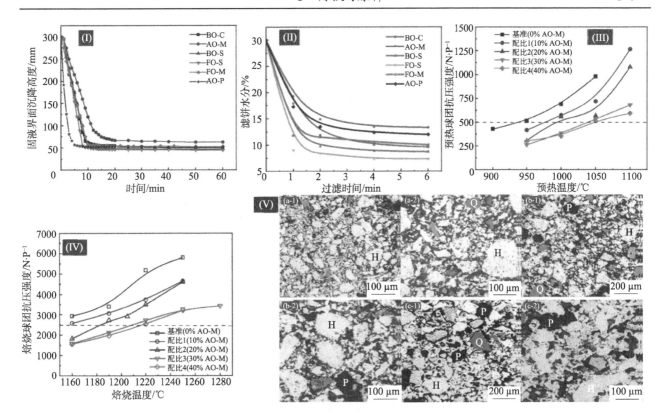

图1 (I) 不同铁矿粉沉降性能对比；(II) 不同铁矿粉过滤性能对比；(III) 不同 AO-M 配比预热球团抗压强度对比；(IV) 不同 AO-M 配比焙烧球团抗压强度对比；(V) 不同 AO-M 配比(a—0%, b—20%, c—40%)条件下的焙烧球团微观结构

改善赤褐铁矿球团生产过程和成品质量的方法

刘胜歌，张俊杰，张巧玉，贾来辉，裴元东，刘 桐，宋亚龙

（中天钢铁集团（南通）有限公司，江苏南通 226100）

摘 要：当前国内大部分球团产线主要以磁铁矿原料为主生产酸性球团。但随着磁精粉价格居高不下，为了降低球团原料成本，部分钢铁厂球团开始以赤、褐铁矿作为原料用于球团生产，进而达到降本增效的目的。但赤褐铁矿尤其是褐铁矿作为球团原料在国内外应用较少，国内也缺乏相关的实践经验。赤褐铁矿由于烧损大、难过滤、含全水量大、无氧化放热过程等特点，导致在焙烧过程难以稳定控制，烧制出来的成品球抗压强度和冶金性能也波动较大，给生产带来困难。本文通过对赤褐铁矿生产过程中原料结构优化，添加一定比例的磁铁矿、硼镁精粉、镜铁矿、高炉返矿、焦化除尘灰以及对过程控制参数进行优化等一系列有效措施，极大改善了焙烧过程控制和成品球的质量，为赤褐铁矿在球团的高比例应用和生产稳定提供了宝贵经验。

关键词：球团；赤褐铁矿；焦化灰；镜铁矿；高炉返矿；硼镁精粉

Abstract: At present, most domestic pelletizing production lines mainly use Magnetite raw materials to produce acid pellets. However, with the high price of magnetite concentrate powder, some iron and steel plants began to use hematite and limonite as raw materials for pelletizing production in order to reduce the cost of pelletizing raw materials. but, hematite Limonite, especially limonite is rarely used as pelletizing raw material at home and abroad before. Because of the characteristics of hematite limonite, such as high burning loss, difficult filtration, high total water content, and no oxidation exothermic process, it is difficult to control the roasting process stably, the compressive strength and metallurgical properties of the fired finished pellets also fluctuate greatly, which brings difficulties to production. Through a series of effective measures are used, such as optimizing the raw material structure, adding a certain proportion of Magnetite, boron magnesium concentrate, specularite, blast furnace return ore, coking ash, and optimizing process control parameters. this paper has greatly improved the roasting process control and the quality of finished pellets, providing valuable experience for the high proportion application of hematite and limonite in pellets.

Key words: pellets; hematite and limonite; coking ash; specularite; return of blast furnace ore; boron magnesium powder

高比例赤褐铁矿碱性球团实践应用与研究

刘胜歌，张俊杰，张巧玉，贾来辉，裴元东，顾建苛，刘 桐，刘建波

（中天钢铁集团（南通）有限公司，江苏南通 226100）

摘 要： 介绍某钢铁公司采用高比例赤、褐铁矿作为主要原料在带式球团焙烧机生产线上开展高硅、含镁碱性球团生产试验，首先将 0~20mm 颗粒状赤、褐铁矿原料通过润磨和压滤，达到粒度-200 目 80%和水分 9.3%左右用于球团生产，然后配加一定比例的磁精粉并在配料系统添加一定比例的焦化灰、消石灰混匀后进行造球，生球筛分后将 8~16mm 左右生球进入焙烧机焙烧。试验通过调整不同碱度情况下的焙烧试验，观测对成品球抗压强度、还原度及还原膨胀影响，尤其是雷区碱度 $R0.3~0.8$ 区间还原膨胀变化趋势，同时也试验了在含镁资源情况下的碱性球的还原膨胀指标情况。最终提出合理的碱度匹配，为高比例赤、褐铁矿球团应用、以及高硅低碱度球团生产模式提供了实践探索，填补了国内在该方向领域研究的空白，该成果也对高炉大比例球团应用、节能减排、绿色炼铁等具有重要意义。

关键词： 球团；赤褐铁矿；焙烧机；碱性球团；还原膨胀

Abstract: Introduction an iron and steel company used a high proportion of hematite and limonite as the main raw materials to carry out the alkaline pellets on belt indurating machine,which contains high silicon and magnesium. Firstly, 0~20mm granular hematite and limonite raw materials were wet ground and filtered to reach a particle size of 0.074mm 80% and a moisture content of 9.3% for pellet production. Then a certain proportion of magnetic powder with coking ash and hydrated lime was added before mixing evenly. After screening the green pellets, approximately 8~16mm of green pellets are fed into the belt indurating machine. The experiment was conducted by adjusting the different alkalinity conditions to observe the impact on the compressive strength, reduction degree, and reduction expansion of the finished pellets, especially the change trend of reduction expansion in the range of alkalinity $R0.3~0.8$ in the minefield. At the same time, the reduction expansion index of the alkaline pellets under magnesium resource conditions was also tested. Finally, a reasonable alkalinity matching is proposed, which provides practical exploration for the application of high proportion hematite and limonite pellets, as well as the production mode of high silicon and low alkalinity pellets, filling the gap in domestic research in this field. This achievement is also the great significance for the large proportion pellets application in blast furnace , energy conservation and emission reduction, green iron smelting, etc.

Key words: pellets; hematite and limonite; belt indurating machine; alkaline pellets; reduction expansion

带式球团配加含铁含碳除尘灰生产实践

胡守景，王业飞，秦立浩

（南钢炼铁事业部球团厂，江苏南京 210035）

摘 要：南钢带式球团产线通过开展含铁、含碳除尘灰小比例配加工业试验及对含碳球团焙烧制度的摸索，已实现含铁、含碳除尘灰稳定配加和焙烧过程控制。通过控制混合料含碳量吨球 5~6kg，焙烧温度可较基准配比低 100℃左右，成品球抗压稳定在 3018N/P，实物质量及冶金性能指标均能满足高炉要求。含铁、含碳除尘灰的稳定配加，年消化除尘灰量 11 万吨，有效回收铁元素 4.7 万吨，利用碳资源 1.3 万吨，吨球煤气消耗下降 0.27GJ，经济效益显著。

关键词：带式焙烧机；含铁含碳除尘灰；热工制度

Production Practice of Adding Iron and Carbon Containing Dust Removal Ash to Belt Pellets

HU Shoujing, WANG Yefei, QIN Lihao

(Iron-making Business Unit Pellet Plant, Nanjing 210035, China)

Abstract: The belt roaster of Nisco has achieved stable addition of iron and carbon containing dust and control of the roasting process by conducting industrial experiments on small proportion mixing of iron and carbon containing dust and exploring the roasting system of carbon containing pellets. By controlling the carbon content of the mixture to 5~6kg per ton of ball, the roasting temperature can be reduced by about 100℃ compared to the benchmark ratio. The compressive strength of the finished ball is stable at 3018N/P. The quality and metallurgical performance indicators of the pellet can meet the requirements of the blast furnace. The stable addition of iron and carbon containing dust removal ash results in an annual digestion of 110000 tons of dust removal ash, effective recovery of 47000 tons of iron elements, utilization of 13000 tons of carbon resources, and a decrease of 0.27GJ in gas consumption per ton of ball. The economic benefits are significant.

Key words: belt roaster; dust removal ash containing iron and carbon; operation method of thermal engineering

高压辊磨预处理国内铁精矿强化球团性能研究

高茂林

（富蕴蒙库铁矿有限责任公司，新疆阿勒泰 836100）

摘 要：为了提高某企业国内铁精矿球团相关性能，本文研究了高压辊磨预处理国内铁精矿对球团性能的影响。结果表明，经过高压辊磨预处理后，铁精矿中小于 0.074mm 的含量和比表面积均得到有效提高；在满足生球落下强度大于 5.0 次条件下，膨润土用量降低了 50%。国内铁精矿经过高压辊磨预处理后，有利于焙烧过程中 Fe_2O_3 再结晶，焙烧球团中赤铁矿晶粒尺寸增大，孔隙率降低，抗压强度得到明显提高。

关键词：国内铁精矿；高压辊磨；预处理；球团性能

Improvement the Pellet Performances of Domestic Iron Concentrate Pretreated by High Pressure Grinding Roller

GAO Maolin

(Fuyun Mengku Iron Ore Co., Ltd., Aletai 836100, China)

Abstract: In order to improve the pellet performances of domestic iron concentrate, the influence of pretreatment of domestic iron concentrate by high pressure grinding roller (HPGR) on pellet performances was studied. The results show that the content of iron concentrate less than 0.074mm and specific surface area are effectively increased after the pretreatment by HPGR. Compared with the iron ore concentrate without pretreated by HPGR, the dosage of bentonite is reduced by 50% under the condition that the drop numbers of green ball meet the requirement of higher than 5.0 times. The compressive strength of pellet is obviously improved because the recrystallization of Fe_2O_3 improves during the roasting process, the grain size of hematite increases and the porosity decreases after the domestic iron concentrate pretreated by HPGR.

Key words: domestic iron ore concentrate; high pressure grinding roller; pretreatment; pellet performances

红磁混合精矿生产低碱度球团生产实践

祁　立

（太钢集团岚县矿业有限公司，山西吕梁　033500）

摘　要： 太钢集团岚县矿业有限公司球团生产线采用以红磁混合精矿粉为原料的链箅机-回转窑生产工艺生产酸性球团。2023 年按集团公司要求，生产低碱度碱性球团。通过不断地实验室研究及对生产现场造球工艺、热工制度等的优化完善，实现了使用红磁混合精矿粉为原料、膨润土与石灰石粉为辅料、煤粉为燃料，长周期稳定生产低碱度氧化球团矿的目标。

关键词： 红磁混合精矿球团；生球；预热球；碱性球团

Abstract: The pellet production line of Lanxian Mining Co., Ltd. of Taiyuan Iron and Steel Group adopts a chain grate rotary kiln production process using red magnetic mixed concentrate powder as raw material to produce acidic pellets. In 2023, according to the requirements of the group company, low alkalinity alkaline pellets will be produced. Through continuous laboratory research and optimization and improvement of production site pelletizing process and thermal system, the goal of using red magnetic mixed concentrate powder as raw material, bentonite and limestone powder as auxiliary materials, and coal powder as fuel to produce low alkalinity oxidized pellets in a long-term and stable manner has been achieved.

Key words: pellet; green pellets; preheat; alkaline pellet

海砂型钒钛磁铁矿颗粒表面性能与球团制备研究

邢振兴[1,2]，刘金生[1,2]，薛向欣[1,2]，程功金[1,2]，杨　合[1,2]

（1. 东北大学冶金学院，辽宁沈阳　110819；
2. 辽宁省冶金资源循环科学重点实验室，辽宁沈阳　110819）

摘　要：针对海砂型钒钛磁铁矿颗粒具有光滑粒粗的独特性，导致其无法大规模高效利用的问题。本文针对海砂矿颗粒的表面性能开展了系列研究，同时提出了多策略强化海砂矿球团制备工艺。研究结果表明：海砂矿的 Zeta 电位为–0.804 mV，颗粒表面带负电荷，与去离子水的接触角为 31.0°，颗粒表面为亲水性矿物。海砂矿的表面自由能数据表明其颗粒表面为低能极性表面，与去离子水之间存在范德华引力和疏水引力。液桥力测试结果揭示了添加膨润土在造球过程中的有益效果，基于此，本文提出了多策略强化海砂矿球团制备工艺。

关键词：海砂型钒钛磁铁矿；表面性能；液桥力；球团

Study on Particle Surface Properties and Pellet Preparation of Sea Sand Type Vanadium Titanomagnetite

XING Zhenxing[1,2], LIU Jinsheng[1,2], XUE Xiangxin[1,2],
CHENG Gongjin[1,2], YANG He[1,2]

(1. School of Metallurgy, Northeastern University, Shenyang 110819, China;
2. Liaoning Key Laboratory of Recycling Science for Metallurgical Resources, Shenyang 110819, China)

Abstract: In response to the unique characteristics of smooth and coarse grain size of sea sand type vanadium titanium magnetite particles, it is difficult to efficiently utilize them on a large scale. In this paper, a series of studies have been carried out on the surface properties of sea sand ore particles, and proposed a multi-strategy to strengthen the preparation process of pellets with sea sand ore. The results show that the Zeta potential of sea sand ore was –0.804 mV, the particle surface was negative charge, the contact angle with deionized water was 31.0°, and the particle surface was hydrophilic mineral. The surface free energy data of sea sand ore indicated that its particle surface was a low energy polar surface, and there was Van der Waals force and hydrophobic gravity between the particle surface and deionized water. The results of the liquid bridge force test revealed the beneficial effect of adding bentonite in the pelletizing process. Based on this, this paper proposed a multi-strategy strengthening process for the preparation of pellets with sea sand ore.

Key words: sea sand type vanadium titanomagnetite; surface properties; liquid bridge force; pellets

循环风流动状态对球团热工制度的影响

张　晨

（宝武集团中钢国际工程技术有限公司，北京　100000）

摘　要：热风循环系统是焙烧球团工艺实现低碳节能的关键技术，由于焙烧机结构复杂、送风支管数量众多，很难避免涡旋流动的产生，这样的流场必然对炉内温度分布、燃料的燃烧过程、以及气流与球床间的传热过程产生决定性的影响。本文采用三维模拟技术对焙烧炉内热风流场仿真，针对循环风流动特性进行研究和分析。结果表明：(1)燃烧室内的旋流和炉膛内的回流对于气体流动和温度分布起到促进作用。(2)隔墙处的回流涡旋虽然无法被消除，但是不影响循环风整体的流动性。(3)热风支管形成的涡旋不仅产生局部阻力损失，对系统的流动状态造成影响；炉膛内高温带前移，形成的温度梯度很小；均热段温度偏低，造成隔墙两侧温差过大，隔墙容易产生错位甚至掉砖；支管处的涡旋使得天然气也发生偏析，形成长火焰燃烧，过程中产生大量的氮氧化物。

关键词：焙烧球团；流场仿真；回流涡旋；温度梯度

The Influence of the Circulating Airflow State on the Thermal System of the Pelletizing Process

ZHANG Chen

(Baowu Group Sinosteel Engineering & Technology Co., Ltd., Beijing 100000, China)

Abstract: The hot air circulation system is a key technology for achieving low carbon and energy saving in the pelletizing process. Due to the complex structure of the pelletizing machine and the large number of air supply ducts, it is difficult to avoid the generation of vortex flow. Such flow field inevitably has a decisive impact on the temperature distribution inside the furnace, the combustion process of the fuel, and the heat transfer between the airflow and the pellet bed. In this study, three-dimensional simulation technology was used to simulate the hot air flow field inside the pelletizing furnace, and the characteristics of the circulating airflow were studied and analyzed. The results show that: (1) The vortex flow in the combustion chamber and the backflow in the furnace promote the gas flow and temperature distribution. (2) Although the backflow vortex at the partition wall cannot be eliminated, it does not affect the overall flow of the circulating air. (3) The vortex formed by the hot air duct not only causes local resistance loss, but also affects the flow state of the system. The high temperature zone in the furnace moves forward, resulting in a small temperature gradient. The temperature in the homogenization section is low, resulting in a large temperature difference between the two sides of the partition wall, which easily leads to displacement or even brick falling off. The vortex at the duct also causes the natural gas to segregate and form a long flame combustion, which produces a large amount of nitrogen oxides during the process.

Key words: gas-based direct reduction; energy conservation and emission reduction; traveling grate process pellet; compressive strength

基于神经网络和 NSGA-Ⅱ 的烧结参数多目标优化

臧疆文

（宝钢集团八钢公司碳中和办公室，新疆乌鲁木齐 830022）

摘　要：为寻求烧结负压、焦粉配比、料层高度等因素对烧结指标影响的内在规律，采用 D-最优设计安排试验；以 D-最优设计的试验数据作为训练样本，建立了以烧结负压、焦粉配比、料层高度为输入，利用系数、转鼓指数、固体燃耗为输出的 BP 神经网络预测模型，经过测试验证了 BP 神经网络预测模型的合理性。将 BP 神经网络预测模型作为 NSGA-Ⅱ 遗传算法的适应度函数进行小球团烧结工艺参数多目标优化，得到 Pareto 最优解集。

关键词：烧结；BP 神经网络；NSGA-Ⅱ；遗传算法；多目标优化；Pareto

Multi-objective Optimization of Sintering Parameters Based on Neural Network and NSGA-Ⅱ

ZANG Jiangwen

(Carbon Neutral Office, Bayi Iron & Steel Co., Baosteel Group, Urumqi 830022, China)

Abstract: In order to explore the internal law of the influence of factors such as sintering negative pressure, coke powder ratio and material layer height on sintering index, D-optimal design was used to arrange the experiments. Using the D-optimal design test data as training samples, a BP neural network prediction model was established with negative sintering pressure, coke powder ratio and material layer height as inputs and utilization coefficient, drum index and solid fuel consumption as outputs. The rationality of the BP neural network prediction model has been verified through testing. The BP neural network prediction model was used as the fitness function of NSGA-II genetic algorithm to optimize the process parameters of pellet sintering, and the Pareto optimal solution set was obtained.

Key words: sintering; BP neural network; NSGA-II; genetic algorithm; multi-objective optimization; Pareto

某进口赤褐铁矿磨矿特性及其氧化球团制备行为研究

杨聪聪，曲士娟，潘 建，朱德庆，郭正启，李思唯

（中南大学资源加工与生物工程学院，低碳与氢冶金研究中心，湖南长沙 410083）

摘 要：在当前"双碳"背景下，大力发展球团矿生产、提高高炉入炉比例是完善高炉长流程，支撑钢铁工业绿色低碳转型的有效手段之一。然而，如何扩大球团原料、降低球团生产成本是钢铁企业当前面临的重要课题之一，也直接影响着球团工艺的发展[1]。赤褐铁矿粉通常用作烧结原料，相比细粒铁精矿具有一定价格优势，对于国内球团厂扩大原料来源、指导现场生产和降低球团生产成本均具有重要的现实意义[2]。

本文以某进口赤褐铁矿(AO-M)为主要原料，其铁品位和SiO_2含量分别为60.31%和4.45%，烧损较高（6.08%），且Al_2O_3含量达到了2.21%，可能对湿式磨矿、生球制备和球团焙烧过程和产品质量产生不利的影响，因而较系统地研究了其湿式磨矿-沉降-过滤特性[3]、原料预处理和球团制备特性，确定了不同AO-M配比条件下适宜的焙烧条件，并考察了对球团冶金性能和矿相特征的影响，为明确适宜的AO-M配比奠定了重要基础。

AO-M的Bond磨矿功指数为12.959kW·h/t，与巴西铁矿粉（BO-C、BO-S）、澳洲铁矿粉（AO-P）和非洲铁矿粉（FO-M、FO-S）的磨矿功指数相近，均属易磨矿石；如图1(I)、(II)可知，单矿矿浆静态沉降及过滤特性试验研究表明，AO-M相比于其他铁矿的沉降和过滤性能均最差，即使过滤时间超过4min，滤饼水分仍>14%，这意味着现场配加AO-M，将带来矿浆沉降慢、过滤和干燥脱水困难的问题，在实际生产中可通过载体强化沉降和过滤过程[4]。

当固定混合精矿比表面积为$1500cm^2/g$左右，AO-M配比为0~40%时，在粘结剂用量0.9%、适宜造球水分（7.9%~8.6%）、造球时间10~12min的条件下，均可制备出落下强度≥20次/(0.5m)、抗压强度≥20N/P、爆裂温度≥330℃的合格生球；但提高AO-M配比趋于增加生球适宜水分，降低生球爆裂温度，对此应适当调整生球干燥制度，避免因生球热稳定性变差而导致球团开裂、粉化和成品率下降等不利问题[5]。

AO-M在配矿条件下（磁铁精矿比例30%~40%）的球团焙烧小型及扩大型试验结果表明：添加AO-M需相应提高适宜的预热和焙烧温度，如图1(III)、(IV)可知，AO-M比例在20%以内时，适宜的焙烧温度为1160~1190℃；当AO-M比例进一步增加到30%~40%时，焙烧温度应相应提高到1220℃以上。通过图1(V)可知，这可能是因为配加AO-M后因结晶水脱除而导致焙烧球团孔隙率及孔径大小有所增加，从而降低了机械性能。同时，配加20%~40% AO-M并未对成品球团的冶金性能造成明显的影响，而是均具有良好的冶金性能，RI≥60%，$RDI_{-3.15}$≥97%，RSI≤20%。综合来看，AO-M在40%以内均可制备出优质高炉用球团；对于采用链箅机-回转窑工艺的球团厂，AO-M的配比控制在20%以内为宜。

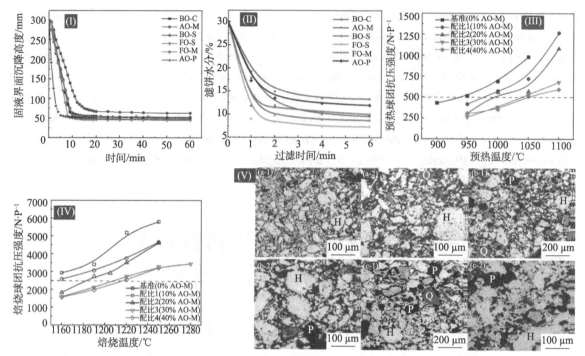

图 1 (I) 不同铁矿粉沉降性能对比；(II) 不同铁矿粉过滤性能对比；(III) 不同 AO-M 配比预热球团抗压强度对比；(IV) 不同 AO-M 配比焙烧球团抗压强度对比；(V) 不同 AO-M 配比（a—0%，b—20%，c—40%）条件下的焙烧球团微观结构

参 考 文 献

[1] 朱德庆, 黄铮静, 杨聪聪, 等. 烧结球团, 2020, 45(6): 61-68.
[2] 彭志坚, 罗浩. 钢铁研究, 2005(6): 1-4.
[3] 毛瑞, 王飞, 金海, 等. 烧结球团, 2022, 47(2): 24-29.
[4] 彭道胜, 潘建, 李军, 等. 烧结球团, 2021, 46(4): 50-57.
[5] 马丽, 青格勒, 田筠清, 等. 烧结球团, 2019, 44(2): 39-41, 49.

MgO 对 CO 分阶段还原赤铁矿过程膨胀性的影响

张 芳，彭 军，常宏涛，王永斌

（内蒙古科技大学材料与冶金学院，内蒙古包头 014010）

摘 要：为了探明 MgO 对铁矿球团逐级还原各阶段膨胀量的影响，首先通过热力学计算确定不同阶段 MgO 的赋存状态，然后将磁铁矿样条在 1250℃氧化焙烧 30min 得到赤铁矿样条，之后用 CO 分阶段还原赤铁矿样条，同时采用线膨胀仪对其线膨胀量进行在线检测，并对还原后试样的矿物组成和微观形貌进行分析，获得 MgO 对赤铁矿还原过程不同阶段体积变化的影响机理。结果表明，在磁铁矿氧化为赤铁矿过程中 MgO 不会进入赤铁矿中；在赤铁矿还原过程中 MgO 主要存在于硅酸盐相中，其次存在于铁氧化物中；并且随着球团矿中 MgO 含量增加，其在硅酸盐和铁氧化物中的含量呈线性增加。对于分向磁铁矿中添加 0、1%、2%和 3%MgO 纯试剂的条形试样，添加 2% MgO 试样的体积变化最小，其 20mm 的长度在 $Fe_2O_3 \rightarrow Fe_3O_4$、$Fe_3O_4 \rightarrow FeO$ 和 $FeO \rightarrow Fe$ 阶段还原过程试样长度的变化分别为 615μm、–25μm 和–378μm，可见赤铁矿还原到磁铁矿的阶段体积膨胀，主要是由 $Fe_2O_3 \rightarrow Fe_3O_4$ 还原阶段铁氧化物的晶型转变，引起赤铁矿颗粒内部产生裂纹引起的。在 $Fe_3O_4 \rightarrow FeO$ 还原阶段，样条产生收缩；MgO 作为高熔点的碱性氧化物，添加量过多造成渣相对铁氧化物颗粒的粘结性变差，抵抗铁氧化物收缩的能力变差。浮士体向金属铁转变阶段，试样长度出现明显的收缩，但是渣相成分适宜的情况下，仍然能够保持最初基本的轮廓。对于条形试样来说，长度增加后缩短的结果体现球团矿上，即表现为由于其还原过程由外向内逐渐膨胀后又收缩，若幅度过大会引起球团表面出现裂纹。添加 0、1%、2%和3%MgO 纯试剂的磁铁矿样条，逐级还原过程由各阶段线膨胀量曲线围成的面积大小分别为 $3.76 \times 10^6 \mu m \cdot s$、$3.23 \times 10^6 \mu m \cdot s$、$3.05 \times 10^6 \mu m \cdot s$ 和 $3.17 \times 10^6 \mu m \cdot s$。

关键词：氧化镁；赤铁矿；还原膨胀性；逐级还原

The Effect of MgO on the Swelling Property of Hematite Reduced by CO Step-by-step

ZHANG Fang, PENG Jun, CHANG Hongtao, WANG Yongbin

(School of Materials and Metallurgy, Inner Mongolia University of Science and Technology, Baotou 014010, China)

Abstract: In order to investigate the influence of MgO on the expansion of iron ore pellets in different stages of progressive reduction, the occurrence states of MgO in different stages were first determined through thermodynamic calculations. Then, the magnetite samples were oxidized and roasted at 1250℃ for 30 minutes to obtain hematite samples. After that, the hematite samples were reduced in stages with CO, and their linear expansion was measured online using a linear dilatometer. And the mineral composition and microstructure of the reduced samples were analyzed to obtain the mechanism of the influence of MgO on the volume changes at different stages of hematite reduction process. The results indicate that MgO does not enter the hematite during the oxidation of magnetite to hematite. During the reduction process of hematite, MgO mainly exists in the silicate phase, followed by iron oxides. And as the MgO content in the pellet increases, its content in silicates and iron oxides increases linearly. For the strip samples with 0, 1%, 2%, and 3% MgO pure reagents added to magnetite, the volume change of the sample with 2% MgO added is the smallest. The length of the 20mm sample during the reduction process of $Fe_2O_3 \rightarrow Fe_3O_4$, $Fe_3O_4 \rightarrow FeO$, and $FeO \rightarrow Fe$ is 615μm, –25μm and –378μm respectively. It can be seen that the volume expansion during the reduction stage from hematite to magnetite is mainly caused by the crystal transformation of iron oxides during the $Fe_2O_3 \rightarrow Fe_3O_4$ reduction stage, which causes cracks to occur inside the hematite particles. During the $Fe_3O_4 \rightarrow FeO$ reduction stage, the spline undergoes shrinkage. As a high melting point alkaline oxide, excessive addition of MgO results in the poor adhesion of slag to iron oxide particles and the poor resistance to iron oxide shrinkage. During the transition phase from the wustite to metallic iron, there is a significant shrinkage in the length of the sample, but when the slag phase composition is suitable, it can still maintain the initial basic contour. For strip samples, the result of shortening after increasing the length is reflected in the pellets, which is manifested that the pellet gradually expand from the outside to the inside and then contracts in the reduction process. If the amplitude is too large, it will cause cracks on the surface of the pellet. For the magnetite samples with 0, 1%, 2%, and 3% MgO pure reagents, the area sizes enclosed by the linear expansion curves of each stage during the step-by-step reduction process were $3.76 \times 10^6 \mu m \cdot s$, $3.23 \times 10^6 \mu m \cdot s$, $3.05 \times 10^6 \mu m \cdot s$ and $3.17 \times 10^6 \mu m \cdot s$, respectively.

Key words: magnesium oxide; hematite; reduction expansibility; step-by-step reduce

3.2 高炉炼铁

高炉开停炉技术创新实践

陈生利

（广东中南股份，广东韶关　512123）

摘　要： 中国宝武集团广东韶钢炼铁厂 6 号、7 号、8 号高炉分别自 2002 年、2005 年、2008 年起陆续投产。近两年来，三座高炉都经历过停炉大修并顺利复产，开炉期间采用了一系列创新技术，最大限度的保证了开炉停炉过程中高炉煤气回收，实现了安全、快速开炉的目标。在历次停炉过程中，通过不断的技术创新，确保了停炉工作的顺利进行。本文重点阐述开停炉过程中关键创新技术及其应用情况，为同行开停炉提供技术支撑。

关键词： 停炉；降料面；开炉；出渣铁；铁水

Technological Innovation Practice of Blast Furnace Opening and Stopping

CHEN Shengli

(Guangdong Zhongnan Stock, Shaoguan 512123, China)

Abstract: The No.6, No.7 and No.8 blast furnaces of Guangdong Shaosteel steelmaking Plant of China baowu Group have been put into operation since 2002, 2005 and 2008 respectively. In the past two years, the three blast furnaces have undergone overhaul and successfully resumed production. A series of innovative technologies have been adopted during the opening of the blast furnace to ensure the recovery of blast furnace gas in the process of opening and closing of the furnace to the maximum extent, and the goal of safe and fast opening of the furnace has been achieved. In the process of stopping the furnace, through continuous technological innovation, to ensure the smooth progress of stopping the furnace. This paper focuses on the key innovation technology and its application in the process of opening and stopping the furnace to provide technical support for the peer.

Key words: the furnace; fall is expected to face; blowing in; slag iron; the molten iron

高炉铁水收得率攻关实践

陈生利，彭　厅，余　骏

（广东中南股份，广东韶关　512123）

摘　要： 近年来，韶钢高炉铁水收得率不足 98%，铁水损失量大，高炉冶炼成本增加。本文针对高炉出铁系统、煤气除尘系统、称量系统、原材料检验系统等进行研究分析，找到了铁水收得率偏低的原因，提出了优化炉前出渣铁、优化炉内气流控制等技术攻关措施,逐步提高铁水收得率至 98%以上,实现了高炉入炉原料理论铁量与实际铁水产量基本吻合的目标。

关键词：高炉；铁水；收得率；技术攻关；高炉炉前

Research Practice of Recovery Rate of Molten Iron in Blast Furnace

CHEN Shengli, PENG Ting, YU Jun

(Guangdong Zhongnan Stock, Shaoguan 512123, China)

Abstract: In recent years, the recovery rate of molten iron in the blast furnace of Shaogang is less than 98%, the loss of molten iron is large, and the smelting cost of blast furnace increases. This paper studies and analyzes the iron extraction system, gas dust removal system, weighing system and raw material inspection system of the blast furnace, and finds out the reasons for the low recovery rate of hot iron. It puts forward technological measures such as optimizing slag iron extraction before the furnace and optimizing air flow control in the furnace, and gradually increases the recovery rate of hot iron to more than 98%. The theoretical iron content of blast furnace feedstock is basically consistent with the actual hot iron production.

Key words: the blast furnace; the molten iron; yield; technology research; before the blast furnace

韶钢 7 号高炉铁口喷溅治理操作实践

匡洪锋

（广东中南钢铁股份有限公司，广东韶关　512123）

摘　要：韶钢 7 号高炉（2200m³）2022 年 3 月大修投产后 3 个铁口出现较为严重的喷溅现象,造成出铁时渣铁流偏小,易导致高炉炉内憋渣铁,影响高炉稳定、顺行。后续采取了降低炉内压差、优化高炉操作制度、铁口压浆、改善炮泥质量、重新制作铁口泥套等措施进行综合治理,取得了良好的效果,使铁口喷溅问题得到了有效的控制,高炉铁产量逐步由 6800t/d 提升至 7200t/d 以上水平,效果显著。

关键词：高炉；铁口；渣铁；压浆

Operation Practice of Iron Spitter Control for No. 7 Blast Furnace at Shaoguan Iron and Steel Group Co., Ltd.

KUANG Hongfeng

(Guangdong Zhongnan Stock, Shaoguan 512123, China)

Abstract: After the overhaul and operation of No. 7 blast furnace (2200m³) in Shaoguan Iron and Steel Group Co., Ltd., in March 2022, serious spatter phenomenon appeared at three iron holes, resulting in small slag flow during iron discharge, which easily leads to slag holding in the blast furnace, affecting the stability and running of the blast furnace. The following measures were taken to reduce the pressure difference in the furnace, optimize the operation system of the blast furnace, press the slurry at the iron mouth, improve the quality of the blast mud, and re-make the mud sleeve at the iron mouth. Good results were obtained, and the spatter problem of the iron mouth was effectively controlled, and the iron output of the blast furnace was gradually increased from 6800t/d to more than 7200t/d, with remarkable results.

Key words: blast furnace; iron mouth; slag iron; grouting

提高铁水硅偏差命中率控制措施

余 骏，陈生利

（广东中南股份，广东韶关 512123）

摘 要： 铁水硅偏差(铁水[Si]含量波动幅度)是衡量高炉铁水质量的重要指标，是代表高炉铁水质量控制水平的重要技术参数。本文重点阐述韶钢 2500m³ 高炉通过提高原料质量、改善炉缸活跃性等技术措施，现实铁水硅偏差稳定率从 50%提升至 80%的良好技术指标，可以为国内高炉铁水质量改善提供技术借鉴。

关键词： 高炉；铁水；硅偏差；命中率

Measures to Improve the Error Hit Ratio of Molten Iron and Silicon

YU Jun, CHEN Shengli

(Guangdong Zhongnan Stock, Shaoguan 512123, China)

Abstract: The molten iron silicon deviation (the fluctuation range of molten iron [Si] content) is an important index to measure the quality of molten iron of blast furnace and an important technical parameter to represent the quality control level of molten iron of blast furnace. This paper focuses on the good technical indicators of the actual molten iron silicon deviation stability rate increasing from 50% to 80% by improving the quality of raw materials and the activity of the hearth of the 2500m³ blast furnace of Shaogang, which can provide technical reference for the improvement of the quality of molten iron in the domestic blast furnace.

Key words: the blast furnace; the molten iron; silicon deviation; shooting

梅钢 4070m³ 高炉炉墙结厚分析与炉况恢复

张光强，韩宏松，占世梅

（上海梅山钢铁股份有限公司，江苏南京 210000）

摘 要： 对梅钢五号高炉炉墙结厚的原因和炉况恢复措施进行总结，认为高炉原燃料质量变差、高炉锌负荷高和高炉频繁崩滑料甚至悬料是炉墙结厚的主要原因。处理炉墙结厚的主要措施是加锰矿洗炉、强力发展边缘气流和保证较高铁水温度等。

关键词： 装料制度；炉墙结厚；炉况异常

昆钢 2500m³ 高炉提高利用系数生产实践

麻德铭，李 淼，李晓东

（武钢集团昆明钢铁股份有限炼铁厂，云南安宁 650309）

摘 要：2021 年，昆钢 2500m³ 高炉通过事前预算、使用跟踪等手段，经济性地提高了入炉品位；通过调整煤枪角度、推行设备包机制，降低了临时休风率；通过"内部精料"、专项攻关等措施，精细化了高炉操作；通过创造富氧条件、"提煤节焦"等方式，提高了高炉的强化冶炼程度。促成高炉利用系数的大幅提高，实现了高效、经济生产。

关键词：高炉；利用系数；品位；休风率；精细化操作管理；强化冶炼

Production Practice of Improving Utilization Factor of Kunming Steel's 2500m³ Blast Furnace

MA Deming, LI Miao, LI Xiaodong

(Wuhan Iron and Steel Group Kunming Iron and Steel Co., Ltd.
Ironmaking Plant, Anning 650309, China)

Abstract: In 2021, the 2500m³ blast furnace of Kunming Iron and Steel Co., Ltd. achieved an economic increase in the incoming grade through pre budgeting, usage tracking, and other means; By adjusting the angle of the coal gun and implementing the equipment package mechanism, the temporary ventilation rate has been reduced; Through measures such as "internal precision material" and special research, the operation of the blast furnace has been refined; By creating oxygen-enriched conditions and "coal extraction and coke saving" methods, the intensity of blast furnace smelting has been improved. This has led to a significant increase in the utilization coefficient of blast furnaces, achieving efficient and economical production.

Key words: blast furnace; utilization coefficient; grade; rest wind rate; refined operation management; enhanced smelting

红钢 3 号高炉安全快速停炉实践

张 波，李晓东

（昆明钢铁股份有限公司炼铁厂，云南昆明 650302）

摘 要：介绍了红钢 3 号高炉在无预休风、未充分洗炉情况下，使用炉顶洒水枪降料面停炉的操作实践。通过"停炉前控制改善用料和精准负荷调控，加盖面焦时采用边缘布焦矩阵并控料线下料，以及降料面过程分阶段调控和合理出铁管理"等技术手段，高炉克服了洒水不均、顶温偏差大、频繁爆震等困难，历时 11h20min 安全降料面至风口区域，实现安全快速降料面停炉。

关键词：高炉；降料面；停炉；安全快速；爆震

Safe and Quick Shutdown Practice of No.3 BF at Honggang

ZHANG Bo, LI Xiaodong

(Ironmaking Plant of Kunming Iron & Steel Co., Ltd., Kunming 650302, China)

Abstract: This paper introduces the operation practice of using the top sprinkler to drop surface to stop the furnace of Honggang No. 3 BF under the condition that there is no pre-rest air and the furnace is not fully washed. Through the technical means of "improving raw materials and precise load control before furnace shutdown, adopting edge coke matrix and controlling the depth of the feed line when covering the surface coke, as well as stage adjustment and control and reasonable iron discharge management during the process of falling the feed surface", the No. 3 BF overcame the difficulties of uneven water spraying, large deviation of top temperature and frequent detonation, and lasted 11h20min to safely drop the material surface to the tuyere area, to achieve a safe and rapid material lowering surface shutdown.

Key words: blast furnace; dropping surface; blowing-out; safety and speediness; knock

配加智利粗粉生产实践

张 月，伊凤永

（承德钒钛，河北承德 067001）

摘 要： 承德地区某钢铁厂为钒钛矿冶炼高炉，在低库存、配加经济料的条件下，通过优化烧结机配料结构，烧结矿配加智利粗粉，改善了烧结、高炉生产条件，达到了生产稳定及提高铁水含钒水的目的，取得了较好的经济技术指标。

关键词： 钒钛矿冶炼；烧结；智利粗粉；钒钛烧结矿

Smelting Practice of Adding Chilean Coarse Powder to Vanadium-titanium Ore Blast Furnace

Abstract: An iron and steel plant in Chengde is smelting vanadium-titanium ore. Under the condition of feeding economic materials, the production conditions of sintering and blast furnace were improved by optimizing the burden structure of sintering machine and adding Chilean coarse powder to the sinter, achieving the goal of stable production and increasing the vanadium content of molten iron, and obtaining better economic and technical indicators.

Key words: Vanadium-titanium ore smelting; sintering; Chilean coarse powder; Vanadium-titanium sinter

ECIA 环保料场创新设计

艾 宇，陈尚伦，王沛庆，刘占稳

（中冶赛迪工程技术股份有限公司钢铁事业本部，重庆 400013）

摘　要： 环保料场是当前原料场技术发展的必然趋势，但现有环保料场无法较好地适应物料多样性要求，厂房内部空间利用率低，资源浪费严重。本文以当前应用最多的 C 型环保料场为研究对象，选择并针对其物料贮量差异适应性差、内部空间利用率低的问题进行系统描述，运用 TRIZ 理论进行系统问题分析，求解得到了一系列方案，并通过方案对比确定了最优方案，该方案提高了料场内部空间利用率，增大了料场贮量，降低了综合投资，经济社会效益显著。

关键词： 散装料；环保料场；贮量；TRIZ

Innovative Design of Eco-friendly Stockyard

AI Yu, CHEN Shanglun, WANG Peiqing, LIU Zhanwen

(CISDI Engineering Co., Ltd., Chongqing 400013, China)

Abstract: The technology of Eco-friendly stockyard is the inevitable trend of the current technical development of stockyard, However, the current technology of Eco-friendly stockyard can not adapt to the requirements of material diversity well and the utilization rate of the internal space of the plant is low, and the waste of resources is serious.In this paper, the most widely used C type stockyard is taken as the research object, and the problems of poor adaptability and low utilization rate of internal space are selected and systematically described, using the theory of TRIZ, get a series of innovative solutions, and through schemes comparison to determine the optimal scheme, the optimal scheme improves the utilization rate of the internal space of the stockyard, increases the storage capacity, reduces the comprehensive investment, and has remarkable economic and social benefits.

Key words: bulk material; eco-friendly stockyard; storage capacity; TRIZ

高块矿比冶炼在兴澄高炉的生产实践

张宏星，陈龙智，史志苗

（江阴兴澄特种钢铁有限公司，江苏江阴　214429）

摘　要： 天然块矿相较于烧结矿和球团有着较高性价比，因此提高块矿比，建立新的经济型炉料结构成为各大钢铁企业降低成本的最直接有效途径。兴澄炼铁通过对块矿性能分析，积极开展高比例块矿的攻关工作，通过优化布料模式、调整送风制度、控制适宜的渣铁热量、强化原燃料控制、加强出铁管理等一系列措施，克服了块矿性能差的"先天缺陷"，成功实现了高块矿比在高炉中的应用，在保持炉况顺行基础上，取得了较好的经济指标，顺利实现了公司铁前降本的任务目标。

关键词： 高炉炼铁；炉料结构；块矿；技术经济指标

Production Practice of High Lump Ore Ratio Smelting in Xingcheng Blast Furnace

ZHANG Hongxing, CHEN Longzhi, SHI Zhimiao

(Jiangyin Xingcheng Special Steel Works Co., Jiangyin 214429, China)

Abstract: Compared with sinter and pellets, natural lump ore has higher cost performance, so increasing the proportion of lump ore in the blast furnace smelting process has become the most direct and effective way for major iron and steel enterprises to reduce costs. Through a series of measures such as optimizing the distribution mode, adjusting the air supply system, controlling the appropriate heat of slag and iron, strengthening the control of raw materials and fuels, and strengthening the management of iron tapping, Xingcheng Ironmaking has overcome the shortcomings of poor lump ore performance and successfully realized the application of high proportion of lump ore in the blast furnace. In addition, based on keeping the furnace running smoothly, the company has achieved good economic indicators in production and successfully achieved the company's task goal of cost reduction.

Key words: blast furnace ironmaking; charge structure; lump ore; technico-economical index

昆钢1号高炉精准快速停炉实践

李晓东，张　波

（武钢集团昆明钢铁股份有限公司炼铁厂，云南昆明　650302）

摘　要：对昆钢1号高炉空料线停炉操作进行了分析总结。通过停炉前方案制定、设备检查确认、充分洗炉作业等充分准备，以及停炉过程对风量、水温差和出铁的精细化操作，高炉历时13h30min成功降料面至风口大套下沿，且降料面过程平稳，停炉炉墙干净、料面疏松无硬块，实现了安全、环保、精准、快速停炉。

关键词：高炉；停炉；降料面；精准快速

Practice of Accurate and Rapid Shutdown of Kunming Iron and Steel No.1 Blast Furnace

LI Xiaodong, ZHANG Bo

(Wuhan Iron and Steel Group Kunming Iron and Steel Co., Ltd. Ironmaking Plant, Kunming 650302, China)

Abstract: An analysis and summary were conducted on the shutdown operation of the empty material line of Kunming Iron and Steel No.1 blast furnace. Through the preparation of pre shutdown plans, equipment inspection and confirmation, and thorough cleaning operations, as well as the refined operation of air volume, water temperature difference, and tapping during the shutdown process, the blast furnace successfully lowered the material surface to the lower edge of the tuyere sleeve for 13 hours and 30 minutes, and the lowering process was smooth. The shutdown furnace wall was clean, and the material surface was loose and free of hard blocks, achieving safe, environmentally friendly, accurate, and rapid shutdown.

Key words: blast furnace; shutdown; material lowering surface; precise and fast

基于高炉透气指数的焦炭质量评价与控制方法

刘起航

（西安建筑科技大学冶金工程学院，陕西西安　710055）

摘　要：高炉强化冶炼、提煤降焦等潜力主要取决于料柱透气性的极限，焦炭作为高炉内唯一的骨架支柱和焦窗，对保证高炉良好的透气透液性起到了关键作用。因此，焦炭质量控制的核心目标是保证焦炭能满足高炉不同区域对透气性的刚性需求，其根本是要控制焦炭在高炉不同区域劣化后的粒度分布（劣化平均粒度和空隙度）。

传统的CRI、CSR等指标没有体现焦炭的梯度劣化特性和劣化后的粒度分布特征（空隙度），因而无法较好的反映焦炭质量与焦床空隙度、高炉透气指数、冶炼指标之间的准确关系，有时甚至会出现较大偏差。焦炭高温溶损劣化过程类似铅笔被削的越来越尖，铅笔(焦炭)断裂劣化后的粒度分布取决于它被削的有多尖（劣化梯度），因而劣化梯度对高炉软熔区焦炭劣化后的粒度分布特征、透气指数等有至关重要甚至决定性的作用。劣化梯度与焦炭结构、粒度和溶损环境有关。

在将CRI和CSR指标内涵扩大为劣化梯度（2个）和劣化强度（3个）基础上，可通过焦炭初始入炉粒度分布、M10、M25或M40、铁矿低温性能评价和控制焦炭在高炉块状带的冷态劣化强度和透气指数，并进一步结合CRI、CSR、k_{rea}/D_{eff}、矿焦耦合高温性能准确模拟和控制焦炭在高炉软熔带的劣化梯度、劣化强度及透气指数，透气性的改善有利于提升高炉的节焦潜力，从而进一步提煤降焦和强化冶炼。研究结果表明，高炉软熔带的透气性受焦炭劣化梯度或 $R_0\sqrt{k_{rea}/D_{eff}}$ 的影响极大，而劣化梯度与焦炭物理化学结构、入炉粒度及高炉气氛环境有关，减小劣化梯度或 k_{rea}/D_{eff} 可明显改善高炉软熔带透气性进而提升高炉节焦潜力。从高炉外部环境看，温度升高、碱金属和水蒸气气氛会明显增大焦炭的 k_{rea}/D_{eff} 和劣化梯度进而恶化高炉透气性，因此高炉应严格控制炉料碱金属和水分含量；从焦炭结构和粒度看，不同焦炭的 k_{rea}/D_{eff} 差异很大，捣固焦的劣化梯度明显高于顶装焦，且焦炭粒度越大，劣化梯度越大，焦炭劣化后的粒度分布就越不均匀（两头多、中间少），对透气性的负面影响也越大，因此入炉粒度并不是越大越好。

k_{rea} 可通过热重分析仪对焦粉气化测定得出，它与焦炭显微组织成分、灰分成分、碳结构等化学结构有关；D_{eff} 可通过气孔结构参数的测定和公式计算得到，它与焦炭孔隙率、孔径等孔结构参数有关，通过炼焦工艺和原料控制合理的 k_{rea} 和 D_{eff} 的值使得 k_{rea}/D_{eff} 降低，是优化焦炭质量和提升高炉透气性的有效方法。因此，将 k_{rea}/D_{eff} 或劣化梯度以及入炉粒度分布指标纳入现有焦炭质量评价体系十分有必要，这是稳步提升高炉节焦潜力和反向制备优质焦炭的有效措施和关键途径。

昆钢新区 1 号高炉降低燃料比实践

段贵军，李　淼，李晓东

（武钢集团昆明钢铁股份有限公司炼铁厂，云南安宁　650300）

摘　要：昆钢新区 1 号高炉有效容积 2500m³，本文对昆钢新区 1 号高炉连续降低燃料比冶炼实践进行了总结。基于昆钢高炉入炉品位低且有害元素高的特点，高炉采取堵风口操作控制适宜鼓风动能，针对昆钢原燃料条件，采用高压差（压差高于顶压 5~10kPa）低顶压操作，再通过提高富氧率，改善焦炭质量、调整布料矩阵提高煤气利用率，控制合理的造渣制度，加强高炉炉体管理等一系列措施，保证了炉况的长周期稳定顺行，燃料比不断降低。
关键词：高炉；燃料比；堵风口；高压差

Practice of Reducing Fuel Ratio for No.1 Blast Furnace in Kunming Iron and Steel New Area

DUAN Guijun, LI Miao, LI Xiaodong

(Wuhan Iron and Steel Corporation Kunming Iron & Steel Co., Ltd.,
Ironmaking Plant, Anning 650300, China)

Abstract: The effective volume of No.1 blast furnace in Kunming Steel New Area is 2500m^3, This paper summarizes the practice of continuously reducing fuel ratio of no. 1 blast furnace in Kunming Steel New Area. Based on the characteristics of low grade and high harmful elements, blast furnace operation control suitable blast kinetic energy, in view of the steel raw fuel conditions, the high pressure difference (pressure difference is higher than the top pressure 5~10 kPa) low top pressure operation, then by improving oxygen rich rate, improve the coke quality, adjust the fabric matrix to improve gas utilization, control reasonable slag system, strengthen the management of blast furnace body and a series of measures, ensure the long cycle of furnace stability, fuel ratio is decreasing.

Key words: blast furnace; fuel ratio; air plugging outlet; high pressure difference

高炉炼铁煤粉制备系统节能探讨

汤楚雄

（中冶南方工程技术有限公司炼铁分公司，湖北武汉　430223）

摘　要： 对高炉喷煤制粉系统节能的各个方面进行了讨论。重点阐述了磨煤机入口正压送烟气新工艺以及减少系统漏风和阻力损失的技术措施。全面系统性地采取减少漏风和降低系统阻力损失的有效措施后，制粉系统节约电能潜力可达 5%~15%。磨煤机设备和操作优化对提高制粉出力和节能尤为重要。主排风机调速形式有多种，应根据原煤供应情况、生产管理要求以及资金和现场条件等方面综合考虑后确定。

关键词： 高炉喷煤；制粉系统；节能；正压；风机调速

Discussion on Energy Saving of Pulverized Coal Preparation System for Blast Furnace Ironmaking

TANG Chuxiong

(WISDRI Iron & Steel Engineering & Research Incorporation Limited, Wuhan 430223, China)

Abstract: Various aspects of energy-saving in blast furnace coal powder injection system were discussed. The new process of positive pressure feeding flue gas at the inlet of coal mill and the technical measures to reduce the system air leakage and resistance loss are emphatically described. After taking effective measures to reduce air leakage and resistance loss of the system comprehensively and systematically, the power saving in potentially of the pulverizing system can reach 5%~15%. The optimization of coal mill equipment and operation is particularly important for improving coal pulverization output and energy conservation. There are various forms of speed regulation for the main exhaust fan, which should be determined based on comprehensive consideration of raw coal supply, production management requirements, funding, and on-site conditions.

Key words: blast furnace coal injection; pulverizing system; energy saving; positive pressure; fan speed regulation

高炉喷吹煤结构优化方法研究与应用

宋灿阳[1]，朱浩民[2]，张晓萍[1]，李　杰[1]，王思维[1]

（1. 马鞍山钢铁股份有限公司技术中心，安徽马鞍山　243000；
2. 重庆大学-辛辛那提大学联合学院，重庆　400030）

摘　要： 本文开发的高炉喷煤结构优化方法将喷吹煤的性价比、喷煤比与喷煤结构相结合起来，采用成本模式、热量模式和混合模式测算相应的高炉喷煤结构，为高炉喷吹煤采购和配煤结构优化提供指导。该方法 2020 年起在某企业实施应用，根据喷吹煤资源条件，按月度测算合理的配煤结构，指导高炉实际生产的喷煤，至 2020 年 12 月份实际烟煤喷吹比例提高至 49.60%，与模型测算值 49.24%相近，较 2019 年提升 5.34 个百分点，有效地降低了炼铁燃料成本。

关键词： 高炉；喷煤；配煤结构；优化方法；研究；应用

Research and Application of Optimization Method of Blast Furnace Coal Injection Structure

SONG Canyang[1], ZHU Haomin[2], ZHANG Xiaoping[1], LI Jie[1], WANG Siwei[1]

(1. Technology Center, Maanshan Iron and Steel Co., Ltd., Maanshan 243000, China;
2. Chongqing University-University of Cincinnati Joint Co-op Institute, Chongqing 400030, China)

Abstract: The optimization method of blast furnace coal injection structure developed in this paper combines the cost performance ratio, coal ratio and injection structure. The cost model, heat model and mixed model are used to calculate the corresponding coal injection structure, which provides guidance for coal procurement and structure optimization. The method has been applied in an enterprise since 2020. According to the injection coal resource conditions, a reasonable coal blending structure is calculated monthly. The actual bituminous coal injection proportion increased to 49.60% in December 2020, which is close to the model calculation value of 49.24%, about 5.34 percentage points higher than that in 2019, effectively reducing the cost of ironmaking fuel.

Key words: blast furnace; coal injection; coal blending structure; optimization method; research; application

宝武多基地高炉使用钢铁循环材料生产实践

王　臣[1]，朱仁良[2]，朱　彤[2]

（1. 宝山钢铁股份有限公司炼铁厂，上海　200941；2. 宝武中央研究院，上海　200941）

摘　要： 随着国家"能耗双控"和"碳排放双控"工作的持续推进，降低钢铁生产化石燃料消耗和 CO_2 排放成为钢铁企业实现可持续发展亟待解决的问题。钢铁循环材料作为一种重要的高载能绿色铁素资源，是唯一可以替代铁矿石的炼钢原料，加强钢铁循环材料综合利用是钢铁行业转型升级、绿色低碳发展的重要措施。目前钢铁循环材料主要应用于电炉和转炉炼钢工序，高炉能否添加部分钢铁循环材料也引起了钢铁生产者的广泛关注和讨论。本文结合宝武集团多基地高炉使用钢铁循环材料的工业试验和生产实践情况，初步探讨了不同钢铁循环材料资源的特点，分析了高炉添加钢铁循环材料对产量和燃料消耗的影响，总结提出了高炉使用钢铁循环材料冶炼需要注意的几个关键问题。

关键词： 高炉炼铁；钢铁循环材料；强化冶炼；低碳

城市碳氢固废高效粒化及粒化产物用于高炉喷吹研究

王广，刘思贤，聂志睿，乔宇峰，王静松，薛庆国

（北京科技大学钢铁冶金新技术国家重点实验室，北京 100083）

摘 要：随着中国经济快速发展，城市生活垃圾产生量巨增，"垃圾围城"现象日益严重。据测算，2亿吨生活垃圾所蕴含的能量约为5000万吨标准煤[1]。随着工业化、城市化进程的加快，我国城市固废的产生量上呈现不断增加趋势，且城市固废种类多样，增加了后期的处理难度[2]。城市固废处理技术主要有（1）填埋法；（2）热转化方法，包括焚烧、热解和气化[3]；（3）冶金过程协同处置城市固废，包括焦化法、高炉喷吹、电炉法以及直接还原法。目前研究较多的是废塑料应用于高炉喷吹的技术，但废塑料的资源量和处理规模有限，而我国城市生活垃圾分类后的其他垃圾产生量大，容易获得，且热值较高，喷入高炉后可取代煤粉和焦炭。其他垃圾中的有机组分（即城市碳氢固废）成分复杂，直接对城市碳氢固废进行破碎难度大、效率低，直接破碎所得到的颗粒都偏大，不适合高炉喷吹。因此，寻求低成本、高效率的破碎工艺使城市碳氢固废粒度达到高炉喷吹的要求成为重点。本文以城市固废中的代表性组分，如PVC、PE、秸秆为原料，将其混合物在不同温度下加热，然后对热处理产物进行破碎，分析热处理产物的粒化性能，最后对粒化产物的热值、粉体物性和燃烧性能进行研究。

从240℃到390℃，热处理产物的颜色逐渐加深，体积逐渐减小，物料结合更加紧密。图1为不同温度下热处理产物形貌。将各温度下的热处理产物经破碎筛分后，通过分析其破碎后的粒度组成，对其破碎性能进行了表征。图2为不同温度下粒化产物的粒度分布情况，得到最佳的热处理温度为330℃。城市碳氢固废粒化产物的工业分析和元素分析如表1所示，爆炸性和着火点如表2所示。

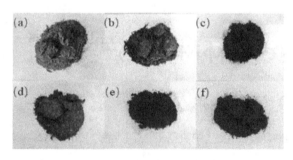

图1 不同温度热处理产物形貌
(a) 240℃; (b) 270℃; (c) 300℃;
(d) 330℃; (e) 360℃; (f) 390℃

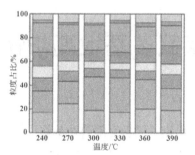

图2 热处理产物破碎后粒度组成

表1 粒化产物的工业分析和元素分析（%）

工业分析			元素分析			
Vd	Ad	FCd	C	H	O	N
75.81	7.58	16.34	65.7	9.70	8.26	0.38

表2 粒化产物的爆炸性和着火点

样品名称	着火点/℃	爆炸火焰长度/mm	挥发分/%
粒化产物	335	>600	75.81

参 考 文 献

[1] 王广，张宏强. 第十三届中国钢铁年会论文集[C]//冶金工业出版社, 2022: 6.
[2] 刘敬武. 中国资源综合利用, 2019, 37(2): 107-109.
[3] Alperen T, Emrah Ö, Ayşegül A. J Renewable and Sustainable Energy Reviews. Journal of Engineering, 2016, 54: 809-815.

鞍钢 11 号高炉风口破损分析

周正义，张立朝，姜庆喜，唐安平，潘世奇

（鞍钢股份有限公司炼铁总厂，辽宁鞍山　114021）

摘　要：对鞍钢 11 号高炉风口寿命及风口破损形式进行了统计研究，重点分析造成风口前端烧损的若干因素，并简要提出改制措施。统计分析表明，鞍钢 11 号高炉风口主要破损形式是前端烧损，造成烧损的主要原因是炉缸活跃状态、有害元素在风口的富集、漏水以及炉前渣铁排放等因素。

关键词：高炉；风口；熔损；炉缸活跃性；碱负荷

Analysis on Damage of No.11 Blast Furnace Tuyere in Ansteel

ZHOU Zhengyi, ZHANG Lichao, JIANG Qingxi, TANG Anping, PAN Shiqi

(Ansteel Iron and Steel Group Co., Ltd., General Ironmaking Plant, Anshan 114021, China)

Abstract: A statistical study was carried out on the life and damage of the tuyere of No.11 blast furnace in Ansteel, with a focus on analyzing several factors that cause burning damage to the front end of the tuyere, and briefly proposing improvement measures. The statistical analysis shows that the main form of damage to the tuyere of No.11 blast furnace in Ansteell is burning damage to the front end. The main reasons for the burning damage are the active state of the furnace body, the accumulation of harmful elements in the tuyere, water leakage, and the discharge of slag iron in front of the furnace.

Key words: blast furnace; tuyere; melting damage; furnace body activity; alkali load

鞍钢 11 号高炉长期闷炉后快速开炉达产实践

周正义，张立朝，姜庆喜，唐安平，潘世奇

（鞍钢股份有限公司炼铁总厂，辽宁鞍山　114021）

摘　要：系统总结了鞍钢 11 号高炉封炉、开炉的实践经验。鞍钢 11 号高炉本次封炉长达 23 天，通过制定详细封炉、开炉方案，实现了安全休风、合理封炉、顺利开炉的目标，并于开炉 3 天后快速达产。

关键词：高炉；闷炉；封炉；开炉；达产

Rapid Blowing-in Production Practice after Long-term Smouldering in No. 11 Blast Furnace of Ansteel

ZHOU Zhengyi, ZHANG Lichao, JIANG Qingxi, TANG Anping, PAN Shiqi

(Ansteel Iron and Steel Group Co., Ltd., General Ironmaking Plant, Anshan 114021, China)

Abstract: This paper systematically summarizes the practical experience of blowing-out and blowing-in No. 11 blast furnace of Ansteel. This time, the No. 11 blast furnace of Ansteel was sealed for 23 days. By formulating a detailed sealing and starting plan, the goal of safe rest, reasonable closure and smooth blow-in was achieved, and rapid production was achieved three days after the blow-in.

Key words: blast furnace; damping down; blow-out; blow-in; production achievement

基于风口图像的高炉状态智能诊断技术及应用

李 强，王子佳，邹宗树

（东北大学冶金学院，辽宁沈阳 110819）

摘 要：《中国制造 2025》明确提出加快钢铁冶金行业智能转型和绿色发展，推动钢铁冶金行业智能制造。高炉炼铁过程存在多相物质共存，同时发生着许多复杂的物理化学现象，生产时具有高温、高压、粉尘、密闭等工艺复杂、条件多变的特点，采用传统的模型方法难以建模和缺少本构关系。但随着大数据驱动的人工智能技术的不断发展，机器/深度学习算法的能力不断增强，使得构建基于大数据驱动的智能模型在高炉过程异常诊断和关键参数预测方面成为可能[1-2]。基于以上，本文围绕着实现炼铁高炉的智能化方向，探索了基于现代机器学习算法结合计算视觉技术的深度卷积神经网络模型，实时智慧感知高炉"心脏"——风口回旋区状态。其核心思想是通过从风口监控系统获取的风口图像，构建风口图像数据库，建立 TIDL-Net（Tuyere Image Deep-Learning Network）模型提取高炉冶炼过程特征，最终利用训练好 TIDL-Net 的模型，基于高炉实时或离线获得的风口监控数据，进行高炉实时在线炉况诊断和监测。

图 1 给出了本研究提出的模型框架和实现思路[3,4]。本研究开发的风口图像深度卷积神经网络（TIDL-Net）的基础架构如图 2 所示。

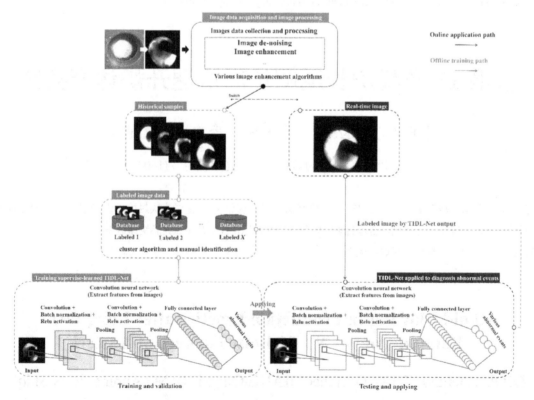

图 1 提出的炼铁高炉异常炉况诊断框架

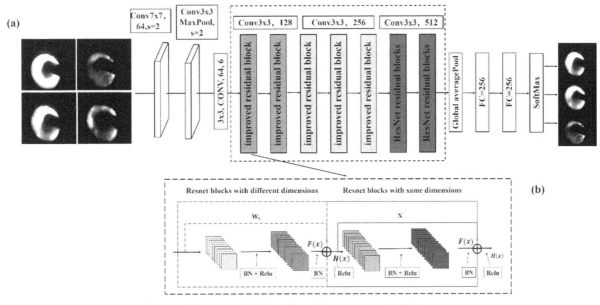

图 2 风口图像深度学习网络 TIDL-Net 示意图
(a) 模型架构示意图；(b) 改进的残差块

该路线图包括通过标记历史数据，驱动深度卷积神经网络监督学习图像内炉况事件特征，从而建立起风口图像深度学习网络（TIDL-Net）模型。模型建立和应用过程可以分为离线训练阶段和基于实时即时数据诊断应用阶段。应用结果表明，TIDL-Net 可成功诊断高炉炉况，训练集和测试集的模型准确率分别为 98% 和 88.58%。此外，将模型性能与经典的深度学习网络进行了比较，包括 AlexNet8、VGG16、Inception-Net10 和 ResNet18，表明 TIDL-Net 易于收敛且具有更高的准确性。经过充分训练的 TIDL-Net 模型建立起风口图像特征和炉况异常事件之间的映射或者高维函数关系，即该模型具有了从标记图像中学习到的特征映射关系后，该模型可在线应用于基于风口监控系统的实时获得的图像上，进行工业高炉的实时智能感知、诊断和监测。

参 考 文 献

[1] 杨尚宝, 杨天钧. 基于神经网络的高炉异常炉况判断专家系统[J]. 北京科技大学学报, 1994, 6: 517-521.
[2] 王帅, 李强. 联合大数据和神经网络的高炉透气性预报模型[J]. 钢铁, 2023, 58(7): 1-10.
[3] 李强, 王子佳, 王帅, 等. 基于风口信息深度学习的监督及预测高炉炉况异常的方法. CN202110187003.0[P]. 2021-05-18.
[4] Li Q, Wang Z J, Wang S, et al. A deep learning-based diagnosis model driven by tuyere images big data for iron-making blast furnaces[J]. Steel Res. Int., 2022, 93: 202100826.

高炉定量化综合操作技术集成创新（上篇）

林安川[1,2]，邱贵宝[1]，刘晓兰[3]，蒋玉波[3]

（1. 重庆大学材料科学与工程学院，重庆 400044；2. 昆明工业职业技术学院，云南昆明 650302；3. 武钢集团昆明钢铁股份有限公司炼铁厂，云南昆明 650302）

摘 要：高炉冶炼过程中操作水平的提升对一定冶炼条件下改善指标具有积极作用。基于炼铁基础理论和生产实践，辅以信息化手段，对原燃料品种及质量管理、高炉炉型管控、有害元素平衡、控硅控料、渣铁成分预测预控、开炉操作及操作制度的互联性等进行量化分析及集成创新，并进行了相关操作范围的定量界定及操作模型的开发，得到在冶炼过程中参数制定与指标控制相关联复杂关系的定量化处理及效果评价模式，得出具体冶炼条件下改善指标的控制措施和新的操作方法，为搞好高炉生产，持续改善指标奠定基础。本文对高炉开炉、日常冶炼过程中矩阵调剂、评价及煤气流在线诊断相关技术及操作模型的建立、实现、效果进行了概述。

关键词：高炉；综合操作；开炉；矩阵评价；煤气流诊断；模型研发；指标改善

Integrated Innovation of Quantitative Integrated Operation Technology of Blast Furnace（Ⅰ）

LIN Anchuan[1,2], QIU Guibao[1], LIU Xiaolan[3], JIANG Yubo[3]

(1. College of Materials Science and Engineering, Chongqing University, Chongqing 400044, China; 2. Kunming Industrial Vocational and Technical College, Kunming 650302, China; 3. Smelting Plant, Kunming Steel & Iron Co., Ltd., China Wuhan Steel Group Co., Ltd., Kunming 650302, China)

Abstract: The improvement of operation level in the process of blast furnace smelting has a positive effect on improving the index under certain smelting conditions. Based on the basic theory and production practice of ironmaking. supplemented by information means, quantitative analysis and integrated innovation are carried out on raw fuel varieties and quality management, blast furnace type control, harmful element balance, silicon control and material control, slag and iron composition prediction and pre-control, furnace opening operation and interconnection of operation system, etc.And carried out the quantitative definition of the relevant operation range and the development of the operation model.Obtained the quantitative treatment and effect evaluation model of the complex relationship between parameter formulation and index control in the smelting process, and the control measures and new operation methods to improve the index under specific smelting conditions. In this paper, Summary the establishment, realization and effect of the relevant technology and operation model of adjustment, evaluation and on-line diagnosis of gas flow in BF's opening and daily smelting.

Key words: blast furnace; comprehensive operation; run the furnace; matrix evaluation; gas flow diagnosis; model development; index improvement

长流程下全氧富氢碳循环低碳炼铁新工艺的可行性分析

李海峰[1,2]，陈靖然[1]，王小艾[3]

（1. 东北大学多金属共生矿生态化冶金教育部重点实验室，辽宁沈阳 110819；2. 东北大学冶金学院，辽宁沈阳 110819；3. 河钢集团有限公司材料技术研究院，河北石家庄 050000）

摘　要：目前高炉炼铁工艺仍为我国主要炼铁流程。随着"双碳"目标时间的临近，钢铁行业亟需绿色低碳转型，国内外均采用氢冶金逐步替代碳冶金的技术路线，以富氧高炉与氢基竖炉为主，但考虑现有高炉存量的惯性，针对中短期降碳目标对高炉工艺进行改革为最佳方案。

经过多年研究，通过高炉下部风口进行富氢喷吹操作的燃料比例仍较低，即使使用全氧高炉通过顶煤气循环技术大幅降低了碳排放，但随之产生的"上冷"和"下热"问题仍无法打破富氢喷吹的上限。作者通过重构工艺方案，提出一种新的全氧富氢低碳还原熔化炉炼铁工艺，打破了传统高炉上部金属化率无法突破70%以上的工艺缺陷，并最终达到至少降低40%的CO_2排放目标。与氢基竖炉的产物固态金属化球团相比，低碳新工艺的冶炼产物仍为铁水，与后续传统流程匹配性更好，显著降低了企业的投资改造成本。

作者基于质量平衡和热平衡原理，开发了一套全氧富氢还原熔化炉炼铁工艺质能分配模型。模型采用局部区域平衡法，上部区域考虑CO、H_2逐级还原铁氧化物及水煤气反应，下部熔化区考虑铁水中元素直接还原反应、补充焦炭燃烧放热反应、成渣反应及碳酸盐分解反应。模型以冶炼1t铁水为基础，以炉身喷吹1000℃还原气CO：H_2=3：7和上部金属化率为90%等为边界条件，以物料平衡、热平衡、碱度平衡及还原势为约束条件，得到低碳炼铁新工艺的物质流分布信息。传统高炉的炉腹煤气量（标态）一般在1400m^3/tHM左右，本工艺中炉身喷吹量（标态）约1000m^3/tHM，底部风口产生的煤气量（标态）约415m^3/tHM，这一变化目的是促进间接还原的发展并能降低下部直接还原的耗热。作者从全炉和高温区分别进行热平衡计算，证明工艺的可行性。全炉平衡来看，碳素氧化热、炉身煤气带入热、氢素燃烧热分别占全炉热收入的58.365%、14.931%、26.336%，全炉热盈余占10.56%；高温区直接还原耗热为390.15MJ/tHM，与传统高炉相比降幅57.8%，这是本工艺节省燃料比的根本原因，高温区热盈余为846.94 MJ/tHM，占比30.53%，仍满足下部渣铁熔化和直接还原所需的热量，炉腹煤气产生热虽较小，但炉身的高温喷吹会弥补这一损失。详细数据可见文献[1]中的表3和表4及相关专利[2]。

作者从限制高炉减排关键参数入手，提出一种全氧富氢低碳还原炼铁的新工艺，通过重新设计炉型、炉身风口喷吹等技术升级实现了上部金属化率达到90%以上的需求，结合顶煤气脱除CO_2等技术实现煤气自循环，实现了与传统长流程相比达到了CO_2排放量降幅44%~69%的目标。

参 考 文 献

[1] 李海峰，陈靖然，王新东，等. 面向长流程的富氢低碳炼铁技术路径分析[J/OL]. 钢铁，2023: 1-14. DOI:10.13228/j.boyuan.issn0449-749x.20230102.

[2] 于勇，王新东，罗志国，等. 全氧富氢低碳还原熔化炼铁系统及炼铁方法. 中国，ZL200210084753.X[P]. 2022-05-26 [2023-01-31].

高炉冶炼喷吹天然气经济性探讨

魏钦帅，罗 凯，潘洪福

（中冶东方工程技术有限公司炼铁与原料事业部，山东青岛 266000）

摘 要：本文从理论计算出发，结合高炉冶炼生产实际，首先对高炉喷吹天然气在生产技术上的可行性进行分析，并拟合出天然气喷吹、煤粉喷吹、富氧等因素变化对风口理论燃烧温度的影响关系式；再利用生产数据验证拟合公式，并对高炉喷吹天然气进行经济分析；本文所得技术和经济分析结论，可为高炉冶炼喷吹提供技术和经济性参考。

关键词：高炉；喷吹天然气；喷煤；煤比；富氧

Discussion on the Economy of Natural Gas Injection in Blast Furnace

WEI Qinshuai, LUO Kai, PAN Hongfu

(Beris Engineering and Research Corporation, Qingdao 266000, China)

Abstract: In this paper, based on the theoretical calculation and the actual production of blast furnace smelting, the feasibility of natural gas injection in blast furnace production technology is analyzed, and the relationship between the changes of natural gas injection, pulverized coal injection, oxygen enrichment and other factors on the theoretical combustion temperature of tuyere is fitted.The fitting formula is verified by the production data, and the economic analysis of gas injection in blast furnace is carried out. The technical and economic analysis results obtained in this paper can provide technical and economic reference for blast furnace smelting injection.

Key words: blast furnace; natural gas injection; coal injection; coal ratio; oxygen-enrichment rate

越南和发榕橘二期绿色智能环保原料场工艺设计及特点

王亚伟，陈尚伦，艾　宇，程　浩，刘　怡，孙贵舟，刘占稳

（中冶赛迪工程技术股份有限公司，重庆　401122）

摘　要： 越南和发榕橘二期原料场主要包含码头输入系统、料场本体系统（含矿石料场和煤场）、混匀系统、熔剂系统以及供料系统等。采用中冶赛迪自主知识产权的上跨式 C 型原料场技术和智能流程管控系统技术，并综合采用大跨度多料条工艺布置技术、超大转运站集中布置技术、数字化和智能化控制系统，实现了二期原料场的在线盘库、智能混匀配料和智能生产应用管理等功能要求。

关键词： C 型原料场；大跨度；数字化；智能化

Process Design and Characteristics of Green Intelligent & Environmental Protection Raw Material Yard of Vietnam's HoaPhat Dung Quat Steel JSC of Phase II

WANG Yawei, CHEN Shanglun, AI Yu, CHENG Hao,
LIU Yi, SUN Guizhou, LIU Zhanwen

(CISDI Engineering Co., Ltd., Chongqing 401122, China)

Abstract: Raw Material Yard of Vietnam's HoaPhat Dung Quat Steel JSC of Phase II includes the dock input system, the material yard system (including ore and coal yards), mixing system, flux system, and supply system. The overhead C-type raw material yard technology and intelligent process control system technology with independent intellectual property rights of CISDI are adopted, and the large-span multi strip process layout technology, super large Transport hub centralized layout technology, digital and intelligent control system are comprehensively adopted to realize the function requirements of the online inventory, intelligent mixing and intelligent production application management of the Phase II raw material yard.

Key words: C-type raw material yard; large span; digitization; intelligence

蓄铁式主沟结构参数优化模拟研究

董晓森，郑　魁

（钒钛资源综合利用国家重点实验室，攀钢集团研究院有限公司，四川攀枝花　617000）

摘　要： 采用数值模拟的方法对攀钢某 1750m³ 高炉配套的蓄铁式主沟进行研究，通过建立主沟流场的流体动力学

模型，从不同角度模拟研究不同结构参数下渣铁分布。结果表明：铁口喷射而出的渣、铁射流在主沟液面的冲击点保持一致，且出铁初期流场波动剧烈；主沟加宽、加长、加深 1/3 或沟底坡度降至 0%后，初期渣层厚度相对减薄，高速流区速度和范围相对降低，流场得到充分发展，且渣中带 Fe 的体积分数最低可控制到 0.32%以下，降幅达 92%，促进了渣铁分离；主沟移动渣口到撇渣器挡墙附近后，虽能降低渣中带铁，但在出铁后期可能出现铁中混渣。

关键词：蓄铁式；主沟；结构优化；数值模拟

Simulation Study on Optimization of Structural Parameters of the Main Channel of Iron Storage Type

DONG Xiaosen, ZHENG Kui

(State Key Laboratory of Vanadium and Titanium Resources Comprehensive Utilization,
Pangang Group Research Institute Co., Ltd., Panzhihua 617000, China)

Abstract: The numerical simulation method was used to study the iron storage type main ditch supporting a 1750m³ blast furnace at Pangang. By establishing a fluid dynamics model of the main ditch flow field, the distribution of slag and iron under different structural parameters was simulated and studied from different perspectives. The results show that the slag and iron jet ejected from the iron mouth remain consistent at the impact point of the main ditch liquid surface, and the flow field fluctuates violently in the early stage of tapping; After the main ditch is widened, lengthened, or deepened by 1/3, or the slope at the bottom of the ditch is reduced to 0%, the initial thickness of the slag layer is relatively thin, and the speed and range of the high-speed flow zone are relatively reduced. The flow field is fully developed, and the minimum volume fraction of Fe in the slag can be controlled below 0.32%, with a decrease of 92%, promoting the separation of slag and iron; After moving the slag mouth of the main ditch to the vicinity of the slag skimmer retaining wall, although it can reduce the iron content in the slag, there may be mixed slag in the iron in the later stage of tapping.

Key words: iron storage type; main channel; structural optimization; numerical simulation

沙钢 2500m³ 高炉用激光料面探测仪研究料面下降速度分布曲线

谭天雷[1]，黄 军[1]，杜友尧[1]，高征铠[2]，高 泰[3]

（1. 江苏省沙钢集团有限公司炼铁厂，江苏张家港 215531；
2. 北京科技大学冶金与生态工程学院，北京 100083；
3. 北京神网创新科技有限公司，北京 100083）

摘 要：2012 年 6 月沙钢在 5800m³ 高炉开始使用在线激光料面探测系统，2019 年 8 月在 2500m³ 高炉使用在线激光料面探测系统，取得了很好的使用效果。鉴于在线激光料面探测系统对高炉生产提供的帮助，现在沙钢 5800m³ 高炉和 4 座 2500m³ 高炉均已应用该料面探测系统。本文主要介绍沙钢 2500m³ 1 号高炉使用在线激光料面探测系统，通过测量同一批炉料不同时刻的料面形状，对两次料面形状的数据进行分析计算，得到料面不同位置下降速度的数值和分布曲线，为高炉布料操作提供精准实时数据，为调整装料制度，改善高炉操作提供了数据支持。

关键词：高炉；激光探测；料面形状；下降速度

Study on the Distribution Curve of Burden Surface Profile Falling Speed Using the Blast Furnace Laser Burden Surface Detector for a 2500m³ Blast Furnace at Shagang

TAN Tianlei[1], HUANG Jun[1], DU Yourao[1], GAO Zhengkai[2], GAO Tai[3]

(1. Jiangsu Shagang Group Co., Ltd., Iron-making Plant, Zhangjiagang 215531, China;
2. University of Science and Technology Beijing, Beijing 100083, China;
3. Shenwang Pioneer Tech. Corporation Beijing, Beijing 100083, China)

Abstract: In June 2012, Shagang began using the Blast Furnace Laser Burden Surface Detector at 5800 m³ blast furnace, and in August 2019, 2500 m³ blast furnace began using the Blast Furnace Laser Burden Surface Detector. Due to the assistance provided by the burden surface profile data for blast furnace production, the Blast Furnace Laser Burden Surface Detector has been applied to both the 5800 m³ blast furnace and four 2500 m³ blast furnaces at Shagang. By measuring the burden surface profile of the same charge material at different times, analyzed and calculated the two times data. and obtained the numerical values and distribution curves of the falling speed at different positions of the burden surface. This provides accurate real-time data for blast furnace material charging operation, and data support for adjusting the material charging and improving blast furnace operation.

Key words: blast furnace; laser measurement; burden surface profile; falling speed

高球团矿配比条件下炉喉矿焦比变化规律及其优化

王 斌[1]，王 炜[1]，郑 魁[1,2]，陈世锦[1]，陈绪亨[1]，赵立辉[1]

（1. 武汉科技大学材料与冶金学院，湖北武汉 430080；
2. 攀钢集团研究院有限公司，四川攀枝花 617000）

摘 要：本文基于1∶3的炉喉截面物理模型，通过布料实验研究了球团矿配比、溜槽倾角、矿石批重对中心矿焦比指数的影响，结果表明：基于响应曲面法构建的多元回归模型，发现溜槽倾角对中心矿焦比指数的影响程度最大，其次是球团矿配比和矿石批重；高炉采用高比例球团矿冶炼时，为了获得较小的中心矿焦比指数，发展中心气流，应适当增加矿石批重，在布料时应采用较大的溜槽倾角以减少球团矿向高炉中心的滚动。

关键词：高炉布料；球团矿配比；矿石批重；溜槽倾角；中心矿焦比指数

The Variation Law of Furnace Throat Ore/Coke under High Pellet Charging Ratio and Its Optimization

WANG Bin[1], WANG Wei[1], ZHENG Kui[1,2], CHEN Shijin[1], CHEN Xuheng[1], ZHAO Lihui[1]

(1. School of Materials and Metallurgy, Wuhan University of Science and Technology, Wuhan 430080, China; 2. Pangang Group Research Institute Co., Ltd., Panzhihua 617000, China)

Abstract: Based on the 1∶3 physical model of furnace throat cross-section, this paper investigates the influence of pellet ratio, chute angle and batch weight on central ore/coke index through fabric experiments, and the results show that: based on the multiple regression model constructed by response surface method, it is found that the chute angle has the greatest influence on central ore/coke index, followed by pellet ratio and batch weight; when blast furnace adopts high proportion of pellet smelting, in order to obtain a smaller central ore/coke index and develop central airflow, the batch weight should be increased appropriately, and a larger chute angle should be used to reduce the rolling of pellet to the center of the blast furnace.

Key words: blast furnace charging; pellet ratio; batch weight; chute angle; central ore/coke index

高炉铜冷却壁渣皮微观结构及导热性能分析

陈世锦[1,2]，王炜[1,2]，陈绪亨[1,2]，王斌[1,2]，宋大为[1,2]

（1. 湖北省冶金二次资源工程技术研究中心，湖北武汉　430081；
2. 武汉科技大学省部共建耐火材料与冶金国家重点实验室，湖北武汉　430081）

摘　要：高炉的强化冶炼导致高炉铜冷却壁过早破损，大量高炉铜冷却壁的寿命仅7~10年，其成为影响高炉寿命的限制性环节之一，高炉铜冷却壁热面形成的渣皮是保障冷却壁寿命的关键。目前对高炉铜冷却壁热面渣皮的物相组成及传热性能研究较少，明确高炉铜冷却壁渣皮的物相组成、微观结构特征，揭示渣皮形成过程，为高炉铜冷却壁形成稳定的渣皮提供理论支撑；获得实际高炉铜冷却壁渣皮的导热系数，修正高炉炉腹炉腰传热模型的关键参数，能更加准确的得到铜冷却壁渣皮厚度，为高炉炉型管理提供操作依据。

本文基于国内某高炉炉身中下部破损调查，以炉腹渣皮为研究对象，通过化学分析、XRD、SEM-EDS、光学数码显微镜等表征方法，对铜冷却壁渣皮的物相组成、微观结构、粘度和导热系数等进行分析，结果表明：渣皮中主要物相为黄长石、尖晶石；渣皮冷面较渣皮热面而言，含有较多的C和Fe，渣皮热面的孔隙呈独立分布状态，而渣皮冷面的孔隙呈相互连通状态，其孔隙率大于渣皮热面；渣皮热面到冷面的微观结构具有明显的分层结构，沿渣皮热面到冷面尖晶石相的数量逐渐减少，且结构由针状逐渐变成柱状，在靠近渣皮冷面时以细颗粒状为主。渣皮黏度较大，在1550℃时黏度就已高达接近1000mPa·s，在温度降低至1470℃时其黏度已超过3000mPa·s，熔化性温度约为1525℃。特别的，黏度实验冷却试样的底部发现有明显的金属铜，表明铜冷却壁上的部分铜进入渣皮内部。通过激光导热系数测量仪得到渣皮的导热系数在1.6~1.8W/(m·K)之间，平均导热系数为1.646W/(m·K)。

关键词：铜冷却壁；渣皮；物相组成；微观结构；导热系数

Microstructure and Thermal Conductivity Analysis of Blast Furnace Copper Stave Slag Crust

CHEN Shijin[1,2], WANG Wei[1,2], CHEN Xuheng[1,2], WANG Bin[1,2], SONG Dawei[1,2]

(1. Hubei Provincial Engineering Technology Research Center of Metallurgical Secondary Resources, Wuhan 430081, China; 2. State Key Laboratory of Refractories and Metallurgy, Wuhan University of Science and Technology, Wuhan 430081, China)

Abstract: The enhanced smelting of the blast furnace leads to premature breakage of the blast furnace copper stave, the life of a large number of blast furnace copper stave is only 7~10 years, it has become one of the limiting links affecting the life of the blast furnace, the formation of slag crust on the hot side of the blast furnace copper stave is the key to ensure the life of the cooling wall. At present, there is less research on the physical composition and heat transfer performance of the slag crust on the hot surface of the blast furnace copper stave, to clarify the physical composition and microstructural characteristics of the slag crust of the blast furnace copper stave, to reveal the formation process of the slag crust, to provide theoretical support for the formation of a stable slag crust for the blast furnace copper stave; to obtain the actual thermal conductivity of the slag crust of the blast furnace copper stave, to correct the key parameters of the heat transfer model of the blast furnace belly furnace waist, to more accurately obtain the slag crust thickness of the copper stave skin thickness, to provide operational basis for blast furnace management.

In this paper, based on the investigation of the damage in the middle and lower part of the furnace body of a domestic blast furnace, the slag crust of the furnace belly was used as the research object, and the physical composition, microstructure, viscosity and thermal conductivity of the slag crust of the copper stave were analysed by chemical analysis, XRD, SEM-EDS, optical digital microscope and other characterisation methods, and the results showed that: the main physical phases in the slag crust are yellow feldspar and spinel; the cold side of the slag crust contains more C and Fe than the hot side of the slag crust C and Fe, the pores on the hot side of the slag crust are independently distributed, while the pores on the cold side of the slag crust are interconnected, and its porosity is greater than that on the hot side of the slag crust; the microstructure from the hot side to the cold side of the slag crust has an obvious layered structure, the number of spinel phases gradually decreases along the hot side to the cold side of the slag crust, and the structure gradually changes from needle-like to columnar, and is mainly fine-grained near the cold side of the slag crust. The viscosity of the slag crust is large, at 1550℃ the viscosity is already up to nearly 1000mPa·s, at a reduced temperature of 1470℃ its viscosity is already more than 3000mPa·s, the melting temperature is about 1525℃. In particular, visible metallic copper was found at the bottom of the viscosity experimentally cooled specimen, indicating that some of the copper from the copper stave had entered the interior of the slag crust. The thermal conductivity of the slag crusts was obtained by laser thermal conductivity measurement between 1.6 and 1.8W/(m·K), with an average thermal conductivity of 1.646W/(m·K).

Key words: copper stave; slag crust; phase structure; microstructure; thermal conductivity

高炉定量化综合操作技术集成创新（下篇）

林安川[1,2]，刘晓兰[3]，蒋玉波[3]

（1. 重庆大学材料科学与工程学院，重庆 400044；2. 昆明工业职业技术学院，云南昆明 650302；3. 武钢集团昆明钢铁股份有限公司炼铁厂，云南昆明 650302）

摘 要：基于高炉炼铁基础理论和生产实践，辅以过计算机信息化手段，对高炉生产过程中涉及的可量化技术及其互联性分析方法进行集成创新，进而得出具体条件下持续改善技术指标的冶炼控制措施和新的操作方法。本文对高炉生产日常冶炼过程中涉及大量原始数据的综合计算及解决各种冶炼参数协调统一发展问题的相关概念、原理、技术、方法进行了概述。包括原燃料经济性校核及品质界定、炉体炉型管控、有害元素平衡、控硅控料、渣铁成分预控等综合操作技术，这为高炉降低冶炼成本、持续改善产量、燃料比等主要技术经济指标奠定基础。

关键词：高炉；综合操作；原燃料评价；冶炼参数综合计算；控硅控料；指标改善

Integrated Innovation of Quantitative Integrated Operation Technology of Blast Furnace (II)

LIN Anchuan[1,2], LIU Xiaolan[3], JIANG Yubo[3]

(1. College of Materials Science and Engineering, Chongqing University, Chongqing 400044, China; 2. Kunming Industrial Vocational and Technical College, Kunming 650302, China; 3.Smelting Plant, Kunming Steel & Iron Co., Ltd., China Wuhan Steel Group Co., Ltd., Kunming 650302, China)

Abstract: Based on the basic theory and production practice of blast furnace ironmaking. supplemented by computer information means, Integrated innovation the quantifiable technology and its interconnection analysis method involved in the BF's production process. and then obtained the smelting control measures and new operation methods for continuous improvement of technical indicators under specific conditions. In this paper. the concepts, principles, techniques and methods involved in the comprehensive calculation of a large number of raw data in the daily smelting process of BF's production and the problem of the coordinated development of various smelting parameters are summarized.Including the original fuel economy verification and quality definition, furnace type control, harmful element balance, silicon control and material control, slag iron composition pre-control and other comprehensive operation technology, This lays a foundation for the reduction of smelting cost, continuous improvement of production, fuel ratio and other main technical and economic indicators of blast furnace.

Key words: blast furnace; integrated operation; raw fuel evaluation; comprehensive calculation of smelting parameters; control silicon control material's descent velocity; moving indicators

金属化炉料模拟高炉条件下的反应行为

闫瑞军[1,2,3]，储满生[1,2,3,4]，柳政根[1,2,3]，刘培军[1,2,3]

（1. 东北大学冶金学院，辽宁沈阳　110819；2. 辽宁省低碳钢铁前沿技术工程研究中心，辽宁沈阳　110819；3. 低碳钢铁前沿技术教育部工程研究中心，辽宁沈阳　110819；4. 重庆大学材料科学与工程学院，重庆　400045）

摘　要： 高炉使用金属化炉料作为一种减排方式已被广泛研究。金属化炉料能否被煤气中 CO_2 氧化成为被关注的问题。金属化炉料对综合炉料熔滴特性有重要影响。本文使用热分析仪研究金属化物料不同 CO_2/CO 气氛非等温反应行为。使用熔滴装置，研究金属化炉料对综合炉料熔滴特性影响。结果表明，在接近炉顶煤气利用率的 CO_2/CO 为 1∶1 气氛中，金属化物料在 720℃ 左右开始发生氧化反应，该温度远高于高炉炉顶温度。随炉料下降，温度升高但还原性气氛增强。所以金属化炉料在高炉中不存在再氧化问题。单烧结矿的熔化开始温度 T_S 和熔化结束温度 T_D 分别为 1283℃ 和 1503℃。随金属化炉料加入且金属化率由 73% 变为 92%，T_S 增加到 1300℃ 和 1316℃，T_D 降低为 1440℃ 和 1455℃。软熔带位置降低，厚度变窄。

关键词： 高炉；金属化炉料；反应行为；软化行为；熔化行为

Reaction Behaviors of Metallized Burden under Simulated Blast Furnace Conditions

YAN Ruijun[1,2,3], CHU Mansheng[1,2,3,4], LIU Zhenggen[1,2,3], LIU Peijun[1,2,3]

(1. School of Metallurgy, Northeastern University, Shenyang 110819, China; 2. Liaoning Low-carbon Steelmaking Technology Engineering Research Center, Shenyang 110819, China; 3. Engineering Research Center of Frontier Technologies for Low-carbon Steelmaking (Ministry of Education), Shenyang 110819, China; 4. College of Materials Science and Engineering, Chongqing University, Chongqing 400045, China)

Abstract: Using the metallized burden in blast furnaces as a means of reducing emissions has been widely studied. Whether the metallized burden can be oxidized by CO_2 has become a concern. Metallized burdens have a significant effect on the softening-melting behaviors of comprehensive burdens. This paper studies the non-isothermal reaction behavior of metallized burdens in different CO_2/CO atmospheres using a thermal analyzer. The effect of metallized burdens on the softening-melting behaviors of comprehensive burdens are studied using a softening-melting-dripping apparatus. The results show that in a CO_2/CO of 1∶1 atmosphere close to the utilization ratio of top gas, the metallized burden can be oxidized at around 720℃. This temperature is much higher than that of the blast furnace top. As the burden descents, the temperature increases but the reducing atmosphere increases. Thus the metallized burden cannot be oxidized in the blast furnace. The melting start temperature T_S and melting end temperature T_D of single sinter are 1283℃ and 1503℃, respectively. With the addition of metallized burdens and the metallization ratio increasing from 73% to 92%, T_S increases to 1300℃ and 1316℃, T_D decreases to 1440℃ and 1455℃. The position of the cohesive zone descents and the thickness narrows.

Key words: blast furnace; metallized burden; reaction behavior; softening behavior; melting behavior

武钢有限8号高炉更换冷却壁快速恢复实践

林巍，陈畏林

（武钢有限炼铁厂生产技术室，湖北武汉 430080）

摘 要：大型高炉采用不挖炉缸半装料的方式开炉的主要难点：（1）对炉缸热状态及所存半凝固渣铁混合物量的估算；（2）如何快速打通风口与铁口的通道，使新产生的渣铁混合物进入炉缸并顺利排出；8号高炉能实现快速恢复也是基于对上述难点予以充分的考虑：通过炉缸热电偶温度分析炉缸热状态，改进炉料结构，并结合恢复进度调整上料，为炉缸补充热量，同时精心设计开炉时风口布局以及开风口的顺序和节奏实现恢复过程中不出现反复。

关键词：风压；料线；风口；物理热

Fast Recovery Practice of Cooling Stave Replacement in No.8 Blast Furnace

LIN Wei, CHEN Weilin

(Production Technology Department, Ironmaking Plant, WISCO, Wuhan 430080, China)

Abstract: The main difficulties of opening large blast furnaces with half-charging method without digging hearth are as follows :(1)How to estimate the thermal state of hearth and the half-solidification stored (2) how to quickly play the channel between the vent and the iron mouth, so that the newly generated iron mixture enters the hearth and is discharged smoothly; The fast recovery of No.8BF is based on full consideration of the above-mentioned difficulties: through the hearth thermocouple temperature analysis hearth thermal state, improve the charge structure。In combination with the recovery schedule, the feeding is adjusted to supplement the heat for the hearth. At the same time, the layout of the tuyere and the sequence and rhythm of the tuyere are carefully designed to achieve recovery. There is no repetition in the complex process.

Key words: wind pressure; feed line; tuyere; physical heat

富氢低碳冶炼工况下高炉热量分配模型

王 刚,许 俊,贺 坤,邹忠平,赵运建,牛 群

(中冶赛迪集团有限公司低碳技术研究院,重庆 401122)

摘 要:高炉的富氢低碳冶炼是当前炼铁领域发展的关键技术方向。本文提出了一种研究高炉炉内热量分配的数学模型,该模型同时考虑了气固换热、渣铁融化及化学反应过程。模型被用于分析了某2300m³高炉的温度分布状态,并与其生产实绩数据对比验证了模型有效性。随后,研究了富氢介质喷吹量、H_2浓度及鼓风富氧率对炉内温度分布的影响。结果表明:富氢介质喷吹量的增加使直接还原度降低,导致炉内温度升高;而富氧率增加导致炉内温度降低,燃料比亦降低;富氢介质中H_2浓度的变化对炉内温度分布的影响较小;在喷吹富氢介质的同时配合提升富氧率,有助于将炉内温度分布调整至与常规高炉操作相同的水平。

关键词:高炉;低碳冶炼;炉温分布;理论计算方法

Heat Distribution Model under Hydrogen-rich Low-carbon Conditions in Blast Furnace

WANG Gang, XU Jun, HE Kun, ZOU Zhongping,
ZHAO Yunjian, NIU Qun

(CISDI Low-Carbon Blast Furnace Research Office, MCC Low-Carbon
Technology Research Institute, Chongqing 401122, China)

Abstract: Low carbon development of blast furnaces is one of the key technological directions in the current development of ironmaking. A heat distribution model was proposed to study the temperature distribution in a blast furnace, simultaneously considering gas-solid heat exchange, slag and iron melting, and chemical reactions. The model was used to analyze the temperature distribution of a 2300 m³ blast furnace and was verified via comparison with actual production data. Subsequently, the effects of the injection rate of hydrogen-rich media, H_2 concentration, and oxygen enrichment rate of the blast on the temperature distribution were investigated. Results indicated that the increase in the injection rate of the hydrogen-rich media decreased the amount of direct reduction and led to an increase in the furnace temperature. Furthermore, an increase in the oxygen enrichment rate led to a decrease in the furnace temperature, but could reduce the solid fuel ratio, while the change in H_2 concentration had less effect on the temperature distribution. The combination of hydrogen-rich media injection and the increase in the oxygen enrichment rate would help to adjust the temperature distribution to the same level as the conventional blast furnace conditions.

Key words: blast furnace; low-carbon condition; heat distribution; calculation model

首钢通钢2号高炉开炉快速达产实践

谢绍飞[1],杨晓辉[2],刘文明[2],于海新[2]

(1. 首钢通钢制造部,吉林通化 134003;2. 首钢通钢炼铁事业部,吉林通化 134003)

摘　要：首钢通钢 2 号高炉（2680m^3）于 2022 年 9 月 30 日降料面停炉检修，停产 171 天，2023 年 3 月 20 日组织点火开炉复产。本次开炉在前期准备、炉前出铁等方面，采用了一系列新技术。通过进行充分的准备工作、制定科学的开炉方案、选择合理的送风参数和装料制度，开炉送风后整体炉况恢复非常顺利。开炉后第 4 天高炉产量达到 5381t，利用系数 2.0t/(m^3·d)，实现了开炉后快速恢复、快速达产的目的。

关键词：大型高炉；开炉；快速达产

Practice of Rapid Production of Shougang Tonggang No. 2 Blast Furnace by Blowing in

XIE Shaofei[1], YANG Xiaohui[2], LIU Wenming[2], YU Haixin[2]

(1. Shougang Tonggang Manufacturing Department, Tonghua 134003, China;
2. Shougang Tonggang Ironmaking Business Unit, Tonghua 134003, China)

Abstract: The No. 2 blast furnace (2680m^3) of Shougang Tonggang was shut down for maintenance on the feeding surface on September 30, 2022, with production suspended for 171 days. On March 20, 2023, ignition and furnace opening were organized to resume production. A series of new technologies have been adopted in the early preparation and pre furnace tapping of this furnace. By conducting sufficient preparation work, developing a scientific blowing plan, selecting reasonable air supply parameters and charging system, the overall furnace condition is restored very smoothly after blowing. On the fourth day after the furnace was blown in, the high furnace output reached 5381 tons, with a utilization coefficient of 2.0t/(m^3·d), achieving the goal of rapid recovery and reaching production quickly after the furnace was blowed in.

Key words: large blast furnace; blow in the furnace; rapid production

武钢有限 6 高炉加废钢的生产实践

余珊珊，李勇波

（武汉钢铁有限公司，湖北武汉　430083）

摘　要：为了提高铁水产量，降低工序能耗，武钢有限 6 号高炉进行了高炉加废钢的工业试验。理论计算每加入 100kg/tFe 的废钢，理论产量的增加率应该为 6.39%，理论燃料消耗为 130kg/t。工业试验表明：高炉加废钢后，有利于高炉降低燃料消耗，提高产量，降低碳排放。

关键词：高炉；燃料比；废钢

Production Practice of Adding Scrap Steel to No.6 Blast Furnace in Wisco

YU Shanshan, LI Yongbo

(Wuhan Iron and Steel Co., Ltd., Wuhan 430083, China)

Abstract: In order to increase the production of molten iron and reduce the energy consumption of the process, the industrial test of adding scrap steel to the No. 6 blast furnace of Wuhan Iron and Steel Co., Ltd. was conducted. Theoretical calculation shows that for every 100kg/tFe of scrap added, the increase rate of theoretical production should be 6.39%, and the theoretical fuel consumption should be 130kg/t. Industrial experiments have shown that adding scrap steel to the blast furnace is beneficial for reducing fuel consumption, increasing production, and reducing carbon emissions.

Key words: blast furnace; fuel ratio; steel scrap

首钢通钢高炉减少风口小套破损技术研究

谢绍飞[1]，郑文书[1]，王德军[2]

（1. 首钢通钢制造部，吉林通化　134003；2. 首钢通钢炼铁事业部，吉林通化　134003）

摘　要：首钢通钢2号、3号高炉设计有效炉容均为2680m^3，分别于2017年、2018年中修完毕后开炉。2019~2021年两座高炉风口小套破损频繁，因换套休风时间年均长达60h。本文介绍了首钢通钢公司根据多年高炉操作经验及结合生产实际，经过2022年1~11月份对风口破损的综合治理，使两座高炉风口小套破损数量大大降低，高炉继续保持安全稳定生产。

关键词：风口破损；锌负荷

Reducing the Damage of Tuyere Small Sleeves in Shougang Tonggang Blast Furnace Technical Study

XIE Shaofei[1], ZHENG Wenshu[1], WANG Dejun[2]

(1. Shougang Tonggang Manufacturing Department, Tonghua 134003, China;
2. Shougang Tonggang Ironmaking Business Unit, Tonghua 134003, China)

Abstract: The effective blast furnace capacity of Shougang Tonggang No. 2 and No. 3 blast furnaces is designed to be 2680m^3, and they were opened after the completion of medium maintenance in 2017 and 2018, respectively. From 2019 to 2021, the small sleeves of the tuyeres of two blast furnaces were frequently damaged, and the average annual shutdown time for replacing the sleeves was 60 hours. This article introduces that Shougang Tonggang Company, based on years of experience in blast furnace operation and combined with production practice, has significantly reduced the number of damaged tuyere sleeves in two blast furnaces through comprehensive treatment of tuyere damage from January to November 2022, and the blast furnace continues to maintain safe and stable production.

Key words: damaged air outlet; zinc load

南钢铁区质量一体化智能管控模型的应用与实践

方　磊，诸　刚，刘永辉

（南京钢铁股份有限公司，江苏南京　210035）

摘 要：南钢基于"5G+工业互联网",融合数据采集、物料跟踪、质量管控等功能板块,覆盖全工序,搭建了行业领先的铁区一体化智能管控平台。本文重点介绍南钢铁区基于平台支撑实现质量一体化智能管控,实现铁区质量一贯制管理和铁钢界面质量一体化精准协同管控,详细介绍了质量一贯制管理、铁水硅预测、铁水磷预测、铁水罐装准率管理和铁水实时产量预测等模型的应用与实践。南钢炼铁事业部基于平台支撑,不断强化"大质量"意识,过程产品质量稳定性和铁水质量不断提升,炉况长周期稳定顺行,燃料比降幅超过 20kg/t,铁水合格率提升 9%,铁钢质量协同效率提高,有力支撑南钢集团高质量发展,对行业智能制造实践具有一定借鉴和示范意义。

关键词：工业互联网；铁区一体化；质量一贯制；硅预测；磷预测

Application and Practice of Quality Integration Intelligent Management Model in Nangang Iron District

FANG Lei, ZHU Gang, LIU Yonghui

(Nanjing Iron & Steel Co., Ltd., Nanjing 210035, China)

Abstract: Based on "5G + Industrial Internet", Nangang integrates data collection, material tracking, quality control, and other functional modules, covering the entire production process, and has constructed an industry-leading integrated intelligent management platform for the iron district. This article focuses on introducing the application and practice of the quality integration intelligent management supported by the platform in the South Steel Iron District, achieving consistent quality management and precise collaborative control of iron and steel interface quality. It provides a detailed introduction to the application and practice of models such as consistent quality management, silicon prediction in molten iron, phosphorus prediction in molten iron, accurate management of molten iron tank loading rates, and real-time production forecasting of molten iron. Supported by the platform, the Ironmaking Division of Nangang continuously strengthens the awareness of "high-quality production". The stability of process product quality improves continuously, the quality of molten iron increases, the stability of furnace conditions during long cycles is enhanced, the fuel ratio decreases by more than 20kg/t, the qualified rate of molten iron increases by 9%, and the synergy efficiency of iron and steel quality improves. It effectively supports the high-quality development of Nangang Group and has certain reference and demonstration significance for intelligent manufacturing practices in the industry.

Key words: industrial internet; iron district integration; consistent quality control; silicon prediction; phosphorus prediction

焦炭在 H_2O+CO_2 气氛中气孔结构的溶损演化行为

孙 章,韩嘉伟,窦明辉,吕叶飞,付佳旭

(华北理工大学化学工程学院,河北唐山 063210)

摘 要：高炉富氢冶炼是一种低碳炼铁技术,"以氢代碳"能够降低碳排放,但亦使炉内 H_2O 含量大幅增加,导致水蒸气与焦炭反应的溶损劣化过程不可忽视。焦炭是一种多孔性的炭材料,溶损过程中气孔结构随之演变,为了深刻理解富氢环境下焦炭溶损行为,研究焦炭在 H_2O+CO_2 气氛中气孔结构的溶损演化规律对于认识焦炭在富氢高炉内溶损劣化行为具有重要意义。

为研究 H_2O 和 CO_2 对焦炭气孔结构溶损演化行为的影响,采用自制连续进水的全自动焦炭反应性测定装置在 1100℃对不同反应气氛下的柱状焦炭进行了溶损实验,运用图像分析法对焦炭不同深度气孔结构进行测试,分析了

H_2O/CO_2 不同气氛下焦炭深度反应与缩核表现,研究了焦炭在 H_2O 和 CO_2 气氛下气孔溶损行为的差异性。研究结果表明:焦炭显微气孔分布呈现偏态分布特征,随反应深度的增加,小孔含量升高,中、大孔含量降低,焦炭气孔结构逐渐趋于未反应核,柱状焦炭在溶损过程中气孔侵蚀呈现"类锥形"(图1)。H_2O 存在能够减弱焦炭深度反应的溶损程度,气孔结构在较浅的反应深度趋于未反应核,焦炭在 CO_2+H_2O 气氛中溶损的各区域气孔变化幅度是 CO_2 气氛溶损变化幅度的 1.2~1.5 倍。基于气孔侵蚀最远距离 l 和焦炭半径 r 提出了溶损过渡层参数 θ 定量表征焦炭溶损缩核行为,焦炭在 CO_2 和 CO_2+H_2O 气氛中溶损的过渡层夹角 θ 分别是 45°的 38°,H_2O 蒸气加入使焦炭溶损过渡层夹角 θ 减小 15.6%,表明反应气氛中 H_2O 的加入增强了焦炭表面反应,气体内部扩散受到阻碍,缩核程度增加。

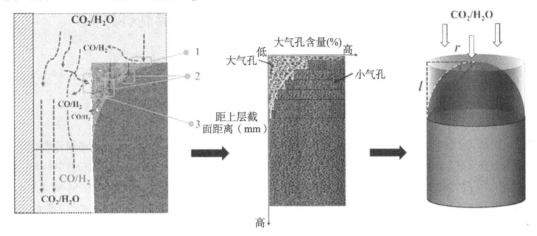

图 1 焦炭溶损过程示意图

参 考 文 献

[1] 高建军, 齐渊洪, 严定鎏, 等. 中国冶金, 2021, 31(9): 64-72.
[2] 张福明, 钢铁. 2022, 57(9): 11-25.
[3] GUO W., XUE Q., LIU Y., et al. Int J Hydrogen Energ, 2015, 40(39): 13306-13313.
[4] WANG P, ZHANG Q Y, LONG M H, et al. ISIJ Int, 2017, 57(4): 643-648.
[5] SHIN S, JEONG I, JUNG S. Ironmak Steelmak, 2017, 45(8): 739-746.

高炉主沟温度运行状态监控及分析

谢 皓,颜秋余,雷 磊,李自帅,姚元辉

(中冶赛迪信息技术(重庆)有限公司,重庆 401122)

摘 要: 高炉主沟运行状态对高炉冶炼效率和生产安全具有重要影响,而主沟温度作为反应主沟运行状态的直接表现,对其进行日常监控和异常报警至关重要。本文从温度趋势、变化速率及绝对值水平等三个维度,对主沟重点区域热电偶数据进行全面跟踪分析,及时识别到主沟异常状态,为高炉操作人员调整操作及现场维护处理争取时间,减少了主沟维护成本,确保主沟安全运行。

关键词: 高炉;主沟;温度监控;分析

Abstract: The operating status of the main channel has a significant impact on the smelting efficiency and production safety of the blast furnace, and the temperature as a direct reflection of the operating status of the main channel, is crucial

煤粉在小套内运动轨迹的模拟研究

孙俊杰

（宝山钢铁股份有限公司中央研究院，上海　201900）

摘　要：高炉风口小套的磨损是高炉生产事故之一，很多高炉在生产过程中会因为喷吹煤粉的原因造成小套内壁的磨损，这类磨损问题是制约喷煤比提升和高炉稳定顺行的限制环节。梅钢五号高炉经常发生煤粉磨穿小套前端内侧焊缝的现象，为了解煤粉磨损小套机理和载气喷吹速度对煤粉轨迹宽度的影响，本文以CFD-DPM模型为基础，对煤粉在小套内运动轨迹开展了模拟研究。模拟结果阐明了煤粉载气量与煤粉迹线宽度的关系，为喷吹系统二次补气量的控制提供参考。

关键词：高炉；小套磨损；煤粉；数值模拟

Simulation Study of Pulverized Coal Movement Trajectory in the Tuyere

SUN Junjie

(Research Institute of Baoshan Iron & Steel Co., Ltd., Shanghai 201900, China)

Abstract: The wear of tuyere is one of the accidents in blast furnace production. The inner wall of tuyere of many blast furnaces is worn because of pulverized coal injection in the production process. This kind of wear problem is the restricted link of increasing the coal injection ratio and running smoothly of blast furnace. The phenomenon of the inner seams of the front end of the tuyere are worn through by pulverized coal often occurs in the No. 5 BF of Meishan Iron and Steel Company. In order to understand the mechanism of wearing and the influence of carrier gas injection velocity on pulverized coal trajectory width the movement trajectory of pulverized coal in the tuyere is simulated based on the CFD-DPM model.The simulation results illustrate the relationship between the gas carrying capacity and the width of the pulverized coal trace, which provides a reference for the control of the secondary gas filling capacity of the injection system.

Key words: blast furnace; wear of tuyere; pulverized coal; numerical simulation

武钢4号高炉降低燃料比操作实践

金　锁，郑剑洪

（宝钢股份武汉钢铁有限公司炼铁厂，湖北武汉　430080）

摘　要：武钢4号高炉通过不断实践与探索，通过使用高风温，高顶压，高富氧，调整风口进风面积、布料角度、布料矩阵等技术手段，维护了合理的操作炉型，提高了高炉煤气利用率，改善了高炉的煤气流分布，降低了高炉的燃料比。

关键词：燃料比；布料制度；高富氧；煤气利用率

Operation Practice of Fuel Ratio Reduction in No. 4 Blast Furnace of WISCO

JIN Suo, ZHENG Jianhong

(Baosteel Wuhan Iron and Steel Co., Ltd., Ironmaking Plant, Wuhan 430080, China)

Abstract: Through continuous practice and exploration, WISCO No. 4 blast furnace maintains reasonable operation furnace type by using high air temperature, high top pressure, high oxygen enrichment, adjusting the air inlet area of tuyere, distributing Angle, distributing matrix and other technical means, improves the gas utilization rate of blast furnace, improves the gas flow distribution of blast furnace, and reduces the fuel ratio of blast furnace.

Key words: fuel ratio; fabric system; high oxygen enrichment; gas utilization rate

高炉风口回旋区传热/传质行为及动力学机理的 DEM-CFD 研究

李　萌，安希忠

（东北大学冶金学院，辽宁沈阳　110819）

摘　要：从风口喷出的热风形成的回旋区为高炉冶炼生产提供热量和能量，对高炉的连续稳定生产至关重要。高炉风口回旋区的形成、形态和燃烧特性直接影响高炉内的煤气成分、高炉炉料的稳定性、软熔带的形成[1-3]。因此，研究高炉风口回旋区的形成、气固运动特性和燃烧行为，对优化高炉生产、提高冶炼效率具有极其重要的意义。然而由于回旋区中的化学反应十分复杂，广大的炼铁工作者在物理实验和数值模拟方面开展了大量工作，但风口回旋区理论尚未形成统一的共识。

物理实验虽然可以从宏观上观察回旋区的演变甚至燃烧行为，但难以有效且全面地描述气体和颗粒的详细信息，包括颗粒速度、力、温度、气体组分分布等，而这些不足可以通过所谓的数值模拟来克服。目前，用于模拟颗粒在回旋区中的运动行为的数值模型主要包括双流体模型（TFM）和离散元-计算流体动力学模型（DEM-CFD）。由于 TFM 模型的计算成本较低，其在模拟大规模颗粒体系方面具有独特的优势，但其计算精度相对较差。然而，DEM-CFD 模型可以解决 TFM 模型中无法考虑颗粒离散特性对流动行为影响的问题，且计算精度较高。目前，利用 DEM-CFD 对风口回旋区内热化学行为的研究比较有限，且处于初步阶段，对鼓风参数和颗粒特性的影响以及及相应的微观机理还不清楚，需要进行全面的研究。

本研究采用 DEM-CFD 方法模拟了高炉风口回旋区中焦炭颗粒的传热、传质行为和动力学特性，其中考虑了焦炭燃烧和碳溶等化学反应。首先，通过将传热和传质行为与之前的工作进行比较，验证了所建立的数值模型和模拟参数的准确性和有效性。然后，分析了工艺参数（包括鼓风速度、鼓风温度和氧气浓度）对回旋区中气体种类分布、反应动力学速率、颗粒温度和颗粒尺寸的影响。此外，还系统研究了风口回旋区中焦炭颗粒的动力学特征（包括回旋区尺寸、颗粒动能）和微观行为（配位数、法向接触力、总力和孔隙分布），并量化了它们与各工艺参数的相关性。最后，建立了 BPNN 模型，用于预测任意工艺参数组合下的风口回旋区尺寸。所获得的结果有望为实际的高炉生产提供深入的见解和有价值的参考。

参 考 文 献

[1] Hilton J, Cleary P. Raceway formation in laterally gas-driven particle beds[J]. Chem. Eng. Sci., 2012, 80: 306-316.
[2] Hou Q, Kuang D E S, et al. DEM-based virtual experimental blast furnace: A quasi-steady state model[J]. Powder Technol., 2017, 314: 557-566.
[3] Hou Q, Kuang D E S, et al. A process scaling approach for CFD-DEM modelling of thermochemical behaviours in moving bed reactors[J]. Fuel Process, Technol, 2020, 202: 106369.

基于高炉风口物料特征及炉缸温度测量判断炉缸活跃性

王元生[1]，张庆喜[2]，郑华伟[2]，鲍俊芳[1]，陈令坤[1]，任玉明[1]

（1. 宝钢股份中央研究院武钢有限技术中心，湖北武汉 430083；
2. 宝钢股份武汉钢铁有限公司炼铁厂，湖北武汉 430083）

摘 要：本文描述了风口焦技术研究进展，对比分析了武钢有限 7、8 号高炉风口平面径向焦炭粒度以及渣铁分布规律，发现风口焦中<3mm 细颗粒焦炭比例由炉缸边缘的 9.28%升高至炉缸中心的 69.95%，不同的炉况引起的炉缸边缘和中心的渣铁分布的变化。进行了高炉休风后风口平面炉缸温度测量，发现休风后 7、8 号高炉风口平面径向炉缸温度最大值位于距离炉墙 5m 处，炉缸边缘温度高于中心温度。有限 7 号高炉采取"平台+漏斗"布料模式，最内侧布料角位较小，矿石布料更加靠近高炉中心，因而，高炉风口平面渣铁从距离炉墙 5m 至炉缸中心占比较高。通过高炉风口焦平均粒度、风口渣铁径向分布规律、渣铁比例并结合风口损坏等，可判断炉缸活跃性，具有与预判性。

关键词：高炉；炉缸温度；风口物料；炉缸活跃性

Judging the Activity of Hearth Based on the Characteristics of Tuyere Material in Blast Furnace and Temperature Measurement of Furnace Hearth

WANG Yuansheng[1], ZHANG Qingxi[2], ZHENG Huawei[2],
BAO Junfang[1], CHEN Lingkun[1], REN Yuming[1]

(1. Research and Development Center of Wuhan Iron and Steel Co., Ltd., Baosteel Central Research Institute, Wuhan 430083, China; 2. Iron Plant, Wuhan Iron and Steel Co., Ltd., Wuhan 430083, China)

Abstract: This article describes the research progress of technology of tuyere coke, and compares and analyzes the particle size of coke and the distribution of slag and iron in the radial tuyere plane of No. 7 and No. 8 blast furnaces of Wuhan Iron and Steel Co., Ltd. It is found that the <3mm fine tuyere coke increases from 9.28% at the edge of the hearth to 69.95% at the center of the hearth. Different conditions of furnace cause changes in the distribution of slag iron at the edge and center of the hearth. The temperature measurement of the flat hearth of the tuyere after the blast furnace has been conducted, It was found that the maximum temperature of the radial hearth of the tuyere plane in No. 7 and No. 8 is located at a distance of 5m from the furnace wall, and the temperature at the edge of the hearth is higher than the center temperature,. The No. 7 blast furnace adopts a "platform+funnel" distribution mode, with a smaller innermost distribution angle and ore distribution closer to the center of the blast furnace. Therefore, the proportion of slag and iron in the blast furnace tuyere plane is relatively high from 5m away from the furnace wall to the center of the hearth. By analyzing the average particle size of blast furnace tuyere coke, the radial distribution pattern of tuyere slag and iron, the ratio of slag and iron, and combining with tuyere damage, the activity of the furnace hearth can be determined, which is predictive.

Key words: blast furnace; temperature of furnace hearth; tuyere material; activity of Hearth

高炉回旋区内煤粉燃烧行为的三维数值模拟研究

吴 丽，温良英，张生富

（重庆大学材料科学与工程学院，重庆 400030）

摘 要： 高炉回旋区内煤粉的燃烧行为对于高炉降低能耗、高炉顺行具有重大影响。煤粉燃烧过程中不可避免出现未燃煤粉，未燃煤粉穿过回旋区进入料层，导致料柱的透气性变差，影响高炉的稳定运行。因此，通过调整某些工艺参数提高煤粉在回旋区内的燃烧率，减少未燃煤粉的含量成为当前研究的热点话题。

以某厂 2500 m^3 高炉喷吹煤粉的实际建立模拟计算风口区几何模型和确定边界条件。利用 FLUENT 软件计算气体流动，ROCKY 软件对单个焦炭颗粒相进行求解，两者通过 Wen&Yu 曳力模型进行 CFD-DEM 双向耦合建立高炉回旋区三维数值模型，同时耦合考虑了多个相间力的作用，包括曳力、虚假质量力、升力和压力梯度力等，研究了回旋区稳定形成的过程。

在此基础上，进一步通过 FLUENT 软件的多孔介质模型模拟焦炭层，将回旋区视为稳定的空腔，加入采用 DPM 模型描述的煤粉，建立描述高炉风口回旋区内煤粉的气固流动、传热和燃烧的数学模型，研究了在煤粉燃烧过程的回旋区温度分布、气体组分分布特征以及煤粉的流动及燃烧行为，并重点考察了不同喷煤量、煤粉粒度和富氧率等工艺条件对煤粉燃尽率的影响，得出煤粉主要在回旋区内进行燃烧反应，沿风口中心线 O_2 含量减少，CO 含量逐渐降低，CO_2 含量增加，温度升高，CO_2 含量在 O_2 基本耗尽时达到峰值，且温度也达到最高值，约为 2573 K；增大喷煤量、减小粒径、提高富氧率都可以有效提高煤粉在回旋区内的燃尽率。

参 考 文 献

[1] 倪澳, 李承志, 张伟, 等. 钢铁研究学报, 2022, 34(5): 41.
[2] 张仕洋, 薛庆国, 刘锦周, 等. 过程工程学报, 2014, 14(2): 273-279.
[3] Miao Zhen, Zhou Zongyan, Yu A B, et al. POWDER TECHNOLOGYELSEVIER SCIENCE BV, 2017, 314: 542-549.
[4] Bambauer F, Wirtz S, Scherer V, et al. POWDER TECHNOLOGYELSEVIER SCIENCE BV, 2018, 334: 53-64.

玉钢高炉提高钒钛铁水质量实践

杨 敬

（玉溪新兴钢铁有限公司炼铁厂，云南玉溪 653100）

摘 要： 通过对玉钢 1080m^3 高炉钒钛矿冶炼实践分析影响钒钛铁水质量原因，制定原燃料质量跟踪管理措施，同时优化高炉四大操作制度改善炉缸工作状态，加强炉前组织管理，强化设备保障，大幅减少了炉况波动，为提升钒钛铁水质量合格率充分发挥钒效益创造了基础条件，保障了高炉钒钛矿长周期冶炼稳定顺行。

关键词：炉况波动；钒钛铁水；脱硫性能

Yugang Blast Furnace to Improve the Quality of Vanadium Iron Quality Practice

YANG Jing

(Yuxi Xinxing Steel Co., Ltd., Iron Plant, Yuxi 653100, China)

Abstract: By analyzing the reasons affecting the quality of vanadium titanium ore in Yugang 1080m3 blast furnace, the quality tracking management measures of raw fuel were formulated, and the four major operating systems of blast furnace were optimized to improve the working state of the furnace cylinder, strengthen the organization and management before the furnace, strengthen the equipment guarantee, greatly reduce the fluctuation of furnace conditions, create basic conditions for improving the quality qualification rate of vanadium titanium molten and give full play to the benefits of vanadium, and enaure the stable and smooth smelting of blast furnace vanadium titanium ore.

Key words: furnace fluctuations; vanadium titanium ferrite; desulfurization performance

玉钢炼铁厂钒钛矿冶炼炉前技术进步

杨 敬

（玉溪新兴钢铁有限公司炼铁厂，云南玉溪 653100）

摘 要：玉钢炼铁厂主要以差异化钒钛矿冶炼为主，自2008年开始相继在两座450m³高炉和1座1080m³高炉开展钒钛矿冶炼，是云南省最大的钒资源综合利用生产基地。以前两座450m³高炉钒钛矿冶炼因炉前技术落后，硬件设备设施不全等因素造成炉前工劳动强度大，生产组织被动，高炉顺行得不到保障，产量低，经济技术指标差，制约公司整体效益的发挥。随着小高炉逐步淘汰，大高炉的投产，炉前通过设备设施升级改造，检修复风操作经验的总结提炼，炉前管理体系的建立等各种设备、技术手段的实施，确保了大高炉冶炼钒钛矿长周期稳定顺行，指标得到提升。

关键词：钒钛矿冶炼；炉前技术；体系

Technical Progress in Front of Vanadium Titanium Ore Smelting Furnace of Yugang Iron Making Plant

YANG Jing

(Yuxi Xinxing Steel Co., Ltd., Iron Plant, Yuxi 653100, China)

Abstract: Yugang iron making plant mainly focuses on differentiated vanadium titanium ore smelting, and has successively carrind out vanadium titanium ore smelting in two 450m³ blast furnaces and one 1080m³ blast furnace 2008, which is the largest comprehensive utilization production base of vanadium resources in Yunnan Province. In the past, the two 450m³ blast furnaces vanadium titanium ore smelting due to backward technology in front of the furnace, incomplete hardware equipment and facilities and other factors caused by the labor intensity of the furnace pre-turnace, passive production organization, the smooth flow of the blast furnace is not guaranteed, the output is low, and the economic and technical indicators are poor, which restricts the overall efficiency of the company. With the gradual elimination of small blast furnaces, the upgrading of equipment and facilities in front of the furnace, the summary and refining of inspection and repair wind operation experience, the establishment of the pre-furnace management system and other equipment and technical means to ensure the long-term stability and smooth progress of the large blast furnace smelting vanadium titanium ore, and the indicators have been improved.

Key words: vanadium titanium ore smelting; pre-furnace technology; system

小颗粒烧结矿在无钟高炉中的优化利用模拟研究

曾　旺[1,2]，滕　飞[1,3]，刘　放[1,2]，倪　俊[1,2]，
周　恒[1,2]，吴胜利[1,2]，寇明银[1,2]

（1. 北京科技大学钢铁冶金新技术国家重点实验室，北京　100083；2. 北京科技大学冶金与生态工程学院，北京　100083；3. 北京钢研新冶科技有限公司，北京　100089）

摘　要： 我国高炉炼铁产量占比达到了 85%以上，在未来几十年高炉仍然是我国炼铁生产的首要工艺[1-2]。为了保证高炉内的透气性，入炉炉料的尺寸一般要求大于 5mm[3]。但是近年来，随着铁矿石资源的不断消耗，各类铁矿石资源呈现劣化趋势，一些小颗粒的矿石会进入高炉，从而影响高炉的透气性，进而影响高炉的煤气利用率以及碳排放[4-5]。因此，研究高炉排料时小颗粒炉料分布对炉喉上部透气性的影响具有重要意义。

本文基于离散单元法建立了 1:1 串罐无钟高炉布料三维模型。该模型模拟了四种炉料多种尺寸组合在一起的实际布料过程，并且研究了直径 4mm 的小粒度烧结矿布料位置的变化对高炉炉喉径向透气性的影响。同时当小颗粒烧结矿在高炉中的使用量达到 100kg/tHM，在考虑炉喉径向透气性的情况下，对装料参数进行了优化。部分结果如图 1 所示，结果表明，当小粒径烧结矿分别位于顶部和底部时，高炉边缘和中部的透气性变差。改变溜槽角度，将底部小粒径烧结矿排至 0.9~1.8m 处压降小的区域，可以发现，高炉炉喉边缘透气性变好，煤气流分布与基准工况一致。基于上述研究本文提出了一种高炉使用小颗粒烧结矿的方法，为优化高炉操作提供了理论基础与依据。

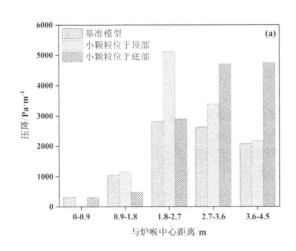 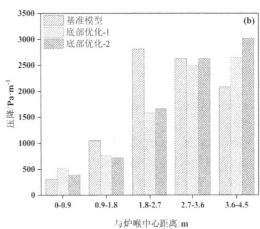

图 1　不同模型在不同区域下的压降

(a) 改变小颗粒位置；(b) 改变小颗粒位置与溜槽角度

参 考 文 献

[1] 潘钊彬, 乔军. 炼铁, 2020, 39(6): 20-26.
[2] Zhao J, Zuo H, Wang Y, et al. Ironmaking & Steelmaking, 2020, 47(3): 296-306.
[3] 吴浩. 中国设备工程, 2018(1): 118-119.
[4] 谢皓, 王劲松, 邹忠平, 等. 炼铁, 2014, 33(4): 55-60.
[5] Zhou Z, Zhu H, Wright B, et al. 2011, 208(1): 72-85.

面向闭环的高炉加减风智能调控系统开发与应用

闫炳基[1], 国宏伟[1], 牟原金[1], 梁合兰[2]

（1. 苏州大学沙钢钢铁学院, 江苏苏州　215137；
2. 苏州大学计算机科学与技术学院, 江苏苏州　215008）

摘　要：为实现高炉稳定顺行, 上下部协调调控是重要的手段, 而通过调整风量来调控高炉冶炼状态是常用的操作手段, 其对稳定炉况和炉温起着重要作用, 但高炉炼铁是一个复杂的工艺过程, 涉及到诸多参数的控制和优化, 具有大滞后、大噪声、多变量和非线性的特点, 且整个内部反应过程无法直接观察到, 因此实现高炉的加减风调控难度较大。

国外部分高炉基于专家系统实现了加减风的闭环控制, 国内部分高炉虽然配备了高炉专家系统, 但专家系统依然还停留在操作指导阶段, 这是由于专家系统给出的操作指导建议与现场工长操作存在较多偏差, 致使专家系统的加减风操作指导未被采纳, 实际对于高炉加减风的操作与未配备高炉专家系统的高炉一样, 依然还是依据高炉工长经验进行操作, 由此导致加减风操作取决于操作工长的水平和主观认识, 存在难以标准化、自动化和智能化的问题。针对国内当前高炉加减风调控操作存在的问题, 基于日本川琦 Go-Stop 专家系统、奥钢联 VAIron 专家系统、芬兰罗德洛基专家系统等的结构设计理念, 结合高炉工长的操作经验, 以专家系统持续循环优化迭代为着力点, 采用模糊数学、专家推理以及深度学习的方法, 将大数据分析的数据挖掘手段同工长制定的规则相结合, 以模糊知识表示、专家系统推理和阈值自适应为三个维度, 实现了面向闭环的高炉加减风智能调控。

面向闭环的高炉加减法智能调控系统自 2023 年 4 月 25 日起在实际生产高炉上投入使用, 通过系统的持续循环优化迭代, 目前真实命中率稳定在 95%以上, 实现加减风的闭环智能调控, 不仅能够准确判断高炉的工作状态和异常情况, 并提供相应的诊断结果和操作建议, 而且可以减少工作人员的主观判断和操作干预, 提高工作效率和操作稳定性, 从而达到稳定高炉状态的效果。

参 考 文 献

[1] Li Hongyang, Li Xin, et al. Industrial internet platforms: applications in BF ironmaking[J]. Ironmaking & Steelmaking, 2022, 49(9): 905-916.
[2] 石泉, 唐珏, 储满生. 基于工业大数据的智能化高炉炼铁技术研究进展[J]. 钢铁研究学报, 2022, 34(12): 11.
[3] 肖鹏. 高炉炼铁技术创新实践及未来展望[J]. 钢铁, 2021, 56(6): 6.
[4] Wang H, Zhao W, Chu M, et al. Current status and development trends of innovative blast furnace ironmaking technologies aimed to environmental harmony and operation intellectualization[J]. Journal of Iron and Steel Research (International), 2017, 24(8): 751-769.
[5] 陈令坤, 左海滨, 于仲洁, 等. 高炉冶炼专家系统的开发研究[J]. 钢铁, 2006, 41(1): 5.

基于智能感知技术的高炉炉前可视化平台开发与搭建

国宏伟[1]，陈令坤[2]，梁合兰[3]，闫炳基[1]，许 浩[1]

（1. 苏州大学沙钢钢铁学院，江苏苏州 215137；2. 宝钢股份中央研究院炼铁所，湖北武汉 430080；3. 苏州大学计算机科学与技术学院，江苏苏州 215008）

摘 要：随着中国经济结构的转型升级，钢铁行业作为传统制造业，生产模式也需向自动化、信息化、智能化转型升级[1-2]。目前，计算机视觉以其优秀可靠的检测算法被广泛应用到工业领域[3-4]，逐渐成为现场操作不可或缺的辅助工具。

高炉炉前出铁是钢铁冶炼流程中的关键环节，随着高炉的大型化发展，高炉炉前出铁对高炉顺行的影响愈加突出[5]。目前国内大部分钢铁企业在高炉炉前出铁过程中仍然面临着一些问题：（1）高炉出铁开始时间、见渣时间、出铁结束时间是高炉渣铁管理的重要信息，当前通过现场工人观测记录，存在记录不及时、不准确的问题；（2）高炉铁口深度和泥炮机打泥量由于高炉炉前出铁环境恶劣，导致关键数据传感器极易发生故障，当前大多数钢铁企业对于铁口深度和泥炮机打泥量依赖现场工人观测记录，也存在观测记录不及时、不准确的问题；（3）出铁铁流的射流形态是反映高炉炉缸状态的重要信息，因此高炉工长需观察铁流射流形态评估高炉状态，该信息存在难以数字化表征的问题。针对高炉炉前出铁存在的上述关键性问题，论文基于高炉炉前出铁现场的视频成像监测数据，以高炉炉前出铁过程视觉深度感知为着力点，采用图像处理技术、目标检测技术、数据拟合分析技术，将高炉炉前开铁口、出铁出渣和堵铁口等工艺过程与深度学习相结合，系统研究开发了高炉铁口开口深度测量模型、出铁出渣时间自动识别与信息存储模型、铁流状态实时监测及关键参数自动提取模型、泥炮机打泥信息测量模型。

经企业实践检验，高炉炉前可视化平台具有如下优势和创新：（1）实现了高炉炉前数据的数字化表征，有效解决了当前高炉炉前出铁状态无法量化的问题；（2）实现了高炉炉前出铁的标准化管理，有效解决了高炉炉前操作人员进行信息采集时存在的作业环境恶劣、作业风险高、测量误差大、数据收集不及时等问题；（3）提高了高炉炉前出铁的智能化，基于智能感知的检测结果为渣铁排放的及时性以及铁口状态维护的有效性提供了有力支撑，从而有效稳定炉况、最大限度地提升铁口运行质量。

参考文献

[1] 王春梅, 周东东, 徐科, 等. 综述钢铁行业智能制造的相关技术[J]. 中国冶金, 2018, 28(7): 1-7.
[2] 李江昀, 杨志方, 郑俊锋, 等. 深度学习技术在钢铁工业中的应用[J]. 钢铁, 2021, 56(9): 43-49.
[3] Dong Pan, Zhaohui Jiang, Zhipeng Chen, et al. IEEE Trans. Instrumentation and Measurement, 2019, 68(10): 3576-3588.
[4] Zhengmi Tang, Kohei Shimasaki, Mingjun Jiang, et al. ISIJ International, 2020, 60(5): 960-970.
[5] 张强, 王迎光, 赵颖, 等. 浅谈安钢高炉炉前出铁操作的若干问题[J]. 炼铁, 2019, 38(4): 56-58.

钢铁企业原料场绿色高效储运技术的研究

毕 琳

（中冶京诚工程技术有限公司冶金公司，北京 100176）

摘 要：在千万吨级的钢铁联合企业中，几十种原料年输入总量近3000万吨，"储"与"运"的高效协同十分重要。

传统原料场储运工艺：来料→卸料→集中储存→混匀和加工→供料→用户[1]，这种原料储运作业模式重"储"轻"运"，大宗原料进企业后先集中输送到有足够储料量的料场"储"，生产需要时再从料场"运"出。既增加"运"的周期、设备和损耗，也增加"储"的规模、占地、投资和成本，并在烧结、球团、焦化等铁前多用户供料相互制约时影响到正常冶炼生产。

近年来，行业内针对原料场环保封闭、智能化提升和数字化料场本体方面有一些技术创新[2-4]，但对原料场高效新工艺系统开发、料场减量化储存、即时供料技术等研究内容较少提及。本文突破传统工艺，从"运"和"储"协同增效，以均衡"储"保障 即时"运"的安全角度，让原料均衡减量储存，节省建设占地和投资，尤其是降低大型原料场环保封闭时的高成本，促可控物料"流动"起来，以一种"即时方式"灵活、高效满足连续、安全供料需求。

通过即时均衡的储运工艺、共用组网的网群输送、自动选线的智能输送和降低损耗的清洁转运等四个方面的研究，探索绿色高效的原料储运新模式。即时均衡的储运工艺以双向直供系统为干线，原料作业工序和原料用户组合的供料系统为多个支线，干支线串并联供料；借助原料信息大数据平台设定即时供料级别，实现"储"和"运"根据需求灵活选取干支线，均衡减量储料。共用组网的网群输送以共用输送路径和组网多通路输送模型实现原料顺畅进厂和铁前多工序一体化生产时有序供料[5]。自动选线的智能输送以物料品种和工艺关联条件的要素模型优选网络线路，辅助输送设备状态全过程可视仿真实现压减供料设备空转和即时发送。降低损耗的清洁转运以转运点新型曲面滑槽、缓冲曲面溜槽和新型密封结构，近距离引导转运改变物料冲击和自由散落，改善输送线作业环境。

工程实践表明，即时均衡的储运工艺节省原料集中储存占地 10%~20%，减少原料重复倒运 10%~15%，降低建设投资 10%~25%，减少原料资金占用成本 10%~15%。共用组网的网群输送减少带式输送机输送线长度 20%~25%，减少输送设备数量 25%~30%。自动选线的智能输送提升供料系统智能化程度，设备利用率提高 10%~15%，设备运行能耗降低 10%。降低损耗的清洁转运可减少单点转运物料破碎率 85%、减少漏撒率 80%，减少单点转运物料扬尘 50%~95%。提升原料物流效率，保障冶炼原料低成本安全运行，推动原料行业向调运量、减储量、少装备、高效率、统管理的绿色、低碳转型和高质量发展。

参 考 文 献

[1] 杨德政，张惠宁，朱金瑞，等. 钢铁企业原料准备设计手册/中国冶金建设协会编[M]. 北京：冶金工业出版社，1997.
[2] 吴旺平. C 型封闭料场在宝钢的投产应用[J]. 烧结球团，2017, 42(2): 53-56.
[3] 康兴东，王东. 现代综合原料场新技术应用实践[J]. 烧结球团，2017, 42(2): 57-61.
[4] 张毅. 宝钢原料场改造新技术的运用[J]. 烧结球团，2017, 42(4): 40-43.
[5] 毕琳，徐培万. 钢铁企业原燃料储运平衡分析及优化设计[C]//中国金属学会 中国金属学会炼铁分会 浙江杭州：2018 年全国炼铁生产技术会暨炼铁学术年会摘要集，2018: 635.

矿焦耦合冶金性能：1 试验及评价方法

汪 琦[1]，宋阳升[2]，李廷乐[1]，李哲熙[1]，EVANS Tim[2]

（1. 辽宁科技大学材料与冶金学院，辽宁鞍山 114051；
2. 力拓集团，澳大利亚西澳大利亚州珀斯 6000）

摘 要：我们发明了采用焦-矿-焦分层试料，模拟矿焦在高炉内还原、溶损、软化和熔融之间关联行为及特性的荷重-热重-气体分析联动试验装置，进行矿焦耦合性能试验及评价的方法（Qisunny 法）。变矿或变焦试料的试验结果示出：矿焦耦合冶金性能按间接还原区、间接还原和直接还原及软化区、熔融直接还原和滴落直接还原四个区域依次演变，性能评价指标包括各区域的温度及区间、间接和直接还原度，总间接还原度、熔融区透气性特征值、上层焦炭溶损率及溶损后强度等。该方法为评价矿焦冶金性能之间的相互作用，优化矿焦炉料结构和质量提供了一种全新的选择。

关键词：高炉；铁矿石；焦炭；耦合冶金性能；评价方法

Ore and Coke Coupling Metallurgical Properties: Part 1 Test and Evaluation Method

WANG Qi[1], SONG Yangsheng[2], LI Tingle[1], LI Zhexi[1], EVANS Tim[2]

(1. School of Materials and Metallurgy, University of Science and Technology Liaoning, Anshan 114051, China; 2. Rio Tinto Group, Perth 6000, West Australia, Australia)

Abstract: We have invented a test device and evaluation methodology (Qisunny method) which deploy samples with coke-ores-coke layer structure in the crucible integrating the reduction under the load-thermogravimetric-gas analysis to simulate coupling ores and coke behaviours and properties, such as iron ore reduction, coke dissolution, burden softening, meting and dropping. The device and methodology can conduct coke and ore coupling behaviours and properties test and evaluate its coupling performance. Changing burden structure and coke type test results showed that coupling coke and ore metallurgical performance evolved with different zones with distinct boundary in an order from indirect reduction zone to indirect and direction reduction co-existing and softening zone, to smelting direct reduction zone and to dripping direct reduction zone. The performance evaluation parameters include the temperature and the temperature interval of different zones, the indirect and direct reduction rate, the total reduction rate, the permeability parameters in melting and smelting zone, the dissolution rate and strength after dissolution of top layer coke. The invented device and methodology can provide a totally new option for evaluating coupling coke-ore behaviours and performances, and optimising burden structure and iron ore and coke quality.

Key words: blast furnace; iron ore; coke; coupling metallurgical properties; evaluation method

昆钢 2 号高炉铁口喷溅治理实践

陈芳林，麻德铭

（武钢集团昆明钢铁股份有限公司炼铁厂，云南安宁 650300）

摘 要：昆钢新区 2500m³ 高炉自 2022 年 2 月 28 日点火投产。投产后铁口喷溅严重，渣铁排放时间短，高炉受渣铁排放不畅影响憋风严重，制约炉况顺行和高炉产量的提升。另一方面，铁口长时间喷溅，产生的大量烟尘，极大增加了除尘压力和环保风险。为解决铁口喷溅问题，高炉采取了压炮处理、铁口孔道压浆、提高炮泥质量、统一炉前操作、优化炉内操作等措施，有效治理了铁口喷溅，保证了高炉出铁顺畅，为高炉提高技术经济指标创造了条件。

关键词：铁口；喷溅；憋风；治理

Practice of Treating Iron Spitter at the Iron Mouth of Kunming Steel's 2# Blast Furnace

CHEN Fanglin, MA Deming

(Wuhan Iron and Steel Corporation Kunming Iron & Steel Co., Ltd., Ironmaking Plant, Anning 650300, China)

Abstract: The 2500m³ blast furnace in Kunming Iron and Steel New Area has been ignited and put into operation since February 28, 2022. After being put into operation, the iron spout is severely splashed, and the discharge time of slag and iron is short. The blast furnace is severely affected by the poor discharge of slag and iron, which restricts the smooth operation of the furnace and the improvement of blast furnace output. On the other hand, the long-term splashing of the iron mouth generates a large amount of smoke and dust, greatly increasing the dust removal pressure and environmental risks. In order to solve the problem of iron spout splashing, the blast furnace has taken measures such as gun pressure treatment, iron spout hole grouting, improving the quality of gun mud, unifying furnace operation, optimizing furnace operation, etc., effectively controlling iron spout splashing, ensuring smooth iron tapping of the blast furnace, and creating conditions for improving technical and economic indicators of the blast furnace.

Key words: iron mouth; splash; holding back the wind; govern

大型高炉低碳冶炼策略之提高顶压

储 健[1,2]，徐 云[1,2]，陈庆林[1,2]，张 南[3]，代 兵[1,2]

（1. 安徽工业大学低碳研究院，安徽马鞍山 243000；2. 冶金减排与资源综合利用教育部重点实验室，安徽马鞍山 243000；3. 鞍钢股份有限公司鲅鱼圈钢铁分公司，辽宁营口 115007）

摘　要：基于高炉炼铁流程的基本工艺及原理，通过分析 2020 年我国 18 座 3200m³ 大型高炉的生产技术数据，系统研究了高炉顶压对高炉内还原、送风、炉顶、能源消耗以及生铁成分的影响。国内某企业新 1 号 3200m³ 高炉提高顶压操作实践表明，提高顶压不仅可以有效改善炉内冶炼进程，改善顺行，减少炉况波动，而且可以减少炉尘吹出量，提高生铁产量和质量，显著降低燃料消耗和工序能耗。在双碳目标的背景下，科学认识高炉冶炼过程的动态运行规律，不断探索极限低碳冶炼方法将成为广大高炉炼铁生产、管理、技术、科研工作者的方向和重点。

关键词：大型高炉；顶压；生产数据；低碳冶炼；高效低耗

Increased Top Pressure as a Low-Carbon Smelting Strategy for Large Blast Furnaces

CHU Jian[1,2], XU Yun[1,2], CHEN Qinglin[1,2], ZHANG Nan[3], DAI Bing[1,2]

(1. Low-Carbon Research Institute, Anhui University of Technology, Maanshan 243000, China; 2. Key Laboratory of Metallurgy Engineering and Resources Recycling of Ministry of Education, Maanshan 243000, China; 3. Bayuquan Branch of Angang Steel Co., Ltd., Yingkou 115007, China)

Abstract: Based on the fundamental processes and principles of ironmaking in a blast furnace, an analysis was conducted on the production technical data from 18 large-scale 3200m³ blast furnaces in China in 2020. The study systematically investigated the influence of top pressure on reduction, air injection, furnace top, energy consumption, and pig iron composition within the blast furnace. Practical operations in a newly built No. 1 3200m³ blast furnace in a domestic company demonstrated that increasing the top pressure not only effectively improved the smelting process, operational stability, and reduced furnace fluctuations but also led to a decrease in dust emissions, increased pig iron yield and quality, and significantly reduced fuel consumption and process energy consumption. Against the backdrop of dual carbon objectives, gaining a scientific understanding of the dynamic operation principles of the blast furnace smelting process and continuously exploring low-carbon ironmaking methods have become the focus for professionals and researchers in the fields of blast furnace ironmaking, production management, technology, and scientific research.

Key words: large blast furnace; top pressure; production data; low-carbon smelting; high efficiency and low consumption

昆钢 1 号高炉开炉及达产生产实践

卢郑汀，胡玉清，王 楠

（昆钢钢铁股份有限公司炼铁厂，云南昆明 650000）

摘 要： 昆钢 1 号高炉 2022 年 12 月 14 日停炉进行为期 24 天中修于 2023 年 1 月 8 日开炉，经过科学谋划、全面统筹、精准管控，实现高炉快速达产，复风后 11h 出铁，36h4min 后喷煤，37h15min 后富氧，48h 后铁水含硅降至 1%，第三天高炉利用系数达到 $2.54t/m^3 \cdot d$ 的成绩。

关键词： 高炉；开炉；生产；降硅；达产

Opening and Production Practice of 1# Blast Furnace in Kunming Steel

LU Zhengting, HU Yuqing, WANG Nan

(Kunming Steel Co., Ltd., Kunming, 650000, China)

Abstract: The 1# blast furnace of Kunming Steel was stopped on December 14, 2022 for 24 days on January 8, 2023, after scientific planning, comprehensive planning, precise control and control, the blast furnace rapid production, 11h after air return, coal injection after 36h after 4 min, oxygen enriched after 37h 15min, molten iron silicon reduced to 1% after 48h, the utilization coefficient of blast furnace reached $2.54t/m^3 \cdot d$ on the third day.

Key words: blast furnace; furnace; production; silicon; production

沙钢高炉经济高效护炉技术研究与应用

雷 鸣[1,2]，杜 屏[1]，魏红超[1]，焦克新[2]，张建良[2]

（1. 江苏沙钢集团有限公司，江苏张家港 215625；
2. 北京科技大学冶金与生态工程学院，北京 100083）

摘 要： 沙钢 2680 高炉炉役中后期炉缸温度相继升高，采用传统的钛矿护炉方法，铁水中的钛含量最高达到了 0.12% 以上，护炉成本达到了 10 元/吨铁，但炉缸安全状态并未得到有效改善，仍需采用休风、堵风口等措施来保证安全生产。除安全方面的问题，高钛铁水还影响下游炼钢工序，不但造成转炉冶炼过程中喷溅，还影响品种钢的生产。高炉安全长寿自修复技术明确了炉缸铁水达到碳饱和，与耐材热面温度低于保护层形成温度是富石墨碳保护层形成的两个必要条件，可以通过优化炉缸传热体系促进铁水析碳和调控炉缸活性促进铁水渗碳两个必要措施形成保护层，实现炉缸自修复。沙钢 2680 m³ 高炉通过提高铁水温度、提高铁水碳含量、活跃炉缸等措施，炉缸温度逐步下降并保持稳定，高炉安全停炉大修。在大修破损调查中发现，炉缸形成了富石墨碳为主的多元综合保护层，厚度普遍在 30cm 以上，在炉役后期起到了至关重要的保护作用。

关键词：护炉；炉缸侵蚀；石墨碳；钛矿

Research and Application of Economical and Efficient Blast Furnace Maintenance Technology of Shasteel

LEI Ming[1,2], DU Ping[1], WEI Hongchao[1], JIAO Kexin[2], ZHANG Jianliang[2]

(1. Jiangsu Shagang Group Co., Ltd., Zhangjiagang 215625, China; 2. School of Metallurgical and Ecological Engineering, University of Science and Technology Beijing, Beijing 100083, China)

Abstract: Hearth temperature of Shasteel 2680m³ blast furnace was increased in the middle and later stages of service. Titanium ore was used as a Traditional maintenance technology. The titanium content in hot metal reached more than 0.12%, and the cost of furnace maintenance reached 10 yuan/ton of iron. But the safety state of the furnace hearth has not been effectively improved, still need to blow-off, tuyere blocking and other measures to ensure safe production. In addition to safety problems, high titanium molten iron also affects the downstream steelmaking process, not only causing spatter in the converter smelting process, but also affecting the production of grade steel. Safety and longevity self-repairing technology of blast furnace specifies that the carbon saturation of the molten iron in the furnace hearth and the formation temperature of the hot surface of the resistant material is lower than that of the protective layer are two necessary conditions for the formation of the graphite rich carbon protective layer. The protective layer can be formed by optimizing the heat transfer system of the furnace hearth to promote the carbon precipitation of molten iron and regulating the activity of the furnace hearth to promote the carburizing of molten iron, so as to realize the self-repair of the furnace hearth. By increasing the hot metal temperature, increasing the carbon content of hot metal and activating the furnace hearth, the furnace hearth temperature gradually decreased and remained stable, and the blast furnace was safely blown down for relining. In the investigation of hearth damage, it was found that the furnace hearth formed a multicomponent comprehensive protective layer mainly rich in graphitic carbon, the thickness of which was generally more than 30cm, which played a crucial protective role in the later stage of service.

Key words: blast furnace maintenance; hearth erosion; graphitic carbon; titanium ore

H_2 对钒钛球团软熔滴落行为的影响研究

马凯辉[1,2]，方云鹏[1,2]，刘凌岭[1,2]，唐文博[1,3]，
张立利[1,3]，蒋胜[1,3]，陈茅[1,2]

（1. 钒钛资源综合利用国家重点实验室，四川攀枝花 617000；2. 攀钢集团研究院有限公司，四川成都 610000；3. 攀钢集团攀枝花钢铁研究院有限公司，四川攀枝花 617000）

摘 要：高炉富氢冶炼目前被认为是钢铁工业降碳非常有效的技术路径之一。目前国内外针对普通矿高炉冶炼已开展了高炉富氢冶炼的实验室研究和工业试验，开展了许多很有意义的探索。但因钒钛磁铁矿高炉冶炼的特殊性，富氢冶炼相关研究尚处于初级阶段，还存在许多基础理论及关键技术问题亟待厘清，如富氢对钒钛矿炉

料透气性影响以及对渣、铁生成机理等。本文研究了 100%CO、50%H$_2$+50%CO、100%H$_2$ 气氛下，钒钛球团矿的软熔滴落行为。

不同还原气氛下钒钛球团矿的软熔滴落结果如表 1 所示。结果表明，在 100%CO 条件下，钒钛球团矿软化开始温度为 1187℃，软化终了温度为 1315℃，软化区间为 128℃。通入 50%H$_2$+50%CO 后，软化开始温度、软化终了温度及软化区间分别降低 21℃、26℃、5℃。100%H$_2$ 条件下，相较于混合气氛，软化开始温度、软化终了温度及软化区间升高。然而，相较于 100%CO，软化开始温度、软化终了温度降低，软化区间增大。

在 100%CO 条件下，钒钛球团矿熔融开始温度、熔融终了温度、熔融区间分别为 1370℃、1514℃、144℃，滴落物质量为 137.3g。通入 H$_2$ 后，熔融开始温度、熔融终了温度升高，熔融区间降低，50%H$_2$+50%CO 条件下滴落物质量为 66g。100%H$_2$ 条件下熔融开始温度、熔融终了温度分别高达 1556℃、1568℃，滴落物质量仅为 8g。这一结果被认为与还原铁的渗碳过程密切相关，因为渗碳过程降低了金属铁的熔点。渗碳过程有两种碳源，一种是布尔多反应（Boudouard）的逆反应，另一种是焦炭。与焦炭相比，布尔多反应的逆反应对渗碳过程的影响不仅更强，而且更早。因此，当用 H$_2$ 代替或部分代替还原气中的 CO 时，产生的铁水量减少[1]。此外，H$_2$ 的加入极大提高了料层的透气性。

本文基于全 H$_2$（100% H$_2$）、全 CO（100% CO）气氛条件下对钒钛炉料软熔滴落过程的深入剖析，进一步加深钒钛炉料与 H$_2$-CO-C 耦合反应过程的理解，为钒钛磁铁矿高炉富氢冶炼提供理论支撑，有利于促进我国多金属复合矿资源利用科研和产业整体水平在低碳目标中的不断提升，助力行业低碳冶炼技术进步、变资源优势为技术和经济优势。

表 1 不同气氛下钒钛球团矿软熔滴落实验结果

还原气氛	软化开始温度/℃	软化终了温度/℃	软化温度区间/℃	熔融开始温度/℃	熔融终了温度/℃	熔融温度区间/℃	ΔP(max)/kPa	S 特征值/kPa·℃
100%CO	1187	1315	128	1370	1514	144	3.49	320.44
50%H$_2$+50%CO	1166	1289	123	1466	1553	87	2.56	112.86
100%H$_2$	1171	1312	141	1556	1568	12	1.33	10.04

参 考 文 献

[1] Ma K, Xu J, Deng J, et al. Hydrogen impact on the shrinkage behaviors of wustite packed beds above 900 ℃[J]. International Journal of Hydrogen Energy, 2019, 44: 19555-19562.

泰钢高炉长寿生产实践与技术探讨

薛玉卿，陈培敦，李 胜，王学军，孙其德，王 涛

（山东泰山钢铁集团有限公司炼铁部，山东济南 271100）

摘 要：本文重点对泰钢两座高炉长寿管理生产实践进行总结，结合高炉大修炉缸解剖调研结果，明确稳定边缘煤气流，减少冷却壁温度波动是实现炉体冷却壁长寿的主要因素；改善炉缸活跃状态，减小铁水边缘环流，优化铁水成分控制，强化炉缸冷却，促使炭砖热面形成稳定的保护层，阻止炭砖继续侵蚀，是有效控制炉缸侧壁温度上升，延长炉缸寿命的重要措施。

关键词：高炉；长寿管理；煤气流调控；铁水环流；铁水成分调控

Discussion on Practice and Technology of Blast Furnace Longevity Production in Taishan Steel

XUE Yuqing, CHEN Peidun, LI Sheng, WANG Xuejun, SUN Qide, WANG Tao

(Shandong Taishan Iron and Steel Group Co., Ltd., Ironmaking Department, Jinan 271100, China)

Abstract: This paper mainly summarizes the practice of longevity management of two blast furnaces in Taisteel. Combined with the investigation results of furnace cylinder overhaul, it is clear that stabilizing the edge gas flow and reducing the temperature fluctuation of the cooling wall are the main factors to achieve the longevity of the furnace cooling wall. Improving the active state of the hearth, reducing the circulation at the edge of the molten iron, optimizing the composition control of the molten iron, strengthening the cooling of the hearth, promoting the formation of a stable protective layer on the hot surface of the carbon brick and preventing the continuous erosion of the carbon brick are important measures to effectively control the temperature rise of the side wall of the hearth and extend the life of the hearth.

Key words: blast furnace; longevity management; gas flow regulation; hot metal circulation; composition control of hot metal

泰钢一号高炉长期焖炉快速恢复实践

薛玉卿，李 胜，孙其德，王 涛，韩传宝

（山东泰山钢铁集团有限公司炼铁部，山东济南 271100）

摘 要： 泰钢1号1780m³高炉根据指令进行为期20d的休风焖炉作业，为保证复风后快速恢复，从焖炉料的制定、休风操作、复风准备、送风初始制度和加风节奏、炉温平衡以及出铁管理等方面进行了严格的控制，实现了安全、顺利、快速达产的目的。

关键词： 高炉；焖炉；复风；达产

The Practice of Fast Recovery of Long - term Stewed Furnace in No. 1 Blast Furnace of Taishan Steel

XUE Yuqing, LI Sheng, SUN Qide, WANG Tao, HAN Chuanbao

(Shandong Taishan Iron and Steel Group Co., Ltd., Ironmaking Department, Jinan 271100, China)

Abstract: The No. 1 blast furnace of Taishan Steel carried out a 20d stop-blast operation according to the instruction. In order to ensure the rapid recovery after the restart, strict control was carried out from the aspects of the formulation of the stop-blast charge, the operation of the restart air, the preparation of the restart air, the initial system of air supply and the rhythm of air addition, the balance of the furnace temperature and the management of iron discharge, so as to achieve the purpose of safe, smooth and fast production.

Key words: blast furnace; stew oven; rewind; reach the production target

MgO 对包钢高炉渣系流变特性及熔化性的影响

冯 聪[1]，王雅军[1]，李玉柱[1]，白晓光[1]，郑占斌[2]，于恒亮[2]

（1. 内蒙古包钢钢联股份有限公司技术中心，内蒙古包头 014010；
2. 内蒙古包钢钢联股份有限公司炼铁厂，内蒙古包头 014010）

摘 要：基于白云鄂博矿的特殊性及造渣制度在高炉炼铁生产过程中的重要作用，本研究以包钢4号高炉现场渣为基础，采用纯化学试剂调制渣样，在氮气气氛条件下运用熔体物性综合测定仪分析了炉渣MgO含量对包钢高炉渣系熔化性温度、黏度及黏流活化能的影响。同时，采用Factsage8.2热力学软件研究了MgO对包钢高炉渣系熔化温度及热焓的影响。结果表明，炉渣MgO含量为9%~12%时，随着炉渣MgO含量的升高，包钢高炉渣系熔化性温度与黏流活化能呈先降低后升高趋势，各高温下黏度呈降低趋势，渣系流动性变好。另外，炉渣熔化温度与热焓呈升高趋势，渣系在高炉炉缸的蓄热能力增强。
关键词：白云鄂博矿；高炉渣；MgO；流变特性；熔化性

Effects of MgO on Viscous and Melting Behaviors of Blast Furnace Slag in Baotou Steel

FENG Cong[1], WANG Yajun[1], LI Yuzhu[1], BAI Xiaoguang[1],
ZHENG Zhanbin[2], YU Hengliang[2]

(1. Technical Center of Inner Mongolia Baotou Steel Union Co., Ltd., Baotou 014010, China;
2. Iron-making Plant of Inner Mongolia Baotou Steel Union Co., Ltd., Baotou 014010, China)

Abstract: Based on the specific mineral characteristics of Bayan Obo ore and the importance of slag system in the blast furnace (BF) ironmaking process, the effects of MgO on the break point temperature, viscosity and activated energy for viscous flow of BF slag in Baotou steel were analyzed. The experimental slag samples were prepared with the pure chemical reagents and the practical production slags in 4# BF of Baotou steel in a certain proportion. The measurements of viscous behaviors of slag were carried out under N_2 atmosphere with a melt property tester. Besides, the influences of MgO on the melting temperature and enthalpy of BF slag were studied by Factsage8.2 thermodynamic software. The results shows that the break point temperature and activated energy for viscous flow of BF slag in Baotou steel are initially decreased and subsequently increased with the increase of MgO content from 9.00wt% to 12.00wt%. At given temperatures, the viscosity of slag decreases when the MgO content increases, the fluidity of slag becomes better. In addition, as the MgO content increases, the melting temperature and enthalpy of BF slag calculated by Factsage8.2 increase, and the heat storage capacity of slag in the BF hearth is enhanced.
Key words: Bayan Obo ore; blast furnace slag; MgO; viscous behaviors; melting behaviors

高炉布料落点精准预测系统的应用实践

李 胜，薛玉卿，杨立春

（山东泰山钢铁集团有限公司炼铁部，山东济南 271100）

摘 要：针对目前原燃料质量波动频繁，尤其是粒度组成的变化导致高炉布料落点差异大，从而影响煤气流分布的问题，本系统将高炉原燃料粒度智能检测与布料落点预测有机融合，实现了高炉在日常布料过程中，根据不同原燃料粒度、料线深度以及溜槽倾角，实时预测布料落点位置，预测结果更加及时、精确，对指导高炉精准布料，控制煤气流合理分布，降低燃料消耗具有重要意义。

关键词：原燃料粒度；布料落点；煤气流；燃料消耗

Application Practice of Accurate Prediction System of Blast-furnace Distribution Drop Point

LI Sheng, XUE Yuqing, YANG Lichun

(Shandong Taishan Iron and Steel Group Co., Ltd., Ironmaking Department, Jinan 271100, China)

Abstract: In view of the frequent fluctuation of raw fuel quality, especially the change of particle size composition leads to a large difference in the drop point of the blast furnace, thus affecting the gas flow distribution, the system organically integrates the intelligent detection of raw fuel particle size of the blast furnace with the prediction of the drop point of the blast furnace. In the daily feeding process of the blast furnace, according to different raw fuel particle size, the depth of the feed line and the dip Angle of the chute, The real-time prediction of the drop point is more timely and accurate, which is of great significance for guiding the accurate distribution of blast furnace, controlling the reasonable distribution of gas flow and reducing fuel consumption.

Key words: size of raw fuel; cloth drop point; gas flow; fuel consumption

高温对焦炭宏观性能和微观结构的影响

付晓微[1]，何志军[1]，高立华[1]，杨立春[2]

（1. 辽宁科技大学材料与冶金学院，辽宁鞍山 114051；
2. 山东泰山钢铁集团有限公司，山东济南 271100）

摘 要：近些年，由于国家大力倡导"碳达峰，碳中和"，使钢铁行业面临更大的挑战。为了减少环境污染，钢铁企业需要降低焦比来减少二氧化碳的排放，这也对焦炭的质量提出了更高的要求。焦炭在高温环境下会产生粉末，导致焦炭强度下降，影响高炉的顺利运行。为了研究高温对宏观性能和微观结构的影响，在模拟高炉加热制度的条件下，将两种焦炭加热到1100~1400℃。结果表明，随着温度从1100℃升至1400℃，捣固焦炭和顶装焦炭的重量分别减少了9.04g和8.7g，反应后焦炭强度分别降低了4.23%和3.86%。此外，从环境温度到1400℃，捣固焦炭和顶装焦炭的电阻都有所下降。电阻变化率与反应后的焦炭强度呈负相关，随着电阻变化率的增加，焦炭强度也随之降低。比较焦炭的光学纹理指数、碳层间距、碳结构参数和大孔隙百分比与加热温度的关系发现，高温导致焦炭的各向异性纹理减少，石墨化程度增加，无定形碳部分转化为石墨碳，大孔隙百分比增加。此外，焦炭的孔隙面积越大，其孔壁硬度和弹性模量就越小。这些结果表明，各向异性纹理的减少、大孔隙和石墨化程度的增加会降低耐温性，导致反应后强度降低。

3.3 非高炉炼铁

HIsmelt 渣铁间硅分配比共存理论预报模型研究

李 林[1]，张福明[2]，郭汉杰[3]

（1. 北京中日联节能环保工程技术有限公司，北京 100040；2. 首钢集团有限公司，
北京 100041；3. 北京科技大学冶金与生态工程学院，北京 100083）

摘 要：为验证基于共存理论（IMCT）硅分配比预报模型的准确性，分析炉渣组元对脱硅的影响及贡献率。基于炉渣离子-分子共存理论，建立预测硅在 HIsmelt 熔融还原 CaO–SiO$_2$–MgO–FeO–Fe$_2$O$_3$–Al$_2$O$_3$–P$_2$O$_5$ 七元渣系与铁水间分配比的热力学模型，即 IMCT–L_{Si} 模型，并用国内某工厂 18 炉工业生产数据进行验证。结论为：首先，所建立的预报模型预测的硅分配比与 18 炉现场生产数据一致，说明 IMCT–L_{Si} 模型可以成功应用于 HIsmelt 工艺工业生产过程。其次，HIsmelt 七元渣系 CaO–SiO$_2$–MgO–FeO–Fe$_2$O$_3$–Al$_2$O$_3$–P$_2$O$_5$ 与铁水间平衡时，随渣中 CaO 和各类氧化铁含量的增加，硅分配比的对数线性增加；而随 SiO$_2$、Al$_2$O$_3$ 含量增加，硅分配比的对数线性减小。最终，炉渣中形成 2CaO·SiO$_2$、CaO·SiO$_2$、CaO·MgO·2SiO$_2$、2CaO·Al$_2$O$_3$·SiO$_2$ 和 CaO·MgO·SiO$_2$ 对脱硅的贡献率分别为 22.81%、19.92%、18.24%、16.22%、12%，是硅的分配比的主要贡献者，而形成其他结构单元对脱硅的贡献率极低，以至于可以忽略。

关键词：HIsmelt；硅分配比；共存理论；预报模型

Study on IMCT Prediction Model of Silicon Distribution Ratio for Slag-metal of HIsmelt

LI Lin[1], ZHANG Fuming[2], GUO Hanjie[3]

(1. Beijing JC Energy and Environment Engineering Co., Ltd., Beijing 100040, China;
2. Shougang Group Co., Ltd., Beijing 100041, China; 3. School of Metallurgical and Ecological
Engineering, University of Science and Technology Beijing, Beijing 100083, China)

Abstract: In order to verify the accuracy of the theoretical silicon distribution ratio prediction model based on IMCT theory, the influence and contribution rate of slag components on desiliconization were analyzed. Based on the IMCT theory, a thermodynamic IMCT–L_{Si} model was established to predict the silicon distribution ratio of slag-metal for CaO–SiO$_2$–MgO–FeO–Fe$_2$O$_3$–Al$_2$O$_3$–P$_2$O$_5$ slag system, this model was verified by industrial production data of 18 furnaces in a domestic plant. The conclusion is as follows: firstly, the predicted silicon distribution ratio of IMCT model is consistent with the industrial production data of 18 furnaces, this shows that IMCT-L_{Si} model can be successfully applied to HIsmelt industrial production process. Secondly, when HIsmelt CaO–SiO$_2$–MgO–FeO–Fe$_2$O$_3$–Al$_2$O$_3$–P$_2$O$_5$ slag system lies in slag-metal equilibrium, the logarithm of distribution ratio increases with the increase of content of CaO and all kinds of iron oxide in the slag, decreases with the increase of content of SiO$_2$ and Al$_2$O$_3$. Finally, the content of 2CaO·SiO$_2$, CaO·SiO$_2$, CaO·MgO·2SiO$_2$, 2CaO·Al$_2$O$_3$·SiO$_2$ and CaO·MgO·SiO$_2$ to desilication is 22.81%, 19.92%, 18.24%, 16.22% and 12% respectively, they are main contributors of desilication, while the contribution of other structural units to desilication is so low that they can be ignored.

Key words: HIsmelt; silicon distribution ratio; IMCT; prediction model

HIsmelt 渣铁间磷分配比共存理论预报模型研究

李 林[1]，张福明[2]，郭汉杰[3]，林 松[1]，徐海如[1]

（1. 北京中日联节能环保工程技术有限公司，北京 100040；2. 首钢集团有限公司，北京 100041；3. 北京科技大学冶金与生态工程学院，北京 100083）

摘 要：为验证基于共存理论（IMCT）的磷分配比预报模型的准确性，分析炉渣组元对脱磷的影响及贡献率。基于炉渣离子–分子共存理论，建立预测磷在 HIsmelt 熔融还原 $CaO-SiO_2-MgO-FeO-Fe_2O_3-Al_2O_3-P_2O_5$ 七元渣系与铁水间分配比的热力学模型，即 $IMCT-L_P$ 模型，并用国内某工厂 18 炉工业生产数据进行验证。结论为：$IMCT-L_P$ 模型可成功应用于 HIsmelt 工业生产过程。基于 HIsmelt 渣金间平衡时，随渣中 CaO、MgO 和各类氧化铁含量的增加，磷分配比的对数线性增加；而随 SiO_2、Al_2O_3 含量增加，磷分配比的对数线性减小。炉渣中生成的 $3CaO \cdot P_2O_5$ 对磷分配比的贡献率为 99.7%，CaO 对脱磷起到极大的协同作用。

关键词：HIsmelt；磷分配比；共存理论；预报模型

Study on IMCT Prediction Model of Phosphorus Distribution Ratio for Slag-metal of HIsmelt

LI Lin[1], ZHANG Fuming[2], GUO Hanjie[3], LIN Song[1], XU Hairu[1]

(1. Beijing JC Energy and Environment Engineering Co., Ltd., Beijing 100040, China; 2. Shougang Group Co., Ltd., Beijing 100041, China; 3. School of Metallurgical and Ecological Engineering, University of Science and Technology Beijing, Beijing 100083, China)

Abstract: In order to verify the accuracy of the theoretical silicon distribution ratio prediction model based on IMCT, the influence and contribution rate of slag components on desiliconization were analyzed. Based on the IMCT, a thermodynamic $IMCT-L_P$ model was established to predict the silicon distribution ratio of slag-metal for $CaO-SiO_2-MgO-FeO-Fe_2O_3-Al_2O_3-P_2O_5$ slag system, this model was verified by industrial production data of 18 furnaces in a domestic plant. The conclusion is as follows: $IMCT-L_P$ model can be successfully applied to HIsmelt industrial process. Based on the HIsmelt slag-metal equilibrium, the logarithmic of phosphorus distribution ratio increases linearly with the increase of CaO, MgO and various iron oxides in the slag. With the increase of SiO_2 and Al_2O_3 content, the logarithmic of phosphorus partition ratio linearly decreases. The contribution rate of $3CaO \cdot P_2O_5$ on the phosphorus distribution ratio is 99.7%, and CaO plays a great effect on dephosphorization.

Key words: HIsmelt; phosphorus distribution ratio; IMCT; prediction model

氨气还原不同种类铁矿石的低温冶金性能研究

刘玥君，李先春，李 丽，林建廷

（辽宁科技大学化学工程学院，辽宁鞍山 114000）

摘　要：随着工业化进程的不断加快，传统高炉炼铁工艺因消耗大量焦炭造成了严重的环境污染问题[1-3]。2020 年我国在联合国大会上提出了"中国的 CO_2 排放量将于 2030 年前达到峰值，于 2060 年前实现碳中和"的重要承诺。钢铁行业的 CO_2 年排放量占全球碳排放的 7%左右[4]，正面临着能源低碳转型的挑战[5]。为了实现氨气在钢铁行业的高效大规模应用，推动氢冶金向氨冶金的转变，不仅要探究还原过程的反应规律，还要从实验室规模拓展到中试规模，为钢铁行业提供相关的理论支撑。氢基竖炉内以球团矿和块矿作为原料，因此考察还原气的改变对铁矿石还原粉化性能的影响是至关重要的。

利用铁矿石冶金综合性能测定装置考察了 900℃时 NH_3 还原不同种类铁矿石的冶金性能。结果表明，当 NH_3 低于 20%时，球团矿还原 3h 后样品内除了 Fe 还有少量 $CaFeSi_2O_6$，30% NH_3 对不同种类赤铁矿、针铁矿、球团矿和烧结矿均具有良好的还原特性，还原度都达到了 97%以上，而用 40%CO 还原球团矿 3h 后还原度仅有 68.6%。氨气浓度越高，还原反应越剧烈，吸热量越大，反应器中心温度下降的越多。用 N_2 和 Ar 作为载气对球团矿还原速率和还原度几乎没有影响。利用偏光显微镜对还原后的矿相组织进行分析，结果表明经过 30% NH_3 还原 3h 后，烧结矿内复合铁酸钙的针状形貌和球团矿内的未反应核形貌几乎完全消失；由于 NH_3 和 H_2 的扩散作用强，样品内产生了大量裂纹且海绵铁大量生成。还原后不同种类的铁矿石的抗压强度都得到了显著提高。在相同温度下利用 NH_3 还原后的铁矿石还原粉化率比 CO 的小，分析因为 NH_3 还原体系内是吸热的，不存在析碳反应，对铁矿石结构的影响没有 CO 的高。NH_3 较比 CO 有更好的还原能力，对铁矿石的品位和种类要求更低，在竖炉内应用氨气作为还原剂具有极高的潜力。

参 考 文 献

[1] Bhaskar A, Assadi M, Somehsaraei H N. Decarbonization of the Iron and Steel Industry with Direct Reduction of Iron Ore with Green Hydrogen[J]. Energies, 2020, 13(758): 1-23.

[2] Yanni X, Yue Q. Retrospective and Prospective Analysis on the Trends of China's Steel Production[J]. Journal of Systems Science and Information, 2016, 4(4): 291-306.

[3] Jialin S, Qi Z, Lisong X, et al. Future CO_2 emission trends and radical decarbonization path of iron and steel industry in China[J]. Journal of Cleaner Production, 2021, 326(11): 1-10.

[4] Alexandra D, Jannik K, Haulwen G. Global green hydrogen-based steel opportunities surrounding high quality renewable energy and iron ore deposits[J]. Nature Communications, 2023, 14: 1-14.

[5] Ming R, Pantao L, Xiaorui L, et al. Decarbonizing China's iron and steel industry from the supply and demand sides for carbon neutrality[J]. Applied Energy, 2021, 298: 1-13.

氨气直接还原铁矿石的冶金性能研究

刘玥君，李先春，李　丽，林建廷

（辽宁科技大学化学工程学院，辽宁鞍山　114000）

摘　要：随着工业化进程的不断加快，传统高炉炼铁工艺因消耗大量焦炭造成了严重的环境污染问题[1-3]。2020 年我国在联合国大会上提出了"中国的 CO_2 排放量将于 2030 年前达到峰值，于 2060 年前实现碳中和"的重要承诺。钢铁行业的 CO_2 年排放量占全球碳排放的 7%左右[4]，正面临着能源低碳转型的挑战[5]。为了实现氨气在钢铁行业的高效大规模应用，推动氢冶金向氨冶金的转变，不仅要探究还原过程的反应规律，还要从实验室规模拓展到中试规模，为钢铁行业提供相关的理论支撑。氢基竖炉内以球团矿和块矿作为原料，因此考察还原气的改变对铁矿石还原粉化性能的影响是至关重要的。

利用铁矿石冶金综合性能测定装置考察了 900℃时 NH_3 还原不同种类铁矿石的冶金性能。结果表明，当 NH_3

低于 20%时，球团矿还原 3h 后样品内除了 Fe 还有少量 CaFeSi$_2$O$_6$，30%NH$_3$ 对不同种类赤铁矿、针铁矿、球团矿和烧结矿均具有良好的还原特性，还原度都达到了 97%以上，而用 40%CO 还原球团矿 3h 后还原度仅有 68.6%。氨气浓度越高，还原反应越剧烈，吸热量越大，反应器中心温度下降的越多。用 N$_2$ 和 Ar 作为载气对球团矿还原速率和还原度几乎没有影响。利用偏光显微镜对还原后的矿相组织进行分析，结果表明经过 30%NH$_3$ 还原 3h 后，烧结矿内复合铁酸钙的针状形貌和球团矿内的未反应核形貌几乎完全消失；由于 NH$_3$ 和 H$_2$ 的扩散作用强，样品内产生了大量裂纹且海绵铁大量生成。还原后不同种类的铁矿石的抗压强度都得到了显著提高。在相同温度下利用 NH$_3$ 还原后的铁矿石还原粉化率比 CO 的小，分析因为 NH$_3$ 还原体系内是吸热的，不存在析碳反应，对铁矿石结构的影响没有 CO 的高。NH$_3$ 较比 CO 有更好的还原能力，对铁矿石的品位和种类要求更低，在竖炉内应用氨气作为还原剂具有极高的潜力。

参 考 文 献

[1] Bhaskar A, Assadi M, Somehsaraei H N. Decarbonization of the iron and steel industry with direct reduction of iron ore with green hydrogen[J]. Energies, 2020, 13(758): 1-23.

[2] Yanni X, Yue Q. Retrospective and prospective analysis on the trends of china's steel production[J]. Journal of Systems Science and Information, 2016, 4(4): 291-306.

[3] Jialin S, Qi Z, Lisong X, et al. Future CO$_2$ emission trends and radical decarbonization path of iron and steel industry in China[J]. Journal of Cleaner Production, 2021, 326(11): 1-10.

[4] Alexandra D, Jannik K, Haulwen G. Global green hydrogen-based steel opportunities surrounding high quality renewable energy and iron ore deposits[J]. Nature Communications, 2023, 14: 1-14.

[5] Ming R, Pantao L, Xiaorui L, et al. Decarbonizing China's iron and steel industry from the supply and demand sides for carbon neutrality[J]. Applied Energy, 2021, 298: 1-13.

绿电低温电解铁

杨海涛 [1,2,3]，王 鑫 [1]，谭钰桦 [1]，程佳鑫 [1]，朱庆山 [1,2,3]

（1. 中国科学院过程工程研究所多相复杂系统国家重点实验室，北京 100190；2. 中国科学院大学化学工程学院，北京 101408；3. 中国科学院绿色过程制造创新研究院，北京 100190）

摘 要：钢铁工业是国民经济的支柱产业，也是二氧化碳排放大户。传统高炉炼铁-转炉炼钢流程中，炼铁工序碳排放约占全流程的 70%。开发中的低碳炼铁技术主要可分为绿氢冶金和绿电冶金。绿氢冶金采用可再生能源电力电解水制氢，然后采用氢气还原铁矿石，本质上也是消耗的绿色电能。绿电冶金采用电解的方法还原铁矿石，跨过制氢工序。另一方面，近些年光伏和风电取得了快速发展，然而风光绿电是间歇和波动的。传统的冶金都是连续生产，为了能够与连续冶金相适应，就需要加配储电装置、或者储氢装置，储电和储氢本身也消耗能量。着眼于未来可再生电力与钢铁减碳需求，本文探索了绿电直接（柔性）电解铁工艺。实验分别采用恒电流、脉冲电流、模拟光伏电流、真实光伏电流等条件进行，获得了平滑光亮的电解铁，纯度≥99.9%，能耗 3.5~4.0kWh/kg-Fe；揭示了非线性电沉积铁的微观形貌及物相变化规律。绿电低温电解铁为钢铁低碳转型提供了一个多元化的技术路线。

关键词：绿电低温电解铁；柔性电还原炼铁；低碳钢铁；光伏电解铁；绿色纯铁

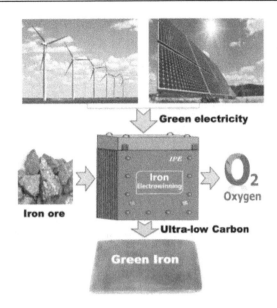

图1 绿电低温电解铁

参 考 文 献

[1] 朱庆山. 化工进展, 2022, 41(3): 1391-1398.
[2] Cheng J X, Yang H T, Wang X, et al. Renewable and Sustainable Energy Reviews, 2023, 183:113487.
[3] Wang X, Yang H T, Yu X H, et al. Journal of Applied Electrochemistry, 2023, 53: 1521-1536.
[4] Tan Y H, Yang H T, Tian G C, et al. Industrial & Engineering Chemistry Research, 2023, 62(12): 4817-4830.
[5] Tan Y H, Yang H T, Tian G C, et al. International Journal of Hydrogen Energy, 2022, 47(90): 38134-38154.

基于 HIsmelt 工艺 $CaO-SiO_2-MgO-Al_2O_3-TiO_2-FetO$ 渣系脱硫能力的研究

刘晏廷[1]，刘 然[1]，高艳甲[1]，王小艾[2]

（1. 华北理工大学，河北唐山 063200；2. 河钢集团材料技术研究院，河北石家庄 050023）

摘 要：炉渣脱硫在炼铁工艺中具有显著的环境重要性。硫是许多工业活动和燃烧过程中产生的主要污染物之一，其排放会导致严重的环境问题。二氧化硫等硫化物不仅对大气质量造成影响，还会形成酸雨，损害土壤和水体，危害植物生长和水生生物。此外，硫化物排放还会对人类健康产生负面影响，引发呼吸道疾病等健康问题。通过炉渣脱硫技术，可以有效减少硫化物的排放。而当今的环保形势日益严峻，这对传统高炉炼铁工艺的冶炼带来了诸多限制。为了应对"双碳"战略压力，钢铁行业急需降低碳排放。因此，非高炉炼铁工艺备受冶金工作者的关注，其中 HIsmelt 工艺以原料适应性强，污染少等特点备受推崇[1,2]。然而，该工艺冶炼钒钛磁铁矿时，熔渣中 FeO 的含量高，熔渣脱硫能力较低。因此，为了充分发挥该工艺的优势，需要对高钛高 FeO 渣系的硫容量进行深入研究[3]。

本研究采用气-渣平衡法，测定了六元渣系 $CaO-SiO_2-Al_2O_3-MgO-FeO-TiO_2$ 熔渣硫容量，研究了炉渣成分和温度对硫容量的影响规律。熔渣硫容量随着碱度、MgO 和 FeO 含量的增加而增加，随着 TiO_2 含量的增加先增加后降低。结果表明：温度 1773K 时，碱度可以提高熔渣硫容量。研究表明随着二元碱度的提高，熔渣硫容量逐渐增加。

当碱度达到 1.3 时硫容量的 log 值达到最大值–3.3699；温度能够显著增加熔渣硫容量，本研究发现熔渣的 $\log C_s$ 值与 $10^4/T$ 成线性关系。其斜率为 2.14 与其他文献相差较大，可能是由于渣中 FeO 含量较高；在 $CaO-SiO_2-Al_2O_3-MgO-FeO-TiO_2$ 渣系中，随着 TiO_2 含量的从 10%增加到 30%，熔渣硫容量先增加，当 TiO_2 含量增加到 15%时达到最大值–3.3611 继续增加 TiO_2 含量硫容量反而降低并趋于稳定；增加 MgO 含量能够提高熔渣硫容量，MgO 含量从 6%增加到 8%时硫容量显著提高，再提高 MgO 含量熔渣硫容量升高趋势变缓。因此，造渣制度中 MgO 含量应该控制在 8%。渣中 FeO 含量对熔渣硫容量起到积极作用，当 FeO 含量由 3%增加到 9%时，硫容量增加的趋势缓慢，当 FeO 含量由 9%增加到 15%时，硫容量显著增大。

参 考 文 献

[1] 贾利军, 汤彦玲. HIsmelt 熔融还原技术现状及生产实践[J]. 工业炉, 2022, 44(1): 20-23.
[2] 张树石, 胡鹏, 饶家庭, 等. 钒钛磁铁矿综合利用现状及 HIsmelt 冶炼可行性分析[J]. 中南大学学报(自然科学版), 2021, 52(9): 3085-3092.
[3] CONDO A F T, ALLERTZ C, SICHEN D. Experimental determination of sulphide capacities of blast furnace slags with higher MgO contents[J]. Ironmaking & Steelmaking, 2017, 46(3): 207-210.

八钢冶金煤气脱碳系统运行总结

田 果[1]，王 鑫[1]，朱利刚[1]，刘西坤[2]

（1. 新疆八一钢铁股份有限公司炼铁厂，新疆乌鲁木齐 830022；
2. 中钢设备有限公司，北京 100000）

摘 要：2019 年八钢公司开展冶金煤气 CO_2 捕集技术科研工作，通过捕集欧冶炉冶金煤气中的 CO_2、水分，进行煤气成分重整。至今系统已经运行 3 年，未进行全面检修，在近几年运行的过程中出现了较多的问题，溶液循环、溶液丢失、设备故障、脱碳效率无法达标等多重问题考验系统设施的稳定运行，文章对八钢冶金煤气脱碳系统运行实践过程进行了分析总结。
关键词：冶金煤气；循环泵；溶液丢失；腐蚀；压缩机

Operation Analysis of Metallurgical Gas Decarbonization System of Bayi Iron and Steel Co.

TIAN Guo[1], WANG Xin[1], ZHU Ligang[1], LIU Xikun[2]

(1. Xinjiang Bayi Iron and Steel Co., Ltd., Iron Making Plant, Urumqi 830022, China;
2. Sinosteel Equipment Co., Ltd., Beijing 100000, China)

Abstract: In 2019, Bayi Steel Co., Ltd. carried out scientific research on CO_2 capture technology of metallurgical gas, and reformed gas components by capturing CO_2 and moisture in metallurgical gas of European metallurgical furnace. So far, the system has been running for 3 years without comprehensive overhaul. In recent years, many problems have appeared in the process of operation, such as solution circulation, solution loss, equipment failure, decarbonization efficiency can not reach the standard and other multiple problems test the stable operation of the system facilities. The paper analyzes and summarizes the operation practice process of metallurgical gas decarbonization system of Basteel.
Key words: metallurgical gas; circulating pump; solution loss; corrosion; compressor

基于焦炉煤气的直接还原技术污染物协同治理生产实践

李晓兵

（河钢张宣高科科技有限责任公司，河北张家口　075000）

摘　要：在我国"双碳"政策背景下，河钢张宣科技采用焦炉煤气零重整工艺生产直接还原铁，可以达到减碳的目的，在整个工艺流程中产生各种污染物，建设过程中实施多种污染物治理技术，实现环境颗粒物排放≤5mg/m³，水中氨氮含量≤10mg/L，CO_2气体脱除后H_2S含量<0.1ppm，焦炉煤气总硫<10mg/m³，不仅满足了工艺生产的需求，更是达到了国家排放标准，实现了焦炉煤气在直接还原技术中成功应用。

关键词：焦炉煤气；直接还原工艺；污染物；治理

Production Practice of Collaborative Treatment of Pollutants Based on Direct Reduction Technology of Coke Oven Gas

LI Xiaobing

(HBIS Zhangxuan high-tech technology Co., Ltd., Zhangjiakou 075000, China)

Abstract: Under the backdrop of China's "Dual Carbon" policy, HBZX use zero reforming process of coke oven gas to produce direct reduced iron, which can achieve the purpose of carbon reduction. Various pollutants are produced in the whole process, and various pollutant treatment technologies are implemented in the construction process to achieve the emission of particulates remains at or below 5mg/m³, the concentration of ammonia nitrogen in water is maintained at or below 10mg/L, the content of H_2S after CO_2 removal remains under 0.1ppm, and the total sulfur presence within the COG is kept under 10mg/m³. This approach not only fulfills the demands of the production process, but also adheres to national emission standards, thereby demonstrating the successful incorporation of COG into direct reduction technologies.

Key words: coke oven gas; direct reduction process; pollutants; treatment

$NaHCO_3$对铌精矿微波碳热还原的影响研究

文　明，刘雅琦

（内蒙古科技大学材料与冶金学院，内蒙古包头　014010）

摘　要：由于白云鄂博铌矿物相互依存、相互包裹以及嵌布复杂的特点，导致选冶工艺复杂，不能得到品位较高的铌富集物。基于此，本文提出了白云鄂博含铌矿物资源选冶利用的新思路-采用冶金手段将多种铌矿相转化为单一的$NaNbO_3$矿相，通过磁选得到含有$NaNbO_3$的富集产物，实现铌的高效分离与富集，便于后期选冶利用。以白云鄂博含铌矿物为研究对象，在微波场下，分别考察添加剂$NaHCO_3$加入量、还原温度以及还原时间对含铌矿物碳热还原反应的影响，获得含铌矿物微波碳热还原的最佳工艺参数。

利用双参数模型计算二元氧化物 $FeNb_2O_6$ 和 $NaNbO_3$ 的熵变、焓变、热容和吉布斯自由能，填补了相关铌矿物热力学参数的空白，加入钠剂后铌矿物的转变趋势为：$NaNbO_3$＞$FeNb_2O_6$＞NbC。通过计算铌矿物物相转变的热力学行为，证明铌矿物中 Nb_2O_5 与 $NaHCO_3$ 反应生成 $NaNbO_3$ 在热力学上是可行的。

在含铌矿物微波碳热还原过程中，还原产物的金属化率以及失重率随着还原温度的升高而增长；随着添加剂 $NaHCO_3$ 加入量的增多，金属化率显著增大，当加入量为 2%时，金属化率达到最大值，之后趋于平缓；金属化率随着还原时间的增加先增大后减小，还原时间在 30min 时，金属化率达到最大。在微波场下，含铌矿物微波碳热还原反应进行研究，得到最佳的工艺参数：还原温度 1100℃，还原时间为 30min，添加剂 $NaHCO_3$ 加入量为 2%，该条件下对应金属化率为 94.80%。

通过热力学分析以及微波碳热还原实验证明白云鄂博含铌矿物资源选冶利用的新思路是可行的，实现了将多种铌矿相转化为单一的 $NaNbO_3$ 矿相的目的，实现了铌的高效分离与富集，便于后期选冶利用，并且得出了含铌矿物加入 $NaHCO_3$ 微波碳热还原的最佳工艺条件，具有实际应用价值。

4 炼钢与连铸

大会特邀报告
第十三届冶金青年科技奖获奖人特邀报告
分会场特邀报告
矿业工程
焦化及节能环保
炼铁与原料
★ 炼钢与连铸
电冶金与废钢铁
轧制与热处理
表面与涂镀
金属材料深加工
粉末冶金
先进钢铁材料及应用
节能与低碳技术
冶金环保与资源利用
冶金设备与工程技术
冶金自动化与智能化
冶金物流
冶金流程工程学
其他

4.1 炼 钢

基于沸腾出钢工艺的 SPHC 钢渣洗脱磷研究

李 超，黄 岩，尚德义，冉茂铎

（鞍钢股份鲅鱼圈钢铁分公司炼钢部，辽宁营口 115007）

摘 要：本文针对沸腾出钢渣洗工艺进行了研究，确定影响脱磷的主要因素为氧含量、白灰加入量、吹氩时间，通过生产数据分析得到：脱磷率随着氧含量增大不断增大；脱磷率随着白灰重量不断增大，但在白灰加入量达到一定范围后，脱磷率维持最大值没有明显变化；随着吹氩时间增加，开始阶段脱磷率没有变化，50s 之后脱磷率增大并在 180s 达到最大值，之后通过延长吹氩无法实现脱磷率的增大。通过统计采用新工艺生产数据，平均脱磷率由试验前 9.4%提高到 16.7%，磷含量超标准的质量事故平均降低 0.3 罐/月，未出现因新工艺应用造成的钢水温度低相关事故。

关键词：渣洗；氧含量；白灰；吹氩

Study on Slag-washing Phosphorus Elution of SPHC Steel Based on Boiling Steel Tapping Process

LI Chao, HUANG Yan, SHANG Deyi, RAN Maoduo

(Steelmaking Department Bayuquan Iron & Steel Subsidiary of
Angang Steel Co., Ltd., Yingkou 115007, China)

Abstract: In this paper, phosphorus elution process of boiling steel slag-washing is Studied, the main factors affecting deshosphating are oxygen content, lime addition weight and argon blowing time, According to the production data analysis, dephosphorization rate is increasing with the increase of the oxygen content, The dephosphorization rate increases with the weight of the lime, but after the addition of the lime reaches a certain range, the maximum rate remains, The dephosphorization rate does not change with the increase of argon blowing time, After 50 seconds later, the dephosphorization rate increases, then the maximum value reaches at 180 seconds, the increase of the dephosphorization rate cannot be achieved by extending the argon blowing time. According to the production data of the new process, the average phosphorus dephosphorization rate increased from 9.4% to 16.7%, and the quality accidents with phosphorus content exceeding the standard decreased by 0.3 tank / month, and there was no accident of low molten steel temperature caused by the application of the new process.

Key words: slag-washing; oxygen content; lime; blowing argon

电感耦合等离子体发射光谱法测定钛铁中微量元素含量

任玲玲，谭胜楠，戚振南

（河钢材料技术研究院，河北石家庄 050000）

摘　要：在测定钛铁合金中硅、锰、磷、铝、铬、锆等微量元素含量时，采用碱熔融溶解样品存在操作复杂、检出限高的问题，故本文采用硝酸-氢氟酸-盐酸常温或冷水浴中溶解样品，降低硅元素、磷元素损失，建立了采用电感耦合等离子体发射光谱法（ICP-AES）测定钛铁合金中硅、锰、磷、铝、铬、锆元素的方法。结果表明，采用基体匹配法配制校准曲线，各元素校准曲线相关系数均大于 0.99994，元素测定下限在 0.0008%~0.015% 之间。根据实验方法测定钛铁标准样品中硅、锰、磷、铝、铬，结果相对标准偏差（RSD, n=8）在 0.76%~4.30%，回收率在 95%~104%，检测结果与 2 个标准样品认定值相吻合。

关键词：电感耦合等离子体发射光谱法；钛铁；硅；铝；磷；锰；铬；锆

Determination of Silicon, Manganese, Phosphorus, Aluminum, Chromium, Zirconium in Ferrotitanium by Inductively Coupled Plasma Atomic Emission Spectrometry

REN Lingling, TAN Shengnan, QI Zhennan

(HBIS Materials Technology Research Institute, Shijiazhuang 050000, China)

Abstract: During the determination of silicon, manganese, phosphorus, aluminum, chromium, zirconium in ferrotitanium, the sample always be melted with alkali, which has problems of complicated operation and high detection limit. Therefore, for reducing the loss of silicon, phosphorus, the samples were decomposed with nitric acid - hydrofluoric acid - hydrochloric acid at room temperature or in a cold bath. The contents of Si, Mn, P, Al and Zr in ferrotitanium were determined by inductively coupled plasma atomic emission spectrometry. The calibration curves were prepared by matrixing matching method. Within linear range of calibration curves of elements, the correlation coefficients are greater than 0.99994. The limit of detection for elements is between 0.0008% and 0.015%. The relative standard deviations (RSD, n=8) of Si, Mn, P, Al and Zr in the standard sample were between 0.76% and 4.30%, and the recoveries were between 95% and 104%. The detection results were consistent with the determination values of the two standard samples.

Key words: inductively coupled plasma atomic emission spectrometry(ICP-AES); ferrotitanium; silicon; manganese; phosphorus; aluminum; chromium; zirconium

真空脱气炉关键设备管理对钢水质量提升浅谈

翟昱明，王玉翠，苏小宇

（包钢钢管有限公司，内蒙古包头 014010）

摘　要：包钢钢管公司现有方圆坯连铸机建设于 1995 年，炼钢厂 2012 年建立干 VD 真空脱气炉。真空脱气炉主要是为了实现钢液中的氮，氢气体及一些杂质的去除。在提高公司产品"品种，质量，效益"等方面发挥重要作用，同时为了保生产，提效率，对泵组进行细致的，周期性维护管理是必不可少的。除了必要的设备维护与管理，泵的使用与现场工况磨合也是泵维护管理关键因素。为了满足现场使用避免故障，维护及使用都制定了严格的检修维护标准。

关键词：连铸设备；VD 真空脱气炉；周期检修；周期管理

Study on Maintenance and Management of VD Vacuum Degasification Furnace for Steelmaking

ZHAI Yuming, WANG Yucui, SU Xiaoyu

(Baotou Steel Tube Co., Ltd., Baotou 014010, China)

Abstract: Baotou steel tube company existing square round casting machine construction in 1995, a dry VD vacuum degasification furnace was established in 2012. The vacuum degassing furnace is mainly used to remove nitrogen, hydrogen gas and some impurities in the steel liquid. Play an important role in improving product "variety, quality and efficiency" at the same time in order to maintain production and improve efficiency, it is necessary to carry out detailed and periodic maintenance and management of the pump group .In addition to the necessary equipment maintenance and management, pump use and site working condition running in is also a key factor of pump maintenance management .In order to meet the requirements of site operation to avoid failure, strict maintenance standards have been established for maintenance and use.

Key words: continuous casting equipment; VD vacuum degassing furnace; periodic maintenance; cycle management

重庆钢铁无取向硅钢 RH 精炼温度控制实践

文　敏，陈露涛，刘向东，姚前光，张　俊，吴　伟

（重庆钢铁股份有限公司，重庆　401258）

摘　要：无取向硅钢 RH 精炼过程温度控制是主要难点，特别是脱氧合金化后温度低，无有效的温度补偿手段，本文主要介绍无取向硅钢 RH 精炼过程温度控制方法。

关键词：无取向硅钢；RH 精炼；温度控制

洁净钢生产现状及钢包引流新工艺研究

李德军[1,2]，杨　光[1,2]，金　喆[1,2]，李博洋[1,2]，张维维[1,2]，
黄玉平[1,2]，栗　红[1,2]，赵成林[1,2]

（1. 海洋装备用金属材料及其应用国家重点实验室，辽宁鞍山　114009；
2. 鞍钢集团钢铁研究院，辽宁鞍山　114009）

摘　要：本文通过对洁净钢和目前国内外洁净钢生产工艺发展现状的介绍，论述了提高钢铁产品洁净度的实际意义，在此基础之上提出了钢包电磁引流工艺技术，并对该技术的开发背景和基本原理进行了系统分析，并通过数值模拟实验和热态实验对该技术的可行性以及该技术中的各参数进行了分析说明，表明该技术是可行的，并能够为洁净钢的生产提供有益的借鉴。

关键词：洁净钢；夹杂物；钢包；电磁引流

Clean Steel Production Status and New Process of Ladle-tapping

LI Dejun[1,2], YANG Guang[1,2], JIN Zhe[1,2], LI Boyang[1,2], ZHANG Weiwei[1,2], HUANG Yuping[1,2], LI Hong[1,2], ZHAO Chenglin[1,2]

(1. State Key Laboratory of Metal Material for Marine Equipment and Application, Anshan 114009, China;
2. Iron & Steel Research Institutes of Ansteel Group Corporation, Anshan 114009, China)

Abstract: The status of clean steel and its production at home and abroad are introduced and the practical significances of improving the steel products' cleanliness are discussed in this article. On this basis, a new process of electromagnetic-tapping of ladle is proposed, of which background and basic principles were systematically analyzed. Therefore through numerical simulation and thermal experiment the feasibility and the various parameters of the technology was analyzed. It shows that the technology is feasible and able to provide a beneficial reference for clean steel production.

Key words: clean steel; inclusion; ladle; electromagnetic drainage

转炉高废钢比冶炼的关键单元技术进展

杨　光[1,2]，李德军[1,2]，李博洋[1,2]，赵成林[1,2]，
金　喆[1,2]，朱晓雷[1,2]

（1. 海洋装备用金属材料及其应用国家重点实验室，辽宁鞍山　114009；
2. 鞍钢集团钢铁研究院，辽宁鞍山　114009）

摘　要：钢铁行业的绿色发展一直备受关注，转炉高废钢比冶炼则是长流程钢铁制造过程中实现降低碳排的重要措施。本文主要对废钢预热、转炉内添加补热剂、二次燃烧和底喷粉等关键单元技术的研究及应用进展进行了介绍。

关键词：高废钢比；转炉；冶炼；低碳

Development of Key Unit Technology with High Scrap Ratio in Converter Steelmaking

YANG Guang[1,2], LI Dejun[1,2], LI Boyang[1,2], ZHAO Chenglin[1,2], JIN Zhe[1,2], ZHU Xiaolei[1,2]

(1. State Key Laboratory of Metal Material for Marine Equipment and Application, Anshan 114009, China; 2. Ansteel Iron & Steel Research Institutes, Anshan 114009, China)

Abstract: The green development of the steel industry has been attracting much attention, and the high scrap ratio smelting of converter is an important measure to reduce carbon emission in the long-process steel manufacturing process. This paper mainly introduces the research and application progress of key unit technologies such as scrap steel preheating, adding heat supplement in the converter, post combustion, and converter bottom powder.

Key words: high scrap ratio; converter; steelmaking; low-carbon

管线钢精炼渣物相及岩相组成对炉渣性能的影响研究

吕春风[1,2]，滕行泽[1,2]，尚德礼[1,2]，廖相巍[1,2]，赵成林[1,2]

（1. 海洋装备金属材料及应用国家重点实验室，辽宁鞍山　114009；
2. 鞍钢集团钢铁研究院，辽宁鞍山　114009）

摘　要：为提高管线钢精炼渣的稳定控制和精炼效果，研究了管线钢 LF 精炼渣的化学组分、结晶矿相和微观结构对炉渣熔化性能和精炼效果的影响，结果表明：管线钢精炼渣结晶矿相主要由作为基质相的低熔点铝酸钙和在基质相上析出的高熔点相 $2CaO \cdot SiO_2$、$MgO \cdot Al_2O_3$ 等高熔点相组成，高熔点相的含量对炉渣的熔化性影响较大。研究表明，为了保证精炼渣的性能，应调整精炼渣成分位于相图中低熔点的钙铝酸盐区域，严格避免精炼渣成分处于相图中的硅酸盐、石灰、尖晶石等高熔点区域范围内，可使精炼渣具有良好的流动性，促进钢渣反应，提高精炼渣吸附夹杂的能力，提高精炼渣的精炼效果。

关键词：物相组成；岩相组成；精炼渣；夹杂物

Study on the Influence of Physical and Petrographic Composition of Slag on Slag Properties in Pipeline Steel Refining

LV Chunfeng[1,2], TENG Xingze[1,2], SHANG Deli[1,2], LIAO Xiangwei[1,2], ZHAO Chenglin[1,2]

(1. Key Laboratory of Metal Materials for Marine Equipment and Application, Anshan 114009, China;
2. Metallurgy Department of Ansteel Iron & Steel Research Institutes, Anshan 114009, China.

Abstract: Abstract: In order to improve the stable control and refining effect of the refining slag for pipeline steel, the influence of the chemical composition, crystalline mineral phase and microstructure of LF refining slag on the melting performance and refining effect of the slag has been investigated for pipeline steel. The results show: The crystalline mineral phase of pipeline steel refining slag is mainly composed of the low melting point calcium aluminate as the matrix phase and the high melting point phase $2CaO \cdot SiO_2$, $MgO-Al_2O_3$ and other high melting point phases precipitated on the matrix phase, the content of the high melting point phase has a large impact on the melting properties of the slag. Research shows that in order to ensure the performance of the refining slag, the refining slag composition should be adjusted to be located in the low melting point calcium aluminate region in the phase diagram, and strictly avoid the refining slag composition in the high melting point region range of silicate, lime and spinel in the phase diagram, which can make the refining slag has good fluidity, promote the steel slag reaction, improve the ability of the refining slag to adsorb inclusions and improve the refining effect of the refining slag.

Key words: physical composition; petrographic composition; refining slag; inclusion

转炉吹炼过程锰元素反应规律的研究

赵成林[1,2]，王　鹏[3]，黄玉平[1,2]，齐志宇[3]，张维维[1,2]，王一名[3]

（1. 海洋装备用金属材料及其应用国家重点实验室，辽宁鞍山　114009；2. 鞍钢集团钢铁研究院，辽宁鞍山　114009；3. 鞍钢股份有限公司炼钢总厂；辽宁鞍山　114021）

摘　要：本文对锰元素在转炉内的反应机理进行了热力学分析，计算出转炉吹炼不同阶段锰元素的平衡浓度，并通过理论计算与实际生产数据进行对比，确定了转炉吹炼终点锰元素的化学平衡形式。利用方差分析、回归分析等方法，建立了转炉单渣冶炼工艺终点锰含量的回归方程；对于转炉双渣冶炼工艺，明确了半钢碳含量及转炉终点碳含量对半钢脱锰率及转炉终点锰含量的影响规律。

关键词：转炉；脱锰；双渣法；终点控制

Study of Manganese Reaction Rule at Different Blowing Stages in Converter

ZHAO Chenglin[1,2], WANG Peng[3], HUANG Yuping[1,2], QI Zhiyu[3], ZHANG Weiwei[1,2], WANG Yiming[3]

(1. State Key Laboratory of Metal Material for Marine Equipment and Application, Anshan 114009, China;
2. Ansteel Iron & Steel Research Institutes, Anshan 114009, China;
3. General Steelmaking Plant of Angang Steel Co., Ltd., Anshan 114021, China)

Abstract: In this article, thermo-mechanical analysis was done on the reaction mechanism of manganese in the converter and equilibrium concentration at different blowing stages in converter was calculated. Through comparison of theoretical calculation and actual manufacturing data, the chemical equilibrium form of manganese was determined at the end-point in converter. Using methods of variance analysis and regression analysis, etc, regression equation was established. And aiming at double-slag smelting technology, affection rule of semi-steel carbon content and converter end-point carbon content on semi-steel manganese removal rate and end-point manganese content was illustrated.

Key words: converter; manganese removal; double-slag method; end-point control

基于出钢冶金与渣冶金的低成本精炼工艺

汪洪峰，李月林，贾军艳

（唐山钢铁集团有限公司质量管理部，河北唐山　063600）

摘　要：利用出钢过程钢水的高冲击、高氧势、高温度的有利条件，将复合脱氧和白渣前置到出钢过程的新型顶渣改质和脱氧造渣出钢冶金工艺和后续在钢包渣面直接加铝粒进行扩散脱氧与控铝的新型渣冶金工艺。能起到脱氧、脱硫、去夹杂，改善钢包渣流动性等作用，代替 LF、RH 等精炼工艺，与连铸快节奏配合，达到去除夹杂物的目的，满足成品钢水的生产要求，获得高洁净度钢，降低冶炼成本。

关键词：钢水精炼；出钢冶金；渣冶金

Low Cost Refining Process Based on Tapping and Slag Metallurgy

WANG Hongfeng, LI Yuelin, JIA Junyan

(Tangshan Iron and Steel Group Co., Ltd., Tangshan 063600, China)

Abstract: Taking advantage of the advantages of high impact, high oxygen potential and high temperature of molten steel in the process of the new tapping metallurgy and slag metallurgy process of composite deoxidation and white slag in the process of steel production and the new slag metallurgy process of deoxidation and aluminum control by adding aluminum particles directly to the ladle slag surface. It can play the role of deoxidation, desulphurization, removal of inclusions, improve the flow of ladle slag and so on. It can replace the refining process of LF and RH and cooperate with the fast rhythm of continuous casting to achieve the purpose of removing inclusions, obtain high cleanliness steel and reduce the smelting cost.

Key words: refining process; tapping metallurgy; slag metallurgy

炼钢厂降低铁合金消耗的工艺措施

李 京

（湘潭钢铁集团有限公司炼钢厂，湖南湘潭 411101）

摘 要：铁合金是炼钢的重要原料之一，是钢的组成成分。铁合金在炼钢过程中有两个主要用途：一是作为脱氧剂，消除钢液中过量的氧；二是作为合金元素添加剂，改善钢的质量与性能。炼钢根据钢种加入各种各样的铁合金，生产各种各样的优质钢、合金钢，以满足国民经济、国防建设的多种需要。

随着我国钢铁行业不断向绿色环保低碳转型升级，对行业内部提质降本提出了更高的要求。铁合金作为炼钢过程的重要原料之一，其消耗对炼钢厂冶炼生产的成本影响很大，约占炼钢成本的20%-40%。降低铁合金消耗不仅在短期内可以明显降低产品生产成本，而且又有利于改善公司其他产品经济技术指标，是促进企业降低产品成本控制的主要有效手段之一。降低对铁合金的消耗还同时可以通过有效控制降低工业各类冶金废弃物的总产生量，为工业开展安全清洁绿色生产、发展新型循环清洁经济体系创造一种良好条件。

本文针对近年来影响炼钢厂铁合金消耗增加的相关因素进行对比分析得出:通过采取选择非直上工艺路线、优化转炉终点控制、滑板挡渣出钢技术、窄成分控制、合金替代、钢轧协同、管理流程细化等措施，铁合金成本得到有效控制，实现了企业降本增效的目的。

关键词：炼钢；铁合金；消耗；措施

Technological Measures for Reducing Ferroalloy Consumption in Steel Making Plant

LI Jing

(Xiangtan Iron and Steel Group Co., Ltd., Xiangtan 411101, China)

Abstract: This article compares and analyzes the relevant factors that have affected the increases in ferroalloy consumption in steelmaking plants in recent years. The technical and economic indicators have been improved by taking non-direct route, optimizing converter end point control, slag blocking technology of slide plate tapping, narrow component control, alloy substitution, production line coordination and refinement of management process, etc. The cost of ferroalloy has been effectively controlled, achieving the goal of reducing cost and increasing efficiency.

Key words: steelmaking; ferroalloy; consumption; measure

富氢熔态还原炼钢工艺路径下钢水渗氢及脱氢研究

李明明，杨 宇，邵 磊，邹宗树

（东北大学冶金学院，辽宁沈阳 110819）

摘 要：在全球脱碳大潮的背景下，构建"以氢代碳"的钢铁用能新体系是中国钢铁行业实现绿色低碳转型发展的主要抓手。富氢气体或富氢等离子体熔态还原炼钢工艺路线是钢铁行业实施"以氢代碳"低碳发展的热点技术路径之一。但作为一种变革性钢铁冶金新技术，该工艺存在如下待解问题：（1）在富氢（尤其是等离子体氢）环境下，铁矿物熔态还原过程产生的铁/钢水氢含量是否超出目前"碳工艺"路线的钢水氢含量，产生高氢甚至超高氢含量的钢水？（2）对于高氢钢水而言，基于现行操作参数和反应器设计的钢包真空精炼脱氢工艺效能是否满足钢水温降和生产节奏对处理时间的约束？（3）如何实现高氢钢水的高效脱氢？

首先，基于热力学原理，定量分析了铁矿物富氢气体或富氢等离子体熔态还原炼钢工艺路径下生产的钢水中氢的饱和溶解度，表明该工艺生产的钢水氢含量可高于传统碳基工艺流程钢水氢含量的数倍，甚至达到数十个 ppm 的超高含量水平。其次，针对高氢钢水的脱氢，以某钢厂真空脱气（Vacuum Degassing, VD）精炼工艺为对象，基于流体力学与脱氢反应热力学原理，利用欧拉-欧拉方法，耦合 VD 钢包内钢水流体力学传输行为与钢包内脱氢机理，构建了 VD 真空脱氢数值仿真模型，研究了 VD 工艺对高氢钢水的脱氢能力。研究表明，基于现行常规 VD 工艺参数（真空度、底吹 Ar 气量、脱氢时间）进行高氢钢水脱氢处理无法达到氢的高效脱除，虽然通过大幅度提高底吹气量可提高脱氢效率，但会引起严重的喷溅等操作难题，因此需改进现行 VD 工艺。最后，探讨了底吹模式（喷孔数量、喷孔角度、喷孔位置、供气模式）对脱氢的影响，表明喷孔数量对脱氢效率影响最大，若实现高氢钢水的高效脱氢，需突破现行 VD 钢包单或双喷孔底吹设计模式，采用四孔及以上底吹布置，或者采用三孔布置并适当延长真空处理时间。

高废钢比对转炉炼钢工艺影响及其对策

李伟东，王金辉，何海龙

（鞍钢股份有限公司炼钢总厂，辽宁鞍山 114021）

摘 要：针对鞍钢股份有限公司炼钢总厂降低铁水单耗衰 提高废钢比后影响转炉炼钢工艺的问题，优化了废钢尺寸和结构、转炉造渣工艺、转炉提温剂的使用及转炉冶炼控制模型，实现了高废钢比情况下转炉生产指标的稳定，并提高了生产效率。

关键词：转炉；废钢比；造渣；提温剂

Effect of High Scrap Ratio on Smelting Process by Converter and Relevant Countermeasures

LI Weidong, WANG Jinhui, HE Hailong

(General Steelmaking Plant of Angang Steel Co., Ltd., Anshan 114021, China)

Abstract: In allusion to the problem that the smelting process by converter was influenced after increasing the scrap ratio and reducing the unit consumption for hot metal in General Steelmaking Plant of Angang Steel Co., Ltd., the relevant countermeasures such as optimizing the size and compositions in terms of scrap, optimizing slagging process for converter, optimizing the addition of the temperature increasing agent for converter and optimizing the control model for smelting by converter were taken. After that, the production indexes for converter were stable and its production efficiency was also improved.

Key words: converter; scrap ratio; slag making; temperature increasing agent

基于高清彩色 CCD 相机的 RH 炉内监控与测温系统

李保华[1]，周东东[1]，徐 科[1]，雷雨田[2]，化越飞[2]

（1. 北京科技大学钢铁共性技术协同创新中心，北京 100083；
2. 苏州宝联重工股份有限公司，江苏苏州 215152）

摘 要： RH 精炼过程中的顶枪火焰检测、精炼过程监测以及钢水温度测量是精炼过程中非常重要的参考因素，对于 RH 精炼炉的正常运行以及钢水的最终质量有着重要的影响。目前，RH 精炼过程的顶枪火焰检测以及精炼过程监测缺乏较成熟的手段，而精炼过程中温度测量存在着不连续、实时性差等问题。本文基于彩色 CCD 相机开发了一套 RH 炉内监控系统，并利用彩色 CCD 图像和比色测温法开发了 RH 精炼炉内钢水温度实时测量方法。系统可满足 RH 精炼生产的炉内监控需求，并为 RH 精炼工艺优化提供重要的数据参考。

关键词： RH 精炼炉；彩色 CCD 相机；比色测温法；炉内监测

Furnace Monitoring and Thermometry System of RH Refining Based on High-definition Color CCD Camera

LI Baohua[1], ZHOU Dongdong[1], XU Ke[1], LEI Yutian[2], HUA Yuefei[2]

(1. Collaborative Innovation Center of Steel Technology, University of Science and Technology Beijing, Beijing 100083, China; 2. Suzhou Baolian Heavy Industry Co., Ltd., Suzhou 215152, China)

Abstract: Flame detection, refining process monitoring and molten steel thermometry are very important reference factors for the working process of RH refining furnace, which have a significant impact on the normal operation of RH refining furnace and the final quality of molten steel. At present, there are still no mature methods for flame detection and monitoring of the RH refining furnace, and there are problems with thermometry during the refining process, such as discontinuity and poor real-time performance. A system based on color CCD is developed to meet the real-time detection of RH refining furnace flame and refining process. Based on color CCD images, a real-time temperature measurement of RH refining furnace using colorimetric thermometry method is proposed. The system can meet the needs of furnace monitoring of RH refining, and provide important data reference for optimizing the process of RH refining.

Key words: RH refining furnace; color CCD camera; colorimetric thermometry; furnace monitoring

基于共存理论的精炼熔渣脱硫热力学模型研究

苏建铭，孙 群，温荣宇

（鞍钢股份有限公司炼钢总厂，辽宁鞍山 114021）

摘 要：针对钢水在 LF 炉精炼过程硫含量难以预测的问题，基于冶金熔渣共存理论建立了精炼熔渣脱硫的热力学模型，系统分析了在不同熔渣碱度、初始硫含量、钢水温度以及氧活度条件下，LF 炉脱硫终点硫含量的变化规律，研究表明：提高熔渣碱度的增大、钢水温度以及降低钢水氧活度有助于促进脱硫反应的进行，在工艺可控制范围内，钢水温度对于促进脱硫反应、降低处理后钢水硫含量的贡献更为突出。

关键词：共存理论；精炼熔渣；熔渣脱硫热力学模型

Thermodynamic Model for Desulfurization of Refining Slag Based on Coexistence Theory

SU Jianming, SUN Qun, WEN Rongyu

(Steelmaking Plant, Anshan Iron and Steel Group Co., Ltd., Anshan 114021, China)

Abstract: In response to the problem of difficult prediction of sulfur content in molten steel during LF furnace refining, a thermodynamic model for refining slag desulfurization was established based on the coexistence theory of metallurgical slag. The changes in sulfur content at the end point of LF furnace desulfurization were systematically analyzed under different slag alkalinity, initial sulfur content, molten steel temperature, and oxygen activity conditions. The results indicate that increasing the alkalinity of the slag, increasing the temperature of the molten steel, and reducing the oxygen activity of the molten steel help promote the desulfurization reaction. Within the controllable range of the process, the contribution of the temperature of the molten steel to promoting the desulfurization reaction and reducing the sulfur content of the treated molten steel is more prominent.

Key words: coexistence theory; refining slag; thermodynamic model for slag desulfurization

稀土 Ce 对 NM450 钢中夹杂物的影响

郭树豪[1]，刘金瑞[1]，高 军[2]，樊立峰[3]，崔 衡[1]

（1. 北京科技大学 钢铁共性技术协同创新中心，北京 100083；2. 内蒙古包钢钢联股份有限公司制造部，内蒙古包头 014010；3. 内蒙古工业大学材料科学与工程学院，内蒙古呼和浩特 010051）

摘 要：耐磨钢具有高强度、高耐磨性以及较高的冲击性能的特点，因而被广泛应用于制作煤炭、采矿、机械设备的零部件。通过提高耐磨钢的洁净度和控制钢中各类非金属夹杂物的尺寸的方式，可以增强耐磨钢的性能，进而有效的延长耐磨钢的使用寿命[1]。稀土处理是控制钢中非金属夹杂物的一种重要方式。近年来，国内外学者的研究表明，稀土处理可将高硬度氧化物改性为低硬度稀土夹杂物，并降低钢中夹杂物数量[2-3]，进而提高钢的性能。稀土是我国的重要资源，研究并优化钢的稀土处理工艺，是目前冶金领域的一个重要方向。本研究以耐磨钢为研究对象，研究了添加稀土后，钢中夹杂物的演变规律。

根据某厂生产的 NM450 钢成分进行热力学计算。结果表明，随着稀土 Ce 的添加量的增多，加稀土后钢中夹杂物的演变顺序为：$CeAlO_3 \to Ce_2O_2S \to Ce_2O_2S + CeS$。当添加的 Ce 质量分数为 0.0015%时，钢液中 $CeAlO_3$ 相比于其他含 Ce 夹杂物具有更低的生成吉布斯自由能。这说明，在该条件下，$CeAlO_3$ 具有更高的热力学稳定性，更容易在钢液中稳定存在。

通过系统取样、SEM-EDS 检测和 AZtecFeature 全自动非金属夹杂物分析得到了添加稀土后对钢中夹杂物形貌、成分和尺寸的影响。加稀土前，钢中 O 含量以及夹杂物的面密度均较高，且典型夹杂物为 Ca-Al-O。添加质量分数为 0.001%~0.002%的 Ce 后，钢中 O 含量大幅降低，夹杂物的面密度降低且平均尺寸增加。EDS 能谱分析结果表明，钢中的 Ca-Al-O 夹杂物被 Ce 改性为 $CeAlO_3+xCaO \cdot yAl_2O_3$ 夹杂物。钙处理后，钢中的 O、Ce 含量降低，生成了 CaS 夹杂物和 Ce-Al-Ca-O-S 夹杂物，而 MnS 夹杂物的面密度降低。大量生成的小尺寸 CaS 夹杂物使得钢中夹杂物的面密度增加而平均尺寸大幅降低。得益于稀土处理工艺，钢中大部分的 Ca-Al-O 夹杂物会被 Ce 改性并被除去，有害夹杂物 MnS 得到了有效控制。综上所述，通过稀土处理工艺，钢中的夹杂物面密度增加，平均尺寸减小，这有利于得到具有更优良性能的耐磨钢。

关键词：耐磨钢；稀土处理；夹杂物演变；夹杂物改性

转炉底吹 O_2-CO_2-石灰粉工艺的应用实践研究

王春阳[1]，董 凯[1]，朱 荣[1]，赵长亮[2]，关顺宽[2]，丁国慧[2]

（1. 北京科技大学碳中和研究院，北京 100083；
2. 首钢京唐钢铁联合有限责任公司，河北唐山 063200）

摘 要：为实现转炉绿色洁净生产并满足高废钢比冶炼的需求，本团队对转炉底吹 O_2-CO_2-石灰粉工艺进行研究并将其应用于首钢京唐 300t 脱磷转炉进行工业试验。试验结果表明，应用底吹 O_2-CO_2-CaO 工艺具有改善熔池的动力学条件，提高石灰利用率，高效稳定脱磷，降低渣中 T.Fe 含量，使碳氧反应更加彻底等优势，从而降低生产成本。具体指标为：吨钢石灰消耗降低 4.94kg；终点磷含量降低 0.00155%；配合顶吹供氧强度的增强，渣中 T.Fe 降低 8.19%；终点碳氧积最低可达 13.4×10^{-4}。

关键词：转炉炼钢；底吹 O_2-CO_2-石灰粉；高效脱磷；渣中 T.Fe；碳氧积

The Application Practice of Bottom-blown O_2-CO_2-CaO Process in Steelmaking

WANG Chunyang[1], DONG Kai[1], ZHU Rong[1], ZHAO Changliang[2], GUAN Shunkuan[2], DING Guohui[2]

(1. Institute for Carbon Neutrality, University of Science and Technology Beijing, Beijing 100083, China;

2. Shougang Jingtang United Iron and Steel Co., Ltd., Tangshan 063200, China)

Abstract: In order to achieve green and clean production in the converter and to meet the needs of high scrap ratio smelting, we studied the bottom-blowing O_2-CO_2-CaO process and applied to the Shougang Jingtang 300t dephosphorization converter for industrial trials. The test results show that the current process can improve the kinetic conditions of the melt pool, increase the lime utilization ratio, efficiently and stably dephosphorization, reduce the T. Fe content in the slag and make the carbon-oxygen reaction more complete, thus reducing production costs. The specific indicators are: 4.94kg/t lower lime consumption; 0.00155% lower phosphorus content; 8.19% lower T. Fe in the slag in combination with the enhanced oxygen supply strength of the top blow; and a minimum carbon and oxygen product of 13.4×10^{-4} at the end-point.

Key words: steelmaking; bottom-blown O_2-CO_2-CaO; high efficiency dephosphorization; T.Fe in slag; the product of carbon and oxygen content

冷硬产品夹杂缺陷分析及质量控制

辛 鑫[1]，付 薇[1]，高洪涛[2]，黄秋菊[1]

（1. 鞍钢股份有限公司制造管理部，辽宁鞍山 114000；
2. 鞍钢股份有限公司炼钢总厂，辽宁鞍山 114000）

摘 要： 针对镀锌用户提出的冷硬原料的缺陷样板，利用SUPRA 55场发射扫描电镜，结合OXFORD能谱仪进行了检测分析，结果表明缺陷位置有聚集的夹杂物，此缺陷产生于钢水冶炼及浇注过程。对夹杂缺陷炼钢过程进行调查与分析，得出炼钢板坯夹杂主要与过程氧、中间包过热度、浇铸拉速、中间包吨位等因素有关。研究发现，炼钢过程氧高，是板坯氧化物夹杂的主要来源。浇铸过程过热度低、中间包吨位低、生产拉速高均不利于夹杂物上浮，从而导致板坯夹杂缺陷增多。通过制定炼钢各工序过程氧相应控制基准、优化浇铸过热度、中间包吨位及浇铸速度等措施，可以有效控制板坯夹杂缺陷。

关键词： 冷硬产品；夹杂；缺陷分析；质量控制

Analysis and Quality Control of Inclusion Defects in Cold Hard Products

XIN Xin[1], FU Wei[1], GAO Hongtao[2], HUANG Qiuju[1]

(1. Manufacture Management Department, Ansteel, Anshan 114000, China;
2. Steel Mill, Ansteel, Anshan 114000, China)

Abstract: In response to the defect sample of cold hard raw materials proposed by galvanized users, scanning electron microscopy was used for inspection and analysis, and the results showed that the defect location was aggregated inclusions, this defect occurs during the smelting and pouring process of molten steel. Investigation and analysis of the steel making process with inclusion defects, resulting in the production of steel plate billets Inclusions are mainly related to factors such as process oxygen, tundish superheat, casting speed, and tundish tonnage. Research has found that the steelmaking process has a high oxygen content, which is attributed to the main source of oxide inclusions in billets; The low superheat during the casting process, low tonnage of the tundish, and high production speed are all unfavorable for the floating of inclusions, resulting in an increase in slab inclusion defects. By establishing corresponding control standards for oxygen in various steelmaking processes, optimizing casting superheat, tundish tonnage, and casting speed measures such as temperature control; It can effectively control slab inclusion defects.

Key words: cold and hard products; inclusion; defect analysis; quality control

浅析我国钢包用耐火材料的品种及应用

李仕龙

（河钢集团承钢公司棒材事业部，河北承德　067102）

摘　要：钢包担负着载运钢水和进行炉外精炼的双重任务，随着炼钢技术的发展，我国的钢包用耐火材料也得到了很好的发展。特别是自20世纪80年代以来，我国的耐火材料科研机构、生产企业和使用厂家，密切配合，结合我国的国情，不断开发出新型的钢包用耐火材料，使我国的钢包用耐火材料以较快的速度向前发展，满足了我国炼钢工业快速发展的需要。

关键词：钢包；耐火材料

60t 脱磷站-AOD 炉-LF-CCM 生产 1Cr13 钢低成本生产实践

冯文甫[1,2,3]，张育明[1,2,3]，范建英[1,2,3]，郭志彬[1,2,3]，阎丽珍[1,2,3]

（1. 河北邢钢科技有限公司，河北邢台　054799；2. 邢台钢铁有限责任公司，河北邢台　054027；
3. 河北线材工程技术创新中心，河北邢台　054027）

摘　要：钢铁行业日益竞争激烈，成本控制显得格外重要。针对"脱磷站-AOD 炉-LF-CCM"一步法工艺生产 1Cr13 钢成本较高的问题，通过对全流程的梳理，采取优化脱磷站吹炼模型、开发 AOD 炉一次还原不加萤石化渣的工艺、提高连铸机连浇炉数等，各项指标降低较明显，累计实现成本降低 145 元/t，经济效益显著；同时通过实践表明：AOD 炉冶炼过程中还原期利用低碱度进行一次还原化渣的工艺可行，但需要增加二次还原；AOD 吹炼过程中高碳区采用增加氧气量、低碳区采用降低氧压比可实现周期缩短，同时随着连铸中包耐材的替换，能够稳定实现连铸机连浇 12 炉。

关键词：不锈钢；成本；生产实践

Low Cost Production Practice of 1Cr13 Steel Produced by 60t DEP-AOD-LF-CCM

FENG Wenfu[1,2,3], ZAHNG Yuming[1,2,3], FAN Jianying[1,2,3], GUO Zhibin[1,2,3], YAN Lizhen[1,2,3]

(1. Hebei Xinggang Technology Co., Ltd., Xingtai 054799, China;

2. Xingtai Iron & Steel Co., Ltd., Xingtai 054027, China; 3. Hebei Wire Engineering
Technology Innovation Center, Xingtai 054027, China)

Abstract: With the increasingly fierce competition in the steel industry, cost control is particularly important. In view of the high cost of the one-step process of "dephosphorization station AOD furnace lf CCM" for the production of 1Cr13 steel, by combing the whole process, optimizing the blowing model of dephosphorization station, developing the process of AOD furnace reduction without adding fluorite slag, increasing the number of continuous casting furnaces of continuous casting machine, etc., the indicators are obviously reduced, the cumulative cost is reduced by 145 yuan / ton, and the economic benefit is significant; at the same time, through practice The results show that the process of primary reduction slag with low basicity in reduction period is feasible, but secondary reduction needs to be increased; the cycle can be shortened by increasing oxygen content in high carbon area and reducing oxygen pressure ratio in low carbon area in AOD blowing process, and the continuous casting of 12 heats can be realized stably with the replacement of refractory in continuous casting.

Key words: stainless steel; cost; productive practice

稀土—碳材质浸入式水口连铸稀土钢的应用实践

谌智勇[1]，王小军[2]，任　强[3]，智建国[1]

（1. 内蒙古包钢钢联股份有限公司，内蒙古包头　014010；2. 青岛正望新材料股份有限公司，山东青岛　266200；3. 燕山大学，河北秦皇岛　066000）

摘　要： 稀土钢连铸絮流、堵塞是行业难题。本文以人工合成 $CeAlO_3$ 为基础，设计开发了采用稀土—碳材质和铝碳材质复合的浸入式水口。通过工业化应用实践，完善了该水口的制造工艺，并实现了连铸生产顺行。通过对残余水口内部材质的分析，表明使用优化后的结构材料，能够避免钢水中稀土与耐火材料反应而导致的水口堵塞问题。

关键词： 稀土钢；稀土—碳材质；中间包功能材料

Application Practice of Rare-earth Carbon Immersion Continuous Casting of Rare-earth Steel

CHEN Zhiyong[1], WANG Xiaojun[2], REN Qiang[3], ZHI Jianguo[1]

(1. Steelmaking Plant, Inner Mongolia Baotou Steel Union Co., Ltd., Baotou 014010, China;
2. Qingdao Zhengwang New Material Limited Liability Company, Qingdao 266200, China;
3. Yanshan University, Qinhuangdao 066000, China)

Abstract: The flow and plugging of rare earth steel continuous casting is a difficult problem in the industry. On the basis of synthetic CeAlO°, the immersion nozzle with rare earth carbon and aluminum carbon composite material was designed and developed. Through the industrial application practice, the two manufacturing techniques of the nozzle were perfected and the continuous casting production was realized. Through the analysis of the inside of the residual nozzle, it can be seen that the use of rare earth-carbonaceous internal materials can avoid the problem of nozzle blockage caused by the reaction of rare earth in the molten steel with refractory materials.

Key words: rare earth steel; rare-earth carbon material; tundish functional material

稀土处理对不同钢种冲击韧性的影响

隋亚飞，赵 如，罗 钢，梁 亮

（湖南华菱涟源钢铁有限公司技术中心，湖南娄底 417000）

摘 要：为了研究稀土处理对品种钢冲击韧性的作用，分别在低合金结构钢、高强钢、热轧汽车用钢、上进行了现场的稀土处理试验。摸索了不同钢种适当的稀土加入时机与加入量，比较了各钢种稀土处理下的夹杂物变化和性能改善情况。稀土处理后钢中小尺寸夹杂物变性为（La-Ce-S）和（La-Ce-Al-S-O）稀土夹杂物，大尺寸夹杂物的变性程度有限。稀土处理改善冲击功的关键是控制了钢中 MnS 夹杂，当 Ti 含量较高时，钢中几乎不存在 MnS 夹杂，此时稀土处理会增加 B 类夹杂物评级，反而恶化冲击功。因此，铝脱氧高钛钢并不适合使用稀土处理工艺。

关键词：稀土处理；夹杂物；冲击功；可浇性

Effect of Rare Earth Treatment on Impact Energy of Different Steel Grades

SUI Yafei, ZHAO Ru, LUO Gang, LIANG Liang

(Hunan Valin Lianyuan Iron and Steel Co., Ltd., Technical Center, Loudi 417000, China)

Abstract: In order to study the effect of rare earth treatment on the impact toughness of various types of steel, on-site rare earth treatment experiments were conducted on low alloy structural steel, high-strength steel, and hot-rolled automotive steel. Explored the appropriate timing and amount of rare earth addition for different steel grades, and compared the changes in inclusions and performance improvement under rare earth treatment for each steel grade. After rare earth treatment, small and medium-sized inclusions in steel undergo denaturation into (La Ce-S) and (La Ce-Al-S-O) rare earth inclusions, while the degree of denaturation of large-sized inclusions is limited. The key to improving impact energy through rare earth treatment is to control the MnS inclusions in the steel. When the Ti content is high, there are almost no MnS inclusions in the steel. At this time, rare earth treatment will increase the rating of B-type inclusions and instead worsen the impact energy. Therefore, aluminum deoxidized high titanium steel is not suitable for the use of rare earth treatment processes.

Key words: rare earth treatment; inclusions; impact energy; castability

转炉熔池火点区传热行为模拟研究

蒋 睿，孙建坤，刘 青

（北京科技大学绿色低碳钢铁冶金全国重点实验室，北京 100083）

摘 要：为探究转炉熔池多相流体的传热行为，本文采用数值模拟的方法构建了 100t 顶吹转炉熔池火点区传热数学模型，研究了工艺操作参数对熔池多相流体传热行为的影响，明晰了熔池温度分布规律。数值模拟结果与经验公式计算的熔池升温速率吻合良好。研究结果表明：火点区与熔池之间的温度梯度以及火点区周围的高温区域均会随着火点区放热时间的增加而增大。此外，氧枪枪位能够显著影响钢液的平均流速，从而影响熔池的升温情况。当氧枪枪位从 1.35m 增加到 1.65m 时，熔池的平均升温速率由 0.161K/s 增加到 0.271K/s，提高了 40.59%。氧枪操作压力对钢液的平均流速影响较小，不能显著改变熔池的平均升温速率，当氧枪操作压力从 0.7MPa 增加到 1.1MPa 时，熔池平均升温速率仅提高了 18.23%。

关键词：转炉炼钢；火点区；流动；温度分布；升温速率

Simulation Study on Heat Transfer Behavior in Hot Spot Zone of Converter Molten Bath

JIANG Rui, SUN Jiankun, LIU Qing

(State Key Laboratory of Advanced Metallurgy, University of Science and Technology Beijing, Beijing 100083, China)

Abstract: In order to investigate the heat transfer behavior of the multiphase fluid inside the converter molten bath, a mathematical model of heat transfer in the hot pot zone of 100t top blown converter is established. The temperature distribution laws of the molten bath are clarified, and the effects of process parameters on the heat transfer behavior of the molten bath are numerically studied. The numerical simulation results are in good agreement with the heating rate calculated by the empirical formulas. The research results show that the temperature gradient between the hot spot zone and the molten bath and the high temperature zone around the hot spot zone will increase with the increase of the hot spot zone heat release time. In addition, the oxygen lance height can significantly affect the average flow velocity of molten steel so as to influence the heat of the molten bath. When the oxygen lance height is increased from 1.35m to 1.65m, the average heating rate of the molten bath is increased from 0.161K/s to 0.271K/s, which is increased by 40.59%. The operating pressure of oxygen lance has little effect on the average flow velocity of the molten steel, and can not significantly change the average heating rate of molten bath. When the operating pressure of oxygen lance is increased from 0.7MPa to 1.1MPa, the average heating rate is only increased by 18.23%.

Key words: converter steelmaking; hot spot zone; flow; temperature distribution; heating rate

120t 转炉高效冶炼提升生产实践

罗焕松，周　甫，曾维侃，吴振坤，刘春辉

（武汉钢铁有限公司条材厂，湖北武汉　430083）

摘　要：为解决宝钢股份武钢有限一炼钢转炉炼钢工序严重限制产能释放的问题，通过对一炼钢 120t 转炉设备、供氧强度、造渣工艺等进行优化，建立转炉热平衡静态模型，推行留渣、少渣冶炼模式，利用投弹副枪指导直接出钢，推进快速烘烤新型补炉方式等，将转炉冶炼周期从 38min/炉缩短至 32min/炉，有效的提高了转炉的生产节奏，实现了转炉的高效冶炼，为武钢有限一炼钢产能提高和效益提升提供了进一步的支撑。

关键词：转炉高效冶炼；热平衡；转炉冶炼周期

Production Practice of Efficient Smelting and Upgrading of 120 tons Converter

LUO Huansong, ZHOU Fu, ZENG Weikan, WU Zhenkun, LIU Chunhui

(Plant of Long Product of Wuhan Iron & Steel Co., Ltd., Wuhan 430083, China)

Abstract: In order to solve the problem of severely limiting the capacity release in the converter steelmaking process of No.1 Steelmaking Plant of Baosteel, the 120t converter equipment, oxygen supply intensity and slagging process were optimized, the static model of converter heat balance was established, the slag-remaining and slag-less smelting mode was carried out, the direct tapping was guided by the ejection sublance, and the new type of fast baking method was promoted. The smelting cycle of the converter was shortened from 38 min/furnace to 32min/furnace, which effectively improved the production rhythm of the converter and realized the efficient smelting of the converter. It provides further support for the improvement of the production capacity and efficiency of No.1 Steelmaking Plant of Baosteel.

Key words: converter efficient smelting; thermal equilibrium; converter smelting cycle

钢包透气砖防钢液渗漏的数值模拟

娄文涛，张加毓，朱苗勇

（东北大学，辽宁沈阳 110819）

摘　要： 钢包底吹氩已成为炉外精炼运用最广泛的工艺，透气砖作为钢包底吹氩的关键功能性元件，其吹氩效果和使用寿命直接关系着钢水质量、节奏、成本及安全。透气砖在服役过程中，常因钢液渗漏进入缝隙内部造成堵塞。为清除缝隙内夹钢，需每次对透气砖砖面进行高温烧氧，进而造成了透气砖的快速损毁。目前，国内钢铁企业透气砖寿命仅为 25 炉次左右，在一个钢包炉役期间需多次下线更换透气砖，不仅造成大量耐材、热量和人力消耗，而且严重影响生产效率和节奏。尤其在精炼过程中，透气砖夹钢堵塞后，极大恶化透气砖吹氩精炼效果，甚至需要中途更换钢包，导致严重的钢水二次污染和温度损失。因此，如何保证透气砖良好的透气性，并大幅提升透气砖工作寿命，成为本行业领域需要解决的共性技术难题。

针对透气砖狭缝钢液渗漏堵塞问题，课题组研发了钢包透气砖防渗漏装备技术，该技术目的是要在不影响现有底吹氩精炼工艺条件下，在钢包底部设置有防渗漏装置设备，可以在钢包非吹氩阶段（如吊运、浇注）时持续为透气转释放氩气，保证透气砖狭缝内长期微正压，防止钢液和熔渣渗漏进入狭缝堵塞透气砖。这样不仅可以消除透气砖的渗钢现象，为底吹精炼效果提供基础保障，还可免氧气清洗工序，大幅提升透气砖工作寿命。目前，课题组已分别在国内某两个钢铁企业 120t 和 260t 钢包实施应用，生产实践表明通过此技术消除了狭缝渗钢现象，透气砖寿命可由 25 炉次提升到 60 炉次以上，寿命提升 1 倍以上。

结合课题组研发的钢包透气砖防渗漏装备，本文采用数值模拟方法对防渗漏所需的安全流量与透气砖狭缝数量、宽度、长度以及钢水液面高度之间的定量关系展开研究。研究表明，狭缝内近壁面函数模型对透气砖内的气流-压损具有重要影响，采用 Non-Equilibrium Wall Function 函数模型可以较好预测透气砖内的流动行为。随着透气砖狭缝宽度和长度的增加，透气砖防渗漏所需安全流量增大，且透气砖宽度对安全流量的影响要大于透气砖长度的影响。随着狭缝条数增加，透气砖防渗漏的安全流量近线性增加，液面高度的对防渗漏安全流量影响较小。最后基于因次分析方法，提出防渗漏装置安全流量的计算式，为钢包透气砖防渗漏技术开发应用提供重要理论基础。

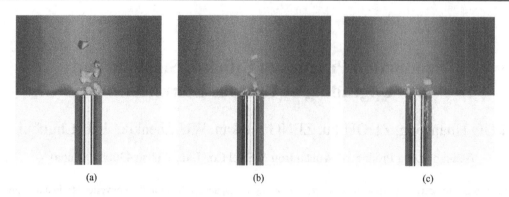

图 1 不同底吹流量下钢包透气砖狭缝处的钢液界面渗漏行为的模拟预测云图

(a) 10NL/min；(b) 8NL/min；(c) 6NL/min

稀土镁处理顺序和钙预处理对钢中稀土夹杂物特征的影响

任泽斌，陈 凯，李 艳，唐 鑫，郑立春，姜周华

（东北大学冶金学院，辽宁沈阳 110000）

摘 要：为细化钢中稀土夹杂物，本文研究了稀土镁处理顺序和钙预处理对稀土夹杂物特征的影响。结果表明：先镁处理、稀土和镁同时处理、先稀土处理三种处理顺序中稀土夹杂物含量均远高于 MgO 夹杂物含量，且先稀土铈处理时稀土夹杂物的含量最高。先进行稀土铈处理时，生成的稀土夹杂物尺寸最大，且有轻微团聚，不利于细化稀土夹杂物；稀土铈和镁同时处理时，产生的稀土夹杂物尺寸和先镁处理相当，但由于受到镁气化产生的强烈搅拌影响，稀土夹杂物形貌严重变形且团聚严重。先镁处理会产生大量 MgO 夹杂物，MgO 夹杂物的尺寸小，且不易团聚。稀土处理会使原始的氧化铝夹杂物和钙铝酸盐夹杂物的尺寸大幅增加。对于未经钙预处理的夹杂物，稀土处理后变为形状不规则的 Ce-Al-O 夹杂物。而钙预处理后，稀土夹杂物形貌呈球形，其成分为 Al_2O_3-CaO-Ce_2O_3，其中 CaO 含量在 13% 以内，Al_2O_3 含量在 20% 左右。

关键词：稀土；镁；钙预处理；非金属夹杂物；形貌；尺寸

Effect of Rare Earth Magnesium Treatment Sequence and Calcium Pretreatment on the Characteristics of Rare Earth Inclusions in Steel

REN Zebin, CHEN Kai, LI Yan, TANG Xin,
ZHENG Lichun, JIANG Zhouhua

(School of Metallurgy, Northeastern University, Shenyang 110000, China)

Abstract: Refinement of rare earth inclusions in steel, this paper studies the rare earth magnesium treatment order and calcium pretreatment on the rare earth inclusions characteristics. The results show that: the first magnesium treatment, rare earth and magnesium treatment at the same time, the first rare earth treatment of the three treatment order of rare earth inclusions are much higher than the content of MgO inclusions, and the first rare earth cerium treatment of the highest content of rare earth inclusions. First rare earth cerium treatment, the generation of rare earth inclusions size is the largest, and there is a slight agglomeration, is not conducive to the refinement of rare earth inclusions; rare earth cerium and magnesium treatment at the same time, the size of the rare earth inclusions and the first magnesium treatment is comparable, but due to the magnesium gasification produced by the impact of strong stirring, rare earth inclusions morphology is seriously deformed and agglomeration is serious. Rare earth magnesium treatment will produce a large number of MgO inclusions, MgO inclusions size is small, and not easy to agglomerate. Rare earth treatment increases the size of the original alumina inclusions and calcium aluminate inclusions significantly. For inclusions without calcium pretreatment, the rare earth treatment changes them into irregularly shaped Ce-Al-O inclusions. In contrast, after calcium pretreatment, the rare-earth inclusions have a spherical shape, with a composition of Al_2O_3-CaO-Ce_2O_3, in which the CaO content is within 13% and the Al_2O_3 content is about 20%.

Key words: rare earth; magnesium; calcium pretreatment; non-metallic inclusions; morphology; size

不同赋存状态下稀土对钢力学性能的影响机理研究

刘香军，杨昌桥，杨吉春，任慧平

（内蒙古科技大学材料与冶金学院，内蒙古包头　014000）

摘　要： 科学技术日新月异，下游制造业对钢品质要求越发严苛，研发高强韧性、高耐腐蚀性钢种将是未来钢铁材料发展的重要方向。微量稀土的添加便可大幅度提升钢材的综合性能，当前的研究以实验研究为主，理论研究较少，尤其缺乏微观尺度上的认识，致使稀土的作用本质及内在机理不明确，缺少理论指导是发展稀土在钢中广泛应用的最大瓶颈。本文采用理论计算结合实验研究的研究方法探索稀土 Ce 在钢中作用的微观机理，以 Ce 在钢中的不同赋存状态，固溶 Ce 及 Ce 夹杂物两个维度对钢力学性能展开研究。本文采用模拟计算的方式开展研究，基于密度泛函理论采用第一性原理计算方法，通过溶解能、几何结构、电子结构、Bader 电荷等维度来揭示 Ce 原子在 α-Fe 中的固溶机理，并对 Fe-Ce 掺杂体系的力学性能进行详细计算。Ce 的掺杂降低了体系的不可压缩性、抗剪切应变能力、刚度以及维氏硬度，但提高了体系的韧性以及可加工性。Ce 掺杂导致体系金属键强度降低，这是 Fe-Ce 掺杂体系不可压缩性、刚性和硬度降低的主要原因；另一方面，Ce 的掺杂增加了体系中电子云的密度，这有利于掺杂体系韧性的提高。

通过计算 Ce 夹杂物自身物理化学性质，分析夹杂物对周围钢基体残余应力的影响。TiN、Al_2O_3 夹杂物的体积模量、剪切模量、杨氏模量以及维氏硬度较大，呈现出较大的刚性和硬度，表现为脆性特征，而 Ce_2O_3 和 Ce_2O_2S 夹杂物表现为韧性特征。与 Al_2O_3 和 TiN 相比，Ce_2O_3、$CeAlO_3$ 夹杂的热膨胀系数与铁基体接近，而 Ce_2O_2S 夹杂物的热膨胀系数比铁基体稍大。Ce 夹杂物与基体在不可压缩性、刚性、硬度、韧脆性及热膨胀性等方面的差异较小，钢基体塑性变形的一致性得到提升，有利于延缓微孔洞微裂纹的萌生，有助于钢材力学性能的提高。

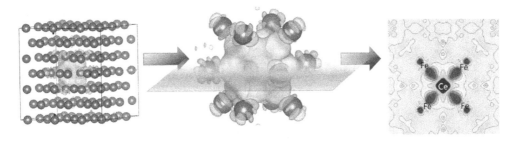

图 1　Fe-Ce 掺杂体系的差分电荷密度图

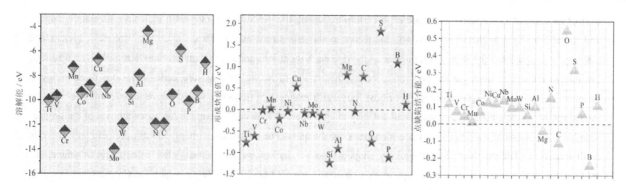

图2 钢中常见20中元素对Ce的相互作用

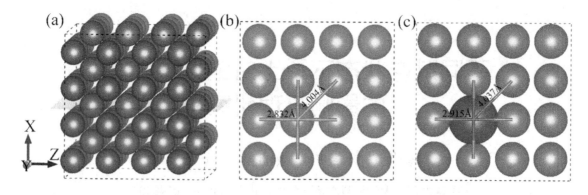

图3 Ce在钢中的几何结构

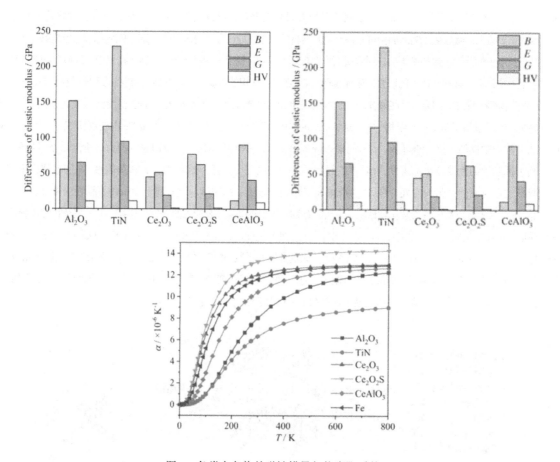

图4 各类夹杂物的弹性模量与热膨胀系数

转炉炉膛结构对熔池流动特性影响研究

刘福海[1,2]，朱 荣[1,2]，董 凯[2,3]，魏光升[2,3]，程 钰[2,3]

（1. 北京科技大学国家材料服役安全科学中心，北京 100083；2. 北京科技大学碳中和研究院，北京 100083；3. 北京科技大学冶金与生态工程学院，北京 100083）

摘 要：转炉炼钢是我国目前最主要的炼钢方法，其以高温铁水与冷态废钢为主要原料，具有冶炼周期短、单炉产量大、热能负载充沛等特点。目前，转炉顶吹供氧的设计原则及过程操作均基于初始理想炉型结构，并通过理论或经验公式计算得到。

但转炉全炉役周期范围内，其"前-中-后"各时期炉膛结构特点具有明显的差异性，单一恒定的顶吹供氧参数加剧了不规则的熔渣乳化流动状态，导致顶吹供氧、底吹供气与高效造渣等单元操作工艺的同步、精准控制难度大。本文针对"前-后"期炉役结构与冶炼工艺特点，利用水力学模拟与数值仿真研究方法，分析各时期熔池流动特性与顶吹供氧参数的作用特点，以提高转炉炼钢的生产技术与经济效益指标，研究结果表明：

（1）初始炉型熔池平均混匀时间为78.8s，中后期炉型熔池平均混匀时间为65.7s，中后期炉型结构下熔池混匀效果得到明显提高。在初始炉型结构下，工况因素对熔池混匀时间影响程度排序为：氧枪类别>顶吹流量>枪位。在中后期炉型结构下，工况因素对熔池混匀时间影响程度排序为：顶吹流量>氧枪类别>枪位。

（2）初始炉形与末期炉形的平均冲击面积分别为 $0.164m^2$ 及 $0.218m^2$。相较于初始炉形结构所用氧枪，末期炉形氧枪角度平均提高 0.5°~1°，且末期炉膛直径较初期直径提高 31.9%，炉壁对熔渣径向扩散阻碍效果相对下降，导致末期炉形熔池冲击面积明显提高。

（3）初始炉形与末期炉形的平均冲击深度分别为 0.130m 及 0.113m。由于转炉中期所用氧枪夹角更大，射流轴向冲击能力下降。同时，末期熔池深度较初期深度下降43.5%，转炉炉底对熔池轴向扩散阻碍效果相应提高，导致末期熔池冲击深度下降。

（4）初始炉型原氧枪形成的速度死区集中在氧枪喷管径向夹角区，且靠近炉壁一带（呈U型分布），及炉底中心区一带。末期炉型速度死区仅分布在氧枪喷管径向夹角区，且靠近炉壁一带，并不规则边型带状分布。

（5）相比于初始炉型，转炉末期结构增大了熔池内"金-渣"接触面积，其接触面积平均提高78.7%，进一步证明了转炉末期结构可显著增大渣中FeO与熔池[C]元素的反应速率，不利于熔池渣层稳定的泡沫化。

板坯中间包用耐火材料长寿技术

董战春，朱静然，白晓卫，王 兴，刘根胜，章荣会

（北京联合荣大工程材料股份有限公司，北京 101400）

摘 要：中间包是钢水连铸工艺中的重要生产设备，包龄高低、寿命长短直接影响着连铸坯的收得率以及生产组织的连续性和稳定性。延长连铸中间包的使用寿命，既可缩短中间包更换的操作时间、减少烤包用煤气的消耗和现场工人砌筑施工的劳动强度，还有利于减少连铸铸坯的重接废品、提高铸坯合格率，以及可有效的降低中间包单包用耐火材料吨钢消耗，提高钢水纯净度。通过对一机一流和二机二流板坯连铸中间包用耐火材料进行工艺调整和优化，

并经过多次工业现场应用实践结果表明,经过调整和优化的各中间包用耐火材料产品均能够满足连铸生产工艺的提寿使用要求,有助于实现降本增效的目标。

关键词:连铸;中间包;耐火材料;长寿

Long-life Technology of Refractory Materials for Slab Tundish

DONG Zhanchun, ZHU Jingran, BAI Xiaowei, WANG Xing,
LIU Gensheng, ZHANG Ronghui

(Beijing Allied Rongda Engineering Material Co., Ltd., Beijng 101400, China)

Abstract: Tundish is an important production equipment in molten steel continuous casting process. The tundish age and service life directly affect the yield of continuous casting slab and the continuity and stability of production organization. Extending the service life of the continuous casting tundish can not only shorten the operation time for tundish replacement, reduce the consumption of gas for baking and the labor intensity of on-site workers in masonry construction, but also help reduce the reconnection waste of continuous casting billets, improve the qualification rate of billets, and effectively reduce the consumption of refractory materials per ton of steel for single tundish, and improve the purity of molten steel. By adjusting and optimizing the process of refractory materials used in the slab continuous casting tundish, and through multiple industrial field applications, the results show that the adjusted and optimized refractory materials used in each tundish can meet the requirements for improving the lifespan of the continuous casting production process, which can help to achieve the goal of cost reduction and efficiency increase.

Key words: continuous casting; tundish; refractory materials; long life

RH 单工位双钢包车工艺布置的工程实践

丁永锷

(宝钢工程技术集团有限公司,上海 201900)

摘 要:多功能高效 RH 精炼技术在中国已有二十多年的应用实践,从中演化出多种工艺布置方案。单工位 RH 主要分为单钢包车、双钢包车等形式,双工位 RH 主要分为两车三位、两车四位、三车五位和四车六位等形式。

追求"性价比"是业主投资项目永恒的主题,即花较少的投资产生较大的效益。那么"单工位双钢包车"这种形式的 RH 因具备这一特点,被广泛采用。从而又演化出多种形式的布置方案,如:垂直交叉轨、45°交叉轨、横向一字轨、纵向一字轨(顶升和提升)、双平移台车等形式。

本文就着重介绍单工位双钢包车 RH 的几种典型的布置方案和特点。

关键词:RH;工艺布置;解决方案

The Layout Solution of Single Station RH with Double Ladle Car in Real Project

DING Devin

(Baosteel Engineering & Technology Group Co., Ltd., Shanghai 201900, China)

Abstract: Multi-functional and efficient RH refining technology has been applied in China for more than 20 years, from which a variety of process layout schemes have evolved. Single station RH is mainly divided into single ladle car, double ladle car and other forms, double station RH is mainly divided into two cars three, two cars four, three cars five and four cars six and other forms.

The pursuit of "cost performance" is the eternal theme of the owner's investment projects, that is, the investment that spends less will produce greater benefits. Then, the RH of "single station double ladle car" is widely used because of this feature. Thus, it has evolved a variety of layout schemes, such as: vertical cross rail, 45° cross rail, horizontal one-frame rail, longitudinal one-frame rail (jacking and lifting), double translation car and other forms.

This paper mainly introduces several typical layout schemes and characteristics of RH in single station double ladle car.

Key words: RH; layout; solution

含 TiO_2 精炼渣对 MgO 耐火材料的侵蚀行为研究

郝广御[1,2]，邓志银[1,2]，魏春新[1,2,3]，朱苗勇[1,2]

（1. 东北大学多金属共生矿生态化冶金教育部重点实验室，辽宁沈阳　110819；
2. 东北大学冶金学院，辽宁沈阳　110819；3. 本钢板材股份有限公司，辽宁本溪　117000）

摘　要： 钢中的钛（Ti）能起到细化晶粒的作用，进而能提高钢的强度和韧性。因此，Ti 被广泛添加在齿轮钢、汽车板钢、不锈钢和其他钢种中。

在含钛钢种精炼的过程中，虽然常使用 $CaO-SiO_2-Al_2O_3-MgO$ 四元渣系进行精炼，但由于钢渣平衡的影响，钢中部分 Ti 元素会向渣中迁移，最终形成 $CaO-SiO_2-Al_2O_3-MgO-TiO_2$ 的终渣，导致钢液中 Ti 收得率降低。考虑向精炼渣中加入一定量的 TiO_2，改变钢-渣平衡，可以提高 Ti 的收得率。然而，TiO_2 的添加会影响炉渣的物理性质，如黏度和熔点等，因此也可能会影响精炼渣对耐火材料的侵蚀行为。

为了解 $CaO-SiO_2-Al_2O_3-MgO-TiO_2$ 精炼渣中的 TiO_2 对 MgO 耐火材料侵蚀的影响，在实验室预熔制成精炼渣，研究了 MgO 耐火材料（致密 MgO 棒和工业镁碳砖）在精炼渣中的溶解行为以及精炼渣对耐火材料的渗透行为。此外，对不同 TiO_2 含量的精炼渣的黏度、表面张力和与 MgO 耐火材料的接触角进行测量，探究了精炼渣物理性质变化对渗透的影响机理。研究发现，向渣中添加 TiO_2（0~10%）后，TiO_2 会降低精炼渣的液相线，进而提高 MgO 在渣中的溶解度，加剧 MgO 耐火材料向渣中溶解。同时，TiO_2 的添加还会影响精炼渣的物理性质。随着精炼渣中 TiO_2 含量的增加，精炼渣的黏度和表面张力先降低后升高，而精炼渣与 MgO 耐火材料的接触角则呈相反的变化趋势，导致精炼渣渗入耐火材料的深度先增加后减少。精炼渣的渗透会进一步加剧耐火材料向渣中溶解。因此，向精炼渣中添加 TiO_2 后，会加剧渣对 MgO 耐火材料的侵蚀。为了减少含 TiO_2 精炼渣对耐火材料的侵蚀，实验进一步尝试提高精炼渣中的 MgO 含量。结果发现，适当增加渣中的 MgO 含量不仅能减弱 MgO 耐火材料的溶解，还能降低精炼渣对耐火材料的渗透。此外，更加致密的耐火材料也有利于抵抗精炼渣对耐火材料的侵蚀。

参 考 文 献

[1] Pak J J, Jo J O, Kim S I, et al. ISIJ Int., 2007, 47(1): 16-24.
[2] Byun J S, Shim J H, Cho Y W, et al. Acta Mater., 2003, 51(6): 1593-1606.
[3] Wang Y, Cho J H, Jeong T S, et al. Mater. Trans. B., 2021, 52(6): 3986-4001.
[4] Zhang X M, Yan Z W, Deng Z Y, et al. Metals, 2023, 13(2): 431.
[5] Gao H B, Tang J, Chu M S, et al. J. Iron Steel Res. Int., 2023, 30(3): 456-464.

钙处理工艺对车轴钢中夹杂物的影响

郭俊波，沈　昶，杨克枝，陆　强

（马鞍山钢铁股份有限公司，轨道交通关键零部件安徽省技术创新中心，安徽马鞍山　243000）

摘　要：利用 ASPEX 对某厂采用 LF 炉钙处理工艺、RH 钙处理工艺生产车轴钢的钙处理前、RH 精炼终点的夹杂物进行了研究。结果表明：LF 炉钙处理工艺精炼终点的 T.O 含量、氮含量均低于 RH 钙处理工艺。与钙处理前相比，LF 炉钙处理工艺的 RH 精炼终点钢中夹杂物数量明显减少，RH 钙处理工艺的精炼终点钢中夹杂物数量增加，且明显高于 LF 炉钙处理工艺。RH 钙处理工艺的 RH 精炼终点 1~5μm、5~15μm、>15μm 的夹杂物数量均高于 LF 炉钙处理工艺。

关键词：车轴钢；钙处理；夹杂物

Effect of Calcium Treatment on Non-metallic Inclusions in Railway Axle Steel

GUO Junbo, SHEN Chang, YANG Kezhi, LU Qiang

(Maanshan Iron & Steel Co., Ltd, Pivotal Components of Rail Transit Technology Innovation Center of Anhui Province, Maanshan 243000, China)

Abstract: The effect of LF Calcium treatment process and RH calcium treatment process on the inclusions were investigated by ASPEX. The results show that the T.O content and nitrogen content at the end of refining of LF calcium treatment process are lower than those of RH calcium treatment process. Compared with before calcium treatment, the number of inclusions in the end of RH refining of LF Calcium treatment process is obviously reduced, while the number of RH calcium treatment process increases. The number of inclusions in RH refining end point steel of RH calcium treatment process is significantly higher than that of LF calcium treatment process. The number of inclusions with 1~5μm, 5~15μm and >15μm at the end of RH refining in RH calcium treatment process is higher than that in LF furnace calcium treatment process.

Key words: railway axle steel; calcium treatment; inclusions

浅谈 RH 液压系统冲洗方案

雷丛卉，宁　博

（中国重型机械研究院股份公司，陕西西安　710018）

摘　要：在对液压系统故障的分析中可以发现，大部分液压系统故障和失效是油液污染导致的。由于新制液压系统都会经历在制造厂内预安装后拆解，运输到现场再重新安装、焊接的过程，管路中不可避免会有杂质存在，所以必须对液压系统进行冲洗，待油液的清洁度达标后才能进入设备调试运行阶段。本文针对某钢厂新投产的 RH 精炼装置液压系统冲洗方案进行探讨。

关键词：液压系统；油液清洁度；RH；冲洗方案

Discussion on the Flushing Plan of RH Hydraulic System

LEI Conghui, NING Bo

(China Heavy Machinery Research Institute Co., Ltd., Xi'an 710018, China)

Abstract: In the analysis of hydraulic system faults, it can be found that most hydraulic system faults and failures are caused by oil contamination. Due to the process of pre installation, disassembly, transportation to the site, reinstallation, and welding in new hydraulic systems, impurities are inevitably present in the pipelines. Therefore, it is necessary to flush the hydraulic system and wait for the cleanliness of the oil to meet the standards before entering the equipment debugging and operation stage. This article explores the flushing plan for the hydraulic system of the RH refining device newly put into operation in a certain steel plant.

Key words: hydraulic system; oil cleanliness; RH; flushing plan

邢钢 50t 转炉改善挡渣效果实践

韩清连[1,2,3]，张育明[1,2,3]，霍志斌[1,2,3]，和红杰[1,2,3]，
李智刚[1,2,3]，曹红波[1,2,3]

（1. 邢台钢铁有限责任公司，河北邢台 054027；2. 河北邢钢科技有限公司，河北邢台 054799；
3. 河北省线材技术创新中心，河北邢台 054027）

摘　要：邢钢 50t 转炉出钢前期使用挡渣塞挡渣，出钢后期使用挡渣锥挡渣，由于挡渣塞尺寸偏小、挡渣锥密度较小、挡渣锥加锥定位不准等因素，全年回磷>0.003%比例为 7.37%。邢钢炼钢厂冶炼车间通过增加挡渣塞尺寸、提高挡渣锥密度、制作挡渣锥定位装置、规范渣料加入量等措施，将精炼过程回磷>0.003%比例控制到 5.0%以下，对降低炼钢综合铝耗、改善钢水质量具有显著的效果。

关键词：50t；密度；回磷；挡渣锥；挡渣塞

Practice of Improving Slag Blocking Effect in 50t Converter of Xingtai Iron and Steel Co., Ltd.

HAN Qinglian[1,2,3], ZHANG Yuming[1,2,3], HUO Zhibin[1,2,3],
HE Hongjie[1,2,3], LI Zhigang[1,2,3], CAO Hongbo[1,2,3]

(1. Xingtai Iron & Steel Co., Ltd., Xingtai 054027, China; 2. Hebei Xinggang Technology Co., Ltd., Xingtai 054799, China; 3. Hebei Wire Engineering Technology Innovation Center, Xingtai 054027, China)

Abstract: In the converter of 50 tons which belongs to Xingtai Iron and Steel Co.. Ltd, the slag blocking plug is used to block the slag in the early stage of tapping and the slag blocking cone is used in the late stage of tapping. Due to the shortage size of the slag blocking plug, the density lack of the slag blocking cone, the inaccurate positioning of the slag block cone and so on, the whole year's rate about the amount of rephosphorization >0.003% was 7.37%. By increasing the size of slag blocking plug, increasing the density of slag blocking cone, making locating device of slag blocking cone and standardizing the amount of slag material, the rate about the amount of rephosphorization >0.003% is controlled to below 5.0% on refining process, which has significant effect on reducing the comprehensive aluminum consumption and improving the quality of molten steel.

Key words: 50 tons; density; rephosphorization; slag blocking cone; slag blocking plug

RH-BPI 钢液流动行为数值分析

刘 壮[1,2]，娄文涛[1,2]，朱苗勇[1,2]

（1. 东北大学多金属共生矿生态化冶金教育部重点实验室，辽宁沈阳 110819；
2. 东北大学冶金学院，辽宁沈阳 110819）

摘 要： 硫在硅钢，IF钢等高端钢材中作为有害元素应该被尽可能地去除。铁水预处理-转炉冶炼-RH精炼-连铸是生产上述钢材的典型流程。其中在转炉冶炼过程中，硫不可避免地从渣中进入钢液，很容易造成钢液内的硫含量超标，需要在后续RH工序进一步脱除。此外，RH工序的整体冶炼时间短，冶金功能多，这要求RH具备快速脱硫能力。因此，通过在钢包底部设置喷粉孔，将脱硫粉剂喷入钢液进行脱硫的RH-BPI (RH-bottom powder injection) 方法被提出。该技术旨在增加RH脱硫精炼效率。事实上，由于RH内特殊的钢液循环流动结构，在钢包底部喷粉将影响钢液的流动行为。需要对RH-BPI过程中的钢液的流动行为进行分析。

针对RH-BPI精炼过程，建立Eulerian-Eulerian耦合DPM数值模型来预测钢液的流动行为和粉剂的运动行为，同时建立了物理模型用来验证数值模型。结果表明，Eulerian-Eulerian耦合DPM模型可以准确地预测RH-BPI过程的钢液流动和粉剂运动。喷粉将影响钢包内气泡羽流的分布进而影响钢液的流动如图1所示。不进行喷粉时，钢包内的部分气泡将会进入上升管和钢包壁的缝隙中，导致钢液直接冲击浸渍管底部，同时也增加了钢包顶部形成"渣眼"的风险；另一部分气泡进入上升管，使得循环流量增加11.5%。当开始喷粉后，钢包内的气泡羽流转向上升管，因而大部分气泡进入上升管，使得循环流量增加31.7%。同时减少了钢液对浸渍管的冲击和钢包顶部形成"渣眼"的风险。

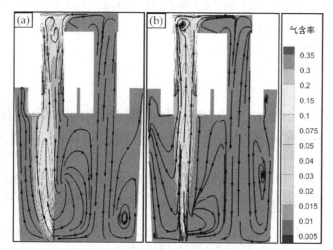

图1 气泡羽流和钢液流动对比

(a) RH-钢包底吹氩；(b) RH-BPI过程

超低碳钢夹杂物的影响因素分析与改进

舒宏富，赵傲南

（马鞍山钢铁股份有限公司技术中心，安徽马鞍山 243000）

摘 要：影响超低碳钢夹杂物的因素很多，本文讨论了钢液活度氧、炉渣、钢液成分[Si]、[Mn]和[P]元素含量对夹杂物的影响，并提出了相应的改进措施。研究结果表明：转炉终点碳氧积越低，钢液 C-O 关系更接近于平衡，并能遗传至 RH 脱碳反应过程，减小了 RH 真空条件下的碳氧反应层厚度，RH 脱碳结束时的活度氧也就越低。降低炉渣氧化性，能显著降低夹杂物数量和尺寸，并明显改变簇状夹杂物的种类及其影响程度。超低碳钢中的 Mn、Si 和 P 元素含量较高时，中包钢液全氧以及冷轧板表面检测仪检测的夹杂物数量较少。

关键词：超低碳钢；夹杂物；钢洁净度；分析与改进

Analysis and Improvement of Influence Factors of Inclusion in Ultra Low Carbon Steel

SHU Hongfu, ZHAO Aonan

(Technology Centre of Maanshan Iron and Steel Co., Ltd., Maanshan 243000, China)

Abstract: There are many factors affecting inclusion of ultra-low carbon steel. The effects of oxygen activity, slag, composition [Si], [Mn] and [P] of liquid steel on inclusion are discussed in this paper, and the corresponding improvement measures are proposed. The results show that the lower the carbon and oxygen volume at the end of the converter, the C-O relationship is closer to the equilibrium, and can be inherited to the RH decarburization reaction process, and the thickness of the carbon and oxygen reaction layer under RH vacuum condition is reduced, and the activity oxygen at the end of RH decarburization is also lower. Reducing the oxidation of slag can significantly reduce the number and size of inclusions, and obviously change the type and influence degree of cluster inclusions. When the content of Mn, Si and P elements in ultra-low carbon steel is high, the total oxygen content of liquid steel and the number of inclusions detected by cold-rolled plate surface detector are small.

Key words: ultra low carbon steel; inclusions; steel cleanliness; analysis and improvement

210t 转炉强供氧冶炼工艺技术研究与应用

王鹏飞

（山东钢铁集团日照有限公司炼钢厂生产技术室，山东日照 276800）

摘 要：针对山东钢铁集团日照有限公司炼钢厂 210t 转炉产能提升，优化脱磷、脱碳、升温效果，解决氧枪喷头使用炉数低、钢铁料消耗高、钢水过氧化导致钢水纯净度差的问题，通过设计新型"5+1"孔氧枪喷头，包括设计喷孔数、马赫数、喉口直径、出口直径、中心倾角，提高供氧强度进而缩短纯冶炼周期，提高产能的同时为设备检修提供充足时间，同时，强供氧冶炼氧气射流冲击深度深、冲击面积大、钢液搅拌充分、爆发性喷溅显著减少、钢铁料消耗降低、炉衬维护提升、氧枪喷头使用炉数提升、钢水过氧化降低、钢水絮流减少、非稳态情况得到有效解决、钢水质量大幅提升，实现年产由 850 万吨钢提升至 920 万吨钢的产能，为新型高级别品种钢的研究开发及生产提供有力保障。

关键词：强供氧强度；新型"5+1"孔氧枪喷头；纯冶炼时间；脱磷；过氧化；钢铁料；氧枪枪龄；炉衬维护

Research and Application of Intense Oxygen Supply Smelting Technology in 210 tons Converter

WANG Pengfei

(Production Technology Room, Shandong Iron and Steel Group Rizhao Co., Ltd., Rizhao 276800, China)

Abstract: Aiming at increasing the capacity of 210 tons of converter in Shandong Iron and Steel Group Rizhao Co., Ltd., optimizing the effect of dephosphorization, decarbonization and heating, and solving the problems of low number of furnaces used by oxygen gun nozzle, high consumption of steel material and poor purity of molten steel caused by peroxide, the new "5+1" hole oxygen gun nozzle is designed. Including the design of nozzle number, Mach number, throat diameter, outlet diameter, center inclination Angle, improve the oxygen supply intensity and shorten the pure smelting cycle, improve production capacity while providing sufficient time for equipment maintenance, at the same time, the strong oxygen supply smelting oxygen jet impact depth, impact area is large, liquid steel mixing is sufficient. Metal spatter and explosive spatter are significantly reduced, steel material consumption is reduced, furnace liner maintenance is improved, the number of furnaces used by oxygen gun nozzle is increased, the peroxide of molten steel is reduced, the flocculation of molten steel is reduced, and the unstable situation is effectively solved, the quality of molten steel is greatly improved, and the annual output capacity is increased from 8.5 million steel to 9.2 million tons of steel. It provides a strong guarantee for the research, development and production of new high-grade steel.

Key words: strong oxygen supply intensity; new "5+1" hole oxygen gun nozzle; pure smelting time; dephosphorization; peroxide; steel material; oxygen gun age; lining maintenance

二次燃烧氧枪下顶吹转炉气相区燃烧特征数值研究

董澎源，郑淑国，朱苗勇

（东北大学冶金学院，辽宁沈阳 110819）

摘　要：在转炉炼钢过程中，氧气通过氧枪形成超音速射流吹向金属熔池，进而对熔池脱碳、脱磷、脱硫、升温，直至冶炼出钢水。冶炼过程的熔池脱碳产生大量的 CO 气体，这些弥散在转炉空间的 CO 气体若能进一步氧化成 CO_2 将会释放出大量的热。这一过程被称为二次燃烧（简称 PC），它是一个高效放热过程，约为[C]氧化生成 CO 气体所释放热量的 3 倍[1-3]。如果这些能量被用于熔化更多的废钢，将大幅降低转炉冶炼过程所需的能耗，这也是 PC 技术得以发展的经济驱动力。值得注意的是，若要更有效地利用这部分能量，则需要实现两点：一是提高转炉内 CO 气体的二次燃烧率；二是要确保二次燃烧反应所释放的能量有效地传递到金属熔池，而不是作为高温气体离开转炉。因此，深入研究转炉内气相区的二次燃烧特性是很有必要的。国内外学者利用 CFD 方法对二次燃烧过程进行了一些探索[4-5]。但在炼钢方面，尤其是转炉炼钢过程，有关气相区燃烧特性的研究比较有限。

本文以国内某钢厂 120t 顶吹转炉为研究对象，建立了一个用于描述顶吹转炉气相区燃烧现象的三维模型，对比了普通四孔氧枪和二次燃烧氧枪下转炉内的速度场、温度分布、各组分质量分数和炉内 CO 的二次燃烧率。结果表明，超音速射流离开氧枪喷嘴后向下流动并撞击熔池表面，而后在液面的反作用下沿炉壁爬升，最终从炉口排出。高速氧射流在撞击熔池表面后会在转炉内形成环流区，该区域会不断地卷吸和夹带弥散在转炉内的 CO 气体。与普通氧枪射流不同的是，分布在二次燃烧氧枪周围的二次射流在轴向流动过程会逐渐与主射流耦合，这会影响主射流的流动特性（速度衰减、流动轨迹）。燃烧反应主要集中在射流边界处，因为该处是 CO 和 O_2 的掺混区。由于二次射流的存在，二次燃烧氧枪具有较大的燃烧区域。在轴向高度 H=0.15 m、0.65 m 及 1.25 m 处，其燃烧区的面积分别是普通氧枪的 1.37 倍、1.52 倍及 1.19 倍。炉内 CO 气体主要沿炉壁分布，且在相同截面高度处，采用二次燃烧氧枪吹炼时 CO 浓度低于普通氧枪。CO_2 气体作为燃烧产物主要在射流边界处产生，且在二次燃烧氧枪下，转炉内 CO_2 浓度高于普通氧枪下的 CO_2 浓度。为了对比两种氧枪下炉内 CO 的二次燃烧率，本文统计了炉口处 CO 和 CO_2 浓度，发现普通氧枪和二次燃烧氧枪吹炼下 CO 的二次燃烧率分别为 15.4%和 20.0%。可见，相比普通氧枪，二次燃烧氧枪使炉内 CO 的二次燃烧率提高了 4.6%。

参 考 文 献

[1] Hirai M, Tsujino R, Mukai T, et al. ISIJ Int., 1987, 27(10): 805-813.
[2] Gou H, Irons G A, Lu W -K. Metall. Mater. Trans. B, 1993, 24B: 179-188.
[3] Li Y, Fruehan R J. Metall. Mater. Trans. B, 2003, 34B: 333-343.
[4] Doh Y, Chapelle P, Jardy A, et al. Metall. Mater. Trans. B, 2013, 44B: 653-670.
[5] Liu F H, Zhu R, Dong K, et al. Metall. Mater. Trans. B, 2016, 47B: 228-243.

水平入料不锈钢-碳钢固液复合工艺技术研发

吴恩旭[1]，刘益民[2]，赵金龙[1]，纪 红[1]

（1. 鞍钢重型机械设计研究院有限公司，辽宁鞍山 114031；
2. 鞍钢重型机械有限责任公司，辽宁鞍山 114031）

摘 要：通过自主研发实验室装置，预见性复合模拟铸造，开展固液复合过程温度场数值模拟研究，明确固液复合主参数，研发不锈钢—碳钢水平固液复合工艺技术。依据不锈钢-碳钢固液复合实验室原理型装备制造出的成品，开展具有保护浇注功能的水平固液复合铸造中试生产线工艺设备设计。

关键词：水平入料；不锈钢-碳钢；固液复合；铸造模拟；复合工艺

Solid-Liquid Composite Process Technological Development of Stainless Steel Clad Carbon Steel

WU Enxu[1], LIU Yimin[2], ZHAO Jinlong[1], JI Hong[1]

(1. Angang Heavy Machinery Designing and Research Institute Co., Ltd., Anshan 114031, China;
2. Angang Heavy Machine Co., Ltd., Anshan 114031, China)

Abstract: By using experimental device which is designed , the numerical simulation of temperature field in solid-liquid composite process is carried out ,the main parameters of solid-liquid composite are determined and solid-liquid composite new technology for continuous casting of stainless steel-carbon steel is developed including predictive compound casting simulation through Huazhu software .According to the composite product produced by the principle equipment of the stainless steel clad carbon steel solid-liquid composite in the laboratory, the horizontal composite process and composite technology of the solid-liquid continuous casting pilot line with protective pouring function is proposed.

Key words: horizontal feeding; stainless steel clad carbon steel;solid-liquid composite; casting simulation; composite process

新型狭缝-弥散透气砖组合高效底吹工艺

李向龙[1]，王德永[1]，屈天鹏[1]，陈 刚[2]

(1. 苏州大学钢铁学院冶金工程系，江苏苏州 215137；
2. 张家港浦项不锈钢股份有限公司，江苏张家港 215621)

摘 要： 针对传统对称底吹模式存在搅拌效率较低，残余夹杂物较多的问题，本项目提出了一种新型非对称底吹工艺。首先通过 C 语言程序编写了异质颗粒（气泡、夹杂物）聚并及破碎行为的数学模型，在此基础上研究了不同透气砖搭配情况下的夹杂物去除情况。狭缝式吹气孔流出的气泡对流场的搅动能力较大，有利于快速混均；弥散式吹气孔流出的气泡细小，比表面积大，有利于吸附大量夹杂物上浮去除。目前常用的双狭缝式透气砖吹出来的气泡冲击顶面动能较大，在钢包内部形成双循环流，渣金界面波动较大，夹杂物去除效率较低；双弥散式透气砖的流场动力学条件较差，除杂效果仅比双狭缝式透气砖略强。狭缝+弥散式透气砖搭配，改善了前两种透气砖的天然弊端，夹杂物去除效率可以分别提高 22.7%和 5.2%。

关键词： 钢包精炼；非对称底吹；数值模拟；物理模型

High-efficiency Bottom Blowing Technology Through Slot-porous Matched Dual Tuyeres

LI Xianglong[1], WANG Deyong[1], QU Tianpeng[1], CHEN Gang[2]

(1. Soochow Univeristy, Suzhou 215137, China;
2. PZSS, Zhangjiagang 215621, China)

Abstract: In this work, a new kind of blowing technology is proposed. The coalescence and breakup between particles with different properties are described through C language. Then based on this mathematical model, the inclusion removements through different tuyere matches are investigated and compared. Slot tuyere is beneficial for fast stirring because the bubbles injected from it is quite large; By comparison, porous tuyere is beneficial for inclusion removal because the bubbles are fine. For double tuyere blowing, argon bubbles injected through two slot tuyeres are large, leading to the phenomenon that the slag-metal interface is turbulent, which is harmful for inclusion removal. Argon bubbles injected from two porous tuyeres are fine, however, dynamic behavior is largely reduced. Taking advantages of these two tuyeres, the inclusion removal ratio is increased to 22.7% and 5.2% of traditional slot-slot and porous-porous tuyeres.

Key words: ladle refining; asymmetry bottom blowing; numerical simulation; physical model

外部加入纳米 TiO₂ 对 HRB400 夹杂物性质的影响研究

程 翔，李 慧，赵煜昊，孔 辉

（安徽工业大学冶金工程学院，安徽马鞍山 243002）

摘 要：1873K，通过真空感应炉将 TiO₂ 纳米粒子直接加入 HRB400 钢液中，通过中间取样，对生成氧化物的性质演变过程进行了研究。结果表明，随着处理时间的增加，钢中钛含量和含钛氧化物的数密度均逐渐增加，并且与空白样相比，钢中氧化物的尺寸减小。这表明了 TiO₂ 外部加入的有效性和优越性，有助于加深对外部加入法的理解。

关键词：氧化物冶金；外部加入法；钛氧化物；夹杂物性质

The Effect of TiO₂ Nanopowder Adding on Inclusion Properties of HRB400

CHENG Xiang, LI Hui, ZHAO Yuhao, KONG Hui

(School of Metallurgical Engineering, Anhui University of Technology, Maanshan 243002, China)

Abstract: The TiO₂ nanopowder was directly added to the HRB400 steel melt at 1873K using a vacuum induction furnace, and the property evolution of introduced oxides was studied through intermediate sampling. The results indicate that with increasing treatment time, both the titanium content in steel and the number density of titanium-bearing oxides increases. Moreover, compared to the blank sample, the oxide diameter is reduced after the external addition of TiO₂. These results demonstrate the effectiveness and superiority of TiO₂ addition, and contribute to the understanding of the external addition method.

Key words: t oxide metallurgy; external addition method; titanium oxide; inclusion properties

转炉冶炼过程增硫现象研究与控制

赵明哲

（天津铁厂有限公司技术中心，河北涉县 056040）

摘 要：转炉炼钢为氧化反应，脱硫能力较弱，约 40%[1]。但实际生产中，时常出现转炉冶炼后钢水硫含量反高于所用铁水的现象，同时由于转炉增硫现象往往在某一炉次突然开始出现，具有相当的不可预见性，往往给后续生产造成极大的处理负担和质量风险。针对此问题，分析转炉增硫现象的主要原因，简要介绍了天津铁厂有限公司（以下简称天铁公司）生产实践中实施的分析及排查过程和采取的控制措施，为钢铁联合企业同类问题的分析与控制提供参考。

关键词：钢铁冶金；转炉炼钢；洁净钢；硫含量控制；炼钢原料管理

Research and Control of Sulfur Increasing Phenomenon in Converter Smelting Process

ZHAO Mingzhe

(Tianjin Iron PLANT Co., Ltd., Shexian 056040, China)

Abstract: Converter steelmaking is an oxidation reaction with weak desulfurization ability, which is generally considered to reach about 40% under ideal conditions. However, in actual production, there is often a phenomenon where the sulfur content in the molten steel after converter smelting is higher than that of the used iron. At the same time, due to the sudden onset of converter sulfur increase in a certain furnace, it is quite unpredictable and often causes great processing burden and quality risk to subsequent production. In response to this issue, the main reasons for the sulfur increase phenomenon in the converter are analyzed, and a brief introduction is given to the analysis and troubleshooting process and control measures implemented in the production practice of Tianjin Iron Works Co., Ltd., providing reference for the analysis and control of similar problems in steel complex enterprises.

Key words: iron and steel metallurgy; converter steelmaking; clean steel production; sulfur content control; management of steelmaking raw materials

钢中硫化物形貌控制

张立峰[1]，周秋月[2]，高小勇[3]

（1. 北方工业大学机械与材料工程学院，北京 101441；2. 北京科技大学冶金与生态工程学院，北京 100083；3. 燕山大学机械工程学院，河北秦皇岛 066004）

摘 要：本文详细对比了钢液钙处理、镁处理、稀土处理和碲处理等方法控制钢中硫化物的效果，提出了钢中硫化物形貌控制的解决方案。实验结果表明，使用精准钙处理可以有效控制钢中硫化物的形貌，并提出了精准钙处理三个区的概念。不同钢种硫含量的目标控制范围如图1所示，其中齿轮钢和易切削钢为高硫钢。钢中硫化锰夹杂物的典型形貌可分为如图2所示的四类，Ⅱ类 MnS 夹杂物有良好的变性能力，在轧制过程中延展成为大尺寸长条状，使钢材力学性能呈现各向异性，降低了材料横向性能，因此，高硫钢中硫化物形貌的控制尤为重要。

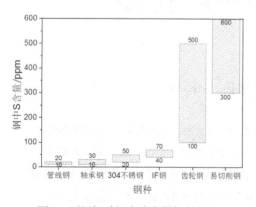

图 1 不同钢种钢中硫含量控制范围

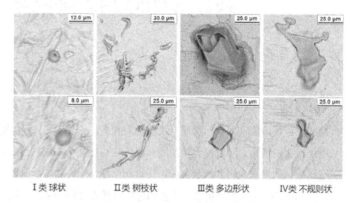

图 2 钢中硫化物形貌分类

钢中硫化物形貌的控制方法及技术总结于图3，在以下方法中，通过钢液精准钙处理，在凝固过程优先生成点状 CaS 夹杂物，可以最经济、有效的控制钢中硫化物形貌。为了更好地控制不同钢种钙处理改性硫化物形貌的目标加入量，如图 4 所示，将精准钙处理分为三个区域：Ⅰ区是 100%液相夹杂物区，该区域可以利用于大型钢锭中防止固态夹杂物碰撞聚合；Ⅱ区是 20%液相夹杂物区，该区域夹杂物造成水口结瘤的概率很低，可以用于实际生产中防止水口结瘤[1]；Ⅲ区是 $M_{CaS}/M_{liquid\ inclusions}=0.2\sim0.3$ 区域，Ⅲ区必须在Ⅱ区范围内，该区域可以用于硫化物形貌控制，且降低连铸过程中水口结瘤的发生率。

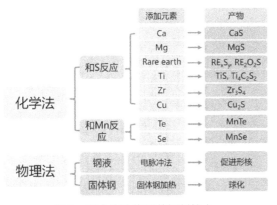 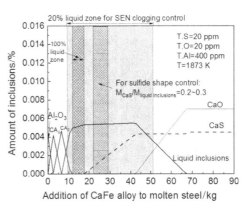

图 3　钢中硫化物形貌控制技术　　　　　　图 4　精准钙处理三个区域的理论基础[1]

参 考 文 献

[1] Li W, Wang Y, Wang W, et al. Dependence of the clogging possibility of the submerged entry nozzle during steel continuous casting process on the liquid fraction of non-metallic inclusions in the molten Al-killed Ca-treated steel[J]. Metals, 2020, 10(9): 1205.

钢液精炼过程钢中夹杂物成分和数量演变

王举金，张立峰

（北方工业大学机械与材料工程学院，北京　100144）

摘　要：本文基于耦合反应模型建立了精炼过程多相多元反应动力学模型，综合考虑了钢-渣反应、钢-夹杂物反应、钢-耐火材料反应、渣-耐火材料反应的影响。自主编写程序求解了该动力学模型。图 1(b)为耐火材料-钢液-夹杂物体系下钢液和夹杂物成分演变的测量结果和动力学模型计算结果的对比。随着保温时间的增加，MgO 质耐火材料被钢中溶解铝还原而不断向钢液中传递镁，而钢中溶解镁[Mg]进一步和 Al_2O_3 夹杂物反应，夹杂物中 Al_2O_3 含量逐渐降低，MgO 含量呈上升趋势。

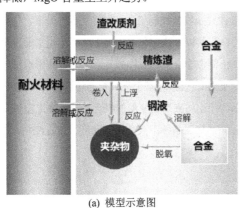 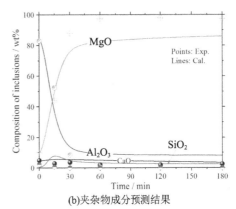

(a) 模型示意图　　　　　　　　　　　　　(b) 夹杂物成分预测结果

图 1　耦合反应模型示意图和计算结果[1]

基于颗粒尺寸分组法（PSG法）求解了钢液中夹杂物的碰撞长大模型。铝脱氧后，钢中夹杂物多是纯 Al_2O_3，Al_2O_3 夹杂物多呈聚合物状或团簇状，如图2所示。碰撞前小夹杂物占多数，碰撞 300 s 时，直径小于 2.5μm 的夹杂物数密度（$d_1 \sim d_4$）有明显下降，从 $10^{13} \sim 10^{15}$ 个/m^3 下降到 10^1 个/m^3，碰撞后出现了直径大于 53.02μm（d_{15} 和 d_{16}）的夹杂物并快速增多。随着碰撞的继续，直径小于 4.61μm（$d_1 \sim d_6$）的夹杂物数密度不断下降。碰撞 1800s 后，$d_1 \sim d_{12}$ 组的夹杂物的数密度为 10^9 个/m^3 量级，而大尺寸夹杂物（如 d_{16} 组夹杂物）则处于很低的水平，如图3所示。

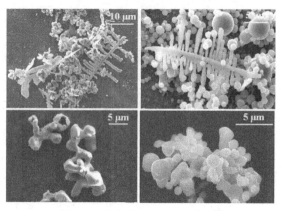

图 2　钢液中典型 Al_2O_3 夹杂物[2]

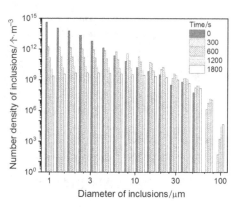

图 3　不同时间夹杂物尺寸分布[2]

参 考 文 献

[1] Wang Jujin, Zhang Lifeng, Cheng Gong, et al. Dynamic mass variation and multiphase interaction among steel, slag, lining refractory and nonmetallic inclusions: Laboratory experiments and mathematical prediction[J]. International Journal of Minerals Metallurgy and Materials, 2021, 28(8): 1298-1308.

[2] Peng Kaiyu, Wang Jujin, Li Qilan, et al. Multiphase simulation on the collision, transport, and removal of non-metallic inclusions in the molten steel during RH refining[J]. Metallurgical and Materials Transactions B, 2023, 54(2): 928-943.

钢中非金属夹杂物变形的新认识

杨　文[1]，张立峰[2]

（1. 北京科技大学冶金与生态工程学院，北京　100083；
2. 北方工业大学机械与材料工程学院，北京　100144）

摘　要：非金属夹杂物是高端钢材高品质发展的瓶颈难题之一。以往夹杂物的控制工作主要针对钢液开展，包括减少夹杂物的生成、控制夹杂物的改性和促进夹杂物的去除等。然而在连铸以及连铸之后的加工过程中由于非金属夹杂物的变温转变和轧制变形会导致夹杂物成分和形态的进一步转变。钢中非金属夹杂物的变形性能直接影响钢的加工和服役性能。本文结合不同钢种夹杂物特征要求以及热加工和冷加工工艺特点，基于夹杂物的随温塑脆转变现象，在非金属夹杂物变形性能的传统认识基础上提出了夹杂物变形的新认识。将夹杂物变形的概念扩展为高温塑性变形和低温脆性变形或脆性破碎，并据此提出夹杂物变形的控制方略，即热加工条件下夹杂物低黏度或熔点控制、冷加工条件下夹杂物低杨氏模量控制，如图1所示。

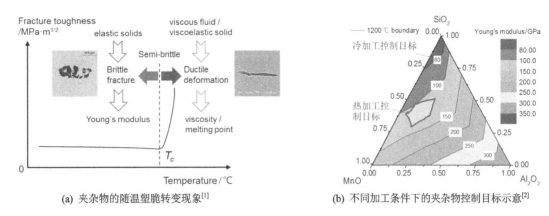

(a) 夹杂物的随温塑脆转变现象[1]　　　　　　(b) 不同加工条件下的夹杂物控制目标示意[2]

图 1　不同加工温度下的夹杂物变形机理及控制目标

基于夹杂物变形的新认识，分别开展了热轧过程和冷拔过程夹杂物变形控制的应用研究。通过增大非金属夹杂物的黏度，抑制了无取向硅钢热轧过程钢中夹杂物的塑性变形，提升了无取向硅钢的磁性能；通过减小非金属夹杂物的杨氏模量，促进了帘线钢冷拉拔过程钢中夹杂物的脆性变形，提升了帘线钢的拉拔性能，结果如图 2 所示。

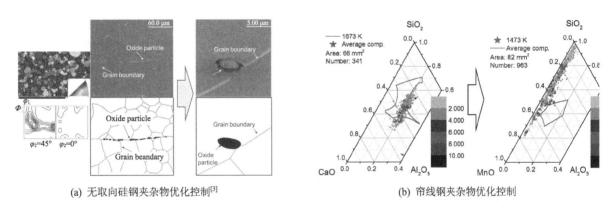

(a) 无取向硅钢夹杂物优化控制[3]　　　　　　(b) 帘线钢夹杂物优化控制

图 2　夹杂物变形控制在无取向硅钢和帘线钢中的应用

参 考 文 献

[1] Yang Wen, Peng Kaiyu, Zhang Lifeng, et al. Deformation and fracture of non-metallic inclusions in steel at different temperatures[J]. Journal of Materials Research and Technology, 2020, 9(6): 15016-15022.

[2] Zhang Lifeng, Guo Changbo, Yang Wen, et al. Deformability of oxide inclusions in tire cord steels[J]. Metallurgical and Materials Transactions B, 2018, 49(2): 803-811.

[3] Ren Qiang, Zhang Lifeng, Yang Wen. Pinning effect of oxide particles on grain boundaries of a low aluminum non-oriented electrical steel[J]. Steel Research International, 2020, 91(1): 1900303.

不锈钢精炼过程中渣-钢-夹杂物反应机理研究

任　英[1]，张立峰[2]

（1. 北京科技大学冶金与生态工程学院，北京　100083；
2. 北方工业大学机械与材料工程学院，北京　100144）

摘　要： 本文通过工业实验和热力学计算研究了精炼过程中精炼渣成分对硅锰脱氧不锈钢中夹杂物的影响。工业实验表明，LF 精炼过程中 CaO 的加入使炉渣碱度由 1.9 提高到 2.2，有利于夹杂物的去除，同时使夹杂物中 Al_2O_3 含量由 20%左右提高到 40%。低矿渣碱度使夹杂物中 Al_2O_3 含量降至 20%左右，而在 LF 重点产生较高的 T.O.。高碱度精炼渣有利于不锈钢的脱硫和脱氧，低碱度精炼渣有利于降低 304 不锈钢中的[Al]含量和夹杂物中 Al_2O_3 含量，结果如图 1[1]所示。

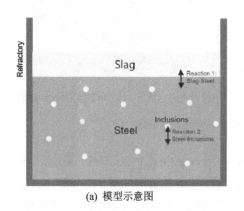
(a) 模型示意图

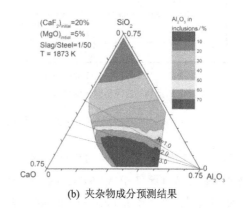
(b) 夹杂物成分预测结果

图 1 精炼渣-钢-夹杂物反应热力学模型和计算结果[1]

建立了渣-钢-夹杂物平衡反应热力学模型，模型考虑了渣-钢反应、钢-夹杂物反应，计算结果可以较好地与实验结果吻合。此模型可应用于不锈钢精炼渣成分的优化设计，预测不同精炼渣成分对钢液成分、脱硫、夹杂物成分、夹杂物熔点等的影响。为了更好了研究精炼渣对钢液成分和夹杂物成分的影响，建立了钢渣-夹杂物反应动力学模型，模型考虑了渣-钢反应、钢-夹杂物反应、钢包吹氩、夹杂物上浮等 LF 精炼参数的影响，模型计算了渣改性对夹杂物组成的影响，对渣改性过程中渣、钢和夹杂物的演变进行了较好的预测，结果如图 2 所示[2]。

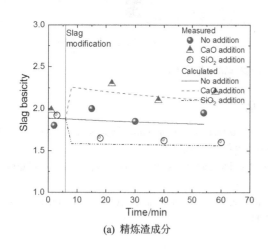
(a) 精炼渣成分

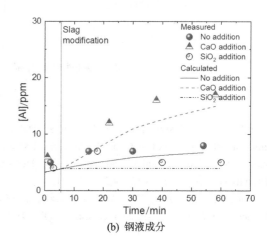
(b) 钢液成分

图 2 精炼渣成分对不锈钢洁净度影响工业试验和动力学预测[2]

参 考 文 献

[1] Ren Ying, Zhang Lifeng. Thermodynamic Model for Prediction of Slag-Steel-Inclusion Reactions of 304 Stainless Steels[J]. ISIJ International, 2017, 57(1): 68-75.

[2] Yuan Tianxiang, Zhang Lifeng, Ren Ying, et al. Effect of Slag Modification on Inclusions in Si-Mn-Killed 304 Stainless Steels[J]. steel research international, 2020: 2000506.

稀土对铝脱氧钙处理低合金耐磨钢中非金属夹杂物的影响

郑海翔[1]，任 强[2]，张立峰[3]

（1. 燕山大学亚稳材料科学与技术国家重点实验室，河北秦皇岛 066004；2. 燕山大学机械工程学院，河北秦皇岛 066004；3. 北方工业大学机械与材料工程学院，北京 100144）

摘 要：本文通过添加稀土合金再熔炼的方式，研究了稀土对初始 Al 脱氧 Ca 处理低合金耐磨钢中非金属夹杂物的影响，分析了稀土对不同类型夹杂物的改性作用规律。结果表明：当初始夹杂物为氧化物和 CaS 复合时，稀土会优先将氧化物改性为$(Ca, La)AlO_3$，CaS 此时并不参与稀土改性的这个过程；当初始夹杂物中含有较少的氧化物时，即主要夹杂物为 CaS、MnS 和 MgS 的三者其一其二或者其三，稀土会直接与硫化物发生反应，形成 CaS-MnS-MgS-CeS-LaS 的多元复合夹杂物。

关键词：低合金耐磨钢；夹杂物；稀土处理

Effect of Rare Earth on Non-metallic Inclusions in Al Deoxidized Ca Treated Low Alloy Wear Resistant Steel

ZHENG Haixiang[1], REN Qiang[2], ZHANG Lifeng[3]

(1. State Key Lab of Metastable Materials Science and Technology, Yanshan University, Qinhuangdao 066004, China; 2. School of Mechanical Engineering, Yanshan University, Qinhuangdao 066004, China; 3. School of Mechanical and Materials Engineering, North China University of Technology, Beijing 100144, China)

Abstract: In this paper, the effect of rare earth on the non-metallic inclusions in Al deoxidization Ca treated low alloy wear-resistant steel was studied by adding rare earth alloy to remelt, and the modification effect of rare earth on different types of inclusions was analyzed. The results show that when the initial inclusion is oxide and CaS, the rare earth will preferentially modify the oxide to (Ca, La) AlO_3, and CaS does not participate in the process of rare earth modification. When the initial inclusions contain less oxides, that is, the main inclusions are one or two or three of CaS, MnS and MgS, rare earth will directly react with sulfide to form CaS-MnS-MgS-CeS-LaS multi-component composite inclusions.

Key words: the low alloy wear resistant steel; inclusions; rare earth treated

第一性原理计算钢中夹杂物腐蚀性能的研究

张 静[1]，张立峰[2]

（1. 燕山大学车辆与能源学院，河北秦皇岛 066004；
2. 北方工业大学机械与材料工程学院，北京 100144）

摘　要：本文采用第一性原理计算研究了钢中夹杂物对钢腐蚀性能的影响。建立了钢中常见夹杂物的晶体结构，对晶体构型进行几何优化，充分弛豫晶格常数及各原子空间位置，使计算体系处于能量最低状态。计算结果表明，MnS 和 TiN 的形成能最大，分别为 –0.869 和 –1.707，相对最难形成；功函数可以很好的反映了钢中夹杂物的腐蚀倾向，计算夹杂物平均功函数大小为 TiN > Fe > MgO > CaO > CaS > MnS，TiN 与 MnS 对点蚀影响相对较大，更易诱导点蚀的生成，结果如图 1[1]所示。

稀土元素铈（Ce）的添加可以有效降低晶面功函数，通过添加 Ce 改性钢中夹杂物时，钢中 S 元素与活泼的 Ce 元素生成了 CeS 和 Ce_2O_2S 夹杂物，减少了凝固过程中 MnS 的生成；而部分氧化物夹杂物改性成 Ce_2O_3、Ce_2O_2 和 $CeAlO_3$ 夹杂物。5 种夹杂物和钢基体的平均功函数大小顺序为 $CeAlO_3$>Fe>MnS>CeS>Ce_2O_2S>Ce_2O_3，结合实验结果可知，Ce_2O_3 诱导点蚀发生的概率最高，$CeAlO_3$ 可以有效提高钢的耐腐蚀性能[2]。

稀土元素镧（La）的添加可以改性 MnS 夹杂物组成成分并形成含 La 夹杂物，含 La 夹杂物电子功函数相比于 MnS 夹杂物有所提高，进而增强 MnS 夹杂物局部腐蚀抗性，结果如图 2 所示[3]。

采用第一性原理计算和实验相结合的方法，计算钢中夹杂物的形成能和功函数，可从微观角度研究分析钢中夹杂物的腐蚀趋势和抗腐蚀性能，有助于理解夹杂物腐蚀机理，进一步控制夹杂物、提高钢的洁净度。

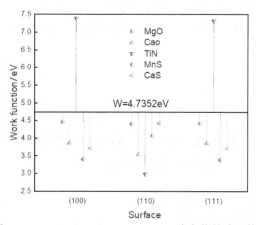
图 1　MgO、CaO、TiN、MnS、CaS 夹杂物的功函数[1]

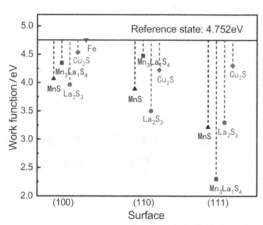
图 2　MnS、$Mn_3La_1S_4$、La_2S_3 和 Cu_2S 夹杂物的电子功函数[3]

参 考 文 献

[1] Liu Hanze, Zhang Shikun, Zhang Jing, et al. Properties of typical non-metallic inclusions in steel:First-principles calculations [J]. Materials Today Communications, 2023, 34: 105118.

[2] 刘瀚泽, 张静, 张继, 等. 稀土元素铈对钢中非金属夹杂物改性和腐蚀影响的第一性原理研究 [J]. 工程科学学报, 2022, 44(9): 1516-1528.

[3] Zhang Jing, Su Chunming, Chen Xianpei, et al. First-principles study on pitting corrosion of Al deoxidation stainless steel with rare earth element (La) treatment[J]. Materials Today Communications, 2021,27: 102204.

含钇不锈钢中 TiN 在 Y_2O_3 表面析出机理研究

徐晓峥[1]，张　静[1]，张立峰[2]

（1. 燕山大学车辆与能源学院，河北秦皇岛　066004；
2. 北方工业大学机械与材料工程学院，北京　100144）

摘　要：钢中非金属夹杂物是影响铁素体不锈钢的主要因素，向钢中添加稀土可以优化夹杂物成分进而提高不锈钢的性能。本文通过实验研究和理论计算对含有 0.014wt%钇(Y)不锈钢中主要夹杂物 Y_2O_3/TiN 界面进行了研究。实验结果表明，Y 金属氧化物发挥氧化物冶金作用形成 Y_2O_3-TiN 复合夹杂物，添加 Y 后钢中夹杂物主要为小尺寸的 TiN[1]和以 Y_2O_3 为核心表面生长 TiN 的复合夹杂物，Y_2O_3/TiN 夹杂物如图 1 所示。

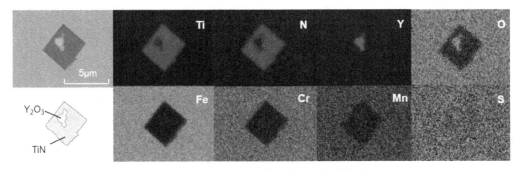

图 1 Y_2O_3/TiN 夹杂物形貌及面扫描结果

通过第一性原理计算对 Y_2O_3(100)/TiN(110)、Y_2O_3(110)/TiN(100)界面体系的界面稳定性和电子结构进行了研究。Y_2O_3 (100)/TiN(110)界面体系的粘附功为 $4.125J/m^2$，TiN 与 Fe 基体[2]形成的 Fe/TiN 界面体系粘附功最大值达到 $3.18J/m^2$。与 Fe 基体相比，TiN 与 Y_2O_3 形成的界面体系结合能力更强，TiN 更容易在 Y_2O_3 基底上成核生长。界面处 Y_2O_3 通过电子相互作用与 TiN 发生化学键合和电子转移，Y_2O_3/TiN 界面体系平面平均差分密度如图 2 所示。电荷的重新分配主要发生在 Y_2O_3/TiN 界面处，电子主要通过界面从 Y_2O_3 侧转移到 TiN 侧。界面处原子轨道发生杂化作用形成 O-Ti、Y-N 共价键，促进了界面的结合。

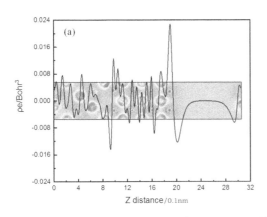 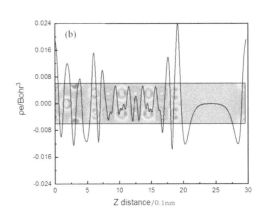

图 2 Y_2O_3/TiN 界面体系平面平均差分密度
(a)Y_2O_3(110)/TiN(100)界面体系；(b)Y_2O_3(100)/TiN(110)界面体系

参 考 文 献

[1] 张静, 马宏博, 张继, 等. 钇含量对铝脱氧含钛不锈钢中夹杂物的影响[J]. 钢铁, 2022, 57(9): 82-94: 0449-749.
[2] Matthias Gsellmann, Daniel Scheiber, Thomas Klünsner, et al. Bond strength between TiN coating and microstructural constituents of a high speed steel determined by first principle calculations[J]. Acta Materialia 2022: 117439.

基于神经网络技术的精炼过程中钢液终点硫含量预测

王伟健，张立峰

（北方工业大学机械与材料工程学院，北京　100144）

摘　要：人工神经网络，是一种在生物神经网络的启发下建立的数据处理模型。神经网络通过不断调整连接神经元的权值对输入的数据进行建模，最终具备解决实际问题的能力。神经网络将一系列具有简单数据处理能力的神经元节点通过权值相互连接，神经元模型如图1所示，权值可以不断进行调整，因此神经网络的可塑性较强。当神经网络的权值及阈值调整至恰当时，即可输出正确的结果。图2为深度神经网络的结构示意图。

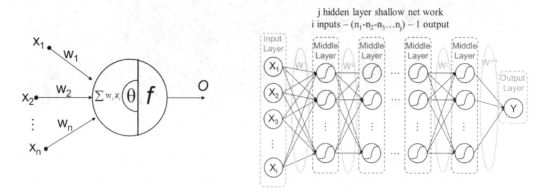

图1　神经元模型示意图　　　　　　图2　深度神经网络结构示意图[1]

神经网络的计算流程主要包括三个步骤：（1）对数据进行预处理，选择合适的神经网络结构参数，构建合适的神经网络，分别选择浅层神经网络、深度神经网络以及遗传算法优化的神经网络进行预测，并挑选出最优的预测模型，计算流程如图3所示；（2）对不同类型的神经网络进行训练；（3）利用训练好的神经网络对测试数据进行预测。图4为通过神经网络模型预测的硫含量与实际精炼过程终点硫含量的对比，预测结果与实际结果较接近。

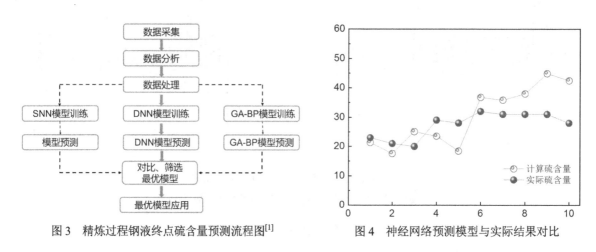

图3　精炼过程钢液终点硫含量预测流程图[1]　　　图4　神经网络预测模型与实际结果对比

参 考 文 献

[1] Wang W, Zhang L, Ren Y, et al. Prediction of Calcium Yield during Calcium Treatment Process Performed in Steelmaking Using Neural Network[J]. Metallurgical and Materials Transactions B, 2021, 53(1): 1-7.

钢中常见元素固溶对钢的力学性能影响研究

孙　莹[1]，张　静[1]，张立峰[2]

（1. 燕山大学车辆与能源学院，河北秦皇岛　066004；
2. 北方工业大学机械与材料工程学院，北京　100144）

摘　要：工程应用中对钢的力学性能要求较严格，需要在钢中添加不同元素调整钢的力学性能从而满足不同应用场景。本研究以 α-Fe 为基础，研究了不同掺杂元素对钢力学性能的影响，并利用局域电荷密度分析了不同元素掺杂的影响机理。结果表明，Cr、Si、N 元素对钢力学性能都有不同程度的降低，其中 Si 和 N 元素对钢力学性能降低的影响较小。Si 和 N 能够形成较强的离子键从而更加稳定的存在于 α-Fe 晶胞中。本研究通过第一性原理计算预测了不同元素对钢力学性能的影响，减少了实验成本，节能环保。

关键词：掺杂体系；弹性模量；第一性原理

Effect of Common Elements in Steel Dissolved in Steel on Mechanical Properties

SUN Ying[1], ZHANG Jing[1], ZHANG Lifeng[2]

(1. School of Vehicle and Energy, Yanshan University, Qinhuangdao 066004, China;
2. School of Mechanical and Material Engineering, North China University of Technology, Beijing 100114, China)

Abstract: The mechanical properties of steel in the process application are more stringent, and it is necessary to add different elements to the steel to adjust the mechanical properties of the steel to meet different application scenarios. Based on α-Fe, the influence of different doped elements on the mechanical properties of steel was studied, and the influence mechanism of doping of different elements was analyzed by using local charge density. The results show that Cr, Si and N elements have different degrees of reduction in the mechanical properties of steel, and Si and N elements have little influence on the reduction of mechanical properties of steel. Si and N can form strong ionic bonds and thus are more stable in α-Fe unit cells. In this study, the influence of different elements on the mechanical properties of steel was predicted through first-principles calculation, which reduced the experimental cost, and saved energy and environmental protection.

Key words: doping systems; elastic modulus; first-principles

钛锆复合处理钢中夹杂物诱导针状铁素形核研究

姚　浩[1]，刘承军[1]，张立峰[2]

（1. 东北大学冶金学院，辽宁沈阳　110819；
2. 北方工业大学机械与材料工程学院，北京　100144）

摘　要：本文结合原位观察结果通过理论计算的方法研究了针状铁素体在夹杂物表面形核行为，原位实验表明，针状铁素体开始在夹杂物表面形核的温度与原奥氏体晶粒尺寸、冷却速率和夹杂物特性有关[1-3]；计算结果表明[1, 2]，在所有实验条件下，随着奥氏体晶粒尺寸降低至 100μm 左右和冷却速率增加至 10℃/s，针状铁素体在夹杂物表面的形核功最大，意味着夹杂物诱导针状铁素体形核的能力最强，结果如图 1 所示。

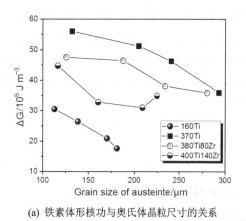 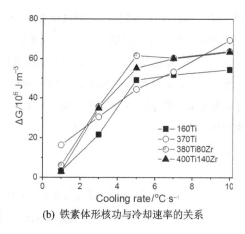

(a) 铁素体形核功与奥氏体晶粒尺寸的关系　　(b) 铁素体形核功与冷却速率的关系

图 1　针状铁素体形核功与奥氏体晶粒尺寸和冷却速率之间的关系

建立了复合夹杂物与奥氏体和铁素体之间界面的应力夹角计算模型[3],模型考虑了单一成分夹杂物与铁素体之间的界面能与奥氏体之间的界面能和夹杂物的尺寸的影响,模型认为复合夹杂物可被视为单一成分夹杂物的加权组合,计算结果可以较好地与实验结果吻合。计算结果表明,夹杂物的应力夹角的余弦大于0.3,夹杂物直径大于1μm时,夹杂物可诱导针状铁素体形核,在此基础上统计不同钢种的夹杂物成分与形核尺寸的关系,随着夹杂物中 TiO_x 和 ZrO_2 含量的增加,诱导针状铁素体形核的夹杂物平均直径降低,结果如图2所示。

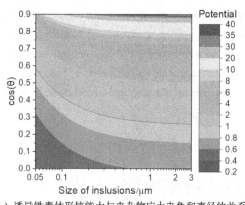 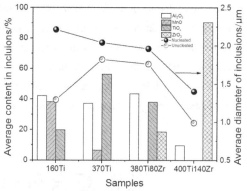

(a) 诱导铁素体形核能力与夹杂物应力夹角和直径的关系　　(b) 诱导针状铁素体形核的夹杂物成分与直径的关系

图 2　诱导针状铁素体形核的夹杂物的夹杂物成分和直径之间的关系

参 考 文 献

[1] Yao Hao, Ren, Qiang Yang Wen, et al. In situ observation and prediction of the transformation of acicular ferrites in Ti-containing HLSA steel[J]. Metallurgical and Materials Transactions B, 2022, 53(3): 1827-1840.
[2] 姚浩, 张立峰, 任强, 等. 冷却速率对 Ti-Zr 处理钢针状铁素体转变的影响[J]. 钢铁, 2021: 1-11.
[3] Yao Hao, Zhang Lifeng, Ren Qiang. Influence of inclusions on the nucleation of acicular ferrites in a Ti-Zr-bearing steel[J]. Steel Research International, 2022, 93(2): 2100468.

高纯铁素体不锈钢 TiN 夹杂物控制研究

周全磊[1]，任　英[2]，杨　文[2]，张立峰[3]

（1. 北京工业大学材料与制造学部，北京　100022；2. 北京科技大学冶金与生态工程学院，

北京 100083；3. 北方工业大学机械与材料工程学院，北京 100144）

摘　要：高纯铁素体不锈钢主要通过向钢中加入一定量的 Ti 进行稳定化处理，Ti 与 C、N 反应会形成细小的化合物，降低钢中 C、N 含量的不利影响，Ti 加入还可以促进晶粒细化和等轴晶粒的形成。但是大尺寸的 TiN 夹杂物会破坏钢基体力学性能的连续性，并在轧制过程中引发裂纹[1]，钢中含有较多的 TiN 还会在后续轧制后引发表面线缺陷[2]。本文通过定向凝固实验精确控制冷却速度以表征冷却速度对连铸坯中 TiN 夹杂物析出时机和析出规律的影响，分析 TiN 夹杂物与凝固组织的形成关系和其聚集导致缺陷的原因。不同冷速下 TiN 夹杂物形貌演变如图 1 所示，在冷却速度小于 1 ℃/s 时，TiN 夹杂物主要呈四面体状，部分以氧化物为核心长大；当冷速大于 5 ℃/s 后，钢中大量出现小尺寸聚集型的 TiN 夹杂物与长条状夹杂物。

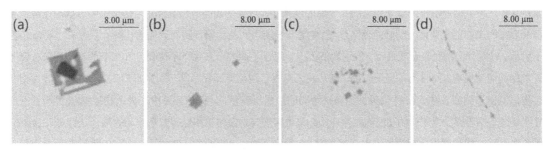

图 1　不同冷速下典型 TiN 夹杂物形貌
（a）0.2 ℃/s；（b）1 ℃/s；（c）5 ℃/s；（d）10 ℃/s

定向凝固实验表明，随着冷却速度的增大，TiN 夹杂物平均尺寸从 3.4 μm 降低到 0.9 μm，TiN 夹杂物平均长宽比从 2.0 增加到 6.4。在冷速大于 5 ℃/s 后钢中开始大量出现小尺寸长条状与聚集型 TiN 夹杂物是导致其长宽比显著增大的原因。

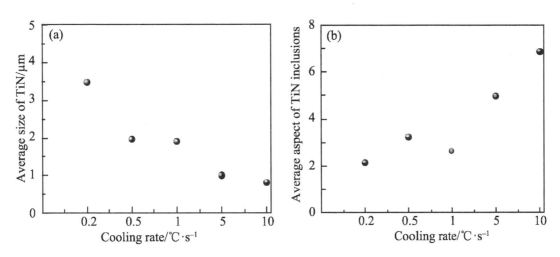

图 2　氮化钛夹杂物平均尺寸和长宽比随冷却速度的变化
（a）平均尺寸；（b）长宽比

参 考 文 献

[1] Liu T, Long J, Chen F, et al. Effect of coarse TiN inclusions and microstructure on impact toughness fluctuation in Ti micro-alloyed steel[J]. Journal of Iron and Steel Research International, 2018, 25(10): 1043-1053.
[2] Duan H, Zhang Y, Ren Y, et al. Distribution of TiN inclusions in Ti-stabilized ultra-pure ferrite stainless steel slab[J]. Journal of Iron and Steel Research International, 2019, 26(9): 962-972.

钢液中夹杂物界面现象研究

段豪剑[1]，张立峰[2]

（1. 北京科技大学冶金与生态工程学院，北京　100083；
2. 北方工业大学机械与材料工程学院，北京　100144）

摘　要：本文介绍了钢液中与夹杂物相关的几个界面现象。针对夹杂物间的相互作用，基于空腔形成理论分析了夹杂物在聚合过程中的能量变化，提出了活化态、稳定态、平衡态和临界聚合距离等概念，揭示了夹杂物聚合机理[1]，如图1所示。针对气泡捕获夹杂物过程，采用数值模拟计算了不同湍流条件下不同尺寸和形貌的气泡浮选去除不同尺寸夹杂物的捕获概率；同时为了便于应用，还开发了用户可视化界面程序[2]，如图2所示。针对钢-渣界面捕获夹杂物过程，提出了湍流条件下在钢-渣界面捕获夹杂物的边界层理论[3]，在该理论中，夹杂物被钢-渣界面捕获的有效边界层厚度是钢渣界面流速和湍动能的函数，该结果可实现与钢液精炼和连铸宏观流动的耦合。针对夹杂物在钢-渣界面分离过程，通过理论分析讨论了夹杂物在钢-渣界面分离的临界条件，建立了夹杂物在钢-渣界面分离模型，推导得到了夹杂物从钢-渣界面分离的临界速度，并且通过与流动参数相结合，建立了分离概率与流速和湍动能间的函数关系[4]，如图4所示。

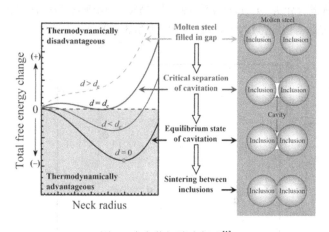

图1　夹杂物间聚合机理[1]

图2　气泡浮选去除夹杂物[2]

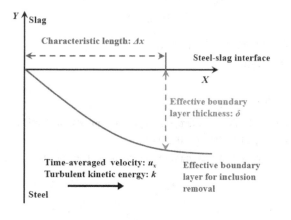

图3　钢-渣界面捕获夹杂物边界层理论[3]

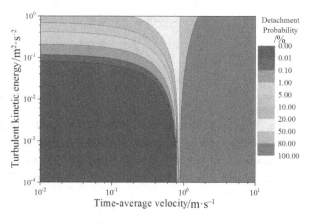

图4　夹杂物在钢-渣界面分离概率云图[4]

参 考 文 献

[1] Duan Haojian, Ren Ying, Thomas Brian G, et al. Agglomeration of Solid Inclusions in Molten Steel[J]. Metallurgical and Materials Transactions B, 2019, 50(1): 36-41.
[2] Duan Haojian, Ren Ying, Zhang Lifeng. Inclusion Capture Probability Prediction Model for Bubble Floatation in Turbulent Steel Flow[J]. Metallurgical and Materials Transactions B, 2019, 50(1): 16-21.
[3] Sun Yipeng, Duan Haojian, Zhang Lifeng. A boundary layer model for capture of inclusions by steel-slag interface in a turbulent flow[J]. Journal of Iron and Steel Research International, 2023, 30(6): 1101-1108.
[4] Hu Zhiyuan, Duan Haojian, Zhang Lifeng. Detachment Mechanism of Inclusions From the Interface Between Molten Steel and Slag[J]. Metallurgical and Materials Transactions B, 2022, 53(3): 1339-1343.

Nb-Ti 元素对铁素体不锈钢中 Fe 基合金性能影响研究

张世锟[1]，张　静[1]，张立峰[2]

（1. 燕山大学车辆与能源学院，河北秦皇岛　066064；
2. 北方工业大学机械与材料工程学院，北京　100144）

摘　要：Fe-Cr 合金作为铁素体不锈钢的主体合金，在众多工业领域应用广泛。其中 Nb、Ti 元素对合金的固溶强化有重要影响，研究 Nb、Ti 元素对合金性能影响，对 Nb、Ti 合金化冶炼具有重要意义。本文基于第一性原理计算的密度泛函理论，构建了不同 Nb、Ti 元素浓度的 Fe-Cr-Nb-Ti 合金模型，研究了合金的形成能与力学性能随合金元素浓度变化情况。结果表明，Cr、Nb 元素浓度的升高使合金的形成能升高，合金稳定降低。而 Ti 元素添加降低了合金形成能，合金稳定性得到改善。当 Cr 浓度为 6.25%，Nb 浓度为 3.125%时，合金形成能达到最低。此外，合金剪切模量随 Nb、Ti 浓度变化并不明显，受 Nb、Ti 元素影响较小。体积模量和杨氏模量随 Cr、Ti 浓度升高逐渐增大，合金的弹性及抗体积形变能力得到改善。

关键词：FeCrNbTi 合金；形成能；弹性模量；铁素体不锈钢；第一性原理

Effect of the Nb-Ti Elements on the Properties of Fe-based Alloys in Ferritic Stainless Steel

ZHANG Shikun[1], ZHANG Jing[1], ZHANG Lifeng[2]

(1. School of Vehicle and Energy, Yanshan University, Qinhuangdao 066064, China;
2. School of Mechanical and Material Engineering, Engineering, North China University of Technology, Beijing 100144, China)

Abstract: Fe-Cr alloys are widely applied in industrial fields as the main ferritic stainless steel alloys. Nb and Ti elements have an important influence on the solid solution strengthening of the alloys. Studying the effect of Nb and Ti elements on the properties of the alloys is of great significance to the alloying of Nb and Ti. Based on the density functional theory of first-principles calculation, the Fe-Cr-Nb-Ti alloy models with different Nb and Ti element concentrations were constructed. The variation of formation energy and mechanical properties of the alloys with the concentration of alloying elements was studied. The results show that the formation energy of the alloys increases and the stability of the alloy decreases with the increase of the concentrations of Cr and Nb. The addition of the Ti element reduces the formation energy of the alloy and improves the stability of the alloys. When the Cr concentration is 6.25% and the Nb concentration is 3.125%, the alloy formation energy reaches the lowest. In addition, the shear modulus of the alloys does not change significantly with the concentration of Nb and Ti. The effect of the Nb and Ti elements on the shear modulus is little. The bulk modulus and Young's modulus increase with the increase of Cr and Ti concentrations. The elasticity and volume deformation resistance of the alloys is improved.

Key words: FeCrNbTi alloys; formation energy; elastic modulus; ferritic stainless steel; first-principles

稀土钆及热处理对 C70S6 钢种拉伸性能的影响

卫红[1]，任强[2]，张立峰[3]

（1.燕山大学亚稳材料科学与技术国家重点实验室，河北秦皇岛 066044；2. 燕山大学机械工程学院，河北秦皇岛 066044；3. 北方工业大学机械与材料工程学院，北京 100144）

摘 要：本研究采用室温拉伸实验，研究了稀土 Gd 含量和热处理对 C70S6 材料拉伸性能的影响，分析对比了不同状态下材料的拉伸性能。结果表明：材料的抗拉强度和屈服强度随着稀土 Gd 含量的增加先降低后增加。在 Gd 含量为 51ppm 时抗拉强度最低，达到 825.48MPa。材料的断面收缩率和延伸率与之成反比，随着稀土 Gd 含量的增加先增加后降低。材料在 950℃下保温 45min 后，材料的抗拉强度、屈服强度、断面收缩率和延伸率均得到大幅提高。未添加稀土试样的抗拉强度由 879.41MPa 提升至 964.62MPa。断面收缩率由 12.48%提升至 35.68%。材料的强度和塑韧性均得到改善。

关键词：C70S6；稀土 Gd；拉伸性能；热处理

Effect of Rare Earth Gd and Heat Treatment on Tensile Properties of C70S6 Steel

WEI Hong[1], REN Qiang[2], ZHANG Lifeng[3]

(1. State Key Lab of Metastable Materials Science and Technology, Yanshan University, Qinhuangdao 066044, China; 2. School of Mechanical Engineering, Yanshan University, Qinhuangdao 066044, China; 3. School of Mechanical and Materials Engineering, North China University of Technology, Beijing 100144, China)

Abstract: The present investigation delves into an examination of the impact of rare earth Gd content and heat treatment on the tensile properties of C70S6 materials. Employing a room temperature tensile test, the study scrutinizes the tensile behaviors of materials in distinct states, subsequently facilitating a comprehensive analysis and comparison. Remarkably, the results unveil a distinct trend: the tensile strength and yield strength of the material exhibit an initial decline followed by a subsequent increase as the rare earth Gd content is incrementally augmented. Particularly noteworthy is the nadir observed at a Gd content of 51ppm, where the tensile strength registers at 825.48MPa. Subsequent to a heat treatment regimen entailing a 950℃ hold period spanning 45 minutes, substantial enhancements in the tensile properties materialize. Specifically, the tensile strength, yield strength, reduction of area, and elongation undergo substantial augmentation. The tensile strength of the rare earth-free specimen undergoes a considerable elevation from 879.41MPa to 964.62MPa, representing a significant improvement. Furthermore, the reduction of section area experiences a substantial rise from 12.48% to 35.68%. These compelling findings collectively underscore the pronounced ameliorative effects of heat treatment on the mechanical attributes of the material, resulting in elevated strength and enhanced plasticity.

Key words: C70S6; rare earth Gd; tensility; heat treatment

钛处理对易切削钢中硫化物的影响

焦亚兴[1]，高小勇[2]，任　强[2]，张立峰[3]

（1. 燕山大学亚稳材料科学与技术国家重点实验室，河北秦皇岛　066044；2. 燕山大学机械工程学院，河北秦皇岛　066044；3. 北方工业大学机械与材料工程学院，北京　100144）

摘　要：对于易切削钢来说，钢中夹杂物的成分、形貌、尺寸及分布都会影响易切削钢的切削性能。本文通过实验室实验，使用场发射电子扫描显微镜研究了不同 Ti 含量对易切削钢中夹杂物成分、形貌和硬度的影响。结果表明：随着 Ti 含量的增加，钢中夹杂物的形貌发生显著变化：在未添加 Ti 的易切削钢中硫化物主要为球形和棒状，而在添加 2000ppm 的钢中，钢中硫化物的成分发生显著变化，钢中硫化物由 MnS 全部转变为 $Ti_4C_2S_2$，硫化物的形貌主要为棒状和树枝状，还有少量的纺锤状。然而，与未添加 Ti 的易切削钢相比，在添加大量 Ti 的易切削钢中钢基体的硬度以及钢中夹杂物的硬度明显增加。

关键词：易切削钢；小样电解；硫化物形貌；维氏硬度

Effect of Titanium Treatment on Sulfides in Free Cutting Steel

JIAO Yaxing[1], GAO Xiaoyong[2], REN Qiang[2], ZHANG Lifeng[3]

(1. State Key Lab of Metastable Materials Science and Technology, Yanshan University, Qinhuangdao 066044, China; 2. School of Mechanical Engineering, Yanshan University, Qinhuangdao 066044, China; 3. School of Mechanical and Materials Engineering, North China University of Technology, Beijing 100144, China)

Abstract: For free cutting steels, the composition, morphology, size and distribution of inclusions in the steel will effect the cutting performance of the free cutting steel. The effects of titanium content on the composition, morphology and hardness of inclusions in free-cutting steel were studied by electron scanning microscope and Vickers hardness tester. The results show that the morphology of inclusions in steel changes significantly with the increase of Ti content; In the free cutting steel without Ti, the sulfide is mainly spherical and rod-like, while in the steel with 2000ppm, the composition of the sulfide in the steel changes significantly, and the sulfide in the steel is changed from MnS to $Ti_4C_2S_2$, and the morphology of the sulfide is mainly rod-like and dendritic, and a small amount of spindle. However, compared with the free cutting steel without Ti, the hardness of the steel matrix and the hardness of inclusions in the free cutting steel with a large amount of Ti are significantly increased.

Key words: free-cutting steel; small sample electrolyzation; sulfide morphology; vickers hardness

界面反应对硅硫钢与氧化钙接触角的影响

马天浩[1]，任　强[2]，张立峰[3]

（1. 燕山大学亚稳材料制备技术与科学国家重点实验室，河北秦皇岛　066044；

2. 燕山大学机械工程学院，河北秦皇岛 066044；

3. 北方工业大学机械与材料工程学院，北京 100144）

摘 要：本文对 Si-S 钢与 CaO 基板的接触角进行测量，对脱硫界面进行了的观察。Si-S 钢与 CaO 基板接触角在 2 min 内稳定在 132°左右，2 min 到 4 min 逐渐降低到 126°，4 min 后由于反应产物溢出到表面，接触角无法测量。钢液表面溢出的反应产物主要为 CaS，脱硫界面的产物主要为 Ca_2SiO_4 与 CaS。脱硫反应分为两个阶段，致密反应层形成后阻碍脱硫反应；致密反应层被破坏后，脱硫反应继续进行，反应产物在界面聚集到上限后溢出表面。文

关键词：界面反应；脱硫；接触角；润湿性；CaO；硅硫钢

Effect of Interfacial Reaction on the Contact Angle Between Si-S Steel and Calcium Oxide

MA Tianhao[1], REN Qiang[1], ZHANG Lifeng[3]

(1. State Key Lab of Metastable Materials Science and Technology, Yanshan University, Qinhuangdao 066044, China; 2. School of Mechanical Engineering, Yanshan University, Qinhuangdao 066044, China; 3. School of Mechanical and Materials Engineering, North China University of Technology, Beijing 100144, China)

Abstract: In this study, the contact angle between a Si-S steel and a CaO substrate was measured, and the desulfurization interface was observed. The contact angle between the Si-S steel and the CaO substrate was 132° within 2 min and gradually decreased to 126° from 2 min to 4 min. After 4 min, the contact angle could not be measured due to the overflow of reaction product to the surface. The reaction product of the liquid steel surface overflow was CaS, and the product of desulfurization interface was Ca_2SiO_4 and CaS. Desulphurization reaction was divided into two stages. The formation of dense reaction layer hindered desulphurization reaction. After the dense reaction layer was destroyed, the desulfurization reaction was continued, and the reaction product was accumulated to the interface and was overflowed the surface.

Key words: interfacial reaction; desulfurization; contact angle; wettability; CaO; Si-S steel

U71Mn 重轨钢中氧化物夹杂物对 MnS 评级结果的影响

高梓淇[1]，任 强[1]，张立峰[2]

（1. 燕山大学机械工程学院，河北秦皇岛 066044；
2. 北方工业大学机械与材料工程学院，北京 100144）

摘 要：非金属夹杂物是重轨钢内部损伤的主要原因，重轨钢采用无铝脱氧工艺，严格控制引入硬质夹杂物。本研究取自国内某钢厂无铝脱氧配合调整精炼渣系生产的重轨钢 MnS 评级合格与不合格样品，通过自动扫描电镜与能谱结合分析了合格与不合格样品夹杂物的形貌、平均成分、数量密度、面积分数对 MnS 评级的影响。发现 U71Mn 重轨钢合格与不合格样品化学成分并无差异，钢轨中夹杂物的平均化学成分为：78.86% Al_2O_3，2.57%SiO_2，1.76%MnO，4.64%CaO，12.17%MgO。夹杂物形貌主要以氧化物硫化物复合夹杂物为主，同时还有一些纯氧化铝夹杂、镁铝尖晶石夹杂以及硅酸盐夹杂。此外，合格样品中的氧化物数量较多，有利于诱导 MnS 非均质形核析出，

促进 MnS 的弥散分布，所以合格样品中 MnS 的数量密度与面积分数更小，级别也较不合格样品中的 MnS 更小。

关键词：重轨钢；非金属夹杂物；夹杂物形貌；非均质形核；MnS 评级

Effect of Non-metallic Inclusions in U71Mn Heavy Rail Steel on MnS Rating Results

GAO Ziqi[1], REN Qiang[1], ZHANG Lifeng[2]

(1. School of Mechanical Engineering, Yanshan University, Qinhuangdao 066044, China;
2. School of Mechanical and Materials Engineering, North China University of Technology, Beijing 100144, China)

Abstract: Non-metallic inclusions are the main cause of internal damage of heavy rail steel. Heavy rail steel adopts aluminum-free deoxidization process and strictly controls the introduction of hard inclusions. In this study, MnS-rated qualified and unqualified samples of heavy rail steel produced by a domestic steel plant without aluminum deoxidization and adjustment of refining slag system were taken. The effects of morphology, average composition, number density and area fraction of inclusions in qualified and unqualified samples on MnS rating were analyzed by automatic scanning electron microscopy and energy spectrum. It was found that there was no difference in chemical composition between qualified and unqualified samples of U71 Mn heavy rail steel. The average chemical composition of inclusions in rail was: 78.86% Al_2O_3, 2.57% SiO_2, 1.76% MnO, 4.64% CaO, 12.17% MgO. The morphology of inclusions is mainly composed of oxide sulfide composite inclusions, as well as some pure alumina inclusions, magnesium aluminate spinel inclusions and silicate inclusions. In addition, the number of oxides in the qualified samples is large, which is beneficial to induce the heterogeneous nucleation and precipitation of MnS and promote the dispersion distribution of MnS.Therefore, the number density and area fraction of MnS in the qualified samples are smaller, and the grade is also smaller than that of MnS in the unqualified samples.

Key words: heavy rail steel; non-metallic inclusions; inclusion morphology; heterogeneous nucleation; MnS rating

304 不锈钢中 $MnO \cdot Cr_2O_3$-MnS 夹杂物耐点蚀性能研究

刘 震[1]，任 强[2]，张立峰[3]

（1. 燕山大学亚稳材料科学与技术国家重点实验室，河北秦皇岛 066044；2. 燕山大学机械工程学院，河北秦皇岛 066044；3. 北方工业大学机械与材料工程学院，北京 100144）

摘 要：本研究采用原位腐蚀实验，研究了 304 不锈钢中 $MnO \cdot Cr_2O_3$ 夹杂物及 $MnO \cdot Cr_2O_3$-MnS 复合夹杂物的点蚀萌生过程，分析对比了两种夹杂物的耐点蚀性能。结果表明：$MnO \cdot Cr_2O_3$ 夹杂物不容易引起点蚀，在 $FeCl_3 \cdot HCl$ 溶液中浸泡 30min 后夹杂物表面及基体界面处没有发生明显变化。$MnO \cdot Cr_2O_3$-MnS 复合夹杂物的耐点蚀性能较差，在 $FeCl_3 \cdot HCl$ 溶液中浸泡 3min 后，夹杂物边缘处 MnS 颜色变深；浸泡 10min 后，MnS 与钢基体界面处萌生点蚀，并向基体扩展；浸泡 30min 后，点蚀坑继续扩展，钢基体及夹杂物边缘处的 MnS 大量溶解，夹杂物心部的 $MnO \cdot Cr_2O_3$ 没有发生腐蚀。

关键词：夹杂物；点蚀；原位腐蚀

Study on Pitting Resistance of MnO·Cr$_2$O$_3$-MnS Inclusions in 304 Stainless Steel

LIU Zhen[1], REN Qiang[2], ZHANG Lifeng[3]

(1. State Key Lab of Metastable Materials Science and Technology, Yanshan University, Qinhuangdao 066044, China; 2. School of Mechanical Engineering, Yanshan University, Qinhuangdao 066044, China; 3. School of Mechanical and Materials Engineering, North China University of Technology, Beijing 100144, China)

Abstract: In this study, the pitting initiation process of single MnO·Cr$_2$O$_3$ inclusions and MnO·Cr$_2$O$_3$-MnS composite inclusions in cast 304 stainless steel was studied by in-situ corrosion experiment, and the pitting corrosion resistance of the two inclusions was analyzed and compared. The results showed that the MnO·Cr$_2$O$_3$ inclusion was not easy to cause pitting corrosion, and there was no change on the surface of the inclusion and the interface of the matrix after soaking in FeCl$_3$·HCl for 30 min. The pitting corrosion resistance of MnO·C$_{r2}$O$_3$-MnS composite inclusions was poor. When immersed in FeCl$_3$·HCl for 3min, the color of MnS at the edge of the inclusions becomed darker. When immersed for 10 min, pitting corrosion initiated at the interface between MnS and steel matrix and expanded to the matrix. When immersed for 30 min, the pitting pits continued to expand, MnS at the steel matrix and the edge of the inclusions dissolved a lot, and MnO·Cr$_2$O$_3$ in the core of the inclusions did not corrode.

Key words: inclusions; pitting corrosion; in-situ corrosion method

304 不锈钢中 Mn-Cr-O 夹杂物的形成机制

齐江涛[1]，王志军[1]，李院高[2]，伊 璞[1]

（1. 山西太钢不锈钢股份有限公司制造部，山西太原 030003；
2. 山西禄纬堡太钢耐火材料有限公司，山西太原 030104）

摘 要： 本文主要研究了 304 不锈钢在连铸和热连轧工序 Cr-Mn-O 夹杂的成因。通过中包覆盖剂、中包涂抹料（耐火材料）、结晶器保护渣、304 不锈钢化学成分，结合 304 钢连铸和热连轧过程冶炼条件，对 304 不锈钢在连铸和热连轧工序进行热力学分析，发现耐火材料和中包覆盖剂对 304 不锈钢成分和夹杂没有影响；在保护渣和 304 钢水共同作用下，形成 MnS 夹杂以及富 Cr 液相，在热应力和表面张力的驱动下聚集于连铸坯的边棱周围，在热连轧弱氧化气氛下，二者氧化形成 Mn-Cr-O 夹杂物；由于三种保护渣在高温下物相组成与物相含量接近，因此三种保护渣均会形成 Mn-Cr-O 夹杂。

关键词： Mn-Cr-O 夹杂；304 奥氏体不锈钢；结晶器保护渣；中包涂抹料；中包覆盖剂

The Formation Mechanism of Mn-Cr-O Inclusions in 304 Stainless Steel

QI Jiangtao[1], WANG ZhiJun[1], LI Yuangao[2], YI Pu[1]

(1. Manufacturing Department of Shanxi Taigang Stainless Steel Co., Ltd., Taiyuan 030003, China;

2. Shanxi Luweibao Taigang Refractory Materials Co., Ltd., Taiyuan 030104, China)

Abstract: This article mainly studies and analyzes the formation mechanism Cr-Mn-O inclusions in 304 stainless steel during continuous casting and hot continuous rolling processes. Thermodynamic analysis was conducted on 304 stainless steel during the continuous casting and hot continuous rolling processes using tundish covering fluxes, tundish magnesite coating (refractories), mold fluxes, and chemical composition of 304 stainless steel, combined with the smelting conditions. It was found that refractories and tundish covering fluxes have no effect on the composition and inclusions of 304 stainless steel. Under the joint action of mold fluxes and 304 molten steel, MnS inclusions and Cr-rich liquid phases are formed, which aggregate around the edges of 304 stainless steel CC slab under the driving force of thermal stress and surface tension. Under the low oxidation atmosphere of hot continuous rolling, MnS inclusions and Cr-rich liquid phases are oxidized to form Mn-Cr-O inclusions. Due to the nearly equal phase composition and content of the three types of mold fluxes at high temperatures, Mn-Cr-O inclusions are formed in all mold fluxes.

Key words: Cr-Mn-O inclusions; 304 austenitic stainless steel; mold fluxes; tundish magnesite coating; tundish covering fluxes

炉壁氧枪垂直角对电弧炉炼钢影响的模拟研究

曾召鹏，王　耀，杜习乾，马建超

（江苏省（沙钢）钢铁研究院炼钢连铸研究室，江苏张家港　215625）

摘　要：以某厂 100t 电弧炉为原型，建立了全尺寸三维几何模型，利用数值模拟的方法研究了炉壁氧枪垂直角对电弧炉炼钢的影响。数值模拟结果表明：随着氧枪垂直角减小，射流到达钢液面的总行程变短，核心区流速增加，有利于氧气射流对熔池的冲击和搅拌，熔池死区比例呈现减小趋势；但伴随而来的是钢液喷溅现象加剧，氧气射流对熔池冲击点距离炉壁更近，随着壁面剪切力增大，炉衬侵蚀更为严重。当氧枪垂直角由 50°减小到 40°时，熔池平均冲击深度增加了 56.7%，熔池死区比例减少了 40.1%，炉壁剪切力增加了 67.2%。因此，调整氧枪垂直角可以增强电弧炉熔池搅拌效率，减少渣线处的炉衬侵蚀，对延长炉衬寿命有着重要意义。

关键词：电弧炉；数值模拟；炉壁氧枪；垂直角；熔池搅拌；炉衬侵蚀

Simulation Research on Effect of Vertical Angle of Wall Oxygen Lance on EAF Steelmaking

ZENG Zhaopeng, WANG Yao, DU Xiqian, MA Jianchao

(Steelmaking-Casting Group, Institute of Research of Iron and Steel, Shagang,
Jiangsu Province, Zhangjiagang 215625, China)

Abstract: Based on a 100t EAF in a factory, a full-scale three-dimensional geometric model was established. The effect of vertical angle of wall oxygen lance on EAF steelmaking is studied by numerical simulation. The numerical simulation results show that with the decrease of the vertical angle in wall oxygen lance, the jet distance to the molten steel surface becomes shorter and the flow velocity in the core area increases, which is conducive to the impact and stirring of the oxygen jet on the molten bath, and the proportion of dead zone in molten bath appeared a decreasing trend. However, the splashing of molten steel is becoming worse, the impingement point of the oxygen jet on the molten bath is closer to the furnace wall, as the wall shear on the furnace wall increases, the lining erosion is becoming worse. When the vertical angle of the oxygen lance decreases from 50° to 40°, the average impact depth of the molten bath increases by 56.7%, the proportion of dead zone of the molten bath decreases by 40.1%, and the wall shear on the furnace wall increases by 67.2%. Therefore, adjusting the vertical angle of oxygen lance can increase the molten bath stirring efficiency of EAF and reduce the loss of wear and tear at the slag line, which is of great significance for prolonging the life of furnace lining.

Key words: electric arc furnace; numerical simulation; wall oxygen lance; vertical angle; molten bath stirring; lining erosion

富锰渣的副产品高效利用的试验研究

吴伟[1], 杨勇[1], 赵斌[2], 赵博[1], 梁强[1], 赵进宣[1]

(1. 钢铁研究总院有限公司, 冶金工艺研究所, 北京 100081;
2. 西昌学院, 机械与电气工程学院, 四川西昌 615000)

摘 要: 为了高效地利用富锰渣副产品高锰生铁, 在500kg感应炉进行了高锰生铁冶炼。把高锰生铁的冶炼过程分为脱锰保碳期和去碳脱磷期。脱锰保碳期通过向熔池内加入铁矿石, 控制铁水温度在1300和1400℃之间。终点得到含氧化锰大于50%的高锰渣和碳含量为3.2%左右的半钢。去碳脱磷过程是向熔池内加入石灰、镁球等造渣材料, 终点得到磷含量<0.015%和硫含量<0.02%的钢水。结果表明, 得到高锰渣的化学成分中FeO在10.8%和15.3%之间, MnO在70%和77%之间, SiO_2在3%和7%之间。高锰渣的主要矿相组成是$2MnO·SiO_2$和$FeO·MnO·SiO_2$, 这些符合制备高碳锰铁原料的条件。得到半钢的化学成分中, C是在3.2%和4.2%之间, Mn是在0.3%和1.0%之间, P是在0.12%和0.22%之间, S是在0.01%和0.024%之间。进一步冶炼, 得到低磷、低硫的洁净钢水。

关键词: 富锰渣; 副产品; 高效利用; 脱锰保碳期; 去碳脱磷期; 洁净钢水

Study on Efficient Use of By-products for Rich Manganese Slag

WU Wei[1], YANG Yong[1], ZHAO Bin[2], ZHAO Bo[1],
LIANG Qiang[1], ZHAO Jinxuan[1]

(1. Metallurgical Technology Research Institute, General Iron and Steel Research Institute,
Beijing 100081, China; 2. School of Mechanical and Electrical Engineering, Xichang
University, Xichang 615000, China)

Abstract: In order to efficiently utilize the high manganese pig iron by-product of manganese-rich slag, the smelting process of high manganese pig iron was studied in a 500kg induction furnace. The smelting process of high manganese pig iron was divided into demanganese and carbon retention period and decarbonization and dephosphorization period. The hot metal temperature is controlled between 1300 and 1400°C by adding iron ore to the molten bath. At the end point, high manganese slag containing more than 50% manganese oxide and semi-steel with carbon content of about 3.2% are obtained. Lime, magnesium balls and other slag-forming materials are added to the molten bath for dephosphorization and desulfurization of hot metal. At the end point, molten steel with a phosphorus content of less than 0.015% and a sulfur content of less than 0.02% is obtained. The results show that in the chemical composition of high manganese slag, the iron oxide content is between 10.8% and 15.3%, the manganese oxide content is between 70% and 77%, and the silicon oxide content is between 3% and 7%. The main mineral phases of the high-manganese slag are $2MnO·SiO_2$ and $FeO·MnO·SiO_2$, which are suitable for the preparation of high-carbon ferromanganese raw materials. In the chemical composition of the semi-steel, the carbon content is between 3.2% and 4.2%, the manganese content is between 0.3% and 1.0%, the phosphorus content is between 0.12% and 0.22%, and the sulfur content is between 0.01% and 0.024%. Further smelting, the chemical composition of the clean molten steel was obtained.

Key words: manganese-rich slag; by-product; efficient utilization; demanganese and carbon retention period; decarbonization and dephosphorization period; clean molten steel

双渣法降低铁钢比的实践探索

王少波[1,2]，鹿 焱[2]，刘 震[3]，谷茂强[2]，朱苗勇[1]

（1. 东北大学冶金学院，辽宁沈阳 110819；2. 宝山钢铁股份有限公司，上海 201900；
3. 重庆钢铁股份有限公司，重庆 401220）

摘 要：针对目前碳中和、碳达峰发展的需要，在转炉生产过程中增加废钢的使用有重要意义，对此重钢进行了低碳冶炼方面的探索，即采用双渣法进行转炉吹炼。根据生产实绩对单渣法、双渣法进行了热量收支理论计算，对比单渣法评估了双渣法对废钢加入量的影响。根据重钢生产实践，针对 210t 转炉板坯产线，不补热情况下，采用双渣法，平衡铁钢比从单渣法 840kg/t 降低至 810kg/t。

关键词：转炉；少渣；热平衡；铁钢比

Exploration of Reducing Iron-to-Steel Ratio Through Double-Slag Method

WANG Shaobo[1,2], LU Yan[2], LIU Zhen[3], GU Maoqiang[2], ZHU Miaoyong[1]

(1. Metallurgy Department of Northeastern University, Shenyang 110819, China; 2. Baosteel, Shanghai 201900, China; 3. Chongqing Iron and Steel Group, Chongqing 401220, China)

Abstract: In response to the current need for carbon neutrality and carbon peaking development, increasing the utilization of scrap in the BOF steelmaking process holds significant importance. In this regard, a low-carbon and low-cost BOF steelmaking exploration has been conducted by Chonggang, employing the "Double-Slag" method for converter refining. Theoretical calculations on heat balance were performed for the Single-Slag method based on production data, and the influence of the "Double-Slag" method on the amount scrap was calculated. According to production practice, for a 210-ton BOF slab line without heat supplementation, the implementation of the "Double-Slag" method led to a reduction in the iron-to steel ratio, dropping from 840kg/t to 810kg/t.

Key words: BOF; less slag; heat balance; scrap ratio

100t 转炉低耗高效长材炼钢技术生产实践

张朝发，李双武，张 垚，李 硕，张宝景

（唐山钢铁集团有限责任公司长材事业部，河北省高品质钢
连铸技术创新中心，河北唐山 063000）

摘　要：100t 转炉通过采取优化喷头参数、转炉终点低氧含量控制、合金料在线烘烤、钢包高效周转和连铸低温快浇等措施，逐步解决了冶炼周期长、生产效率低、原辅料消耗高等问题，主要技术指标明显提升，实现了直轧工艺下炉机匹配的高效化稳定生产。转炉冶炼周期缩短 1.3min，稳定控制在 29min 左右，连铸拉速稳定在 3.8~4.1m/min，浇注周期逐步控制到 30min 以内，中包过热度稳定在 22~24℃，转炉钢铁料消耗降低 8kg/t 左右，低合金转炉出钢温度降低 11℃左右。转炉-连铸生产效率显著提升，炼钢成本明显降低。

关键词：100t 转炉；钢铁料消耗；冶炼周期；出钢温度；生产实践

Production Practice of Steelmaking Technology with Low Consumption and High Efficiency for Long Product in 100 ton Converter

ZHANG Chaofa, LI Shuangwu, ZHANG Yao, LI Shuo, ZHANG Baojing

(Long Product Department of HBIS Group Tangsteel Company, Tangshan 063000, China)

Abstract: The problems of long smelting cycle, low production efficiency and high raw materials consumption were gradually solved by optimizing nozzle parameters, low oxygen content at the end of the converter, alloy materials on-line baking, efficient ladle turnover and low-temperature fast casting of continuous casting in 100t converter. The main technical indicators improved significantly. The smelting time of the converter was stably at about 29min by decreasing 1.3min, the continuous casting pulling speed was stably at 3.8~4.1m/min, the pouring cycle was within 30min, the superheat of the tundish was about 22~24℃, the steel consumption of the converter reduced by about 8kg/t, and the tap steel temperature of converter reduced by about 11℃. The production efficiency of converter and continuous casting improved significantly, and the cost of steelmaking reduced significantly.

Key words: 100t converter; steel consumption; smelting time; tap steel temperature; production practice

转炉炼钢渣中微观渣相对低磷钢冶炼的影响

吕延春[1,2]，李海波[1]，刘　洋[1]，王海宝[1]，孔祥涛[1]

（1. 首钢集团有限公司技术研究院，北京　100043；
2. 北京绿色可循环钢铁流程北京市重点实验室，北京　100043）

摘　要：随着钢铁工业快速发展及结构升级，先进钢铁材料在制造、装备、重大工程等领域不断突破，对钢水超高洁净度提出更高要求，对于临氢钢、抗酸容器钢、低温容器钢等钢种要求钢水磷含量极其严格，生产中遇到了诸多技术瓶颈。与底吹效果好的常规大型转炉不同，公称容量小的弱底吹转炉在冶炼拉碳后至出钢前的这个过程段易发生转炉内的回磷，尤其是低碳钢冶炼炉次。

以首钢首秦 100t 转炉为例，底吹最大供气强度 0.03Nm³/min，采用留渣+双渣冶炼工艺，在转炉冶炼后期可以将钢水磷降低至 30~40ppm，但是极易发生出钢前转炉内钢水中磷含量升高 10~30ppm 的现象，导致出钢时钢水 P 含量升高，成品磷最低 70ppm。研究发现高 FeO 含量的钢渣富磷相 $2CaO·SiO_2$ 的尺寸相对较小，渣中 P 含量较低，转炉终渣高 FeO 条件下产生回磷[1-4]，其原因是渣中自由 FeO 与富磷相 $2CaO·SiO_2-3CaO·P_2O_5$ 相反应，生成了低熔点的液态渣相，磷进入液相后参与了液相渣与钢水间磷的重新分配，产生了回磷的现象。通过试验与计算表明，钢渣中存在固态富磷相 C_2S-C_3P 对渣-钢磷分配比有影响，渣中存在固态富磷相可以提高渣-钢间的磷分配比。转炉入炉铁水 Si 含量变化对转炉终点磷含量产生一定影响，双渣法冶炼低磷钢，在转炉入炉铁水 Si 量较低及炼钢辅料加入量稳定的情况下，入炉铁水每减少 0.1%的 Si，终点钢中 P 增加 9~12ppm。另外顶吹供氧量过高，也容易导致转炉终点磷的升高，主要原因是由于供氧量高的炉次渣中 FeO 易升高致使渣中固态 C_2S-C_3P 减少，导致磷在液渣中磷含量升高，进而产生磷在渣-钢间的重新分配。超低磷钢冶炼拉碳时低温控制 P≤40ppm，继续吹炼至终点过程中温度继续提高，该过程易发生回磷，对于该关键点的控制，为了防止回磷，必须控制升温过程中新产生的 FeO 与富磷相 $2CaO·SiO_2$ 的反应。对于发生回磷炉次进行分析发现，拉碳渣中富磷相 $2CaO·SiO_2$ 颗粒尺寸在 5×5μm-10μm，而对应的终点渣的渣相 $2CaO·SiO_2$ 颗粒变的更加细小，其尺寸在 2×5μm-5μm。未发生回磷炉次拉碳渣与终点渣的 $2CaO·SiO_2$ 颗粒尺寸没有发生明显变化。

参 考 文 献

[1] 王新华, 朱国森, 李海波, 等. 氧气转炉"留渣＋双渣"炼钢工艺技术研究[J]. 中国冶金, 2013, 23(4): 41-45.
[2] 吕延春, 王新华, 秦登平, 等. 转炉炼钢渣磷元素的富集及影响因素[J]. 钢铁, 2017, 52(10): 29-36.
[3] Nurse R W, Welch J H, Gutt W, et al. High-temperature Phase Equilibria in the System Dicalcium Silicate-tricalcium Phosphate[J]. Chem. Sco, 1959(63): 1080-1083.
[4] 周寒梅, 包燕平, 林路. P_2O_5 对 CaO-SiO_2-$FetO$-P_2O_5 渣中富磷相的影响[J]. 中国冶金, 2013, 23(1): 45-49.

冶金模型定义的数字化炼钢连铸工艺

李立勋，田 伟，李 杨，刘善喜，孙 博，徐 政

（河钢集团唐钢公司，河北唐山 063000）

摘 要：本文系统研究炼钢连铸区域应用冶金工艺模型研发高端钢种的实践及炼钢区域冶金工艺模型的集群应用。阐明转炉炼钢冶金工艺模型在低成本洁净钢冶炼过程、连铸冶金工艺模型在均质化铸坯生产过程发挥的关键核心作用。应用转炉炼钢数字化冶金模型实现熔炼过程稳定低耗洁净钢生产；应用系统级的连铸数字化冶金工艺模型有效控制不同类型铸坯中心偏析，实现连铸均质化铸坯生产；系统解决炼钢连铸数字化、模型化生产用工业软件问题。详细比对分析板坯连铸应用数字化工艺模型改进前后中碳类钢种铸坯中心偏析及其如何遗传影响热轧带钢产品的组织及夹杂物分布的工程应用及冶金效果。

关键词：炼钢；连铸；冶金模型；高端钢种；洁净钢；均质化

Digital Steelmaking and Continuous Casting Process Defined by Metallurgical Model

LI Lixun, TIAN Wei, LI Yang, LIU Shanxi, SUN Bo, XU Zheng

(HBIS Group Tangsteel Company, Tangshan 063000, China)

Abstract: This paper systematically studies the practice of applying metallurgical process model to research and develop high-end steel grades in the steelmaking and continuous casting area of Hesteel Group Tangshan Iron and Steel Co., Ltd. and the cluster application of metallurgical process model in the steelmaking area. Clarify the key core role of the converter steelmaking metallurgical process model in the low-cost clean steel smelting process, and the continuous casting metallurgical process model in the homogenized billet production process. Applying the converter steelmaking metallurgical model to achieve stable and low consumption digital model control production during the smelting process; Applying a system level continuous casting digital metallurgical process model to achieve homogenized billet production and effectively control center segregation of different types of billets; Solve the industrial software problem of digitalization and modeling production in steelmaking and continuous casting. Detailed comparison and analysis of the engineering application effect of the center segregation of medium carbon steel billets and its genetic impact on the microstructure and inclusion distribution of hot rolled strip products before and after the improvement of the digital process model for slab continuous casting.

Key words: steelmaking; continuous casting; metallurgical model; high end steel grade; clean steel; homogenization

炼钢炉渣总量对稀土收得率影响

张胤[1,2]，张顺[3]，刁望才[1,2]，张怀军[1,2]，
曹瑞峰[3]，韩春鹏[1,2]

（1. 内蒙古包钢钢联股份有限公司技术中心，内蒙古包头 014010；
2. 内蒙古自治区稀土钢产品研发企业重点实验室，内蒙古包头 014010；
3. 内蒙古包钢钢联股份有限公司炼钢厂，内蒙古包头 014010）

摘　要： 炉外精炼过程中，炉渣中富集氧、硫元素。在稀土钢冶炼过程中，随合金加入的稀土Ce元素与炉渣中氧、硫元素作用形成化合物而大量留存于炉渣内，只有少部分稀土元素随钢水进入浇注环节，影响稀土收得率。本文通过对稀土元素在精炼炉渣中存在形式进行分析，以对比试验结果分析了精炼炉渣量与稀土元素Ce收得率之间关系，结果显示精炼炉渣量对稀土Ce收得率产生影响。

关键词： 稀土；炉渣；收得率

Effect of Total Amount of Steelmaking Slag on Yield of Rare Earth

ZHANG Yin[1,2], ZHANG Shun[3], DIAO Wangcai[1,2], ZHANG Huaijun[1,2],
CAO Ruifeng[3], HAN Chunpeng[1]

(1. Technical Center of Inner Mongolia Baotou Steel Union Co., Ltd., Baotou 014010, China;
2. Key Laboratory of Rare Earth Steel Products R&D Enterprise of Inner Mongolia Autonomous Region, Baotou 014010, China; 3. Steel Plant of Inner Mongolia Baotou Union Co., Ltd., Baotou 014010, China)

Abstract: Oxygen and sulfur are enriched in the slag in the process of refining outside the furnace. In the process of rare earth steel smelting, the rare earth Ce element added with the alloy and the oxygen and sulfur elements in the slag to form compounds and a large number of retained in the slag, only a small part of rare earth elements with the molten steel into the pouring link, affecting the rare earth recovery rate. In this paper, the existence form of rare earth elements in the refining slag is analyzed, and the relationship between the refining slag and the yield of rare earth elements Ce is analyzed by comparing the test results. The results show that the amount of refining slag has an effect on the yield of rare earth Ce.

Key words: rare earth; refining slag; yield

基于时空分辨成像光谱分析的炼钢供氧火点区在线监测技术与研究

孟令辉,董 凯

(北京科技大学碳中和研究院,北京 100083)

摘 要:本文根据炉口火焰分析技术现存的一些问题,提出了一种新的基于时空分辨光谱的炼钢火点区在线监测技术,该技术选择了一条全新的光谱测量途径,可以为进一步探索炼钢火点区冶金反应特性提供有利条件,同时,有效避免了炉口火焰分析技术所存在的缺陷。本文通过热态实验对该技术的实现进行了初步研究,并对该技术的光谱测量和数据分析方法,以及与转炉配合的控制系统和运行逻辑进行了简要介绍。

关键词:炉口火焰分析;火点区;光谱分析;自动控制;转炉炼钢

Research and Development of Oxygen Injection Fire Zone in Steelmaking Online Monitoring Technology Based on the Spatio-temporal Resolution Imaging and Spectroscopy Analysis

MENG Linghui, DONG Kai

(Carbon Neutral Innovation Institute, University of Science and Technology Beijing, Beijing 100083, China)

Abstract: In this paper, a new oxygen injection fire zone in steelmaking online monitoring technology based on the spatio-temporal resolution imaging and spectroscopy analysis is proposed, aiming to address some existing issues in the field of tuyere flame analysis. This technology adopts a novel approach for spectral measurements, providing favorable conditions for further exploration of the metallurgical reaction characteristics in the oxygen injection fire zone in steelmaking, while effectively avoiding the drawbacks of traditional tuyere flame analysis techniques. Preliminary research on the implementation of this technology is conducted through thermal experiments. The spectral measurement and data analysis methods, as well as the control system and operating logic in coordination with the converter, are briefly introduced.

Key words: tuyere flame analysis; fire zone; spectral analysis; automatic control; converter steelmaking

Mg 脱氧的氧化物冶金技术

杨 健,李婷婷,徐龙云,潘晓倩,张银辉

(上海大学材料科学与工程学院,上海 200444)

摘 要:本文研究了 Mg 脱氧对船板钢第二相纳米颗粒,以及 Nb 和 B 含量对 Mg 脱氧船板钢 HAZ 组织和韧性的影响。结果表明，Mg 的强还原性使 TiN 纳米颗粒发生增量细化析出，Mg 脱氧有利于改善船板钢 HAZ 韧性。Mg 脱氧船板钢中 Nb 含量从 0.002 增加到 0.016wt.%引起 TiN 转变为(Ti, Nb)(C, N)，颗粒稳定性降低，加速了焊接过程中小尺寸颗粒溶解和大尺寸颗粒粗化，引起 PAG 尺寸增大。大尺寸 PAG 中铁素体形核密度较低以及生长速率较高，促进脆性组织生长，因此 –40℃ HAZ 韧性从 127 降低到 58J。Mg 脱氧船板钢中添加 0.0022wt.% B 时，B 以 4at.%浓度在晶界偏析，使得晶界能降低 0.13J/m^2，这将通过提高铁素体形核的临界活化能来降低边界铁素体形核密度。因此，GBF 和 FSP 降低，IAF 增加，–40℃ HAZ 韧性从 40 提高至 141J。

关键词：船板钢；Mg 脱氧；Nb；B；HAZ；组织；韧性

Oxide Metallurgy Technology with Mg Deoxidization

YANG Jian, LI Tingting, XU Longyun, PAN Xiaoqian, ZHANG Yinhui

(School of Materials Science and Technology, Shanghai University, Shanghai 200444, China)

Abstract: The effect of Mg deoxidization on the secondary phase nano particles in shipbuilding steel plates and the effects of Nb and B contents on the microstructures and toughness of HAZ for Mg-deoxidized shipbuilding steels were studied. The results show that the strong reducibility of Mg makes TiN nano particles in Mg-deoxidized shipbuilding steel plates be incrementally refined. Increasing Nb content in Mg-deoxidized shipbuilding steels from 0.002 to 0.016wt.% makes TiN change into (Ti, Nb)(C, N) with the particle stability decreasing, which enhances the small-size particle dissolution and the large-size particle coarsening during welding, and further increases the PAG size. Accordingly, the ferrite nucleation density is lowered and the ferrite growth rate is increased, which coarsens the microstructures and reduces the HAZ toughness at –40℃ from 127 to 58J. With adding 0.0022wt.% B to Mg-deoxidized shipbuilding steels, B is segregated at the austenite grain boundary with the concentration of 4at.%. This leads to the grain boundary energy decreasing by 0.13J/m^2, which further reduces the boundary ferrite nucleation density. Therefore, GBFs and FSPs decrease, and IAFs increase, resulting in the HAZ toughness at –40℃ increasing from 40 to 141J.

Key words: shipbuilding steel; Mg deoxidization; Nb; B; HAZ; microstructure; toughness

降低 120t 转炉炉渣铁含量的研究

乔西亚[1,2]，韩啸[1,2]，何志军[1,2]，杨鑫[1,2]

（1. 辽宁省绿色低碳与智能冶金重点实验室，辽宁鞍山　114051；
2. 辽宁科技大学材料与冶金学院，辽宁鞍山　114051）

摘　要：转炉炉渣中氧化铁含量是炼钢生产的一项重要技术经济指标，转炉吹炼过程中炉渣中铁元素的消耗约占整个铁损失的 50%，增加了转炉炼钢成本[1,2]。降低终渣（T.Fe）含量可以提高转炉吹炼终点的金属收得率，降低吨钢生产成本，还有利于降低钢液氧含量，实现降低脱氧剂及合金消耗、减少钢液中夹杂物的产生量[3]。现场生产中的转炉终渣粒铁质量偏高，造成铁损较多、成本增加。

本研究先是通过开展工业试验转炉终点炉渣取样分析，工业试验终渣出现夹带无法打磨碎的铁粒情况，终渣（SiO$_2$）含量和出钢温度对终渣粒铁的影响如图 1 所示，发现将终渣（SiO$_2$）含量控制在 16%以内，出钢温度高于 1630℃，可明显降低炉渣铁粒占比，并提出了控制终渣（SiO$_2$）含量、提高出钢温度、分批次加入铝矾土等措施，转炉炉渣的铁粒含量由平均 8%降至 0.5%，吨钢成本降低 25 元；之后通过对试验炉次的对比分析，揭示了转炉渣（T.Fe）含量与底吹强度、含铁冷料的加入量和加入时机、脱碳速度和终点碳含量的关系，提出（T.Fe）含量控制措施：120t 转炉吹炼末期底吹流量应控制在 300m³/h（标态）左右，增大钢液和炉渣接触面积，促进[C]-(FeO)反应的进行，（T.Fe）含量可降低 3.5%；采用后搅工艺，可继续降低（T.Fe）含量 3.3%；含铁冷料应在吹氧 90%之前全部加完，减少补吹和过吹；保证终点[C]含量在 0.06%以上，可将转炉终渣（T.Fe）含量稳定控制在 15%以下；出钢碳含量控制在 0.1%以上，可将炉渣（T.Fe）含量控制在 12.5%以下；对于出钢磷含量≥0.025%的钢种，终渣（T.Fe）最低降至 14%以下，吨钢成本减少 3.6 元。

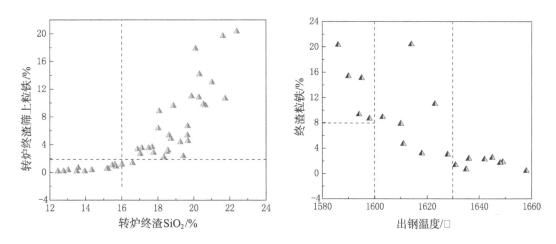

图 1　工业试验终渣(SiO₂)、出钢温度含量对终渣粒铁的影响

参 考 文 献

[1] 赵志超, 孙彦辉, 罗磊, 等. 300t 顶底复吹转炉炉渣 FeO 动态预测模型[J]. 炼钢, 2015, 31(6): 13-16, 22.
[2] 王旭. 降低转炉炉渣中氧化铁含量的研究[J]. 山西冶金, 2023, 46(2): 154-156.
[3] 刘飞, 官挺, 孙凤梅, 等. 转炉吹炼过程炉渣的变化及其对脱磷的影响[J]. 炼钢, 2013, 29(6): 23-27.

高品质特殊钢冶炼恒稳定工艺技术研究与应用

杨 平，王海达，陈 列，李 亮，胡金海

（建龙北满特殊钢有限责任公司，黑龙江齐齐哈尔　161041）

摘　要：针对高品质特殊钢炼钢工艺不稳定的问题，采用跟踪排查、定点监测、规范操作、控制物料成分、优化炼钢工艺等手段系统研究了不同流程中钢水成分、温度和周期的变化规律，确定了影响高品质特殊钢加工性能和产品质量波动的关键工艺参数，建立了以高拉碳、强脱氧、均搅拌、精操作为工艺特色的钢-渣-气-料恒通量技术体系，提出了 LD→LF→VD→CC 炼钢全流程、多维度、跨界面的控制策略。现场成品钢碳、锰、铬含量与目标值偏差分别不超过±0.01%、±0.02%、±0.02%的比例高于 99.5%、99.3%、99.3%；精炼钢水连铸上台过热度在（25±5）℃之间的比例超过 96.2%，冶炼-精炼-连铸周期匹配炉次比例超过 95.0%，奠定了连铸恒稳控制的基础。
关键词：高品质特殊钢；工艺稳定；成分控制；工艺协同

Research and Application of Constant Stability Production Technology for High Carbon Alloy Steel Smelting

YANG Ping, WANG Haida, CHEN Lie, LI Liang, HU Jinhai

(Jianlong Manchuria Special Steel Co., Ltd., Qiqihar 161041, China)

Abstract: In view of the problem of unstable steelmaking process of high carbon alloy steel, by means of tracking and checking, site monitoring, standard operation, control of material composition and optimization of steelmaking process, the variation of liquid steel composition, temperature and period in different processes were systematically studied, the key process parameters affecting the machining properties and product quality fluctuation of high carbon wear-resistant steel were determined, the technology system of constant flux of steel-slag-gas-material, which is characterized by high carbon content, strong deoxidation, uniform stirring and fine operation, has been established, the control strategy of LD→LF→VD→CC steelmaking process is put forward. The deviation of carbon, manganese and chromium contents from the target values of finished steel in the field was less than ±0.01%, ±0.02% and ±0.02%, respectively, and the ratio was higher than 99.5%, 99.3% and 99.3%. The proportion of superheat between (25±5)℃ and 95.0% of matching furnaces of smelting-refining-continuous casting period were established, which laid the foundation for constant control of continuous casting.

Key words: high carbon alloy steel; stable process; composition control; process collaboration

高碳合金钢矩形坯均质化技术研究与应用

王海达，陈 列，张亚楠，杨 平，范世强，杨 铭

（建龙北满特殊钢有限责任公司，黑龙江齐齐哈尔 161041）

摘 要：通过对 250mm×280mm 矩形坯连铸末端电磁弱搅拌配合轻重混合压下协同的作用下，所得高碳钢连铸坯中心碳偏析 1.05、碳极差不大于 0.08% 的比例达到 98.4%，中心缩孔不超过 0.5 级的比例达到 99.5%。同时基于铸坯内部质量的提升，实现了低压缩比大规格棒材轧制，其内部超声波探伤使用 GB/T 4162 A 级合格率高于 99.95%。大幅度降低工序成本提高了连铸连轧的生产效率。

关键词：末端电磁搅拌；轻压下；重压下；偏析；缩孔

Research and Application of Homogenization Technology for High Carbon Alloy Steel Rectangular Billet

WANG Haida, CHEN Lie, ZHANG Ya'nan, YANG Ping, FAN Shiqiang, YANG Ming

(Jianlong Manchuria Special Steel Co., Ltd., Qiqihar 161041, China)

Abstract: The main reason of matrix spalling and dropping is that the segregation and shrinkage cavity of continuous casting billet can be inherited into the product, which leads to abnormal structure and discontinuous structure. The central segregation ratio of high carbon steel billet is less than 1.05 and the carbon gap is less than 0.08%, which is 98.4%, 99.5%

of the bars have no central shrinkage of more than 0.5 grade, and 99.5% of the bars have no "White circle" or "Black Center" macro-segregation defects. At the same time, based on the improvement of slab internal quality, the rolling internal qualified rate of large-size bar with low pressure reduction ratio is higher than 99.95%. The production efficiency of continuous casting and rolling is improved by greatly reducing the process cost.

Key words: electromagnetic stirring; low pressure; high pressure; segregation; shrinkage cavity

钢中夹杂物稳定区域图的绘制方法、典型案例与重要问题

王宏坡，康 健，王 雨

（重庆大学材料科学与工程学院，重庆 400044）

摘 要：夹杂物控制是钢液洁净度控制的核心内容之一，稳定区域图（优势区域图）是研究夹杂物演化和制定夹杂物调控策略的重要依据。基于主流脱氧、脱硫和夹杂物变质工艺，稳定区域图主要涉及 Al-O 系、Al-Mg-O 系和 Al-Ca-O 系、Si-Mn-O 系、Al-Ti-O 系、Al-Ca-O-S 等夹杂物体系。近年来随着稀土在钢中应用的大力发展，研究者也尝试计算绘制了 Al-La-O 系、Al-Ce-O 系和 Al-Y-O 系等夹杂物的稳定区域图。稳定区域图的绘制方法主要包括边界线法、等氧浓度线法和计算相图法（CALPHAD）等。20 世纪末 Itoh H 教授[1]绘制了 Al-Mg-O 系夹杂物的稳定区域图，张立峰教授[2]和 Kang Y B 教授[3]等对二维稳定区域图的绘制进行了系统描述和讨论，本文作者[4]在此基础上将基于瓦格纳活度模型绘制的稳定区域图拓展至三维空间，更直观地阐述了稳定区域图的实际意义。

稳定区域图的绘制当前主要存在几个突出问题：一是不同学者报道的相关热力学数据波动较大，绘制的稳定区域图相应相差较大；二是无论基于瓦格纳活度模型计算还是 FactSage 软件直接绘制稳定区域图均存在很多难点，具有一定的技术门槛；三是绘制结果的准确性如何进行判断。本文针对上述问题，以采用 FactSage 软件绘制典型 Al-Mg-O 系夹杂物稳定区域图为案例，详细阐述了其绘制方法和必须注意的问题，绘制过程如图 1 所示。并就其绘图结果与和瓦格纳活度模型自编程计算绘图结果间的差异进行了对比分析，阐述了导致两种结果存在差异的主要原因。研究结果对准确判断夹杂物的生成和制定相应调控策略具有较好的指导意义。

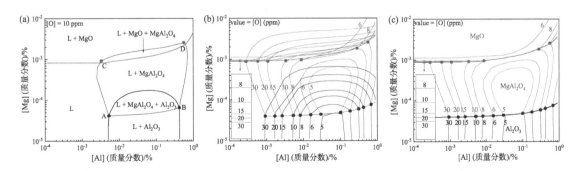

图 1 1600℃下铁液中 Al-Mg-O 系夹杂物稳定区域图绘制过程
(a) 相界的判定；(b) 多氧浓度下相界的确定；(c) 等氧浓度图

参 考 文 献

[1] Itoh H, Hino M, Ban-Ya S. Metallurgical and Materials Transactions B, 1997, 28(5): 953-956.
[2] Zhang L, Ren Y, Duan H, et al. Metallurgical and Materials Transactions B, 2015, 46(4): 1809-1825.
[3] Kang Y B, Jung S H. ISIJ International, 2018, 58(8): 1371-1382.
[4] Wang H, Yu P, Zhou X, et al. Journal of Materials Research and Technology, 2021, 12: 43-52.

基于 Python 的转炉合金化控制系统开发

胡倩倩[1,2]，董晓雪[1,2]，朱海琳[1,2]，施树蓉[1,2]，
韩　啸[1,2]，张友建[1,2]，何志军[1,2]

（1. 辽宁科技大学，辽宁鞍山　114051；
2. 辽宁省绿色低碳与智能冶金重点实验室，辽宁鞍山　114051）

摘　要：转炉炼钢是钢铁企业产业链的核心工艺，炼钢成本关系整个钢铁产品成本。针对不同钢种，炼钢成本变化最大的为合金成本，其占比最低在4%左右，最大可达50%。特别针对高合金钢种，成本占比较大，由此需考虑在满足产品需求的同时降低合金成本，从而降低炼钢工序成本。实现炼钢工序中合金成本最优，一方面需要控制物料的使用成本，采用综合成本较低的合金，精准控制合金加入量，另一方面还需控制合金物料的采购成本以及合理的库存。针对以上问题，采用 Python 开发了转炉合金化控制系统。控制系统通过线性规划算法，综合分析实际生产参数、合金采购成本、合金元素成分、合金收得率、钢种元素上下限、所需元素以及有害元素限制、冶炼终点残余元素含量等因素后构建合金成本优化算法模型，同时，控制系统采用 Web 和桌面应用程序两个方向进行开发，其中 Web 主要是针对不同数据进行管理，而桌面应用程序主要提供给现场操作人员直接运算应用。开发的转炉合金化控制系统具有以下特点和功能：利用线性规划算法，将计算结果再次作为条件进行优化计算，保证结果最优；同时直接关联生产系统，提高数据准确性同时，能很好地符合现场实际操作要求；系统应用过程自动采集生产数据，无需人工干预，可以降低操作人员劳动强度，降低岗位轮换限制要求，有助于生产稳定运行；针对合金市场价格波动，系统能够有效为合金物料采购及库存的合理控制提供指导。

精炼渣氧化性对 Y15 硫易切削钢中夹杂物的影响

王桂林[1,2]，刘丽霞[1,2,3,4]，彭　军[1,2,3,4]，
张　芳[1,2]，刘　爽[1,2,3]，安胜利[1,2,3]

（1. 内蒙古科技大学材料与冶金学院，内蒙古包头　014000；2. 内蒙古自治区先进陶瓷材料与器件重点实验室（内蒙古科技大学），内蒙古包头　014010；3. 教育部轻稀土资源绿色提取与高效利用重点实验室（内蒙古科技大学），内蒙古包头　014010；
4. 内蒙古科技大学稀土产业学院，内蒙古包头　014010）

摘　要：本文通过 FactSage 热力学计算与碳硫分析仪、场发射扫描电镜、夹杂物统计等方面结合，研究了 Y15 硫系易切削钢增硫后和增硫后添加精炼渣对钢中硫含量和夹杂物的影响。研究表明，只增硫的钢中，夹杂物析出温度为 1500℃，且硫的收得率较低，为 50%，钢样中的夹杂物呈链状沿晶界聚集分布，对钢的性能不利。采用 FeO-CaO-Al_2O_3-SiO_2-MgO 精炼渣冶炼后，夹杂物析出温度降低至 1200℃，与只增硫的钢样相比，钢中夹杂物的析出温度降低，析出量减少，从 Ⅱ 类链状分布向 Ⅰ 类弥散分布转变。随着渣中 FeO 含量的增加，钢样中硫含量增加，硫的收

得率升高，夹杂物尺寸增大，群聚程度降低。当渣中 FeO 含量，达到 5%时，硫的收得率最高，并且此时夹杂物为块状和球状，呈弥散分布。

关键词：增硫；精炼渣；硫易切削钢；夹杂物

稀土 Ce 对 20CrMnTi 钢中 TiN 夹杂析出影响热力学计算及实验研究

彭 军[1,2,3,4]，王 健[1,2]，张 芳[1,2]，刘 爽[1,2,3]，安胜利[1,2,3]

（1. 内蒙古科技大学材料与冶金学院，内蒙古包头 014010；2. 内蒙古自治区先进陶瓷材料与器件重点实验室（内蒙古科技大学），内蒙古包头 014010；3. 教育部轻稀土资源绿色提取与高效利用重点实验室（内蒙古科技大学），内蒙古包头 014010；4. 内蒙古科技大学稀土产业学院，内蒙古包头 014010）

摘 要：TiN 夹杂物尺寸和形貌对 20CrMnTi 齿轮钢产品质量、服役性能均有影响，控制 TiN 析出尺寸及形貌是提高 20CrMnTi 钢服役稳定性的关键。本文对添加稀土 Ce 元素前后 20CrMnTi 钢中 TiN 夹杂物形核、析出行为及演变进行了研究。结果表明：1873K 下，钢液中无法析出 TiN 等夹杂物，即 TiN 只能在钢液凝固过程中析出长大。加入稀土 Ce 元素后，钢中稀土 Ce 与 Al_2O_3 反应生成 $CeAlO_3$ 可作为 TiN 夹杂物的形核核心，形成椭球型的 $CeAlO_3$-TiN 复合夹杂物。对 Ce 元素加入后，20CrMnTi 钢中夹杂物尺寸进行统计，因形成尺寸更小的复合夹杂物 $CeAlO_3$-TiN，使钢中尺寸＞5μm 夹杂物降低了 17.2%，有效减小了钢中 TiN 夹杂物尺寸，从形貌和尺寸上，达到改质钢中 TiN 的效果。

关键词：稀土 Ce；微合金化；夹杂物；形核核心

Thermodynamic Calculation and Experimental Study on the Effects of Rare Earth Ce on TiN Inclusion Precipitation in 20CrMnTi Steel

PENG Jun[1,2,3,4], WANG Jian[1,2], ZHANG Fang[1,2], LIU Shuang[1,2,3], AN Shengli[1,2,3]

(1. School of Materials and Metallurgy, Inner Mongolia University of Science and Technology, Baotou 014010, China; 2. Inner Mongolia Key Laboratory of Advanced Ceramic Materials and Devices (Inner Mongolia University of Science and Technology), Baotou 014010, China; 3. Key Laboratory of Green Extraction & Efficient Utilization of Light Rare-Earth Resources, Ministry of Education (Inner Mongolia University of Science and Technology), Baotou 014010, China; 4. School of Rare Earth Research and Development, Inner Mongolia University of Science and Technology, Baotou 014010, China)

Abstract: The size and morphology of TiN inclusions have an effect on the quality and service performance of gear steel 20CrMnTi. Controlling the size and morphology of TiN precipitation is the key to improving the service stability of 20CrMnTi. In this paper, the nucleation behaviors of TiN inclusions in 20CrMnTi before and after the addition of rare earth Ce are studied. The results show that TiN inclusions cannot precipitate in the molten steel at 1873K. TiN can only precipitate and grow during the solidification of the steel. After adding rare earth Ce, Ce reacts with Al_2O_3 in steel to generate $CeAlO_3$, which can serve as the nucleation cores of TiN inclusions to form ellipsoidal $CeAlO_3$-TiN composite inclusions. The size of inclusions in 20CrMnTiCe steel was statistically analyzed. Due to the formations of the smaller composite inclusions of $CeAlO_3$-TiN, the number of inclusions with a size larger than 5μm reduced by 17.2%. The addition of Ce effectively reduces the size of TiN inclusions in steel, and achieves the effect of improving TiN in terms of morphology and size.

Key words: rare earth Ce; microalloying; inclusions; nucleation core

Si 添加对高 Mn-N 型 TWIP 不锈钢力学性能的影响

曾泽瑶

（钒钛资源综合利用国家重点实验室，材料工程技术研究所，四川攀枝花　617000）

摘　要：利用 SEM、TEM 等研究了不同热处理工艺制度对 1.5%（质量分数）Si 添加后的高 Mn-N TWIP 不锈钢显微组织和力学性能的影响，结果发现，试验钢退火孪晶厚度随着固溶温度的升高而增大，在高的固溶温度下，Si 添加以提高合金的应变硬化率，拉伸变形过程中生成大量细长的形变孪晶可以有效阻碍位错滑移，提高材料强度与塑性。时效处理对拉伸性能影响较小，在 700~800℃时效过程中形成晶界胞状形式析出，胞状析出内以短棒状 Cr_2N 为主，造成沿晶脆性断裂，冲击吸收功快速降低。

关键词：不锈钢；析出相；沿晶断裂；冲击韧性

Effect of Si Addition on Mechanical Properties of High Mn-N Type TWIP Stainless Steel

ZENG Zeyao

(State Key Laboratory of Vanadium and Titanium Resources Comprehensive Utilization, Material Engineering Technology Research Institute, Panzhihua 617000, China)

Abstract: The effects of different heat treatment processes on the microstructure and mechanical properties of high Mn-N type of TWIP stainless steel with 1.5(wt.%) Si element addition were studied by SEM and TEM. The results showed that the annealing twin thickness of the experimental steel increases with the increase of solution temperature,. At high solution temperature, Si was added to improve the strain hardening rate of the alloy, and a large number of slender deformation twins were generated during tensile deformation, which can effectively prevent dislocation slip. Improve material strength and plasticity. Aging treatment has little effect on tensile properties. During aging at 700~800℃, grain boundary cellular precipitation is formed, and the cellular precipitation is mainly short rod Cr_2N, resulting in intergranular brittle fracture and rapid reduction of impact absorption energy.

Key words: stainless steel; precipitated phase; intergranular fracture; impact toughness

水分子在稀土元素掺杂铁表面的吸附研究

雷 超

（钒钛资源综合利用国家重点实验室，材料工程技术研究所，
四川攀枝花 617000）

摘 要：利用第一性原理计算的方法，对水分子在有稀土元素掺杂的铁晶格表面的吸附作用进行了研究，分析了水分子在 La、Y、Ce 三种稀土元素掺杂的体心立方、面心立方、密排六方铁表面的吸附构型以及吸附能大小。研究发现，掺杂 La、Y、Ce 三种稀土元素的铁晶格表面更容易吸附水分子，其中 Y 的促进作用最强，Ce 次之，La 最弱。

关键词：第一性原理计算；表面吸附；稀土元素；吸附能

A Study about the Adsorption of H_2O on the Surface of Fe with Rare Earth Doping

LEI Chao

(State Key Laboratory of Vanadium and Titanium Resources Comprehensive Utilization,
Material Engineering Technology Research Institute, Panzhihua 617000, China)

Abstract: The adsorption of water molecules on the surface of iron lattice doped with rare earth elements was studied by the method of first principles calculation. The adsorption configuration and adsorption energy of water molecules on the body-centered vertical, face-centered cubic and close-packed hexagonal iron surface doped with La, Y and Ce were analyzed. It is found that the surface of iron lattice doped with La, Y and Ce is more likely to adsorb water molecules, in which Y has the strongest promoting effect, Ce is the second, La is the weakest.

Key words: first principles calculation; surface adsorption; rare earth elements; adsorption energy

4.2 连 铸

基于铸坯质量提升关键设备精度校准浅谈

翟昱明，王玉翠，何文浩

（包钢钢管有限公司，内蒙古包头 014010）

摘 要：连铸机在设备安装阶段，往往会选择几个关键区域埋下永久设备基准点，为后续大修过程中参考使用。但连铸机连续生产 5~6 年后，基准点可能因基础下沉、漂移、锈蚀等原因失去基准精度；或者因为技术革新改造，基

准点被其他设备遮挡,很难参考。连铸坯质量与连铸机精度息息相关,为生产特殊品质钢种,往往提前进行连铸机大修,而我们往往纠结于原有永久标点的寻找和可信度参考。需要探究的就是,当设备精度下降,根据现场生产如何找回关键的基准点,来恢复现场设备运行精度,确保生产稳定性及产品的质量。

关键词:基准点;设备精度;设备运行;产品质量

Discussion on Accuracy Maintenance of Continuous Caster

ZHAI Yuming, WANG Yucui, HE Wenhao

(Seamless Tube Plant of Steel Union Co., Ltd. of Baotou Steel (Group) Corp.,
Baotou 014010, China)

Abstract: In the equipment installation stage of continuous caster, several key areas are often selected to bury permanent equipment reference points for reference in the subsequent overhaul process. After 5 or 6 years of continuous production, the datum point may lose its accuracy due to foundation sinking, drift, corrosion and other reasons; or because of technological innovation and transformation the datum point is blocked by other equipment, which is difficult to refer to. The quality of continuous casting slab is closely related to the accuracy of continuous casting machine, in order to produce special quality steel, the overhaul of continuous casting machine is often carried out ahead of time. However, we are often involved in the search for the original permanent punctuation and reliability reference. What needs to be explored is that when the accuracy of the equipment decreases, how to find the key reference point according to the on site production to restore the accuracy of the on site equipment operation and ensure the stability of production and product quality.

Key words: datum point; equipment accuracy; equipment operation; product quality

方坯定重剪切工艺优化与应用

胡铁军,贺保堂,李光辉,李子辉

(河钢集团承钢分公司棒材事业部,河北承德 067002)

摘 要:目前大部分连铸采用的是定尺切割,受限于连铸温度及拉速等因素影响,相同定尺下铸坯重量波动范围较大,轧钢在轧制过程中,因铸坯重量波动,造成通尺材长度变化较大,影响轧钢成材率。通过铸坯定重切割系统,根据轧钢需求,通过调整铸坯定尺,使铸坯在小范围内波动,稳定通尺材长度,提高成材率。

关键词:铸坯定尺;定重切割

中厚船板拉伸试样断口分层缺陷分析

康 伟[1,2],田永久[3],赵启斌[3],苏小利[3],方恩俊[3],廖相巍[1,2]

(1. 海洋装备金属材料及应用国家重点实验室,辽宁鞍山 114009;2. 鞍钢钢铁研究院炼钢技术研究所,辽宁鞍山 114009;3. 鞍钢股份鲅鱼圈分公司,辽宁营口 115007)

摘　要：为明确中厚船板拉伸试样断口分层的成因，对中厚船板不同厚度、不同牌号的拉伸试样进行了化学成分、断口形貌、金相组织、中心微区偏析、断口能谱的检验分析。结果表明：厚规格、锰含量高的拉伸试样带状组织严重，严重的带状组织导致拉伸试样出现断口分层；所有拉伸试样中心微区都存在不同程度的锰偏析，中心微区锰偏析程度与带状组织严重程度无关联性；出现分层的拉伸试样断口存在大量片层状硫化锰、中心偏析带、微裂纹、大量细小的铌钛析出相；热送铸坯加热炉加热时间过长、轧制冷却强度大会导致轧板异常金相组织产生，也会导致拉伸试样断口分层的出现。

关键词：船板分层；断口分层；锰偏析；带状组织；异常金相组织

Analysis on Fracture Delamination Tensile Test of Medium Thick Ship Plate Steel

KANG Wei[1,2], TIAN Yongjiu[3], ZHAO Qibin[3], SU Xiaoli[3],
FANG Enjun[3], LIAO Xiangwei[1,2]

(1. Key Laboratory of Metal Materials for Marine Equipment and Application, Anshan 114009, China;
2. Metallurgy Department of Ansteel Iron & Steel Research Institutes, Anshan 114009, China;
3. Bayuquan Iron & Steel Subsidiary, Ansteel Co., Ltd., Yingkou 115007, China)

Abstract: To clarify the fracture delamination in tensile specimens of medium thick ship plates, chemical composition, fracture morphology, metallographic structure, central micro zone segregation, and fracture energy spectrum were analyzed for tensile specimens of different thicknesses and grades of medium thick ship plates. The results showed that the thick and high manganese content tensile specimens had severe banded structure, and the severe banded structure led to fracture delamination in the tensile specimens. Manganese segregation exist in the central micro area of all tensile specimens, and there is no correlation between the degree of manganese segregation in the central micro area and the severity of the banded structure. There are a large number of layered manganese sulfide, central segregation bands, microcracks, and a large number of fine niobium titanium precipitates on the fracture surface of the tensile specimen with delamination. The hot delivery slab overstays in heating furnace and the rolling cooling overpowered can lead to abnormal metallographic structure in the rolled plate, which can also lead to the occurrence of fracture delamination in the tensile specimen.

Key words: layering of ship plates; fracture stratification; manganese segregation; banded structure; abnormal metallographic structure

连铸结晶器智能预报系统开发

张维维[1,2]，王丽娟[1,2]，赵成林[1,2]，陈　东[1,2]，
栗　红[1,2]，黄玉平[1,2]，李德军[1,2]

（1. 海洋装备用金属材料及其应用国家重点实验室，辽宁鞍山　114009；
2. 鞍钢集团钢铁研究院，辽宁鞍山　114009）

摘　要：本文采用数值模拟的方法，将结晶器内部温度场、流场、凝固与结晶器铜板温度场进行耦合计算，得到三维坐标系下对应的速度、温度数据及对应的云图。模拟计算得到的结晶器热电偶温度数据与实时采集数据的误差率小于3%。采用.Net软件建立耦合数据库，编制应用操作界面，开发了结晶器智能预报系统软件。该软件离线命中

率可达到95%。

关键词：数值模拟；结晶器温度场；预报软件

Development of Intelligent Prediction System for Continuous Casting Mold

ZHANG Weiwei[1,2], WANG Lijuan[1,2], ZHAO Chenglin[1,2], CHEN Dong[1,2], LI Hong[1,2], HUANG Yuping[1,2], LI Dejun[1,2]

(1. State Key Laboratory of Metal Material for Marine Equipment and Application, Anshan 114009, China; 2. Ansteel Iron & Steel Research Institutes, Anshan 114009, China)

Abstract: In this article, numerical simulation method was used to couple the temperature field, flow field, solidification inside the mold, and temperature field of the copper plate. And the corresponding velocity, temperature data, and cloud map in a three-dimensional coordinate system were obtained. The error rate between the simulated temperature data of the mold thermocouple and the real-time collected data is less than 3%. A coupling database was established using the .Net software, and an application operation interface was developed to monitor the intelligent prediction system software for mold. The offline hit rate of this software can reach 95%.

Key words: simulation; mould temperature field; forecast software

150方弹簧钢连铸坯负偏析的研究与控制

孙俊喜，靳国兵，陈 涛，李 杰

（河钢集团邯郸分公司，河北邯郸 056015）

摘 要： 为改善弹簧钢铸坯中心偏析问题，邯钢一炼钢厂150方连铸采用凝固末端电磁搅拌来改善铸坯的中心偏析等缺陷。凝固末端电磁搅拌技术就是通过安装在连铸坯凝固末端的电磁搅拌器产生电磁力来打断两相区内相互交错的枝晶并促使富集的浓缩钢液流动，从而达到均匀化钢液、消除铸坯中心偏析的目的。弹簧钢液由于其糊状区直径小而且黏度高，必须有足够的搅拌强度，才能真正达到改善中心偏析的作用。然而当末搅电流过大，感应电磁力较大，搅拌过于剧烈，对凝固前沿固液两相区产生强烈冲刷，未凝固的枝晶间富含溶质的液体向中心区域流动，造成搅拌区域的负偏析和更严重的中心偏析。

关键词： 中心偏析；负偏析；末端电磁搅拌；硬度

Research and Control of Negative Segregation in 150 Square Spring Steel Continuous Casting Billets

SUN Junxi, JIN Guobing, CHEN Tao, LI Jie

(Handan Iron & Steel Group Co., Ltd., Handan 056015, China)

Abstract: In order to improve the central segregation problem of spring steel billets, the 150 square meter continuous casting at the first steelmaking plant of Han Steel adopts electromagnetic stirring at the solidification end to improve the central segregation and other defects of the billets. The electromagnetic stirring technology at the solidification end of the continuous casting billet is achieved by generating electromagnetic force through an electromagnetic stirrer installed at the solidification end of the billet, which interrupts the intersecting dendrites in the two-phase zone and promotes the flow of enriched concentrated steel liquid, thereby achieving the goal of homogenizing the steel liquid and eliminating the center segregation of the billet. Spring steel liquid, due to its small diameter and high viscosity in the paste zone, must have sufficient stirring strength to truly achieve the effect of improving center segregation. However, when the final stirring current is too high, the induced electromagnetic force is large, and the stirring is too intense, causing strong erosion of the solid-liquid two-phase zone at the solidification front. The solute rich liquid between the non solidified dendrites flows towards the central region, causing negative segregation and more severe central segregation in the stirring region.

Key words: center segregation; negative segregation; end electromagnetic stirring; hardness

C-Mn-Nb-Cr 管线钢高温力学性能研究

张英慧[1,2]，黄明浩[1,2]，孔祥磊[1,2]，王　杨[1,2]，刘　干[1,2]

（1. 海洋装备用金属材料及其应用国家重点实验室，辽宁鞍山　114009；
2. 鞍钢集团钢铁研究院，辽宁鞍山　114009）

摘　要：为制定切实可行的管线钢连铸及热轧工艺，采用 Gleeble-3800 试验机对 C-Mn-Nb-Cr 管线钢进行高温力学性能的测试研究，确定 C-Mn-Nb-Cr 管线钢的零塑性温度 ZDT 和零强度温度 ZST；测得在 2×10^{-3}/s 应变速率下，试验钢有两个脆性温度区，从凝固温度至 1350℃为第Ⅰ脆性区，650~900℃为第Ⅲ脆性区；其脆性温度区间小，裂纹敏感性较低，在高温脆性区抵抗裂纹的能力较强。这些参数对实际连铸生产具有较高的参考价值。

关键词：C-Mn-Nb-Cr 管线钢；零塑性温度；零强度温度；连铸；热轧

Study on Mechanical Properties of C-Mn-Nb-Cr Pipeline Steel at High Temperatures

ZHANG Yinghui[1,2], HUANG Minghao[1,2], KONG Xianglei[1,2],
WANG Yang,[1,2] LIU Gan[1,2]

(1.State Key Laboratory of Metal Materials and Application for Marine Equipment, Anshan 114009, China;
2. The Iron and Steel Research Institute of Ansteel Group, Anshan 114009, China)

Abstract: In order to set down the practical processes of continuous casting and hot rolling for C-Mn-Nb-Cr pipeline steel, its mechanical properties at high temperatures were measured by the Gleeble-3800 tester to determine the Zero Ductility Temperature (ZDT) and the Zero Strength Temperature (ZST). The experimental results showed that there exists two brittle temperature regions for C-Mn-Nb-Cr testing pipeline steel under the strain rate of 2×10^{-3}/s, that is brittle region Ⅰ from the melting point to 1350℃, and brittle region Ⅲ between 650℃ and 900℃. The brittle temperature range for the steel is small, the crack susceptibility for the steel is lower, and the ability for resisting cracking at high temperature brittle region is stronger. These parameters are much valuable for setting down the real processes of continuous casting and hot rolling of the steel.

Key words: C-Mn-Nb-Cr pipeline steel; zero ductility temperature; zero strength temperature; continuous casting; hot rolling

RAMON 智能出坯系统

谢卫东，王 彬，蒋 琼

（衡阳镭目科技有限责任公司，湖南衡阳 421200）

摘 要：介绍了一种基于机器视觉、智能传感器、AI 智能控制的智能出坯系统。该系统通过建立铸坯履历、图像识别和智能分析决策，能够在集控中心对铸坯实现精准定位和控制。该系统可根据连铸后区实时工况进行智能分段控制，实现节能降耗，全流程实现自动化控制，并具备自诊断功能，可实现智能闭环控制。系统还可离线对连铸后区进行沙盘演练，是设备维护、流程优化、产能提升的一大利器，也是智能连铸不可或缺的一环。

关键词：出坯系统；连铸；铸坯跟踪；智能控制

RAMON Intelligent Casting System

XIE Weidong, WANG Bin, JIANG Qiong

(Ramon Science and Technology Co., Ltd., Hengyang 421200, China)

Abstract: This article introduces an intelligent slabbing system based on machine vision, smart sensors, and AI intelligent control. By establishing a slab resume, image recognition, and intelligent analysis and decision-making, the system can achieve accurate positioning and control of the slab in the control center. The system can perform intelligent segmented control based on real-time conditions in the continuous casting area, achieving energy-saving and consumption reduction throughout the process and automatic control, and having self-diagnostic capabilities to achieve intelligent closed-loop control. The system can also perform offline sandbox simulation of the continuous casting area, making it a powerful tool for equipment maintenance, process optimization, and capacity improvement, and an essential part of intelligent continuous casting.

Key words: casting system; continuous casting; slab tracking; intelligent control

基于小波分析的结晶器液面波动特征解析与控制

崔 衡，王振东，高 煜

（北京科技大学钢铁共性技术协同创新中心，北京 100083）

摘 要：连铸坯的质量控制一直是冶金行业关注的重点，铸坯质量等级直接影响到后续热、冷轧过程中缺陷产生的占比。在双碳目标和绿色钢铁的大背景下，企业对铸坯质量的控制提出了更为严格的要求。铸坯缺陷的产生与结晶器保护渣的卷入有着密切关系，不合理的结晶器液面波动会直接增加了保护渣卷入的频率；还会影响亚包晶钢初生凝固坯壳的均匀生长，从而会导致铸坯表面纵裂纹、凹陷等缺陷发生。因此，将液面波动进行解析并提出控制对铸坯质量的提升有着十分重要的作用。

目前，一般以结晶器液面波动的波高范围作为判据之一对铸坯质量进行判定，大于某一范围铸坯会被降级。然而，结晶器液位传感器监测到的波形是由许多波叠加而成的，只用波幅评价，虽能在宏观上大致表征液面波动且方便应用，但易忽略很多细节，难以适应精度要求越来越高的结晶器液面波动研究。F 数[1]和快速傅里叶变换(FFT)[2]也被用于分析液面波动，但它们存在一定的局限性，均无法在时间尺度做出分析，对现场情况不能快速做出反应。

小波分析被广泛应用在信号分解、降噪和机械故障检测[3]等，对非周期类信号分解有着独特的优势。采用小波分析能较好地对影响因素多、瞬间变化复杂的液面波动进行精确分析。采用小波分析方法对液面波动解析，可同时在时域和频域进行分析，对液面波动情况做出准确判断。课题组已将小波分析结合信息熵引入水模型研究，对结晶器液面波动进行了精确表征并取得良好效果[4]。在现场生产中，浸入式水口发生结瘤，结晶器内钢液发生偏流，铸坯发生"鼓肚"等均会引起结晶器液面异常波动，应用小波分析方法对液面波动解析，根据解析结果可对液面异常波动做出准确判断，同时可结合相应的工艺参数如塞棒位置变化幅度，拉速变化等进行对比，可准确追溯异常波动产生原因，从而对异常波动进行控制。同时基于小波分析方法的液面波动特征解析与控制也有望用于结晶器液面波动情况的在线预报、预测等，可对铸坯质量进行判定，减少结晶器液面波动异常对铸坯质量带来的影响，从而提高连铸生产效益。

关键词：结晶器；液面波动；小波分析；卷渣；铸坯质量

参 考 文 献

[1] Teshima T, Kubota J, Suzuki M et al. Tetsu-to-Hagané, 1993, 79: 576-582.
[2] Jiang Z K,. Su Z J, Xu C Q, et al. J Iron Steel Res Int, 2020, 27: 160-168.
[3] Silva K M, Souza B A, Brito N S D. Ieee T Power Deliver, 2006: 2058-2063.
[4] Zhang K T, Liu J H, Cui H, et al. Metall Mater Trans B, 2018, 49: 1174-1184.

湘钢特厚板坯连铸技术研究进展

徐龙云，蒙　刚，师　文，马仲新，邢银超，张明志，杨建华

（湖南华菱湘潭钢铁有限公司，湖南湘潭　411101）

摘　要：随着重型装备向超大型化方向发展，电力、石化、船舶、海洋工程、机械制造、模具等行业对高品质特厚板（80mm 以上）的需求日益增加。如何解决高质量低成本特厚板坯料问题一直是钢铁冶金材料领域关注的焦点。利用特厚板坯连铸机直接生产轧制特厚板所需的特厚板坯（400mm 以上），为绿色、优质、高端特厚板钢铁产品的制备开辟了新的途径。目前，南钢、湘钢、营口、兴澄特钢、南阳汉冶、新余钢铁、首钢等先后建设了特厚板坯连铸机[1]。湘钢特厚板坯连铸机于 2021 年 7 月 31 日一次热试成功，目前该铸机生产的铸坯最大厚度为 452mm，最大宽度为 2500mm，涵盖了海上风电用钢、容器钢、高强海工钢、模具钢、高强度桥梁结构钢等高端产品，满足了特厚板轧制（特别是 200mm 以上）的要求，为湘钢创造了良好的经济效益。湘钢在特厚板品种开发过程中进行了系列特厚板坯连铸技术研究，为企业高质量发展提供了技术支撑。

（1）高裂纹敏感性钢种特厚板坯表面裂纹控制技术。

高端容器用钢 13MnNiMoR 属于典型的高裂纹敏感性钢种，易在矫直过程中产生表面横裂纹。通过降低二冷喷嘴雾化压缩空气压力、提拉速、优化二冷配水（弱冷→弱弱冷）等系列措施，将矫直段铸坯表面温度由 850℃提高至 915℃以上，避开了第Ⅲ脆性区，成功开发了高裂纹敏感性钢种特厚板坯表面裂纹控制技术。该技术在后续生产中有效地解决了 07MnMoVR、07MnNiMoDR 等高裂纹敏感性钢种的铸坯表面裂纹问题。

（2）特厚板坯非均匀凝固控制技术。

特厚板坯断面宽、厚度大，非均匀凝固现象愈加显著，中心凝固终点与两侧凝固终点距离可达 2~3m，难以通过连

铸压下技术解决板坯中心偏析问题[2]。湘钢针对450断面低碳钢种铸坯非均匀凝固问题,通过调整二冷配水,通过减弱铸坯宽向中部冷却水量减小中心凝固终点与两侧凝固终点距离,有效解决宽度方向1/8位置局部中心偏析严重问题,全宽度低倍评级均为C0.5。

(3)特厚板坯凝固末端凸形辊大压下技术。

将扇形段平辊改凸形辊可显著提升大压下扇形段的压下能力[3]。湘钢通过凸形辊压下技术将450断面凝固末端单段压下量增加3倍以上,实现了单段压下10mm以上的能力,提升了铸坯致密度;与动态轻压下技术结合,在有效保证铸坯致密度的前提下,减少凝固终点所在扇形段的压下量,降低铸坯低倍中间裂纹发生率。此外,凸形辊压下的铸坯表面峰值应力小,减少了辊套磨损,可增加大压下扇形段使用寿命,降低设备维护成本[4]。

参 考 文 献

[1] 徐龙云, 佐龙. 第23届全国炼钢学术会议论文集, 2022, B卷, 5-03.
[2] 王臻明, 赵晶, 王玉龙, 等. 连铸, 2019, 44(6): 47-50.
[3] 逯志方, 赵昊乾. 第十二届中国钢铁年会论文集, 2019: 1-6.
[4] 杜一哲, 李丽, 汪勤政, 等. 连铸, 2021(1): 47-54.

鞍钢高品质汽车用钢板坯新缺陷的研究

金宏斌,陈 宇,刘佳铭,宋 宇,高洪涛,杜 林

(鞍钢股份有限公司炼钢总厂,辽宁鞍山 114021)

摘 要:鞍钢股份有限公司炼钢总厂三分厂主要生产超低碳汽车用钢,年产量预计在2023年增长为180万吨。三分厂生产的超低碳汽车钢均为IF钢,因其具有良好的深冲和拉延性能,在各种复杂的汽车难冲件中脱颖而出。但随着汽车行业的飞速发展,客户对优质汽车钢提出了极为严苛的表面质量要求。炼钢厂为了提高铸坯表面质量引进了火焰清理机。随着新设备的投入与使用,新型缺陷也随之产生。本文阐述了2023年冷轧厂在生产炼钢总厂三分厂超低碳IF钢时出现的新型缺陷,针对新缺陷进行分析、成因排查及整改措施,有效地杜绝了日后类似缺陷的产生。

关键词:角部熔渣;火焰清理机;冲渣粒化高压水;后夹辊;氧化铁

Research of New Defects of Ansteel High Quality Automobile Steel Slab

JIN Hongbin, CHEN Yu, LIU Jiaming, SONG Yu, GAO Hongtao, DU Lin

(General Steelmaking Plant of Angang Steel Co., Ltd., Anshan 114021, China)

Abstract: The third branch of the General steelmaking plant of Anshan Iron and Steel Co., Ltd. mainly produces ultra-low carbon automobile steel with an annual output of about 1.8 million tons. Ultra-low carbon automobile steel of the third branch are all Interstitial-Free Steel, which emerges in a variety of complex automotive parts due to its good deep drawing and ductility. But with the rapid development of the automobile industry, customers have put forward more strict requirement on the surface quality of high-quality automobile steel. In order to improve the surface quality of the casting billet, the third branch have introduced the automatic scarfing machine. With the introduction and use of new equipments, new defects are also produced. In this paper, the new defects appeared in the cold-rolling plant by producting ultra-low carbon IF steel which are produced in the third branch of the general steel-making plant in 2023. Eliminating the occurrence of similar defects effectively in the future by conducting analysis, causing investigation and rectifying measures for new detects.

Key words: edge defect; automatic scarfing machine; slag flushing and granulating water; back pinch roll; iron oxide

减少冷轧低碳钢轧材表面夹杂缺陷的工业实践

杜 林，孙 群，苏建铭，王一名

（鞍钢股份有限公司炼钢总厂，辽宁鞍山 114021）

摘 要： 客户对冷轧低碳钢产品的表面质量提出了更高的要求，技术人员对低碳钢表面缺陷进行检验，电镜扫描结果显示为结晶器保护渣卷入钢水占较大比例。通过对结晶器保护渣卷入原因分析，发现连铸的拉速、浸入式水口的进入深度、氩气流量、保护渣的理化性能对结晶器中保护渣卷入钢水中有着影响。通过合理匹配拉速、水口浸入深度、氩气流量，调整优化保护渣的理化性能，减少了低碳钢结晶器保护渣在轧材表面出现夹杂缺陷的情况。

关键词： 夹杂；拉速；水口浸入深度；氩气流量；保护渣

Industrial Practice of Reducing Shell Defects on Surface of Cold Rolled Low Carbon Steel

DU Lin, SUN Qun, SU Jianming, WANG Yiming

(General Steelmaking Plant of Angang Steel Co., Ltd., Anshan 114021, China)

Abstract: Customers put forward higher requirements for surface quality of cold rolled low carbon steel products, the surface defects of low carbon steel were examined by technicians, sem results showed that the mold slag powder involved in molten steel accounted for a large proportion. Through analyzing the reason why slag powder is involved, found that the continuous casting speed, the nozzle immersed depth, the argon flow rate, the physical and chemical properties of slag powder have an effect on slag powder involved in molten steel. By reasonable matching of the casting speed, the nozzle immersed depth and the argon flow rate, adjust and optimize the physical and chemical properties of slag powder, the shell defect of slag powder on the surface of rolled steel was reduced.

Key words: shell; casting speed; nozzle immersed depth; argon flow rate; slag powder

轮辐钢中心裂纹原因分析及工艺改进

宋 宇，杜 林，金宏斌，陈柏宇

（鞍钢股份有限公司炼钢总厂，辽宁鞍山 114021）

摘　要：LQ590 钢为高级别汽车轮辐钢，轮辐是汽车车轮的重要组成部分，起支撑轮钢的作用，决定了车轮的寿命。2022 年炼钢四分厂在生产 LQ590 时在铸坯断口处发现中心裂纹，导致多块铸坯判废，损失较大。本文对 LQ590 中心裂纹的产生原因进行了分析，并结合鞍钢炼钢厂炼钢工序的实际工艺过程，提出了 LQ590 中心裂纹的工艺改进办法，实施后取得了良好的效果。

关键词：轮辐钢；中心裂纹；工艺改进；轻压下

Cause Analysis of Central Cracks in Wheel Arm Steel and Craft Improvement

SONG Yu, DU Lin, JIN Hongbin, CHEN Baiyu

(General Steelmaking Plant of Angang Steel Co., Ltd., Anshan 114021, China)

Abstract: LQ590 steel is the top of automobile wheel arm steel. Wheel arm is an important component of commercial automobile wheel, supporting felly, which influences the life of a whole wheel. The fourth branch of the general steel making plant of Anshan Iron and Steel Co., Ltd.found central cracks when mainly produces LQ590 steel in 2022, a lot of casting blank became waste, suffered a great loss. In this paper,author anslysis the cause of central cracks, and in the view of the actual craft in the smelting process, the method to improve the process to control central cracks developing is proposed. Good effect was achieced after implementation.

Key words: wheel arm steel; central cracks; craft improvement; soft reduction

连铸中间包冲击区不同控流装置的对比研究

杨成虎，张江山，刘　青，秦邦明

（北京科技大学钢铁冶金新技术国家重点实验室，北京　100083）

摘　要：中间包是钢液凝固前的最后一个冶金容器，在钢铁生产过程中发挥着重要的作用。在早期的发展阶段，中间包主要用于缓冲、稳流、连浇和分流等方面。然而，随着对钢水质量要求的提高，尤其是洁净钢生产技术的快速发展，中间包的冶金功能变得越来越重要。中间包不仅能够用于调控钢液的温度和化学成分，还能够去除其中的非金属夹杂物。中间包冶金过程中，湍流抑制器和长水口是重要的耐材装置，是影响中间包冲击区钢液流动方式的主要湍流控制装置，合理的结构设计与使用可以减少冲击器的卷渣、吸气等现象。

冲击区通常是连铸中间包内部湍流强度最大的区域，最易发生卷渣和二次氧化等行为，是设计和优化中间包流场的重点。长水口和稳流器是控流中间包冲击区主要的两个控流装置，本文采用物理和数值模拟相结合的方法对比研究了不同长水口、稳流器及其组合对冲击区流动特征的影响规律。本研究以单流中间包为原型，结构如图 1 所示。原型中间包上液面长 936 mm，下液面长 826 mm，高 300 mm，长水口浸入深度 90 mm；铸坯断面尺寸是 110 mm×900 mm，拉速是 4.0 m/min。结果表明，所构建的中间包流动数学模型与物理模型的对比验证结果较好，喇叭型水口的出口直径和容积较大，出口速度较小，能够减轻冲击区耐材的冲刷腐蚀，改善中间包内钢液的流动。与直筒型长水口方案相比，喇叭型水口方案的死区比例相差不大，表面最大流速从 0.078 m/s 降低至 0.065 m/s，中间包冲击区最大湍动能从 1×10^{-2} m^2/s^2 降低到 1×10^{-3} m^2/s^2。湍流抑制器能够改变钢液的运动方向，并且会减小死区体积，湍流抑制器型与喇叭型水口组合方案中间包表面最大流速从 0.078 m/s 降低至 0.032 m/s，死区比例从 16.48%降低至 9.46%。当长水口位置发生偏移时，死区比例从 16.48%增加至 19.53%，最大表面流速从 0.078 m/s 增加至 0.1 m/s，冲击器流动发生恶化。

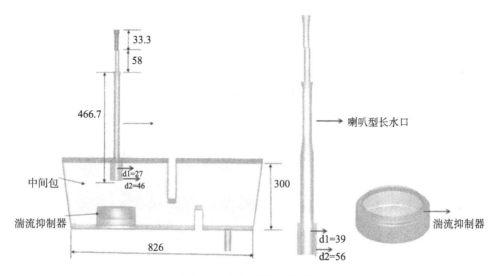

图 1　中间包三维几何模型

Fig.1　3D geometry of modeled tundish

基于机器视觉的异形坯表面缺陷在线检测系统

秦　观[1]，杜昕弢[1]，李保华[1]，张学民[2]，赵立峰[2]，徐　科[1]

（1. 北京科技大学钢铁共性技术协同创新中心，北京　100083；

2. 山东钢铁集团研究院，山东济南　250101）

摘　要：异形坯表面缺陷检测目前基本采用人工目视检测的方式，该方法存在主观性强、漏检率高、工人劳动强度大等问题。本文开发了一套高温异形坯表面缺陷在线检测系统，该系统安装在异形坯火焰切割后，采用蓝色激光对高温异形坯进行表面照明，并通过高分辨率线阵相机结合窄带滤波方式获取异形坯表面高清图像。开发了基于 YOLOV5 的异形坯表面缺陷检测算法，可在线检测裂纹缺陷。系统运行后，异形坯的精整量降低 2803.2t/月，后道工序轧材修磨量降低了 55.6%。

关键词：异形坯；表面缺陷；机器视觉；表面检测

Online Surface Defect Detection System of Hot Profiled Billets Based on Machine Vision

QIN Guan[1], DU Xintao[1], LI Baohua[1], ZHANG Xuemin[2], ZHAO Lifeng[2], XU Ke[1]

(1. Collaborative Innovation Center of Steel Technology, University of Science and Technology Beijing, Beijing 100083, China; 2. Research Institute of Shandong Iron and Steel Company Ltd., Ji'nan 250101, China)

Abstract: At present, surface defect detection of shaped billets is basically carried out by manual visual inspection, which has problems such as strong subjectivity, high missing rate, high labor intensity, and so on. An online surface defect detection system of hot profiled billets is developed. The system is installed after flame cutting of profiled billets. The profiled billet is illuminated with blue lasers, and high-resolution line scan CCD cameras with narrow-band filtering are used to capture HD images of the red-hot profiled billet. A surface defect detection algorithm based on YOLOV5 is developed to detect cracks. After working of the system, the finishing amount of the profiled billets is reduced by 2803.2 tons/month, and the grinding amount of the subsequent rolled steels is reduced by 55.6%.

Key words: profiled billet; surface defect; machine vision; surface inspection

中间包内衬冲蚀及夹杂物生成和演化的研究

王　强[1,2]，谭　憧[1,2]，刘　畅[1,2]，贺　铸[1,2]，李光强[1,2]

（1. 武汉科技大学省部共建耐火材料与冶金国家重点实验室，湖北武汉　430081；
2. 武汉科技大学钢铁冶金及资源利用省部共建教育部重点实验室，湖北武汉　430081）

摘　要：为了探究高温钢液对中间包内衬耐火材料的冲刷及夹杂物的生成与去除规律，本文建立了三维非稳态流固耦合数学模型，解析钢液在中间包内的流动和传热，以及中间包内衬耐火材料的温度分布。利用壁面切应力和总压分布计算夹杂物的生成位置、粒径和质量流量，然后采用欧拉-拉格朗日方法追踪夹杂物在中间包内的运动轨迹。结果表明，钢液对中间包内衬耐火材料的物理损毁可以分为冲击和摩擦两种来源。对于中间包而言，由于钢液冲击作用造成的耐火材料损毁要比由于钢液摩擦作用造成的耐火材料损毁更为严重。在长水口内壁产生的新生夹杂物数量占比38%，在湍流抑制器底部产生的新生夹杂物数量占比49%，而其余部位只产生了13%的新生夹杂物。

关键词：中间包；耐火材料损毁；夹杂物；流固耦合；数值模拟

Numerical Simulation on Refractory Flow-Induced Erosion and Inclusion Formation in Continuous Casting Tundish

WANG Qiang[1,2], TAN Chong[1,2], LIU Chang[1,2], HE Zhu[1,2], LI Guangqiang[1,2]

(1. The State Key Laboratory of Refractories and Metallurgy, Wuhan University of Science and Technology, Wuhan 430081, China; 2. Key Laboratory for Ferrous Metallurgy and Resources Utilization of Ministry of Education, Wuhan University of Science and Technology, Wuhan 430081, China)

Abstract: In order to understand the flow-induced erosion on the refractory lining and the formation and removal of inclusion in a continuous casting tundish, an unsteady 3D comprehensive numerical model of the respective fluid-structure interaction. The flow and heat transfer of the molten steel, as well as the refractory temperature profile was numerically clarified. The formation position, initial diameter, and mass flow rate of the inclusion were then determined by the wall shear stress and total pressure. Euler-Lagrange approach was then adopted to estimate the detachment and motion of the inclusion. The flow-induced erosion on the refractory lining could be divided into impacting and washing effects. As for the tundish, the damage caused by the impacting effect is more serious than that caused by the washing effect. At a 1.2 m/min casting speed, 49% and 38% of inclusions are created at the turbulent inhibitor inner bottom and long nozzle inner wall, respectively. In contrast, only 13% of new inclusions are produced at all other inner walls.

Key words: tundish; refractory damage; inclusion; flow-structure coupled; CFD

板坯连铸结晶器内气泡捕获夹杂物的数值模拟

刘 畅[1,2]，张立峰[3]，DAS Lipas[4]，THOMAS Brian G.[4]，
王 强[1,2]，李光强[1,2]

(1. 武汉科技大学省部共建耐火材料与冶金国家重点实验室，湖北武汉 430081；
2. 武汉科技大学钢铁冶金及资源利用省部共建教育部重点实验室，湖北武汉 430081；
3. 北方工业大学机械与材料工程学院，北京 100144；
4. 科罗拉多矿业大学机械工程系，美国科罗拉多州丹佛 80401)

摘 要：本研究建立了结晶器中注入的氩气泡的分布及其与夹杂物之间的相互作用，包括它们在湍流中的传输，以及被凝固坯壳捕获或运动到顶面渣层去除。研究了不同直径的气泡对夹杂物的捕获效率。结果表明，直径大于0.488mm的气泡进入结晶器后很容易上浮到表面去除，几乎所有大于1mm的气泡均从上表面逃逸，停留时间不超过10s。单个气泡的尺寸越大，其在运动到边界时所捕获到的夹杂物数量越多。小尺寸的气泡在捕获夹杂物之后会导致整体的直径和密度发生改变，两者之间的作用更接近碰撞现象，大尺寸气泡在捕获夹杂物后与原本的密度和直径差别不大。注入氩气对10μm夹杂物总的去除效率为61%。
关键词：结晶器流动；非金属夹杂物；气泡浮选；夹杂物去除率

Numerical Simulation on Bubbles Capturing Inclusions in Slab Continuous Casting Mold

LIU Chang[1,2], ZHANG Lifeng[3], DAS Lipsa[4], THOMAS Brian G.[4],
WANG Qiang[1,2], LI Guangqiang[1,2]

(1. The State Key Laboratory of Refractories and Metallurgy, Wuhan University of Science and Technology, Wuhan 430081, China; 2. Key Laboratory for Ferrous Metallurgy and Resources Utilization of Ministry of Education, Wuhan University of Science and Technology, Wuhan 430081, China; 3. School of Mechanical and Materials Engineering, North China University of Technology, Beijing 100144, China; 4. Mechanical Engineering, Colorado School of Mines, Golden, CO 80401, USA)

Abstract: A mathematical model of the interaction between a realistic distribution of injected bubbles and inclusions in the molten steel was established, including their transport in the turbulent flow field, and their capture into the solidifying steel shell or removal into the top surface slag layer. The removal efficiency of inclusions by bubbles with different bubble diameters was studied. The results show that bubbles with a diameter greater than 0.488 mm can easily float up to the top surface. Almost all bubbles with a diameter greater than 1mm escape from the top surface, and the retention time is not more than 10s. The larger the size of a single bubble, the more inclusions it captures when moving to the boundary. The diameter and density of small-size bubbles may change due to their capturing inclusions, and the interaction between them is similar to a collision. While large-size bubbles have little difference in density and diameter after capturing inclusions. The total removal efficiency of 10μm inclusions by gas injection is 61%.
Key words: mold flow; non-metallic inclusions; bubbles floatation; inclusions removal efficiency

中间包开浇阶段湍流抑制器应力变化行为的数值模拟研究

林 鹏[1]，金 焱[1]，甘菲芳[2]，刘子钰[1]，李晓婷[1]，吴健舟[1]

(1. 武汉科技大学耐火材料与冶金国家重点实验室，湖北 武汉 430081；
2. 宝山钢铁股份有限公司，上海 430083)

摘 要： 为了研究开浇过程中湍流抑制器的应力变化行为，本文采用流-热-固耦合模型对两种方案的湍流抑制器进行数值模拟研究。结果发现，开浇阶段湍流抑制器所受应力主要集中于导流孔附近，导流孔为湍流抑制器的结构薄弱点。开浇阶段湍流抑制器所受应力主要来源于钢液的冲击作用，其中横向的拉应力和垂直方向的拉应力为主要应力来源，导流孔作为结构薄弱点，是应力集中的区域，湍流抑制器的侧壁面承受较大的垂直方向的拉应力；导流孔上表面外部所受的横向拉应力最大，导流孔中部两侧所受垂直方向的拉应力最大。

关键词： 湍流抑制器；流固耦合；应力变化；振动

Numerical Simulation Study on Stress Change Behavior of the Turbulent Inhibitor during the Opening Stage

LIN Peng[1], JIN Yan[1], GAN Feifang[2], LIU Ziyu[1], LI Xiaoting[1], WU Jianzhou[1]

(1. State Key Laboratory of Refractory Materials and Metallurgy, Wuhan University of Science and Technology, Wuhan 430081, China;
2. Baoshan Iron and Steel Co., Ltd., Shanghai 430083, China)

Abstract: In order to study the stress variation behavior of turbulence suppressors during the pouring process, this paper uses a fluid thermal solid coupling model to numerically simulate the two schemes of turbulence suppressors. The results showed that the stress on the turbulence suppressor during the pouring stage was mainly concentrated near the guide hole, which was the structural weakness of the turbulence suppressor. The stress on the turbulence suppressor during the pouring stage mainly comes from the impact of the molten steel, with transverse tensile stress and vertical tensile stress as the main stress sources. The diversion hole, as a structural weakness, is a stress concentration area, and the side wall of the turbulence suppressor bears a large vertical tensile stress; The lateral tensile stress on the outer surface of the diversion hole is the highest, while the vertical tensile stress on both sides of the middle of the diversion hole is the highest.

Key words: turbulence inhibitor; fluid-structure interaction; stress; vibration

连铸工艺机理数字化及其发展趋势

刘和平，仇 灏，杨 丽

(中国钢研科技集团有限公司数字化研发中心，北京 100081)

摘 要： 连铸是钢铁制造流程中衔接炼钢与轧钢过程的关键工序。加快推进传统连铸的数字化转型，以数字化赋能

"双碳"目标和智能制造的实现，既是时代的命题也是必然要求。传统上，连铸工艺数字化主要是借助于实时监测或采集的数据来进行预测与控制，如结晶器铜板测温、液面波动监测数据、铸坯射钉法的坯壳厚度、红外测温法的铸坯表面温度、铸坯组织照片等。受限于连铸高温环境、缺少全流程检测手段、多变量影响及连铸拉坯大滞后等和经济成本因素，这些监测数据点或检测量极为有限且准确性差。此外，这些数据大多属于"表象"或"结果"，而非连铸工艺过程的 "内在"信息，所以难以完全认识连铸工艺过程本质并实现稳定性地控制，通过传统"试错法"的研究，周期长、成本高，已成为连铸生产过程数字化转型所面临的重大挑战。因此，发展连铸工艺机理数字化，即通过计算机数值模拟方法来研究连铸过程中多相流动、传热与凝固、应力应变、外场作用等复杂现象与机理，加速实现连铸工艺过程中"黑箱"工艺机理的全面感知、促进连铸工艺大数据形成和高效智能优化就成为连铸生产中的一种经济有效的选择。

为此，本文特别提出了基于计算机数值模拟技术的连铸工艺机理数字化基本概念；阐明了工艺机理数字化与传统自动化概念中的机理仿真模型之间的区别与联系；分析了工艺机理数字化在连铸过程数字孪生、智能制造中的作用与意义；介绍了基于 B/S 架构的工艺机理数字化的云计算系统、材料 APP 系统及数据资源系统等构建及其应用；并以电磁制动作用下 CSP 薄板坯漏斗形结晶器传热和凝固过程、板坯电磁感应加热过程中的磁-热现象分析、电磁搅拌作用下圆坯连铸过程的流动、传热和凝固行为机理类模型为例，讨论了工艺机理数字化在连铸生产过程中的应用方向及前景。最后，提出了连铸工艺机理数字化的未来发展趋势。总体上，连铸工艺机理的数字化朝着更复杂的物理模型描述、更全尺度/多场耦合、更准确的预测数据、更智能化的在线应用等方向发展。

参 考 文 献

[1] 朱苗勇, 罗森. 冶金自动化, 2023, 47(1): 68-85.
[2] 梅康元, 米进周, 郭岩嵩, 等. 工业控制计算机, 2021, 34(7): 139-141.
[3] 朱苗勇, 娄文涛, 王卫领. 金属学报(中文版), 2018, 54(2): 131-150.
[4] Liu H, Zhang J, Tao H, et al. Metallurgical Research & Technology, 2020, 117(602): 1-16.
[5] Liu H, Wang Z, Qiu H. ISIJ International, 2020, 60(9): 1924-1937.
[6] Liu H, Wang X, Si L, et al. Journal of Iron and Steel Research International, 2020, 27(7): 420-432.

关于低频电磁场下低反应性连铸保护渣吸收 Al_2O_3 夹杂的研究

白 博，王 雨，丁昭霖

（重庆大学材料科学与工程学院，重庆　400044）

摘　要：随着电磁搅拌在连铸邻域的广泛应用，以及低反应性保护渣的快速发展，关于电磁场下低反应性保护渣与 Al_2O_3 夹杂的研究具有重要意义。本文采用旋转圆柱法，以吸收速率和比溶解量作为表征方法，探究了电磁场环境对低反应性保护渣同化吸收 Al_2O_3 夹杂的影响和作用机理。研究发现，电磁场环境可以明显促进低反应性连铸保护渣同化吸收 Al_2O_3 夹杂，这主要与焦耳热效应和熔渣离子在电磁场下受到电磁力有关，并得到了吸收速率和比溶解量与磁场强度、频率之间的关系。

关键词：低频电磁场；低反应性连铸保护渣；Al_2O_3 夹杂；同化吸收

Study of the Absorption of Al₂O₃ Inclusion by Low Reactivity Mold Flux under Low Frequency Electromagnetic Field

BAI Bo, WANG Yu, DING Zhaolin

(College of Materials Science and Engineering, Chongqing University, Chongqing 400044, China)

Abstract: The study of low reactivity mold flux and Al_2O_3 inclusion under electromagnetic field is very important because of the widespread use of electromagnetic stirring in continuous casting and the quick development of low reactivity mold flux. In this paper, the rotating cylinder method with the absorption rate and specific dissolution amount as characterization methods were used to investigate the effect of electromagnetic field environment on the assimilation and absorption of Al_2O_3 inclusion in low reactivity mold flux and its mechanism. The assimilation and absorption of Al_2O_3 inclusion in the low-reactivity mold flux is found to be significantly facilitated by the electromagnetic field environment. This is primarily due to the Joule thermal effect and the electromagnetic force of the slag ions under the electromagnetic field, and a relationship between the absorption rate and specific dissolution amount, as well as the magnetic field intensity and frequency, is discovered.

Key words: low frequency electromagnetic field; low reactivity mold flux; Al_2O_3 inclusion; assimilation and absorption

20CrMnTi 连铸凝固条件下的 TiN 析出行为研究

王卫领，商庭瑞，罗 森，朱苗勇

（东北大学冶金学院，辽宁沈阳 110819）

摘 要： 20CrMnTi 属 Ti 微合金钢，在我国被广泛加工成汽车齿轮，应用于汽车变速箱、减速器以及差速器等部件。20CrMnTi 连铸坯中微米级大尺寸 TiN 析出物是齿轮服役过程中磨损、开裂的关键诱因之一，且难以通过后续的热处理有效消除。因此，研究揭示 20CrMnTi 连铸凝固条件下枝晶间溶质偏析与 TiN 析出耦合作用机理对于从源头控制 TiN 析出至关重要。本研究以国内某钢厂 20CrMnTi 160mm×160mm 小方坯连铸凝固过程为研究对象，建立了凝固传热-溶质微观偏析-TiN 析出的宏-微观跨尺度组合模型，将连铸工艺条件与 TiN 析出有效联系起来，揭示了连铸凝固条件下 20CrMnTi 的溶质微观偏析和 TiN 析出行为，阐明了 N 含量和二次冷却比水量对 TiN 析出的影响规律。

由于凝固潜热的补偿，20CrMnTi 连铸坯初始凝固阶段冷却速率较小，凝固中后期逐渐增大。铸坯内弧侧皮下 20mm、40mm、60mm 处包晶反应发生的固相率为 0.81。越靠近铸坯中心，溶质微观偏析程度越大，60mm 处粗大的等轴晶间尤甚。TiN 的析出造成凝固前沿 Ti 元素的富集速度减缓，但是降低 N 元素偏析。20mm、40mm、60mm 处 TiN 最大析出尺寸的预测值分别为 6.128μm、7.836μm、7.593μm，实测值分别为 4.758μm、7.552μm、6.679μm。随着初始 N 含量的提升，可在较小固相率下达到 TiN 的析出热力学条件，因而促进 TiN 的生长。40mm、60mm 处 TiN 尺寸的决定性因素从生长时间转变为 TiN 前沿的浓度差。将初始 N 含量（质量分数）降低到 0.002% 以下，40mm、60mm 处 TiN 尺寸可有效控制在 6μm 以下。随着二冷比水量的提升，凝固组织得到细化，TiN 生长时间缩短，TiN 尺寸减小。将二冷比水量提升到 0.75L/kg，40mm、60mm 处 TiN 尺寸分别降低到 7.053μm 和 6.574μm。

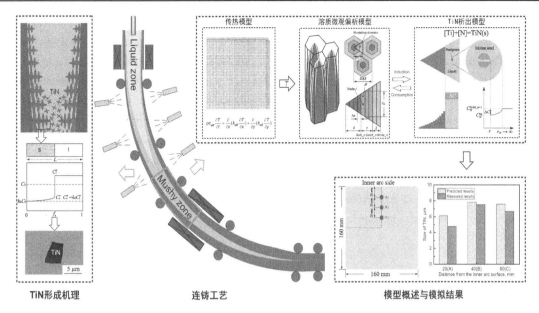

图 1　20CrMnTi 连铸凝固条件下 TiN 析出行为

参 考 文 献

[1] 商庭瑞, 王卫领, 康吉柏, 等. 钢铁, 2022, 57(7): 73-85.
[2] Shang T R, Wang W L, Kang J B, et al. Journal of Materials Research and Technology, 2023, 24: 3608-3627.

重型异型坯铸机漏钢原因与解决对策

李静文，王金坤，吴耀光，刘建坤，朱卫群，张学森

（马钢股份特钢公司，安徽马鞍山　243000）

摘　要：马钢于 2020 年正式投产当前世界上最大断面重型异型坯，但是在 1030mm×440mm×130mm 断面上在翼缘顶部位置频繁发生漏钢事故，严重影响了铸机的正常生产。为了分析该断面漏钢的主要原因，通过对保护渣理化性能以及结晶器锥度适应性进行研究，结果表明结晶器保护渣理化性能与结晶器锥度设计不合理是漏钢的主要原因。通过研究结晶器保护渣中 Na_2O 质量分数对保护渣黏度、熔点与钢之间润湿性能影响，将保护渣中 Na_2O 质量分数由 4%提高至 8%，并且将不适用于当前生产结晶器锥度进行了调整，将结晶器液面区的锥度由 2.4%/m 减小至 1.1%/m。通过以上措施实施解决了 1030mm×440mm×130mm 断面翼缘顶部漏钢问题。

关键词：重型异型坯；翼缘顶部；漏钢；保护渣；锥度

Measures and Practices of Breakout Prevention of Heavy Beam Blank

LI Jingwen, WANG Jinkun, WU Yaoguang, LIU Jiankun, ZHU Weiqun, ZHANG Xuesen

(Steelmaking and Rolling Plant of Number 1, Maanshan Iron and Steel Co., Ltd., Maanshan 243000, China)

Abstract: Maanshan Iron & Steel Co., Ltd. officially put the world largest beam blank caster into production in 2020. However, breakout accidents frequently occur at the top of flange on the 1030mm×440mm×130mm cross-section, which seriously affects the normal production of the production line. In order to study the main reason of breakout in this section, through the study on the physical and chemical properties of the mold powder and the adaptability of the mold taper, the research results show that the main reason of breakout is that the physical and chemical properties of the mold powder and the mold taper design are unreasonable. By studying the mass fraction of Na_2O of mold powder on the viscosity、melting point and the wettability between steel, increase mass fraction of Na_2O conten of mold powder from 4% to 8%, and modified the unsuitable mold taper which is not adapt to current production, decrease the mold taper from 2.4%/m to 1.1%/m around the level domain. Through the above improvement measures eventually solved the top flange breakout problem of 1030mm×440mm×130mm section.

Key words: heavy beam blank; flange top; breakout; mold powder; mold taper

连铸非稳态过程板坯高洁净度综合控制技术

刘 洋[1]，朱志远[2]，谢翠红[2]，王卫华[1]，季晨曦[1]，赵 晶[2]

（1. 首钢技术研究院，北京 100043；2. 首钢京唐钢铁联合有限责任公司，河北曹妃甸 063200）

摘 要：板坯连铸过程不可避免存在启车及封顶、异钢种混浇、换水口等各种非稳态情况，非稳态过程意味着拉速、浇铸条件、液面高度等波动变化，对钢水洁净度及铸坯质量有很大影响，针对问题，首钢京唐二期中厚板产线开展了连铸非稳态过程洁净度控制相关研究，近期开发了三项主要技术。

（1）新型五孔多股控流水口开发。

针对结晶器表面流速差较大、液渣层厚度较薄且全宽度方向厚度不均匀的问题，开发了新型五孔控流水口。图 1 为新型水口和原水口结构对比，根据伯努利原理，将水口上部设计为阶梯状；水口底部由 112mm 减薄至 102mm，水口底部和结晶器之间的空间大；水口出口设计为"工"字结构，增加流场束流功能；缓解侧流股冲击力，水口底部设计为通孔。由于水口底部和结晶器之间的空间更大，钢液更新快，水口周围的温度也相应提高。

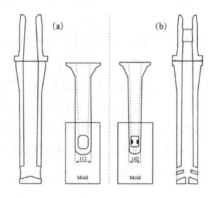

图 1 新型水口和原水口结构对比

基于水模拟和数值模拟的方法研究新型水口和原水口对结晶器流场的影响规律，见图 2。新水口实现了结晶器宽度方向表面流速均匀性大幅提高（0.25~0.4m/s），液渣厚度均匀性±2mm、液渣厚度 10~15mm，传热稳定性改善，纵裂纹发生率由 3.8%降至 0.3%。

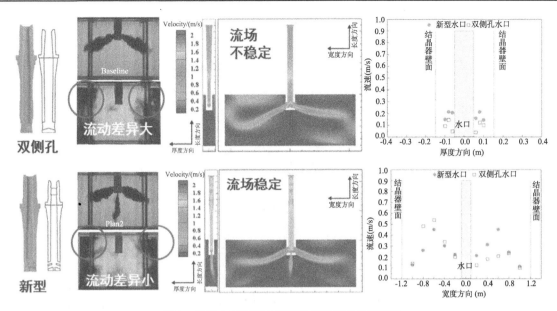

图 2 新型水口和原水口对结晶器流场影响研究

（2）浇铸末期下渣及尾坯质量控制技术。

连铸末期中间包低液位浇铸是减少尾坯长度的主要手段，低液位情况的中包下渣控制是技术的关键，需要中间包具有钢液停留时间长、稳流防末期旋涡的特点，针对此问题，开发了方形湍流抑制器、近堰坝、深平台的新型结构中间包，如图 3 所示。

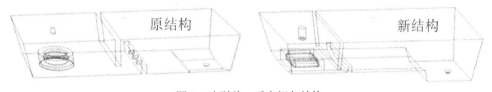

图 3 改型前、后中间包结构

新结构中间包钢液平均停留时间增加 15%、死区面积减少 13%，对于夹杂物的进一步上浮去除及稳定流股起积极作用，图 4 为新旧两种结构中间包流场数值模拟（20μm 粒子轨迹仿真）。

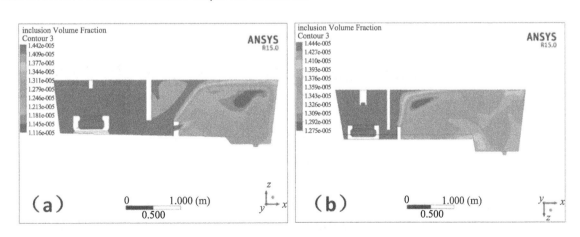

图 4 中间包流场仿真
(a) 原结构；(b) 新结构

基于中间包卷渣规律研究，开发了连铸封顶分级降速工艺，提高尾坯表面温度、改善尾坯质量。实现铸余由 10t 减少至 7t，尾坯长度减至 900mm。

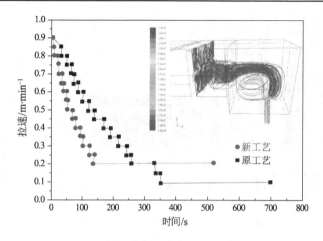

图 5 尾坯分级降速工艺

（3）异钢种混浇坯智能预测系统。

异钢种混浇是中厚板产品连铸生产的组织方式，同浇次内不同钢种混浇存在明显的成分过渡坯，通常采用铸坯下线切角成分复检的方式进行判别，效率低、成本高。

图 6 混浇坯智能预测系统

针对以上问题，开发了异钢种混浇坯成分智能预测系统，模型首次考虑了异钢种结晶器内钢液的滞后混合行为以及多元素成分综合判断，实现混浇坯元素成分在线快速判断，减少复检量 90%，误判率≤1%。

板坯连铸结晶器铜板热/力学行为数值模拟研究

姜毅银，蔡兆镇，朱苗勇

（东北大学冶金学院，辽宁沈阳　110819）

摘　要：基于国内某钢厂 1600mm×230mm 断面板坯连铸机生产实际，建立"铸坯-保护渣膜-铜板-冷却水"体系的三维多场耦合计算模型，研究分析了结晶器铜板服役过程的温度场与应力场分布规律。在此基础上，讨论了镀层厚度及其分布对铜板传热与受力行为的影响。结果表明：某钢厂典型连铸工艺下的铜板最高温度和最大应力出现在弯月面下约 20mm 处，宽、窄面侧的最高温度分别约为 214.5℃和 259.2℃，对应的最大应力分别约为 918.2MPa 和 941.6MPa。铜板热面温度和应力均随着镀层厚度增加而增大。在兼顾磨损的前提下，适当减小镀层厚度有利于降低镀层剥落和裂纹风险。

关键词：板坯连铸；结晶器；铜板；热/力学行为；镀层；数值模拟

Numerical Simulation of Thermal-mechanical Behavior in Slab Continuous Casting Mold Copper Plate

JIANG Jiyin, CAI Zhaozhen, ZHU Miaoyong

(School of Metallurgy, Northeastern University, Shengyang 110819, China)

Abstract: Based on the actual production of 1600mm×230mm continuous casting machine for a domestic steel plant, a three-dimensional multi-field coupling calculation model was developed to the system of "slab - mold flux film - copper plate - cooling water". The temperature and stress field distribution during the service process of the mold copper plate were studied and analyzed. On this basis, the influence of coating thickness and its distribution on the thermal-mechanical behavior of the copper plate was discussed. The results show that, the mold temperatures and stresses reached their maximum at 20mm below the meniscus for wide face and narrow face, their maximum temperatures were 214.5℃ and 259.2℃, their maximum stresses were 918.2MPa and 941.6MPa, respectively. The hot face temperature and stress of copper plates increase with the increase of coating thickness. Reducing the coating thickness appropriately while considering wear is beneficial for reducing the risk of coating peeling and cracking.

Key words: slab continuous casting; mold; copper plates; thermal-mechanical behavior; coating; numerical simulation

中间包功能稀土材料的开发

谌智勇[1]，任 强[2]，王小军[3]，智建国[1]，任 磊[4]

（1. 内蒙古包钢钢联股份有限公司，内蒙古包头 014010；
2. 燕山大学，河北秦皇岛 066000；3. 青岛正望新材料股份有限公司，山东青岛 266200；
4. 内蒙古科技大学，内蒙古包头 014010）

摘 要： 稀土钢连铸是行业难题，尤其稀土含量增加到500ppm以后，稀土收得率不稳定和钢水絮堵的现象将严重制约稀土钢的工业化进程和产品质量。针对如何解决稀土钢浇注过程中，由于钢液中的稀土与中间包功能耐火材料反应，导致的控流失控以及稀土收得率波动的问题，本文提供了通过在中间包功能耐火材料的工作表面复合稀土氧化物材料，实现隔断稀土与镁铝质耐火材料反应的思路。通过人工合成$CeAlO_3$并作为原料，采用常规检验方法检验了材料性能，按照一般的加工工艺检测了其加工及应用的可行性。对比来看，$CeAlO_3$作为耐火材料的原料其物性指标优于现有常规的镁质和铝质材料。本文通过实验室耐侵蚀对比试验，证明其具有优越的抗稀土侵蚀的作用。

关键词： 稀土钢；$CeAlO_3$；中间包功能材料

Development of Rare Earth Oxides in Tundish Functional Material

CHEN Zhiyong[1], REN Qiang[2], WANG Xiaojun[3], ZHI Jianguo[1], REN Lei[4]

(1. Inner Mongolia Baotou Steel Union Co., Ltd., Baotou 014010, China; 2. Yanshan University, Qinhuangdao 066000, China; 3. Qingdao Zhengwang New Material Limited Liability Company, Qingdao 266200, China; 4. Inner Mongolia University of Science and Technology, Baotou 014010, China)

Abstract: Continuous casting of rare earth steel is a difficult problem in the industry, especially when the rare earth content increases to about 500ppm, the phenomenon of unstable rare earth yield and steel flocculation plugging is more serious. On the basis of synthetic CeAlO₃ as tundish functional material, the physical properties of this material are analyzed to be better than the existing conventional materials, and through the laboratory erosion resistance test, it is proved that it has superior effect of rare earth erosion resistance.

Key words: rare earth steel; CeAlO₃; tundish functional materials

低碳钢板坯连铸结晶器液位瞬时异常波动时频特征解析

孟晓亮[1]，罗 森[1]，周业连[2]，王卫领[1]，朱苗勇[1]

（1. 东北大学冶金学院，辽宁沈阳 110819；2. 上海梅山钢铁股份有限公司，江苏南京 210039）

摘 要： 结晶器是连铸生产的核心环节，是确保铸坯质量的起点。根据冶金大数据统计结果可知，大部分的连铸坯表面缺陷来源于结晶器。而结晶器中的钢液液位波动作为连铸过程中重要的物理现象，对于最终连铸坯的质量会造成很大影响。结晶器液位异常波动不仅容易造成卷渣现象，使得初生凝固坯壳附近夹杂物含量增加，影响初生坯壳的均匀生长，同时还会影响保护渣的润滑性能，严重时将会造成连铸坯表面裂纹甚至漏钢事故。

而发展以高拉速、无缺陷为核心内涵的高效连铸是各工厂努力实现的目标，但是随着拉速的提升，结晶器液位异常波动现象会更加严重，势必影响拉速的提高，从而极大地限制了连铸的高效生产，这为结晶器液位波动的稳定性控制带来了前所未有的挑战。结晶器液位瞬时异常波动相较于大范围周期型异常波动，并不是由于某种物理现象所引起的，而主要是由结晶器中各复杂工艺所造成的。目前大部分研究从不同角度对结晶器液面异常波动进行了分析，但研究对象主要集中在周期型液面异常波动，对瞬时液面异常波动的研究并不多。并且研究方法主要集中在时域范围内，而许多研究表明由于结晶器中各工艺之间的复杂关系，结晶器中很多重要的现象在时域中获得信息所受的局限性很大，因此有许多的专家学者将研究的目光放在了频域的研究上。结果表明，相较于时域，频域分析可以获得更多的信息。与时域分析方法相比，频域分析方法可以确定感兴趣的频带，可以实现感兴趣信号频率的分离和处理。而时频域分析方法可以同时保留时域和频域的特征，并且针对非平稳、非线性的信号具有更好的分析能力。

根据现场数据对比发现，低碳钢出现瞬时异常波动的现象更加严重。因此本文利用快速傅里叶变换（fast Fourier transform，FFT）和连续小波变换（continuous wavelet transform，CWT）分析数据特征，进而研究工艺参数对低碳钢结晶器液位瞬时异常波动的影响。FFT 分析结果表明，鼓肚对于结晶器液面瞬时异常波动并无明显影响。通过 CWT 分析了结晶器液位瞬时异常波动和塞棒位置的时频特性，结果表明，低碳钢在不同拉速下，结晶器液位瞬时异常波动发生之前塞棒位置高频区 CWT 系数都呈线性增加趋势。当拉速保持恒定时，其增加率在 0.1~0.2/s 的范围内；对于由拉速急剧变化引起的瞬时异常结晶器液位波动，增加率可达 0.58/s。对于由换包引起的结晶器瞬时异常波动，增加率可达 0.82/s。因此，通过对塞棒位置高频区的 CWT 分析，可以预测结晶器液位瞬时异常波动。

板坯连铸结晶器电磁搅拌工艺参数优化研究

卢海彪[1,2]，雷作胜[1,2]，钟云波[1,2]，程常桂[3]，任维丽[1,2]

（1. 上海大学省部共建高品质特殊钢冶金与制备国家重点实验室，上海 200444；
2. 上海大学上海市钢铁冶金新技术应用重点实验室，上海 200444；
3. 武汉科技大学钢铁冶金及资源利用省部共建教育部重点实验室，湖北武汉 430081）

摘 要：板坯连铸结晶器电磁搅拌技术，作为生产高质量、高附加值钢铁产品的重要技术手段之一，已在国内外诸多钢铁企业中得到了广泛应用[1,2]。其原理是通过调控结晶器内钢水的流动状态，从而实现铸坯质量的提升。然而，在生产实践中发现，当搅拌电流选择不合理时，电磁搅拌会诱发卷渣、坯壳重熔和夹杂物聚集等负面效果，并造成铸坯质量缺陷[3,4]。因此，在板坯连铸结晶器内，如何电磁搅拌成为了冶金工作者关注的问题之一。结晶器是一个涉及多相流、传热、凝固、夹杂物传输等宏观传输行为的复杂体系[5]。要回答这个问题，需要从根本上厘清电磁搅拌对结晶器内宏观传输行为的影响机理，才能更加有效地发挥电磁搅拌的冶金效果。因此，本文主要采用数值模拟的方法，综合研究了电磁搅拌对钢渣界面卷渣、坯壳生长均匀性、凝固前沿夹杂物冲刷和分布的作用机理，并利用现场数据和低熔点物理模拟实验验证了数学模型的准确性。

基于大涡模拟、VOF多相流模型，建立了三维钢液-保护渣-空气的磁场、流场双向耦合数学模型，探究了电磁搅拌下钢渣界面卷渣的机理，构建了净卷渣速率与工艺参数之间的关系。研究表明：当水平环流未形成(弱搅拌电流)时，卷渣类型主要是漩涡卷渣，形成水平环流(强搅拌电流)后，卷渣类型主要是剪切卷渣，因此可以通过水平环流与否判断卷渣的类型。

在上述模型的基础上耦合了凝固传热模型和结晶器振动，探究了电磁搅拌对保护渣渣耗和初始凝固行为的影响机理，发现结晶器内初始凝固坯壳行为由渣膜导热和钢水对流换热共同作用，电磁搅拌通过改善结晶器上部温度场，使得沿铸坯宽度方向液态渣膜厚度分布更均匀，进而有助于坯壳的均匀生长，而在射流区域，由于电磁驱动流与射流的共同作用，坯壳均匀性有所降低。

构建了流动-传热-凝固-夹杂物传输的数学模型和凝固前沿夹杂物捕获模型，探究了电磁搅拌对夹杂物在垂直段和弧形段的捕获规律，随后系统地研究了不同工艺参数下夹杂物的去除和其在凝固坯壳内的分布规律。研究发现，夹杂物去除和捕获率随结晶器流场的变化呈瞬态波动，电磁搅拌有助于夹杂物的上浮去除和减小夹杂物在弯曲段被捕获的几率随着搅拌电流的增大，夹杂物分布均匀性和铸坯表层洁净度均先增大后减小。

参考文献

[1] Lu H, Li B, Li J, et al. ISIJ Int., 2021, 61: 1860-1871.
[2] Li B, Lu H, Shen Z, et al. ISIJ Int., 2019, 59: 2264-2271.
[3] Li B, Lu H, Zhong Y, et al. ISIJ Int., 2020, 60: 1204-1212.
[4] Lu H, Zhong Y, Ren Z, et al. J. Iron. Steel Res. Int., 2022, 29: 1807-1822.
[5] 朱苗勇, 钢铁, 2019: 21-36.

湍流抑制器对两流中间包流场影响的模拟研究

刘崇[1]，刘纲[2,3]，张彩东[1]，丁志军[4]，李杰[1]，田志强[1]

（1. 河钢材料技术研究院，河北石家庄 050023；2. 清华大学信息国家研究中心，北京　100084；3. 江阴兴澄特种钢铁有限公司，江苏无锡　214429；

4. 石家庄钢铁有限责任公司，河北石家庄　050031）

摘　要：以双流对称中间包为研究对象，在稳态浇注下，对三种不同的湍流抑制器中间包钢水的流动状态进行数值模拟研究，并从不同截面的速度云图、湍动能云图进行定性分析中间包内钢液的流动行为，结果表明：原中间包和采用圆形加檐湍流抑制器的中间包长水口附近具有较大的流速，钢液表面的速度分布不均匀，且在钢水冲击区和浇注区内存在较大的死区比例，而采用方形加檐湍流抑制器能改善中间包在冲击区和浇注区的流动状态，在稳态浇筑状态下可以形成理想的大环流，有助于中间包内钢液的传质与传热的顺利进行。

关键词：数值模拟；中间包；湍流抑制器

Simulation Study on the Effect of Turbulence Suppressor on Flow Field in Two-strand Tundish

LIU Chong[1], LIU Gang[2,3], ZHANG Caidong[1], DING Zhijun[4],
LI Jie[1], TIAN Zhiqiang[1]

(1. HBIS Material Technology Research Institute, Shijiazhuang 050023, China; 2. National Information Research Center of Tsinghua University, Beijing 100084, China; 3. Jiangyin Xingcheng Special Steel Works Co., Ltd., Wuxi 214429, China; 4. Shijiazhuang Iron & Steel Co., Ltd., Shijiazhuang 050031, China)

Abstract: Taking the two-strand symmetrical tundish as the research object, the flow state of molten steel in the tundish with three different turbulence inhibitors was studied by numerical simulation under stable casting, and the flow behavior of molten steel in the tundish was qualitatively analyzed from the velocity nephogram and turbulent energy nephogram of different sections. The results show that the flow velocity near the long nozzle of the original tundish and the tundish with circular eaves turbulence inhibitor is higher, the velocity distribution on the surface of molten steel is not uniform, and there is a large proportion of dead zone in the impact zone and the casting zone of molten steel, while the square eaves turbulence inhibitor can improve the flow state in the impact zone and the casting zone of tundish. An ideal large circulation can be formed in the steady pouring state, which is helpful for the smooth progress of mass and heat transfer of molten steel in the tundish.

Key words: numerical simulation; tundish; turbulence inhibitor

连铸大方坯点状偏析的轻压下调控机制

王海杰[1]，王　璞[1]，汤群伟[2]，陈　刚[2]，李伟涛[2]，张家泉[1]

（1. 北京科技大学冶金与生态工程学院，北京　100083；
2. 潍坊特钢集团有限公司，山东潍坊　262400）

摘　要：考虑到凝固组织形态和尺寸对溶质分布的影响时，基体成分不均匀性可达到半宏观或宏观尺度，形成点状偏析乃至宏观偏析，加热炉中难以彻底去除，遗传到轧材中会形成带状组织[1-3]，不利于钢材服役。近年来，作者团队对连铸半宏观偏析的研究成果[1-5]表明，基于凝固组织调控的连铸工艺(高过热度、弱结晶器电磁搅拌)是降低铸坯点状偏析和轧材带状缺陷的有效控制思路。然而，轻压下作用于凝固末端等轴晶间浓化钢液进而改善点状偏析鲜有研究。

因此，本研究以国内某厂 42CrMoA 钢种为例，分析了 200mm×240mm 矩形坯不同轻压下参数连铸坯凝固组织和点状偏析特征，创新提出了轻压下对点状偏析的影响机制。实验结果表明，相比于未使用轻压下，轻压下 10mm 时，连铸坯中心疏松和缩孔明显改善；疏松型点状偏析数量降幅为 29.3%，尺寸在 400μm 以上的斑块型点状偏析完全消失。压下量由 10mm 提高至 12mm 时，在不产生中间裂纹且满足后续轧制工序的前提下可令疏松型点状偏析数量继续减少 18 个，但斑块型点状偏析数量和尺寸变化幅度较小。根据本实验对齿轮钢连铸坯凝固组织和点状偏析研究发现，连铸坯凝固末端轻压下可补偿体积收缩进而减轻对残留钢液抽吸，对固定等轴晶区域的点状偏析形成具有直接的抑制作用。

综上所述，凝固末端轻压下可以明显改善连铸坯中心疏松、缩孔和点状偏析，为提高棒材乃至产品组织和性能均匀性提供了一种新的控制思路。

参 考 文 献

[1] 张壮, 李海洋, 周蕾, 等. 齿轮钢铸态点状偏析及其在热轧棒材中的演变[J]. 金属学报, 2021, 57(10): 1281-1290.
[2] 李博, 张忠铧, 刘华松, 等. 高强耐蚀管钢点状偏析及带状缺陷的特征与演变[J]. 金属学报, 2019, 55(6): 762-772.
[3] 李博, 杨飞飞, 刘华松, 等. 凝固组织对高强耐蚀管带状缺陷的影响[J]. 钢铁, 2020, 55(3): 87-95.
[4] 兰鹏, 铁占鹏, 张伟, 等. 连铸坯点状偏析缺陷研究进展[J]. 钢铁, 2020, 55(2): 11-22.
[5] 耿豪, 张忠铧, 唐海燕, 等. 高强度油井套管钢偏析与含 Nb 析出相研究[J]. 钢铁研究学报, 2019, 31(4): 387-393.

重轨钢氢含量控制研究

韩元庭[1]，潘艳华[1]，徐国涛[2]，白 静[1]，刘 鹏[1]，王彦林[1]

（1. 宝钢股份武汉钢铁有限公司条材厂，湖北武汉　430083；
2. 宝钢股份中央研究院（青山），湖北武汉　430083）

摘 要：重轨钢是一种高强度、细珠光体钢，它对钢中的氢有较高的敏感性。过量的氢在应力作用下会使钢轨产生白点，引发氢脆，恶化钢的物理特性。本文对重轨生产过程进行研究，目标是增加真空脱氢效率，减少冶炼过程和连铸过程增氢以及促进铸坯氢的释放，使得重轨氢含量得到了有效控制。
关键词：重轨；氢；真空；中包；缓冷

Study on Hydrogen Content Control of Heavy Rail Steel

HAN Yuanting[1], PAN Yanhua[1], XU Guotao[2], BAI Jing[1],
LIU Peng[1], WANG Yanlin[1]

(1. Plant of Long Product, Wuhan Iron & Steel Co., Ltd., Wuhan 430083, China;
2. Central Research Institute of Baosteel (Qingshan), Wuhan 430083, China)

Abstract: Heavy rail steel is a kind of high-strength, fine pearlite steel, which is highly sensitive to hydrogen in steel. Excessive hydrogen under the action of stress will cause white spots on the rail, cause hydrogen embrittlement and deteriorate the physical properties of the steel. In this paper, the production process of heavy rail is studied. The goal is to increase the efficiency of vacuum dehydrogenation, reduce the increase of hydrogen in smelting process and continuous casting process, and promote the release of billet hydrogen. Effective hydrogen control measures were obtained.
Key words: heavy rail; hydrogen; vacuum; medium package; slow cooling

小方坯电缆钢浸入式水口侵蚀机理分析及改善措施

刘 鹏，杨文清，韩元庭，张昌宁，白 静

（武钢有限条材厂一炼钢分厂，湖北武汉 430083）

摘 要：简述了浸入式水口的侵蚀机理。针对浇铸电缆钢时浸入式水口不耐侵蚀的原因进行分析，采取优化保护渣、增加水口壁厚等方法来延长浸入式水口的使用寿命，电缆钢的单中包连浇炉数也有了一定的提高，满足了多炉连浇的生产要求。

关键词：小方坯连铸；电缆钢；浸入式水口；侵蚀机理

Analysis and Countermeasures on Corrosion Mechanism of SEN in Cable Steel of Small Section Billet

LIU Peng, YANG Wenqing, HAN Yuanting, ZHANG Changning, BAI Jing

(Plant of Long Product, Wuhan Iron & Steel Co., Ltd., Wuhan 430083, China)

Abstract: The erosion mechanism of submerged nozzle was described in the paper. It aimed at the factors analysis about the non-corrosive submerged nozzle when pouring cable steel. Through analysis, it extended the useful life of submerged nozzle by optimizing casting powder, increasing thickness of nozzle and so on. Moreover, it also improved the continuous casting heats of tundish in cable steel so that to satisfy the production requirement on tundish's life.

Key words: small section billet; cable steel; submerged nozzle; erosion mechanism

H13 多合金钢凝固热力学实验和模拟研究

罗腾飞，王卫领，商庭瑞，罗 森，朱苗勇

（东北大学冶金学院，辽宁沈阳 110819）

摘 要：为实现双碳目标，钢铁企业越来越多地采用效率高、成本低的连铸生产 H13 合金钢。H13 属于过包晶钢，凝固过程中发生包晶相变，导致连铸坯凝固坯壳不均匀生长，诱发表面纵裂纹。另外，合金元素的偏析会产生大尺寸碳化物。因此，研究揭示 H13 多合金钢凝固相变热力学对其连铸凝固质量控制具有重要意义。随着合金元素的种类和含量的提升，传统 Ueshima 模型采用 Fe-X 二元平衡相图热力学数据判定相转变已无法适用。为此，本文首先采用差示扫描量热仪，开展了 H13 凝固热力学实验研究。其次，借助热力学软件 Thermo-Calc 计算了不同 C、Si、Mn、P、S、Cr、Mo、V 含量下的伪二元相图，获得了 δ/γ 相转变温度数据集，建立了 δ/γ 相转变温度 Bp 神经网络预测模型，并对液相线温度求解公式进行了优化，进而耦合至 Ueshima 模型，并采用实测热力学数据进行了验证。在此基础上，揭示了不同冷却速率下溶质元素的微观偏析行为，阐明了溶质元素对凝固相变特征温度和溶质微观偏析的影响规律。

在冷却速率 10℃/min、15℃/min 和 20℃/min 条件下，液相线温度 T_L 的实测值分别为(1475.94±2.88)℃、(1476.94±3.22)℃和(1476.07±2.13)℃，固相线温度 T_S 的实测值分别为(1316.79±28.36)℃、(1307.83±22.60)℃和(1305.48±3.91)℃。Bp-Ueshima 模型预测的 T_L 为 1473.35℃；T_S 分别为 1306.12℃、1306.89℃ 和 1307.43℃，与实测结果吻合。另外，预测的包晶相变起始温度 T_P 分别为 1420.31℃、1420.35℃ 和 1420.37℃；铁素体消失温度 $T_δ$ 分别为 1376.21℃、1376.40℃ 和 1376.52℃。另外，凝固末期溶质元素的偏析程度由高到低依次为 S、P、Si、Mo、C、V、Mn 和 Cr。T_L 和 T_S 随溶质初始含量的提升而降低，T_L 随溶质 C、Si、Mn、P、S、Cr、Mo 和 V 含量（质量分数）的变化规律分别为–83.15℃/%、–14.89℃/%、–5.82℃/%、–37.38℃/%、–38.72℃/%、–0.90℃/%、–2.02℃/%和–2.03℃/%；T_S 随溶质初始含量（质量分数）的变化规律分别为–397.14℃/%、–87.74℃/%、–15.40℃/%、–1044.76℃/%、–1739.99℃/%、–1.77℃/%、–1.23℃/%和–7.20℃/%。T_P 和 $T_δ$ 随 C 和 Mn 含量的增加而增大，其余溶质元素则反之，T_P 随溶质初始含量（质量分数）的变化规律分别为 23.90℃/%、–9.38℃/%、7.65℃/%、–90.57℃/%、–117.07℃/%、–3.86℃/%、–16.82℃/%和–8.01℃/%。$T_δ$ 随溶质初始含量（质量分数）的变化规律分别为 28.22℃/%、–51.75℃/%、1.81℃/%、–100.43℃/%、–53.58℃/%、–4.63℃/%、–45.21℃/%和–58.51℃/%。

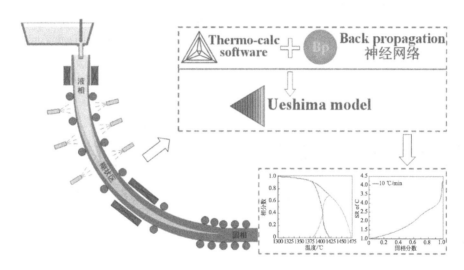

图 1　H13 多合金钢连铸凝固过程热力学行为

结晶器微区电磁搅拌对大方坯内部质量的影响

佘勇军，牛　帅

（南京钢铁股份有限公司特钢事业部，江苏南京　210000）

摘　要：本文通过对比在常规结晶器电磁搅拌和微区电磁搅拌两种工况下大方坯的凝固组织和宏观碳偏析的差异，研究结晶器微区电磁搅拌对 250mm×300mm 断面的齿轮钢（20MnCr5M）铸坯成分均匀性的影响，并从轧材的凝固组织、原位分析以及宏观碳偏析等方面对其进行了验证。结果表明，相比常规的结晶器电磁搅拌，在连铸过程中使用结晶器微区电磁搅拌，铸坯内部的成分均匀性得到了一定程度的提高：铸坯的等轴晶率增大，混晶区宽度明显变窄，混晶区枝晶尺寸得到明显细化，铸坯的凝固中心和几何中心的偏离度得到减小，铸坯内外弧及侧弧方向的成分均匀性得到改善。

关键词：结晶器微区电磁搅拌；大方坯；铸态组织；宏观碳偏析；成分均匀性

Effect of Mold Modulated Wave Stirring on Internal Quality of Bloom

SHE Yongjun, NIU Shuai

(Nanjing Iron & Steel Co., Ltd., Nanjing 210000, China)

Abstract: By comparing the difference of solidification structure and macro segregation of carbon of the bloom under conditions of conventional mold electromagnetic stirring (M-EMS) and mold modulated wave stirring (M-MWS), the influence of M-MWS on the composition uniformity of the gear steel (20MnCr5M) bloom with the section of 250mm×300mm was studied in the present paper. The influence was further verified by the analysis of the solidification structure, in situ analysis and macro segregation of carbon of the subsequently rolled bar. It was found that, compared with the conventional M-EMS, the composition uniformity of the bloom had a certain extent of improvement by applying the M-MWS during continuous casting: The equiaxed grain rate of the bloom was increased, the width of the columnar to equiaxed transition (CET) region was obviously decreased, the dendrites were refined in the CET region, the deviation degree of the solidification center and geometric center of the bloom was reduced, and the composition uniformity along the direction of loose side to fixed side and narrow side of the bloom was also improved.

Key words: mold modulated wave stirring; bloom; as cast structure; macro segregation of carbon; composition uniformity

基于大区域碳浓度分布预测的连铸坯低倍组织精细化评级系统

侯自兵[1,2]，郭坤辉[1,2]，易 凯[1,2]，彭治强[1,2]，郭中傲[1,2]，文光华[1,2]

（1. 重庆大学材料科学与工程学院，重庆 400044；
2. 重庆大学钒钛冶金及新材料重庆市重点实验室，重庆 400044）

摘 要： 针对目前连铸坯评级结果误差大且难以满足生产需求等问题，本文尝试构建连铸坯低倍组织精细化自动评判系统。首先，根据连铸坯酸洗后的低倍组织形貌与碳浓度分布的对应关系建立了基于图像灰度的大区域碳浓度分布预测模型；然后以中心偏析与疏松缺陷为例，定量化分析其评级标准，并得到定量化标准参考曲线；同时对实际连铸坯大区域碳元素浓度分布进行预测并参照定量化标准参考曲线确定实际铸坯中心偏析与中心疏松的浓度及面积阈值；最后根据阈值参数对连铸坯中心偏析与疏松识别并提取评级参数，参照国家标准定量化参考曲线构建精细化评级系统，且评级精度至少提高一个数量级。利用该评级方法举例应用分析过热度对中心偏析和中心疏松等级的影响，发现对于所研究铸坯而言，相比于低过热度（29℃），高过热度（34℃）时铸坯中心偏析等级较低（由1.62级变为0.83级，中心疏松等级较高（由0.78级变为1.77级）；故针对钢种性能要求，基于精细化低倍组织量化评价从而获得合适的过热度至关重要。以上方法及举例对于其他类型缺陷精细化评判、提高现场工艺优化精度以及数字化智能化控制水平具有重要参考意义。

关键词： 连铸；偏析；碳浓度分布；疏松；低倍组织；精细化评级

A Fine Grading System of Macrostructure of Continuous Casting Billet Based on Prediction of Carbon Concentration Distribution in Large Area

HOU Zibing[1,2], GUO Kunhui[1,2], YI Kai[1,2], PENG Zhiqiang[1,2], GUO Zhongao[1,2], WEN Guanghua[1,2]

(1. College of Materials Science and Engineering, Chongqing University, Chongqing 400044, China;
2. Chongqing Key Laboratory of Vanadium-Titanium Metallurgy and New Materials,
Chongqing University, Chongqing 400044, China)

Abstract: According to the corresponding relationship between macrostructure morphology and carbon concentration distribution after pickling of continuous casting slab, a prediction model of carbon concentration distribution in large area based on image gray level was established. Then, the central segregation and porosity rating standards in the national standard were quantitatively analyzed, and the quantitative standard reference curve was obtained. Then, the carbon concentration distribution in the large area of the actual continuous casting slab was predicted, and the concentration and area threshold of the central segregation and central porosity of the actual slab were determined by referring to the quantitative standard reference curve. Finally, according to the threshold parameters, the central segregation and porosity of the continuous casting billet were identified and the rating parameters were extracted. According to the national standard quantitative reference curve, the fine rating was realized and the rating method was applied to the influence of superheat on the central segregation and central porosity grade. It was found that compared with low superheat (29℃), the central segregation grade of the billet was lower and the central porosity grade was higher at high superheat (34℃).

Key words: continuous casting; segregation ; carbon concentration distribution; porosity; macrostructure; refined rating

原位观察下等温冷却对包晶钢凝固行为的影响

王伟安，李小明，杨永坤，朱佳雨，王阳，王建立

（西安建筑科技大学冶金工程学院，陕西西安 710055）

摘 要： 利用原位观察实验研究了等温冷却条件对包晶钢凝固行为的影响，通过样品表面粗糙度反映了包晶转变收缩程度。结果表明，等温冷却使钢液凝固过程不经历包晶转变，凝固路径为 L→L+δ-Fe→δ-Fe→δ-Fe+γ-Fe→γ-Fe。等温冷却样品表面不再呈粒状凹凸不平，而是表面平整无粒状形貌。在 10、50 和 70℃/min 非等温冷却速率下，样品表面粗糙度分别为 8.59μm、6.72μm 和 5.27μm；等温冷却使得样品表面粗糙度大幅降低，显著改善了钢液凝固收缩，为包晶钢连铸降低裂纹风险提供理论依据。

关键词： 包晶钢；原位观察；等温冷却；凝固行为；连铸

Effect of Isothermal Cooling on Solidification Behavior of Peritectic Steel under In-situ Observation

WANG Weian, LI Xiaoming, YANG Yongkun, ZHU Jiayu, WANG Yang, WANG Jianli

(College of Metallurgical Engineering, Xi'an University of Architecture and Technoloyy, Xi'an 710055, China)

Abstract: The solidification behavior of peritectic steel under isothermal cooling conditions was studied by in-situ observation experiment, and the transformation shrinkage degree of peritectic steel was reflected by sample surface roughness. The results show that the solidification path is L→L+δ-Fe→δ-Fe→δ-Fe+γ-Fe→γ-Fe without peritectic transformation. The surface of the sample is no longer granular uneven under isothermal cooling, but flat without granular morphology. At the cooling rates of 10, 50 and 70℃/min, the surface roughness of the samples was 8.59μm, 6.72μm and 5.27μm, respectively. The surface roughness was significantly reduced during isothermal cooling, which significantly improved the solidification shrinkage of the steel and provided theoretical basis for the reduction of cracking risk in the continuous casting of clad crystal steel.

Key words: peritectic steel; in-situ observation; isothermal cooling; solidification behavior; continuous casting

钢的热裂机理及扩展条件研究

仲红刚，林增煌，赵 宇，李天宇，翟启杰

（上海大学先进凝固技术中心，上海 200444）

摘 要： 热裂（亦称凝固裂纹）是连铸坯及铸件凝固过程常见缺陷，但目前对钢热裂机理的认识仍有很大分歧，并且缺乏准确的判据来确定热裂纹的扩展条件。我们提出一种模拟实际凝固条件的热裂热模拟方法，并设计了相应装置，研究了低碳钢和高碳钢等材料的热裂形成过程。结果表明，晶间搭桥及搭桥的形式对热裂纹的扩展影响显著。机械搭桥对裂纹的扩展影响很小，呈脆性断裂特征，而冶金搭桥会阻碍裂纹扩展，从而呈现出塑性断裂的特征。同时，碳钢热裂纹是否发生不可逆的迅速扩展，与裂纹功增长率是否达到临界值有关。据此，我们提出了基于能量平衡的裂纹扩展模型，模型考虑了裂纹在纯液膜中和晶间搭桥（冶金搭桥）中扩展机制的差异，同时提出了基于裂纹功增长率临界值的热裂纹扩展判据，判据考虑了应变速率与冷却速率对热裂的耦合作用，有望定量地预测热裂纹的形成扩展，从而为优化生产工艺参数提供指导。

关键词： 凝固；热裂；裂纹扩展；晶间搭桥；连铸工艺

结晶器液压伺服调宽系统可靠性验证

雷丛卉，宁 博，刘 洪，徐学华，樊小强

（中国重型机械研究院股份公司，陕西西安 710018）

摘　要：针对河北某钢厂已投产的板坯连铸机结晶器液压伺服在线调宽系统，通过在结晶器窄边铜板四个调宽油缸与铜板连接点附近安装顶针式位移传感器，对 iba 数据采集系统采集的数据进行分析计算，验证结晶器液压伺服在线调宽系统的可靠性。

关键词：结晶器调宽；液压伺服；iba 数据采集；可靠性

Reliability Verification of Hydraulic Servo Width Adjustment System for Crystallizers

LEI Conghui, NING Bo, LIU Hong, XU Xuehua, FAN Xiaoqiang

(China Heavy Machinery Research Institute Co., Ltd., Xi'an 710018, China)

Abstract: For the hydraulic servo online width adjustment system of the crystallizer of a slab continuous casting machine that has been put into operation in a certain steel plant in Hebei, a pin type displacement sensor is installed near the connection point between the four width adjustment cylinders and the copper plate on the narrow edge of the crystallizer. The data collected by the iba data collection system is analyzed and calculated to verify the reliability of the hydraulic servo online width adjustment system of the crystallizer.

Key words: mold width adjustment; hydraulic servo; iba data collection; reliability

水口结构改善大圆坯结晶器电磁偏心搅拌的数值模拟

朱佳雨，杨永坤，王建立，王伟安，王　阳，李小明

（西安建筑科技大学冶金工程学院，陕西西安　710055）

摘　要：为改善大圆坯结晶器电磁偏心搅拌对 Φ500mm 和 Φ650mm 圆坯的负面影响，本文建立了电磁、流动、传热和凝固相耦合的三维数学模型，研究了电磁偏心搅拌对不同断面圆坯内钢液流动和传热的影响。通过将直通型水口改变为四孔水口，从而改善了电磁偏心造成的流场分布不均等现象，并引入对称指数 S 定量评价了电磁偏心搅拌对不同断面圆坯钢液流场的影响。电磁偏心搅拌通过将来自直通型水口的高速射流发生偏移，进一步影响内外弧流场分布，改用四孔水口后，钢液无法通过水口正下方，有效避免了电磁偏心对射流发生偏移的影响。在电磁偏心搅拌作用下，Φ500 mm 和 Φ650 mm 圆坯内外弧最大温差分别出现在距弯月面 1.35m 和 1.73m 处，为 63K 和 26K；凝固坯壳最大差值分别出现在距弯月面 1.03m 和 1.78m 处，为 11.5mm 和 5.3mm。改用四孔水口后内外弧坯壳分布几乎一致。Φ500mm 圆坯的对称性指数由 0.55 提升至 0.77，极大地提高了钢液流场的均匀性。四孔水口可用于改善电磁偏心搅拌对钢液流场产生的负面影响。

关键词：水口结构；电磁偏心搅拌；对称性指数；钢液流动

Numerical Simulation of Nozzle Structure to Improve Eccentric Mold Electromagnetic Stirring in a Round Bloom

ZHU Jiayu, YANG Yongkun, WANG Jianli, WANG Weian, WANG Yang, LI Xiaoming

(School of Metallurgical Engineering, Xi'an University of Architecture and Technology, Xi'an 710055, China)

Abstract: To mitigate the adverse impact of eccentric mold electromagnetic stirring (EM-EMS) on Φ500 mm and Φ650 mm round blooms, a three-dimensional mathematical model coupling electromagnetic, flow, heat transfer, and solidification was established in this paper. The effect of EM-EMS on the flow and heat transfer of molten steel in round blooms with different cross sections was studied. The uneven distribution of the flow field caused by EM-EMS was improved by changing the straight submerged entry nozzle (SEN) to a four-port SEN. The symmetry index, S, was used to quantitatively assess the effect of EM-EMS on the flow field of round blooms with different sections. EM-EMS can affect the flow field distribution by shifting the high-speed jet from the straight SEN. After changing to a four-port SEN, the molten steel cannot pass right under the nozzle, which effectively avoided the impact of EM-EMS on the jet offset. In the presence of EM-EMS, the maximum temperature difference between the inner and outer arcs of Φ500mm and Φ650mm round bloom appears at 1.35m and 1.73m on the meniscus, respectively, 63K and 26K; the maximum difference of solidified shell appears at 1.03m and 1.78m according to the meniscus, which is 11.5mm and 5.3mm, respectively. After using the four-port SEN, the temperature and the shell distribution on the inner and outer curves for the Φ500mm round bloom were almost the same. The symmetry index of Φ500mm round bloom was increased from 0.55 to 0.77, resulting in a significant enhancement in the uniformity of the flow field. The four-hole nozzle can be used to mitigate the negative impact of EM-EMS on the steel flow field.

Key words: nozzle structure; electromagnetic eccentric stirring; symmetry index; steel flow

基于棒材直接轧制工艺的铸坯温度精准控制研究

方 明[1]，张朝晖[1]，习晓峰[2]，郭红民[2]，
梁少鹏[2]，薛 魁[1]，王贺龙[3]，吕 明[1]

（1. 西安建筑科技大学冶金工程学院，陕西西安 710055；2. 陕西龙门钢铁有限责任公司，陕西韩城 715405；3. 陕西钢铁集团有限公司，陕西西安 710055）

摘 要： 连铸坯温度精准控制是实现棒材直接轧制工艺的关键，由于未经加热炉加热和不同流之间拉速差异，导致铸坯温度场分布不均匀，温差较大，不利于轧材性能的稳定。本研究基于传热理论，建立方坯凝固传热模型，研究不同参数对铸坯温度的影响规律，通过提高拉速、优化冷却水量来提高出坯温度。同时探究铸坯输送过程温降和头尾温差变化，保证温度分布均匀性。结果表明，对于170mm×170mm方坯连铸，拉速在2.8m/min的条件下，铸坯到达轧机前平均温度保持在900℃以上，满足直接轧制工艺温度要求；通过优化二冷配水实现铸坯温度窄范围控制，在2.8~3.5m/min内，切断处坯头温度保持在930~950℃内；铸坯输送过程中，表面中心温降速率在0.381~0.643℃/s，随输送时间的增加，头尾温差逐渐缩小。通过对铸坯温度精准控制研究，使头尾温差在轧制过程中降低到30℃以

内，得到的棒材力学性能保持均衡稳定。

关键词：直接轧制；连铸；温度场；二冷强度；温度均匀性

Research on Precise Control of Billet Temperature Based on Direct Rolling Technology of Bars

FANG Ming[1], ZHANG Zhaohui[1], XI Xiaofeng[2], GUO Hongmin[2],
LIANG Shaopeng[2], XUE Kui[1], WANG Helong[3], LV Ming[1]

(1. School of Metallurgical Engineering, Xi'an University of Architecture and Technology, Xi'an 710055, China; 2. Shaanxi Longmen Iron and Steel Co., Ltd., Hancheng 715405, China; 3. Shaanxi Iron and Steel (Group) Co., Ltd., Xi'an 710055, China)

Abstract: Accurate temperature control of continuous casting billets is the key to achieving the direct rolling process of bars. Due to the lack of heating in the heating furnace and the difference in casting speed between different streams, the temperature distribution of the billets is uneven and the temperature difference is large, which is not conducive to the stability of the rolling performance. This study is based on heat transfer theory and establishes a solidification heat transfer model for square billets, studying the influence of different parameters on the temperature of the billets. By increasing the casting speed and optimizing the cooling water volume, the billet temperature can be increased. Simultaneously explore the temperature drop and temperature difference changes between the head and tail during the billet transportation process to ensure uniform temperature distribution. The results indicate that for 170mm × 170mm square billet continuous casting, with a casting speed of 2.8m/min, the average temperature of the billet before reaching the rolling mill remains above 900℃, meeting the temperature requirements of the direct rolling process; By optimizing the secondary cooling water distribution, the narrow range control of casting billet temperature is achieved. Within 2.8~3.5m/min, the temperature of the billet head at the cutting point is maintained within 930~950℃; During the transportation process of the billet, the temperature drop rate of the surface center is between 0.381~0.643℃/s. As the transportation time increases, the temperature difference between the head and tail gradually decreases. Through precise control research on the temperature of the casting billet, the temperature difference between the head and tail is reduced to within 30℃ during the rolling process, and the mechanical properties of the obtained bar remain balanced and stable.

Key words: direct rolling; continuous casting; temperature field; secondary cooling strength; temperature uniformity

316L 奥氏体不锈钢方坯残留铁素体特征研究

王　洋[1]，张政睿[1]，陈　超[1]，刘　涛[1]，闫琦吉[1]，林万明[1,2]

（1. 太原理工大学材料科学与工程学院，山西太原　030024；
2. 山西电子科技学院新能源与材料工程学院，山西临汾　041075）

摘　要：316L 奥氏体不锈钢磁性来源为残留铁素体相。本文对 AISI316L 奥氏体不锈钢连铸方坯的残留铁素体特征进行了研究。通过低倍组织观察发现，铸坯可划分为表层细晶区、柱状晶区和中心等轴晶区三个区域，各区域的厚度和晶粒结构有所差异，其中表面细晶区厚度为 20mm，柱状晶区厚度为 50mm，等轴晶区厚度为 40mm。在细晶区，残留铁素体呈颗粒状和板条状分布于奥氏体晶界处。柱状晶区的残留铁素体主要以骨骼状存在，并向内逐渐变为不连续的网状结构。铁素体含量表面细晶区最少，中心等轴晶区最多。铁素体中 Cr 元素富集、Ni 元素贫乏，奥氏体中 Ni 元素富集、Cr 元素贫乏，且不同区域铁素体和奥氏体化学成分有区别。

关键词：316L 奥氏体不锈钢；宏观组织；残余铁素体形貌；残余铁素体成分

Characteristics of Residual Ferrite in 316L Austenitic Stainless Steel Billet

WANG Yang[1], ZHANG Zhengrui[1], CHEN Chao[1],
LIU Tao[1], YAN Qiji[1], LIN Wanming[1,2]

(1. College of Materials Science and Engineering, Taiyuan University of Technology, Taiyuan 030024, China; 2. College of New Energy and Materials Engineering, Shanxi Electronic Science and Technology Institute, Linfen 041075, China)

Abstract: The magnetic source of 316L austenitic stainless steel is the residual ferrite phase. In this paper, the residual ferrite characteristics of AISI316L austenitic stainless steel continuous casting billet were studied. Through metallographic observation, it is found that the casting billet can be divided into three regions: surface fine crystal region, columnar crystal region and central equiaxed crystal region. The thickness and grain structure of each region are different. The thickness of surface fine crystal region is 20mm, the thickness of columnar crystal region is 50mm, and the thickness of equiaxed crystal region is 40mm. In the fine grained region, the residual ferrite is distributed in granular and lathy form at the boundaries of austenite. The residual ferrite in the columnar crystal area mainly exists in skeletal form, and gradually changes inward into discontinuous network structure. The content of ferrite is the least in the surface fine crystal region and the most in the central equiaxed crystal region. The ferrite is rich in Cr and poor in Ni, and the austenite is rich in Ni and poor in Cr, and the chemical composition of ferrite and austenite is different in different regions.

Key words: 316L austenitic stainless steel; macrostructure; residual ferrite morphology; residual ferrite composition

异型坯单水口浇铸结晶器流场温度场特征及优化

彭治强[1,2]，侯自兵[1,2]，唐 萍[1,2]，李 昊[1,2]

（1. 重庆大学材料科学与工程学院，，重庆 400044；
2. 重庆大学钒钛冶金及新材料重庆市重点实验室，重庆 400044）

摘 要：H 型钢作为重要的钢铁产品，被广泛用于建筑、桥梁和铁路等领域。异型坯连铸热轧因能耗较低成为 H 型钢生产的主要手段。随着对节能降耗要求的提高，近终型薄腹板异型坯成为未来的发展方向。出于对力学性能的要求，H 型钢成分多设计为碳含量在 0.1%~0.2%的包晶微合金钢，加上异型坯断面复杂性，在其表面容易产生纵裂纹[1]，这限制了异型坯连铸技术的进一步发展。浇铸方式的选择对于异型坯缺陷的形成至关重要。翼缘双定径水口半保护浇铸的方式在生产中被广泛采用，但这种方式难以适应氧含量要求高以及包含 Cr、Ni 等合金元素的钢种。对于双水口保护浇铸[2]方式而言，仍然需要采用拉矫机控制结晶器液位，但拉速频繁波动不利于纵裂纹的控制。目前也有采用腹板中心扁平式水口的浇铸方式[3]，但由于水口寿命较短难以提高连浇炉数。故对于包晶微合金钢薄腹板异型坯的浇铸，翼缘单水口保护浇铸的方式[4]往往是更为合适的选择，但其结晶器内流场和温度场不均匀，易在腹板不同位置产生纵裂纹。当前缺乏对于这种浇铸方式条件下初始凝固特征的研究，也没有针对同一铸坯不同位置纵裂纹发生机理进行解释，不利于纵裂纹的控制。

通常认为，弯月面强冷是导致包晶钢纵裂纹产生的重要因素，而这又取决于结晶器内的流场和温度场。故本文采用物理模拟和数值模拟的方法，对某厂尺寸为 730mm×370mm×90mm，拉速为 0.8m/min 的异型坯直通型单水口（内径为 30mm）条件下铸坯初始凝固过程中流场及温度场特征进行研究，并提出优化措施。首先，采用物理模拟方法利用示踪剂显示流场形态，并模拟液渣在钢液面上的分布；其次构建考虑凝固坯壳的流温耦合模型对流场和温度场进行研究。

研究发现，模拟钢液在表面腹板位置处存在从远离水口（南侧）至靠近水口（北侧）的单向流动，这使得液渣在钢液黏性力的作用下从南侧往北侧流动；受液渣黏度影响，南侧液渣层厚度较薄，不利于该位置下弯月面的控热及润滑，在实际生产铸坯对应位置也发现纵裂纹及渣沟缺陷；除液渣黏度以外，水口插入深度对液渣分布存在较大影响，过浅的插入深度易导致液渣分布不均。同时，弯月面温度场分布也不均匀；在循环流的作用下，腹板北侧的弯月面区域钢液温度较低，铸坯初始冷却强度较大，易引起不均匀凝固，是北侧腹板纵裂纹产生的主要原因。最后，为减少纵裂纹的发生，从控制弯月面缓冷的角度，应提高液渣黏度以保证南侧液渣层厚度；另一方面，应当避免北侧腹板过低的弯月面温度。为此，本研究得到了最佳的水口插入深度及对应连铸工艺，并且现场生产统计数据表明水口插入深度从 65mm 增至 75mm 后铸坯纵裂纹率相对降低了 39%。

参 考 文 献

[1] 陈伟, 刘明华, 付振宇, 等. 连铸, 2020, 45(3): 52-56.
[2] Wang Wei, Su Wang, Liu Jianhua, et al. Ironmaking & Steelmaking, 2020, 47(8): 929-941.
[3] 王学新, 于铭杰, 石磊, 等. 冶金设备, 2010, 4: 38-40, 43.
[4] 李四军, 杜金科, 彭永香, 等. 连铸, 2020, 45(3): 1-9.

基于铜浴法模拟料层厚度对保护渣熔化行为的影响

陈富杭，文光华，唐 萍，张 义

（重庆大学材料科学与工程学院，重庆 400044）

摘 要：结晶器保护渣是连铸过程重要的功能材料，用以保障连铸进程顺行和改善铸坯表面质量。随着保护渣技术的发展，如今已经可以满足不同钢种、断面及拉速的连铸生产。然而，目前仍存在问题需要解决。Mills[1]指出，保护渣在不同连铸机上使用效果存在差异，是连铸的关键问题。Wolf[2]认为这种波动与连铸结晶器内保护渣喂入工艺密切相关。研究表明[3]，采用间歇式保护渣喂入工艺，结晶器内熔渣池深度随保护渣总料层厚度呈周期性变化。这种周期性变化可能与保护渣的熔化行为有关[4]。因此，探究料层厚度对保护渣熔化行为的影响，对指导连铸结晶器内保护渣喂入工艺，稳定保护渣的使用效果具有重要意义。

基于上述背景，本研究采用基于铜浴的保护渣熔化行为测试方法，如图 1 所示，模拟渣层厚度对保护渣熔化行为的影响。结果表明，随保护渣料层厚度变化，保护渣的熔化速率表现出"M 曲线"规律。"M 曲线"主要与料层中炭的燃烧行为有关。如图 2 所示，随料层厚度上升，保护渣熔化速率的变化分为四个阶段。在 I 阶段，料层过薄，炭质材料与氧气充分燃烧。此时温度是限制保护渣熔化的主要因素。随着料层厚度上升，保温性能改善，保护渣熔化速率上升。在 II 阶段，氧气向颗粒料层中的扩散受限，炭质材料的燃烧成为保护渣熔化的限制性环节。随料层厚度上升，保护渣熔化速率降低。在 III 阶段，料层下部发生 Boundouard 反应，消耗炭质材料，保护渣熔化速率升高。在 IV 阶段，由于料层上部温度过低，炭质材料无法燃烧。此时，缺乏 CO_2 参与渣层下部的 Boundouard 反应，保护渣熔化速率降低。

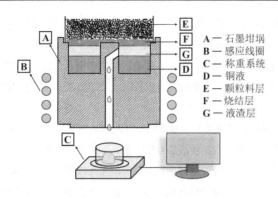
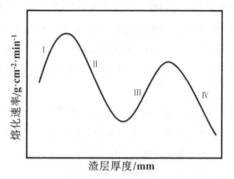

图1 基于铜浴的保护渣熔化行为测试方法　　图2 渣层厚度对保护渣熔化速率的影响

参 考 文 献

[1] Mills K C. Ironmaking & Steelmaking, 2017, 44(5): 326-332.
[2] Wolf M M. 81st Steelmaking conference proceedings, Tronto, Iron and Steel Society, 1998: 53-62.
[3] Wolf M M. The Making, Shaping, and Treating of Steel, 11th Edition, Casting Volume, Section Two, Warrendale, AIST, 2003: 21.
[4] Supradist M, Cramb A W, Schwerdtfeger K, ISIJ Int., 2004, 44(5): 817-826.

连铸保护渣中炭黑在浆料中的分散行为研究

许彦可，唐 萍，郭 俊，文光华，李 昊，黄晓燕

（重庆大学材料科学与工程学院，重庆 400044）

摘 要：结晶器保护渣在连铸生产过程中起着控热和润滑等作用，是十分重要的功能材料，它主要以硅酸盐为基料、多种助熔剂和炭质材料三大部分组成[1]。其中炭质材料中的炭黑在保护渣中起着隔离其他物料相互接触，进而控制熔化速度的作用[2]，但炭黑具有比表面积大、高表面能等特点，使炭黑很容易相互吸引形成聚集体，导致对熔化速度的控制不稳定。最初的粉状态保护渣为使炭黑分散，需采用球磨碾混。为了实现保护渣加渣的自动化和环保的目标，现如今都采用喷雾造粒技术来制备空心颗粒保护渣。在喷雾造粒制浆过程中需要在干料中加入水制成流动性较好的浆料[3]，而介质水的加入会对炭黑的分散性产生新的影响，因为炭黑的表面官能团含有羧基、羟基等，与水分子发生反应使炭黑表面带有电荷，形成稳定的电荷分散体系。由于不同类型炭黑表面官能团的种类和数量不相同，所以在浆料中的分散程度有区别。

针对此问题，本文将对炭黑在浆料中的分散行为展开研究。研究对保护渣原料中会参与水化反应的物质进行计算。结果表明，原料中起熔剂作用的 Na_2CO_3 与水会发生水化反应，改变浆料的 pH 值，进而影响炭黑的表面官能团结构，改变分散性；其他物质则以絮凝物（硅酸盐）或不溶于水的形式存在。另外，向浆料中添加微量影响羧基、羟基平衡的木质素，是否对炭黑的分散性产生影响，值得研究。由此，开展了炭黑类型、木质素分散剂、Na_2CO_3 溶液浓度对炭黑分散行为的研究。实验采用机械球磨机（转速：1000r/min，时间：30min）对炭黑在浆料中进行分散，用动态光散射（DSL）对浆料中炭黑的分散程度进行表征。研究结果表明：（1）炭黑官能团数量越多，pH 值越大，表面活性越大与溶液作用越强，分散程度更好。pH 值分别为 7.84、9.26、9.45 的 N220、PC、H30 三种炭黑，随 pH 值的增大，炭黑官能团数量增多，DSL 测试得到的炭黑在浆料中聚集体粒径减小，分别为 2.72μm、2.05μm、1.27μm。（2）阴离子木质素分散剂的加入，对炭黑表面官能团与溶液之间进行了接枝改性，减弱了炭黑之间的聚集。研究的三种炭黑在添加 2%（质量分数）的木质素分散剂后，粒径分别为 2.20μm、1.87μm、1.02μm，聚集体粒径整

体减小 20%左右。(3) 不同 Na_2CO_3 浓度对炭黑的分散性存在影响，随浆料中 Na_2CO_3 浓度的增加，炭黑聚集体增大。以 N220 炭黑为基，Na_2CO_3 含量（质量分数）分别为 0、1%、3%、5%、7%、10%，炭黑聚集体变化分别为 0.4μm、1.56μm、2.27μm、2.49μm、2.63μm、2.85μm。添加 1% Na_2CO_3 时，聚集体粒径由 0.4μm 增大至 1.56μm，对炭黑分散性有大的影响，添加 3% Na_2CO_3 后炭黑聚集体增加趋势变小。综上研究，可通过调整炭黑表面官能团与浆料之间的相互作用来促进炭黑的分散。

参 考 文 献

[1] 唐萍, 高金星, 文光华. 炼钢, 2017, 33(3): 1-19, 30.
[2] 颜慧成, 杨春梅, 席常锁, 等. 钢铁研究学报, 2011, 23(4): 28-31.
[3] 谢智勇. 新疆有色金属, 2020, 43(1): 95-96.

钢液中氢含量对连铸结晶器传热影响的机理研究

郭 俊[1,2]，唐 萍[1,2]，李 昊[1,2]，黄晓燕[1,2]，付 濯[1,2]，文光华[1,2]

（1. 重庆大学材料科学与工程学院，重庆 400044；
2. 重庆大学钒钛冶金及新材料重庆市重点实验室，重庆 400044）

摘 要：在空气潮湿的时节，空气中湿度较高，会导致最终钢液中氢含量增加，而氢在钢中的溶解度随温度下降而降低，过饱和的氢在连铸过程中会从钢液中析出。在连铸生产过程中，发现钢液中[H]含量与结晶器传热热流密度曲线存在一定的相关性，当钢液中[H]含量升高，保护渣渣条厚度增加，传热热流密度降低，热流密度曲线波动增大，有产生粘结报警甚至粘结漏钢等问题的风险[1]。

针对此问题，本文在碳含量为 0.03%~0.05%的低碳钢上开展了在钢液氢含量不同的条件下，保护渣渣条与结晶器传热性能之间的关系研究。生产工艺：转炉炼钢—LF 钢包精炼—（VD）—连铸。首先，对于进行真空 VD 及不真空的不同氢含量（连铸中间包钢液定氢）的钢液，在进行浇铸时的热流密度和热电偶温度进行数据采集，然后，分别对钢液中[H]=1.72ppm 和[H]=6.70ppm 所使用的保护渣渣条进行了取样，通过影像仪、台式电镜、XRD 对渣条进行宏观形貌和微观组织的分析。得出了以下结论：（1）钢液中氢含量上升时，结晶器热流密度和热电偶温度呈下降趋势。结晶器热流密度和热电偶值的降低，反映铸坯传递到结晶器铜板上的热量减少，结晶器内凝固坯壳厚度变薄，高温强度降低，有产生粘结报警甚至粘结漏钢等问题的风险。钢液中[H]含量分别为 3.2~3.5ppm、4.9~5.9ppm、6.5~8.2ppm 时，对应的热流密度分别为 2.6~2.7MW/m^2、2.5~2.6MW/m^2、2.4~2.5MW/m^2；在钢液中氢含量由 1.72ppm 增加为 6.70ppm 时，结晶器前三排热电偶温度降低，最大温差达到约 20℃。（2）钢液中氢含量更高的保护渣渣条更为厚大，过大的渣条会堵塞通道，液渣流入受阻，影响保护渣的润滑功能，同时过大的渣条形成也会消耗液渣，使得液渣层厚度变薄。氢含量为 1.72ppm 时，渣条厚度为 5.5mm，长度为 22mm，而氢含量为 6.70ppm 时，渣条厚度为 13.5mm，长度为 35mm。（3）钢液中氢含量升高，保护渣渣条结晶物相未发生改变，物相主要为枪晶石和霞石，但晶粒尺寸变大。这是由于钢液中析出的 H_2 进入液渣中，作为非均质形核的核心促进了熔渣的结晶和晶粒的长大[2]。按米氏散射理论，晶粒尺寸的增大，渣膜的辐射热阻增加，辐射传热量变小，结晶器热电偶温度降低，热流密度降低。同时，钢液中氢含量增加，渣膜结晶性能增强，渣膜易脱落，使得热电偶温度产生波动。钢液氢含量由 1.72ppm 增加为 6.70ppm，渣条的细晶粒区、柱状晶区和等轴晶区晶粒的当量直径分别由 5.1μm、8.3μm、13.2μm 提高至 8.7μm、16.9μm、15.7μm。

综上所述，钢液中氢含量增加，析出的 H_2 诱导了保护渣结晶性能增强，导致结晶器传热减弱。可从两个方面解决，一是控制生产流程中钢液的氢含量；二是采用结晶性能更弱的保护渣。

参 考 文 献

[1] 蒋文波, 文光华, 江婧, 等. 保护渣渣圈形成长大机理及影响因素研究现状[J]. 连铸, 2021(2): 7-13.
[2] BAEK J-Y, CHO J-W, KIM S-H. Effect of Water Vapor on Evaporation and Melt Crystallization of Mold Fluxes[J]. Metallurgical and Materials Transactions B, 2016, 47: 32-36.

镀锡板 T5 钢结晶器液面波动控制实践

肖同达[1]，毛会营[1]，张洪雷[2]

（1. 宝钢股份武钢有限炼钢厂，湖北武汉　430080；2. 宝钢股份中央研究院，上海　201900）

摘　要：镀锡板 T5 钢浇铸过程中出现结晶器液面波动，被迫降低铸机拉速，但效果不明显，导致铸坯裂纹及漏钢事故发生，结晶器液面波动机理分析及过程控制愈加迫切。分析发现钢水在结晶器内凝固时发生包晶反应，导致坯壳生长不均匀，铸坯在扇形段内发生鼓肚是导致结晶器液面波动的主要原因。通过采取结晶器水、二冷水调整，钢水成分优化、铸机精度控制等措施，结晶器液面波动得到控制，铸坯质量得到改善。

关键词：包晶反应；结晶器；二冷水；液面波动；镀锡板

Control Practice of Mould Liquid Level Fluctuation of T5 Tinplate

XIAO Tongda[1], MAO Huiying[1], ZHANG Honglei[2]

(1. Steelmaking Plant, WISCO Co., Ltd., Wuhan 430080, China;
2. Baoshan Iron and Steel, Shanghai 201900, China)

Abstract: Tinplate T5 appeared in the process of casting mould steel liquid level fluctuation, forced to reduce casting machine speed, but the effect is not obvious, lead to slab cracks and steel leakage accidents, mold level fluctuation mechanism analysis and process control of more and more urgent. It was found that the main factor for mold level fluctuation was the unevenness of strand shell caused by peritectic reaction during solidifying of molten steel and the billet is bulged in the fan-shaped segment. By the adjustment of second cooling water and mold water, the optimization of molten steel composition, accuracy control of casting machine, the problem had been settled and slab quantity was improved.

Key words: peritectic reaction; mold; second cooling water; mold level fluctuation; tinplate

电流相位分布对克莱姆式电磁搅拌器搅拌效率的影响

孟纯涛[1]，张　静[2]，张立峰[3]

（1. 东北大学冶金学院，辽宁沈阳　110819；2. 燕山大学车辆与能源学院，河北秦皇岛　066000；
3. 北方工业大学机械与材料工程学院，北京　100144）

摘 要：本文通过数值模拟的方式研究了三相交变电流的相位分布对克莱姆式电磁搅拌器搅拌效率的影响。以缠绕了 12 组线圈的克莱姆式电磁搅拌器为对象重点研究了三种交变电流相位分布形式。如图 1 所示，图中 A、B、C 为三相交变电流的三相，A 与–A 分别表示同一相的不同绕线方向。该三种电流相位分布形式皆可以在连铸坯内部产生旋转的电磁体积力。为了探究该三种电流相位分布情况下的最优搅拌效率，建立了磁场、电磁体积力的计算模型。利用 ANSYS 16.0 软件对电磁模型进行计算，通过对计算模型中线圈施加图 1 所示的三种相位分布形式的三相交变电流得到连铸坯内部时均电磁体积力的分布情况。

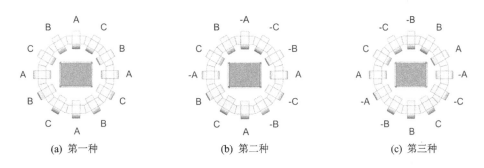

(a) 第一种　　　　(b) 第二种　　　　(c) 第三种

图 1　克莱姆式电磁搅拌器电流三种相位分布示意图

电流参数为 200 A、2.5Hz 情况下电磁模型在搅拌器中心截面的计算结果数据如图 2 所示，对比了搅拌器中心截面 Z=0m 线上三种相位布置下的时均电磁体积力计算结果，统计了三种电流相位布置情况下搅拌器中心截面的时均电磁体积力的平均值与时均电磁体积力平均值同最大值的比值两种数据用来评判克莱姆式电磁搅拌器对内部连铸坯的搅拌效率。经过对比发现，在三种相位布置方式中，第二种为最优的相位布置方式，利用相同电流参数能够得到最大的搅拌效率。

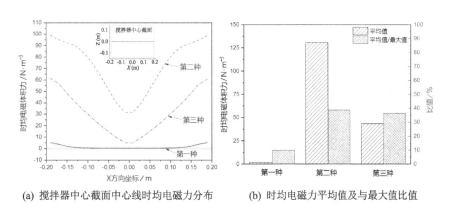

(a) 搅拌器中心截面中心线时均电磁力分布　　(b) 时均电磁力平均值及与最大值比值

图 2　搅拌器中心截面计算结果

连铸坯全凝固长度内夹杂物碰撞聚合与上浮去除的数值模拟研究

李琪蓝[1]，王举金[2]，张立峰[2]

（1. 北京科技大学冶金与生态工程学院，北京　100083；
2. 北方工业大学机械与材料工程学院，北京　100144）

摘　要：本文建立了断面为 380mm×280mm 的大方坯的全凝固长度的连铸模型,在考虑结晶器电磁搅拌的条件下,基于计算稳定的流场、温度场与凝固的工况,将中间包出口取样扫描后通过 PSG 分组法考虑了从 1.00μm 到 71.95μm 共计 15 组夹杂物的湍流碰撞与 Stokes 碰撞[1],应用 Eulerian-Eulerian 模型结合自定义函数模拟夹杂物的碰撞聚合、上浮与去除行为。图 1 和图 2 分别展示了第 1 组 1.00μm 和第 14 组 53.02μm 的夹杂物在连铸坯宽面中心截面和连铸终点断面上的无量纲浓度（无量纲浓度=该组夹杂物数密度/第一组夹杂物数密度）分布。可以看出,夹杂物在结晶器范围内有明显的断层分布,这是由于结晶器电磁搅拌造成的。另外,小尺寸夹杂物呈现出中心浓度低、表面浓度高的趋势,大尺寸夹杂物与之相反。

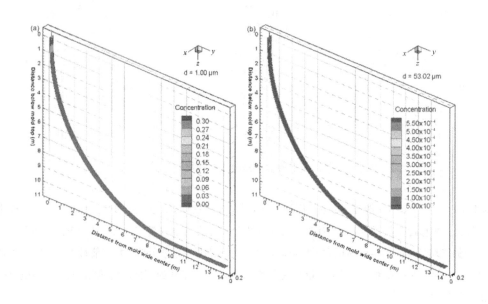

图 1　连铸坯宽面中心截面夹杂物无量纲浓度分布

(a) 1.00μm; (b) 53.02μm

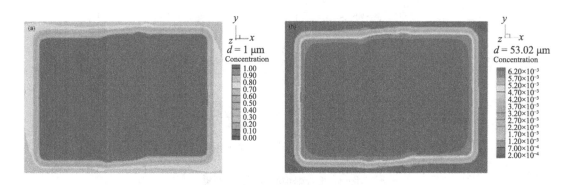

图 2　连铸终点处连铸坯断面夹杂物无量纲浓度分布

(a) 1.00μm; (b) 53.02μm

图 3 显示了全凝固长度范围内 15 组夹杂物平均直径的变化,前期小尺寸夹杂物迅速碰撞聚合导致平均尺寸急剧增大,而后逐渐增大并稳定在 3.946μm。此外,夹杂物的总氧变化如图 4 所示,整个过程中夹杂物总氧从 30.36ppm（1ppm=10^{-6}）先降低再升高,最后稳定在 31.65ppm。小尺寸夹杂物进入结晶器后迅速碰撞聚合生成大尺寸夹杂物,大尺寸夹杂物迅速上浮至结晶器表面被钢渣去除导致夹杂物总氧降低；而后夹杂物进入到连铸坯深处进行碰撞聚合,但不容易上浮到结晶器表面,导致夹杂物难以通过上浮去除,从而总氧升高。

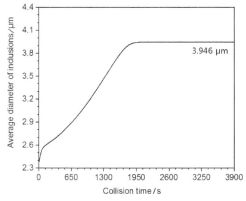
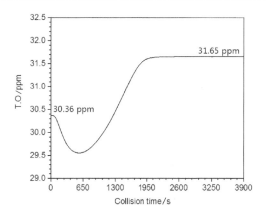

图 3　连铸坯全凝固长度内夹杂物平均直径的变化　　　图 4　连铸坯全凝固长度内夹杂物总氧的变化

参 考 文 献

[1] Peng Kaiyu, Wang Jujin, Li Qilan, et al. Multiphase Simulation on the Collision, Transport, and Removal of Non-metallic Inclusions in the Molten Steel During RH Refining[J]. Metallurgical and Materials Transactions, B. Process metallurgy and materials processing science, 2023.

连铸坯全断面夹杂物成分分布耦合模型

张月鑫[1]，王举金[2]，张立峰[2]

（1. 北京科技大学冶金与生态工程学院，北京　100083；
2. 北方工业大学机械与材料工程学院，北京　100144）

摘　要：本文通过建立钢在凝固冷却过程中夹杂物成分变化模型，预报了连铸坯全断面夹杂物成分的分布。图 1[1]为连铸坯全断面夹杂物成分分布预测模型，该模型考虑了流体流动、传热、凝固和偏析；考虑了夹杂物在冷却凝固过程中成分转变热力学计算；考虑了工业参数，例如拉速和过热度的影响；考虑了夹杂物成分转变动力学的影响。图 2[2]为通过模型（不考虑偏析）计算结果和实验结果，验证了该模型的正确性，预测值和测量值都显示了沿着连铸坯内弧到外弧，夹杂物中 CaO 和 CaS 的含量变化是相反的。

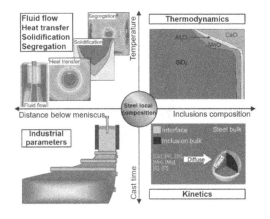
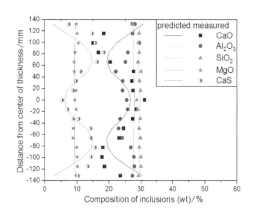

图 1　连铸坯中夹杂物成分分布预测模型[1]　　　图 2　模型（不考虑偏析）重轨钢连铸坯内弧到外弧夹杂物成分预测值和测量值的对比[2]

图 3[2]为模型（不考虑偏析）预测重轨钢连铸坯全断面 CaO 夹杂物成分的分布，在连铸坯表层 CaO 含量分布最大，含量（质量分数）大于 29%。并且在连铸坯表层夹杂物的成分与中间包中的初始成分接近，这是由于快速冷却，从而导致表层夹杂物成分来不及发生转变。图 4[3]为对比模型（考虑偏析）计算得到重轨钢连铸坯内弧到外弧夹杂物成分变化的预测值和测量值。通过对比图 2[2]和图 4[3]，可以得知，考虑了元素偏析的计算模型得到的结果与实测结果更加吻合。

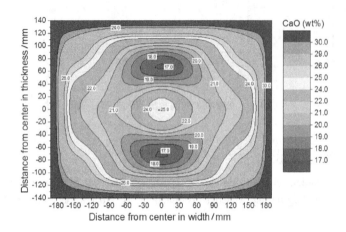 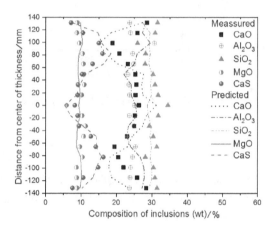

图 3　模型（不考虑偏析）计算得到重轨钢连铸坯全断面中 CaO 夹杂物含量分布[2]

图 4　模型（考虑偏析）计算重轨钢连铸坯内弧到外弧夹杂物成分预测值和测量值的对比[3]

参 考 文 献

[1] 张月鑫, 张立峰, 王举金, 等. 钢液凝固与冷却过程及固体钢加热过程钢中非金属夹杂物成分动力学转变的几个概念和特征曲线[J]. 工程科学学报, 2023, 45(3): 369-379.

[2] Ren Qiang, Zhang Yuexin, Zhang Lifeng, et al. Prediction on the spatial distribution of the composition of inclusions in a heavy rail steel continuous casting bloom[J]. Journal of Materials Research and Technology, 2020, 9(3): 5648-5665.

[3] Zhang Yuexin, Chen Wei, Wang Jujin, et al. Prediction of Distribution of Composition of Inclusion in Continuous Casting Bloom of the Heavy Rail Steel Coupling Element Segregation, Heat Transfer, and Kinetics[C]. TMS 2022 151st Annual Meeting & Exhibition Supplemental Proceedings. Springer, 2022: 87-94.

钢包底部改造对浇注末期下渣的影响研究

吕彬玉[1]，张立峰[2]

（1. 北京科技大学冶金与生态工程学院，北京　100083；
2. 北方工业大学机械与材料工程学院，北京　100144）

摘　要：钢包浇注过程中，对包底添加台阶进行改造会影响着出钢过程中的流体流动，加剧出钢末期钢包流体在径向的流动，从而减缓下渣的发生。本文通过研究不同底部台阶总高度的影响，有助于选取更为合适的生产参数，延缓钢包出钢末期下渣的发生。如图 1 所示，钢包浇注末期会在水口上方形成高速旋流区域，本研究通过改变生产现场常用的包底台阶总高度，研究其对于钢包下渣临界参数的影响。

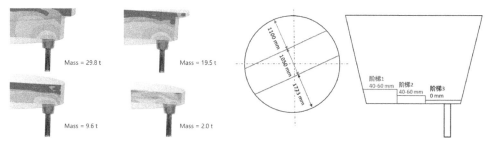

(a) 钢包改造前浇注末期流场变化图　　(b) 钢包底部改造示意图

图 1　钢包底部改造前后示意图

浇注末期钢包发生下渣时，有无台阶以及台阶的高度对临界下渣参数均有影响，如图 2 所示，当钢包底部无台阶时，发生下渣时钢液的临界高度为 122mm，剩余钢液质量为 7.852t；当钢包底部台阶增大到 120mm 时，发生下渣时钢液的临界高度为 114mm，剩余钢液质量为 4.741t。由于台阶的设置位置与形成漩涡的高速区有一定距离，因此在钢包末期下渣过程中，底部改造对发生下渣时的钢液临界高度影响不大，但随着底部台阶的升高，漩涡发生时的临界余钢质量逐渐减小，这主要是由于在现场实际生产中，改造包底台阶会使钢包容量减少，在临界高度变化不大的基础上，发生漩涡下渣的临界钢液质量也相应减少。因此，生产现场钢包包底改造工作需要根据漩涡形成的临界时刻综合考虑钢液洁净度及成本控制，选择合适的台阶高度。

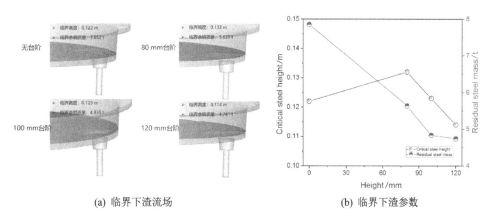

(a) 临界下渣流场　　(b) 临界下渣参数

图 2　钢包底部改造后浇注末期临界下渣参数

板坯结晶器开浇过程多相流动及凝固的数值模拟研究

李丰辰[1]，陈　威[1]，张立峰[2]

（1. 燕山大学机械工程学院，河北秦皇岛　066004；
2. 北方工业大学机械与材料工程学院，北京　100144）

摘　要：本文建立了 1100mm×230mm 断面板坯结晶器三维数值模型，耦合了 Realizable k-ε 湍流模型与 VOF 多相流模型，对结晶器开浇过程"钢-气"两相流体流动现象以及钢液凝固现象进行数值模拟研究。开浇初期钢液与空气混合剧烈，两条钢液射流直接冲击结晶器底部区域，随着液位逐渐上升，在开浇 30s 时完成 1/2 钢液浇注质量，此刻开始射流冲击区域逐渐从结晶器底部上抬至窄面,速度分布如图1(a)所示。随着液位继续上升下环流逐渐发展,钢液面完全没过浸入式水口出口后,下环流发展完全,上环流逐渐发展并于结晶器浇满至工作液位后发展完全,三维流态如图1(b)所示。

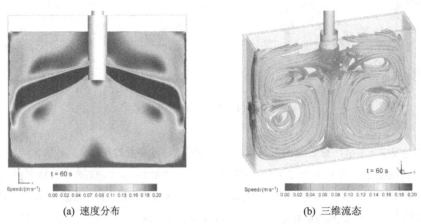

(a) 速度分布　　　　　　　　　　(b) 三维流态

图1　结晶器开浇过程钢液速度分布与三维流态

开浇过程中流动-多相-传热-凝固的耦合计算结果表明,随着射流冲击区从结晶器底部逐渐上抬至窄面,在35~36s时底部生成完整凝固坯壳,靠近窄面射流冲击区凝固坯壳最薄约3mm,底部中心非冲击区凝固坯壳最厚约18mm,结晶器开浇过程凝固坯壳厚度分布如图2所示。

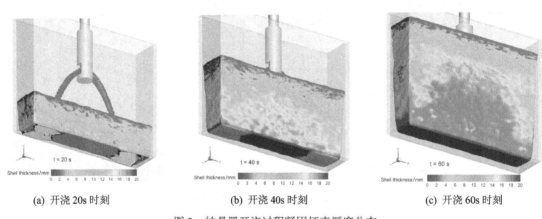

(a) 开浇20s时刻　　　　(b) 开浇40s时刻　　　　(c) 开浇60s时刻

图2　结晶器开浇过程凝固坯壳厚度分布

板坯结晶器电磁搅拌对卷渣的影响

郑　甫[1],陈　威[1],张立峰[2]

(1. 燕山大学机械工程学院,河北秦皇岛　066004;
2. 北方工业大学机械与材料工程学院,北京　100144)

摘　要:板坯连铸生产拉速变化、水口堵塞、吹氩以及电磁搅拌都会影响结晶器多相流场和卷渣,控制结晶器多相流场和卷渣对于控制连铸坯质量至关重要[1,2],本文建立了钢-渣-气三相模型,模拟了结晶器电磁搅拌(M-EMS, mold electromagnetic stirring)对弯月面流动速度的影响,如图1所示。模拟结果表明,不加载M-EMS时结晶器上环流冲击区域流速较大,在左右两侧对称分布,水口附近流速较小;加载M-EMS后,弯月面整体速度分布更加均匀,但宽面两侧速度明显增加，使卷渣发生概率提升，可能对铸坯质量控制产生不利影响。

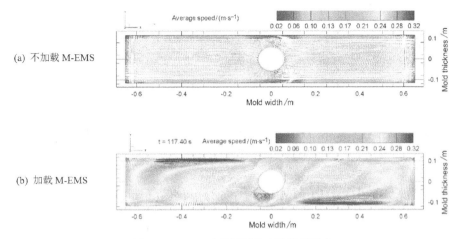

图 1 M-EMS 对弯月面流动速度的影响

不合理的 M-EMS 参数可能提升卷渣的发生概率，本文研究了不同电流强度对结晶器净卷渣速率的影响，结果如图 2 所示。模拟结果表明，200A、400A 和 600A 电流强度下净卷渣速率较不加载 M-EMS 情况下分别提升了 1.51 倍、5.51 倍和 8.27 倍，说明 400A 和 600A 电流强度下，净卷渣速率大幅提升，对铸坯的表面及中心质量产生不利影响，而 200A 电流参数下净卷渣速率提升较小，有利于控制结晶器卷渣。

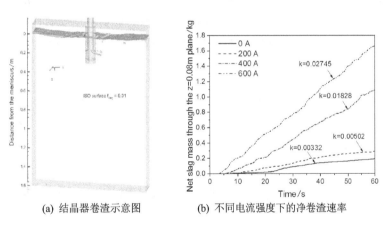

(a) 结晶器卷渣示意图　　(b) 不同电流强度下的净卷渣速率

图 2 M-EMS 对净卷渣速率的影响

参 考 文 献

[1] Zhang L, Yang S, Cai K, et al. Thomas. Investigation of fluid flow and steel cleanliness in the continuous casting strand[J]. Metallurgical and Materials Transactions B, 2007, 38: 63-83.
[2] Chen W, Zhang L, Wang Y, et al. Mathematical simulation of two-phase flow and slag entrainment during steel bloom continuous casting[J]. Powder Technology, 2021, 390: 539-554.

电磁制动技术研究进展与应用

肖玉宝[1]，秦　哲[1]，金百刚[2]，郭庆涛[1]，张凯伦[1]，潘　栋[1]

（1. 鞍钢集团北京研究院有限公司钢铁研究院分院，北京　102211；
2. 鞍钢股份鲅鱼圈钢铁分公司，辽宁营口　115007）

摘 要：本文对连铸过程中结晶器流场控制的电磁制动技术的研究进展与应用情况进行了综合评述。介绍了电磁制动技术主要分类及其特点、数值模拟以及物理模拟的新进展，并给出了电磁制动技术在连铸结晶器中的应用实例。分析了电磁制动技术的冶金效果及存在的不足，指出新型电磁制动技术磁场分布的合理化建议、钢液流场优化方案。最后，提出了有助于改善连铸工艺的结晶器电磁制动技术的未来发展趋势。

关键词：电磁制动；电磁场；电磁力；流场；连铸；结晶器

Research Progress and Application of the Electromagnetic Brake Technique

XIAO Yubao[1], QIN Zhe[1], JIN Baigang[2], GUO Qingtao[1], ZHANG Kailun[1], PAN Dong[1]

(1. Iron and Steel Research Institute, Ansteel Beijing Research Institute Co., Ltd., Beijing 102211, China;
2. Ansteel Bayuquan Iron and Steel Subsidiary, Yingkou 115007, China)

Abstract: In this paper, the research process of the electromagnetic brake technology and its application situation in terms of controlling the flow field of the continuous casting mold was comprehensively reviewed. The main classification and characteristics of electromagnetic brake technology, new progress in numerical simulation and physical simulation, and the presentative application of electromagnetic brake technique in continuous casting mold was introduced. Furthermore, the metallurgical effect of electromagnetic brake and its deficiencies was analyzed, and the reasonable suggestion for magnetic field distribution of new type of electromagnetic brake, as well as the optimized scheme for the molten steel flow field was pointed out. Finally, the future development trend of the electromagnetic brake technology contributing to the improvement of the continuous casting technology was put forward.

Key words: electromagnetic brake; electromagnetic field; electromagnetic force; flow field; continuous casting; casting mold

超大型异形坯连铸结晶器内凝固过程仿真

刘 艳，雷少武，李 伟，耿明山，王 颖

（中冶京诚工程技术有限公司，北京 100176）

摘 要：本文采用有限差分法模拟仿真了超大型异形坯在结晶器内凝固过程，建立了结晶器、铸坯、冷却水以及保护渣三维组合模型，创建了移动边界条件，既考虑了铜板与冷却介质对铸坯横向冷却，也考虑了气隙对铸坯纵向凝固的影响，较为全面、真实地仿真了结晶器内腔在有无锥度、不同水孔布置方案下异形坯和结晶器铜板三维传热耦合温度场。在此基础上针对其中水孔布置方案，在接近结晶器铜板内壁设置了测温热电偶，计算了各特征点的温度，揭示了铜板热面温度的分布规律，为超大型异形坯结晶器设计提供了理论依据。

关键词：异型坯；结晶器；水冷；仿真

Simulation of Solidification in Mold for Large Beam Blank's Continuous Casting Process

LIU Yan, LEI Shaowu, LI Wei, GENG Mingshan, WANG Ying

(MCC Capital Engineering & Research Incorporation Limited, Beijing 100176, China)

Abstract: The solidification process of large beam blank in mold is simulated with finite difference method. A 3D combined model with mold, blank, cooling water and covering slag is established, and the moving boundary conditions are created in this paper. Both the transverse cooling effect of copper plate and cooling medium on the blank, as well as the influence of air gap on the longitudinal solidification of the blank are considered. The 3D heat transfer coupled temperature field between the blank and mold copper plate is simulated comprehensively and truly under with and without taper on the inner wall of the copper plate and different cooling water holes arrangements. On this basis, the thermocouple points are set close to the inner wall of the copper plate when the cooling water hole diameter is $\phi 9mm$, and the temperature of each characteristic point is calculated, the distribution pattern of the temperature on thermal surface of the copper plate are revealed. The simulation results provide theoretical foundation for the mold design of large beam blank.

Key words: beam blank; mold; water cooling; simulation

钢铁冶金过程的水跃行为研究

杨　泽，许浩峰，贾志扬，杨文城，兰文泽，王彦强，陈　进

（东北大学 EPM 实验室，辽宁沈阳　100819）

摘　要： 钢铁冶金是在高温下进行相关物理变化、化学变化的复杂过程。它以及辅助工艺和设备等涉及物质的传输、热能和动量的传输、固体的运动、流体的流动、化学反应和相转变等基本现象，其中，水跃是此过程中一个普遍和重要的现象。不同的水跃行为在中间包、结晶器钢液浇注拉坯、二冷冷却喷淋等工艺中发生，包括环形水跃、经典水跃及不同水跃之间的转变等。揭示水跃的机理和变化规律从而调控好水跃行为，可以实现对铸坯质量的优化和控制。

关键词： 水跃；钢铁冶金；流场控制

Hydraulic Jump Behavior in Iron and Steel Metallurgical Process

YANG Ze, XU Haofeng, JIA Zhiyang, YANG Wencheng,
LAN Wenze, WANG Yanqiang, CHEN Jin

(Key Laboratory of Electromagnetic Processing of Materials (Ministry of Education),
Northeastern University, Shenyang 110819, China)

Abstract: The intricate process of iron and steel metallurgy involves high-temperature physical and chemical changes. It comprises basic phenomena like the transfer of mass, heat, and momentum, the movement of solids, the flow of fluids, chemical reactions, and phase transitions, along with auxiliary processes and equipment. The hydraulic jump is a frequent and important phenomenon in these processes. Different hydraulic jump behaviors occur in processes such as tundish, casting in mold, and secondary cooling spray, including circular hydraulic jump, classical hydraulic jump, and transition between different hydraulic jumps. Revealing the mechanism and variation law of hydraulic jump and regulating the behavior of hydraulic jump can realize the optimization and control of slab quality.

Key words: hydraulic jump; iron and steel metallurgy; flow field control

脉冲电流对 321 不锈钢中 TiN 析出的影响

梁尚东，魏光旭，肖紫轩，杨成成，朱正海，黄冬建，常立忠

（安徽工业大学冶金工程学院，安徽马鞍山 243000）

摘　要：脉冲电流处理可以有效改善金属凝固组织和析出物尺寸，本文通过热态实验及理论分析研究了脉冲电流强度对 TiN 析出的影响机理。实验结果表明：当未通脉冲电流时试样含 TiN 析出物分布密度为 555 个/mm^2，施加 30A、50A、70A 脉冲电流时分布密度变为 600 个/mm^2、644 个/mm^2、689 个/mm^2。理论分析表明：脉冲电流能显著增大钢液中 Ti 原子扩散通量，促进 TiN 粒子形核和析出，电流越大，效果越明显。研究结果为通过脉冲电流处理控制 TiN 粒子的析出提供了理论依据。

关键词：脉冲电流；不锈钢；TiN；扩散通量

Effect of Pulsed Current on TiN Precipitation in 321 Stainless Steel

LIANG Shangdong, WEI Guangxu, XIAO Zixuan, YANG Chengcheng,
ZHU Zhenghai, HUANG Dongjian, CHANG Lizhong

(School of Metallurgical Engineering, Anhui University of Technology, Maanshan 243000, China)

Abstract: Pulsed current treatment can effectively improve the solidification organization of metal and the size of precipitates. In this paper, the influence mechanism of pulsed current intensity on TiN precipitation is investigated through thermal state experiments and theoretical analysis. The research shows that: the experimental results show that: when the specimen contains TiN precipitates without pulse current, the distribution density is 555/mm^2, and the distribution density becomes 600/mm^2, 644/mm^2 and 689/mm^2 when the pulse current is applied at 30A, 50A and 70A; the theoretical analysis shows that: the pulse current can significantly increase the diffusion flux of Ti atoms in the liquid steel and promote the nucleation and precipitation of TiN particles, and the larger the current, the more obvious the effect. The results provide a theoretical basis for controlling the precipitation of TiN particles by pulsed current treatment.

Key words: pulsed current; stainless steel; TiN; diffusive fluxes

五机五流非对称连铸中间包流场优化及特性研究及优化设计

王海达[1]，陈　列[1]，成国光[2]，杨　平[1]，范世强[1]

（1. 建龙北满特殊钢有限责任公司，黑龙江齐齐哈尔　161041；2. 北京科技大学，北京　100083）

摘　要：通过对五机五流非对称中间包的三维模拟、水模型物理模拟和 ANSYS Fluent 有限元计算软件的数值模拟手段，提出中间包钢液流动行为的优化方案，逐步应用于现场实际生产并达到工艺和操作稳定化。验证结果表明：优化后的中间包平均停留时间显著延长，最高达到 467.47s，死区比例最低降至 22.22%，RTD 曲线由陡峭、分散转变为平缓、一致。铸坯内夹杂物尺寸以 1~2.5μm 为主，尺寸大于 5μm 以上夹杂物数量较少，夹杂物平均尺寸在 2μm 左右，各流钢水停留时间差异小于 20%，多流之间过热度实测差异小于 2℃。各流一致性、钢水纯净度得到了明显改善。

关键词：非对称中间包；水模拟；流场；三维模拟；夹杂物

Optimization of Flow Field and Characteristics of Five-machine and Five-strand Unsymmetrical Continuous Casting Tundish

WANG Haida[1], CHEN Lie[1], CHENG Guoguang[2], YANG Ping[1], FAN Shiqiang[1]

(1. Jianlong Manchuria Special Steel Co., Ltd., Qiqihar 161041, China;
2. University of Science and Technology Beijing, Beijing 100083, China)

Abstract: By means of three-dimensional simulation of five-machine and five-stream asymmetric tundish, physical simulation of water model and numerical simulation of ANSYS Fluent finite element software, the optimization scheme of molten steel flow behavior in Tundish is put forward, gradually applied to the actual production site and achieve process and operation stability. The results show that the optimized Tundish's average residence time is prolonged to 467.47s and the ratio of dead zone is reduced to 22.22%. The RTD curve changes from steep and dispersive to smooth and consistent. The size of inclusions in billet is mainly 1~2.5μm, and the number of inclusions is less when the size is larger than 5μm. The average size of inclusions is about 2μm, and the residence time difference of molten steel is less than 20%, the measured difference of superheat between multi-flow is less than 2℃. The consistency of each flow and the purity of molten steel were improved obviously.

Key words: asymmetric tundish; water simulation; flow field; three-dimensional simulation; inclusions

高锰奥氏体低温钢凝固特性与微观裂纹敏感性

何　杨，徐　浩，刘建华

（北京科技大学工程技术研究院，北京　100086）

摘　要：高锰奥氏体低温钢具有高强度、低成本、优良焊接性等优势，是生产 LNG 储罐极具竞争力的理想材料。然而，高锰奥氏体低温钢在连铸过程中面临铸坯裂纹、漏钢等问题，严重制约高锰奥氏体低温钢连铸工业化生产。为此，本研究结合 DSC 和定向凝固实验方法，探究高锰奥氏体低温钢凝固特性及微观裂纹敏感性，对解决高锰奥氏体低温钢连铸过程遇到的凝固裂纹问题有积极作用。

本研究实验用钢以工业纯铁、电解锰、工业硅、电解铬、石墨等为原料，采用 50kg 真空感应炉熔炼，并在真空条件下浇铸成锭，检测成分为（质量分数）0.362% C、0.224% Si、26% Mn、0.0047% P、3.35% Cr、0.395% Cu、0.052%V。实验铸锭用于 DSC 和定向凝固实验取样。DSC 实验采用高纯氩保护，试样先以 30℃/min 升温速率加热至 1200℃，再分别以 25℃/min、10℃/min、5℃/min 升温速率加热至 1500℃后完全熔化。定向凝固实验同样采用高纯氩保护，试样先以 10℃/min 升温速率加热至 1500℃后完全熔化，再分别以 5μm/s、50μm/s、100μm/s、200μm/s、

300μm/s 抽拉速率进行冷却，并在 Ga-In-Sn 合金液中淬火。

DSC 实验表明，高锰奥氏体低温钢升温过程仅发生固-液相转变，升温速率对固相线温度影响较小，但对液相线温度影响较大。在平衡凝固条件下，实验用钢的固相线温度约为 1342℃，液相线温度约为 1400℃。在连铸生产过程中，可根据高锰奥氏体低温钢的固、液相线温度，选择合理的过热度及浇铸制度，以保证连铸坯质量。

定向凝固实验表明，高锰奥氏体低温钢在不同抽拉速率下均呈现树状晶组织，且二次枝晶间距与抽拉速率存在指数关系。结合定向凝固两相区组织和固、液相线温度，进一步确定高锰奥氏体低温钢固相率与温度的关系式。随着抽拉速率增加，高锰奥氏体低温钢的平均二次枝晶间距逐渐减小，即提高抽拉速率（冷却速率）可细化二次枝晶。

通过扫描电镜观察高锰奥氏体低温钢定向凝固两相区组织微观裂纹，发现微观裂纹与冷却速率存在直接关系，越大的冷却速率更易产生微观裂纹，且裂纹总长度、平均长度均与冷却速率正相关。统计结果分析表明，裂纹平均长度可作为反映高锰奥氏体低温钢微观裂纹敏感性的一种判定标准。

参 考 文 献

[1] Tang J, He S, Mo Y, et al. Materials Characterization, 2022, 190: 112024.
[2] Liu Z G, Gao X H, Xiong M, et al. Materials Science and Engineering: A, 2021, 807: 140881.
[3] Shen Y, Liu J, Xu H. Materials Science and Technology, 2021, 37(4): 446-457.
[4] Wang X, Sun X, Song C, et al. Materials Characterization, 2018, 135: 287-294.
[5] Shen Y, Liu J, Xu H, et al. Metallurgical and Materials Transactions B, 2020, 51(6): 2963-2975.

重轨钢连铸大方坯 V 型偏析探究

李红光[1,2]，徐明丽[1]

（1. 昆明理工大学冶金与能源工程学院，云南昆明 650093；2. 攀钢集团攀枝花钢铁研究院有限公司，钒钛资源综合利用国家重点实验室，四川攀枝花 617000）

摘 要：针对重轨钢连铸大方坯 V 型偏析，检测了 V 型偏析的宏观形貌特征和凝固组织，计算对比了理论收缩体积与实际 V 型体积。通过凝固末端压下工艺对比、钢种对比及凝固组织调控试验，探究了 V 型偏析形成的影响因素。结果表明，重轨钢大方坯 V 型偏析分两步形成，首先是凝固前沿捕捉的晶核充分长大成粗大球化等轴晶组织后发生搭接，搭接产生封闭的富集溶质微区而形成初始 V 型偏析，进一步在钢液静压力和凝固收缩的负压抽吸等综合作用下发生组织滑移产生中心凹陷，导致 V 型偏析的宏观形貌进一步改变（减小 V 型角）而形成最终的 V 型偏析，研究制定了改善重轨钢大方坯 V 型偏析的关键工艺。

关键词：重轨钢；凝固组织；V 型偏析

Study on the V-shape Segregation of Rail Steel Bloom

LI Hongguang[1,2], XU Mingli[1]

(1. Faculty of Metallurgical and Energy Engineering, Kunming University of Science and Technology, Kunming 650093, China; 2. Pangang Group Research Institute Co., Ltd., State Key Laboratory of Vanadium and Titanium Resource Comprehensive Utilization, Panzhihua 617000, China)

Abstract: The feature of macro-profile and microstructure were tested, also the theoretical shrinkage value and actual volume of V-shape was compared for the formation of V-shape segregation in rail steel bloom. The influencing factors including dynamic soft reduction processes, grade of steel and technology of micro-structure control were researched by carried out experiment. Result shows the V-shape segregation of rail steel bloom tend to formed from two steps. the first step is "initial V-shape segregation formed", the specific process is the crystal nucleus which caught by solidification front of bloom were fully developing to be the coarse equiaxed crystals and connected, this lead the closed micro-regions produced, the molten steel in micro-regions are tend to be solute rich during the solidification process. And then the second step is V-shape changed, which is formed from the microstructure subside and slip by the effect of the static pressure and the negative pressure suction by shrinkage. The key processes were established to improve V-shape segregation of rail steel bloom.

Key words: rail steel; solidification structure; V-shape segregation

5 电冶金与废钢铁

大会特邀报告
第十三届冶金青年科技奖获奖人特邀报告
分会场特邀报告
矿业工程
焦化及节能环保
炼铁与原料
炼钢与连铸
★ 电冶金与废钢铁
轧制与热处理
表面与涂镀
金属材料深加工
粉末冶金
先进钢铁材料及应用
节能与低碳技术
冶金环保与资源利用
冶金设备与工程技术
冶金自动化与智能化
冶金物流
冶金流程工程学
其他

5.1 电冶金

电渣重熔精炼过程数值模拟研究进展

王 强[1,2]，刘 畅[1,2]，王 芳[3]，贺 铸[1,2]，

李光强[1,2]，李宝宽[3]

（1. 武汉科技大学省部共建耐火材料与冶金国家重点实验室，湖北武汉　430081；
2. 武汉科技大学钢铁冶金及资源利用省部共建教育部重点实验室，湖北武汉　430081；
3. 东北大学冶金学院，辽宁沈阳　110819）

摘　要：电渣重熔技术在高品质特殊钢的生产中占据重要地位。由于电渣重熔过程的复杂性以及现场测量条件的限制，数值模拟已成为解析其过程现象和机理的必要手段。本文主要综述了电渣重熔过程数值模拟的研究进展，包括电磁场、流场、传热、凝固、宏观偏析、精炼反应和夹杂物去除等，还涉及到了导电结晶器、旋转电极、加压电渣重熔、电渣接续制备双合金和电渣重熔回收不合格电解金属锰等电渣冶金新技术的数值模拟研究。利用数值模拟可以准确掌握重熔过程中多物理场的耦合作用规律，进而预测电渣锭成分分布，优化电渣锭质量，推动电渣冶金技术的发展。

关键词：电渣重熔；高品质特殊钢；传输现象；数值模拟

Development of Numerical Simulation in Electroslag Metallurgy

WANG Qiang[1,2], LIU Chang[1,2], WANG Fang[3], HE Zhu[1,2],
LI Guangqiang[1,2], LI Baokuan[3]

(1. The State Key Laboratory of Refractories and Metallurgy, Wuhan University of Science and Technology, Wuhan 430081, China; 2. Key Laboratory for Ferrous Metallurgy and Resources Utilization of Ministry of Education, Wuhan University of Science and Technology, Wuhan 430081, China; 3. School of Metallurgy, Northeastern University, Shenyang 110819, China)

Abstract: Electroslag remelting (ESR) technology creates a great contribution in the production of high quality special steel. Due to the process complication and measuring limitation, numerical simulation has become a necessary approach for understanding the transport phenomena and mechanism in the ESR process, which includes electromagnetism, flow, heat transfer, solidification, macrosegregation, refining reaction, and inclusion removal, etc. Besides, the numerical simulations of novel ESR technologies such as conducting mold, rotating electrode, pressurized ESR, ESR connecting preparation, and ESR recycling have also been reviewed. The numerical simulation is able to clarify the coupled physical phenomena during the ESR process and predicate the composition profile, which contribute to the improvement of the ESR ingot performance. It is believed that the ESR technology development could be promoted by the numerical simulation.

Key words: electroslag remelting; high quality special steel; transport phenomena; numerical simulation

碳粉喷吹过程气-固混合喷吹动力学研究

刘崇[1]，刘纲[2,3]，王强[1]，李杰[1]，张彩东[1]，丁志军[4]

（1. 河钢材料技术研究院，河北石家庄　050023；2. 清华大学信息国家研究中心，北京　100084；3. 江阴兴澄特种钢铁有限公司，江苏无锡　214429；4. 石家庄钢铁有限责任公司，河北石家庄　050031）

摘　要：为了进一步提升电炉冶炼过程碳粉的利用率，建立了气-固混合喷吹数值模拟模型，利用数值模拟分析了气-固混合喷吹过程气体和粉剂的动力学行为特性以及两者间的相互作用。结果表明：纯气相喷吹过程，气相射流核心段长度约为250mm，采用气-固混合喷吹后，由于受到固相颗粒的影响，气相的速度核心段消失，出口处的气相与固相速度一致，且随着碳粉粒度或碳枪直径的增加，粉气流的最大速度以及出口速度均呈现增加的趋势；环境温度对纯气相和粉气流的最大速度影响不大，但是对于纯气相和粉气流的出口速度影响较大，高温环境有助于延迟气-固混合射流的衰减，增大气-固混合喷吹粉气流的出口速度。

关键词：数值模拟；气-固混合喷吹；电炉炼钢

Study on the Dynamic Characteristics of Gas-solid Mixed Injection in Carbon Powder Blowing Process

LIU Chong[1], LIU Gang[2,3], WANG Qiang[1], LI Jie[1], ZHANG Caidong[1], DING Zhijun[4]

(1. HBIS Material Technology Research Institute, Shijiazhuang 050023, China;
2. National Information Research Center of Tsinghua University, Beijing 100084, China;
3. Jiangyin Xingcheng Special Steel Works Co., Ltd., Wuxi 214429, China;
4. Shijiazhuang Iron & Steel Co., Ltd., Shijiazhuang 050031, China)

Abstract: In order to further improve the utilization rate of carbon powder in the process of electric furnace smelting, a numerical simulation model of gas-solid mixed injection was established, and the dynamic behavior characteristics of gas and powder and the interaction between them in the process of gas-solid mixed injection were analyzed by numerical simulation. The results show that the core length of gas jet is about 250mm in the process of pure gas-phase injection. After gas-solid mixing injection, the core length of gas velocity disappears due to the influence of solid particles, and the velocity of gas and solid at the outlet is the same. With the increase of carbon particle size or carbon gun diameter, the maximum velocity of powder gas flow and the outlet velocity show an increasing trend; Ambient temperature has little effect on the maximum velocity of pure gas and powder flow, but has great effect on the outlet velocity of pure gas and powder flow. High ambient temperature is helpful to delay the attenuation of gas-solid mixing jet and increase the outlet velocity of gas-solid mixing blowing powder flow.

Key words: numerical simulation; gas-solid mixed injection; EAF steelmaking

Ce-Mg 协同处理对 M50 航空轴承钢中夹杂物演变的影响

王礼超，田家龙，蒋成钢，任　吉，姜周华

(东北大学冶金学院，辽宁沈阳　110819)

摘　要：本文采用真空感应熔炼工艺冶炼航空轴承钢 M50，对比分析 Ce-Mg 协同处理对氧、硫含量的影响，并分析了 Ce-Mg 处理对夹杂物组成、尺寸分布特征的影响。结果表明，Ce-Mg 协同处理具有强大的脱氧和脱硫能力。添加质量分数为 0.026% 的 Ce 和 0.002% 的 Mg 后，实验钢中氧、硫质量分数分别降低至 0.00075% 和 0.001%；夹杂物的数量密度、面积分数和平均直径显著降低；夹杂物类型从 Al_2O_3 和 MnS 变成 Ce_2O_2S、Ce_2O_3、Ce_xS_y，以及含 As 的稀土复合夹杂物。热力学计算结果表明 Ce-Mg 处理后最稳定的夹杂物为 Ce_2O_2S。

关键词：稀土；镁；航空轴承钢；夹杂物；热力学

Effect of Ce-Mg Synergistic Treatment on the Inclusion Evolution in M50 Aerospace Bearing Steel

WANG Lichao, TIAN Jialong, JIANG Chenggang, REN Ji, JIANG Zhouhua

(School of Metallurgy, Northeastern University, Shenyang 110819, China)

Abstract: In this work, two heats of the M50 aircraft bearing steel were prepared using vacuum induction melting process. The effects of Ce-Mg synergistic treatment on oxygen content, sulfur content, and inclusion distribution characteristics were analyzed comparatively. The results showed that Ce-Mg treatment could significantly improve the deoxidation and desulfurization efficiency. After adding Ce with mass fraction of 0.026% and Mg with mass fraction of 0.002%, the mass fraction of oxygen and sulfur were reduced to 0.00075% and 0.001%, respectively. The area fraction, number density and average diameter of inclusions in Ce-Mg treated steel noticeably decreased, and the Al_2O_3 and MnS phases were modified into Ce_2O_2S、Ce_2O_3 and Ce_xS_y phases together with complex inclusions containing As element. The thermodynamic analysis results showed that Ce_2O_2S phase was the most stable product.

Key words: rare earth; magnesium; aerospace bearing steel; inclusions; thermodynamics

D2 冷作模具钢中的碳化物形成机理及控制

李万明[1]，刘峻辰[1,2]，臧喜民[1]，姜周华[2]

(1. 辽宁科技大学材料与冶金学院，辽宁鞍山　114051；
2. 东北大学材料与冶金学院，辽宁沈阳　110819)

摘　要：D2 冷作模具钢因具有良好的淬硬性、优异的耐磨性和抗高温氧化性，被广泛应用在冷加工模具领域。由于 D2 冷作模具钢高碳高合金成分，其铸态组织内部容易析出聚集大尺寸 M_7C_3 碳化物。本文研究了凝固冷速对 D2 冷作模具钢碳化物特征的影响，并从三维层面研究了冷却速率对棒状 M_7C_3 碳化物生长机制的影响。研究了 D2 冷作模具钢高温处理过程碳化物的转变行为，并计算出等温热变形时 M_7C_3 碳化物相变激活能。研究结果表明，碳化物三维形貌主要为六边形空心棒状，存在少量的片层状、弯曲团簇棒状。六边形空心棒状 M_7C_3 碳化物以侧向包抄方式螺旋位错生长。经加热和热变形处理后，D2 钢基体中大尺寸 M_7C_3 碳化物得到细化。D2 钢基体中的 M_7C_3 碳化物会在热加工锻造过程发生断裂，产生相变的激活能为 226.974kJ/mol。

关键词：D2 冷作模具钢；冷却速率；碳化物；热加工

Formation Mechanism and Control of Carbides in D2 Cold Work Die Steel

LI Wanming[1], LIU Junchen[1,2], ZANG Ximin[1], JIANG Zhouhua[2]

(1. School of Materials and Metallurgy, University of Science and Technology Liaoning, Anshan 114051, China; 2. School of Materials and Metallurgy, Northeastern University, Shenyang 110819, China)

Abstract: D2 steel is widely used in the field of cold working molds because of its good hardening, excellent wear resistance and high temperature oxidation resistance. Due to the high carbon and alloy composition of D2 steel, it is easy to precipitate and accumulate large size M_7C_3 carbides in the as-cast structure. In this paper, the effect of solidification cooling rate on the carbide characteristics of D2 steel is studied, and the effect of cooling rate on the growth mechanism of rod-like M_7C_3 carbide is studied from the three-dimensional level. The transformation behavior of carbide in D2 steel was studied, and the activation energy of M_7C_3 carbide during isothermal deformation was calculated. The results show that the three-dimensional morphology of carbides is mainly hexagonal hollow rod, and there are a few lamellar and curved cluster rod. Hexagonal hollow rod-shaped M_7C_3 carbides grow in helical dislocations by lateral encapsulation. After high temperature treatment, the large size M_7C_3 carbides of D2 steel matrix are refined. The M_7C_3 carbide in D2 steel matrix is fractured during hot working forging process, and the activation energy of phase transition is 226.974kJ/mol.

Key words: D2 steel; cooling rate; carbide; hot working

新型 15Cr-30Ni 气阀合金电渣锭均质化控制

史成斌，张 淮，王世舟，李 晶

（北京科技大学，北京 100083）

摘 要：气阀合金用于生产汽车内燃机进气阀和排气阀。进气阀工作温度通常在 400℃以下，排气阀的工作温度通常在 500～850℃之间，这要求气阀材料需要具有良好的高温强度和抗高温废气腐蚀性能，同时不得有宏观偏析、缩孔、疏松、内裂和杂质富集等非均质化问题。涡轮增压技术是近年来广泛应用且经济有效的节能减排技术，提高了内燃机的燃烧效率。排气阀的服役温度随着缸内燃烧温度的提高而不断提高，这对气阀合金的性能提出了更高的要求。因此，研发了一种兼顾高性能和低成本的新型的铁镍基型的气阀合金材料以满足内燃机排气阀的材料需求。

合金元素分布均匀、析出相尺寸及其分布以及晶粒尺寸等微观特征对于保障气阀合金机械性能至关重要。本研究开发了新型 15Cr-30Ni-Fe 沉淀强化气阀合金成分和生产工艺参数，有望在 700～760℃下长期服役。开发了电渣重熔生产气阀合金专用五元渣系，实现电渣重熔过程中 Ti 烧损量<0.2%（质量分数），增 Al 量<0.04%（质量分数），电渣锭中 Ti、Al 含量均匀性稳定控制，电渣锭全氧含量<13×10⁻⁶，电渣后夹杂物数量密度也显著降低，尺寸

均小于 4μm。夹杂物类型为单独 TiN 夹杂物和以 MgO•Al$_2$O$_3$ 为核心外层包裹 TiN 的复合夹杂物。通过电渣重熔工艺参数优化，电渣锭元素偏析显著降低，大尺寸共晶碳化物和金属间化合物减少了 40%，提高了铸锭组织均匀性。通过建立合金的本构方程以及热加工图，确定了合金在热加工过程中应变速率、变形温度和变形程度的最优匹配参数，避免了锻造开裂和组织不均匀问题，锻后方坯的平均晶粒尺寸为 21.4μm±9.6μm，实现了锻后晶粒尺寸的均匀控制。

参 考 文 献

[1] Antonialli A Í S, Carvalho M R D D, Diniz A E. J. Braz. Soc. Mech. Sci., 2020, 42(10).
[2] 王俏俏，龚志华，李康雪. 特钢技术, 2021, 27(3): 30-32, 36.
[3] 嵇爽，赵英利，陈文，等. 热加工工艺, 2023, 1-5.
[4] Shi C B, Wang S J, Li J et al. J. Iron Steel Res. Int., 2021, 28(12): 1483-1503.
[5] Wang S Z, Shi C B, Li J. et al. J. Alloy. Compd., 2023.

电炉回收废弃镁碳耐火材料用于造泡沫渣研究

宋生强

（武汉科技大学省部共建耐火材料与冶金国家重点实验室，湖北武汉　430081）

摘　要：在 EAF 炼钢中，白云石和镁砂通常用于电炉造含 MgO 饱和炉渣以保护镁质耐火材料不受侵蚀。考虑到废弃镁碳耐火材料中 MgO 含量较高，废弃 MgO-C 耐火材料可以作为炉渣调渣剂回收资源化利用，再者其中的碳也可作为造泡沫渣碳源。基于 FactSage 软件模拟计算典型 EAF 炉渣中 MgO 溶解度的等温溶解度图（ISD），结果表明：电炉渣 CaO-SiO$_2$-FeO-MgO-Al$_2$O$_3$ 体系中 MgO 溶解度随着 CaO/SiO$_2$ 碱度的增加而显著降低，电炉渣中其他组元如 Al$_2$O$_3$、P$_2$O$_5$ 等会导致 MgO 饱和度缓慢增加。针对某电炉厂炉渣成分表征分析，基于 ISD 图分析可知适宜理想造泡沫渣的目标 MgO 含量为 7.5%，因此理论上每吨钢最多可回收 13.2kg 废弃 MgO-C 耐火材料。废弃 MgO-C 耐火材料在电炉渣中的溶解实验表明：镁碳颗粒在熔渣中溶解限制性环节为熔渣中 FeO 传质，原位产生的 CO 气体大大加速了氧化镁晶粒的溶解过程，并促进熔渣泡沫化。

关键词：电炉炼钢；泡沫渣；废弃镁碳砖；ISD 图

Recycling of Spent MgO-C Refractories as Slag Foaming Additives in EAF

SONG Shengqiang

(Wuhan University of Science and Technology, the State Key Laboratory of Refractories and Metallurgy, Wuhan 430081, China)

Abstract: In EAF steelmaking, dolomite and magnesite are commonly used to saturate a slag with MgO for refractory protection. Spent MgO-C refractories can be recycled as slag conditioners considering the high concentration of MgO. Isothermal Solubility Diagrams (ISD) to represent MgO solubility in EAF slag were calculated using FactSage. MgO solubility significantly decreases with increasing the basicity of CaO/SiO$_2$ in the CaO-SiO$_2$-FeO-MgO-Al$_2$O$_3$ system. Other components in the electric furnace slag such as Al$_2$O$_3$, P$_2$O$_5$, etc. will cause the MgO saturation to increase slowly. For the EAF slag compositions of steel plant X, ISD shows the optimized MgO content is 7.5% for ideal slag foaming. Theoretically it can recycle a maximum amount of 13.2kg of used MgO-C per ton steel in plant X. Dissolution experiments of waste MgO-C refractory materials in electric furnace slag show that the limited step in the dissolution of magnesium carbon particles in slag is the mass transfer of FeO in slag, and the CO gas generated in situ greatly accelerates the dissolution process of magnesia grains, and promote slag foaming.

Key words: EAF; slag foaming; spent MgO-C refractories; ISD diagram

Ce-Mg 处理对 H13 钢洁净度、组织和性能的影响

任 吉，田家龙，王礼超，蒋成钢，姜周华

（东北大学冶金学院，辽宁沈阳 110819）

摘 要：本文采用真空感应炉冶炼 H13 热作模具钢，研究了 Ce-Mg 处理对 H13 钢洁净度、组织和性能的影响。结果表明，Ce 和 Mg 都可以改性 H13 钢中夹杂物，并且夹杂物平均尺寸都有所减小，Ce 处理夹杂物数量密度最大，Ce-Mg 处理夹杂物平均尺寸最小，由 1.94μm 降至 1.41μm；Ce-Mg 复合处理具有更强的脱氧和脱硫能力，Ce-Mg 处理后钢中氧含量 5.4ppm，硫含量 9ppm；稀土 Ce 可以改善 H13 钢的铸态组织，明显减小最后凝固区域尺寸，从而抑制液析碳化物长大；Mg 显著改善了二次碳化物的形貌、尺寸和分布状态，添加 Mg 之后碳化物更加细小弥散；Ce-Mg 处理后 H13 钢的综合力学性能最好，Mg 处理和 Ce-Mg 处理使钢冲击功分别增加了 16% 和 23%。

关键词：H13 钢；铈镁处理；洁净度；组织；性能

Effect of Ce-Mg Treatment on Cleanliness, Microstructure and Properties of H13 Steel

REN Ji, TIAN Jialong, WANG Lichao, JIANG Chenggang, JIANG Zhouhua

(School of Metallurgy, Northeastern University, Shenyang 110819, China)

Abstract: In this work, the vacuum induction furnace was used to smelt H13 hot work die steel, and the effects of Ce-Mg treatment on the cleanliness, microstructure and properties of H13 steel were studied. The results show that Ce and Mg can change the type of inclusions in H13 steel and reduce the average size of inclusions; Ce-Mg composite treatment has stronger deoxidation and desulfurization ability, and the oxygen content of the steel after Ce-Mg treatment is 5.4ppm and the sulfur content is 9ppm; Ce can improve the as-cast microstructure of H13 steel, significantly reduce the size of the final condensation zone, thereby inhibiting the growth of primary carbides; Mg can significantly improve the secondary carbide morphology, size and distribution of the state, and the carbides become smaller and more diffuse after the addition of Mg; The mechanical properties of Ce-Mg treated H13 steel were the best, and the impact energy of Mg treated and Ce-Mg treated steels increased by 16% and 23%, respectively.

Key words: H13 steel; Ce-Mg treatment; cleanliness; microstructure; properties

基于随机森林的电弧炉终点参数预报研究

陆泓彬，朱红春，姜周华，吴泽文，艾翔宇，杨 策，冯 浩，张树才

（东北大学冶金学院，辽宁沈阳 110819）

摘　要：电弧炉终点控制主要包括终点碳含量和终点温度控制。实现终点参数的有效预测可以提前调整冶炼工艺，从而提高终点控制精度。现有的电弧炉终点预测模型预测精度较低，难以实现终点成分和温度的有效预测。本文针对电弧炉终点参数预测模型的不足，提出了一种基于冶金机理和随机森林算法的预报模型，经实际生产数据测试，模型具有较高的适用性。

关键词：随机森林；电弧炉；终点预报；集成学习；机器学习

Random Forest-based Prediction of Electric Arc Furnace Endpoint Parameters

LU Hongbin, ZHU Hongchun, JIANG Zhouhua, WU Zewen, AI Xiangyu, YANG Ce, FENG Hao, ZHANG Shucai

(School of Metallurgy, Northeastern University, Shenyang 110819, China)

Abstract: Electric arc furnace endpoint control mainly includes endpoint carbon content and endpoint temperature control. Effective prediction of the end point parameters can adjust the smelting process in advance, thus improving the accuracy of the end point control. The existing electric arc furnace endpoint prediction model has low prediction accuracy, and it is difficult to realize the effective prediction of endpoint composition and temperature. In this paper, for the shortcomings of the electric arc furnace endpoint parameter prediction model, a prediction model based on metallurgical mechanism and random forest algorithm is proposed, and the model has high applicability after the actual production data test.

Key words: random forest; electric arc furnace; endpoint prediction; integrated learning; machine learning

加压电渣重熔高氮不锈钢洁净度控制研究

杨守星，李花兵，冯　浩，姜周华，陈　敏

（东北大学冶金学院，辽宁沈阳　110819）

摘　要：本文探究了加压电渣重熔（PSER）高氮不锈钢用炉渣成分优化与钢质洁净度控制。结果表明，炉渣中添加适量 Na_2O 可促进渣-钢脱硫反应，提高 PSER 脱硫率，减少钢中含硫夹杂物。炉渣的熔化温度和黏度随 Na_2O 含量的增高而降低，2%~4%Na_2O 有助于促进夹杂物的吸附去除、减小平均尺寸；但 Na_2O 过高时挥发损失严重，引起轻微增氧和氧化物夹杂增多。高氮不锈轴承钢电渣锭中铝含量由[Si]+(Al_2O_3)反应平衡控制，通过向炉渣中添加适量 SiO_2 和降低电极硅含量，实现了零增铝和无 AlN 夹杂物。

关键词：加压电渣重熔；高氮不锈钢；洁净度；炉渣成分；夹杂物

Cleanliness Control of High Nitrogen Stainless Steels during Pressurized Electroslag Remelting

YANG Shouxing, LI Huabing, FENG Hao, JIANG Zhouhua, CHEN Min

(School of Metallurgy, Northeastern University, Shenyang 110819, China)

Abstract: This paper investigated the composition optimization of slags for pressurized electroslag remelting (PSER) high nitrogen stainless steels and cleanliness control. The results show that proper Na_2O addition in slag improved slag-steel desulfurization reactions, increased PSER desulfurization degree and reduced sulfur-containing inclusions in steel. The melting temperature and viscosity of slag decreased with the increase of Na_2O content. 2%~4%Na_2O addition promoted the adsorption and removal of inclusions, and reduced the average size. However, excessive Na_2O exacerbated volatilization loss, leading to slight increase of oxygen content and oxide inclusions. The aluminum content of remelted HNSBS ingot was determined by the equilibrium of $[Si]+(Al_2O_3)$ reaction. By proper SiO_2 addition in slag and decreasing the silicon content of electrode, aluminum pickup was avoided and AlN inclusion disappeared.

Key words: pressurized electroslag remelting; high nitrogen stainless steel; cleanliness; slag composition; inclusion

CERI S-Arc™ 复合型废钢预热电炉技术的可行性研究

张宇航，王　佳，王少臣，李佳辉

（中冶京诚工程技术有限公司，北京　100176）

摘　要：全废钢电炉短流程炼钢以其绿色低碳、生态环保、节能低耗等优点，成为当前我国钢铁工业实现"碳达峰、碳中和"发展战略的重要途径。与长流程相比，全废钢电炉可实现吨钢降碳约 75%，是我国钢铁行业绿色低碳转型的主力军。但现有全废钢电炉技术存在生产成本高、竖式废钢预热电炉设备结构复杂和水平加料电炉预热效果差等局限，限制了我国电炉炼钢的发展。因此，开发高效低耗的新型废钢预热电炉技术对于我国钢铁行业绿色低碳转型发展具有重要意义。

本文针对当前电炉废钢预热技术的问题，自主开发了 CERI S-Arc™ 复合型废钢预热电炉技术，并对技术的可行性和可靠性进行研究分析、模拟仿真和实验验证。结合现有废钢预热技术竖井与水平预热的优缺点，CERI S-Arc™ 开发了新型竖井+水平加料的复合型废钢预热和加料技术，可使废钢预热竖井远离炉体高温区布置，降低了竖井的热负荷和漏水爆炸的风险，提高了生产安全性。竖井和水平复合预热增强了烟气与废钢的对流换热时间和效率，降低电炉烟尘量的同时提高金属收得率约 2%，吨钢电耗同比降低 30~50kW·h/t，吨钢综合能耗同比降低 7kgce/t 以上。预热后的废钢采用水平加料入炉可实现电炉不开盖、不断电加料，全程平熔池冶炼，可缩短电炉冶炼周期 6min，同时电炉的冶炼噪声可控制在 85~90dB（A）。采用柔性供电技术替代传统变压器供电方式，提高电弧稳定性，将功率因数从 0.8 提高至 0.95 以上，供电线路损耗降低 15~22kW·h/t。开发"全程高温烟气+急冷余热锅炉+活性炭吸附"二噁英超净消除技术，充分利用水平加料段进行烟气二次燃烧，保证废钢预热后烟气温度达到二噁英裂解的温度，无需烟气再升温的燃气消耗，确保二噁英排放浓度（标态）≤$0.1ng/m^3$，同时回收蒸汽≥100kg/t，实现了以较低的能源消耗达到更高的排放标准。为保证新型电炉对于原料的适用能力，CERI S-Arc™ 采用经典的 EBT 炉型，可采用当前电炉炼钢所应用的各种炼钢原料，保证冶炼工艺成熟可靠，原料和钢种适应性更高，且维护简单。

CERI S-Arc™ 电炉技术在基于复合型废钢预热和加料技术、柔性供电技术、i-Arc 智能电极调节技术、急冷余热锅炉技术以及二噁英消除与粉尘超净排放的烟气净化技术的基础上，与当前应用较多的水平加料电炉技术相比，吨钢生产成本可降低约 39 元，吨钢碳排放可间接降低约 36kg。

面向"碳中和"的中国电炉流程发展战略研究

上官方钦[1]，崔志峰[2]，周继程[1]，倪 冰[1]，李 涛[1]

（1. 中国钢研科技集团有限公司钢铁绿色化智能化技术中心，北京 100081；
2. 北京科技大学冶金与生态工程学院，北京 100083）

摘 要： 钢铁工业是资源、能源密集型行业，是国民经济的重要基础产业，也是实现绿色低碳发展的重要领域。当前，我国钢铁工业的流程结构以高炉-转炉长流程为主，能源结构以煤炭为主，加上大规模的粗钢生产使得钢铁行业成为 CO_2 排放大户，约占全国碳排放总量的 15%~17%，钢铁工业脱碳化发展任重道远。2020 年 9 月，习近平总书记在第 75 届联合国大会上提出"双碳目标"；为积极响应党中央号召，各部委陆续发布《工业领域碳达峰实施方案》《关于促进钢铁工业高质量发展的指导意见》等一系列政策文件，并明确指出要有序合理引导全废钢电炉流程发展，深入推进钢铁工业绿色低碳转型。

结合当前钢铁行业发展形势，我国电炉流程在发展过程中困难重重，如：电炉钢生产成本高，企业无利可图；电炉钢比偏低，发展道路漫长；废钢资源短缺，废钢流向不合理；诸多电炉流程优化技术及低碳制造技术待研发。为应对上述各类挑战，利用钢铁生命周期法对未来我国废钢资源量[1, 2]进行预测；并构建钢铁行业能源结构预测模型，对双碳进程中钢铁行业电力能源需求量进行测算分析。研究发现：未来我国电炉流程发展具备坚实的废钢资源及电力能源保障，其产量占比将逐年提高，预计 2030 年达到 15%以上，2035 年进一步提升至 30%以上，2040 年后，有望成为我国主要钢铁制造流程。

在推动电炉流程发展的进程中，需做到有序推进、循序渐进，适当布局以"两链一流"系统为核心的"城市钢厂"[3]，并根据产品需求形成以下三类主要的发展模式：

（1）全废钢电炉流程替代中小高炉，生产螺纹钢等长材；

（2）返回法与近终型制造技术结合，生产优特钢、不锈钢、硅钢、热轧薄板带等；

（3）氢还原-电炉流程生产国防军工、高新技术产业等所需的特殊钢材。

中国电炉流程的技术结构从只注重产品制造功能逐步拓展到"三个功能"[4]上，在重点关注单体技术研发的同时，兼顾电炉流程优化及设计方面的研究。

最后，在上述研究基础上，构建了双碳分析模型，分别从工序及时间两个层面绘制了中国电炉流程低碳发展路线图。分析发现：电炉工序和轧钢工序的降碳潜力最大，分别占全流程降碳潜力的 40.7%和 36.7%；随着各工序低碳技术的应用，2030 年电炉流程的吨钢碳排放强度有望在 2020 年的基础上降低 35.1%，2040 年降低 74.7%，2050 年实现"近零碳"冶炼。

参 考 文 献

[1] 上官方钦, 郦秀萍, 周继程, 等. 中国废钢资源发展战略研究[J]. 钢铁, 2020, 55(6): 8-14.
[2] 崔志峰, 上官方钦, 王方杰, 等. 2022—2060 年中国废钢资源量分析预测[J]. 钢铁, 2023, 58(6): 126-133.
[3] 上官方钦. 全废钢电炉流程优化研究——城市钢厂与钢铁工业脱碳化[D]. 北京: 钢铁研究总院, 2021.
[4] 殷瑞钰. 冶金流程工程学[M]. 北京: 冶金工业出版社, 2004.

全废钢电炉-ESP 产线绿色高效冶炼关键技术研究实践

魏光升[1]，李 挺[2]，朱 荣[1]，廖交志[2]，李 欣[1]

（1. 北京科技大学，北京 100083；2. 福建鼎盛钢铁有限公司，福建宁德 352000）

摘 要：以再生钢铁原料（社会废钢）低碳地生产高性能钢铁材料是未来钢铁工业绿色低碳发展的重要方向。其中，"全废钢电炉-ESP 近终形无头轧制"制造流程简约高效，大幅降低了冶炼和轧制环节的 CO_2 排放，近几年在业内备受关注。本文主要结合鼎盛钢铁公司国内首条量子电炉-ESP 绿钢产线的生产实践，对影响"全废钢电炉-ESP 近终形无头轧制"流程高效稳定运行的关键难点进行了分析，围绕量子电炉高效节能冶炼、全流程钢液洁净度控制等方面介绍了相关单元技术的研究及应用情况。同时，针对"全废钢电炉-ESP 近终形无头轧制"流程面临的有害残余/杂质元素稳定控制难题及其进一步节能降碳生产近零碳绿钢的需求，开展了深度脱磷、稳定控氮、钢液脱铜、能效评价及生物质炭利用等诸多新技术的前期研究与探索。

关键词：电炉；ESP；低碳生产；洁净冶炼

Research and Application on Green and Efficient Smelting Technology for EAF-ESP Production Process with 100% Scrap Steel

WEI Guangsheng[1], LI Ting[2], ZHU Rong[1], LIAO Jiaozhi[2], LI Xin[1]

(1. University of Science and Technology Beijing, Beijing 100083, China;
2. Fujian Dingsheng Iron and Steel Company, Ningde 352000, China)

Abstract: The low-carbon production of high-performance steel materials using recycled steel raw materials (social scrap) is an important direction for the green and low-carbon development of the future steel industry. In recent years, the manufacturing process of "Scrap-EAF-ESP" has attracted much attention in the steel industry, which is simple and efficient, significantly with lower CO_2 emissions in the smelting and rolling processes. Combining the production practice of the first domestic Quantum EAF-ESP green steel production line in Dingsheng Iron and Steel Company, this report mainly analyzes the key difficulties that affect the efficient and stable operation of the EAF-ESP process. It introduces the research and application of relevant unit technologies in efficient EAF energy-saving smelting and molten steel cleanliness control. At the same time, in response to the problem of stable control of harmful residual/impurity elements and the need for further energy-saving and carbon reduction production of near zero carbon green steel, preliminary research and exploration have been carried out on many new technologies such as deep dephosphorization, stable nitrogen control, steel liquid copper removal, energy efficiency evaluation and biomass carbon utilization.

Key words: EAF; ESP; low carbon production; clean smelting

全废钢电弧炉尾渣回用造渣工艺研究

王 岩[1]，李 晶[1]，李 波[1]，刘 燊[2]

（1. 北京科技大学钢铁冶金绿色低碳全国重点实验室，北京 100083；
2. 台鑫钢铁有限公司，江西上饶 334602）

摘 要：钢渣是炼钢过程的副产物，具有产量大、难处理的特点。电弧炉炼钢过程钢渣排放量为 100~150kg/t 钢，经磁选提铁约 10%后剩余的废渣称为电炉尾渣，国内仅有约 30%的炼钢尾渣通过水泥掺合料等方式低值利用，其余堆存抛撒，造成了严重的环境污染和土地占用。国内外亦有企业将尾渣用于高炉减少外排，但此方式对短流程全废钢电炉炼钢企业并不适用。

电弧炉尾渣为碱性氧化渣，碱度在 2.0~2.2，以电弧炉尾渣资源化循环利用为目标，进行了全废钢电弧炉尾渣回用造渣工艺研究。明确尾渣加入量对电炉终渣碱度的影响，最大限度减少炼钢过程石灰消耗，从而降低石灰生产过程的碳排放，同时，减少钢渣产出量，降低钢渣对环境的影响。根据不同碱度下渣中(MgO)饱和溶解度随(FeO)含量的变化可知，碱度低于 1.8 时，MgO 饱和溶解度在 9%以上，这将导致炉衬侵蚀加剧或含镁渣料加入量增加，成本增加。若渣中 MgO 不饱和，则渣中不存在(Mg,Fe)O 第二相，渣有效黏度低，发泡性能降低[1]。因此，在保证终渣碱度不低于 1.8 的前提下，采用理论计算和现场工业试验相结合的方法，明确了尾渣与石灰质量的优化配比。尾渣单次回用时，采用 13.0kg/t 尾渣、28.5kg/t 石灰和 7.5kg/t 镁球的造渣工艺；连续多炉回用时，第一炉采用 13.0kg/t 尾渣、28.5kg/t 石灰和 7.5kg/t 镁球的渣料组合，后续炉次采用 13.0kg/t 尾渣、33.4kg/t 石灰和 7.7kg/t 镁球的渣料组合进行造渣。后续回用炉次要多加 4.9kg/t 石灰和 0.2kg/t 镁球，原因是补偿尾渣带来的碱度损失和渣量增加导致的 MgO 含量的降低。连续 20 炉回用，石灰与镁球平均消耗分别降低 8.45 kg/t 钢和 0.6kg/t 钢。

电炉渣发泡性能与渣有效黏度有关，有效黏度与渣中悬浮第二相含量（通常为(Mg,Fe)O）及黏度有关[2]。经磁选后的尾渣 FeO 含量为 17%~19%，MgO 含量为 6%~7%，为探明尾渣加入对炉渣泡沫化的影响，对比研究了前期加入石灰和加入等质量尾渣时，渣中第二相的变化，发现加入石灰的渣因过剩的自由 CaO 与(Mg,Fe)O 反应生成了低熔点的 CaO·FeO 及难熔 MgO 固相，导致渣中(Mg,Fe)O 含量降低。加入尾渣的渣中(Mg,Fe)O 悬浮第二相含量基本不变，炉渣黏度较大，泡沫化效果更好，电能利用效率更高。与采用 41.6kg/t 石灰和 8.3kg/t 镁球的造渣工艺相比，优化后的尾渣连续回用造渣工艺，氧气消耗降低 2.68Nm³/t 钢、碳粉消耗降低 1.12kg/t 钢，废钢加入速度提高 0.14t/min，电耗平均降低 10.12kW·h/t 钢。

参 考 文 献

[1] Pretorius E B, Carlisle R C. Foamy slag fundamentals and their practical application to electric furnace steelmaking[J]. Iron & Steelmaker, 1999, 26(10): 79-88.
[2] Almeida R A M, Vieira D, Bielefeldt W V, et al. Slag foaming fundamentals-a critical assessment[J]. Materials Research, 2017, 20: 474-480.

5.2 废钢铁预处理及应用

100t 全废钢 Consteel 电弧炉冶炼能量平衡与能效分析

李龙飞，胡砚斌，张曦东，徐立军，孟华栋，王 明

（钢铁研究总院有限公司冶金工艺研究所，北京 100081）

摘 要：本文利用能量平衡原则计算分析了 100t 全废钢 Consteel 电弧炉各部分能量输入、能量消耗、能量损失以及能量效率。计算结果表明，冶炼前期化渣阶段碳粉利用率低，随着冶炼时间延长，废钢和渣料熔化升温消耗能量占比升高，其他热损失降低，总能量效率升高。建议适当提高留钢量，调整冶炼供氧喷碳制度，通过改变料型等方式提高刨花料废钢进料速度，以达到缩短冶炼周期及降低电耗和综合成本的目的。

关键词：全废钢；Consteel 电弧炉；化学能；能效分析

Energy Balance and Energy Efficiency Analysis of 100t Consteel Electric Arc Furnace with All Scrap

LI Longfei, HU Yanbin, ZHANG Xidong, XU Lijun,
MENG Huadong, WANG Ming

(Institute for Metallurgical Process, Central Iron and Steel Research Institute, Beijing 100081, China)

Abstract: In this paper, the energy input, energy consumption, energy loss, and energy efficiency of the 100 ton Consteel electric arc furnace with all scrap were calculated and analyzed using energy balance principle. The calculation results show that the utilization rate of carbon powder is lower during the slag melting period. As the smelting time prolongs, the proportion of energy consumption for melting and heating of scrap steel and slag increases, other heat loss reduces and overall energy efficiency increases. It is recommended to appropriately increase hot heel, adjust the oxygen supplying and carbon spraying system, and increase the feeding speed of shaving scrap steel by changing the shape, in order to shorten the smelting cycle and reduce electricity consumption and comprehensive costs.

Key words: all scrap; Consteel electric arc furnace; chemical energy; energy efficiency analysis

面向废钢高效高质利用的智能验质技术研究

刘丽冉，杨冬靓

（河钢数字技术股份有限公司，河北石家庄 053000）

摘 要：废钢是炼钢流程工艺中的主要原料之一，在炼钢工艺中，不仅可以大量代替铁矿石，还是回收率最高的再

生资源，是中国钢铁工业实现"碳达峰"和"碳中和"的重点之一。因此，合理充分利用好废钢资源，不仅可以达到减少铁矿石依赖和减少碳排放的双重效果，还对整个行业乃至国家的绿色化发展进程有着积极的推动作用。本文基于深度学习技术和神经网络算法研发的废钢智能验质系统，可通过高效的图像处理、人机交互等技术大幅提升废钢利用效率，在节能降碳的同时为废钢流程的优化路径提供思路。

关键词：废钢应用；废钢判级；机器学习；机器视觉

Research on Intelligent Quality Inspection Technology for High-efficiency and High-quality Utilisation of Steel Scrap

LIU Liran, YANG Dongliang

(HBIS Digital Technology Co., Ltd., Shijiazhuang 053000, China)

Abstract: Scrap is one of the main raw materials in the steel making process, which can not only replace iron ore in large quantities, but also is the renewable resource with the highest recycling rate, which is one of the key points for China's iron and steel industry to realise "Carbon Peak" and "Carbon Neutral". Therefore, making full use of scrap resources can not only achieve the double effect of reducing the dependence on iron ore and reducing carbon emissions, but also have a positive impact on the whole industry and even the country's green development process. This paper based on deep learning technology and neural network algorithms to develop an intelligent quality inspection system for scrap steel, through efficient image processing, human-computer interaction and other technologies to significantly improve the efficiency of scrap steel, energy saving and carbon reduction at the same time as the optimisation of the scrap steel process to provide ideas for the path.

Key words: scrap applications; scrap grading; machine learning; machine vision

铁碳熔池中废钢熔化行为的数值模拟研究

刘孟珂[1,2]，马国军[1,2]，徐 菊[1,2]，张 翔[1,2]，郑顶立[1,2]

（1. 武汉科技大学钢铁冶金及资源利用省部共建教育部重点实验室，湖北武汉 430081；
2. 武汉科技大学钢铁冶金新工艺湖北省重点实验室，湖北武汉 430081）

摘 要：在"双碳"目标的政策背景下，绿色低碳是钢铁行业向高质量发展的关键[1-5]。国务院印发的《2030年前碳达峰行动方案》中指出，钢铁行业应促进结构优化和清洁能源替代，提升废钢资源回收利用水平。当前，废钢作为金属料被广泛应用在转炉、电弧炉和铁水包等设备中。废钢的熔化行为是影响转炉废钢比和炼钢过程温度轨迹的关键因素，同时也会影响电弧炉炼钢的能耗和产能。因此，有必要探究废钢在铁碳熔池中的熔化行为及机理。

前人已对废钢在熔池中的熔化行为进行了多项研究，主要包括热模实验、水模实验、数值计算和数值模拟[6-13]。相对于物理模拟、热模实验和工业试验而言，数值模拟实验周期较短，且具有模拟真实条件和理想条件的能力，已成为冶金工作者解决实际生产问题的有效手段。为探究实际条件下废钢的熔化过程，并为转炉和电弧炉熔炼废钢工艺建模提供依据，本研究使用ANSYS Fluent软件开展了废钢熔化行为的数值模拟，研究了熔池温度、熔池碳含量、废钢特征长度、废钢预热温度、底吹氩气量和底吹元件分布对废钢在铁碳熔池中熔化行为的影响。研究表明：（1）当废钢浸入液态熔体中后，废钢表面形成固化层。随着浸泡时间的增加，固化层的厚度先增加后减小，然后废钢本体开始熔化。同时，在废钢熔化过程中，废钢中心的温度始终最低，且随着浸泡时间的增加而升高。（2）随着熔池温度、熔池碳含量和废钢预热温度的升高，废钢特征长度的减小，固化层的形成时间逐渐缩短，最大厚度逐渐减小，废钢的熔化速度逐渐加快，废钢的完全熔化时间逐渐缩短，废钢中心的升温速率逐渐提高。在废钢本体的熔化过程中，当熔池温度为 1873K 时，废钢中心的升温速率为 28.6K/s；当熔池碳含量（质量分数）为 4%时，废钢中心的升温速率为 25.0K/s；当废钢特征长度为 10mm 时，废钢中心的升温速率约为 45K/s；当废钢预热温度为 1100K 时，废钢中心的升温速率约为 18.5K/s。（3）底吹氩气可以起到搅拌熔池的作用，适当增大熔池的底吹气量，选择合适的底吹元件分布方式可以加强熔体向废钢的传热与传质过程，有利于废钢的熔化，缩短其完全熔化时间。当底吹氩气量为 0.5L/min 时，固化层在 63s 时达到最大厚度，废钢的完全熔化时间为 170s。但当底吹气量过大时，搅拌效率的提升程度有效，废钢完全熔化时间的缩短程度减缓。其中，当底吹氩气量增大至 1L/min 时，废钢的完全熔化时间为 149s，这与水模实验的结果较为一致。

关键词：废钢；铁碳熔池；熔化机理；数值模拟；吹氩搅拌

Numerical Simulation of Melting Behavior of Steel Scrap in Iron-carbon Bath

LIU Mengke[1,2], MA Guojun[1,2], XU Ju[1,2], ZHANG Xiang[1,2], ZHENG Dingli[1,2]

(1. Key Laboratory for Ferrous Metallurgy and Resources Utilization of Ministry of Education, Wuhan University of Science and Technology, Wuhan 430081, China; 2. Hubei Provincial Key Laboratory of New Processes of Ironmaking and Steelmaking, Wuhan University of Science and Technology, Wuhan 430081, China)

Key words: steel scrap; iron-carbon bath; melting mechanism; numerical simulation; argon stirring

参 考 文 献

[1] 王永中. 碳达峰、碳中和目标与中国的新能源革命[J]. 人民论坛·学术前沿, 2021, 14: 88-96.

[2] 徐匡迪. 低碳经济与钢铁工业[J]. 钢铁, 2010, 45(3): 1-12.

[3] 刘树洲, 张建涛. 中国废钢铁的应用现状及发展趋势[J]. 钢铁, 2016, 51(6): 1-9.

[4] 单亦和. 废钢作为可持续发展资源在钢铁工业中的应用[J]. 钢铁, 2001, 36(10): 6-11.

[5] 上官方钦, 郦秀萍, 周继程, 等. 中国废钢资源发展战略研究[J]. 钢铁, 2020, 55(6): 8-14.

[6] Shukla A, Deo B, Robertson D. Scrap dissolution in molten iron containing carbon for the case of coupled heat and mass transfer control[J]. Metallurgical and Materials Transactions B, 2013, 44(6): 1407-1427.

[7] Li J, Provatas N, Irons G. Modeling of late melting of scrap in an EAF[J]. Iron and Steel Technology, 2008, 5(8): 216-223.

[8] Shukla A, Deo B, Robertson D. Role of air gap in scrap dissolution process[J]. Metallurgical and Materials Transactions B, 2013, 44(6): 1398-1406.

[9] Shukla A, Dmitry R, Volkova O, et al. Cold model investigations of melting of ice in a gas-stirred vessel[J]. Metallurgical and Materials Transactions B, 2011, 42(1): 224-235.

[10] Li J, Brooks G, Provatas N. Kinetics of scrap melting in liquid steel[J]. Metallurgical and Materials Transactions B, 2005, 36(2): 293-302.

[11] Gao Ming, Gao Jintao, Zhang Yanling, et al. Two-dimensional temperature distribution and heat transfer during scrap melting[J]. JOM, 2020.

[12] Kruskopf A. A model for scrap melting in steel converter[J]. Metallurgical and Materials Transactions B, 2015, 46(3): 1195-1206.

[13] Kruskopf A, Holappa L. Scrap melting model for steel converter founded on interfacial solid/liquid phenomena[J]. Metallurgical Research and Technology, 2018, 115(2): 1-7.

废钢预热在提高转炉废钢比中的应用与发展

胡砚斌，黄 芳，侯中晓，何 赛，贾志立

（钢铁研究总院有限公司冶金工艺研究所，北京 100081）

摘 要：氧气转炉冶炼过程中通过供氧操作，使铁水中的碳、硅等元素被氧化放热从而提供冶炼需要的热量，受铁水温度、成分和装入量等因素限制，转炉炼钢富余热量有限，废钢加入比例受到限制。对废钢进行预热可提高废钢自身的物理热，增加熔池容纳废钢能力，是提高转炉废钢比的重要手段，本文对目前主要的废钢预热技术进行了介绍及对存在的问题进行了分析，并探索了废钢预热技术下一步的发展方向，对钢铁企业废钢预热技术的选择和发展具有一定的参考意义。

关键词：废钢预热；高炉-转炉流程；高废钢比；高温气体循环

Application and Development of Scrap Preheating Technology in Improving Scrap Ratio of Converter

HU Yanbin, HUANG Fang, HOU Zhongxiao, HE Sai, JIA Zhili

(Metallurgical Technology Institute, Central Iron and Steel Research Institute, Beijing 100081, China)

Abstract: In the process of oxygen converter smelting, the elements such as carbon and silicon in hot metal are oxidized and exothermic to provide the heat required for smelting. Due to the limitation of hot metal temperature, composition and its amount, the surplus heat of converter steelmaking is limited, and the proportion of scrap added is limited. Preheating of scrap can improve the physical heat of scrap itself and increase the capacity of molten pool to accommodate scrap, which is an important means to improve the scrap ratio of converter. This paper introduces the main preheating technology of scrap and analyzes the existing problems, and explores the development direction of scrap preheating technology in the next step. It has a certain reference significance for the selection and development of scrap preheating technology in iron and steel enterprises.

Key words: scrap steel preheating; BF-BOF process; high scrap ratio; high temperature gas circulation

废钢智能判级系统发展现状及未来趋势

赵东伟

（用友网络科技股份有限公司钢铁冶金行业事业部，北京 100094）

摘 要：废钢质量检验一直是困扰各大钢铁企业重要问题。为解决废钢验质过程中存在的主观因素影响大、判级过程难追溯、质量异议无法妥善解决等问题，基于人工智能技术的废钢智能判级系统近年来应运而生，并且在钢铁行业得到了极大的关注。作为新鲜事物，很多企业对废钢智能判级系统的理解不全面、不科学，甚至存在误解。本文基于对废钢智能判级系统的多年实践经验，重点阐述废钢智能判级系统的原理、技术难点、局限性以及给企业带来的价值和未来的发展趋势，希望为各位读者解开废钢智能判级系统的神秘面纱。

关键词：废钢；智能判级；发展现状；未来趋势

Development Status and Future Trend of Scrap AI Grading System

ZHAO Dongwei

(Yongyou Network Technology Co., Ltd. Steel Metallurgy Industry Division, Beijing 100094, China)

Abstract: The quality inspection of scrap steel has always been an important problem that troubles the major iron and steel enterprises. In order to solve the problems, including large subjective factors in the process of scrap quality inspection, difficult trace of the grading process, quality objections. The scrap AI grading system based on artificial intelligence technology has emerged in recent years, and has received great attention in the steel industry. As a new thing, many companies have an incomplete, unscientific and even misunderstood understanding of the scrap AI grading system. Based on several years of practical experience of scrap AI grading system, this paper focuses on the principle, technical difficulties, limitations, value brought to enterprises and future development trend. We hope to unlock the mystery of scrap AI grading system for readers.

Key words: scrap steel; AI grading; current status; development trend

基于机器视觉的废钢智能验质方法研究

徐文广[1]，肖鹏程[1,2,3]，王杏娟[1]

（1. 华北理工大学冶金与能源学院，河北唐山 063210；2. 北京科技大学冶金与生态学院，北京 100083；3. 河钢集团有限公司，河北石家庄 053000）

摘 要：废钢是一种可替代铁矿石的绿色资源，也是现代钢铁行业的重要原料。近年来，由于地条钢的取缔、社会钢铁蓄积量的增加以及再生钢铁原料进口政策的调整，我国废钢供给量逐年增多，为钢铁工业的低碳发展提供重要的资源保障。炼钢过程中，铁元素的收得率与废钢质量等级有着密切的关系。例如，在电炉冶炼中6mm以上的重型废钢收得率可达98%~99%，而3mm以下薄料型废钢收得率仅为90%左右。目前钢企对于废钢的分类评级仍以人工查验为主，主要由质量管理人员目测和卡尺测量共同进行判定，存在着危险性高、评级精度低、公正性易受到质疑等诸多困难。

随着未来我国废钢回收量和进口量的逐步提升，钢企废钢验收过程中质量评价难题将会进一步加剧。因此，开发基于机器视觉的废钢智能验质系统，是钢铁工业面对原料采购难题的迫切需要，是钢企原料管理数字化、智能化转型的必然途径。

本研究针对废钢种类多、实际检测情景复杂、人工系统衔接难度大等问题，以废钢验质过程中的公开、公正、公平为目标，提出基于机器视觉的CSBFNet废钢智能验质深度学习模型。在本模型中，将SE注意力机制加入特征提取网络，并将特征融合网络更换为双向特征金字塔模块BiFPN，使得在废钢类别特征提取和多尺度特征融合上具有显著优势，提升了网络的整体性能。实验先后在实验室数据集HK_L和现场采集的数据集HK_T上进行模型训练和优化。首先，搭建1:3废钢质量查验物理模型，采用高分辨率视觉传感器模拟采集货车卸载废钢作业场景下不同废钢的形貌特征。然后，利用该数据训练CSBFNet模型，获取特征信息，用于对各类废钢进行分类评级。最后，在中国某钢厂进行测试和改进废钢智能验质深度学习模型CSBFNet。结果表明，CSBFNet模型的各类废钢全类别平均准确率达到92.4%，mAP达到了90.7%，且各评价指标高于未加入SE注意力机制和BiFPN模块的模型，以及主流的目标检测模型Faster R-CNN、YOLOv4、YOLOv5系列、YOLOv7等。本模型不仅准确率高，而且检测速度高，对单张图片的检测时间仅需 11ms，能够满足实际生产场景下废钢分类评级的实时性要求，且具有较高的性能和优秀的验质效果。废钢智能验质模型 CSBFNet 与传统人工废钢验质相比，在准确性、安全性和公正性上有着明显优势。此外，废钢智能分类评级不仅影响废钢回收成本结算，更在对接电炉加料，改善电炉工艺制度方面有着积极影响。

参 考 文 献

[1] 上官方钦, 殷瑞钰, 李煜, 等. 论中国发展全废钢电炉流程的战略意义[J]. 钢铁, 2021, 56(8): 86-92.
[2] 朱立光, 陈泊羽, 肖鹏程, 等. 废钢料智能识别的语义分割模型选择与实现[J]. 冶金自动化, 2023, 47(3): 81-92.
[3] 肖鹏程, 徐文广, 常金宝, 等. 基于深度学习的废钢分类评级方法研究[J]. 工程科学与技术, 2023, 55(2): 184-193.
[4] 肖鹏程, 徐文广, 张妍, 等. 基于 SE 注意力机制的废钢分类评级方法[J]. 工程科学学报, 2023, 45(8): 1342-1352.
[5] XU W, XIAO P, ZHU L, et al. Classification and rating of steel scrap using deep learning[J]. Engineering Applications of Artificial Intelligence, 2023, 123: 106241.

镀锌废钢锌铁分离机理研究

朱正海[1]，孙前进[2]，骆小刚[2]，李 勇[2]，
邹海历[2]，殷志宏[1]，周 俐[1]

（1. 安徽工业大学冶金工程学院，安徽马鞍山 243002；
2. 欧冶链金再生资源有限公司，安徽马鞍山 243011）

摘　要：随着双碳工作的持续推进和国内钢铁蓄积量的不断上升，废钢在炼钢过程的使用占比不断增高，废钢相关的研究持续加强，其中镀锌废钢更是逐渐受到关注。

镀锌废钢在冶炼后形成的炉尘通常被送去烧结，但进入高炉后锌易在高炉富集，对高炉生产过程有着严重危害[1]。而随着镀锌钢板产量的逐年升高，金属锌的需求同时也不断提高，镀锌废钢中的锌具有巨大的回收循环价值。因此，镀锌废钢的锌铁分离研究具有重要意义。一些学者对炼钢厂含锌尘泥进行过研究，探索竖炉还原脱锌[2]、烧结过程锌还原[3]、还原焙烧再选[4]等工艺实现锌分离，也有湿法脱锌工艺的研究[5]，但在冶金火法过程对镀锌废钢直接进行锌铁分离的研究报道较少。

本研究针对镀锌废钢板，探索升温过程中表面锌层的形态及废钢基体的变化。在强氧化性条件下，镀锌板表面锌氧化严重，有明显鼓泡状的黄色氧化锌生成，随着温度上升氧化加剧；在弱氧化性条件下，升温过程镀锌板仍会被氧化，氧化锌出现较晚，后随温度升高逐渐减少，表面锌层减少；在真空条件下，随着温度升高表面锌含量不断降低，脱锌效果与一定温度下锌的蒸气压有关；在弱还原性条件下，随温度升高脱锌效果显著；在强还原性条件下，升温过程锌层液化后以蒸发方式逐步脱除。基于不同条件下的试验结果，分析得到废钢表面锌层和基体的变化机理，结合进一步试验对火法工艺锌铁分离的热力学和动力学影响因素进行了研究。由此，可根据不同现场情况，设计火法条件下不同的工艺，创造热力学和动力学条件，实现镀锌废钢的锌铁分离，提高镀锌废钢的资源利用效率。

参 考 文 献

[1] 朱斌, 周进东, 赵龚池, 等. 钢铁, 2020, 55(8): 130-139.
[2] Palimaka P, Pierazyk S, Stepien M. TMS: The Minerals, Metals & Materials Society, 2017: 181-189.
[3] 春铁军. 长沙：中南大学, 2014.
[4] 李辽沙, 李开元. 过程工程学报, 2009, 9(3): 468-473.
[5] Schönfelder I, Gock E, Vogt V, et al. Chemie Ingenieur Technik, 2012, 10(84): 1749-1756.

6 轧制与热处理

大会特邀报告
第十三届冶金青年科技奖获奖人特邀报告
分会场特邀报告
矿业工程
焦化及节能环保
炼铁与原料
炼钢与连铸
电冶金与废钢铁
★ 轧制与热处理
表面与涂镀
金属材料深加工
粉末冶金
先进钢铁材料及应用
节能与低碳技术
冶金环保与资源利用
冶金设备与工程技术
冶金自动化与智能化
冶金物流
冶金流程工程学
其他

6.1 板　　材

固溶处理工艺对奥氏体不锈钢 S31603 组织性能的影响

邢梦楠[1,2]，胡昕明[1,2]，姚　震[2]，欧阳鑫[1,2]，刘晨希[1,2]，王　储[1,2]

（1. 海洋装备用金属材料及其应用国家重点实验室，辽宁鞍山　114009；
2. 鞍钢股份有限公司，辽宁鞍山　114021）

摘　要：为了进一步研究固溶热处理与钢板轧制工艺与性能，采用连铸坯四阶段加热轧制工艺结合固溶热处理制备 316 系列 S31603 奥氏体不锈钢试样，采用光学显微镜和扫描电子显微镜（SEM）、透射电子显微镜（TEM）表征制备试样金相组织和微观形貌，用电子万能试验机测试试样力学性能。研究发现，增加钢板固溶温度和时间钢板强度下降，硬度值变化不大。因此采用连铸坯预热段 880℃，一加热段 1130℃，二加热段 1235℃，均热段 1195℃，总加热时间 6.5~7h 的四阶段加热，结合 1060℃×60min 固溶热处理的方式，钢板综合性能优异。试验钢板屈服强度为 264MPa，抗拉强度为 553MPa，延伸率为 61%，布氏硬度值为 141HBW。按照 GB/T 4334《金属和合金的腐蚀 不锈钢晶间腐蚀试验方法》中的 E 法对试验钢板进行晶间腐蚀性能试验，结果表明实验参数下得到的钢板晶间腐蚀性能优异。

关键词：S31603；奥氏体不锈钢；工艺；组织与性能

Effect of Solution Treatment on Microstructure and Properties of Austenitic Stainless Steel S31603

XING Mengnan[1,2], HU Xinming[1,2], YAO Zhen[2], OUYANG Xin[1,2], LIU Chenxi[1,2], WANG Chu[1,2]

(1. State Key Laboratory of Metal Material for Marine Equipment and Application, Anshan 114009, China;
2. Angang Steel Company Limited, Anshan 114021, China)

Abstract: In order to further study the process and properties of solid solution heat treatment and steel plate rolling, the 316 series of S31603 low carbon austenitic stainless samples were prepared by the four-stage heating rolling process of continuous casting billet combined with solid solution heat treatment. The microstructure and morphology of the samples were characterized by optical microscopy, scanning electron microscopy (SEM) and transmission electron microscopy (TEM). The mechanical properties of sample were tested by electronic universal testing machine. The results show that with the extension of the heating time of continuous casting billet, the steel plate strength increases slightly at first and then decreases. With the increase of solution temperature and time of steel plate, the strength of steel plate decreases and the hardness value changes little. Therefore, the three stages of continuous casting preheating stage 880℃, the first heating stage 1130℃, the second heating stage 1235℃, soaking stage 1195℃, total heating time 6.5~7h combined with 1060℃×60min solution heat treatment method, the comprehensive performance of the steel plate is excellent. The yield strength, tensile strength, elongation and Brinell hardness of the test steel plate are 264MPa, 553MPa, 61%, 141HBW and 306J of impact energy at 20℃. According to Method E in GB/T 4334 "Test Method for intergranular Corrosion of Stainless steel on Corrosion of Metals and Alloys", the intergranular corrosion properties of test plates were tested. The results showed that the steel plates obtained under the experimental parameters had excellent intergranular corrosion properties.

Key words: S31603; austenitic stainless steel; technology; organization and performance

提高重卷线薄规格运行速度和产能

黄海生

（新余钢铁集团有限公司，江西新余　338001）

摘　要： 从重卷线薄规格生产量增大的实际需求出发，解决重卷线薄规格产能不足的问题。对重卷线的设备运行状态、设备参数进行分析，在确保设备安全运行的情况下，不增加设备投入，在原有设备状态下，提出一种最大限度挖掘设备潜能的速度提升方案。通过消化吸收重卷线 PLC 控制和变频传动控制程序，优化 PLC 控制程序，优化变频传动控制参数，实现了机组最大运行速度从 300m/min 提升到 390m/min。经过 3 年多的运行跟踪，设备运行安全稳定。

关键词： 重卷线；速度；PLC；变频传动

热处理工艺对 690MPa 级容器钢组织转变的影响

刘晨希[1,2]，胡昕明[1,2]，欧阳鑫[1,2]，邢梦楠[1,2]，王　储[1,2]

（1. 海洋装备用金属材料及其应用国家重点实验室，辽宁鞍山　114009；
2. 鞍钢集团钢铁研究院，辽宁鞍山　114009）

摘　要： 随着大型乙烯球罐朝着大型化、高参数化和减量化发展，因此解决国家大型炼化乙烯项目重大技术装备材料供给难题。本文通过对实验钢进行淬火温度 820~970℃、保温时间 20min 的淬火工艺和回火温度 610℃、保温时间 20min 的回火工艺，并采用光学显微镜、扫描电镜研究实验钢在不同热处理工艺条件下的组织演变。结果表明，实验钢在进行 820~970℃ 的淬火热处理工艺，组织均由贝氏体和铁素体组成；实验钢经回火热处理工艺后，受组织遗传性影响，部分淬火组织未分解被保留下来。

关键词： 热处理；容器钢；贝氏体；力学性能；组织遗传性

Effect of Heat Treatment Process on Microstructure Transformation of 690MPa Vessel Steel

LIU Chenxi[1,2], HU Xinming[1,2], OUYANG Xin[1,2], XING Mengnan[1,2], WANG Chu[1,2]

(1. State Key Laboratory of Metal Material for Marine Equipment and Application, Anshan 114009, China;
2. Iron & Steel Research Institute of Angang Group, Anshan 114009, China)

Abstract: With the development of large ethylene spherical tanks towards large scale, high parameter and reduction, the problem of supplying major technical equipment materials for national large ethylene refining and chemical projects has been solved. In this paper,the microstructure evolution of experimental steel under different heat treatment conditions was studied by means of quenching process at 820℃ to 970℃ and holding time for 20min,tempering process at 610℃ and holding time for 20min.The results show that the microstructure of the experimental steel is composed of bainite and ferrite after quenching heat treatment at 820℃ to 970℃.After the tempering heat treatment,part of the quenched microstructure is retained due to the genetic effect of the microstructure.

Key words: heat treatment;container steel;bainite;mechanical property;tissue heritability

宝钢 5m 厚板轧机展宽 PVPC 设定模型优化

张敏文

（宝钢股份厚板事业，上海 201900）

摘 要：本文通过调查厚板轧制在无展宽 PVPC 条件下头部形状与过程参数的相关性，并在原有静态表的基础上建立外挂模型，纳入目标轧制厚度等必要变量，最终设计输出展宽 PVPC 设定深度，为减轻轧钢操作工手动设定工作量与成材率稳步提升准备先期工作。

关键词：厚板轧制；展宽 PVPC；深度

Design of Broadsizing PVPC Model Defect in 5m Heavy Plate Mill Baosteel

ZHANG Minwen

(Heavy Plate Mill Division Baosteel, Shanghai 201900, China)

Abstract: This paper investigates the correlation between the head shape and the process parameters of heavy plate rolling without broadsizing PVPC, and establishes a plug-in model on the basis of the original static table, including the necessary variables such as the target rolling thickness, and finally designs the setting depth of broadsizing PVPC, which is the preparatory work for reducing the manual setting workload of the rolling operator and steadily improving the yield.

Key words: heavy plate rolling; broadsizing PVPC; depth

S6 轧机工作辊常见失效分析与再生技术研究

张 青，侯兴慧，罗 昌

（宝钢轧辊科技有限责任公司，江苏常州 213019）

摘 要：本文通过对轧制高强钢的 S6 单机架轧机的国内外工作辊的不同失效方式进行分析和总结，检测对比了材质、硬度和金相等，成功开发和应用了新型高速钢材质适用于此轧机；同时针对主要失效方式，对工作辊结构进行优化设计，实现此类工作辊的绿色再生制造，大大降低了使用成本，获得了极高的性价比，彻底取代了进口轧辊。

关键词：高强钢；S6 轧机；失效分析；高速钢轧辊；再生技术

Research on Common Failure Analysis and Regeneration Technology of S6 Rolling Mill Work Rolls

ZHANG Qing, HOU Xinghui, LUO Chang

(Baosteel Roll Technology Co., Ltd., Changzhou 213019, China)

Abstract: In this paper, by analyzing and summarizing the different failure methods of domestic and foreign working rolls of the S6 single cold rolling mill for rolling high-strength steel, this paper detects and compares the material, hardness and metallographic phase, and successfully develops and applies a new high-speed steel material suitable for this rolling mill. At the same time, according to the main failure mode, the structure of the working rolls is optimized to realize the green recycling and manufacturing of such working rolls, which greatly reduces the cost of use, obtains a very high cost performance, and completely replaces the imported rolls.

Key words: high strength steel; S6 rolling mill; failure analysis; high speed steel rolls; regenerative technology

降低高强度热轧酸洗板材降级品率

李子文，杜宏松，卢　振，刘云飞，李艳杰，姬明利

（承德钒钛新材料有限公司板带事业部，河北承德　067002）

摘　要： 优质高强度的热轧酸洗板因从规格上可部分替代冷轧板，通过酸洗、镀锌、以热代冷实现效益最大化，受到广大客户青睐，市场前景广阔。但由于屈服强度高，热轧过程中轧制稳定性较差，带钢头尾跑偏刮撕甩尾就会造成后续带钢出现辊印、压痕缺陷，降级品率高严重影响产品效益提高。所以本文将系统总结高强度热轧酸洗板材产生辊印与压痕缺陷的原因，同时结合本单位轧制高强度酸洗系列产品降级品率的历史最好数据和行业先进水平进行比对，分析缺陷高发原因、制定解决措施，以降低缺陷发生率，进一步提高高强酸洗板材热卷表面质量，实现产品效益最大化。

关键词： 高强度；热轧酸洗板；辊印；压痕；轧制稳定性

Reduce the Degradation Rate of High Strength Hot Rolled Pickling Sheet

LI Ziwen, DU Hongsong, LU Zhen, LIU Yunfei, LI Yanjie, JI Mingli

(Chengde Vanadium Titanium New Material Co., Ltd., Chengde 067002, China)

Abstract: High quality and high strength hot rolled pickling plate can partially replace cold rolled plate from the specifications, through pickling, galvanized, hot instead of cold to maximize the benefit, favored by the majority of customers, broad market prospects. However, due to high yield strength and poor rolling stability in hot rolling process, strip head and tail deviation, scraping, tearing and throwing tail will lead to subsequent roll marks and indentation defects of strip steel, and high degraded product rate seriously affects the improvement of product benefits. Therefore, this paper will systematically summarize the causes of roll marks and indentation defects in high-strength hot rolling and pickling plates, and combine the best historical data of degraded product rate of high-strength pickling series products in our unit with the advanced level of the industry to compare, analyze the causes of high incidence of defects, and formulate solutions to reduce the incidence of defects and further improve the surface quality of high-strength pickling plates. Maximize product benefits.

Key words: high strength; hot-rolled pickling plate; roll printing; indentation; rolling stability

镀锌极限规格产品生产稳定性的研究与改进

张振峰，谢苏源，刘海超

（承德钒钛冷轧薄板有限公司，河北承德 067000）

摘　要： 根据镀锌产线极限规格产品需保证焊缝质量和钢带表面完整易镀锌的生产要求，对现场带钢跑偏、刮伤、卡阻等现象进行分析并改进，同时增加活套套量，用于增加入、出口段重焊和穿带时间，增强产线生产稳定性，满足产品的质量与产量要求。

关键词： 镀锌；极限规格；稳定性；活套套量

The Research and Improvement about the Production Stability of Galvanized Limit Specification Products

ZHANG Zhenfeng, XIE Suyuan, LIU Haichao

(Chengde Xinxin Vanadium and Titanium Co., Ltd., Chengde 067000, China)

Abstract: The products shall meet the production requirements of high welding quality and surface integrity of the steel strip for standard-compliant galvanization. Matters such as deviation, scuffing and obstruction during the production need to be solved and analyzed in line with the limit specification of the galvanizing production line, meanwhile to ensure the product quality and production requirements, the company has to increase the looper capacity and the stability of the production line as well as to free up additional time for the production processes of re-welding and coiling at the entry section and exit section.

Key words: galvanization; limit specification; stability; looper capacity

针对无头轧制产线的轧制模型开发和辊形配置研究

周冠禹[1]，何安瑞[2]，郭蓝田[1]，李朝阳[3]，李宏军[3]，郭智勇[3]

（1. 北京科技大学国家板带生产先进装备工程技术研究中心，北京 100083；
2. 北京科技大学钢铁共性技术协同创新中心，北京 100083；
3. 唐山东华钢铁企业集团有限公司钢轧厂，河北唐山 063300）

摘　要： 在国家"双碳"战略背景下，具有流程更加简约高效，资源利用率更高的无头轧制工艺受到了广泛关注。

无头轧制技术能够实现 1.0mm 以下超薄规格热轧带钢稳定生产，具有部分产品"以热代冷"的优势。连浇连铸连轧是无头轧制产线的鲜明特征，工序间紧密协同是保证无头轧制稳定进行的前提。就连轧部分而言，轧制模型参数精确设定和延长轧辊服役公里数是核心难点。因此，针对无头轧制产线开发了具备工业在线计算能力的辊系-轧件耦合计算模型和兼顾多目标控制的非对称工作辊辊形及其配套支撑辊辊形方案。具体研究结果如下：

（1）无头轧制产线具有大压下率、极薄终轧厚度等轧制工艺特点，使得其对数值仿真模型的求解能力提出了更高要求。基于此，通过构建轧制过程轧辊带钢接触界面控制方程，组成轧辊轧件变形的全局方程组，实现了辊系轧件模型的同步迭代，有效提升了模型的稳定性和计算速度。新模型的单工况求解耗时 40ms 左右，能够快速准确对无头轧制极薄规格轧制过程进行仿真模拟并达到了工业在线计算速度。

（2）为延长轧辊服役公里数，开发了具有多目标协同控制能力的非对称工作辊辊形（MOC-ATR）。MOC-ATR 辊形可分为边部直线段、抛物线辊形段以及一次曲线锥区段，其中边部直线段防止极薄规格的工作辊辊端压靠，抛物线段的辊形凸度设计值综合考虑了轧辊热膨胀和带钢目标凸度等指标，一次曲线锥区段配合在线窜辊实现了打开磨损凹槽功能。窜辊策略根据矩形磨损假设和轧辊半径磨损量确定，保证了磨损辊形的平滑控制，改善了轧制中后期的板形问题，工作辊轧制公里数平均提升了 15.7%。

（3）为了减小有害接触区，提高辊间接触压力均匀性，设计了与 MOC-ATR 辊形相配套的支撑辊辊形。基于智能寻优算法，联合辊系-轧件耦合模型，得到了不同工作辊辊形、不同窜辊位置下的综合辊间接触压力分布平滑性最优的 VCR+支撑辊辊形系数，上机后支撑辊磨损辊形更加平滑，并提升了轧机弯辊调控能力，支撑辊服役周期延长了 9.5%。

参 考 文 献

[1] 康永林. 轧钢, 2015, 32(1): 7-11.
[2] 田维兵. 轧钢, 2020, 37(2): 69-72.
[3] Yao C, He A, Shao J, et al. International Journal of Mechanical sciences, 2019, 159: 91-102.
[4] Li Z, Qi Z, Guo L, et al. Journal of Iron and Steel Research International, 2023, 30(2): 267-276.

冷轧板边部表面色差原因分析及改进措施

何士国，李江委，王 存，王 杰，刘旺臣，宋运涛

（鞍钢股份有限公司鲅鱼圈钢铁分公司，辽宁营口 115007）

摘 要：针对生产 st13 镀锌用冷轧板边部表面色差问题，通过热轧板和酸洗板表面粗糙度检测分析结果表明，热轧板边部与中间部位的表面粗糙度差异大是边部表面色差的主要原因。结合热轧板横断面的铁皮厚度，分析了钢板表面粗糙度差异大的原因。根据实际生产情况，调整冷轧末机架轧机负荷和优化板形，改善钢板表面的色差缺陷，通过合理匹配热轧卷取温度和层流冷却模式，解决了 st13 钢板表面的边部色差问题。

关键词：层流冷却；卷取温度；表面色差；粗糙度

Cause Analysis and Improvement Measures of Edge Surface Color Difference of Cold-rolled Sheet

HE Shiguo, LI Jiangwei, WANG Cun, WANG Jie, LIU Wangchen, SONG Yuntao

(Bayuquan Iron and Steel Branch Company of Angang Steel Co., Ltd., Yingkou 115007, China)

Abstract: Aiming at the problem of edge surface color difference of cold-rolled sheet used in the production of st13 galvanizing, the results of surface roughness detection and analysis of hot-rolled sheet and pickling sheet show that the big difference between edge surface roughness and middle surface roughness of hot-rolled sheet is the main reason for edge surface color difference. Combined with the thickness of iron sheet in the cross section of hot rolled sheet, the reasons for the large difference of surface roughness of steel sheet are analyzed. According to the actual production situation, adjusting the load of the cold rolling mill at the end of the cold rolling mill, optimizing the shape and improving the color difference defects on the surface of the steel plate, and solving the edge color difference problem on the surface of st13 steel plate by reasonably matching the hot rolling coiling temperature and laminar cooling mode.

Key words: laminar cooling; coiling temperature; surface color difference; roughness

550MPa级热轧高强集装箱板形翘曲的控制研究

李江委，王 刚，王 杰，任俊威，王 存，宋运涛

（鞍钢股份有限公司鲅鱼圈钢铁分公司，辽宁营口 115007）

摘 要：本文针对用户在使用550MPa级高强集装箱板过程中存在的翘曲问题，通过研究分析出了带钢内应力是导致钢板翘曲的主要原因。结合热轧产线工艺和设备装备的现状，通过增加立辊切水板、优化在线板形、改进层流冷却方式、开发高强集装箱的平整工艺等措施，高强集装箱板的翘曲问题得到了有效解决，满足了用户需求。

关键词：高强集装箱；板形；层流冷却；平整工艺

Cause Analysis and Control Measures of Flatness Warping of 550MPa Grade High Strength Container Steel

LI Jiangwei, WANG Gang, WANG Jie, REN Junwei, WANG Cun, SONG Yuntao

(Bayuquan Iron and Steel Branch Company of Angang Steel Co., Ltd., Yingkou 115007, China)

Abstract: In this paper, aiming at the warping problem existing in the process of using 550MPa grade high strength container plate, it is pointed out that the internal stress of strip steel is the main cause of steel plate warping. Combined with the present situation of the process and equipment of the hot rolling production line, by increasing the vertical roll water cutting plate, optimizing the on-line shape, improving the laminar cooling mode, developing the leveling process of high-strength container, and so on, the warping problem of high-strength container plate has been effectively solved to meet the needs of users.

Key words: high strength container steel; plate shape; laminar cooling; leveling process

冷轧某产品的缺陷发生率差异研究

谷开峰，郑 涛

（宝钢股份冷轧厂，上海 200941）

摘 要：近年来一些汽车板在最终工序完成后出现某种缺陷，这个缺陷是热轧产生的，受制于热轧工序质量改进的

限制，希望通过冷轧工序来消除或者减少这个缺陷。冷轧工序存在不同产线方向，缺陷发生率差异很大。本文根据不同方向产线和缺陷的对应关系进行研究和分析，梳理出不同产线之间的差异。对比不同方向产线酸洗使用盐酸腐蚀速率，发现了不同再生工艺去除废酸中金属离子能力不同，进而发现废酸中有铜离子并通过化学分析得出铜离子的来源。通过对比添加缓蚀剂后盐酸的腐蚀速率，来确定酸中的铜离子是否是不同方向产线缺陷发生率差异的主要原因。

关键词：热轧缺陷；酸洗；盐酸再生；铜离子

Analysis of Difference between Defect Occurrence Rate of a Cold-rolled Product

GU Kaifeng, ZHENG Tao

(Baoshan Iron & Steel Co., Ltd., Cold Rolling Plant, Shanghai 200941, China)

Abstract: In recent years, some car panels have a certain defect after the final production process, which is caused by hot rolling. Due to the limitations of quality improvement in the hot rolling process, this defect can be eliminated or reduced by the cold rolling process. The cold rolling process has different production line directions, and the defect occurrence rate varies greatly. This article analyzes the relationship between different production lines and defects, and summarizes the differences between production lines. By comparing the corrosion rate of hydrochloric acid used in different directional lines, we found that the ability of different recycling processes to remove metal ions in waste acid is various, and thus discovered the source of copper ions in waste acid through chemical analysis. By comparing corrosion rate of hydrochloric acid after adding corrosion inhibitors, we conclude that copper ions in acid are the main reason for the variation in defect occurrence rate among several directional lines.

Key words: hot rolling defects; pickling; HCL regeneration; Copper ions

浅谈砷元素对酸洗板表面的影响及改善措施

陈 锦

（宝武集团欧冶云商股份有限公司，上海 201900）

摘 要：砷（As）元素除了对钢材性能有影响，生产中还发现 As 对热轧酸洗带钢表面质量也有很大影响，会造成酸洗板表面出现明显的氧化铁皮斑迹缺陷。通过对斑迹部位进行能谱分析，发现 As 富集是造成斑迹缺陷的主要原因。为此，制定针对性的措施并实施之后，酸洗板表面斑迹缺陷大幅度改善。

关键词：砷（As）元素；热轧酸洗板；氧化铁皮斑迹；酸洗板表面质量

冷轧机组入口卸套筒功能改进

姜 雨，段晓溪，张旭亮，吴建学，祁 静

（首钢京唐钢铁联合有限责任公司冷轧作业部，河北唐山 063200）

摘　要：某冷轧厂机组入口生产 508mm 内径钢卷，执行卸尾卷或卸套筒自动步时，经常把开卷机芯轴上的橡胶套筒带出，导致卸套筒时间过长，对入口上卷时间产生严重影响，间接影响工艺能否正常运行。本文主要针对卸套筒功能进行分析研究，对该过程进行改进，使整个卸套筒过程稳定高效。

关键词：入口段；套筒；运卷小车

热镀锌光整机高压清洗系统改造与应用

姜　雨，陈　瑶，王　良，闫彤彤，
李文波，张森建

（首钢京唐钢铁联合有限责任公司冷轧作业部，河北唐山　063200）

摘　要：热镀锌机组光整机高压清洗对于带钢质量有着十分重要的作用，是保证生产质量不可或缺的一部分。某冷轧厂高压清洗泵为进口设备，维护成本高，并且使用时间较长状态呈现不稳定趋势，频繁出现泵体漏水、压力不稳等运行问题，严重影响产品质量。本文主要针对高压清洗系统进行分析研究，对该系统进行国产化改造，使整个系统稳定得了有效提升。

关键词：高压清洗；电气控制；压力

基于全面自动控制提升的连续退火模型功能优化研究

宋利伟[1,2]，孙荣生[1,2]，蔡顺达[1,2]，
刘军友[3]，洪天生[3]，姜丽丽[3]

（1. 海洋装备用金属材料及其应用国家重点实验室，辽宁鞍山　114009；
2. 鞍钢集团钢铁研究院，辽宁鞍山　114009；3. 鞍钢股份冷轧厂，辽宁鞍山　114021）

摘　要：连续退火机组退火数学模型是连退机组关键控制系统，承担着连退机组生产过程控制指挥中枢的作用。连退数学模型的自动化程度及控制精度，对机组生产自动化率、生产效率及产品质量稳定性均有重要影响。现有连续机组退火数学模型功能主要集中在对炉区加热段、冷却段带钢温度和区域温度的控制方面，而对于连退炉相关的光整机区域等缺少系统的、高精度高自动化水平的控制方案。同时，因机组产品品种结构的不断升级，新钢质、新工艺的不断推出，既有数学模型在退火温度及速度控制精度及响应灵敏度方面已经逐渐显现出不足。本文通过基于全面自动控制水平提升的研究，对鞍钢既有连退机组数学模型功能进行了优化，显著提升了退火模型自动化率及控制精度，提升机组生产效率的同时，实现产品性能实物质量的稳步提高。

关键词：连续退火；数学模型；光整自动化；效率提升

Research on Function Optimization of Continuous Annealing Model Based on Full Automatic Control Lifting

SONG Liwei[1,2], SUN Rongsheng[1,2], CAI Shunda[1,2],
LIU Junyou[3], HONG Tiansheng[3], JIANG Lili[3]

(1. State Key Laboratory of Metal Material for Marine Equipment and Application, Anshan 114009, China;
2. Iron and Steel Research Institute of Angang Group, Anshan 114009, China;
3. Cold Rolling Plant of Ansteel, Anshan 114021, China)

Abstract: The annealing mathematical model of continuous annealing unit is the key control system of continuous annealing unit, and it plays the role of the command center of continuous annealing unit's production process control. The automation degree and control precision of the mathematical model have an important influence on the automation rate, production efficiency and product quality stability of the unit. The functions of the existing annealing mathematical model of continuous unit mainly focus on the control of the temperature of the strip steel in the heating section and the cooling section of the furnace area and the regional temperature, while there is a lack of systematic, high-precision and high-automation control scheme for the annealing mill area related to the continuous annealing furnace. At the same time, due to the continuous upgrading of the product variety and structure of the unit, and the continuous introduction of new steel and new technology, the existing mathematical models have gradually shown deficiencies in the control accuracy and response sensitivity of annealing temperature and speed. In this paper, based on the research of overall automatic control level improvement, the mathematical model function of Angang's existing continuous annealing unit was optimized, which significantly improved the automation rate and control precision of annealing model, improved the production efficiency of the unit, and realized the steady improvement of product performance and physical quality.

Key words: continuous annealing; mathematical model; automation of finishing; efficiency improvement

低合金钢带钢辊弯开裂原因分析与热轧工艺控制

江 傲，程志诚，余 伟

（北京科技大学工程技术研究院，北京 102206）

摘 要： 辊压或冲压成形过程中材料断裂形成的裂纹是严重的加工缺陷之一，不仅影响材料的利用率，还影响制造产品服役的安全性。提高钢材的深加工性能是结构材料性能控制的重要方向。控轧控冷工艺是提高钢材强度和韧性的重要手段，但如何同时控制钢材的成形性能仍需要探索。

针对采用控轧控冷工艺生产的屈服295MPa级低合金钢板在辊弯成形过程中出现边部开裂现象，采用金相分析、弯曲试验、力学性能测试和扫描电镜能谱（EDS）分析等方式，对冷弯钢板母材不同部位的组织性能进行测试分析，以确定导致钢材辊弯成形开裂的原因。

相对钢板中部区域而言，钢板边部的纵向抗拉强度和抗弯强度高出4%和6%，平均硬度较其他位置硬度提高约15%，低应变区（0.025）边部加工硬化率提高约12%。令人惊奇的是其断后伸长率却提高了13%。从微观组织看，钢板均由铁素体相和珠光体相构成。经测量统计，边部区域铁素体平均晶粒尺寸约为2.61μm，钢板横向其余部位约为5.35μm。在珠光体形态上，边部珠光体内部渗碳体颗粒呈现球状或小块状，且分布较为弥散，而其余部位珠光体多呈现片层状分布。采用图像法统计珠光体面积百分数,边部区域约为 51%,其余部位面积约为 25%,边部珠光体较中部明显增多。钢板边部区域强度偏高的原因是铁素体晶粒细小、珠光体面积百分比高;而细小的铁素体晶粒与球化渗碳体形态的珠光体,增加了钢的塑性变形能力,导致边部区域延伸率增加。进一步对辊弯开裂部位的裂纹源及裂纹扩展路径进行形貌观察,发现裂纹扩展路径总是起源于晶界渗碳体,尤其是多沿着珠光体相与铁素体基体相的交界处萌生。主裂纹附近的微裂纹内部还发现存在细小的 MnS 夹杂。

边部呈粒状和块状的异常珠光体通常称为退化珠光体。热轧带钢层流冷却过程中,边部温度较中部区域低,快速进入低于 A_{r1} 温度,抑制了奥氏体向先共析铁素体的转变,大量过冷残余奥氏体最终形成了后续的退化珠光体,导致了钢板边部的高强度与高硬度。虽然边部区域的塑性更高,但是晶粒细化和珠光体面积百分数增加,提高了钢材的加工硬化率,铁素体晶界的位错塞积更容易导致裂纹沿铁素体晶界的渗碳体界面扩展,使得钢板边部区域更易出现成形裂纹。因此,要提高钢板整体的成形性能,可以考虑适当增大晶粒尺寸、改变渗碳体形态,降低屈服强度和抗拉强度。在热轧带钢生产加工过程中,提高热轧和层流过程钢板的温度均匀性和卷取温度是可供选择的控制策略。

Graphic abstract:

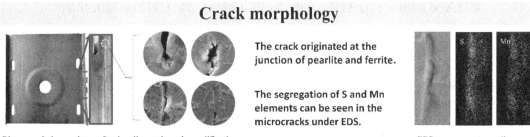

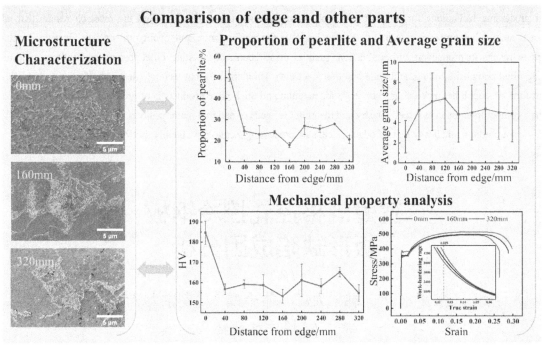

中厚板产品钢坯处理技术的研究与应用实践

张 健,黄 健,孙殿东,陈军平

(鞍钢股份有限公司,辽宁鞍山 114000)

摘　要： 随着中国冶金工业的发展，中厚板产品的需求和质量要求越来越高，特别是对于一些特殊使用环境下的关键装备材料，提出更严苛的要求，钢坯处理技术是保证中厚板产品品质和性能的关键环节，本文主要阐述了钢坯处理技术在中厚板产品中的重要性，并以鞍钢为例，介绍了其在中厚板产品钢坯处理方面的研究与应用。例如采用连铸坯锻轧、连铸坯复合轧制以及宽薄钢板叠轧等技术，从研究成果中可以看出，采用不同的钢坯处理技术可以提高钢铁产品的性能和质量，并在生产和应用中取得优异的效果。因此，研究和应用中厚板产品钢坯处理技术对于提升我国钢铁行业竞争力和满足国家重大项目需求具有重要的意义。

关键词： 中厚板产品；钢坯处理技术；品质；性能

Research and Application of Steel Slab Processing Technology for Medium and Heavy Plate Products

ZHANG Jian, HUANG Jian, SUN Diandong, CHEN Junping

(Angang Steel Co., Ltd., Anshan 114000, China)

Abstract: With the development of China's metallurgical industry, the demand and quality requirements for medium and thick plate products are increasing, especially for some key equipment materials under special usage environments that require stricter requirements. The steel billet processing technology is a key link to ensure the quality and performance of medium and thick plate products. This article mainly expounds the importance of steel billet processing technology in medium and thick plate products, and takes Angang Steel as an example to introduce its research and application in steel billet processing technology for medium and thick plate products. It can be seen from the research results that using different steel billet processing technologies can improve the performance and quality of steel products and achieve excellent results in production and application, such as using continuous casting billet forging and rolling, continuous casting billet composite rolling, and wide and thin steel plate laminating rolling technologies. Therefore, the research and application of steel billet processing technology for medium and thick plate products is of great significance for enhancing China's competitiveness in the steel industry and meeting the needs of national major projects.

Key words: medium and heavy plate products; steel slab processing technology; quality; performance

宽薄规格控轧控冷钢板板形缺陷成因分析

张建平[1,2]，韩　旭[3]，韩千鹏[3]，庞宗旭[1,2]，
黄　健[1,2]，李新玲[3]

（1. 海洋装备用金属材料及其应用国家重点实验室，辽宁鞍山　114009；
2. 鞍钢集团钢铁研究院，辽宁鞍山　114009；
3. 鞍钢股份公司鲅鱼圈分公司，辽宁营口　115007）

摘　要： 为探明宽薄规格控轧控冷态钢板轧制、冷却和剪切后板形平直度较差的原因，通过生产跟踪和数据分析，明确了板形缺陷主要产生工序。利用手持式热成像仪测量了钢板冷却和剪切过程表面温度分布，验证了温度分布与板形缺陷的对应关系。结果表明，轧机刚度影响轧制板形控制，辊道冷却水使钢板产生周期性垂直于轧制方向的低温带，层流冷却水残留容易使钢板表面产生局部大温差区，这些局部大温差区是钢板在冷床冷却过程变形的主要发

生位置，精整剪切位置温差较大是钢板剪切后发生瓢曲变形的主要原因。根据分析结果，提出了相关控制和优化措施，相关措施实施后，板形合格率大幅提升。

关键词：宽薄规格中厚板；板形；层流冷却；温度分布

Analysis of Shape Defects of Wide and Thin Gauge Controlled Rolling and Controlled Cooling Steel Plate

ZHANG Jianping[1,2], HAN Xu[3], HAN Qianpeng[3], PANG Zongxu[1,2], HUANG Jian[1,2], LI Xinling[3]

(1. State Key Laboratory of Metal Material for Marine Equipment and Application, Anshan 114009, China; 2. Ansteel Group Iron and Steel Research Institute, Anshan 114009, China; 3. Angang Steel Company Limited, Bayuquan Iron & Steel Subsidiary, Yingkou 115007, China)

Abstract: In order to find out the reason of poor flatness after rolling, cooling and shearing of wide and thin gauge controlled rolling and controlled cooling steel plate, the main process of flatness defect was defined through production tracking and data analysis. The temperature distribution on the surface of steel plate during cooling and shearing process was measured by a hand-held thermal imager, and the corresponding relationship between temperature distribution and shape defect was verified. The results show that the mill stiffness affects the shape control of the rolling plate, the cooling water of the roller table causes the steel plate to produce the low temperature strip which is periodically perpendicular to the rolling direction, and the residual laminar cooling water easily causes the local large temperature difference on the surface of the steel plate, these local large temperature difference areas are the main places where the steel plate deforms during the cooling process in the cooling bed. According to the analysis results, the relevant control and optimization measures are put forward. After the relevant measures are implemented, the qualified rate of flatness is greatly increased.

Key words: wide and thin gauge plate; plate shape; laminar cooling; temperature distribution

临界退火温度对中锰钢奥氏体稳定性和力学性能的影响

张哲睿[1,2]，任毅[1,2]

（1. 海洋装备用金属材料及其应用国家重点实验室，辽宁鞍山 114009；
2. 鞍钢钢铁研究院，辽宁鞍山 114009）

摘 要：本文设计了一种新型 Fe-6Mn-4Al-0.4C 高强度中锰热轧钢板。采用 SEM、XRD、EBSD、TEM 研究了临界间退火工艺对实验钢组织演变和力学性能的影响。实验钢分别在 640℃、680℃、720℃、760℃、800℃下退火 30min。当临界间退火温度为 640℃时，渗碳体颗粒在 γ/α 相界面间大量析出。随着退火温度的升高，拉伸强度降低，伸长率提高。渗碳体逐渐溶解消失，片层奥氏体含量显著增加。实验钢在 760℃保温 30min 后表现出优异的综合力学性能，抗拉强度为 978MPa，屈服强度为 743MPa，总伸长率为 73.5%，强度与伸长率之积为 71.9GPa·%。合理的退火温度与时间可以有效调控奥氏体含量、稳定性与形貌从而改善力学性能。

关键词：Fe-Mn-Al-C；临界退火；拉伸性能；奥氏体稳定性

Effect of Critical Annealing Temperature on Austenite Stability and Mechanical Properties of Medium Manganese Steel

ZHANG Zherui[1,2], REN Yi[1,2]

(1. State Key Laboratory of Metal Material for Marine Equipment and Application, Anshan 114009, China;
2. Ansteel Iron & Steel Research Institute, Anshan 114009, China)

Abstract: In this study, a new Fe-6Mn-4Al-0.4C high strength medium manganese hot rolled steel sheet was designed. The effects of the intercritical annealing process on microstructure evolution and mechanical properties of experimental steel were studied by SEM, XRD, EBSD and TEM. The experimental steel was annealed for 30 minutes at 640℃, 680℃, 720℃, 760℃, 800℃, respectively. When the intercritical annealing temperature was 640℃, cementite particles substantially precipitated between the γ/α phase interface. As the annealing temperature increased, the tensile strength decreased and the elongation increased. The cementite gradually dissolved and disappeared, the fraction of lamellar austenite increased significantly. The experimental steel exhibited excellent comprehensive mechanical properties after held at 760℃ for 30min. The ultimate tensile strength was 978MPa, the yield strength was 743MPa, and the total elongation was 73.5%, the product of strength and elongation was 71.9GPa•%. Reasonable annealing temperature and time can effectively control austenite content, stability and morphology to improve mechanical properties.

Key words: Fe-Mn-Al-C; intercritical annealing; tensile property; austenite stability

酸洗段圆盘剪剪切质量的控制

毛玉川[1], 孙荣生[2], 刘英明[1], 张一凡[1], 蔡顺达[2], 王金星[1]

（1. 鞍钢股份有限公司冷轧厂, 辽宁鞍山 114021; 2. 鞍钢集团钢铁研究院, 辽宁鞍山 114009）

摘　要： 本文主要在详细分析酸轧机组工艺段圆盘剪切变形过程及剪切断面的基础上，对圆盘剪剪切工艺参数进行了深入的研究，并通过对现场实际情况的跟踪，最终提出了不同材质厚度带钢对应的工艺参数，提高了酸洗段圆盘剪剪切质量，保证了轧后带钢的边部质量。

关键词： 圆盘剪；剪切质量；工艺参数

The Control of the Cutting Quality of Rotary Shear on the Pickling Section

MAO Yuchuan[1], SUN Rongsheng[2], LIU Yingming[1], ZHANG Yifan[1], CAI Shunda[2], WANG Jinxing[1]

(1. Cold Rolled Strip Steel Mill of Angang Steel Co., Ltd., Anshan 114021, China;
2. Ansteel Iron and Steel Research Institutes, Anshan 114009, China)

Abstract: The deforming process and cutting profile of the rotary shear on the pickling line was analyzed detailed in this paper, and the rotary shear cutting technology parameter was researched deeply, through tracking the fact on the workshop, the optimal technology parameter of the different thickness and grade strip for control the cutting quality was brought at last, the cutting quality of the pickling line was heightened, the border quality was ensured after the rolling.

Key words: rotary shear; cutting quality; technology parameter

不锈钢复合板 316L+Q500qE 的组织与性能的研究

李中平[1]，周文浩[1]，史术华[1]，王智聪[2]，
罗　登[1]，熊祥江[1]，范　明[1]

（1. 湖南华菱湘潭钢铁有限公司，湖南湘潭　411101；
2. 湖南浮胜金属新材有限公司，湖南湘潭　411101）

摘　要： 采用真空复合方法制复合板坯，利用 TMCP+T 工艺生产 316L+Q500qE 复合板，通过光学显微镜、力学性能测试、晶间腐蚀、热矫试验和焊接接头试验等手段研究了复合板的组织、性能、可焊性等。结果表明真空复合的不锈钢 316+LQ500qE 桥梁板冶金结合完整，复合板的界面结合牢固，剪切强度达到 430MPa；复合钢板 316+LQ500qE 性能均匀稳定，低温冲击韧性优良；复合钢板具有良好的可焊性，焊缝外观质量、内部质量和接头力学性能均满足 GB/T 13148—2008《不锈钢复合钢板焊接技术要求》、GB/T 16957—2012《复合钢板焊接接头力学性能试验方法》标准的要求。

关键词： 复合板；316L；Q500qE；真空复合

Microstructure and Properties of 316L + Q500qE Stainless Steel Clad Plate Researching

LI Zhongping[1], ZHOU Wenhao[1], SHI Shuhua[1], WANG Zhicong[2],
LUO Deng[1], XIONG Xiangjiang[1], FAN Ming[1]

(1. Xiangtan Iron & Steel Group Co., Ltd., Xiangtan 411101, China;
2. Hunan Fusheng Metal New Material Co., Ltd., Xiangtan 411101, China)

Abstract: The 316L+Q500qE composite plate was produced by vacuum composite method and TMCP + tempering process. The microstructure, properties and weldability of the composite plate were studied by optical microscope, mechanical property test, intergranular corrosion test, hot straightening test and welding joint test. The results show that the metallurgical bonding of 316L+Q500qE stainless steel bridge plate in vacuum is complete, the interface bonding of the composite plate is firm, and the shear strength reaches 430MPa. The composite steel plate Q500qE+316L has uniform and stable performance and excellent low-temperature impact toughness; The clad steel plate has good weldability, and the appearance quality, internal quality and mechanical properties of the weld meet the requirements of GB/T 13148—2008 Technical Requirements for Welding of Stainless Clad Steel Plates and GB/T16957—2012 Test Method for Mechanical Properties of Clad Steel Plate Welded Joints.

Key words: clad plate; 316L; Q500qE; vacuum recombination

退火温度对 1500MPa 冷轧 Q&P 钢组织性能的影响

蔡顺达[1,2]，孙荣生[1,2]，宋利伟[1,2]，刘军友[3]，阮国庆[3]，钟莉莉[1,2]

（1. 海洋装备用金属材料及其应用国家重点实验室，辽宁鞍山 114009；2. 鞍钢集团钢铁研究院，辽宁鞍山 114009；3. 鞍钢股份冷轧厂，辽宁鞍山 114021）

摘　要：通过三种不同的退火工艺参数的变化研究了 1500MPa 级淬火配分（Quenching and Partitioning，简称 Q&P）钢的组织、残余奥氏体含量及力学性能的变化规律，不同退火工艺参数下的试验钢均由铁素体、马氏体和残余奥氏体组成，但各相的含量和晶粒尺寸存在较大差异。结果表明随着退火温度的提升，延伸率和屈服强度随之逐步下降，但抗拉强度在临界区退火温度下逐步降低，在全奥氏体化退火温度得到提升。力学性能主要受相比例和细晶强化作用共同影响。

关键词：退火温度；1500MPa 级 Q&P 钢；组织性能

Effects of Annealing Temperature on Microstructure and Property of Cold-rolled Q&P Steel with Tensile Strength of 1500MPa

CAI Shunda[1,2], SUN Rongsheng[1,2], SONG Liwei[1,2], LIU Junyou[3], RUAN Guoqing[3], ZHONG Lili[1,2]

(1. State Key Laboratory of Metal Material for Marine Equipment and Application, Anshan 114009, China; 2. Ansteel Iron & Steel Research Insititute, Anshan 114009, China; 3. Ansteel Coldronnling Plant, Anshan 114021, China)

Abstract: Three different annealing processes were applied to the quenching and partitioning (Q&P) steel with tensile strength of 1500MPa, the microstructure, retained austenite fraction and mechanical property were analyzed. The microstructure composed of ferrite, martensite and retained austenite was obtained, the content and grain size of each phase were completely different. Consequently, the elongation decreased, yield strength increased, the tensile strength first invarinted and then increased with the increasement of annealing temperature. The cooperation of phase fraction and refined strengthening have primary influence on mechanical property.

Key words: annealing temperature; 1500MPa Q&P steel; microstructure and properties

回火工艺对低合金高强耐磨钢组织与性能的影响

王万慧，夏强强，韩会全，辜蕾钢，刘显军

（中冶赛迪工程技术股份有限公司钢铁事业本部，重庆 400013）

摘　要：本文研究了回火温度和回火时间对低合金高强耐磨钢组织与性能的影响。通过拉伸性能、显微组织、布氏硬度分析，建立了回火温度和回火时间对耐磨钢组织和性能的影响规律。结果表明，回火温度对耐磨钢的性能影响远远高于回火保温时间；当回火温度为200℃时，回火保温时间对实验钢组织性能的影响不大。回火温度较低时（≤400℃），显微组织均以回火马氏体为主，而当回火温度较高时（>400℃），显微组织出现少量的再结晶铁素体，并形成拉伸屈服平台。随着回火温度的升高，耐磨钢的强度和硬度逐渐降低，伸长率变化不大，回火温度高于350℃，其布氏硬度值下降明显。随着回火保温时间的延长，耐磨钢强度和延长率小范围波动，变化不明显。

关键词：耐磨钢；回火温度；回火时间；马氏体；拉伸性能

Effect of Tempering Process on Microstructure and Properties of Low-alloy High-strength Wear-resistant Steel

WANG Wanhui, XIA Qiangqiang, HAN Huiquan, GU Leigang, LIU Xianjun

(CISDI Engineering Co., Ltd., Chongqing 400013, China)

Abstract: The influence of tempering temperature and time on the microstructure and properties of low-alloy high-strength wear-resistant steel was studied in this paper. Besides, the relationship between tempering temperature and time on the microstructure and properties of wear-resistant steel was established by analyzing the tensile properties, microstructure, hardness. The results show that the influence of tempering temperature on the properties of wear-resistant steel was much higher than that of tempering time. The influence of tempering holding time on the microstructure and properties of experimental steel was little when the tempering temperature was 200℃. In addition, when the tempering temperature was low (≤400℃), the microstructure was mainly tempered martensite, while a small amount of recrystallized ferrite appeared and formed a tensile yield platform when the tempering temperature was high (>400℃). With the increase of tempering temperature, the strength and hardness decreased gradually, and the elongation feebly changed. The hardness decreased obviously when tempering temperature was higher than 350℃. The strength and elongation of wear-resistant steel fluctuated in a small range with the tempering time, but the change was not obvious.

Key words: wear-resistant steel; tempering temperature; tempering time; martensite; tensile property

家电彩涂板表面麻点缺陷产生原因分析及预防措施

汤晓东，谢义康，朱彰荣，吴银义，郑 磊

（马鞍山钢铁股份有限公司冷轧总厂，安徽马鞍山 243000）

摘 要：本文主要针对马钢2号彩涂线在生产一种抗刮面家电彩涂板过程中出现的周期性麻点缺陷，通过微观形貌、能谱、红外及结合现场实际分析了其产生的原因，并通过底漆稀释剂型号更改、现场黏度、辊速以及过滤袋目数调整，使麻点缺陷得到了有效改善。

关键词：家电彩涂板；麻点；产生原因；预防措施

Analysis of the Causes and Preventive Measures for the Surface Pitting Defects of Household Appliances

TANG Xiaodong, XIE Yikang, ZHU Zhangrong, WU Yinyi, ZHENG Lei

(Cold Rolling Mill of Maanshan Iron & Steel Co., Ltd., Maanshan 243000, China)

Abstract: This paper mainly aims at the periodic pitting defects of Masteel 2# color coating line in the process of producing a kind of anti-scratch color coating board for home appliance. The causes are analyzed through the microscopic morphology, energy spectrum, infrared and combined with the actual situation on the site, and through the change of primer dilution dosage form number, field viscosity, roll speed and filter bag mesh adjustment, so that the pitting defects have been effectively improved.

Key words: household appliances; pitting; cause; preventive measure

可实现轴向变张力轧制的张力辊系统

马晓宝，彭永恒，李 鹏，王 涛，陈 鹏

（太原理工大学机械与运载工程学院，山西太原 030024）

摘 要： 精密薄带属于板带材的高端产品，是航空航天、军工核电、电力电子、能源医疗等领域的关键基础材料。目前精密薄带轧制过程总是存在四分浪等复杂板形问题，带材越薄，板形问题越严重，成为制约精密薄带产业发展的重要瓶颈。

不同类型的轧机主要通过调整弯辊力、轧辊横移、倾辊、轧辊分段冷却、大张力轧制等进行板形控制，然而弯辊力、轧辊横移对板形控制作用主要体现在边部，倾辊主要控制板带楔形，分段冷却主要控制高次浪形，且属于近稳态调控过程，实时性差，大张力有利于减小轧制力，但对局部板形的精细控制能力差，因而各个板形控制手段在板带宽度方向的作用域有限。为解决冷轧精密薄带轧制中存在的板形问题以及平直度和横断面轮廓调控的耦合现象，迫切需要开发新型板形控制手段。

张力是影响轧制力分布的重要工艺参数，是板形控制最直接、最有效的控制手段之一。常规轧制过程张力沿宽度方向分布均匀，但带材变形形成内部残余应力，造成轧制过程张应力沿横向分布是不同的，对精密薄带的纵向拉伸及产生的横向拉压作用不同，对四分浪等复杂板形的形成有重要影响，构造横向变张力，有望校正变形区轧制压力，影响金属塑性变形。可实现轴向变张力轧制的张力辊系统，在张力辊芯轴外侧设有轴向分段布置的可膨胀囊以支撑板带，通过在张力辊芯轴两端的回转接头连接张力辊芯轴内部多条孔道，实现与可膨胀囊连通，使用液压控制阀组控制各段囊体内部压力。工作时，通过调控各段可膨胀囊内部压力，实现沿张力辊轴向施加不同支撑力，进而在板带宽度方向构造不同张应力。该装置可以用作卷取导向辊、张力辊组，通过对板带构造不同支撑反力，实现轧制过程板带宽度方向变张力轧制，提高张力对板形的控制能力和灵活度，有望为精密带高效轧制提供更加强大的板形控制手段。

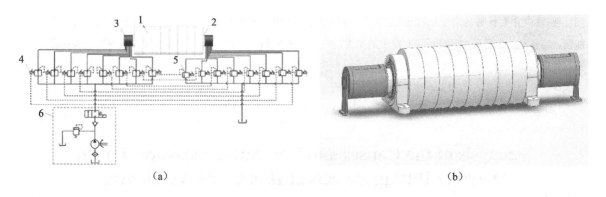

图1 张力辊系统（a）和张力辊系统的三维结构（b）

1—可膨胀囊；2—张力辊芯轴；3—回转接头；4，5—控制阀组；6—压力源

1420UCM 五机架连轧机振动抑制措施研究

黄 勇，龚 艺，张东方，胡岸松，陈立钢

（武钢日铁镀锡板(武汉)有限公司，湖北武汉 430083）

摘 要：通过对 FAT 期间轧机振动工艺调整过程的总结，发现 1420UCM 连轧机振动的特点，其振动主要发生在第四机架的辊缝处；并从通过提高辊缝润滑，降低轧制力可以抑制轧机振动，使得轧制镀锡板 T5 牌号速度最高可达 2000m/min。

关键词：轧机振动；辊缝润滑；UCM 轧机

Analyzing Reason and Controlling Measure of the Vibration in 1420 UCM Cold Continuous Rolling Mill

HUANG Yong, GONG Yi, ZHANG Dongfang, HU Ansong, CHEN Ligang

(Baowu Steel Group WISCO-NIPPON Steel Tinplate Co., Ltd., Wuhan 430083, China)

Abstract: by summarizing the process of rolling mill vibration process adjustment during fat, it is found that the vibration characteristics of 1420UCM continuous rolling mill mainly occur at the roll gap of the fourth stand; The rolling mill vibration can be restrained by improving the roll gap lubrication and reducing the rolling force, so that the maximum speed of rolling T5 tinplate can reach 2000m/min.

Key words: rolling mill vibration; roll gap lubrication; UCM rolling mill

减少热轧带钢头尾不规则形状延伸技术研究与实践

曾龙华，赵金凯，王 波

（湛江钢铁有限公司热轧厂，广东湛江 524072）

摘 要：本文阐述湛钢热轧厂通过研究热轧带钢轧制过程头尾不规则形状形成机理，试验不同工艺参数验证头尾不规则延伸长度，得出最优化参数比优化头尾不规则形状，改善了带钢轧制过程轧制稳定性，减少头尾不规则形状延伸切损量 20%，使湛钢热轧轧线切损率在侧压量超 150mm 情况下"破四进三"实现 0.38%水平。

关键词：热轧带钢；不规则形状；切损量

Research and Practice on Reducing Irregular Shape Extension of Hot Rolled Strip

ZENG Longhua, ZHAO Jinkai, WANG Bo

(Hot Rolling Mill of Zhanjiang Iron & Steel Co., Ltd., Zhanjiang 524072, China)

Abstract: By studying the formation mechanism of the irregular shape of the head and tail during hot rolled strip rolling and testing different process parameters to verify the irregular extension length of the head and tail, it is concluded that the optimized parameters can improve the rolling stability of the strip rollingprocess andreducethecutloss oftheirregularshapeof theheadandtailby 20%. The cuttingrate of hot rolling line of Zhan steel is broken into four and three to achieve 0.38% level.

Key words: hot rolled strip steel; irregular shape; cut loss

热轧钢卷端部缺陷自动识别技术研究与实践

曾龙华,高佳瑞

(湛江钢铁有限公司热轧厂,广东湛江　524072)

摘　要: 本文阐述湛钢热轧厂为突破带钢端部缺陷全自动识别技术难题,通过对热轧带钢端面缺陷特征进行分析,建立适应的模型优化迭代,攻克因缺陷缺陷小、种类多、光线暗等难识别技术难点,并构建端部缺陷全自动识别流程框架,实现了缺陷低时延识别判定技术,使得热轧带钢端部缺陷全自动识别技术成功在湛钢热轧厂两条产线应用落地,取代原有人工实时监控作业,提高缺陷识别准确率至98%以上,降低人工劳动负荷70%以上。

关键词: 带钢端面；模型；自动识别

Research and Practice of Automatic Defect Identification Technology of Hot Rolled Steel Coil End

ZENG Longhua, GAO Jiarui

(Hot Rolling Mill of Zhanjiang Iron & Steel Co., Ltd., Zhanjiang 524072, China)

Abstract: In order to break through the technical problems of automatic identification of strip end defects in Zhan Steel hot rolling plant, by analyzing the defect characteristics of hot rolled strip end faces, an adaptive model optimization iteration is established to overcome the difficult technical difficulties of identification due to small defects, many types and dark light, and the automatic identification process framework of end defects is constructed to realize the low-delay identification and judgment technology of defects. The automatic defect identification technology of hot rolled strip end has been successfully applied in two production lines of Zhan Steel hot rolling plant, replacing the original manual real-time monitoring operation, improving the accuracy rate of defect identification to more than 98%, and reducing the manual labor load by more than 70%.

Key words: strip end face; the model; automatic recognition

退火工艺对高强 DR 材组织性能的影响

孙超凡[1,2]，方 圆[1,2]，孙 晴[3]，王雅晴[1,2]

（1. 首钢集团有限公司技术研究院，北京 100043；
2. 绿色可循环钢铁流程北京市重点实验室，北京 100043；
3. 首钢京唐钢铁联合有限责任公司镀锡板事业部，河北唐山 063200）

摘 要：镀锡板是一类重要的金属包装材料，在食品、饮料、化工以及电子元件加工制造领域均有着广泛应用，其中 DR 材采用二次冷轧工艺，具有厚度更薄、强度更高的特点，可以有效降低包装用户生产成本，因此受到终端用户和市场的广泛关注。

传统 DR 材大多采用铁素体再结晶区退火工艺组织生产，为了提高成品的强度级别，往往在退火后的二次冷轧工序采用较大的二次冷轧压下率，但是较大的二次冷轧压下率往往导致材料加工硬化严重，材料断后伸长率尤其是 TD 方向断后伸长率急剧下降，材料塑性显著恶化。因此，如何采用成分设计和退火工艺，开发强度级别更高，同时兼具更好塑性的 DR 材，满足下游制罐加工企业的技术需求成为镀锡板产品开发需要重点解决的问题。

本研究采用 α+γ 两相区均热退火和快速冷却工艺开发了一种兼具高强度、高塑性的 DR 材，重点分析了两相区退火和快速冷却工艺条件下材料的显微组织特点，研究了二次冷轧压下率对 DR 材成品力学性能的影响。结果表明，两相区退火和快速冷却工艺条件下，带钢显微组织为铁素体+岛状珠光体+渗碳体，其中铁素体平均晶粒尺寸 5.0μm，珠光体呈岛状分布在铁素体晶界位置，片层间距 100nm 以下，渗碳体数量较少，弥散分布在铁素体晶粒内部。退火带钢二次冷轧压下率控制在 16%左右时，成品调质度达到 DR-8M 级别，产品洛氏硬度值 73.3，屈服强度 $Rp_{0.2}$ 达到 580MPa 以上，RD、45°方向和 TD 三向断后伸长率 A_{50} 均达到 5%以上，材料塑性获得显著改善。

回火温度对 500MPa 级高强钢屈强比和内应力的影响

武凤娟，曲锦波，杨 浩

（江苏省（沙钢）钢铁研究院，江苏张家港 215625）

摘 要：420MPa 及以上级别高强度结构钢广泛应用于桥梁、建筑、风电等钢结构领域，随着钢结构安全性和装配精度要求的不断提高，设计单位对钢板的屈强比和不平度也提出了越来越高的要求，例如建筑结构用钢要求屈强比≤0.83（420MPa 级），桥梁钢板普遍要求不平度≤3mm/m。高强钢生产过程中因轧制变形大、冷却强度高等原因，容易产生较大的内应力，在后续切割、焊接等加工过程中可能会产生局部变形，造成不平度超标，增加了钢结构制造难度并带来安全隐患。因此，如何降低高强钢的内应力，成为其质量改进的一个重要方向。回火热处理可以在一定程度上降低钢板的内应力，且回火温度越高，内应力释放越彻底，但同时可能会对其力学性能，尤其是屈强比带来一定的影响。

本研究以 500MPa 级高强钢为对象，对比分析了热机械轧制+回火（TMCP+T）、淬火+回火（Q+T）、亚温淬火+回火（L+T）等三种工艺条件下，回火温度对钢板屈强比和内应力的影响规律。试验钢的主要化学成分为

0.06%C-1.65%Mn-0.035%Nb-0.25%Ni-0.3%Cr-0.25%Mo-0.25%Cu（质量分数），试验钢板厚度为 16mm，对热机械轧制态、淬火态（淬火温度 920℃）、亚温淬火态（淬火温度 840℃）钢板分别进行 300~600℃的回火热处理，然后分别测量钢板的屈强比和内应力。内应力采用 HK21A 应力检测仪进行测量，根据 GB/T 31310—2014，利用钻孔应变法，在钢板上钻孔，孔的区域由于部分应力释放产生相应位移形变，经换算得到孔处原有内应力。当内应力低于 5MPa 时可视为无内应力，即内应力已完全释放。

结果表明，随着回火温度的升高，内应力逐渐降低，但三种工艺的降低程度不同：在 500℃回火后，TMCP+T 的内应力降低到 29MPa，仍较高；Q+T 的内应力降低到 18MPa，尚未释放完全；L+T 的内应力降低到 2MPa，已完全释放。随着回火温度的升高，屈强比升高，但三种工艺的升高幅度不同：TMCP+T 的升高幅度最大，L+T 的升高幅度最小，在 500℃回火后，TMCP+T 工艺由 TMCP 态的 0.75 升高到 0.91，Q+T 工艺由 Q 态的 0.73 升高到 0.87，L+T 工艺由 L 态的 0.69 仅升高到 0.81。在本研究试验参数范围内，采用 TMCP+T 和 Q+T 工艺无法获得低屈强比且低内应力的钢板；L+T 工艺在经过 500℃回火后屈强比仍保持较低水平，且内应力已完全释放，因此是生产 500MPa 级高强钢的理想工艺。

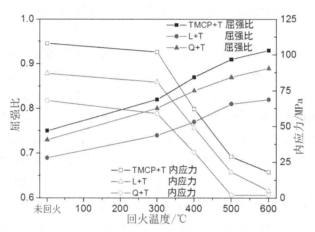

图 1　回火温度对试验钢板屈强比和内应力的影响

影响冷轧高强钢板形质量的性能因素

陈　飞，张　栋，孙翼洲，李高峰

（首钢集团有限公司技术研究院，北京　100043）

摘　要：冷轧产品以其优异的尺寸精度广泛应用于各类制造部门。随着制造业转型升级的深入，客户对板形质量提出了更高的要求，这给冷轧高强钢的生产带来了新的挑战。传统观点下，冷轧过程板形问题仅是机械自动化意义上的控制问题，板形控制不稳或归因于轧机板形调控能力不够，或归因于控制系统的固有缺陷。

受制于检测手段的不足，目前还无法对来料性能进行在线测量，通常冷轧过程只能假设来料性能是均匀的，并在此基础上进行板形设定。因此，一旦来料性能实际值与目标值之间出现较大偏差，板形设定也会随之出现偏差，导致控制失稳，最终产出板形不合的产品。冷轧高强钢板形问题不仅是控制问题，还是与来料性能密切相关的材料问题。

本文针对 590~980MPa 强度范围高强钢在冷轧生产过程及成品中存在的带中板形突变、带头尾厚度波动、通卷轧制力差异过大等问题进行全流程溯源，通过工艺曲线对比、温度跟踪测量、仿真模拟及力学性能检测，建立了冷轧过程板形与热卷性能之间的联系，并从工艺角度给出了解决方案。综合研究结果表明，热卷通卷性能均匀性与冷轧过程板形稳定性之间有着极强的遗传关系，而轧后冷却过程是影响热卷室温性能的关键。

通过高强钢性能均匀性与冷轧过程板形控制的研究，对当前轧制技术面临的问题提出以下几点思考：

（1）性能差异归根结底是组织差异，而组织差异的根源在冷速，想要避免性能差异过大，需要关注热卷的实际热经历，当前仅关注中心线温度的现状无法满足该需求，热轧工序需要对热卷建立起冷速场的概念。

（2）传统力学性能检测手段已无法满足当前对进一步提高带钢性能均匀性的高要求，实践证明，温度、厚度、轧制力等参数可以辅助分析热卷性能变化趋势，对提高产品开发过程中工艺制度的改进效率有着极大的促进作用。

（3）冷轧机组是整个钢铁生产流程中自动化水平最高的单元，但性能在线检测技术的缺失制约着其控制能力的进一步提升，这可能是未来轧制技术实现跃迁式突破的重要方向。

硅对超高强轻质钢热变形行为影响的研究

张振山，宋成浩，吴文源，王皓亮，孙振忠

（东莞理工学院机械工程学院，广东东莞　523808）

摘　要：气候变化，特别是全球变暖，在过去几十年里已经成为世界范围内的一个严重问题，温室气体一直被认为是气候变化的主要原因。在环境问题的驱使下，汽车工业的首要要求是环保减排和提高安全性。先进高强钢是开发下一代节能环保汽车的关键结构材料，中锰钢作为最新一代先进高强钢之一，以其优异的力学性能以及低合金化成本在过去十年中备受关注。近期课题组研发了一种新型高 Si 中锰钢[1-2]，其未经任何热处理的热轧板即具有超高抗拉强度（2163MPa）和良好的断后伸长率（11.7%），打破了长期以来 Si 使钢变脆的观念，并且同无 Si 中锰钢相比降低了一定的密度。该热轧板的显微组织是由未再结晶的原奥氏体转变成的马氏体组成，其为获得良好强塑性的关键。因此在热变形过程中，奥氏体晶粒的演变行为是影响此钢种力学性能的主要因素。

本文以不同 Si 含量（0~4wt.%，质量分数）的中锰钢为研究对象，在温度为 950~1150℃、应变速率为 0.01~10s^{-1} 范围内，通过热模拟试验研究了 Si 对中锰钢热变形行为的影响机理。通过数据分析绘制了材料的真应力-应变曲线和加工硬化速率曲线，确定了发生动态再结晶（DRX）的临界变形条件，并建立 Zener-Hollomon 模型。基于动态材料模型（DMM），绘制了热加工图，研究分析了稳定区和不稳定区的 DRX 显微组织特征，分析并获得材料的耗能效率（η）、微观结构与软化机理之间的关系：当峰值 $\eta(\eta_p)<0.3$ 时，随着真应变的增加，η 逐渐减小，软化主要表现为动态回复；当 $\eta_p>0.3$ 时，η 随真应变的增大而增大，软化以 DRX 为主；η 在变形后期达到峰值后逐渐降低，表明其为晶粒粗化过程。最后，建立了动态再结晶模型，描述了 Si 含量对 DRX 行为的影响。通过以上研究，表明 Si 具有显著提高加工硬化能力、降低热变形激活能及促进奥氏体形核的作用，极大调控了传统中锰钢的动态再结晶行为。本研究可为实际生产提供理论支撑，并对未来超高强塑钢的发展提供了一定的借鉴。

参 考 文 献

[1] Song C, Wang H, Sun Z, et al. Scripta Materialia, 2022, 212: 114583.
[2] Song C, Zhang Z, Wu W, et al. Materials Science & Engineering A, 2023, 869: 144825.

热轧高强钢平整过程中残余应力的分析及应用

刘天武[1]，张彩东[1]，孙　力[1]，潘　进[1]，张志强[2]，丁文红[3]

（1. 河钢材料技术研究院，河北石家庄　050023；2. 河钢集团邯钢公司，河北邯郸　056002；
3. 武汉科技大学材料与冶金学院，湖北武汉　430081）

摘 要：高强钢具有良好的成形性以及碰撞吸能效果，是汽车等领域的重要基础材料，也是衡量国家冶金技术水平和国力的重要标志之一。热轧过程中，在不均匀温度场、应力场和复杂相变耦合作用下，高强钢组织内部形成了较高的残余应力，恶化了带钢的平直度，因此制约了高强钢应用领域的扩展。如何准确测量高强钢的残余应力，并基于此分析残余应力改善方法是个难点问题。本文针对某热轧厂平整机生产高强钢的情况，测量了平整前后钢板残余应力，分析了残余应力的影响因素，研究了改善措施，并取得了明显的应用效果。

关键词：高强钢；平整；残余应力；裂纹柔度法

Analysis and Application of Residual Stress on Hot Rolled High Strength Steel during Temper Passing Process

LIU Tianwu[1], ZHANG Caidong[1], SUN Li[1], PAN Jin[1], ZHANG Zhiqiang[2], DING Wenhong[3]

(1. HBIS Material Technology Research Institute, Shijiazhuang 050023, China; 2. HBIS Group Hansteel Company, Handan 056002, China; 3. School of Materials and Metallurgy, Wuhan University of Science and Technology, Wuhan 430081, China)

Abstract: High strength steel has good formability and impact energy absorption effect. It is an important basic material in automobile and other fields. It is also one of the important symbols to measure the national metallurgical technology level and national strength. In the process of hot rolling, under the coupling effect of heterogeneous temperature field, stress field and complex phase transition, high residual stress is formed in the microstructure of high strength steel, which worsens the flatness of the strip, and therefore restricts the extension of the application field of high strength steel. It is very difficult to accurately measure the residual stress of high strength steel and analyze the improvement method based on it. In this paper, the residual stresses of hot rolled high strength steel before and after temper passing are measured, and the influence factors of the residual stress are analyzed. The improvement measures are studied, and the obvious application effect is obtained.

Key words: high strength steel; temper passing; residual stress; crack compliance method

高强度宽幅中厚钢板残余应力的低热-磁-振复合时效消减调控方法与机理研究

宋和川[1,2]，张清东[1,2]，周晓敏[1,2]，张勃洋[1,2]

（1. 北京科技大学机械工程学院，北京 100083；
2. 北京科技大学顺德创新学院，广东佛山 528399）

摘 要：残余应力是指在没有对物体施加外力时，物体内部存在的保持自相平衡的应力系统，它是固有应力或内应力的一种。轧制过程中金属的不均匀变形导致了板带中残余应力的存在，残余应力的不均匀分布又是板形不良的根本原因。

板形问题实质是板带材在生产过程中发生了屈曲变形，使得板带材在纵向方向产生了瓢曲浪形。其产生的力学机理是板带材内部残余应力大于了维持板带材平衡的临界应力，而使得板带材发生失稳现象。

高强度宽幅中厚钢板作为板带钢中的一种新型产品，具有显著的减重节材和节能减排的效果，为进一步落实国家"碳达峰、碳中和"的重要战略部署，提前布局绿色转型，在后续竞争中占据有利地位，课题组提出一种适用于高强度宽幅中厚钢板的低热-磁-振复合时效残余应力精准调控方法（时效温度大大低于现有热处理温度），基于此，以改善高强度宽幅中厚钢板的残余应力分布和力学性能为目标，借助低热-磁-振复合时效装置的搭建与实验，验证该工艺的残余应力消减与均化效果以及力学性能的强化效应；依据实验结果，结合弹塑性力学、位错理论及强化机制等，系统揭示钢板在低热-磁-振复合时效作用下的残余应力与力学性能的演变规律及其调控机理；最终利用机器学习优化算法寻求高强度宽幅中厚钢板低热-磁-振复合时效残余应力最佳调控方案。该方法可有效替代现有高强度宽幅中厚钢板高耗能高污染的机械或热处理去应力方法，同时可延长钢铁产品服役寿命，降低部件更换频次，从长远角度降低原材料消耗，实现源头减排。

关键词：高强度钢；残余应力；带钢板形；时效处理；调控机理

传统热连轧变厚轧制功能开发与应用

单旭沂，张　勇，刘　苗

（宝山钢铁股份有限公司热轧厂，上海　200948）

摘　要：对用户的变厚产品的需求进行了分析，根据变厚范围广、过渡段长度短的特点，针对使用调整 AGC 厚度目标值无法实现相关产品开发的问题，提出了模型设定、L1 动态控制以及辅助测厚仪功能设计新的方法，在传统热连轧生产线上成功开发了大范围厚度变化及短过渡长度的不同类型的变厚轧制功能，实现了大批量稳定生产。

关键词：热连轧；变厚轧制；模型

基于三维有限元的 DP590 冷连轧中间辊窜辊板形控制分析

刘　需，李连杰，刘天武，马　成，杨　婷，
赵轶哲，刘宏强，姜正义

（河北大河材料科技有限公司，河北石家庄　050000）

摘　要：板形作为宽幅高强冷轧板带钢的重要指标，多种板形影响因素和不同调控手段的调控特性影响着最终的板形指标，其中，中间辊窜辊被广泛用于改善带钢板形。以河钢某厂 2180mm 酸轧联合生产线连轧机组轧制 590DP 的板形控制工艺为研究对象，建立了考虑材料加工硬化效应和板形继承的冷连轧多机架弹塑性有限元模型，中间辊窜辊的调控特性进行了定量研究和应用。

建立了一种考虑材料加工硬化效应和板形继承的冷连轧多机架有限元模型[1-5]。利用有限元软件建立了多机架冷连轧三维有限元模型（3D-FEM），并通过 S1~S5 机架的实测轧制力和 S5 机架出口的实测板凸度对比，验证了模型的准确性。该模型将冷连轧机分为 5 个单机架模型，各模型分别作为一个单独的有限元模型进行计算。利用相邻模型间的总等效塑性应变和板凸度遗传，将单机架模型连接成一个整体的多机架冷连轧系统模型。大大减少了系统中单次计算的单元数量，并在不改变板带单元数的基础上解决了变形较大时板带单元的畸变问题。

利用多机架冷连轧有限元模型，研究了 S1~S5 机架在金属材料 590DP 轧制过程中，中间辊窜辊量变化对本机

架轧后带钢各项板凸度、平直度指标的影响关系，并定量计算了影响功效系数，总结出影响功效系数的变化规律如下：S1~S5 的中间辊窜辊量对板凸度指标均是反向影响，并大体呈现出逐渐减小的趋势，这取决于带钢的塑性刚度；对平直度指标则均是正向影响，从 S1 到 S4 其对二次平直度的调节能力明显增强，然后在 S5 处下降；这些发现有助于更好地理解在冷连轧过程中 IRS 在控制板形中的作用。

依托多机架冷连轧有限元分析模型分析计算得到的中间辊窜辊量对最终板形指标影响的功效系数表，对 2180mm 酸轧联合生产线的不同钢卷生产工艺参数差值进行分析和计算，验证了功效系数对 2180mm 产线的可用性。

参 考 文 献

[1] 时旭, 刘相华, 王国栋. 薄板轧制的接触摩擦及其对轧制力的影响[J]. 塑性工程学报, 31-34.
[2] 张清东, 李博, 郑武, 等. 冷带轧机板形闭环反馈控制策略及模型研究[J]. 系统仿真学报, 2009, 21(24): 7858-7862.
[3] Li Lianjie, Xie Haibo, Liu Xu, et al. Numerical simulation of strip shape of high-strength steel during hot rolling process[J]. EI: Key Engineering Materials, 2020, 2.
[4] Xie Haibo, Li Lianjie, Liu Tianwu, et al. Three dimensional finite element simulation of strip shape and flatness of high strength steel [J] . Key Engineering Materials, 2020, 2.
[5] 陈火红, 尹伟奇, 薛小香. MSC.Marc 二次开发指南[M]. 北京: 科学出版社, 2004.

高性能耐候桥梁钢 Q500qNH 的开发

彭宁琦，高 擎，史术华，范 明，熊祥江，李中平，陈奇明

(湘潭钢铁集团有限公司技术中心，湖南湘潭 411101)

摘 要：通过优化成分设计和采用 TMCP+T 工艺，开发了 Q500qNH 高性能耐候桥梁钢。检测开发钢的拉伸性能、冲击性能、冷弯性能、抗层状撕裂性能和应变时效冲击性能，并进行斜 Y 坡口焊接裂纹试验、焊接热影响区最高硬度试验、焊接工艺评定试验、无塑性转变温度落锤试验、等温型双重拉伸止裂试验、中性盐雾试验和大气暴露试验。结果表明，开发钢的综合性能良好，其强韧性、可焊性、止裂性、耐候性等与国外同类产品水平相当，并具有低屈强比。该项工作有助于高性能耐候桥梁钢的国产化及其推广应用。

关键词：耐候钢；Q500qNH；力学性能；焊接性能；止裂性能；耐候性能

Development of Q500qNH High-performance Atmospheric Corrosion Resisting Steel for Bridge

PENG Ningqi, GAO Qing, SHI Shuhua, FAN Ming,
XIONG Xiangjiang, LI Zhongping, CHEN Qiming

(Technology Center, Xiangtan Iron and Steel Group Co., Ltd., Xiangtan 411101, China)

Abstract: By optimized composition and TMCP+T process, the high-performance atmospheric corrosion resisting steel for bridge of Q500qNH was successfully developed. Then the properties of the produced steel plates on tensile, impact, bending, lamellar tear resistance and strain aging impact were measured, and some experiments were carried out that include Y-groove cracking test, maximum hardness test in weld heat affected zone, welding procedure qualification test, drop-weight test to determine nil-ductility transition temperature, isothermal double tensile crack arrest test, neutral salt spray test, and atmospheric corrosion test. Results show that the comprehensive properties of the developed steel plates are good, including obdurability, weldability, anti-crack and weathering of the steel plates are equivalent to those of foreign like products. So this work is helpful to localization of high-performance bridge structural steel with weather resistant and its popularization and application.

Key words: atmospheric corrosion resisting steel; Q500qNH; mechanical property; welding performance; crack-arrest performance; weathering performance

960MPa级工程机械用钢的连续冷却转变规律

惠亚军[1,2]，田志红[1]，肖宝亮[1]，邱　宇[3]，冯　超[1]，黄天华[4]

（1. 首钢集团有限公司技术研究院，北京　100043；2. 绿色可循环钢铁流程北京市重点实验室，北京　100043；3. 首钢集团有限公司技术研究院迁顺技术中心，河北迁安　064404；4. 北京首钢股份有限公司制造部，河北迁安　064404）

摘　要：利用扫描电子显微镜（SEM）研究了不同冷却速度对960MPa级工程机械用钢连续冷却转变规律。结果表明：实验钢在加热过程中铁素体向奥氏体转变开始点和结束点分别为725℃和865℃，动静态CCT曲线均由三部分组成：高温铁素体和珠光体转变区，中温贝氏体转变区，低温马氏体转变区。与静态CCT曲线相比，动态CCT曲线中奥氏体向铁素体转变开始曲线向高温区移动，奥氏体向铁素体转变结束曲线移动不明显，贝氏体开始转变温度与结束温度都有所升高，在冷速较低时变化幅度较小，冷速较高时变化幅度较大。验钢在5~50℃/s的较大冷速范围内获得贝氏体组织，随着冷却速度的增加，组织中粒状贝氏体的量下降，板条贝氏体的量增加，在轧制过程中可以通过增加冷却速度来获得大量板条贝氏体组织，进而减少Mo与B合金元素的含量。

关键词：960MPa；工程机械用钢；连续冷却转变；贝氏体

Continuous Cooling Transformation Law of 960MPa Grade Construction Machinery Steel

HUI Yajun[1,2], TIAN Zhihong[1], XIAO Baoliang[1],
QIU Yu[3], FENG Chao[1], HUANG Tianhua[4]

(1. Shougang Group Co., Ltd., Shougang Research Institute of Technology, Beijing 100043, China;
2. Beijing Key Laboratory of Green Recyclable Process for Iron & Steel Production, Beijing 100043, China;
3. Qianshun Technology Center, Shougang Group Co., Ltd., Qian'an 064404, China;
4. Beijing Shougang Co., Ltd., Manufacturing Department, Qian'an 064404, China)

Abstract: Scanning electron microscope (SEM) was used to study the continuous cooling transformation law of 960MPa grade engineering machinery steel with different cooling rates. The results show that the starting point and ending point of the ferrite to austenite transformation of the experimental steel are 725℃ and 865℃ respectively during the heating process. The dynamic and static CCT curves are composed of three parts: high temperature ferrite and pearlite transformation zone, medium temperature bainite transformation zone, and low temperature martensitic transformation zone. Compared with the static CCT curve, in the dynamic CCT curve, the austenite to ferrite transformation start curve moves to the high temperature region, the austenite to ferrite transformation end curve does not move significantly, and both the bainite transformation temperature and the end temperature are increases, and the change range is small when the cooling speed is low, and the change range is large when the cooling speed is high. The bainite structure was obtained in the large cooling rate range of 5~50℃/s. With the increase of cooling rate, the amount of granular bainite decreased, and the amount of lath bainite increased. During the manufacturing process, a large amount of lath bainite structure can be obtained by increasing the cooling rate, thereby reducing the content of Mo and B alloying elements.

Key words: 960MPa; construction machinery steel; continuous cooling transformation; bainite

热轧板带钢轧制残余应力的研究

王卫卫[1]，胡霄雨[2]，肖金福[1]，张宏亮[1]，丛菁华[1]

（1. 钢铁研究总院有限公司冶金工艺研究所，北京 100081；
2. 中国金属学会，北京 100081）

摘 要：本文在理论分析和假设的基础上，应用有限元模拟软件对带钢轧制过程建模，计算分析了带钢宽度、压下率、工作辊径、开轧温度对带钢轧后残余应力的分布和大小的影响。并通过轧制实验和残余应力测量实验，对结果进行了验证。结果表明，轧制过程中，带钢的宽度、道次压下率及工作辊径对轧制后带钢残余应力的分布和大小有一定的影响。在一定的条件下，采用最优的工作辊辊径和压下率进行轧制，能够使带钢轧后残余应力分布最均匀。

关键词：工艺参数；轧制；残余应力；有限元；热轧板带

Research on Residual Stress in Hot Rolled Plate and Strip

WANG Weiwei[1], HU Xiaoyu[2], XIAO Jinfu[1],
ZHANG Hongliang[1], CONG Jinghua[1]

(1. Metallurgical Technology Institute of Central Iron & Steel Research Institute, Beijing 100081, China;
2. The Chinese Society for Metals, Beijing 100081, China)

Abstract: On the basis of theoretical analysis and assumptions, this paper uses finite element simulation software to model the rolling process of strip steel, and calculates and analyzes the influence of strip width, reduction rate, work roll diameter and rolling temperature on the distribution and size of residual stress after rolling. The results were verified by rolling experiments and residual stress measurement experiments. The results show that the strip width, pass reduction and work roll diameter have certain effects on the distribution and magnitude of residual stress in the rolled strip. Under certain conditions, using the optimal work roll diameter and reduction rate for rolling can make the residual stress distribution of the rolled strip the most uniform.

Key words: process parameters; rolling; residual stress; FEM; hot rolled strip

浅谈轧制油对不锈钢表面质量的影响

吴雪，刘金元，陈普

（山东泰嘉新材料科技有限公司，山东济南 271100）

摘　要： 不锈钢冷轧产品的轧制过程中轧制油对不锈钢表面质量有着极大的影响，因此要求不锈钢冷轧轧制油应具备优异的润滑、冷却及清洁性能，本论文旨在简要探讨分析轧制油各种性能指标对不锈钢表面质量的影响因素，为之后提高不锈钢表面质量提供理论依据。

关键词： 轧制油；性能指标；不锈钢；表面质量

Discussion on the Effect of Rolling Oil on the Surface Quality of Stainless Steel

WU Xue, LIU Jinyuan, CHEN Pu

(Shandong Taijia New Material Technology Co., Ltd., Jinan 271100, China)

Abstract: Rolling oil in the rolling process of stainless steel cold rolled products has a great impact on the surface quality of stainless steel, so stainless steel cold rolled rolling oil should have excellent lubrication, cooling and cleaning performance, this paper aims to briefly discuss and analyze the various performance indicators of rolling oil on the surface quality of stainless steel influencing factors, for the later improvement of stainless steel surface quality to provide a theoretical basis.

Key words: rolling oil; performance indicators; stainless steel; surface

基于理论预测模型的轧机单调自动控制系统开发及其不锈钢轧制应用实践

李慧峰[1]，王中杰[2]，李建民[1]，王　忠[1]，杨永杰[1]，郭新峰[2]

（1. 宝武集团太原钢铁集团有限公司技术中心，山西太原　030003；

2. 山西大学物理电子工程学院，山西太原　030006）

摘　要： 山西大学和太原钢铁集团有限公司组建联合团队，联合开发了太钢2250热连轧线轧机单调自动控制系统，以通过优化轧机单调控制来改善400系不锈钢热轧卷的卷形质量。所开发的轧机单调自动控制系统是基于现场获得的必要的工艺参数，通过自主研究形成的轧机单调理论模型进行计算获得轧机单调组合预测值，组批发送后在现场轧机信号触发下依次通过轧机液压单调机构执行，目标是消除轧件在轧制过程中的楔形波动，同时获得良好的直线度。该系统在400系不锈钢热轧中得以批量应用，并在生产稳定性、卷形质量和表面缺陷控制等方面取得良好效果。

关键词： 热连轧；不锈钢带钢；直线度自动调平；系统

提高410S不锈钢2B产品光泽度工艺研究与实践

刘金元，张爱建，陈　普

（山东泰嘉新材料科技有限公司，山东济南　271100）

摘 要： 410S 不锈钢 2B 产品表面发白，光泽度低，无法满足客户对亮度的要求，针对 410S 不锈钢 2B 产品表面发白不亮问题，本文通过从白皮原料、冷轧、退火、酸洗、平整工艺技术研究，成功生产出 410s 不锈钢 2B 表面白光度低，光泽度高的产品，形成了完整的生产工艺，本工艺生产的产品满足客户对表面质量光泽度的要求。

关键词： 410S 不锈钢 2B 面；光泽度；工艺研究

Research on the Process of Improving the Gloss of 410S Stainless Steel 2B Products

LIU Jinyuan, ZHANG Aijian, CHEN Pu

(Shandong Taijia New Material Technology Co., Ltd., Jinan 271100, China)

Abstract: 410S stainless steel 2B product surface white, low gloss, can not meet customer requirements for brightness, for 410S stainless steel 2B product surface white not bright problem, this paper through white skin raw materials, cold rolling, annealing, pickling, leveling process technology research, successfully produced 410s stainless steel 2B surface low white luminosity, high gloss products, formed a complete production process, this process production products to meet customer requirements for surface quality gloss.

Key words: 410S stainless steel 2B side; gloss; process research

10Cr17 不锈钢冷轧轧制力优化设计

张爱建，刘金元，陈 普

（山东泰嘉新材料科技有限公司，山东济南 271100）

摘 要： 不锈钢产品作为现代工业中一种重要的金属材料。具有高强度、高韧性、耐腐蚀性等许多优异的特性。为提高市场竞争力，我公司按照控轧、控冷的方式来进行优化。生产 10Cr17 冷轧不锈钢的轧制力模型，对轧制力，变形量和轧制油的使用进行分析和优化。文章的实验钢卷为目标厚度 0.25mm 的 10Cr17 不锈钢薄板轧制情况。

关键词： 不锈钢；冷轧；变形量；优化

Optimized Design of 10Cr17 Stainless Steel Cold Rolling Force

ZHANG Aijian, LIU Jinyuan, CHEN Pu

(Shandong Taijia New Material Technology Co., Ltd., Jinan 271100, China)

Abstract: Stainless steel products are an important metal material in modern industry. It has many excellent characteristics such as high strength, high toughness, corrosion resistance and so on. In order to improve market competitiveness, our company optimizes according to the way of controlled rolling and cold control. Production of rolling force models of 10Cr17 cold-rolled stainless steel, analysis and optimization of rolling force, deformation and use of rolling oil. The experimental steel coil in this article is the rolling of 10Cr17 stainless steel sheet with a target thickness of 0.25mm.

Key words: stainless steel; cold rolling; amount of deformation; optimized

回火工艺对热处理 X80 管线钢组织性能的影响

刘文月[1,2]，李天怡[1,2]，安 涛[1,2]，臧 岩[1,2]，王传军[1,2]，王超逸[1,2]

（1. 海洋装备用金属材料及其应用国家重点实验室，辽宁鞍山 114009；
2. 鞍钢集团北京研究院有限公司，北京 102200）

摘 要：本文研究了回火工艺对 X80 管线钢组织性能的影响，并利用 SEM、EBSD 和 TEM 研究了材料在回火后的微观组织特征与性能的关系。X80 管线钢经过调质处理后组织为贝氏体铁素体，且有亮白色的 Fe_3C 颗粒物分布在基体上。回火温度提高后，材料大角晶界占比提高，而此时实验钢实验钢强度下降，低温韧性上升。大角晶界的存在阻碍了裂纹扩展路径，从而提高了材料低温韧性。回火时间延长后，实验钢大角度晶界占比基本不变，等效晶粒尺寸略微增加，材料冲击韧性略微增加。然而有效晶粒尺寸及大角度晶界密度在 40~80min 的回火时间范围内的变化并不十分明显，相应地性能也未发生明显波动。。

关键词：X80 管线钢；回火工艺；TEM 分析；强化机制

Influence of Tempering Process on Microstructure and Mechanical Properties of Heat-treated X80 Pipeline Steel

LIU Wenyue[1,2], LI Tianyi[1,2], AN Tao[1,2], ZANG Yan[1,2], WANG Chuanjun[1,2], WANG Chaoyi[1,2]

(1. State Key Laboratory of Metal Materials for Marine Equipment and Application, Anshan 114009, China;
2. Ansteel Beijing Research Institute Co., Ltd., Beijing 102200, China)

Abstract: The effect of tempering process on the microstructure and properties of X80 pipeline steel is studied, and the relationship between the microstructure characteristics and properties of the material after tempering is discussed by using SEM, EBSD and TEM. The structure of X80 pipeline steel is bainitic ferrite after quenching and tempering treatment, and there are bright white F_3C particles distributed on the ferrite matrix. With increasing tempering temperature, the proportion of large-angle grain boundaries increases, while the strength of the experimental steel decreases and the low-temperature toughness increases. The existence of high-angle grain boundaries hinders the crack propagation path, thereby improving the low-temperature toughness of the material. After prolonging the tempering time, the proportion of high-angle grain boundaries in the experimental steel remained unchanged, the equivalent grain size increased slightly, and the impact toughness of the material increased slightly. However, the change of effective grain size and large-angle grain boundary density in the tempering time range of 40-80 minutes is not very obvious, and the mechanical fluctuation is not obvious.
Key words: pipeline steel; tempering process; TEM analysis; strengthening mechanism

热处理工艺对吉帕级无缝管用钢组织及性能的影响

安 涛[1,2]，刘文月[1,2]，李天怡[1,2]，臧 岩[1,2]，李江文[1,2]，薛 峰[2]

（1. 海洋装备用金属材料及其应用国家重点实验室，辽宁鞍山 114009；
2. 鞍钢集团北京研究院有限公司，北京 102200）

摘　要：采用光学显微镜，扫描电子显微镜，透射电子显微镜和拉伸试验机等试验设备研究不同的热处理工艺对吉帕级无缝管用钢的微观组织和力学性能的影响。结果表明：热轧态时，钢的显微组织为晶粒粗大的贝氏体和珠光体，强度和韧性最差；淬火态时，钢的显微组织为高位错密度的板条马氏体和少量残余奥氏体，晶粒得到细化，因此强度较高，冲击功为24J；调质态时，钢显微组织为回火索氏体，晶粒得到进一步细化，位错密度降低，因此钢的强度下降，韧性进一步提高。

关键词：无缝管用钢；热处理；横向组织；TEM 分析

Effect of Heat Treatment on Microstructure and Mechanical Property of GPa Seamless Pipe Steel

AN Tao[1,2], LIU Wenyue[1,2], LI Tianyi[1,2], ZANG Yan[1,2], LI Jiangwen[1,2], XUE Feng[2]

(1. State Key Laboratory of Metal Materials for Marine Equipment and Application, Anshan 114009, China; 2. Ansteel Beijing Research Institute Co., Ltd., Beijing 102200, China)

Abstract: Optical microscopy, scanning electron microscopy, transmission electron microscopy and tensile test machine are used to study the effects of different heat treatment processes on the microstructure and mechanical properties of GPa grade seamless pipe steel. The results show that in the hot rolled state, the microstructure of steel is coarse bainite and pearlite, with the worst strength and toughness; After quenching, the microstructure of steel is lath martensite with high dislocation density and a small amount of residual austenite. Due to the grain refinement, the strength increases, the charpy impact absorbed energy work increases to 24J; After quenching and tempering treatment, the microstructure turns into tempered sorbite, the grains are further refined and the dislocation density is reduced, so the strength of the steel is reduced and the low temperature toughness is further improved.

Key words: seamless pipe steel; heat treatment; horizontal microstructure; TEM analysis

6.2　长材和钢管

棒材生产线定尺剪切系统的提速改造

吴光行，庞博文

（中冶南方武汉钢铁设计研究院有限公司，湖北武汉　430081）

摘　要：针对棒材线在增产提速改造时精整收集区产能限制的情况，分析定尺剪切系统的产能限制的原因，研究定尺剪切系统优化的方案，通过一系列的工艺调整、设备改造、新设备创新及应用等措施，提高了棒材线定尺剪切的效率、提高棒材质量和精度，解决棒材线在增产提速改造时精整收集区产能受限的问题。

关键词：棒材提速；定尺剪切改造

Speed-up Refurbishment of Fixed-length Shearing System in Bar Production Line

WU Guangxing, PANG Bowen

(WISDRI South Wuhan Iron and Steel Design and Research Institute Co., Ltd., Wuhan 430081, China)

Abstract: For the situation where the production capacity of the finishing and collecting area of the bar line was limited during the speed-up refurbishment for increased production, we analyzed the reasons for the production capacity limitation of the fixed-length shearing system. We researched optimization plans for the fixed-length shearing system, and through a combination of process adjustments, equipment refurbishment, innovation with new equipment applications, and other measures, we improve the efficiency of the fixed-length shearing process in the bar line, enhance the quality and precision of the bars, and solve the problem of limited production capacity in the finishing and collecting area during the speed-up refurbishment of the bar line.

Key words: bar speed increase; fixed length cutting transformation

高碳钢车轮踏面近表层异常组织控制

国新春，邓荣杰，刘 智，黄孝卿，翟 龙，陶盈龙，张意哲

（宝武集团马钢轨交材料科技有限公司技术中心，安徽马鞍山 243000）

摘 要： 传统CL60材质货运车轮已不能适用当前铁路货运重载的发展需求，国铁集团审时度势适时开发了以碳、硅、锰为强化元素的CL65、CL70重载车轮用钢，但新材质的应用也给车轮生产实现带来一定的难度。本文基于试验，制定热处理工艺参数，通过工业化生产结果和服役表现，证明了制定的热处理工艺参数稳定可靠。

关键词： 重载；热处理工艺；工业化；运用

Study on Heat Treatment Process of Heavy-duty Wheel

GUO Xinchun, DENG Rongjie, LIU Zhi, HUANG Xiaoqing,
ZHAI Long, TAO Yinglong, ZHANG Yizhe

(Baowu Group Masteel Transit Materials Techonlogy Co., Ltd., Maanshan 243000, China)

Abstract: The traditional CL60 material freight wheel can not meet the development needs of the current railway freight heavy haul. According to the situation, the State Railway Group timely developed CL65 and CL70 heavy haul wheel steel with carbon, silicon and manganese as strengthening elements. However, the application of new materials also brings some difficulties to the wheel production. In this paper, based on the test, the heat treatment process parameters are formulated. Through the industrial production results and service performance, it is proved that the heat treatment process parameters are stable and reliable.

Key words: heavy load; heat treatment process; industrialization; application

慢应变速率对 110S 油井管拉伸性能和断口形貌的影响

钟 彬[1,2]，陈义庆[1,2]，艾芳芳[1,2]，高 鹏[1,2]，李 琳[1,2]，
张圣洁[1,2]，伞宏宇[1,2]，苏显栋[1,2]，沙楷智[1,2]

（1. 海洋装备用金属材料及其应用国家重点实验室，辽宁鞍山 114009；
2. 鞍钢集团钢铁研究院，辽宁鞍山 114009）

摘 要：通过扫描电镜（SEM）、慢速率拉伸试验机和金相显微镜等测试手段，研究了 110S 油井管在室温空气介质中的慢拉伸性能和断口形貌。结果表明：应变速率对 110S 油井管拉伸性能影响显著，随着应变速率的降低，延伸量越来越小，断裂时间越来越长，断面收缩率逐渐降低并趋于稳定。同时，随着应变速率的降低，断口形貌中放射区的面积逐渐减少，剪切唇的面积逐渐增加，二次裂纹逐渐减少，形成较大的撕裂楞和多源的断口形貌。应变速率为 5×10^{-6} mm/s 时，晶内位错开动少，增殖效应较弱，使得屈服强度和抗拉强度最低，延伸量最小，是研究 H 与位错交互作用的合适应变速率。

关键词：油井管；应变速率；慢拉伸；断口形貌

Effect of Slow Strain Rate on Tensile Properties and Fracture Morphology of 110S Oil Well Pipe

ZHONG Bin[1,2], CHEN Yiqing[1,2], AI Fangfang[1,2], GAO Peng[1,2], LI Lin[1,2],
ZHANG Shengjie[1,2], SAN Hongyu[1,2], SU Xiandong[1,2], SHA Kaizhi[1,2]

(1. State Key Laboratory of Metal Material for Marine Equipment and Application, Anshan 114009, China;
2. Anshan Iron & Steel Institute, Anshan 114009, China)

Abstract: The slow tensile properties and fracture morphology of 110S sulfur-resistant pipe in room temperature air was carried out by means of scanning electron microscope (SEM), slow rate tensile testing machine and metallographic microscope and other testing methods. The results show that strain rate has a significant effect on the tensile performance of 110S sulfur-resistant pipe, as the strain rate decreases, the amount of extension becomes smaller and smaller, the fracture time gets longer and longer, and the rate of reduction in area gradually decreases and tends to be stable. At the same time, as the strain rate decreases, the area of the radiation zone in the fracture morphology gradually decreases, the area of the shear lip gradually increases, and the number of secondary cracks gradually decreases, forming larger tearing flutes and a kind of multi-source fracture morphology. When the strain rate is 5×10^{-6} mm/s, there are fewer active dislocations, the proliferation effect is weak, and the stress and strain are synchronized, resulting in the lowest yield strength and tensile strength, and the smallest elongation. Therefore, 5×10^{-6} mm/s is the best strain rate for studying the interaction between hydrogen and dislocations.

Key words: sulfur-resistant pipe; strain rate; slow tensile; fracture morphology

智能高速棒材生产线建设的创新与关键技术

徐言东[1,2]，王晓晨[1,2]，程知松[1,2]，余 伟[1,2]，
张勇军[1,2]，刘 洋[1,2]，白金龙[1,2]

(1. 北京科技大学国家板带生产先进装备工程技术研究中心，北京 100083；
2. 北京科技大学高效轧制与智能制造国家工程研究中心，北京 100083)

摘 要：近年来传统棒材生产线在市场的压力下，大多进行了升级改造，正在由局部设备的智能化、少人化向着整体智能化高速棒材生产线迈进。发现其一般都是按照生产过程智能控制、生产管理智能控制的"双智控"架构来进行建设的。北科工研、中冶京诚、中钢设备等设计供货单位通过对长治钢厂高线改高棒、福建三钢一棒改造、济源钢厂二棒改造、珠海粤裕丰棒材成品库、昆钢4580棒线材、建龙西钢双高棒等项目进行了智能化创新建设，都取得了较好的效果。这些信息化和智能化手段的应用可促进高速棒材生产线升级，增强市场竞争力。

关键词：高速棒材；智能化；双智控；工业机器人；智能装备

Innovation and Key Technologies of Intelligent High Speed Bar Production Line Construction

XU Yandong[1,2], WANG Xiaochen[1,2], CHENG Zhisong[1,2], YU Wei[1,2],
ZHANG Yongjun[1,2], LIU Yang[1,2], BAI Jinlong[1,2]

(1. National Engineering Technology Research Center of Flat Rolling Equipment, University of Science and Technology Beijing, Beijing 100083, China;
2. National Engineering Research Center for Advanced Rolling and Intelligent Manufacturing, University of Science and Technology Beijing, Beijing 100083, China)

Abstract: In recent years, under the pressure of the market, most of the traditional bar production lines have been upgraded and transformed, and are stepping from the intellectualization and Flexible Manpower Line construction of local equipment to the intellectualization of high-speed bar production line as a whole.It is found that they are generally constructed according to the "Dual Intelligence Control" architecture of intelligent control of production process and production management.Design and supply units such as Institute of Engineering Technology of USTB, Capital Engineering & Research Incorporation, and SINOSTEEL MECC have carried out intelligent innovation construction through such projects as Changzhi Steel Plant's high speed bar production line transformation, Fujian Sangang's No.1 bar production line transformation, Jiyuan Steel Plant's No.2 bar production line transformation, Zhuhai Guangdong Yufen Bar Product Library, Kungang 4580 bar production line and Jianlongxi Steel's double high speed bar production lines, all of which have achieved good results. The application of these informational and intelligent methods can promote the upgrade of high-speed bar production line and enhance the market competitiveness.

Key words: high-speed bar; intelligence; dual intelligence control; industrial robot; intelligent equipment

B500B 英标钢筋疲劳断裂失效分析

高 敏，贾元海，刘可可，刘效云

（承德钒钛技术中心，河北承德 067102）

摘 要：公司生产的公称直径为 20mm B500B 英标螺纹钢筋在疲劳测试过程中未达到 5×10^6 次应力循环的要求，经对试样断口的形貌、化学成分及金相组织进行分析，发现疲劳试验过程中钢筋过早断裂是由表面缺陷引起的。

关键词：B500B；英标钢筋；疲劳测试；失效分析

Fatigue Fracture Failure Analysis of B500B British Standard Steel Bars

GAO Min, JIA Yuanhai, LIU Keke, LIU Xiaoyun

(V-Ti Technology Center of Chengde, Chengde 067102, China)

Abstract: The nominal diameter of B500B British standard threaded steel bars produced by the company did not reach 5×10^6 stress cycles, after analyzing the morphology, chemical composition, and metallographic structure of the fracture surface of the sample, it was found that the premature fracture of the steel bar during the fatigue test was caused by surface defects.

Key words: B500B; British standard steel bars; fatigue test; failure analysis

铁路车轮踏面 C 型剥离缺陷分析

庞晋龙，李 翔，马俊松，李自钢，陈 刚，徐金辉

（宝武集团马钢轨交材料科技有限公司，安徽马鞍山 243000）

摘 要：随着铁路车辆高速化的飞速发展，近年来车轮踏面 C 型剥离频繁发生，已严重影响到列车安全运行，并造成重大经济损失，因此，本文以车轮 C 型剥离缺陷为研究对象，通过实物解剖及损伤机理分析等手段开展深入研究，寻找 C 型剥离成因，并提出处置和改进防范措施。研究结果表明：通过车轮材质优化，提高车轮强硬度和抗变形能力，可有效减少车轮剥离发生；通过缩短镟修周期和预防性镟修，尽早发现和消除 C 型剥离裂纹，可防止裂纹向深处扩展，有效缓解车轮损伤。从而为铁路部门的车轮使用、检修和更换等方面提供一定的指导意见。

关键词：铁路车轮；C 型剥离；接触疲劳；裂纹

Defects Analysis of C-type Shelling of Railway Wheel Tread

PANG Jinlong, LI Xiang, MA Junsong, LI Zigang, CHEN Gang, XU Jinhui

(Baowu Group Masteel Rail Transit Materials Technology Co., Ltd., Maanshan 243000, China)

Abstract: With the rapid development of high-speed railway vehicles, C-type shelling of wheel tread has occurred frequently in recent years, which has seriously affected the safe operation of trains and caused significant economic losses. Therefore, this paper takes the defects of C-type shelling of wheels as the research object, conducts in-depth research through physical anatomy and damage mechanism analysis, finds out the causes of C-type shelling, and proposes disposal and improvement preventive measures. The results show that the wheel shelling can be effectively reduced by improving the strength, hardness and deformation resistance of the wheel through the optimization of wheel material. By shortening the cycle of rotation repair and preventive rotation repair, early detection and elimination of C-type peel cracks can prevent cracks from spreading deep and effectively alleviate wheel damage. It provides some guidance for railway wheel use, maintenance and replacement.

Key words: railway wheels; C-type shelling; contact fatigue; crack

一种低碳当量易焊接热轧钢轨的开发

冯立果[1], 崔 恺[2], 李钧正[2],
杨正宗[2], 郭朝军[2]

（1. 河钢材料院先进材料研发中心，河北石家庄 052160；2. 邯钢技术中心，河北邯郸 056015）

摘 要： 目前铁路广泛采用无缝线路，由于钢轨多为高碳钢，焊接难度系数极大，焊接接头是整条线路的薄弱地带。依据碳当量计算公式 $Ceq=C+1/6Mn+1/5(Cr+Mo+V)+1/15(Ni+Cu)$，碳是影响钢轨焊接性能的最主要元素。通过生产大数据分析、合理优化成分设计和轧制工艺，开发出一种低碳当量易焊接热轧钢轨。通过小批量生产验证，表明钢轨化学成分设计合理、工艺可行，抗拉强度≥990MPa，延伸率≥12%，踏面中心线硬度范围≥268HB，氧化物洁净度 $K3<2$，产品性能稳定，在降低碳当量的同时，达到国标和铁标同等强度级别钢轨质量水平。

关键词： 低碳当量；易焊接；热轧；钢轨

Development of a Low Carbon Equivalent Hot-rolling Rail for Easy Welding

FENG Liguo[1], CUI Kai[2], LI Junzheng[2],
YANG Zhengzong[2], GUO Chaojun[2]

Abstract: At present, the railway is widely used seamless line, as the rail is mostly high carbon steel, it is difficult to weld, and the welding joint is the weak area of the whole line. According to the carbon equivalent calculation formula $Ceq=C+1/6Mn+1/5(Cr+Mo+V)+1/15(Ni+Cu)$, carbon is the most important element which affects the weldability of rail. A hot rolling rail with low carbon equivalent and good weldability was developed by analyzing big production data, optimizing composition design and rolling process. Through the small batch production verification, it is shown that the rail chemical composition design is reasonable and the process is feasible, the tensile strength is ≥990MPa, the elongation is ≥12%, hardness of the centre line rail running surface is ≥268HB, the oxide cleanness $K3<2$, the product performance is stable, and reaches the same level of rail quality as the national standard and the railway industry standard while reducing the carbon equivalent.

Key words: carbon equivalent; easy to weld; hot rolling; rail

大口径 UNS N08810 无缝管冷扩内壁横裂原因分析及优化

黎 毅，郭盈春，杨 帆，刘 宇，郭建邦，康喜唐

（山西太钢不锈钢钢管有限公司，山西太原 030008）

摘 要：随着光伏产业的全球化发展，其上游多晶硅项目建设迅速扩增。大口径 UNS N08810 无缝管由于其良好的腐蚀特性成为多晶硅项目建设的首选管材，其生产工艺以冷扩+冷拔为主，而在实际生产中，冷扩横裂尤其是内壁横裂，成为 UNS N08810 无缝钢管冷扩常见缺陷，从而导致产生大量废品。本论文利用金相显微镜和扫描电镜对 UNS N08810 内壁横裂试样进行了检验与分析，结合生成工艺分析，结果表明：冷扩内壁横裂裂纹主要方式是沿晶断裂，断裂处富集 Cr 的氧化物，外表面晶粒大小一致，靠近内表面出现混晶现象，使得内表面产生了裂纹。进而分析晶粒形成过程，对热穿孔、冷扩、酸洗和固溶处理工艺进一步优化，解决了内壁横裂的质量问题。

关键词：横裂；沿晶断裂；混晶；冷扩

Analysis and Optimization of Transverse Cracks on the Inner Wall of Large Diameter UNS N08810 Seamless Pipe during Cold Expansion

LI Yi, GUO Yingchun, YANG Fan, LIU Yu, GUO Jianbang, KANG Xitang

(Shanxi Taigang Stainless Steel Pipe Co., Ltd., Taiyuan 030008, China)

Abstract: With the global development of the photovoltaic industry, the construction of upstream polycrystalline silicon projects is rapidly expanding. Large diameter UNS N08810 seamless pipe has become the preferred pipe for polycrystalline silicon project construction due to its excellent corrosion characteristics, its production process mainly involves cold expansion and cold drawing. However, in actual production, cold expansion transverse cracks, especially inner wall transverse cracks, have become common defects in cold expansion of UNS N08810 seamless steel pipes, leading to the production of a large number of waste products. This paper uses a metallographic microscope and scanning electron microscope to inspect and analyze the inner wall transverse crack specimen of UNS N08810. Combined with the analysis of the generation process, the results show that the main mode of transverse crack on the cold expanded inner wall is intergranular fracture, and the fracture site is enriched with Cr oxide. The grain size on the outer surface is uniform, and mixed crystal phenomenon occurs near the inner surface, resulting in cracks on the inner surface. Furthermore, the process of grain formation was analyzed, and the processes of hot perforation, cold expansion, acid washing, and solid solution treatment were further optimized to solve the quality problem of internal wall transverse cracking.

Key words: transverse fissure; intergranular fracture; mixed crystal; cold expansion

起重机钢轨 QU80 开发实践

余选才，郑赟峰，董茂松

（武汉钢铁有限公司条材厂，湖北武汉 430083）

摘　要：通过对吊车轨 QU80 U71Mn 实际开发过程中遇到的主要问题进行分析，阐述了生产质量问题产生原因，并针对性的提出改进措施，稳定生产工艺，实现批量生产及供货。

关键词：吊车轨；QU80；U71Mn；轧制工艺

Development Practice of QU80 Crane Rail

YU Xuancai, ZHENG Yunfeng, DONG Maosong

(Plant of Long Product, Wuhan Iron & steel Co., Ltd., Wuhan 430083, China)

Abstract: Through the analysis of the main problems in the actual development of QU80 U71Mn, the causes of the production quality problems of QU80 was expounded in this paper, and some improving measures aimed at stabilizing the production process so as to realize the batch production and supply were proposed.

Key words: crane rail; QU80; U71Mn; rolling process

X80 管线钢高温热形变行为及析出研究

武晓龙，王娇娇，周玉青，高云哲，弓俊杰，梁爱国，赵林林

（河钢材料技术研究院能源材料研发中心，河北石家庄　050000）

摘　要：以 X80 管线钢为研究对象，采用 Gleeble3500 热模拟试验机检测其高温力学性能，运用金相显微镜、扫描电镜表征拉伸断口的微观组织、形貌及析出相。结果表明：600~1050℃拉伸断口为韧性断裂。在 750~775℃拉伸时发生显著的形变诱导相变，拉伸断口的室温组织为铁素体及粒状贝氏体组织；750℃拉伸断口的室温组织中观察到数量较多 Nb、Ti 的碳氮化物，表现出较强的形变诱导析出效应。因此，750℃高温热塑性最差是形变诱导铁素体转变及析出相共同作用的结果。800~1200℃时拉伸断口室温组织为粒状贝氏体及板条贝氏体，未观察到明显的形变诱导相变的痕迹。当拉伸温度超过 900℃后析出相数量明显增多且尺寸变大。

关键词：管线钢；热塑性；析出相；形变诱导相变；韧性断裂

Study on High Temperature Thermal Deformation Behavior and Precipitation of X80 Pipeline Steel

WU Xiaolong, WANG Jiaojiao, ZHOU Yuqing, GAO Yunzhe,
GONG Junjie, LIANG Aiguo, ZHAO Linlin

(HBIS Group Material Technology Research Institue, Energy Materials Research and Development Centre, Shijiazhuang 050000, China)

Abstract: The high temperature mechanical properties of X80 pipeline steel were tested by Gleeble 3500 thermal simulation testing machine, and the microstructure, fracture morphology and precipitates after tensile fracture were characterized by metallographic microscopy and scanning electron microscopy. The results show that the tensile fracture at 600~1050℃ belongs to ductile fracture. The deformation induced phase transformation occurs at 750~775℃, and the microstructure of tensile fracture at room temperature consists of ferrite and granular bainite. Besides, more Ni-Ti carbonitride are observed in the microstructure of tensile fracture at 750℃, which shows a strong deformation-induced precipitation effect. Therefore, the worst thermoplasticity at 750℃ can be attributed to the combination of deformation induced ferrite transformation and precipitates. By contrast, the microstructure at room temperature of tensile fracture from 800℃ to 1200℃ is composed of granular bainite and lath bainite, and no obvious traces of deformation induced phase transformation are observed. Moreover, when the tensile temperature exceeds 900℃, the number and the size of precipitates increase obviously.

Key words: pipeline steel; thermoplasticity; precipitate; deformation induced phase transformation; ductile fracture

合金成分对 10.9 级高强螺栓钢抗延迟断裂性能影响的研究

任学堂[1]，肖丙政[1]，王青峰[2]，张欢欢[2]

（1. 南京钢铁股份有限公司特钢事业部，江苏南京　210000；
2. 燕山大学材料科学与工程学院，河北秦皇岛　066000）

摘　要： 本文通过恒载荷试验、慢应变拉伸试验、氢渗透试验、TDS 氢热脱附试验研究合金成分对 10.9 级高强螺栓钢抗延迟断裂性能影响。研究结果表明：Nb、V、Mo 合金成分的添加一方面细化了晶粒组织，降低沿晶断裂的倾向，另一方面 V-Nb 形成的碳化物氢陷阱，减少导致延迟断裂的可扩散氢含量并使氢分布均匀，两方面共同作用提高了高强螺栓钢的抗延迟断裂性能。

关键词： 合金成分；高强螺栓钢；抗延迟断裂

Study on the Effect of Alloy Composition on the Delayed Fracture Resistance of 10.9 Grade High Strength Bolt Steel

REN Xuetang[1], XIAO Bingzheng[1], WANG Qingfeng[2], ZHANG Huanhuan[2]

(1. Nanjing Iron & Steel Co., Ltd., Nanjing 210000, China;
2. School of Materials Science and Engineering, Yanshan University, Qinhuangdao 066000, China)

Abstract: This paper studies the effect of alloy composition on the delayed fracture resistance of 10.9 grade high-strength bolt steel through constant load tests, slow strain tensile tests, hydrogen permeation tests, and TDS hydrogen thermal desorption tests. The research results show that the addition of Nb, V, and Mo alloy components not only refines the grain structure and reduces the tendency of intergranular fracture but also forms carbide hydrogen traps of V-Nb, which decrease the diffusible hydrogen content that causes delayed fracture and make the hydrogen distribution uniform. The combination of these two factors improves the delayed fracture resistance of high-strength bolt steel.

Key words: alloy composition; high-strength bolt steel; delayed fracture resistance

高温超导感应加热技术在热挤压钢管生产中的应用前景

张国栋,商庆华

(宝武特种冶金有限公司,上海 200940)

摘 要: 热挤压钢管主要生产流程包括:环形炉预热,一次感应加热,穿(扩)孔,二次感应加热,挤压机挤压等过程,感应加热炉是热挤压机组的核心设备,功能是将管坯加热至穿(扩)孔/挤压工艺所需温度,感应加热后管坯温度的均匀性决定了热挤压钢管的质量。管坯温度不均匀可引起热挤压荒管单边、夹层、微裂纹等缺陷。由于传统工频感应加热炉透热深度有限,只能对管坯表面有限的深度进行同步加热,管坯中心区域的加热靠材料自身热传导实现,所以管坯在直径方向温差较大,在挤压变形过程中局部软硬不一,这是当前热挤压钢管产线所面临的共性问题。

为满足市场对高端无缝钢管的需要,2009年宝钢引进建设的60MN热挤压管生产线投入生产。之后太钢不锈、新兴铸管、内蒙古北方重工、常熟华新特钢等也相继建成了热挤压管生产线[1]。随着我国热挤压机组的快速增加和高端无缝钢管的需求量上升,提升管坯感应加热质量水平满足高端无缝钢管品种的质量要求尤为迫切。

随着超导技术的发展和进步,高温超导感应加热技术逐步进入商业运营阶段,超导感应加热解决了传统感应加热温度不均匀的难题,且对比传统工频感应加热具有以下优势:(1)高透热深度,超导感应加热将透热深度从工频感应加热炉的50mm增加到200mm,能够实现管坯整体同步加热,温度均匀性极高;(2)高效节能,超导线圈在超导态时运行电阻为零,超导直流系统的理论损耗接近于0,只产生低温制冷损耗、机械传动损耗,可以将加热能效提升到80%以上;(3)加热速度快,超导磁体建立直流磁场约0.5~1T,是传统工频炉的100倍以上,能够将加热功率提高一倍以上,加热时间减少一半,提高生产效率[2]。

自2008年德国投产全球首台高温超导感应加热炉以来,韩国、中国相继建成高温超导感应加热炉用于有色金属热挤压产线。实验证明超导感应炉既能加热铁合金等黑色金属物体,也能加热有色金属物体[3],因此超导感应加热炉完全可应用于热挤压钢管产线,能够有效提升高端热挤压钢管的品质。此外,超导感应加热技术凭借大于80%的加热能效和加热速度快的优势,可以代替传统的环形预热炉,实现以电代燃的节能方式。在"双碳"背景的推动下,未来高温超导感应炉在高端的热挤压钢管产线中替代传统工频感应加热炉是必然趋势。

参 考 文 献

[1] 詹才俊. 钢管热挤压机组感应加热设备及其工艺操作的控制[J]. 钢管, 2013, 42(5): 55-60.
[2] 张东, 肖立业, 林良真. 高温超导直流感应加热技术研究综述[J]. 电工电能新技术, 2020, 39(7): 45-53.
[3] Jongho C, Sung-Kyu K, Kwangmin K, et al. Design and Performance Evaluation of a Multi-Purpose HTS DC Induction Heating Machine for Industrial Applications. IEEE Transactions on Applied Superconductivity, 2015, 25(3): 1-5.

PRS轧机应用特点分析

孔令波,马洪磊,王云涛,薛正国,郑佳星,张 旭

(邢台钢铁有限责任公司线材厂,河北邢台 054000)

摘　要：PRS 为新型的四辊轧制新型轧机，在线材厂进行了初步调试应用，通过试用初步取得了提高尺寸精度的效果，但也存在着过钢支数偏少和表面状况不佳的问题。通过总结以往 PRS 轧机生产调试的经验与数据，分析了 PRS 轧机孔型参数结构特点和车削加工参数特点、变形宽展特点、尺寸精度等方面的情况。对实际生产应用中存在孔型问题、轧机匹配问题以及操作繁琐等特点进行了简要分析，初步分析了 PRS 轧机目前的应用状况和推广使用的可行性。

关键词：PRS 轧机；孔型；宽展；尺寸

Application Characteristics Analysis of PRS Rolling Mill

KONG Lingbo, MA Honglei, WANG Yuntao, XUE Zhengguo,
ZHENG Jiaxing, Zhang Xu

(Wire Factory, Xingtai Iron and Steel Co., Ltd., Xingtai 054000, China)

Abstract: PRS is a new type of four-high rolling mill, which has been preliminarily tested and applied in the wire mill, and the effect of improving the dimensional accuracy has been preliminarily obtained through the trial, but there are still some problems such as too few overpasses and poor surface condition. Through summing up the experience and data of PRS rolling mill production and debugging, the characteristics of PRS rolling mill pass parameter structure, turning parameter, deformation width and dimension accuracy are analyzed. This paper briefly analyzes the problems of pass pattern, mill matching and complicated operation in practical production and application, and preliminarily analyzes the current application status of PRS mill and the feasibility of its popularization.

Key words: PRS mill; pass; spread; size

攀钢高强耐磨过共析钢轨的研制及应用

韩振宇[1]，邹　明[1]，刘丰收[2]，
祭　程[3]，王文健[4]

（1. 攀钢集团攀枝花钢铁研究院有限公司轨道交通用钢技术研究所，四川攀枝花　617000；
2. 中国铁道科学研究院集团有限公司金属及化学研究所，北京　100081；
3. 东北大学低碳钢铁前沿技术教育部工程研究中心，辽宁沈阳　110819；
4. 西南交通大学摩擦学研究所，四川成都　610031）

摘　要：针对高强耐磨过共析钢轨研制技术难题，开展了过共析重轨钢化学成分设计、高纯净度冶炼和高均质度连铸、高精度万能轧制、精准在线热处理、低残余应力矫直以及多工况焊接工艺，形成了高碳钒铬复合微合金化成分体系，钢中 T[O]平均 0.0012%，铸坯中心 C 偏析度极大值降至≤1.10；钢轨腹腔尺寸和轨底尺寸精度显著提高，轨底中心残余应力从 274MPa 降低至 184MPa。阐明了低熔点下接头过烧缺陷形成机制，探索出固定及移动闪光焊和铝热焊接工艺。攀钢在国内首家完成 U95Cr 高强耐磨过共析钢轨的研制并实现产业化应用。朔黄铁路、大秦铁路、美国 TTCI 应用表明，U95Cr 钢轨具有优异的耐磨损性能和优良的抗接触疲劳性能，母材及焊接接头服役性能稳定，适宜于重载铁路应用。

关键词：金属材料；过共析钢轨；焊接性；钢轨服役

The Development and Application of Pangang High-Strength & High Wear-Resistance Hypereutectoid Rail

HAN Zhenyu[1], ZOU Ming[1], LIU Fengshou[2], JI Cheng[3], WANG Wenjian[4]

(1. Department of Rail Transit Steels Technology, Pangang Group Panzhihua Iron and Steel Research Institute, Panzhihua 617000, China; 2. Institute of Metals and Chemistry, China Railway Science Research Institute Group Co., Ltd., Beijing 100081, China; 3. Engineering Research Center of Frontier Technologies for Low-Carbon Steelmaking (Ministry of Education), Northeastern University, Shenyang 110819, China; 4. Tribology Research Institute, Southwest Jiaotong University, Chengdu 610031, China)

Abstract: Focused on the technical difficulties in the development of high-strength & high wear-resistance hypereutectoid rail, the chemical composition design, high cleanliness and uniformity smelting, high precision universal rolling and heat-treatment, low residual stress control straightening, and mutli-condition welding technology of hypereutectoid rail were studied. The high-carbon Cr-V multi-microalloying composition system of U95Cr hypereutectoid rail was built, T[O] in rail steel was reduced to an average content of 0.0012%, and maximum C segregation degree in the center of rail steel casting bloom decreases to ≤1.10. The web and base dimensional accuracy of U95Cr hypereutectoid rail has been significantly improved, the residual stress of rail base center decreased from 274MPa to 184MPa. The forming mechanism of overburn defects in welded joint at low melting point was explained, and the technology of fixed and mobile flash welding and thermite welding were established. Pangang has completed the development of U95Cr high-strength & high wear-resistance hypereutectoid rail and achieved its industrialization popularization and application for the first time in China. The trial tests and service application results of Shuo-Huang Railway, DA-Qin Railway and TTCI in America show that U95Cr rail has excellent wear resistance and contact fatigue resistance, both base metal and welded joints have stable service performance, suitable for heavy-haul railway applications.

Key words: metallic materials; hypereutectoid rails; weldability; rail service

棒材精轧前后控冷工艺设备研究与改进

戴川涵，陈育生，黄建辉

（武钢集团昆明钢铁股份有限公司轧钢厂，云南昆明　650302）

摘　要： 针对轧钢厂九年前建成的棒材生产线控冷设备不能满足现合金成分条件下的钢材性能稳定性需要，以及轧件上冷床后头尾弯曲的问题，通过对18H出口眼镜管至3号剪前空过辊道的控冷工艺设备和增加精轧机组前控冷设备研究改进。实现了棒材成品冷却均匀，性能稳定，通过优化精轧前后控冷工艺，在保证钢材性能的前提下，减少炼钢合金的使用量，降低公司生产成本。

关键词： 棒材；精轧；控冷；工艺；设备

Research and Improvement of Controlled Cooling Process Equipment Before and after Bar Finishing Rolling

DAI Chuanhan, CHEN Yusheng, HUANG Jianhui

(Wuhan Iron and Steel Group Kunming Iron and Steel Co., Ltd.,
Rolling Mill, Kunming 650302, China)

Abstract: In order to solve the problem that the control cooling equipment of bar-1 production line built nine years ago in rolling mill can not meet the performance stability of steel under the condition of current alloy composition, and the bending of the end and end after the cooling bed on the rolled piece, the control cooling process equipment of the 18H exit glass tube to the 3# before cutting empty over the roll table and the addition of the front control cooling equipment of the finishing mill group were studied and improved. By optimizing the controlled cooling process before and after finishing rolling, under the premise of ensuring steel properties, the use of steel alloy is reduced, and the production cost of the company is reduced.

Key words: bar; finishing rolling; cold control; technology; equipment

钢结构建筑用高强度厚规格耐候抗震热轧 H 型钢研制

卜向东[1]，涛 雅[1]，郭利宏[2]，王永明[3]，卢雄慧[4]，宋振东[1]

（1. 内蒙古包钢钢联股份有限公司技术中心，内蒙古包头 014010；
2. 内蒙古包钢钢联股份有限公司轨梁厂，内蒙古包头 014010；
3. 内蒙古包钢钢联股份有限公司总工室，内蒙古包头 014010；
4. 内蒙古包钢钢联股份有限公司炼钢厂，内蒙古包头 014010）

摘 要： 根据装配式钢结构住宅使用需求，采用低碳铜铬镍微合金成分体系包晶钢设计，近终型异型坯连铸技术，型钢万能轧制技术，自主开发了 420MPa 级高强度厚规格耐候抗震专用热轧 H 型钢，该产品主要应用于钢结构住宅框架结构的立柱，通过连铸工艺及轧制工艺优化，解决了铸坯翼缘内侧皮下裂纹及中间裂纹，轧制过程中氧化铁皮压入带来的成品表面缺陷，成品各项力学性能均满足技术协议要求，外形尺寸及表面质量满足 GB/T 11263 标准要求，最终实现了厚规格高强耐候抗震热轧 H 型钢批量稳定供货。

关键词： 高强度；厚规格；耐候；抗震；H 型钢

Development of High Strength Thick Gauge Weather-resistant and Anti-seismic Hot-rolled H-Beam

BU Xiangdong[1], TAO Ya[1], GUO Lihong[2], WANG Yongming[3],
LU Xionghui[4], SONG Zhendong[1]

(1. Technical Center of Inner Mongolia Baotou Steel Union Co., Ltd., Baotou 014010, China;

2. Rail and Beam Plant of Inner Mongolia Baotou Steel Union Co., Ltd., Baotou 014010, China;
3. Chief Engineer Office of Inner Mongolia Baotou Steel Union Co., Ltd., Baotou 014010, China;
4. Steel-making Plant of Inner Mongolia Baotou Steel Union Co., Ltd., Baotou 014010, China)

Abstract: According to the use requirements of prefabricated steel structure residential buildings, 420MPa grade high-strength thick specification weather-resistant and anti-seismic special hot-rolled H-shaped steel has been designed and developed, and all properties of the designed products meet the requirements of the technical agreement; Through the optimization of continuous casting process and rolling process, the subcutaneous cracks and intermediate cracks on the inner side of the slab flange and the surface defects of the finished product caused by the pressing of iron oxide scale during the rolling process have been solved.

Key words: high strength; thick specification; weather resistance; anti-seismic; H-shaped steel

60～50kg/m U75V 热轧异型钢轨跟端热处理工艺研究

张凤明，梁正伟，薛虎东

（内蒙古包钢钢联股份有限公司技术中心，内蒙古包头 014010）

摘 要：本文对 60～50kg/m U75V 热轧异型钢轨跟端锻压区域，进行了不同终冷温度及冷却速度的热处理工艺试验研究，分析了异型钢轨成型段、过渡段及热影响区的拉伸性能、轨顶面硬度、横断面硬度、显微组织、脱碳层。结果表明，异型钢轨成型段和过渡段终冷温度控制在 650℃左右，冷却速度控制在 1.3～1.7℃/s 之间，热影响区终冷温度控制在 550℃以下，冷却速度大于等于 2℃/s，可有效地解决异型钢轨成型段及过度段硬度偏高、热影响区硬度偏低、软化区宽度超标的问题。

关键词：U75V 钢轨；异型钢轨；热处理工艺

Study on Heat Treatment Process of 60~50kg/m U75V Hot Rolled Special-shaped Rail Heel End

ZHANG Fengming, LIANG Zhengwei, XUE Hudong

(Technical Center of Inner Mongolia Baotou Steel Union Co., Ltd., Baotou 014010, China)

Abstract: In this paper, the heat treatment process of 60～50kg/m U75V hot rolled special-shaped rail with different final cooling temperatures and cooling rates is experimentally studied, and the tensile properties of the forming section, transition section and heat-affected zone of the special-shaped rail, the hardness of the top surface of the rail, the hardness of the cross section, the microstructure and the decarburization layer are analyzed. The results show that the final cooling temperature of the forming section and the transition section of the special-shaped rail is controlled at about 650℃, the cooling speed is controlled between 1.3～1.7℃/s, the final cooling temperature of the heat affected zone is controlled below 550℃, and the cooling speed is greater than or equal to 2℃/s. It can effectively solve the problems of high hardness in the forming section and the excessive section of the special-shaped rail, low hardness in the heat affected zone and excessive width in the softening zone.

Key words: U75V rail; special-shaped rail; heat treatment process

江西台鑫钢铁有限公司双高速棒材轧钢生产线新工艺设备技术应用

陈立胜

(安徽诺泰工程技术有限公司，重庆 401120)

摘　要：2018 年热轧带肋钢筋新国标颁布后，热机轧制工艺技术得到了广泛应用，成功通过轧钢工序降低生产成本，提高产品性能。经过近几年不断发展和进步，与热机轧制配套的高强度模块轧机也日趋成熟完善，形成了完整的系列化产品。江西台鑫钢铁有限公司双高速棒材轧钢生产线，采用全新一代热机轧制工艺技术（多级控轧控冷技术）、免加热直轧工艺、柔性控制冷却，高速区采用重型 RVM330 系列模块轧机。配合低成本生产工艺技术，实现低合金成分生产，精确负公差轧制，轧材晶粒细小均匀，产品性能稳定，生产成本大幅降低。

关键词：热机轧制；柔性控制冷却；免加热直轧工艺；低成本生产技术；模块轧机；低合金成分

Application of New Technology and Equipment in Double High Speed Bar Production Line of Jiangxi Taixin Steel Co., Ltd.

CHEN Lisheng

(Anhui Rolltai Engineering Co., Ltd., Chongqing 401120, China)

Abstract: The double high-speed bar rolling production line of Jiangxi taixi steel Co., Ltd. adopts a new generation of hot rolling technology (multi-stage controlled rolling and controlled cooling technology), and the heavy-duty RVM330 Series modular rolling mill is used in the high-speed area.

Key words: TMCP; flexible controlled cooling; non-heating direct rolling process; low cost production technology; modular mill; low alloy composition

稀土轨研究发展与现状

薛虎东，赵桂英，王慧军，郑　瑞，刘　阳

(内蒙古包钢钢联股份有限公司技术中心，内蒙古包头 014010)

摘　要：近年来，稀土在工业化生产中加入和利用已取得较大突破，利用稀土净化钢水和微合金化，起到了改善钢的纯净度和提高韧性等作用。稀土在薄板、无缝钢管等产品中已进行工业应用，并且部分产品实现了规模化生产。钢轨作为铁路用重要产品，稀土在重轨的应用研究也一直在进行，利用稀土对钢轨内部夹杂物控制和力学性能综合提高一直是服役长寿化研究的重要方向。本文简述我国稀土轨发展历程，总结了各个阶段稀土轨研究主要工作，重点介绍稀土轨研究现状和产业化应用，提出了我国铁路用稀土轨发展建议。

关键词：钢轨；稀土；长寿化

Development and Current Situation of Rare Earth Orbital Research

XUE Hudong, ZHAO Guiying, WANG Huijun,
ZHENG Rui, LIU Yang

(Inner Mongolia Baotou Steel Union Co., Ltd., Technical Center, Baotou 014010, China)

Abstract: In recent years, significant breakthroughs have been made in the addition and utilization of rare earth elements in industrial production. The use of rare earth elements to purify molten steel and microalloy has played a role in improving the purity and toughness of steel. Rare earth elements have been applied industrially in products such as thin plates and seamless steel pipes, and some products have achieved large-scale production. As an important product for railways, research on the application of rare earths in heavy rail has also been ongoing. The use of rare earths to control internal inclusions and comprehensively improve mechanical properties of rails has always been an important direction for research on service life. This article briefly describes the development process of rare earth rails in China, summarizes the main research work on rare earth rails at various stages, focuses on the current research status and industrial applications of rare earth rails, and puts forward suggestions for the development of rare earth rails for railways in China.

Key words: steel rail; rare earth; longevity

美标、欧标与铁标钢轨标准差异分析

边　影，薛虎东，王嘉伟，彭　飞

（内蒙古包钢钢联股份有限公司技术中心，化检验中心，内蒙古包头　014010）

摘　要：介绍了美标 AREMA—2019、欧标 EN 13674.1—2011 与铁标 TB/T 2344—2012 钢轨标准的差异及应用现状，对比分析了不同国家钢轨标准的断面、成分、性能检验方法。结果表明，虽然我国铁路标准在制定过程中借鉴了国外不少先进标准制定准则，但由于国内外地理环境、标准层次、体例的差异，导致国内钢轨准方面与欧标、美标标准存在一些差异。通过对比这些标准差异，充分了解国内钢轨出口水平，将国内钢轨更广泛的推向国际市场。

关键词：技术标准；成分标准；检验方法

Analysis of American Standard, European Standard and Iron Standard Rail Standard Difference

BIAN Ying, XUE Hudong, WANG Jiawei, PENG Fei

(Technical Center of Inner Mongolia Baotou Steel Union Co., Ltd., Baotou 014010, China)

Abstract: This paper introduces the difference and application status of American standard AREMA—2019, European standard EN 13674.1—2011 and iron standard TB/T 2344—2012 rail standards, and analyzes the section, composition and performance test methods of different national rail standards. The results show that although China's railway standards draw lessons from many foreign advanced standards in the formulation process, there are some differences in domestic European rail standards and European standards and American standards. By comparing these standard differences, fully understand the domestic rail export level, the domestic rail more widely to the international market.

Key words: technical standard; composition standard; inspection method

轨梁厂二号线码垛台架运输链测距方法改进

刘晓宇

（内蒙古包钢钢联股份有限公司轨梁轧钢厂，内蒙古包头 014010）

摘　要： 在钢铁企业中，码垛台架运输链大多是通过码盘测量距离进行反馈，经过计算来实现台架自动上钢。通过码盘对运输链进行控制，虽然控制精准，但是这种方法有以下缺点：传感元件所需的电源及信号线路较多，造成码垛台架下管线路由十分复杂，加之台架下环境恶劣，在使用过程中，不但易造成码盘等紧密传感仪器和网线的损坏，而且容易引起网络故障，维护起来十分麻烦。由于路由复杂，线路较多，一旦发生故障，查找时间较长，由此引发的事故时间较多。维护费用高，码盘价格高昂，一旦损坏，备件费用极高。

关键词： 矢量控制；模拟步距控制；通讯字；模拟位置

万能轧机装配精度对钢轨万能轧法的影响

郭大林，陈　琳，段永强，王　琨

（内蒙古包钢钢联股份有限公司轨梁轧钢厂，内蒙古包头 014010）

摘　要： 万能轧机对于热轧产品质量影响的因素主要包含水平辊、立辊以及导卫系统。装配精度对于钢轨产品，会产生头偏、上下腿厚度不均、出钢上下弯以及对称性不合等问题。通过对装配数据的优化，可以有效提高装配精度，从而减少废品量，提升钢轨成材率。

关键词： 万能轧机；装配精度；水平辊；立辊；导卫

国内棒材高速上钢系统的技术特点及发展趋势

张俊义，方针正，董红卫，何巍巍

（中钢设备有限公司，北京 100080）

摘　要： 介绍了棒材高速上钢系统的发展历程，对棒材高速上钢系统的性能特点，特有技术的进行了分析和探讨，对未来的发展趋势进行了预测，对棒材高速上钢的系统性了解和设计具有重要指导意义。

关键词： 棒材高速上钢系统；工艺特点；高速上钢主要技术

Technical Overview and Development Trend of Domestic High-speed Bar System

ZHANG Junyi, FANG Zhenzheng,
DONG Hongwei, HE Weiwei

(Sinosteel Equipment & Engineering Co., Ltd., Beijing 100080, China)

Abstract: This article introduces the development history of the high-speed bar system, analyzes and discusses the performance characteristics and unique technologies of the high-speed bar system, predicts the future development trend, and has important guiding significance for the systematic understanding and design of high-speed bar system.

Key words: high speed bar system; process characteristics; main technology of high-speed bar system

高速棒材连续工作制飞剪的力学性能和参数分析

张俊义[1]，方针正[1]，惠秦川[2]，董红卫[1]，何巍巍[1]

（1. 中钢设备有限公司，北京 100080；2. 北京佰能盈天科技股份有限公司，北京 100080）

摘　要：综合分析了连续工作制飞剪的工作状态，对其启动性能，加减速性能，剪切能力等力学性能参数计算和分析，讨论了连续工作制飞剪的设计要点，对最高轧制速度可达45m/s的高速棒材生产中的连续工作制飞剪的设计具有重要指导意义。

关键词：高速棒材；连续工作制飞剪；力学性能

The Mechanical Performance Analysis of Continuous Working Flying Shear in High-speed Bar Line

ZHANG Junyi[1], FANG Zhenzheng[1], HUI Qinchuan[2],
DONG Hongwei[1], HE Weiwei[1]

(1. Sinosteel Equipment & Engineering Co., Ltd., Beijing 100080, China;
2. Beijing Bestpower Intelcontrol Co., Ltd., Beijing 100080, China)

Abstract: In this paper continuous working flying shear is comprehensively analyzed in starting performance, acceleration and deceleration performance, shear capacity and other mechanical performance parameters, and its design points are discussed, which has important guiding significance for the design of continuous working flying shear in high-speed bar line with maximum rolling speed of 45m/s.

Key words: high-speed bar line; continuous working flying shear; mechanical properties

高速线材升级改造探索与实践

荣 崎,唐心语

(安徽诺泰工程技术有限公司,重庆 401120)

摘 要:为实现轧钢生产的绿色环保化,势必引进不断迭代的先进轧钢技术,保证钢铁工业的健康发展,提高产品高附加值。本文介绍了全新一代热机轧制工艺技术(多级控轧控冷技术)及其核心装备,以及成功应用于宝武集团鄂城钢铁有限公司高线热机轧制改造项目的经验,成功实现降低合金使用量、提高产品性能、降低生产成本,逐渐转型为优质工业用材生产线,轧线装备水平得到质的提升。对于落后高速线材项目改造较强的借鉴意义。

关键词:热机轧制;核心装备;低成本;技术改造

Exploration and Practice of Upgrading and Transforming High Speed Wire Rod

RONG Qi, TANG Xinyu

(Anhui Rolltai Engineering Co., Ltd., Chongqing 401120, China)

Abstract: In order to realize the green environmental protection of steel rolling production, it is necessary to introduce the continuous iteration of advanced steel rolling technology to ensure the healthy development of iron and steel industry and improve the high added value of products. This paper introduces the new-generation hot mill rolling technology (multi-stage controlled rolling and controlled cooling technology) and its core equipment, as well as the successful application experience in the high-speed wire mill rolling transformation project of Baowu Group E Cheng Iron and Steel Co., Ltd., we have successfully reduced the amount of alloy used, improved product performance, reduced production costs, and gradually transformed into high-quality industrial production lines, rolling line equipment has been upgraded. It has strong reference significance for the renovation of outdated high-speed wire rod projects.

Key words: TMCP; core equipment; low cost; technical transformation

矿用锚索用高强度Φ14mm 82B盘条网状渗碳体成因分析

陈 列[1],朱子轩[2,3],郭 鑫[1],陶立志[1],
潘晓坤[2,3],张朝磊[2,3]

(1. 建龙北满特殊钢有限责任公司,黑龙江齐齐哈尔 161041;2. 北京科技大碳中和研究院,北京 100083;3. 辽宁材料实验室钢铁再生技术研究所,辽宁沈阳 110169)

摘 要:斯太尔摩风冷线生产的大规格82B盘条易出现网状渗碳体,从而使盘条塑性变差,导致后续拉拔过程中出现断裂，严重影响产品质量和生产效率。本文通过场发射扫描电镜（SEM）、碳硫分析仪以及等离子体光谱仪等实验手段观察了盘条横截面的金相组织并测定了盘条横截面的元素分布情况。结果表明：82B 盘条心部存在严重的网状渗碳体，是因为斯太尔摩风冷线上冷速过慢导致的。经风冷工艺优化，网状渗碳体问题得到很好解决。

关键词：斯泰尔摩风冷线；大规格 82B 盘条；网状渗碳体；偏析；冷却速度

Cause Analysis of Mesh Cementite of High Strength Φ14 mm 82B Wire Rod for Mine Anchor Cable

CHEN Lie[1], ZHU Zixuan[2,3], GUO Xin[1], TAO Lizhi[1], PAN Xiaokun[2,3], ZHANG Chaolei[2,3]

(1. Jianlong Beiman Special Steel Co., Ltd., Qiqihar 161041, China; 2. Institute for Carbon Neutrality, University of Science and Technology Beijing, Beijing 100083, China; 3. Institute of Steel Sustainable Technology, Liaoning Academy of Materials, Shenyang 110169, China)

Abstract: The large-size 82B wire rod produced by Stelmor air cooling line is prone to network cementite, which makes the plasticity of the wire rod worse, resulting in fracture in the subsequent drawing process, seriously affecting product quality and production efficiency. In this paper, the metallographic structure of the cross section of the wire rod was observed and the element distribution of the cross section of the wire rod was measured by means of field emission scanning electron microscopy (SEM), carbon sulfur analyzer, and plasma spectrometer. The results show that there is a serious network cementite in the core of 82B wire rod, which is caused by the slow cooling rate on the Stelmor air cooling line. After the optimization of air cooling process, the problem of network cementite is well solved.

Key words: Stelmor line; large size 82B wire rod; network cementite; segregation; cooling rate

热轧钢筋分段气雾冷却工艺及控制模型研究

赵舸[1]，王卫卫[1]，丁哲毅[2]，潘富强[2]，范仲继[2]，
宋为[2]，严雪蕾[2]，苏进彬[2]，唐建生[2]，周道树[2]，
李丰德[2]，林华春[2]，魏勇[2]，黄标彩[2]

（1. 钢铁研究总院冶金工艺研究所，北京 100081；
2. 福建省三钢集团有限责任公司，福建三明 365000）

摘 要：通过对热轧钢筋分级气雾冷却设备、工艺和控制模型的系统研究，获得了气雾冷却工艺对显微组织和力学性能的影响规律，为今后的热轧钢筋降本增效提供技术参考。研究表明，气雾冷却工艺能细化边部到心部的显微组织，可以细化珠光体团块尺寸及片层间距，强化效果明显，同时组织性能预报模型（屈服强度、抗拉强度、伸长率）精度达到 98% 以上。

关键词：高性能钢筋；分级气雾冷却；工艺；珠光体细化；组织性能预报模型

Research on the Staged Spray Evaporative Cooling Process and Control Model for Hot-rolled Rebar

ZHAO Ge[1], WANG Weiwei[1], DING Zheyi[2], PAN Fuqiang[2], FAN Zhongji[2], SONG Wei[2], YAN Xuelei[2], SU Jinbin[2], TANG Jiansheng[2], ZHOU Daoshu[2], LI Fengde[2], LIN Huachun[2], WEI Yong[2], HUANG Biaocai[2]

(1. Metallurgical Technology Institute of Central Iron & Steel Research Institute, Beijing 100081, China;
2. Fujian Sanming Iron and Steel Group Co., Ltd., Sanming 365000, China)

Abstract: Through systematic research on the equipment, process, and control model of the staged spray evaporative cooling process for hot-rolled steel rebar, the influence of spray evaporative cooling process on microstructure and mechanical properties was obtained, providing technical reference for future cost reduction and efficiency improvement of hot-rolled steel bars. The results show that spray evaporative cooling process can refine the microstructure from the edge to the center, refine the size and interlayer spacing of pearlite agglomerates, and achieve significant strengthening effects. At the same time, the accuracy of the microstructure and properties prediction model (yield strength, tensile strength, elongation) can reach over 98%.

Key words: high performance rebar; staged spray evaporative cooling; process; pearlite refinement; microstructure and properties prediction model

热处理工艺对超深井用非标油套管组织性能的影响

臧 岩[1,2]，李天怡[1,2]，刘文月[1,2]，安 涛[1,2]，李江文[1,2]

（1. 海洋装备用金属材料及其应用国家重点实验室，辽宁鞍山 114009；
2. 鞍钢集团北京研究院有限公司，北京 102200）

摘 要：本文研究了不同热处理工艺对非标油套管组织和性能的影响，通过采用扫描电镜（SEM）、电子背散射衍射（EBSD）、透射电子显微镜（TEM）对试验钢的显微组织进行了观察和表征，并测定了实验钢的拉伸和冲击性能。结果表明，随着奥氏体化温度的升高，材料强度和冲击功均呈先上升后下降的趋势；随着回火温度的升高，材料的抗拉强度呈现降低的趋势，冲击功一直升高。与目标性能相比，在860~900℃奥氏体化+620℃回火，或900℃奥氏体化+610~640℃回火，材料强韧性匹配满足力学性能要求。

关键词：石油套管；热处理工艺；微观组织；强韧性

Effect of Heat Treatment Technology on Microstructure and Properties of Non-standard Oil Casing Used in Ultra-deep Wells

ZANG Yan[1,2], LI Tianyi[1,2], LIU Wenyue[1,2], AN Tao[1,2], LI Jiangwen[1,2]

(1. State Key Laboratory of Metal Materials for Marine Equipment and Application, Anshan 114009, China; 2. Ansteel Beijing Research Institute Co., Ltd., Beijing 102200, China)

Abstract: In this paper, the effects of different heat treatment processes on the microstructure and properties of the non-standard oil casing were studied. The microstructure of the test steel was observed and characterized by scanning electron microscopy (SEM), electron back scattering diffraction (EBSD) and transmission electron microscopy (TEM), and the tensile and impact properties of the test steel were measured. The results show that with the increase of austenitizing temperature, the strength and impact energy of the materials increase first and then decrease. With the increase of tempering temperature, the tensile strength of the material decreases and the impact energy increases. Compared with the target properties, austenitizing at 860~900℃ +620℃ tempering, or austenitizing at 900℃ +610~640℃ tempering, the material strength and toughness match to meet the mechanical properties requirements.

Key words: casing and tubing; heat treatment process; microstructure; strength and toughness

2000MPa 级以上超高强钢增塑/韧新机制及其应用

李云杰

（东北大学，辽宁沈阳 110167）

摘 要：针对超高强钢塑韧性低的共性难题，创新提出"马氏体拓扑学结构设计+亚稳相调控"协同增塑新机制，采用绿色化 C-Mn 合金设计和简单热加工工艺构筑出一种全新的拓扑学有序排列的"马氏体+多尺度亚稳奥氏体"纳米级多层次结构。该结构可以诱发板条面位错滑移、界面滑移以及 TRIP 效应等多种增强、增塑、增韧机制，突破了 2000MPa 级超高强钢的性能边界。报告介绍了该机制在热轧、冷轧及锻造等领域的应用，可制备热轧薄板/厚板、冷轧板以及轴类锻件等，在工程机械、汽车工业以及国防军工等领域具备重大应用潜力。针对超高强钢成形难题，提出了"低强度成形+超高强度使用"新策略，实现了 2.2GPa 超高屈服强度和 10%以上均匀延伸率的零件使用性能，大幅度提升材料服役安全性。

关键词：超高强钢；增塑/韧新机制

7 表面与涂镀

大会特邀报告
第十三届冶金青年科技奖获奖人特邀报告
分会场特邀报告
矿业工程
焦化及节能环保
炼铁与原料
炼钢与连铸
电冶金与废钢铁
轧制与热处理
★ 表面与涂镀
金属材料深加工
粉末冶金
先进钢铁材料及应用
节能与低碳技术
冶金环保与资源利用
冶金设备与工程技术
冶金自动化与智能化
冶金物流
冶金流程工程学
其他

工业化生产中热镀锌板表面小渣点缺陷原因研究

宋青松，马幸江，黄 宾，陈 彪，段晓溪

（首钢京唐钢铁公司，河北唐山 063200）

摘 要：本文主要利用扫描电镜（SEM）进行表面微观形貌与 EDS 能谱分析，分析了汽车外板用热镀锌板表面渣点缺陷的形成原因，并提出了有效的控制措施，通过研究表明：造成渣点的形成原因有四种，通过控制带钢生产前产线的清洁度，调整炉鼻子内氧含量以及炉鼻子加湿量，调节锌锅工艺参数，减少镀层凝固过程中的脏污的黏附，可有效控制小渣点缺陷的产生。

关键词：热镀锌；黑点；灰点；翘皮

Study on the Causes of Small Slag Defects on the Surface of Hot Galvanized Sheet in Industrial Production

SONG Qingsong, MA Xingjiang, HUANG Bin, CHEN Biao, DUAN Xiaoxi

(Jingtang Iron and Steel Union Company Limited, Capital Iron and Steel Company, Tangshan 063200, China)

Abstract: This paper mainly uses scanning electron microscope (SEM) to analyze the surface morphology and EDS energy dispersive spectrum, analyzes the causes of slag point defects on the surface of hot galvanized sheet for automobile outer plate, and puts forward effective control measures. The research shows that: There are four reasons for the formation of slag spots. The small slag spots can be effectively controlled by controlling the cleanliness of the production line before strip production, adjusting the oxygen content in the furnace nose and the humidification of the furnace nose, adjusting the technological parameters of the zinc pot, and reducing the adhesion of dirt in the solidification process of the coating.

Key words: hot dip galvanizing; black dot; gray dot; warped skin

高耐蚀锌铝合金镀层 Al 含量选择研究

金永清，黄 建，吴林倬

（攀枝花钢钒有限责任公司板材厂，四川攀枝花 617000）

摘 要：通过改良森吉米尔法热镀锌机组试制了不同 Al 含量的锌铝合金镀层钢板，对镀层组织、耐蚀性、加工性和锌花状态进行了研究。结果表明，锌铝合金镀层由先共晶富 Al 相和 Zn-Al 二元共晶组织组成，随着 Al 含量增加，镀层中先共晶富 Al 相的比例增加，枝晶变粗变多，Zn-Al 二元共晶组织比例减少，形状由网格状变为细条状结构；同时镀层加工性下降，锌花立体感增强。镀层耐蚀性优秀，随 Al 含量升高平板 NSST 有增加趋势，杯突和 T 弯 NSST

变化趋势不明显。

关键词：Al 含量；锌铝合金；耐蚀性；凝固组织；加工性

Study on the Selection of Al Content in High Corrosion Resistance Zinc Aluminum Alloy Coatings

JIN Yongqing, HUANG Jian, WU Linzhuo

(Plate Factory of Panzhihua Steel Vanadium Co., Ltd., Panzhihua 617000, China)

Abstract: The Zn-Al coating steel sheets with with different Al contents were trial produced by Modified Semgemir hot dip gAlvanizing line,and the coating structure, corrosion resistance, workability, and zinc spray state were studied.The results show that the Zn-Al alloy coating consists of pre-eutectic rich Al phase and Zn-Al binary eutectic structure. With the increase of Al content, the proportion of pre-eutectic rich Al phase in the coating increases, dendrites become thicker and more abundant, and the proportion of Zn-Al binary eutectic structure decreases. The shape changes from a grid like structure to a fine strip like structure; At the same time, the processability of the coating decreases, the three-dimensional sense of zinc Spanglesis enhanced. The coating has excellent corrosion resistance, and the NSST of the flat plate shows an increasing trend with the increase of Al content, while the change trend after Cupping test and T-bend NSST is not significant.

Key words: Al content; Zn-Al alloy; corrosion resistance; solidified structure; processability

浅析传统镀锌机组集约化智能化的优化设计

董 斌[1]，杨柏松[1]，李元亭[1]，叶 波[2]，向双发[2]

（1. 赛迪热工环保工程技术有限公司，重庆 401122；
2. 深圳市希洛奥德科技有限公司，广东深圳 518000）

摘 要：本文以传统镀锌机组为背景，按照工艺段分布逐梳理了影响传统机组智能化整体水平的各个因素；提出了实现抢救组的集中智能控制的基本框架；以及影响集中智能化的关键环节及智能化子系统实施方案；为传统机组实现集约化、智能化提供了切实可行的优化设计及解决方案。

关键词：镀锌机组；智能化；集中控制；工业机器人

Preliminary Analysis of Intelligent System Design for a Traditional Galvanizing Line

DONG Bin[1], YANG Baisong[1], LI Yuanting[1], YE Bo[2], XIANG Shuangfa[2]

(1. Chongqing CISDI Thermal & Environmental Engineering Co., Ltd., Chongqing 401122, China;
2. Generica Strapping Machine, Shenzhen 518000, China)

Abstract: Based on the traditional galvanizing line as the background, this article preliminary sorts out the factors that affect the overall level of intelligence of the traditional machine set one by one. The overall centralized control scheme and the implementation scheme of intelligent subsystem of key links are proposed, which provides a practical overall solution for the overall intelligent of traditional units.

Key words: galvanizing line; intelligent equipment; centralized control; industrial robots

连续锌铝镁热涂镀机组针对不同锌铝镁配方的设计优化

杨柏松[1]，环素学[2]

（1. 赛迪热工环保工程技术有限公司，重庆　401122；
2. 河北敬业高品钢科技有限公司河北省，河北石家庄　050400）

摘　要：本文浅析了锌铝镁机组在涂镀区的共性要求；提论述了对锌鼻子、三辊六臂、气刀、镀后冷却器的设计原则；论述了清洗段采用电解清洗结合磁过滤器、炉内高氢冷却器、水冷塔顶辊、消除蒸汽水淬槽、大辊径光整机的选型优化设计；论述了超融服务器结合工业机器人为机组解决智能化、集约化控制创造条件。

关键词：镀后上行冷却器；磁过滤器；超融服务器；机器人

Strip Coating Line Optimal Design for Different ZAM Coating Contents

YANG Baisong[1], HUAN Suxue[2]

(1. Chongqing CISDI Thermal & Environmental Engineering Co., Ltd., Chongqing 401122, China;
2. Hebei Jingye High-Quality Steel Technology Co., Ltd., Shijiazhuang 050400, China)

Abstract: This paper briefly analyzes the common requirements of ZAM line equipment in the coating area. The design principles of zinc snout、sink roller、air knife and up-stroke post air　cooler were proposed. It is proposed that the cleaning section adopts electrolytic cleaning combined with magnetic filter, high hydrogen RJC cooler inside the furnace, water-cooled tower top roller, steam free water quenching tank, and large roller diameter skin pass machine optimization design; It is proposed that the hyperfusion server combined with industrial robots create conditions for the ZAM line to solve intelligent and intensive control.

Key words: multi partition up-leg air jet cooler; magnetic filter; hyperfusion server; industrial robots

高强厚规格热基镀锌板生产工艺研究

贾海超

（河钢集团承钢公司冷轧事业部，河北承德　067000）

摘　要：在生产高强度厚规格镀锌板的生产过程中，初步摸索出了生产此类镀锌产品的相关工艺参数。通过对退火炉温度、锌液温度及成分、气刀、产线张力等工艺参数的现场观察调整，有效改善了高强度厚规格镀锌板的麻面、针孔漏镀、划伤、卷形质量等问题。

关键词：高强度；厚规格；麻面；针孔漏镀；划伤；卷形质量

Study on Production Technology of High Strength and Thick Specification Hot Base Galvanized Sheet

JIA Haichao

(Cold Rolling Division, Chengsteel Company, Hesteel Group, Chengde 067000, China)

Abstract: In the process of producing high strength thick specification galvanized sheet, the relevant technological parameters of producing this kind of galvanized product were initially explored. Through the field observation and adjustment of annealing furnace temperature, zinc liquid temperature and composition, gas knife, production line tension and other process parameters, the problems of pockmarked surface, pinhole leakage plating, scratch and roll shape quality of high strength thick gauge galvanized sheet are effectively improved.

Key words: high strength; thick specifications; pits; the pinhole leak plating; scratch; roll quality

镀锌板上制备氮化钛镀层实验研究

徐闻慧，杨洪刚，金　勇，吕家舜，徐承明，王永明

（海洋装备用金属材料及其应用国家重点实验室，
鞍钢股份有限公司技术中心，辽宁鞍山　114009）

摘　要：本文主要研究采用不同工艺在镀锌板上制备氮化钛镀层的可行性以及镀层的性能，采用电弧离子镀及磁控溅射的方式在镀锌板上制备氮化钛镀层，利用辉光放电发射光谱仪（GDS）分析了镀层中各元素沿深度方向的分布，利用扫描电镜（SEM）及能谱分析（EDX）观察了镀层截面的微观形貌、合金层的形貌和镀层中各元素分布组成，利用电子探针（EPMA）分析了镀层表面各元素的分布，利用了冷弯实验及共聚焦显微镜（OLS）分析了镀层的粘附性，采用盐雾腐蚀试验方法检测了镀层的耐蚀性。结果表明，采用磁控溅射的方式在镀锌板上制备氮化钛镀层，其表面均匀致密，氮化钛层与锌层结合良好，盐雾试验24h无白锈，具备良好的耐腐蚀性能。

关键词：氮化钛镀层；电弧离子镀；磁控溅射

Experomental Study on Preparation of TiN Coating on Galvanized Sheet

XU Wenhui, YANG Honggang, JIN Yong, LV Jiashun, XU Chengming, WANG Yongming

(Ansteel Company Limited Technology Center, Anshan 114009, China)

Abstract: In this paper, the feasibility of preparing titanium TiN coating on galvanized sheet by different technologies and the performance of the coating were studied. TiN was prepared by arc ion plating and magnetron sputtering on galvanized sheet. The distribution of elements in the coating along the depth direction was analyzed by GDS. SEM and EDX were used to observe the microstructure of the cross section of the coating, the morphology of the alloy layer and the distribution of elements in the coating. The distribution of elements on the surface of the coating was analyzed by EPMA. The adhesion of the coating was analyzed by cold bending experiment and OLS. The corrosion resistance of the coating was tested by salt spray corrosion test. The results show that the TiN coating was prepared on the galvanized sheet by magnetron sputtering. The surface of the coating is uniform and compact, and it is well combined with the zinc coating. The salt spray test has no white rust for 24 hours, and has good corrosion resistance.

Key words: TiN coating; arc ion plating; magnetron sputtering

表面粗糙度对全无铬钝化热镀锌板耐腐蚀性能的影响

黎 敏[1]，张 晨[2]，王长成[2]，鲁洋泽[2]，
任新意[2]，邵 蓉[1]，刘永壮[1]

（1. 首钢集团有限公司技术研究院，北京 100041；
2. 首钢京唐钢铁联合有限责任公司，河北唐山 063200）

摘 要：通过三维形貌分析、GDS测试、电化学分析、润湿角测试、中性盐雾实验、循环盐雾实验研究了表面粗糙度对全无铬钝化热镀锌板耐腐蚀性能的影响，结果表明：随着粗糙度增加，其润湿角变小，疏水性变差，其容抗弧变小。这可能是由于在钝化膜厚度一定时，当材料表面粗糙度较小时，其光整坑波峰和波谷的差值较小，有利于钝化膜的完整和连续，当材料表面粗糙度较大时，其钝化膜连续性较差，波峰位置整体覆盖较差。SST和CCT结果表明：在钝化膜重量一定的情况下，粗糙度越大，钝化膜在波峰位置覆盖的均匀性越差，耐蚀性越差。

关键词：GI板；钝化膜；循环腐蚀；表面粗糙度

Effect of Surface Roughness on Corrosion Resistance of Chromium-free Passivated Hot Dip Galvanized Sheet

LI Min[1], ZHANG Chen[2], WANG Changcheng[2], LU Yangze[2],
REN Xinyi[2], SHAO Rong[1], LIU Yongzhuang[1]

(1. Shougang Research Institute of Technology, Beijing 100041, China;
2. Shougang Jingtang United Iron and Steel Co., Ltd., Tangshan 063200, China)

Abstract: The influence of surface roughness on the corrosion resistance of chromium-free passivated hot dip galvanized sheet was studied by three-dimensional morphology analysis, GDS test, electrochemical analysis, wetting angle test, neutral salt spray test and cyclic salt spray test. The results showed that with the increase of roughness, its wetting angle became smaller, its hydrophobicity became worse, and its capacitive reactance arc became smaller. This may be due to the fact that when the thickness of the passivation film is constant, when the surface roughness of the material is small, the difference between the peaks and valleys of the finishing pit is small, which is beneficial to the integrity and continuity of the passivation film. When the surface roughness of the material is large, the continuity of the passivation film is poor and the overall coverage of the peak position is poor. The results of SST and CCT show that the greater the roughness, the worse the uniformity of the passivation film at the peak position and the worse the corrosion resistance.

Key words: GI board; passivation film; cyclic corrosion; surface roughness

通过数值模拟方法研究带钢入锌锅速度对产渣量的影响

刘 尧，赵敬华，吴广新

（上海大学材料科学与工程学院省部共建高品质特殊钢冶金与制备国家重点实验室，上海 200444）

摘 要： 钢铁是日常生活中最常见、应用最广泛的金属材料。由于钢铁自身的理化特性和使用环境，腐蚀一直是制约其使用寿命的重要因素[1-2]。用于汽车、家电领域的带钢通常使用热镀锌手段控制腐蚀。热镀锌过程是指轧制后的带钢退火后通过炉鼻子浸入到熔融的锌液，使带钢表面形成镀锌层[3]。锌液中添加了少量合金元素铝，使带钢与镀层之间形成 Fe-Al 抑制层来提高镀锌带钢的成形性与镀层附着力，此外铝元素的加入还可以有效降低底渣量，甚至实现无底渣作业[4]。但是铝的增加会产生更多浮渣，浮渣易跟随流场运动并附着在带钢表面产生缺陷，造成产品降级和经济损失。尤其是在生产高级别的汽车板时，为了确保表面质量稳定，通常只在较低的工艺速度下生产。如何在较高的工艺速度条件下，生产出表面渣量少、级别高的热镀锌带钢，一直是困扰着各企业的难题。

针对上述难题，本团队通过数值模拟方法研究了不同带速对锌渣产量的影响，研究过程和成果如下：

（1）以国内某钢厂带钢连续热镀锌机组为基础构建了锌锅三维几何模型，提出了一种基于热力学、计算流体动力学（CFD）和离散相模型（DPM）耦合的数值模拟方法，建立了多物理场耦合下的锌锅数值模型和锌锅内流场的瞬态三维数学模型，通过用户自定义函数（UDF）实现界面反应的优化模型，包括带钢表面铁的溶解（见图1）和铝富集的瞬时速率，发现在抑制层形成之前铁的溶解和铝的富集速率都较快，当抑制层产生之后界面反应速率非常小，铁的溶解和铝的富集速率可以忽略不计。

（2）数值模拟了 Zn-0.3Al 锌锅中 η-Fe_2Al_5 悬浮渣的生成分布规律（见图 2）、分析不同粒径的悬浮渣颗粒在 Zn-0.3Al 镀锌锅中的运动轨迹和壁面沉积情况，结果表明悬浮渣主要在加锌区域温度较低的区域产生，锌渣颗粒的运动轨迹会受锌液的流动和颗粒尺寸的显著影响，较小尺寸的锌渣（20μm）易沉积于带钢表面，较大尺寸的锌渣（300μm）更易沉积于沉没辊表面。

（3）最后通过数值模拟研究了不同带速情况下，带钢表面和 V 型区内的平均产渣量（见表1），发现带钢速度越大带钢表面和 V 型区内的产渣量越多。

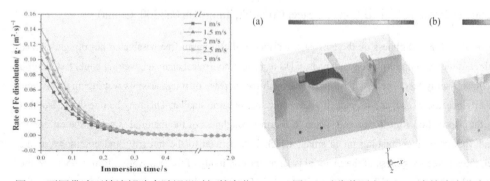

图 1 不同带速下铁溶解速率随浸没时间的变化　　图 2 对称截面上 Fe_2Al_5 渣的浓度分布(a)和锌液温度分布(b)

表 1 不同带速下带钢表面和 V 型区内平均产渣量

带钢速度/m·s^{-1}	带钢表面 Fe$_2$Al$_5$(质量分数)/%	V 型区 Fe$_2$Al$_5$(质量分数)/%
1	1.47142×10^{-5}	1.14785×10^{-5}
2	1.47184×10^{-5}	1.14795×10^{-5}
3	1.47243×10^{-5}	1.14808×10^{-5}

参 考 文 献

[1] 张启富, 郝晓东. 钢结构腐蚀防护现状和发展[J]. 中国建筑金属结构, 2006, 5(9): 22-26.
[2] 陈菌. 镀锌防腐工艺[J]. 钢铁, 2007, 42(7): 38-42.
[3] 张杰, 江社明, 张启富. 热成形钢镀层研究进展[J]. 金属热处理, 2015, 40(3): 169-172.
[4] 周国平, 于磊, 谷田, 等. 带钢表面锌渣产生的机理及控制措施[J]. 冶金管理, 2020, 3(15): 2.

一种锆化条纹缺陷及其改进

陈 光

(宝钢股份有限公司制造管理部,上海 201900)

摘 要: 汽车车身在电泳前一般需要进行磷化处理,以提高汽车板与油漆层之间的结合力,以及整个涂层体系的耐蚀性。磷化处理一般需要在加热条件下进行,能耗较大,工序冗长复杂[1-2]。新型的绿色无磷转化膜技术逐渐成熟并快速发展,无磷前处理技术替代磷化技术已成为趋势[1]。锆化前处理作为新型的无磷前处理工艺,可实现无磷酸盐排放、常温反应成膜、不需要表调工艺[3],工艺流程短,因此更加环保、节能、降低车厂制造成本。锆化工艺形成的薄膜厚度一般在 20~50nm,而磷化工艺的薄膜在 2~3μm[4]。锆化处理膜更加薄,对基材遮盖性较差,对基板的微观均匀性要求更加苛刻,对带钢表面细微缺陷的容忍度下降。

本文研究了一种锆化条纹缺陷,缺陷平行于带钢轧制方向,在带钢宽度方向几乎整板面分布。该缺陷仅在油漆清洗槽出来后可见,在冲压零件上目视和打磨都难以发现。对带钢表面进行硝酸酒精擦拭,擦拭后可以明显看到条纹缺陷,形貌与电泳和锆化处理后的条纹缺陷形貌对应。该缺陷产生在酸洗前的热轧工序,为带钢表面的"山峰状"缺陷所致。"山峰状"缺陷常见于酸洗和电镀锌产品,在普冷和热镀锌产品上几乎不可见。"山峰状"缺陷本质上是热轧的氧化铁皮压入。其成因主要是带钢由于板坯加热温度过高,热轧除鳞不充分,氧化铁皮过厚,导致氧化铁皮轧制进入带钢表面。热轧粗轧为多道次往复轧制,因此缺陷延轧制方向呈正反两个方向的"山峰状"。氧化铁皮与基体相互嵌入、轧合,并随钢板的减薄延轧制方向逐渐拉长形成山水画的形貌[5]。此缺陷的深度一般为十几微米到几十微米不等,主要取决于化学成分、板坯加热和除鳞工艺、除鳞设备能力等。经酸洗后部分氧化铁皮被清除,但带钢表面会残存深度十几微米到几十微米的小凹坑,经冷轧轧制和退火后,部分缺陷被碾压轧合,缺陷不明显。但在电镀过程中,随着锌离子在带钢表面的结晶和沉积,缺陷容易显露。这也是"山峰状"缺陷常见于酸洗和电镀锌产品,而在普冷板上目视难以发现的原因。通过优化热轧工艺使缺陷的发生率从 3.5% 降低到了 0.1%以下。

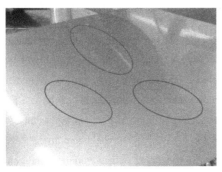

图 1 钢板电泳涂覆后的条纹缺陷

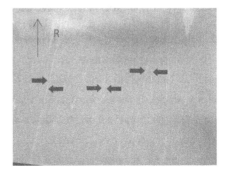

图 2 锆化处理后钢板表面的缺陷

参 考 文 献

[5] 单铭棋. 锆化薄膜前处理工艺环境下的镀锌板缺陷分析[J]. 现代涂料与涂装, 2019, 22(11): 32-34.
[6] Terrence R, Bruce H, William E. An Update of New Conversion Coating for the Automotive Industry[C]//2008 World Congress, 2008: 11441-11449.
[7] 李治国, 杨安庆, 杨书杰, 等. 锆化前处理工艺在汽车涂装中的应用[J]. 现代涂料与涂装, 2017, 20(11): 47-49.
[8] 郑福斌, 苏和, 梁炳华. 锆化前处理工艺的应用研究[J]. 现代涂料与涂装, 2016, 19(4): 15-18.
[9] 齐慧滨, 何晓明, 钱余海, 等. 热轧带钢的氧化皮缺陷类型与成因分析[C]//第七届中国钢铁年会论文集(中), 北京: 冶金工业出版社, 2009: 631-636.

冷轧汽车板缺陷在锆化前处理工艺中的传递性及其对表面耐蚀性的影响

方百友

（宝钢日铁汽车板有限公司技术质量管理部，上海　201941）

摘　要：随着绿色环保概念在全球的推广，汽车生产过程中的前处理工艺也在逐渐实现从磷化向锆化的切换。与传统的磷酸盐薄膜相比，锆系薄膜前处理工艺不需要表调和钝化步骤，而且成膜反应可在室温条件下进行，污水排放也对环境更加友好。但是，由于锆酸盐膜非常薄（20~200nm），其对微小缺陷的遮盖力逊于磷酸盐膜（约 3μm），以此替代传统的磷化层可能会对后序汽车板的使用性能造成潜在的影响。本研究模拟实际工况条件，对冷轧汽车板进行锆酸盐前处理，探究了汽车板表面缺陷在锆化工艺中的传递性及其对耐蚀性的影响。结果表明，尽管锆化处理后冷轧板和镀锌板缺陷处均覆盖有均匀的锆化膜，表面缺陷仍具有很强的传递性，即绝大多数缺陷无法被遮盖。相较于裸板，锆酸盐膜的存在可有效提高缺陷处的耐蚀性能；锆化后部分缺陷表面与完好表面耐蚀性能基本相当。

关键词：汽车板前处理；锆化工艺；磷酸盐转化膜；耐蚀性能；表面传递性

The Transferability of Surface Defects and Its Influence on the Corrosion Resistance of Zirconizing Pretreatment for Cold Automotive Sheets

FANG Baiyou

(Baosteel-Nippon Steel Automotive Steel Sheets Co., Ltd., Shanghai 201941, China)

Abstract: With the rising consciousness of green concept, phosphating pretreatment of automobile production is gradually being replaced by zirconizing pretreatment. The zirconizing pretreatment has a lot of advantages over the phosphating pretreatment such as requiring no surface passivation procedure, being able to carry out at room temperature and discharging greener sewage. However, the ultrathin zirconium conversion coatings (20~200nm) have a poorer shielding ability to surface defects compared with the phosphate conversion coatings (~3μm), making the replacement of phosphating pretreatment with zirconium conversion coating exerting potential adverse effects on the automotive sheets. In this work, zirconizing pretreatment for automotive sheets was carried out in simulating actual working condition to study the transferability of surface defects and its influence on the corrosion resistance. Results shows that the surface defects on cold rolled sheets and galvanized sheets have strong transferability regardless the formation of zirconium conversion coatings, indicating that the zirconium conversion coatings are unable to shield most defects. Nevertheless, the presence of zirconium conversion coatings can significantly improve the corrosion resistance of defects and some surfaces bearing defects have comparable corrosion resistance with perfect surfaces.

Key words: automotive pretreatment; zirconizing process; phosphate conversion coating; corrosion resistance; transferability

光整液残留对无铬耐指纹钝化板耐蚀性影响

邵 蓉，黎 敏，刘永壮，曹建平，李学涛

（首钢集团有限公司技术研究院，北京 100041）

摘 要： 本文通过 SEM、GDS、XPS 等测试对锌铝镁（ZM）空白板、表面涂覆光整液、钝化液以及光整+钝化液样板进行表面微观形貌、元素分布、价态成分等分析，结果表明，ZM 板表面光整液残留会对后续钝化液在钢板表面成膜性具有明显影响。表面存在光整液残留样板其钝化膜厚度较薄，且致密性变差。中性盐雾试验及电化学测试结果表明，表面光整液残留样品耐蚀性变差。因此，在工业生产过程中，光整工序后需清洗彻底，避免光整液残留。

关键词： 光整液；钝化膜；锌铝镁板；耐蚀性

Effect of Finishing Fluid Residue on Corrosion Resistance of Chromium-free Fingerprint Passivation Film

SHAO Rong, LI Min, LIU Yongzhuang, CAO Jianping, LI Xuetao

(Shougang Research Institute of Technology, Beijing 100041, China)

Abstract: In this paper, the surface microstructure, element distribution and valence composition of Zn-Al-Mg (ZM) blank plate, ZM with finishing film, ZM with passivation film and ZM with finish and passivation film were analyzed by SEM, GDS, XPS tests. The results show that the residue of finishing liquid on ZM surface has a significant impact on the passivation film formation on the surface of ZM plate. The thickness of the passivation film is thinner and the densification is worse. The results of neutral salt spray test and electrochemical test show that corrosion resistance of ZM with finish and passivation film is poor. Therefore, in the industrial production process, it is necessary to clean thoroughly after the finishing process to avoid the finishing liquid residue.

Key words: finishing fluid; passivation film; Zn-Al-Mg plate; corrosion resistance

锌铝镁镀层电化学腐蚀行为及其耐蚀机理研究

刘 洁，阎元媛，戴竞舸

（宝山钢铁股份有限公司中央研究院，上海 201900）

摘 要： 为了探讨锌铝镁镀层的腐蚀行为及其耐腐蚀机制，本文采用 SEM、XRD 等表征手段对锌铝镁镀层的微观结构和物相组成进行了分析，采用开路电位、动电位极化曲线、交流阻抗谱技术对锌铝镁镀层的电化学腐蚀行为进行了研究。结果表明，同样地经过腐蚀液浸泡 30h 后，锌铝镁镀层的腐蚀电流密度比镀锌层的腐蚀电流密度小 2 个数量级，比镀铝锌层的腐蚀电流密度小 1 个数量级，锌铝镁镀层表现出明显的腐蚀抑制作用。而起到腐蚀抑制作用的主要因素为锌铝镁镀层在腐蚀液中浸泡之后表面生成的以 $Zn_5(OH)_8Cl_2$ 和 $Zn_4Al_2(OH)_{12}CO_3$ 为主的难溶腐蚀产物，这些腐蚀产物在镀层表面堆积形成了一层缺陷密度较低的致密的钝化膜，能够有效地阻止腐蚀液向镀层内部渗透，从而抑制腐蚀行为的进一步扩散。

关键词：锌铝镁镀层；腐蚀产物；钝化膜；抑制腐蚀；电化学腐蚀行为；耐腐蚀机理

Study of Electrochemical Corrosion Behavior and Corrosion Resistant Mechanism of Zn-6Al-3Mg Coating

LIU Jie, YAN Yuanyuan, DAI Jingge

(Research Institute, Baoshan Iron & Steel Co., Ltd., Shanghai 201900, China)

Abstract: In order to discuss the corrosion behavior and corrosion resistance mechanism of Zn-6Al-3Mg coating, the microstructure and phase composition of Zn-6Al-3Mg coating were analyzed by SEM and XRD, and the electrochemical corrosion behavior of Zn-6Al-3Mg coating was studied by open-circuit potential, potentiodynamic polarization curve and AC impedance spectroscopy. Results show that the corrosion current density of Zn-6Al-3Mg coating decreases by 2 orders of magnitude compared with that of Zn coating, and decreases by 1 order of magnitude compared with that of Al-Zn coating in the same environment, showing obvious corrosion inhibition effect. After analysis, $Zn_5(OH)_8Cl_2$ and $Zn_4Al_2(OH)_{12}CO_3$ corrosion products are mainly generated on the surface of the Zn-6Al-3Mg coating. These corrosion products accumulate on the surface of the coating and form a passivation film with low defect density. The passivation film with less defects can effectively prevent the corrosion liquid from penetrating into the coating, thus inhibiting the diffusion of corrosion behavior.

Key words: Zn-6Al-3Mg coating; corrosion products; passivation film; corrosion inhibition; electrochemical corrosion behavior; corrosion resistant mechanism

乳化液斑对冷轧带钢表面质量的影响及控制措施

关洪星[1]，陈　平[1]，梁　直[2]，罗　军[1]，瞿作为[1]，唐　华[1]

（1. 宝钢股份武汉钢铁有限公司冷轧厂，湖北武汉　430083；
2. 宝钢股份中央研究院武汉分院（武钢有限技术中心），湖北武汉　430083）

摘　要：本文结合某冷轧厂轧钢机组的生产工艺及设备，对乳化液斑缺陷的形成机理进行分析研究，探究在实际生产中乳化液斑对板面质量的影响和优化工艺进行控制，为冷轧带钢表面乳化液斑迹缺陷的控制提供指导，制定出改善冷轧带钢乳化液斑迹缺陷的控制手段。

关键词：乳化液斑；带钢；吹扫；工艺优化

Effect of Emulsion Spot on Surfacein Quality of Cold Rolled Strip and Its Purification

GUAN Hongxing, CHEN Ping, LIANG Zhi, LUO Jun, QU Zuowei, TANG Hua

(Baosteel Wuhan Iron and Steel Company Limited, Wuhan 430083, China)

Abstract: Combined with the production process and equipment of cold rolling, The formation mechanism of emulsion spot defect was analyzed and studied, The influence of emulsion spot on the quality of plate surface in actual production and the control of optimization process are studied, It provides guidance for the control of emulsion spot defect on the surface of cold rolled strip, It was worked out that the control method to improve the emulsion spot defect of cold rolled strip steel.

Key words: emulsion spot; strip steel; purge; process optimization

连续退火露点对高强钢表面可镀性和镀层附着性的影响

董志磊，杨向鹏，阎元媛

（宝山钢铁股份有限公司研究院，上海 201900）

摘 要：本文以抗拉强度 590MPa 的高强钢为基板，利用 Iwatani HDPS 设备，开展连续退火工艺以及热镀锌铝镁工艺的模拟试验，分别在不同退火露点的 N_2-5%（体积分数）H_2 气氛中制备了连续退火样板和热镀锌铝镁样板。利用 GDS 分析了退火露点对 Mn、Si 合金元素在基板表面富集的影响，利用扫描电镜对热镀样板表面和抑制层进行了微观形貌表征和成分分析，评估了热镀样板镀层附着力。结果显示，随退火露点升高至 −10℃ 以上，Mn、Si 元素在基板表面富集趋势显著减轻，界面抑制层的完整性和致密性提高。同时热镀样板表面漏镀缺陷减少，基板的可镀性和镀层附着性大大提高。为大生产中提高高强钢热镀产品的可镀性和镀层附着性提供了退火工艺优化思路。

关键词：高强钢；热镀；露点；元素富集；抑制层；可镀性；附着性

Effect of Continuous Annealing Dew Point on the Surface Patability and Coating Adhesion of High-strength Steel

DONG Zhilei, YANG Xiangpeng, YAN Yuanyuan

(Research Institute, Baoshan Iron and Steel Co., Ltd., Shanghai 201900, China)

Abstract: In this paper, high-strength steel with tensile strength of 590 MPa was used as the substrate. The Iwatani HDPS equipment was used to simulate continuous annealing process and hot dipping galvanized aluminum-magnesium process. Continuous annealing samples and hot dipping galvanized aluminum-magnesium samples were prepared respectively in N_2-5 vol% H_2 atmosphere at different annealing dew points. The influence of annealing dew point on the surface enrichment of Mn and Si on the substrate was analyzed by GDS. The surface morphology and composition of the hot dipped templates and the interfacial inhibition layers were characterized by SEM. The adhesion of the hot dipped templates was also evaluated. The results show that with the increase of annealing dew point to −10℃, the surface enrichment of Mn and Si elements on the substrate is significantly reduced, and the integrity and density of the interfacial inhibition layer are improved. At the same time, the surface skip plating defect of the hot dipped plate is reduced, and the platability and coating adhesion of the substrate are greatly improved. The optimization idea of annealing process is provided for improving the platability and coating adhesion of hot dipping products of high strength steel in large-scale production.

Key words: high-strength steel; hot dipping; dew point; enrichment; inhibition layer; platability; adhesion

55%Al-Zn-1.6%Si-1.2%Mg 合金镀层的弯曲裂纹分析与改善

李 超[1]，王 滕[1]，刘茂林[2]，杨 平[1]

（1. 马钢股份技术中心，安徽马鞍山 243000；2. 马钢股份冷轧总厂，安徽马鞍山 243000）

摘 要： 55%Al-Zn-1.6%Si-1.2%Mg 四元合金镀层在凝固过程中生成的富 Al 枝晶相与分布在枝晶间的 Si 相、Mg-Si 相或 Mg-Zn 相之间的微塑性差异，导致镀层在弯曲加工过程中容易产生裂纹。本文通过对 55%Al-Zn-1.6%Si-1.2%Mg 合金镀层钢板进行弯曲变形以获得微小尺寸的裂纹，利用扫面电子显微镜及能谱仪分析了镀层的组织结构及开裂位置与镀层中合金相的关系，并基于镀层中各相的弹性各向异性浅要解释了高铝锌铝锌镁镀层弯曲变形后容易开裂的原因。

关键词： 四元合金镀层；微塑性；弯曲变形；合金相

Analysis and Improvement of Bending Cracks of 55%Al-Zn-1.6%Si-1.2%Mg Alloy Coating

LI Chao[1], WANG Teng[1], LIU Maolin[2], YANG Ping

(1. Technology Center, Maanshan Iron & Steel Co., Ltd., Maanshan 243000, China;
2. Cold Rolling Plant, Maanshan Iron & Steel Co., Ltd., Maanshan 243000, China)

Abstract: The microplastic difference between the Al rich dendrite phase generated during the solidification process of the 55Al-Zn-1.6Si-xMg quaternary alloy coating and the Si phase, Mg-Si phase or Mg-Zn phase distributed between the dendrites results in the coating being prone to cracking during bending processing.In this paper, bending deformation is applied to the 55%Al-Zn-1.6%Si-1.2%Mg alloy coated steel plate to obtain tiny cracks, and the relationship between the crack position of the coating and the alloy phase in the coating is analyzed by scanning electron microscope and energy spectrum, and based on the Young's modulus of the alloy phase in the coating, the reason why the high aluminum zinc aluminum magnesium coating is prone to cracking after bending deformation is explained briefly.

Key words: quaternary alloy coating; microplastic; bending deformation; alloy phase

预氧化技术在镀锌高强钢表面质量控制中的应用

孙卫华，谢英秀，陈庆军，查 凯，宇 凡，刘万春

（山东钢铁集团日照有限公司，山东日照 276800）

摘　要：先进高强钢（AHSS）在大幅减轻汽车车身质量的同时保证其安全性，使用热镀锌高强钢可以同时提高车身结构的安全性和耐腐蚀性。为了获得高强度和延展性，通常第三代 AHSS 中会添加较高的 Si、Mn 等合金元素，但 Si、Mn 在退火过程中扩散到钢板表面形成氧化物，影响热镀锌的可镀性，容易产生漏镀缺陷。特别是 Si、Mn 含量越高的热镀锌高强钢，产生漏镀缺陷的风险越高。

预氧化技术作为改善高强钢表面质量的典型技术，主要通过预氧化后钢板表面形成 Fe 的氧化物，而合金元素的氧化物较少。之后钢板进入退火炉均热段，表面 Fe 氧化物被还原成海绵铁，而 Mn 等合金元素氧化物被覆盖于海绵铁之下，产生"内氧化"现象，阻止了合金元素的析出，从而改善带钢可镀性。

从预氧化室的设备功能优化及工艺参数优化等方面优化控制工艺，提出改进措施，可大幅提高热镀锌高强钢表面质量控制水平。主要措施包括：保持退火炉预氧化室氧化气氛和以外区域还原气氛；提前 6 卷开始稳定炉内气氛，手动开启均热段放散 30%，加热段放散 40%，预热段放散 50%，提高氮氢保护气阀门开度和注入流量，稳定加热段炉压 2.0~3.0hPa，均热段氢含量 4%~7%，露点-35℃左右。准确控制氧化室内部露点：改造 CAA02 分析柜取样管路，将炉内气氛取样点由氧化室密封辊外部，改造为氧化室腔体内部，提高氧化室内露点测量准确度。提高氧化室内气氛均匀性：通过清理流量孔板，将预氧化室循环风机电机由 1900rpm 型号更换为 3000rpm 型号，循环风机流量由 260m³/h 提高到 360m³/h，提高氧化室气体循环和气氛均匀性。控制炉鼻子露点：提前关闭炉鼻子氮气加湿，避免二次氧化。同时为避免锌渣过多附着在带钢表面，对炉鼻子溜槽实施改造，通过氮气吹扫置换，使其具备净化炉鼻子气氛的功能。在关闭氮气加湿时进行一次炉鼻子气氛净化，并且在生产过程中每 3 卷再投用净化一次，控制锌灰缺陷。

通过预氧化技术优化，目前镀锌双相钢 590DHD 和 780DHD 表面漏镀问题均得到解决，整体表面质量良好，产品规格覆盖 0.7~2.5mm×(1000~1650)mm，月产可达 5000t 以上。

厚规格高附着量锌铝镁镀层板用气刀结构与工艺参数的研究

许秀飞[1]，许　正[2]，张　瑶[1]，廖嘉玮[1]

（1. 中冶赛迪集团有限公司，重庆　401122；
2. 上海交通大学船舶学院，上海　200030）

摘　要：在"双碳"经济的新形势下，以光伏支架为代表的厚规格大附着量锌铝镁镀层产品需求量越来越多，成为当前涂镀产品生产技术研究的热点。由于液态锌铝镁合金表面氧化膜较厚，厚规格大附着量锌铝镁镀层产品生产时，对气刀气流的稳定性要求非常高。但是，厚板生产时生产线速度很低，带钢从镀浴内带出来的液态金属较少，生产高附着量时气刀压力只能调整到很低的水平；加上厚板往往板形较差，气刀与带钢的距离不得不加大；因此，保证气流的稳定性，生产表面质量优良、镀层均匀的产品，成为急需解决的难题。现有各种品牌的气刀及相应的工艺参数，都是以生产薄板、低附着量产品为主设计和制定的，不能满足厚板生产 600g/m² 以上的镀锌板或 450g/m² 以上的镀锌铝镁板的需要。为此，进行了气刀结构与工艺参数的改进研究。

采用 fluent 软件可以仿真计算不同工况下气流运动场的平均特征参数，包括平均速度和压力等，但无法研究气流的稳定性问题。采用 fluent 软件的 les 大涡模型可以模拟气体流动随着时间的动态变化情况，计算瞬时速度和压力，但只能以动画的形式展示出涡流的形态，或以曲线表达参数随着时间的变化情况，无法进行不同工况的对比。

本研究创造性地引进了数理统计的方法，对不同工况下 les 大涡模型仿真计算出的气体流动随时间变化的瞬时速度和压力数据进行统计分析，使问题迎刃而解。通过对不同方案瞬时速度和压力的平均值和极差进行对比优选，设计出了适应厚规格高附着量锌铝镁镀层板用气刀的结构，并优选出了最佳的工艺参数组合，颠覆了现有的部分认知，提出了全新的系统性的解决方案，对解决锌流纹、锌起伏等难题有着积极的指导意义。

锌铝镁产品表面锈蚀缺陷原因分析与控制

李志庆，刘茂林，陈德春，章一樊，
钱　鑫，阮康康

（马鞍山钢铁股份有限公司冷轧总厂，安徽马鞍山　243000）

摘　要：针对客户反馈厚规格锌铝镁镀层光伏产品表层锈蚀斑迹缺陷，从生产工艺、原卷存储、物流运输、分条加工、分条卷仓储以及客户使用跟踪分析原因，得出锌铝镁光伏产品表层锈蚀斑迹主要为分条卷层间发生电化学腐蚀导致。通过下线卷包装温度、后处理工艺、钝化药剂改良及分条后的运输、仓储环境方面改进后，厚规格锌铝镁镀层光伏产品表层锈蚀缺陷得到解决。

关键词：锌铝镁；间隙锈蚀；控制方法

Analysis and Control of Corrosion Defects on Surface of Zinc-aluminum-magnesium Products

LI Zhiqing, LIU Maolin, CHEN Dechun, ZHANG Yifan,
QIAN Xin, RUAN Kangkang

(Maanshan Iron and Steel Co., Ltd., Maanshan 243000, China)

Abstract: Aiming at corrosion defects on surface of thick zinc-aluminum-magnesium photovoltaic products feedback by customers. From the aspects of production technology, original roll storage, logistics transportation, slitting processing, slitting roll storage and customer use, tracking and analyzes the reasons. It is concluded that the corrosion defects on the surface of Zn-Al-Mg photovoltaic products are mainly caused by electrochemical corrosion between steel coils and layers. The corrosion defects of thick Zn-Al-Mg photovoltaic products have been solved through the improvement of packaging temperature, post-treatment process, improvement of chemical treatment, logistics transportation and storage environment of slitting roll.

Key words: zinc-aluminum-magnesium; gap corrosion defects; control method

涂覆钢板产品研发现状与进展

曹宏玮[1]，孙　力[1]，张　鹏[1]，杨士弘[1]，邹炎斌[2]，赵秀娟[2]

（1. 河钢材料技术研究院，河北石家庄　050023；

2. 青岛河钢新材有限公司，山东青岛　266000）

摘　要：相较传统的木材、石材、陶瓷、玻璃等常规装饰材料，涂覆钢板具有更加优良的装饰性、耐蚀性和成形性，同时，涂覆钢板可实现定制化设计，适用于多种场景和环境，满足用户个性化需求，现已广泛应用于室内外装饰、家电等行业。本文概述了涂覆产品发展历程和最新研发进展，对涂覆钢板产品特性进行了介绍，重点分析了覆膜彩涂板、印刷彩板、数码喷绘打印板等产品的技术工艺、产品功能和结构特点。最后对涂覆钢板未来发展方向进行了展望，以期为市场提供高品质、功能性、个性化的优质涂覆产品。

关键词：涂覆钢板；覆膜彩涂板；印刷彩板；数码喷绘打印板

Research and Development Direction of Coated Steel Plate Technology

CAO Hongwei[1], SUN Li[1], ZHANG Peng[1], YANG Shihong[1],
ZOU Yanbin[2], ZHAO Xiujuan[2]

(1. HBIS Materials Technology Research Institute, Shijiazhuang 050023, China;
2. HBIS New Material, Qingdao 266000, China)

Abstract: Compared with traditional wood, stone, ceramic, glass and other conventional decorative materials, coated steel plate has better decorative, corrosion resistance and formability. At the same time, the coated steel plate can achieve customized design, suitable for a variety of scenarios and environments, to meet the individual needs of users. It has been widely used in indoor and outdoor decoration, home appliances and other industries. The Latest research and development progress of coated products were reviewed in the paper. The performance of coated sheet were introduced, and the property characteristics of coated color metal sheet, printed metal sheet, digital visual metal sheet and other products were summarized. Finally, the future development direction of coated steel plate is prospected, in order to provide the market with high-quality, functional and personalized high-quality coated products.

Key words: coated steel sheet; coated color metal sheet; printed metal sheet; digital visual metal sheet

电磁驱动下锌锅表面锌液流动行为及其扒渣效果分析

雷作胜[1,2]，卢海彪[1,2]，罗贤文[1,2]，
钟云波[1,2]，任维丽[1,2]

（1. 上海大学省部共建高品质特殊钢冶金与制备国家重点实验室，上海　200444；
2. 上海大学上海市钢铁冶金新技术应用重点实验室，上海　200444）

摘　要：带钢连续热镀锌锌锅表面流场的运动行为是决定锌锅面渣分布，影响带钢表面质量的重要因素[1-2]。随着带钢连续热镀锌生产的要求提高，传统的优化方法如锌锅结构改进、锌锅组分控制和气刀吹气等方法已经不能满足改善锌锅表面流场以提高带钢生产质量的要求[3]。因此，电磁驱动技术以其非直接接触、高效率、便于实现自动化和安全系数高的优点，逐渐被应用在热镀锌锌锅表面流场优化中。但是目前电磁驱动对锌锅流场影响的研究相对较

少，并缺乏系统的研究。通过锌锅内多场耦合数值模拟的方式研究不同优化技术对锌锅流场的影响，定义相关技术对锌锅流场优化的定量评价指标，成为研究人员急需解决的问题。

为此，本文以某热镀锌厂锌锅设备参数为原型，建立了锌锅内磁场-流场耦合的数学模型，数值模拟了电磁驱动作用时锌锅内锌液的流动，重点考察带钢速度不同电磁场条件对锌锅表面流场和内流场的影响，定量评价了气刀喷吹作用及电磁驱动技术对锌锅表面流场的优化作用。研究发现：电磁驱动技术显著的改变了锌锅上层区域的锌锅流速而对锌锅底层区域锌液运动影响较小。相较于仅有气刀喷吹的情况，电磁驱动对锌锅表面扒渣指数的提高更加显著。当带钢速度和气刀喷吹压力不变时，随着驱动电流的增加，锌锅表面扒渣指数显著增加，但锌锅表面扒渣指数的增幅逐渐下降。当驱动电流和气刀喷吹压力不变时，带钢速度增加使锌锅表面扒渣指数有所增加。但当驱动电流较低时，带钢速度对锌锅表面扒渣指数的增幅比较明显，而在驱动电流较高的情况下，带钢速度对锌锅表面扒渣指数贡献较小。当保持驱动电流、气刀喷吹压力和带钢速度不变时，不同电磁驱动装置分布方式对锌锅表面扒渣指数均有显著提升，且不同分布方式对锌锅表面流场的影响有显著不同。结果说明，锌锅表面电磁驱动技术能够在原有气刀喷吹技术上进一步提高锌锅表面锌液运动速度以改善锌锅表面流场。

参 考 文 献

[1] 侯晓光, 钱洪卫, 陆勇. 轧钢, 2022, 39(4): 87-96.
[2] Xiao Y B, Guo Q T, Liu T, et al. Metals, 2022, 12(10): 1-12.
[3] 李嘉勋, 卢海彪, 王昇军, 等. 钢铁研究学报, 2022, 34(1): 64-74.

无铬钝化镀锌板粉末喷涂脱漆的原因分析及应对措施

徐卫国，董立华，刘 闯，宇 凡，吕明鹏

（山东钢铁集团日照有限公司，山东日照　276800）

摘　要：山钢无铬钝化板在某家电工厂零件制造过程中经粉末喷涂后，进行百格实验时出现脱漆问题，影响使用。本文通过现场排查、微观分析、实验室模拟等方式解决了该问题。

对山钢与竞品钢厂原材料表面钝化膜类型进行排查，检测发现山钢镀锌板为无铬钝化，钝化膜厚1.02g/m²，竞品样板实际为三价铬钝化后，钝化膜厚40mg/m²。初步判断基板采用的钝化工艺差异，可能是该家电制造厂家此批次产品百格脱漆不合的主要原因。

选取山钢缺陷样板与竞品合格样板一同在扫描电镜下进行高倍率观察分析，将样板表面皮膜放大到200倍观察钝化皮膜整体情况，发现缺陷样板表面凸起，边缘处发生皮膜疏松，严重位置甚至有孔洞和裂纹，而合格样板表面良好，说明缺陷样板钝化皮膜发生了损坏，影响了涂装效果造成脱漆。

将无铬钝化样板拿到实验室采用不同的脱脂时间，硅烷溶液处理时间，烘干温度，烘干时间观察钝化表面皮膜情况，并结合百格实验，杯凸实验观察脱漆情况。根据实验结果可以发现，脱脂10min+硅烷6min处理工艺下的4个样板表面皮膜开裂分布密集，损伤比较严重，脱漆面积最大。脱脂6min+硅烷3min处理工艺下皮膜损伤程度有明显减轻，脱漆面积明显减轻。脱脂5min+硅烷2min处理工艺下，皮膜几乎无损伤，几乎无脱漆现象。可以得出以下结论：（1）当脱脂时间越长，硅烷处理时间越长，无铬钝化皮膜损伤越严重，脱漆面积就越大。（2）当烘干温度越高，无铬钝化皮膜损伤越严重，脱漆面积就越大。（3）当烘干时间越长，无铬钝化皮膜损伤越严重，脱漆面积就越大。（4）山钢无铬钝化板与硅烷溶液具有兼容性，本次脱漆的原因为脱脂时间长，硅烷时间长，烘干温度高，烘干时间长造成无铬钝化皮膜损伤引起。

最后通过合理调整脱脂时间、硅烷处理时间、烘干时间和烘干温度，能够有效避免表面钝化膜损伤引起钢板的涂装性能降低。

稀土元素对锌铝镁镀层组织及耐腐蚀性能的影响

宋　帅，弓俊杰，张　青，薛仁杰，王立辉，陈屹松

（河钢材料技术研究院，河北石家庄　050023）

摘　要：通过向 Zn-2%Al-1.5%Mg 镀液中添加微量稀土元素（La、Ce），制备了不同稀土含量（0.05%、0.10%、0.15%）的锌铝镁镀层钢板。使用扫描电镜、X 射线衍射仪、电化学工作站、盐雾试验箱等设备研究了不同稀土元素含量对锌铝镁镀层组织及耐腐蚀性能的影响。结果表明：稀土的添加有利于锌铝镁镀层组织的细化，随着稀土元素的增加，镀层组织中富锌相晶粒尺寸逐渐减小；当添加稀土含量在 0.10%时，锌铝镁镀层的耐腐蚀性能最好。

关键词：锌铝镁；镀层；稀土；组织；耐蚀性

Effect of Rare Earth Elements on the Microstructure and Corrosion Resistance of Zinc-Aluminum-Magnesium Coating

SONG Shuai, GONG Junjie, ZHANG Qing, XUE Renjie,
WANG Lihui, CHEN Yisong

(HBIS Materials Technology Research Institute, Shijiazhuang 050023, China)

Abstract: Zinc-aluminum-magnesium coated steel plates with different rare earth contents(0.05%, 0.10%, 0.15%) were prepared by adding trace rare earth elements to the plating solution. The effects of different rare earth elements on the microstructure and corrosion resistance of Zn-Al-Mg coated steel plate were studied by means of SEM, X-ray diffractometer, electrochemical workstation and salt spray test chamber. The results show that the addition of rare earth is beneficial to the refinement of the Zn-Al-Mg coating microstructure, and the grain size of zinc-rich phase decreases with the increase of rare earth elements. When the content of rare earth is 0.10%, the corrosion resistance of Zn-Al-Mg coating is the best.

Key words: zinc-aluminum-magnesium; coating; rare earth; organization; corrosion resistance

冷轧 IF 带钢表面清洁度的控制实践

薛仁杰[1]，李岚涛[2]，杨士弘[1]

（1. 河钢材料技术研究院，河北石家庄　052165；
2. 河钢集团河钢股份有限公司，河北石家庄　052165）

摘　要：随着车企对带钢表面清洁度要求的提升，汽车板表面清洁度已成为钢企的重要供货指标之一。但随着冷轧厂汽车板的上量，冷轧 IF 钢表面清洁度波动问题严重影响产品的正常供货。本文结合某钢铁冷轧厂现有设备介绍了冷轧 IF 带钢表面清洁度的影响因素，并从酸洗工艺、乳化液、轧辊及轧制速度等方面进行工艺优化，冷轧 IF 钢的清洁度得到显著提升。

关键词：表面清洁度；IF 钢；乳化液；轧制

The Practice of IF Strip Surface Cleanliness Control in Cold Rolling Line

XUE Renjie[1], LI Lantao[2], YANG Shihong[1]

(1. Materials Technology Research Institute, HBIS Group, Shijiazhuang 052165, China;
2. HBIS Company Limited, HBIS Group, Shijiazhuang 052165, China)

Abstract: With the improvement of strip surface cleanliness requirements of automobile enterprises, the surface cleanliness of automobile plate has become one of the important supply indicators of steel enterprises. However, with the volume of automobile plate in cold rolling plant, the fluctuation of surface cleanliness of cold rolled IF steel seriously affects the normal supply of products. This paper introduces the factors affecting the surface cleanliness of IF strip steel in cold rolling combined with the existing equipment of a steel cold rolling plant, and the cleanliness of cold-rolled IF steel has been significantly improved by optimization of pickling process, roll, emulsion and rolling speed.

Key words: surface cleanliness; IF steel; pickling emulsion; roll

时效对镀锡板钝化膜结构影响的研究

张　鹏[1,2]，李　虎[3]，孙　力[1]，史文礼[4]，张　诚[3]，曹宏玮[1]

（1. 河钢材料技术研究院，河北石家庄　050000；2. 钢铁研究总院先进金属材料涂镀国家工程实验室，北京　100081；3. 河钢集团衡水薄板有限责任公司，河北衡水　053000；
4. 河钢集团塞钢公司，塞尔维亚斯梅代雷沃　11300）

摘　要：针对某厂家反映其镀锡板钝化膜中铬含量测试结果随时间变化的现象，选取两个不同厂家 311 钝化后的镀锡板，使用 XPS 方法测试两种产品钝化膜结构随时间变化的规律，结合镀锡板电解钝化反应原理以及现有 Cr 含量测试设备特点，提出 $Cr(OH)_3$ 随时间的水解是检测中 Cr 含量变化的主要原因，通过钝化工艺的优化，降低 $Cr(OH)_3$ 的含量可有效改善该现象。

关键词：镀锡板；钝化膜；水解；XPS

Study on the Effect of Aging on the Structure of Passivation Film of Tinplate

ZHANG Peng[1,2], LI Hu[3], SUN Li[1], SHI Wenli[4], ZHANG Cheng[3], CAO Hongwei[1]

(1. HBIS Materials Technology Research Institute, Shijiazhuang 050000, China;
2. National Engineering Laboratory of Advanced Coating Technology for Materials, Central Iron and Steel Research Institute, Beijing 100081, China; 3. Hengshui Sheet Co., Ltd., Hesteel Group,

Hengshui 053000, China; 4. HBIS Group Serbia Company, Smederevo 11300, Serbia)

Abstract: In view of the phenomenon that a manufacturer reflects the change of the test results of chromium content in the passivation film of its tinplate with time, two different manufacturers of 311 passivated tinplate were selected, and the XPS method was used to test the change of the passivation film structure of the two products with time. Combined with the principle of electrolytic passivation reaction of tin plate and the characteristics of existing Cr content testing equipment, it is proposed that the hydrolysis of Cr(OH)$_3$ over time is the main reason for the change of Cr content in the detection. Through the optimization of the passivation process, the reduction of the content of Cr (OH)$_3$ can effectively improve this phenomenon.

Key words: tin plate; passivation film; hydrolysis; XPS

合金化镀锌板表面白点缺陷原因分析

王 川[1], 刘李斌[1], 李 勇[2], 史 昌[3], 黄学启[3]

(1. 首钢集团技术研究院, 北京 100043; 2. 北京首钢冷轧薄板有限公司, 北京 101304;
3. 北京首钢股份有限公司, 北京 100043)

摘 要: 使用扫描电子显微镜、能谱仪、白光干涉仪等设备,对合金化镀锌板表面白点缺陷进行分析,明确了其产生机理,并通过工艺优化解决此类缺陷。结果表明此类芯部含有黑点的白点缺陷与底渣相关,其形成过程可被描述为:被锌液搅动的底渣与锌液中的 Al 反应形成外层低 Al 内部高 Al 的复合锌渣形态并被裹于锌层中带出锌锅,在随后的合金化加热过程中与周围纯锌相发生锌铁扩散,锌渣外层首先与纯锌相反应形成爆发组织,同时由于锌渣的裹入镀层进一步增厚,而内部由于高 Al 的存在阻碍了 Zn 向内扩散的进程,导致内部相对周围产生凹陷,再经过光整后,芯部保持凹陷而周边被压平,最终形成芯部含有黑点的白点缺陷的宏观形貌。

关键词: 合金化镀锌板;底渣;锌铁合金化;白点缺陷

Causes of White-spot Defects on the Surface of Galvannealed Sheets

WANG Chuan[1], LIU Libin[1], Li Yong[2], SHI Chang[3], HUANG Xueqi[3]

(1. Research Institute of Technology, Shougang Group Co., Ltd., Beijing 100043, China;
2. Beijing Shougang Cold Rolled Sheet Co., Ltd., Beijing 101304, China;
3. Beijing Shougang Co., Ltd., Beijing 100043, China)

Abstract: The white-spot defects on the surface of galvannealed sheets were analyzed by using scanning electron micrscope, energy dispersive spectrometer and white light interferometer, and the cause of the defects was found out and solve such defects through process optimization. The results show that this type of white spot defect with black spots in the core is related to the bottom dross. The bottom dross reacts with Al in the zinc liquid to form a composite zinc dross which having the bottom dross component wrapping the surface dross component and then carried out of the zinc pot. The composite zinc dross reacts with surrounding pure zinc during the galvannealing process. The outer layer of dross first reacts with pure zinc to form an explosive structure. At the same time, the coating is further thickened due to the inclusion of the dross, while the presence of high Al inside hinders the inward diffusion of Zn, resulting in a depression relative to the surrounding area. After skin pass, the core remains depressed while the sorrounding area is flattened, ultimately forming a macroscopic morphology of white spot defects wrapped in black spots in the core.

Key words: hot-dip galvannealed steel; bottom dross; galvannealing; white-spot defects

铸管外喷工艺的应用

姚震宇，张海艳

（安钢集团永通球墨铸铁管有限责任公司，河南安阳 455133）

摘　要：以国内某球墨铸管企业生产项目为例，主要讲述了铸管外喷工序的工艺流程、设备参数；喷涂工作前的设备检查和喷涂过程中的安全工作；喷涂完成后的质量自检工作；外喷工序的质量处理的方式，对同行业的对应喷涂工作具有一定的指导意义。

关键词：球墨铸管；外喷工序；设备参数；质量自检

Application of Casting Pipe External Spray Process

YAO Zhenyu, ZHANG Haiyan

(Angang Group Yongtong Ductile Iron Pipe Co., Ltd., Anyang 455133, China)

Abstract: Taking the production project of a ductile pipe enterprise in China as an example, it mainly describes the process flow and equipment parameters of the external spraying process of casting pipe; Equipment inspection before spraying work and safe work during spraying; Quality self-inspection work after spraying; The quality treatment of the external spraying process has certain guiding significance for the corresponding spraying work in the same industry.

Key words: ductile cast pipe; external spraying process; equipment parameters; quality self-inspection

镀锡板表面特性评价技术与应用

李建中[1]，李　雪[1]，刘包发[2]，王悦鼎[3]

（1. 东北大学，辽宁沈阳　110000；2. 宝钢股份有限公司，上海　200941；
3. 沙钢钢铁研究院有限公司，江苏张家港　215000）

摘　要：镀锡板是一种生产技术复杂、制造流程长且产品质量要求高的产品。从镀锡技术来看，目前主要体现在镀液体系差异，酸性镀锡液通常由三部分组成：一种酸类、可溶性锡盐、一种或几种有机添加剂。按酸和锡盐的种类不同，又可分为卤素型、氟硼酸盐型、硫酸盐型、苯酚磺酸盐型（PSA）、氨基磺酸盐型和甲基磺酸盐型（MSA）等。目前国内 70%以上的生产线均采用以 PSA 镀液体系为主的镀锡工艺，这主要是由于 PSA 电镀液具有电流密度范围宽、沉积速度快、电流效率高、易于维护和操作等优点，然而 PSA 镀液体系同时也存在镀液导电性能低、镀层晶粒粗大、孔隙率高、以及环境污染严重等问题。随着环保压力的日益增大和电镀锡市场的竞争日益激烈，环境友好型低成本镀液体系的开发逐渐成为电镀锡工业发展的热点之一，如 MSA 镀液体系在镀锡板总产量占比例逐年提升，2017 年 MSA 镀液体系的镀锡板产量占比 32%。因此，MSA 镀液体系也在此背景下得到了大量的研究。MSA 镀锡液组成比较简单，对 Sn^{2+} 溶解性极高，锡利用率高，且镀液废水处理简单，所以 MSA 镀液体系可以被定义为环保型镀液。可以说，电镀锡行业正处在 PSA 和 MSA 镀液体系并行的状态。

本文从镀锡板表面质量的实际问题出发，结合电镀产业生产特点，通过铁溶出值法和电化学方法对基板和不同电镀成品的耐腐蚀性进行分析，应在不同情况下对溶液浓度和反应时间进行优化设计，建立了耐蚀性量化评价方法；采用 XPS 检测技术对镀锡板和基板表面元素含量相对比例、不同深度的物质成分组成比例以及镀锡板表面钝化膜成分进行分析，并添加硫酸、过硫酸铵和硝酸银溶液分别对溶液进行中和、氧化和稳定处理后，通过吸光度测试对镀锡板钝化膜成分进行有效表征。在镀锡板表面氧化膜成分进行分析时，建立了重现率较高的电化学表征方法，以氢氧化钠体系下镀锡板 E-t 曲线的拐点作为反应的起始点，以反应的时间差对镀锡板表面锡氧化物含量进行定性分析。上述表征方法为镀锡产业的实际生产和工艺改进提供了理论支撑。

关键词：镀锡板；表面特性；耐腐蚀；表征方法

镀锌模拟实验机锌灰收集过滤装置设计及应用

田绍鹏，崔忠信，冯小雷，王学慧，石玉龙，苏立冬

（河钢材料技术研究院中试试验中心，河北石家庄 050023）

摘 要：多功能镀锌和合金化模拟机试验过程中产生的锌灰，不仅会对环境造成严重的污染，还会影响电控阀的动作以及锌金属的流失增加镀锌成本。为了消除这种弊端，设计了过滤装置加装在锌锅排气的出口位置。该装置不需要能源，而是通过粗、精过滤滤芯进行过滤。该装置投用后，过滤效果明显，电控阀堵塞问题彻底消除，降低了工人操作难度和镀锌成本费用。对锌灰的排放也起到了过滤回收作用，减少了环境污染。

关键词：镀锌模拟机；过滤装置；锌灰的回收

Design and Application of Zinc Ash Collecting and Filtering Device for Galvanizing Simulator

TIAN Shaopeng, CUI Zhongxin, FENG Xiaolei,
WANG Xuehui, SHI Yulong, SU Lidong

(Hegang Material Technology Research Institute Pilot Test Center, Shijiazhuang 050023, China)

Abstract: The zinc ash generated during the testing process of multifunctional galvanizing and alloying simulators not only causes serious environmental pollution, but also affects the operation of electric control valves and increases the cost of galvanizing due to the loss of zinc metal. In order to eliminate this drawback, a filtering device was designed and installed at the outlet of the zinc pot exhaust. This device does not require energy, but is filtered through coarse and fine filter elements. After the device is put into use, the filtering effect is obvious, and the problem of blockage in the electric control valve is completely eliminated, reducing the difficulty of workers' operation and the cost of galvanizing. It also plays a filtering and recycling role in the discharge of zinc ash, reducing environmental pollution.

Key words: galvanizing simulator; filtering device; recovery of zinc ash

8 金属材料深加工

大会特邀报告
第十三届冶金青年科技奖获奖人特邀报告
分会场特邀报告
矿业工程
焦化及节能环保
炼铁与原料
炼钢与连铸
电冶金与废钢铁
轧制与热处理
表面与涂镀
★ 金属材料深加工
粉末冶金
先进钢铁材料及应用
节能与低碳技术
冶金环保与资源利用
冶金设备与工程技术
冶金自动化与智能化
冶金物流
冶金流程工程学
其他

韩国光伏发电装置钢梁基础模块化工艺

罗 晔

（宝钢中央研究院武钢有限技术中心，湖北武汉 430080）

摘 要：光伏发电装置主要包括光伏模块、逆变器和支架结构。为了实现长期稳定运行，支架结构的形态和施工方法的选取都显得非常重要。本文重点介绍了韩国ISON公司开发的光伏发电装置钢梁基础模块化工艺，其全新的光伏支架结构采用了耐腐蚀钢板PosMAC和锁紧螺栓MacBOLT，在制作和施工时，不仅可以减少碳排放和土壤污染，还可以进行回收再利用。

关键词：光伏发电；模块化；支架；耐腐蚀钢板

Modular Process for Steel Beam Foundations for Photovoltaic Power Installations in Korea

LUO Ye

(R&D Center of Wuhan Iron & Steel Co., Ltd., Baosteel Central Research Institute, Wuhan 430080, China)

Abstract: Photovoltaic power installations mainly consist of photovoltaic modules, inverters and racking structures. In order to achieve long-term stable operation, the form of the racking structure and the selection of the construction method are of great importance. This paper focuses on the modular process for steel beam foundations for photovoltaic power generation installations developed by ISON in South Korea, whose new PV racking structure uses corrosion-resistant steel plates PosMAC and locking bolts MacBOLT, which can be fabricated and constructed in a way that reduces carbon emissions and soil contamination, and can also be recycled.

Key words: photovoltaics; modularity; racking; corrosion-resistant steel plates

铁锰因瓦合金的强化研究

刘津伊[1]，严 玲[2,3]，刘文月[1]，齐祥羽[2,3]，张 鹏[2,3]，李广龙[2,3]

（1. 鞍钢集团北京研究院，北京 102299；2. 海洋装备用金属材料及其应用国家重点实验室，辽宁鞍山 114021；3. 鞍钢集团钢铁研究院，辽宁鞍山 114009）

摘 要：铁锰因瓦合金具有较低的热膨胀系数与反铁磁性，在高磁场环境下的精密仪器中具有潜在应用。由于铁锰因瓦合金屈服强度较低导致其在高新技术领域的应用受到限制。为了提高这类合金的综合性能，我们通过大塑性冷拔丝变形与热处理的方法，调控合金的微观结构，分别构建了双峰晶粒异质结构、纳米孪晶与再结晶晶粒混合组织两种微观结构的铁锰合金，其屈服强度在1~1.3GPa之间，延伸率在20%~34%之间，并且在–100~0℃展现出8.4×10^{-6}/℃的低热膨胀系数。结果表明该方法在不损失合金低热膨胀特性的同时达到强化力学性能的效果。

关键词：铁锰合金；因瓦效应；异质结构；纳米孪晶

Study on Strengthening of Fe-Mn Invar Alloy

LIU Jinyi[1], YAN Ling[2,3], LIU Wenyue[1], QI Xiangyu[2,3], ZHANG Peng[2,3], LI Guanglong[2,3]

(1. Ansteel Beijing Research Institute Co., Ltd., Beijing 102299, China; 2. State Key Laboratory of Metal Material for Marine Equipment and Application, Anshan 114021, China; 3. Iron and Steel Research Institute, Ansteel Group, Anshan 114009, China)

Abstract: Fe-Mn Invar alloy shows low thermal expansion coefficient and antiferromagnetism, which has potential applications in precision instruments under high magnetic field environments. Due to the low yield strength of Fe-Mn Invar alloy, their application in high-tech fields is limited. In order to improve the comprehensive performance of this alloy, we controlled the microstructure of the alloy through large plastic deformation by cold draw and heat treatment. We constructed two types of microstructure Fe-Mn alloy, namely bimodal grain heterostructure and nano twins and recrystallized grains mixed structure, with yield strength between 1 and 1.3GPa, elongation between 20% and 34%, and low thermal expansion coefficient of 8.4×10^{-6}/℃ at -100℃ to 0℃. The results indicate that this method achieves the effect of strengthening mechanical properties without losing the low thermal expansion characteristics of Fe-Mn Invar alloy.

Key words: Fe-Mn alloy; invar effect; heterogeneous structure; nano twins

粗轧微张力控制方法与实践

刘 伟，闫 鹏

（武钢热轧厂1580产线，湖北武汉 430083）

摘 要： 热连轧中粗轧平辊与立辊之间的微张力是通过两辊之间的速差调节和立辊传动系统的力矩补偿两个方面来进行控制的，将张力值的控制在一个较小的范围之内，力求使平辊与立辊之间的微张力保持恒定，通过优化微张力可以适当改善中间坯宽度和平直度，尤其对粗轧轧辊打滑的改善有一定的帮助作用，提高变频器对速度和转矩的匹配控制运算的精度，改善电机电流平滑度，消除速度波动，提高转矩能力，减少主传动变频器跳闸造成卡钢损辊，降低生产中异常辊耗的成本。

关键词： 微张力；恒定；控制；打滑；平滑

Optimization of Micro Tension Controlling in Roughing Mill Area

LIU Wei, YAN Peng

(WISCO HSM Ltd., HSM3, Wuhan 430083, China)

Abstract: The micro tension is between the flat roll and edger roll in the rough mill, which is based on the speed difference between the rolls and edger roll torque compensation in main drive system, the micro tension is controlled though these two method. The tension value is controlled within a smaller range, it strives to make the micro tension constant. The intermediate slab width and straightness can be appropriately improved by optimizing the micro tension. It is especially some helpful for roll slip improvement in roughing. It improves drive speed and torque matching control calculation precision in converter. While improving the smoothness of the motor current, eliminating motor speed fluctuation, increasing the motor torque capacity. It reduces the loss of roll, that is because the main drive converter trip resulting the strip stopped in the middle of two rolls. And it ultimately reduces the production cost of abnormal consumption rolls.

Key words: micro tension; constant; control; slip; smooth

低应力热轧高强钢薄板矫正策略

高智平，张鹏武

（武汉钢铁有限公司热轧厂，湖北武汉 430083）

摘 要： 本文介绍了热轧高强薄板板形及应力消减常用的矫正手段，提出了各工序的主要作用及控制要点。明确了要实现热轧高强薄板残余应力小加工使用不出现变形翘曲，需多工序协作多手段同时使用的策略。

关键词： 应力；平整；矫直；去应力退火

热连轧卷取机镰刀弯防夹死功能开发及应用

梅荣利，袁 金，李君波

（宝钢股份武钢有限热轧厂，湖北武汉 430080）

摘 要： 针对热轧带钢头部严重镰刀弯引发的卷取机入口受阻堆钢问题，通过创建卷取侧导板防夹死及二次防夹死自动控制程序，自动识别钢卷头部镰刀弯特征值并反馈修正侧导板控制输出，实现不同来料头部镰刀弯动态自动调整控制，避免人工未干预、误干预而引发的堆钢事故并兼顾卷形控制，有效减少废钢事故时间及侧导板异常损耗。

关键词： 卷取机；镰刀弯；防夹死

Abstract: In order to solve the problem of blocked pile steel at the entrance of coiler caused by severe scythe bending of hot rolled strip head, the automatic control program of anti-clamping and secondary anti-clamping of coiling side guide plate was created to automatically identify the characteristic value of scythe bending of steel coil head and feedback and correct the control output of side guide plate, so as to realize dynamic automatic adjustment and control of scythe bending of different incoming material heads. Avoid the pile steel accidents caused by manual intervention and misintervention and take into account the coil shape control, effectively reduce the scrap accident time and side guide plate abnormal loss.

Key words: coiler; camber; prevent-clamping

酸洗钢辊系铁皮缺陷分析及改进

周 坤，梅荣利，袁 金，徐 浩，李君波

（武钢有限公司热轧厂，湖北武汉 430083）

摘　要：对酸洗钢辊系铁皮缺陷的微观组织及成分分析，结合轧辊氧化膜的生成机理，通过提高工作辊冷却水能力、降低轧制温度、优化精轧负荷分配等措施，有效降低了该缺陷的发生率，提高了实物质量。

关键词：酸洗钢；辊系铁皮

Research and Improve of Roll System Algam Defects on Pickling Steel

ZHOU Kun, MEI Rongli, YUAN Jin, XU Hao, LI Junbo

(WISCO, Wuhan 430083, China)

Abstract: Based on the analysis of the microstructure and composition of the iron sheet defect the pickling steel roll system, combined with the formation mechanism of the oxide film on the roll, the occurrence rate of the defect was effectively reduced and the actual material quantity was increased by improving the cooling water capacity of the work roll, reducing the rolling temperature and optimizing the distribution of the finishing rolling.

Key words: pickling steel; roll system algam defects

微合金化高强钢热轧卷尾部性能研究

王　成[1]，童善康[2]，陈志辉[2]，袁　清[2]，甘晓龙[2]，丁　茹[1]

（1. 武汉钢铁有限公司热轧厂，湖北武汉　430083；2. 武汉科技大学，湖北武汉　430080）

摘　要：钢卷尾部的强度通常是非均匀的，本文详细研究了钢卷尾部最外圈、第3圈和第5圈的显微组织和力学性能。采用光学显微镜(OM)和电子背散射衍射(EBSD)技术对铁素体晶粒和晶界进行表征，通过高分辨率透射电子显微镜(HR-TEM)和能谱仪(EDS)对析出物进行了表征。研究发现，钢卷尾部不同位置的性能差异主要是由纳米尺度(Ti, Nb, Mo)C粒子产生的析出强化变化所引起。纳米级(Ti, Nb, Mo)C颗粒主要在卷取和冷却过程中析出。不同冷却速率对铁素体晶粒尺寸影响不大。

关键词：微合金化高强钢；铁素体钢；性能；析出物

Investigating the Properties of Coil Tail in Microalloyed Hot-rolled Strip

WANG Cheng[1], TONG Shankang[2], CHEN Zhihui[2],
YUAN Qing[2], GAN Xiaolong[2], DING Ru[1]

(1. Wuhan Iron and Steel Co., Ltd., Hot Rolling Plant, Wuhan 430083, China;
2. Wuhan University of Science and Technology, Wuhan 430080, China)

Abstract: In the present study, The strength at the tail of a strip coil is often found to be non-uniform. The microstructural and the mechanical properties of the specimens in the outermost circle as well as in the third circle and the fifth circle from the outermost circle were analyzed in detail. The optical microscopy (OM) and the electron backscatter diffraction (EBSD) techniques were employed for the characterization of ferrite grains and grain boundaries. The precipitates were revealed by high-resolution transmission electron microscopy (HR-TEM) and energy dispersive spectroscopy (EDS). It was found that the precipitation strengthening induced by nano-scale (Ti, Nb, Mo)C particles was the dominant factor for the disparity in properties in different positions of the coil tail. Nano-scale (Ti, Nb, Mo) C particles mainly precipitated in ferrite grains during coiling and subsequent cooling. Furthermore, different cooling rates along the strip had inconspicuous influences on the grain size and lengths of ferrite grain boundaries.

Key words: microalloy; ferritic steel; property; precipitates

主提升钢丝绳断丝失效分析

赵宪海

（鞍钢股份有限公司，辽宁鞍山　114021）

摘　要：矿井提升用钢丝绳是提升机关键部件，异形股钢丝绳是主要产品品种，断丝是其失效的主要表现形式。北方某矿务局提升用 6V×37S+FC-1670MPa 异形股钢丝绳使用 12 个月后陆续发生断丝，45 天后断丝超标造成整组钢丝绳报废。本文从钢丝化学成分、金相组织、力学性能、断口形貌和钢丝绳结构参数等方面对断丝原因进行了分析，认为钢丝绳放绳时出现局部扭结，该部位后续使用过程中与滑轮高速接触时产生类马氏体组织，并形成裂纹乃至断丝。钢丝表面的缺陷造成断丝是造成这次钢丝绳报废的主要原因。

关键词：煤矿；钢丝绳；断丝

Failure Analysis on Wire Breaking in Wire Rope for Mine Hoisting Purposes

ZHAO Xianhai

(Angang Steel Company Limited, Anshan 114021, China)

Abstract: The wire rope used in mine hoist is the key part of hoist, the shaped strand wire rope is the main product, and the broken wire is the main form of its failure. Broken wires in a 6V×37S+FC-1670MPa shaped strand wire rope for mine hoisting purposes used for 12 months were found and failure of the set of ropes 45 days later. This article analyzes the chemical composition, metallograpllic, mechanical properties of the broken wires, the appearance of the facet breaking and the parameters of construction of the wire rope. It is believed that the wire rope has local kinks when the wire rope is laid. This part would be in contact with the pulley at high speed during subsequent work. Martensite-like structure is produced, and cracks and even broken wires are formed. Defects on the surface of the steel wire cause broken wires are the main reason for the failure of the wire rope.

Key words: coal mine; wire rope; broken wires

一种矿用钢丝绳插接新技术

赵宪海[1]，王靖涵[2]，王国辉[3]，王洪亮[1]，刘思洋[1]，李广宇[2]

（1. 鞍钢股份有限公司，辽宁鞍山　114021；2. 辽宁对外经贸学院，辽宁大连　116052；
3. 鞍钢钢绳有限责任公司，辽宁鞍山　114021）

摘　要：矿用提升装置主要包括斜井提升、立井提升、带式输送机提升等几种方式，与之配套的钢丝绳也品种繁多。其中，带式输送机是以钢丝绳为牵引主体，通过驱动轮带动对称布置的两根钢丝绳，将皮带及其上面的物料运送到指定位置，钢丝绳在整个运行轨迹上形成闭环，同时还要经过若干次弯折，对疲劳性能和插接部位的质量要求很好，历来受到使用单位的重视，但一直没有形成规范的方法。本文在生产实践的基础上，综合其他索道行业的插接经验，使用了新的复合填充材料，总结出一套新的钢丝绳插接技术。该插接新技术可有效地解决插接部位运行过程中不牢固、易断丝等缺点，可推广使用。

关键词：钢丝绳；插接；工艺技术

The New Technology Relates to Inserting the Steel Wire Rope

ZHAO Xianhai[1], WANG Jinghan[2], WANG Guohui[3],
WANG Hongliang[1], LIU Siyang[1], LI Guangyu[2]

(1. Ansteel Iron and Steel Co., Ltd., Anshan 114021, China; 2. Liaoning Institute of Foreign Trade and Economics, Dalian 116052, China; 3. Ansteel Wire Rope Co., Ltd., Anshan 114021, China)

Abstract: Mine hoisting device mainly includes inclined shaft hoisting, vertical shaft hoisting, belt conveyor hoisting and so on. Among them, the belt conveyor is to take the steel wire rope as the traction main body, through the driving wheel to drive the symmetrical arrangement of two steel wire rope, the belt and the above material transport to the designated position, the steel wire rope forms a closed loop on the whole running track, at the same time, it has to go through several bends, but there has been no way to form a standard. On the basis of production practice, this paper summarizes a set of new technology of wire rope splicing by using new composite filling material and combining with the experience of other ropeway industry. The new technology can effectively solve the defects such as weak and easy to break wires in the process of operation of the plug-in part, and can be popularized and used

Key words: wire rod; deep processing; sewage treatment

耐热套管特殊螺纹接头 BG-TH 开发理论与试验研究

詹先觉，董晓明，高　展

(宝山钢铁股份有限公司研究院，上海　201900)

摘　要：钢铁材料在热采井等350℃超高温度条件下，不仅材料屈服强度和抗拉强度下降，材料的其它物性参数也发生了很大变化，热应变从次要矛盾上升为主要矛盾。本文对P110和110H两种材质进行了室温与不同高温等级下的试验研究，基于高温下材料的"弹性应力"与"塑性应变"设计理念开发了BG-TH（Thermal）耐热套管，使用有限元分析与实物试验对设计结构进行了验证，结果表明，将耐热管材与耐热螺纹接头两者配合使用可提高套管在高温中的耐用性。

关键词：套管；热采井；高温

Research on Thermal Well Premium Casing Connection BG-TH

ZHAN Xianjue, DONG Xiaoming, GAO Zhan

(Research Institute, Baoshan Iron & Steel Co., Ltd., Shanghai 201900, China)

Abstract: The yield strength and tensile strength of iron and steel materials not only decrease under the ultra-high temperature of 350℃ such as thermal recovery Wells, but also the other physical parameters of the materials change greatly, and the thermal strain rises from the secondary contradiction to the main contradiction. In this paper, two kinds of materials P110 and 110H were tested at room temperature and different high temperature levels. Based on the design concept of "elastic stress" and "plastic strain", BG-TH (Thermal) casing was developed. The design structure was verified by finite element analysis and physical test. The combination of heat-resistant pipe and heat-resistant threaded joint can improve the durability of casing at high temperature.

Key words: casing; thermal well; high temperature

DIL 805 热膨胀仪在先进高强钢中的典型应用

李春诚，郭晓静，杨 波，海 超，李科龙，姜乐朋

（本钢集团有限公司技术中心先进汽车用钢开发与应用技术
国家地方联合工程实验室，辽宁本溪 117000）

摘 要：DIL 805 是一台高度自动化的淬火相变热膨胀仪，可以在极端的可控升温以及冷却条件下检测试样尺寸的变化。为更好地了解和应用 DIL 805 热膨胀仪设备，服务于先进高强钢产品的研制与开发，本文主要介绍了 DIL 805 热膨胀仪的设备型号、常用功能、测量原理、技术参数及其典型应用，包括钢的临界点测定、CCT 曲线测定、TTT 曲线测定和其他模拟实验等。

关键词：热膨胀仪；先进高强钢；CCT；TTT

Typical Application of the DIL 805 Thermal Dilatometer in Advanced High Strength Steel

LI Chuncheng, GUO Xiaojing, YANG Bo, HAI Chao,
LI Kelong, JIANG Lepeng

(National and Local Joint Engineering Laboratory of Advanced Automotive Steel
Development and Application Technology of Technical Center of Bengang Group Corporation,
Benxi 117000, China)

Abstract: The DIL 805 is a fully automated self-contained quenching dilatometer used to observe dimensional changes under extreme conditions of controlled heating and cooling. In order to better understand and apply the DIL 805 thermal dilatometer and serve the research and development of advanced high strength steel products, the equipment model, common functions, measurement principles, technical parameters, and typical applications of the DIL 805 thermal dilatometer are introduced, including the determination of critical points, CCT curve, TTT curve and other simulation experiments of steel.

Key words: thermal dilatometer; advanced high strength steel; CCT; TTT

Cr 对 22MnB5 热成形钢高温氧化行为的影响

朱蓉，张淇，米振莉，吴彦欣，江海涛

（北京科技大学高效轧制与智能制造国家工程研究中心，北京 102206）

摘 要：本文以 22MnB5 和 22MnB5Cr3 两种热成形钢为研究对象，对其进行高温氧化试验，并结合场发射扫描电镜、能谱仪和 X 射线衍射仪研究了 Cr 元素对 22MnB5 热成形钢高温氧化行为的影响。结果表明，在初始氧化过程中，Cr 元素会加速氧化过程并形成含 Cr 的保护性氧化层，随着氧化时间的延长，22MnB5Cr3 钢表现出更优异的抗氧化性能。Cr 元素含量提高后，热成形钢氧化层表面的起泡和脱落情况得到明显改善。氧化后生成从外到内分别为 Fe_2O_3、Fe_3O_4、FeO 和 AB_2O_4 尖晶石的氧化层，添加 Cr 元素后，氧化层中 FeO 占比明显减小，AB_2O_4 尖晶石氧化层的占比明显提高，AB_2O_4 尖晶石氧化层可以对基体起到保护作用。

关键词：热成形钢；Cr 元素；高温氧化；氧化层

Effects of Cr on High Temperature Oxidation Behavior of 22MnB5 Hot Stamping Steel

ZHU Rong, ZHANG Qi, MI Zhenli, WU Yanxin, JIANG Haitao

(National Engineering Research Center for Advanced Rolling and Intelligent Manufacturing, University of Science and Technology Beijing, Beijing 102206, China)

Abstract: In this paper, the high-temperature oxidation tests of the 22MnB5 and 22MnB5Cr3 steels were conducted. The effects of Cr element on the high-temperature oxidation behavior of the 22MnB5 hot stamping steel were studied by means of field emission scanning electron microscopy, energy dispersive spectrometer and X-ray diffraction. The results indicate that during the initial oxidation process, chromium element accelerates the oxidation process and forms a protective oxide layer of $FeCr_2O_4$. The 22MnB5Cr3 steel exhibits better oxidation resistance than 22MnB5 steel with time. After adding the chromium element, the bubbling and peeling of the oxide layers on the hot stamping steels are significantly improved. After oxidation, an oxide layer consisting of Fe_2O_3, Fe_3O_4, FeO, and AB_2O_4 spinels is generated from the outside to the inside. After adding the chromium element, the proportion of FeO in the oxide layer is significantly reduced, while the proportion of AB_2O_4 spinel oxide layer is significantly increased. The AB_2O_4 spinel oxide layer can protect the substrate.

Key words: hot stamping steel; Cr element; high-temperature oxidation; oxide layer

低密度中锰钢脆性断裂研究

左汪楠[1,2]，杨永刚[1,2]，宿泽轩[1,2]，袁长辉[1,2]，吴彦欣[1,2]，米振莉[1,2]

（1. 高效轧制与智能制造国家工程研究中心，北京 100083；
2. 北京科技大学工程技术研究院，北京 100083）

摘 要：近年来，中锰钢中添加 Al 元素降低其密度已成为重要发展趋势，但其安全性仍然需要重点关注。本文针对临界退火后中锰钢热轧板出现脆性断裂的现象,采用 SEM、EDS 等实验手段对材料的组织、断口形貌等进行分析。结果表明:低密度中锰钢脆性断裂的裂纹源主要集中在孔洞和夹杂物处,沿着(γ+α)/δ边界和夹杂物的方向快速扩展。

关键词:中锰钢;脆性断裂;临界退火;断口形貌

Study on Brittle Fracture of Low Density Medium Manganese Steel

ZUO Wangnan[1,2], YANG Yonggang[1,2], SU Zexuan[1,2],
YUAN Changhui[1,2], WU Yanxin[1,2], MI Zhenli[1,2]

(1. National Engineering Research Center for Advanced Rolling and Intelligent Manufacturing, University of Science and Technology Beijing, Beijing 100083, China; 2. Institute of Engineering Technology, University of Science & Technology Beijing, Beijing 100083, China)

Abstract: In recent years, adding Al element to reduce the density of medium manganese steel has become an important development trend, but its safety still needs to be focused on. This article focuses on the phenomenon of brittle fracture in hot-rolled plates of medium manganese steel after critical annealing. The microstructure and fracture morphology of the material were analyzed using experimental methods such as SEM and EDS. The results indicate that the crack source of brittle fracture in low density medium manganese steel is mainly nucleation at holes and inclusions, along the (γ+α)/δ The boundary and inclusion direction rapidly expand.

V 对 Q355 钢 CCT 曲线的影响

刘欣悦[1],方幸[1],左汪楠[1],吴彦欣[1],米振莉[1],潘欣[2]

(1. 北京科技大学 高效轧制与智能制造国家工程研究中心,北京 100083;
2. 吉林华信新型结构科技有限公司,吉林磐石 132300)

摘 要:利用热膨胀仪、显微硬度计等实验设备,测定了 Q355 钢和含 V 的 Q355 实验钢的连续冷却转变(CCT)曲线,研究了 V 对 CCT 曲线的影响。实验结果表明,V 元素一方面可以降低奥氏体临界转变温度,并推迟奥氏体向珠光体转变,使 CCT 曲线中珠光体区右移减小;另一方面可以降低贝氏体相变临界冷却速度,扩大实验钢的贝氏体转变区,从而有利于贝氏体相变;另外发现,两个实验钢硬度均随着冷速的增大而变大,且相同冷却速度下,添加微合金元素 V 的实验钢的硬度始终高于未添加元素 V 的实验钢。

关键词:V 元素;Q355 实验钢;CCT 曲线

The Influence of V on the CCT Curve of Q355 Steel

LIU Xinyue[1], FANG Xing[1], ZUO Wangnan[1], WU Yanxin[1], MI Zhenli[1], PAN Xin[2]

(1. National Engineering Research Center for Efficient Rolling and Intelligent Manufacturing, University of Science & Technology Beijing, Beijing 100083, China;
2. Jilin Huaxin New Structure Technology Co., Ltd, Panshi 132300, China)

Abstract: The continuous cooling transformation (CCT) curves of Q355 steel and experimental steel containing V were measured using experimental equipment such as a thermal dilatometer and a microhardness tester, and the effect of V on the CCT curves was studied. The experimental results indicate that V element can reduce the critical transformation temperature of austenite and delay the transformation of austenite to pearlite, resulting in a decrease in the right shift of the pearlite zone in the CCT curve; On the other hand, it can reduce the critical cooling rate of Bainite transformation and expand the Bainite transformation zone of the experimental steel, which is conducive to Bainite transformation; Furthermore, it was found that the hardness of both experimental steels increased with the increase of cooling rate, and at the same cooling rate, the hardness of the experimental steel with the addition of microalloyed element V was consistently higher than that of the experimental steel without the addition of element V.

Key words: V element; Q355 experimental steel; CCT curve

连铸连轧工艺下 SPHC 热变形行为模拟

常　江，米振莉，苏　岚，杨永刚，张茂才，刘心爽

（北京科技大学工程技术研究院，北京　100083）

摘　要：目前计算轧制过程的力学场、温度场已是现代材质预报的一个重要手段。研究轧件在轧制过程中的应力、温度变化，以及轧件内部应变场、温度场具有十分重要的意义。本文通过数值模拟技术，借助 ANSYS 有限元分析软件，采用弹塑性有限元法，对 SPHC 钢在热轧全过程中的应力场、应变场、温度场进行三维热力耦合有限元仿真，包括粗轧、精轧及道次间空冷。利用 CSP 产线的实测数据与仿真结果进行对比验证，规格为 7 道次精轧且轧后厚度为 3.03mm 的带钢，计算的精轧出口上表面温度为 790℃，实测温度为 798.9℃，计算值与实测值相差 8.9℃，相对误差 1.12%，且 7 道次的轧制力相对误差在 10%以内，板坯温度与轧制力的实际测量结果与仿真结果吻合，验证了有限元模型的有效性，对轧制产线的工艺制定和组织性能预测提供参考。

关键词：SPHC 钢；热变形；热力耦合模拟；有限元仿真

Simulation of Hot Deformation Behavior of SPHC during Continuous Casting and Rolling

CHANG Jiang, MI Zhenli, SU Lan, YANG Yonggang,
ZHANG Maocai, LIU Xinshuang

(Institute of Engineering Technology, University of Science & Technology Beijing, Beijing 100083 China)

Abstract: At present, calculating the mechanical field and temperature field of the rolling process has become an important means of modern material prediction. It is of great significance to study the changes of stress and temperature in rolling process, as well as the internal strain field and temperature field of rolled parts. In this paper, the stress field, strain field and temperature field of SPHC steel in the whole process of hot rolling are simulated by using three-dimensional thermo-mechanical coupling finite element, which using elastoplastic finite element method with numerical simulation technology and ANSYS finite element analysis software, and the whole process include rough rolling, finishing rolling and air cooling between passes. The measured data of CSP production line are compared with the simulation results. The calculated surface temperature of the slab with a thickness of 3.03mm after 7 passes of precision rolling is 790℃, while the measured temperature is 798.9℃. The difference between them is 8.9℃, with a relative error of 1.12%. Moreover, the relative error of the rolling force for 7 passes is within 10%. The actual measurement results of the slab temperature and rolling force are consistent with the simulation results, verifying the effectiveness of the finite element model. The finite element model provides a reference for the process formulation and prediction of microstructure and property of rolling production line.

Key words: SPHC steel; thermal deformation; thermo-mechanical coupling simulation; finite element simulation

超临界二氧化碳输送用 HFW 焊管性能研究

黄晓辉[1,2]，张锦刚[1,2]，毛浓召[1,2]，韦 奉[1,2]，王博玉[1,2]，赵红波[1,2]

（1. 宝鸡石油钢管有限责任公司，陕西宝鸡 721008；
2. 中油国家石油天然气管材工程技术研究中心有限公司，陕西西安 710018）

摘 要： 目前，全球已建二氧化碳管道近万公里，其中美国约有 6000km 二氧化碳管道，其余二氧化碳管道主要分布在加拿大、挪威等国家，二氧化碳多采用超临界态高频电阻（HFW）焊管输送。而当前，国内已建成的二氧化碳管道大都为气相输送，尚未开始采用 HFW 焊管输送超临界二氧化碳。由于二氧化碳管道泄漏或放空时，密度发生剧烈突变，易于在强节流效应下形成-78～-30℃低温，并可能有干冰生成，钢管需要具备一定的低温抗脆性断裂能力和抵抗长程扩展的止裂能力，因此输送超临界二氧化碳的 HFW 焊管需要管体和焊缝都需具有低温下优异的韧性。

为了满足高压超临界二氧化碳及含少量杂质载荷条件下管材低温高韧性，联合钢厂投料 200t 开发出抗低温性能优异的低碳中锰高铌热轧板卷，卷板微观组织为细长扁平的细小多边形铁素体（PF）+少量珠光体（P），控制 Mn/Si 比 5～7 以减少减小后期 HFW 制管时焊缝中高熔点 Mn、Si、O 夹杂物的数量和尺寸。经过控制 HFW 焊管机组挤压成型量到 5.5mm，优化高频焊接工艺，在生产线上开发出超临界输送用 X60MC/X65MC 钢级 ϕ219.1mm/ϕ323.9mm×10mmHFW 电阻焊管。第三方力学性能检测表明，试制的 HFW 焊管新产品完全符合 API 5L、GB/T 9711 标准要求，管环压扁到平板相贴时候，焊缝区域未见任何裂纹，表明母材和焊缝都具有高的强塑形。母材和焊接接头硬度相近，最高硬度仅 224HV10，表明母材和焊缝都具有一定抗应力腐蚀性。特别是母材和焊缝具有优异的焊缝低温韧性，在技术规格书要求的-45℃下 X60MC 焊缝冲击功 292～351J，X65MC 焊缝冲击功 177～414J。当试验温度下降到-78℃下，X60MC/X65MC 母材冲击功、焊缝冲击功、热影响区冲击功大数据平均值都在 204～350J 范围内，远远高于常规 X60M/X65M 钢级 HFW 焊管焊缝在-45℃下冲击单值≥60J、冲击平均值≥80J 的低温管要求。

开发出超临界输送用 X60MC/X65MC 钢级 ϕ219.1mm/ϕ323.9mm×10mmHFW 电阻焊管，具有优异的性价比，特别是在-78℃下 HFW 管体和焊缝的高韧性确保了管材具有高的抗裂纹启裂和抗延性扩展止裂能力，完全满足超临界二氧化碳管安全输送服役要求，高的低温韧性是 HFW 焊管制造技术的又一新的突破，今后将该技术制作的 HFW 焊管作为油气站场用管，严寒和极地地带油气输送管得到大量应用。

参 考 文 献

[1] 张强, 杨玉锋, 张学鹏, 等. 超临界二氧化碳管道完整性管理技术发展现状与挑战[J]. 油气储运, 2023, 42(2): 152-160.
[2] 徐源. 含杂质超临界 CO_2 管道裂纹延性扩展研究[D]. 陕西:西安石油大学, 2021.
[3] 吕家兴, 侯磊, 王昕, 等. 超临界 CO_2 管道输送管径技术经济性评价[J]. 油气储运, 2022, 41(1): 114-120.

冷轧薄板激光切割翘曲变形分析研究

曹 政[1,2]，冉茂宇[1,2]，吕 冬[1,2]，芦延鹏[1,2]，徐 鑫[1,2]，林森木[1,2]

（1. 海洋装备用金属材料及其应用国家重点实验室，辽宁鞍山 114001；
2. 鞍钢股份技术中心，辽宁鞍山 114001）

摘 要：钢板在冷轧过程中受到的轧制力不均导致内部板带中存在残余应力，板带在激光切割过程中进一步引入热输入，引起不同区域不均匀的热胀冷缩导致激光切割后翘曲变形明显。文章研究了不同激光切割尺寸、不同切割方法对翘曲变形的影响，同时对比分析了落料与激光切割的翘曲变形分布规律，并进一步通过冷轧板带轧制模型说明试板上下表面残余应力分布不均是引起激光切割翘曲变形的主要原因，该研究对控制冷轧薄板激光切割翘曲变形有重要意义。

关键词：激光切割；冷轧薄板；翘曲变形；残余应力

Analysis and Research on Warpage Deformation of Cold Rolled Plate Laser Cutting

CAO Zheng[1,2], RAN Maoyu[1,2], LV Dong[1,2], LU Yanpeng[1,2], XU Xin[1,2], LIN Senmu[1,2]

(1. State Key Laboratory of Metal Materials for Marine Equipment and Application, Anshan 114001, China; 2. Ansteel Iron and Steel Researc Institute, Anshan 114001, China)

Abstract: The uneven rolling force of the steel plate during the cold rolling process leads to residual stress in the internal strip, and the strip further introduces heat input during the laser cutting process, causing uneven thermal expansion and contraction in different areas, resulting in warping after laser cutting The deformation is obvious. This article studies the effects of different laser cutting sizes and different cutting methods on warping deformation, and compares and analyzes the distribution of warping deformation between blanking and laser cutting, and further illustrates the remaining surface residual on the upper and lower surfaces of the test plate through the cold-rolled strip rolling model. Uneven stress distribution is the main cause of warpage in laser cutting. This research is of great significance for controlling warpage in laser cutting of cold-rolled sheet.

Key words: laser cutting; cold rolled sheet; warping deformation; residual stress

网格应变分析在冲压成形质量检测中的应用

赵轶哲[1]，孙 力[1]，韩世绪[1]，杨 婷[1]，吕 浩[2]

（1. 河钢材料技术研究院，河北石家庄 050023；2. 河钢集团唐钢公司，河北唐山 063000）

摘 要：通过网格应变分析技术检测了河钢 DX56D+Z 冲压汽车内板成形质量，通过网格印制-零件拉延冲压-应变数据采集与光学图像处理，得到主次应变和厚度减薄率分布云图，结合成形极限值云图和成形极限图进行分析，零件厚度减薄率大部分低于 16%，最大厚度减薄率为 17.21%，板料变形充分，成形性较好。应变主要分布在拉-压区域，应变点完全落在安全区域内，安全裕度为 19.09%。冲压零件无起皱开裂现象，零件安全裕度高，使用安全性高。网格应变分析技术可准确测量分析零件冲压成形质量，适用于汽车冲压行业进行安全生产监控及辅助工艺优化。

关键词：网格应变分析；冲压；减薄率；安全裕度；成形极限图

A Application of Grid Strain Analysis in the Quality Analysis of Stamping

ZHAO Yizhe[1], SUN Li[1], HAN Shixu[1], YANG Ting[1], LV Hao[2]

(1. HBIS materials Technology Research Institute, Shijiazhuang 050023, China;
2. HBIS Group Tangsteel Company, Tangshan 063000, China)

Abstract: The forming quality of the inner plate of HBIS DX56D+Z stamping truck was measured by grid strain analysis technology. Through grid print-part drawing and stamping-strain data acquisition and optical image processing, the distribution cloud map of primary and secondary strain and thickness thinning rate was obtained. The analysis was combined with forming limit value cloud map and forming limit diagram. The maximum thickness thinning rate is 17.21%, and the sheet material has full deformation and good formability. The strain is mainly distributed in the tension and pressure region, and the strain point completely falls within the safety zone, and the safety margin is 19.09%. Stamping parts without wrinkling and cracking phenomenon, parts safety margin is high, high safety in use. Mesh strain analysis technology can accurately measure and analyze the forming quality of parts, which is suitable for safety production monitoring and auxiliary process optimization in automotive stamping industry.

Key words: grid strain analysis; stamping; thickness reduction percent; safety margin; forming limit diagram

热卷 Q355B 低合金钢表面裂纹研究与应用

赵启帆，向浪涛，王　灿，戴　林，何　璋，陈启发

（重庆钢铁股份有限公司，重庆　400080）

摘　要：通过对结构钢热卷 Q355B 成品表面出现的裂纹问题进行分析，表明部分铸坯在高温装炉后，由于铸坯初始晶粒粗大，氮化铝相在晶界形核，在加热过程进一步长大，晶间容易出现微裂纹，后续轧制过程塑性降低，裂纹扩展，导致表面裂纹。同时，针对表面裂纹的改善，进行了热装工艺调整，试验结果显示，新工艺下的 Q355B 卷板表面裂纹缺陷得到明显改善，性能合格率进一步提高约 0.2 个百分点，成品卷板的组织、性能指标均满足设计要求。

关键词：热卷 Q355B；氮化铝；晶界形核；表面裂纹

Research and Application of Surface Crack in Hot Coil Q355B Low Alloy Steel

ZHAO Qifan, XIANG Langtao, WANG Can, DAI Lin, HE Zhang, CHEN Qifa

(Chongqing Iron & Steel Co., Ltd., Chongqing 400080, China)

Abstract: The cracks on the surface of structural steel hot coil Q355B are analyzed. The results show that the aluminum nitride phase nucleates at the grain boundary and grows further during heating, and micro-cracks are easy to appear between the grains. In the subsequent rolling process, the shape decreases and cracks expand, resulting in surface cracks. At the same time, chemical composition optimization and hot-loading process adjustment were carried out to improve the surface crack. The test results showed that the surface crack defects of Q355B coil under the new process were significantly improved, the performance qualification rate was further improved, and the structure and performance indexes of the finished coil met the design requirements.

Key words: hot coil Q355B; aluminum nitride; grain boundary nucleation; surface crack

双相钢铁素体相变动力学模型及预测

雷明钢[1]，常 江[2]，李嘉旋[2]，何 方[1]，
李守华[1]，杨永刚[2]，米振莉[2]

（1. 邯郸钢铁集团有限责任公司，河北邯郸 056015；
2. 北京科技大学高效轧制与智能制造国家工程研究中心，北京 100083）

摘 要：在钢铁材料数字化研发的背景下，建立物理冶金机理模型成为实现组织性能预测及智能研发的迫切需要。本文以双相钢为研究对象，通过 DIL805A/D 热膨胀试验测得了其在不同冷却速率下的膨胀曲线；结合热膨胀曲线和金相组织分析，确定了临界区退火工艺下的铁素体体积分数和晶粒尺寸；并基于 Avrami 方程建立了铁素体相变动力学模型和晶粒尺寸预测模型。结果表明：铁素体体积分数和晶粒尺寸随着保温时间的提高而不断增大，基于实验结果确定了相变动力学参数 n 和 K 分别为 0.47 和 1.29，经过验证模型预测精度高达 95%，为准确预测铁素体相变的体积分数和晶粒尺寸提供指导。

关键词：双相钢；组织预测；相变动力学；铁素体；晶粒尺寸

Construction of a Kinetic Model of Ferrite Phase Transition in Dual Phase Steel

LEI Minggang[1], CHANG Jiang[2], LI Jiaxuan[2], HE Fang[1],
LI Shouhua[1], YANG Yonggang[2], MI Zhenli[2]

(1. Handan Iron and Steel Group Co., Ltd., Handan 056015, China;
2. National Engineering Research Center for Advanced Rolling and Intelligent Manufacturing, University of Science & Technology Beijing, Beijing 100083, China)

Abstract: In the context of digital research and development of steel materials, establishing physical metallurgical mechanism models has become an urgent need to achieve intelligent development and prediction of microstructures and properties. In this paper, the DIL805A/D thermal expansion test was used to measure the expansion curves of dual Phase steel at different cooling rates. By combining the thermal expansion curve and metallographic analysis, the ferrite volume fraction and grain size under the annealing process in the critical region were determined. Based on Avrami equation, a dynamic model of ferrite transformation and a prediction model of grain size were established. The results show that the volume fraction and grain size of ferrite continue to increase with the increase of holding time. Based on the experimental results, the phase transformation kinetic parameters n and K were determined to be 0.47 and 1.29. After verification, the prediction accuracy of the model is as high as 95%, providing guidance for accurately predicting the volume fraction and grain size of ferrite phase transformation.

Key words: dual phase steel; prediction of microstructures; the phase transformation kinetic; ferrite; grain size

Fe-Mn 系列汽车用钢拉深成型性能分析

代永娟，武祥祥，陈佳佳，周 涛

（天津职业技术师范大学，天津 300222）

摘 要： 各类材料在被发现和使用的历史进程中，虽然新材料层出不穷，但是钢铁材料以其多样的性能在经济社会发展中的作用仍然不可替代。在碳中和，碳达峰的环境要求下，钢铁行业"绿色低碳，高质量发展"促进了先进高强钢的研发和应用。应绿色经济的发展，汽车工业提出汽车轻量化创新发展理念。汽车行业的迅速发展使减轻汽车自重和提高安全性能成为了行业发展的关键。以其优异的强塑性在保证汽车用钢的性能达到安全标准的前提下可以实现汽车的轻量化。研究学者已进行了大量的 Fe-Mn 系列钢微观组织性能方面的研究工作[1-2]，但其成型性能研究较少[3-4]。

利用冲压专业分析 Dynaform 软件对 Fe-0.58C-23Mn、Fe-0.44C-23Mn、Fe-25Mn-3Al-3Si、Fe-0.6C-23Mn-1.37Al、Fe-0.6C-23Mn-4.72Al、Fe-0.6C-23Mn-8.31Al 六种成分 TWIP 钢进行板料拉深成型性能模拟计算，六种材料的屈服强度均在 550MPa 左右，抗拉强度在 890～1140MPa，延伸率均大于 36.5%。使用 UG 三维建模软件建立模型，主要组成部分有凸模、凹模、压边圈和不同厚度的板料，拉深成型三维计算模型如图 1 所示。通过控制变形板料厚度、压边力、冲压速度三个工艺参数，分析其成型极限图与厚度变化图，获得不同成分 Fe-Mn 系列汽车用钢成型性能。模拟结果表明，材料 Fe-25Mn-3Al-3Si 在本模拟计算中拉深成型性能较差，只有冲压速度为 5000mm/s 没有出现破裂缺陷。Fe-0.6C-23Mn-8.31Al 整体的成型性能良好，未出现破裂缺陷。其拉深工艺参数为拉深厚度 1.0mm，7000mm/s 拉深速度在本模拟中从未出现拉裂缺陷，拉深件成型无缺陷。本模拟计算过程中，六种成分材料的成型性能按照从优到差顺序依次为：Fe-0.6C-23Mn-8.31Al>Fe-0.6C-23Mn-4.72Al>Fe-0.6C-23Mn-1.37Al>Fe-0.44C-23Mn>Fe-0.58C-23Mn>Fe-25Mn-3Al-3Si。

6 种不同成分、相同强度级别的 Fe-Mn 系列 TWIP 钢成型性能模拟计算结果分析对比结果，为其工业化应用提供理论参考。

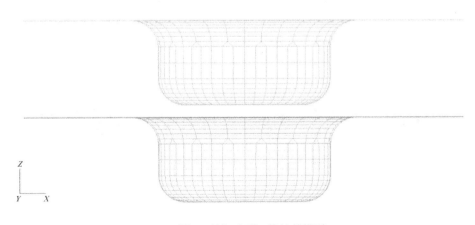

图 1 拉深成型三维计算模型

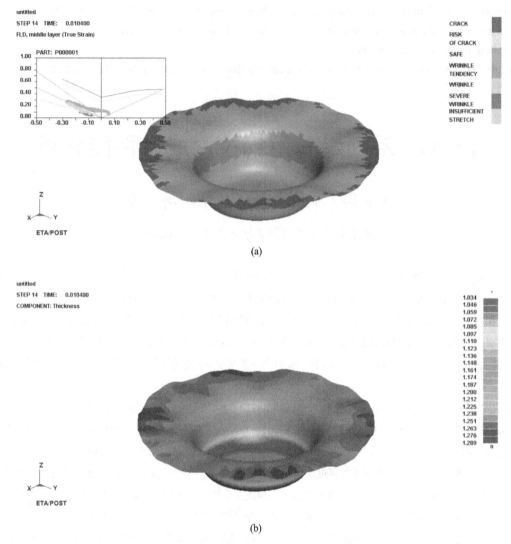

图 2　Fe-0.6C-23Mn-8.31Al 成形极限图（a）和厚度变化过程（b）

参 考 文 献

[1] 米振莉, 唐荻, 陈雨来, 等. 高强度高塑性 TWIP 钢的开发研究[J]. 钢铁, 2005(1): 58-60.
[2] Bouaziz O, Allain S, Scott C P, et al. High manganese austenitic twinning induced plasticity steels: A review of the microstructure properties relationships[J]. Current Opinion in Solid State and Materials Science, 2011, 15(4): 141-168.
[3] 段吉超, 杨冰. TWIP950 在某新车型前纵梁的应用可行性研究[J]. 材料与冶金学报, 2018, 17(1): 62-68.
[4] 米振莉, 薛瑶, 吴彦欣, 等. 基于 Dynaform 软件的 TWIP 钢延迟断裂研究[J]. 金属热处理, 2014, 39(6): 126-129.

35MnB 钢高强韧耐磨件的热处理技术

罗　扬[1], 夏占雪[2], 李福勇[3], 段路昭[1], 宋　月[1],
白丽娟[1], 赵　博[2], 冯立果[1]

（1. 河钢材料技术研究院, 河北石家庄　052160; 2. 第一拖拉机股份有限公司,
河南洛阳　471000; 3. 河钢集团石钢公司, 河北石家庄　050100）

摘 要：研究并确定了适用于 35MnB 钢高强韧耐磨件的淬火和低温回火工艺，并与进口同类产品进行了性能比较。结果表明，采用本研究热处理技术制备的 35MnB 钢高强韧耐磨件的硬度达到 50HRC 以上，常温冲击功达到 30J 以上，抗拉强度达到 1700MPa 以上，断后伸长率达到 12%以上，综合力学性能和耐磨性能均达到了与进口同类产品相当的水平。

关键词：35MnB 钢；热处理；力学性能；耐磨性能

Heat Treatment Technology for 35MnB Steel Parts with High Strength, Toughness and Wear Resistance

LUO Yang[1], XIA Zhanxue[2], LI Fuyong[3], DUAN Luzhao[1], SONG Yue[1],
BAI Lijuan[1], ZHAO Bo[2], FENG Liguo[1]

(1. HBIS Group Material Technology Research Institute, Shijiazhuang 052160, China;
2. First Tractor Company Limited, Luoyang 471000, China;
3. HBIS Group Shisteel Company, Shijiazhuang 050100, China)

Abstract: The quenching and low temperature tempering processes suitable for high strength, toughness and wear-resisting parts of 35MnB steel were studied and determined, and the properties were compared with those of imported products. The results indicate that the hardness of the high strength, toughness and wear-resistant parts of 35MnB steel prepared by the heat treatment technology in this study can reach more than 50HRC, the impact energy at room temperature can reach more than 30J, the tensile strength can reach more than 1700MPa, and the elongation after breaking can reach more than 12%. The comprehensive mechanical properties and wear resistance have reached the same level as imported similar products.

Key words: 35MnB steel; heat treatment; mechanical property; wear resistance property

高强度桥梁缆索用盘条的开发

王 雷，李月云，张 宇

（江苏省沙钢钢铁研究院有限公司，江苏张家港 215625）

摘 要：桥梁缆索是斜拉桥和悬索桥等大跨度桥梁的核心承重构件[1]。伴随着桥梁大跨度、超重载的需求，桥梁缆索钢丝的也在向大规格、高强度方向发展。高强度桥梁缆索钢丝已经成为特大型桥梁建设的材料瓶颈。使用更高强度的钢丝不仅可以提高桥梁缆索的安全系数，还可以减少材料用量，提高桥梁活载，降低建设成本。近十年来桥梁缆索钢丝行业进入快速发展期，桥梁缆索钢丝的强度快速从 1770MPa 提高至 2100MPa 级。新开工建设的桥梁缆索钢丝的强度还在不断提升，如已开工的张皋长江大桥主缆钢丝的设计强度已提高至 2200MPa 级。

我国正处于基础设施建设高峰期，高强度桥梁缆索用钢的用量还在逐年上升。据统计，2023~2027 年 2000MPa 及以上级别的高强度桥梁缆索钢丝的需求量就超过了 40 万吨。桥梁缆索用盘条作为缆索钢丝的关键原材料，其成分设计、组织均匀性、力学性能等对最终成品钢丝的组织和性能均有显著的影响[2-4]。近些年我国在高强度桥梁缆索用钢方面取得了快速发展，整体水平已经达到世界前列。

如何实现桥梁缆索钢丝的高扭转是高强度桥梁缆索用钢开发的关键与难点。钢丝强度越高，作为塑性指标的扭转性能下降越严重。为解决世界在建桥梁中最高强度等级的 2100MPa 级桥梁缆索的扭转问题，我们以扭转恶化的机理为出发点，在系统研究了拉拔及热镀工艺的基础上形成了以扭转为导向的成分设计方案。进而通过深入分析中心偏析、夹杂物、组织均匀性以及钢丝生产工艺对扭转合格率的影响，形成了包括结晶钢冶炼、高均匀性热处理、低损伤拉拔等在内的一整套超高强度桥梁缆索用钢生产控制技术，实现了 7mm 2100MPa 级桥梁缆索用钢 SWRS96Si-TP 的成功开发与批量供货，并取得了 2200MPa 级桥梁缆索用钢开发应用的初步成功。

关键词：桥梁缆索；抗拉强度；扭转性能；热处理

参 考 文 献

[1] Kang J R. Global attractor for suspension bridge equations with memory[J]. Mathematical Methods in the Applied Sciences, 2016, 39(4): 762-775.
[2] 王雷，李月云，胡磊，等. 6.0mm 2060MPa 级桥梁缆索镀锌铝钢丝用盘条的研制[J]. 金属材料与冶金工程, 2019(5): 9-15.
[3] Konno S, Eguchi T, Maruyama N, et al. High strength galvanized steel wire for bridge cables[J]. Structural Engineering International, 2002, 12(3): 209-213.
[4] Zhou L, Fang F, Wang L, et al. Torsion performance of pearlitic steel wires: Effects of morphology and crystallinity of cementite[J]. Materials Science and Engineering: A, 2019,743:425-435.

压边力及拉延筋对 DC03 板材拉深影响的有限元模拟研究

陈 健，方进秀，钱健清

（安徽工业大学冶金工程学院，安徽马鞍山 243032）

摘 要：采用有限元模拟的方法对筒形件拉深成形工艺参数进行模拟，研究了筒形件拉深成形过程中压边力以及拉延筋布置对成形的影响。结果发现：在其他工艺参数相同的情况下，增加压边力对于 DC03 板材拉深成形壁厚分布影响不大；布置拉延筋对于 DC03 板材拉深成形壁厚分布有明显的影响，并且不同阻力的拉延筋对最终成形壁厚分布影响差别较大。

关键词：压边力；拉延筋；拉深成形；Dynaform；有限元模拟

Finite Element Simulation Study on the Effects of Blank Holder Force and Drawing Ribs on the Deep Drawing of DC03 Plate

CHEN Jian, FANG Jinxiu, QIAN Jianqing

(School of Metallurgical Engineering, Anhui University of Technology, Maanshan 243032, China)

Abstract: The finite element simulation method was used to simulate the process parameters of deep drawing of cylindrical parts, and the effects of edge force and layout of drawing ribs on the forming process of cylindrical parts were studied. The results showed that under the same other process parameters, increasing the blank holder force had little effect on the wall thickness distribution of DC03 sheet during deep drawing; The arrangement of stretching ribs has a significant impact on the wall thickness distribution of DC03 sheet during deep drawing, and the influence of stretching ribs with different resistance on the final formed wall thickness distribution varies greatly.

Key words: blank holding force; stretch reinforcement; deep drawing forming; Dynaform; finite element simulation

钢铁工艺知识图谱的构建与应用研究

米振莉，袁长辉，杨永刚，宋　勇，倪道军，常　江

（北京科技大学，工程技术研究院，北京　100083）

摘　要：传统的钢铁成分和工艺参数的推荐方法主要依赖于人工经验和试错，生产效率低下且容易导致成品质量不稳定等问题。知识图谱以图数据库作为存储引擎，对海量钢铁材料数据信息进行智能化处理，可以形成大规模的工艺知识库并进而支撑业务应用。本研究以钢铁生产全生命周期工艺数据与相关知识为背景，首先进行类、对象属性和数据属性三方面的抽象，完成知识图谱模式层的交互式构建，其次按照模式层需要将钢铁工艺数据存入关系型数据库，通过数据与预设模式层映射构建形成钢铁工艺知识图谱，最后结合深度学习或图神经网络的方法能够自动、智能地在现有的工艺知识数据和信息中发掘、推导隐性信息，并实现钢铁成分和工艺参数的推荐。通过钢铁工艺知识图谱的构建与应用，实现对钢铁工艺知识沉淀和知识重用，对于降低生产成本和质量风险，提高企业生产的效率和稳定性具有重要意义，为钢铁数字化转型提供了重要途径。

关键词：钢铁工业；知识图谱；深度学习；工艺参数推荐；数字化

Research on the Construction and Application of Knowledge Graph for Iron and Steel Processes

MI Zhenli, YUAN Changhui, YANG Yonggang, SONG Yong,
NI Daojun, CHANG Jiang

(Institute of Engineering Technology, University of Science and Technology Beijing, Beijing 100083, China)

Abstract: The traditional method of recommending steel composition and process parameters mainly relies on manual experience and trial-and-error, which is inefficient and prone to lead to unstable quality of finished products. Knowledge graph uses a graph database as a storage engine to intelligently process massive steel material data information, which can form a large-scale process knowledge base and support business applications. The research is based on process data and knowledge of the whole life cycle of iron and steel production. First of all, the abstraction of the three aspects of class, object attribute and data attribute were carried out, the interactive construction of the knowledge graph model layer was completed, and then the iron and steel process data were stored in the relational database according to the needs of the model layer, forming the knowledge graph of iron and steel process through the construction of the mapping of the data and the preset model layer. Finally, the steel knowledge graph was combined with deep learning or graph neural network can automatically and intelligently discover and derive implicit information from the existing process knowledge data and information, and realize the recommendation of steel composition and process parameters. Through the construction and application of steel process knowledge graph, it realizes the knowledge precipitation and knowledge reuse, which is of great significance for reducing production cost and quality risk, improving the efficiency and stability of enterprise production, and providing an important way for the digital transformation of iron and steel industry.

Key words: iron and steel industry; knowledge graph; deep learning; process parameter recommendation; digitization

稀土对 Mn13 组织及性能影响

石 帅[1,2]，赵燕青[1,2]，高云哲[1,2]，弓俊杰[1,2]

（1. 河钢材料技术研究院，河北石家庄　050023；
2. 河北大河材料科技有限公司，河北石家庄　050023）

摘　要：采用微量稀土 La-Ce 微合金化高锰铸钢，借助金相显微镜、扫描电镜、能谱仪、冲击试验机、显微硬度计等检测手段，对添加 La-Ce 混合稀土的铸造高锰钢显微组织、夹杂物和力学性能进行了表征和分析。结果表明：高锰钢中加入微量的稀土 La-Ce 使铸态组织柱状晶区缩短，等轴晶区扩大，二次枝晶间距减小，晶界碳化物细化，夹杂物变性。混合稀土 La-Ce 改善铸态高锰钢的冲击韧性及显微硬度，0℃、-20℃、-40℃横向冲击功分别提升 32.3%、67.5%、170.0%，低温韧性提升更加显著；显微硬度 HV10 由 221 提升至 238。

关键词：高锰钢；稀土；夹杂物；冲击功

Effect of Mixed Rare Earth La、Ce on the Microstructure and Inclusion Modification of Cast High Manganese Steel

SHI Shuai[1,2], ZHAO Yanqing[1,2], GAO Yunzhe[1,2], GONG Junjie[1,2]

(1. Material Technology Research Institute of HBIS Group, Shijiazhuang 050023, China;
2. Hebei Dahe Material Technology Co., Ltd., Shijiazhuang 050023, China)

Abstract: High manganese cast steel was microalloyed with trace rare earth elements La Ce. The microstructure, inclusions, and mechanical properties of the cast high manganese steel with La Ce mixed rare earth elements were characterized and analyzed using metallographic microscopy, scanning electron microscopy, energy dispersive spectroscopy, impact testing machine, and microhardness tester. The results show that the addition of trace amounts of rare earth La-Ce in high manganese steel shortens the columnar crystal zone of the as-cast structure, expands the equiaxed crystal zone, reduces the spacing between secondary dendrites, refines grain boundary carbides, and denatures inclusions. The mixed rare earth La Ce improves the impact toughness and microhardness of as cast high manganese steel, with a transverse impact energy increase of 32.3%, 67.5%, and 170.0% at 0℃, -20℃, and -40℃, respectively. The low-temperature toughness improvement is more significant; The microhardness HV10 has been increased from 221 to 238.

Key words: high manganese steel; rare earth; inclusions; impact energy

金属管材热气胀成形技术发展及应用

张　猛[1,2]，阮尚文[1,3]，成　刚[1]，程鹏志[1]，
吴彦欣[3]，张　淇[3]，米振莉[3]

（1. 航宇智造（北京）工程技术有限公司，北京　100080；2. 北京航空航天大学能源动力与工程学院，北京　100191；3. 北京科技大学工程技术研究院，北京　100083）

摘　要： 金属管材热气胀（Hot Metal Gas Forming，HMGF）是一种高温状态下借助高压气体迫使管坯发生塑性变形直至其与模具型面完全贴合的净成形技术。相较于液压成形技术，HMGF 技术具有更高的温度，进而使材料具有更好的延展性，从而实现难变形材料的精确成形。

HMGF 技术始于上世纪末，是美国 NIST 所支持的先进技术研究项目所提出的。其研究结果表明，采取该技术能将钢的铝的变形膨胀率分别提高至 50%和 150%，并通过两端进给实现了更小的减薄[1]。欧洲煤炭和钢铁研究基金（RFCS）于 2004 年起开始进行一项为期三年的科研项目资助，并成功实现了 140%变形量的自由胀形[2]。自 2006 年开始，HEAform 公司报告了一种新的研究方法，并生成在 550℃状态下仅用 20s 时间内就实现铝合金高达 270%的变形量。值得注意的是，加热方式及温度还会对材料的组织产生影响。特别是对于高强钢来讲，对其进行加热保温可以实现其奥氏体转换，从而进一步提升成形后的材料力学性能[3]。这使得 HMGF 更加适用于超高强钢的成形，且该技术已在部分高端车型的 A 柱和 B 柱上开始应用。公开资料显示，采取 HMGF 技术成形的高强度 A 柱可以带来 25%偏置碰撞性能，并带来 10%~20%的结构减重效果。此外，相较于传统液压成形技术，材料在 HMGF 技术中的流动应力较低，进而降低高强钢成形后零件的回弹量，从而实现精确成形[4,5]。其中，德国的 HEAform、加拿大的 Multimatic 以及挪威的 Hydro Extruded Solutions 均在 HMGF 领域取得了大量的技术积累与实践进展。

相较于国外，国内的 HMGF 发展技术起步较晚，相关工艺、设备及行业标准还鲜有报道。其中，具有代表性的应用成果为航宇智造为福特某型号汽车生产的 A 柱总成。该项目已成功实现了 HMGF 技术的全国产应用，具有完全自主知识产权，并成功进行了 OEM 装车。

参 考 文 献

[1] Winter, Sven W, Markus Haase, et al. Processing Q&P steels by hot-metal gas forming: Influence of local cooling rates on the properties and microstructure of a 3rd generation AHSS[J]. Journal of Materials Processing Technology, 2021, 293(1).
[2] A K D, B K W A, A W C, et al. Study on fast gas forming with in-die quenching for titanium alloys and the strengthening mechanisms of the components[J]. Journal of Materials Research and Technology, 2022.
[3] Zheng K, Zheng J H, He Z, et al. Fundamentals, processes and equipment for hot medium pressure forming of light material tubular components[J]. International Journal of Lightweight Materials and Manufacture, 2019, 3(1).DOI:10.1016/j.ijlmm.2019.10.003.
[4] Martin S, Wolf S, Martin U, et al. Deformation mechanisms in austenitic TRIP/TWIP steel as a function of Temperature[J]. Metallurgical and Materials Transactions A, 2014. DOI:10.1007/s11661-014-2684-4.
[5] 张国进. 超高强度钢板 BR1500HS 热成形连续冲压热平衡工艺参数研究[D]. 重庆：重庆理工大学, 2017: 15-20.

定制辊压成型技术与中国制造业的高质量发展

晏培杰 [1,2]

（1. 南京工业大学先进轻质高性能材料研究中心，江苏南京　210037；
2. 苏州亿创特智能制造有限公司，江苏昆山　215300）

摘　要： 定制辊压成型技术是钢铁材料深加工技术的重要分支，如今正面临着制造业发展带来的严峻挑战。定制辊压成型技术可以实现轻量化、节能减排与智能制造等先进制造理念，满足制造业转型升级的迫切需求。本文对比了定制辊压成型技术在国内外的发展现状，阐述其对我国制造业发展的重大意义，同时针对国内定制辊压成型技术发展过程的主要难点，提出了相应的解决方案和实施路径。最后对定制辊压成型技术在重点行业的应用状况以及市场规模做出展望。

国内外发展现状对比：（1）辊压型材的设计和应用技术。欧洲国家对辊压型材的设计和应用已经形成体系，一方面,产品设计工程师将形式各异的型材作为常规结构件进行设计;另一方面,型材供应商对型材的性能及应用工况了解深入,能够指导用户使用。(2)辊压为平台的多工艺集成技术。辊压技术作为一种平台技术,可以与各种不同工艺相集成,以满足产品需要。(3)先进的模具设计技术。(4)轻质材料辊压成型技术。

对我国制造业发展的意义:(1)引领国内辊压行业技术升级。(2)轻质材料辊压成型技术的系统开发。定制辊压成型技术可以根据不同轻质材料的成型性能,给出辊压技术的解决方案,形成量产能力,推动轻质材料的应用和发展。(3)先进型材对智能制造的推动。(4)钢材-定制辊压-产品供应链上下游关系中,定制辊压成型技术起到从材料到产品的桥梁作用,它识别产品的需求,选择超高强钢等合适材料,经过辊压成型技术,满足产品的结构功能一体化,应用于下游产品。

定制辊压成型技术的解决方案:面对制造业升级所带来的诸多新需求,定制辊压型材通过材料优化设计和结构整合优化两种手段改良辊压型材产品。(1)材料优化设计。(2)结构优化设计。(3)产品模块化。

本文围绕定制辊压成型技术发展的背景、定义、方案和市场进行了系统综述。定制辊压成型技术未来发展的关键点在于:(1)辊压成型创新技术的研发。(2)定制辊压成型技术的社会重视。(3)产业链上下游的共同价值链形成。也将定制辊压成型技术的典型应用进行了阐述,比如客车车身骨架航空领域。汽车领域建筑领域物流领域。

高强度链条用 20Mn2 拉力不足原因分析

王冬晨[2,3],李永超[1,2,3],樊亚鹏[1,2,3],李亚硕[1,2,3],韩光洋[1,2]

(1. 河北邢钢科技有限公司,河北邢台 054799;2. 邢台钢铁有限责任公司,河北邢台 054027;
3. 河北省线材工程技术创新中心,河北邢台 054027)

摘 要:本文利用光学显微镜检验组织、扫描电镜观察微观组织及断口、洛氏硬度计等手段研究了组织类型对 20Mn2 链条性能的影响。分析结果认为,当链条显微组织中残留较高比例的块状铁素体时,导致链条强度、硬度不足且韧性降低,原因主要是链条受力时组织中的块状铁素体处产生应力集中,从而导致链条整体性能下降。

关键词:20Mn2;链条;组织;性能

Analysis of Causes of Insufficient Tensile Force of 20Mn2 for High Strength Chain

WANG Dongchen[2,3], LI Yongchao[1,2,3], FAN Yapeng[1,2,3],
LI Yashuo[1,2,3], HAN Guangyang[1,2]

(1. Hebei Xinggang Technology Co., Ltd., Xingtai 054799, China; 2. Xingtai Iron and Steel Corp Ltd., Xingtai 054027, China; 3. Hebei Province Engineering and Technology Innovation Center for Wire Rod, Xingtai 054027, China)

Abstract: In this paper, the effect of structure type on the properties of 20Mn2 chain was studied by means of optical microscope, scanning electron microscope and Rockwell hardness tester. The results show that when a high proportion of massive ferrite remains in the chain microstructure, the strength and hardness of the chain are insufficient and the toughness is reduced. The main reason is that the stress concentration occurs at the massive ferrite in the structure when the chain is stressed, resulting in the decline of the overall performance of the chain.

Key words: 20Mn2; chain; microstructure; properties

酸洗冷连轧机组薄规格启动断带问题研究

刘前亮，李玉冰，张 倩，杨雪松

（河钢集团邯钢公司冷轧厂，河北邯郸 056015）

摘 要：分析研究了酸洗冷连轧机组薄规格启动断带的原因。重点研究了轧机入口张力建立及波动问题，轧机静态张力过大，停车启动轧机建张原理和顺序，附加速度和张力控制及厚度控制的关系，生产过程中张力波动的原因和张力控制调节器的工作原理等。本文对各项问题进行了深入研究，建立了厚度与张力附加速度数学模型。成功解决了薄规格启动断带问题。

关键词：薄规格；启动断带；张力控制；附加速度

Research on Start-up Strip Break Issues in the Continuous Pickling Line-Tandem Cold Rolling Mill for Thin Specifications

LIU Qianliang, LI Yubing, ZHANG Qian, YANG Xuesong

(HBIS Group Hansteel Company Cold Rolling Mill Plant, Handan 056015, China)

Abstract: Analyzed and researched the causes of strip break during the start-up of the acid pickling cold rolling mill for thin specifications. Emphasized the investigation of issues related to establishing and fluctuating tension at the mill entrance, excessive static tension in the rolling mill, the principles and sequence of tension establishment during mill stoppage and restart, the relationship between additional acceleration and tension control, as well as thickness control. Explored the reasons for tension fluctuations during the production process and the operating principles of tension control regulators. This paper conducted in-depth research on these issues, provided solutions, and established a mathematical model for thickness and tension with additional acceleration. Successfully resolved the problem of strip break during the start-up of thin specifications.

Key words: thin specifications; start-up brokennin; tension control; additional speed

轨梁厂CCS工作辊道减速机失效分析及结构优化

贾 欣，冯指名，刘志强，张 骞，刘剑锋

（包钢（集团）公司轨梁轧钢厂，内蒙古包头 014000）

摘 要：根据CCS工作辊道减速机的结构，并结合坯料撞击工作辊道的原因及其冲击载荷，校核了减速机齿轮的弯曲疲劳强度，推导出在坯料对工作辊道的冲击载荷下，齿轮弯曲疲劳强度严重偏低，从而导致经常断齿和定位键切键问题。在原工况条件下，通过使用热处理调质后的双键槽定位键，增加减速机输出轴长度，提高减速机输出轴高度，改变减速机底座尺寸，增大输出最大扭矩和输入最大功率等参数，得出工作辊道减速机的合理结构，优化后减速机的齿根弯曲疲劳强度和齿面接触疲劳强度均能满足现场工况要求。

关键词：定位键；扭矩；功率；结构优化；减速机；疲劳强度

Failure Analysis and Structure Optimization of Rail and Beam Factory CCS Work Roll Table Reducer

JIA Xin, FENG Zhiming, LIU Zhiqiang, ZHANG Qian, LIU Jianfeng

(Baogang(Group) Company Rail and Beam Factory, Baotou 014000, China)

Abstract: According to the structure of CCS working table reducer, combined with the reason of billet impacting the working table and its impact load, the bending fatigue strength of high-speed shaft pinion of reducer is checked, and it is deduced that the bending fatigue strength of high-speed shaft pinion is seriously low under the impact load of billet on the working table, which leads to the problems of frequent tooth breakage and positioning key cutting. Under the original working conditions, the reasonable structure of the working roller reducer was obtained by using the double-key slot positioning key after heat reatment and tempering, increasing the length and height of the output shaft of the reducer, changing the size of the reducer base, increasing the maximum output torque and the maximum input power, etc. After optimization, the tooth root bending fatigue strength and tooth surface contact fatigue strength of the reducer can meet the requirements of the field working conditions.

Key words: locatung key; twist moment; power; structure optimization; reducer; fatigue strength

辊弯成型过渡区长度计算方法分析

刘　鹏，李宏图，王海隆，龚学堂

（中冶赛迪工程技术股份有限公司低碳技术研究院，重庆　401122）

摘　要： 辊弯成型工艺设计中，过渡区长度是一个十分重要的设计参数，与机架间距、弯曲角度以及弯曲力的计算均有关联。本文对辊弯成型的钢板咬入情况、几何法求解和平衡方程求解过渡区长度的情况进行分析，得到影响过渡区的一些因素。结合方形管设计参数，对影响过渡区长度计算的其它因素进行了简单分析，得到了部分有益结论。

关键词： 辊弯；过渡区长度；机架间距；几何法；平衡方程法

Analysis of the Calculation Method for the Length of the Transition Zone in Roll Forming

LIU Peng, LI Hongtu, WANG Hailong, GONG Xuetang

(CISDI Group, Co., Ltd., Chongqing 401122, China)

Abstract: In the design of roll bending forming process, the length of the transition zone is a very important design parameter, which is related to the calculation of frame spacing, bending angle, and bending force. This article analyzes the biting situation of steel plates formed by roll bending, geometric method solving, and balance equation solving for the length of the transition zone, and obtains some factors that affect the transition zone. Combining square tube design parameters, a simple analysis was conducted on other factors that affect the calculation of the length of the transition zone, and some beneficial conclusions were obtained.

Key words: roll bending; length of transition zone; frame spacing; geometric method; balance equation method

9　粉末冶金

大会特邀报告

第十三届冶金青年科技奖获奖人特邀报告

分会场特邀报告

矿业工程

焦化及节能环保

炼铁与原料

炼钢与连铸

电冶金与废钢铁

轧制与热处理

表面与涂镀

金属材料深加工

★ 粉末冶金

先进钢铁材料及应用

节能与低碳技术

冶金环保与资源利用

冶金设备与工程技术

冶金自动化与智能化

冶金物流

冶金流程工程学

其他

基于选区激光熔化的镍基单晶高温合金修复研究

张 鹏[1,2]，沈 超[1]，张 宇[1]，章 林[1,2]，曲选辉[1,2]

（1. 北京科技大学新材料技术研究院，北京 100083；
2. 北京材料基因组工程高精尖创新中心，北京 100083）

摘 要：单晶镍基高温合金叶片修复可以节约昂贵的高温合金材料，延长叶片的使用寿命。选区激光熔化光斑小精度高，热影响更小，几乎不会造成变形，是一种重要的单晶叶片修复方法之一。本工作研究了高温合金沉积层在单晶 CM247LC 基板上沉积过程中的显微组织演变规律以及力学性能，主要结论如下：

激光功率和扫描速度对修复组织有显著影响。随着激光功率的降低以及扫描速度的增高，沉积区裂纹密度减少，而修复界面处深宽比小的熔池结构使得晶粒取向趋于一致，深宽比大的熔池因晶体垂直于熔合线生长使得在熔池中心取向差较大。获得了宽深比为 6 的熔池结构，此时晶粒取向垂直于熔池底部沿构建方向生长，沉积组织与基体的取向差较小。

修复界面处杂晶的产生与共晶和碳化物强烈相关，这主要是由于局部的元素偏析影响成分过冷，从而导致固液界面形状的改变，以及低熔点相重熔导致界面下沉变形。同时共晶和碳化物会成为界面处的裂纹源；热处理后加剧界面 Ti、Al 等元素扩散，界面熔池形貌逐渐消失，界面逐渐变平整。在 1140℃下固溶温度能够大幅减少界面处的共晶组织，而当固溶温度达到 1230℃发生了明显再结晶，枝晶转变为胞状晶。时效温度升高会导致共晶尺寸会增加，使组织性能恶化。

最终通过调控参数，当激光功率为 180W、扫描速度为 900mm/s、激光能量密度为 111.11J/mm^3，得到了裂纹密度少（26μm/mm^2），晶粒取向单一，修复界面结合优异的单晶沉积层。沉积区的抗拉强度达到 1094MPa，延伸率达 22%。经过 1140℃/2h+850℃/4h+760℃/16h 热处理工艺后，γ'析出相呈双峰分布，抗拉强度较打印态提升 33%，达到 1459MPa，超过铸态 CM247LC 的室温力学性能。

关键词：单晶高温合金；激光修复；熔池；元素分布；力学性能

注射成形高强钛合金研究

魏敬浩，陈 刚，秦明礼，曲选辉

（北京科技大学新材料技术研究院，北京 100083）

摘 要：钛合金具有低密度、高比强度、耐腐蚀等优点，广泛应用于航空航天、汽车、电子等领域。近年来，航空航天以及军工行业对钛合金性能尤其是强度要求越来越高，不仅如此消费电子、电子穿戴领域也需要高强钛合金来代替传统高强材料，实现整体轻量化。钛合金很难通过传统的加工方式稳定高效的制备出形状复杂的钛合金零件，并且加工过程中会造成资源和能源浪费。金属粉末注射成形技术（MIM）可以大批量生产形状复杂的小型零部件，可以解决钛合金难加工的问题。然而，当前仍缺乏注射成形高强钛合金的成分设计准则，以至于注射成形制备的钛合金强度仍低于 1000MPa，远未达到高强钛合金（强度高于 1100MPa）的性能要求。因此，亟需开发一种注射成形高强钛合金成分及注射成形工艺。本课题创新性的设计了一款注射成形用高强钛合金，通过研究其注射成形工艺，使得钛合金样品具有很好的力学性能和尺寸精度，烧结后样品抗拉强度>1200MPa，屈服强度>1100MPa，延伸率>4%，产品性能稳定满足需求。

关键词：高强钛合金；注射成形；尺寸精度

增材制造 NiTi-Nb 医用多孔支架孔径策略设计及性能研究

刘国浩，王　瑞，谢锦丽，毕中南，孙广宝

（北京钢研高纳科技股份有限公司，北京　100081）

摘　要：随着以 SLM 为代表的增材制造技术的发展，研究人员可以设计更为精细的金属多孔支架更高效的治疗骨缺损[1]。通过多孔结构设计不仅能够降低支架的弹性模量，避免"应力遮蔽"效应，还能够促进细胞的增殖分化和骨组织的长入。但较大的孔隙率会导致支架强度下降，支架承载能力不足，较低的孔隙率则会导致支架渗透性下降，骨组织长入效果不理想，因此获得强度-孔隙率的更优匹配一直是医用多孔支架研究的重点[2]。NiTi-Nb 合金具有超弹性、形状记忆功能、良好的生物相容性和耐腐蚀性能，是一种良好的生物医学金属材料[3]。本文基于极小曲面（TPMS），通过改变结构单元尺寸和孔径分布函数的思路设计了相同孔隙率（70%）的不同孔径大小策略多孔支架（500μm、750μm 和 900um）及径向成梯度分布的孔径分布策略多孔支架（500~900μm 和 900~500μm），通过选区激光熔融（SLM）制备了多孔支架，研究了孔径策略对支架力学性能、渗透性和生物相容性的影响。

本文通过 SLM 制备的新型 NiTi-Nb 多孔支架具有良好的通透性，尺寸和实际孔隙率与设计模型基本保持一致。压缩实验和有限元模拟结果表明，NiTi-Nb 多孔支架的弹性模量在 0.80~1.05GPa 范围内，与松质骨相匹配，抗压强度和最大应变分别在 42.3~65.8MPa 和 15.3%~19.1%之间，与相近孔隙率的其他材料支架相比具有较好的力学性能组合；对于不同孔径大小支架，力学性能没有显著差异；对于不同孔径分布径向梯度支架，弹性模量与均匀支架相近，但通过改变孔径分布，提高了支架的局部孔隙率，提升了支架的承载能力，抗压强度较均匀支架得到了显著提高。渗透性实验结果显示：所有支架在体液环境中的渗透率均与松质骨相当，保证了营养物质的运输和细胞的新陈代谢；对于孔径大小策略多孔支架，增大结构单元的策略降低了结构单元数，减小了支架的表面积，进而降低了对流体的摩擦阻力，支架的渗透率随孔径的增大而减小；对于孔径分布策略梯度支架局部较大的孔隙率提高了局部流量，进而提高了支架整体的渗透性。生物相容性实验结果表明：NiTi-Nb 多孔支架在 BMSC 细胞接种 1 天后表现出良好的生物相容性和细胞黏附能力，共培养 4 天和 7 天的细胞 OD 值显示，BMSC 细胞能够支架表面增殖生长；小孔径支架由于较大的表面积，有利于细胞的黏附，大孔径支架更高的渗透性有利于细胞的增殖。本研究为获得更优力学性能、渗透性和生物性能组合的医用多孔植入物提供了新的思路。

参 考 文 献

[1] Lv Y T, Wang B H, Liu G H, et al. Metal material, properties and design methods of porous biomedical scaffolds for additive manufacturing: A review[J]. Front. Bioeng. Biotechnol., 2021, 9: 641130.
[2] Murphy V, Atala A. 3D bioprinting of tissues and organs[J]. Nat. Biotechnol., 2014, 32(8): 773-785.
[3] Guo S, Zhang H, Chen Y, et al. A sandwich-structured Nb/NiTi composite with good bio-compatibility, near-linear-elastic deformation and large elastic admissible strain[J]. Compos. Part B Eng., 2021, 207: 108586.

"核壳"结构特种功能粉体及其潜在工程应用

杨亚锋，李少夫

（中国科学院过程工程研究所，北京 100190）

摘 要：粉末冶金和金属 3D 打印近净成形技术能够实现粉体到高性能复杂构件的快速加工制造，具有节约原材料、缩短生产周期、绿色环保等优点，其发展符合"双碳"背景下我国绿色制造技术发展战略。粉末颗粒是 3D 打印和粉末冶金技术的原材料，也是获得高性能、高质量、高精度零部件的基础。针对当前市场上存在的粉体种类单一、复合粉体缺乏以及功能粉体制备困难等问题，本报告重点介绍一种特殊"壳核"结构包覆型粉体的流态化制备原理和弘量化生产技术与配套装备，重点阐述不同种类包覆改性粉体的当前工程实践探索。主要涉及领域包括：通过在钛粉表面均匀包覆烧结辅助剂，实现了大尺寸、复杂结构粉末冶金钛合金制品的控形控性；通过粉体表面功能改性实现了铝、铜等高激光反射率金属粉体的高精度、高效率、高性能打印成形；通过包覆晶粒细化剂，消除了 3D 打印钛合金的粗大柱状晶组织，并解决了组织性能各向异性问题，获得了性能超过 ASTM 铸锻钛合金标准的综合性能；创制了高品质金属基复合材料 3D 打印专用粉体以及符合 3D 打印需求的新型陶瓷粉体，实现了高性能、复杂结构复合材料和陶瓷器件的高精度一次打印成形。本报告还将深入探讨其他新型复合粉末的设计需求、可行性研制以及它们潜在的工程应用。

基于粉末 3D 打印原理的材料高通量制备实验方法与应用

张百成

（北京科技大学，北京 100083）

摘 要：材料基因工程是材料领域的颠覆性前沿技术，将对材料研发模式产生革命性的变革，全面加速材料从设计到工程化应用的进程，大幅度提升新材料的研发效率，缩短研发周期，降低研发成本，促进工程化应用。本文从基础理论与方法、关键技术与装备、新材料研发与工程化应用、人才培养以及材料基因工程新理念的形成和推广等方面，综述了中国材料基因工程的研究进展，并提出了未来发展方向建议。

紧耦合气雾化技术制备铁基非晶粉末的数值模拟与试验研究

刘佳奇[1]，王璞[1]，董延楠[1]，赵欢[2]，庞靖[2]，张家泉[1]

（1. 北京科技大学冶金与生态工程学院，北京 100083；

2. 青岛云路先进材料技术股份有限公司，山东青岛　266232）

摘　要：Fe 基非晶合金因具有低矫顽力、高饱和磁感、高电阻率与低磁晶各向异性等优秀的软磁性能，可用于制备变压器、扼流圈与电感等电子零部件[1]。在当下电子元件小型化与高频化的发展趋势下，新一代非晶电子零部件对高端非晶原材料，尤其是 Fe 基非晶粉末及其高效制备工艺提出了迫切需求[2]。紧耦合气雾化技术因具备成本低、冷却能力足够、粉末球形度高以及大规模生产能力等优势而成为工业制备 Fe 基非晶粉末的主要方法[3]。但现有研究中针对雾化法生产非晶粉末的相关研究较为匮乏，粉末缺陷的控制与液滴冷却速率的提高仍然是限制粉末性能的关键问题。

本研究基于计算流体力学方法，首先通过 VOF(volume of fluid)模型模拟一次雾化过程，详细研究了金属液质量流速与雾化压力对于破碎过程与粉末特性的影响。然后，进一步使用双向耦合 DPM 模型(discrete particle model)计算二次雾化过程，针对不同工艺参数对颗粒飞行与冷却过程进行了详细研究，并分析了热气体雾化技术对于制备微细非晶粉末的应用前景。

一次雾化计算结果表明，金属液流速大于 0.075kg/s 时，一次雾化破碎不充分；金属液质量流速过低（0.025kg/s）时，发生易导致堵钢的液膜破碎。过低雾化压力（1.0MPa）下液滴逆流，而 3.0MPa 以上的雾化压力则导致大量卫星粉、空心粉与针状粉等缺陷粉的产生。对于本研究中分析的雾化工艺，发现气液比可作为工业生产中预测一次雾化破碎模式的简易标准，当气液比≥4.4 时发生液膜破碎，而气液比≤4.3 时则发生"微型喷泉"破碎。二次雾化结果表明，颗粒的 d_{50} 同样随气液比的增加；颗粒粒度分布标准差 d_{84}/d_{50} 随金属液质量流速的增加而增加，过高与过低雾化压力均会导致较高的 d_{84}/d_{50}；降低金属液质量流速、增加雾化气压会加速液滴与气体间的传热，进而增加颗粒平均冷却速率。适当增加气体温度至 400K，可以大幅降低 d_{50} 并可将颗粒冷却速率提高近一倍。但过高的气体温度不仅对颗粒冷却速率改善有限，而且易导致大量缺陷粉末的产生。不同雾化压力下的工业试验结果与上述模拟分析大致吻合，即在 2.0MPa 与 0.050kg/s 下制备的粉末 d_{50} 为 58.9μm，缺陷粉数量较少，且粒径小于 50μm 的粉末非晶度可达 96.6%。

本工作通过建立雾化全过程数值模型，针对缺陷粉控制、雾化冷却能力的提高与粉末综合性能的改善提供了理论与方法指导，以期实现高性能 Fe 基非晶粉末的批量生产。

参 考 文 献

[1] Silveyra J M, Ferrara E, Huber D L, et al. Science, 2018, 6413(362): eaao0195.
[2] Liu J Q, Wang P, Zhang J Q, et al. Journal of Iron and Steel Research International, 2023, 30: 1-14.
[3] 刘佳奇, 庞靖, 王璞, 等. 中国冶金, 2022, 32(2): 1-14.

铜基 CBN 超硬复合材料界面反应行为研究

范永刚，王　聪

（东北大学，辽宁沈阳　110819）

摘　要：围绕国内航天、汽车和船舶等高精密机械加工领域对立方氮化硼（Cubic Boron Nitride，CBN）复合材料的加工性能要求越来越高。传统 CBN 复合材料中，CBN 颗粒只是被机械包埋在基体中，两者之间无冶金结合，界面结合强度低，导致加工中立方氮化硼(CBN)容易整体脱落和砂轮堵塞，进而严重影响其加工性能这一关键共性问题。因此，开发成本低、耐磨性优良、能长时间进行稳定加工的 CBN 复合材料是必然的发展趋势。

本论文采用先冷压成型后高温真空活性烧结方法制备 CBN/Cu-Sn-Ti 复合材料，通过原料配比和工艺优化—精确调控界面反应层厚度—调控超硬复合材料耐磨性。通过对不同 TiH_2 含量下烧结后 CBN/Cu-Sn-Ti 复合材料的润湿性、界面反应层厚度和耐磨性等系统的研究，确定最佳 TiH$_2$ 含量。进一步探究不同烧结温度下 CBN/Cu-Sn-Ti 界面产物种类、反应层厚度和生成物形貌的系统变化规律，揭示活性元素 Ti 在界面和基体的分布情况；随后深入分析不同保温时间下金属基体强度和界面反应层生长的规律性变化，量化保温时间与反应层厚度之间的关系，确定最优工艺参数，实现对反应层厚度的精确调控。利用 FIB-TEM 对界面反应层结构和界面元素的分布进行精确表征，通过选区衍射确定界面反应产物种类，明确 Ti 的全流程扩散路径，阐明界面反应层的形成机理和生长过程，使界面形成高强度的冶金结合。此外，对界面产物形核能力进行动力学计算，确定界面产物的生成顺序。研究不同烧结条件下复合材料失重和孔隙率的变化规律，评价复合材料的耐磨性能。同时对磨损后的形貌进行分析，揭示磨损过程中失重机理，为制备加工性能优良的 CBN/Cu-Sn-Ti 复合材料提供坚实的理论依据。

参 考 文 献

[4] 王乾, 薛茂权, 董笑瑜. 硬质合金, 2005, 22(4): 216-220.
[5] Faran E, Gotman I, Gutmanas E Y. Mater Lett, 2000, 43(4): 192-196.
[6] 卢金斌, 贺亚勋, 穆云超, 等. 金刚石与磨料磨具工程, 2015, 5(35): 51-54.
[7] Naidich Y V, Zhuravlev V S, Gab I I, et al. J Eur. Ceram. Soc, 2008, 28(4):717-728.

耐高温金属多孔材料的高温防护策略

张惠斌，万磊磊，郭 菲，马骏梁

（浙江工业大学材料科学与工程学院，浙江杭州 310014）

摘 要：近年来，金属多孔材料作为核心过滤元件或高比表面支撑体在高温过滤、化工合成气工序如蒸汽甲烷重整（SMR）、自热重整（ATR）、催化部分氧化、高温电化学如固体氧化物燃料电池(SOFC)、气体渗透膜等应用场合展现出广阔的应用前景。金属多孔材料的力学性能优异，可以承受更高的温度和压力波动；其工艺性良好，易加工和组合成各种异形元件和模块；同时具有理想的导热和传质特性，在作为催化剂载体时可以避免大的温度梯度和局部热点，从而提升转化率和选择性。然而，上述应用场合要求金属多孔材料能够在高温腐蚀性环境下稳定服役，对材料的高温力学性能、耐腐蚀性、孔结构稳定性等提出了极高的要求。

值得注意的是，当前研究人员主要基于致密体合金的研究结果，用于耐高温金属多孔材料的开发；却很少有人从金属多孔材料的结构特征出发，"自下而上"地进行金属多孔材料的合金成分设计和性能研究。金属多孔材料的结构特征可以归结为：（1）具有高度连通的孔结构和较高的比表面积，在形成连续氧化膜过程中会消耗更多的成膜元素和活性元素；（2）多孔骨架曲率半径小（数微米至几十微米），孔结构曲折复杂，氧化膜与基体应力状态匹配难度大；（3）高温烧结过程晶粒过分长大，具有大量高度裸露的大角度晶界，容易成为氧、硫、碳等原子向合金基体扩散的通道。因此，相较于致密体材料，金属多孔材料的结构特征决定了对其进行高温防护需要全新的策略。

作者提出从金属多孔材料的结构特征出发，"自下而上"地进行金属多孔材料合金成分和组织设计。基于孔结构特点开展成膜元素的临界含量研究，发现高比表面积的多孔材料在构建稳定氧化膜时需要更高的临界成膜元素含量。引入活性元素硼和钇，利用硼强化晶界并抑制晶界腐蚀，利用大原子半径元素钇形成的中间钇铁相抑制烧结晶粒长大，并抑制富铬相调幅分解，提高了材料的高温力学性能稳定性。同时，硼的表面偏聚以及钇在氧化膜与金属基体间的偏聚，可以起到降低氧分压以及抑制铝离子扩散等效果，促进了致密 α-Al$_2$O$_3$ 膜在较低温度和短时间内形成，从而显著提高了金属多孔材料在高温下的稳定服役性能。

参 考 文 献

[1] Zhang H B, Gao H Y, Liu X L, et al. Separation and Purification Technology, 2019, 220: 152-161.

[2] Zhang H B, Guo F, Wan L L, et al. Journal of Alloys and Compounds, 2023, 930:167437.
[3] Zhang H B, Yu H, Ma J L, et al. Transactions of Nonferrous Metals Society of China, 2022,32:2620-2633.
[4] Zhang H B, Ma J L, Gao Z C, et al. Materials, 2022, 15: 3718.
[5] Zhang H B, Wang L F, Ma J L, et al. Materials Chemistry and Physics, 2021,273:124929.

微纳钨极的近终形制备研究

吴昊阳，王 杰，董宏月，秦明礼，曲选辉

（北京科技大学，北京 100083）

摘 要：随着空间技术的飞速发展，高精度微牛级推力系统已成为引力波探测、重力场测量、微纳卫星等重大空间工程的研究热点。场发射推进器（Field Emission Electric Propulsion，FEEP）是一种新型微牛级电推进系统，具有推力小且大范围精确可调、比冲高、效率高、功耗成本低、结构紧凑、重量轻等特点，被"天琴计划"列为精确调节编队卫星相对位置的首选推进器，也是微纳卫星完成高精度复杂飞行任务的最优推进器之一。

微纳发射极是含精细尖端且直径微米级、内部孔隙纳米级的多孔针状物，其作为 FEEP 推进器结构中的核心组件，对场发射器推力的大小及精度具有重要的影响。金属钨耐高温、耐腐蚀、与推进剂润湿性好且不互溶，是发射极的理想材料。但由于硬度高，脆性大，加工性能差，其复杂构件无法直接通过机械加工制备，传统工艺是先采用粉末冶金工艺制备出可加工的钨铜假合金，加工成所需形状构件后，再将铜去除。这种方法不仅工艺复杂，材料利用率低（通常低于20%），特别是受加工工艺限制，难以制备出细晶粒、纳米孔隙结构的微纳钨极，造成推进器的启动电压高、推进力精度低、质量效率低、寿命短，已成为高性能场发射推进器的"瓶颈"。

基于高精度微纳发射极的迫切需求及制备难题，采用气流磨技术与微注射近净成形相结合的新方法，具体路径为：将钨粉经气流分散处理后与有机粘结剂均匀混合，以有机粘结剂为载体，采用塑料注射的方法成形，再将成形坯中的粘结剂脱除，经烧结得到微纳钨极。研究了粉体颗粒整形改性对注射成形微纳钨极的影响，对气流磨处理前后粉末特性、装载量、喂料均匀性、流变性能以及孔隙结构进行了表征。结果表明：将费氏粒度为 $0.8\mu m$ 的商业钨粉经气流分散处理后，团聚消失，粉末分散性显著提高，粉末颗粒的 D50 由 $1.41\mu m$ 转变为 $1.09\mu m$，比表面积由 $1.06m^2/g$ 增大为 $1.58m^2/g$，喂料装载量（体积分数）由 40%提高到了 50%，经 1100℃烧结处理后，微纳钨极的平均孔径为 401.37nm，孔隙率为 33.5%。

参考文献

[1] Holste K, Dietz P, et al. Review of Scientific Instruments, 2020, 91(6): 1-55.
[2] Hu W R, Wu Y L. National Science Review Siemes, 2017, 4(5): 685-686.
[3] Muller A, Ewert D, Galatanu A, et al. Fusion Engineering and Design, 2017, 124: 455-459.
[4] Choi J P, Park J S, Hong E J, et al. Powder Technology, 2017, 319: 253-260.

晶界自净化钨镁合金的制备与性能研究

王一甲，张大越，李彬周

（鞍钢集团北京研究院有限公司，北京 102200）

摘　要：将纯钨粉与镁粉进行机械混合，在真空下进行分段热压烧结，制备出钨镁合金。在高温和压力作用下，镁发生熔化和流动，通过毛细作用，充分润湿钨晶界，并与钨基体中的残余氧发生反应，达到净化晶界，生成强化相的目的。通过调整镁含量，制备了一系列晶界净化钨镁合金，对合金进行性能测试，结果显示：相比纯钨，钨镁合金的抗弯强度最高提高了 300%，硬度最高提高了 50%。对合金的组织和断口进行分析，发现镁的加入，有效吸收了钨基体中的残余氧，并生成以氧化镁和钨酸镁为主的第二相。这些第二相大部分均匀分布于钨基体晶界上，也有少量存在于晶内。钨镁合金通过晶界净化和弥散强化的共同作用，大幅度提高了钨材料的力学性能。

关键词：钨镁合金；晶界强化；断裂形貌；力学性能

Preparation of Grain Boundary Strengthened W-Mg Alloy Via Powder Metallurgy

WANG Yijia, ZHANG Dayue, LI Binzhou

(Ansteel Beijing Research Institute Co., Ltd., Beijing 102200, China)

Abstract: After mechanical mixing of pure tungsten (W) powder and magnesium (Mg) powder, W-Mg alloys were prepared by subsection hot pressing sintering under vacuum. Under the action of high temperature and pressure, Mg melted and flowed. By capillary action, W grain boundary was fully wetted. Mg reacted with residual oxygen in W matrix to purify grain boundaries and generate strengthening phase. A series of W-Mg alloys were prepared by adjusting Mg content, and the mechanical properties of the alloys were tested. Compared with pure W, W-Mg alloys had a maximum flexural strength increase of 300% and a maximum hardness increase of 50%. It was found that the addition of Mg effectively absorbed the residual oxygen in the tungsten matrix and formed a second phase dominated by MgO and $MgWO_4$. Most of these second phases were uniformly distributed in the grain boundaries and the rest existed inside the grains. The mechanical properties of W-Mg alloys were significantly improved through the combined action of grain boundary purification and dispersion strengthening.

Key words: W-Mg alloy; grain boundary strengthening; fracture morphology; mechanical properties

激光熔覆铁基涂层强化热轧导板技术应用

王一甲，李彬周，张大越

（鞍钢集团北京研究院有限公司，北京　102200）

摘　要：生产热轧钢球时，需利用上下导板对钢棒进行引导和夹持。现有导板由 45 号钢制备，材质偏软，极易磨损，需定期下线并采用手工堆焊工艺修复。受技能水平和材料匹配因素的影响，采用高硬度焊条易开裂，低硬度焊条则耐磨性差，目前产线一个生产周期需更换多块导板，影响生产效率，迫切需求新的修复工艺与材料。

为改善上述问题，本团队采用激光熔覆技术在导板表面熔覆铁基耐磨涂层。相比手工堆焊，激光熔覆过程热输入小，工件变形忽略不计，母材稀释率和热影响区小；Fe 基合金（Cr 19%、C 0.15%、Si 0.8%、Mn 0.4%、Mo 1%、Ni 2.5%、V 0.1%）可与基体进行良好的冶金结合且价格低廉，有助于降低成本[1]。

热轧导板上表面为长条楔形，首先开展精准测绘并使用 solidworks 建模，形成上表面数字模型，利用切片软件对其分割，规划熔覆路径[2]；之后，采用激光同轴送粉增材设备开展熔覆，迭代工艺参数，获得最优工艺；最后，成功在导板表面形成熔覆层，厚度从 3mm 至 20mm 可调。如图 1 所示，熔覆层均匀平整，未发现开裂现象，熔覆层与基体之间亮白色的熔合线，表明二者形成了牢固的冶金结合。可观察到激光熔覆较高冷速产生的细小的枝晶组织。用维氏显微硬度测试仪对基体与熔覆层进行硬度测试，熔覆层部位硬度较高，约为基体的 2.5 倍。经鞍钢矿山机械制造有限公司上线测试，证明 3mm 熔覆层即可使导板服役寿命提高两倍以上，强化效果显著。

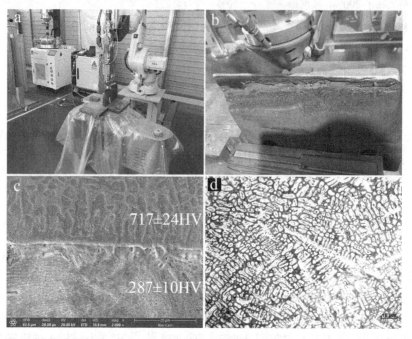

图 1　熔覆过程（a）、修复面形貌（b）、熔覆界面形貌（c）和熔覆层组织（d）

参 考 文 献

[1] 徐一飞，孙耀宁，王国建，等. 高速激光熔覆铁基合金涂层的组织及性能研究[J]. 中国激光, 2021, 48(10): 9. DOI:10.3788/CJL202148.1002122.
[2] 王杨霄，孙文磊，刘金朵，等. 复杂曲面零件激光随形熔覆轨迹规划研究[J]. 机械设计, 2022(2): 39.

不同粒度 Cu 粉的力学性能研究

徐　强[1]，刘一波[1,2]，杨志威[2]

（1. 钢铁研究总院，北京　100081；
2. 北京安泰钢研超硬材料制品有限责任公司，北京　102200）

摘　要：本文对四种不同粒度 Cu 粉进行了研究，对其粉末颗粒的形貌和氧含量进行对比分析，并分别经过 700℃、750℃、800℃和 850℃ 4 种烧结温度下制备了烧结试样，利用阿基米德排水法测试样密度，HR-150A 型洛氏硬度计测量试样的 HRB 值，三点弯曲法测定试样的抗弯强度和 SEM 观察试样断口形貌。试验结果表明：随着烧结温度的升高，四种测试块的密度都呈增加趋势，当温度超过 800℃后，密度逐渐趋于平缓，而且粉末粒度越粗，密度随温度增加的趋势越大；粉末越细，致密化的温度越高；Cu-325 的抗弯强度随着烧结温度的提高而增加，世佳威尔 Cu 的抗弯强度在 800℃以前在缓慢下降，超过 800℃有开始增加，而精研 Cu-400 和 Cu-P3 的抗弯强度随着烧结温度的升高在逐渐下降；随着温度的升高，晶粒都在不断地长大，而 Cu-325 的对温度敏感性较小，当温度达到 800℃以后，晶粒几乎不再长大；从断口形貌上得到，四种铜粉的断裂方式均为韧性断裂，粉末越细，韧窝越细小。

关键词：铜粉；粉末形貌；烧结性能；断口形貌

参 考 文 献

[3] 李占荣,周友智,张敬国,等. 中国铜基粉末产业发展现状及展望[J]. 粉末冶金工业, 2021, 31(2): 1-11.
[4] 徐浩翔,麻洪秋,罗锡裕,等. 雾化预合金胎体粉末的制备及其在金刚石工具中的应用[J]. 金刚石与磨料磨具工程, 2004, 2(1): 45-48.
[5] 刘维平. 利用铜精矿直接制备超细铜粉[J]. 矿冶工程, 2007(6): 41-43.
[6] 刘文胜,李淑娟,周殿春. 雾化铜粉制备技术的研究进展[J]. 粉末冶金工业, 2021, 21(3): 53.
[7] 谭芳香,黄以伟. 水雾化法制备铜及铜合金粉[J]. 金属功能材料, 2021, 28(3): 18-22.
[8] 舒适. 耦合压力—气体雾化制粉工艺及制备粉末性能的研究[D]. 北京:北京有色金属研究总院, 2018.

PVP 含量对纯钛注射成型的影响

张韦晨[1], 李 璐[1,2], 李传勇[1], Muhammad Dilawer Hayat[3]

（1. 昆明理工大学材料科学与工程学院,云南昆明 650093；
2. 昆明理工大学云南省分析测试中心,云南昆明 650093；
3. 怀卡托大学工程学院,新西兰汉密尔顿 3240）

摘 要：金属注射成型（MIM）是一种将传统粉末冶金与塑料注射成型相结合的近净成形技术。它具有成本低、可加工复杂形状的零件、生产效率高、材料利用率高等优点,这些优点使其成为昂贵金属（如钛）的理想加工技术。

水溶性粘结剂因其环保性质,近年来受到广泛关注。聚乙二醇（PEG）因其易水溶性、无毒和稳定性好等优点,通常用作水溶性粘合剂体系中的主要成分。聚甲基丙烯酸甲酯（PMMA）与 PEG 具有良好的相容性和高拉伸强度,常被选为 PEG 基粘合剂的主链聚合物。此外,PMMA 是一种非常干净的聚合物,在分解过程中留下的残留物很少。其热解产物相对清洁,90%的热解产物由纯气态甲基丙烯酸甲酯（MMA）单体组成。因此,考虑到钛的活泼性,PEG/PMMA 体系可能是纯钛注射成型的较好选择。

尽管具有清洁的性质,但 PEG/PMMA 粘结剂体系在 MIM 行业中的使用仍然相当有限。PEG 的结晶温度与 PMMA 的玻璃化转变温度之间的差异引起的空隙仍然是一个大问题。本文选择了 PEG（北京索莱宝科技有限公司）、PMMA（台湾奇美）和 SA（东莞市宏兴新材料有限公司）作为粘结剂组分,与球形纯钛粉（平均粒径为 34 μm）混合,在该体系中分别加入不同含量的结晶抑制剂 PVP,以减少冷却过程中产生的孔隙,提高样品的力学性能。通过流变仪测定 PVP 的加入对喂料黏度的影响；采用差示扫描量热法（DSC）研究了 PVP 对 PEG 结晶的影响；热重（TGA）用于分析粘结剂的热分解行为；借助扫描电镜（SEM）观察样品内部水脱前后的结合情况以及样品的断裂面形貌；采用显微 CT 观察样品内部的三维结构及形貌；使用 MTS 拉伸试验机测定烧结样品的力学性能。

参 考 文 献

[1] Dehghan-Manshadi A, Bermingham M J, Dargusch M S, et al, Powder Technology, 2017, 319: 289-301.
[2] Hayat M D, Zhang H, Karumbaiah K M, et al. Powder Technology, 2021, 382: 431-440.
[3] Wen J, Liu W, Xie Z, et al. Ceramic International, 2018, 44: 2718-2726.
[4] Hayat M D, Li T, Wen G, et al. The International Journal of Advanced Manufacturing Technology, 2015, 80: 1665-1671.
[5] Sidambe A T, Figueroa I A, Hamilton H G, et al. Journal of Materials Processing Technology, 2012, 212: 1591-1597.

10 先进钢铁材料及应用

大会特邀报告
第十三届冶金青年科技奖获奖人特邀报告
分会场特邀报告
矿业工程
焦化及节能环保
炼铁与原料
炼钢与连铸
电冶金与废钢铁
轧制与热处理
表面与涂镀
金属材料深加工
粉末冶金
★ 先进钢铁材料及应用
节能与低碳技术
冶金环保与资源利用
冶金设备与工程技术
冶金自动化与智能化
冶金物流
冶金流程工程学
其他

10.1 钢铁材料数字化

管线钢中夹杂物追踪及成因分析

王复越[1]，任　毅[1]，田永久[2]，崔福祥[2]，赵钰琛[3]，张万顺[3]

（1. 海洋装备用金属材料及其应用国家重点实验室，鞍钢集团钢铁研究院，辽宁鞍山　114009；
2. 鞍钢股份有限公司鲅鱼圈分公司，辽宁营口　115007；3. 辽宁科技大学，辽宁鞍山　114051）

摘　要： 本研究针对高钢级管线钢超纯净度的要求，采用计算机数值模拟与采样实测统计的办法对管线钢中夹杂物从轧板到连铸坯进行追溯、表征与分析。研究借助 ANSYS 有限元分析软件，建立结晶器内多物理场耦合的数值模拟模型。模拟结果表明铸造速度与铸造温度对熔体流动行为影响显著，进而改变夹杂物在铸坯中的数量与分布。此外，对高速低温与低速高温铸造条件下生产的连铸坯断面夹杂物分布情况进行统计。结合坯/板中夹杂物的成分、形貌、位置等特征分析夹杂物来源与成因，研究结果对于提升设备运行水平、提高产品质量以及降低不合格品率等方面都具有重要的意义。

关键词： 连铸；夹杂物；数值模拟；结晶器；超声波探伤

Inclusion Tracking and Cause Analysis of Inclusions in Pipeline Steel

WANG Fuyue[1], REN Yi[1], TIAN Yongjiu[2], CUI Fuxiang[2],
ZHAO Yuchen[3], ZHANG Wanshun[3]

(1. State Key Laboratory of Metal Material for Marine Equipment and Application, Anshan 114009, China
2. Bayuquan Branch of Angang Steel Co., Ltd., Yingkou, 115007, China; 3. School of Applied Technology,
University of Science and Technology Liaoning, Anshan 114051, China)

Abstract: In this study, computer numerical simulation and sampling statistics are used to trace the inclusions from the rolled plate to the continuous casting billet in a reverse direction for the requirement of ultra-purity of high-grade pipeline steel. The simulation results show that the casting speed and casting temperature significantly affect the behavior of inclusions' movement, which in turn changes the quantity and distribution of inclusions in the billet. In addition, microscopic observation and characterization of inclusions in the cross-section of the continuous casting billet were carried out. The distribution of inclusions in the cross-section of the straight casting billet was counted under high-speed, low-temperature, and low-speed, high-temperature casting conditions. The sources and causes of inclusions were analyzed according to the composition, morphology, and location of inclusions in the billet/plate. The study results are of great significance for improving equipment operation, improving product quality, and reducing the rate of rejects.

Key words: continuous casting; inclusions; numerical simulation; crystallizer; ultrasonic flaw detection

统计分析技术在 1580 热轧线生产质量改进中的应用

刘旺臣，王　存，王　杰，李江委，任俊威，刘　磊

(鞍钢股份有限公司鲅鱼圈钢铁分公司，辽宁营口　115007)

摘　要：利用统计分析技术对热轧带钢生产过程中的工艺参数和质量结果进行分析，可以提高过程控制与产品质量的控制稳定性。通过利用均值-极差控制图对 FDT 进行控制分析，可以及时发现 FDT 控制过程中的异常问题，通过及时改进，可以提高控制过程能力。通过利用多元回归分析技术，对带钢的屈服强度和工艺因子进行建模分析，可以建立回归分析模型，对带钢的屈服强度进行提前预测，并对工艺参数进行优化调整。

关键词：统计分析技术；热轧；控制图；多元回归分析

Application of Statistical Analysis Technology in 1580 Hot Rolling Line

LIU Wangchen, WANG Cun, WANG Jie, LI Jiangwei, REN Junwei, LIU Lei

(Bayuquan Branch of Angang Steel Co., Ltd., Yingkou 115007, China)

Abstract: Using statistical analysis technology to analyze the process control and performance in the production process of hot rolled strip can improve the stability of process control and quality results. By using the Mean-range control charts to control and analyze the FDT, the abnormal problems in the FDT control process can be found in time. Through timely improvement, the control process capability index can be improved. By using multiple regression analysis technology to model and analyze the yield strength of strip steel, the regression analysis model can be established to predict the yield strength of strip steel in advance and optimize and adjust the process.

Key words: statistical analysis technology; hot rolling; mean-range control charts; multiple regression analysis technology

点缺陷在 Cr_2O_3/Fe_2O_3 界面氢损伤的作用

米志杉

(中国钢研科技集团有限公司　数字化研发中心，北京　100081)

摘　要：不锈钢在生产生活中应用广泛，主要是因为其表面形成的致密的钝化膜结构。但是，不锈钢中的氢会影响钝化膜的组成和结构，进一步会影响不锈钢的抗腐蚀能力。因此，研究氢对不锈钢钝化膜的影响，对于理解氢损伤机理具有重要意义。不锈钢钝化膜为外层富铁，内层富铬的氧化物结构。对此建立了不锈钢钝化膜的简化模型，Cr_2O_3/Fe_2O_3 界面结构，利用第一性原理密度泛函理论 LDA+U 方法研究了 Cr_2O_3/Fe_2O_3 界面处不同类型的点缺陷(主要是空位缺陷和合金元素)与 H 原子的相互作用，厘清了这些点缺陷在界面处结合 H 原子的规律和机理。我们研

究发现在没有其他缺陷存在时，钝化膜内部的 Cr_2O_3/Fe_2O_3 界面很难结合 H 原子；当界面处存在 Cr 空位、Fe 空位以及合金元素 Zn、Ni 和 Cu 时，H 结合能为负值，范围在-1.3 eV 至-2.9 eV，此时 H 原子很容易与界面结合，这些点缺陷容易导致钝化膜界面发生氢损伤；当界面处存在合金元素 Mn，Mo，Ti，Al 和 V 时，H 的结合能为正，说明此时 H 不容易存在于界面处，这些点缺陷会减少钝化膜界面发生氢损伤的可能性。我们的研究结果对选择合适的合金元素获得钝化膜具有低氢脆敏感性的耐蚀不锈钢具有指导意义。

关键词：钝化膜；氢损伤；点缺陷；密度泛函理论；合金元素

高硅高铝耐磨钢在中应变速率下的本构模型

杨哲懿[1,2]，高磊[1,2]，王尊呈[1,2]，王帅[1,2]，刘威[3]，董洋[3]

（1. 海洋装备用金属材料及其应用国家重点实验室，辽宁鞍山 114009；2. 鞍钢集团钢铁研究院热轧产品研究所，辽宁鞍山 114009；3. 鞍钢股份有限公司热轧带钢厂，辽宁鞍山 114009）

摘 要：使用 Gleeble-3800 对热轧态耐磨钢在中水平应变速率下，以 800~1200℃的变形温度进行单轴热压缩试验，压缩量为 60%。根据变形过程中的真实应力-应变数据，以 Arrhenius 方程为基准建立带有变形程度参量的高硅高铝耐磨钢本构模型，研究其高温流变形为。研究结果表明：在处于中应变速率（$\dot{\varepsilon} \leqslant 1 \times 10^2$）变形时，材料的变形抗力仅在高温区呈现动态再结晶软化的特点，而在低温区尤其当应变量累积超过 0.4 时，因加工硬化导致应力逐渐增大。在对传统 Arrhenius 方程改进后的能够准确的计算高硅高铝耐磨钢在 800~1200℃和 1~10s^{-1} 的流变行为，预测误差约为 3.79%。

关键词：耐磨钢；本构模型；Arrhenius 方程；非线性拟合

The Mid-range Strain Rate of Constitutive Relationship about Wear-resistant Steel with High Silicon and High Aluminum

YANG Zheyi[1,2], GAO Lei[1,2], WANG Zuncheng[1,2], WANG Shuai[1,2], LIU Wei[3], DONG Yang[3]

(1. State Key Laboratory of Metal Material for Marine Equipment and Application, Anshan 114009, China; 2. Ansteel Group Iron and Steel Research Institute Hot Rolling Product Research Institute, Anshan 114009, China; 3. Ansteel Co., Ltd., Hot-rolled Steel Factory, Anshan 114009, China)

Abstract: Using Gleeble-3800 to perform a single-axis hot compression test. At the medium level of hot-rolled wear-resistant steel, the deformation temperature is 800~1200℃, at the same time, the deformation should be 60%. Based on the true stress-strain data in the test, establish a constitutive relationship about wear-resistant steel with high silicon and high aluminum with the degree of deformation with the Arrhenius equation. Research indicates: when deformation at the mid-range strain rate($\dot{\varepsilon} \leqslant 1 \times 10^2$), the deformation resistance of the material only presents the characteristics of dynamic crystallization softening at high temperature. While at low temperature especially when the true-strain reach to 0.4, stress gradually increases due to processing hardening. After refine the Arrhenius equation, its calculations about flow-stress are more accurate for wear-resistant steel with high silicon and high aluminum, at the temperature range of 800~1200℃ and the strain-rate range of 1~10s^{-1}, the prediction error is about 3.79%.

Key words: wear-resistant steel; constitutive relationship; Arrhenius equation; nonlinear fitting

ICME 框架下汽车用高强钢的设计研发

郑伟森，赵 宁，何燕霖，李 麟，鲁晓刚

（上海大学材料科学与工程学院，上海 200444）

摘　要：随着集成计算材料工程（ICME）的发展，多种材料计算方法应用于高性能材料的设计研发，不仅可以轻松地探究大量的合金体系和工艺参数，有效地探索宽广范围的成分空间，而且突破了传统试错法的成本和时间限制[1]。计算热力学和动力学作为重要的计算方法之一，通过与专有数据库的结合，使得多元多相材料的相变过程和显微组织演变模拟成为可能。而专有数据库的可靠性直接决定了相变行为模拟预测的准确度。然而，针对汽车用高强钢设计的热力学和扩散动力学数据库仍需进行完善。同时，如何确定影响汽车用高强钢特定性能的热力学或动力学特征变量，仍然是 ICME 框架下材料设计面临的首要问题。以相变诱发塑性（TRIP）钢为例，其力学性能由包含铁素体、贝氏体、残余奥氏体或马氏体的复杂组织所决定，但组织占比无法确定。Huang 等人[2]报道当低合金 TRIP 钢含有 53%铁素体、31%贝氏体和 16%残余奥氏体时，强塑积达到 23.6GPa%，表现出优异的力学性能。Dong 等人[3]指出当铁素体、贝氏体、残余奥氏体及马氏体的比例为 1.8∶4.9∶3.4∶1 时，TRIP 钢的力学性能得到显著提升，屈服强度、抗拉强度、断后延伸率分别达到 890MPa、1220MPa 和 44%。Zhu 等人[4]根据预测的热力学相图确定了临界退火温度，以获得占比约 33%的铁素体含量，由此设计的 TRIP 钢热处理后抗拉强度高达 1300MPa，均匀延伸率约为 22%。然而，Zhu 等人[4]的设计方法只有在化学成分或热处理条件已知的情况下才能通过热力学计算粗略预测实验钢的相分数。目前，在 ICME 框架下通过预测不同合金成分在各种加工条件下可能的微观组织演化来设计 TRIP 钢的报道仍然很少。鉴于此，本工作采用相图计算（CALPHAD）方法开展热力学和扩散动力学建模研究，完善汽车用高强钢设计的基础数据库。在此基础上，通过探究材料组织结构与性能之间的关系来确定影响材料特定性能的热力学和动力学特征变量，并通过使用特征工程从影响目标性能的许多原始特征中选择合适的特征参量，提高新材料设计的准确性，成功设计出不同强度级别的汽车用高强钢。

参 考 文 献

[1] 鲁晓刚, 王卓, 崔予文, 等. 科学通报, 2013, 58(35): 3656-3664.
[2] Huang J N, Tang Z Y, Ding H, et al. Mater. Sci. Eng. A, 2019, 759: 40-46.
[3] Dong X X, Shen Y F. Mater. Sci. Eng A, 2022, 852: 143737.
[4] Zhu R, Li S, Karaman I, et al. Acta Mater. 2012, 60: 3022-3033.

钒微合金化 TRIP 钢的高通量计算设计

梅　宇，吴　通，郑伟森，何燕霖，鲁晓刚，李　麟

（上海大学，材料科学与工程学院，上海 200444）

摘　要：中锰 TRIP 钢作为第三代先进高强钢因其优异的强塑性而受到广泛关注。然而，大量文献报道主要集中在中锰钢微观组织和性能调控的实验研究，仍然缺少针对特定目标性能的中锰钢逆向设计探索。针对抗拉强度≥

980MPa、断后伸长率≥30%、密度≤7.65g/cm³ 的目标性能，本工作基于高通量计算方法通过引入钒微合金化提升强度并保持合适的塑性，同时添加轻质元素 Al 以降低实验钢密度，从而获得理想性能的钒微合金化中锰钢。首先，基于 CALPHAD 方法建立马氏体转变起始温度预测的热力学模型，同时结合热力学计算软件 Thermo-Calc 软件的应用编程接口，通过自制的 MATLAB 代码对钒微合金化中锰 TRIP 钢成分进行自动化高通量计算筛选，根据高通量计算优选出钒微合金化中锰 TRIP 的最佳成分范围。经过后续实验验证，实验钢具有良好的综合力学性能，其抗拉强度和断后延伸率分别达到 1010MPa 和 37.0%，强塑积为 37.7GPa，同时密度为 7.51g/cm³，成功实现了针对特定目标性能中锰钢的逆向设计。

关键词：高通量计算；中锰 TRIP 钢；材料逆向设计

高炉数字孪生驾驶舱与操作评价平台的建设

孟昕阳[1]，徐 震[2]，王宝海[3]，郝云东[4]，孙敏敏[1]，庞克亮[1]

（1. 鞍钢集团北京研究院有限公司，北京 102209；2. 鞍钢股份鲅鱼圈钢铁分公司炼铁部，辽宁营口 115000；3. 鞍钢集团工程技术有限公司，辽宁鞍山 114021；4. 北京北科亿力科技有限公司，北京 100041）

摘 要：随着物联网、大数据人工智能等新一代数字化技术的快速发展，数字化转型已经成为钢铁企业重塑竞争优势的关键举措。针对高炉各类数据信息孤岛，信息价值化不足和高炉操作标准化管控体系缺失等现象，研发了高炉数字孪生驾驶舱与操作评价平台。借助已有高炉数字化监测系统，完善数据采集，构建实时数据驱动的高炉 3D 数字孪生。利用"专家经验+冶炼机理+智能算法"，开发操作评价模型，定期对高炉进行多维度的解析与优化。同时与人工操作相互评价对比，将操作经验和冶炼机理充分结合。保证高炉长期稳定顺行，提高高炉寿命，降低燃料消耗，践行智能化操作。

关键词：数字孪生；智能制造；高炉；操作评价模型；互评价

Construction of Digital Twin Cockpit and Operation Evaluation Platform for Blast Furnace

MENG Xinyang[1], XU Zhen[2], WANG Baohai[3], HAO Yundong[4], Sun Minmin[1], PANG Keliang[1]

(1. Ansteel Beijing Research Institute Co., Ltd., Beijing 102209, China; 2. Bayuquan Branch Company of Angang Steel Co., Ltd., Yingkou 115000, China; 3. Ansteel Engineering Technology Co., Ltd., Anshan 114021, China; 4. Beijing Beike Yili Technology Co., Ltd., Beijing 100041, China)

Abstract: With the rapid development of new digital technologies such as the Internet of Things, big data and artificial intelligence, digital transformation has become a key initiative for steel companies to reshape competitive advantage. The blast furnace digital twin cockpit and operation evaluation platform were developed to address the phenomena of various data information silos, insufficient information valorization and lack of standardized control system for blast furnace operation. With the existing blast furnace digital monitoring system, the data acquisition is improved, and the real-time data-driven blast furnace 3D digital twin is constructed. Using "expert experience+smelting mechanism+intelligent algorithm", the operation evaluation model is developed. The multi-dimensional analysis and optimization of the blast furnace is carried out in a fixed period. At the same time, it is compared with manual operation to fully integrate operation experience and smelting mechanism. It ensures the long-term stable operation of the blast furnace, improves the life of the blast furnace, reduces fuel consumption, and implements intelligent operation.

Key words: digital twin; intelligent manufacturing; blast furnace; operational evaluation mode; mutual evaluation

结合物理冶金原理和迁移学习的耐热钢蠕变寿命预测及设计

魏晓蓉，王晨充，徐 伟

（东北大学，辽宁沈阳 110819）

摘 要： 蠕变性能等长时服役性能是钢铁材料最重要的力学性能之一，尤其是对于高温下使用的钢种如火力发电机组用耐热钢。因此，蠕变性能的评估和预测对于新材料研发尤为重要。然而，蠕变试验昂贵且耗时，传统的以蠕变性能为导向的合金设计试错法成本高昂。尽管数据挖掘和机器学习已被广泛用于材料设计以取代试错法，但其成功实现需要足够的数据积累和优质的数据库。然而对于新材料，高质量及大规模蠕变数据库的建立显然是困难的，基于机器学习方法的蠕变寿命预测及合金设计面临小样本问题。同时蠕变复杂的成分/工艺-组织-性能关系进一步增加了建模难度。综上，数据稀缺与机理复杂成为了制约机器学习方法在金属结构材料设计领域进一步发展的核心瓶颈问题。

为了解决上述小样本下基于机器学习方法难以实现复杂蠕变性能可靠评估预测及合金优化设计的问题，提出了结合物理冶金原理和迁移学习的耐热钢蠕变寿命不确定性预测及设计框架 (PM-TR-BCNN)。具体而言，该框架中，依据耐热钢多种析出相粗化相关的物理冶金参数指导(PM)以及高温短时拉伸性能与蠕变性能间相关性提供的迁移学习指导(TR)，并与贝叶斯卷积神经网络(BCNN)相结合，形成了小样本下对蠕变寿命的不确定性预测框架。该框架实现耐热钢蠕变寿命准确评估的同时可获得模型预测的不确定性，实现了模型可靠性评估。将该框架进一步与多目标遗传算法相结合，成功实现了蠕变寿命-不确定性双目标合金优化设计，所获合金更符合物理冶金原理。当前工作进一步揭示了不确定性在蠕变寿命预测及设计中的作用，并评估了预测框架的外推能力，其对于未知合金体系及测试条件均实现了相比传统机器学习方法更为准确的评估。以上结果证实了不确定性迁移框架在蠕变寿命预测及设计中的可靠性，其为机器学习和物理冶金学的整合提供了一个范例，有潜力移植到其他材料体系中。

助力热轧棒线材工程精准精细化——中冶南方数字预测系统

李杨齐，骆艳萍，李沐泽，柯衡珍，闵 威

（中冶南方工程技术有限公司，湖北武汉 430223）

摘 要： 随着现代钢铁企业追求降低研发、生产成本的同时，确保产品性能稳定并实现高效批量生产的需求日益凸显，预测钢材组织性能的技术变得越来越重要。热轧棒线材的组织性能预测是实现这一目标的关键环节。本文介绍了一种基于物理冶金模型的组织性能预测技术，该技术由中冶南方开发，已经在武钢高线、高棒上得到验证与试用，预测精度较高，轧制过程温度预测的误差在±10℃以内，产品强度预测的误差在±10MPa 以内。

首先，该系统基于有限差分法建立了热轧全流程各环节棒线材轧件的二维温度场数值计算模型。相关模型考虑了热轧棒线材与环境、轧辊、除鳞水箱和水冷箱的关系，以及摩擦热、变形热的影响，从而能够获得准确的模拟温度场。

根据模拟的温度场，该系统建立了一系列组织性能预测模型，包括加热模型、再结晶模型、相变模型和组织-性能关系模型。这些模型可以模拟小方坯在加热炉中的晶粒长大过程，奥氏体在热轧过程中的再结晶行为，以及在冷床或吐丝机后的风冷阶段奥氏体的相变规律，还可以预测基于化学成分和组织结构参数的钢材力学性能。

此外，该系统还可以根据目标温度反算水冷阶段的水箱个数、水量、水箱间距和轧件速度，满足用户多样化的需求。本系统还引入了人工智能技术，为水冷过程提供自学习的在线和离线调试功能。用户可以根据实际情况选择在线检测或离线导入数据的方式，以水冷参数和实测温度等数据作为输入值，对软件的水冷阶段换热系数进行调试。这种结合了物理冶金模型与人工智能的方法，有效解决了传统物理冶金模型预测技术适应性不强、依赖模型参数精度的问题，实现了产品性能的高精度预测。

总的来说，中冶南方开发的热轧棒线材数字预测系统实现了对钢材组织性能的高精度预测，有助于降低研发、生产成本，保证产品性能稳定，且支持高效批量化生产。此外，该系统为用户提供了全过程的温度场、变形、组织和性能预测，以及局部的水箱参数调整，为满足用户多样化的需求提供了可能。

中冶南方致力于技术创新和应用研究，以实际行动为钢材组织性能预测这一关键共性技术提供强有力的支撑和保障，满足企业的市场个性化需求和现场共性化生产要求，同时秉持新时代绿色钢铁、智慧钢铁的发展理念实现了对钢材组织性能的准确预测，从而有效助力热轧棒线材工程的精准精细化发展。

材料和工艺设计多尺度仿真方法及应用

杨 丽[1,2]，苏 航[1]，刘和平[1]，孙 旭[1]，米志杉[1]，程 挺[1]，李 通[1]

(1. 中国钢研科技集团有限公司，北京　100081；
2. 南京钢铁股份有限公司，江苏南京　211500)

摘　要：近年来，随着材料基因工程、材料集成计算等技术的发展，多尺度仿真计算方法被越来越广泛地运用到材料的研发和工业生产中，为材料和工艺的设计提供技术支撑。

在纳观尺度，第一性原理计算基于量子力学理论，可以计算材料结构、能量、基础性能等；分子动力学基于牛顿力学理论，可模拟分子、原子体系的运动。在微观尺度，材料热动力学计算基于热力学原理，可计算材料的相变信息，计算元素扩散过程；相场法用于材料微观组织演变模拟，可以还原凝固树枝晶生长、奥氏体-铁素体固态相变、晶粒长大和再结晶过程。在宏观尺度，计算流体用于分析工艺过程的流场分布情况；宏观有限元主要用于结构力学、热传递、疲劳、裂纹扩展等场景。

通过将纳观、微观、宏观等多尺度的仿真方法结合起来，可更好地分析和预测材料在生产和服役过程中的晶体结构、微观组织、基础性能，从而指导材料和工艺的正向设计。晶体相场法融合了密度泛函与相场理论的理念，可应用于分析纳观尺度的团簇、缺陷、有序-无序、形核长大等演变过程；将第一性原理计算与有限元计算相结合，可计算金属材料氢致裂纹扩展等过程；将热动力学计算、相场模拟和有限元相结合，可模拟钢铁材料在加工及服役过程中的温度场-应力场-组织场分布和演变情况。

中国钢研建立了面向行业级的金属材料数字化研发平台 Material-DLab。通过该平台的多尺度公网云计算技术，为行业用户提供 10 余款国产化多尺度仿真软件在线使用、计算的机会，致力于推动国产化材料和工艺设计多尺度仿真软件在行业的应用，助力软件服务模式从 License 向 Service 的转变。此外，通过开发和上线典型应用场景的材料 APP，为用户提供了材料和工艺设计各种应用场景的计算案例，包括材料设计、材料加工、结构力学、服役腐蚀等，极大地降低了国产化软件的使用门槛。

低压脉冲渗碳组织及变形的预测

谢东兴[1]，贺笃鹏[2]，韩 瑞[1]，王泽辉[1]，秦湘阁[1]

（1. 佳木斯大学，黑龙江佳木斯 154007；2. 中国钢研集团有限公司，北京 100081）

摘 要：低压脉冲渗碳是在 70~3000Pa 的低压真空状态下，由交替的渗碳和扩散组成的脉冲式渗碳工艺过程[1-2]。强渗碳阶段，通入乙炔作为渗碳气体，并使之在炉内充分裂解，吸附并渗入零件表面；扩散阶段，在炉内通入高纯惰性气体，确保工件表面的碳向内部扩散。经过多次渗碳与扩散的脉冲循环，最终获得工件的渗层要求。低压脉冲渗碳可以精确控制零件表层的碳化物形态、尺寸和分布[1]，不产生表面氧化和晶间碳化物，极大改善了航空齿轮和重载齿轮疲劳性能和服役寿命，已在航空发动机行业等领域得到应用[2]。

目前，低压脉冲渗碳工艺的设计主要是基于一维有限差分法来计算渗碳工艺曲线[3]，只能预测零件表面碳浓度、有效渗层深度和表层碳浓度分布，关于低压渗碳工艺过程的有限元模拟很少报道[4,5]。本文建立了一个有限元模型来研究低压脉冲渗碳 C 型缺口试样的加热、渗碳、扩散和淬火过程中的碳浓度分布、显微组织和淬火变形。在计算碳浓度分布时，考虑了合金成分和渗碳温度的影响，为基于碳浓度场精确计算显微组织场、应力-应变场和淬火变形提供了基础。在计算应力-应变分布时，考虑了奥氏体、马氏体和贝氏体的碳和温度相关的材料特性，包括相变引起的体积变化、相变潜热和相变塑性，主要取得了以下主要结果。

（1）考虑了渗碳过程中扩散系数的变化和渗碳-扩散阶段不同的边界条件，利用有限元法求解 Fick 第二定律预测了低压脉冲渗碳过程中试样的表面碳浓度变化和渗碳层的碳浓度分布。

（2）二维有限元法模拟预测的表面碳浓度不同于文献中用一维有限差分法计算的表面碳浓度，可以敏感反映零件形状对表面局部碳含量的影响，为基于有限元法优化低压脉冲渗碳工艺提供了可靠的基础。

（3）二维有限元法模拟预测的表面碳浓度不同于三维有限元模拟的表面碳浓度。

（4）考虑了低压脉冲渗碳淬火过程中温度场、浓度场、应力场和组织场之间的耦合，预测了渗碳淬火后 C 型试样的组织分布与变形。

参 考 文 献

[1] Wang H, Wang B, Wang Z, et al. Journal of Materials Science & Technology, 2019, 35(7): 1218-1227.
[2] 王斌，何燕萍，王昊杰，等. 材料研究学报, 2020, 34(1): 35-42.
[3] Woowiec-Korecka E . Engineering Structures, 2018, 177: 489-505.
[4] Zajusz M, Tkacz-Miech K, Danielewski M. Surface & Coatings Technology, 2014, 258: 646-651.
[5] Kim D W, Cho H H, Lee W B, et al. Materials & Design, 2016, 99: 243-253.

InterMat：一种基于区块链的材料数据共享基础设施

王畅畅[1,2]，苏 航[1,3]，段琳娜[2]，李 灏[2]

（1. 钢铁研究总院，北京 100081；2. 北京钢研新材料科技有限公司，北京 100081；
3. 中国钢研科技集团有限公司数字化研发中心，北京 100081）

摘　要：数据驱动的材料研发需要大量的数据，传统材料数据共享平台间难以进行数据跨平台共享与整合。本文提出了一种基于区块链的材料数据共享基础设施—InterMat，它拥有去中心化结构，系统架构包括数据层、支撑层、网络层、应用层，通过智能合约实现材料数据存证、材料数据发现与共享、材料数据版权追溯、材料数据价值评估等功能。InterMat 结合了联盟链技术、材料模式识别、材料区块链等技术，解决了不同主体间材料共享的信任问题，服务于材料行业从研发、生产到服役全生命周期的数据管理与安全共享，形成不同组织间数据开放、协作的新生态。

关键词：材料大数据；区块链；数据共享；异构数据

InterMat: A Blockchain Based Material Data Sharing Infrastructure

WANG Changchang[1,2], SU Hang[1,3], DUAN Linna[2], LI Hao[2]

(1. Central Iron & Steel Research Institute, Beijing 100081, China; 2. Beijing MatDao Technology Co., Ltd., Beijing 100081, China; 3. Material Digital R&D Center, China Iron & Steel Research Institute Group, Beijing 100081, China)

Abstract: Data driven material research and development requires a large amount of data, making it difficult for traditional material data sharing platforms to share and integrate data across platforms. This paper proposes a blockchain based material data sharing infrastructure-InterMat, which has a decentralized structure. The system architecture includes a data layer, support layer, network layer, and application layer. Through smart contracts, functions such as material data certification, material data discovery and sharing, material data copyright tracing, and material data value evaluation are achieved. InterMat combines consortium blockchain technology, material pattern recognition and material blockchain technology to solve the trust problem of material sharing between different entities. It provides data management and security sharing services for the entire lifecycle of the material industry, thereby establishing a new ecosystem of data collaboration.

Key words: material big data; blockchain; data sharing; heterogeneous data

质量分级方法在食品接触用不锈钢使用安全性评价方面的应用

王世宏[1]，苏　航[1,2]，李　灏[1]，段琳娜[1]

（1. 北京钢研新材科技有限公司，北京　100081；
2. 中国钢研科技集团有限公司数字化研发中心，北京　100081）

摘　要：不锈钢材料由于其不锈、美观等优点，被大量用来制造食品容器、厨房设备等[1]。但是由于不锈钢中含有大量铬（Cr）、镍（Ni）、锰（Mn）等元素，在使用时可能会发生重金属离子溶出，当食用者摄入过量时，会引发胃溃疡、肌肉痉挛和精神萎靡等症状[2]。因此食品接触用材料的安全问题引起了广泛关注。为了检验食品接触用钢的安全性，GB 31604.1 规定了相应的重金属离子迁移实验方法，即采用 4%醋酸溶液作为食品模拟液，将食品接触用钢在溶液中煮沸，并检测溶液中的 Mn、Cr、Ni 等重金属离子溶出值。GB 4806.9 则对相应重金属离子的溶出值上限做了规定。然而在日常生活中，不锈钢锅需经历蒸、煮、炒等烹饪过程，会经常暴露在含盐环境中，而调研发现菜肴中的含盐量普遍在 1%~4%之间[3]，氯离子含量较高，导致金属表面钝化膜被破坏，极易加剧重金属离子溶出。为了更加全面、综合地评价食品接触用不锈钢的使用安全性，本研究首先对不同材质的不锈钢锅具进行了重金属离子溶出试验；进一步，采用了质量分级方法对不锈钢锅具的使用安全性进行了连续的差异化评价，得出了食品接触用不锈钢使用安全性的分级排名。

（1）重金属离子溶出试验。

本研究中，从市场上选择了非标 200 系、304、430 和 410S 共 4 种类型的不锈钢锅具若干个，参照 GB 31604.1 和 GB 4806.9，首先设计了两组重金属离子迁移试验，模拟不同的烹饪环境，分别在 1）4%醋酸溶液；2）3%NaCl+4%醋酸溶液中煮沸 30min，检验了两种食品模拟液中的重金属离子溶出情况，如表 1 所示。可以看出，与 304 和 430 不锈钢相比，非标 200 系和 410S 不锈钢在两种溶液中的重金属离子溶出值变化较大，其中 410S 在 3%NaCl+4%醋酸溶液中的 Cr 溶出值达到了 209 mg/kg，远超 GB 4806.9 规定的 2 mg/kg 的溶出上限[4]。

表 1 不同食品模拟液中的重金属离子溶出情况 （mg/kg）

类别	Mn		Cr		Ni	
	4%醋酸	3%NaCl+4%醋酸	4%醋酸	3%NaCl+4%醋酸	4%醋酸	3%NaCl+4%醋酸
非标 200 系	0.032	0.98	0.015	1.24	0.068	0.08
304	0.0027	0.002	0.013	0.02	0.034	0.014
430	0.0024	0.002	0.019	0.034	0.273	0.0076
410S	0.0058	5.13	0.21	209	0.068	0.13
GB 4806.9 规定溶出上限	/		2		0.5	

（2）食品接触用不锈钢使用安全性分级排名。

依据 T/CISA 008.1[5]，从材料的化学成分（S、P、Ni）、重金属离子溶出情况（Cr、Mn、Ni）、耐蚀性（点蚀当量）、结构刚度和单位容积材料用量等方面，对从市场上随机购买的若干个不锈钢锅具的使用安全性进行了分级评价。根据分级评价得分，将其分为了 A+、A、B 和 C 级，如图 1 所示。可以看出，304 不锈钢的使用安全性较高，430 和非标 200 系不锈钢次之，410S 不锈钢的使用安全性则较差。因此，在选择锅具时，应尽量选择 304 或更高牌号的不锈钢产品，避免使用 410S 不锈钢锅具进行烹、煮等。同时还可看出，国产品牌的不锈钢锅具也具有相当的竞争力，可比肩某些国际品牌。

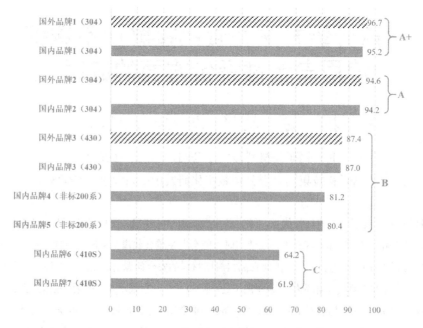

图 1 食品接触用不锈钢使用安全性分级排名

（剖线条形柱为国外企业）

参 考 文 献

[1] 邱宝渭, 陈东华. 不锈钢餐具在不同介质中金属溶出的研究[J]. 应用化工, 2014, 43(10): 1938-1940.
[2] 彭韶华. 不锈钢餐、厨具中铬离子迁移量的影响因素及质量控制[J]. 南方农机, 2017, 48(11): 84-85.
[3] 刘海波, 张尧, 李晓红, 等. 食品中的食用盐含量分级方法[J]. 公共卫生与预防医学, 2017, 28(1): 32-34.
[4] 中华人民共和国国家卫生和计划生育委员会. GB 4806.9—2016 食品安全国家标准 食品接触用金属材料及制品[S]. 北京: 中国标准出版社, 2016.
[5] 中国钢铁工业协会. T/CISA 008.1—2019 钢铁产品质量能力分级规范-第 1 部分：通则[S]. 北京: 冶金工业出版社, 2019.

基于深度学习和专家知识的材料图谱识别

王炫东[1,2], 苏 航[1], 李 南[3], 孟惠民[2]

（1. 中国钢研科技集团有限公司，数字化研发中心，北京　100081；
2. 北京科技大学，新材料技术研究院，北京　100083；
3. 钢铁研究总院，中心实验室，北京　100081）

摘　要：钢铁材料金相组织复杂，种类繁多，通常由实验人员或专家进行观察与分析，但效率有限，同时难以做到定量化统计。近年来，基于卷积神经网络的深度学习方法凭借其独特的特征提取能力，在图像识别、语义分割等方向取得了许多进展，为材料的图谱识别提供了新的有力工具。现有基于深度学习的材料图谱识别工作通常只针对单一场景、单一体系进行建模，尚未形成体系化工作，且针对组织图谱的专家知识未能有效利用。本工作开发了一套金相组织照片特征识别与分析方法。针对图谱中的材料组织，聚焦于不同组织类型具有的独特纹理特征，以框选标记方法取代繁重低效的手动像素标记，采用构建组织纹理库并生成仿真数据集的方式训练模型；同时，开发了图像数字化方法，可将组织照片处理为固定维度的特征向量，通过对比向量在特征空间中的距离可判断待识别照片与数据库中组织照片的相似度，进而判断组织类型、材料体系、热处理状态等信息。将上述功能封装后，形成一套可拓展、可迭代的钢铁材料金相照片识别分析系统，可实现对照片的组织分析与定量化统计等功能。

Cr5 型模具钢水-空交替控冷数值模拟

李 栋[1,4], 梁敬斌[1,4], 尤晓东[1,3], 贺笃鹏[2], 谢志彬[1,3], 邵青立[1,3]

（1. 河冶科技股份有限公司，河北石家庄　052165；2. 中国钢研科技集团数字化研发中心，北京　100081；3. 河北省高速工具钢技术创新中心，河北石家庄　052165；
4. 河北省高品质工模具材料重点实验室，河北石家庄　052165）

摘　要：某厂下游客户要求规格为 $\phi 245mm$ 的 Cr5 型模具钢的退火组织达到 SEP1614 标准中的合格级别或更高级别，统计了之前的工艺情况和组织检测情况，全年退火组织不合格率为 50%。针对 Cr5 型模具钢退火组织合格率低，难以满足标准要求的现实技术难题，需要在 Cr5 型模具钢热加工成型后进行热处理以优化球化退火前的组织。Cr5 型模具钢热处理工艺主要包括高温加热和水空交替冷却，其中水空交替冷却是决定退火组织的最关键工艺[1-2]。利用 SYSWELD 对现有工艺进行数值模拟[3-4]，采用等效热熔的方法来定义相变潜热，建立温度场数学模型；通过 KM 方程定义马氏体相变过程，通过 leblond 方程定义铁素体、珠光体、贝氏体以及奥氏体的相转变过程，建立组织场数学模型；用 SYSWELD 软件对 Cr5 型模具钢材料模型的温度、组织、应力进行数值模拟计算。如图 2 所示，方案 M2 的组织已全部转变为马氏体，未见贝氏体产生，得到合格的热加工后组织。对优化方案进行试验验证，退火组织 GA2，达到 SEP1614 合格级别。根据优化后的方案 2，进行了大批量的工业化生产，全年生产 807t，退火组织合格率经检测达到 100%。

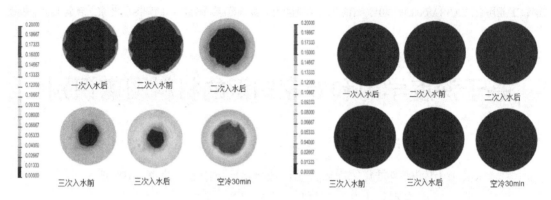

图 1　M1 方案剖面在不同工艺节点的贝氏体云图　　图 2　M2 方案剖面在不同工艺节点的贝氏体云图

参 考 文 献

[1] K farivar H, Bleck W. Microstructural adjustment of carburized steel components towards reducing the quenching-induced distortion[J]. Journal of Materials Processing and Technology, 2019, 264: 313.
[2] 贺笃鹏，张国强，王毛球，等. 18CrNiMo7-6 齿轮钢高温渗碳淬火变形数值模拟[J]. 金属功能材料, 2021, 10(5): 13.
[3] 刘桢，代朋超. 两种不同热变形方式对于 GH4169 合金锻棒组织的影响[J]. 金属功能材料, 2020, 27(3): 65.
[4] 蔡梦茹，王瑞珍，何宜柱，等. 淬火方式对含铜时效钢组织和性能的影响[J]. 金属功能材料, 2019, 26(6): 7.
[5] 袁丽，贺笃鹏，何欣，等. 16Cr3NiWMoVNbE 钢 C 型环真空低压渗碳及淬火有限元模拟[J]. 金属热处理, 2022, 47(9): 257.

基于高通量实验的金属材料腐蚀行为研究

宋有朋[1]，颜鲁春[2]，庞晓露[2]，宿彦京[3]，乔利杰[3]，高克玮[2]

（1. 中国钢研科技集团有限公司数字化研发中心，北京　100081；2. 北京科技大学材料科学与工程学院，北京　100083；3. 北京科技大学新材料技术研究院，北京　100083）

摘　要：海洋经济大发展、双碳目标的绿色发展之路，亟待研发出先进的耐蚀材料。然而，材料成分选择空间大，以试错为特征的传统材料研究方法存在耗时长、成本高、效率低等问题，严重制约了新材料的研发速度。特别是对于多组元合金，其广阔的成分空间为新材料的开发构成了巨大挑战。

旨在加速新材料研发进程、降低研发成本的材料基因工程理念的提出，为新材料的研发提供了一种崭新的模式。近年来，作为"材料基因组技术"三大组成要素之一的高通量实验在加速新材料研发和应用进程方面展现出巨大的潜力。磁控共溅射便是常用的多组元合金材料库制备方法之一。通过该方法，研究者经一次实验即可制备出包含大量合金成分的组合材料库样品。然而，需要指出的是，对于组合材料库中某一特定的合金成分来说，其元素分布并不均匀。而且，组合材料库中的大量合金相互连接，没有明显的界限。在难以建立成分与位置对应关系的同时，对组合材料库样品进行腐蚀筛选时，通常会发生电偶腐蚀问题。此外，通过共溅射方法所制备的合金材料库样品并不适用于传统相对成熟的耐蚀性表征方式，而需要开发相应的高通量微区表征装置。

针对上述问题，我们提出了一种高通量制备多组元均匀合金材料库的新策略，并研制了可一次性制备 97 种不同合金成分的多工位同步旋转样品台，成功解决了合金材料库中所存在的元素分布不均匀及相互连通问题。通过该策略研究了一系列纳米晶 FeCrNi 合金在 3.5%（质量分数）NaCl 溶液环境中的腐蚀行为，从而快速建立了合金成分与耐蚀性之间的关系，为高性能不锈钢的开发提供了指导。此外，在 FeCrNi 合金体系基础上，为进一步探索高性能材料，高通量合成了一系列纳米晶 $Al_xCu_y(FeCrNiCo)_{100-x-y}$（$x$=6.22%~50.51%，$y$=5.58%~40.31%，原子数分数）合金。在二维成分空间内探究了共合金化 Al 和 Cu 对 FeCrNiCo 高熵合金在 3.5%（质量分数）NaCl 溶液环境中耐蚀性以及力学性能的影响，从而实现了高熵合金耐蚀性和力学性能的协同设计，获得了综合性能优异的合金成分，为极端服役环境提供了高性能材料储备。此外，还系统分析了共合金化 Al 和 Cu 对 FeCrNiCo 高熵合金耐蚀性和力学性能的内在作用机制。

参 考 文 献

[1] Song Y P, Qiao L J, Gao K W, et al. Corrosion Science, 2022, 196: 110028.
[2] Song Y P, Qiao L J, Gao K W, et al. Corrosion Science, 2023, 213: 110983.
[3] 高克玮, 宋有朋, 乔利杰, 等. 一种高通量制备多组分均匀薄膜材料的装置和方法[P]. 中国: ZL201910722742.8, 2020-10-09.

高分辨率数字钢卷及应用

刘安平，彭燕华

（中冶赛迪信息技术（重庆）有限公司，重庆 401122）

摘 要：数字钢卷就是在实物钢卷上附有相关的生产数字信息，是一系列数据集合，实现钢卷"全程可视化"和"数字化"，为后续的大数据分析与挖掘，提供完善、准确、可靠的数据基础，是实现智能工厂最关键的一步，然而由于技术及实现方案等原因大多长度方向只能做到米级，同时各工艺段数据对齐复杂，大大限制了实际应用效果，甚至形同虚设。高分辨率数字钢卷采用 PDA(Process Data Acquisition)高速数据采集分析系统内核，非常适合处理庞大的数据，真正精确到厘米级、毫秒级、微秒级，系统理念非常有特点。

关键词：高分辨率；数字钢卷；分组；PDA

High Resolution Digital Steel Coil and Its Application

LIU Anping, PENG Yanhua

(CISDI Information Technology (Chongqing) Co., Ltd., Chongqing 401122, China)

Abstract: The digital steel coil is a series of data sets attached with relevant production digital information on the physical steel coil to realize the "whole process visualization" and "digitalization" of the steel coil and provide a complete, accurate and reliable data basis for subsequent Big data analysis and mining. It is the most critical step to realize the intelligent factory. However, due to technology and implementation schemes, most of the length directions can only reach the meter level, and the data alignment of each process section is complex, It greatly limits the practical application effect, and even appears to be virtual. The high-resolution digital steel coil adopts a PDA high-speed data acquisition and analysis system core, which is very suitable for processing large amounts of data, truly accurate to the centimeter level, millisecond level, and microsecond level. The system concept is very unique.

Key words: high resolution; digital steel coil; grouping; PDA

Al 对 Fe-Cr-Al 合金变形行为影响机制的数值模拟研究

王 鹏，乔英杰，戚 文，都时禹，李学达

（中国石油大学（华东）材料科学与工程学院，山东青岛 266580）

摘 要： 为进一步提高核能系统的综合效益，世界各国于 2000 年启动第四代核能系统的研发计划，要求组成材料将在保持经济性的同时能够在更加严苛的工作环境服役较长时间。2011 年日本福岛核事故之后，要求新型包壳材料具有事故容错能力，即更好的抗氧化性、高温稳定性和抗辐照性能，在高温失水事故下仍然能够在足够长的时间内安全运行。Fe-Cr-Al 包壳材料机械性能良好、组织稳定、抗辐照性能优良，尤其是抗高温水蒸气氧化性能良好，在新型核燃料包壳材料候选材料中受到重点关注。

通过系统开展 Fe-Cr-Al 包壳材料力学性能与抗高温水蒸气氧化性能的组分调控实验研究，建立组分含量对 Fe-Cr-Al 包壳材料微观组织、力学性能的影响规律。在实验基础上通过分子动力学模拟推演，揭示 Al 含量对 Fe-Cr-Al 包壳材料在微观尺度上的组分影响机制，为 Fe-Cr-Al 包壳材料的组分优化提供理论参考。

首先建立不同 Al 含量的 Fe-Cr-Al 包壳材料分子动力学拉伸演化模型，分析合金元素成分对晶体结构、滑移面、位错和形变传播行为的影响。通过理想单晶拉伸演化分析发现理想单晶 Fe-Cr-Al 包壳材料的形变机制为孪生形变，其滑移系启动能量较高，温度升高及 Al 含量增加可以降低滑移系启动所需要的能量。通过对形变多晶 Fe-Cr-Al 包壳材料和理想单晶 Fe-Cr-Al 包壳材料分子动力学模拟的拉伸行为对比可以发现，晶界和位错畸变区域使得其影响范围内的滑移系启动所需能量大幅降低，使得形变多晶 Fe-Cr-Al 包壳材料主要形变机制为滑移。

对形变多晶 Fe-Cr-Al 包壳材料模拟拉伸演化过程进行分析，发现温度升高 Fe-Cr-Al 包壳材料滑移系启动所需能量降低，Al 含量增加也使得滑移启动能量进一步降低，但会增大晶界与畸变区上的阻力。高温时，Al 含量的增加往往使得滑移系启动对塑性的增强影响大于在晶界与畸变区上的阻碍导致的塑性降低影响，即滑移阻力增大了但启动的滑移系更多了，因此可以发现在高温下 Al 含量增加使得 Fe-Cr-Al 包壳材料增强增韧。

由于模拟尺寸过小，不能很好的体现滑移系启动传递数量急剧衰减过程，材料仍处于韧性区间。因此结合实验结果，基于理论分析进行逆向推测。在韧脆转变温度以下时，滑移系启动所需能量增大，Fe-Cr-Al 包壳材料塑性形变能力降低，Al 含量增加所带来的滑移系启动能量减小对滑移系启动的影响并不显著，相反，Al 含量增加对滑移面上的滑移产生阻力增大的影响十分显著，滑移面启动所需临界分切应力较大，滑移系启动传递数量降低，导致常温塑性能力降低，甚至导致脆性断裂行为。

钢铁材料数字化研发应用实践

梁 坤，王 卓

（成都材智科技有限公司，四川成都 640041）

摘　要： 目前国内钢铁企业的生产信息化建设已基本完成，多以产线自动化系统、过程控制系统和生产管理系统三级系统的方式进行精细的生产计划管控与控制管理。随着能源、制造、汽车、航空航天等钢铁材料应用领域的发展对钢铁材料的性能提出更高要求，对钢铁企业的现代化、精细化的研发也提出了新要求，尤其是对工艺参数、实验检测数据等推动材料高质量研发根本所在的研发数据的利用提出了更高的要求。

但在钢铁企业中，工艺参数、实验检测数据等研发数据零散分布于生产控制系统、实验室检测系统、研发人员个人电脑中，数据多源异构的特性导致其利用不通畅，难以满足各应用领域对钢铁材料所需性能提升带来的钢铁材料快速开发要求，拖累了企业的经营效率。

本文以钢铁材料研发数字平台为例介绍钢铁材料数字化研发应用实践，探讨数字化技术在钢铁材料研发中的应用及其未来发展路径。钢铁材料数字化研发平台通过将分散在检化验系统、产供销系统中的检化验数据、产品标准、质量异议、用户档案、质量要点、技术协议、冶金规范等研发相关数据进行同步集成，便于研发人员进行数据查询及利用。搭建用户档案管理系统和研发项目管理系统，形成面向用户的品种开发工作及研发项目管理；搭建铁基材料数据库、材料研发知识管理系统，整合构建研究院内外部知识管理体系；构建基于物理冶金模型的可视化分析、工艺参数管理系统，大幅提高数据可用率与利用效率，提高研发人员研发效率。

钢铁材料数字化研发平台能够将研发人员从分散在过程控制系统、生产控制系统中提取研发数据的工作中解放出来，极大地提高数据处理能力与研发效率，快速响应高端制造业对钢铁性能需求的变化。通过这种方式集中汇聚的研发数据也可用于发展基于机器学习的材料性能预测和优化设计，推动钢铁材料研发向第四范式进行转变。

南钢宽厚板全轧程数值模拟系统开发

王凌宇，邱保文

（南京钢铁股份有限公司新材料研究院新材料研究所，江苏南京　210035）

摘　要： 根据现场实际，构建4700mm轧机辊系三维模型，在几何模型的基础上，针对轧机实际划分网格，建立轧机辊系的模型库。通过典型钢种物性参数的研究，确定材料模型，建立对应的材料数据库。借助"三维热力耦合""弹性辊轧制""全轧程三维组织演变""结构-温度-组织-性能多场耦合"四大关键科研成果，可实现对南钢宽厚板轧制过程进行全方位的模拟。

关键词： 宽厚板；全轧程；三维；网格；数值模拟；系统

Development of Numerical Simulation System for Wide and Thick Plate Rolling in Nangang

WANG Lingyu, QIU Baowen

(Institute of New Materials, Research Institute of New Materials, Nanjing Iron and Steel Co., Ltd., Nanjing 210035, China)

Abstract: According to the actual situation, the three-dimensional model of 4700mm rolling mill system is constructed. On the basis of geometric model, according to the actual grid division of rolling mill, the model library of rolling mill system is established. Through the study of the physical parameters of typical steel, the material model is determined and the corresponding material database is established. By four key scientific research achievements: three-dimensional thermodynamic coupling, elastic roll rolling, three-dimensional microstructure evolution of full rolling process, and multi-field coupling of structure-temperature-structure-property, the simulation of wide and thick plate rolling process of Nangang is realized.

Key words: wide and thick plate; full roll; three-dimensional; gridding; numerical simulation; system

基于相场-神经网络的镍基高温合金析出相演化行为研究

郑国才[1]，秦海龙[2]，毕中南[2]，施荣沛[3]，李东风[1]

（1. 哈尔滨工业大学（深圳），理学院，广东深圳 518055；2. 北京钢研高纳科技股份有限公司，北京 100081；3. 哈尔滨工业大学（深圳），材料科学与工程学院，广东深圳 518055）

摘 要： 镍基高温合金 GH4169 具有出色的力学性能，良好的加工性能，以及较好的经济性，因此在航空航天、能源等领域有着广泛的应用。GH4169 是一种析出相强化型高温合金，其力学性能与析出相的尺寸、形态、体积分数以及空间分布密切相关。近来的研究[1]发现，在热力耦合作用下 GH4169 的主要强化相 γ" 相会发生变体选择，这对合金的力学性能有显著影响，因此厘清强化相的析出和演化机制对优化合金的性能具有重要意义。

本文采用相场模拟与循环神经网络（RNN）相结合的方法来研究 GH4169 中三种 γ" 变体在不同外加载荷下的析出演化过程。如图 1（a）所示，本文首先应用相场法[2]模拟了时效过程中 γ" 相三种变体的析出及粗化过程，给出了无外载时效 2h 后 γ" 相的形貌。本文进一步考虑了外加载荷对时效过程的影响，如图 1（b）所示在[001]方向施加拉应力后变体 1、2（如图 1（a）中黑色和灰色所示）逐渐消失，发生了变体选择，与实验结果一致。本文最后开展了大量的不同外载作用下 γ" 相演化过程的相场模拟，构建了变体体积分数和平均尺寸的时间序列数据集，通过训练 RNN 神经网络（如图 1（c）所示）并进行验证，该网络可以对复杂外载作用下 γ" 相三种变体体积分数和平均尺寸的演化过程进行预测。

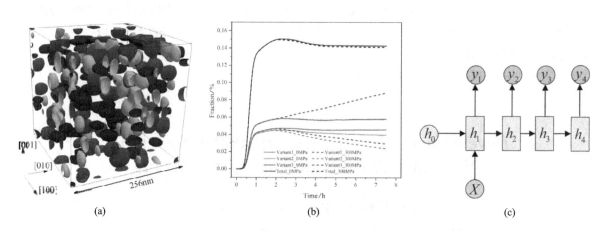

图 1 无外载时效 2h 后的析出相形貌（a）、不同外载下的析出相体积分数变化曲线（b）和 RNN 网络结构示意图（c）

参 考 文 献

[1] 秦海龙. GH4169 合金中 γ″相的变体选择行为与机制研究[D]. 北京：钢铁研究总院，2019.
[2] Zhou N, Lv D C, Zhang H L, et al. Acta Materialia, 2014, 65, 270-286.

表征深冲钢织构演变特征规律的分析测试方法

张 娜，赵美英，史文义，李智丽

（内蒙古包钢钢联股份有限公司技术中心，内蒙古包头 014010）

摘 要：针对一些典型深冲钢产品现阶段的发展要求，通过分析处理近几年 DC 系列深冲钢板材质量异议的案例表明，开发研究有利于深冲性能优化的技术手段，对于提高深冲钢产品质量，降低产品的赔付具有长远的经济效益和社会效益。综合力学性能分析测试选择实验研究的材料，利用传统金相显微分析技术以及宏观、微观表征材料晶体结构及晶粒取向的研究方法，实现表征不同工序深冲钢织构演变特征规律的分析测试方法研究和应用。

关键词：深冲钢；力学性能；织构；EBSD；XRD

Analysis and Test Method for Characterizing of Texture Evolution on Deep Drawing Sheet Steel

ZHANG Na, ZHAO Meiying, SHI Wenyi, LI Zhili

(Technical Center of Inner Mongolia Baotou Steel Union Co., Ltd., Baotou 014010, China)

Abstract: According to the development requirements of some typical deep-drawing sheet steel products at the present stage, through the analysis and treatment of the quality objections of DC series deep-drawing sheet steel in recent years, the development and research of technical means conducive to the optimization of deep-drawing performance has long-term economic and social benefits for improving the quality of deep-drawing sheet steel products and reducing the product compensation. The materials for experimental study were selected by comprehensive mechanical properties analysis and testing, and the traditional metallographic microanalysis technology and the macroscopic and microscopic research methods to characterize the crystal structure and grain orientation of materials were used to realize the research and application of analysis and testing methods to characterize the characteristics of texture evolution of deep-drawing sheet steel in different processes.

Key words: deep drawing sheet steel; mechanical property; texture; EBSD; XRD

耐热钢中 Laves 相抗蠕变机制的分子动力学模拟

王传军[1]，孙 旭[2]，刘文月[1]，刘 鑫[1]，秦 哲[1]

（1. 鞍钢集团北京研究院有限公司，北京 102209；
2. 中国钢研科技集团数字化研发中心，北京 100081）

摘 要：Laves 相是一种化学式主要为 AB2 型的密排立方或六方结构的金属间化合物。Laves 相中原子半径比 rA/rB 在 1.1~1.6 之间。在许多 Laves 相（AB2）中，过渡族金属一般为组元 B，但有时也可以起组元 A 的作用。Laves 相的晶体结构有三种类型：$MgCu_2$ 型、$MgZn_2$ 型、$MgNi_2$ 型。

利用弥散分布的Laves相强化的耐热钢通过结合固溶强化相和弥散分布的Laves相的析出强化可提供在高温服役环境下优秀的抗蠕变性能。而弥散分布的Laves相颗粒与基体相、位错运动关系等机制尚不完全明确。

本文研究体系为17Cr4W，其中Fe、Cr、W的质量分数分别为79%、17%、4%。通过Thermo-Calc软件计算17Cr4W体系在常压650℃下的相和成分分布，然后根据热力学计算结果，构建17Cr4W双相体系的原子结构模型，其中Laves相颗粒用不同晶粒取向的球体代替，Laves相的结构用（Fe,Cr）2W表示，为$MgZn_2$型结构。铁素体相直接将BCC结构的Fe中的部分Fe原子分别按比例随机替换成Cr和W。根据Laves相的粒径大小设计了三种模型：100%小粒径、100%大粒径以及混合粒径（按体积比30%小粒径+70%大粒径），通过分子动力学模拟对这些模型实施单轴加载，研究分析了应变过程中，Laves相与位错的交互作用，考虑了位错的产生、移动、消失等过程，以及其抗蠕变性能的影响规律，分析了不同尺寸、粒径配比的Laves相对上述规律的影响。发现了如下规律：耐热钢中的Laves相可以提高其高温抗蠕变性能；在一定范围内，Laves相的粒径越小，其抗蠕变效果越好，这是由于Laves相体积分数一定时，小粒径的Laves相不仅不容易促进位错形核，其对位错滑移的阻碍作用也更高；对于有一定尺寸分布的Laves相的耐热钢体系，位错形核取决于其中尺寸最大的Laves颗粒，而位错滑移阻力介于纯大粒径和纯小粒径体系之间。以上发现为实际制备高抗蠕变性能的耐热钢提供了一定的理论指导。

参 考 文 献

[1] Bonny G, et al. J. Phys.: Condens. Matter, 2013, 25: 315401.
[2] Hu P, Yan W, Sha W, et al. Study on Laves phase in an advanced heat-resistant steel[J]. Front. Mater. Sci. China 3, 2009: 434-441.
[3] 刘鸿国, 李生志, 熊伟, 等. 材料热处理学报, 2015, 36(9): 109-113.

10.2　汽车用钢

铝硅镀层厚度对热成形钢极限冷弯性能影响

徐德超[1,2]，张博明[1,2]，张士杰[3]，黄　俊[4]，滕华湘[1,2]，韩　赟[1,2]

（1. 首钢集团有限公司技术研究院薄板研究所，北京　100043；2. 绿色可循环钢铁流程北京市重点实验室，北京　100043；3. 首钢京唐钢铁联合有限责任公司，河北唐山　063200；4. 北京首钢股份有限公司，北京　100043）

摘　要：铝硅镀层热成形钢热冲压淬火后，利用金相显微镜、扫描电镜、电子探针、室温拉伸检测等方法研究了显微组织和力学性能变化。极限冷弯力-位移曲线显示，不同镀层厚度对最大力无显著影响，薄镀层较厚镀层实验样力略有下降，占比5%。最大力位移薄镀层较厚镀层实验样由7.1mm增加到8.1mm左右，极限冷弯角度由65°提高到75°左右，提升约14%。镀层厚度较薄时，铝元素和基体铁元素扩散较充分，较软的相互扩散层和富铝铁素体层厚度达到8~10μm。软相层厚度增加，有效缓解了裂纹的萌生和扩展，提高了热成形钢的极限冷弯性能。
关键词：热成形钢；铝硅镀层；极限冷弯性能；微观组织

Influence of Aluminum-silicon Coating Thickness on the Cold Bending Performance of Hot-formed Steel

XU Dechao[1,2], ZHANG Boming[1,2], ZHANG Shijie[3], HUANG Jun[4], TENG Huaxiang[1,2], HAN Yun[1,2]

(1. Research Institute of Technology, Shougang Group Co., Ltd., Beijing 100043, China; 2. Beijing Key Laboratory of Green Recyclable Process for Iron & Steel Production Technology, Beijing 100043, China; 3. Shougang Jingtang United Iron & Steel Co., Ltd., Tangshan 063200, China; 4. Beijing Shougang Co., Ltd., Beijing 100043, China)

Abstract: The microstructure and mechanical properties were studied by microscope, scanning electron microscope(SEM), electron probe(EPMA) and tensile test. The cold bending force-displacement curves show that different coating thickness has no significant effect on the maximum force, and the experimental sample force of thin coating has decreased slightly compared with the thick coating, just about 500-700N, accounting for 5%. The maximum force displacement of thin coating increased from 7.1mm to about 8.1mm compared with the thick coating, and the extreme cold bending angle increased from 65 degree to about 75 degree, with an increment of 14%. When the coating thickness is thin, the aluminum element and the matrix iron element diffuse more fully, and the soft mutual diffusion layer and the aluminum-rich ferrite layer reach 8-10μm thickness. The thickness of soft phase layer effectively alleviates the initiation and expansion of crack and improves the extreme cold bending performance of hot-formed steel.

Key words: hot-forming steel; aluminum-silicon coating; bending performance; microstructure

钼对无碳贝氏体非调质钢动态连续冷却转变的影响

陈曦[1,2]，王福明[2]，代文彬[1]，祁永峰[1]，陈学刚[1]

（1. 中国恩菲工程技术有限公司恩菲研究院，北京 100038；
2. 北京科技大学冶金与生态工程学院，北京 100083）

摘 要： 利用Gleeble3800热模拟机，测定了新型高强韧无碳贝氏体钢过冷奥氏体动态连续冷却转变曲线，分析了Mo含量对实验钢组织的影响。通过FESEM、XRD等方法对组织特性进行了表征，结果表明：Mo含量的增加，推迟了高温扩散型转变，降低了贝氏体转变开始温度，促进了贝氏体转变，使得板条状无碳贝氏体组织在较低的冷速下生成，提高了显微组织的硬度。同时，在一定轧后冷速范围内，Mo的增加可获得更多的残余奥氏体，有利于实验钢获得更理想的强韧性匹配。

关键词： 无碳贝氏体；非调质钢；Mo；组织；动态CCT

Effect of Mo Content on Dynamic Continuous Cooling Transformation of Carbide-free Bainitic Non-quenched and Tempered Steel

CHEN Xi[1,2], WANG Fuming[2], DAI Wenbin[1], QI Yongfeng[1], CHEN Xuegang[1]

(1. ENFI R&D Institute, China ENFI Engineering Co., Ltd., Beijing 100038, China;
2. School of Metallurgical and Ecological Engineering, University of Science and Technology Beijing, Beijing 100083, China)

Abstract: Dynamic continuous cooling transformation curves of newly carbide-free bainitic steels with high strength and toughness were measured based on Gleeble3800 thermal mechanical simulator. The effect Mo content on microstructure of experimental steels has been investigated by FESEM and XRD measurements. The results show that the addition of Mo can effectively delay high-temperature ferrite transformation and decrease bainite start temperature, which means medium-temperature bainite transformation is promoted. Furthermore, with the increase of Mo content, the carbide-free lath bainite was formed at a lower cooling rate, the hardness of the microstructure was improved, more retained austenite can be obtained within a certain range of cooling rate after rolling, which is beneficial to obtain more ideal matching of strength and toughness of experimental steel.

Key words: carbide-free bainite; non-quenched and tempered steel; Mo; microstructure; dynamic CCT

低成本高表面质量600MPa级热轧双相钢研制开发

董毅[1,2]，时晓光[1,2]，刘仁东[1,2]，孙成钱[1,2]，韩楚菲[1,2]，王俊雄[1,2]

（1. 海洋装备用金属材料及其应用国家重点实验室，辽宁鞍山 114009；
2. 鞍钢集团钢铁研究院，辽宁鞍山 114009）

摘 要：基于鞍钢股份热轧带钢厂连轧生产线超快冷系统工艺特点，进行了600MPa级低成本高表面质量热轧双相钢研制开发。通过低Si成分设计和超快冷+空冷+超快冷的三段式冷却工艺制定，开发了具有优良力学性能匹配的600MPa级热轧双相钢板，钢板组织为典型的双相组织，其中铁素体百分含量约为85%，铁素体晶粒尺寸约为6.0μm；钢板屈服强度为360~390MPa，抗拉强度达到600~630MPa，屈强比为0.57~0.63，延伸率大于27%。钢板在具有优良力学性能的同时具有良好的表面质量和较低的成本，满足汽车车轮等部件制造要求。

关键词：热轧双相钢；表面质量；低成本；超快冷

Evelopment of 600MPa Grade Hot-rolled Dual-phase Steel with Low Cost and High Surface Quality

DONG Yi[1,2], SHI Xiaoguang[1,2], LIU Rendong[1,2], SUN Chengqian[1,2], HAN Chufei[1,2], WANG Junxiong[1,2]

(1. State Key Laboratory of Metal Material for Marine Equipment and Application, Anshan 114009, China; 2. Ansteel Iron & steel Research Insititute, Anshan 114009, China)

Abstract: Based on the process characteristics of ultra-fast cooling system in the continuous rolling production line of Anshan Iron and Steel Co., Ltd., 600MPa grade dual-phase steel with low cost and high surface quality was developed. A 600MPa grade hot-rolled dual-phase steel plate with excellent mechanical properties was developed by low Si composition design and three-stage cooling process of ultra-fast cooling+air cooling+ultra-fast cooling, the yield strength of the steel plate is 360-390 MPa, the tensile strength is 600-630 MPa, the yield strength ratio is 0.57-0.63, and the elongation is more than 27%. The steel plate has good mechanical properties, good surface quality and low cost, and meets the manufacturing requirements of automobile wheels and other parts.

不同应变速率下超高强热成形钢的动态变形行为

马为涛，王鑫玮，陈星翰，宋仁伯

（北京科技大学材料科学与工程学院，北京 100083）

摘 要：热成形钢在汽车制造中扮演着至关重要的角色，被广泛应用于汽车的抗碰撞关键部件，如A柱、B柱、C柱和防撞梁等，以确保乘客在碰撞事故中得到足够的保护。随着汽车轻量化减重的趋势不断发展，我们也必须同时关注汽车的服役安全性。对于热成形零件来说，不仅需要具备足够的强度和刚性，还必须能够在碰撞时有效地吸能和变形，以减轻碰撞对乘客的影响。因此在汽车用钢的强度不断提升的大背景下，有必要研究超高强热成形钢在高应变速率下的动态变形行为，为工业生产提供指导[1-4]。

本文以1800MPa级热成形钢为研究对象，通过进行淬火-回火热处理（奥氏体化温度880℃，保温后低温回火）来模拟工业生产，对实验钢进行不同应变速率（$1s^{-1}$、$10s^{-1}$、$100s^{-1}$、$300s^{-1}$）下的拉伸实验，研究1800MPa级别超高强热成形钢的动态变形行为，深入分析实验钢在高应变速率下的服役性能和断裂机制。

实验结果表明，随着应变速率的提升，实验钢的强度有明显提高，屈服强度$R_{p0.2}$和抗拉强度R_m分别提升267MPa和88MPa，如图1(b)所示；加工硬化曲线和绝热温升曲线均有明显提高，加工硬化行为和绝热软化行为同时发生，特别是在$300s^{-1}$应变速率下绝热温升达到42.3℃，此应变速率下获得最高的断后延伸率，如图1(c)(d)和(e)所示；随着应变速率的增加，实验钢吸收的能量不断增加，在$300s^{-1}$应变速率下，其吸能能量最大，为10668MPa%，如图1(f)所示。在高应变速率下，实验钢的拉伸呈现沿45°方向断口且为典型的韧性断裂，不同应变速率下拉伸断口都由较多韧窝组成，随着应变速率的提升（$1\sim300s^{-1}$），韧窝数量和深度都在逐渐提升。高倍SEM观察韧窝内部发现有较多析出，对其进行了EDS能谱分析，发现析出物多为TiO、Fe-Si-O等氧化物。

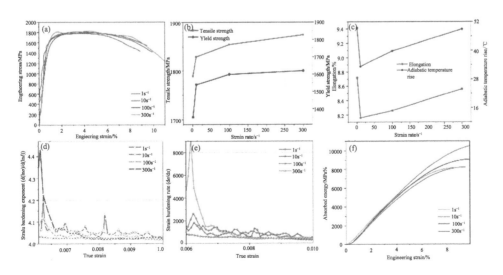

图1 实验钢动态力学性能

(a)应力应变曲线；(b)强度变化；(c)绝热温升和塑性变化；(d)加工硬化曲线；(e)加工硬化指数曲线；(f)吸能曲线

参 考 文 献

[1] 鲁春艳. 上海汽车, 2007(6): 28-31.
[2] 林建平, 田浩彬, 张燕, 等. 制造业高端技术系列[M]. 北京: 机械工业出版社, 2017, 159.
[3] 王小兰. 汽车文摘, 2021(2): 30-34.
[4] Karbasian H, Tekkaya A E. Journal of Materials Processing Technology, 2010, 210(15): 2103-2118.

基于脉冲处理的中锰钢高温力学性能研究

赵文皓，宋仁伯，王永金，赵 帅

（北京科技大学材料科学与工程学院，北京 100083）

摘 要：中锰钢的高温力学性能在生产流程中至关重要，其在连铸生产中高温铸态下的塑性与强度将影响热轧加工的产品质量[1]；若对中锰钢使用脉冲处理这一新型加工工艺，脉冲处理对晶粒的焦耳热效应与电子风力效应[2]使组织产生的晶粒生长、晶粒细化或裂纹修复行为都可能会改善中锰钢的高温力学性能[3]。因此本试验基于脉冲处理对中锰钢的高温力学性能进行研究，以期改善传统的加工工艺，获得更优良的组织性能。

本实验以一种中锰钢为研究对象，从温度和脉冲处理时间两个条件对中锰钢的高温力学性能进行研究。实验设计进行了600~1000℃下的拉伸试验，对700℃下的组织在拉伸前进行0~30min不同时间的脉冲处理，应变速率均为$0.004s^{-1}$；试验后利用 SEM 对断口和断口附近截面组织进行观察与分析。该中锰钢在不同温度下拉伸的应力应变曲线见图1（a），拉伸后的断面收缩率-温度曲线见图1（b）；结果表明当该中锰钢组织温度为600℃时，回复和再结晶软化不足以进行，又由于残余奥氏体[4]与形变强化共同作用使材料在抗拉强度较高（438MPa）的同时保留部分塑性，致使断裂与颈缩部位不一致；650℃时组织在强度较高(369MPa)的同时具备较优的塑形，断面收缩率达55.4%；800℃下的组织具有优良的热塑性，断面收缩率达62.1%；当温度为1000℃时组织晶粒粗大，强度较低（41MPa），断口表现出脆性，裂纹沿晶界扩展，导致沿晶断裂；施加脉冲处理的样品均表现出抗拉强度降低；如图2所示，进行较长时间脉冲处理（30min）的试样断口观察到等轴状颗粒，具有平整的边缘和表面，具有明显的脆断特征，但总应变相比无脉冲样品大幅提升（0.992/0.483），表现出较低的强度和较高的塑性。

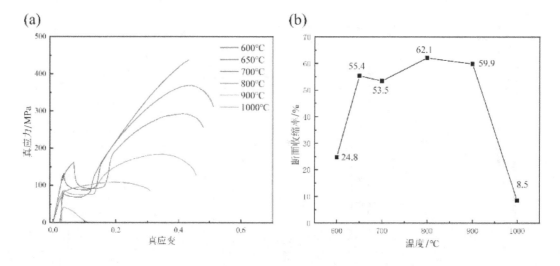

图1 600~1000℃拉伸力学性能曲线

(a) 不同温度下拉伸的应力应变曲线；(b) 断面收缩率-温度曲线

图 2 脉冲 30min 后的中锰钢试样在 700℃下的拉伸断口形貌

(a) 20 倍；(b) 96 倍；(c) 424 倍

参 考 文 献

[1] 雷志国. 江西理工大学, 2020.
[2] 王睿楠. 武汉理工大学, 2019.
[3] 唐国翌, 姜雁斌, 崔敬泉. 冶金设备, 2008(3): 63-66.
[4] 刘春泉, 彭其春, 邓明明, 等. 钢铁研究学报, 2017, 29(6): 431-440.

退火工艺对高成形钢组织和力学性能的影响

张　峰[1,2]，郭金宇[1,2]，孟静竹[1,2]，王科强[1,2]，靳友鹏[1,2]，张瑞坤[1,2]

（1. 海洋装备用金属材料及其应用国家重点实验室，辽宁鞍山　114009；
2. 鞍钢股份有限公司技术中心，辽宁鞍山　114009）

摘　要：应用连退热模拟、扫描电镜（SEM）分析、力学性能分析等实验方法，研究了高成形钢在不同的退火制度下的组织和力学性能的变化规律。结果表明，实验钢在一步退火工艺中，随着退火温度的升高，其抗拉强度增大，延伸率升高，残余奥氏体的含量增多；随着配分温度的升高，抗拉强度先降低后增大，延伸率先增大后降低，残余奥氏体的含量先增多后降低。在预淬火退火工艺中，实验钢组织中的奥氏体化程度更高，马奥岛晶粒尺寸更加的细小且均匀，残余奥氏体含量更高，组织均匀性更高，高含量的残余奥氏体和细小且分布均匀的马奥岛晶粒是提升高成形钢力学性能的关键。

关键词：高成形钢；退火工艺；残余奥氏体；马奥岛

Effect of Continuous Annealing Process on Microstructure and Mechanical Properties of High Forming Steel

ZHANG Feng[1,2], GUO Jinyu[1,2], MENG Jingzhu[1,2], WANG Keqiang[1,2],
JIN Youpeng[1,2], ZHANG Ruikun[1,2]

(1. State Key Laboratory of Metal Material for Marine Equipment and Application, Anshan 114009, China;
2. Automobile and Home Application Steel Insititute, Technology Center of Angang Steel
Company Limited, Anshan 114009, China)

Abstract: The microstructure and mechanical properties of high forming steel under different annealing regimes were studied by means of continuous annealing hot simulation, SEM analysis and mechanical properties analysis.The results

show that in the one-step annealing process of the experimental steel, as the annealing temperature increases, its tensile strength increases, elongation increases, and the content of retained austenite increases. With the increase of partitioning temperature, the tensile strength decreases first and then increases, the elongation increases first and then decreases, and the content of retained austenite increases first and then decreases. In the pre-quenching annealing process, the degree of austenitization in the experimental steel structure is higher, the grain size of MA island is finer and more uniform, the content of retained austenite is higher, the structure uniformity is higher, and the high content of residual austenite Austenite and fine and evenly distributed MA island grains are the key to improving the mechanical properties of high forming steels.
Key words: high forming steel; continuous annealing process; retained austenite; MA island

热处理工艺对中锰钢组织性能的影响

敬树坤，宋仁伯，霍巍丰，赵 帅，王鑫玮，陈星翰

（北京科技大学材料科学与工程学院，北京 100083）

摘 要：随着汽车产业的不断发展，人均汽车保有量的不断提高，人们对车身高强度、轻量化的需求也不断增长，追求尾气减排、燃油经济性和车辆安全性。中锰钢是第三代先进高强钢的代表产品之一，凭借较低的成本，且兼具高强度、高塑性的特点，受到了汽车生产行业的广泛关注[1-2]。中锰钢的性能受到残余奥氏体积分数和稳定性的显著影响，这些残余奥氏体通常是通过 IA（临界退火）[3]和 Q&P（淬火和分配）工艺[4]获得的，其在拉伸变形过程中的 TRIP 和 TWIP 效应是获得优异性能的关键机制[5]。

本文以 Fe-0.26C-2.06Mn-0.54Al 为研究对象，对热轧及冷轧后的实验钢进行了不同温度（700~860℃）、不同时间（10~30min）的临界退火实验，并引入 Q&P（淬火和分配）工艺，进一步提高残余奥氏体的稳定性，从而提升实验钢的各项力学性能。并通过 SEM、XRD、EBSD 等表征手段探究残余奥氏体含量、形态、尺寸、化学成分等因素对于实验钢力学性能的影响。

实验结果表明，在保温时间为 30min 时，随着临界退火温度的升高，实验钢内残余奥氏体的含量比较波动，整体呈先上升后下降的趋势，在 840℃时达到最高值，此时试样的抗拉强度也最高，达到了 1850MPa。通过对比不同退火时间下实验钢内的奥氏体含量，随着退火时间的延长，试样内残余奥氏体的含量逐渐减少，且整体含量普遍偏低，在退火温度为 740℃时，退火时间由 10min 增加到 30min，残余奥氏体含量也从 14.61%下降到 1.9%。选取热轧后 720℃临界退火、退火时间 30min 的实验钢进行冷轧，并对冷轧后实验钢进行不同温度的临界退火实验，实验发现冷轧后的实验钢随退火温度的升高，抗拉强度由 1200MPa 上升到 1800MPa，而断后延伸率由 10.5%下降到 6.8%。对部分实验钢进行 IQ&P（临界退火后配分）实验，发现残余奥氏体含量显著提高，且实验钢各项力学性能也大幅提升。通过各项表征实验，并对比力学性能数据，发现不同热处理条件下，残余奥氏体含量接近的实验钢也拥有不同的力学性能，是由于其中残余奥氏体的形态、尺寸也影响了实验钢的力学性能。

参 考 文 献

[1] Cai Z H, Ding H, Misra R D K, et al. Acta Mater, 2015, 84: 229-236.
[2] Yang D P, Du P J, Wu D, et al. Mater. Sci. Technol, 2021, 75: 205-215.
[3] Li X, Song R B, Zhou N P, et al. Scr. Mater. 2018, 154: 30-33.
[4] Latypov M I, Shin S, De Cooman B C, et al. Acta Mater. 2016, 108: 219-228.
[5] Li J J, Song R B, Li X, et al. Mater. Sci. Eng. A, 2019, 745: 212-220.

超高强度热成形钢的强化机制及氢致延迟开裂行为

陈伟健，章顺虎

（苏州大学沙钢钢铁学院，江苏苏州 215021）

摘 要：在"碳达峰"和"碳中和"的时代背景下，汽车轻量化的发展迫切需要开发超高强度汽车用钢并提高其服役性能。热成形工艺有效解决了高强度汽车用钢不易冷冲压成形的问题，所制备的热成形钢零部件已经成为汽车中强度级别最高的安全部件之一。然而，随着强度的提高，热成形钢在服役过程中不可避免地面临塑性差和氢脆问题，且强度级别越高，氢致延迟开裂问题越严重。因此，揭示超高强度热成形钢的强化机制和氢脆机理，实现其具有高强塑性的同时拥有优异的抗氢脆敏感性，对实现汽车轻量化具有重要的工程意义。

新型试验钢（40MnCr2NbV）完全奥氏体化后向马氏体转变的临界冷速为 0.7℃/s，表现出优异的淬透性。将试验钢冷轧板完全奥氏体化后直接油淬至室温，微观组织由大量回火马氏体和少量淬火马氏体构成，抗拉强度达到 2428MPa，屈服强度为 1457MPa，总延伸率达到 8.4%。对比研究商用 2000MPa 级（34MnBV）和 40MnCr2NbV 热成形钢的强化机制，结果表明，位错强化占屈服强度的比例分别达到 41%（34MnBV）和 62%（40MnCr2NbV）而成为主要强化机制。此外，当慢应变拉伸速率为 $10^{-5}s^{-1}$ 时，计算得出热成形钢中位错的移动速率远小于氢原子的扩散速率，表明位错可束缚氢原子移动以降低氢原子的扩散速率。然而热成形钢的氢脆敏感性系数随着位错密度的增加而增高，并且氢致裂纹易在晶界处萌生并沿着晶界扩展，因此，位错会携带氢原子（有目的性）在晶界处聚集，从而促进氢致裂纹的产生。可通过纳米级碳化物细化组织增加晶界面积提高氢原子分布的均匀性、钉扎位错阻碍"氢-位错柯氏气团"的移动防止氢原子在晶界处聚集、作为不可逆氢陷阱直接束缚氢原子的扩散和聚集，使热成形钢具有高强塑性的同时拥有优异的抗氢致延迟开裂性能。

关键词：热成形钢；强化机制；氢致延迟开裂；组织性能

高成形性 980MPa 级复相钢组织性能研究

谢春乾[1,2]，刘华赛[1,2]，韩赟[1,2]，刘李斌[1,2]，王川[1,2]，李明远[1,2]

（1. 首钢集团有限公司技术研究院，北京 100043；
2. 绿色可循环钢铁流程北京市重点实验室，北京 100043）

摘 要：利用连续退火模拟试验机、光学显微镜、扫描电镜等设备，研究了退火工艺对高成形性 980MPa 级复相钢组织性能的影响。结果表明，试验钢中的组织为铁素体-贝氏体-马氏体+残余奥氏体组织。随着过时效温度的增加，试验钢中马氏体含量逐渐增加，残余奥氏体含量逐渐减少，同时材料抗拉强度降低而延伸率增加，体现了残余奥氏体的 TRIP 效应，采用低温过时效具有更好的性能表现。

关键词：复相钢；连续退火；残余奥氏体；组织性能

Research on the Microstructure and Mechanical Properties of 980MPa Complex Phase Steel with High Formability

XIE Chunqian[1,2], LIU Huasai[1,2], HAN Yun[1,2], LIU Libin[1,2], WANG Chuan[1,2], LI Mingyuan[1,2]

(1. Shougang Group Co., Ltd. Research Institute of Technology, Beijing 100043, China; 2. Beijing Key Laboratory of Green Recyclable Process for Iron & Steel Production Technology, Beijing 100043, China)

Abstract: Effect of annealing process on microstructure and mechanical properties of 980MPa complex phase steel with high formability was studied by means of annealing simulator, optical microscope and SEM. The results show that the microstructure of test steel consists of ferrite, bainite, martensite and retained austenite after annealing progress. The martensite increase with annealing temperature rises, and at the same time, the tensile strength decrease while the elogation increase, and the best properties can be obtained at the low-temperature over aging.

Key words: complex phase steel; annealing; retained austenite; microstructure and mechanical properties

汽车用镀锌热成形钢胶接性能的研究

陈屹松，李志昂，王立辉，张馨月，刘丽君

（河钢材料院汽车材料研发中心，河北石家庄 050000）

摘　要： 现代汽车生产工艺中，为保证车身强度并减少应力集中，往往采用胶接结构，来提升车身的强度与稳定性。镀锌热成形钢通过高淬透性、微合金化等成分设计，并采用热成形的思路，近年来在车身上的应用越来越广泛。为满足镀锌热成形钢在车身零部件的顺利应用，本文以镀锌热成形钢、结构胶为基础材料，针对结构胶处于不同测试条件和服役环境的形态以及拉伸力学性能进行初步的研究，为后续汽车用镀锌热成形钢的胶接性能中需要进一步研究的工作提出了展望。

关键词： 胶接结构；镀锌热成形钢；结构胶；拉伸力学性能

Study on the Bonding Performance of Galvanized Hot Formed Steel for Automobile

CHEN Yisong, LI Zhi'ang, WANG Lihui, ZHANG Xinyue, LIU Lijun

(Department of Automotive Material Research Center, HBIS Group Material Technology Research Institute, Shijiazhuang 050000, China)

Abstract: In the production process of Modern Motor Company, in order to ensure the body strength and reduce stress concentration, adhesive structures are often used to improve the strength and stability of the body. Galvanized hot-formed steel is designed with high hardenability, microalloying and other components, and adopts the idea of hot forming. In recent years, its application in car bodies has become increasingly widespread. In order to meet the smooth application of galvanized hot-formed steel in car body components, this article takes galvanized hot-formed steel and structural adhesive as the basic materials, conducts preliminary research on the morphology and tensile mechanical properties of structural adhesive under different testing conditions and service environments, and proposes prospects for further research on the bonding performance of galvanized hot-formed steel for automotive use.

Key words: adhesive structure; galvanized hot formed steel; structural adhesive; tensile mechanical properties

30MnCrB5 连续冷却组织转变研究

张　青，薛仁杰，宋　帅，王立辉，陈屹松，陈　晨

（河钢材料技术研究院，河北石家庄　050023）

摘　要： 利用 Gleeble-3800 型热模拟试验机研究分析 30MnCrB5 动态连续冷却过程组织转变，与静态热膨胀相比，动态热膨胀过程晶粒经过压缩变形，存在大量形变储能导致相变温度提高。模拟热轧工艺过程研究轧制对显微组织的影响，结果表明，冷速为 0.5℃/s 时，组织为铁素体和珠光体，硬度值为 210HV；冷速为 2℃/s 时，组织为针状铁素体、珠光体和少量贝氏体，硬度值为 258HV 和 290HV；冷速为 5℃/s 时，组织为贝氏体占主体和少量马氏体，硬度值在 327HV 之间；冷速为 10℃/s、15℃/s 时，组织中板条马氏体束增多，贝氏体逐渐减少，硬度值在 437～487HV 之间；冷速为大于 25℃/s 时，组织全为马氏体，硬度达到 500HV 以上。

关键词： 30MnCrB5；热模拟；动态连续冷却；金相组织

Study on Transformation of 30MnCrB5 after Continuous Cooling

ZHANG Qing, XUE Renjie, SONG Shuai, WANG Lihui, CHEN Yisong, CHEN Chen

(HBIS Materials Technology Research Institute, Shijiazhuang 050023, China)

Abstract: The microstructure transformation of 30MnCrB5 during dynamic continuous cooling process was studied by Gleeble-3800 thermal simulation test machine. Compared with the static thermal expansion, the grain in the dynamic thermal expansion process is compressed and deformed, and there is a large amount of deformation energy storage, which leads to the increase of phase transition temperature. The effect of rolling on the microstructure was studied by simulating the hot rolling process. The results show that when the cooling rate is 0.5℃/s, the microstructure is ferrite and pearlite, and the hardness is 210HV. When the cooling rate is 2℃/s, the microstructure is acicular ferrite, pearlite and a small amount of bainite, and the hardness values are 258HV and 290HV. When the cooling rate is 5℃/s, the microstructure is bainite and a small amount of martensite, and the hardness value is between 327HV. When the cooling rate is 10℃/s and 15℃/s, the lath martensite bundle in the microstructure increases, the bainite gradually decreases, and the hardness value is between 437-487HV; when the cooling rate is greater than 25℃/s, the microstructure is all martensite, and the hardness reaches more than 500HV.

Key words: 30MnCrB5; thermal simulation; dynamic continuous cooling; metallographic organization

2000MPa级超高强热成形钢退火–淬火–回火工艺及组织性能研究

王鑫玮,宋仁伯,陈星翰,霍魏丰,赵 帅

(北京科技大学材料科学与工程学院,北京 100083)

摘 要:本文在1500MPa级热成形钢22MnB5成分基础上,适当增加碳含量,同时添加0.04%的微合金元素Nb设计一种2000MPa级超高强度热成形钢,试验钢成分体系为C 0.34%~0.36%,Mn 1.4%~1.6%,Nb 0.03%~0.05%,Si 0.3%~0.5%,Cr 0.15%~0.25%,Ti 0.03%~0.05%,B 0.004%~0.006%。通过设置不同的退火温度、奥氏体化保温温度、回火温度、回火时间等工艺参数以获得一种抗拉强度不低于2000MPa,总伸长率不低于7%的超高强度热成形钢。退火方式采取连续退火方式,退火温度范围设置为740~820℃,奥氏体化保温温度范围设置为860~920℃,回火温度范围设置为170~200℃,回火时间设置为10~20min。分析了不同热处理工艺参数对组织及性能的影响规律,探究了试验钢在不同热处理工艺参数下的组织演变和性能特征。

结果表明,试验钢退火温度在740~760℃之间时,退火组织为铁素体+碳化物,当退火温度在780~820℃之间时,退火组织为铁素体、碳化物和马氏体,当退火温度为800℃时,退火后试验钢综合力学性能达到最佳:屈服强度为782MPa,抗拉强度为1062MPa,总伸长率为8.0%。试验钢奥氏体化后进行水冷淬火,组织均为马氏体。随着奥氏体化保温温度升高,奥氏体晶粒逐渐变大,淬火后相对应的马氏体板条也变得粗大,试验钢强度先升高后降低,总伸长率逐渐升高,奥氏体化温度为900℃时,综合力学性能最佳,故后续回火实验选择900℃奥氏体化保温温度基础上进行。随着回火温度的升高以及回火时间的增加,试验钢强度逐渐降低,塑性逐渐增加。为满足试验钢的设计要求,选择淬火-回火工艺为900℃淬火 5min+170℃回火 20min,试验钢组织为回火马氏体;力学性能为:屈服强度1365MPa,抗拉强度2012MPa,总伸长率7.3%,满足了2000MPa级超高强度热成形钢的性能要求。

Nb含量对Fe-4Mn-2Al-0.2C钢组织性能及屈服行为的影响

霍魏丰,宋仁伯,王永金,赵 帅,苏盛睿,张应超,王鑫玮

(北京科技大学材料科学与工程学院,北京 100083)

摘 要:随着社会的发展,能源消耗和环境污染问题促进了汽车用先进高强钢(AHSSs)的发展。中锰钢(Mn含量(质量分数)在3%~12%之间)作为第三代汽车用先进高强钢的代表产品,利用临界退火工艺获得极细晶的铁素体+奥氏体组织,可以获得优异的综合力学性能[1,2]。Nb常作为微合金元素被考虑到中锰钢的成分设计中,但有关于Nb添加量对中锰钢组织性能的影响却鲜有研究。

因此,本文通过对热轧后的试验钢进行临界退火处理,研究了铌含量分别为0.02%和0.1%的两种Fe-4Mn-2Al-0.2C钢的显微组织、力学性能和屈服行为。结果表明,Nb含量的增加可带来50~100MPa的强化效果,主要来源于更为明显的细晶强化和析出强化效果。但另一方面,更多Nb的加入会促进渗碳体的析出,阻碍或延缓

再结晶的发生，恶化奥氏体的稳定性。Nb 含量为 0.1%的试验钢，由于临界退火中的静态再结晶受到抑制，组织呈现出单一的板条形态。板条状奥氏体在冷却过程中发生部分马氏体转变，碳原子从已转变的马氏体向残余奥氏体扩散，并使碳在奥氏体相界处偏聚，导致了吕德斯效应的出现。当 Nb 含量为 0.02%时，组织中的奥氏体分数较高，并存在等轴状和板条状两种形态，使其在变形过程中的 TRIP 效应持续性更强，获得了更优的综合力学性能。在 725℃临界退火 30min 后，Nb 含量为 0.02%的试验钢屈服强度、抗拉强度和断后伸长率分别为 570MPa、1011MPa 和 35.6%，强塑积达到了 36GPa·%，并有效避免了吕德斯效应的出现。与之对应，Nb 添加量更高的试验钢（0.1%）的各项力学性能为：屈服强度 687MPa、抗拉强度 1071MPa、断后伸长率 29.8%、强塑积 31.9GPa·%，强韧性相对较低，并伴随着屈服平台的出现。

参 考 文 献

[1] Suh D W, Kim S J. Scr. Mater., 2017, 126: 63-67.
[2] Wang M M, Tasan C C, Ponge D, et al. Acta Mater., 2016, 111: 262-272.

温轧-热处理耦合工艺对热轧 3Mn 钢力学性能的影响研究

赵　帅，宋仁伯，张应超，霍巍丰，苏盛睿，王永金

（北京科技大学材料科学与工程学院，北京　100083）

摘　要：近年来，针对中锰钢的强化工艺研究逐渐由单一的热处理向轧制-热处理方向发展。温轧作为一种有效提高中锰钢力学性能的新型强化工艺备受关注。其中，温轧的保温和轧制温度选取以及与热处理工艺之间的耦合机制一直是温轧工艺的研究要点。

本文针对热轧 3Mn 钢分别进行两相区保温-两相区轧制，两相区保温-铁素体区轧制和铁素体区保温-铁素体区轧制的温轧实验，并对温轧后的 3Mn 钢分别进行了退火/回火处理，探究了温轧-热处理耦合工艺对力学性能的影响机制。温轧实验结果表明，轧制温度是影响组织和力学性能的关键因素，两相区轧制使 3Mn 钢获得板条状的马奥结构，导致其呈现连续的加工硬化行为。当轧制温度处于铁素体区时，大量渗碳体析出且无奥氏体保留至室温，导致 TRIP 效应消失和加工硬化能力降低。温轧促进大量具有位错胞的板条晶粒产生，晶粒内部虽然位错发生缠结但并没有演化为完整的晶界。另外，在两相区保温-铁素体区轧制处理过程中，两相区保温形成的奥氏体在铁素体区轧制时会转变为铁素体，这种形变诱导铁素体相变对晶粒回复长大有抑制效果，晶粒尺寸更为细小。

温轧后的热处理（退火/回火）有效提升 3Mn 钢的塑性。其中，两相区轧制-回火工艺产生的异构多相组织使力学性能的提升最为明显。回火后 3Mn 钢的塑性失稳行为（屈服平台和应力锯齿）对回火温度具有强烈的依赖性。高温回火状态下，晶粒粗化和位错湮灭比较明显，屈服后位错增殖和滑移的等待时间增加，导致屈服平台出现。低温回火抑制了温轧细晶的长大，使得组织为板条晶粒+等轴晶粒混合的异构细晶组织，强化了加工硬化能力的同时也延缓了塑性失效。同时，回火温度降低导致 M_7C_3 的析出增多，促进了塑性变形过程中碳化物周围位错的增殖，提高了加工硬化能力。

最终，两相区温轧的 3Mn 钢在 300℃回火 10min 时获得了最佳力学性能（抗拉强度为 1591MPa，断后伸长率为 24.5%）且无塑性失稳行为产生。

参 考 文 献

[1] Zou Y, Ding H, Zhang Y, et al. International Journal of Plasticity, 2022, 151: 103212.

[2] 赵晓丽，张永健，黄海涛，等. 钢铁研究学报, 2018, 30(8): 642-649.
[3] 田亚强，曹仲乾，毕文强，等. 热加工工艺, 2022, 51(2): 133-137.
[4] Song C, Wang H, Sun Z, et al. Materials Characterization, 2021, 181: 111486.

中锰钢中 VC 析出物对奥氏体逆转变行为的影响——纳米尺度原位观察

张应超，宋仁伯，赵　帅，苏盛睿，霍巍丰，王海波，王永金

（北京科技大学 材料科学与工程学院，北京　100083）

摘　要：纳米析出物对钢铁材料中界面迁移过程的作用机制是重要的科学问题，本研究通过原位加热球差矫正高分辨透射电镜研究了中锰钢加热时初始相界面处纳米 VC 析出物对奥氏体逆转变行为的影响规律，发现了析出物对界面的迁移存在显著的温度效应。

首先，通过 FIB 提取温轧中锰钢中板条状间隔分布的奥氏体-铁素体双相组织，在局部相界面处获得直径约 20nm 纳米 VC 析出物。然后通过原位加热装置将试样加热至 720℃保温，发现在奥氏体中发生了明显的界面迁移，其方向沿着板条组织的长轴方向。当界面迁移至界面纳米析出时，其靠近析出物的位置被明显地拖拽，而远离析出物的位置在跨过析出物后生长发生停滞。随着保温的进行，靠近析出物的新界面逐渐跨过界面析出物，但远离析出物的新界面不再发生任何变化。最终在 720℃下，系统保持稳定，新界面在析出物处停滞生长，但保持着平直的状态。但是当继续加热至 800℃时，奥氏体逆转变表现得更加活跃，同时析出物的稳定性降低，发生分解。温度升高直接导致新界面的进一步活跃，这促进了界面析出物的分解。同时，初始相界面在析出与新界面的交汇处发生迁移，直接导致界面析出物被分割为两部分。随着保温的进行，新界面和初始界面在板条内发生活跃的迁移，最终获得完全转变的奥氏体。

本研究首次观察了了纳米析出物与奥氏体逆转变行为相互作用行为，明确了二者相互影响的温度效应。该研究结果对于新一代汽车用中锰钢的组织设计和性能优化提供了重要的基础理论支撑。

参 考 文 献

[1] Tomota Y, Gong W, Harjo S, et al. Scripta Materialia, 2017, 133(5): 79-82.
[2] Cai Z H, Ding H, Misra R D K, et al. Acta Materialia, 2015, 84(2): 229-236.

热轧中锰钢对超快加热和低温回火的力学性能敏感性

苏盛睿，宋仁伯，全书仪，张应超，霍巍丰，赵　帅，王永金

（北京科技大学材料科学与工程学院，北京　100083）

摘　要：为了响应钢铁行业对能源效率和提高生产效率的需求，最近，一种超快加热工艺被提出以满足要求[1]。超

快加热可以在合适的条件下可以大幅提升钢的屈服强度同时不牺牲过多塑性[2]，已成为解决上述挑战的一种有前途的技术。

在这项工作中，我们以约 100℃/s 的加热速度将热轧 7%（质量分数）Mn 钢加热至 700℃，随后立即冷却至室温或保温 20s 左右后冷却至室温，接下来两种试样均在 170℃保温 20min。我们发现热轧中锰钢的机械性能对超快速加热的持续时间有很强的敏感性。这些试样对随后的低温回火的敏感性也有着显著差异。借助多尺度的实验表征、屈服强度的模型计算和元素扩散的数值模拟，我们讨论了这些敏感性。结果表明，经过无等温过程的超快加热处理的试样的力学性能对随后的低温回火处理不敏感。这是由于在超快加热过程中，C 原子已经在位错处偏聚，导致低温回火中的 C 原子偏聚作用不明显，因此低温回火前后的力学性能变化不大。超快加热后的短时间的等温过程导致钢中大量纳米级碳化物的快速析出。但析出强化的贡献小于固溶强化，导致屈服强度明显下降。同时，低温回火过程使 C 原子向位错、晶界或相界面偏析，显著改善了试样的 Portevin-Le Chatelier (PLC)效应，提高了钢的延伸率（从 10% 到 19%）。此外，Cottrell 气氛诱导的强化和相边界偏析诱导的强化共同作用有助于提高屈服强度（112MPa）。

参 考 文 献

[1] Wen P, Hu B, Han J, et al. A strong and ductile medium Mn steel manufactured via ultrafast heating process[J]. J. Mater. Sci. Technol, 2022, 97: 54-68. 10.1016/j.jmst.2021.04.035.
[2] Wan X, Liu G, Yang Z, et al. Flash annealing yields a strong and ductile medium Mn steel with heterogeneous microstructure[J]. Scr. Mater, 2021, 198: 113819. 10.1016/j.scriptamat.2021.113819.

淬火温度对 Fe-0.2C-2.1Si-2.5Mn 淬火配分钢组织性能的影响

高鹏飞[1,2]，刘旭明[1]，赵征志[2]

（1. 鞍钢集团北京研究院有限公司，北京　102211；
2. 北京科技大学钢铁共性技术协同创新中心，北京　100083）

摘　要：本研究采用完全奥氏体化淬火配分工艺对 Fe-0.2C-2.1Si-2.5Mn 钢进行热处理，研究了淬火温度对试验钢微观组织结构及力学性能的影响规律。实验结果表明，当淬火温度从 165℃提升至 315℃时，试验钢抗拉强度在 1298~1400MPa 范围内波动，屈服强度由 1186MPa 逐渐下降至 779MPa，延伸率随淬火温度先升高后降低。当淬火温度为 265℃时，试验钢延伸率达到最大值 18.3%，此时试验钢的强塑积亦达到 23.8GPa·%。当淬火温度偏低时，试验钢的性能由残余奥氏体的体积分数决定；当淬火温度偏高时，新生马氏体体积分数增加可导致力学性能下降。

关键词：淬火配分钢；淬火温度；残余奥氏体；力学性能

The Effect of Quenching Temperature on the Microstructure and Properties of Fe-0.2C-2.1Si-2.5Mn Quenched and Partitioning Steel

GAO Pengfei[1,2], LIU Xuming[1], ZHAO Zhengzhi[2]

(1. Ansteel Beijing Research Institute Co., Ltd., Beijing 102211, China; 2. Collaborative Innovation Center of Steel Technology, University of Science and Technology Beijing, Beijing 100083, China)

Abstract: A fully austenitized quenching & partitioning heat treatment of Fe-0.2C-2.1Si-2.5Mn was carried out. The effects of quenching temperature on the microstructure and mechanical properties of the test steel were studied. The experimental results show that when the quenching temperature is raised from 165℃ to 315℃, the tensile strength of the test steel fluctuates in the range of 1298-1400MPa, the yield strength gradually decreases from 1186MPa to 779MPa, and the elongation first increases and then decreases with the quenching temperature. When the quenching temperature is 265℃, the elongation of the test steel reaches the maximum value of 18.3%, and the product of strength and plasticity of the test steel reaches 23.8GPa·%. When the quenching temperature is low, the properties of the test steel are determined by the volume fraction of retained austenite; when the quenching temperature is high, the increase in the volume fraction of new martensite leads to a decrease in mechanical properties.

Key words: quenched and partitioning steel; quenching temperature; retained austenite; mechanical properties

含 V 奥氏体低密度钢退火时的组织性能演变

谢志奇，惠卫军，张永健，赵晓丽

（北京交通大学机械与电子控制工程学院，北京 100044）

摘　要：对于高 Mn、高 Al、高 C 含量的奥氏体低密度钢，其常见的热处理制度包括固溶、时效和退火。固溶处理的温度范围较高，可以使奥氏体低密度钢获得良好的塑性，但强度水平较低；低温时效处理会大幅提高奥氏体低密度钢的强度，但会造成塑性急剧下降；退火处理的温度一般介于固溶和时效之间，通过退火得到的奥氏体低密度钢的强度和塑性均可达到优异的水平。奥氏体低密度钢的常规成分包括 Mn、Al、C、Si 等元素，在这种成分体系下，奥氏体低密度钢的组成相包括少量奥氏体、铁素体和 κ-碳化物，κ-碳化物既包括晶内的纳米级 κ-碳化物，也包括晶界处粗大的 κ-碳化物。

以可被剪切变形的纳米级 κ-碳化物强化的奥氏体低密度钢的屈服强度仍不是很理想，虽然通过时效处理增大 κ-碳化物的尺寸和数量后屈服强度可以进一步提升，但塑性的大幅下降以及综合性能的恶化使得时效处理的应用场景较窄。因此，借助传统钢铁材料的设计理念，在奥氏体低密度钢中引入不可以剪切变形的硬质粒子是提升其屈服强度的另一种思路。基于此，一些研究人员在奥氏体低密度钢中添加了 V、Nb 等微合金元素，并成功地获得了极为优异的力学性能[1,2]。但诸如 V 等微合金元素添加后除了产生强化效果外，还会对奥氏体低密度钢的微观组织产生多方面的影响。

本研究在前期工作的基础上[3]，以 Fe-25Mn-10Al-1.1C-(0/0.5)V 奥氏体低密度钢为研究对象，对含 V 与不含 V 两种实验钢进行奥氏体共析反应温度范围内的退火处理，诱导实验钢中的奥氏体发生共析反应，对比探究 V 元素对奥氏体的共析反应及其对实验钢微观组织和力学性能的影响。结果表明，含 V 与无 V 两种实验钢在 800℃退火时会发生共析反应（γ→α+κ），组织由奥氏体基体、晶界铁素体、κ-碳化物三相组成，此外，含 V 实验钢基体中不仅存在纳米级 κ-碳化物，还存在大量 VC 析出相。由于更细的锻态晶粒提供了更多的形核点以及奥氏体稳定性的下降，退火过程中含 V 实验钢中奥氏体的共析反应受到显著促进，晶界铁素体和 κ-碳化物含量上升。V 微合金化细化了奥氏体晶粒尺寸、降低了奥氏体再结晶程度、抑制了孪晶发育。同时，晶粒细化和共析反应是一个相互促进的动态过程。与无 V 钢相比，含 V 钢强度大幅提高，塑性有所降低，应变硬化率随应变增加而下降的速度增大。本研究有望为含 V 奥氏体低密度钢的设计与开发提供实验参考。

参 考 文 献

[1] Liu M, Li X, Zhang Y, et al. Intermetallics, 2021, 134: 107179.
[2] Ma T, Gao J, Li H, et al. Metals, 2021, 11: 345.
[3] Xie Z, Hui W, Zhang Y, et al. Mater Sci Eng A, 2022, 861: 144306.

IF 钢成形后波纹度增量的研究

张露星，张　军，葛　浩，施刘健，崔　磊

(马鞍山钢铁股份有限公司，安徽马鞍山　243000)

摘　要：研究不同 IF 钢成形前与 5%成形后的波纹度。成形后钢板的波纹度由原板波纹度与成形过程中波纹度增量 ΔW_{sa} 叠加决定。钢板成形后 ΔW_{sa} 与钢板表层晶粒形态相关，当表层晶粒出现未回复再结晶会导致 ΔW_{sa} 异常偏高。在晶粒形态正常情况下 ΔW_{sa} 与表层组织大小呈正相关趋势，晶粒越粗大则相应的 ΔW_{sa} 越高。在晶粒尺寸相近的情况下，原板的波纹度较高，则相应的 ΔW_{sa} 会略微降低。

关键词：IF 钢；波纹度；ΔW_{sa}

Study on Waviness Increment of IF Steel after Forming

ZHANG Luxing, ZHANG Jun, GE Hao, SHI Liujian, CUI Lei

(Maanshan Iron and Steel Company, Maanshan 243000, China)

Abstract: The waviness of different IF steel before and after forming is studied. The ΔW_{sa} is closely related to the surface grain morphology of the steel plate. When the surface grain is not recrystallized, the ΔW_{sa} is abnormally high. Under normal grain morphology, the ΔW_{sa} is positively correlated with the surface microstructure size, the larger the grain is, the higher the ΔW_{sa} is. With similar grain sizes, if the waviness of the original plate is higher, the ΔW_{sa} will be slightly reduced.

Key words: IF steel; waviness; ΔW_{sa}

低密度高强塑 Fe-Mn-Al-Cr 系奥氏体不锈钢的研究进展

周彦君，肖　蕾，邓想涛

(轧制技术及连轧自动化国家重点实验室，辽宁沈阳　114000)

摘　要：目前，高强度、高韧/塑性是钢铁材料发展的重要方向，尤其是在汽车行业，开发出更安全更轻便的汽车用材是汽车研究人员一直追求的目标。汽车大梁是载货车的主要承载部件，梁板用钢需要具有足够的强韧性、冷成型性、抗疲劳等良好的综合性能。目前汽车大梁钢板的主要以 510L 和 610L 钢为主，它的屈服强度和抗拉强度只有 355MPa 和 510~650MPa，力学性能处于一个较低的水平。本课题组开发出一种新型的 Fe-Mn-Al-Cr 汽车用钢，并对新钢种进行了一系列的测试表征，结果表明，经过比重计测量，随着 Al 含量的增加钢的密度降低，密度可达 6.8g/cm³，比普通碳素钢密度降低了 13%。经过力学性能测试以及 SEM 和 TEM 等表征，发现 Cr 和 Al 的添加一方面可以在奥氏体基体中形成纳米级 κ-碳化物有效提高钢的强度，强度可达 1000MPa 以上，另一方面还能提高钢的塑韧性，延伸率可达 40%以上，这种钢板与普通的碳素钢相比具有优异的轻量化、高强度高塑性、可成形性以及其他许多优点，符合汽车大梁钢板等汽车零部件的各项性能指标，是结构材料的潜在替代品，在汽车行业具有非常大的潜力和应用价值，并为未来汽车用钢的研究方向提供了新思路。

关键词：Fe-Mn-Al-Cr 系钢；轻量化；高强塑；汽车用钢

微合金化 35MnB5 热成形钢在不同热处理状态下的氢脆敏感性研究

林 超[1]，郭晓菲[1,2]，韦习成[1,2]

（1. 上海大学材料科学与工程学院，上海 200444；
2. 上海大学（浙江）高端装备基础件材料研究院，浙江嘉兴 314100）

摘 要：本文比较了含铌微合金化 35MnB5 热成形钢在罩退状态，淬火状态，以及淬火+回火状态下的组织和力学性能演变，以及预充氢条件下的氢脆敏感性。通过结合电化学氢渗透实验，进一步阐释组织结构对其表观扩散速率的影响。研究表明退火态 35MnB 在在同样充氢条件下获得更高的氢含量，具有较高的容氢能力和低氢脆敏感性。该材料在淬火态强度达到 1800MPa，具有高氢脆敏感性和较退火状态低氢表观扩散速率的特征，含有大量位错的板条状马氏体组织使其具有较低容氢能力。在淬火+300℃回火状态，位错密度显著降低，氢脆敏感性较淬火状态降低。

关键词：热成形钢；氢脆；氢扩散；热处理

Effect of Heat Treatment on Hydrogen Embrittlement Sensitivity of 35MnB5 Hot Press Forming Steel

LIN Chao[1], GUO Xiaofei[1,2], WEI Xicheng[1,2]

(1. School of Materials Science and Engineering, Shanghai University, Shanghai 200444, China;
2. Zhejiang Institute of Advanced Materials, Jiaxing 314100, China)

Abstract: This work compares the evolution of microstructure and mechanical properties of niobium-containing micro-alloyed 35MnB5 hot press forming steel in annealed, quenched, and quench & tempered states, as well as their susceptibility to hydrogen embrittlement under pre-hydrogen-charged conditions. The influence of microstructure on the apparent diffusion rate is further elucidated by incorporating electrochemical hydrogen permeation experiments. It is shown that 35MnB in the annealed state obtains higher hydrogen content under the same hydrogen charging conditions, with higher hydrogen tolerance and low hydrogen embrittlement susceptibility. The material reaches a strength of 1800MPa in the quenched state, which is characterized by a high hydrogen embrittlement susceptibility and a lower apparent hydrogen diffusion rate than in the annealed state. The lath martensitic with high dislocation density leads to low hydrogen tolerance. In the quench & tempered (300 ℃) condition, the dislocation density decreases significantly, and the hydrogen embrittlement susceptibility is reduced compared to the quenched states

Key words: hot press forming steel; hydrogen embrittlement; hydrogen diffusion; heat treatment

10.3 特殊钢

碲极微合金化对非调质钢组织与性能影响

徐翔宇，刘年富，王梓菲，付建勋

（上海大学，上海 200444）

摘　要：碲在钢铁材料领域属于小众元素，主要作为调控硫化物形态元素使用。碲改质技术最初应用于易切削钢，近年来国内开始应用于非调质钢[1-3]，以改善硫化物形态，提高产品品质，减少成品磁痕缺陷率。然而，碲改质技术对非调质钢组织与性能影响鲜有研究。本文将 Te/S 质量比小于 0.07 的碲处理技术称为碲极微合金化，此时钢中无碲化锰析出，碲主要以取代硫原子的形式掺杂于硫化锰中。

本工作依托国内某特钢厂生产的 38MnVS 工业棒材（直径 60mm），碲改质钢的碲含量为 23ppm，Te/S 质量比为 0.043。材料生产工艺为：BOF-LF-VD-CC-CR。碲以包芯线形式添加至钢液中，工序位于 VD 和 CC 之间，以提高碲的收得率。未改质材料与碲改制材料的连铸与热轧工艺完全一致。再加热温度为 1160℃，初轧和精轧开轧温度分别为 1050℃ 和 990℃，终轧尺寸为直径 60 mm 圆棒。热轧后，冷床空冷至室温。

试验钢的显微组织由铁素体、珠光体和 MnS 夹杂物组成。未改质钢的和碲改质钢的铁素体面积百分比分别为 30.6±4.0% 和 32.0±2.7%。碲改质对铁素体和珠光体百分含量改变不明显，但对组织有着一定细化作用。碲对硫化物有着细化作用，夹杂物数量由 230 个/mm² 增加至 261 个/mm²，夹杂物长宽比小于 4 的百分比明显多于未改质钢。极微量的碲也可有效改善热轧态硫化物的形态。

碲极微合金化后，材料的屈服强度略有提升，抗拉强度变化不明显，延伸率横纵向差异得到一定程度改善。屈服强度的提升主要与显微组织的细化有关。横向和纵向力学性能的差异主要与硫化物的形态有关。碲改质后，硫化物的长宽比减小，降低了力学性能的各向异性。

一般来说，材料的疲劳强度取决于抗拉强度或者硬度，尽管碲极微合金化对材料的抗拉强度改善不明显，但碲改质钢的疲劳强度略高于未改质钢，这也与显微组织的细化有关，提高了滑移带开动初始应力，从而延迟裂纹萌生[4]。疲劳裂纹始终开始于试样表面，疲劳裂纹的扩展机制为准解理断裂机制，碲改质钢的韧窝数量明显多于未改质钢，疲劳裂纹主要沿着铁素体/珠光体界面和铁素体晶界传播。

碲改质后，材料的切削性能得到有效改善，切削合力降低，表面光洁度提高。极微碲改质条件下，碲主要掺杂于 MnS 中，提高了 MnS 硬度，提高了 MnS 作为应力集中源的作用。

参 考 文 献

[1] 胡涛, 钟亮美, 周蕾, 等. 炼钢, 2022, 38(1): 63-67.
[2] 刘年富, 沈伟, 田钱仁, 等. 钢铁研究学报, 2023, 35(1): 71-78.
[3] 轩康乐, 刘栋林, 俞杰, 等. 特殊钢, 2022, 43(6): 79-82.
[4] Hui W, Chen S, Zhang Y, et al. Materials & Design (1980-2015), 2015, 66: 227-234.

生活垃圾焚烧炉用耐热钢炉排铸件对比分析及优化设计

王春奕，欧红燕，张全新，胡 毅

（宝武特冶航研科技有限公司，重庆 400084）

摘 要：本文通过对生活垃圾焚烧炉常用的五种炉排进行化学成分分析、力学性能试验、显微组织的观察和检测，结合垃圾焚烧炉排的服役环境以及理化检测分析结果对炉排性能优劣进行了对比分析，并根据分析结果提出了旨在提高垃圾焚烧炉排使用寿命的优化设计建议方案。

关键词：垃圾焚烧炉；耐热铸钢；腐蚀

Comparative Analysis and Optimum Design of Heat-resistant Cast Steel for MSW Incinerator Grate

WANG Chunyi, OU Hongyan, ZHANG Quanxin, HU Yi

(Baowu Specialmetallurgy Aeronautical Research Technology Co.,Ltd., Chongqing 400084, China)

Abstract: In this paper, the chemical composition, mechanical properties and microstructure of five kinds of grates commonly used in waste incineration project are analyzed, combined with the service environment of waste incinerator grate and the results of physical and chemical detection and analysis, the performance of grate was compared and analyzed, based on the analysis results, the optimal design proposal for improving the service life of waste incinerator grate is put forward.

Key words: waste incinerator; heat-resistant cast steel; corrosion

激光切割对弹簧钢 50CrV4 冷成型影响的研究

杨 玉[1,2]，王英海[1,2]，许 成[1,2]，王润琦[1,2]，张吉富[3]，任俊威[3]

（1. 海洋装备用金属材料及其应用国家重点实验室，辽宁鞍山 114009；
2. 鞍钢集团钢铁研究院，辽宁鞍山 114009；
3. 鞍钢股份有限公司鲅鱼圈钢铁分公司，辽宁营口 115007）

摘 要：本文主要分析了 50CrV4 钢热轧板激光切割后折弯开裂的原因及基体冷成型性能。结果表明，激光切割端面存在厚度 0.11～0.25mm 的马氏体层及微裂纹，折弯时外表面受拉应力，马氏体层内微裂纹处应力集中成为裂纹源引起微裂纹扩展是折弯开裂的主要原因；50CrV4 钢板基体 90°冷弯未开裂，冷成型性能优异。

关键词：激光切割；冷弯性能；马氏体；折弯开裂

Effect of Laser Cutting on Cold Forming of 50CrV4 Steel

YANG Yu[1,2], WANG Yinghai[1,2], XU Cheng[1,2],
WANG Runqi[1,2], ZHANG Jifu[3], REN Junwei[3]

(1. State Key Laboratory of Metal Materials and Application for Marine Equipment, Anshan 114009, China;
2. The Iron and Steel Research Institute of Ansteel Group, Anshan 114009, China;
3. Bayuquan Branch of Angang Steel Co., Ltd., Yingkou 115007, China)

Abstract: In this paper, the causes of bending cracking of 50CrV4 steel hot-rolled plate after laser cutting and the cold forming properties of the substrate were analyzed. The results showed that there were martensite layers and micro-cracks with thickness of 0.11-0.25mm on the laser cutting surface, and the external surface was subjected to tensile stress when bending. The main reason of bending crack was that the stress concentration at the micro-crack in martensite layer became the crack source and caused the micro-crack propagation. The 50CrV4 steel plate substrate 90 degrees cold bending without cracking, cold forming performance is excellent.

Key words: laser cutting; cold forming performance; martensite; bending cracking

热轧卷取温度对弹簧钢氧化铁皮结构及酸洗特性影响研究

李雯，裴新华

（宝钢研究院梅钢技术中心，江苏南京 210039）

摘 要： 本文以弹簧钢 60Si2Mn 为对象，研究不同卷取温度下氧化铁皮结构特点及酸洗特性，结果表明：热轧板的边部及中部氧化铁皮结构存在显著差异，边部氧化铁皮厚度均匀，以致密的 Fe_3O_4 及表层 Fe_2O_3 为主，中部氧化铁皮厚度极不均匀，除 Fe_3O_4 外，存在 60%以上的 Fe，其中，620℃低温卷取样板的 Fe 析出物主要分布在上表面，而 680℃高温卷取的样板除上表面外，基底层也有明显的聚集状铁析出。经酸洗，两种样板表层以 Fe 为主的膜状氧化铁皮呈现整片剥离，剩余部分氧化铁皮结构极为致密，酸洗以匀速腐蚀溶解为主，其中高温卷取样板由于基底层更多的 Si 聚集，酸洗后期反应速度下降，完全酸洗用时更长。

关键词： 弹簧钢；氧化铁皮；酸洗

Effect of Rolling Temperature on the Iron Oxide Scale Characteristics and Pickling Characteristics of the Spring Steel

LI Wen, PEI Xinhua

(R&D Center for Meisteel of Baosteel Research Institution, Nanjing 210039, China)

Abstract: In this paper, the structural characteristics and pickling characteristics of iron oxide scale at different winding temperatures are studied by taking spring steel 60Si2Mn as the object, and the results show that there are significant differences in the structure of the edge and middle iron oxide scale of the hot-rolled plate, the thickness of the edge iron oxide scale is uniform, mainly dense Fe_3O_4 and surface Fe_2O_3, the thickness of the middle iron oxide scale is extremely uneven, except for Fe_3O_4, there are more than 60% eutectoid iron, among which, the eutectoid iron of the 620℃ coil sample plate is mainly distributed on the upper surface, while eutectoid iron of the 680℃ coil sample plate not only appeared on the surface, but in the basal layer. After pickling, the surface layer, mainly Fe-based film-like iron oxide sheet, peeled off the whole piece, the remaining part of the iron oxide scale structure is extremely dense, pickling is mainly based on uniform dissolution, among which the 680℃ rolling sample plate shows the lower reaction speed in the later stage of pickling due to more Si aggregation in the basal layer, consequently, the complete pickling takes longer.

Key words: spring steel; iron oxide scale; pickling

轴承钢热轧盘条表面缺陷的研究

巩延杰[1,2,3]，张育明[1,2,3]，范振霞[1,2,3]，董 庆[1,2,3]，赵昊乾[1,2,3]

(1. 河北邢钢科技有限公司，河北邢台 054799；2. 邢台钢铁有限责任公司，河北邢台 054027；3. 河北省线材工程技术创新中心，河北邢台 054027)

摘 要：针对轴承钢生产过程产生的裂纹、结疤及折叠等各种表面缺陷，利用光学及电子显微镜进行系统分析，根据缺陷的宏观、微观形貌及成分分析，准确地判断出造成缺陷的原因及改进方向，从而有效地避免了该类缺陷的再次产生。

关键词：轴承钢；裂纹；结疤；折叠；表面缺陷

Study on Surface Defects of Bearing Steel Hot-rolled Rod

GONG Yanjie[1,2,3], ZHANG Yuming[1,2,3], FAN Zhenxia[1,2,3], DONG Qing[1,2,3], ZHAO Haoqian[1,2,3]

(1. Hebei Xinggang Technology Co., Ltd., Xingtai 054799, China; 2. Xingtai Iron and Steel Co., Ltd., Xingtai 054027, China; 3. Hebei Wire Engineering Technology Innovation Center, Xingtai 054027, China)

Abstract: In view of the bearing steel production process of various surface defects such as cracks, scarring and folding, using system with optical and electron microscope analysis, according to the defects of macro and micro morphology and composition analysis, accurately judge the causes of the defects and the direction of improvement, effectively avoids the production of this kind of defect again.

Key words: bearing steel; crack; scar; fold; surface defects

N 合金化对辙叉用高锰奥氏体钢组织和性能的影响

齐向阳，陈 晨

(燕山大学材料科学与工程学院，河北秦皇岛 066004)

摘　要： Hadfield 钢是目前应用最为广泛的辙叉用钢，经水韧处理后，其在室温条件下即可获得全奥氏体组织，这种成分和组织特点使其具有优异的塑韧性和加工硬化能力，在冲击载荷的作用下能够产生快速的加工硬化从而获得很高的冲击耐磨性[1,2]。然而，高速重载铁路的发展也对铁路辙叉提出了更高的要求。一方面随着列车运行速度的提高和运输重量的增大，列车运行的动态环境急剧恶化，轮轨间的相互作用更加复杂和剧烈，由此引发的磨损、疲劳和失效问题也更加剧烈；另一方面，Hadfield 钢初始强度低的缺点常常导致辙叉部件在服役初期发生严重塑性变形而提前失效，严重影响使用寿命[3]。除了改进辙叉结构来改善局部受力状态以外，对辙叉材料本身的强韧性、抗疲劳性能和耐磨性能也提出了更高的要求。

通过再合金化处理能够有效提高高锰奥氏体钢的基体强度和加工硬化能力，从而改善辙叉整体的性能状态。近年来，随着冶金科技的进步，高氮钢以其优异的综合性能越来越受钢铁行业和科研学者的高度关注。氮是一种强奥氏体稳定化元素，能够通过间隙固溶强化奥氏体，增加钢的强度，同时不会损害塑性和韧性[4,5]。因此，将氮元素的积极作用引入到辙叉用高锰钢中，开发高氮的高锰奥氏体辙叉钢，有望显著提升辙叉钢的综合性能。

通过增锰添铬的合金化设计，成功制备了氮含量为 0.2% 的高锰奥氏体钢。与不含氮钢相比，含氮钢具有更高的强度和延伸率，这是由于氮元素提高了高锰钢的层错能，导致变形过程中形变孪晶缓慢生成，最终导致高锰钢具有更高的强塑性配合；此外，由于氮元素促进了位错的平面滑移，含氮钢还表现出更高的疲劳寿命。

参 考 文 献

[1] Frank E E. Evolution of the rail-bound manganese frog[J]. Transportation Research Record, 1986.
[2] 张福成. 辙叉钢及其热加工技术[M]. 北京：机械工业出版社, 2009: 29-44.
[3] 马广清, 于文馨, 翟启杰, 等. 水韧处理工艺对合金高锰钢组织和性能的影响[J]. 铸造, 2004, 53(4): 308-309.
[4] Hanniene M, Romu J, Ilola R, et al. Effect of processing and manufacturing of high nitrogen-containing stainless steels on their mechanical, corrosion and wear properties[J]. Journal of Materials Processing and Technology, 2001, 117: 424-430.
[5] Kubota S, Xia Y, Tomota Y. Work-hardening behavior and evolution of dislocation-microstructures in high-nitrogen bearing austenitic steels[J]. ISIJ international, 1998, 38(5): 474-481.

S30408 奥氏体不锈钢不同轧制与固溶工艺条件下的组织演变与再结晶行为

黄　健[1,2]，庞宗旭[1,2]，管吉春[3]，范刘群[1,2]，
张建平[1,2]，孙殿东[1,2]，王　勇[1,2]

（1. 海洋装备用金属材料及其应用国家重点实验室，辽宁鞍山　114009；
2. 鞍钢集团钢铁研究院，辽宁鞍山　114009；3. 鞍钢股份有限公司，辽宁鞍山　114009）

摘　要： 采用 OM、SEM 等方法对高温控轧、高温热轧、低温控轧以及不同温度（950~1100℃）及时间（10~40min）固溶处理后的 S30408 奥氏体不锈钢的微观组织进行了研究。结果表明高温控轧与高温热轧试样固溶后上下表面依然存在混晶组织，而低温控轧态试样经 1050℃净保温 20min 后可得到平均晶粒尺寸为 42.5μm 的均匀组织；EBSD 实验结果表明，轧后材料内部储能不均匀，混晶组织中大尺寸晶粒内部小角度晶界占比高于小尺寸晶粒导致固溶时再结晶驱动力强，并且随固溶温度越高再结晶速率越快。

关键词： 奥氏体不锈钢；晶粒度；TMCP；固溶处理；微观组织；再结晶

Microstructure Evolution and Recrystallization Behavior of S30408 Austenitic Stainless Steel under Different Rolling and Solution Treatment Conditions

HUANG Jian[1,2], PANG Zongxu[1,2], GUAN Jichun[3], FAN Liuqun[1,2], ZHANG Jianping[1,2], SUN Diandong[1,2], WANG Yong[1,2]

(1. State Key Laboratory of Metal Material for Marine Equipment and Application, Anshan 114009, China;
2. Iron & Steel Research Institutes of Ansteel Group Corporation, Anshan 114009, China;
3. Angang Steel Company Limited, Anshan 114009, China)

Abstract: The microstructure and recrystallization behavior of S30408 austenitic stainless steel after high-temperature controlled rolling, high-temperature hot rolling, low-temperature controlled rolling and solution treatment at different temperatures (950-1100℃) and holding (10-40min) was studied. The results that a coarse-grained structure was still present on the upper and lower surfaces of the high-temperature controlled rolling and high-temperature hot rolling samples after solution treatment, while a uniform structure with an average grain size of 42.5μm was obtained after the low-temperature controlled rolling sample was treated at 1050℃ for 20min. The microstructure characterization results showed that the internal stored energy of the rolled material was not uniform, and the proportion of low-angle grain boundaries in large-size grains was higher than that in small-size grains, which lead to the recrystallization behavior first in the large-size grains during solution treatment. The grains grew by means of grain boundary migration and grain merging. We found that the higher the solution temperature, the faster the rate of recrystallization.

Key words: austenitic stainless steel; grain size; thermo-mechanical control process; solution treatment; microstructure; recrystallization

消应力处理对双相不锈钢 S32205 组织及力学性能的影响

张瀚宁[1,2]，王 勇[1,2]，李黎明[3]，孙殿东[1,2]，颜秉宇[1,2]

（1. 海洋装备用金属材料及其应用国家重点实验室，辽宁鞍山 114009；2. 鞍钢集团钢铁研究院，辽宁鞍山 114009；3. 鞍钢集团有限公司科技发展部，辽宁鞍山 114009）

摘 要： 对 S32205 双相不锈钢进行 300~400℃消应力处理组织及性能变化行为进行研究，采用光学显微镜（OM）、扫描电子显微镜（SEM）等显微组织分析手段，分析在不同消应力处理制度下材料组织的变化行为，并对其进行力学性能分析，研究消应力处理对强度和韧性的影响。实验结果表明，消应力热处理会促进铁素体组织中第二相析出，使不锈钢强度提高，韧性下降。经过消应力处理后力学性能仍能保证较大余量，可满足工业应用要求。

关键词： 双相不锈钢；消应力处理；金相组织；低温韧性

Effect of Stabilization Treatment on Microstructure and Mechanical Properties of S32205 Duplex Stainless Steel

ZHANG Hanning[1,2], WANG Yong[1,2], LI Liming[3], SUN Diandong[1,2], YAN Bingyu[1,2]

(1. State Key Laboratory of Metal Material for Marine Equipment and Application, Anshan 114009, China;
2. Iron & Steel Research Institute of Angang Group, Anshan 114009, China; 3. Science and Technology Development Department of Ansteel Group Corporation Limited, Anshan 114009, China)

Abstract: The microstructure and properties of S32205 duplex stainless steel were studied by stabilization treatment on 300-400℃. The microstructure changes of the material under different stabilization treatment methods were analyzed by means of optical microscope (OM), scanning electron microscope (SEM), etc. The mechanical properties were analyzed to study the influence of stabilization treatment on strength and toughness. The experimental results show that the stabilization treatment can promote the second phase precipitation in the ferrite structure, improve the strength of stainless steel and reduce the toughness, but the mechanical properties can still ensure a large margin after the stabilization treatment, which can satisfy the requirements of industrial application.

Key words: duplex stainless steel; heat treatment; microstructure; low temperature iMPact toughness

汽车悬架弹簧钢表面脱碳及对疲劳性能的影响

王程明[1]，孙晓冉[1]，赵 楠[1]，赵中昱[1]，海 岩[2]

（1. 河钢材料技术研究院理化检测中心，河北石家庄 052000；
2. 和芯星通科技（北京）有限公司，北京 100083）

摘 要：对汽车悬架用55Cr3弹簧钢进行不同热处理制度下的脱碳试验，用金相法测量脱碳层厚度，研究了加热温度和保温时间对试验钢脱碳的影响规律，分析了脱碳组织形貌规律及脱碳对疲劳性能的影响。结果表明：700～750℃试验钢为全脱碳层组织，850℃以上为部分脱碳组织，且脱碳层厚度随加热温度的增加而增加；随保温时间的延长，脱碳层厚度逐渐增加；脱碳大大降低了弹簧钢的疲劳寿命。为减小脱碳对试验钢性能的影响，试验钢热处理制度应控制为850℃，保温时间小于60min。

关键词：55Cr3弹簧钢；脱碳；脱碳层；组织形貌

Surface Decarbonization of Spring Steel for Automobile Suspension and Influence on Fatigue Properties

WANG Chengming[1], SUN Xiaoran[1], ZHAO Nan[1], ZHAO Zhongyu[1], HAI Yan[2]

(1. HBIS Materials Technology Research Institute, Shijiazhuang 052000, China;
2. Unicorn Communications Technology (Beijing) Co., Ltd., Beijing 100083, China)

Abstract: The decarbonization test of 55Cr3 spring steel used for automobile suspension under different heat treatment systems was carried out. The thickness of decarbonization layer was measured by metallographic method. The influence of heating temperature and holding time on the decarbonization of test steel was studied, the decarbonized structure morphology and the effect of decarburization on fatigue properties were analyzed. The results show that the test steel at 700-750℃ is fully decarbonized layer structure, and above 850℃ is partially decarbonized structure. the thickness of the decarbonized layer increases with the increase of heating temperature. With the extension of holding time, the thickness of decarbonized layer increases gradually. Decarbonization greatly reduces the fatigue life of spring steel. In order to reduce the influence of decarbonization on the performance of test steel, the heat treatment system of test steel should be controlled at 850℃ and holding time less than 60min.

Key words: 55Cr3 spring steel; decarbonization; decarbonization layer; structure morphology

开坯加热工艺对硅脱氧弹簧钢盘条夹杂物尺寸控制影响

孟耀青,李建立

(武汉科技大学材料与冶金学院,湖北武汉 430081)

摘 要: 在节能减排理念不断深入的过程中,汽车材料轻量化生产技术与材料开发稳步前进。汽车弹簧作为汽车重要零部件,降低其重量的主要措施是提升钢材许用应力强度级别。但随着弹簧钢丝抗拉强度的提高,钢中夹杂物对弹簧疲劳性能的影响愈加显著,夹杂物临界尺寸越来越小。当55SiCr(德标54SiCr6)热处理弹簧钢丝抗拉强度由1600MPa逐步提升至2100MPa,引起悬架簧疲劳断裂的夹杂物尺寸由200μm以上降至50μm以下,如图1所示(弹簧制造商德国慕贝尔内部检测数据)。

螺旋弹簧疲劳断裂总是发生在沿垂直于线材轴向的半径方向上或与轴线呈45°断面上扩展[1],夹杂物的特征尺寸可用半径方向上夹杂物的厚度来表征。通过采用弹簧钢夹杂物塑性化生产工艺,可以使夹杂物在轧制、拉拔阶段变成沿加工方向伸长的形状,减小其在垂直于最大主应力方向平面上的投影面积,有效提高弹簧的疲劳性能[2]。汽车弹簧钢用热轧盘条一般采用两火成材工艺,即轧制盘条前铸坯需经过开坯加热和轧制加热两次加热处理,但铸坯加热过程影响夹杂物的玻璃化稳定,会导致夹杂物轧制塑性变形性发生恶化[3-4]。通过多炉次工业生产试验,借助FEI Explorer 4 自动扫描电镜和极值统计法,对比研究不同开坯加热工艺对硅脱氧弹簧钢热轧坯夹杂物成分、形貌及成品线材夹杂物特征尺寸的影响。结果表明,开坯加热工艺对硅脱氧弹簧钢盘条夹杂物尺寸控制有着重要影响,如图2所示,这为改善硅锰脱氧钢夹杂物控制提供了新的指导思路。后续拟对夹杂物结晶转变 TTT 曲线进一步研究,揭示 $CaO-SiO_2-Al_2O_3$ 体系夹杂物结晶行为演变及动力学调控机理。

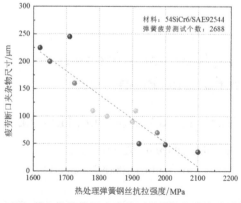

图1 悬架簧夹杂物临界尺寸与原材料抗拉强度关系

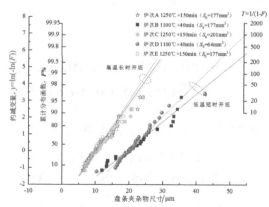

图2 开坯加热工艺对盘条夹杂物特征尺寸影响

参 考 文 献

[1] 王仁智, 汝继来, 中国表面工程, 2016, 29(4): 1-9.
[2] Hu Y, Chen W Q, Wan C J, et al. Metallurgical and Materials Transactions B, 2018, 49: 569-580.
[3] Rocabois P, Pontoire J N, Lehmann J, et al. Journal of Non-Crystalline Solids, 2001, 282(1): 98-109.
[4] Zhao C, Jung S M, Kashiwaya Y, et al. ISIJ International, 2008, 48(6): 747-754.

高均质特殊钢大构件的微观机制与工程应用

曹艳飞，李殿中，刘宏伟，傅排先

（中国科学院金属研究所，辽宁沈阳 110016）

摘 要：宏观和微观成分/组织均匀性是核电、航空航天、军工等高端装备用先进钢铁材料和关键构件高品质制备的关键，对服役寿命和安全具有重要影响。本研究基于多尺度模拟计算、实验表征和工程应用，系统揭示了特殊钢氧致偏析的微观机制，提出了高均质钢的氧含量模型和临界判据，拓展了经典偏析理论。开发了低氧高纯净高均质技术，成功应用于核电压力容器、核电低压转子、核用筒体等大构件的高端制造；与法国电力集团合作，为核电大构件偏析问题追根溯源。在宏观均质性基础上，阐明了微观成分不均匀导致的粗大碳化物的形成条件、演化规律和影响因素，开发了稀土钢高均质技术、全流程热加工碳化物细化技术、精细组织调控技术，实现了航空发动机主轴承长寿命高可靠制备，在多种型号实现应用。

GCr15 轴承钢钢坯加热开裂原因分折及防止措施

完颜卫国，张树山，许 兴，王福礼，李 勇

（马鞍山钢铁股份有限公司特钢公司，安徽马鞍山 243000）

摘 要：在用 ϕ450mm GCr15 钢连铸圆坯轧制生产 ϕ150mm 圆钢时，多支钢坯在开坯时出现了开裂的现象，钢坯开裂均为横裂。

理化检验结果：主要化学成分并无异常；从宏观断口形貌可见，裂纹源在钢坯心部近中心位置，裂纹是从内部向外扩展的；断口附近显微组织未见过热。过烧迹象，钢坯表面脱碳层厚度约为 633.8μm，断口面脱碳层厚度约为 83.66μm，两者的差异较大，表明钢坯出炉前，裂纹处并未完全与外部联通；对钢坯上开裂处的位置进行了测量，表明钢坯上对应两侧水冷梁的位置最易发生开裂，中间开裂处不一定全在中间水冷梁的位置，但对应中间水冷梁的位置处是容易发生开裂的位置。

冷坯在加热过程中,热量的传导是由表及里的,从钢坯的表面到中心存在着温度梯度，相应地，在钢坯内部产生热应力。热应力作用的结果，使钢坯的表层受到压应力，而钢坯中心会受到拉应力。首先钢坯断面越大、加热速度越快则表里温差也越大，其二水冷梁位置处会加大这种温度差，尤其是靠近火焰的两端水冷梁位置，增加热应力，从而增加开裂的倾向；其三钢坯中心部位是低溶点组分和杂质富集的地方，强度最低，甚至还存在间断性的缩孔，钢坯芯部的缺陷还会造成应力集中；其四钢坯芯部残余应力与热应力同向；其五加热时相变组织应力，附加了心部拉应力；其六冷连铸坯入炉加热初期时，钢坯的塑性差，加热过快极易产生裂纹。

当 GCr15 钢坯心部温度在约 500℃以下（对应大规格坯料表面温度 647～683℃以下，对应炉温约 1026℃以下，

参见图1），钢仍处于低塑性区，500℃附近而钢的强度已经明显下降。如果温差造成的热应力过大，加上铸坯残余应力导致芯部叠加拉应力高于材料的强度，就会产生芯部裂。当钢坯心温度升到500℃以上以后，塑性会明显提高，芯部的拉应力通过塑性变形可以得到松弛。因此红送钢坯可以直接装入高温炉内快速加热而不易产生芯部开裂。

防止钢坯加热开裂的措施有：647～683℃（表面温度）以下心部低塑性温度区加热要缓慢，力求铸坯内外升温均匀，缩小铸坯内外温差，降低热应力；760～810℃（表面温度）区间加热也要缓慢，以免热应力和组织应力叠加，促进裂纹、或其它连铸坯缺陷的扩展；连铸坯入坑缓冷应充分，尽量降低连铸坯心部残余拉应力；提高连铸坯芯部的质量；"红送"装炉很有效，但要避开700～900℃温度范围装炉，以免产生"红送裂"。

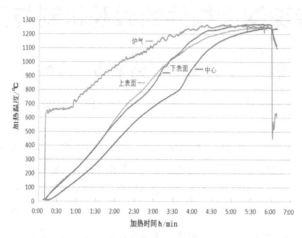

图 1　φ450mm 钢坯加热温度曲线

Fe-Ni-Cr 系中熵合金拉拔过程中性能及微观组织演变

孙智妍[1]，任　帅[1]，赵英利[1]，张志旺[1]，吴迎飞[1]，孙彩凤[1]，张中武[2]

（1. 河钢材料技术研究院特种材料研发中心，河北石家庄　050000；
2. 哈尔滨工程大学材料科学与化学工程学院，黑龙江哈尔滨　150001）

摘　要：本文将 φ6mmFe-Ni-Cr 系中熵合金盘条通过9道次拉拔、2次氢退工艺加工成 φ2.45mm 线材，采用拉伸实验、显微维氏硬度计、光学显微镜、扫描电子显微镜及电子探针对 Fe-Ni-Cr 系中熵合金冷拉拔过程中的力学性能及微观组织演变进行分析。实验结果表明：随着累计变形量的增加，屈服强度、抗拉强度、显微维氏硬度明显增加，断面收缩率、延伸率明显降低；本文在累计变形量达到20%和40%进行了氢退处理，发现其能明显改善 Fe、Ni、Cr、Mo、Al 元素的偏析情况，降低材料内应力，短时高效地实现普通固溶退火的效果，这极大地提高了 Fe-Ni-Cr 系中熵合金的冷拉拔效率。另外，随着拉拔道次的增加，粗大的析出相逐渐破碎，使得退火后达到相同变形量的试样具有更高延伸率和断面收缩率。

关键词：中熵合金；拉拔；力学性能；微观组织

Mechanical Properties and Microstructure Evolution of Fe-Ni-Cr System Medium Entropy Alloy during Drawing Process

SUN Zhiyan[1], REN Shuai[1], ZHAO Yingli[1], ZHANG Zhiwang[1],
WU Yingfei[1], SUN Caifeng[1], ZHANG Zhongwu[2]

(1. HBIS Materials Technology Research Institute, Shijiazhuang 050000, China; 2. School of Materials Science and Chemical Engineering, Harbin Engineering University, Harbin 150001, China)

Abstract: In this work, a ϕ2.45mm Fe-Ni-Cr system medium entropy alloy wire was obtained by 9 passes of cold drawing and 2 hydrogen annealing processes, starting with an untreated alloy of ϕ6mm. The mechanical properties and microstructure evolution during cold drawing were analyzed by a tensile test, a micro vickers hardness tester, an optical microscope, a scanning electron microscope and an electron probe microanalysis. The results indicated that with an increase in cumulative deformation, the yield strength, tensile strength, and the vickers microhardness are significantly enhanced, while the reduction of area and elongation are decreased; In addition, online hydrogen annealing was carried out at the cumulative deformation of 20% and 40%. It was found that online hydrogen annealing significantly improved the segregation of Fe, Ni, Cr, Mo, and Al elements, reduced the internal stress, and achieved similar annealing effects as conventional solution treatment. These improvements greatly enhanced the cold drawing efficiency of the Fe-Ni-Cr system medium entropy alloy. Moreover, as the number of drawing passes increased, the coarse precipitates were broken, resulting in higher elongation and reduction of area for samples with the same deformation after annealing.

Key words: medium entropy alloy; cold drawing; mechanical properties; microstructure evolution

热轧不锈钢板氧化层贫铬层的微观结构特征

岳莹莹[1]，李晓亮[1]，刘承军[2,3]

（1. 辽宁科技学院，辽宁本溪　117004；2. 东北大学多金属共生矿生态化冶金教育部重点实验室，辽宁沈阳　110819；3. 东北大学冶金学院，辽宁沈阳　110819）

摘　要： 热轧不锈钢板工序短，能耗低，"以热代冷"逐渐成为不锈钢市场的消费趋势。在热轧不锈钢板的生产过程中，需经热轧退火，调整晶粒度，提高不锈钢塑性，在该过程中，不锈钢表面会形成成分、结构复杂的氧化层，同时伴生贫铬层[1]，严重影响不锈钢的表面质量，需通过后续酸洗工艺去除。以高温氧化层为核心的表面控制技术以及环保高效的酸洗技术成为高表面质量不锈钢产品开发的关键。

明确不锈钢表面氧化层和贫铬层的组成、分布特征及形成机理是优化酸洗工艺的先决条件。本文通过高温氧化动力学实验，研究了430铁素体不锈钢和304奥氏体不锈钢的高温氧化过程，对比分析了两种不锈钢板的氧化层微观结构及贫铬层的分布特征。结果显示，430不锈钢与304不锈钢氧化层的组成物相类似，自外而内呈如下分布：Fe_2O_3为主的富铁外氧化层，中层为尖晶石相及铁铬复合氧化物（$FeFe_{2-x}Cr_xO_4$），内层为以Cr_2O_3为主的富铬氧化层。430不锈钢氧化前沿均匀向内生长，未形成氧化结节，304不锈钢会形成多个氧化结节，氧化结节内富铬相氧化层与富Ni尖晶石相交错分布，其氧化机制如图1所示。不锈钢的贫铬层始终位于内氧化层前沿，厚度为1.5~3 μm，其分布不受氧化时间与气氛的影响。高温氧化过程中，430铁素体不锈钢的氧化前沿均匀前进，无明显氧化结节。430不锈钢表面贫铬层分布较均匀。而304奥氏体不锈钢氧化前沿形成多个氧化结节，氧化结节的半径及深度不一，造成贫铬层的不均匀分布。因此，相较于铁素体不锈钢，奥氏体不锈钢的酸洗难度更大，在进行酸洗体系设计时，要提高酸洗液向氧化层中的渗透能力，加大酸洗液对贫铬层的溶解，以促进表面氧化层的剥离脱落。

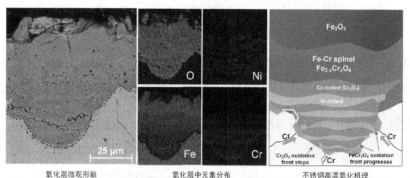

图1 304奥氏体不锈钢高温氧化机制

参 考 文 献

[1] 刘承军, 岳莹莹, 史培阳, 等. 中国冶金, 2016, 26(8): 39-44.

异质超细晶/细晶奥氏体不锈钢的显微组织和力学性能

全书仪，宋仁伯，苏盛睿，王永金，王开坤

（北京科技大学材料科学与工程学院，北京 100083）

摘 要：奥氏体不锈钢因具有优异的力学性能和耐蚀性能广泛应用于航空航天，石油化工、生物医疗等领域。但奥氏体不锈钢具有屈服强度低的共性问题，严重制约了其在要求具有高强度的结构材料领域的应用。细晶强化是提高奥氏体不锈钢力学性能的最有效手段。随着晶粒尺寸的减小，材料的强度将按 Hall-Petch 关系增加。但当晶粒尺寸达到亚微米级别时，强度的提升会伴随着塑性的显著下降。异质结构将具有低强度、高应变硬化能力的细晶嵌入到具有高强度、低应变硬化能力的超细晶组织中，超细晶提升强度，细晶保证塑性，二者的相互协调可以使材料获得优异的强塑性匹配。本文通过430℃温轧+冷轧的两步轧制工艺对 200 系奥氏体不锈钢 Fe-18Cr-6Mn-3Ni-0.24N 施加 96%的严重塑性变形，结合轧后的逆相变退火处理，获得异质超细晶/细晶奥氏体不锈钢。通过力学性能测试和微观组织表征，研究了退火时间对奥氏体不锈钢组织演变和力学性能的影响。退火时间延长，材料的强度降低塑性升高。在770℃退火30s的样品获得了最佳的综合力学性能，抗拉强度为1161MPa，屈服强度为1056MPa，断后伸长率为31.1%。实验结果表明，奥氏体不锈钢经大变形轧制+逆相变退火后，通过应变诱导马氏体的逆转变和形变奥氏体的再结晶，获得了异质超细晶/细晶组织。异质变形诱导强化使材料获得了明显高于传统的奥氏体不锈钢的屈服强度；孪晶诱导塑性和异质变形诱导硬化效应的协同作用，使奥氏体不锈钢具有良好的塑性。

镍对 Cr-Mo-V 系热作模具钢淬透性及碳化物析出的影响

岳建博[1]，田家龙[1]，龚 伟[1]，廖 俊[2]，于 洋[2]，姜周华[1]

（1. 东北大学冶金学院，辽宁沈阳 110819；
2. 江苏天工工具新材料股份有限公司，江苏丹阳 212312）

摘　要：已发现在热作模具钢中添加一定量的镍有利于提高淬透性。本文向 Cr-Mo-V 系热作模具钢中添加镍元素改善淬透性，并在此基础上重点探讨其对碳化物析出行为的影响机理。经研究结果表明：Ni 元素作为非碳化物形成元素不改变平衡时碳化物析出类型及总量，但会溶于基体与 Fe 原子形成置换型晶格畸变，促进回火过程碳化物在基体的弥散析出过程。这一机制会加快在长时间回火保温过程中二次碳化物的聚集长大，对模具钢的高温抗软化性能产生不利影响。通过热力学和动力学计算揭示了 Ni 的作用机理。计算结果表明，Ni 显著提高了回火过程面心立方晶格结构中 Cr 原子的扩散系数，并且对各类析出相的相变驱动力也有明显的提升。这项工作对 Ni 在热作模具钢中的作用机制有了新的认识，为以碳化物作为主要强化相的二次硬化钢的成分设计提供参考依据。

关键词：热作模具钢；镍；淬透性；热稳定性；热力学和动力学

Effect of Ni on Hardenability and Carbide Precipitation in Cr-Mo-V Hot-work Die Steel

YUE Jianbo[1], TIAN Jialong[1], GONG Wei[1], LIAO Jun[2], YU Yang[2], JIANG Zhouhua[1]

(1. School of Metallurgy, Northeastern University, Shenyang 110819, China;
2. Jiangsu Tiangong Tools New Materials Co., Ltd., Danyang 212312, China)

Abstract: The addition of a certain amount of nickel in hot work tool steels has been found to improve their hardenability. This article investigates the effect of adding nickel to Cr-Mo-V hot-work die steels to improve their hardenability, with a focus on exploring the impact on carbide precipitation behavior. The research results show that as a non-carbide-forming element, nickel does not alter the type and overall amount of carbide precipitation at equilibrium. However, it dissolves in the matrix and forms lattice distortion with Fe atoms, promoting the diffusion precipitation of carbides during the tempering process. This mechanism accelerates the growth and aggregation of secondary carbides during prolonged tempering, thereby adversely affecting the high-temperature anti-softening performance of the tool steels. The mechanism of nickel's effect is revealed through thermodynamic and kinetic calculations. The calculations demonstrate that nickel significantly enhances the diffusion coefficient of Cr atoms in the face-centered cubic lattice structure during tempering, and also increases the phase transformation driving force for various types of precipitates. This work provides new insights into the mechanism of nickel in hot work tool steels and serves as a reference for the design of secondary hardening steels with carbides as the main strengthening phase.

Key words: hot-work die steel; Nickel; quenching and tempering ability; thermal stability; thermodynamics and kinetics

高温快速回火对中碳超细贝氏体钢轨钢组织、性能和残余奥氏体稳定性的影响

王建军[1]，李宏光[1]，王庆超[1]，贾德诚[1]，杨志南[1,2]，张福成[3]

（1. 燕山大学国家冷轧板带装备及工艺工程技术研究中心，河北秦皇岛　066004；
2. 燕山大学亚稳材料制备技术与科学国家重点实验室，河北秦皇岛　066004；
3. 华北理工大学冶金与能源学院，河北唐山　063210）

摘　要：提高回火温度有利于降低钢轨中的残余应力，然而，高温回火后超细贝氏体钢轨钢组织、性能的演变规律

尚不明晰。因此，本研究使用高温盐浴炉，将等温淬火生成的超细贝氏体钢以20℃/s的速度加热至550℃，分别保温2min、5min和10min。利用系列微结构表征及性能测试手段，详细研究了高温快速回火对中碳超细贝氏体钢微观组织、力学性能和残余奥氏体在拉伸过程中机械稳定性的影响。

随着回火时间的延长，钢的体积先快速收缩后缓慢收缩。回火2min时，碳原子向位错处偏聚[1]；回火5min时，部分残余奥氏体分解，有纳米碳化物析出；回火10min时，碳化物析出数量和残余奥氏体分解量均增多，贝氏体板条粗化，在冷却过程中有少量马氏体生成。得益于团簇强化，回火2min试样的屈服强度最高，但随着回火时间的延长，由于碳化物析出、板条粗化，屈服强度逐渐降低。纳米碳化物的析出，提高了贝氏体基体的变形能力，同时促进位错在拉伸过程中增值，提高加工硬化能力。随着回火时间的延长，试样的抗拉强度逐渐提高，延伸率先升高后降低[2]。回火2min时，由于贝氏体基体包壳效应的提升，残余奥氏体的稳定性增强[3]；回火5min时，由于贝氏体基体强度降低以及残余奥氏体中碳含量降低，残余奥氏体稳定性下降，但是，纳米碳化物的析出使得残余奥氏体的稳定性表现出动态强化现象。本结果可为贝氏体轨道钢快速回火工艺的开发提供理论支持。

关键词：超细贝氏体钢；快速回火；组织；性能；残奥稳定性

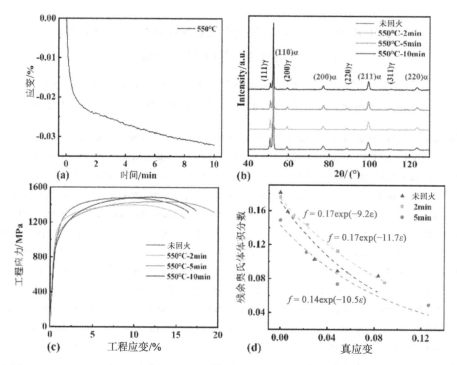

图1 550℃保温时的应变曲(a)、XRD图谱(b)、工程应力应变曲线(c)和残余奥氏体体积分数相对于真应变的变化(d)

参 考 文 献

[1] Cheng L, Brakman C M, Korevaar B M, et al, Tempering of iron-carbon martensite; dilatometric and calorimetric analysis[J]. Metall. Trans. A, Phys. Metall. Mater. Sci. 1988 (A 19): 2415-2426.

[2] Yang Z N, Liu C B, Zhang C Y, et al. Microplasticity behavior of multiphase high-strength nanobainitic steel based on a modified law of mixtures[J].Mater. Sci. Eng., 2021 (A 825): 141848.

[3] Wang K, Gui X L, Bai B Z, et al. Effect of tempering on the stability of retained austenite in carbide-free bainitic steel[J]. Mater. Sci. Eng., 2022 (A 850): 143525.

预冷变形对 GCr15Si1Mo 轴承钢相变、微结构和强韧性的影响机制

贾德诚[1]，张春生[1]，董润洲[1]，王贺霖[1]，杨志南[1,2]，张福成[3]

（1. 燕山大学国家冷轧板带装备及工艺工程技术研究中心，河北秦皇岛 066004；
2. 燕山大学亚稳材料制备技术与科学国家重点实验室，河北秦皇岛 066004；
3. 华北理工大学冶金与能源学院，河北唐山 063210）

摘　要：轴承钢在全部合金钢中是品质要求最高、生产制造技术最难、质检项目最多的钢种之一，长久以来都有"钢中之王"之称[1]。近年来，随着对轴承钢研究的深入，尤其是对贝氏体轴承钢的研究，国内外开发出了新一代的轴承用钢。我国开发的 G23Cr2Ni2Si1Mo 钢和 GCr15Si1Mo 钢已经成功纳入到国家标准《渗碳轴承钢》和冶金行业标准《轴承钢辗轧环件及毛坯》中，受到了国内广泛关注。一般来说在轴承制造过程中，轴承钢在热处理前都会通过机加工或成型加工来生产轴承部件，在某些情况下用于进一步的冷成型过程，如冷镦和冷锻。而其中冷辗扩技术是制造高性能轴承套圈的发展方向之一，目前针对轴承套圈冷辗扩的研究主要集中在工艺参数控制及材料参数选择上[2]。然而，冷辗扩对轴承套圈热处理过程中微观组织的演变影响显著，直接影响后期服役性能。

本研究通过对 GCr15Si1Mo 轴承钢进行不同变形量的冷变形，利用动态淬火相变仪、电子探针、扫描电子显微镜、透射电子显微镜、EBSD、力学性能测试等系列研究方法系统研究了预变形对 GCr15Si1Mo 轴承钢的相变、微结构及性能的影响。结果表明，预冷变形处理能降低试验钢 A_{c1} 温度，小变形会减少碳化物溶解，大变形使得碳化物析出。奥氏体化过程中碳化物的析出导致经预变形处理后试样的 M_s 温度升高，组织中渗碳体尺寸更加均匀。此外，通过相变动力学曲线发现小变形会延缓贝氏体转变，但随预变形量增大，贝氏体峰值转变速率出现的时间逐渐缩短，峰值转变速率逐渐增大，这表明预变形处理会加速贝氏体转变。同时预变形处理会细化原始奥氏体晶粒尺寸，当预变形量达到 50%时，原始奥氏体晶粒尺寸细化到 3.3 μm，减小幅度达 50%。对于贝氏体轴承钢，奥氏体晶粒的细化和过冷奥氏体中碳含量的增加能显著减低贝氏体板条厚度。此外随着预冷变形的增加，原奥氏体晶粒的减小会使贝氏体会发生明显的变体选择倾向。同时晶粒与板条尺寸的细化导致超细贝氏体轴承钢的冲击韧性从 67.8 J/cm^2 提高到 114.3 J/cm^2。且随预变形量增大，贝氏体轴承钢的硬度呈小幅度提升的趋势。对马氏体轴承钢而言，预冷变形处理在保持试验钢韧性的同时，大量孪晶结构的形成会使轴承钢硬度大幅度提高。本研究表明了预冷变形可以使轴承钢的性能显著提高，期待为高碳超细贝氏体轴承钢的热加工工艺优化提供理论支持。

关键词：轴承钢；预冷变形；微结构；强韧性

紧固件用钢的高性能化

陆恒昌[1,2]，胡杰[1]，杨现亮[2]，胡春东[1,2]，史文[1,2]，董瀚[1,2]

（1. 上海大学材料科学与工程学院，上海 200444；
2. 上海大学（浙江）高端装备基础件材料研究院，浙江嘉兴 314113）

摘　要：中国紧固件每年用钢量约 900 万吨。高强化是紧固件高性能化的主要方向。伴随强度的提高，需要解决疲

劳失效和延迟断裂趋势增大的问题，以及相应的加工工艺变化。8.8~12.9 级是目前高强紧固件用量最大的品种，要求强度和疲劳性能相匹配。我们研制了超窄成分（ΔC≤0.01%）及较小硬度散差（ΔHRC≤2）的发动机关键紧固件用 SCM435 钢线材；开发出风塔用大规格、低成本、高淬透性的 10.9 和 12.9 级紧固件用钢；研制了满足 650℃ 持久性能的 A286 合金线材。考虑到减省热处理，研制出免退火冷镦钢和细长杆紧固件用非调质钢线材。装备轻量化需要进一步使用 13.9 级及以上超高强紧固件，需要解决延迟断裂的问题。我们研制出调质型 14.9 级紧固件用钢；利用等温淬火工艺制备 14.9 级超高强紧固件，可以减少合金的使用。19.8 级以上的超高强紧固件需要探索新的强化方式。我们利用二次硬化原理制备出具有耐疲劳和延迟断裂特性的 19.8 级紧固件用钢。

关键词：紧固件；钢；高强度；疲劳失效；延迟断裂

On High Performance Steels for Fasteners

LU Hengchang[1,2], HU Jie[1], YANG Xianliang[2], HU Chundong[1,2], SHI Wen[1,2], DONG Han[1,2]

(1. School of Materials and Science Engineering, Shanghai University, Shanghai 200444, China;
2. Zhejiang Institute of Advanced Materials, Jiaxing 314113, China)

Abstract: The consumption of steel for fasteners in China is about 9 million tons annually, accounting for approximately one-third of worldwide usage. High strengthening is the main development trend of high performance of fasteners. As strength increase, the sensitivity of both fatigue failure and delayed fracture increases, and the processing technology varies. These are the main problems need to be addressed. Fasteners of property classes 8.8-12.9 are currently the most widely used high-strength fasteners. The match between fatigue life with strength is commonly the paramount consideration. We have developed a series of steels involving (1) SCM435 steel wire for key fasteners of engine with ultra-narrow composition (ΔC ≤0.01%), slight hardness variation (ΔHRC≤2) and a qualified fatigue performance of 1×10^7 cycles; (2) 10.9 and 12.9 fastener steels with large size, low cost and high hardenability for wind tower; (3) A286 alloy wire with desirable 650 °C durability. In addition, we developed ML35 wire free of spheroidizing and microalloyed steels for slender bolts free of quenching & tempering process. The use of ultra-high strength fasteners ≥14.9 is necessary to achieve equipment light weighting. We have developed steels for 14.9 fasteners with an excellent resistance to delayed fracture. An austempering process was applied to ultra-high strength fasteners, which enables the reduction of the alloy content. A novel strengthening method needs to be introduced to develop fasteners in excess of class 19.8 fastener. We have developed 19.8 fastener steel with good fatigue and delayed fracture resistance. Secondary hardening plays a key role in the improvement of strength.

Key words: fasteners; steels; high strength; fatigue failure; delayed fracture

加压熔炼条件下钙处理对高铝钢微观组织的影响

李文宇，李 阳，杨 皓，夏 铃，陈浩嘉

（东北大学冶金学院，辽宁沈阳 110819）

摘 要：在 2MPa 气相压力条件下，对 38CrMoAl 进行钙处理精炼实验，获得了 Ca 质量分数最高为 0.01% 的 4 组实验钢。对其不同热处理态微观组织进行研究发现，加压熔炼条件下，随着钙质量分数的增加，实验钢的退火态珠光体组织由片层状变为纳米级的粒状珠光体；同时回火态钢中残余奥氏体含量有一定程度的下降，从 27.92% 降低至 12.36%，同一原始奥氏体晶粒内的马氏体取向呈均一化趋势，且趋向于细密且大小均匀的针状马氏体。

关键词：高铝钢；钙处理；加压；奥氏体

Effect of Calcium Treatment on Microstructure of High-alumina Steel under Pressurized Melting Conditions

LI Wenyu, LI Yang, YANG Hao, XIA Ling, CHEN Haojia

(Northeastern University Institute of Metallurgy, Shenyang 110819, China)

Abstract: Ca treatment refining experiments were carried out on 38CrMoAl under 2 MPa gas-phase pressure conditions, and four groups of experimental steels with Ca mass fraction up to 0.01% were obtained. The microstructure of its different heat treatment states was investigated and found that under pressurized melting conditions. The pearlitic organization of the experimental steels in the annealed state changed from lamellar to nano-sized granular pearlitic as the calcium mass fraction increased. The residual austenite content in the steel in the tempered state decreased to a certain extent, from 27.92% to 12.36%, and the martensite orientation within the same original austenite grains showed a tendency to homogenization and Tend to be fine and uniform size of the needle martensite.

Key words: high-aluminum steel; calcium treatment; pressurization; austenitic

海洋工程用高耐蚀钢筋开发

陈焕德，周　云，杨晓伟，张　宇

（江苏沙钢集团有限公司，江苏张家港　215625）

摘　要：钢筋是桥梁、码头等海洋工程基础设施建设中应用最广泛的建筑材料；钢筋锈蚀导致海洋工程腐蚀失效，严重阻碍了海洋工程长寿命化进程；随着海洋强国战略的实施，海洋工程腐蚀失效问题日趋严峻；研制和使用高耐蚀钢筋是解决钢筋锈蚀问题的有效方法，对确保海洋工程耐久性具有重要意义。

钢筋锈蚀引发的海洋工程腐蚀失效已成为建筑工程的关键技术难题，国内外学者对此开展了许多研究，也有一些产品或技术试图来解决该问题，但均无法大面积推广应用；如不锈钢耐蚀性好，但价格很高；耐候钢成本低，但耐蚀性不足；涂层钢筋不易加工，且与混凝土的握裹力差、施工难度大；现有产品及防护技术尚不能完全满足海洋工程的服役需求，特别是在南海区域的高盐雾、高湿热、强辐射等严酷环境下，钢筋的锈蚀问题更为突出；因此，亟待开发出具有更高品质的耐蚀钢筋，为我国海洋战略的顺利实施提供材料保障。

为解决钢筋锈蚀问题，系统研究了碳、铬、氮等元素对钢筋耐蚀性能及组织形态的影响规律，设计了超低碳铬钼系多元合金成分体系，通过高铬含量实现了钢筋表面钝化行为的增强，而超低碳含量的设计有效减少碳化物的形成，防止了晶界贫铬现象发生；同时通过耐点蚀元素的添加，以及洁净冶炼技术的开发，解决了钢筋耐腐蚀性能不足的技术难题。此外，为提高超低碳钢的力学性能，在成分设计时添加了一定量的合金强化元素，并开发了控轧控冷技术，获得了铁素体/贝氏体双相组织，解决了钢筋强度-塑性-耐蚀性协调匹配难题。

为解决耐蚀钢筋产业化问题，开发了转炉高效冶炼和精炼深脱碳技术，解决了采用"转炉-LF-RH"工艺冶炼时出现的热量平衡难控制、冶炼周期长等技术难题；配套开发了专用固态连接技术和高可靠度套筒连接工艺，解决了接头与母材性能差异性难题；构建了钢筋耐久性评价、服役寿命预测平台，解决了服役行为难以定量表征难题。

技术成果已形成国家标准和施工规范，并实现了重大工程应用；工业化产品屈服强度≥400MPa，强屈比≥1.30，最大力总延伸率≥9%；实际服役环境下的腐蚀数据表明，与普通钢筋相比，耐蚀钢筋的腐蚀速率下降5~10倍，能够满足海洋工程百年服役需求。

新型低密度高强耐热不锈钢的设计研究

陈 卉，江志华，倪志铭，金建军

（中国航发北京航空材料研究院钢与稀贵金属研究所，北京 100095）

摘 要：追求高的比强度是推动航空材料发展的源动力。钢铁材料由于密度大，制约了其在航空上的应用。降低钢铁材料的密度，是提高钢铁材料比强度的有效技术途径。低密度高强钢具有低的密度，与 Al、Mg 合金等轻质材料相比又具有高的弹性模量，在保证强度的同时可直接降低重量，是目前结构件材料轻量化发展的重要方向。本文以研发具有自主知识产权、密度≤7.0g/cm³、综合性能不低于 1Cr11Ni2W2MoV 的低密度高强度耐热不锈钢为目标，突破成分设计、熔炼、开坯锻造、热处理等关键技术，取代目前发动机上使用的 1Cr11Ni2W2MoV 不锈钢，满足发动机对减重的需求，并抢占低密度高强度不锈钢研究领域的先机，为研制性能与 GH4169 相当的低密度不锈钢打下基础。

关键词：低密度；高强度；不锈钢；高温拉伸性能

Research of Low-density High-strength Heat Resistant Stainless Steels

CHEN Hui, JIANG Zhihua, NI Zhiming, JIN Jianjun

(AECC Beijing Institute of Aeronautical Materials, Beijing 100095, China)

Abstract: The high strength-to-weight ratio is motive force of aeronautical material development. Owing to the high density of steel materials, it restricts aeronautical applications. The high strength-to-weight ratio of these steels is due to reduce the density of steel materials. The low-density high-strength steel possesses the low density and higher elastic modulus compared with magnesium and aluminum. The development direction of the lightweight materials is decreasing weight and maintaining strength. In low-density steels area, low-density high-strength stainless steels are rarely reported. This paper aims at developing low-density high-strength heat resistant stainless steels with the density no larger than 7.0g/cm³, performance exceeding to 1Cr11Ni2W2MoV and independent property rights. It breaks through the key technique such as composition design, melting, forging, and heat treatment to satisfy requirement of lightweight engines. It provides basis of low-density stainless steels with performance similar to GH4169.

Key words: low-density; high-strength; stainless steels; tensile properties at high temperature

减振降噪耐蚀阻尼钢的开发

李江文[1]，侯华兴[1]，李大航[2]，孙美慧[1]，郭呈宇[1]，张 弛[1]

（1. 鞍钢集团北京研究院有限公司，北京 102299；2. 鞍钢集团钢铁研究院，辽宁鞍山 114000）

摘 要：Fe-Mn 型阻尼钢作为一种减振降噪的高阻尼金属材料，由于屈服强度低，耐蚀性差，极大地限制了应用。为此，我们开发了一种具备高屈服强度、高塑韧性、高阻尼性能以及良好耐蚀性能的 Fe-Mn 阻尼钢，可以满足结构力学性能和减振降噪功能性能需求。目前屈服强度可覆盖 345~500MPa 级别，屈强比小于 0.65，断后伸长率大于 30%，-20℃冲击功大于 180J，实际使用测试中的减振降噪效果显著。

关键词：减振降噪；阻尼钢；ε马氏体；高强韧；高阻尼

Development of Vibration and Noise Reduction Corrosion-resistant Damping Steel

LI Jiangwen[1], HOU Huaxing[1], LI Dahang[2], SUN Meihui[1],
GUO Chengyu[1], ZHANG Chi[1]

(1. Ansteel Beijing Research Institute Co., Ltd., Beijing 102299, China;
2. Iron and Steel Research Institute, Ansteel Group, Anshan 114000, China)

Abstract: As a high-damping metal material with vibration and noise reduction, Fe-Mn type damping steel greatly limits its application due to its low yield strength and poor corrosion resistance. To this end, we have developed a Fe-Mn damping steel with high yield strength, high plastic toughness, high damping performance and good corrosion resistance, which can meet the structural mechanical properties and vibration and noise reduction function performance requirements. At present, the yield strength can cover 345-500MPa level, the yield ratio is less than 0.65, the elongation is greater than 30%, the impact power of –20℃ is greater than 180J, and the vibration and noise reduction effect in the actual use test is remarkable.

Key words: vibration and noise reduction; damping steel; ε martensite; high strength and toughness; high damping

Φ800mm 大断面连铸圆坯 42CrMo4 凝固传热过程模拟

任春节[1]，冷永磊[2]，尹修刚[2]，任立坤[2]，邢 伟[2]，
张英男[2]，刘福斌[1]，姜周华[1]

（1. 东北大学冶金学院，辽宁沈阳　110000；2. 承德建龙特殊钢有限公司，河北承德　067000）

摘　要：为研究工艺参数对连铸坯的凝固传热过程的影响，本文以断面尺寸为 Φ800mm 的连铸圆坯 42CrMo4 钢为研究对象，建立凝固传热模型，研究过热度、拉速和二冷区比水量对凝固传热规律的影响。模拟结果表明：铸坯表面温度模拟结果和工业测量结果吻合良好，此模型可以很好反映实际情况。浇铸温度从 1504℃增加到 1549℃，铸坯表面温度增加 27℃，铸坯中心温度增加 42℃，凝固终点后移 1.5m，出结晶器坯壳厚度减少 5mm；拉坯速度从 0.16m·min^{-1} 增加到 0.22m·min^{-1}，铸坯表面温度增加 100℃，铸坯中心温度增加 293℃，凝固终点后移 8.4m，出结晶器坯壳厚度减少 13mm；二冷比水量从 0.16L/kg 增加到 0.22L/kg，二冷区表面温度约减少 130℃，铸坯中心温度减少 66℃，凝固终点前移 2m。模拟结果为现场连铸工艺参数的优化提供了理论指导。

关键词：42CrMo4；凝固传热；过热度；拉速；比水量

Simulation of Solidification and Heat Transfer Process of Φ800mm Large Cross Section Continuous Casting Round Billet 42CrMo4

REN Chunjie[1], LENG Yonglei[2], YIN Xiugang[2], REN Likun[2], XING Wei[2], ZHANG Yingnan[2], LIU Fubin[1], JIANG Zhouhua[1]

(1. School of Metallurgy, Northeastern University, Shenyang 110000, China;
2. Chengde Jianlong Special Steel Col., Ltd., Chengde 067000, China)

Abstract: To study the influence of process parameters on the solidification and heat transfer process of continuous casting billets, The article takes the cross-sectional size as solidification heat transfer model was established for the Φ800 mm continuous casting round billet 42CrMo4 steel, and the effects of superheat, casting speed, and specific water content in the secondary cooling zone on the solidification heat transfer law were studied. The simulation results indicate that the surface temperature simulation results of the casting billet are in good agreement with the industrial measurement results, and this model can well reflect the actual situation. The casting temperature increased from 1504℃ to 1549℃, the surface temperature of the billet increased by 27℃, the center temperature of the billet increased by 42℃, the length of the liquid core increased by 1.5m, and the thickness of the crystallizer shell decreased by 5mm; The casting speed increased from 0.16m·min^{-1} to 0.22m·min^{-1}, the surface temperature of the casting billet increased by 100℃, the center temperature of the casting billet increased by 293℃, the length of the liquid core increased by 8.4m, and the thickness of the crystallizer shell decreased by 13mm; The specific water content of the secondary cooling zone increased from 0.16L/kg to 0.22L/kg, resulting in a decrease of approximately 130℃ in the surface temperature of the secondary cooling zone, a decrease of 66℃ in the center temperature of the billet, and a decrease of approximately 2min the length of the liquid core. The simulation results provide theoretical guidance for the optimization of on-site continuous casting process parameters and the determination of the end electromagnetic stirring position.

Key words: 42CrMo4; solidification heat transfer; degree of superheat; pulling speed; specific water volume

高温扩散时间对 GCr15SiMn 轴承钢碳化物带状的影响

王 强，陈 列，谢奎龙，夏伟鹤

（建龙北满特殊钢有限责任公司，黑龙江齐齐哈尔 161000）

摘 要： GCr15SiMn 轴承钢 Φ60mm 规格棒材的生产流程为 EAF-LF-RH-CC（250mm×280mm）-连轧。生产试验了连铸方坯采用 2.5~4.5h 不同的高温扩散时间、在 1200~1250℃ 的高温扩散温度下的热轧材碳化物带状的影响。结果表明，随着高温扩散时间的增加，热轧圆钢心部和 1/3R 位置的碳化物带状评级、碳化物带状宽度呈明显下降趋势，碳化物更易呈弥散形貌分布。

关键词： GCr15SiMn；碳化物带状；高温扩散

Effect of High Temperature Diffusion Time on the Banded Structure of GCr15SiMn Bearing Steel

WANG Qiang, CHEN Lie, XIE Kuilong, XIA Weihe

(Jianlong Beiman Special Steel Co., Ltd., Qiqihar 161000, China)

Abstract: GCr15SiMn bearing steel the production process of 60mm specification bars is EAF-LF-RH-CC (250mm× 280mm)-continuous rolling. Production experiments were conducted to investigate the influence of different high-temperature diffusion times of 2.5-4.5h on the carbide banding of hot-rolled steel billets at high-temperature diffusion temperatures of 1200-1250℃. The results show that with the increase of high-temperature diffusion time, the band level and width at the center and 1/3R position of the hot rolled round steel show a significant downward trend, and the carbides are more likely to exhibit a dispersed morphology distribution.

Key words: GCr15SiMn; high-temperature diffusion; carbide banded structure

感应加热下非均匀温度场对高锰钢硬化机理的探索

房启文[1]，闫学峰[2]，杨志南[1,2]，张福成[3]

（1. 燕山大学亚稳材料制备技术与科学国家重点实验室，河北秦皇岛　066004；
2. 燕山大学国家冷轧板带装备及工艺工程技术研究中心，河北秦皇岛　066004；
3. 华北理工大学冶金与能源学院，河北唐山　063210）

摘　要： 高锰钢由于其良好的韧性与优异的加工硬化能力被广泛应用于铁路、矿山、电力、建材、农机等工程领域。然而，高锰钢水韧处理后其强度和硬度低的缺点严重影响了其使用寿命。目前，预硬化是提高高锰钢使用寿命主要手段之一，高锰钢的加工硬化能力也是工程领域研究学者们关注的重点。利用高锰奥氏体钢热导率小、线膨胀系数大的特点，本文提出了一种针对高锰奥氏体钢的高效预硬化方法，探索了感应加热过程产生的非均匀温度场对高锰奥氏体钢组织和性能的影响，并利用 ANSYS 有限元分析软件的 APDL 语言建立了电磁-热-结构多场耦合的三维有限元模型，通过建立各个物理场的物理环境文件，并采用顺序耦合物理文件法实现了高碳高锰奥氏体钢感应加热到淬火过程的数值模拟。研究结果表明：感应加热过程中高锰奥氏体钢感应加热表面最高温度可达940℃，最低温度为500℃；不均匀温度场使高锰奥氏体钢内部产生了不均匀的应力应变场（峰值等效应力可达185MPa），从而使高锰奥氏体钢产生了不均匀的微观应变，最终导致高锰奥氏体钢的硬度提高，峰值硬度点硬度提高37%。由于感应加热过程时间较短，高锰奥氏体钢中并未发现碳化物，但仍可能产生脱溶分解，形成 C 原子团簇对高锰奥氏体钢起到强化作用，这也是下一步继续研究的方向。

10.4 高温合金及信息化技术

基于反向传播神经网络的高温合金蠕变曲线预测

何金珊，马博浩，王西涛

（北京科技大学，北京 100083）

摘 要：高温合金在服役过程中常因发生蠕变变形而失效，从而造成重大服役安全隐患和经济损失问题。为了解决该问题，亟需一种准确预测高温合金蠕变时间和蠕变变形的方法，避免达到蠕变失效之前对高温合金最大程度地使用。

1985 年，Evans 提出了 θ 投影法，通过描述蠕变曲线来预测材料的蠕变寿命[1]。具体来说，该方法将蠕变应变 ε 以时间 t 的函数形式描述为：$\varepsilon = \theta_1(1-\exp(-\theta_2 t)) + \theta_3(\exp(\theta_4 t) -1)$，其中 θ_1 和 θ_3 分别表示第一阶段和第三阶段蠕变的幅值，θ_2 和 θ_4 分别表示这两个阶段的拐点。尽管近年来出现了多种预测蠕变曲线的数值方法[2,3]。但 θ 投影法因其原理简单而被广泛应用。然而，θ 投影法对蠕变稳态阶段没有明确的描述，而蠕变稳态阶段在蠕变过程占据重要部分。此外，这些方法较难考虑微观组织退化对蠕变曲线的影响。因此，通过构建唯象模型很难满足蠕变曲线预测的精度要求。

本文结合 Larson-Miller 参数和反向传播神经网络（BPNN）模型构建了蠕变曲线的预测模型。研究发现，仅采用有限的蠕变曲线数据，经过 4000 次迭代训练以后，对未参与训练的蠕变曲线预测误差在±20%以内。另外，发现使用该方法对单晶、定向和等轴高温合金的蠕变曲线均获得较准确的预测，说明该模型可应用于不同类型的高温合金，具有广泛的适用性。

参 考 文 献

[1] Evans R W, Wilshire B. Creep of Metals and Alloys[C]//The Institute of Metals: London, UK, 1985.
[2] Kulkarni S S, Tabarraei A. An ordinary state based peridynamic correspondence model for metal creep[J]. Eng. Fract. Mech, 2020, 233: 107042.
[3] Li K-S, Wang R Z, Yuan G J, et al. A crystal plasticity-based approach for creep-fatigue life prediction and damage evaluation in a nickel-based superalloy[J]. Int. J. Fatigue, 2021, 143: 106031.

MES 系统在鞍钢热轧 EPS 生产线的应用

高 松，车志良，黄玉彬，张 喆，孙上海

（鞍钢集团信息产业有限公司，辽宁鞍山 114051）

摘 要：鞍钢热轧 EPS 生产线 MES 系统主要负责 EPS 生产线的基础数据管理、订单管理、质量管理、库房管理、生产管理、储运管理，同时与公司 ERP 系统、热轧 1700、1780、2150MES 系统、一级 PLC 系统进行通讯，收集和传递生产数据，保证各级系统信息的数据一致性。鞍钢热轧 EPS 线是国内第四条 EPS 生产线，生产速度快、产品规格覆盖面广、保供能力强。EPS 生产是纯物理过程，不产生有害废料，无尘、无化学残留、无有害物排放，更加环保。与传统的酸洗法相比，通过 EPS 工艺处理钢材，占地面积能够减少 50%、原材料、设备损耗折旧、人工等工作成本降低 30%，符合行业"绿色低碳 高质量发展"的主旨。

关键词：鞍钢；EPS；MES；绿色低碳

The Application of MES in the EPS Production Line of Ansteel Hot Strip Mill

GAO Song, CHE Zhiliang, HUANG Yubin, ZHANG Zhe, SUN Shanghai

(Ansteel Information Industry Co., Ltd., Anshan 114051, China)

Abstract: The MES system of Ansteel Hot Rolling EPS Production Line mainly takes charge of the basic data management, order management, quality management, warehouse management, production management, transportation management of the EPS production line. Meanwhile, it communicates with the company's ERP system, 1700, 1780, 2150 MES, and L1 PLC system, to transfers and gathers various production data, keep the information data consistency at all levels system. Ansteel hot rolled EPS line is the fourth EPS production line in China, with fast production speed, wide product specification coverage, and strong supply guarantee capacity. EPS production is a pure physical process that does not generate hazardous waste, dustless, no chemical residues, and no harmful emissions, making it more conducive to environmental protection. Compared with the traditional acid pickling method, using EPS process to treat steel can reduce the floor area by 50%, reduce the depreciation of raw materials and equipment, and reduce labor costs by 30%, which is in line with the industry's theme of "green, low-carbon, and high-quality development".

Key words: Ansteel; EPS; MES; green and low-carbon

基于神经网络的烧结工艺设计系统

陈 浩

（鞍钢集团信息产业有限公司，辽宁鞍山 114051）

摘 要：提出使用 BP 神经网络算法建立烧结工艺终点预测模型，并使用仿真手段对网络进行验证。BP 神经网络预测模型可以在较短的时间内完成训练，预测结果与期望值误差小。仿真结果表明该方法可以用于解决预测烧结终点问题，对烧结生产过程有良好的指导意义。

关键词：烧结工艺；神经网络；终点预测

Sintering Process Design Syetem Based on Neural Network

CHEN Hao

(Ansteel Information Industry Co., Ltd., Anshan 114051, China)

Abstract: This paper manages to set up a predicting model of the sintering BTP basing on the BP neural network to solve the problem of sintering BTP, and then verifies the network by simulation method. BP neural network can be trained within a relatively short period, and the error between the prediction value and the expected value is small. The simulation results show that the method has good effect on solving the problem of the prediction of BTP, therefore, it can be used for the guidance of sintering production process.

Key words: ore sintering; neural network; prediction of sintering end point

镍钴基变形高温合金电子束焊接接头的微观组织演变及力学性能的高通量表征

周海晶[1,2]，谢锦丽[2,3]，蔡世平[1,2]，董金鑫[1,2]，于鸿垚[1,2]，毕中南[2,3]

（1. 钢铁研究总院有限公司高温材料研究院，北京 100081；2. 高温合金新材料北京市重点实验室，北京 100081；3. 北京钢研高纳科技股份有限公司，北京 100081）

摘　要： 新型镍钴基高温合金是在商业合金 U720Li 基础上提高 Co、Ti、Nb 等元素含量，使其具备更高的承温能力，从而成为新一代高性能航空发动机涡轮盘材料[1]。熔焊是一种已实现商业化的构件修复技术，具有良好的应用前景。已有报道表明电子束焊接能够应用于高温合金，但是焊接过程中极快的冷却速度及温度梯度能够导致该区域内微观组织及力学性能产生较大的变化[2]。因此，系统分析该区域微观组织及力学性能的分布规律，精准定位影响接头力学性能的关键区域及薄弱环节，能够为后续热处理工艺的制定提供理论依据及数据支撑。

纳米压痕和微柱压缩是目前广泛应用的高通量力学性能分布规律测试方法[3]。然而，上述两种方法均仅能够分析合金的压缩性能，并且分析区域的变形是孤立的，与合金的实际服役条件差别较大。接近服役条件下力学性能的高通量测试仍未见报道。

本文以镍钴基变形合金 GH4251 为研究对象，利用原位扫描电镜系统（in-situ SEM），结合数字图像相关法（Digital Image Correlation，DIC）技术，定量分析了电子束焊接接头中微观变形的分布规律以及焊缝至母材区域微观组织的演变对力学性能的影响。本研究提供了一种更接近实际服役条件的高通量力学性能测试方法，并为后续热处理工艺制定提供指导。

焊缝区组织为包含枝晶组织的柱状晶，晶粒较大且无 γ′ 相析出。在热影响区，晶粒转变为等轴晶且尺寸降低，该区域能够观察到 γ′ 相和 MC 碳化物，且含量随与焊缝中心距离增加而提高（图 1（a））。母材区域为等轴晶、γ′ 相及 MC 碳化物的混合组织，析出相的尺寸及含量保持稳定。焊接接头的高通量力学性能测试表明（图 1（b）），焊缝区及母材区的杨氏模量分别为 216GPa 和 267GPa；热影响区的力学性能介于两者之间。由于热影响区内晶粒度及析出相含量等组织参数变化复杂，该区域存在与组织参数相关的力学性能变化。

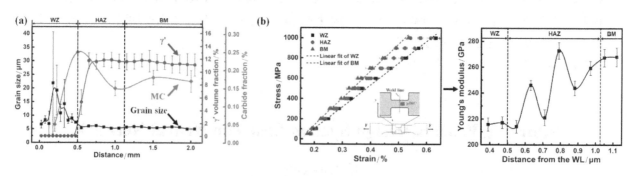

图 1　高温合金电子束焊接接头
（a）微观组织参数演变；（b）力学性能演变

参 考 文 献

[1] Zhang R, Liu P, Cui C, et al, Acta Metallurgica Sinica, 2021, 57(10): 1215-1228.

[2] Han K, Wang H, Zhan g, B et al, Journal of Materials Research and Technology, 2022(17): 1030-1042.
[3] Zhao J C. Pogress in Materials Science, 2006, 51(5): 557-631.

P92 钢高温蠕变损伤模拟试验研究

宋宏峰

（钢研纳克检测技术股份有限公司，北京 100081）

摘 要：通过 P92 钢的在 650℃不同应力水平下单轴蠕变试验，基于连续损伤力学理论，建立了 P92 钢非均匀损伤的 K-R 本构方程并通过非线性拟合的方法获得了其参数。通过自行编制用户子程序 UMAT，实现了在 ABAQUS 中对非均匀损伤的 K-R 本构的模拟，并计算了单轴蠕变试样的损伤发展情况。

关键词：单轴蠕变试验；K-R 本构方程；UMAT；损伤

The Study on the High Temperature Creep Damage of the P92 Steel

SONG Hongfeng

(NCS Testing Technology Co., Ltd., Beijing 100081, China)

Abstract: Based on the uniaxial creep tests of the P92 steel at 650℃ under various stress levels and the continuum damage mechanics, the non-uniform damage K-R constitutive equations are established and the parameters are obtained by non linear fitting. The K-R constitutive equation has been incorporated into finite element program ABAQUS through its user subroutine-UMAT and the creep damage of uniaxial creep sample is calculated.

Key words: uniaxial creep test; K-R constitutive equations; UMAT; creep damage

GH4169 合金涡轮盘残余应力演化的全过程模拟仿真

宋润华[1]，李东风[1]，秦海龙[2,3]，毕中南[2,3]，张继[2,3]

（1. 哈尔滨工业大学（深圳），广东深圳 518055；
2. 钢铁研究总院，高温合金新材料北京市重点实验室，北京 100081；
3. 北京钢研高纳科技股份有限公司，北京 100081）

摘 要：高温合金涡轮盘作为航空发动机中最重要的热端核心部件之一，在工作时承受巨大的离心惯性力和热应力的耦合作用，涡轮盘的力学性能与其制备工艺水平密切相关。GH4169 合金涡轮盘在生产过程中需要经过严格的热处理工艺处理，热处理过程中涡轮盘内部发生着复杂的温度场变化和材料微观组织演化，这会使涡轮盘内部引入不可忽略且难以预测的残余应力[1]。并且，残余应力会像涡轮盘的"遗传基因"一样，影响后续服役阶段的性能表现。因此开展涡轮盘制造-服役全过程模拟仿真研究，对阐明工艺-性能之间的映射关系、实现制备工艺优化具有重要意义。

针对这些问题，本文基于顺序热力耦合有限元方法同，开发了 GH4169 盘件制备-服役跨阶段残余应力仿真有限元模型。该模型涵盖了盘件淬火、时效、机加工和旋转蠕变四个阶段，在不同阶段分别对盘件模型施加相对应的热机械边界条件，并结合作者前期发展的材料本构模型[2]，实现了对淬火残余应力产生、时效残余应力演化、机加工残余应力再平衡、抗蠕变性能的预测。四阶段的仿真结果均与实验结果进行了验证，结果显示模型具有理想的模拟预测精度。图1为模拟方法和实验方法获得的涡轮盘时效后残余应力分布的对比。基于上述跨阶段涡轮盘残余应力仿真预测方法，本文进一步探究了不同热处理工艺对抗蠕变性能的影响。结果表明，通过有针对性的热处理工艺设计，可以起到反向施加有益的残余应力，从而改善涡轮盘抗蠕变性能的效果。

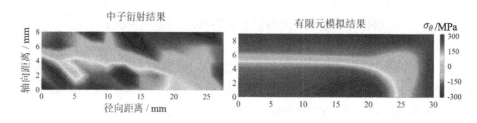

图 1 盘件时效残余应力的实验与模拟结果对比

参 考 文 献

[1] 秦海龙, 张瑞尧, 毕中南, 等, 金属学报, 2019, 55(8): 997-1007.
[2] Song R, Qin H, Li D, et al. ENG COMPUTATION, 2020, 38(1): 19-35.

选区激光熔化 TiC 增强 CM247LC 高温合金基复合材料的组织与性能

王 瑞，刘国浩，谢锦丽，毕中南

（北京钢研高纳科技股份有限公司，北京 100081）

摘　要：以选区激光熔化为代表的增材制造技术作为一种迅速发展的先进技术在精密成型镍基高温合金构件方面展现出巨大潜力，已经成为镍基高温合金复杂结构部件精密成形的一种重要的技术方案。本文针对高 γ′含量镍基高温合金 CM247LC 增材制造工艺性能差的瓶颈难题，提出控制 γ′相形成元素含量，通过外加法引入纳米碳化物颗粒，利用纳米碳化物颗粒细化晶粒并改善晶界处应力集中的作用，发展了一种既能满足选区激光熔化工艺要求又能保证力学性能的高性能碳化物增强镍基复合材料。本文通过选区激光熔化技术成功制备了高致密度的碳化物增强镍基复合材料，研究了镍基复合材料的不同状态下的微观组织及力学性能。

本文经过优化选区激光熔化工艺的参数，成功制备了以 CM247LC 合金为基体，缺陷占比可低至 1.95% 以下，内部极少裂纹等危害性较大的缺陷的碳化物增强镍基复合材料，最优的工艺参数为扫描速度 1000mm/s，激光功率 200W，层厚 30μm，扫描间距 80μm。EBSD 结果显示，TiC/CM247LC 复合材料的晶粒形态与 CM247LC 类似，都是柱状晶，其平均晶粒尺寸为 70μm，略小于 CM247LC（77μm）。同时，KAM 结果显示，TiC/CM247LC 复合材料的局域应变程度比 CM247LC，表明 TiC 具有明显的消除应力集中的作用。SEM、TEM 及 EDS 结果表明，TiC/CM247LC 复合材料的晶粒内部由柱状组织与胞状组织组成，柱状组织与胞状组织边缘为缠结的位错，大量的 TiC 颗粒沿着柱状组织与胞状组织的边缘分布。拉伸实验结果显示 TiC/CM247LC 复合材料的屈服强度比 CM247LC 高出 93MPa，表征结果证实 TiC 可以起到钉扎位错、承载应力的效果，并且由于 TiC 与基体线膨胀系数的差异提高了材料的位错密度进一步强化了合金。本研究发展了抑制高 γ′ 含量镍基高温合金开裂的新方法，为研制增材制造专用高性能高温合金提供了有益的研究思路。

机器学习和三目标优化算法辅助调控高温合金 γ/γ′ 微观组织

刘 沛 [1,2,3,4]，黄海友 [3,4]，文 成 [5]，Turab Lookman [3,6]，
宿彦京 [3,4]，秦海龙 [1,2]，谢锦丽 [1,2]

（1. 北京钢研高纳科技股份有限公司，北京 100081；
2. 高温合金新材料北京市重点实验室，钢铁研究总院，北京 100081；
3. 北京材料基因工程高精尖创新中心，北京科技大学，北京 100083；
4. 腐蚀与防护中心，新材料技术研究院，北京科技大学，北京 100083；
5. 机械工程学院，广东海洋大学，广东湛江 524000；
6. AiMaterials Research LLC，Santa Fe，New Mexico 87501，USA）

摘 要：凭借着高效的拟合和预测能力，数据驱动的机器学习技术已经广泛用于高温合金新材料的发现与改进。在材料数据量有限的客观条件下，算法在很大程度上决定了机器学习模型的收敛和寻优能力，进而影响到高性能新材料发现的效率。

在多个材料数据量都很少的情况下，多目标优化算法结合机器学习模型预测可以对高温合金的多种性能同时进行优化提升，经过少量的实验即可发现综合性能优良的新合金。目前主流的双目标优化算法有筛选法、概率提升以及基于帕累托前沿的连续改进算法等，至于同时优化更多材料性能的高效算法，目前鲜有报道。为此，有必要发展一种寻优效率更高的通用型可拓展式多目标主动学习技术用于指导高温合金的多目标性能优化。本研究中，多目标性能优化的焦点问题选择了单晶高温合金多个关键组织参量的调控。在学术界和工程界中，具有较高的 γ′ 相体积分数、较小的 γ′ 相尺寸和较为立方的 γ′ 相形貌的单晶高温合金其蠕变寿命往往较长，AM1、CMSX-4、CMSX-10 和 TMS-238 等合金都具备此类组织，而如何对此三个关键组织参量同时进行调控是单晶高温合金问世 40 年以来一直悬而未决的问题。

基于多维高斯分布，我们提出了一种多目标主动学习技术（ED-MCI），结合测试函数和机器学习模型，与现有的两种多目标优化方法进行了寻优效率的比较，结果表明 ED-MCI 算法迭代效率最高。进一步，我们将其用于高温合金 γ′ 相体积分数、尺寸和形貌的同时优化，经过 4 轮实验，合成了 3 种具有高 γ′ 相体积分数（> 54%）、小 γ′ 相尺寸（< 480 nm）和 γ′ 相形貌立方的多组元高温合金新材料（已授权发明专利）。此外，对该方法的协同优化能力、全局寻优能力和捕捉高温合金耐火元素强化机制等方面进行了有效性的论证。该方法在数学推导阶段不受材料特定问题的具体束缚，原则上可用于其他材料领域的多目标优化问题，为材料工作者从算法开发的角度解决材料问题提供了新的思路。

参 考 文 献

[1] Liu P, Huang H Y, Antonov S, et al. NPJ Comput. Mater., 2020, 6(62).
[2] Liu P, Huang H Y, Jiang X, et al. Acta. Mater., 2022, 235: 118101.
[3] Liu P, Huang H Y, Wen C, et al. NPJ Comput. Mater., 2023, accept.

基于数据与计算智能的高温合金数字化研发平台

梁 坤，王 卓

（成都材智科技有限公司，四川成都 640041）

摘 要：高温合金是航空航天领域的关键材料，随着其应用领域逐步扩展到电力、汽车等工业领域，市场对新型高温合金的需求进一步扩大。然而传统的高温合金研发以试验和经验优化，研发耗时久、费用高、效率低，如何加速高温合金材料的性能优化、设计和研发成为一个关键问题。在此背景下，利用先进的数据和计算智能技术，建立高温合金数字化研发平台来支持高温合金的研发具有重要的意义。

基于数据与计算智能的高温合金数字化研发平台集成材料数据库、高通量计算仿真、机器学习等工具和技术，以实现高温合金材料的快速设计、优化和开发。通过材料数据库技术，收集并整合来自材料手册、实验室测试、模拟计算及文献研究等多来源的高温合金成分、物理性能、化学性质、工艺参数、服役评价等高温合金全生命周期数据。

高温合金具有可通过调整热变形工艺参数和热处理工艺获得具有不同晶粒尺寸和不同性能水平的最终产品的特点，利用高通量的热力学、动力学集成计算工具，开展计算相图（CALPHAD）方法，对高温合金的组织控制以及高温服役状态下的组织演变开展机理研究，将计算结果用于优化合金配方、控制工艺窗口，并指导实验设计和测试计划。

高温合金的性能与显微组织高度相关，尤其是组织形貌、合金的成分分布、相分布。平台以数据融合手段，耦合高温合金实验数据、相场计算数据，并以机器学习手段的多尺度设计方法，建立描述高温合金性能与显微组织间的关系模型，如高温合金 γ' 相稳定性与 γ' 相固溶温度和成分、工艺定量关系模型。通过机器学习手段，利用融合数据集提取高温合金关键性能影响因子，实现新型高温合金中研发中热处理工艺和热变形工艺的高效设计。

基于数据与计算智能的高温合金数字化研发平台通过机器学习耦合高温合金热力学、动力学数据库和高通量计算相图等工具及数据，有望实现基于高温合金的组织控制以及高温服役状态下的组织演变机理研究的新型高温合金开发，最终发展基于高温合金特点的分子动力学、相图计算、相场模拟和有限元方法等多尺度计算模型耦合的方法，推动高温合金领域的创新与发展。

参 考 文 献

[1] 谢建新, 宿彦京, 薛德祯, 等. 机器学习在材料研发中的应用[J]. 金属学报, 2021, 57(11): 1343-1361.
[2] 刘兴军, 陈悦超, 卢勇, 等. 新型钴基高温合金多尺度设计的研究现状与展望[J]. 金属学报, 2020, 56(1): 1-20.

GH4151 合金高温流变行为的研究

黄科杰，黄 彬，马丹蕊，刘慧鑫，张麦仓

（北京科技大学材料科学与工程学院，北京 100083）

摘 要：GH4151 合金是典型的难变形镍基高温合金，强化相 γ' 含量接近 60%，变形阻力大，加工窗口较窄，在变形过程中发生了具有复杂力学性能和显微组织演变的塑性变形。系统研究了不同一次 γ' 相分布及不同初始晶粒度 GH4151 合金的高温流变特性及组织演化机理，进而建立了 GH4151 合金动态再结晶动力学模型。研究结果对该合金热加工过程工艺设计及进一步的组织演化动力学分析具有重要意义。

本研究分为两部分，一部分对锻态 GH4151 合金进行循环热处理使二次 γ'完全回溶，一次 γ'均匀分布，然后进行热压缩试验：变形温度（1110℃、1130℃），应变速率（0.001s^{-1}、0.01s^{-1} 和 0.1s^{-1}），变形量为 50%。另一部分对锻态 GH4151 合金进行热处理使其具有不同晶粒度，随后进行热压缩试验：变形温度（1110℃、1130℃、1150℃、1170℃），应变速率（0.01s^{-1}、0.1s^{-1}、1s^{-1}），变形量（15%、30%、50%），对两部分热压试样进行 EBSD、SEM、TEM 分析，并建立动态再结晶模型。

对一次 γ'分布与高温塑性关系的研究发现：（1）循环热处理次数增加，再结晶程度更完全，高温流变应力更小，这是由于循环热处理次数增加，一次 γ'细化，对位错的钉扎阻碍作用增强，阻碍再结晶的发生，导致合金流变应力提高[1]。（2）循环次数增加，热压缩晶粒尺寸减小。这是由于循环热处理次数增加，一次 γ'细化且钉扎于晶界，阻碍晶界运动，抑制晶粒长大。（3）相同条件下，高应变速率时，合金以非连续动态再结晶为主；应变速率降低，合金发展为 γ'相诱发动态再结晶机制；当应变速率降低到 0.001s^{-1} 时，进一步发展为亚晶或者动态再结晶。

对不同晶粒度合金高温流变行为的研究发现：（1）原始晶粒尺寸增大，合金流变应力提高，这是因为晶粒尺寸大，晶界占比下降，晶界滑移减少，应力增加；（2）应变速率提高，合金流变应力增大，是因为速率增加，晶粒来不及再结晶，加工硬化占主要作用，流变应力提高；（3）变形温度升高，流变应力降低，是因为温度升高，原子移动能力提高，位错增多，再结晶形核率提高，降低流变应力；（4）建立了动态再结晶模型如下所示：

$$\varepsilon_c = 0.58423\varepsilon_p$$

$$X_{drx} = 1 - \exp\left(-1.067\left(\frac{\varepsilon - \varepsilon_c}{\varepsilon_{0.5}}\right)^{2.552}\right)$$

$$\varepsilon_{0.5} = 0.434 d_0^{1.635} \dot{\varepsilon}^{0.07789} \exp\left(\frac{78925.47}{RT}\right)$$

$$D_{drx} = 2.66 \times 10^{12} \dot{\varepsilon}^{-0.149} \exp\left(-\frac{307663.42}{RT}\right)$$

参 考 文 献

[1] 李福林, 付锐, 白云瑞, 等. 初始晶粒尺寸和强化相对 GH4096 高温合金热变形行为和再结晶的影响[J]. 金属学报, 2023, 59(7): 855-870.

恢复热处理对单晶高温合金组织演变的影响

谭科杰[1,2,3]，王新广[3]，孟 杰[3]，梁静静[3]，周亦胄[3]，孙晓峰[3]

（1. 北京钢研高纳科技股份有限公司，北京 100081；2. 高温合金新材料北京市重点实验室，钢铁研究总院，北京 100081；3. 中国科学院金属研究所，辽宁沈阳 110016）

摘 要：单晶高温合金涡轮叶片在航空发动机中扮演了至关重要的角色，其服役环境为高温和高压的环境，并在服役过程中承受复杂载荷的作用。在叶片服役过程中，叶片的组织性能会不可避免的发生退化，最终导致叶片失效。随着叶片性能的不断提高，叶片的制备成本也越来越高。因此，人们更加注重零部件的修复技术。常用的修复技术主要分为热等静压（HIP）技术和恢复热处理（RHT）技术。HIP 可以有效修复合金内出现的蠕变孔洞、裂纹等缺陷，但对于处于损伤初期无明显蠕变孔洞的合金，RHT 技术具有热等静压技术具有工艺简单、成本低的优势。因此，对单晶高温合金的恢复热处理工艺的研究也具有十分重要的指导作用。

本课题研究了恢复热处理工艺对蠕变损伤后的二代单晶高温合金表面及其内部组织的影响。经蠕变变形后，样品内部可以观察到筏化不同程度的 γ'相以及位错网，其表面出现氧化层、γ'贫化层和胞状再结晶；经恢复热处理后，样品内部 γ'相转变为规则的立方状，其表面氧化层和 γ'贫化层依旧存在，但胞状再结晶内 γ'相转变为立方状，且在 γ'贫化层与规则的 γ'相之间出现了过渡层。恢复热处理影响区包括 γ'贫化层、再结晶和过渡区，其深度在 20μm 以内，对零件的性能影响很小。恢复热处理后样品内部具有与标准热处理后的样品相同的 γ'相体积分数和尺寸大小。进一步研究表明，组织的元素组成在蠕变-恢复热处理过程中起到了很重要的作用。不同的元素比例会导致恢复热处理后不同的组织形态，由于扩散作用形成的元素梯度导致了过渡层的形成。这一结论同时适用于有再结晶和无再结晶的样品内表面区域。

对比蠕变寿命和内部组织，证明了恢复热处理对修复单晶高温合金蠕变损伤的可行性。经恢复热处理后的样品具有与原始样品相近的蠕变寿命，样品总寿命提高了 28%~59%，样品内部损伤后的 γ'相恢复，位错网消失，证明恢复热处理工艺是有效可行的。即在合适的时机进行恢复热处理，可以将样品内部组织恢复至原始状态，从而恢复样品性能，且样品表面影响区较小，不会影响性能。

参 考 文 献

[1] Tan K J, Wang X G, Liang J J, et al. JMST, 2021 (60).

磷和硼对优质 GH4738 合金偏析特征的影响

马亚芬，马丹蕊，刘慧鑫，黄 彬，张麦仓

（北京科技大学材料科学与工程学院，北京 100083）

摘 要：GH4738 合金是一种典型的镍基时效强化型高温合金，服役温度可达 600℃以上，因其具有良好的强韧化匹配性及较低的裂纹扩展速率而被广泛应用于航空航天领域[1]。

近年来，我国针对 GH4738 合金已经做了很多方面的研究，在合金冶炼、开坯、均匀化及固溶热处理参数优化等方面均有突破，但在成分优化方向仍缺乏更深入的研究。磷和硼元素作为高温合金中常见的微合金化元素，其可以通过改善晶界析出相分布、提高晶界结合力、延缓晶界氧原子扩散以及影响 γ'强化相等方式，在很多牌号的高温合金中都发挥着重要的作用。Zhang 等人[2]研究磷对 IN706 合金力学性能的影响，发现磷以固溶态存在时可以强化晶界，提高合金的晶间抗氧化性。陈国胜等人[3]研究了 B 含量对 GH4049 晶界析出相和高温性能的影响，发现随着 B 含量升高，合金中一次 M_3B_2 增多，晶粒细化，合金的高温性能大幅上升，但 B 含量过高，M_3B_2 密集析出会导致高温性能下降。因此，针对 GH4738 合金中的磷、硼元素进行成分优化，将改善合金的微观组织，进而提高合金的综合力学性能。

本研究采用三联冶炼技术制备了不同磷（0.0040%~0.0655%，质量分数）和硼（0.0010%~0.0059%，同上）含量的优质 GH4738 铸锭，采用光学显微镜、扫描电镜及电子探针研究了磷和硼含量对优质 GH4738 铸态合金偏析特征的影响。实验结果表明，随着磷含量的增加，枝晶间距先增大后减小。磷含量在 0.0091%时对合金的铸态组织影响最大，不仅枝晶间距明显增加，枝晶组织的形貌也有显著变化。磷元素能与合金中的金属原子形成共偏析原子对，从而加重 Ti、Al、Mo 元素的偏析程度，使枝晶间的 Mo、Ti 浓度升高，因此随着磷含量的增加，碳化物的含量增多。枝晶间距随着硼含量的增加先减小后增大，硼同样会促进铸态组织中碳化物的析出，且随着硼含量的增加，块状碳化物有转变为网状碳化物的趋势。

参 考 文 献

[1] 毕中南, 曲敬龙, 邓群等. 钢铁研究学报, 2011, 23(S2): 290-293.

[2] Zhang S, Zhang A, Chang L, et al. Materials Science and Engineering: A, 2019, 761: 137981.
[3] 陈国胜, 金鑫, 周奠华, 等. 金属学报, 2005(6): 622-626.

抽拉速率及铸件结构对单晶高温合金雀斑形成的影响

王志成[1,2]，李嘉荣[1]，刘世忠[1]，王效光[1]，杨万鹏[1]

（1. 北京航空材料研究院先进高温结构材料重点实验室，北京 100095；
2. 北京钢研高纳科技股份有限公司，北京 100081）

摘 要：单晶高温合金雀斑的形成显著破坏了涡轮叶片的单晶完整性，直接导致叶片的报废。本文设计了沿定向凝固方向截面突变的阶梯铸件，通过试验和定向凝固数值模拟相结合的方法，分析了单晶高温合金铸件雀斑形成机制，研究了定向凝固抽拉速率及铸件结构对一种三代单晶高温合金雀斑形成的影响。基于本论文研究结果，完善了目前广泛使用的单晶高温合金雀斑形成瑞利数判据。

将试验结果与数值模拟结果相结合，分析了单晶高温合金雀斑形成机制，结果如图1所示。雀斑易于形成在铸件上等温线较高位置的表面，这是由于侧向散热带来糊状区等温面的倾斜，使糊状区偏析的密度较轻的液相向等温面较高位置倾斜流动，从而增加了等温面较高位置表层液相的不稳定性，使其易于发生强烈的流动，导致枝晶破碎或重熔而形成雀斑。

分析了抽拉速率对单晶高温合金雀斑形成的影响。抽拉速率的增大，一方面使一次枝晶间距减小，增大了糊状区液相的流动阻力，降低雀斑形成倾向；另一方面，如图2所示，使糊状区的倾斜程度和宽度增大，增加液相倾斜流动倾向，增大了雀斑形成倾向。抽拉速率的上述两种作用综合影响着雀斑的形成。

分析了铸件结构对单晶高温合金雀斑形成的影响。结果表明，铸件横截面尺寸的减小，减小了糊状区的倾斜程度和整体宽度，增加了糊状区液相的稳定性，从而降低了雀斑形成倾向。完善了单晶高温合金雀斑形成的瑞利数判据。在瑞利数判据中，增加了表征糊状区倾斜程度的影响因子 θ 和表征铸件尺寸的影响因子 P，弥补了未考虑糊状区形貌及铸件结构来预测雀斑形成的不足。

关键词：第三代单晶高温合金；雀斑；抽拉速率；铸件结构；雀斑形成判据

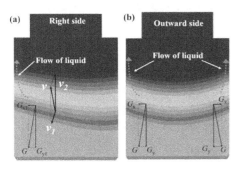

图1 倾斜糊状区液相流动分析

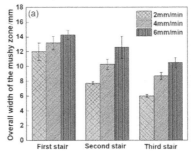 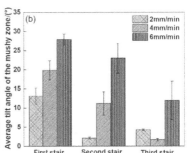

图2 抽拉速率对糊状区形貌的影响
（a）宽度；（b）倾斜程度

基于理论模型和高通量计算的镍基高温合金设计与优化

程懿歆，徐　斌，谢锦丽

（北京钢研高纳科技股份有限公司，北京　100089）

摘　要：镍基高温合金因在高温下具有优异的强度、抗疲劳、抗蠕变和抗氧化性能而广泛应用于航空航天领域[1]，近年来其增材制造技术是研究热点[2]。然而，开裂是镍基高温合金打印过程的主要问题。由于多组元化学成分的特点，传统实验方法设计新型增材制造用镍基高温合金面临着成本高、周期长和资源限制等挑战。理论模型和高通量热力学计算相结合已成为一种高效筛选和优化镍基高温合金的方法。本文采用结合理论模型和高通量热力学计算的分步式筛选方法，探索增材制造用γ′沉淀强化镍基高温合金的优化方案，综合强化效应、可焊接性、组织稳定性和密度的要求，筛选有潜力的增材制造用镍基高温合金成分。

成分空间要求 $w(Ni)\geq40\%$ 以确保 Ni 作为基体；氧化抗腐蚀性要求 $12\%\leq w(Cr)\leq20\%$；其他元素下限为 1%，上限分别为 $w(Ti)\leq7\%$、$w(Al)\leq5\%$、$w(Co)\leq20\%$、$w(Nb)\leq5\%$、$w(Mo)\leq4\%$、$w(Ta)\leq5\%$ 和 $w(W)\leq4\%$。考虑计算效率而采用 1% 的步长，成分空间共 2985735 种合金。为保证镍基高温合金的可焊接性，考虑三个指标：应变失效开裂风险系数（M_{SAC}）、凝固区间（ΔT）和裂纹敏感指数（M_{cs}）。首先，需要降低应变失效开裂风险以保证可焊接性，对关键元素 Al、Ti、Nb 和 Ta 的含量约束条件为 $M_{SAC} = [Al] + 0.5[Ti] + 0.3[Nb] + 0.15[Ta]\leq4\%$[2]；通过 Scheil 模型计算的凝固区间要求小于 260K；M_{cs} 应小于 680。组织稳定性保证 γ + γ′ 的相组成以避免 TCP 相和 η 相产生，分别要求 d 轨道能级 $M_d\leq0.915eV$ 且 $M_{γ'} = (Ti + Nb + Ta) / Al$（原子数分数/%）$\leq1$；然后基于热力学计算进行精细分析，650℃ 时要求 γ + γ′ $\geq$ 99.99%（体积分数）。镍基高温合金的应用要求是轻质高强，因此要求密度设计条件为 $\rho\leq8.40g/cm^3$；并且要求 750℃ 时 γ′体积分数（$V_{γ'}$）大于 20%（体积分数）以保证沉淀强化。

基于上述限制条件对成分空间 2985735 种合金成分进行分步筛选，过程如图 1 所示。最终保留 15 个合金，选择 Ni-12Cr-1Co-1Mo-2W-3Al-1Ti-1Nb-1Ta 作为候选合金，其具有优异的综合性能，具体表现为 $M_{SAC} = 3.95\%$（质量分数）、$M_{γ'} = 0.33$、$M_d = 0.892$ eV、$\rho = 8.24$ g/cm³、650℃ 下 γ + γ′ = 100%（体积分数）、750℃ 下 $V_{γ'} = 27.2\%$（体积分数）、凝固区间 $\Delta T = 128$ K、$M_{cs} = 627$。理论模型与热力学计算相结合的筛选方法展示出精确且高效的优势，可以弥补传统统计方法存在的明显误差，降低单一热力学计算筛选所产生的高计算成本，加速增材制造用镍基高温合金的开发。

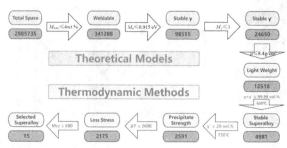

图 1　分步筛选过程

参 考 文 献

[1] Zhou Z, Zhang R, Cui C, et al. J. Mater. Sci. Technol., 2022, 115: 10-18.
[2] Tang Y T, Chinnapat P, Joseph N G, et al. Acta Mater, 2021, 202: 417-436.

基于机器学习的镍基高温合金热变形行为建模与分析

徐 斌[1]，尹海清[2]，谢锦丽[1]

（1. 北京钢研高纳科技股份有限公司，北京 100081；
2. 北京科技大学，钢铁共性技术协同创新中心，北京 100083）

摘 要：镍基高温合金因具有高强度、抗蠕变、抗疲劳和耐腐蚀的优异综合性能，成为航空发动机高温结构部件的关键材料[1]。增加航空发动机推重比要求更高的合金化以提升高温性能，然而高合金化会增加热变形加工时的失效风险。传统的材料物理模型与唯象理论模型在描述热变形行为时需要大量的实验测量参数，反复实验测试会消耗较多的时间与成本。基于数据驱动的机器学习为加速镍基高温合金研发提供了新思路。本论文利用机器学习与理论模型相结合的方法研究了镍基高温合金的热变形行为，可以快速且精确地描述不同热变形条件的流变曲线，并实现外推预测。

本论文以加速镍基高温合金热变形行为研究为目标，结合机器学习模型与流变曲线数学模型预测镍基高温合金流变曲线。通过不同热变形条件实验数据，外推预测四种已开发新型合金的流变曲线，基于预测流变曲线进一步构建四种合金的本构关系模型与动态再结晶体积分数预测模型，加速研究镍基高温合金的热变形行为。图1为基于机器学习的镍基高温合金热变形行为研究过程示意图。利用遗传算法（GA）对镍基高温合金流变曲线数学模型的关键参数进行优化，建立了流变曲线机器学习预测模型，精确获取描述流变曲线的数学模型关键参数。使用少量流变曲线数据实现了对四种合金的流变曲线外推预测。基于获取的关键参数，比较机器建模方式的精度，独立参数梯度提升树模型相较于多目标输出神经网络模型的精度提升近一倍，梯度提升树预测的型判定系数 R^2 为0.986，预测的平均绝对误差为37MPa。基于预测参数构建了四种合金的流变曲线模型，进而构建双曲正弦型Arrhenius本构关系模型以描述热变形行为。结合动态再结晶理论模型与机器学习流变曲线所获得的参数，对四种合金不同热变形条件的动态再结晶体积分数进行预测，预测结果分别为81.1%、79.5%、32.3%和92.1%，而实验结果分别为67.5%、74.6%、25.0%和89.1%，预测的平均绝对误差为7.2%，证明基于机器学习的流变曲线可以预测镍基高温合金动态再结晶。将机器学习模型与有限元模型进行对比，证明了机器学习在流变曲线外推预测上具有优势。机器学习可以为镍基高温合金的热加工提供参考。

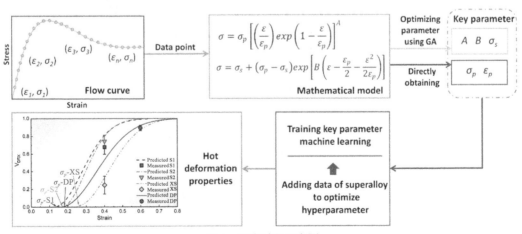

图1 研究过程示意图

参考文献

[1] Xu B, Yin H Q, Jiang X, et al. J. Mater. Sci. Technol, 2023, 155: 175-191.

选区激光熔化高温合金构件变形和内部残余应力的数值模拟

孙志民

(北京钢研高纳科技股份有限公司，北京 100081)

摘 要：选区激光熔化（SLM）是目前金属增材制造技术中发展最成熟、应用最广泛的技术之一。在选区激光熔化过程中，粉末在高能激光束作用下快速熔化后快速凝固，温度梯度极高，且局部热输入造成温度场分布不均，沉积层不同部位物相变化存在先后差异，造成选区激光熔化成形件内部存在高残余应力。SLM 过程中形成的残余应力会极大地影响成形件的力学性能，导致成形件产生几何变形，甚至造成成形件的开裂。对于增材制造构件的变形和应力的准确预测有助于增材制造工艺的优化、后续工艺的选择以及成本的降低。

本文对材质为 IN738LC 的薄壁（外径 240mm，壁厚 5mm，高 70mm）和厚壁（外径 240mm，壁厚 50mm，高 70mm）圆环的 SLM 成形和去基板过程进行了研究。以热-结构耦合的方式建立了描述 SLM 成形→去基板过程的数学模型。通过数值模拟，获得了两个构件在不同工艺参数下成形和去基板后的变形数据以及应力分布情况。为了验证模型的有效性，实际打印了一组工艺参数下的薄壁和厚壁两个构件，对两个构件不同高度处的径向变形情况进行了测量，并通过轮廓法对厚壁构件去基板后的中心纵剖面上的残余应力分布进行了测量。将模拟值与实验值进行了比对，发现变形的模拟值与实验值符合的较好。模拟的残余应力分布与实验得到的分布情况近似，最大压应力和最大拉应力符合的较好，证明了模型的有效性。通过对比不同打印参数下的构件变形数据和残余应力的分布变化，揭示了打印参数对几何变形和残余应力的影响规律。对于不同的扫描功率，随着扫描功率的增加，变形量增大。对于薄壁构件，随着扫描功率的增大，应力逐渐增大。对于厚壁构件，其应力并不与扫描功率成正比，外层的拉应力随着扫描功率的增大而增大，但内部的压应力随着扫描功率的增大而减小。

γ'相与孪晶交互作用对新型 Ni-Co 基高温合金的拉伸性能影响规律

段继萱[1,2]，安 腾[1,2]，谷 雨[1,2]，于鸿垚[1,2]，杜金辉[1,2]，毕中南[1,2]

(1. 高温合金新材料北京市重点实验室钢铁研究总院，北京 100081；
2. 北京钢研高纳科技股份有限公司，北京 100081)

摘 要：镍基高温合金由于其优异的高温强度、抗蠕变、疲劳寿命和耐腐蚀[1-2]性能，已广泛应用于航空发动机热端部件的制造，比如涡轮盘和叶片等。目前有研究表明微纳孪晶可以显著提高超合金的强度，不影响延性。目前基于低 SFE 设计一种新型 Ni-Co 基高温合金，该合金在变形过程中容易产生微孪晶，实现了合金在高温条件下的高强度和延展性。涡轮盘高温合金的拉伸性能是其最重要的性能之一，晶粒尺寸和 γ'相影响镍基高温合金在高温条件下的拉伸性能。因此，可以通过调控不同的热处理条件来优化合金的拉伸性能。本研究主要关注 1090~1130℃ 的 6 个固溶温度对该合金微观结构的影响，以及拉伸性能改变的主要原因。

随着固溶温度的升高，合金微观组织从以一次 γ′ 相为主的细晶粒组织演变到一次 γ′ 相和微孪晶共存的过渡区组织，再到以微孪晶为主的粗晶粒组织。随着固溶温度增加，晶粒尺寸增加，一次 γ′ 相逐渐溶解，二次 γ′ 相尺寸增加。

如图 1 所示，随着固溶温度增加，合金的室温拉伸强度逐渐下降，但 750℃ 拉伸强度保持稳定。该合金的室温抗拉强度随固溶温度的升高而降低。该合金的室温抗拉强度主要来自位错剪切 γ′ 沉淀物（APB 机制）。与室温拉伸试验相比，在 1120℃ 下处理的高温合金在 750℃ 拉伸试验中拉伸强度最高，一次 γ′ 沉淀物溶解，高温合金晶粒度增大，这表明，在 750℃ 下进行的拉伸试验中，高温合金的微观变形机制发生了变化，以补偿合金晶粒生长导致的强度降低。在 750℃ 下进行的拉伸试验中，高温合金的微结构中存在大量的微孪晶和层错，表明，高温合金的变形机理从 APB 剪变为微晶和 SFs 剪变。

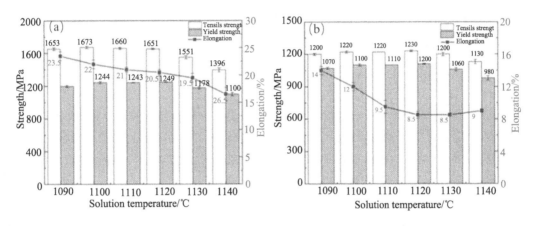

图 1 合金拉伸性能随固溶温度变化趋势
(a) 室温拉伸；(b) 750℃ 拉伸

参 考 文 献

[1] Reed R C. The Superalloys: Fundamentals and Applications[M]. Cambridge University, 2006: 123-136.
[2] Gu Y, Harada H, Cui C, et al. New Ni—Co-base disk superalloys with higher strength and creep resistance[J]. Scripta Mater, 2006, 55(9): 815-818.

航空航天用高温钛合金及钛基（Ti-Al）材料研究概况及应用

张天馨[1]，岳 颖[2]，于继洋[1]，黄召阔[1]，滕艾均[1]

(1. 鞍钢集团北京研究院有限公司钒钛研究院，北京 102200；
2. 成都先进金属材料产业技术研究院股份有限公司，四川成都 610300)

摘 要：概述了航空航天用高温钛合金（Ti-Al-Sn-Zr-Mo-Si 系）及钛基材料（Ti-Al 系）发展历程，分析其工艺和应用技术特点，概述高温钛合金及高温 Ti-Al 基材料优势及应用局限性，并对目前高温钛合金及 Ti-Al 基材料未来研究趋势进行研判。

关键词：航空航天；高温材料；Ti-Al-Sn-Zr-Mo-Si；Ti-Al

Abstract: The development history of high temperature titanium alloys (Ti-Al-Sn-Zr-Mo-Si) and titanium base materials (Ti-Al) for aerospace applications was summarized, and their process and application characteristics were analyzed. The advantages and application limitations of high temperature titanium alloys and high temperature Ti-Al base materials were summarized, and the current future research trends of high temperature titanium alloys and Ti-Al base materials were evaluated.

Key words: aerospace field; high temperature material; Ti-Al-Sn-Zr-Mo-Si; Ti-Al

选区激光熔化制备宏观金属构件应力演化研究

孙志民[1]，孔豪豪[1,2,3]，秦海龙[1]，谢锦丽[1]，毕中南[1]

（1. 北京钢研高纳科技股份有限公司，北京 100081；2. 钢铁研究总院有限公司，北京 100081；
3. 中国钢研科技集团有限公司数字化研发中心，北京 100081）

摘　要： 选区激光熔化工艺作为一种先进成形方法，因可利用金属粉末直接制备得结构复杂的金属构件而备受关注[1]。然而，该技术在工业应用时存在较大挑战，主要体现在材料-工艺-结构适配不合理时，构件内部易温度分布不均匀而累积较高水平的残余应力，这将大大增加构件发生变形和开裂的风险，且风险随构件尺寸增加而非线性递增，严重限制选区激光熔化技术的发展与应用[2]。

高水平的残余应力主要与材料特性、扫描策略、构件结构等要素相关[3]，若通过传统的"试验法"构建各要素与残余应力的映射关系时，需要进行高成本且费时费力的试错实验，难以明晰要素对应力演化的影响，并较难形成通用性的解决规范，而采用数值仿真手段可高保真的复现工艺过程关键控制参数与应力场之间的映射关系，有利于实现残余应力的预测与控制，因而在学术界与工业界备受关注。

目前国内外研究者主要通过固有应变法和超层模拟法对宏观构件应力演化进行模拟，借助这两种方法研究者们在残余应力预测与控制方面取得了较多的研究成果[4]，但是这两种方法仍存在一定的局限性，如固有应变法需通过实验数据驱动获得等效应变因子，在应用等效应变因子对构件整体应力分布预测时存在一定的误差且无法捕捉熔化、凝固等细节对应力分布的影响，而超层模拟虽能够以多个物理层组成的超层为基本单元，考虑顺序激活超层时结构效应对构件应力分布的影响，但忽略了扫描策略对超层温度均匀性的影响，这在一定程度上影响了残余应力的预测效果。

基于此背景，本文在考虑扫描策略、结构效应叠加影响的基础上，通过耦合有限元技术、生死单元法和等效微区热源模型，对In718高温合金"L"型构件制备过程中温度场及应力场进行预测，明晰了L型构件打印过程等效应力及三向应力的演化规律，为保障金属构件材料-工艺-结构的合理适配，提高构件的成型率提供参考。其中，模型的可靠性通过文献报道的中子衍射数据进行了验证[5]。

参 考 文 献

[1] McNamara K, Ji Y, Lia F, et al. Addit Manuf, 2022, 49: 102478.
[2] Zhang B, Li Y, Bai Q, Chin. J. Mech. Eng, 2017, 30: 515-527.
[3] Dive V, Lakade S. Mater. Today, 2021, 47: 3282-3287.
[4] Bayat M, Klingaa C G, Mohanty S, et al. Addit Manuf, 2020, 36: 101508.
[5] Pant P, Proper S, Luzin V, et al. Addit Manuf, 2020, 36: 101501.

VIM+VAR 双联工艺熔炼 GH2901 合金的控制实践

柳金瑞,马志伟,苏爱民

(张宣科技特材公司,河北宣化 075100)

摘 要:特材公司采用真空感应炉(VIM)+真空自耗炉(VAR)双联工艺开发生产 GH2901 合金,熔炼过程工艺参数运行稳定,铸锭化学成分精准内控,高熔点元素 Mo 偏析改善,Pb、Sn、Sb 等低熔点易挥发有害杂质元素和夹杂物数量减少,氧、氢、氮平均质量分数分别降至 5ppm、0.45ppm、19.5ppm,去气效果非常好。成功试制出 6 支成分均匀、纯净度高、宏微组织良好的高温合金铸锭,通过了用户的严格评价。

关键词:VIM;VAR;成分控制;气体

Control Practice of Smelting GH2901 Alloy by VIM+VAR Duplex Process

LIU Jinrui, MA Zhiwei, SU Aimin

(Zhang xuan Technology Special Materials Company, Xuanhua 075100, China)

Abstract: Special Materials Company develops and produces GH2901 alloy by the dual process of vacuum induction furnace (VIM) and vacuum consumable furnace (VAR), and the melting process is a process.The operation of process parameters is stable, the chemical composition of ingot is accurately controlled, the segregation of high melting point element Mo is improved, the number of volatile and harmful impurity elements and inclusions with low melting points such as Pb, Sn and Sb is reduced, and the average mass fractions of oxygen, hydrogen and nitrogen are reduced to 5ppm, 0.45ppm and 19.5ppm respectively, and the degassing effect is very good.Six high-temperature alloy ingots with uniform composition, high purity and good macro-and micro-structure were successfully trial-produced, which passed the strict evaluation of users.

Key words: VIM; VAR; composition control; gas

基于孪晶强化机理的镍钴基高温合金晶体塑性有限元模型

邓锐杰[1],甘 斌[2],李东风[1]

(1. 哈尔滨工业大学(深圳)理学院,广东深圳 518055;
2. 钢铁研究总院高温材料研究所,北京 100081)

摘 要:高温合金作为航空涡轮发动机和燃气轮机等动力系统热端部件的不可替代材料,在使役过程中处于复杂的热力耦合环境,具有复杂的多场耦合特征,其宏观力学行为与材料微纳尺度的缺陷组织紧密相关。研究表明[1],微孪晶化是 FCC 结构合金中一种重要的变形机制，具有低层错能的镍钴基高温合金在加工硬化过程中会形成大量的纳米孪晶，在合金基体中形成大量孪晶界阻碍位错运动，从而达到合金强化的目的。因此，发展滑移和孪生协同变形的力学本构模型，可以帮助研究者们深入理解低能合金纳米缺陷的跨尺度力学调控机制。

晶体塑性有限元法（CPFEM）在近几十年得到飞速发展，其中一个重要优势在于可以求解复杂的内外部边界条件下的细观力学响应，处理晶粒间和晶粒内部的边界效应，合理解释组织结构界面上的界面力学和尺度效应[2]。本文基于有限变形率相关粘弹塑性理论[3]，从位错动力学的角度阐述低能界面的强化机制，在晶体塑性本构模型中引入滑移和孪生的热激活模型来描述位错在滑移面内的连续剪切导致的非弹性变形和孪晶形核生长引起的塑性变形，发展变形阻力等内变量的演化模型形成完备自洽的本构模型。通过隐式离散格式的本构方程建立用户材料子程序来实现所构建的力学模型，如图 1(a)所示，使用三维重构技术建立了能表征材料晶粒性质的代表体积元模型（RVE），借助商业有限元软件求解不同边界条件下的控制方程，可以合理预测低能合金高温合金在不同加载条件下的力学响应（如图 1(b)所示，典型低能合金 MP159 的数值模拟结果与实验数据高度吻合）和缺陷演化（图 1(c)为MP159 常温拉伸下的孪晶体积分数演化的预测结果）。

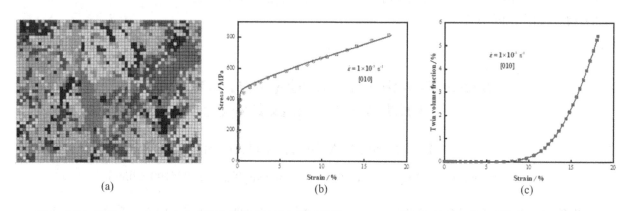

图 1　RVE 模型(a); MP159 合金室温拉伸实验数据与模拟结果(b); MP159 室温拉伸孪晶体积分数变化模拟结果(c)

参 考 文 献

[1] Barba D, Alabort E, Garcia-Gonzalez D, et al. Int. J. Plast., 2018, 105: 74-89.
[2] Roters F, Eisenlohr P, Hantcherli L, et al. Acta Mater., 2010, 58(4): 1152-1211.
[3] Li D F, O'Dowd N P. Mech J. Phys. Solids., 2011, 59(12): 2421-2441.

机器学习辅助增材制造合金内部冶金缺陷调控研究

贵云玮，付华栋

（北京科技大学材料科学与工程学院，北京　100083）

摘　要：在粉末床熔融增材制造过程中，影响合金成形质量的工艺参数众多，内部冶金缺陷时有发生，并成为拉伸和疲劳裂纹源区，降低服役性能。如何对增材制造合金内部冶金缺陷进行精准调控，进而精准指导高力学性能合金的制备已成为目前急需解决的科学难题。在这项研究中，我们根据表面粗糙度确定了表面质量分类的量化标准，并发现不同的表面质量（均匀、凸起和多孔）对应不同类型的内部缺陷。表面凸起的合金由于能量输入过多而产生许多球形气孔，而表面多孔的合金由于能量输入不足而产生大量形状不规则的缺陷和未熔化的粉末。此外，多种机器学习技术被应用于合金工艺参数优化，其中，支持向量机的模型性能最高，我们利用它构建了与内部冶金缺陷相对应的加工图，并确定了无缺陷样品的工艺窗口。本研究还提出了构建粉末床熔融增材制造合金工艺窗口的新框架，可有效加速制造无内部冶金缺陷合金。本研究提出的方法可以作为一种强有力的工具，有效优化粉末床熔融增材制造工艺的高维参数，从而生产出具有优异力学性能的高温合金、碳钢等材料。

关键词：增材制造；机器学习；内部冶金缺陷

Machine Learning-assisted Regulation of Internal Metallurgical Defects in Additively Manufactured Alloys

GUI Yunwei, FU Huadong

(School of Materials Science and Engineering, University of Science and Technology Beijing, Beijing 100083, China)

Abstract: In the process of powder bed fusion additive manufacturing, there are many process parameters affecting the forming quality of alloys, and internal metallurgical defects occur from time to time and become the source area of tensile and fatigue cracks, which reduce the service performance. How to accurately regulate the internal metallurgical defects of additively manufactured alloys, and then accurately guide the preparation of alloys with high mechanical properties has become an urgent scientific challenge. In this study, we determined quantitative criteria for surface quality classification based on surface roughness and found that different surface qualities (even, uneven and porous) corresponded to different types of internal defects. Alloys with uneven surfaces produce many spherical pores due to excessive energy input, while alloys with porous surfaces produce many irregularly shaped defects and unmelted powder due to insufficient energy input. In addition, a variety of machine learning techniques have been applied to optimize the alloy process parameters, among which, support vector machine has the highest model performance, and we have used it to construct processing maps corresponding to internal metallurgical defects and to determine the process window for defect-free samples. This study also proposes a new framework for constructing process windows for powder bed melt additive manufacturing alloys, which can effectively accelerate the fabrication of alloys without internal metallurgical defects. The method proposed in this study can be used as a powerful tool to effectively optimize the high-dimensional parameters of the powder bed melt additive manufacturing process to produce high-temperature alloys, carbon steels and other materials with excellent mechanical properties.

Key words: additive manufacturing; machine learning; internal metallurgical defects

新型高强殷钢材料的创新研发及应用

孙中华

(河钢材料技术研究院中试试验中心，河北石家庄　052165)

摘　要：高强殷钢丝新材料是制作倍容量输电导线的关键核心材料。本文简要介绍了高强殷钢新材料的发展历程，并以倍容量输电导线为例，详细阐述了殷钢材料的强化措施。进一步介绍了河钢集团利用自主创新技术，在行业内首次提出"纳米尺度第二相颗粒复合析出强化殷钢"新思路，研制出了具有自主知识产权的新型高强殷钢丝材料。最后，展望了本专业领域亟需解决的重要难题和未来的重要发展方向。

关键词：殷钢；倍容量输电导线；高强度；低膨胀

Innovative Research and Application of New-typed High-strength Invar Alloy

SUN Zhonghua

(Pilot Test Center, HBIS Group Materials Technology Research Institute, Shijiazhuang 052165, China)

Abstract: The high-strength invar wire is the key core material that is made of the double-capacity transmission wire. This article briefly introduce the developed progress of high-strength invar materials and take the double-capacity transmission wire as an example to elaborate the strengthened measurement of invar materials. Furthermore, it is introduced that the HBIS Group make use of the self-innovative technology that the firstly proposed "nanosized second-phase particles composite precipitation strength invar" new method in the field to produce the novel high-strength invar wire with the full intellectual property. Finally, it is expected in the area of high-strength invar materials that the important puzzled problem that an urgent need to solve and the developing direction in the future.

Key words: invar alloy; double-capacity transmission wire; high-strength; low thermal expansion

10.5 低合金钢

Ce 对高品质风电用钢组织性能的影响

宋成浩[1]，张振山[1]，于 浩[2]

（1. 东莞理工学院，广东东莞 523808；2. 北京科技大学，北京 100083）

摘 要：大力发展清洁能源可以逐步改变传统能源消费结构，减小对能源进口的依赖度，降低环境代价，对保障国家能源安全具有重大意义。风电已成为国际上公认的技术最成熟、开发成本最低、最具发展前景的清洁能源，大力发展风电是推进我国能源结构调整的重要举措。随着风电机组朝着大型化、高效率的方向发展，对风电用钢的力学性能提出了更高的要求。本工作主要围绕微量的稀土 Ce 对低碳低合金钢夹杂物状态、组织状态及位错状态的影响来展开。加入 50ppm 的稀土 Ce 使钢中长条状 MnS、块状 Al_2O_3 和棱角状尖晶石转变为细小、弥散分布的粒状 Ce_2O_2S。50ppm 的稀土 Ce 可使晶粒尺寸从 17 μm 降至 10 μm。加入 50ppm 的稀土 Ce，可使铁素体晶界发生 Ce 偏聚，其为基体 Ce 含量的 30-40 倍，且珠光体内部也存在 Ce 的富集。然后通过 XRD 和中子衍射技术分析 Ce 对 BCC 中位错状态的影响，可以看出 Ce 的添加增加了 BCC 中的位错密度及螺位错含量。然而 Ce 对位错密度的影响很小，因此三种钢的屈服强度没有因为位错密度的差别而有显著影响。但是，50Ce 的试样比不加稀土的螺位错提高了近一倍。螺/刃位错的迁移速率比是控制韧脆转变及提升低温韧性的决定性因素，稀土除了净化钢液、改性夹杂物状态、优化组织外，另一重要作用就是极微量的稀土即可改变位错类型，其对低温韧性具有至关重要的影响。

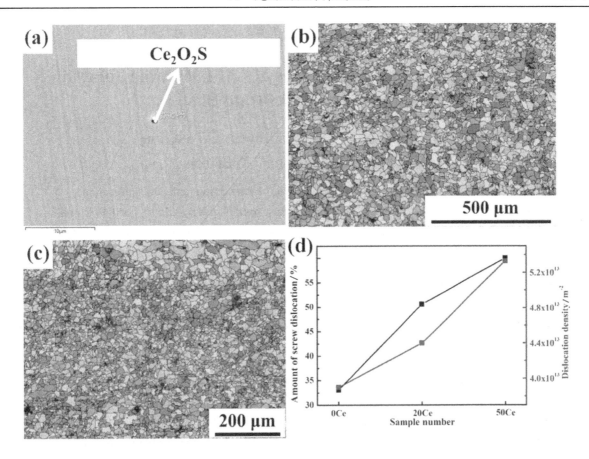

图1 代表性夹杂物图像（a）；不加稀土的 EBSD 图像（b）；加入 50ppm 稀土 Ce 的 EBSD 图像（c）；
通过 XRD 及中子衍射得到的不同稀土含量钢的位错状态曲线（d）

NM450 耐磨钢 CMT 焊接接头微观组织与力学性能

张志强[1,2]，张露云[3]，王佳骥[1,2]，胡奉雅[1,2]，苗 隽[4]，魏 勇[1,2]

（1. 海洋装备用金属材料及其应用国家重点试验室，辽宁鞍山 114009；2. 鞍钢集团钢铁研究院，辽宁鞍山 114009；3. 辽宁科技大学材料与冶金学院，辽宁鞍山 114051；
4. 本钢技术研究院，辽宁本溪 117000）

摘 要：本文采用 ER50-6 焊丝对 6mm 的 NM450 钢板分别进行冷金属过渡焊接（CMT）和熔化极混合气体保护焊（MAG），并对上述两种工艺获得的焊接接头的微观组织和力学性能进行了系统分析。结果表明：在 CMT 和 MAG 模式下均可获得成形良好的接头，CMT 和 MAG 接头热影响区（HAZ）经历不同程度的热循环，HAZ 均可分为粗晶区（CGHAZ）、细晶区（FGHAZ）、临界区（ICHAZ）和亚临界区（SCHAZ），CMT 和 MAG 接头的焊缝（WM）均由针状铁素体和块状铁素体组成，CMT 模式下焊缝的组织更为细小。CMT 双面焊、MAG 双面焊和 MAG 单面焊接头的抗拉强度分别为 720MPa、664MPa 和 717MPa，接头的断裂位置均位于 WM 处。CMT 模式下的熔滴温度较低，降低了 WM 的温度梯度，形成了更多细小等轴的铁素体组织，进而使 CMT 接头力学性能提高。

关键词：NM450 耐磨钢；CMT 焊接；MAG 焊接；微观组织；力学性能

Microstructure and Properties of CMT Welding Joint of NM450 Wear-resistant Steel

ZHANG Zhiqiang[1,2], ZHANG Luyun[3], WANG Jiaji[1,2], HU Fengya[1,2], MIAO Jun[4], WEI Yong[1,2]

(1. State Key Laboratory of Metal Material for Marine Equipment and Application, Anshan 114009, China; 2. Ansteel Iron & Steel Research Institutes, Anshan 114009, China; 3. School of Materials and Metallurgy, University of Science and Technology, Anshan 114051, China; 4. Technology Research Institutes of Benxi Iron & Steel Group Co., Ltd., Benxi 117000, China)

Abstract: In this paper, cold metal transition (CMT) and metal active gas (MAG) arc welding technology were performed on the 6mm NM450 plates with ER50-6 welding wire, and the microstructure and mechanical properties of the welded joints obtained by the above two welding processes were systematically analyzed. The results indicate that well-formed joints can be obtained under both CMT and MAG modes. The heat-affected zone (HAZ) of CMT and MAG joints experiences different degrees of thermal cycling, and the HAZ can be divided into coarse-grained HAZ (CGHAZ), fine-grained HAZ (FGHAZ), inter-critical HAZ (ICHAZ) and sub-critical HAZ (SCHAZ). The welding metal (WM) of CMT and MAG joints were composed of acicular ferrite and massive ferrite, and the microstructure of WM in the CMT joint is finer. The tensile strength of CMT double-sided welding joint, MAG double-sided welding joint, and MAG single-sided welding joints were 762MPa, 664MPa, and 717MPa, respectively. The fracture position of joints were all located at the WM. The droplet temperature in CMT mode is lower, which reduces the temperature gradient of WM and forms more fine equiaxed ferrite structures, thereby improving the mechanical properties of CMT joints.

Key words: NM450 wear-resistant steel; cold metal transfer welding; metal active gas arc welding; microstructure; mechanical property

1400MPa级在线淬火型热轧超高强钢的研究及应用

胡学文，彭 欢，王承剑，王海波，石东亚

（马鞍山钢铁股份有限公司技术中心，安徽马鞍山 243000）

摘 要： 利用金相显微镜、扫描电镜、透射电镜、电子拉伸试验机、布氏硬度计以及湿式橡胶轮试验机等，研究了1400MPa级在线淬火型热轧超高强钢的组织、力学性能、冷弯性能、布氏硬度和耐磨性能。结果表明，1400MPa级热轧超高强钢的显微组织为马氏体+铁素体，马氏体体积分数≥95%，铁素体体积分数≤5%，典型力学性能为屈服强度≥1100MPa，抗拉强度≥1450MPa，屈强比≥0.80，延伸率 A_{50}≥10%，布氏硬度≥450HBW，冷弯性能90°，D=2a 合格，具有高强度、高耐磨、高屈强比、易成形特点；同时，1400MPa热轧超高强钢的相对耐磨性分别是500MPa、700MPa、1000MPa级高强钢的3.32倍、2.67倍、1.61倍，是耐磨钢NM450的0.97倍，耐磨性能和NM450基本相当，适用于对材料耐磨性能要求较高的承载件制造。

关键词： 1400MPa；在线淬火；热轧超高强钢

Research and Application on 1400MPa Gade Hot-rolled Utra-high Strength Seel Poduced by On-line Qenched

HU Xuewen, PENG Huan, WANG Chengjian, WANG Haibo, SHI Dongya

(Technology Center, Maanshan Iron and Steel Co., Ltd., Maanshan 243000, China)

Abstract: Microstructure, mechanical properties, cold bending performance, Brinell hardness and abrasive wear behavior of 1400MPa grade hot-rolled ultra-high strength steel, which produced by on-line quenched, were studied by means of OM, SEM, TEM, electronic tensile tester, Brinell hardness tester and wet rubber rimmed wheel tester. The results show that microstructure of 1400MPa grade hot-rolled ultra-high strength steel is martensite and ferrite, with the volume fraction of martensite and ferrite is more than 95% and less than 5% respectively. It has high strength, good wear resistance, high yield ratio, well cold bending performance , and its typical mechanical properties are yield strength ≥1100MPa, tensile strength ≥1450Mpa, yield strength ratio≥0.80, elongation A_{50} ≥ 10%, Brinell hardness≥450 HBW, cold bending performance under the condition of 90°, D=2a is qualified. Meanwhile, the relative wear resistance of 1400MPa hot-rolled ultea-high strength steel is 3.32 times, 2.67 times, and 1.61 times of that of 500MPa, 700MPa, and 1000MPa grade high-strength steel, respectively, and 0.97 times of that of wear-resistant steel NM450. The wear resistance of 1400MPa hot-rolled ultra-high strength steel is basically equivalent to NM450, making it suitable for the manufacturing of load-bearing parts which require better wear resistance of materials.

Key words: 1400MPa; on-line quenched; hot-rolled ultra-high strength steel

Q345qENH 耐候桥梁钢在模拟不同服役环境下的腐蚀行为

高 鹏[1,2]，陈义庆[1,2]，李 琳[1,2]，钟 彬[1,2]，艾芳芳[1,2]，
苏显栋[1,2]，伞宏宇[1,2]，沙楷智[1,2]，张圣洁[1,2]

（1. 海洋装备用金属材料及其应用国家重点实验室，辽宁鞍山 114009；
2. 鞍钢集团钢铁研究院 焊接与腐蚀研究所，辽宁鞍山 114009）

摘 要：通过周期浸润腐蚀实验、X 射线衍射、超景深显微镜等手段研究了 Q345qENH 耐候桥梁钢在模拟工业大气和田园大气服役环境条件下的腐蚀行为。研究结果表明，Q345qENH 耐候桥梁钢在模拟乡村大气条件下腐蚀速率比模拟工业大气条件下的腐蚀速率低一个数量级左右；在模拟工业大气和乡村大气条件下 Q345qENH 耐候桥梁钢腐蚀动力学符合"双对数规律"，在模拟乡村大气条件下 Q345qENH 耐候桥梁钢表面更早生成保护性锈层；Q345qENH 耐候桥梁钢腐蚀产物 Fe_3O_4 在模拟工业大气下更易向其他物相转化，而在模拟乡村大气条件下 γ-FeOOH 更早生成；腐蚀过程中后期，模拟两种服役环境下腐蚀产物均为 α-FeOOH 和 γ-FeOOH。

关键词：耐候桥梁钢；工业大气；田园大气；腐蚀行为

Corrosion Behavior of Q345qENH Weathering Bridge Steel in Different Simulated Service Environments

GAO Peng[1,2], CHEN Yiqing[1,2], LI Lin[1,2], ZHONG Bin[1,2], AI Fangfang[1,2], SU Xiandong[1,2], SAN Hongyu[1,2], SHA Kaizhi[1,2], ZHANG Shengjie[1,2]

(1. State Key Laboratory of Metal Material for Marine Equipment and Application, Anshan 114009, China;
2. Iron & Steel Research Institute of Angang Group, Anshan 114009, China)

Abstract: The corrosion behavior of Q345qENH weathering bridge steel in a laboratory simulated industrial and rural atmosphere was investigated via means of cyclically alternative immersion-drying accelerated corrosion test, as well as X-ray diffraction, ultra depth of field microscope. The results show that the corrosion rate of Q345qENH weathering bridge steel in simulated rural atmospheric environment was about one order of magnitude smaller than that for the corrosion rate in simulated industrial atmospheric environment. The corrosion kinetics of Q345qENH weathering bridge steel accorded with a double logarithmic law. The protective rust layer on the surface of Q345qENH weathering bridge steel was produced earlier in rural atmospheric conditions. The corrosion product Fe_3O_4 was more easily converted to other phases in simulated industrial atmospheric environment. In simulated industrial atmospheric environment γ-FeOOH was generated earlier. At the later middle and stage of corrosion in the simulated two service environment corrosion products were compoced of α-FeOOH and γ-FeOOH.

Key words: weathering bridge steel; industrial atmosphere; rural atmosphere; corrosion behavior

海洋环境下 Q690 高强钢对接焊缝疲劳性能试验研究

魏欢欢 [1,2,3]

（1. 东南大学土木工程学院，江苏南京 211189；2. 东南大学江苏省工程力学分析重点实验室，江苏南京 211189；3. 西安理工大学土木建筑工程学院，陕西西安 710048）

摘 要： 相比普通建筑钢材而言，高强钢（指名义屈服极限不低于460MPa的钢材）具有承载力高、韧性好、可焊接性和耐腐蚀性等特点，在近海及海岸建、构筑物中，采用高强度钢材能够有效减小构件截面与焊缝尺寸、降低成本，提高海浪作用下结构的疲劳强度，广泛应用于海洋工程结构领域内[1,2]。长期服役于浪花飞溅区环境下，面临高氯离子、干湿循环下的海水冲蚀影响，加之海上波浪荷载的往复作用，材料内部存在较大的累积损伤量，当损伤累积超过临界状态时，材料表面开始萌生裂纹，承重钢构件最终将会引发腐蚀疲劳断裂，失效行为具有危害性和普遍性[3]。为此，开展海洋环境浪溅区下钢材腐蚀机理研究，分析疲劳性能退化规律尤为重要。

通过模拟海洋浪溅区环境，进行了锈蚀 Q690 高强钢对接焊缝疲劳荷载试验，对比不同周期 S-N 曲线，分析了腐蚀损伤对高强钢对接焊缝疲劳性能影响，基于损伤理论研究了试件的力学性能退化规律。结果表明：随着腐蚀周期增加，对接焊缝处分布较多锈蚀产物，金属光泽逐渐变暗，腐蚀 100d 的锈蚀率为 8.46%，焊缝区和热影响区平均腐蚀深度速率分别为 1.34mm/a、1.46mm/a，在高周循环加载时，疲劳极限值降低 32.7%。根据疲劳断口扫描结果可知，焊缝缺陷与腐蚀损伤对于裂纹扩展影响较大。此外，损伤曲线能够反映断口形成机理，处于高应力水平下，疲劳条带数量减少，损伤累积量逐渐提高。研究结果对 Q690 高强钢的推广及实际工程应用具有重要意义。

关键词：Q690 高强钢；海洋环境；对接焊缝；疲劳性能；腐蚀损伤；疲劳断口

参 考 文 献

[1] Gong K, Wu M, Liu G X. Constr. Build. Mater., 2020, 235: 1-12.
[2] Guo H C, Wan J H, Liu Y H, et al. Thin Wall. Struct., 2018, 131: 45-54.
[3] Gkatzogiannis S, Weinert J, Engelhardt I, et al. Int. J. Fatigue, 2019, 126: 90-102.

易焊接煤矿液压支架用钢的研发及推广应用

衣海龙[1]，刘少华[1]，郑中洪[1]，黄 重[2]，陈尹泽[2]

（1. 东北大学，辽宁沈阳 110819；2. 河南钢铁集团有限公司，河南郑州 450046）

摘 要： 煤炭作为主要能源基础和工业原料，高效推动了我国经济快速发展，其国内探明储量达到了 1145 亿吨，位居世界第三位，但其储采比远低于世界平均水平。根据能源发展规划要求，煤炭仍是能源安全和经济安全的基础，具有不可替代的地位，应推进煤炭高效开发利用工作，因此，对煤矿开采中关键的液压支架产品提出了高效、超大及低成本等综合性能要求，以使煤炭开采更加安全化、高效化与绿色化。

在煤矿开采过程中，煤矿液压支架属于整个生产工序中的关键装备之一。液压支架产品需要长期在高载荷、高潮湿环境下使用，服役周期达到 10 年以上，但从其生产及使用需求来看，还存在一些问题。首先，由于煤机行业的工况条件恶劣，除了对钢板本身强韧性指标有要求外，还对钢板整体的质量稳定性与低温韧性等均有更高的要求。其次，高端液压支架用钢已经属于高强钢产品，当采用现有焊接方法焊接时，其接头易产生冷裂纹，为防止焊接冷裂纹的产生，焊前必须进行预热处理，对应预热温度取决于高强钢的化学成分、强度等级、母材厚度以及焊接结构等。尽管焊前预热处理能改善焊接接头的冷裂纹敏感性，但是仍不能完全保证最终产品的质量，同时，考虑到液压支架产品形状复杂、尺寸较大，焊前、焊后预热处理为焊接工作带来了较大的难度，而且焊前、焊后预热处理必须采用人工操作，在增加生产成本的同时，生产效率大大降低。随着高效机器人焊接技术的广泛应用，迫切需要解决高端煤矿液压支架用钢在使用过程中的焊接性能差与焊接效率低等问题。

基于以上研究背景，聚焦高强煤矿液压支架用钢焊接热影响区组织调控关键技术，实现了屈服强度为 690MPa 级和 890MPa 级钢板的批量稳定生产，其中，免预热焊接 690MPa 级钢板实现了厚度 30mm 以下、零下 10℃ 以上无预热焊接，突破了同类产品 100～150℃ 预热焊接工艺的制约；开发了易焊接 890MPa 级高强韧煤矿机械用钢板的生产技术，实际产品焊接热输入量由 12 kJ/cm 提升至 15～20kJ/cm，显著提高焊接效率。采用易焊接 890MPa 级高强韧煤矿机械用钢板，制备了工作阻力为 26000kN、支护高度达 8.8m 的高端液压支架产品。采用高端液压支架产品后，下游煤矿企业的工作效率得到了提升，生产成本降低，资源回采率得到了提高，为企业带来了显著的经济效益，使煤炭开采向安全化、高效化和绿色化迈进了一大步。

回火过程精细组织调控对 P690QL2 钢低温韧性的影响

杜 林[1,2]，王堤鹤[3]，庞启航[3]，张宏亮[1,2]，
王亮亮[4]，朱莹光[1,2]，李相禹[3]，侯家平[1,2]

（1. 海洋装备用金属材料及其应用国家重点实验室，辽宁鞍山 114009；2. 鞍钢集团钢铁研究院，辽宁鞍山，114009；3. 辽宁科技大学材料与冶金学院，辽宁鞍山 114051；4. 鞍钢股份鲅鱼圈分公司厚板部，辽宁营口 115007）

摘 要：海上运输成为促进各国经济文化发展的新方式。随着碳捕集、利用与封存（Carbon Capture Utilization and Storage，CCUS）技术有效促进了二氧化碳的循环利用，减少了温室效应对生态环境的破坏。近年来设计液化二氧化碳运输船与 CCUS 技术密切相关，储存液化二氧化碳新型高强度 P690QL2 钢，需要解决初期研发生产中遇到的低温冲击性能不稳定问题。本文提出了基于人为判定的且优于其的一种图像识别方法精准判断口形貌。借助光学显微镜（OM）、透射电镜（SEM）和大数据分析等表征手段，使用图像识别研究调质处理后 P690QL2 钢的低温韧性与断口形貌的关系。结果表明：试验钢经–60℃夏比冲击试验后断口形貌为结晶状和纤维状共存；借助 Image Pro Plus 6.0 判别后分别计算得出试验钢断口纤维区占比约 80%时，影响试验钢韧脆性的多边形块状回火索氏体占比约 70%，冲击功可达 180J 以上。基于计算结果差异明显而导致试验钢性能不稳定的现象，主要是因为回火过程中产生的"硬相"回火马氏体和回火屈氏体。此方法应用于实际生产既能够合理分析断口特征，相比于人为判断方法精准度提升显著，精准调控回火马氏体和回火索氏体占比对 P690QL2 钢塑韧性的稳定提升具有重要意义。该研究方法广泛推广可为 P690QL2 钢的高效高质量生产提供理论基础。

关键词：P690QL2 钢；断口形貌；图像识别；回火索氏体；低温韧性

Effect of Fine Microstructure Control on Low Temperature Toughness of P690QL2 Steel During Tempering Process

DU Lin[1,2], WANG Dihe[3], PANG Qihang[3], ZHANG Hongliang[1,2],
WANG Liangliang[4], ZHU Yingguang[1,2], LI Xiangyu[3], HOU Jiaping[1,2]

(1. State Key Laboratory of Metal Material for Marine Equipment and Application, Anshan 114009, China; 2. Ansteel Group Iron and Steel Research Institute, Anshan 114009, China; 3. School of Materials and Metallurgy, University of Science and Technology Liaoning, Anshan 114051, China; 4. Bayuquan Iron and Steel Subsidiary Company of Angang Steel Co., Ltd., Yingkou 115007, China)

Abstract: Maritime transport has become a new way to promote the economic and cultural development of various countries. Carbon Capture Utilization and Storage (CCUS) technology has effectively promoted the recycling of carbon dioxide and reduced the damage to the ecological environment caused by the greenhouse effect. In recent years, the design of liquefied carbon dioxide carrier is closely related to CCUS technology, and the storage of liquefied carbon dioxide new high-strength P690QL2 steel needs to solve the problem of low temperature impact performance instability encountered in the initial research and development production. In this paper, an image recognition method based on human judgment is proposed, which is superior to human judgment. The relationship between low temperature toughness and fracture morphology of tempered P690QL2 steel was studied by means of optical microscopy (OM), transmission electron microscopy (SEM) and big data analysis. The results show that the fracture morphology of the test steel after –60℃ Charpy impact test is both crystalline and fibrous. After judging by Image Pro Plus 6.0, it is calculated that when the proportion of fiber area in the fracture of the test steel is about 80%, the proportion of polygonal block tempered sorbite affecting the toughness and brittleness of the test steel is about 70%, and the impact energy can reach more than 180J. Based on the obvious difference of calculation results, the phenomenon of unstable properties of test steel is mainly due to the "hard phase" tempered martensite and tempered troostite produced during the tempering process. When applied to the actual production, this method can not only reasonably analyze the fracture characteristics, but also significantly improve the accuracy compared with the artificial judgment method. Precisely adjusting the proportion of tempered martensite and tempered sotensite is of great significance for the stable improvement of the toughness of P690QL2 steel. This research method can provide a theoretical basis for the high efficiency and high quality production of P690QL2 steel.

Key words: P690QL2 steel; fracture; image recognition; tempered sorbite; cryogenic toughness

低合金高强度极地船舶用钢低温断裂韧性及海冰磨损性能

王超逸[1,2]，严 玲[1]，常雪婷[3]，安文瑞[2]

（1. 海洋装备用金属材料及其应用国家重点实验室，辽宁鞍山 114009；2. 鞍钢集团北京研究院有限公司，北京 102200；3. 上海海事大学海洋科学与工程学院，上海 201306）

摘 要： 为了研究低温环境对低合金高强度极地船舶用钢的影响，使用 TMCP 工艺生产了 460MPa 级别的极地船舶用钢，测试分析了极地船舶用钢在低温环境下的强度和断裂韧性，进一步开展极地船舶用钢低温海冰介质磨损试验，得到了极地船舶用钢在低温海冰介质下的厚度磨损参考数据。结果表明，随着温度的下降，极地船舶用钢的强度上升而断裂韧性下降；试验温度高于–40℃，材料脆性失稳特征值 δu 的平均值在 0.5mm 以上；而在–70℃极端低温环境下，材料的脆性失稳特征值 δu 的平均值约为 0.17mm，其中最小脆性起裂特征值 δc 为 0.076mm。在–40℃模拟低温海冰介质磨损环境下，极地船舶用钢因为低温海冰磨损导致的每海里表面厚度损失约为 0.082mm，海冰磨损会对无保护钢板表面造成明显损伤。研究极地船舶用钢的低温力学及海冰磨损性能，对于开发新型国产极地破冰船用材料，加快我国极地船舶事业有积极的推动作用。

关键词： 极地船舶用钢；TMCP；低温环境；断裂韧性；海冰磨损

Low Temperature Mechanics and Sea-ice Wear Properties of Low Alloy High Strength Polar Ship Steel

WANG Chaoyi[1,2], YAN Ling[1], CHANG Xueting[3], AN Wenrui[2]

(1. State Key Laboratory of Metal Material for Marine Equipment and Application, Anshan 114009, China; 2. Ansteel Beijing Research Institute Co., Ltd., Beijing 102200, China; 3. College of Ocean Science and Engineering, Shanghai Maritime University, Shanghai 201306, China)

Abstract: In order to study the effect of polar low temperature environment on low alloy high strength polar ship steel, 460MPa grade high strength polar ship steel has been produced via thermome-chanical controlled processing(TMCP) route.The strength and fracture toughness of polar ship steel at low temperature were tested and analyzed,The reference data of wear thickness loss of polar ship steel under low temperature sea ice medium were obtained. The results show that with the decrease of temperature, the strength of polar ship steel increases and the fracture toughness decreases.When the test temperature is higher than −40℃, the mean value of the brittle instability characteristic value δu is above 0.5mm. In the extreme low temperature environment of −70℃, the average brittle instability characteristic value δu is about 0.17mm, and the minimum brittle cracking characteristic value δc is 0.076mm. In the −40℃ simulated low-temperature sea-ice wear environment, the surface thickness loss of polar ship steel due to low-temperature sea ice wear is about 0.082mm per nautical mile.The study of the low temperature mechanics and sea ice wear properties of polar ship steel has a positive role in promoting the development of new domestic polar icebreaker materials.

Key words: polar ship steel; TMCP; low temperature; fracture toughness; sea-ice wear

电感耦合等离子体发射光谱法测定水电钢中酸溶硼元素含量

谭胜楠，葛晶晶，任玲玲，戚振南，禹青霄

（河钢材料技术研究院理化检测中心，河北石家庄 050000）

摘 要：本文研究了电感耦合等离子体发射光谱仪测定水电钢中酸溶硼含量的方法。试样借助电热板，采用一定浓度硫酸进行溶解处理，选用慢速滤纸过滤不溶物，取滤液待测。采用高纯物质进行基体匹配后，配制标准溶液系列，选择合适的分析谱线和优化的实验条件，完成使用电感耦合等离子体发射光谱(ICP-OES)法分析其中酸溶硼的含量。酸溶硼的质量分数为 0.0005%～0.010%，范围内校准曲线呈线性，线性相关系数 r 不小于 0.990，结果的相对标准偏差（RSD，n=10）小于 5%，加标回收率为 95%～105%，建立的方法快速、准确，测定结果较为满意。

关键词：电感耦合等离子体发射光谱（ICP-OES）法；酸溶硼；水电钢；分析谱线

Determination of Acid Solvable Boron in Hydroelectric Steel by Inductively Coupled Plasma Emission Spectrometry

TAN Shengnan, Ge Jingjing, REN Lingling, Qi Zhennan, YU Qingxiao

(HBIS Materials Technology Research Institute, Shijiazhuang 050000, China)

Abstract: The determination of acid-soluble boron in hydroelectric steel by inductively coupled plasma emission spectrometer is studied in this paper. The sample was dissolved by a certain concentration of sulfuric acid with the help of an electric heating plate, the insoluble matter was filtered by a slow filter paper, and the filtrate was taken to be measured. After the matrix matching with high purity material, the standard solution series was prepared, the suitable analytical spectrum line and the optimized experimental conditions were selected, and the content of acid-soluble boron was analyzed by inductively coupled plasma emission spectrometry (ICP-OES). The mass fraction of acid-soluble boron is 0.0005% ~ 0.010%, the calibration curve is linear in the range, the linear correlation coefficient r is not less than 0.990, the relative standard deviation (RSD, n=10) of the results is less than 5%, and the standard recovery rate is 95%~105%. The established method is fast and accurate, the results are satisfactory.

Key words: inductively coupled plasma emission spectrometry (ICP-OES); acid solvable boron; hydroelectric steel; analytical spectral lines

组织对船板钢在模拟热带海洋大气中的耐蚀性影响

高峰，李健，周乃鹏，罗小兵

（钢铁研究总院工程用钢研究院，北京 100081）

摘　要：近年来，海洋资源的开发利用不断引起重视，而钢铁材料在海洋中的腐蚀情况十分严重，尤其是船舶腐蚀的问题越来越突出。随着船舶发展的大型化、绿色化和智能化，船体钢的腐蚀问题亟需解决。通过实验室内干湿循环腐蚀试验模拟热带海域高湿热的腐蚀环境，经金相分析、锈层形貌和成分表征以及电化学实验等分析表征方法，研究不同热处理工艺下的组织演变对440MPa船板钢的耐蚀性影响，从而为改善船板钢的腐蚀性能提供新思路。

关键词：船板钢；耐蚀性；组织；热带海洋大气

Effect of Microstructure on Corrosion Resistance of Hull Steel in Simulated Tropical Marine Atmosphere

GAO Feng, LI Jian, ZHOU Naipeng, LUO Xiaobing

(Department of Structural Steels, Central Iron and Steel Research Institute, Beijing 100081, China)

Abstract: In recent years, the development and utilization of marine resources have been continuously attracting attention, while the steel corrosion in the ocean is very serious, especially the corrosion of ships is becoming more and more prominent. With the large-scale, green, and intelligent ship development, the corrosion problem of hull steel needs to be solved. The effect of microstructure evolution under different heat treatment processes on the corrosion resistance of 440 MPa hull steel was investigated by dry/wet cyclic corrosion test, morphology and composition analysis, and electrochemical characterization. It provides a new idea for improving the corrosion performance of hull steel.

Key words: hull steel; corrosion resistance; structure; tropical marine atmosphere

390MPa级抗碰撞船体钢动态再结晶行为研究

陈作宁[1,2]，师仲然[2]，胡骞[1]，展之德[2,3]，罗小兵[2]

（1. 武汉科技大学省部共建耐火材料与冶金国家重点实验室，湖北武汉　430081；
2. 钢铁研究总院有限公司工程用钢研究院，北京　100081；
3. 上海应用技术大学材料与科学工程学院，上海　201418）

摘　要：本文利用Gleeble-3800热模拟试验机对390MPa级抗碰撞钢进行单道次压缩试验，模拟其轧制变形过程，研究试验钢在应变速率为$0.01\sim10s^{-1}$，变形温度为850~1100℃条件下的动态再结晶行为，并建立试验钢的应力应变曲线，绘制动态再结晶图并进行动力学分析。实验结果表明390MPa级船体钢在变形过程中发生了动态再结晶，且随变形温度的升高和变形速率的降低，应力应变曲线逐渐向再结晶型过渡，即动态再结晶更容易进行；当变形温度升高时，试验钢发生动态再结晶所需的临界应变会逐渐减小，动态再结晶开始时间也会缩短；通过计算试验钢动态再结晶激活能为 424.62kJ/mol，实验结果为抗碰撞钢的热变形工艺参数制定提供了理论支持。

关键词：抗碰撞钢；变形温度；应变速率；动态再结晶行为；激活能

Study on Dynamic Recrystallization Behavior of High Toughness Steel

CHEN Zuoning[1,2], SHI Zhongran[2], HU Qian[1], ZHAN Zhide[2,3], LUO Xiaobing[2]

(1. The State Key Laboratory of Refractories and Metallurgy, Wuhan University of Science and Technology, Wuhan 430081, China; 2. Division of Structurale Steels, Central Iron and Steel Research Institute, Beijing 100081, China; 3. Shanghai Institute of Technology School of Materials Science and Engineering, Shanghai 201418, China)

Abstract: This article uses a Gleeble-3800 thermal simulation testing machine to conduct a single pass compression test on 390MPa grade impact resistant steel, simulate its rolling deformation process, study the dynamic recrystallization behavior of the test steel under strain rates of 0.01~10s^{-1} and deformation temperatures of 850~1100℃, and establish a stress-strain curve of the test steel, draw a dynamic recrystallization diagram, and perform dynamic analysis. The experimental results show that 390MPa class ship plate steel undergoes dynamic recrystallization during deformation, and with the increase of deformation temperature and the decrease of deformation rate, the stress-strain curve gradually transitions to recrystallization type, which means that dynamic recrystallization is easier to occur; When the deformation temperature increases, the critical strain required for dynamic recrystallization of the test steel will gradually decrease, and the start time of dynamic recrystallization will also be shortened; By calculating the dynamic recrystallization activation energy of the experimental steel to be 424.62kJ/mol, the experimental results provide theoretical support for the formulation of hot deformation process parameters for collision resistant steel.

Key words: anti expansion steel; deformation temperature; strain rate; dynamic recrystallization behavior; activation energy

20CrMoH 齿轮钢棒材的全尺寸成分偏析研究

任 帅，孙智妍，年保国，段路昭，罗 扬，冯立果

（河钢材料技术研究院先进材料研发中心，河北石家庄 053000）

摘 要：利用电子探针 EPMA 设备对 20CrMoH 齿轮钢棒材进行了全尺寸、各主要元素的线扫描及面扫描分析研究。线扫描结果表明，C 元素几乎不受位置影响，含量（质量分数）在 0.15%~0.30%之间稳定波动；Si 元素在棒材芯部 1/4D-3/4D 位置范围内的偏析略严重于棒材边部；S 元素几乎不受位置影响，含量在 0.006%~0.010%之间稳定波动；P、Mn、Cr、Mo 元素表现出相同的偏析情况，即在棒材 1/4D~3/4D 位置范围内的偏析情况明显比棒材边部严重。面扫描结果表明，Mn、Cr 元素在钢中形成了明显的带状组织，且越往芯部带状组织越严重，在 1/2D 位置处的带状组织宽度达到 100μm 以上；Mo、Si 元素未形成明显的带状组织。

关键词：齿轮钢；成分偏析；线扫描；面扫描

Study on Full Size Composition Segregation of 20CrMoH Gear Steel Bar

REN Shuai, SUN Zhiyan, NIAN Baoguo, DUAN Luzhao,
LUO Yang, FENG Liguo

(HBIS Materials Technology Research Institute, Shijiazhuang 053000, China)

Abstract: The full-scale line and area scanning analysis of 20CrMoH gear steel bar were carried out by electron probe microanalyzer. The results of line scanning showed that the content of C was almost independent of the position, and the content fluctuated stably between 0.15-0.30wt%. The segregation of Si element at the position of 1/4D-3/4D in the core of the bar is slightly more serious than that at the edge of the bar. The S element is hardly affected by its position, and its content fluctuates between 0.006-0.010wt% stably. The element of P, Mn, Cr and Mo show the same segregation. The segregation at the position of 1/4D-3/4D in the core of the bar is obviously more serious than that at the edge of the bar. The area scanning results show that Mn and Cr element forms obvious banded structure in the steel. The more toward the core, the more serious the banded structure is. The width of banded structure at 1/2D position is more than 100μm. The Mo and Si element does not form obvious banded structure.

Key words: gear steel; composition segregation; line scanning; area scanning

U71MnH 钢轨成分、组织及力学性能优化研究

于海鑫[1,2]，金纪勇[1,2]，廖德勇[1,2]，陈 昕[1,2]，王 冬[1,2]，张 瑜[1,2]

（1. 海洋装备用金属材料及其应用国家重点实验室，辽宁鞍山 114021；
2. 鞍钢集团钢铁研究院，辽宁鞍山 114009）

摘 要：试验开展 U71MnH 钢轨提高 C 含量、降低 Mn 含量成分优化及提高冷速的在线热处理工艺优化研究。采用光学显微镜、扫描电镜对优化前后钢轨试样显微组织进行观察，并开展试样力学性能检验。结果表明：优化后钢轨实验试样显微组织仍由珠光体及少量铁素体组成，钢轨试样平均片层间距由 0.208μm 降到 0.180μm，抗拉强度由 1173MPa 提高到 1221MPa，踏面平均硬度由 341HBW 提高到 347HBW，断裂韧性较优化前提高，裂纹扩展速率处于优化前性能波动范围下线。实验开展优化后钢轨工业生产检验，结果表明：成分优化后生产过程成分控制稳定，优化后在线热处理工艺适用于工业生产，钢轨力学性能符合标准要求。

关键词：成分优化；在线热处理工艺；显微组织；力学性能

Optimization of Composition, Microstructure and Mechanical Properties of U71MnH Rail

YU Haixin[1,2], JIN Jiyong[1,2], LIAO Deyong[1,2], CHEN Xin[1,2],
WANG Dong[1,2], ZHANG Yu[1,2]

(1. State Key Laboratory of Metal Material for Marine Equipment and Application, Anshan 114021, China;
2. Ansteel Iron & Steel Research Institutes, Anshan 114009, China)

Abstract: The optimization of increasing C content and reducing Mn content in U71MnH rail and the optimization of on-line heat treatment process to improve cooling rate were studied. The microstructure of rail samples before and after optimization was observed by optical microscope and scanning electron microscope, and the mechanical properties of the samples were tested. The results show that the microstructure of the optimized rail sample is still composed of pearlite and a small amount of ferrite, the average lamellar spacing of the rail sample is reduced from 0.208μm to 0.180μm, and the tensile strength is increased from 1173MPa to 1221MPa, the average hardness is increased from 341HBW to 347HBW, the fracture toughness is higher than before optimization, and the crack growth rate is below the performance fluctuation range before optimization. The industrial production test of the optimized rail was carried out. The results show that the composition control of the production process is stable after the composition optimization. The optimized online heat treatment process is suitable for industrial production, and the mechanical properties of the rail meet the standard requirements.

Key words: component optimization; online heat treatment process; microstructure; mechanical property

冷却方式对耐海水腐蚀钢组织与腐蚀性能的影响

解德刚[1,2]，赵　波[1,2]，吴　红[2]，王善宝[2]，袁　琴[2]

（1. 海洋装备用金属材料及其应用国家重点实验室，辽宁鞍山　114021；
2. 鞍钢集团钢铁研究院，辽宁鞍山　114021）

摘　要： 本文对一种海洋用低合金钢采取控制相变温度的方法，得到了不同金相组织、不同晶粒尺寸和组织偏析程度的试验钢。试验分析了不同金相组织状态的该种低合金钢在模拟海水环境中的盐雾腐蚀速率的变化情况，并对钢的盐雾腐蚀性能与金相组织组成、晶粒尺寸及组织偏析的关联性进行了讨论。试验结果显示，冷却方式对试验钢的盐雾腐蚀性能有一定影响。

关键词： 低合金钢；金相组织；海水；腐蚀速率；抗腐蚀性能

The Effect of Cooling Methods on the Structure and Corrosion Performance of Seawater Corrosion Resistant Steel

XIE Degang[1,2], ZHAO Bo[1,2], WU Hong[2], WANG Shanbao[2], YUAN Qin[2]

(1. State Key Laboratory of Metal Material for Marine Equipment and Application, Anshan 114021, China;
2. Anshan Iron and Steel Group Co., Ltd., Anshan 114021, China)

Abstract: In this paper, the experimental steels with different microstructure, grain size and segregation degree were obtained by controlling the transformation temperature of a marine low alloy steel. The change of corrosion rate of the low alloy steel with different microstructure in simulated seawater environment was analyzed. The correlation between seawater corrosion resistance and microstructure composition, grain size and microstructure segregation was discussed. The experimental results show that the cooling method has a certain impact on the seawater corrosion resistance of the test steel.

Key words: low alloy steel; microstructure; seawater; corrosion rate; corrosion resistance

铝含量对铁素体-贝氏体热轧双相钢的组织性能及磨损行为的影响

陈昊天，宋仁伯，任书濠，王永金

（北京科技大学材料科学与工程学院，北京 100083）

摘 要：热轧双相耐磨钢具有生产流程短、耗能低、抗拉强度高、屈强比低、连续屈服等特点，常应用于搅拌车、自卸车厢体制造。通过控轧控冷工艺制备了高铝（质量分数0.62%）和低铝（质量分数0.35%）两种成分体系的铁素体-贝氏体热轧双相钢，结合多种手段对不同铝含量实验钢的显微组织及力学性能进行表征，并通过环块磨损实验对其磨损行为进行研究。

结果表明，不同Al含量的实验钢均呈多边形铁素体+下贝氏体层状组织，由于Al元素扩大了铁素体相区因此相同工艺下高Al实验钢中的铁素体量更高。同时，高Al实验钢表面由于热轧过程中脱碳还产生了富铁素体层。高Al实验钢中存在稳定的{112}<110>取向织构，有利于提高材料轧制方向上的韧性。而相同工艺下的低Al实验钢中则存在{110}<001>方向上的高斯织构。两种实验钢拉伸曲线均呈连续屈服，高Al实验钢的屈服强度为497MPa，抗拉强度为1088MPa，屈强比为0.45，断后伸长率为15.8%，硬度为285HBW；低Al实验钢的屈服强度为561MPa，抗拉强度为1161MPa，屈强比为0.48，断后伸长率为15.0%，硬度为301HBW。富铁素体层的存在使得高Al实验钢表面硬度更低。由于高Al实验钢中铁素体含量较高从而使其强度更低而塑性更高。

环块磨损实验中高Al实验钢的失重量高于低Al实验钢。通过对其磨损轨迹进行观察发现，高Al实验钢磨损机理主要为犁沟+分层剥落，而低Al实验钢的磨损机理以微切削为主。富铁素体层的存在一方面使得材料的表面硬度下降，另一方面使得相邻的贝氏体具有更高的碳含量及硬度。磨损初期压头能够较轻易的将高Al实验钢表面的软质铁素体层去除，从而产生犁沟。随后压头与硬质贝氏体+铁素体基体发生相对运动时，则通过疲劳裂纹的萌生与扩展使基体发生分层剥落。低Al实验钢的贝氏体+铁素体基体硬度低于高Al实验钢，因此磨损初期产生的金属屑会作为磨粒对新生表面进行切削，从而使其磨损轨迹产生微切削的特点。

关键词：热轧双相钢；显微组织；力学性能；磨损行为

1000MPa级高强度水电用钢板Q890SF的研制与开发

潘中德[1,2]，刘心阳[1]，武会宾[2]，王思聪[1]

（1. 南京钢铁股份有限公司板材事业部，江苏南京 210035；
2. 北京科技大学钢铁共性技术协同创新中心，北京 100083）

摘 要：随着水电工程建设向高水头、大容量等方向发展，水电压力钢管设计对钢板的高强度、低温断裂韧性和焊接性能提出了更为严格的要求。本文阐述了南钢通过低碳、低Pcm成分设计以及淬火+回火热处理工艺设计，成功研发了1000MPa级高强度水电用钢板Q890SF，钢板经检测各项性能指标满足T/SSEA 0225标准要求，其中钢板屈服强度≥890MPa，抗拉强度≥950MPa，–60℃低温冲击功平均值大于100J，–40℃应变时效冲击功平均值大于100J，钢板探伤质量满足 NB/T47013.3 标准 T1 级要求；钢板通过焊条电弧焊和埋弧焊的焊接工艺评定，焊接接头具有优异的综合性能，能够满足水电项目设计及服役要求。

关键词：水电用钢；Q890SF；成分设计；工艺设计；焊接性能

Research and Development of 1000MPa High Strength Steel Plate Q890SF for Hydropower

PAN Zhongde[1,2], LIU Xinyang[1], WU Huibin[2], WANG Sicong[1]

(1. Plate Business Unit of Nanjing Iron & Steel Co., Ltd., Nanjing 210035, China;
2. Collaborative Innovation Center of Steel Technology of University of Science and Technology Beijing, Beijing 100083, China)

Abstract: With the development of hydropower engineering towards high HD and large capacity, the design of hydropower pressure steel pipes has put forward stricter requirements for the high strength, low-temperature fracture toughness, and welding performance of steel plates. This article elaborates on the design of low-carbon, low Pcm composition, and quenching and tempering heat treatment process, NISCO has successfully developed 1000MPa high-strength steel plate Q890SF for hydropower. The steel plate has been tested and all performance indicators meet the requirements of T/SSEA 0225 standard. The yield strength of the steel plate is not less than 890MPa, the tensile strength is not less than 950MPa, the average low-temperature impact energy at –60℃ is greater than 100J, and the average strain aging impact energy at –40℃ is greater than 100J. The quality of steel plate inspection meets the T1 level requirements of NB/T47013.3 standard; The steel plate has passed the SMAW and SAW welding process evaluation, and the welded joints have excellent comprehensive performance, which can meet the design and service requirements of hydropower projects.

Key words: hydropower steel; Q890SF; composition design; process design; weldability

高钢级输氢管线钢 L360MH 的开发及应用

李少坡，张　镇，张彩霞，王志勇，马长文，白学军

（首钢集团有限公司，北京　100043）

摘　要：高压输氢管道在整个氢能产业链中处于"卡脖子"地位，最主要原因在于输氢管道材料存在"氢脆"问题。本文介绍了首钢高钢级输氢管线钢 L360MH 开发和应用情况。输氢管线钢 L360MH 的成分设计重点关注 C、Mn、P、S 等易偏析元素，其含量的控制是关键问题；其微观组织为多边形铁素体组织，大角度晶界比例≥40%，心部带状组织≤1.5 级，组织要足够均匀化。力学性能屈强比控制在 0.88 以内，–30℃夏比冲击韧性在 300J 以上，–15℃落锤性能达到 100%。在 6.3MPa 高压氢环境下进行慢应变速率拉伸试验，材料的断后延伸率呈微弱的下降趋势，氢脆敏感指数 HEIδ 控制在 10%以内。2022 年首钢实现高钢级 L360MH 输氢管线钢批量生产，并在我国第一条内蒙古高压长输掺氢管道项目上得到示范应用。

关键词：输氢管线；L360MH；微观组织；氢脆敏感指数

Development and Application of High Grade Hydrogen Pipeline Steel L360MH at Shougang Steel

LI Shaopo, ZHANG Zhen, ZHANG Caixia, WANG Zhiyong, MA Changwen, BAI Xuejun

(Shougang Group Co., Ltd., Beijing 100043, China)

Abstract: The high-pressure hydrogen pipeline is in the "stuck neck" position in the whole hydrogen energy industry chain, the main reason is that the hydrogen pipeline material has the problem of "hydrogen embrittlement". This paper introduces the development and application of Shougang high grade hydrogen pipeline steel L360MH. The composition design of hydrogen pipeline steel L360MH focuses on C, Mn, P, S and other easily separated elements, and the control of their content is the key problem. The microstructure is polygonal ferrite structure, the proportion of large Angle grain boundary is ≥40%, the core banding structure is ≤1.5, and the structure should be homogenized enough. The mechanical properties shows that yield ratio is controlled within 0.88, the Charpy impact toughness is above 300J at –30℃, and the DWTT performance reaches 100% at –15℃. In the slow strain rate tensile test under 6.3MPa high pressure hydrogen environment, the elongation of the material after fracture shows a weak decreasing trend, and the hydrogen embrittlement sensitivity index (HEIδ) is controlled within 10%. In 2022, Shougang achieves mass production of high-grade L360MH hydrogen pipeline steel, and has been demonstrated in China's first Inner Mongolia high pressure long distance Hydrogen mixing in natural gas pipeline project.

Key words: hydrogen pipeline; L360MH; microstructure; hydrogen embrittlement sensitivity index

690MPa级超低碳贝氏体钢控轧控冷工艺与组织性能研究

梅 涛[1,2]，何有洪[1,2]，柴希阳[1]，罗小兵[1]，刘 静[2]，杨才福[1]

（1. 钢铁研究总院有限公司 工程用钢研究院，北京 100081；
2. 武汉科技大学 省部共建耐火材料与冶金国家重点实验室，湖北武汉 430081）

摘 要： 随着海洋资源竞争愈发激烈，高端船舶与海工装备的需求正在不断牵引着新材料的快速发展与进步。传统Ni-Cr-Mo-V型马氏体钢由于焊接难度大、屈强比高的问题推动了船用高强韧、易焊接、低屈强比超低碳贝氏体钢的发展需求。本文围绕690MPa级超低碳贝氏体易焊接钢研制目标，设计了3炉基于Mn-Mo-Nb成分的试验钢，研究了其相变行为以及实验室控轧控冷工艺，主要研究结果如下：

测定了Mn-Mo-Nb、Mn-Mo-Nb-B、Mn-Mo-Nb-B-Cu三个合金体系的试验钢的静态CCT曲线，分析了不同冷速条件下相变组织与硬度演变规律。结果表明，三个成分体系的试验钢均能在较宽的冷速范围内均获得组织类型一致与硬度波动小的超低碳贝氏体组织，且相变开始温度与终了温度对冷速不敏感，表现出良好的相变稳定性和组织均匀性。本试验的超低碳贝氏体钢在0.03~44℃/s的冷速范围内相变开始温度均低于520℃，只发生中温贝氏体相变。随着冷速的增加，Mn-Mo-Nb钢组织由粒状贝氏体向板条贝氏体过渡，显微硬度逐渐增加。加入16ppm的B能显著提高贝氏体淬透性，促进板条贝氏体的形成，降低贝氏体转变开始温度，并提高了组织的硬度。增B+Cu后进一步降低贝氏体转变温度，得到板条贝氏体倾向进一步增加，显著提高了组织的硬度。

对 Mn-Mo-Nb、Mn-Mo-Nb-B、Mn-Mo-Nb-B-Cu 钢进行了实验室控轧控冷工艺试验，研究了不同终轧温度与冷却速度下的组织与性能。结果表明，Mn-Mo-Nb-B-Cu 钢在埋沙、空冷以及风冷条件下获得了良好的强韧性匹配，屈强比低。在 900~750℃范围内，终轧温度对试验钢的强韧性以及显微组织影响较小。在 0.2~2.3℃/s 范围内，轧后冷速对超低碳贝氏体钢的强度影响较小；当冷速加快到 13.8℃/s 时，强度明显增加，韧性降低，其中 Mn-Mo-Nb-B-Cu 钢屈服强度达到 852MPa，抗拉强度达到 1033.5MPa。随着冷速的提高，试验钢中大小角度晶界密度提高，晶粒细化，贝氏体板条变窄。

参 考 文 献

[1] 贾希光, 刘宁. 控制冷却工艺对 DH36 钢板质量影响的分析研究[J]. 宽厚板, 2016, 22(4): 6-9.
[2] 罗志俊, 沈俊昶, 苏航, 等. 10CrNi5Mo 船体高强韧特厚钢板的研究[J]. 材料开发与应用, 2009, 24(5): 1-6.
[3] 严翔. 船体结构用 HSLA100 钢热处理工艺、组织及性能研究[D]. 武汉: 武汉科技大学, 2015.
[4] 李继红, 徐蔼彦, 李露露, 等. ULCB 钢的研究开发现状与发展前景探讨[J]. 热加工工艺, 2015, 44(16): 12-14, 18.

电磁搅拌对 ML40Cr 合金冷镦钢盘条锭型偏析的影响

马立国[1,2]，郭大勇[1,2]，王秉喜[1,2]，高 航[1,2]，张 博[1,2]，潘 阳[1,2]

（1. 海洋装备用金属材料及其应用国家重点实验室，辽宁鞍山 114009；
2. 鞍钢集团钢铁研究院，辽宁鞍山 114009）

摘 要：采用正交试验设计方法，研究结晶器电磁搅拌工艺参数对连铸坯等轴晶比率的影响，结果表明，结晶器电磁搅拌各参数对铸坯等轴晶率影响的主次顺序为：电流强度＞电压＞频率。优化结晶器电磁搅拌参数，进行 ML40Cr 合金冷镦钢的工业试验。结果表明，优化后，连铸坯等轴晶比率下降到 22%，A、B、C、D 方向上 C 元素偏析成波浪形态均匀分布，最高点小于 1.05，最低点大于 0.95，C 元素分布均匀；盘条表面无明显的框形，颜色较浅，偏析处白亮色铁素体组织较多；优化前盘条横截面硬度差为 20，优化后硬度差大幅降低到 10。从试验结果看，通过优化结晶器电磁搅拌参数，能有效减轻 ML40Cr 合金冷镦钢盘条的锭型偏析缺陷。
关键词：ML40Cr；电磁搅拌；C 偏析；锭型偏析

Effects of Electromagnetic Stirring on ML40Cr Cold Heading Steel Ingot Segregation

MA Liguo[1,2], GUO Dayong[1,2], WANG Bingxi[1,2], GAO Hang[1,2], ZHANG Bo[1,2], PAN Yang[1,2]

(1. State Key Laboratory of Metal Material for Marine Equipment and Application, Anshan 114009, China;
2. Ansteel Group Iron and Steel Research Institue, Anshan 114009, China)

Abstract: By orthogonal experiment design method, this paper examines mold electromagnetic stirring process parameters effect on slab equiaxed crystal ratio, the analysis show that ,primary and secondary order of equiaxed crystal ratio for each parameter of mould electromagnetic stirring: current strength, voltage and frequency. The test results show that, after the optimization of ML40Cr cold heading steel, billet equiaxed crystal ratio reduce to 22%, A, B, C, D directions of Carbon element segregation index into a wave shapeevenly distributed, wave form peak is less than 1.05, the lowest point is greater than 0.95, the Carbon element distribution was more uniform; Wire rod surface have no obvious frame, color shallow and segregation white light ferrite organization more; cross section maximum, minimum hardness difference is 20, the optimized hardness difference greatly reduced, about 10. The test results show, by optimizing the parameters of electromagnetic stirring, ingot segregation defects of ML40Cr cold heading steel can effectively reduce.

Key words: ML40Cr; EMS; carbon segregation; ingot segregation

含 V 低碳超细贝氏体钢在中温回火过程中的微观结构演变及强韧化机制

梁转琴[1]，王建军[1]，李宏光[1]，张向泽[2]，刘长波[2]，
孙东云[3]，杨志南[1,2]，张福成[3]

（1. 燕山大学国家冷轧板带装备及工艺工程技术研究中心，河北秦皇岛 066004；
2. 燕山大学亚稳材料制备技术与科学国家重点实验室，河北秦皇岛 066004；
3. 华北理工大学冶金与能源学院，河北唐山 063210）

摘 要： 低碳超细贝氏体钢因其优异的综合性能被广泛应用在钢轨、桥梁和船舶等领域。然而，贝氏体组织是一种非平衡组织，在热扰动或加热条件下，它可以转变为平衡组织，从而导致力学性能的变化。为了更好的利用低碳超细贝氏体钢，研究其在回火过程中组织性能的变化尤为重要[1]。

研究发现，随着回火温度的升高，残余奥氏体的体积分数以及碳的扩散系数显著增加，有利于提高其稳定性，保证了低碳超细贝氏体钢优异的力学性能。但过高的回火温度又会使残余奥氏体稳定性的降低、贝氏体板条粗化及位错密度降低等有损于塑韧性[2]。因此，大多数的低碳超细贝氏体钢均采用中温回火以增强其强韧性能。此外，在低碳超细贝氏体钢的相变过程中，延长回火时间可释放残余应力、促进碳元素均匀分布以及纳米级碳化物析出等进一步提高其力学性能[3]。然而目前对低碳超细贝氏体钢的中温回火行为进行系统研究很少，长时间回火对低碳超细贝氏体钢的微观结构演变及性能的影响机制尚不清楚。

因此，本研究通过提高含 V 低碳超细贝氏体钢的回火温度并延长回火时间，以促进纳米级 VC 析出，来提高贝氏体钢的综合力学性能。结果表明，回火前期，由于贝氏体转变的驱动力高，抑制碳化物形成元素的扩散可以忽略不计，其堆积在碳化物-基质界面，成为进一步生长的障碍，使得贝氏体转变在回火过程中占主导地位。随着回火时间的延长，元素扩散速率都变高，组织中析出纳米级碳化物。但是，过长的回火时间，会导致残余奥氏体中析出渗碳体，降低其化学稳定性。且在随后的冷却过程中，部分残余奥氏体发生了马氏体相变，因此需要合理调控回火时间。拉伸实验结果表明，随着回火时间的延长，弥散分布纳米级 VC 能够有效地阻止位错和晶界的运动，其强化作用大于贝氏体板条粗化和位错密度降低所造成的损失而且随着偏聚于位错等缺陷处的碳含量的增加，位错进一步回复重排，这可能会提高试样中位错运动的临界应力，进而提高低碳超细贝氏体钢的屈服强度。与此同时，我们发现，在回火过程中，虽然基体中析出的渗碳体会在尖端产生应力集中和裂纹，损害塑性，但 VC 的析出使晶格畸变缓慢下降，结合组织中位错密度的降低，使得试样的变形能力得到提高，而且不稳定的残余奥氏体充分转变或分解，可以延缓塑性不稳定，从而显著提高了延伸率。此外，组织中析出纳米级 VC 还可以提高贝氏体钢在析出相与基体界面处的裂纹扩展应力，从而提高低碳超细贝氏体钢的强韧性能。

关键词： 含 V 低碳超细贝氏体钢；微观组织；力学性能；中温回火

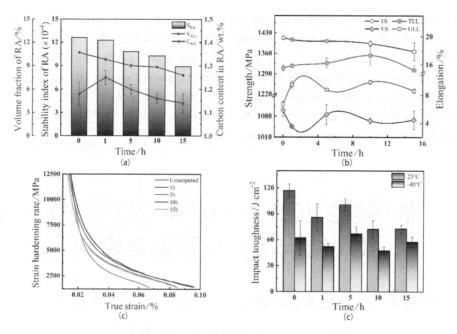

图 1 （a）残余奥氏体中各参数在回火过程中的变化规律；（b）相应的力学性能；
（c）加工硬化率曲线；（d）冲击韧性变化规律曲线

参 考 文 献

[1] Liu X, Han Y, Wei J, et al. Effect of tempering temperature on microstructure and mechanical properties of a low carbon bainitic steel treated by quenching-partitioning-tempering (QPT) process[J]. Journal of Materials Research and Technology, 2023, 23: 911-918.

[2] Królicka A, Żak A M, Caballero F G. Enhancing technological prospect of nanostructured bainitic steels by the control of thermal stability of austenite[J]. Materials & Design, 2021, 211.

[3] Shen Y, Zhou X, Huang X, et al. Identification of precipitate phases in CLAM steel[J]. Fusion Engineering and Design, 2021, 163.

板坯表面热喷涂对 45Mn 钢冷弯性能的影响

杨 玉[1,2]，王英海[1,2]，许 成[1,2]，王 刚[3]，张吉富[3]

（1. 海洋装备用金属材料及其应用国家重点实验室, 辽宁鞍山 114009；2. 鞍钢集团钢铁研究院，辽宁鞍山 114009；3. 鞍钢股份有限公司鲅鱼圈钢铁分公司，辽宁营口 115007）

摘 要：针对 45Mn 钢加工零件过程中折弯开裂的问题，本文主要分析了折弯开裂的原因，并介绍了通过板坯热喷涂工艺改善 45Mn 钢冷成型性能的生产实践。结果表明，钢板表面存在晶间氧化是 45Mn 钢折弯开裂和冷弯性能差的主要原因；采用板坯热喷涂防氧化涂料工艺可以有效抑制钢板表面晶间氧化的形成，改善 45Mn 钢冷成型性能，180°弯曲不开裂。

关键词：晶间氧化；冷弯性能；全脱碳层；部分脱碳层

Effect of Thermal Spraying on Slab Surface on Cold Bending Properties of 45Mn Steel

YANG Yu[1,2], WANG Yinghai[1,2], XU Cheng[1,2], WANG Gang[3], ZHANG Jifu[3]

(1. State Key Laboratory of Metal Materials and Application for Marine Equipment, Anshan 114009, China;
2. The Iron and Steel Research Institute of Ansteel Group, Anshan 114009, China;
3. Bayuquan Branch of Angang Steel Co., Ltd., Yingkou 115007, China)

Abstract: In view of the problem of 45Mn steel bending cracking in the process of machining parts, this paper mainly analyzed the causes of bending cracking, and introduced the production practice of improving the cold forming performance of 45Mn steel through slab Thermal spraying process. The results indicated that the presence of the grain boundary oxidation on the surface of the steel plate was the main reason for the bending cracking and poor cold bending performance of 45Mn steel; The formation of grain boundary oxidation on the surface of steel plate was restrained by using slab Thermal spraying coating process, and the cold bending formability of hot rolled 45Mn steel was improved, and there was no cracking when bending 180° degrees.

Key words: grain boundary oxidation; cold forming performance; fully decarburized layer; half decarburized layer

先进核电设备关键用钢的研制和应用

刘慧斌，侯　洪，张汉谦

（宝山钢铁股份有限公司中央研究院，上海　201900）

摘　要： 近年来，宝山钢铁股份有限公司研制开发了核安全壳用 SA738Gr.B 钢等一系列高难度的三代和四代核电工程关键设备用核心材料，攻克了成分优化设计、高精度工艺控制和高温长时服役性能评价等重大难关。这些关键核电用钢成功在我国多个核电示范工程实现了首次工程应用，添补了国内空白，打破了国外的垄断，甚至有些是世界上首次研制应用。它们满足了我国能源行业对关键材料的需求，提高了国家重点工程用材料的自主化水平，保障了国家重大工程顺利建设，为我国能源结构优化和节能减排做出了贡献，创造了显著的经济和社会效益。

关键词： 核电；示范工程；钢板；研制；应用

含氮微合金钢中氮化物的生成行为研究

杨　睿，李　阳，姜周华，孙　萌，马　帅，马彦硕，李天赐

（东北大学冶金学院，辽宁沈阳　110004）

摘　要： 本研究在实验室条件下分别制备了含氮钛微合金钢和含氮硼微合金钢，对钢中氮化物的生成行为进行了研究，通过扫描电镜对析出相的形貌、尺寸和成分进行了分析，并通过 FactSage 8.2 进行了平衡热力学计算。结果表

明，含氮钛微合金钢中生成了大量立方状的大尺寸 TiN 夹杂（≥5μm）和 TiN-AlN 复合夹杂（≥5μm），含氮硼微合金钢中生成了椭球状（1~2μm）、棒状（≥5μm）的 BN 夹杂以及方形的 Al_2O_3-AlN-MnS 大尺寸复合夹杂（≥10μm），近圆形的 Al_2O_3-BN（2~3μm）复合夹杂。实验观测结果同平衡热力学计算基本吻合，但在含氮硼合金钢中观测到的 AlN 析出相，在热力学计算结果中的生成量明显低于其他相，这可能是受偏析等因素影响所致。

关键词：氮化物；析出相；热力学计算；微合金钢

Study on Nitride Formation Behavior in Nitrogen-containing Microalloyed Steel

YANG Rui, LI Yang, JIANG Zhouhua, SUN Meng,
MA Shuai, MA Yanshuo, LI Tianci

(School of Metallurgy, Northeastern University, Shenyang 110004, China)

Abstract: Nitrogen-containing titanium microalloyed steels and nitrogen-containing boron microalloyed steels were prepared under laboratory conditions. The formation behavior of nitrides in the steels was studied. The morphology, size and composition of precipitated phases were analyzed by SEM, and the equilibrium thermodynamic calculation was carried out by FactSage 8.2. The results show that a large number of cubic TiN inclusions (≥5μm) and TiN-AlN composite inclusions (≥5μm) are precipitated in titanium microalloyed steel containing nitrogen, and ellipsoidal (1~2μm), rod-shaped (≥5μm) BN inclusions, square Al_2O_3-AlN-MnS large-size composite inclusions (≥10μm) and nearly round Al_2O_3-BN (2~3 μm) are precipitated in boron microalloyed steel containing nitrogen. The experimental results are basically consistent with the equilibrium thermodynamic calculation, but the precipitation amount of AlN phase observed in nitrogen-boron alloy steel is obviously lower than that of other phases in thermodynamic calculation, which may be caused by segregation and other factors.

Key words: nitride; precipitated phase; thermodynamic calculation; low carbon microalloyed steel

热轧工艺对 X80 变形奥氏体和落锤性能的影响

熊雪刚[1]，张开华[1]，陈 述[1]，田光林[2]，任守斌[2]，龚 慧[2]

（1. 攀钢集团研究院有限公司 钒钛资源综合利用国家重点实验室，四川攀枝花 617000；
2. 攀钢集团西昌钢钒有限公司 产品发展部，四川西昌 615000）

摘 要：通过工业试制，采用金相显微镜、扫描电镜、落锤试验机等检测手段，研究了不同热轧工艺对 X80 变形奥氏体尺寸及落锤性能的影响。结果表明，随着变形奥氏体组织的细化，相变储能增加，晶粒细小的贝氏体铁素体（BF）、粒状贝氏体（GB）等组织比例增加，其中的小角度晶界比例更高、位错密度更高，促进材料的落锤性能升高。反之，变形奥氏体组织粗大，会增加奥氏体的稳定性，促进马氏体奥氏体岛（M/A 岛）数量增多，导致钢的落锤性能降低。同时，随着板坯加热温度和加热时间降低、粗轧末道次变形量和中间坯厚度增加，变形奥氏体带宽度和成品晶粒尺寸降低，准多边形铁素体（QF）比例降低，BF、GB 比例增加，M/A 岛数量减少、尺寸降低，落锤性能升高。

关键词：变形奥氏体；加热温度；加热时间；粗轧变形量；中间坯厚度；落锤性能

Effects of Hot Rolling Process on Deformed Austenite and DWTT Properties of X80

XIONG Xuegang[1], ZHANG Kaihua[1], CHEN Shu[1], TIAN Guanglin[2], REN Shoubin[2], GONG Hui[2]

(1. Labloratory of Vanadium and Titanium Resource Comprehensive Utilization, Pangang Group Research Institute Co., Ltd., Panzhihua 617000, China; 2. Pangang Group Xichang Steel & Vanadium Co., Ltd., Xichang 615000, China)

Abstract: An industrial production test of X80 pipeline steel was carried out, to study the influence of hot rolling process on the size of deformed austenite and drop-weight tear test (DWTT) properties, through the research means of optical microscope, scanning electron microscope and drop-weight tear testing machine. The results showed that with the refinement of deformed austenite, the phase transformation energy storage increased, so that the formation of acicular ferrite such as banitic ferrite (BF) and granular banite (GB) was promoted. And the microstructure of BF and GB had fine grains, high density of dislocation and low angle boundries, thus increased the crack growth inhibition ability and the DWTT property of the test steel. On the contrary, the coarsening of the deformed austenite would improve the stability of itself, increase the number of M / a islands, and reduce the DWTT properties. At the same time, with the decrease of slab heating temperature and heating time, the increase of deformation of the last pass of rough rolling and the thickness of intermediate billet, the grain size of deformed austenite and the finally microstructure were refined, the number of QF and M/A islands decreased, the number of BF and GB proportion and the DWTT properties increased.

Key words: deformed austenite; slab heating temperature; slab heating time; deformation of rough rolling; thickness of intermediate billet; DWTT properties

稀土元素 Ce 对大型矿用车高强钢板 Q620E 的影响研究

卢晓禹[1,2]，袁晓鸣[1]，黄 利[1]，王少炳[1]

（1. 内蒙古包钢钢联股份有限公司技术中心，内蒙古包头 014010；
2. 内蒙古自治区稀土钢产品研发企业重点实验室，内蒙古包头 014010）

摘 要：本文采用中试试验平台完成不同稀土含量对大型矿用车高强钢 Q620E 的影响研究，分析不同稀土含量对 Q620E 力学性能的影响，利用蔡司显微镜、扫描电镜对组织、夹杂物、冲击断口进行分析。研究结果表明：适当的稀土含量通过细化晶粒、降低夹杂物尺寸和增大韧窝断口尺寸等作用，可使 Q620E 强度增加 30MPa 以上，低温冲击功提高 30J 以上。优化后的 Q620E 既降低了合金成本，又保证了强度和韧性。

关键词：稀土；高强钢；Q620E；显微组织；夹杂物

Effect of Rare Earth Element Ce on High Strength Steel Plate Q620E of Large Mining Vehicle

LU Xiaoyu[1,2], YUAN Xiaoming[1], HUANG Li[1], WANG Shaobing[1]

(1. School of Material and Metallurgy(Rare Earth Institute), Inner Mongolia University of Science and Technology, Baotou 014010, China; 2. Rare Rarth Steel Plate Co., Ltd. of Inner Mongolia Baotou Steel Union Co., Ltd., Baotou 014010, China)

Abstract: In this paper, the pilot test platform was used to complete the study on the effects of different rare earth contents on the mechanical properties of Q620E, and the microstructure, inclusions and impact fracture were analyzed by Zeiss microscope and scanning electron microscope.The results show that the strength of Q620E can be increased by more than 30MPa and the low-temperature impact energy can be increased by more than 30J with appropriate rare earth content by refining the grain, reducing the size of inclusion and increasing the size of dimple.The optimized Q620E not only reduces the alloy cost, but also ensures the strength and toughness.

Key words: rare earth; identity; Q620E; microstructure; inclusions

耐候无缝钢管 Q355 的开发与应用

姚晓乐[1]，张学颖[1]，米永峰[1]，姜海龙[1]，孙文秀[1]，何建中[2]

（1. 内蒙古包钢钢管有限公司，内蒙古包头　014010；
2. 内蒙古包钢钢联股份有限公司，内蒙古包头　014010）

摘　要： 针对装备制造和耐蚀结构钢市场需求及社会发展需要，根据各元素在钢中作用特点及耐候无缝钢管的使用需求，制定出适用于无缝钢管的耐候钢成分及生产工艺流程，开发出耐候无缝钢管 Q355。结果表明，耐候无缝钢管 Q355 成分设计合理，生产工艺控制得当，各项性能满足设计要求。产品内外表面质量良好，具有高强、高韧、耐低温、易焊接、抗震性好、良好的耐大气腐蚀性能等特点，完全满足用户使用要求。

关键词： 耐候无缝钢管；耐低温；易焊接；抗震性；耐大气腐蚀

Development and Application of Weathering Seamless Steel Tube Q355

YAO Xiaole[1], ZHANG Xueying[1], MI Yongfeng[1], JIANG Hailong[1], SUN Wenxiu[1], HE Jianzhong[2]

(1. Inner Mongolia Baotou Steel Pipe Co., Ltd., Baotou 014010, China;
2. Inner Mongolia Baotou Steel Union Co., Ltd., Baotou 014010, China)

Abstract: According to the demand of equipment manufacturing industry and corrosion resistant construction steel market and needs of social development, based on the characteristics of each element in steel and the using needs of weathering seamless steel tube, the composition and production process of weathering steel suitable for seamless steel tube was produced, and we developed the weathering seamless steel tube Q355. The results showed that the composition design of Q355 was reasonable, the production process was properly controlled, and the properties met the design requirements. The internal and external surface quality of the product was well, with high strength, high toughness, low temperature resistance, easily welding, good anti-seismic, good atmospheric corrosion resistance and characteristics, fully met the requirements of users.

Key words: weathering seamless steel tube; low temperature resistance; easily welding; anti-seismic; atmospheric corrosion resistance

Nb、V 对热煨弯管用 X80 管线钢组织性能的影响研究

李天怡[1,2]，刘文月[1,2]，安 涛[1,2]，臧 岩[1,2]，刘津伊[1,2]，薛 峰[2]

（1. 海洋装备用金属材料及其应用国家重点实验室，辽宁鞍山 114009；
2. 鞍钢集团北京研究院有限公司，北京 102200）

摘 要：本文研究了 V 微合金化和 Nb-V 微合金化 X80 管线钢 TMCP 态、回火态和调质态的组织性能，并利用 SEM、EBSD 和 TEM 研究了两种材料在调质处理后的微观组织特征与性能的关系。对于 V 微合金化管线钢来说，相比 TMCP 态，回火和调质热处理均能提高其强度、硬度以及低温韧性。回火后，钢板强度、韧性均有所提高，并具备最佳的强韧性匹配。而 Nb-V 微合金化钢板在回火、调质处理后强度、硬度上升，低温韧性下降。在调质处理后，基体组织由针状铁素体转变为贝氏体铁素体，通过晶粒细化、析出强化同时提高钢板强度，调质态 Nb-V 微合金化钢板具有最佳的强韧性匹配。与调质态 V 微合金化管线钢相比，调质态 Nb-V 微合金化管线钢大角度晶界比例更低、等效晶粒尺寸更小，力学性能上表现为强度降低，低温韧性提高。两种管线钢等效晶粒尺寸的变化规律与冲击韧性的变化规律一致，而大角度晶界比例的变化规律与强度的变化规律一致。

关键词：X80 管线钢；微合金化；EBSD 分析；热处理工艺；强化机制

Influence of Nb、V Microalloying on Microstructure and Mechanical Properties of X80 Grade Hot Bent Pipeline Steel

LI Tianyi[1,2], LIU Wenyue[1,2], AN Tao[1,2], ZANG Yan[1,2], LIU Jinyi[1,2], XUE Feng[2]

(1. State Key Laboratory of Metal Materials for Marine Equipment and Application, Anshan 114009, China;
2. Ansteel Beijing Research Institute Co., Ltd., Beijing 102200, China)

Abstract: The microstructures and properties of V microalloyed and Nb-V microalloyed X80 pipeline steels after TMCP process, tempering process and QT process were studied, and the relationship between microstructure and mechanical property of two materials after QT process was studied by SEM, EBSD and TEM. Compared with TMCP process, both tempering and QT heat treatment can increase the strength, hardness and low temperature toughness of V microalloyed steel. Tempering process leads to an increase in the strength and toughness of the steel plate, and the steel has the best strength and toughness matching. However, after tempering and QT process, the strength and hardness of the Nb-V microalloyed steel plate increase, but the low-temperature toughness decreases. After QT treatment, the matrix structure changes from acicular ferrite to bainitic ferrite, and the strength of the steel plate is simultaneously improved through grain refinement and precipitation strengthening. The QT state Nb-V microalloyed steel plate has the best matching of strength and toughness.

Compared with the V microalloyed pipeline steel, the proportion of high-angle grain boundaries and the equivalent grain size of QT state Nb-V microalloyed pipeline steel are lower. In terms of mechanical properties, the strength is reduced and the low temperature toughness is improved.

Key words: pipeline steel; microalloying; EBSD analysis; heat treatment process; strengthening mechanism

V、Nb 加入量对微合金化钢高温热塑性的影响

连小金，陈雪慧，罗小兵，杨才福

（钢铁研究总院有限公司，北京　100081）

摘　要：充分认识 V、Nb、N 等合金元素在微合金化钢中对高温热塑性的影响规律是改善连铸连轧生产中表面质量问题的重要方法。使用 Gleeble 热模拟机模拟实验钢连铸生产的变形过程，结合金相显微镜、扫描电子显微镜等研究了添加不同含量 V、Nb 对 V-N 微合金化钢的高温热塑性变化规律。结果表明，在 V 含量为 0.06% 的 V-N 微合金化钢中，额外添加 0.024%Nb 或添加 0.06%V 均会提高微合金化钢的强度，但会降低第三脆性区的热塑性，使得易开裂临界区间（断面收缩率<40%）的宽度和深度增加。对比三者易开裂临界区间，0.06V 钢为 862~713℃，0.12V 钢为 887~700℃，0.06V-0.024Nb 钢为 903~700℃。薄膜铁素体的形成和析出相数量增加是额外增加微合金化元素后高温热塑性降低的主要原因。

关键词：V-N 微合金化钢；Nb 微合金化钢；高温热塑性；力学性能；显微组织

Influence of V、Nb Addition on High Temperature Thermoplasticity of Microalloyed Steel

LIAN Xiaojin, CHEN Xuehui, LUO Xiaobing, YANG Caifu

(Central Iron & Steel Research Institute Co., Ltd. Beijing 100081, China)

Abstract: An important method to improve surface quality issues in continuous casting and rolling production is to fully understand the influence of alloy elements such as V, Nb, and N on high-temperature thermoplastic properties in microalloyed steel. The deformation process of experimental steel continuous casting production was studied by using Gleeble thermal simulator. The high-temperature thermoplastic behavior of V-N microalloyed steel with different contents of V and Nb was studied by combining techniques such as metallographic microscope and scanning electron microscope. The results show that In V-N microalloyed steel with a V content of 0.06%, the addition of 0.024% Nb or 0.06% V can increase the strength of the microalloyed steel, but it will reduce the high temperature thermoplasticity of the third brittle zone, and makes the plastic trough zone (reduction of area<40%) wider and deeper. Comparing the plastic trough zone, 0.06V steel is 862-713 ℃, 0.12V steel is 887-700 ℃, and 0.06V-0.024Nb steel is 903-700 ℃. The synthesis of thin film ferrite and carbonitride precipitation is the main reason for the decrease in high temperature thermoplasticity after adding additional microalloying elements.

Key words: V-N microalloyed steel; Nb microalloyed steel; high temperature thermoplasticity; mechanical properties; microscopic structure

大型桥梁用高品质钢材的研发与应用

麻晗，杨浩，王雷，陈焕德

（江苏省沙钢钢铁研究院，江苏张家港　215625）

摘　要：近些年，随着桥梁结构向大跨度、重载荷、轻量化、长寿命等方向发展，沙钢开展了大型桥梁用大厚度、高强度、高韧性、耐腐蚀、易焊接、抗震、抗疲劳的高品质钢材的研发工作，形成了相对完善的桥梁用钢产品体系，成为国内为数不多的同时具备桥梁用板材、线材、棒材产品供货能力的钢厂。

（1）沙钢大型桥梁用高品质钢材产品开发。

1）高强桥梁钢板：开发了500MPa级、690MPa级高强度桥梁钢产品，采用低C成分设计，TMCP工艺生产，屈强比≤0.86，韧脆转变温度≤-50℃，具有良好的强韧性、抗震性及焊接性能，其中Q500qE最大厚度96mm，Q690qE最大厚度60mm。

2）耐候桥梁钢板：开发了Q345/370/420/500qENH系列耐候桥梁钢，产品实物质量优异，经川藏铁路沿线实地挂片试验，耐大气腐蚀性能较普通桥梁钢提升3~5倍。此外，针对耐候桥梁钢用户使用难题，开展了耐候桥梁钢配套焊接工艺、锈层稳定化工艺的研究，形成了产品+使用的完善的应用技术体系。

3）易焊接桥梁钢板：开发了可适用于大线能量焊接的桥梁钢，具有焊接质量好、焊接效率高的特点，相比传统焊接工艺，焊接入量提高3~5倍，可进行双丝或三丝焊接。产品经桥梁制造厂实际焊接，接头热影响区-40℃冲击功≥100J，表现出良好的焊接性能。

4）桥梁钢复合板：开发了Q345/370/420/500qE+316L的系列桥梁钢复合板，形成了高质量复合板坯制备技术、复合板轧制技术、复合板内应力及板形控制技术等关键生产技术。产品具有复层结合紧密、性能稳定、板形优良等特点。

5）桥梁缆索用钢：系统研究了中心偏析、夹杂物、组织均匀性以及钢丝生产工艺对桥梁缆索性能的影响，形成了包括洁净钢冶炼、高均匀性热处理、低损伤拉拔等在内的一整套超高强度桥梁缆索用钢生产控制技术。迄今已成功开发7mm 2100MPa级桥梁缆索用钢SWRS96Si-TP并实现批量供货，同时取得2200MPa级桥梁缆索用钢开发应用的初步成功。

6）耐蚀钢筋：提出了超低碳铬钼多元素复合耐蚀合金成分体系，开发了洁净钢冶炼及棒材复相组织调控技术，获得具有复合钝化膜+铁素体/贝氏体双相组织的高耐蚀钢筋，其屈服强度为400~600MPa，最大力总延伸率≥9%，耐腐蚀性能较普通钢筋提升10倍以上，为解决钢筋混凝土结构的耐久性不足问题提供新的方向。

（2）应用业绩。

目前沙钢高品质桥梁钢材已应用于沪苏通公铁两用长江大桥、平潭铁路跨海大桥、马鞍山公铁两用长江大桥、张靖皋长江大桥、川藏铁路大渡河大桥、胶州湾跨海大桥等十几座世界级桥梁工程，多项产品填补国内空白，满足了国家大型桥梁用钢新需求。

Nb含量和变形量对高强度调质钢板淬火再加热奥氏体晶粒尺寸及其分布的影响

邹 扬，张苏渊，秦丽晔

（首钢集团有限公司技术研究院，北京 100043）

摘 要：利用 Gleeble 3500 热/力模拟试验机，通过 1000℃+820℃两阶段热变形+900℃淬火再加热联合模拟试验，研究了不同 Nb 含量和不同热变形量对水电站用 800 MPa 级高强度试验钢淬火再加热晶粒尺寸及其分布的影响规律，并通过透射电镜（TEM）对形变诱导析出的 Nb(C,N)的粒子尺寸、分布进行了试验和观测。结果表明，热变形态奥氏体晶粒尺寸（D）对于淬火再加热态奥氏体晶粒尺寸（D'）具有重要遗传性影响，二者以及 900℃再加热保温时间 t 之间存在函数关系 $D' = (1.0057D - 6.9785) \times (t/300)^{0.215}$，用于预测 800MPa 高强钢淬火再加热态晶粒尺寸具有较高精度，见图 1。

增加 Nb 含量同时细化晶粒尺寸 D 和 D'，并改善晶粒尺寸分布、显著降低个别粗大晶粒出现的概率。在常用的淬火加热制度下，添加 0.03Nb%和 0.05Nb%的晶粒细化效果基本相当，见图 2。

TEM 观测结果表明，含 Nb 变形态试样中存在大量 10~30 nm 尺寸的 Nb(C,N) 粒子，其数量和密度随 Nb 含量增加而增加，但粒子尺寸并未随之明显增大，见图 3。通过热力学计算并综合粒子尺寸和形成时间推断[1]，这些 Nb(C,N)粒子在 1000℃ 形变诱导析出，在后续 900℃淬火再加热温度下亦无法完全回溶，通过晶界钉扎效应致使再加热态奥氏体晶粒组织获得进一步细化。

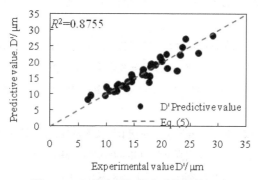

图 1 900℃淬火再加热奥氏体晶粒尺寸预测值和试验值的对比图

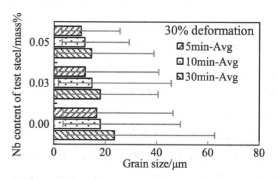

图 2 900℃再加热不同保温时间试样奥氏体晶粒尺寸分布与 Nb 含量的对应关系

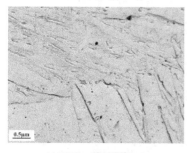

(a) 0%Nb，粒子密度 1.35

(b) 0.03%Nb，粒子密度 14.41

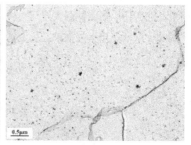

(c) 0.05%Nb 粒子密度 20.25

图 3 不同 Nb 含量试验钢 30%变形量试样的析出物 TEM 显微照片和粒子密度（单位，10^7 个/cm^2）

参考文献

[1] 雍岐龙. 钢铁材料中的第二相[M]. 北京: 冶金工业出版社, 2006.

10.6 电工钢

不同 Si、Al 含量对无取向硅钢 50W800 的夹杂物、微观组织及电磁性能的影响研究

高兴健，王思琴

（宝钢湛江钢铁有限公司制造管理部，广东湛江 524076）

摘 要：对高硅高铝（1.1% Si、0.60% Al）和低硅低铝（0.55% Si、0.30% Al）两种成分的无取向硅钢 50W800 的电磁性能进行了研究，分析了其微观组织、织构及夹杂物的变化规律。结果表明，随 Si+Al 的含量升高，晶粒平均尺寸增大，铁损值降低，但磁感强度也降低。高硅高铝试样的磁滞损耗占比更高，主要为高硅高铝试样中的 0.42~2μm 尺寸的 AlN 数量异常增多导致。由此可见，提高 Al 含量可提高电阻率，从而降低涡流损耗，但 Al 含量较高时，易与 N 元素结合生产大量的 AlN，使磁滞损耗升高，因此 Al 含量的添加量需选择合适的范围。(Si+Al)含量的增加，不利的{111}和{112}面织构强度增加，造成磁感强度降低。

关键词：无取向硅钢；50W800；电磁性能；微观组织；夹杂物

Study on the Influence of Different Si and Al Contents on Inclusions, Microstructure and Electromagnetic Properties of 50W800 Non-oriented Silicon Steel

GAO Xingjian, WANG Siqin

(Products and Techniques Management Division of Baosteel Zhanjiang Iron & Steel Co., Ltd., Zhanjiang 524076, China)

Abstract: The electromagnetic properties of unoriented silicon steel 50W800 with high silicon and high Aluminum (1.1%Si, 0.60%Al) and low silicon and low Aluminum (0.55%Si, 0.30%Al) were studied, and the microstructure, texture and inclusions of the steel were analyzed. The results show that with the increase of Si+Al content, the average grain size increases, the iron loss value decreases, but the magnetic induction strength also decreases. The proportion of hysteresis loss in high-Si and high-aluminum samples is higher, which is mainly caused by the abnormal increase in the amount of 0.42-2μm AlN in high-Si and high-aluminum samples. It can be seen that increasing the Al content can improve the resistivity and thus reduce the eddy current loss. However, when the Al content is high, it is easy to combine with N element to produce a large amount of AlN, which increases the hysteresis loss. Therefore, the addition amount of Al content needs to be selected in an appropriate range. With the increase of (Si+Al) content, the intensity of the unfavourable texture of {111} and {112} surface increases, leading to the decrease of the magnetic induction intensity.

Key words: non-oriented silicon steel; 50W800; magnetic; microstructure; occluded foreign substance

取向硅钢环保型表面除鳞技术的生产应用

夏强强[1]，刘玉堂[1]，何国芳[2]，周建平[2]

(1. 中冶赛迪工程技术股份有限公司，重庆 401122；
2. 赐宝新型薄板（江苏）有限公司，江苏南通 226100)

摘 要：取向硅钢 Si 含量较高，在热轧过程中形成的氧化皮附着力强导致酸洗效率低和表面质量差，同时在酸洗过程中产生"硅泥"并对酸洗系统运行稳定性造成影响。中冶赛迪自主研发的 STD（Surface Treatment Device，简称"STD"）环保型表面除鳞技术，是将含水磨料高速抛射至金属表面均匀去除氧化层。热轧取向硅钢（CGO）经 STD 技术处理后可以完全去除表面氧化层，经 STD 工艺除鳞的钢卷满足一次轧制、脱碳退火、二次轧制、高温退火、热拉伸平整和激光刻痕等下游工序要求，最终产品的磁感（B_8=1.89T）和铁损（$P_{17/50}$=0.98W/kg）与酸洗板相当，均完全达标。STD 技术完全可以替代传统酸洗用于取向硅钢工业生产。

关键词：中冶赛迪；取向硅钢；STD；酸洗；除鳞

Production Application of Environmentally Friendly Descaling Technology for Oriented Silicon Steel

XIA Qiangqiang[1], LIU Yutang[1], HE Guofang[2], ZHOU Jianping[2]

(1. CISDI Engineering Co., Ltd., Chongqing 401122, China;
2. Cibao New Type Thin Plate (Jiangsu) Co., Ltd., Nantong 226100, China)

Abstract: Oriented silicon steel has a high Si content, and the strong adhesion of the oxide skin formed during the hot rolling process leads to low pickling efficiency and poor surface quality. At the same time, "silicon sludge" is generated during the pickling process, which affects the stability operation of the pickling system. Surface Treatment Device (STD), an environmentally Protection descaling technology independently developed by CISDI, can uniformly remove the iron oxide scale by high-speed jetting of water-based abrasives onto the metal surface. Hot rolled oriented silicon steel (CGO) can completely remove surface iron oxide scale after STD technology treatment. The STD plate meets downstream process requirements such as primary rolling, decarburization annealing, secondary rolling, secondary recrystallization annealing, hot stretching leveling, and laser scribing. The magnetic induction (B_8=1.92T) and iron loss ($P_{17/50}$=0.98W/kg) of the final product are equivalent to those of the pickling plate, both of which fully meet the standards. The results indicate that STD technology is feasible for replacing pickling technology in the production of oriented silicon steel.

Key words: CISDI; oriented silicon steel; STD; pickling; descaling

退火工艺对 27AHSW450 冷轧无取向硅钢组织性能的影响

饶彦俊，宋仁伯，赵知洋，张应超，马为涛

（北京科技大学材料科学与工程学院，北京　100083）

摘　要：冷轧无取向硅钢作为功率与能量转化的关键软磁材料，主要用于制造新能源汽车驱动电机的定子与转子铁芯。对感应电机而言，需使用高磁感硅钢材料来提高电机的效率；对永磁同步电机而言，车辆起步和低速爬坡阶段要求电机输出足以带动汽车启动的巨大扭矩，因此同样需要硅钢具有尽可能高的磁感；随着汽车马达转速的提升，磁场频率的加大，铁损值升高，因而要求在高频下的铁损小。此外，汽车驱动电机在高速运行时会产生极大的离心力，并且定子与转子的间隙设计较为严苛，因此要求硅钢片有较高的屈服强度和抗拉强度。总体而言，追求更高的磁感、更低的铁损，同时能满足较高的机械强度需求是驱动电机用硅钢材料的发展趋势[1]。

高牌号冷轧无取向硅钢制造工艺相对复杂、成分控制严格、制造工序长，影响其组织性能的因素有很多，除了要考虑冶炼的纯净度以及热轧工艺等以外，冷轧后的最终退火工艺也是一个非常重要的环节[2]。退火工艺对冷轧无取向硅钢的组织和性能影响较大，合适的退火工艺参数，可使晶粒尺寸达到临界尺寸、粗化夹杂物，改善硅钢片的组织性能[3]。

本文以 27AHSW450（其中 27 代表成品厚度 0.27mm、A 为企业代号、HS 代表高强度、W 表示无取向、450 表示成品屈服强度高于 450MPa）为研究对象，重点研究了退火工艺对冷轧无取向硅钢组织、磁性能以及力学性能的影响，研究结果将为大规模工业生产时退火工艺的制定提供理论依据。在退火温度为 950℃时，对试样分别退火 3min、6min、10min，然后对退火后试样进行组织观察和分析，并测量铁损值与磁感应强度，最后对试样进行拉伸试验测定其力学性能参数，从而发现了退火时间对 27AHSW450 冷轧无取向硅钢组织与性能的影响规律。结果表明：随退火时间从 3min 延长至 10min，试样晶粒尺寸从 27μm 增长到 36μm，晶粒均匀度降低；试样铁损 $P_{10/400}$ 从 22.5W/kg 升至 23.6W/kg，磁感 B_{50} 从 1.57T 升至 1.64T；试样屈服强度从 493MPa 升至 503MPa，抗拉强度从 624MPa 升至 647MPa，伸长率从 21%升至 22%。

参 考 文 献

[1] 龚坚, 罗海文. 材料工程, 2015, 43(6): 102-112.
[2] 王媛, 孙竹. 电工材料, 2019(6): 38-40.
[3] 陈春梅, 高振宇, 李亚东, 等. 鞍钢技术, 2019(4): 28-32.

新一代高技术冷连轧机电工钢同板差板形控制研究进展与创新发展趋势

曹建国[1,2,3,4,5]，宋纯宁[1,2,3,4,5]，孙 磊[1,2,3,4,5]，
赵秋芳[1,2,3,4,5]，夏文辉[1,2,3]，孙双涛[1,2,3]

（1. 北京科技大学机械工程学院，北京 100083；2. 北京科技大学人工智能研究院，北京 100083；3. 北京科技大学顺德创新学院，广东佛山 528399；4. 北京科技大学国家板带生产先进装备工程技术研究中心，北京 100083；5. 北京科技大学高效轧制与智能制造国家工程研究中心，北京 100083）

摘 要：针对新一代高技术冷连轧机电工钢"Dead Flat"矩形断面同板差高精度板形控制问题，比较分析了国际上不断探索的新一代高技术国际主流的典型四辊冷连轧机（含CVC-4、SmartCrown、T-WRS和T-WRS&C等）、无/有工作辊液压窜辊系统的六辊冷连轧机（含UCM、CVC-6、VCMS/UCMW、CVC-6+EDC等）机型与同板差板形控制研究进展，结合我国电工钢骨干工业生产线研发具备工作辊液压窜辊系统的ECC-4四辊边降与凸度紧凑冷连轧机和适应无工作辊液压窜辊系统的ECC-6六辊边降与凸度紧凑冷连轧机及其电工钢同板差板形控制实践。近年来为了支撑实现电工钢高精度同板差板形控制，宽带钢冷连轧机在线升级改造为有工作辊液压窜辊系统的5机架甚或新建6机架"SUPER MILL"全六辊生产线的冷连轧机组极致追求具有重要探索实践价值，但成本高且难度大，更应重点关注结合我国自主研发的新一代高技术热连轧全板形融合π（PCFC All-in-one Integrated，PAI）机型的宽带钢热、冷连轧机"矩形断面"同板差全流程一体化板形控制创新方案，为从根本上突破解决新一代高技术冷连轧机矩形断面同板差高精度板形控制瓶颈难题提供创新实践路径，并展望了新一代高技术轧机电工钢同板差板形控制创新发展趋势。

关键词：冷连轧机；电工钢；轧机机型；同板差；板形控制

Research Progress and Innovative Development Trends of Transverse Thickness Difference for Shape Control of Electric Steel in the New-generation High-tech Tandem Cold Rolling Mills

CAO Jianguo[1,2,3,4,5], SONG Chunning[1,2,3,4,5], SUN Lei[1,2,3,4,5],
ZHAO Qiufang[1,2,3,4,5], XIA Wenhui[1,2,3], SUN Shuangtao[1,2,3]

(1. School of Mechanical Engineering, University of Science and Technology Beijing, Beijing 100083, China; 2. Institute of Artificial Intelligence, University of Science and Technology Beijing, Beijing 100083, China; 3. Shunde Innovation School, University of Science and Technology Beijing, Foshan 528399, China; 4. National Engineering Research Center of Flat Rolling Equipment, University of Science and Technology Beijing, Beijing 100083, China; 5. National Engineering Research Center for Efficient Rolling and Intelligent Manufacturing, University of Science and Technology Beijing, Beijing 100083, China)

Abstract: This text discusses the high-precision shape control problem of the "Dead Flat" rectangular section of the new generation high-tech tandem cold rolling mills (TCMs) for electrical steel. It compares and analyzes the international mainstream new generation high-tech 4-high TCMs (including CVC-4, SmartCrown, T-WRS, and T-WRS&C, etc.) and 6-high TCMs (including UCM, CVC-6, VCMS/UCMW, CVC-6+EDC, etc.) with or without work roll shifting hydraulic system, as well as the research progress and increasingly stringent and complex bottleneck issues of transverse thickness difference for shape control. In combination with the research and development of China's backbone industry production lines for electrical steel, the text also introduces ECC-4 (4-high Edge drop and Crown Compact) TCMs with work roll hydraulic shifting system, and ECC-6 (6-high Edge drop and Crown Compact) TCMs suitable for without work roll hydraulic shifting systems, as well as their practical application for transverse thickness difference for shape control of electrical steel. In recent years, to support the transverse thickness difference for high-precision shape control for electrical steel, upgrading and transforming wide-strip TCMs into 5-stand or even newly built 6-stand 6-high "SUPER MILL" TCMs production lines for the ultimate pursuit with work roll hydraulic shifting system have important practical value for exploration, but they are very costly and difficult. Therefore, more attention should be paid to the innovative solution of the integrated shape control of the "rectangular section" transverse thickness difference for the hot/cold rolling mill of wide strip steel with the new generation high-tech hot rolling π (PCFC All-in-one Integrated, PAI) mill. This provides an innovative practical path to fundamentally break through the bottleneck of shape control for the new generation high-tech TCMs for electrical steel, and anticipates the innovative development trend of transverse thickness difference for shape control with high-tech rolling mills of electrical steel.

Key words: cold tandem rolling mill; electrical steel; mill type; transverse thickness difference; shape control

无取向硅钢热轧常化工艺研究

张 健[1,2]，张智义[1,2]，罗 理[1,2]，胡万卿[1,2]，姜福建[1,2]，张仁波[1,2]

（1. 海洋装备用金属材料及其应用国家重点实验室，辽宁鞍山 114009；
2. 鞍钢集团钢铁研究院，辽宁鞍山 114009）

摘 要：为了响应国家能源战略部署，家电行业节能增效尤为重要，作为家电压缩机主要材料的中低牌号的无取向硅钢电磁性能提升成为研究热点，为满足用户对性能的需求，中低牌号无取向硅钢开始考虑进行常化处理，本文以50W800作为研究对象实验钢，在实验室进行热轧，加热温度为1150℃，热轧厚度为2.3mm，利用VL2000DX高温激光共聚焦显微镜模拟常化工艺，常化温度设定分别为840℃、880℃、920℃、960℃、1000℃，1100℃，在线观察其组织演变过程。同时设定不同加热速率，分析加热速率对组织演变的影响。最终通过实验发现，常化温度为920℃，加热速率为10℃/s时，平均晶粒尺寸可达148μm。综合考虑根据此工艺在实验室模拟常化、酸洗、冷轧、退火后测其性能为$P_{1.5}$=3.47W/kg，B_{50}=1.772T，再结合现场实际工况为最合理工艺。

关键词：无取向硅钢；激光共聚焦；常化；再结晶

Reserch on Hot Rolling Normalization Process of Nonoriented

ZHANG Jian[1,2], ZHANG Zhiyi[1,2], LUO Li[1,2], HU Wanqing[1,2],
JIANG Fujian[1,2], ZHANG Renbo[1,2]

(1. State Key Laboratory of Metal Materials for Marine Equipment and their Applications, Anshan 114009, China; 2. Ansteel Group Iron and Steel Research Institute Liaoning, Anshan 114009, China)

Abstract: In response to the national energy strategy, home appliance industry is particularly important, energy saving and efficiency increasing as main materials, household appliances compressor without orientation of low and medium grade silicon steel electromagnetic performance improvement become the research hot spot, in order to meet user demand for performance, low grades often treated without oriented silicon steel began to consider, in this paper, the experimental steel 50W800 as a research object, In the laboratory, the hot rolling was carried out at 1150℃ and the thickness of the hot rolling was 2.3mm. The normalization process was simulated by VL2000DX high temperature confocal laser microscope. The normalization temperature was set at 840℃, 880℃, 920℃, 960℃, 1000℃ and 1100℃, respectively, and the tissue evolution process was observed online. At the same time, different heating rates were set to analyze the effects of heating rates on the tissue evolution. The results show that the average grain size can reach 148μm when the normalization temperature is 920℃ and the heating rate is 10℃/s. According to the comprehensive consideration of this process, the performance of P1.5=3.47W/kg and B50=1.772T were measured after the simulation of normalization, pickling, cold rolling and annealing in the laboratory, which is the best performance under the current process.

Key words: non-oriented silicon steel; laser confocal; normalize; recrystallization

高牌号无取向电工钢 50W270 工艺优化研究

刘文鹏[1,2]，李亚东[1,2]，李 浩[3]，陈春梅[1,2]，孙 超[1,2]，王祥辉[3]

（1. 海洋装备用金属材料及其应用国家重点实验室，辽宁鞍山 114009；2. 鞍钢集团钢铁研究院，辽宁鞍山 114009；3. 鞍钢股份冷轧硅钢厂，辽宁鞍山 114009）

摘 要： 对高牌号无取向硅钢 50W270 进行在线工艺调整试验，依次针对常化、连退工艺和气氛进行单工艺调整。结果表明，当常化工艺在试验工艺区间内变化时，常化板晶粒尺寸组织遗传性弱，低温中速（950℃-28m/min）常化工艺对应的成品板铁损值最低。当连退工艺在试验工艺区间内变化时，铁损随温度升高呈单调降低趋势。50W270 由于硅铝含量和连退温度高，易产生内氧化现象。连退生产时需要采用干气氛，避免氧化问题发生。

关键词： 无取向硅钢；常化工艺；连退温度；内氧化

The Study on Process Optimization of High Grade Non-oriented Electrical Steel 50W270

LIU Wenpeng[1,2], LI Yadong[1,2], LI Hao[3], CHEN Chunmei[1,2], SUN Chao[1,2], WANG Xianghui[3]

(1. State Key Laboratory of Metal Materials and Applications for Marine Equipment, Anshan 114009, China; 2. Ansteel Group Iron and Steel Research Institute, Anshan 114009, China; 3. Cold Rolled Silicon Steel Plant of Angang Steel Co., Ltd., Anshan 114009, China)

Abstract: By conducting online process adjustment tests on high grade non-oriented silicon steel 50W270, we got following results: When the normalization process changes within the experimental process range, the heritability of the grain size structure of the normalization plate is weak, and the iron loss value of the finished plate corresponding to the low-temperature medium speed (950℃-28m/min) normalization process is the lowest. When the continuous annealing process changes within the experimental process range, the iron loss shows a monotonic decreasing trend with increasing temperature. During continuous production, a dry atmosphere is required to avoid oxidation issues.

Key words: non oriented silicon steel; normalization process; continuous annealing temperature; internal oxidation

薄规格取向硅钢脱碳退火工艺研究

王艺橦，刘旭明，郭 函，耿志宇，薛 峰

（鞍钢集团北京研究院有限公司，北京 102211）

摘 要：取向硅钢由于取向硅钢工艺窗口较窄，因此被认为是钢铁工业的明珠[1-3]。脱碳退火工艺对其组织和织构的研究至关重要，可为后续的高斯晶粒异常长大提供基础。本文采用薄规格取向硅钢进行脱碳退火工艺，对退火温度及退火时间进行研究，探究其对织构的影响。

本文采用商用 0.27mm 冷轧取向硅钢进行脱碳退火，脱碳退火工艺为 1 号试样 820℃ 150s，2 号试样 840℃ 120s，3 号试样 860℃ 120s，4 号试样 820℃ 120s。

由图 1 可知，脱碳退火后，试样中均含有(114)[841]、(111)[11$\bar{2}$]等有利织构和(001)[120]、(001)[520]等不利织构及少量的 Goss 织构(011)[100]。3 号试样中具有较强的有利织构，其平均晶粒尺寸较小，为 20.59μm。退火过程

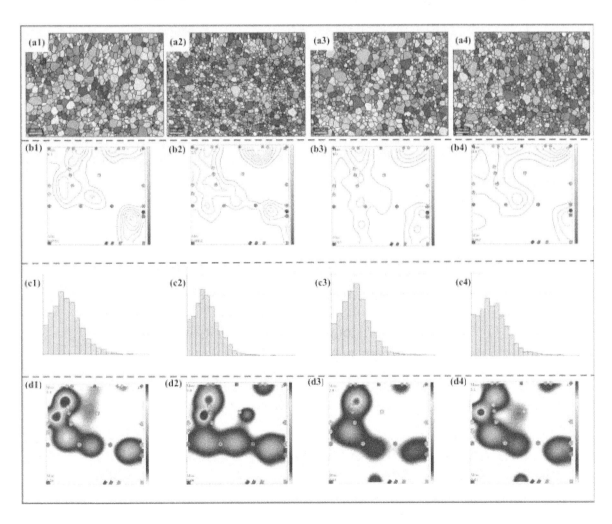

图1 1~4 号试样的 IPF(a)、ODF(b)、尺寸分布(c)及宏观织构(d)

（a1 为 1 号试样，以此类推）

中存在(114)[841]织构，即 α*织构，这是由于在轧制变形时会形成类似于剪切带的变形，从而在再结晶过程中形成(114)[841]晶核，在大角晶界迁移作用下形成(114)[841]织构。初次再结晶过程中形成的 α*织构可促进二次再结晶过程中 Goss 织构的异常长大。

退火时间相同时，随着退火温度升高，试样的晶粒尺寸先增大后减小，有利织构含量先减少后增加。3 号试样具有较好的脱碳效果，可促进二次再结晶过程中 Goss 织构的异常长大。

参 考 文 献

[1] 何忠治，赵宇，罗海文. 电工钢[M]. 北京: 冶金工业出版社, 2012.
[2] 毛卫民，杨平. 电工钢的材料学基础[M]. 北京: 高等教育出版社, 2013.
[3] 王洋. 薄带连铸取向硅钢成分设计与组织织构调控机理的研究[D]. 沈阳: 东北大学, 2017.

薄规格无取向电工钢冷轧工艺优化

耿志宇[1]，李亚东[2]，孙 超[2]，刘旭明[1]，董林硕[1]，郭 函[1]

（1. 鞍钢集团北京研究院有限公司硅钢组，北京 102211；
2. 鞍钢集团钢铁研究院硅钢所，辽宁鞍山 114000）

摘 要：利用钢铁企业生产的常化板为原材料，研究了在不同二次冷轧工艺和中间退火温度下 0.2mm 规格新能源电工钢的力学性能、磁性能、晶粒尺寸和织构的变化规律，以优化 0.2mm 新能源电工钢的冷轧工艺。结果表明，冷轧工艺和中间退火温度对 0.2mm 新能源电工钢的织构和磁性能有较大影响。当一次冷轧压下率小于 60%（即压下至小于 0.8mm）时，最终产品的织构以 γ 纤维和 α*纤维织构为主。当一次冷轧压下至 0.4mm 时，最终产品织构以高斯织构和立方织构为主，此时的磁极化强度达到最大值 1.65T。当一次冷轧压下至 0.8mm 且 950℃中间退火时，最终产品的磁性能最优。

关键词：无取向硅钢；新能源电工钢；二次冷轧；织构

Optimization of Cold Rolling Process of Thin Gauge Non-oriented Electrical Steel

GENG Zhiyu[1], LI Yadong[2], SUN Chao[2], LIU Xuming[1], Dong Linshuo[1], GUO Han[1]

(1. Ansteel Beijing Research Institute Co.,Ltd., Silicon Steel Group, Beijing 102211, China;
2. Ansteel Group Iron and Steel Research Institute, Silicon steel group, Anshan 114000, China)

Abstract: The mechanical properties, magnetic properties, grain size and texture of 0.2mm new energy electrical steel under different cold rolling process and intermediate annealing temperature were studied by using normalized plate produced by industrial production, to optimize the cold rolling process of 0.2mm new energy electrical steel. The results show that the cold rolling process and the intermediate annealing temperature have great influence on the texture and magnetic properties of 0.2mm new energy electrical steel. When the reduction rate of the first cold rolling is less than 60% (the reduction is less than 0.8mm), the texture of the final product is dominated by γ fiber and α* fiber texture. When the first cold rolling is reduced to 0.4mm, the texture of the final product is mainly goss texture and cubic texture, and the magnetic polarization intensity reaches the maximum value of 1.65T. When the first cold rolling is reduced to 0.8mm and intermediate annealing reach to 950℃，The magnetic polarization intensity reached the maximum 1.65T.

Key words: non-oriented silicon steel; new energy electrical steel; two-stage cold rolling; texture

取向硅钢硅酸镁层质量的影响因素及其控制研究

张　刚[1,2]，胡　煜[3]，赵天亮[1,2]，胡丞杨[1,2]，宗　俊[4]，吴开明[1,2]

（1. 武汉科技大学高性能钢铁材料及其应用省部共建协同创新中心，湖北武汉　430081；
2. 武汉科技大学耐火材料和冶金国家重点实验室，湖北武汉　430081；3. 中科信工程咨询（北京）有限公司，北京　100037；4. 上海实业振泰化工有限公司，上海　201424）

摘　要： 取向硅钢轧制成成品厚度后，经脱碳退火、渗氮并在表面涂布氧化镁涂层等工序，随后进入高温罩式炉内退火，使高斯方位晶粒发生二次再结晶，形成{110}<001>高斯织构。氧化镁涂层作为高温退火隔离剂，不仅可以防止在高温退火时钢带发生粘结，更重要的是与钢带表面的氧化膜层发生固相反应，生成玻璃质的硅酸镁层（镁橄榄石层）。硅酸镁层作为取向硅钢的重要组成部分，不仅能够增加取向硅钢表面的绝缘电阻，而且对钢基体产生张应力，能有效提升取向硅钢产品的磁性能。获得无缺陷、结构致密的硅酸镁层是高性能取向硅钢的追求目标，同时也是取向硅钢生产工艺控制的技术难点。本文通过对硅酸镁层质量的三个影响因素的分析研究，提出不同影响因素的工艺控制方案，为取向硅钢在生产过程中形成稳定连续的致密化硅酸镁层提供理论和数据支撑，满足高性能取向硅钢对硅酸镁层的性能要求。

关键词： 硅酸镁层；质量控制；取向硅钢；脱碳退火；高温罩式炉退火

Study on the Influence Factors and Control of Quality of Mg_2SiO_4 Layer on the Surface of Grain-oriented Electrical Steel

ZHANG Gang[1,2], HU Yu[3], ZHAO Tianliang[1,2],
HU Chengyang[1,2], ZONG Jun[4], WU Kaiming[1,2]

(1. Collaborative Innovation Center for Advanced Steels, Wuhan University of Science and Technology, Wuhan 430081, China; 2. The State Key Laboratory of Refractories and Metallurgy, Wuhan University of Science and Technology, Wuhan 430081, China; 3. Zhongkexin Engineering Consulting (Beijing) Co., Ltd., Beijing 100037, China; 4. Shanghai SIIC Zhentai Chemical Co., Ltd., Shanghai 201424, China)

Abstract: After decarburization annealing and nitriding treatment, grain-oriented electrical steel strip was first coated with magnesia coating and then followed the final box annealing process for the Goss grains secondary recrystallisation. Magnesia coating, as a high-temperature annealing separator, can not only prevents the bonding of steel strips during the final box annealing process, but more importantly, generates a glass film that is Mg_2SiO_4 (forsterite) layer by the solid-state chemical reactions with the oxide layer on the surface of the steel strip under high-temperature. The formed Mg_2SiO_4 layer is an important component of grain-oriented electrical steel which on one hand provides the insulation resistance, and on the other hand generates tensile stresses on the steel which effectively improving the magnetic performance of the grain-oriented electrical steel. Obtaining an excellent Mg_2SiO_4 layer is a technical difficulty and also a pursuing goal in the production process of high-grade grain-oriented electrical steel. This article analyzes the three factors affecting the quality of Mg_2SiO_4 layer an provide corresponding control solutions for different factors. All these studies provide theoretical and data support for the production of high-quality grain-oriented electrical steel on site.

Key words: Mg$_2$SiO$_4$ layer; quality control; grain-oriented electrical steel; decarburization annealing; final box annealing

30W1500 无取向硅钢磁性能优化研究

郭飞虎[1,2]，刘 磊[2]，乔家龙[2]，廖德桥[3]，付 兵[3]，仇圣桃[2]

（1. 东北大学冶金学院，辽宁沈阳　110819；2. 钢铁研究总院有限公司连铸技术国家工程研究中心，北京　100081；3. 新余钢铁股份有限公司硅钢事业部，江西新余　338001）

摘　要： 结合工业试验、ASPEX、EBSD 和 XRD 等检测方法研究 30W1500 无取向硅钢中残余元素及常化温度对成品板组织、织构及对磁性能的影响。结果表明：钢中残余元素（Ti 和 S）的降低，夹杂物的分布密度由 155.3 减少到 127.3 个/mm^2，对晶粒长大的抑制能力 F_{Zener} 由 2.08×10^4 降低到 1.52×10^{4}Pa，成品板晶粒尺寸增大，$P_{1.0/400}$ 由 14.85 降到 14.62W/kg。常化温度的升高，常化板晶粒尺寸的增加，形成强{100}<012>和{114}<418>织构。随着常化板晶粒的增加，成品板中 γ 纤维织构含量降低，{114}<481>织构的含量增加，产品磁性能提高。

关键词： 无取向硅钢；残余元素；常化；组织；织构；磁性能

Optimization of Magnetic Properties of 30W1500 Non Oriented Silicon Steel

GUO Feihu[1,2], LIU Lei[2], QIAO Jialong[2], LIAO Deqiao[3], FU Bing[3], QIU Shengtao[2]

(1. School of Metallurgy, Northeastern University, Shenyang 110819, China; 2. National Engineering Research Center of Continuous Casting Technology, Central Iron and Steel Research Institute Co., Ltd., Beijing 100081, China; 3. Silicon Steel Division, Xinyu Iron and Steel Co., Ltd., Xinyu 338001, China)

Abstract: By combining industrial experiments, ASPEX, EBSD, and XRD detection methods, the effects of residual elements and normalizing temperature on the microstructure, texture, and magnetic properties of 30W1500 non oriented silicon steel were studied. The results show that the reduction of residual elements (Ti and S) in the steel results in a decrease in the distribution density of inclusions from 155.3 to 127.3pieces/mm^2, and a decrease in F_{Zeners} ability to suppress grain growth from 2.08 ×10^4 reduced to 1.52 ×10^{4}Pa, the grain size of the finished plate increased, and $P_{1.0/400}$ decreased from 14.85 to 14.62W/kg. The increase in normalizing temperature increases the grain size of the normalizing plate, resulting in the formation of strong {100}<012>and {114}<418>textures. As the grain size of the normalized plate increases, the content of fiber γ texture in the finished plate decreases, the content of {114}<481>texture increases, and the magnetic properties of the product improve.

Key words: non oriented silicon steel; residual elements; normalization; organization; texture; magnetic properties

硅钢连续退火炉炭套结瘤机理及防控措施

何明生，张　敬，龚学成，丁　勇

（武汉钢铁有限公司，湖北武汉　430080）

摘　要：石墨具有独特的叠层结构，硬度低，摩擦系数小，石墨炭套是硅钢连续退火炉中用于钢带支撑和传输最好的炉辊。在高温和 H_2-N_2-H_2O 弱氧化或 H_2-N_2 强还原气氛条件下，炭套使用一段时间后，其表面可能形成粘附瘤子，而表层孔洞内可能形成嵌入式瘤子。尤其生产一些高牌号无取向硅钢和含低熔点金属的低、中牌号无取向硅钢时更容易形成嵌入式瘤子。一旦在炭套表面或表层形成瘤子，容易造成压印、擦伤和划伤，严重影响带钢表面质量，甚至造成改判或废品。近年来，炭套结瘤的原因和机理引起了人们的极大关注，炭套结瘤的真正原因仍然不清楚。

根据对各种瘤子的形貌、微观结构和成分分析，结果表明瘤子的主要成分是 Fe，还有少量 Fe、Mn、Al、Si 的氧化物或复合氧化物。一些中、低温炭套表面的粘附瘤子或表层的嵌入式瘤子中 P 含量较高。嵌入式瘤子形成经历了"成核-长大"过程，由明暗相间的同心圆环状浅色带和深色带组成，其形成有两个必要条件：（1）炭套表层有较大孔洞；（2）有丰富的可形成瘤子的物料来源。在炭套生产过程中形成的孔洞以及在硅钢连续退火炉内因为氧化而形成的孔洞为顽固的嵌入式瘤子形成提供了必要条件。当炭套表层有较大孔洞和钢带表面携带丰富的氧化铁皮、铁锈、油污、粉尘等形成瘤子的物质时就可能在较短时间内形成顽固的瘤子。影响炭套结瘤的主要因素包括炭套质量、炉内气氛、露点、碱洗不干净、炉内不干净等。

瘤子中 70%~95%都是铁，而且所有瘤子由微米级甚至纳米级晶粒组成，但在 800~950℃而没有外力的条件下，坚硬致密的瘤子很难形成。通过在线模拟试验证明在 750℃上，纳米 Fe_2O_3 和 Fe_3O_4 粉末在还原气氛下可以形成坚硬的烧结体。公所周知，纳米金属颗粒的熔点与颗粒大小有关，颗粒越小，其熔点越低。因此，瘤子的形成经历了"成核-长大"的过程，通过纳米液相烧结而成。从现有生产工艺、设备和条件来看，要完全消除结瘤还不可能，但采取合适的控制技术和预防措施可以减少结瘤，甚至杜绝瘤子的形成，延长炭套的使用寿命，提高硅钢的生产效率和产品质量，降低硅钢生产成本。防控炭套结瘤的主要方法和措施：

（1）进一步改善炭套的表面质量；
（2）改进除鳞设备和工艺，加强酸洗；
（3）在线严格监控炉内气氛和露点；
（4）强化钢带清洁，加强碱洗；
（5）定期清理炉内粉尘、异物；
（6）不要使用质量较差的调整材；
（7）合理安排生产，避免频繁进行工艺调整。

电工钢板形控制与优化实践

张　庭，钱震茂，傅潇然

（武汉钢铁有限公司，湖北武汉　430080）

摘　要：电工钢板形控制一直是热轧板形控制的难点。通过分析某热轧厂电工钢生产中出现的板形问题，找出影响板形的主要因素，提出相应的改进措施。通过不断优化模型和电气控制精度，确保活套稳定动作，提高轧制稳定性，降低了板形缺陷，减少了因为板形不良而导致的废钢。

关键词：热轧；板形；控制；活套

Control and Optimization of Electrical Steel Plate Shape

ZHANG Ting, QIAN Zhenmao, FU Xiaoran

(Wuhan Iron and Steel Co., Ltd., Wuhan 430080, China)

Abstract: The electrical shape control is always the difficulty of hot rolling shape control. By analyzing the shape problems of Electrical steel produced in a hot rolling plant, the main factors affecting the shape are found out, and the corresponding improvement measures are proposed. By continuously optimizing the model and electrical control accuracy, the stable operation of the loop is ensured, rolling stability is improved, plate shape defects are reduced, and scrap caused by poor plate shape is reduced.

Key words: hot rolling; plate shape; control; looper

不同加工方式对新能源驱动电机用无取向电工钢磁性能的影响

吴圣杰，岳重祥，钱红伟，麻 晗

（江苏省（沙钢）钢铁研究院，江苏张家港 215600）

摘 要： 系统对比研究机械加工和激光切割两种加工方式对新能源驱动电机用无取向电工钢磁性能的影响。研究结果表明，在不同励磁条件下对比，激光切割样条的矫顽力和铁损均高于机械剪板加工。通过电子背散射衍射（EBSD）技术对晶粒内应力分析发现，剪板加工样条切断面存在应力集中区，而激光切割样条切断面附近则无明显应力集中区。但通过切断面能谱（EDS）分析，发现激光切割样条在切断面处包裹一层厚度为 1μm 的氧化层。该氧化层导致激光切割样条在磁感应强度为 1.0T 条件下，磁滞损耗增加，且随加载频率的升高，影响呈放大趋势。但当磁感应强度增加至 1.5T 后，氧化层对磁滞损耗的影响降低，反常损耗影响增加。

关键词： 无取向电工钢；激光切割；磁滞损耗；反常损耗

Effect of Processing Methods on the Magnetic Properties of Non-oriented Electrical Steel for Electrical Vehicle

WU Shengjie, YUE Chongxiang, QIAN Hongwei, MA Han

(Institute of Research of Iron and Steel, Shagang, Jiangsu Province, Zhangjiagang 215600, China)

Abstract: Effects of two sample processing methods, mechanical punching and laser cutting, on the magnetic properties of non oriented electrical steel for electrical vehicle were studied. The results show that the coercivity and iron loss of laser cutting samples are higher than those of mechanical punching, regardless of the comparison of 20 groups of parallel data or a single group of samples under different excitation conditions. Through the analysis of the internal stress in the grain by electron backscatter diffraction (EBSD) technique, it is found that there is a stress concentration area in the cutting section of punching sample, while there is no obvious stress concentration area near the cutting section of laser cutting sample. However, through the energy dispersive spectrum (EDS) analysis of the cutting surface, it is found that the laser cutting sample wraps an oxide layer with thickness of 1μm. The oxide layer causes the hysteresis loss of laser cutting sample to increase when the magnetic induction intensity is 1.0T, and the influence increases with the increase of loading frequency. However, when the magnetic induction intensity increases to 1.5T, the influence of oxide layer on hysteresis loss decreases and the influence of abnormal loss increases.

Key words: non-oriented electrical steel; laser cutting; hysteresis loss; abnormal loss

10.7 非晶合金

Ga 对 $Fe_{83.3}Si_2B_{10}P_4Cu_{0.7}$ 纳米晶合金结构及软磁性能的影响

韩成府，李福山

(郑州大学材料科学与工程学院，河南郑州　450000)

摘　要：Nanomet 系列（FeSiBPCu）纳米晶软磁合金具有高磁饱和、低矫顽力及高磁导率等优异的软磁性能，且未含有贵重元素，成本较低，因此成为具有发展潜力的软磁合金体系。但由于 Nanomeet 合金非晶形成能力低，较难制得非晶前驱体，且在纳米晶化过程中需要极快的加热速率，制约了其工业化进程。本文基于 Ga 与 Fe 具有负的混合焓，且 Ga 可提高 Fe 的原子磁矩等特点，研究了 Ga 替代 Si 对于 $Fe_{83.3}Si_{2-x}B_{10}P_4Cu_{0.7}Ga_x$（x=0, 0.5, 1, 1.5, 2 at.%）结构及软磁性能的影响，同时探索常规退火和快速退火两种不同处理工艺对于纳米晶软磁合金的影响。

研究结果表明，Ga0.5 和 Ga1 能够获得非晶带材，未添加或者添加较多 Ga 时淬态带材自由面存在表面晶化现象（图1）；通过对其热物性进行分析可知，随着 Ga 含量的增加，初晶相 α-Fe 的形成温度 T_{x1} 和 Fe(B, P) 相的形成温度 T_{x2} 均有所降低，但是 T_{x1} 降低较多（图2），使得 ΔT ($T_{x2}-T_{x1}$) 增加，有利于拓宽获得优异纳米晶结构的退火温度区间。采用常规退火（加热速率约为 2K/s），其矫顽力 Hc 呈现"W"变化趋势，且去应力阶段的矫顽力低于纳米晶化阶段（图3），而采用快速退火（加热速率约为 590K/s），其矫顽力较淬态先下降后上升，主要呈现"√"变化趋势（图4）。

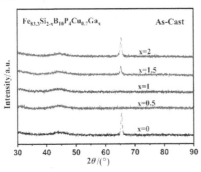

图 1　淬态 XRD 结果

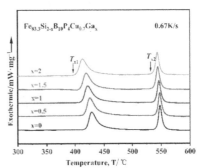

图 2　淬态 DSC 结果

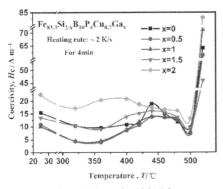

图 3　常规退火后矫顽力

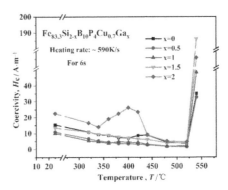

图 4　快速退火后矫顽力

因此，采用 Ga（0～1at.%）替代 Si 对于 $Fe_{83.3}Si_{2-x}B_{10}P_4Cu_{0.7}Ga_x$ 纳米晶软磁合金在一定程度上提高其非晶形成能，降低初晶相 α-Fe 的形成温度，增加 ΔT 区间，同时采用快速退火工艺能够有效地细化晶粒，进一步提升软磁性能。

$FeCrNi_2Nb_{0.1}$ 中熵合金的微观结构调控及其对力学性能的影响

李福山，刘孝威

（郑州大学材料科学与工程学院，河南郑州　450000）

摘　要：由于多种元素的混合引起的晶格畸变和扩散缓慢，使得中/高熵合金具有稳定的组织和优异的力学性能，尤其是单相 FCC 中/高熵合金具有极高的低温断裂韧性。然而，由于 FCC 晶体结构的特点，造成合金的屈服强度较低。因此，研究人员致力于研究如何获得高强、高韧的中/高熵合金。一方面，可通过成分设计获得双相结构以提高强度，如片层组织的共晶中/高熵合金兼具良好的力学性能和优异的铸造性能；另一方面，可通过晶粒细化等组织调控，从而显著提高强度。但传统的强化方式往往伴随着明显的塑性降低，能否同时调控相结构和晶粒尺寸匹配以同时获得高强度、大塑性的中/高熵合金？

本研究从共晶系 FeCrNi2Nbx 中熵合金选取具有优异变形能力的 $FeCrNi_2Nb_{0.1}$ 中熵合金作为研究对象，其具有 FCC 基体+Laves 相的结构特点。其后，采用多种热机械处理工艺（调控变形量、变形温度、轧制工艺和退火温度等）调控 Laves 相分布和晶粒组织，系统研究不同微观结构对力学性能的影响规律。研究结果表明，深冷轧制+退火处理的工艺设计有助于获得具有多级异构组织的中熵合金，其结构组成为大量具有退火孪晶的完全再结晶晶粒、少量具有高密度位错密度的未再结晶晶粒以及小尺寸弥散分布的 Laves 相。相比于其他具有均匀尺寸晶粒结构或大量未再结晶区域结构的合金，具有多级异构组织的合金表现出最佳的屈服强度和均匀延伸率（约720MPa 和约23%），且具有较好的加工硬化能力。这种良好的强度-塑性匹配可归因于多种的强化机制贡献，包括细晶强化、位错强化、变形孪晶强化和异质结构强化。本研究为设计高强、高韧的双相中/高熵合金提供了一种可行的方法和思路。

金属玻璃力学谱的考察以及粘弹性模型的构造

王　昊[1]，张广强[2]，董帮少[2]

（1. 日本东北大学未来科学技术共同研究中心，日本宫城仙台　980-8577；
2. 江苏集萃安泰创明先进能源材料研究院有限公司，江苏常州　213032）

摘　要：我们在等温条件下测量了代表性合金成分 $Zr_{55}Cu_{30}Ni_5Al_{10}$ 金属玻璃的动态模量，并按照温度-时间叠加法创建了储能和损耗模量构成的主曲线长度为 10 位的角频率 $10^{-4}<ω<10^4 rad/s$。此外，根据 Arrhenius 方程确定了各种弛豫模式下的 transfer factor 与温度的相关性。α-, slow β-, 和 fast β-弛豫的 pre-exponential factors 与表观活化能分别得到为 5.1eV 与 $5.1×10^{41} rad/s$，1.6eV 与 $1.1×10^{12} rad/s$，和 0.04eV 与 $1×10^{14} rad/s$。此外，动态粘弹性 Maxwell 模型定量再现了储能模量和损耗模量的力学谱。

关键词：金属玻璃；动态粘弹性；力学谱；拟和

Investigation of the Mechanical Spectrum of Metallic Glass and the Construction of a Viscoelastic Model

WANG Hao[1], ZHANG Guangqiang[2], DONG Bangshao[2]

(1. New Industry Creation Hatchery Center, Tohoku University, Sendai 980-8577, Japan; 2. Jiangsu JITRI Advance Energy Materials Research Institute Co., Ltd., Changzhou 213032, China)

Abstract: The dynamic modulus of the representative alloy composition $Zr_{55}Cu_{30}Ni_5Al_{10}$ metallic glass was measured under isothermal conditions, and a master curve of storage and loss modulus at an angular frequency of 10 digits in length of $10^{-4}<\omega<10^4$ rad/s was created based on the temperature-time superposition rule. Further, various relaxation modes from the temperature dependence of the transfer factors were identified according to the Arrhenius equation. The pre-exponential factors and apparent activation energies for α-, slow β-, and fast β-relaxations were derived as 5.1eV and 5.1×10^{41} rad/s, 1.6eV and 1.1×10^{12} rad/s, and 0.04eV and 1×10^{14} rad/s, respectively. In addition, the dynamic viscoelastic Maxwell model quantitatively reproduced the mechanical spectra of the storage and loss moduli.

Key words: metallic glass; dynamic viscoelasticity; mechanical spectrum; fitting

FeSiBC 混合 FeSiCr 或 CIP 复合磁粉芯软磁性能改善及机理分析

王璞[1]，朱争取[1]，庞靖[2]，张家泉[1]

（1. 北京科技大学冶金与生态工程学院，北京 100083；
2. 青岛云路先进材料技术股份有限公司，山东青岛 266232）

摘 要： 小型化、集成化与高功率化是当前电子终端产品的发展趋势[1]。Fe 基非晶粉是近几十年的一种兼具高饱和磁化强度、低矫顽力和低损耗的新型软磁粉末，是制备高性能的小型电感的理想原料[2]。然而，非晶粉因结构的特殊而硬度极大，导致制备的磁粉芯强度差、磁导率较低，如何在提高非晶磁粉芯的密度和磁导率的同时不恶化其总损耗，是阻碍非晶磁粉芯软磁性能进一步跨越的关键性难题。在非晶粉中混入一定比例的细粉，可有效提高磁粉芯的压制性和致密性，是改善非晶磁粉芯的有效手段[3]。

本研究分别以三种不同粒度的 FeSiCr 水雾化粉和固定粒度的 CIP 为原料，探究细粉粒度和种类对大粒径的 FeSiBC SMPCs 软磁性能的影响。结果表明，由 SEM、EDS、XPS 和 FTIR 的结果分析可知，粉末包覆情况良好，表面均匀的包覆了 SiO_2 绝缘涂层，且三种粉均具有极高的 Ms(163.11~199.96 emu/g)，其中，FeSiBC 的 Hc 最低(0.88 Oe)，表明该粉末具有很高的非晶度。在对应的 SMPCs 中，通过加入一定比例的 FeSiCr 细粉，可以有效填充大粒径 FeSiBC 非晶粉之间的孔隙，使 SMPCs 的密度和磁导率随着 FeSiCr 含量的增加而单调增加，而孔隙率和直流偏置性能则单调减小。此外，FeSiCr 的粒度越小，越有利于降低 SMPCs 的损耗，这主要源于粉末的小粒径减小了涡流损耗。在混 FeSiCr 方案中，加入 10%（质量分数）平均粒径为 6 μm 的 FeSiCr 最优，此时 SMPCs 的磁导率为 18.01，损耗为 239.15 mW/cm(100kHz, 50mT)。FeSiBC、FeSiCr 和 CIP 三种粉末的硬度分别为（1171.2±101.6）HV、（838.8±97.7）HV 和（88.1±35.2）HV。与较硬的 FeSiCr 相比，CIP 因硬度极低导致其在压制过程中变形较大，能够更有效地填充 SMPCs 的孔隙，从而大幅度提高其密度、强度、磁导率并显著降低损耗。

本研究制备的 FeSiBC/30%CIP SMPCs 具有优异的软磁性能（磁导率为 26.26，直流偏置能力为 94.58% at

100 Oe，总损耗为 192.10mW/cm³，100kHz，50mT），较高的强度（(6.54±0.08) MPa）和较低的成本（13.7$/kg），综合性能显著优于粒配 FeSiCr。可见，通过在 FeSiBC 非晶粉中少量混合一种与其在硬度、粒度方面差距大且软磁性能各有互补的粉末，是提高 SMPCs 磁性能更直接有效的手段。

参 考 文 献

[1] Woo H J, Ahn J H, Kim C P, et al. Effect of the particle size classification of FeSiCrB amorphous soft magnetic composites to improve magnetic properties of power inductors[J]. J. Non Cryst. Solids, 2022(577): 121309.

[2] Zhou B, Dong Y Q, Liu L, et al. Enhanced soft magnetic properties of the Fe-based amorphous powder cores with novel TiO$_2$ insulation coating layer[J]. J. Magn. Magn Mater. 2019(474): 1-8.

[3] Chan C T, Guo J J, Li Q, et al. Improvement of soft magnetic properties of FeSiBPNb amorphous powder cores by addition of FeSi powder[J]. J. Alloys Compd, 2019(788): 1177-1181.

Si/B 比对高 B_s-FeSiBNbCu 纳米晶合金组织结构及磁性能的影响

王超人，李艳辉，张 伟

（大连理工大学材料科学与工程学院，辽宁大连 116024）

摘 要：Finemet 系纳米晶软磁合金因其极低的矫顽力（H_c）、高磁导率和低损耗等特性，而广泛地应用于高频变压器、电感器等磁性元器件中。随着电子电力器件向小型化和高效化发展，对软磁材料的性能提出了更高要求，Finemet 系合金较低的饱和磁感应强度（B_s）限制了其应用范围，因此研制具有优异软磁性的高 B_s 铁基纳米晶合金具有重要意义。本工作在 Finemet 系合金的基础上，提高 Fe 元素、降低 Nb 元素的含量，研究了 Si/B 比对 Fe$_{78}$Si$_x$B$_{19.5-x}$Nb$_{1.5}$Cu$_1$ (x = 9.5~13.5)急冷合金组织结构、热稳定性、结晶化组织以及磁性能的影响。

结果表明：x = 9.5~10.5 急冷合金为非晶态，而 x = 11.5~13.5 急冷合金在条带表面析出了 α-Fe 相织构。经热处理结晶化后，合金均得到了纳米 α-Fe/非晶双相结构。随着 Si/B 比的增加，合金的 B_s 逐渐降低，α-Fe 相平均晶粒尺寸（$\bar{D}_{\alpha\text{-Fe}}$）和 H_c 均呈先降低后上升的趋势；其中，x = 11.5 合金具有较微细的组织和良好的软磁性能，其 $\bar{D}_{\alpha\text{-Fe}}$、$B_s$ 和 H_c 分别为 26.8nm、1.53T 和 35.6A/m。适当提高 Cu 含量可使合金急冷态结构均形成非晶态，且热处理后形成的纳米晶组织更加均匀、微细，而显示优异的软磁性能。

关键词：Fe 基纳米晶软磁合金；Fe-Si-B-Nb-Cu 合金；Si/B 比；表面晶化；软磁性能

元素添加对 Finemet 系纳米晶合金高频软磁性能影响研究

郭树深，李艳辉，张 伟

（大连理工大学材料科学与工程学院，辽宁大连 116024）

摘 要：Finemet 系（Fe-Si-B-Cu-Nb）纳米晶软磁合金具有低矫顽力、高磁导率、低铁心损耗和极低的磁致伸缩系数,相比于传统晶态和非晶态软磁合金,其在高频下的综合软磁性能最佳,也是目前唯一实现大规模工业化应用的纳米晶软磁合金体系。随着电力电子行业向高频化、高效化和节能化方向发展,进一步提高 Finemet 系纳米晶合金高频下的软磁性能、降低损耗具有十分重要的意义。高频下,铁基纳米晶合金带材的磁滞损耗可忽略不计,总损耗主要由涡流损耗和剩余损耗组成,故降低带材涡流损耗对于降低总损耗至关重要。涡流损耗与合金带材电阻率密切相关,提高铁基纳米晶合金带材中非晶相或 α-Fe 纳米晶相的电阻率可能是降低涡流损耗的有效途径。本研究以 Fe-Si-B-Cu-Nb 纳米晶软磁合金为基础,添加适量可固溶于 α-Fe 相的 Al、Co 元素以及不固溶于 α-Fe 相的 Mo 元素,系统研究了元素添加对纳米晶合金带材的组织结构、电阻率、静态磁性能、高频磁导率和损耗的影响,并探讨了添加元素改善纳米晶合金高频软磁性能的机理。

结果表明:磁性能方面,适量添加 Mo 和 Al 元素可显著降低纳米晶合金带材的矫顽力(H_c),略微降低饱和磁感应强度(B_s),而添加 Co 元素未大幅改变 H_c,但使 B_s 明显提升;添加三种元素均可有效提高纳米晶合金的电阻率和高频磁导率,显著降低高频损耗,添加 Al、Co 和 Mo 元素分别使纳米晶合金在 100 kHz/0.2 T 下的总损耗降低 29.6%、10.9%和 26.8%,损耗分离结果显示元素添加均降低了纳米晶合金的涡流损耗,但对于剩余损耗的改变不大。组织结构方面,添加 Al 和 Mo 元素细化了纳米晶合金中 α-Fe 相平均晶粒尺寸,而添加 Co 元素未明显改变晶粒尺寸;添加三种元素未明显改变 α-Fe 晶粒的体积分数;添加 Co 和 Al 元素使得 α-Fe$_{(110)}$ 晶面间距变大。

关键词:铁基纳米晶软磁合金;高频损耗;磁导率;电阻率

难熔金属添加对 FeCoNiSiB 高熵块体非晶合金的形成和性能影响研究

李艳辉,张 伟

(大连理工大学材料科学与工程学院,辽宁大连 116024)

摘 要:块体非晶合金具有优异力学、物理和化学性能,特别是高强度、高弹性和优异耐腐蚀性能是其作为新型结构或功能材料应用的独特优势。但非晶合金为亚稳材料,当温度高于其结晶化温度时即发生结晶而失效,这限制了非晶合金高温下应用。高熵非晶合金因其多主元高熵效应,表现出比传统单一主元的非晶合金更为优异的热稳定性和抗结晶化能力,增加合金熵值是提高非晶合金热稳定性、减缓结晶速率的有效手段。此外,在合金中添加难熔金属(RM)或高熔点类金属元素也被证实可提高非晶合金的热稳定性和强度。前期我们成功研制出金属–类金属型 FeCoNi(B, Si, C, P)和 FeNiCrMo(P, C, B)高熵块体非晶合金体系,其具有远高于金属–金属型高熵非晶合金的热稳定性和强度,同时软磁性能/耐腐蚀性能优异。最近,为进一步提高非晶合金热稳定性,拓展其在高温领域应用,我们在 FeCoNi(B, Si)高熵非晶合金中加入 W、Mo、Nb、Ta 等 RM 元素,研究了 RM 对高熵非晶合金的热性能、非晶形成能力(GFA)和力学性能的影响。

结果表明,添加 RM 元素可以有效提高基础高熵合金的热稳定性和 GFA,合金的玻璃化转变温度和初始结晶化温度可分别由 772K 和 805K 提升至 829K 和 879K,形成非晶的临界直径也从 1.0mm 提高至 2.5mm;FeCoNiRM(B, Si)块体高熵非晶合金的强度也得到显著提高,其屈服强度可达 4.2GPa,并有一定的压缩塑性变形能力。我们通过第一性原理计算模拟了 RM 元素对基础高熵非晶合金局域结构的影响,讨论了添加 RM 元素提高合金热稳定性和 GFA 的机理。

Cr 添加对 FeBCu 纳米晶合金组织结构及磁性能的影响

郭明月，薛钧升，李艳辉，张 伟

（大连理工大学材料科学与工程学院，辽宁大连 116024）

摘 要：Fe 基纳米晶软磁合金具有低矫顽力（H_c）、高磁导率、低铁损、低磁致伸缩系数等特性，尤其是高频软磁性能优异，广泛应用于变压器、电动机、互感器等电能转换和传输器件中。随着电子电力器件向小型化和高效化发展，对 Fe 基纳米晶合金的性能提出了更高要求。最近研究报道了高饱和磁感应强度（B_s）的 Fe-B-Cu 纳米晶合金急冷带材非晶基体中存在高数密度的纳米 α-Fe 晶粒，在工况热处理条件下即可形成微细纳米 α-Fe 晶组织，显示出良好的软磁性能[1]。但该系合金的急冷前驱体结构不易控制且耐腐蚀性能不高，影响了其工程应用。本工作研究了 Cr 添加量对 $Fe_{85.5-x}B_{13}Cu_{1.5}Cr_x$ (x = 0~6) 系合金急冷带材的结构、结晶化组织和磁性能的影响，并评价了合金的耐腐蚀性能。

结果表明，$Fe_{85.5-x}B_{13}Cu_{1.5}Cr_x$ (x = 0~6) 急冷合金均为非晶基体中析出纳米 α-Fe 晶粒的复相结构，随 Cr 含量的增加，其 α-Fe 晶粒的数密度和平均尺寸逐渐减小。经热处理后，合金均形成了微细的纳米 α-Fe 晶组织结构；且随 Cr 含量的增加，其平均晶粒尺寸（$\bar{D}_{α-Fe}$）和 H_c 逐渐减小，B_s 值则逐渐降低，其中 x = 3 合金的 $\bar{D}_{α-Fe}$、H_c 和 B_s 分别为 14.9nm、13.8A/m 和 1.74T。添加 Cr 后合金的耐腐蚀性得到了明显改善。此外，对 Cr 含量、非晶前驱体结构、晶化组织和性能之间的关系也进行了探讨。

关键词：Fe 基纳米晶合金；Cr 添加；预存 α-Fe 晶粒；软磁性能；耐腐蚀性能

参 考 文 献

[1] 朱春健, 李艳辉, 朱正旺, 等. 金属学报, on line.

高 B_s 纳米晶合金非晶前驱体结构调控及其晶化组织和磁性能研究

张 伟

（大连理工大学材料科学与工程学院，辽宁大连 116024）

摘 要：Fe-B-Cu、Fe-(B, Si)-Cu、Fe-(B, Si, P)-Cu 等纳米晶合金具有 1.80 T 以上的饱和磁感应强度（B_s），低矫顽力、高磁导率、低铁损等优异的软磁特性，是用于电动机、变压器、互感器等电能转换和传输器件的新一代软磁材料。为获得优异软磁性能，这些合金通常需借助高升温速率（H_r）热处理使其非晶前驱体结晶化，形成微细的纳米晶组织。由于高 H_r 热处理在工况条件下难以实施，限制了其工业化应用。

本工作系统研究了 Fe-B-Cu 和 Fe-(B, Si)-Cu 系纳米晶合金急冷带材的结构、结晶化组织和磁性能，探讨了合金成分-非晶前驱体结构-晶化组织-磁性能间的关系。结果表明：在 Cu 含量（原子数分数）较低（1.0%~1.5%）的成分范围内，急冷合金带材可形成非晶态结构；经低 H_r 热处理形成的纳米 α-Fe 晶组织较粗大，致使软磁性不佳。在 Cu 含量较高（1.5%~2.0%）的成分范围内，急冷合金带材的非晶基体中存在高数密度、尺寸<10nm 的 α-Fe 晶粒（晶核），经低 H_r 热处理也可形成微细的纳米 α-Fe 晶组织，而显示良好的软磁性能。通过研究合金成分及非晶前驱体结构对晶化行为的影响，阐明其影响机制，建立了不同前驱体结构 Fe-B-Cu 合金的纳米晶化模型。此外，研究了添加合金元素对合金非晶前驱体结构及其晶化组织和磁性能的影响，探讨了相应的结晶化机制。

关键词：Fe 基纳米晶合金；Fe-B-Cu 合金；非晶前驱体结构；结晶化行为；软磁性能

P 与 Hf 元素置换对 $Fe_{84}Nb_7B_9$ 纳米晶合金组织结构及软磁性能的影响

李特尼格尔，薛钧升，李艳辉，张 伟

（大连理工大学材料科学与工程学院，辽宁大连 116024）

摘 要：Fe-Si-B-Cu-Nb 纳米晶合金（FINEMET）因具有低矫顽力（H_c）、高磁导率、低损耗和磁致伸缩等特性，已被广泛应用于变压器、电感器、电动机等电力电子器件中，但较低的饱和磁感应强度（B_s，约 1.24T）不利于器件和设备的小型化。相较之下，Fe-M-B (M = Zr, Nb, Hf) 系纳米晶合金（NANOPERM）具有高的 B_s（>1.5T），但其急冷带材表面易析出 α-Fe 相织构，导致热处理结晶化形成的 α-Fe 晶粒异常长大，恶化软磁性能。因此，提高该合金非晶形成能力（AFA），抑制 α-Fe 相织构的产生，是改善其软磁性能的途径之一。

本工作中，以 $Fe_{84}Nb_7B_9$ 合金为对象，分别以 P 和 Hf 对合金中的 B 和 Nb 进行部分置换，以在确保其高 B_s 的前提下，提高合金的 AFA，达到改善纳米晶合金组织和软磁性能的目的。结果表明，适量的 P 置换 B 可消除 $Fe_{84}Nb_7B_9$ 合金急冷带材表面 α-Fe 相织构，形成完全非晶态；并细化了热处理结晶化形成的 α-Fe 相，提高合金的软磁性能。$Fe_{84}Nb_7B_9$ 和 $Fe_{84}Nb_7B_3P_4$ 纳米晶合金的 α-Fe 相平均粒径尺寸（$\bar{D}_{\alpha\text{-Fe}}$）、$H_c$ 与 B_s 分别为 14nm、20.0A/m 与 1.50T 和 12nm、15.3A/m 与 1.57T。Hf 对 $Fe_{84}Nb_7B_5P_4$ 合金中的 Nb 进行置换，可进一步细化纳米晶组织，改善软磁性能，其中 $Fe_{84}Nb_{3.5}Hf_{3.5}B_5P_4$ 纳米晶合金的 $\bar{D}_{\alpha\text{-Fe}}$、$H_c$ 和 B_s 分别为 9nm、11.2A/m 与 1.55T。此外，还探讨了 P、Hf 置换提高 AFA，细化纳米晶组织及改善软磁性能的机制。

关键词：Fe 基纳米晶合金；Fe-Nb-B 合金；相似元素置换；非晶形成能力；软磁性能

Fe-Si-B-Cu-Al 系纳米晶合金组织结构与软磁性能的研究

薛钧升，李艳辉，张 伟

（大连理工大学材料科学与工程学院，辽宁大连 116024）

摘 要：Fe 基纳米晶合金由于兼备低矫顽力（H_c）、高磁导率、高饱和磁感应强度（B_s）、低损耗等优异的软磁特性，使其在电力、电子领域展现出广阔的应用前景。近年来，成功研发出诸多 B_s 超过 1.7T 的 Fe-(Si, B, P, C)-Cu 系纳米晶合金，其对电力、电子设备及器件的小型化具有重要的工程应用价值。由于合金体系中不含抑制 α-Fe 晶粒生长的前过渡族金属元素，通常需借助高升温速率的热处理工艺获得微细的纳米晶组织，而这在工况条件下难以实现。前期工作中发现，高 Cu 含量的 Fe-B-Cu[1]、Fe-Si-B-Cu[2]、Fe-Si-B-P-Cu[3]等纳米晶合金前驱体中预存有高数密度的 α-Fe 纳米晶，即使在低升温速率的热处理条件下仍能获得微细的纳米晶组织及良好的软磁性能。但这些合金的急冷前驱体微结构不易控制，不利于其软磁性能的进一步改善。尽管通过添加合金元素的方式可实现对合金前驱体的微结构、纳米晶组织及软磁性能的调控，但会使合金的 B_s 值降低[4,5]。在确保高 B_s 的同时，为实现对高 Cu 含量纳米晶合金前驱体微结构及其纳米晶化组织的调控，本工作研究了微量 Al 添加对 Fe-Si-B-Cu 合金急冷结构、晶化组织及软磁性能的影响。

结果表明，添加微量的 Al 可降低 Fe-Si-B-Cu 合金的非晶形成能力，使其急冷合金的非晶基体中析出高数密度的 α-Fe 纳米晶粒；经低升温速率热处理后，获得了微细的 α-Fe 纳米晶结构和良好的软磁性能；其中添加 0.6 at.% Al 合金的 α-Fe 晶平均粒径、B_s 和 H_c 分别为 21 nm、1.77 T 和 13.2 A/m。此外，根据合金热处理前后的组织结构，探讨了 Al 含量影响合金晶化行为及软磁性能的机制，提出了相应的结晶化模型。

关键词：Fe 基纳米晶合金；Al 含量；预存 α-Fe 相；软磁性能；晶化行为

参 考 文 献

[1] Wu L C, Li Y H, Yubuta K, et al. J. Magn. Magn. Mater., 2020, 497: 166001.
[2] Li Y H, Jia X J, Xu Y Q, et al. J. Alloy. Compd., 2017, 722: 859.
[3] Jia X J, Li Y H, Wu L C, et al. J. Mater. Sci., 2019, 54: 4400-4408.
[4] Wu L C, Li Y H, Jia X J, et al. Acta Metall. Sin.(English Letters), 2022, 35: 235-242.
[5] Li Y H, Jia X J, Zhang W, et al. J. Mater. Sci. Technol., 2021, 65: 171-181.

非晶合金的流动模型和临界阈值准则

乔珺威，张 浩，王 重

（太原理工大学，山西太原　030024）

摘　要：非晶合金是一种具有优异力学和物理性能的新型合金，有望成为结构承重材料。在远低于玻璃化转变温度下，几乎所有的塑性都被限制在纳米跨越剪切带内，导致不均匀的塑性流动，表现为应力-应变曲线上的锯齿。

（1）非晶合金的流动模型[1]。

非晶合金在单轴压缩下的剪切带具有两种传播模式：渐进式增长和同时滑动。研究人员总结了一系列具有扩展可塑性的非晶合金，并揭示了遵循阿伦尼乌斯定律的剪切带机制。

通过适用于非晶合金的微观机制，如脆性固体断裂理论和自由体积理论，建立了剪切带类裂纹传播速度的本构方程。进而推导出非晶合金在单轴压缩下的流动模型，$\ln v_{SB} = \ln(v_0 a_0) + E_S/(kT_1) - E_S/(kT)$，定义了阿伦尼乌斯定律并描述了非晶合金的材料依赖性。

与之前工作报道的许多实验相比，目前的模型不仅可以准确地预测出非晶合金中锯齿流和非锯齿状流的过渡，而且大多数非晶合金确定的阿伦尼乌斯方程和活化能与实验结果惊奇地一致。该研究为理解金属玻璃力学行为随外部条件的演化提供了新的视角。

（2）非晶合金的临界阈值准则[2]。

剪切转变区对于理解大块金属玻璃的不可逆变形具有至关重要的意义。通过统计最大剪应力的累积分布可以估计剪切转变区的激活体积大小。为了更好地理解大块金属玻璃的微观塑性变形行为，迫切需要通过适当的方法获得这些参数之间的相关性。

首先，通过降噪处理和统计分析得到通过不同的低温热循环获得三个样本的激活体积值。然后，根据平均场理论给出了最大剪应力临界阈值的判据。最后，结合前人关于多种典型金属玻璃最大剪应力和激活体积的关联，$(\tau_{max})_C = V^{*-\lambda}$，并总结出该准则的三个适用条件。

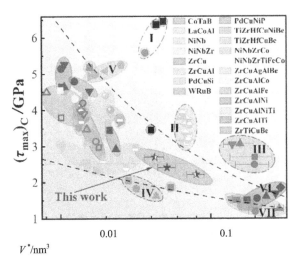

图 1　非晶合金的最大剪切应力临界阈值作为激活体积的函数

参考文献

[1] Zhang H, Wang Z, Yang H J, et al. Scripta Materialia, 2023, 222: 115047.
[2] H Zhang, Wang Z, Liaw P K, et al. Materials Science and Engineering: A, 2023, 873: 145031.

成型压力对 FeSiBCuNb 纳米晶磁粉芯电磁性能的影响

史贵丙，王　丽，王明旭

（机电与信息工程学院山东大学，山东威海　264209）

摘　要：铁基纳米晶软磁复合材料由于具有独特的纳米晶和非晶双相结构而展现出优异的软磁性能，因而引起了广泛的关注。但由于其具有较高的硬度，通常在较高的压力条件下成型制备，这进一步提高了对成型设备的需求。因此，在保证磁粉芯电磁性能的条件下，选择合适的成型工艺也成为了磁粉芯制备过程中必不可少的环节。本文基于未退火的部分晶化的 FeSiBCuNb 纳米晶气雾化粉展开研究，详细探讨了成型压力（800~1600MPa）对于磁粉芯电磁性能的优化程度。发现成型压力的增加能够有效降低磁粉芯中颗粒之间的距离，从而造成磁导率的上升，因此认为成型压力对磁粉芯磁导率的提升存在上限。如图 1 所示，随着成型压力的增加，磁导率呈现单调递增趋势，然而其相对增长比却呈现单调下降趋势，因此在后续的研究中将进一步确定成型压力的阈值区间。此外，在当前的研究压力范围内，磁粉芯的品质因数与交流损耗均随着成型压力的增加有所改善，在 1600MPa 时，品质因数可达到 120 以上，交流损耗可降低至 760mW/cm³（B_m=20mT、f=1MHz）。本文的研究结果对于高收率的部分晶化的 FeSiBCuNb 纳米晶气雾化粉的利用具有重要的指导意义。

关键词：FeSiBCuNb；成型压力；磁粉芯

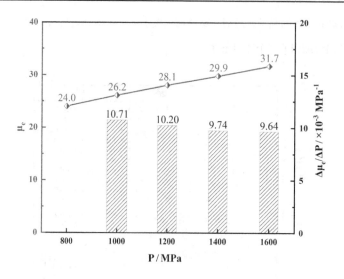

图 1 磁粉芯有效磁导率及相对增长比

参 考 文 献

[1] Tian M Y, Xu J, Yang S, et al. Journal of Alloys and Compounds, 2022, 923: 166394
[2] Sun H B, Wang C, Chen W H, et al. Metals, 2019, 9(3): 381-390

中柱可替换型一体成型电感的仿真性能分析

史贵丙,苗泓毅,王 丽,王明旭

(机电与信息工程学院山东大学,山东威海 264209)

摘 要:电子产品、通信网络的快速发展不断促使电子元器件朝着高频化、小型化、节能化、集成化的方向发展。一体成型电感由于绕组线圈内置而具有良好的磁屏蔽和 EMI 抑制效果,通常被应用在大电流的工况下。然而其磁导率相对较低,当尺寸固定且需要较大感量时需通过增加绕组实现,这不可避免的造成了绕组损耗的增加。为进一步优化一体成型电感的感量,本文基于 Ansys Maxwell 建立了图 1(b)所示的中柱可替换型一体成型电感模型,并详细分析了固定规格尺寸条件下中柱磁芯材料磁导率变化对一体成型电感整体感量的影响。发现在保证电感的感量一定时,中柱磁芯材料磁导率的增加能够有效降低边缘材料磁导率的需求。此外当边缘材料磁导率一定时,中柱磁芯材料磁导率的增加能够有效提升一体成型电感的感量。值得注意的是,中柱磁芯材料磁导率的提升对于一体成型电感的优化效果并非无限,而是随着中柱磁芯材料磁导率的增加,这种优化效果逐渐被弱化。当边缘材料磁导率固定为 30 时,随着中柱磁芯材料磁导率提升 1 倍以及 2 倍,一体成型电感的整体感量则同比增加 14.2%和 22.7%。当一体成型电感的感量固定为 1.41μH 时,则使边缘材料磁导率同比降低 16.7%和 23.3%。本文的分析结果对一体成型电感性能的进一步优化具有重要的指导意义。

关键词:一体成型电感;电磁仿真;中柱磁芯

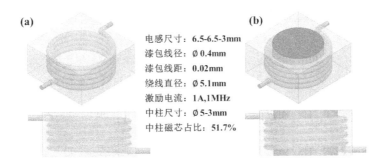

图 1 一体成型电感模型

(a) 常规一体成型电感；(b) 中柱可替换型一体成型电感

参 考 文 献

[1] 聂敏, 谈敏, 黄静, 等. 磁性材料及器件, 2020, 51(2): 47-50.
[2] 董博儒, 覃维, 王鲜, 等. 磁性材料及器件, 2021, 52(5): 28-33.

玻璃态物质弛豫基本单元的探测

宋丽建，霍军涛，王军强

（中国科学院宁波材料技术与工程研究所，浙江宁波 315000）

摘 要：退火是调控金属玻璃等玻璃态材料的重要方法。然而，由于玻璃态材料能量状态丰富，退火弛豫演化规律非常复杂，存在多种弛豫模式，而且不同弛豫模式之间存在耦合和记忆效应，缺乏精准调控的理论和方法。亚稳特征对物理化学性能的影响规律和机制仍然不清楚，极大限制了高性能金属玻璃的研发进程。

一种经典理论认为，玻璃表现出的宽广的弛豫峰源自一系列具有指数特征的弛豫基本单元叠加而成，不同能量的弛豫基元反映了玻璃结构的不均匀性。但目前仍然缺少弛豫基元谱的实验证据。研究探测玻璃弛豫子谱对理解玻璃态本质、精准调控退火工艺改善性能具有重要意义。

最近中科院宁波材料所王军强研究员团队从玻璃态物质弛豫过程中能量变化角度出发，利用高精度闪速差示扫描量热仪研究了金属玻璃、高分子玻璃和小分子玻璃等不同玻璃态材料，在不同退火温度和退火时间下的热流变化。通过精准控制退火温度和时间，他们测量了热流弛豫峰。发现不同温度或弛豫时间的热流弛豫峰的谱线与力学弛豫谱具有一致性，表明宽泛的谱峰来自独立弛豫单元谱的叠加。

为了进一步验证探测到的热流弛豫谱是否是弛豫基元，他们使用 Debye 模型拟合弛豫峰，得到了合理的激活能或特征时间，说明这些弛豫基元与晶体中声子的动力学相似，并据此提出了"弛豫子"概念（relaxation unit, 简称弛豫子, relaxun）。

他们通过控制多步退火中的温度和时间可以实现了对特定弛豫子的激活、湮灭或编程，证实了宏观弛豫源于具有指数特征弛豫谱的非均匀性叠加假说。此外，激活能随退火温度和退火时间存在从 γ/β' 弛豫向 β 弛豫，并最终进入到 α 弛豫的转变动力学行为，在焓空间中实现了对不同弛豫模式含量的定量表征。相关结果为理解玻璃态本质提供了重要实验证据，也为精准调控退火工艺提供了重要理论指导。

研究成果以"玻璃态物质指数弛豫谱的探测"（Detecting the exponential relaxation spectrum in glasses by high-precision nanocalorimetry）为题发表在《美国国家科学院院刊》（Proceedings of the National Academy of Sciences of the United States of America, PNAS 120, e2302776120（2023））上。该研究工作得到了国家自然科学基金委、浙

江省和宁波市的资助。

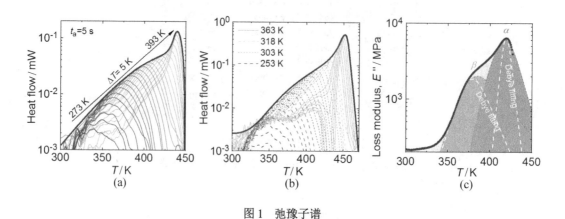

图 1 弛豫子谱

(a) Au 基金属玻璃在退火温度 Ta=273~393 K 下退火 5 s 的热流弛豫峰；(b) Au 基金属玻璃在退火温度 T_a=253K、303K、318K、363K 下退火不同时间的热流弛豫峰；(c) Au 基金属玻璃的力学弛豫谱

11 节能与低碳技术

大会特邀报告
第十三届冶金青年科技奖获奖人特邀报告
分会场特邀报告
矿业工程
焦化及节能环保
炼铁与原料
炼钢与连铸
电冶金与废钢铁
轧制与热处理
表面与涂镀
金属材料深加工
粉末冶金
先进钢铁材料及应用
★ 节能与低碳技术
冶金环保与资源利用
冶金设备与工程技术
冶金自动化与智能化
冶金物流
冶金流程工程学
其他

大型制氧机精氩塔氩固化在线处理方法的研究

董昕宏，舒　畅，王政林

（鞍钢能源科技有限公司，辽宁鞍山　114000）

摘　要：简介某钢厂 35000m³/h 制氧机组内压缩工艺流程及装置产能，通过分析精氩塔氩固化原因及采取的冷态处理方法，保证了空分工况的稳定运行，有效降低制氧机启停能耗。

关键词：大型空分；内压缩；精氩塔；氩固化；原因；方法

Research on Online Processing Method for Argon Solidification in the Precision Argon Tower of Large Oxygen Generators

DONG Xinhong, SHU Chang, WANG Zhenglin

(Ansteel Energy Technology Co., Ltd., Anshan 114000, China)

Abstract: The process flow and unit capacity of 35000m³/h oxygen making unit in Ansteel are briefly introduced. The reasons for freezing and blocking of plate heat exchanger and the treatment measures adopted are analyzed. Some experiences of air separation operation are expounded.

Key words: large scale air separation; internal compression; refined argon tower; argon solidification; reason; method

关于提升炼钢余热蒸汽过热度增加发电量的措施

郑振海，薛晓金，张明扬，张元华

（鞍钢股份有限公司鲅鱼圈钢铁分公司，辽宁营口　115007）

摘　要：本文针对钢厂转炉余热饱和蒸汽发电系统，由于炼钢饱和蒸汽的间歇性、周期性和波动性，造成饱和蒸汽发电机组无法稳定运行，于是炼钢转炉余热蒸汽的不得不进行放散，造成余热蒸汽的极大浪费。结合厂区蒸汽管网的布置情况，合理利用厂区蒸汽管网的便利条件，解决了饱和蒸汽发电机组无法连续运行以及能源利用率低的难题，以最少投资换来最大的利润。

关键词：微过热蒸汽；蒸汽管网；炼钢余热蒸汽

Measures for Increasing the Superheat of Steam from Waste Heat of Steelmaking to Increase Power Generation

ZHENG Zhenhai, XUE Xiaojin, ZHANG Mingyang, ZHANG Yuanhua

(Bayuquan Steel Branch of Angang Steel Co., Ltd., Yingkou 115007, China)

Abstract: This paper aims at the saturated steam power generation system of steel converter waste heat. Because of the intermittently, periodicity and fluctuation of the saturated steam, the saturated steam generator set can not run stably, so the waste heat steam of the converter has to be released, resulting in a great waste of waste heat steam. Combined with the layout of the steam pipe network in the plant area, the convenient conditions of the steam pipe network in the plant area are rationally utilized to solve the problem that the saturated steam generator set cannot run continuously and the energy utilization rate is low, so as to obtain the maximum profit with the least investment.

Key words: slightly superheated steam; steam pipe network; waste heat steam from steelmaking

包钢厂内鱼雷罐车加盖装置的研究与应用

王少龙，刘　强，刘俊杰

（内蒙古包钢钢联股份有限公司运输部，内蒙古包头　014000）

摘　要： 由于包钢厂内鱼雷罐车在运输过程中罐口是敞开的，该过程中伴随着铁水的温降和大量废气排放，造成在炼钢工艺中能源的高消耗及厂内环境污染。包钢运输部为解决此类问题，根据现场实际生产环境，研究开发了鱼雷罐加盖装置，并在包钢厂内得以应用，此加盖装置可以有效防止烟尘外溢，实现超低排放改造目标，同时减少铁水温降，降低生产成本。

关键词： 鱼雷罐车；加盖装置；超低排放改造；铁水温降

Research and Application of the Covered Technology of Torpedo Cars in Baotou Steel Factory

WANG Shaolong, LIU Qiang, LIU Junjie

(Inner Mongolia Baotou Steel Union Co., Ltd., Baotou 014000, China)

Abstract: Because the torpedo cars in the transportation process are lidless in Baotou Steel factory. The process is accompanied by the temperature drop of molten iron and exhaust gas emission. Resulting the high energy consumption in the steelmaking process and environmental pollution. In order to solve such problems, the Transportation Department of Baotou Steel studied and developed a covered technology of torpedo cars according to the actual production environment, which was applied in the Baotou Steel factory. The covered technology can effectively prevent the overflow of soot and achieve the goal of ultra-low emission. At the same time, reducing the temperature drop of molten iron and production cost.

Key words: torpedo car; covered technology; ultra-low emission; temperature drop of molten iron

高炉煤气干煤气直送管网技术研究

吕 勇[1]，贾冬颖[2]，胥中平[1]

（1. 攀钢钒能源动力分公司，四川攀枝花 617000；
2. 攀钢集团设计研究院有限公司，四川攀枝花 617062）

摘 要：2009年9月攀钢5座高炉实现了高炉煤气全干式除尘工艺，有效利用高炉冶炼产生的二次能源，同时减少环境污染，满足国家环保要求。由于高炉煤气主管网设计温度60℃和焦炉使用的高炉煤气换向阀密封等要求，经全干式布袋除尘发电后干煤气不得不设置喷水装置降低煤气温度，同时由于煤气中含有Cl^-，煤气含有饱和水后Cl^-溶解，形成酸性冷凝水腐蚀管网，带来需喷碱脱氯的问题，以满足高炉煤气主管网和用户安全运行。这种运行方式严重浪费高炉煤气干煤气的焓值，增加了煤气的含水量，影响到煤气使用效率，导致冶金炉窑理论燃烧温度下降、热效率降低、冶炼成本上升。

针对上述问题，开展了高炉煤气干煤气直送管网的含水量对比技术分析，进行了干煤气焓值变化对比技术分析，可以大幅度减轻高炉煤气管网的腐蚀和减少喷碱脱氯成本，认为高炉煤气干煤气直送管网具有巨大的经济效益和社会效益。

攀钢高炉煤气干煤气直送利用已在高炉热风炉成功应用十年，为其他高炉煤气用户输送干煤气是完全可行的。第一步，可通过散热器出口温度控制，以保证透平机组排气温度在65℃以内；停运喷碱脱氯装置；1号高炉、4号高炉和新3号高炉的高炉煤气直接送主管网。第二步，一期电鼓和第二台100MW发电机组建设完成后，建设2号、3号高炉透平机组后煤气间接喷水降温装置和管网，保证2号、3号高炉进入高炉煤气主管网煤气温度在60℃以内。

该技术的应用可以充分利用高炉煤气干煤气的焓值，降低煤气含水量，提高冶金炉窑理论燃烧温度、热效率，降低生产成本，提升煤气使用效率。不需要设置喷碱脱氯的装置，减轻煤气主管网腐蚀，降低生产成本。

可推广应用到其他钢铁行业，特别是新建钢铁行业可以将煤气管网设计温度提高到100~150℃，以进一步提高高炉煤气干煤气的应用效果，实现利用效率的提升。

关键词：高炉煤气；干煤气；措施；研究

Research on the Technology of Dry Gas Direct Delivery Pipeline Network for Blast Furnace Gas

LV Yong[1], JIA Dongying[2], XU Zhongping[1]

(1. Panzhihua Steel Vanadium Co., Ltd., Energy Power Branch, Panzhihua 617000, China;
2. Panzhihua Pangang Group Design and Research Institute Co., Ltd., Panzhihua 617062, China)

Abstract: In September 2009, five blast furnaces at Panzhihua Iron and Steel Group achieved a fully dry dust removal process for blast furnace gas, effectively utilizing the secondary energy generated by blast furnace smelting, while reducing environmental pollution and meeting national environmental protection requirements. Due to the design temperature of the blast furnace gas main network of 60℃ and the sealing requirements of the blast furnace gas directional valve used in the coke oven, after fully dry bag dust removal and power generation, the dry gas has to be equipped with a water spraying device to reduce the gas temperature. At the same time, due to the presence of Cl- in the gas, which dissolves in saturated water, forming acidic condensate water that corrodes the pipeline network, the problem of alkali spraying and dechlorination is brought about to meet the safe operation of the blast furnace gas main network and users. This operation mode seriously wastes the enthalpy value of dry blast furnace gas, increases the water content of the gas, affects the use efficiency of the gas, and causes the theoretical combustion temperature of the metallurgical furnace to drop, the thermal efficiency to drop, and the smelting cost to rise.

In response to the above issues, a comparative technical analysis was conducted on the water content of the blast furnace gas dry gas direct delivery pipeline network, and a comparative technical analysis was conducted on the changes in dry gas enthalpy value. This can significantly reduce the corrosion of the blast furnace gas pipeline network and reduce the cost of alkali spraying and dechlorination. It is believed that the blast furnace gas dry gas direct delivery pipeline network has enormous economic and social benefits.

The direct utilization of dry gas from Panzhihua Iron and Steel Group's blast furnace has been successfully applied in hot blast furnaces for ten years, and it is completely feasible to transport dry gas to other blast furnace gas users. The first step is to control the outlet temperature of the radiator to ensure that the exhaust temperature of the turbine unit is within 65℃; Stop the alkali spraying and dechlorination device; The blast furnace gas from the 1 # blast furnace, 4 # blast furnace, and the new 3 # blast furnace is directly sent to the main network. The second step is to complete the construction of the first phase electric drum and the second 100MW generator unit. After the construction of the 2 # and 3 # blast furnace turbine units, the indirect gas spray cooling device and pipeline network will be installed to ensure that the gas temperature of the 2 # and 3 # blast furnaces entering the blast furnace gas main network is within 60℃.

The application of this technology can make full use of the enthalpy value of dry gas of blast furnace gas, reduce the water content of gas, improve the theoretical combustion temperature and thermal efficiency of metallurgical furnace, reduce production costs and improve the efficiency of gas use. There is no need to install a device for alkali spraying and dechlorination to reduce corrosion of the gas main network and reduce production costs.

It can be promoted and applied to other steel industries, especially in the newly built steel industry. The design temperature of the gas pipeline network can be increased to 100-150℃ to further improve the application effect of blast furnace gas dry gas and achieve an improvement in utilization efficiency.

Key words: blast furnace gas; dry gas; measures; research

减温减压器（RTP）减温水系统改造

张腾飞，王 晨

（河钢承钢，河北承德　067000）

摘　要： 河钢承钢能源事业部供风作业区前置背压式发电机组减温减压器（RTP）在运行热备过程中，发生漏量大导致高品位能源损失、发电量减少的故障，经过精确的热力计算找到问题存在的基础数据，并对减温减压器（RTP）的减温水系统进行改造，改造后机组汽耗率降低 8.52kg/kWh，并进一步发展和完善了减温减压器（RTP）小流量调节技术。

关键词： 减温减压器；减温水；小流量调节

Retrofit of Temperature and Pressure Reducing Device (RTP) Desuperheating Water System

ZHANG Tengfei, WANG Chen

(HBIS Chenggang, Chengde 067000, China)

Abstract: During the operation and hot standby process of the front back pressure generator unit temperature and pressure reducer (RTP) in the air supply operation area of Hegang Chenggang Energy Business Unit, a fault occurred that caused high grade energy loss and reduced power generation due to large leakage. After accurate thermal calculation, the basic data of the problem was found, and the temperature and pressure reducer (RTP)'s temperature and pressure reducing water system was modified. After the modification, the steam consumption rate of the unit was reduced by 8.52kg/kWh. And further developed and improved the temperature and pressure reducer (RTP) small flow regulation technology.

Key words: temperature and pressure reducer; desuperheating water; low flow regulation

基于煤气调配途径的轧钢加热炉烟气污染物达标排放对策研究

周劲军，刘自民，唐嘉瑞，周 栩

（马鞍山钢铁股份有限公司，安徽马鞍山 243000）

摘 要： 本文介绍了通过采用煤气调配及燃烧优化控制等手段实现轧钢加热炉烟气污染物达标排放的技术路径，其中包括煤气全硫含量实际值及最大允许值测算、满足热值及排放要求的煤气调配、NO_x 的分段监测预警控制等技术。旨在为企业应对当下"钢铁企业超低排放"及自身降本增效要求提供参考，同时对富氧助燃带来的烟气排放监测值偏高问题进行了探讨。

关键词： 轧钢加热炉；烟气污染物；达标排放；对策研究

Research on Emission Countermeasures of Flue Gas Pollutants in Steel Rolling Heating Furnace Based on Gas Allocation Method

ZHOU Jinjun, LIU Zimin, TANG Jiarui, ZHOU Xu

(Maanshan Iron & Steel Co., Ltd., Maanshan 243000, China)

Abstract: This paper introduces the technical path of achieving the emission of flue gas pollutants from rolling furnaces by means of gas allocation and combustion optimization control, including the calculation of the actual and maximum allowable value of total sulfur content of gas, gas allocation that meets the requirements of calorific value and emissions, and segmented monitoring and early warning control of NO_x. It aims to provide reference for enterprises to cope with the current "ultra-low emissions of steel enterprises" and their own cost reduction and efficiency improvement requirements, and at the same time discuss the problem of high flue gas emission monitoring value caused by oxygen-enriched combustion and boosting.

Key words: rolling furnace; flue gas pollutants; compliance with emissions; countermeasure research

基于低碳冶金的碳捕集工艺

潘 阳，马方曙，陈 誉，陈露露，马志强

（南京胺云化学科技有限公司，江苏南京 210000）

摘　要：碳捕集是助力钢铁行业实现低碳冶炼的重要技术手段。已有碳捕集技术主要以有机胺溶剂吸收-再生法为主，不同的工艺要求会导致酸性气去除装置的工艺配置差异大，能耗与投资差异同样大。在相同原料气条件下，对不同工艺流程的酸性气脱除装置进行比较，发现低投资成本的工艺流程通常能耗更高，导致运营成本高。大型工厂，使用流程相对复杂的工艺流程从长远看更加经济。

关键词：碳捕集；低碳冶金；有机胺；工艺配置；能耗；投资

Comparison of Carbon Capture Process Configurations for Low Carbon Metallurgy

PAN Yang, MA Fangshu, CHEN Yu, CHEN Lulu, MA Zhiqiang

(Nanjing Anyun Chemical Technology Co., Ltd., Nanjing 210000, China)

Abstract: Carbon capture technology is an important stage in achieving low-carbon metallurgy for steel industry. The existing industrialized carbon capture technologies are mainly amine solvent based absorption and regeneration process, however different process schemes of acid gas removal unit, result in different energy consumption and capital investment. Comparing different configurations of acid gas removal unitswith the same feed gas conditions, the lower investment cost process schemes normally with higher energy consumption, and higher operating cost. For large plants, build with complex process configurations will be more cost-efficient from long term aspect.

Key words: carbon capture; low carbon metallurgy; amine; process scheme; energy consumption; investment

煤气加压机的振动分析和处理

高子丰，郑　晨，李星星，王龙锋

（中冶南方工程技术有限公司公用设计所，湖北武汉　430223）

摘　要：在当前经济快速化发展的时代背景下，对于能源的需求量也在持续增加。离心式煤气加压机作为冶金、煤化等领域中一种常用的能源转换设备，在企业的生产过程中发挥着重要的作用。但是其在使用时振动故障问题频发，严重影响了企业的经济效益最大化，阻碍了企业的进一步发展壮大，限制了企业无人化、远程操控的普及，也为企业员工的生命和财产埋下了很大的安全隐患。鉴于此，本文基于离心式煤气加压机工作原理，对其振动故障现象及原因进行了分析，进而从煤气加压机的安装、维护、检修和运行方式等方面提出了相应的措施，降低其振动故障发生率，保证设备能够高效稳定、安全长久地运行。

关键词：离心式煤气加压机；振动原因；解决方法

Vibration Analysis and Treatment of Gas Compressor

GAO Zifeng, ZHENG Chen, LI Xingxing, WANG Longfeng

(WISDRI Iron & Steel Engineering & Research Incorporation Limited, Wuhan 430223, China)

Abstract: In the current era of rapid economic development, the demand for energy is also increasing. As a common energy conversion equipment in the fields of metallurgy and coal gasification, centrifugal gas compressor plays an important role in the production process of enterprises. However, its frequent vibration failure in use has seriously affected the maximization of the economic benefits of the enterprise, hindered the further development and expansion of the enterprise, restricted the popularization of unmanned and remote control of the enterprise, and also buried a great safety hazard for the life and property of the enterprise staff. In view of this, based on the working principle of centrifugal gas compressor, this paper analyzes its vibration failure phenomenon and causes, and then puts forward corresponding measures from the aspects of installation, maintenance, overhaul and operation mode of gas compressor to reduce its vibration failure rate and ensure that the equipment can operate efficiently, stably and safely for a long time.

Key words: centrifugal gas compressor; vibration reasons; solution

连续退火机组平整液循环利用技术开发与实践

何建锋，潘勋平，胡 斌

（宝钢日铁汽车板有限公司，上海 201900）

摘 要：为了配合平整液循环回用工艺的开发，本文在实验室中搭建了模拟平整装置，研究并获取了平整液循环使用过程中的各种平整液性能变化规律，及对生产带钢的产品质量的可能影响。将本项目研究结果运用到现场后，支撑平整液循环回用工艺顺利投产；对生产的带钢进行取样分析，通过各项技术评估，发现平整液循环回用工艺生产的带钢能满足各项要求。

关键词：冷轧；平整；平整液循环回用；技术开发

Research and Application of Recycling Technology for Cold-rolling Temper Lubricant

HE Jianfeng, PAN Xunping, HU Bin

(Baosteel-Nippon Steel Automotive Steel Sheets Co., Ltd., Shanghai 201900, China)

Abstract: In order to cooperate with the research and development of recycling technology for cold-rolling temper lubricant, a testing equipment for simulating temper rolling was built and put to use in the laboratory, to study the changes of various temper lubricant's properties in the process of recycling flow, and also, the possible influence of recycling flow of temper lubricant on the quality of temper-rolled strips were obtained. After applying the research results of this paper to the on-site production, the recycling technology of cold-rolling temper lubricant was successfully put into production. Moreover, the temper-rolled strips have been sampled and analyzed. Through various technical evaluations, it is found that the cold-rolling strips produced by the recycling temper lubricant could meet all kinds of product requirements.

Key words: cold rolling; temper rolling; recycling temper lubricant; technical research and development

钢铁流程能-碳解耦与碳排放解析

蒋滨繁，熊 瑞，程春云，田卓航，夏德宏

（北京科技大学，北京 100083）

摘　要：钢铁是国民经济、社会发展和国防建设的支柱性产业，同时也是我国最大的终端用能和 CO_2 排放行业。传统钢铁生产遵循以焦炭为燃料/还原剂的碳基还原炼铁技术路线，生产过程高度依赖含能源与资源、能/质利用率低，年 CO_2 排放占我国总排放约 16%[1]。钢铁低碳路径选择和节能降碳技术实施依赖于合理的碳素流追踪、分配、测算和评价。然而，由于碳氢耦合热还原炼铁过程是由从微观层面 Fe_xO_y 与 CO/C 等分子/原子反应、到中观层面工序装备热/质传递、再到宏观层面流程系统等组成的多层级能/质耦合复杂网络[2]，存在碳排放界限不清、碳排放测算方法难统一、碳减排潜力难评价等问题。

为此，本文基于物质守恒和"载能体"理论，提出"载碳体"概念——将一切物质和能源介质都视为"载碳体"，用载碳值描述潜在碳排放强度[3]。直接含碳资源或能源中包含的碳元素量为直接载碳值，非含碳能/资源由于生产、运输等过程引发的碳排放为间接载碳值。以此为基础，利用半生命周期溯源计算钢铁全流程直接碳排放、间接碳排放；利用能值对多产品输出工序进行主/副产品的直/间接"载碳值"分配，评估由于余能或废弃物循环利用等引发的负碳排放效益。以高炉-转炉传统流程为例，进行能-碳解耦与碳素流解析，由于化石燃料、含碳资源等利用造成的直接碳排放为 $1.67 tCO_2/tCS$，如图 1 所示；由于纯氧、电力等二次非含碳资源/能源利用引发的间接碳排放为 $0.4\sim0.6 tCO_2/tCS$；由于高温固体散料或烟气等余能回收、炉渣或废料等资源再利用产生的负碳排放效益理论上超过约 $0.6 tCO_2/tCS$。因此，若长流程余能回收利用效率达到 70%，则吨钢 CO_2 排放强度将降低 15%~20%。

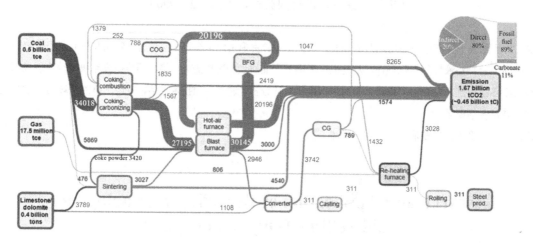

图 1　钢铁全流程碳素流解析与直接 CO_2 排放

参　考　文　献

[1] 国家统计局. 中国能源统计年鉴[M]. 2021.
[2] 殷瑞钰. "流"、流程网络与耗散结构——关于流程制造型制造流程物理系统的认识[J]. 中国科学: 技术科学, 2018, 48(2): 136-142.
[3] Jiang Binfan, Xia Dehong. Toward carbon neutrality in China: A national wide carbon flow tracing and the CO_2 emission control strategies for CO_2-intensive industries[J]. Sci. Total Environ, 2023, 879:163009.

鞍钢 2、3 高炉水冲渣节能减排生产实践

王佳熙，孙成国

（鞍钢股份有限公司炼铁总厂，辽宁鞍山　114021）

摘　要：鞍钢炼铁总厂 2、3 高炉水冲渣采用 INBA 工艺，由于水冲渣工艺特殊性，用水、耗水量较大，且实现水冲渣系统节能减排一直是炼铁总厂致力解决的难题。本文针对鞍钢炼铁总厂 2、3 高炉水冲渣 INBA 工艺系统实施节能减排工作进行阐述和分析，2、3 高炉水冲渣工艺系统实施冗余水回收利用系统、事故渣池水回收利用及各水池液位变频调控等措施，通过生产实践，2、3 高炉水冲渣仅能减排工作效果显著，具有推广应用价值和意义。

关键词：INBA；水平衡；冗余水；节能减排

Energy Saving and Emission Reduction Production Practice of Water Flushing Slag for Angang 2 and 3 Blast Furnaces

WANG Jiaxi, SUN Chengguo

(General Ironmaking Plant of Angang Steel Co., Ltd., Anshan 114021, China)

Abstract: The INBA process is used for the water flushing of blast furnaces 2 and 3 at Angang Iron and Steel Plant. Due to the unique nature of the water flushing process, water consumption and consumption are high, and achieving energy conservation and emission reduction in the water flushing system has always been a challenge that the Iron and Steel Plant is committed to solving. This article elaborates and analyzes the implementation of energy conservation and emission reduction in the INBA process system of blast furnace water flushing slag at Angang Iron and Steel General Plant. The 2 and 3 blast furnace water flushing slag process system implements measures such as redundant water recovery and utilization system, accident slag pool water recovery and utilization, and variable frequency control of water levels in each pool. Through production practice, the 2 and 3 blast furnace water flushing slag can only achieve significant emission reduction work, which has promotion and application value and significance.

Key words: INBA ; water balance; redundant water; energy conservation and emission reduction

冲渣外排水回收利用

王　维[1]，滕福亮[1]，刘　壮[1]，罗　涵[1]，
刘百志[1]，王希坤[1]，栗　艳[2]

（1. 鞍钢股份炼铁总厂，辽宁鞍山　114000；
2. 鞍钢集团工程技术有限公司，辽宁鞍山　114000）

摘　要：某高炉冲渣工艺采用茵芭法冲渣，运行中存在外来水源进入循环系统中，导致系统补充水量大于消耗水量，造成渣池溢流排水，后续环保排水处理压力较大，现把排水收集后循环回收利用，满足环保排放要求。

关键词：冲渣；回收利用

高炉区域工业新水回收利用实践

王 维[1]，王丽芳[2]，罗 涵[1]，王希坤[1]，刘百志[1]，周 鑫[1]，刘鹏羽[1]

（1. 鞍钢股份炼铁总厂，辽宁 鞍山 114000；
2. 鞍山市水务集团有限公司立山自来水公司，辽宁 鞍山 114000）

摘 要： 某高炉液压站、炉顶水冷站等设备在生产过程中需要冷却，冷却水水质采用工业新水，原冷却水直排浪费能源，新建水泵、冷却塔等设施回收利用，降低能源消耗。

关键词： 工业新水；回收利用

Abstract: Some equipments such as the hydraulic pressure station and the water cooling station on the top of a blast furnace need cooling in the production process. The quality of the cooling water adopts the new industrial water, the original cooling water is discharged directly to waste energy, and the newly-built water pump and cooling tower are recycled to reduce energy consumption.

高炉煤气醇胺法 CO_2 吸收塔设计

王博林，李 伟，韩渝京，李 鹏

（北京首钢国际工程技术有限公司能源环境分公司技术部，
北京市冶金三维仿真设计工程技术研究中心，北京 100043）

摘 要： 高炉煤气碳捕集是钢铁行业减排 CO_2 的重要手段之一。本文以 3000m³/h（标态）的工业试验装置为例，以 30%的单乙醇胺（MEA）溶液为吸收剂，进行高炉煤气的 CO_2 吸收塔设计，包括吸收方案的确定、填料的选择、物料衡算、塔径计算、填料层高度计算等一系列设计。

关键词： 高炉煤气；碳捕集；填料吸收塔；设计

Design of CO_2 Absorption Tower with the Method of MEA for Blast Furnace Gas

WANG Bolin, LI Wei, HAN Yujing, LI Peng

(Technology Division of Energy & Environment Branch, BSIET, Beijing Metallurgical 3-D Simulation Design Engineering Technology Research Center, Beijing 100043, China)

Abstract: Carbon capture of blast furnace gas is one of the most important means to reduce CO_2 emission in iron and steel industry. In this paper, a BFG handling capacity of 3000Nm³/h industrial side-stream test is taken as an example, and 30% MEA aqueous solution is used as absorbent to design the CO_2 absorption tower of blast furnace gas. According to the determination of absorption scheme, selection of packing, material balance, calculation of tower diameter and height of packing layer, series of steps are taken.

Key words: blast furnace gas; carbon dioxide capture; packed absorption tower; design

钢铁企业余热余能回收利用措施研究

范金龙

（中冶南方工程技术有限公司，湖北武汉　430080）

摘　要：随着我国经济的飞速发展，各类能源消耗也日益增加，能源供需的矛盾也与日俱增，节能降耗成为钢铁企业未来发展的重要方向。在全球经济增长放缓、环保政策收紧等一系列条件影响下，钢铁企业需要通过对余能和余热进行回收利用以达到节能降耗和降本增效的目的，为实现钢铁生产综合效益最大化。本文以炼铁工艺中的各种余热余能回收利用方式为例，以 TRT 发电，热风炉余热利用，高炉煤气发电为切入点，探究钢铁企业余热余能回收利用措施，并结合 XX 钢厂三种余热利用方式的案例对节能降耗效果及综合效益进行了对比分析，为工程实践提供了指导意义和借鉴价值。

关键词：节能降耗；余热余能；回收利用；综合效益

Research on Measures for the Recovery and Utilization of Residual Heat and Energy in Iron and Steel Enterprises

FAN Jinlong

(Wisdri Engineering and Research Incorporation Limited, Wuhan 430080, China)

Abstract: With the rapid development of China's economy, various types of energy consumption are also increasing, and the contradiction between energy supply and demand is also increasing. Energy conservation and consumption reduction has become an important direction for the future development of iron and steel enterprises. Under the influence of a series of conditions such as the slowdown of global economic growth and the tightening of environmental protection policies, iron and steel enterprises need to achieve the purpose of energy conservation and cost reduction and efficiency increase through the recovery and utilization of residual energy and waste heat, in order to maximize the comprehensive benefits of iron and steel production. Taking various waste heat and energy recovery and utilization methods in ironmaking process as an example, taking TRT power generation, hot blast furnace waste heat utilization, and blast furnace gas power generation as the starting point, this paper explores the measures of waste heat and energy recovery and utilization in iron and steel enterprises, and compares and analyzes the energy-saving and consumption reduction effects and comprehensive benefits with the case of three waste heat utilization methods in XX Steel Plant, providing guidance and reference value for engineering practice.

Key words: energy saving and consumption reduction; residual heat and energy; recycling; comprehensive benefits

钢铁行业中能源管理系统应用分析

范金龙

（中冶南方工程技术有限公司，湖北武汉　430080）

摘　要：能源管理系统通过先进的通信和控制基础设施，或在高峰需求时，或在系统可靠性受到威胁时，降低能源的需求响应，通过持续监控能源调度，在非高峰期等待时间来运行高需求负载，从而有效地管理和减少能源消耗。因此，在钢铁行业中应用能源管理系统，既是预防能源枯竭战略的体现，也是钢铁行业可持续发展的应对。根据钢铁行业的生产特点，论述钢铁行业中能源管理系统的构成，分析钢铁行业中能源管理系统的功能，为钢铁行业中能源管理系统的应用提供理论基础和经验借鉴。

关键词：能源管理系统；钢铁行业；架构；功能；应用

Application Analysis of Energy Management System in the Steel Industry

FAN Jinlong

(Wisdri Engineering and Research Incorporation Limited, Wuhan 430080, China)

Abstract: The energy management system reduces the energy demand response through advanced communication and control infrastructure, or in peak demand, or when the system reliability is threatened, and runs high demand loads through continuous monitoring of energy scheduling and waiting time in off-peak hours, thus effectively managing and reducing energy consumption. Therefore, the application of energy management system in the steel industry is not only the embodiment of the strategy of preventing energy depletion, but also the response to the sustainable development of the steel industry. According to the production characteristics of the steel industry, this paper discusses the composition of the energy management system in the steel industry, analyzes the functions of the energy management system in the steel industry, and provides theoretical basis and experience for the application of the energy management system in the steel industry.

Key words: energy management system; steel industry; framework; function; application

高温高压煤气发电的应用

姚震宇，张海艳

（安钢集团永通球墨铸铁管有限责任公司，河南安阳　455133）

摘　要：当前钢铁企业的重大战略任务即是节能减排以及能源的综合利用。通过企业能源结构战略调整，采用高温高压与煤气联合再热发电技术，实现能源的综合利用率的提高。通过分析公司当前多种能级高低不同的高炉煤气、环冷机余热锅炉产生的蒸汽等能源利用情况，按照"能质对口、梯级利用"的原则，实现余能余热的高效梯级利用，降低能耗。

关键词：节能减排；高温高压；煤气；蒸汽；余能余热

Application of High Temperature and High Pressure Gas Power Generation

YAO Zhenyu, ZHANG Haiyan

(Angang Group Yongtong Ductile Iron Pipe Co., Ltd., Anyang 455133, China)

Abstract: At present, the major strategic task of steel enterprises is energy conservation, emission reduction and comprehensive utilization of energy. Through the strategic adjustment of the energy structure of the enterprise, the combined reheat power generation technology of high temperature and high pressure and gas is adopted to achieve the improvement of the comprehensive utilization rate of energy. By analyzing the current energy utilization of blast furnace gas and steam generated by the waste heat boiler of the ring cooler with different energy levels, the company realizes the efficient cascade utilization of residual energy and waste heat and reduces energy consumption in accordance with the principle of "energy quality matching and cascade utilization".

Key words: energy conservation and emission reduction; high temperature and high pressure; coal gas; steam; residual energy and residual heat

氨气还原铁氧化物的反应特性及动力学研究

刘玥君，李先春，李 丽，林建廷

（辽宁科技大学化学工程学院，辽宁 鞍山 114000）

摘 要： 2022 年 2 月，全球粗钢产量为 1.427 亿吨。钢铁行业能源活动中 CO_2 排放量占全国 CO_2 排放量的 15%左右，仅次于电力行业[1]。在应对全球气候变化和双碳要求的背景下，将低碳能源应用于绿色冶金产业链中是必然趋势。H_2 是目前应用较为广泛的无碳清洁能源，但制氢和储运成本昂贵，应当利用其他氢载体继续开展炼铁脱碳研究。氨在常压下极易液化，易于安全运输，合成氨技术已较为成熟。氨作为一种极具潜力的氢载体，近些年来得到了广泛的关注[2]。相关研究发现，NH_3 和 H_2 同时吸附在赤铁矿表面上会对还原反应产生抑制作用[3]，还原后期主要由 NH_3 分解产生的 H_2 起主导作用[4]。为了钢铁企业能安全高效地利用 NH_3，需要就 NH_3 直接还原铁矿石的还原特性、反应动力学及机理等问题进行研究。

利用 TG-MS 在 700~900℃下对 40%NH_3-Ar 还原 Fe_2O_3 粉末进行了等温动力学研究。结果表明 40%NH_3 还原 Fe_2O_3 的还原度略高于理论上分解最大量的 H_2 还原 Fe_2O_3 的还原度，且 SEM 结果显示 NH_3 还原后的铁相较 H_2 还原后的形貌更为清晰，EDS 结果显示 NH_3 还原后的 O 含量更低，说明还原过程中部分 NH_3 被铁氧化物催化分解成 H_2，大量 NH_3 参与了还原反应。NH_3 还原体系内不仅存在着 Fe_4N 的产生与分解，还有 NH_3 和 H_2 的竞争吸附，因此还原速率比 H_2 还原速率略慢，NH_3 还原 Fe_2O_3 的总包反应活化能 E = 57.66 kJ/mol。利用卧式高温电炉在 400~900℃下探究了不同浓度 NH_3 还原球团矿粉末的反应特性。900℃时，20%以上的 NH_3 可以将球团矿粉末在 3h 内完全还原成单质铁；当 NH_3 浓度低于 20%时部分 Fe 会被 NH_3 氮化产生 Fe_4N。探究了温度对 40%NH_3 还原球团矿粉末的物相演变规律，结果表明随着还原温度的升高会产生 Fe_4N，当温度升高至 700℃ Fe_4N 会被分解，产物内只有单质 Fe。利用偏光显微镜对还原后的矿相组织进行分析，结果表明随着还原温度的升高，赤铁矿外层被磁铁矿层覆盖，完全还原后未反应核形貌几乎完全消失。随着还原时间的推移，由于单质 Fe 有较高的自由能，样品烧结程度逐渐增强。

参 考 文 献

[1] Jian Z, Jingyang L, Li D, et al. Influencing factors and scenario analysis of China's energy consumption CO_2 emissions[J]. Journal of Environmental Engineering Technology, 2021: 1-11.

[2] Wu C, Zheng S, Wang Z, et al. Discussion on ammonia as one of the energy storage media of solar energy in China[J]. Energy Strategy Reviews, 2021, 38: 1-12.

[3] Hosokai S, Kasiwaya Y, Matsui K, et al. Ironmaking with ammonia at low temperature[J]. Environmental Science & Technology, 2011, 45(2): 821-826.

[4] Yan Ma, Jae Wung Bae, Se-Ho Kim, et al. Reducing iron oxide with ammonia: A sustainable path to green steel[J]. Advanced Science, 2023: 1-7.

简述钢铁企业能源管理及智能化的发展现状与展望

陈 琛，倪书权，徐继法，邱明英，李加旺，史 光

（五矿集团中冶京诚工程技术有限公司，北京 100176）

摘 要：钢铁产业是我国国民经济的重要支柱型产业，也是能源消耗和环境污染的主要来源之一。为了应对日益严峻的就业、能源和环境挑战，钢铁企业需要加强能源管理和智能化建设，提高能源效率和资源利用率，降低碳排放和生产成本，增强市场竞争力和可持续发展能力。本文综述了国内外钢铁企业在能源管理及智能化方向的发展现状与展望，重点介绍了一些知名钢铁企业在该方向的代表性技术、成果和应用，并矩阵化的分析了存在的问题和未来的发展趋势形成了分析能源管理及智能化发展的研究框架。

首先，基于矩阵化研究框架，本文介绍了钢铁企业在能源管理及智能化方向的主要内容和目标，包括能源数据采集、分析、优化、控制、节约、利用等方面，以及实现智能化生产、运营、管理、服务等方面。其次，在上述应用内容和目标的基础上本文举例分析介绍了一些知名钢铁企业在各自发展方向上的代表性技术、成果和应用。例如我国宝武集团着力发展基于数据驱动的能源智能管控系统；印度塔塔钢铁和韩国浦项钢铁则基于自动化系统升级迭代发展高级自动化技术、人工智能技术的智能能源控制系统等，并说明了这些技术对于提高钢铁企业的能源效率、质量水平、生产效率、经济效益等方面的作用和价值。第三，本文分析了钢铁企业在"能源管理及智能化"方向存在的问题和挑战，如技术创新不足、标准规范缺失、数据安全风险等。最后，本文提出了未来的能源管理和智能化发展趋势和建议，指出通过加强技术研发和推广、建立统一的标准体系、加强数据保护和共享的发展战略为钢铁企业提供参考和借鉴，并促进钢铁产业的转型升级和绿色发展。

关键词：能源管理；人工智能；钢铁企业；工业应用；ESG

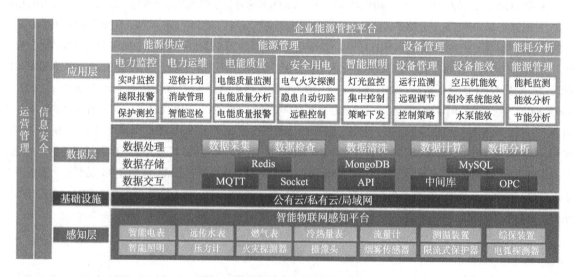

图 1 典型钢铁企业能源管控架构

参 考 文 献

[1] 刘昱辰. 能源管理系统在钢铁行业的应用[J]. 冶金管理, 2022, 443(9): 1-3.
[2] 吴群. 基于物联网的钢铁行业节能检测信息系统[D]. 浙江理工大学, 2022.
[3] 曲泰安, 高大鹏, 白雪, 等. 钢铁行业智能化能源管控平台的建设与应用[J]. 鞍钢技术, 2020, 424(4): 65-70.

[4] 何维. 钢铁企业能源管理系统智能优化调度研究[D]. 湘潭大学, 2010. DOI:10.7666/d.D384289.
[5] 邓万里. 智能制造视野下钢铁企业能源管控系统展望[J]. 钢铁, 2020, 55(11): 9.
[6] 许逵. 云平台下钢铁企业能源管理系统设计与实现[J]. 冶金与材料, 2022, 42(6): 4.
[7] 任国明. 浅谈能源管控系统在钢铁行业中的应用[J]. 工程技术研究, 2022, 4(8): 61-65.
[8] 孙涛. 能源管理系统在钢铁企业上的应用[J]. 数字技术与应用, 2021(039-010).
[9] 张福明. 智能化钢铁制造流程信息物理系统的设计研究[J]. 钢铁, 2021, 56(6): 9.

低碳发展背景下炼铁工艺革新的新思考

金永龙，孙宇佳，郝良元，田京雷

（河钢集团有限公司，河北石家庄 050023）

摘　要：低碳发展是应对全球气候变暖的重要战略性举措，低碳冶金是钢铁行业未来可持续发展的必然选择，低碳冶金工艺和技术革新是行业发展的重要措施。废钢-电炉短流程冶金是无可争议的低碳发展路径，但经典的长流程工艺，即造块-炼铁-炼钢工艺，尤其是高炉炼铁工艺在国内仍将占据重要的发展位置，在中长期内仍会有长足的发展，长流程的优化和低碳发展技术仍是业界研究的重点方向。而对于氢冶金等替代传统长流程工艺的技术发展，将是未来极致低碳的重要技术路线。本文针对企业或行业发展过程中，对新的工艺技术研发和关键技术选择等事关企业生存和发展大计的问题进行深入分析和探讨，为脚踏实地的低碳发展提供借鉴。

关键词：炼铁技术革新；低碳发展；创效能力；竞争力

New Thinking on Ironmaking Process Innovation under the Background of Low Carbon Development

JIN Yonglong, SUN Yujia, HAO Liangyuan, TIAN Jinglei

(HBIS Group Co., Ltd., Shijiazhuang 050023, China)

Abstract: Low carbon development is an important strategic measure to address global climate change. Low carbon metallurgy is an inevitable choice for the future sustainable development of the steel industry, and low-carbon metallurgical processes and technological innovation are important measures for industry development. Scrap steel electric furnace short process metallurgy is an undisputed low-carbon development path, but the classic long process process, namely block making iron making steel making process, especially the blast furnace iron making process, will still occupy an important development position in China, and there will be significant development in the medium to long term. The optimization of long process and low-carbon development technology are still the key directions of industry research. For the development of hydrogen metallurgy and other technologies to replace the traditional long process, it will be an important technology roadmap of extreme low carbon in the future. This article conducts in-depth analysis and exploration of issues related to the survival and development of enterprises or industries, such as the research and development of new process technologies and the selection of key technologies, in order to provide reference for down-to-earth low-carbon development.

Key words: innovation in ironmaking technology; low carbon development; ability to create efficiency; competitive power

基于生命周期理论的汽车钢碳足迹评价

房晓晴[1,2]，孙文强[1,2]

（1. 东北大学热能工程系，辽宁沈阳　110819；
2. 国家环境保护生态工业重点实验室（东北大学），辽宁沈阳　110819）

摘　要：随着2021年以来全球汽车业快速复苏，"双碳"目标对汽车产业发展提出新要求[1]，汽车可持续发展面临的问题更加严峻，汽车行业对于结构材料的低碳化要求也更为严格。汽车钢的碳足迹是产品生命周期中温室气体排放的量化值。基于生命周期理论分析碳足迹可以对汽车钢生产各环节的环境影响有更清晰的认识[2]，可以从中识别出环境影响的重要因素，以便进行有针对性的改进。通过开展汽车钢生命周期碳足迹评价，可有效指导并帮助企业真正了解汽车钢生产对气候变化的影响[3]。因此，钢铁企业更要积极开展汽车钢的相关碳足迹评价工作以应对越来越严格的结构材料需求。

研究针对汽车钢开展了生命周期碳足迹评估，以生产单位汽车钢为功能单位，建立了"摇篮到大门"的汽车钢生产生命周期碳足迹评价模型。该模型从生命周期的视角出发，将铁矿石、煤炭等原料、燃料开采，经过焦化、烧结等原料加工工序到炼铁、炼钢等制造工序纳入系统，形成汽车钢产品的系统边界。并通过物质流与能量流将生产工序、上游资源能源开采加工工序、运输等建立联系[4]。最后将采集的单元过程输入输出数据在所定义的生命周期系统边界范围内依据工艺过程形成汽车钢生命周期碳足迹清单模型。模型中涵盖了上游过程、运输过程生产系统与能源系统等不同环节，便于后续区分不同来源的环境影响。基于所建立的生命周期评价模型，从上下游环节、工序、物质流、能量流等多种层面分析对碳足迹贡献较大的因素。

在汽车钢"摇篮到大门"的生命周期过程中，洗精煤作为能源的使用及石灰石作为原材料的输入将带来较大的碳足迹贡献，高炉煤气作为副产品的回收将产生最大碳足迹抵扣。选取其中对碳足迹影响较大的资源结构、厂内能源等影响因素类型，分析其在实际生产参数范围内变化的敏感度。最后，对各因素敏感度分析进行了结果排序。敏感度分析结果表明，加强高炉煤气回收利用、选择更清洁的能源、降低铁钢比等措施可作为降低汽车钢生命周期碳足迹的可靠措施。

参 考 文 献

[1] Xu, H, Chen Y. Energy Reports, 2022, 8(1): 315-320.
[2] Yang H, Ma L, Li Z. Renewable and Sustainable Energy Reviews, 2023, **172**: 113040.
[3] 刘涛, 刘颖昊, 周烨. 中国冶金, 2021, 31(9): 130-134.
[4] 刘书含, 孙文强, 石晓星, 等. 中国冶金, 2022, 33(2): 77-83.

大型钢铁企业焦炉煤气柜全离线情况下的煤气平衡实践

吕　军，黄　腾

（武钢有限制造管理部，湖北武汉　430080）

摘　要：本文介绍了武钢有限焦炉煤气柜全离线情况下的煤气平衡实践情况。实践表明：在没有焦炉煤气柜的情况下，通过系统策划，采取一系列有效的管控措施也可实现焦炉煤气平衡。此次焦炉煤气柜离线运行期间圆满实现了安全事故为零、用户生产中断事件为零、煤气放散为零等目标，为以后公司焦炉煤气平衡积累了宝贵经验，相关经验也可推广至其他大型钢铁企业。

关键词：大型钢铁企业；焦炉煤气柜；离线；煤气平衡

Gas Balance Practice of Large Iron and Steel Enterprise under Coke Oven Gas Tanks are All Offline

LV Jun, HUANG Teng

(Products and Technique Management Department of WISCO, Wuhan 430080, China)

Abstract: This paper introduces the practice of large iron and steel enterprise under coke oven gas tanks are all offline. The practice shows that the balance of coke oven gas can be achieved by systematic planning and a series of effective control measures. During the off-line operation of the coke oven gas tank successfully achieved the goal of zero safety accidents, zero user production interruption events, and zero gas emission, which accumulated valuable experience for the balance of coke oven gas in the future, and relevant experience can also be extended to other large iron and steel enterprises.

Key words: large iron and steel enterprise; coke oven gas tank; offline; gas balance

转底炉系统风机配置与节能优化分析

朱红兵

（上海宝钢节能环保技术有限公司，上海　200999）

摘　要：转底炉技术已经成为钢铁企业固废尤其是含铁含锌尘泥处置的优选工艺。在"双碳"背景下，如何降低转底炉能耗是其发展的研究方向[1]。本文结合已经实施的典型节能案例，为转底炉工艺系统中的风机节能降耗指明方向。

（1）存在问题溯源及对策分析。

1）风机设备选型和设计参数偏差问题。

①系统阻力偏差问题。管网系统阻力与实际值偏差过大是导致风机运行效率偏低的一个主要因素。存在差异的主要原因归结为阻力系数的试验数据所依赖的条件与实际情况有偏差造成。而文献[2]中则指出，采用计算流体力学的方法更适用于管网形状复杂、流动变化的情况；同时指出经验公式法计算管网系统的整体阻力时为所有沿程阻力和局部阻力的线性叠加，而计算流体力学的方法则可以较好地反映各管件之间的相互影响，计算精度更高更准确。

②进气条件偏差问题。风机进气条件一般包括：当地大气压或海拔高度、进气端压力（通常意义上为进气负压）、进气温度、进气介质成分和含量百分比、进气介质密度、进气介质含尘量（包括颗粒物成分、粒径等）。

在很多的设备技术规格书中，进气负压都没有特别注明，并且进气温度也和实际监测的温度存在一定的差距，这在转底炉污泥烘干系统和生球烘干系统中表现明显。

③模型机风机性能真实性的论证。原始数据库中风机模型的性能必须真实可靠，并且考到尺寸效应影响。尤其是涉及到模型机的效率确定，必须用扭矩仪实测轴功率。

2）管道布置问题。

系统效应的问题目前已经得到足够重视，但就转底炉项目中需要特别提出的就是转底炉助燃风机的出口布置问题，虽然风机的出风口条件不能像进口涡流一样改变风机的特性曲线，但出口条件不良会造成一定量的能量损失，其实这个问题在很多的助燃风机系统中是个共性问题。

由图1可以看出，这是很明显的出口管道逆向布置。从流场模拟分析可以明显看出，风机出口的气流是没有充满整个出口截面、各向异性的气流，见图2，气流没有足够的距离进行充分的再分布，在拐弯之前，没有能够呈现典型的全紊流特征，气流速度最高点总是在出口弯管的对侧，气流再分布会造成很大的动能能耗。在初始设计时将原来的左（右）90°修改成右（左）135°的旋向布置压力损失最小，同时对机壳进行合理的剖分，检修时吊装同样方便。

 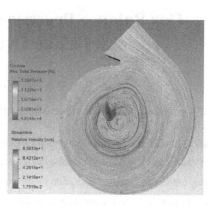

图1　助燃风机出口管道布置图（均为逆向布置）　　　图2　助燃风机机壳内部流线图

3）变频方式的适用性。

以生球烘干除尘风机采用变频调速改造为例，在管网系统中运行时，风机处于30~35Hz区间内运行。由于原设计点与实际需求偏差太大，变频改造后的节电率虽然高达60%，但机组运行效率只有55.1%，风机内效率只有65.7%，与风机最高效率点83.1%相差甚远。以此案例说明变频改造方式的局限性[3]。

因此必须明确变频在风机设计中的地位，变频调速只是一种调节手段，必须从工艺需求方面明确变频改造的必要性。

（2）小结。

由于转底炉工艺为近几年发展起来的固废处置新工艺，其中的风机配置和技术参数的确定需要一个积累和摸索的过程。从几条生产线风机节能项目的实施效果来看，平均节电率达到30%以上，对降低转底炉吨原料电耗贡献度比较明显。随着节能工作的深度推进，转底炉产线的整体能耗有望得到进一步降低。

参 考 文 献

[1] 刘长正. 转底炉处理冶金固废的节能研究与应用[J]. 中国资源综合利用, 2022(6): 180-184.
[2] 王枭, 饶杰, 王弼. 管道阻力计算[J]. 压缩机技术, 2018(6): 26-29.
[3] 吴民强. 泵与风机节能技术问答[M]. 北京: 中国电力出版社, 1998: 267-271.

面向转炉炼钢的能源管控系统设计与实现

李霜青[1]，赵　菲[1]，谷　牧[2]，王　宏[1]，郑　磊[2]

（1. 包钢集团新联信息产业有限公司，内蒙古包头　014010；
2. 北京航天智造科技发展有限公司，北京　100144）

摘　要：本文针对炼钢过程的高能耗、高污染问题，研究并设计了一种用于转炉炼钢过程的能源管控系统，可以实现对能源系统综合管理、能源监控预警、综合能耗监测、能耗成本监测等一体化智能监测管理，并构建了能耗分析知识库系统，可以有效实现能源系统化监控与管理优化的目标。

关键词：转炉炼钢；能源系统；能源管控；能耗分析

Design and Implementation of Energy Control System for Converter Steelmaking

LI Shuangqing[1], ZHAO Fei[1], GU Mu[2],
WANG Hong[1], ZHENG Lei[2]

(1. Baogang Group Xinlian Information Industry Co., Ltd., Baotou 014010, China; 2. Beijing Aerospace Intelligent Manufacturing Technology Development Co., Ltd., Beijing 100144, China)

Abstract: Focuses on the high energy consumption and pollution issues in the steelmaking process, this paper studies and designs an energy control system for the converter steelmaking process. It can achieve integrated intelligent monitoring and management of energy system comprehensive management, energy monitoring and warning, comprehensive energy consumption monitoring, energy cost monitoring, and other aspects. It also constructs an energy analysis knowledge base system, which can effectively achieve the goal of energy systematic monitoring and management optimization.

Key words: converter steelmaking; energy system; energy control; energy consumption analysis

自备电厂煤粉锅炉掺烧生物质试验

郁 雷[1]，邱全山[1]，刘自民[1]，曹曲泉[2]，顾 骁[2]，赵 华[2]

（1. 马鞍山钢铁股份有限公司技术中心，安徽马鞍山 243000；
2. 马鞍山钢铁股份有限公司能环部，安徽马鞍山 243000）

摘 要：生物质资源是"零碳"燃料，在我国能源转型以及"碳中和"战略中处于极其重要的地位。本文在自备电厂开展煤粉锅炉掺烧生物质工业试验，研究了生物质燃料基本性能及掺烧对煤粉锅炉运行状态的影响，对直燃耦合掺烧的安全性及可行性进行了分析，试验结果表明，在230MW发电机组的220t煤粉锅炉上生物质掺烧质量比可到达4%~6%，掺烧对制粉储料系统、入炉煤煤质、锅炉运行、烟气排放影响甚微，CO_2减排作用明显，通过措施优化可改善颗粒粉化现象。

关键词：自备电厂；煤粉锅炉；生物质；CO_2减排

Experiment on Blending Biomass in Pulverized Coal Boiler of Self-provided Power Plant

YU Lei[1], QIU Quanshan[1], LIU Zimin[1], CAO Ququan[2],
GU Xiao[2], ZHAO Hua[2]

(1. Maanshan Iron & Steel Co., Ltd., Technology Center, Maanshan 243000, China; 2. Maanshan Iron & Steel Co., Ltd., Energy and Environmental Protection Department, Maanshan 243000, China)

Abstract: Biomass resource is a "zero-carbon" fuel, and plays an extremely important role in China's energy transformation and "carbon neutral" strategy. In this paper, an industrial experiment on blending biomass in pulverized coal boiler was carried out in self-owned power plant, and the basic performance of biomass fuel and its influence on the operation of pulverized coal boiler were studied. The safety and feasibility of direct-combustion coupled blending was analyzed. The test results show that the mass ratio of biomass blending can reach 4%~6% on 220 tons pulverized coal boiler and 230MW generator set. Mixed firing has little effect on pulverizing system, coal composition, boiler operation and flue gas emission, but has obvious effect on CO_2 emission reduction.

Key words: self-contained power plant; pulverized coal boiler; biomass; CO_2 emission reduction

复杂工况下钢包全程保温技术装备研发与应用

赵金龙[1]，杨轶龙[2]，吴恩旭[1]，张洪亮[1]

（1. 鞍钢重型机械设计研究院有限公司，辽宁鞍山 114031；
2. 鞍钢股份公司设备工程部，辽宁鞍山 114021）

摘　要： 在炼钢工序实施钢包全程保温技术对降低转炉出钢温度、减低钢水夹杂率具有重要意义，但在紧凑型炼钢工序复杂工况下实施钢包全程保温技术始终是一项亟待解决的瓶颈性技术难题，本文围绕鞍钢炼钢四分厂紧凑型工艺布局，研发出国内首套正反双向插齿式钢包全程保温技术装备，打破了紧凑型炼钢工序空间受限、工艺顺行性差的技术瓶颈，解决了钢水夹杂缺陷率居高不下的技术难题，实现了炼钢工序提质增效、节能低碳等一系列技术创新成果。

关键词： 钢包保温；复杂工况；节能降碳；插齿式；加揭盖设备

Research and Development and Application of Ladle Insulation Technology and Equipment under Complex Working Conditions

ZHAO Jinlong[1], YANG Yilong[1], WU Enxu[1], ZHANG Hongliang[1]

(1. Angang Heavy Machinery Design & Research Institute Co., Ltd., Anshan 114031, China;
2. Equipment Engineering Department of Ansteel Co., Ltd., Anshan 114021, China)

Abstract: The implementation of ladle insulation technology in the steelmaking process is of great significance to reduce the converter steel output temperature and reduce the molten steel inclusion rate，under the complex working conditions of compact steelmaking processes, the implementation of ladle insulation technology has always been a bottleneck technical problem that needs to be solved urgently, this paper focuses on the compact process layout of the fourth branch plant of Angang steelmaking, developed the first set of front and reverse two-way shaping steel ladle full insulation technology and equipment in China, it breaks the technical bottleneck of limited space in the compact steelmaking process and poor process prograde,it solves the technical problem of high molten steel inclusion defect rate, it has realized a series of technological innovation achievements such as improving quality and efficiency of steelmaking process, energy saving and low carbon.

Key words: iadle insulation; complex operating conditions; energy saving and carbon reduction; gear shaping; add the uncover device

朝阳钢铁智慧水务管理探索与研究

黄永梁[1]，潘大龙[1]，王念军[2]，鲁 璐[1]，刘 洋[1]，王艳霞[2]

(1. 鞍钢集团自动化有限公司，辽宁鞍山 114000；
2. 鞍钢集团朝阳钢铁有限公司，辽宁朝阳 122000)

摘 要：朝阳钢铁基于自身水系统管理特点，建立全水种、全工序水系统运行图及各工序水耗、公司水效分析模型，对购水、制水、配水、用水、排水等全流程和水温、水压、水量、水质等全要素进行运行管理的综合分析，建立了智慧水务平台，打造全工序、全过程、全要素的极致水效管理模式。

关键词：全工序；全过程；全要素；极致水效

Experience Summary of Smart Water Management of Chaoyang Iron and Steel

HUANG Yongliang[1], PAN Dalong[1], WANG Nianjun[2],
LU Lu[1], LIU Yang[1], WANG Yanxia[2]

(1. Automation Co., Ltd. of Ansteel Group Corporation Ltd., Anshan 114000, China;
2. Ansteel Group Chaoyang Steel & Iron Co., Ltd., Chaoyang 122000, China)

Abstract: Based on its own water system management characteristics, Chaoyang iron and steel has established a water system operation diagram for all kinds of water and all working procedure, as well as a water consumption and company water efficiency analysis model for each process. It has comprehensively analyzed the whole process of water purchase, water production, water distribution, water use, water drainage, and all factors such as water temperature, water pressure, water yield and water quality. It has established an intelligence water service platform to create an extreme water efficiency management mode for total working procedure, total processes and total factors.

Key words: total working procedure; total processes; total factors; extreme water efficiency management

钢铁企业高炉氧气消耗预测模型开发及应用

刘 凯[1]，刘磊刚[2]，宋岩峰[1]，覃天强[1]，董惠文[1]，王亚腾[1]

(1. 鞍钢集团自动化有限公司，辽宁鞍山 114000；2. 鞍钢股份有限公司，辽宁鞍山 114000)

摘 要：高炉炼铁是现代钢铁生产的主要工艺过程，其生产过程中需要大量的氧气，在钢铁企业氧气消耗总量中高炉耗氧占比较大，氧气发生量与消耗量的供需平衡对管网稳定运行有着至关重要的影响，因此，预测高炉氧气发生量对氧气调度平衡优化有着重要意义。近年来，随着计算机技术和人工智能的快速发展，越来越多的数学模型和算法被应用于钢铁领域。本模型依托现场实时数据以及生产计划、生产实绩数据进行分析，对未来一段时期内高炉氧气消耗量进行预测,为氧气调度人员提供决策参考信息。

关键词:高炉氧气预测;调度平衡优化;预测模型

Development and Application of Oxygen Consumption Prediction Model for Blast Furnace in Iron and Steel Enterprises

LIU Kai[1], LIU Leigang[2], SONG Yanfeng[1], QIN Tianqiang[1], DONG Huiwen[1], WANG Yateng[1]

(1. Ansteel Group Information Industry Co., Ltd., Anshan 114000, China;
2. Angang Steel Company Limited, Anshan 114000, China)

Abstract: Blast furnace ironmaking is the main technological process in modern steel production, which requires a large amount of oxygen. In the total oxygen consumption of steel enterprises, blast furnace oxygen consumption accounts for a relatively large proportion. The supply and demand balance between oxygen generation and consumption has a crucial impact on the stable operation of the pipeline network. Therefore, predicting blast furnace oxygen generation is of great significance for optimizing oxygen scheduling balance. In recent years, with the rapid development of computer technology and artificial intelligence, more and more mathematical models and algorithms have been applied in the field of steel. This model relies on real-time on-site data, as well as production plans and performance data for analysis, to predict the oxygen consumption of blast furnaces in the future, providing decision-making reference information for oxygen scheduling personnel.

Key words: blast furnace oxygen prediction; optimization of scheduling balance; prediction model

绿色低碳,高质量发展-磁悬浮风机在焦化生化站应用

木塔力甫·玉苏甫,韩 亮

(宝武水务科技有限公司新疆分公司,新疆乌鲁木齐 830022)

摘 要:随着环境保护和可持续发展的重要性日益突出,推动绿色低碳、高质量的发展已成为各行各业的共同目标。在这个背景下,磁悬浮风机作为一种创新的曝气设备,正在逐渐取代传统的离心风机,为环境保护和可持续发展做出卓越贡献。

宝武水务科技有限公司新疆分公司焦化生化站好氧池曝气原使用 5 台 160kW 离心式鼓风机,随着使用年限增加,离心式鼓风机弊端不断显现。本文以宝武水务新疆分公司焦化生化站的磁悬浮风机代替离心风机曝气的案例为例,探讨磁悬浮风机在绿色低碳发展中的优势和应用。

首先,磁悬浮风机通过磁悬浮轴承技术取代了传统的机械轴承,减少了摩擦损耗,提高了能效。根据实际测试数据,原离心风机的功率为 3×160kW(开三备二),而将离心风机替换为磁悬浮风机,新的风机功率为 2×150kW(开二备一)。这意味着在相同的曝气量下,磁悬浮风机的能耗降低了约 180kW,实现了显著的节能效果。

其次,磁悬浮风机还具有噪声低、振动小的优点。根据测量数据,原离心风机的噪声超过 100dB,对工作环境和员工健康带来负面影响。而通过磁悬浮风机的应用,风机噪声得到了大幅度的减少,不超过 80dB。这种噪声减少的优势,提高了工作环境的舒适度,也符合绿色低碳发展的要求。

此外，磁悬浮风机还通过智能化控制系统实现了自动化运行，大大提高了运行的稳定性和可靠性。在原有的离心风机中，人工干预调节和维护频繁，而磁悬浮风机通过自动调整风量和转子位置，减少了人为干预的需求。这在案例中也得到了验证，磁悬浮风机的故障率明显降低，维修成本显著减少。

根据经济效益分析，通过磁悬浮风机的应用，公司预计年度节约电量2277600kW，节约费用将达到58.58万元。这些数据显示了磁悬浮风机在节能降耗方面的显著效果，为企业的可持续发展和经济效益带来了实实在在的益处。

磁悬浮曝气风机代替离心式曝气风机在绿色低碳发展中具有重要意义。通过宝武水务新疆分公司的实际探索，我们可以清楚地看到磁悬浮风机在能耗降低、噪声减少、运行稳定性提高和经济效益增加等方面的优势。相信随着磁悬浮曝气风机的不断改进和推广应用，其在绿色低碳、高质量发展中的作用将更加重要，为我们的环境保护和可持续发展做出更大的贡献。

浅谈BTG亚临界发电机组煤气-烟气换热器应用问题解析

郝 鑫，王 浩

（南钢能源动力事业部发电厂，江苏南京 210000）

摘 要：近年来，随着技术进步和钢铁企业对能源利用水平提高的强烈需求，高参数小型化煤气发电机组投产案例逐步增加，其中根据部分钢企生产情况，超高温亚临界技术跟具有优势。在燃烧系统方面，为降低排烟热损失，提高整体系统热效率，新建机组时设计院通常会增加一套煤气-烟气换热器，回收烟气余热，预热煤气，提高锅炉效率，但是煤气-烟气换热器在实际运行中存在一定问题，文章主要讨论该装置运行中的问题和探索解决方法的过程，为后续同类型项目实施提供参考意见。

关键词：BTG亚临界发电机组；煤气-烟气换热器；流场分析

Analysis of Application Problems of Gas Flue-Gas Heat Exchanger in BTG Subcritical Power Generation Units

HAO Xin, WANG Hao

(Nanjing Nangang Industrial Development Co., Ltd., Nanjing 210000, China)

Abstract: In recent years, with the advancement of technology and the strong demand for energy utilization by steel enterprises, there has been a gradual increase in the number of high-parameter small-scale gas generator units put into operation. Among them, according to the production situation of some steel enterprises, ultra-high temperature subcritical technology has advantages. In terms of the combustion system, in order to reduce exhaust heat loss and improve the overall system thermal efficiency, the design institute usually adds a set of gas flue gas heat exchangers when building a new unit to recover waste heat from the flue gas, preheat the gas, and improve the boiler efficiency. However, there are certain problems in the actual operation of the gas flue gas heat exchanger. The article mainly discusses the problems in the operation of the device and the process of exploring solutions, Provide reference opinions for the implementation of similar projects in the future.

Key words: BTG subcritical power generation unit; gas flue-gas heat exchanger; flow field analysis

焦炉煤气能源绿色转换技术开发与应用

刁呈振

（河钢集团张宜科技，河北张家口　075100）

摘　要：焦炉煤气能源绿色转换技术开发与应用项目属于化工与能源环境技术领域。2020年9月，在联合国大会上，我国宣布了二氧化碳排放力争于2030年前达到峰值，努力争取2060年前实现碳中和的宏伟目标。随着京津冀区域雾霾问题日趋严峻和北京携手张家口举办2022冬奥会举行，位于首都北京上风向的河北省张家口市，锅炉的燃料结构低碳化、排放洁净化势在必行。为积极相应国家环保政策，应对能源结构变化，更大量消化焦炉煤气，实现烟气超低排放，经过一系列对全燃气锅炉进行低碳燃料结构改造、锅炉受热面结构改造、烟气净化改造，最终实现燃气锅炉超绿色洁净低碳生产。

关键词：焦炉煤气；能源；绿色；转换技术

Development and Application of Green Conversion Technology for Coke Oven Gas Energy

DIAO Chengzhen

(Zhangxuan Tech, Hebei Iron and Steel Group, Zhangjiakou 075100, China)

Abstract: The development and application project of green conversion technology for coke oven gas energy belongs to the field of chemical engineering and energy environmental technology. In September 2020, at the United Nations General Assembly, China announced its goal of peaking carbon dioxide emissions by 2030 and striving to achieve carbon neutrality by 2060. With the increasingly serious haze problem in Beijing Tianjin Hebei region and the 2022 Winter Olympic Games held by Beijing and Zhangjiakou, Zhangjiakou City, Hebei Province, located in the upwind direction of Beijing, the fuel structure of boilers is low-carbon and emission cleaning is imperative. In order to actively respond to national environmental policies, respond to changes in energy structure, digest more coke oven gas, and achieve ultra-low flue gas emissions, a series of low-carbon fuel structure transformation, boiler heating surface structure transformation, and flue gas purification transformation have been carried out on all gas boilers, ultimately achieving ultra-green, clean, and low-carbon production of gas boilers.

Key words: coke oven gas; energy; green; conversion technology

芜湖新兴烧结工序能效提升生产实践

魏瑞瑞[1,2]，楚天福[1]，段龙龙[1]，于恩斌[1]，朱玉停[1]，武彪[1]

（1. 芜湖新兴铸管有限责任公司铁前事业部，安徽芜湖　241002；
2. 芜湖新兴铸管有限责任公司研究院，安徽芜湖　241002）

摘　要：钢铁行业是耗能大户，我国钢铁行业的用能量约占全社会总能耗的10%。钢铁行业已经度过了粗放发展、无序耗能的时期。随着钢铁行业竞争不断加剧，企业间整合不断加速，市场需求不断萎缩，能效指标将成为钢铁企业能否生存的重要评价指标。芜湖新兴针对烧结工序能耗较高的问题，对烧结环冷系统进行了改造提升，将传统环冷机改造为新型水密封环冷机，极大的减小了环冷机漏风率和用电消耗，漏风率从40%降至5%，环冷风机用电消耗从每小时2520kW·h降低至630kW·h，降低了75%，余热回收蒸汽量从20 t/h增加到28 t/h，增加了40%。对烧结点火炉进行升级改造，使用帘式点火器，改变烧嘴结构，增加烧嘴数量，并对助燃空气在炉膛内进行预热，大幅降低了煤气消耗量，吨矿煤气消耗量由$6.03 m^3/t$降低至$4.2 m^3/t$，降低了30.3%。此外，通过对烧结烟道弯头及竖管进行改造，有效降低了漏风率，烟气中氧含量从17.6%降至少16%。在主抽后端增加冷风阀，从而在烧结停机时避免使用主抽风机为脱硫脱硝工序的活性炭冷却，有效降低了停机时主抽的无效运转时间，每次停机约节约电量27000kW·h。

关键词：能效提升；水密封环冷机；点火炉；冷风阀

Wuhu Xinxing Sintering Process Energy Efficiency Improvement Production Practice

WEI Ruirui[1,2], CHU Tianfu[1], DUAN Longlong[1],
YU Enbin[1], ZHU Yuting[1], WU Biao[1]

(1. Ironmaking Department, Wuhu Xinxing Ductile Iron Pipes Co., Ltd., Wuhu 241002, China;
2. Research Institute, Wuhu Xinxing Ductile Iron Pipes Co., Ltd., Wuhu 241002, China)

Abstract: The iron and steel industry is a large energy consumer, the energy used in China's iron and steel industry accounts for about 10% of the total energy consumption of the whole society. The iron and steel industry has gone through the period of extensive development and disorderly energy consumption. With the fierce competition in the iron and steel industry, the continuous acceleration of the integration between enterprises, and the continuous shrinking of market demand, the energy efficiency index will become an important evaluation index for the survival of iron and steel enterprises. In view of the high energy consumption of the sintering process, Wuhu Xinxing has carried out the transformation and upgrading of the sintering loop cooling system, and transformed the traditional loop cooling machine into a new water-sealed loop cooling machine, which greatly reduces the air leakage rate and electricity consumption of the loop cooling machine, the air leakage rate decreases from 40% to 5%, the electricity consumption of the loop cooling fan decreases from 2520 kW·h per hour to 630 kW·h, which reduces by 75%, and the amount of waste heat recovery steam increases from 20 t/h to 28 t/h, which increases by 40%. The sintering ignition furnace is upgraded and transformed, the curtain ignition is used, the burner structure is changed, the number of burners is increased, and the combustion air is preheated in the furnace, which greatly reduces the gas consumption, the consumption of tonnage coal gas decreases from $6.03 m^3/t$ to $4.2 m^3/t$, which reduces by 30.3%. In addition, through the transformation of the sintering flue elbow and vertical pipe, the air leakage rate is effectively reduced, and the oxygen content in the flue gas decreases from 17.6% to at least 16%. A cold air valve is added at the rear end of the main suction, so as to avoid using the main suction fan to cool the activated carbon for the desulfurization and denitration process during the sintering shutdown, effectively reducing the invalid operation time of the main suction during the shutdown, and saving about 27000 kW·h of electricity each time.

Key words: energy efficiency improvement; water sealed ring cooler; ignition furnace; cold air valve

转炉余热蒸汽蓄热系统的技术改造实践

程 鑫[1],饶江平[2],朱善合[2],王 军[1],汪文婷[3]

(1. 武钢有限炼钢厂,湖北武汉 430080; 2. 武钢有限技术中心,湖北武汉 430080;
3. 武钢有限能源环保部,湖北武汉 430080)

摘 要:介绍了武钢有限三炼钢厂转炉余热蒸汽系统的现状及问题,结合炼钢厂实际运行情况,通过转炉蒸汽蓄热器及蒸汽管道的技术改造,提升了蓄热器容量,降低了蒸汽管网的压力波动,减少了RH使用电厂过热蒸汽量,取得了良好的节能与降本效果。

关键词:转炉;余热蒸汽;蓄热器;节能

Utilization of Technical Transformation on Storage System for Saturated Steam in Steel-making

CHENG Xin[1], RAO Jiangping[2], ZHU Shanhe[2], WANG Jun[1], WANG Wenting[3]

(1. Steel-making Plant of WISCO Ltd., Wuhan 430080, China; 2. Research and Development Center of WISCO Ltd., Wuhan 430080, China; 3. Department of Energy and Environment Protection of WISCO Ltd., Wuhan 430080, China)

Abstract: This paper briefly introduces the present situation and problems of the converter waste heat recovery system in the third steel-making Branch Mill of WISCO, Ltd. The relevant technical transformation on heat accumulator and pipeline of saturated steam system in Steel-making was put forward. Based on running record, energy-saving and cost-reducing has been obtained by improving the volume of heat accumulator, reducing the pressure oscillation of the steam system and the using level of superheated steam from electricity generating station for RH.

Key words: converter; waste heat steam; heat accumulator; energy-saving

供热电站蒸汽管网优化应用分析

操加元[1],程 海[1],严 亮[1],朱善合[2]

(1. 武钢有限能源环保部,湖北武汉 430080; 2. 武钢有限技术中心,湖北武汉 430080)

摘 要:本文对供热电站送出蒸汽管网进行系统分析,结合供热电站蒸汽管网近年来发生的事故及供热电站发展阶段,系统阐述蒸汽管网存在的问题与不足,并对系统缺陷进行分析,提出切实有效的优化方案,提高供热电站蒸汽管网的安全保产能力,同时从经济运行角度有效提高蒸汽使用效率。

关键词:供热电站;蒸汽管道;蒸汽流量;高效利用

Analysis on Optimum Usage of Steam Pipeline in Heat Power Station

CAO Jiayuan[1], CHENG Hai[1], YAN Liang[1], ZHU Shanhe[2]

(1. Department of Energy and Environment Protection of WISCO Ltd., Wuhan 430080, China;
2. Research and Development Center of WISCO Ltd., Wuhan 430080, China)

Abstract: According the accident of steam pipeline in heat power station recent years and the stage of heat power station development, the problem and the shortage of the steam pipeline has been comprehensively discussed. A series of optimum proposal has been put forward to improve the ability of safety in production of the steam pipeline in heat power station, meanwhile improving the efficiency of steam utilization from economical aspect.

Key words: heat power station; steam pipeline; steam flow; efficient utilization

转炉汽化冷却蒸汽平衡与高效利用技术研究

饶江平[1]，朱善合[1]，付刚强[2]，王 军[2]，彭有杰[2]，程 鑫[2]

（1. 武钢有限技术中心，湖北武汉 430080；2. 武钢有限炼钢厂，湖北武汉 430080）

摘 要： 针对武钢有限三炼钢厂转炉汽化冷却蒸汽利用现状与存在问题，全面介绍了转炉汽化饱和蒸汽高效利用与炼钢区域蒸汽自平衡的技术实践、实施效果，对推动钢铁行业转炉蒸汽自用率提升、工序成本减降具有巨大参考价值。

关键词： 转炉；汽化蒸汽；蒸汽平衡；高效利用

Research on Steam Balance of Vaporization Cooling and Efficient Utilization in Steel-making Converter

RAO Jiangping[1], ZHU Shanhe[1], FU Gangqiang[2],
WANG Jun[2], PENG Youjie[2], CHENG Xin[2]

(1. Research and Development Center of WISCO Ltd., Wuhan 430080, China;
2. Steel-making Plant of WISCO Ltd., Wuhan 430080, China)

Abstract: According the present situation and problems of the vaporization cooling steam in steel-making converter in the third steel-making Branch Mill of WISCO, Ltd., the efficient utilization of vaporization cooling saturated steam and the application of steam balance technology practice is comprehensively discussed. The research of this paper makes contributions to the promotion of steam self-use rate and the cost reduction in steel- making converter.

Key words: converter; vaporized steam; steam balance; efficient utilization

多变内外部环境下的能源高效利用，降低外购能源费用

王 剑

（邯郸钢铁有限公司，河北邯郸 056015）

摘 要：钢铁属高耗能行业，尤其是长流程炼钢，能源成本较高。生产连续性较强，各工序之间紧密衔接，设备必须要确保稳定、长周期、经济运行。河钢集团邯钢公司近年来快速稳定发展，但受所在地域限制，环保因素影响，停限产较为频繁；市场行情变化较大，公司谋求效益最大化，更好服务新老客户，产品结构需动态调整；设备服役时间较长，事故停机及设备升级改造增多；能源介质实时动态平衡等等，都会导致设备运行方式的频繁变化，容易出现低效率运行甚至设备空耗、介质放散等问题，影响能源利用效率，致使能源成本提高。在如此复杂多变的内外部环境下，综合考虑生产组织、设备状态、环保、安全要求等因素，针对各单位及各工序制定了能源指标，实现能源成本多维度分层管理，建立了完整的能源指标管理体系。按照全员参与、集中一贯的管理思路，打造全面的能源根因分析为基础的能源管控系统，支撑企业目前的节能需求，达到国内一流水平。

关键词：能源费用；节能；能源管控；创新技术；发电效率

微波场下 H_2 和 CO 还原磁铁矿的微观结构比较

洪陆阔，周美洁，艾立群，孙彩娇，佟 帅

（华北理工大学冶金与能源学院，河北唐山 063210）

摘 要：比较了微波加热技术在 H_2 和 CO 气氛下对磁铁矿的还原效果。比较了微波场和常规场对 H_2+CO 混合气氛中磁铁矿的还原效果。用电子显微镜观察微观结构变化。结果表明：在 900~1100℃范围内，H_2 原磁铁矿的金属化速率和还原程度远高于 CO 还原磁铁矿；H_2 还原生成的 Fe 相形成空腔结构，CO 还原生成的 Fe 相形成致密块状。在常规加热条件下，纯 CO 气氛中 H_2 含量越高，还原效果越好。在微波场作用下，还原率最高的气氛为 50% H_2 + 50% CO，与常规加热相比，相同条件下微波场中还原铁形成的气泡更大。

关键词：H_2 还原；磁铁矿；微波加热；微观结构表征

Comparison of Microstructures of Magnetite Reduced by H_2 and CO under Microwave Field

HONG Lukuo, ZHOU Meijie, AI Liqun, SUN Caijiao, TONG Shuai

(College of Metallurgy and Energy, North China University of Science and Technology, Tangshan 063210, China)

Abstract: The reduction of magnetite in H_2 and CO atmospheres was compared using a microwave-heating technique. The reduction of magnetite in a mixed H_2 + CO atmosphere was compared with respect to the effects of a microwave field and a conventional field. Microstructural changes were observed using an electron microscope. The results show that the metallization rate and reduction degree of the H_2-reduced magnetite are much higher than those of the magnetite reduced by CO at 900–1100 ℃. The Fe phase generated by H_2 reduction forms a cavity structure, and the Fe phase generated by CO reduction forms a dense block. Under conventional heating conditions, the higher the H_2 content in a pure CO atmosphere, the better the reduction effect. Under the effect of a microwave field, the atmosphere with the highest reduction rate was 50% H_2 + 50% CO. Compared with conventional heating, the bubble holes formed by reduced iron in microwave field are larger under the same conditions.

Key words: H_2 reduction; magnetite; microwave heating; microstructural characterization

浅谈氢冶金工程低压供配电系统

王 伟

（中冶京诚工程技术有限公司电气所，北京 100176）

摘 要：在中国提出碳达峰和碳中和目标的背景下，国内开始实验新的氢冶金工艺。本文基于已经投产的某钢厂氢冶金工程的电气设计基础上，对氢冶金工程负荷分级及供电要求做了介绍，较为简要的阐述了氢冶金供配电的范围、特点，柴油发电机的负荷计算和选型过程以及应急 MCC 互投断路器的工作过程和保护参数配合的选择，也对智能 MCC 的架构做了具体分析。

关键词：氢冶金；供配电；智能 MCC；柴油发电机

12 冶金环保与资源利用

大会特邀报告

第十三届冶金青年科技奖获奖人特邀报告

分会场特邀报告

矿业工程

焦化及节能环保

炼铁与原料

炼钢与连铸

电冶金与废钢铁

轧制与热处理

表面与涂镀

金属材料深加工

粉末冶金

先进钢铁材料及应用

节能与低碳技术

★ 冶金环保与资源利用

冶金设备与工程技术

冶金自动化与智能化

冶金物流

冶金流程工程学

其他

12.1 冶金环保

复合菌剂增效处理焦化废水的研究与应用

王 永[1,2]，王 飞[1,2]，刘 芳[1,2]，安 宁[1,2]，李函霏[1,2]，张天赋[1,2]

（1. 海洋装备用金属材料及其应用国家重点实验室，辽宁鞍山 114009；
2. 鞍钢集团钢铁研究院 环境与资源研究所，辽宁鞍山 114009）

摘 要： 本文首先对焦化废水的处理工艺进行简介，并简述了生物强化技术的发展历史和最佳应用条件。针对厂内某焦化厂好氧池出水的苯酚，苯甲醇指标较高的特点，向好氧池生化池中加入源位提取筛选出的菌剂，强化苯酚、苯甲醇的去除效果，在不改变整个工艺的情况 35 下提高 COD 降解能力。结果表明，添加菌剂以后，好氧池中 COD 去除率提高了 52.8%，降低后续的处理焦化废水的费用。

关键词： 芽孢杆菌；焦化废水；生物强化；降解 COD

The Research and Application on Coking Waste-water Treatment Through Combined Bacteria

WANG Yong[1,2], WANG Fei[1,2], LIU Fang[1,2],
AN Ning[1,2], LI Hanfei[1,2], ZHANG Tianfu[1,2]

(1. State Key Laboratory of Metal Material for Marine Equipment and Application,
Anshan 114009, China; 2. Iron & Steel Research Institute of Ansteel Group,
Institute of Environment and Resources, Anshan 114009, China)

Abstract: Firstly, the coking waste-water treatment processes are introduced in this paper, also, including the brief history of bioaugmentation technology and its optimum application scenario. Based on the characteristic of high index of phenol and benzyl alcohol in effluent of the aerobic tank in a plant, with addition of the selected bacteria agent to intensified the removal effect of phenol and benzyl alcohol, biodegradation capacity of COD is improved while the whole processes remain unchanged. The results show that with the addition of bacteria agent, the removal rate of COD is improved by 52.8% which could lead to lower costs in the following coking waste-water treatment process.

Key words: bacillus; coking wastewater; phenol; degradation COD

钢企高炉煤气前端脱硫与煤气用户末端脱硫的技术经济比较

任海霞，王志荃，文浩锦

（中冶南方工程技术有限公司，湖北武汉 430223）

摘 要：在钢铁企业超低排放的背景下，高炉煤气精脱硫的选择是一个现实难题。根据某钢企实际运行数据，分析了高炉煤气硫分及用户烟气排放特征。分别对前端、末端脱硫多种典型的工艺技术路线及技术经济进行比较，分析前端、末端脱硫的应用现状及各自优缺点。最终总结影响企业工艺路线选择的关键因素，并对脱硫技术选择及技术改进方向提出建议。

关键词：高炉煤气；脱硫；技术经济

Technological and Economic Comparison between Front Fine Desulphurization of Blast Furnace Gas and Terminal Desulphurization of Gas User

REN Haixia, WANG Zhiquan, WEN Haojin

(WISDRI Engineering & Research Incorporation Limited, Wuhan 430223, China)

Abstract: Under the background of ultra-low emission of steel enterprises, the selection of blast furnace gas desulfurization is a practical issue.Basedonactualoperatingdata of a steelenterprises, sulfur content of blast furnace gas and emission characteristics of exhaust gas wereanalyzed. Afterwards, several typical technological routes and technological economy of front and terminal desulfurization were compared respectively. Then, application status and technical economy of front and terminal desulphurization were compared. Next, key factors affecting the choice of enterprise process route were summarized. Atlast, some suggestions were given for the choice of desulphurization technology and the direction of technology improvement.

Key words: blast furnace gas; fine desulphurization; technological economy

焦炉煤气精脱硫工艺分析及设计优化

邓万里[1]，杨 静[2]，李 霁[1]

（1. 宝山钢铁股份有限公司能源环保部，上海 201900；
2. 宝钢工程技术集团有限公司，上海 201999）

摘 要：为实现多个使用混合煤气的轧钢加热炉超低排放，通过烟气的含硫量分析，以及反算煤气含硫量，确定了首先应当实施焦炉煤气的精脱硫改造。在分析和吸取煤化工行业水解法脱硫技术的基础上，宝钢结合自身特点，对原料气净化、反吹气选用、再生气去向、水解的分段，设计中进行了适应性优化，并充分考虑了与煤气柜和管网的合理连接。项目实施后，煤气总硫脱除率在90%以上，烟气 SO_2 排放减少50%以上，初步达到预期效果。

关键词：焦炉煤气；精脱硫；二氧化硫；煤气柜

Process Analysis and Design Points of Fine Desulfurization of Coke Oven Gas

DENG Wanli[1], YANG Jing[2], LI Ji[1]

(1. Energy & Environment Department, Baoshan Iron & Steel Co., Ltd., Shanghai 201900, China;

2. Baosteel Engineering & Technology Group Corporation, Shanghai 201999, China)

Abstract: In order to realize the ultra-low emission of several steel rolling furnaces using mixed gas, through the analysis of the sulfur content of flue gas and the reverse calculation of the sulfur content of gas, it is determined that the refined desulfurization transformation of coke oven gas should be implemented first. Based on the analysis and absorption of the hydrolysis desulfurization technology in the coal chemical industry, combined with its own characteristics of Baosteel, the design has carried out adaptive optimization of feed gas purification, back blowing gas selection, regeneration gas direction, hydrolysis section, etc., and considered the reasonable connection with the gas holder and pipe network. After the implementation of the project, the total sulfur removal rate of gas is more than 90%, and the SO_2 emission of flue gas is reduced by 50%, which has initially achieved the expected effect.

Key words: coke oven gas; fine desulfurization; SO_2; gas holder

高炉煤气精脱硫系统工艺研究

胡 伟[1]，董映红[1]，耿云梅[1]，郭利朋[2]

（1. 佰利天控制设备（北京）股份有限公司研发部，北京 100020；
2. 安阳钢铁集团有限责任公司，河南安阳 455004）

摘 要： 近两年中国粗钢产量突破10亿吨，多个大型钢铁基地落成，造成环境承载能力减弱，大气污染防治重点区域内的钢铁企业为达到超低排放限值，投入运行了多套高炉煤气精脱硫系统。本文阐述了高炉煤气硫化物来源及去向，分析了高炉煤气含硫量变化、高炉煤气系统对精脱硫系统的要求、精脱硫系统对高炉煤气工艺条件的要求，总结了高炉煤气精脱硫技术路线，包括"转化吸收法""转化吸附法"和"直接吸附法"，三种技术路线都已投入使用。最后展望了污染物超低排放和双碳目标下高炉煤气精脱硫系统发展方向。

关键词： 高炉煤气；煤气脱硫；超低排放；羰基硫；硫化氢

Research on the Process of Blast Furnace Gas Fine Desulfurization System

HU Wei[1], DONG Yinghong[1], GENG Yunmei[1], GUO Lipeng[2]

(1. Bailitian Control Equipment (Beijing) Co., Ltd., Beijing 100020, China;
2. Anyang Iron & Steel Group Co., Ltd., Anyang 455004, China)

Abstract: In the past two years, China's crude steel output has exceeded 1 billion tons, and a number of large-scale steel bases have been completed, resulting in a weakening of environmental carrying capacity, and steel enterprises in key areas for air pollution prevention and control have put into operation a number of blast furnace gas refined desulfurization systems in order to reach ultra-low emission limits. This paper expounds the source and destination of blast furnace gas sulfide, analyzes the changes of sulfur content of blast furnace gas, the requirements of blast furnace gas system for fine desulfurization system, and the requirements of fine desulfurization system for blast furnace gas process conditions, and summarizes the technical route of refined desulfurization of blast furnace gas, including "conversion absorption method", "conversion adsorption method" and "direct adsorption method", all three technical routes have been put into use. Finally, the development direction of blast furnace gas fine desulfurization system under the goal of ultra-low pollutant emission and double carbon is prospected.

Key words: blast furnace gas; gas desulfurization; ultra-low emissions; carbonyl sulfur; hydrogen sulfide

一条线材深加工企业工业污水处理线的设计与实践

赵宪海[1]，金晓谦[2]，祝 贺[2]，唐 盈[2]，范思豪[2]，谈志昊[3]

（1. 鞍钢股份有限公司，辽宁鞍山 114021；2. 鞍钢钢绳有限责任公司，辽宁鞍山 114021；3. 无锡艾德环境科技有限公司，江苏无锡 214000）

摘　要：本文介绍了一条新建的线材深加工企业污水处理线的设计与实践情况。本文简要描述了工程项目总体设计思路、工艺原理和工艺流程，提出了污水处理线进水水量、进水水质指标。在实践过程中，结合出现的问题从工艺流程完善、设备补充、管理优化等方面也进行了实事求是的阐述，这些都对线材深加工企业新建及改扩建污水处理系统提供参考。

关键词：线材；深加工；污水处理

Design and Practice of an Industrial Sewage Treatment Line for Wire Rod Deep Processing Enterprises

ZHAO Xianhai[1], JIN Xiaoqian[2], ZHU He[2],
TANG Ying[2], FAN Sihao[2], TAN Zhihao[3]

(1. Ansteel Iron and Steel Co., Ltd., Anshan 114021, China; 2. Ansteel Wire Rope Co., Ltd., Anshan 114021, China; 3. Wuxi Ade Environment Technology Co., Ltd., Wuxi 214000, China)

Abstract: This paper introduces the design and practice of sewage treatment line in a newly-built wire rod deep processing enterprise. This paper briefly describes the general design idea, process principle and process flow of the project, and puts forward the water intake quantity and quality index of the sewage treatment line. In practice, problems were Seek truth from facts in terms of process improvement, equipment replenishment and management optimization, all of these provide references for the new and expanded sewage treatment system of wire rod deep processing enterprises.

Key words: wire rod; deep processing; sewage treatment

高线精轧机辊箱稀油污染源的检测方案

卢永清，吴春东，张建业

（承德钒钛新材料有限公司棒材事业部，河北承德 067002）

摘　要：为快速准确判断高线精轧机辊箱稀油污染源，设计并制定此实施检测方案，达到预知检修，提升轴承寿命的目的，为轧钢生产提供了可靠保证。

关键词：高线精轧机辊箱；稀油污染源；检测方案

酸再生系统常见故障及改进措施

党文文

（承德钒钛冷轧薄板有限公司，河北承德 067000）

摘　要：为了保证酸再生机组稳定性，对酸再生机组常见故障的关键点进行了分析。以酸再生文丘里浓缩酸液杂质、酸再生废气排放及管道堵塞为重点，分析了故障原因。通过改进设备、调整工艺参数及优化操作，保证了酸再生系统稳定性，并对酸再生系统的一些设备改造建议和操作方法进行了详细的介绍。

关键词：酸再生；焙烧炉；稳定性；管道堵塞

Common Faults and Improvement Measures of Acid Regeneration System

DANG Wenwen

(Chengde Vanadium and Titanium Co., Ltd., Chengde 067000, China)

Abstract: In order to ensure the stability of acid regeneration unit, the key points of common faults of acid regeneration unit were analyzed. The causes of the failure were analyzed by focusing on the concentration of impurities in acid regenerated venturi solution, acid regenerated exhaust gas discharge and pipeline blockage. By improving equipment and adjusting process method, the stability of acid regeneration unit is ensured, and some improvement suggestions and operation methods of acid regeneration are introduced in detail.

Key words: ACID regeneration plant; baking furnace; acid regeneration unit stability; pipeline blockage

循环冷却水应急除氰可行性研究

蒋青利

（武汉钢电股份有限公司生产技术室，湖北武汉 430000）

摘　要：武汉钢电股份有限公司循环冷却水系统的补水来自净化水和中水，其中中水中含总氰化物，当中水作为补水进系统后，随着系统不断浓缩，极易造成循环冷却水系统水质中总氰化物指标超标，超出环保要求的外排管制标准，从而影响到整个厂区的废水外排。本文通过密切监视中水总氰化物含量，充分研究总氰化物脱除原理，对比市面几种药剂筛选并选中药剂进行试验，最终得出当中水作为补水进系统后，极端情况下，按 10ppm 投加除氰剂，在保证水质其他指标不受影响的前提下，循环冷却水系统水质中总氰化物快速脱除是可行的。

关键词：总氰化物；氧化剂；次氯酸钠；循环水；中水

聚合硅酸铝钙处理焦化废水中氟化物的实验研究

王 飞[1,2]，王 永[1,2]，刘 芳[1,2]，安 宁[1,2]，李函霏[1,2]

（1. 海洋装备用金属材料及其应用国家重点实验室，辽宁鞍山 114009；
2. 鞍钢集团钢铁研究院，辽宁鞍山 114009）

摘 要：以含有氟化物的焦化废水为研究对象，利用新型聚合硅酸铝钙除氟剂进行了氟化物去除试验研究，考察了除氟剂用量、pH、水温、搅拌时间、搅拌速度以及沉降时间等因素对氟化物去除效果的影响，确定了最佳除氟条件。

关键词：聚合硅酸铝钙；除氟剂；焦化废水；含氟废水

Experimental Study on the Treatment of the Wastewater Containing Fluoride by Using Poly-silicate-aluminum-calcium

WANG Fei[1,2], WANG Yong[1,2], LIU Fang[1,2], AN Ning[1,2], LI Hanfei[1,2]

(1. State Key Laboratory of Metal Material for Marine Equipment and Application, Anshan 114009, China;
2. Ansteel Iron & Steel Research Institutes, Anshan 114009, China)

Abstract: An experimental study was conducted on the removal of fluoride from coking wastewater containing fluoride using a new type of polymeric aluminum calcium silicate fluoride removal agent. The effects of factors such as the dosage of agent, pH, water temperature, stirring time, stirring speed, and settling time on fluoride removal efficiency were investigated, and the optimal experimental conditions were determined.

Key words: poly-silicate-aluminum-calcium; fluorine-removal agent; coking wastewater; fluoride wastewater

烧结烟气错流式活性焦脱硫脱硝技术特征与工程实践

李 林[1]，辻孝典[1]，初 霖[2]，李 健[2]，张 涛[2]，尚久石[1]，李彦哲[1]

（1. 北京中日联节能环保工程技术有限公司，北京 100040；
2. 北京首钢国际工程技术有限公司，北京 100043）

摘 要：烧结烟气治理是我国钢铁行业超低排放需彻底解决的重大课题之一。首先，总结烧结烟气各治理工艺的优劣势、分析不同工艺反应原理。其次，重点介绍错流式活性焦干法工艺的试验方案，开展与工艺设计密切相关的脱硝性能试验，以确定吸附塔配置、活性焦循环量、空速 SV 等主要技术参数。该试验研究结论为：烟气温度 120~140℃时，装置在此温度范围内的较高温度下运行，可提高脱硝性能；较高温度且 SV 较低时，能够有效抑制氨逃逸；活性焦滞留时间 RT 低于 60 时，对脱硝性能影响较大。最后，以河南某钢铁烧结烟气错流式活性焦干法净化工程为例，阐述错流式工艺的主要技术特征。该工程投产至今的运行实践表明，错流式工艺系统运行的整体稳定性高，对钢铁行业烧结烟气治理具有重要意义。

关键词：烧结烟气；活性焦；错流式；脱硫脱硝；超低排放

Technical Characteristics and Engineering Practice of Cross-flow Activated Coke Desulfurization and Denitrification for Sintering Flue Gas

LI Lin[1], TSUJI Takanori[1], CHU Lin[2], LI Jian[2], ZHANG Tao[2], SHANG Jiushi[1], LI Yanzhe[1]

(1. Beijing JC Energy & Environment Engineering Co., Ltd., Beijing 100040, China;
2. Beijing Shougang International Engineering Technology Co., Ltd., Beijing 100043, China)

Abstract: Sintering Sintering flue gas treatment is one of the major issues that need to be thoroughly solved in Chinese steel industry ultra-low emissions. Firstly, the advantages and disadvantages of sintering flue gas treatment processes are summarized, the principle of desulfurization and denitrification reaction for different processes are analyzed. Secondly, the test scheme of activated coke cross-flow process is proposed to determine adsorption tower configuration, active coke cycle amount, SV and so on, and the denitrification performance that is closely related to the process design is carried out. It is concluded that when the flue gas temperature is 120~140℃, it can improve the denitrification performance when operating at a higher temperature in this temperature range. At higher temperatures and low SV, ammonia escape can be inhibited. When RT is lower than 60, RT has a greater influence on the denitrification performance. Lastly, taking a sintering flue gas active coke cross-flow purification project in Henan as an example, the main technical characteristics of the cross-flow process are expounded. The operation practice shows that overall stability of the cross-flow process system is high, it plays an important role for the sintering flue gas treatment in steel industry.

Key words: sintering flue gas; activated coke; cross-flow; desulfurization and denitrification; ultra-low emission

烧结烟气错流式活性焦干法净化工艺的构建

李　林[1]，初　霖[2]，辻孝典[1]，李　健[2]，张　涛[2]，尚久石[1]，李彦哲[1]

（1. 北京中日联节能环保工程技术有限公司，北京　100040；
2. 北京首钢国际工程技术有限公司，北京　100043）

摘　要：构建烧结烟气错流式活性焦干法净化工艺，阐述整套装置的工艺流程和系统组成，介绍装置运行参数和烟气排放指标，为我国钢铁企业烧结烟气超低排放提供思路与借鉴。

关键词：烧结烟气；活性焦；错流式；脱硫脱硝；超低排放

Construction of Activated Coke Desulfurization and Denitrification Process for Sintering Flue Gas

LI Lin[1], CHU Lin[2], TSUJI Takanori[1], LI Jian[2], ZHANG Tao[2],
SHANG Jiushi[1], LI Yanzhe[1]

(1. Beijing JC Energy & Environment Engineering Co., Ltd., Beijing 100040, China;
2. Beijing Shougang International Engineering Technology Co., Ltd., Beijing 100043, China)

Abstract: Cross-flow activated coke dry purification process for sintering flue gas is constructed, the process flow and system composition of the entire device are explained, the operating parameters and flue gas emission indicators of the device are introduced, ideas and references for ultra-low emission of sintering flue gas in Chinese steel enterprises are provided.
Key words: sintering flue gas; activated coke; cross-flow; desulfurization and denitrification; ultra-low emission

探地雷达技术在地下供水管道漏损检测中的应用研究

陈 鹏[1,2]，安 宁[1,2]，胡绍伟[1,2]，李函霏[1,2]，王 飞[1,2]，刘 芳[1,2]

（1. 海洋装备用金属材料及其应用国家重点实验室，辽宁鞍山 114009；
2. 鞍钢集团钢铁研究院环境与资源研究所，辽宁鞍山 114009）

摘 要： 供水管道漏损一直是企业节水工作中的突出问题。在各种漏损检测方法中，探地雷达具有高效、无损、准确的优点。本文介绍了探地雷达在鞍钢供水管道漏损检测中的实际应用，证明了探地雷达在钢铁企业供水管道漏损检测中的适用性。结果表明，湿润土壤会破坏地层和管道线性信号的连续性。此外，漏损区域反射回波有明显的下降趋势，接收电磁波的回波时间增长，反射强度增大，并呈现多个垂向分布的双曲线信号。
关键词： 供水管道；漏损检测；探地雷达；图像识别

Application Research of Ground Penetrating Radar Technology in Leakage Detection of Underground Water Pipeline

CHEN Peng[1,2], AN Ning[1,2], HU Shaowei[1,2], LI Hanfei[1,2],
WANG Fei[1,2], LIU Fang[1,2]

(1. State Key Laboratory of Metal Material for Marine Equipment and Application, Anshan 114009, China; 2. Iron & Steel Research Institute of Ansteel Group, Institute of Environment and Resources, Anshan 114009, China)

Abstract: Leakage of water pipeline is always a prominent problem for water saving in steel enterprises. Among various leak detection methods, ground penetrating radar (GPR) has the advantages of high efficiency, nondestructive and accuracy. In this paper, the application of GPR in water pipeline of Ansteel was introduced. The work proved the applicability of GPR in water pipeline leakage detection. The results showed that wetting soil disrupted signal continuity of formation signal and pipeline signal. Moreover, the reflected wave in the wet region had a obvious downward trend, the time of the received reflected wave increased, and the intensity of reflection increased. Meanwhile, multiple hyperbolic signals distributed vertically were presented in the image.

Key words: water pipeline; leakage detection; ground penetrating radar; image identification

外旋式大型旋流沉淀池在热轧水处理中的改进探讨

赵治国，李 湧，杨富刚，蔡 毅，谭 焕

（中国重型机械研究院股份公司陕西冶金设计研究院有限公司，陕西西安 710018）

摘 要： 旋流沉淀池被广泛应用于热轧水处理中，本文章针对乌兹别克斯坦某1450热连轧浊环水处理系统，探讨了关于外旋式大型旋流沉淀池在热轧水处理中的应用和改进措施。通过改进进水方式、增加除油设施、改进吸水井、改进稳流板等措施优化了外旋式旋流沉淀池。

关键词： 旋流沉淀池；热轧水处理；设计；优化

Discussion on the Improvement of External Spinning Extra-large Cyclone Well in Hot Rolled Water

ZHAO Zhiguo, LI Yong, YANG Fugang, CAI Yi, TAN Huan

(China National Heavy Machinery Research Institute Co., Ltd., Shaanxi Metallurgical Design and Research Institute, Xi'an 710018, China)

Abstract: rotational flow sedimentation wells are widely used in hot rolled water treatment, This paper discusses the application and improvement measures of the extra-large rotational flow sedimentation well in hot water judgment for a Uzbek 1450mm project. The external rotary cyclone well was optimized by improving the water inlet method, increasing the oil removal facilities, transforming the water absorption well, and transforming the stabilizer plate.

Key words: rotational flow sedimentation well; hot rolled water treatment; design; optimization

基于智慧取水的钢厂纯水制备系统废水减排改造实践

丁宗琪，王慧军，曹小兵，李 璐

（宝钢股份能源环保部，上海 201999）

摘 要： 目前国内各钢厂均在大力推进节水减排工作。宝钢股份宝山基地结合自身水库取水的独特优势，开发了避咸取淡智慧化取水技术取到优质水源，在此前提下对纯水制备系统进行改造，将原有离子交换工艺变成为"双膜"

工艺，并根据水质水量平衡关系，实现纯水制备系统废水梯级利用，取得了良好的节水效果同时，也大幅度减少了原有离子交换系统的废水排放量。

关键词：钢厂；取水模型；纯水制备；梯级利用

Renovation Practice of Water Extraction System in Steel Plants Based on Intelligent Water Extraction

DING Zongqi, WANG Huijun, CAO Xiaobing, LI Lu

(Baosteel Energy and Environmental Protection Department, Shanghai 201999, China)

Abstract: Currently, various steel mills in China are vigorously promoting water-saving and emission reduction work. Baosteel Baoshan Base, taking into account its unique advantages in water intake from its own reservoirs, has developed a smart water intake technology that avoids saltiness and takes light water to obtain high-quality water sources. Under this premise, the pure water preparation system has been transformed from the original ion exchange process to a "double membrane" process. Based on the balance between water quality and quantity, the pure water preparation system has achieved cascade utilization of wastewater, achieving good water-saving effects, It also significantly reduces the wastewater discharge of the original ion exchange system.

Key words: steel mill; intake model; preparation of pure water; step utilization

转炉烟气余热、化学能回收及超低排放技术综述与展望

陈 琛，倪书权，徐继法，李加旺，郝景章，徐 蕾，史 光

（中国五矿集团中冶京诚工程技术有限公司，北京 100176）

摘 要：随着国家环保部在 2018 年 5 月 7 日关于征求《钢铁企业超低排放改造工作方案(征求意见稿)》意见函的发布，随着生态环境部、市场监督总局和国家能源局在 2021 年 11 月 1 日联合发布《关于严格能效约束推动重点领域节能降碳的若干意见》的发布，钢铁行业烟气治理同时兼顾能效和能源利用进一步引起了环保行业和从业人员的高度关注。于此同时，钢铁行业烟气超低排放和极致能效也逐渐成为热点研究课题和项目。本文论述了转炉烟气的余热和化学能回收技术的历史发展沿革并对主流除尘技术做了工艺流程和设备组成的综述，并结合核心问题的解决，探寻了适合中国本土化超低排放和极致能效要求的转炉一次烟气净化技术的新思路。

关键词：超低排放；极致能效；转炉冶炼；余热回收；化学余能回收

钢铁行业 CO_2 资源化利用技术应用现状及发展趋势

王雪琦，王改荣，李鹏阳，张彩东，李兰杰，田志强

（河钢材料院前沿技术创新中心，河北石家庄 050023）

摘 要：本文对钢铁行业烟气 CO_2 资源化利用现状进行了分析，并在此基础上，着重综述 CO_2 化学利用、生物利用、矿化利用等关键技术的工艺流程及国内外研究应用现状，初步探讨了碳中和目标下钢铁企业 CO_2 资源化利用方向，分析了我国钢铁企业 CO_2 资源化利用需要研究解决的问题，继而凸显钢化联产技术应用优势，提高钢厂烟气资源利用率，促进可持续发展，为企业创造更多经济效益，并对我国钢铁行业煤气资源化利用及低碳转型提出了对策意见。

关键词：钢铁工业；CO_2；资源化利用；碳减排

Application Status and Development Trend of CO_2 Resource Utilization Technology in the Steel Industry

WANG Xueqi, WANG Gairong, LI Pengyang,
ZHANG Caidong, LI Lanjie, TIAN Zhiqiang

(HBIS Material Institute Frontier Technology Innovation Center, Shijiazhuang 050023, China)

Abstract: This paper analyzes the current situation of CO_2 utilization in the steel industry and focuses on the process of key technologies such as CO_2 chemical utilization, CO_2 biological utilization and mineralization, as well as the research and application status at home and abroad. Meanwhile, the directions of CO_2 resource utilization of steel enterprises under the goal of carbon neutrality are discussed, the problems of CO_2 resource utilization of steel enterprises in China are analyzed. Moreover, the application advantages of co-production technology are emphasized, and this will improve the utilization of flue gas resources in steel mills and create more economic benefits for enterprises. And then, the suggestions for the low-carbon transformation of CO_2 utilization in the steel industry are presented.

Key words: steel industry; CO_2; resource utilization; carbon emission reduction

八钢烧结机头烟气脱硫脱硝超低排放改造浅析

王 宁

（新疆八一钢铁股份炼铁厂，新疆乌鲁木齐 830022）

摘 要：文章以原烧结机头烟气脱硫系统（CFB 工艺）设计参数为依据，结合近年来烧结机头烟气入口 SO_2 浓度变化实绩，对照钢铁行业超低排放改造要求，分析了烧结机头烟气脱硫脱硝改造的必要性；同时对当前在烧结机头烟气治理超低排放改造中几种主流的工艺路线进行了对比分析，阐述了八钢烧结机头烟气超低排放改造两种工艺路线确定的基本原则。

关键词：烧结机头烟气；脱硫脱硝；超低排放改造；工艺路线

Analysis on Ultra-low Emission Transformation of Flue Gas Desulfurization and Denitration in the Sintering Machine Head of Bagang

WANG Ning

(Xinjiang Bayi Iron and Steel Co., Ltd., Ironmaking Plant, Urumqi 830022, China)

Abstract: Based on the design parameters of the original sintering and sintering head flue gas desulfurization system (CFB process), combined with the actual changes in SO_2 concentration at the inlet of the sintering head flue gas in recent years, and in accordance with the requirements for ultra-low emission transformation in the steel industry, the article analyzes the necessity of reforming the sintering head flue gas desulfurization and denitrification; At the same time, several commonly used process routes in the current ultra-low emission transformation of sintering machine head flue gas treatment were compared and analyzed, and the basic principles for determining the two process routes for the ultra-low emission transformation of sintering machine head flue gas at Bagang were expounded.

Key words: flue gas from sintering machine head; desulfurization and denitrification; ultra low emission transformation; process route

混凝-浸没式超滤短流程工艺处理钢铁综合废水的实验研究

杨建峡，余云飞，陈思雨，邱利祥

（中冶赛迪工程技术股份有限公司，重庆　401147）

摘　要： 以某钢铁厂综合废水为处理对象，进行混凝-浸没式超滤短流程工艺实验研究。以 PFS 作为混凝剂，PAM 作为絮凝剂，利用矾花视境®智能加药系统监测不同投药量下絮体粒径、分形维数等絮体结构特性的同时，考察不同混凝剂投加量、絮凝剂投加量、运行通量等条件下膜通量、跨膜压差的变化趋势，并在最佳运行参数下进行连续实验。结果表明，混凝-浸没式超滤工艺不建议在混凝基础上增加 PAM 进行助凝，并且通过对混凝剂的投加量加以控制，确保絮体粒径位于合理范围内可以有效地缓解膜污染的程度，确保短流程工艺长期稳定运行。

关键词： 混凝-浸没式超滤工艺；絮体特性；膜通量；跨膜压差

Experimental Study on the Treatment of the Iron and Steel Plant Wastewater by Using Coagulation-Immersion Ultrafiltration Technology

YANG Jianxia, YU Yunfei, CHEN Siyu, QIU Lixiang

(MCC CISDI Engineering Technology Co., Ltd., Chongqing 401147, China)

Abstract: Taking a comprehensive wastewater from a certain iron and steel plant as the research object, a coagulation-immersion ultrafiltration short process experiment was conducted. PFS was used as the coagulant, PAM was used as the coagulant aid, and the Floc-Focus® intelligent dosing system was used to monitor the floc structure characteristics such as floc particle size and fractal dimension at different dosage levels. At the same time, the changing trend of membrane flux and transmembrane pressure under different coagulant dosage, coagulant aid dosage, and operating flux conditions was investigated, and a continuous coagulation-immersion ultrafiltration experiment under the optimal operating parameters was conducted. The results indicate that the coagulation-immersion ultrafiltration process does not recommend adding PAM as coagulant aid, and it can effectively alleviate the degree of membrane pollution and ensure long-term stable operation of the short process process by controlling the dosage of coagulants to ensure that the floc particle size is within a reasonable range.

Key words: coagulation-immersion ultrafiltration; floc structure characteristics; membrane flux; transmembrane pressure

水质悬浮物测定影响因素分析及流程改进

付志军

（中国宝武新余钢铁集团有限公司技术中心，江西新余 338028）

摘 要：系统分析水质悬浮物检测流程，确定影响检测结果准确性的各个因素，经改进的分析流程和重新选定试剂材料后，滤膜初重和过水后湿滤膜恒重均达到国标规定要求，初重恒重≤0.2mg，过水后湿滤膜恒重在0.4mg以内，单个样品检测四次的绝对误差在0.4mg以内，检测时间在3h内，长期抽查比对符合率达到99.5%以上。

关键词：水质；悬浮物；影响因素

Analysis and Improvement of Influencing Factors for Determination of Suspended Solids in Water Quality

FU Zhijun

(Test Centre, Xinyu Iron & Steel Co., Ltd., Xinyu 338028, China)

Abstract: System analyses the determination of suspended water quality inspection process, determine the various factors influencing the accuracy of the test results, the steps of analysis and the improvement to the selected reagents material, at the beginning of a membrane after weight and water wet membrane filter constant weight all reach national standard regulations, at the beginning of heavy constant weight not more than 0.2 mg, after water wet membrane filter constant is 0.4 mg of less than, The absolute error of single sample was less than 4mg for four times, and the detection time was less than 4 hours. The coincidence rate of long-term spot check was more than 99.5%.

Key words: determine; water quality; interfering factor

环保抑尘处理技术在钢铁生产全流程的应用和发展

徐言东[1]，韩 爽[1]，白金龙[1]，王占坡[2]，张瑞新[3]

（1. 北京科技大学国家板带生产先进装备工程技术研究中心，北京 100083；2. 冶金自动化研究设计院，北京 100071；3. 柯美瑞（唐山）环保科技有限公司，河北唐山 063021）

摘 要：在钢铁生产全流程中伴随着诸多氧化粉尘、油脂烟气、废酸气体的生成，对产品质量、操作人员职业健康、设备及环境造成损害，环保抑尘处理十分必要。针对当前普通水喷淋+高压水清洗除尘方法对粉尘的捕捉率较低，抑尘率只有20%~30%，且由于喷流量较大，降尘罩+风扇通风除尘方法由于其结构特点，负压吸入口离尘源点有一定距离，无法做到密封，不能形成负压，无法把粉尘全部抽干净而出现部分粉尘外逸现象等问题，提出了采用喷雾湿式除尘器+降尘罩+管道通风除尘系统的方案。能够实现高效喷雾覆盖，达到一定的环保抑尘要求，但因吸尘罩口离轧件过高和过远，导致降尘罩对烟气的吸纳有限，有少数烟气逃逸；虽然操作简单，运行维护方便，但降尘功耗较高。介绍了美国某品牌喷雾降尘系统、日本某品牌干雾抑尘处理系统的特点及其应用情况,通过对比,提出了对于一些没有环保降尘措施的料场、带式运输机、钢铁行业全流程,采用喷雾湿式除尘器+降尘罩+管道通风多级尘雾分离系统与美国某品牌喷雾降尘系统、日本某品牌干雾抑尘处理系统相配合应用,可以取得更好的效果。

关键词:料场;带式运输机;钢铁行业全流程;抑尘;氧化粉尘;喷雾除尘;干雾除尘

活性焦脱硫脱硝生产实践及优化建议

邵久刚,毛 瑞,苏 航,李 涛

(江苏省(沙钢)钢铁研究院,江苏张家港 215625)

摘 要:在活性焦脱硫脱硝技术应用过程中,由于参数控制、烟气条件甚至初始设计等原因往往会存在一些共性问题。本文主要围绕沙钢生产实践,对活性焦脱硫脱硝系统中存在的部分关键问题进行分析并提出优化建议。如通过降低进入系统的粉尘、减少热风炉异常停机、优化氮气和压力控制等措施,可以有效解决解析腐蚀和改善解析效果;通过长期温度监测,发现模块烟气流量或温度分布对脱硝有重要影响,通过采取措施促进烟气在模块间的稳定分布,可以改善系统的脱硝效果;对比了不同活性焦装填量对系统运行的影响,在装填量一定的前提下,适当提高脱硫层高度有利于在烟气进入脱硝层或喷氨之前将 SO_2 和 HCl 充分脱除,对于系统的长期稳定运行更为重要。

关键词:活性焦;钢铁烟气;脱硫脱硝;关键技术

Practical Application of Flue Gas Desulfurization and Denitrification by Activated Coke and Optimal Suggestions

SHAO Jiugang, MAO Rui, SU Hang, LI Tao

(Institute of Research of Iron and Steel, Jiangsu Province (Shasteel), Zhangjiagang 215625, China)

Abstract: In the practical application of flue gas desulfurization and denitrification by activated coke, due to different parameters control, various flue gas conditions, as well as initial design, there are some common problems. Based on practical production of Shagang, analysis and optimization suggestions about some key problems for activated coke desulfurization and denitrification system were proposed. For example, the problems of erosion and insufficient desorption of stripper process could be solved through measures of reducing the dust entering stripper, reducing abnormal shutdown of hot stove, and optimizing nitrogen and pressure control. From long term monitoring, it could be found that the volume distribution of flue gas or temperatures of different modules have significant influence on denitrification efficiency, and through measures to hold steady distribution of flue gas in different modules the denitrification efficiency could be improved. The effects of different initial activated coke loading weight on operation of the system were compared, and appropriately increasing the height of desulfurization lawyer under the premise of a certain loading is conducive to fully removal of SO_2 and HCl before entering denitrification lawer or ammonia spraying, which is more important for long term stable operation of the system.

Key words: activated coke; flue gas from steel industry; desulfurization and denitrification; key technology

烧结烟气挥发性有机物及氮氧化物协同减排研究

吕大友，申明锐，王毅璠，龙红明

（安徽工业大学冶金工程学院，安徽马鞍山　243032）

摘　要：烧结烟气中除粉尘、NO_x、SO_2、CO_x、HCl、氟化物、二噁英类（PCDD/Fs）等污染物外，还含有大量的 VOCs (volatile organic compounds)[1]。随着技术迭代和思路创新，在短流程、高效低耗脱除污染物的基础上实现多污染因子达标排放已成为当下的研究热点。在过去的几十年，针对烧结烟气污染物的净化开展了大量的研究工作，已经逐渐形成了"除尘-脱硫-SCR 脱硝"或"除尘+活性焦"的主流工艺路线[2,3]。近年来，基于 SCR 技术开发各种催化剂应用于烧结过程氮氧化物减排已成为行业研究热点。现行的烧结烟气的脱硝工艺一般采用 NH_3-SCR 技术，需要在催化反应器中通入 NH_3。历史上，Iwamoto 等根据 Toyota 和 Volkswagen 的专利首先报道了碳氢化合物在富氧条件下选择性催化还原 NO_x，该方法可同时脱除两种污染物以达到净化的目的[4]。而烧结过程中，化石燃料在预热层分解产生了大量的碳氢化合物，以这些碳氢化合物作为还原剂，选择性催化还原 NO_x，为烧结烟气 VOCs 与氮氧化物的协同减排提供了新的思路[5]。本文采用溶胶凝胶法合成了一系列 La 基钙钛矿型催化剂，初步探讨了烧结烟气 VOCs 与 NO_x 协同减排的可行性，为烧结烟气一体净化提供理论依据。

如图 1 所示，经过严格配比后，对制备的 $LaBO_3$ 催化剂进行了 XRD 表征分析，其中 $LaMnO_3$ 和 $LaCoO_3$ 分别与其标准样品衍射峰（PDF#86-1229）和（PDF#84-0848）完全匹配，其余四种均展现了完整的钙钛矿晶相特征峰。

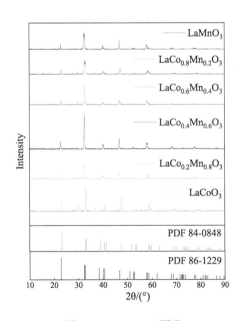

图 1　$LaBO_3$ XRD 图谱

前期研究发现，$LaMn_{1-y}Co_yO_3$（y=0、0.2、0.4、0.6、0.8、1）钙钛矿催化剂在甲苯减排中效果显著，如图 2（a）所示。其中，$LaMn_{0.2}Co_{0.8}O_3$ 钙钛矿的催化反应在 257℃的条件下实现了 90%的甲苯转化率。通过一系列表征手段，揭示了掺杂改性对钙钛矿型催化剂活性提升的机制。少量的元素掺杂，可以在不破坏原本钙钛矿结构的基础上，激发氧物种活性、增加表面氧浓度、增加氧空位或制造晶格缺陷。然而，在氮氧化物减排过程中，$LaMnO_3$ 上 NO_x 的转化率在 160℃时达到最佳，但在继续升温后出现了明显的下降。尤其是温度高于 260℃时，反应尾气中氮氧化物含量急剧升高，并超过入口氮氧化物的浓度，原因是气体组分中的 NH_3 被氧化。钙钛矿型催化剂在催化氧化 VOCs 时性能优异，且在低温条件下对 NO_x 具有高度的选择性，但其高效的氧化性在降解 NO_x 的过程可能存在"反向催化"的弊端。大量研究证实，钙钛矿型催化剂具有高度调节性，可灵活优化，其氧化还原性能因 B 位活性元素的不同，存在较大差异。因此，选择合适的金属元素，合理控制氧化态分布，设计可控的氧化还原性，将有望实现烧结烟气挥发性有机物与氮氧化物协同减排。

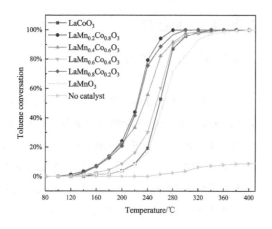
图 2　$LaMn_{1-y}Co_yO_3$ 催化氧化甲苯曲线

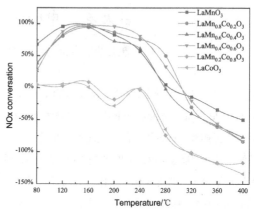
图 3　$LaMn_{1-y}Co_yO_3$ 催化氧化 NO_x 曲线

参 考 文 献

[1] Wang Y F, Qian L X, Yu Z W, et al. Inhibition Behavior of PCDD/Fs Congeners by Addition of N-containing Compound in the Iron Ore Sintering[J]. Aerosol and Air Quality Research, 2020, 20(11): 2568-2579.

[2] 闫伯骏, 邢奕, 路培, 等. 钢铁行业烧结烟气多污染物协同净化技术研究进展[J]. 工程科学学报, 2018, 40(7): 767-775.

[3] 龙红明, 丁龙, 钱立新, 等. 烧结烟气中 NO_x 和二噁英的减排现状及发展趋势[J]. 化工进展, 2022, 41(7): 3865-3876.

[4] Iwamoto M, Yahiro H. Novel catalytic decomposition and reduction of NO[J]. Catalysis Today, 1994, 22(1): 5-18.

[5] Shin D N, Koh D J, Kim K T, et al. Insight into the chemical reaction of CH_2OH with NO: Formation of N-hydroxy formamide as an isocyanic acid precursor[J]. Catalysis Today, 2007, 119(1-4): 209-212.

八钢炼铁厂翻车机负压收尘技术运用实践

王　宁

（新疆八一钢铁股份炼铁厂，新疆乌鲁木齐　830022）

摘　要：文章介绍了八钢炼铁厂翻车机卸料过程中粉尘治理采取的两种方式：抑尘及收尘两种工艺路线对粉尘抑制的作用效果对比。结合钢铁行业超低排放改造对厂房内无组织可视化粉尘的治理要求，重点对负压除尘技术在新建翻车机除尘项目中的运用进行分析总结。

关键词：翻车机；无组织扬尘治理；抑尘；负压收尘

Application Practice of Negative Pressure Dust Collection Technology for Dumper at Bayi Iron and Steel Ironmaking Plant

WANG Ning

(Xinjiang Bayi Iron and Steel Co., Ltd., Ironmaking Plant, Urumqi 830022, China)

Abstract: The article introduces two methods for dust control during the unloading process of the tippler at the Iron and Steel Plant of Baosteel: dust suppression and dust collection. The comparison of the effects of two process routes on dust suppression is presented. Based on the requirements of ultra-low emission transformation in the steel industry for the treatment of unorganized visual dust in the factory building, the application of negative pressure dust removal technology in the newly built dumper dust removal project is analyzed and summarized.

Key words: tippler; unorganized dust control; dust suppression; negative pressure dust collection

283.15K 和 353.15K 下 Na_2SO_4-$Na_2S_2O_3$-NaSCN-H_2O 四元相平衡及其在焦化脱硫废液中的应用

阳露波，王建山，张 衡，张小龙，于云涛

（攀钢集团研究院有限公司环保技术研究所，
钒钛资源综合利用国家重点实验室，四川攀枝花 617000）

摘 要： 利用湿渣法绘制了 283.15K 和 353.15K 下 Na_2SO_4-$Na_2S_2O_3$-NaSCN-H_2O 四元相图。在三元体系中，$Na_2S_2O_3$ 和 NaSCN 对 Na_2SO_4 均有较强的盐析作用，在四元体系中，353.15K 下 $Na_2S_2O_3$ 和 NaSCN 对 Na_2SO_4 有极强的盐析作用，母液能完全分离 Na_2SO_4 晶体。利用相图计算，焦化脱硫废液可通过分步结晶分离 Na_2SO_4、$Na_2S_2O_3$ 和 NaSCN，单程收率分别为 100%、55%和 55%。

关键词： 硫酸钠；硫氰酸钠；硫代硫酸钠；相平衡；焦化脱硫废液

Quaternary Phase Equilibria of Na_2SO_4-$Na_2S_2O_3$-NaSCN-H_2O at 283.15K and 353.15K and Its Application in Coking Desulfurization Waste Water

YANG Lubo, WANG Jianshan, ZHANG Heng, ZHANG Xiaolong, YU Yuntao

(Pangang Group Research Institute Co., Ltd., State Key Laboratory of Vanadium and Titanium Comprehensive Utilization, Panzhihua 617000, China)

Abstract: The quaternary phase diagram of Na_2SO_4-$Na_2S_2O_3$-NaSCN-H_2O at 283.15K and 353.15K was obtained by wet slag method. In the ternary system, $Na_2S_2O_3$ and NaSCN both have strong salting out effect on Na_2SO_4. In the quaternary system, $Na_2S_2O_3$ and NaSCN have a strong salting effect on Na_2SO_4 at 353.15K, and the mother liquor can completely separate Na₂SO₄ crystals. According to the phase diagram, Na₂SO4, Na₂S₂O₃ and NaSCN can be separated by step-by-step crystallization with the yield of 100%, 55% and 55%, respectively.

Key words: Na₂SO₄; NaSCN; Na₂S₂O₃; phase equilibrium; coking desulfurization waste water

12.2 冶金固废资源综合利用

火电厂掺烧固废热力计算前提条件确定应注意的问题分析与解决建议

张鲁峻，贺 宇，徐新文

（武汉钢电股份有限公司生产技术室，湖北武汉 430080）

摘 要：目前火电厂掺烧固体废弃物已越来越广泛。事先进行热力计算，对预判锅炉参数的变化趋势、估算经济效益、环境评价、安全性评估都是很有必要的基础工作。本文依据煤质化验的标准，针对不同固废的特点进行了分析，指出事先热力计算过程中容易忽视的问题，并提出了解决建议。

关键词：火电厂；掺烧；固废；热力计算；前提条件；问题；分析；建议

钢铁行业历史固废堆场系统性环境深度治理研究与实践

刘剑平，王如意，高 亮

（宝山钢铁股份有限公司能源环保部，上海 201900）

摘 要：钢铁生产每生产 1t 钢会产生 0.6t 固废，历史上因固废加工利用产业不成熟时会有大量的冶金固废堆存，固废堆存一是会造成环境污染，二是占用宝贵的土地资源。在国家长江大保护及双碳发展的大背景下，解决这个历史问题更为迫切。近年来国内也有固废堆场治理案例，但多以固废清理复绿为主，大规模系统性深度治理缺乏先例。宝钢股份以高度的责任感，率先在行业内开展以风险管控、清理利用、生态绿化、地下水大规模深度治理的多层次综合治理研究及探索，为行业内固废堆场深度综合治理及利用提供可供借鉴的经验，成为钢铁行业固废堆场环境整治正面典型案例。

关键词：钢铁；固废堆场；风险管控；治理

Research and Practice to Depth Treatment the Solid Wastes Landfill in Steel Industry

LIU Jianping, WANG Ruyi, GAO Liang

(Energy and Environmental Department of Baoshan Iron and Steel Co., Ltd., Shanghai 201900, China)

Abstract: Every ton of steel production will generate 0.6 tons of solid waste. Historically, a large amount of steel solid waste has been piled up for lack of rational disposal and utilization, which has caused serious environmental pollution, and large areas of valuable lands have been occupied. With implementation of the Yangtze River protection strategy and the Double-Carbon policy in China, it is more urgent to solve this kind historical problem, especially along the sides of Yangtze River. In recent years, there have been some cases of solid waste landfill treatment, but most of them focus on solid waste cleaning and re-greening, but not systematic treatment and environmental risk elimination. Baowu Group is the first to practice the comprehensive improvement of steel slag landfill problems in China, which contains risk management and control, clean-up and utilization, ecological restoration, and remediation of groundwater contamination. The elimination of environmental risks in steel slag landfills has provided valuable experience and knowledge to treatment and utilization of industry solid wastes in steel industry.

Key words: steel; solid waste; risk management and control; treatment

球磨粒子钢热压块出钢水率检测方法研究

窦 猛[1]，戴 伟[1]，聂文金[1]，皋 萍[1]，王晓东[2]

（1. 江苏沙钢集团有限公司，江苏苏州 215000；2. 江苏省沙钢钢铁研究院有限公司，江苏苏州 215000）

摘　要： 采用石墨黏土坩埚熔化球磨粒子钢易导致钢水增碳增重，影响出钢水检测的准确度。本研究采用在石墨黏土坩埚底部铺硼砂的方法，增强坩埚的耐侵蚀性能，同时钢水与坩埚内壁之间形成的大量含硼熔渣层造成 C、Si 的淀析，C、Si 扩散受阻，钢水增碳、增硅减少，出钢水率检测偏差降低 2.0%。对采用石墨黏土坩埚与石英坩埚，是否加硼砂四种熔样方式熔后钢锭做光镜金相分析，发现石墨黏土坩埚熔后钢锭中有大量大粒度杂质，以含 Mn、Ca、Al、Si 的氧硫化物为主，加入一定量硼砂后，MnS 颗粒明显减小，金属氧化物数量减少。改用石英坩埚掺入 8.5%硼砂熔样，铝钙类氧化物及含硅杂质大幅减少，MnS 颗粒几乎消失，杂质量极少，出钢水率检测偏差再次降低 5%。研究也发现，需要向粒子钢中加入 3%~4%的 Al 粒以还原其中的铁氧化合物。最终，研究发现采用石英坩埚掺入 8.5%硼砂的熔样方法比仅采用石墨黏土坩埚熔化检测粒子钢出钢水率偏差降低 7%。

关键词： 出钢水率；石墨黏土坩埚；含硼渣层；硼砂；光镜金相分析；石英坩埚

Study on the Method of Measuring the Rate of Molten Steel Output of Hot Briquetting Iron of Ball Mill Particle Steel

DOU Meng[1], DAI Wei[1], NIE Wenjin[1], GAO Ping[1], WANG Xiaodong[2]

(1. Jiangsu Shagang Group Co., Ltd., Suzhou 215000, China;

2. Jiangsu Shagang Iron and Steel Research Institute Co., Ltd., Suzhou 215000, China)

Abstract: Using a graphite clay crucible to melt ball milled particle steel can easily cause carbonization and the increase of weight in liquid steel, which affects the accuracy of detection of the rate of molten steel output .This study adopts a method of laying borax at the bottom of the graphite clay crucible to enhance the corrosion resistance of the crucible. At the same time, a large amount of boron containing slag layer between the steel in liquid and the inner wall of the crucible causes the precipitation of C and Si, and prevents the diffusion ,which reduces the addition of carbon and silicon of steel in liquid, the detecting deviation of the rate of molten steel output decrease 2.0%. Optical microscope metallographic analysis was conducted on the steel ingots molted by graphite clay crucibles or quartz crucibles, with or without the addition of borax. It was found that there were a large number of large particle impurities in the steel ingots melted by graphite clay ,major component is oxygensulfides containing Mn, Ca, Al, Si. After adding a certain amount of borax, the particle size of MnS impurities significantly become smaller, the number of metal oxides also decreases. When using a quartz crucible and adding 8.5% borax for sample melting significantly reduces aluminum calcium oxide and silicon containing impurities, MnS inclusions almost eliminates, impurities is little. The detection deviation of molten steel output rate reducs 5% again. In addition,a research is also found that it is necessary to add 3%-4% Al particles to the particle steel to reduct iron oxides in it. Finally. it is found that the deviation of the rate of molten steel output d molted by quartz crucible mixed with 8.5% borax is 7% lower than that of graphite clay crucible only.

Key words: rate of molten steel output; graphite clay crucible; boron bearing slag layer; borax; optical microscope metallographic analysis; quartz crucible

炼钢中低价物料使用与替代的生产实践

李　超，尚德义，何　冲，冉茂铎

（鞍钢股份鲅鱼圈钢铁分公司炼钢部，辽宁营口　115007）

摘　要： 本文介绍了 260t 转炉中使用脱硫渣铁、石灰石替代白灰、转炉尾渣造渣应用的实践情况，进行了相关的理论分析，并介绍了几种经济物料使用的具体方法，生产数据表明：脱硫渣铁的使用，对现有工艺没有较大影响，据统计回硫量为 0.001%/t，同时提高了其他经济指标；石灰石替代白灰提高了脱磷水平，终点磷较实施前降低 0.009%，煤气回收量增加 1.1m³/t；转炉尾渣使用提高脱磷率 3.35%，同时熔剂消耗等生产指标也得到改善。

关键词： 石灰石；脱硫渣铁；尾渣；脱磷

Production Practice of the Economic Materials Using in Steelmaking

LI Chao, SHANG Deyi, HE Chong, RAN Maoduo

(Steelmaking Department Bayuquan Iron & Steel Subsidiary of Angang Steel Co., Ltd., Yingkou 115007, China)

Abstract: This paper introduces the practice of using desulfurization slag iron and limestone replacing lime and wasting-slag slagging in 260t converter, A relevant theoretical analysis was performed and it introduces several using specific methods of economic materials . The production data show that the utilization of desulfurization slag iron has no great impact on the existing process,statistics indaicates the amount of sulfur coming-back is 0.001%/t，it improves other economic indicators; limestone replacing lime improves dephosphorization level. The endpoint phosphorus decreased by

0.009%, and the gas recovery increased by 1.1m³/t; the phosphorus removal rate of converterincreased by 3.35% by using tail slag, and the production indexes such as flux consumption also are improved.

Key words: limestone; desulfurization slag iron; wasting slag; dephophorization

宝钢二次资源消纳实践

吴旺平，王跃飞，鲁 健

（宝山钢铁股份有限公司炼铁厂，上海 201900）

摘 要：随着新固废法及超低排放等政策出台，钢铁企业开始逐步减少二次资源的出厂。对于如何全量消化回收二次资源，宝钢统筹规划，利用原料均质化系统、烧结粉尘制粒、转底炉、固废处理中心等处理工艺消纳厂内产生的二次资源。但是实际应用过程中，发现部分物料过于湿粘，导致部分物料无法利用上述处理工艺消化。此外超低排放要求减少卡车运输过程中扬尘污染路面。针对此，宝钢在原有基础上，开发了新工艺，诸如低锌干灰直接进入烧结、高炉二次湿灰进入转底炉、含杂物料的处理等，确保了固废不出厂。

关键词：转底炉；固废；锌负荷；二次资源

Application of Water-saving Utilization for Raw Material Yard in Baosteel

WU Wangping, WANG Yuefei, LU Jian

(Ironmaking Plant, Baoshan Iron & Steel Co., Ltd., Shanghai 201900, China)

Abstract: With the release of the new solid waste law、ultra-low emission requirements for iron and steel enterprises，Steel enterprises began to reduce the secondary resources to off-site processing. As for how to fully digest the recovered secondary resources, Baosteel planned to use homogenized treatment system、sintering dust granulation, rotary hearth furnace and solid waste treatment center to dispose of the secondary resources. However, in the actual application process, it was found that some materials were too wet and sticky, which caused some materials could not be used in the original way. In addition, ultra-low emissions require less dust pollution during truck transport,so Baosteel has to develop new method to dispose these materials, such as low zinc dry ash directly into the sinter、blast furnace secondary wet ash into the rotary hearth furnace、treatment process for miscellaneous materials.

Key words: rotary hearth furnace; solid waste; zinc load; secondary resources

冷轧磁过滤油泥处理技术实验与思考

徐鹏飞[1,2]，杨大正[1,2]，耿继双[1,2]，王 飞[1,2]，吴文浩[1,2]

（1. 海洋装备用金属材料及其应用国家重点实验室，辽宁鞍山 114009；

2. 鞍钢集团钢铁研究院，辽宁鞍山 114009）

摘　要：冷轧磁过滤油泥是钢企冷轧过程中产生的危险废物，需要进行无害化处理和资源化利用。文章对比分析了离心法、酸破乳法、萃取法、热解法、调质-机械分离法的技术特点并进行处理实验，结果表明，离心速率达到5000r/min以上、盐酸浓度为1~2mol/L、萃取温度80℃以上、热解温度大于350℃时，都可较好地处理冷轧磁过滤油泥，结合上述方法的优缺点，提出了油泥综合处理技术，可实现油泥中油和铁屑的资源化回收，为钢企冷轧磁过滤油泥处理提供技术参考。

关键词：冷轧磁过滤油泥；危险废物；酸破乳；萃取；热解；调质分离

Experiment and Thought on Magnetic Filtered Oily Cold-rolling Mill Sludge Treatment Technology

XU Pengfei[1,2], YANG Dazheng[1,2], GENG Jishuang[1,2], WANG Fei[1,2], WU Wenhao[1,2]

(1. State Key Laboratory of Metal Material for Marine Equipment and Application, Anshan 114009, China;
2. Ansteel Iron & Steel Research Institute, Anshan 114009, China)

Abstract: Magnetic filtered oily cold rolling mill sludge is a hazardous waste generated during the cold-rolling process of steel enterprises, which required harmless treatment and resource utilization. The characteristics of centrifugation, acid demulsification, extraction, pyrolysis, and tempering-mechanical separation technologies were compared and analyzed, and experiments were carried out, respectively. The results showed that when the centrifugal rate was above 5000r/min, the hydrochloric acid concentration was 1-2mol/L, the extraction temperature was over 80℃, and the pyrolysis temperature was higher than 350℃, the sludge could be treated, incompletely. Combining the advantages and disadvantages of the above methods, a comprehensive treatment method for sludge was proposed, which could recycle the resource of oil and iron in the sludge and the reference for the treatment of magnetic filtered oily cold rolling mill sludge in steel enterprises was provided.

Key words: magnetic filtered oily cold rolling mill sludge; hazardous waste; acid demulsification; extraction method; pyrolysis method; tempering-mechanical separation technology

电炉灰循环减量及资源化利用技术探索与实践

饶　磊，刘自民，桂满城，张耀辉，马孟臣，刘英才

（马鞍山钢铁股份有限公司技术中心，安徽马鞍山　243000）

摘　要：本文分析了某钢铁企业电炉灰特性，通过电炉灰返回电炉内部循环利用实现减量化，同时采用转底炉造球焙烧还原工艺实现电炉灰中锌资源的回收，并将脱锌后的转底炉球团用于转炉冷却剂，实现电炉灰的减量化和资源化利用。试验结果表明：电炉灰按13kg/t钢返回电炉循环利用，电炉灰产生量由13kg/t钢降至5.7kg/t钢；转底炉造球配用电炉灰，随着电炉灰比例提升，生球落下强度下降，含粉率上升，使用声波除尘器使换热器运行周期时长延长一倍；转底炉金属化球团按照100kg/t钢的比例返回转炉利用，转炉终点S含量增加0.005%，转底炉球团冷却效应是铁矿石的0.7倍。

关键词：电炉灰；电炉；转底炉；转炉；减量化；资源化

Exploration and Practice of Electric Furnace Ash Recycling Reduction and Resource Utilization Technology

RAO Lei, LIU Zimin, GUI Mancheng, ZHANG Yaohui, MA Mengchen, LIU Yingcai

(Technology Center of Maanshan Iron & Steel Co., Ltd., Maanshan 243000, China)

Abstract: In this paper, the characteristics of electric furnace ash in an iron and steel enterprise are analyzed, and the reduction of zinc resources in electric furnace ash is realized through the recycling of electric furnace ash back to the electric furnace. Meanwhile, the recovery of zinc resources in electric furnace ash is realized by the process of ball roasting and reduction in rotary hearth furnace. The dezincized rotary hearth furnace pellets are used as the coolant of converter to achieve the reduction and resource utilization of electric furnace ash. The test results show that the electric furnace ash is recycled with 13kg/t steel, and the output of electric furnace ash decreases from 13kg/t steel to 5.7kg/t steel. With the increase of the proportion of electric furnace ash, the falling strength of green ball decreases and the powder content increases. The operation cycle time of heat exchanger is doubled by using acoustic dust collector. The metallized pellets of rotary hearth furnace are returned to the converter according to the ratio of 100kg/ ton steel, the content of S at the end of the converter is increased by 0.005%, and the cooling effect of rotary hearth furnace pellets is 0.7 times that of iron ore.

Key words: electric furnace ash; electric furnace; rotary hearth furnace; converter; reduction; resource

烧结-高炉法协同处置多源铬渣和有机固废新技术

涂义康，张元波，苏子键，姜　涛

（中南大学资源加工与生物工程学院，湖南长沙　410083）

摘　要：中国每年产排大量铬渣（铬盐渣、酸洗泥等）和有机固废（钢铁厂含碳粉尘、废活性炭等），其成分复杂、无害化处理难度大。若采用堆存或填埋处理，其中的有价组分无法得到回收利用，且潜在极大环境隐患。本研究提出利用红土镍矿烧结机协同处置多源铬渣和有机固废技术路线：首先将多源铬渣和有机固废制成复合球团，而红土镍矿、熔剂、燃料等混匀后制成基体料，再将复合球团和基体料布料到烧结机上联合焙烧，生产含铬复合烧结矿并用于高炉冶炼不锈钢母液。烧结过程中复合球团中的Cr(VI)被有机质还原成金属铬或Cr(III)，复合烧结矿满足毒性浸出要求。该方法可实现有毒铬渣和有机固废的无害化处理和资源化利用。以年产量50万吨红土镍矿烧结矿的生产线为例，粗略估计，本技术的成功实施可为企业带来超3亿元的年收益，减少CO_2排放3.85万吨。

关键词：铬渣；有机固废；烧结；解毒；资源化

Collaborative Disposal of Chromium Slag and Organic Solid Waste Via a Novel Technology of Composite Agglomeration Process (CAP)

TU Yikang, ZHANG Yuanbo, SU Zijian, JIANG Tao

(School of Minerals Processing and Bioengineering, Central South University, Changsha 410083, China)

Abstract: Large amount of chromium slag (chromium salt slag, pickling mud, etc.) and organic solid waste (C-bearing dust from steel industry, waste activated carbon, etc.) were produced in China annual. and its composition is complex and it is difficult to treat it harmlessly. As the complex components, it is difficult to treat them harmlessly. If it is stored or landfilled, the valuable components in it cannot be recycled, and there is a great potential environmental hazard. In this study, the technical route of collaborative disposal of multi-source chromium slag and carbon-containing solid waste by limonitic laterite sintering machine is put forward: Chromium slag and organic solid waste were made into composite pellets first, and limonitic laterite, flux, fuel, etc. were evenly mixed to prepare matrix materials. The composite pellets and matrix materials were then distributed on sintering machine for co-roasting to produce Cr-bearing composite sinters and used for smelting stainless mother liquor in blast furnace. Cr(VI) in composite pellets was reduced to metal chromium or Cr(III) by organic matter, and the composite sinters met the requirements of toxic leaching. This method could realize the harmless treatment and resource utilization of toxic chromium slag and organic solid waste. Taking the limonitic laterite sinter production line with an annual output of 500000 tons as an example, it was roughly estimated that the successful implementation of this technology could bring enterprises more than 300 million yuan of annual income and reduce CO_2 emissions by 38500 tons.

Key words: chromium slag; organic solid waste; sintering; detoxification; resource utilization

电炉粉尘资源化转底炉工艺试验研究

李 林[1]，野田悦郎[2]，徐 萌[3]，中山俊孝[2]，辻孝典[1]，李彦哲[1]

（1. 北京中日联节能环保工程技术有限公司，北京 100040；2. 日铁工程技术株式会社，日本福冈 804-8505；3. 首钢集团有限公司技术研究院，北京 100043）

摘 要： 含锌电炉粉尘资源化是我国钢铁工业践行绿色制造的关键课题。在分析转底炉工艺技术优势和反应原理的基础上，进行与工艺设计紧密相关的落下强度和管式炉还原试验。试验得出：原料配加水量3%、粘结剂用量0.5%时，球团成型率满足工业生产要求；还原温度1300℃时，球团的铁金属化率86.2%，脱锌率93.8%。本试验研究对含锌电炉粉尘资源化的转底炉工艺设计具有重要指导意义。

关键词： 电炉粉尘；转底炉；脱锌率；次氧化锌

Process Test Research of Rotary Hearth Furnace for Resource Utilization of Electric Arc Furnace Dust

LI Lin[1], NODA Etsuro[2], XU Meng[3], NAKAYAMA Toshitaka[2], TSUJI Takanori[1], LI Yanzhe[1]

(1. Beijing JC Energy & Environment Engineering Co., Ltd., Beijing 100040, China; 2. Nippon Steel Engineering Co., Ltd., Fukuoka 804-8505, Japan; 3. Technology Research Institute of Shougang Group Co., Ltd., Beijing 100043, China)

Abstract: Recycling of electric arc furnace (EAF) dust containing elements such as zinc are important issue for the iron and steel industry to promote green manufacturing. On the basis of summarizing the technological advantages and the reduction reaction principles of rotary hearth furnace (RHF), drop strength and reduction tests closely related to the process design were carried out. Test results were as follows, when the water amount was 3% and the binder amount was 0.5%, the drop strength and the molding rate of the molded pellets all met the actual production requirements.The best tube furnace reduction test condition was at the reduction temperature of 1300℃, the Fe metallization rate was 86.2%, and the dezincification rate was 93.8%. Test research is of great significance to RHF for the zinc-containing resource utilization of EAF furnace dust.

Key words: EAF dust; rotary hearth furnace; dezincification rate; zinc suboxide

高效处置钢铁企业含锌固废技术实践

孙宇佳，金永龙，田京雷，王　倩

（河钢集团有限公司，河北石家庄　050023）

摘　要： 针对钢铁企业含锌固废的处置与资源化的难点问题，开发改进型回转窑技术工艺并实现产线落地。本文主要从原料的环保处理及合理配料、回转窑运行参数的实时精准操控、产品的高附加值回收三个方面进行阐述，技术具有能耗优、运行可控、脱锌效率高、无二次固废污染等优势，在钢铁行业内具有全面推广的可能性，资源综合利用效率高，附加值高，对无废工厂、无废城市的建设都具有深远意义。

关键词： 含锌固废；回转窑；脱锌率；环保

Practice on Efficient Disposal of Zinc Containing Solid Waste from Iron and Steel Enterprises

SUN Yujia, JIN Yonglong, TIAN Jinglei, WANG Qian

(HBIS Group Co., Ltd., Shijiazhuang 050023, China)

Abstract: In view of the difficult problems in the disposal and recycling of zinc containing solid waste in iron and steel enterprises, the improved Rotary kiln technology process was developed and the production line was put into practice. This article mainly elaborates from three aspects: environmental protection treatment and reasonable batching of raw materials, real-time accurate control of Rotary kiln operating parameters, and high value-added recovery of products. The technology has the advantages of excellent energy consumption, controllable operation, high dezincification efficiency, and no secondary solid waste pollution. It has the possibility of comprehensive promotion in the iron and steel industry. It has high comprehensive utilization efficiency of resources, high added value, and no waste plants The construction of waste free cities has profound significance.

Key words: zinc containing solid waste; rotary kiln; dezincification rate; environment protection

含铁锌固废资源化再生关键工艺技术研究

陈方元，李菊艳，徐永斌

（中冶南方工程技术有限公司技术研究院，湖北武汉　430223）

摘　要： 主要论述研究一种属于冶金工程及安全环保技术领域，资源化处理含铁锌固废关键技术、工艺、设备及辅助配套设施，验证含铁锌固废还原回收金属锌工艺原理和生产流程。本论文采用外热式还原炉工艺，回收金属锌，显著缩短锌回收工艺流程，提升产品附加值。该技术可用于处理高炉除尘灰、转炉除尘灰、电炉除尘灰、锌浸出渣等含铁锌固废。

关键词：铁锌；固废；资源化再生；工艺技术

Research on Key Process Technologies for Resource Recycling of Solid Waste Containing Iron and Zinc

CHEN Fangyuan, LI Juyan, XU Yongbin

(WISDRI Engineering & Research Incorporation Limited, Wuhan 430223, China)

Abstract: This article mainly discusses and studies a key technology, process, equipmentand auxiliary facilities for the resource treatment of solid waste containingiron and zinc, whichbelongs to the field of metallurgical engineering and safety and environmental protection technology. It verifies the principle and production process of the reduction and recovery of metal zinc from solid wastecontainingiron and zinc. Compared with the existing process of recovering secondary zinc oxide from solid waste, this paper adopts an external heating reduction furnace to directly recover metal zinc, significantly shortening the zinc recovery process and improving product added value. This technology can be used to treat iron and zinc containing solid waste such as blast furnace dust, converter dust, electric furnace dust, and zinc leaching slag.

Key words: iron and zinc; solid waste; resource based regeneration; process technology

矿渣-钢尾渣基充填材料配比优化研究

代梦博[1]，顾宝澍[1]，刘 斌[1]，孙业长[2]，春铁军[1]

（1. 安徽工业大学冶金工程学院，安徽马鞍山 243002；
2. 马钢集团设计研究院有限责任公司，安徽马鞍山 243000）

摘 要：为实现多种固废的协同利用并降低充填成本，以矿渣为基准，引入脱硫灰和钢尾渣等固废，通过研究矿渣与钢尾渣的复合胶凝材料、添加剂组分、灰砂比等因素的影响，采用正交实验探寻调配规律，对固体充填料的配比进行了优化。随着钢尾渣的增加，料浆的流动性得到改善，但充填体的抗压强度降低。在灰砂比1∶6、钢尾渣-矿渣比为1∶4条件下，优化的添加剂组分为脱硫灰∶425硅酸盐水泥∶原添加剂=3∶2∶5。此配方充填料可替代现场的矿渣基材料，并满足C2级充填强度的要求。此外，在最优添加剂组分基础上对固体料的全因素进行正交优化，开发出C3级的矿渣-钢尾渣充填料，不仅可大宗量处理钢尾渣和脱硫灰，且显著降低了充填成本。

关键词：矿山胶结充填；流动性；矿渣—钢尾渣；配比优化；充填体强度

Optimization of the Ratio of Steel Slag Tailings-GGBS Based Backfill Material

DAI Mengbo[1], GU Baoshu[1], LIU Bin[1], SUN Yechang[2], CHUN Tiejun[1]

(1. School of Metallurgical Engineering, Anhui University of Technology, Maanshan 243002, China;

2. Magang Group Design and Research Institute, Masteel Group Holding
Co., Ltd., Maanshan 243000, China)

Abstract: In order to achieve synergistic utilization of multiple solid wastes and reduce the cost of backfill, a composite cementitious material consisting of GGBS, steel tail slag and desulfurization ash was investigated. Through orthogonal experiments, the optimal ratio of admixtures, ash-to-sand ratio and the proportion of steel tail slag to GGBS were determined. Results showed that with increasing steel tail slag content, the fluidity of the slurry was improved, but the compressive strength of the filling body decreased. It was found that a formulation of desulfurization ash: cement clinker: original additive =3 : 2 : 5, with an ash-to-sand ratio of 1 : 6 and 20% steel tail slag replacing GGBS, could meet the requirement of C2 filling strength. Moreover, based on the optimal additive components, an optimized slag-steel tailing slag backfill material of C3 grade was developed, which can process steel tailing slag and desulfurization ash in large quantities and significantly reduced the cost of solid material.

Key words: mine cemented backfill; slurry flow; GGBS-steel slag tailings; formulation optimization; backfill strength

从高炉瓦斯灰中综合回收碳和铁试验研究

刘兴华

（长沙矿冶研究院有限责任公司，湖南长沙　410012）

摘　要： 瓦斯灰中常含有铁、碳和少量有色金属，属宝贵的二次资源，潜在利用价值很高。本研究采用浮选-弱磁选-强磁选工艺对某高炉瓦斯灰中的碳和铁进行综合回收，可获得 C 品位 81.00%、回收率 95.47%的碳精矿，TFe 品位 56.78%、回收率 20.14%的弱磁铁精矿和 TFe 品位 49.02%、回收率 35.04%的强磁铁精矿，合计铁精矿的 TFe 品位为 51.59%、回收率为 55.18%。取得了较好的回收指标。

关键词： 高炉瓦斯灰；浮选；磁选；综合回收

Experimental Study on Comprehensive Recovery of Carbon and Iron from Blast Furnace Gas Ash

LIU Xinghua

(Changsha Research Institute of Mining and Metallurgy Co., Ltd., Changsha 410012, China)

Abstract: Gas ash contains iron, carbon and a small amount of non-ferrous metals, which is a valuable secondary resource with high potential utilization value. In this study, carbon and iron in the gas ash of a blast furnace were comprehensively recovered by floatation-low intensity magnetic separation (LIMS)-high intensity magnetic separation (HIMS) process. Carbon concentrate with C grade of 81.00% and recovery of 95.47%, LIMS iron concentrate with TFe grade of 56.78% and recovery of 20.14% and HIMS iron concentrate with TFe grade of 49.02% and recovery of 35.04% were obtained. The TFe grade of total iron concentrate is 51.59% and the recovery is 55.18%. A good recovery index was obtained.

Key words: blast furnace gas ash; flotation; magnetic separation; comprehensive recovery

钢渣粗、细集料制备沥青混合料及施工特性研究综述

王子鹏[1]，马宏伟[1]，孙宇佳[2]，周永祥[3]

（1. 河北通华公路材料有限公司，河北石家庄 050081；2. 河钢集团有限公司，河北石家庄 050023；3. 北京工业大学，北京 100124）

摘　要：在双碳背景与大宗固废资源化的背景下，以钢铁行业产出量巨大的钢渣为研究对象，结合交通领域建设，针对钢渣沥青混合料的应用前景，重点对钢渣沥青混合料的配合比设计、体积安定性检测、重金属浸出风险、路用性能做了大量总结分析，钢渣沥青混合料在道路工程的应用还有很大的发展空间，最后提出了其在公路行业内存在的应用问题和钢渣沥青混合料的安全问题，并给出相应的建议。

关键词：钢渣沥青混合料；配合比设计；体积安定性；重金属检测；路用性能

Summary of Research on the Preparation and Construction Characteristics of Asphalt Mixture with Steel Slag Coarse and Fine Aggregates

WANG Zipeng[1], Ma Hongwei[1], SUN Yujia[2], ZHOU Yongxiang[3]

(1. Hebei Tonghua Highway Materials Co., Ltd., Shijiazhuang 050081, China; 2. HBIS Group Co., Ltd., Shijiazhuang 050023, China; 3. Beijing University of Technology, Beijing 100124, China)

Abstract: In the context of dual carbon background and the resource utilization of bulk solid waste, taking the steel slag with a huge output in the steel industry as the research object, combined with the construction of the transportation field, and focusing on the application prospects of steel slag asphalt mixture, a large amount of summary and analysis have been conducted on the mix design, volume stability testing, heavy metal leaching risk, and road performance of steel slag asphalt mixture. There is still great development space for the application of steel slag asphalt mixture in road engineering, Finally, the application problems and safety issues of steel slag asphalt mixture in the highway industry were proposed, and corresponding suggestions were given.

Key words: steel slag asphalt mixture; mix design; volume stability; heavy metal detection; road performance

热镀锌渣原位回收技术研究

初仁生[1,2]，李　研[1]，王　哲[2]，李翔宇[1]，李战军[1,3,4]，郭占成[2]

（1. 首钢集团有限公司技术研究院，北京 100043；2. 北京科技大学钢铁冶金新技术国家重点实验室，北京 100041；3. 北京市能源用钢工程技术研究中心，北京 100043；4. 绿色可循环钢铁流程北京市重点实验室，北京 100043）

摘　要：热镀锌工业是金属锌最大的消费领域，全世界每年金属锌总产量的40%以上用于钢材的热镀锌。热镀锌过程中，锌渣的存在对热镀锌板的表面质量造成严重影响，形成锌渣缺陷。为减小锌渣影响，每隔一段时间，锌池表面的锌渣需由人工或机械捞出去除，但在捞渣过程中不可避免会夹带出大量的金属锌，导致锌的利用率较低，锌的损失率较高，目前生产热镀锌汽车钢板时，锌损可以达到17%以上。随着热镀锌工业的快速发展，每年会有大量的热镀锌渣产生，每年产出的热镀锌渣在10万吨以上，锌渣的形成不仅消耗大量的锌，增加了镀锌成本，但同时热镀锌渣中含有大量的锌（90%以上），是一种宝贵的二次锌资源，对热镀锌渣进行回收利用，不仅能缓解资源紧缺的现状，锌渣的去除和再回收利用将进一步降低企业全流程的CO_2排放量，同时提高能耗强度与材料效率。

因此本文提出了一种更有效的、低成本的热镀锌渣收利用技术，即采用超重力冶金技术在线原位回收热镀锌渣，热镀锌渣中金属原位返回锌锅中，而极少部分渣留在超重力设备中，通过开展的实验研究表明：采用超重力技术回收热镀锌渣，可以降低锌渣量60%以上，实现锌损降低由17%降低到7%以下，大大提高锌资源的有效利用，主要技术路线如图1所示。

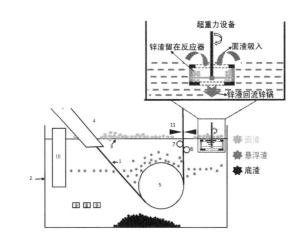

图1　锌锅内面渣在线回收利用示意图

1—带钢；2—锌锅；3—锌液液面；4—炉鼻子；5—沉没辊；6—纠正辊；7—稳定辊；8—感应加热器出口；
9—感应加热器入口；10—锌锭；11—气刀

参 考 文 献

[1] 黄孟阳, 邓志敢, 李兴彬, 等. 热镀锌渣综合回收锌的工艺研究进展及新工艺[J]. 矿冶, 2019, 28(4): 70-74.
[2] 卜二军. 第四届冶金渣固废回收、节能减排及资源综合利用高峰论坛论文集, 2019: 22-27.
[3] Wang Z, Gao J T, Meng L, et al. ISIJ International, 2018, 58: 1175-1177.
[4] Meng L, Guo L, Guo Z C. Waste Management, 2019, 84: 227-234.

钢渣磨料制备与研磨机理仿真研究

裴晶晶，邢宏伟，张玉柱，吴金虎，林文龙，霍文青

（华北理工大学冶金与能源学院，河北唐山　063210）

摘　要：在炼钢工业中，钢渣的产量约为粗钢产量15%左右[1-2]，如果钢渣得不到及时的处理，钢渣的堆积将会占用较大的土地资源，同时对土壤和地下水产生污染[3-4]。因此，利用钢渣生产高附加值产品既是对生态环境的保护，也是钢铁行业发展的新趋势，对废弃钢渣的综合利用具有重要的意义。

采用课题组自制的熔渣气淬实验装置，将液态钢渣利用高压高速气体射流击碎成液态渣滴，液态渣滴飞行过程中快速凝固，得到分散的球形钢渣磨料，钢渣磨料内部呈多孔结构分布，具有较好热稳定性、化学稳定性和力学性能。

利用 ABAQUS 有限元分析软件建立钢渣磨料研磨铝工件的有限元分析模型，钢渣磨料的微观研磨过程分为滑擦、耕犁和成屑过程，随着研磨速度的增加，滑动和滚动钢渣磨料磨削力呈递增的趋势，使磨料研磨部分的稳定性下降且耐磨性降低；改变摩擦系数和研磨深度对滑动磨料的稳定性和耐磨性影响较小，且摩擦系数越大、研磨深度越小滚动磨料的稳定越好、磨损程度越小。

通过实验研究钢渣磨料对铝合金表面质量的影响，磨料粒径为 40~80μm、80~150μm、150~300μm，初始工件表面划痕明显，表面粗糙度 Ra 为 3.2μm，表面轮廓集中在−11.5~8.0μm 之间。不同粒径钢渣磨料研磨 10min 后工件表面粗糙度 Ra 为 0.7μm、1.4μm、2.2μm，表面轮廓主要集中在±2.0μm、−2.8~4.0μm、−9.0~7.0μm 范围内，减小磨料的粒径有利于降低工件表面粗糙度和改善工件表面质量，且使用后的钢渣磨料破损率较低，所以钢渣磨料对工件能形成较好的研磨效果，是一种非常有潜力的研磨材料。

参 考 文 献

[1] 蒋亮, 李佳欣, 吴婷, 等. 材料导报, 2019, 33(15): 2490-2496.
[2] 张浩, 韩伟胜, 程峥明, 等. 光谱学与光谱分析, 2022, 42(12): 3906-3912.
[3] 任旭, 王会刚, 吴跃东, 等. 环境工程, 2022, 40(8): 220-224.
[4] 龙红明, 郑伟成, 裴元东, 等. 钢铁研究学报, 2021, 33(10): 1076-1083

梯级分离钢铁粉尘中 Zn-Pb-In 的工艺与展望

张良进[1,2,3]，张玉柱[1]，龙 跃[1]，杜培培[1]，王贵华[1]

（1. 华北理工大学冶金与能源学院，河北唐山 063009；2. 中冶建筑研究总院有限公司，北京 100088；3. 中冶检测认证有限公司，北京 100088）

摘 要：近年来对钢铁粉尘中有价元素的提取利用已经成为固废领域的研究热点，根据我国目前钢铁粉尘处理现状，在借鉴铅锌冶炼废渣处理技术的基础上，对钢铁粉尘经回转窑焙烧后的氧化锌烟尘进行湿法处理。首先对氧化锌烟尘进行酸浸处理，酸浸后液进行净化、电解得到锌锭，Pb、In 则富集在酸浸渣中；其次对酸浸渣进行二次酸浸，使得 Pb、In 分离，对二次酸浸渣进行氯盐提 Pb，完成 Pb 的靶向提取；最后采用 P_2O_4 对二次酸浸后液进行萃取、洗涤、反萃，再经中和、锌板置换得到海绵铟，从而完成 Zn-Pb-In 的梯级分离、靶向提取。此外，微波、超声等外部强化或联合强化、研发新的萃取体系、开发绿色离子溶剂等都是未来发展方向。

关键词：铅锌冶炼渣；钢铁粉尘；梯级分离；Zn；Pb；In

Process and Prospect of Cascade Swparating Zn-Pb-In in Steel Dust Based

ZHANG Liangjin[1,2,3], ZHANG Yuzhu[1], LONG Yue[1], DU Peipei[1], WANG Guihua[1]

(1. College of Metallurgy and Energy, North China University of Science and Technology, Tangshan

063009, China; 2. Central Research Institute of Building and Construction Co., Ltd., Mcc Group, Beijing 100088, China; 3. Inspection and Certification Co., Ltd., Mcc, Beijing 100088, China)

Abstract: In recent years, the extraction and utilization of valuable elements from iron and steel dust has become a hot research topic in the field of solid waste. According to the current situation of steel dust treatment and utilization in China, on the basis of the lead-zinc smelting slag treatment technology, hydrometallurgical process was adopted to handle the zinc oxide dust in iron and steel dust which was roasted by rotary kiln. After acid leaching treatment of zinc oxide dust, the acid leaching solution was purified and electrolyzed to yield the zinc ingot, while lead and indium were enriched in the acid dip slag. Subsequently, the acid leaching residue was subjected to secondary acid leaching treatment to sperate lead and indium, and then the chlorine salt technology was used to extract lead from the secondary acid leaching residue to complete targeted extraction of lead.The secondary acid leaching solution were extracted, washed and reverse extracted by P_2O_4 technology, and the sponge indium was obtained by the neutralization and replacement of zinc plate. so as to complete the cascade separation and targeted extraction of Pb-Zn-In. In addition, microwave, ultrasound and other external strengthening or combined strengthening, the research and development of new extraction systems, and the development of green ionic solvents are the future significant research filed.

Key words: lead and zinc smelting slag; iron and steel dust; ladder separation; Zn; Pb; In

Cr_2O_3对含铬熔分钛渣黏流特性的影响

曾睿琪[1,2]，王　楠[2,3]，李　尉[1,2]

（1. 东北大学多金属共生矿生态化冶金教育部重点实验室，辽宁沈阳　110819；2. 东北大学冶金学院，辽宁沈阳　110819；3. 东北大学低碳钢铁前沿技术研究院，辽宁沈阳　110819）

摘　要： 为使红格钒钛磁铁矿金属化球团熔分过程中的含铬熔分钛渣具有良好流动性，从而有利于渣金分离的顺利进行，本文研究了Cr_2O_3对含铬熔分钛渣黏流特性的影响。采用半球法和旋转柱体法分别测量了不同Cr_2O_3含量下含铬熔分钛渣的熔点和黏度。可以发现，随着Cr_2O_3含量的增加，含铬熔分钛渣的熔点和黏度逐渐增加。此外，含铬熔分钛渣黏流活化能的变化趋势与其黏度的变化趋势一致。

关键词： 含铬熔分钛渣；熔点；黏度；活化能

Influence of Cr_2O_3 on the Viscous Flow Behavior of Cr-Containing Molten Titanium Slag

ZENG Ruiqi[1,2], WANG Nan[2,3], LI Wei[1,2]

(1. Key Laboratory for Ecological Metallurgy of Multimetallic Mineral (Ministry of Education), Northeastern University, Shenyang 110819, China; 2. School of Metallurgy, Northeastern University, Shenyang 110819, China; 3. Institute for Frontier Technologies of Low-Carbon Steelmaking, Northeastern University, Shenyang 110819, China)

Abstract: In order to obtain the good liquidity of Cr-containing molten titanium slag and optimize the slag-metal separation in melting separation process of Hongge vanadium titanomagnetite metallized pellet, the influence of Cr_2O_3 on the viscous flow behavior of Cr-containing molten titanium slag was investigated in this work. The melting and viscosity properties were studied by hemisphere method and rotating cylinder method, respectively. The results showed that the melting temperature and viscosity both increased with the increase of Cr₂O₃ content. The trend of activation energy for viscous flow and the measured values of viscosity were identical.

Key words: Cr-containing molten titanium slags; melting temperature; viscosity; activation energy

脱磷渣和转炉渣协同处理与磷的浸出回收

于耀辉，杜传明

（东北大学冶金学院，辽宁沈阳　110819）

摘　要：转炉炼钢产生的普通转炉渣及脱磷渣中含有大量的有价组分，如 FeO_x、CaO、MnO 等，其最佳利用方式是作为炼钢原料或熔剂重新回用于冶金流程；然而，渣中 P_2O_5 的存在限制了其循环利用。若能将 P_2O_5 从钢渣中分离，可实现钢渣在冶金流程内的再利用，此外，分离提取的磷酸盐可作为磷肥使用，这对推动钢渣的减量化、无害化、高效资源化利用具有十分重要的意义。

本研究提出利用选择性浸出-沉淀法回收钢渣中的磷，流程如图 1 所示。由于脱磷渣和转炉钢渣成分的互补性（主要指碱度及 P_2O_5 含量），将两种钢渣进行协同改质处理，获得具有适当碱度的改质渣[1]。利用稀 HCl 溶液选择性溶解改质渣中的富磷相，剩余富铁残渣可作为冶金熔剂使用；酸浸后的含磷浸出液，添加 $Ca(OH)_2$ 调整 pH 值后得到磷酸钙沉淀，可作为磷肥使用；提磷后的浸出液可返回浸出工序再利用。本文研究了改质过程矿物相的转变和浸出参数对改质渣中各元素溶出行为的影响，并研究了浸出液中磷的沉淀规律。

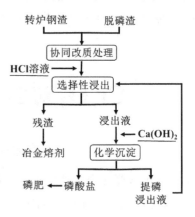

图 1　工艺流程图

浸出结果如图 2 所示。与转炉渣和脱磷渣相比，改质渣中不仅 P 的溶出率最高，且 Fe 几乎不溶解，说明协同改质处理能显著促进 P 的选择性浸出。降低 pH 值不仅能显著提高改质渣中 P 的溶出率，还能提高其溶出速率；降低固液比及钢渣粒度仅能提高 P 的溶出速率，对 P 的最终溶出率影响较小；浸出温度变化对溶出率及溶出速率影响均较小。在最佳条件下，P 的溶出率超过 90%，Fe 和 Mn 几乎不溶解，实现了磷的选择性浸出。浸出后残渣中 Fe_2O_3、MgO、MnO 含量分别为 65.86%、8.64%、8.5%，而 P_2O_5 含量仅有 0.32%。动力学研究表明改质渣中磷的溶出受内扩散控制，活化能为 5.02 kJ/mol，半经验公式可由下式表示：

$$1-3(1-X)^{2/3}+2(1-X) = \left[k_0 \cdot \exp(-2.172\text{pH}) \cdot (S/L)^{-1.031} \cdot (PZ)^{-1.598} \cdot \exp\left(\frac{-0.604}{T}\right)\right]t$$

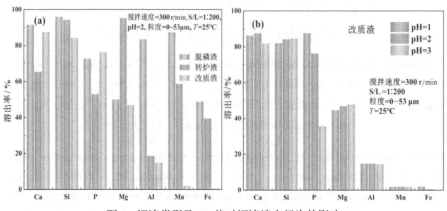

图 2　钢渣类型及 pH 值对钢渣溶出行为的影响

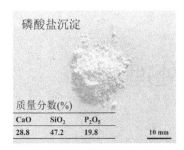

添加 Ca(OH)$_2$ 悬浊液提高含磷浸出液 pH 值至 7.0,浸出液中磷酸根离子沉淀析出。得到的白色磷酸钙盐沉淀如图 3 所示,其中 P$_2$O$_5$ 接近 20%,可作为缓释型磷肥使用,同时也能为农作物补充 Si 元素,增产促收。经过该工艺,含磷钢渣可转变为富铁残渣和磷酸盐肥料两种产品,从而实现了钢渣的高效资源化利用,有助于绿色低碳钢铁冶金流程的构建。

图 3 磷酸盐图片及成分

参 考 文 献

[1] Du C M, Xu G, Ueda S, ISIJ Int, 2018, 58(9): 1659-1668.

120t 转炉钢渣热熔压块使用效果研究

李廷刚,赵和明,甄新刚,张志强,何世长,范祚驿

(日钢营口中板有限公司冶金技术处,辽宁营口 115005)

摘 要:针对 120t 转炉使用的钢渣热熔压块使用效果进行分析研究。通过分析发现,钢渣热熔压块密度越大,融化速度越快,当目标温度设定至 1600℃时,热熔压块融化所用时间较废钢长约 190s。同时发现,达到同等温度时热熔压块吸收的热量是纯净废钢的 1.37 倍。依据转炉生产模式,当转炉采取高铁耗装入制度时,钢渣热熔压块性价比高于废钢,采取低铁耗模式生产时,钢渣热熔压块性价比低于废钢。

关键词:钢渣热熔压块;转炉;性价比;融化速度

The Using Effect Analysis of 120t LD Using Steel Slag Melt Pressure Block

LI Tinggang, ZHAO Heming, ZHEN Xingang, ZHANG Zhiqiang, HE Shichang, FAN Zuoyi

(Rizhao Steel Yingkou Medium Plate Co., Ltd. Metallurgical Technology Department, Yingkou 115005, China)

Abstract: For 120 t LD using steel slag melt pressure block using effect analysis, through the analysis found that steel slag melt pressure block density, the greater the melting faster; Heated to 1600℃ with heating parameters, hot melt press block used for a scrap of about 190 seconds long; Hot melt pressure block absorption of heat is 1.37 times that of the pure steel scrap; On the basis of converter production mode, when to take high iron loss of converter load system, frit briquetting ratio higher than scrap steel, steel slag to adopt low iron loss model production, steel slag melt pressure piece price lower than that of scrap steel.

Key words: steel slag melt press block; LD; cost performance; melting speed

烟气脱硫制酸废液返回烧结处理对活性炭性能的影响

李俊杰[1,2,3]，陶家杰[4]，齐鹏宇[4]，丁 龙[4]，
钱立新[4]，春铁军[4]，龙红明[4]

（1. 中冶长天国际工程有限责任公司，湖南长沙 410205；2. 中国科学院过程工程研究所，多相复杂系统国家重点实验室，北京 100190；3. 中国科学院大学化学工程学院，北京 100049；4. 安徽工业大学冶金工程学院，安徽马鞍山 243032）

摘 要：为探究烧结烟气脱硫制酸废液返回烧结处理对活性炭性能的影响，开展了制酸废液对活性炭的强制中毒实验，并通过固定床反应装置测试了中毒前后活性炭的脱硫性能变化；利用 BET、SEM、XPS、FTIR 和 NH_3-TPD 等表征手段揭示了活性炭制的中毒机理。结果表明：制酸废液中无机成分包括 NH_4Cl、$(NH_4)_2SO_4$ 和 NH_4HSO_4，有机物成分有主要以酚类、酸酯类和烷烃类为主。活性炭中毒后，SO_2 吸附量由 29.7mg/g 降低至 1.1mg/g，穿透时间由 12min 缩短到 1min。制酸废液对活性炭的表面产生了堵塞，活性炭的比表面积、孔容积大幅降低，微孔结构几乎消失。孔道堵塞物质为 $(NH_4)_2SO_4$ 和 NH_4Cl 晶体颗粒。此外制酸废液使活性炭表面碱性官能团酮羰基含量由 6.42%降低到 4.83%。制酸废液中无机盐在有机物组分的作用下，堵塞了活性炭的孔结构，破坏了表面官能团，使得活性炭对 SO_2 的吸附转化性能降低。

关键词：烧结烟气；活性炭；脱硫；制酸废液；中毒

Effect of Return Iron Ore Sintering of Acid Waste Water from Flue Gas Desulfurization on Properties of Activated Carbon

LI Junjie[1,2,3], TAO Jiajie[4], QI Pengyu[4], DING Long[4],
QIAN Lixin[4], CHUN Tiejun[4], LONG Hongming[4]

(1. MCC Changtian International Engineering Co., Ltd., Changsha 410205, China;
2. State Key Laboratory of Multi-phase Complex Systems, Institute of Process Engineering, CAS, Beijing 100190, China; 3. School of Chemical Engineering, University of Chinese Academy of Sciences, Beijing 100049, China; 4. School of Metallurgical Engineering, Anhui University of Technology, Maanshan 243032, China)

Abstract: In order to investigate the effects of returning the desulfurization acid waste water from flue gas to the sintering process on the performance of activated carbon, forced poisoning experiments were conducted using fresh activated carbon and acid waste water obtained from the sintering industrial site. The changes in the desulfurization performance of activated carbon before and after poisoning were tested using a fixed-bed reactor. Characterization techniques such as BET, SEM, XPS, FTIR, and NH_3-TPD were employed to reveal the poisoning mechanism of activated carbon. The results show that the inorganic components in the acid waste water include NH_4Cl, $(NH_4)_2SO_4$, and NH_4HSO_4, while the organic components mainly consist of phenols, acid esters, and alkanes. After the poisoning, the adsorption capacity of SO_2 on the activated carbon decreased from 29.7mg/g to 1.1mg/g, and the breakthrough time reduced from 12 minutes to 1 minute. The physical and chemical properties of activated carbon showed that the acid waste water caused blockage on the surface of activated carbon, resulting in a significant decrease in specific surface area and pore volume, and the microporous structure almost disappeared. Additionally, the acid waste water reduced the content of ketonic carbonyl groups on the surface of activated carbon from 6.42% to 4.83%. The inorganic salts in the acid waste water, in the presence of organic components, blocked the pore structure of activated carbon and damaged the surface functional groups, leading to a decrease in the adsorption and conversion performance of activated carbon for SO_2.

Key words: sintering flue gas; activated carbon; desulfurization; acid waste water; poisoning

赤泥微波碳热还原熔分提铁工艺技术研究

施树蓉[1,2]，胡倩倩[1,2]，丛云伶[1,2]，张军红[1,2]，高立华[1,2]，何志军[1,2]

（1. 辽宁科技大学，辽宁鞍山 114051；2. 辽宁省绿色低碳与智能冶金重点实验室，辽宁鞍山 114051）

摘 要： 赤泥是氧化铝厂生产氧化铝时排出的污染性废渣，赤泥的主要污染物为碱、氟化物、钠和铝等，赤泥一般以堆放处理为主，这不仅对环境造成严重危害，同时也对企业生产经营带来了巨大压力。目前国内外工作者针对从赤泥中提取有价金属元素使得赤泥得到妥善，并在消除对环境污染的同时实现赤泥资源化利用开展了许多相关研究试验。作者团队基于微波能是一种高效的清洁能源利用微波碳热还原技术对赤泥进行了配碳直接还原，同时基于响应曲面法对碳氧比、微波功率、加热时间和添加剂配比四个因素进行了优化实验研究，分析了不同因素对铁回收效果的影响。由于赤泥本身是一种弱吸波物质，它在微波功率1kW下进行齐最高温度只能达到240℃；实验中用到的碳热还原剂无烟煤具有良好的吸波特性，它可以在微波功率500W下快速升温到760℃以上，因此在赤泥微波碳热还原中无烟煤是影响混合原料温度的主要升温物质。通过响应曲面法研究得到，对铁回收率混和铁品位影响影响大小依次为：微波功率>加热时间>碳氧比>添加剂配比。通过 XRD 和 SEM 等手段对还原产物的分析得知，赤泥中 Fe_2O_3 几乎全部还原成 Fe_3O_4 和铁单质颗粒。运用 Box-Behnken 以铁回收率响应值（Y）为因变量，以配碳比（X_1）、微波功率（X_2）、加热时间（X_3）为自变量构建出二次多项式回归模型：$Y=0.72732+30.55835X_1+0.021277X_2+1.26920X_3+1.38353\times10^{-3}X_1X_2-0.056677X_1X_3+1.32520\times10^{-5}X_2X_3-8.38621X_1^2-4.34235\times10^{-6}X_2^2-0.011832X_3^2$。得到最佳工艺参数为碳氧比为1.7、微波功率为2800W、加热时间为40min时，赤泥微波碳热还原处理后可获得金属铁和 Fe_3O_4，铁回收率可达88.9%。当碳氧比为1.7、微波功率为2800W、加热时间40min、添加剂配比为5%时，铁的品位和回收率分别达到了64.58%和90.64%。根据 XRD、SEM 与 EDS 综合分析得知，还原出的铁主要以 Fe_3O_4 为主，并有少量的金属铁生成，模型预测与实验之间的误差为1.65%。

关键词： 赤泥；微波碳热还原；响应曲面法；铁回收率；预测模型

Research on Microwave Carbon-Thermal Reduction Smelting and Iron Extraction from Red Mud.

SHI Shurong[1,2], HU Qianqian[1,2], CONG Yunling[1,2], ZHANG Junhong[1,2], GAO Lihua[1,2], HE Zhijun[1,2]

(1. College of Materials and Metallurgy, University of Science and Technology Liaoning, Anshan 114051, China; 2. Key Laboratory of Green Low-Carbon and Intelligent Metallurgy in Liaoning Province, Anshan 114051, China)

Abstract: Red mud is a pollutant residue generated during the production of alumina in alumina plants. The main pollutants in red mud include alkali, fluoride, sodium and aluminum. Red mud is typically managed through disposal by stacking, which not only poses serious environmental hazards but also creates significant pressure on the operation of the companies. Currently, researchers both domestically and internationally have conducted numerous studies and experiments on the extraction of valuable metallic elements from red mud, aiming to achieve proper treatment of red mud and its utilization as a resource while eliminating environmental pollution. In this study, the authors' team utilized microwave carbon-thermal reduction technology, which makes use of microwave energy as an efficient and clean energy source, to directly reduce red mud through carbon addition. Additionally, the authors conducted optimization experiments based on response surface methodology, focusing on four factors: carbon-to-oxygen ratio, microwave power, heating time, and additive ratio. They analyzed the effects of these factors on iron recovery efficiency. Due to red mud being a weak microwave absorber, its temperature can only reach a maximum of 240℃ under a microwave power of 1 kW. In contrast, the carbon-thermal reduction agent used in the experiment, namely anthracite coal, exhibits good microwave absorption properties and can quickly reach temperatures above 760℃ under a microwave power of 500 W. Therefore, anthracite coal is the main heating material affecting the temperature of the mixed raw materials in the microwave carbon-thermal reduction of red mud. Based on the response surface methodology, the study found that the factors influencing iron recovery efficiency and iron grade in descending order of magnitude are: microwave power > heating time > carbon-to-oxygen ratio > additive ratio. The analysis of the reduction products through XRD and SEM revealed that almost all of the Fe_2O_3 in the red mud was reduced to Fe_3O_4 and iron particles. A Box-Behnken design was employed to construct a quadratic polynomial regression model, with iron recovery rate (Y) as the response variable and carbon-to-iron ratio (X_1), microwave power (X_2), and heating time (X_3) as the independent variables. $Y=0.72732+30.55835X_1+0.021277X_2+1.26920X_3+1.38353\times10^{-3}X_1X_2-0.056677X_1X_3+1.32520\times10^{-5}X_2X_3-8.38621X_1^2-4.34235\times10^{-6}X_2^2-0.011832X_3^2$。 The optimal process parameters were determined to be a carbon-to-oxygen ratio of 1.7, microwave power of 2800W, and heating time of 40 minutes, resulting in the production of metallic iron and Fe_3O_4 through microwave carbon-thermal reduction treatment of red mud, with an iron recovery rate reaching 88.9 %. When the carbon-to-oxygen ratio was 1.7, microwave power was 2800W, heating time was 40 minutes, and the additive ratio was 5%, the iron grade and recovery rate reached 64.58 % and 90.64 %, respectively. Comprehensive analysis using XRD, SEM, and EDS indicated that the reduced iron was primarily Fe_3O_4, with a small amount of metallic iron formation. The model prediction had an error of 1.65 % compared to the experimental results.

Key words: red mud; microwave carbon-thermal reduction; response surface methodology; iron recovery rate; predictive model

13　冶金设备与工程技术

大会特邀报告
第十三届冶金青年科技奖获奖人特邀报告
分会场特邀报告
矿业工程
焦化及节能环保
炼铁与原料
炼钢与连铸
电冶金与废钢铁
轧制与热处理
表面与涂镀
金属材料深加工
粉末冶金
先进钢铁材料及应用
节能与低碳技术
冶金环保与资源利用
★ 冶金设备与工程技术
冶金自动化与智能化
冶金物流
冶金流程工程学
其他

X 射线荧光光谱法分析热压铁块中主次成分

禹青霄，杨晓倩，孙少海，杨慧贤，牛威斌

（河钢材料院理化检测中心，河北石家庄 050000）

摘　要：应用熔融制样-X 射线荧光光谱法测定热压铁块中各主次元素，探讨了不同熔剂、稀释比例、熔融温度和熔融时间对检测结果的影响，解决了热压铁块中金属铁腐蚀铂金坩埚的难题，实现了对热压铁块中的全铁 TFe、SiO_2、Al_2O_3、CaO、MgO、P_2O_5 等元素的快速检测。试验首先对样品进行预处理，然后采用 1100℃熔融温度下熔融 7min，稀释比为 1∶10，以 LiBr 为脱模剂熔解试样，样品熔解完全，且对铂金坩埚无明显腐蚀。试验结果表明该方法精密度为 RSD<3%，正确度结果表明，标准样片和实际样品的测定值分别与认定值及国标湿法值相一致。

关键词：X 射线荧光光谱法；熔融制样；热压铁块；主次成分

Determination of Major and Minor Components of Hot Briquette Iron by X-ray Fluorescence Spectrometry

YU Qingxiao, YANG Xiaoqian, SUN Shaohai, YANG Huixian, NIU Weibin

(HBIS Materials Institute Physical and Chemical Testing Center, Shijiazhuang 050000, China)

Abstract: The main and minor components in hot briquette iron were determined by X-ray fluorescence spectrometry with melting sample preparation, and the effects of different flux, dilution ratio, melting temperature and melting time on the detection results were discussed, the problem of metal iron corroding platinum crucible in hot briquette iron was solved, and the fast detection of TFe, SiO_2, Al_2O_3, CaO, MgO, P_2O_5 in hot briquette iron was realized. The samples were pretreated first, then melted for 7 min at 1100 ℃, the dilution ratio was 1:10, and the samples were melted with LiBr as release agent. The results show that the precision of the method is RSD<3%. The accuracy of the method shows that the measured values of the standard sample and the actual sample are consistent with the certified value and the national standard wet method value respectively.

Key words: X-ray fluorescence spectrometry; fused preparation; hot briquette iron; major and minor components

氢化物发生-电感耦合等离子体原子发射光谱法测定钼铁中砷锡锑铋

葛晶晶

（河钢材料技术研究总院，河北石家庄 050000）

摘　要：建立了一种简单、灵敏度高的测定钼铁中砷、锡、锑、铋的方法，样品经盐酸和硝酸处理后，在10%（体积分数）盐酸介质中，用 10g/L 的硼氢化钾-氢氧化钠作为还原剂，将氢化物发生器与电感耦合等离子体原子发射光谱仪（ICP-AES）联用，可有效将待测元素与基体进行分离，消除基体效应，提高检测灵敏度，方法检出限(3σ)

为砷 1.2μg/L，锡 0.8μg/L，锑 0.76μg/L，铋 0.41μg/L，相对标准偏差（RSD）砷小于 4.1%，锡小于 3.3%，锑小于 2.7%，铋小于 3.2%，加标回收率为 96%~109%。

关键词：氢化物发生；电感耦合等离子体原子发射光谱法（ICP-AES）；钼铁；砷；锡；锑；铋

Determination of Arsenic, Tin, Antimony, Bismuth in Molybdenum Iron by Hydride Generation-inductively Coupled Plasma Atomic Emission Spectrometry

GE Jingjing

(HBIS Materials Institute, Shijiazhuang 050000, China)

Abstract: A simple and high sensitivity method for the determination of arsenic, tin, antimony, and bismuth in molybdenum iron was established. The sample is treated with hydrochloric acid and nitric acid, and in a 10% (volume fraction) hydrochloric acid medium, 10g/L potassium borohydride sodium hydroxide is used as a reducing agent. The hydride generator is combined with inductively coupled plasma atomic emission spectrometer (ICP-AES) to effectively separate the tested element from the matrix and eliminate matrix effects, Improve detection sensitivity, method detection limit (3σ) Arsenic 1.2μG/L, tin 0.8μG/L, antimony 0.76μG/L, bismuth 0.41μG/L, relative standard deviation (RSD) of arsenic is less than 4.1%, tin is less than 2.4%, antimony is less than 2.6%, bismuth is less than 3.2%, and the recovery rate of added standard is 96%~109%.

Key words: hydride generation;inductively coupled plasma atomic emission spectrometry (ICP-AES);molybdenum iron; arsenic; tin; antimony; bismuth

Python 与设备点检管理相结合的实践与运用

周子珍[1]，茹剑波[2]

(1. 宝武钢铁集团宝钢股份武钢有限热轧厂，湖北武汉　430081；
2. 宝武钢铁集团宝钢股份武钢有限设备管理部，湖北武汉　430081)

摘　要：点检是设备管理的一项基本制度，通过人的五感或者借助工器具，按照预先设定的周期和方法，对设备上的规定部位（点）进行有无异常的预防性周密检查的过程。作为设备点检业务管理人员，需要将现场点检人员点巡检的实绩汇总作初步分析，目前统计主要以人工为主。在现有信息平台不能实现快速数据统计的情况下，本文将介绍如何通过结合互联网中的 python 工具及相关第三方库来简化这个数据收集过程，实现数据自动采集，保证准确性的同时，提升工作效率。

关键词：设备点检管理；点检实绩；python；selenium；自动化

The Practice and Application of Python and Equipment Inspection Management

ZHOU Zizhen[1], RU Jianbo[2]

(1. Hot Rolling Mill of Wuhan Iron & Steel Co., Ltd., Baosteel Group, Wuhan 430081, China;

2. Equipment Management Department of Wuhan Iron & Steel Co., Ltd.,
Baosteel Group, Wuhan 430081, China)

Abstract: Spot inspection is a basic system for equipment management. With the five senses of human or the use of equipment, according to the pre-set cycle and method, undertaking an abnormal preventive and thorough inspection process on the specified parts (points) on the device. As a equipment manager of equipment inspection, the actual performance of the on-site inspection personnel needs to be preliminarily analyzed. At present, the statistics are mainly manual. In the case where the existing information platform cannot achieve fast data statistics, this article will introduce how to simplify this data collection process by combining the Python tools and related third-party libraries in the Internet, automated the data to ensure accuracy while improving work efficiency Essence.
Key words: spot inspection; inspection data; python; selenium; automation

浅析大型液压传动系统温度高的问题

梁 婷

（武汉钢铁有限公司，湖北 武汉 430070）

摘 要： 液压源回路是向液压传动系统提供满足执行机构需要的压力和流量。冶金行业轧钢片区机械设备有大量的液压控制回路，液压传动系统在此工厂内占有量较大，液压传动系统运行状态与大量机械设备动作有着密切的关系。液压传动系统在长期运行过程中，会出现各种各样的问题，本文针对某厂大型液压传动系统存在的温度高的问题，提出了对应的解决方案，并且效果良好。

关键词： 液压传动；温度；循环

Abstract: The hydraulic source circuit is to provide the hydraulic transmission system with the pressure and flow required by the actuator. There are a large number of hydraulic control circuits for the mechanical equipment in the steel rolling area of the metallurgical industry. The hydraulic transmission system occupies a large proportion in the factory. The operation status of the hydraulic transmission system is closely related to the action of a large number of mechanical equipment. After the long-term operation, system various problems of the hydraulic transmission system will occur. Aiming at the problem of high temperature in a large hydraulic system of a factory, this paper puts forward corresponding solutions, and the results are good.
Key words: hydraulic transmission; temperature; cycle

炼钢连铸新型高温复合辊断裂失效机理分析

廖礼宝

（宝钢股份设备部，上海 201900）

摘 要： 炼钢连铸工艺装备要承受高温工况条件的考验，特别是连铸扇形段（SEG）的高温连铸辊受高温工况及附加交变载荷的同时作用条件更为苛刻，从而导致该类连铸辊寿命颇受影响。随着技术集成度越来越高，失效分析的难度越来越大。针对某炼钢厂试用的新型双重复合材料制成的高温 SEG 连铸辊所形成的大量裂纹并出现异常断裂现象，从材料和装备技术专业角度，对该类高温加交变复合辊进行了失效专业研究。基于某新型双重复合高温辊的断裂失效，分别从辊子的断口宏观形貌特征、双重成分、硬度梯度分布、力学性能、金相组织、高温氧化及扫描电镜（SEM）等多方面进行了失效机理分析并得出了结论。为避免某新型双重复合高温连铸辊表面形成大量裂纹并出现异常断裂现象，建议分别将使用工况温度、复合连铸辊的性能指标及本身冷却效果、最终焊接质量和最终热处理质量（金相）构成作为关键判定与验收考核指标。

关键词：新型连铸复合辊；扇形段；断裂失效；机理分析

Mechanism Analysis on Fracture-failure of High Temperature and New Type Composite Roller in Continuous-casting Segment

LIAO Libao

(Equipment Department, Baoshan Iron & Steel Co., Ltd., Shanghai 201900, China)

Abstract: Continuous-casting equipment must bear the test of high temperature conditions. In particular, the high temperature continuous-casting rollers of segment influenced by more severe working condition, thus affecting its service life. However, as technology integration becomes higher, failure analysis becomes more difficult. This article is aimed at a large number of cracks and abnormal fractures formed by high temperature segment continuous-casting rollers made of double composite materials, and from the perspective of materials and equipment technology, failure research and analysis of this type of compound rollers are carried out. Based on fracture failure of double composite roller and sum up the research results of the predecessors, we analyzed the failure mechanism of the roller from the macro morphology, dual composition, hardness gradient distribution, mechanical properties, metallurgical structure, high temperature oxidation and scanning electron microscope (SEM). In order to prevent the formation of cracks on the roller surface and abnormal fractures, it is suggested that the working condition temperature, the performance index of the composite caster, the final welding state and the final heat treatment state are considered as the key criteria and evaluation indexes.

Key words: continuous-casting roller; segment; high temperature-thermal fatigue failure; mechanism analysis

冶金旋流井中水流作用下抓渣斗的受力分析与防缠绕防护罩装置

崔忠信，孙立红，冯小雷，田绍鹏，张大伟，田　兴

（河钢材料技术研究院，河北石家庄　050000）

摘　要：为了降低抓渣斗旋转角度，详细分析了冶金旋流井内水流对抓渣斗的冲击作用。基于天车抓渣斗旋转及水平位移产生的根本原因，设计相应的防护结构。经过实践验证，进行二次优化，最终提供了一套可以降低抓渣斗旋转的防护装置。

关键词：旋流井；抓渣斗；防护装置；机械设计

Force Analysis about Water Flow in a Metallurgical Cyclone Well on the Grab and Its Protective Device

CUI Zhongxin, SUN Lihong, FENG Xiaolei, TIAN Shaopeng,
ZHANG Dawei, TIAN Xing

(HBIS Materials Technology Research Institute, Shijiazhuang 050000, China)

Abstract: In order to reduce the rotation angle of the grab, the impcut of water flow in a metallurgical swirl well on the grab is analyzed in detail. Based on the root cause of rotation and horizontal displacement of the crown block grab, We design corresponding protective structures. Through practice, we conducted secondary optimization. Eventually, a set of protective devices was provided that could reduce the rotation of the grab.

Key words: swirl well; grab; protective devices; machine design

全站仪在激光焊机小车行走直线度检测中的应用

赵新宝，高　超，兰晓栋，玄利剑，唐晓宇，
孟召锦，刘　昊，司　泽

（首钢智新迁安电磁材料有限公司，河北迁安　064400）

摘　要：激光焊机处于带钢热处理产线的关键部位，焊机的机械精度直接影响带钢的焊机质量，是保障产线连续顺稳运行的重要因素。本文介绍一种应用全站仪测量焊机小车行走直线度的方法，多点拟合直线法，拟合方法采用最小二乘法，阐述此方法的测量原理及步骤，并通过冷轧现场具体实例来验证此方法的可行性，为其在冷轧工厂广泛应用提供理论及数据支撑。

关键词：激光焊机；全站仪；直线度；最小二乘法

The Application of Total Station in Detecting the Straightness of Laser Welding Machine Trolley

ZHAO Xinbao, GAO Chao, LAN Xiaodong, XUAN Lijian, TANG Xiaoyu,
MENG Zhaojin, LIU Hao, SI Ze

(Shougang Zhixin Qian'an Electromagnetic Materials Co., Ltd., Qian'an 064400, China)

Abstract: Laser welding machine is in the key position of heat treatment production line of strip steel. The mechanical precision of welding machine directly affects the quality of strip steel welding machine and is an important factor to ensure the smooth and stable operation of production line. This paper introduces a method of using total station to measure the walking straightness of welding machine trolley, multi-point fitting straight line method, and the OLS method is used for fitting method. The measurement principle and steps of this method are described, and the feasibility of this method is verified by the specific cold rolling field examples, which provides data support for its wide application in cold rolling plants.

Key words: laser welder; total station; straightness; OLS method

结合视觉与图谱分析的带钢表面质量判定研究

吴昆鹏[1,2]，邓能辉[1,2]，郭亚男[1,2]，郭 爽[1,2]，石 杰[1,2]，杨朝霖[1,2]

(1. 北京科技大学国家板带生产先进装备工程技术研究中心，北京 100083；
2. 北京科技大学设计研究院有限公司，北京 100083)

摘 要：基于视觉的带钢表面缺陷检测系统虽然提供了表面缺陷的统计信息，但仍然存在识别数据不准确、分析层次低等问题，无法形成综合性的表面质量判定结果。通过对人工质量判定过程的分析，提出采用级联式模型强化对缺陷的识别能力，利用图谱分析技术提升特征抽取层次，设计多模态依赖的缺陷严重度分析模型，充分效仿人工高维层次的质量判定规律，形成全新质量判定模式，并以知识图谱方式形成可解释性的带钢判定数据。通过对表面检测系统输出数据结果的再分析，完成对带钢质量的综合判定，能够直接进行封锁、报警等操作，有效降低人工成本，提升生产效率，减少客户质量异议。

关键词：缺陷检测；分布图谱；质量判定；知识图谱

Research on Surface Quality Evaluation of Strip Steel by Combining Visual Inspection and Graph Analysis

WU Kunpeng[1,2], DENG Nenghui[1,2], GUO Ya'nan[1,2],
GUO Shuang[1,2], SHI Jie[1,2], YANG Chaolin[1,2]

(1. National Engineering Technology Research Center of Flat Rolling Equipment, Beijing 100083, China;
2. Design and Research Institute Co., Ltd., University of Science and
Technology Beijing, Beijing 100083, China)

Abstract: Although the visual based strip steel surface defect detection system provides statistical information on surface defects, there are still problems such as inaccurate recognition data and low analysis level, which cannot form comprehensive surface quality judgment results. By analyzing the process of manual quality judgment, a cascaded model is proposed to enhance the recognition ability of defects, and graph analysis technology is used to enhance the level of feature extraction. A multi-modal dependency defect severity analysis model is designed, fully imitating the high-dimensional quality judgment rules of manual work, forming a new quality judgment mode, and forming interpretable strip steel judgment data through knowledge graph. By reanalyzing the output data results of the surface inspection system, a comprehensive judgment of the quality of the strip steel can be completed, and operations such as blocking and alarming can be directly carried out, effectively reducing labor costs, improving production efficiency, and reducing customer quality objections.

Key words: defect detection; distribution map; quality judgment; knowledge graph

高压自密封阀门四开环改进

赵 宇，贾树勋，侯 新，李雪松

（鞍钢股份有限公司能源管控中心，辽宁鞍山 114021）

摘 要：高压自密封阀门更换楔形密封工作中由于四开环移位，造成预紧不足现象经常发生，给后期工作带来隐患，针对这一现象，改进四开环结构，避免预紧不足现象，以供相关人员参考和应用。

关键词：自密封阀；四开环

全氢罩式炉加热罩的使用与维修

吴立军

（鞍钢集团鞍山钢铁冷轧四分厂热处理作业区，辽宁鞍山 114000）

摘 要：全氢罩式炉是再充分利用普通罩式炉优点的基础上，又克服了普通罩式退火炉及连续退火炉不可避免的缺陷而发展起来的。20世纪70年代初，奥地利埃伯纳（Ebner）开发了用于黄铜工业的强对流罩式退火炉，德国LOI工业炉公司在20世纪80年代初研制开发了现已被广泛用于钢铁工业的氢罩式退火炉，这种退火炉称为HICON/H，即强对流全氢罩式退火炉。由于H_2优越的传热性能（其传导系数为N_2的7倍），使罩式炉的生产效率大大提高；采用了金属封闭炉台、气/水组合冷却罩及优化设计的扩散器和对流板等重大改进，降低了露点，使钢板表面不因酸洗后C^{4+}、Cr^{2+}等离子吸附而发黑，这样一来改善了钢板表面质量，使表面光洁而发亮，同时提高了钢板的力学性能和深冲性，降低了能耗，节约了成本。所以用全氢罩式退火炉取代常规的罩式退火炉是罩式退火技术的重大进步，而其中的加热罩更是重中之重，其加热过程直接决定钢卷的品质，不仅影响冷轧板的表面光亮程度，而且更重要的是影响钢板的性能及其均匀性。随着冷轧钢板要求的不断提高和冷轧技术的发展，对现场生产、维护也提出了更高的要求。为提高生产作业率和产品质量以及加热精度，对加热罩的工作精度及稳定性提出了更高的要求。本文结合现场实际应用场景以及日常工作当中的探索，总结出一些经验和具体方案。

关键词：罩式炉；加热罩；使用；维修

The Use and Maintenance of Heating Hood of Full Hydrogen Bell Furnace

WU Lijun

(Heat Treatment Operation Area of Anshan Steel Cold Rolling No.4 Plant of Anshan Steel Group, Anshan 114000, China)

Abstract: Full hydrogen bell furnace is developed on the basis of fully utilizing the advantages of ordinary bell furnace and overcoming the inevitable defects of ordinary bell annealing furnace and continuous annealing furnace. In the early 1970s,

Ebner of Austria developed a strong convection bell-type annealing furnace for brass industry, and LOI Industrial Furnace Company of Germany developed a hydrogen bell-type annealing furnace which has been widely used in steel industry in the early 1980s. This kind of annealing furnace is called HICON/H, that is, a strong convection full hydrogen bell-type annealing furnace. Due to the superior heat transfer performance of H_2 (its conductivity is 7 times that of N_2), the production efficiency of the bell furnace is greatly improved. Major improvements, such as metal closed hearth, gas/water combined cooling cover, optimized diffuser and convection plate, were adopted, which reduced the dew point and prevented the surface of steel plate from blackening due to C^{4+} and Cr^{2+} plasma adsorption after pickling, thus improving the surface quality of steel plate, making the surface smooth and shiny, improving the mechanical properties and deep drawability of steel plate, reducing energy consumption and saving costs. Therefore, replacing the conventional bell-type annealing furnace with hydrogen bell-type annealing furnace is a great progress in bell-type annealing technology, and the heating bell is the most important one. Its heating process directly determines the quality of steel coil, which not only affects the surface brightness of cold-rolled plate, but also affects the performance and uniformity of steel plate. With the continuous improvement of cold-rolled steel plate requirements and the development of cold-rolled technology, higher requirements are put forward for on-site production and maintenance. In order to improve the production rate, product quality and heating accuracy, higher requirements are put forward for the working accuracy and stability of the heating cover. In this paper, some experiences and specific schemes are summarized based on the practical application scenarios and the exploration in daily work.

Key words: hood furnace; heating cover; use; maintenance

基于三维建模分析的热轧飞剪前侧导板改造

王骏超，吴长杰，张会明，屈二龙，醴亚辉，尹贵祖，郭维进，东占萃

（北京首钢股份有限公司，河北迁安　064406）

摘　要： 飞剪侧导板安装热轧产线飞剪设备入口位置，是精轧机组入口区域重要的对中导向设备。传统的飞剪前侧导板采用同步曲柄转轴设计，单液压缸驱动，存在对中力小，侧导板中心线调整困难的缺点，且导向采用固定滑板，在对中力作用下对带钢表面划伤严重。本文新设计了一种侧导板驱动导向装置，采用双液压缸对两侧导板进行分别驱动，增加了对中力，同时实现了侧导板中心线的在线调整。通过对侧导板增加了导轮装置，将对板坯的导向方式由滑动摩擦改为滚动摩擦，改善了带钢边部质量。

关键词： 热连轧机；飞剪；侧导板；导轮

Transformation of Entry Guide of Corp-shear on HSM Line Based on 3D Modeling Analysis

WANG Junchao, WU Changjie, ZHANG Huiming, QU Erlong, LI Yahui, YIN Guizu, GUO Weijin, DONG Zhancui

(Beijing Shougang Co., Ltd., Qian'an 064406, China)

Abstract: The side guide of corp shear was installed on flying shear equipment entrance position of hot rolling line, it is an important centering guide equipment in the entrance area of the finishing mill. The traditional flying shear front side guide plate adopts synchronous crank shaft design, single hydraulic cylinder drive, which has the disadvantages of small centering force and difficult adjustment of the centerline of the side guide, and the guide adopts fixed skateboard, which seriously scratches the surface of the strip under the action of the centering force. In this paper, a new side guide drive guide is designed, which uses double hydraulic cylinders to drive the two side guide plates separately, which increases the centering force and realizes the online adjustment of the center line of the side guide plate. By adding a guide roller device to the opposite side guide plate, the guiding method of the slab is changed from sliding friction to rolling friction, and the quality of the strip edge is improved.

Key words: hot strip mill; corp shear; side guide; guide roller

精轧机增加小立辊的设计与实施

吴长杰，张会明，东占萃，郭维进，张志桥，王艺霖

（北京首钢股份有限公司，河北迁安 064406）

摘 要：精轧机入口对中及微减宽功能对于精轧机轧制稳定性及产品宽度控制有着重要的作用，国内某2250产线由于投产时未设计精轧前立辊设备，随着产品升级及质量需求增加，需要在现有设备空间内增加设计一套立辊设备。本文主要描述了一种在线增加精轧小立辊对中设备的设计及实施方案，充分利用现有的空间结构，并且利用生产及检修时间进行安装调试投入运行，为同类产线升级改造提供了借鉴意义。

关键词：热连轧机；立辊；对中；减宽；跑偏

Design and Application of Adding Edge Roll on Finish Mill

WU Changjie, ZHANG Huiming, DONG Zhancui, GUO Weijin,
ZHANG Zhiqiao, WANG Yilin

(Beijing Shougang Co., Ltd., Qian'an 064406, China)

Abstract: Finishing mill inlet and micro-width reduction function for the finishing mill rolling stability and product width control plays an important role, a domestic 2250 production line due to the commissioning of the finished rolling front vertical roll equipment, with the product upgrade and quality demand increase, need to design a set of vertical roll equipment in the existing equipment space. This paper mainly describes the design and implementation scheme of an online increase of finishing small vertical roll centering equipment, making full use of the existing space structure, and using the production and maintenance time for installation, commissioning and operation, which provides reference significance for the upgrading and transformation of similar production lines.

Key words: hot strip mill; edge roll; centring; reduce width; deviation

热轧钢板高速飞剪过程的有限元模拟

李 轲，陈 敏，高爱民

（中冶赛迪技术研究中心有限公司，冶金智能装备重庆市重点实验室，重庆 401122）

摘　要：建立了高速飞剪剪切过程的有限元模型，对 40Mn 带钢剪切飞剪剪切过程进行了模拟，分析了飞剪的关键参数对剪切的影响规律。首先，通过对飞剪机构进行运动学分析，获得剪切过程中剪刃运动轨迹。接着，在 DEFORM 软件上建立了 40Mn 带钢飞剪过程模拟的平面应变有限元模型，材料采用了考虑应变、应变速率硬化和温度软化的 Johnson-Cook 塑性本构模型及损伤模型。通过所建立的有限元模型的计算分析，得到了剪刃圆角半径及剪刃相对间隙对剪切功和剪切断面质量影响规律。在给定的带钢尺寸和飞剪条件下，确定了最佳的剪刃间隙及圆角半径。

关键词：高速飞剪；参数分析；有限元；DEFORM

Finite Element Simulation of Hot Rolled Steel Plate Flying Shear Process at High Speed

LI Ke, CHEN Min, GAO Aimin

(CISDI Research & Development Co., Ltd., Chongqing Key Laboratory of Metallurgical Intelligent Equipment, Chongqing 401122, China)

Abstract: A finite element model was established to simulate the flying shear process of 40Mn strip steel, and the influence of key parameters of the flying shear on the shear was analyzed. Firstly, the kinematic analysis of the flying shear mechanism was carried out to obtain the motion path of the cutting edge in the shearing process. Then, a plane strain finite element model was established on DEFORM software to simulate the flying shear process of 40Mn strip steel. Johnson-Cook plastic constitutive model and damage model considering strain, strain rate hardening and temperature softening were adopted. Through the calculation and analysis of the finite element model, the influence law of the fillet radius of the cutting edge and the relative clearance of the cutting edge on the shearing work and the quality of the cutting section is obtained. Under the given strip size and flying shear conditions, the optimal cutting edge clearance and fillet radius are determined.

Key words: high speed flying shear; parameter analysis; finite element; DEFORM

轧机油膜轴承润滑系统稳定性控制

武明明，王金星，宋　波

(鞍钢股份有限公司冷轧厂，辽宁鞍山　114027)

摘　要：鞍钢冷轧厂联合机组轧机支持辊润滑系统为静-动压润滑系统。油膜轴承价格昂贵，而润滑系统的稳定性对油膜轴承起到至关重要的作用，冷轧厂轧机由于工艺需要，轧制时需要喷射乳化液进行润滑及冷却，这就造成支持辊油膜轴承进水，影响油脂精度，降低系统稳定性，本文讨论的是通过保证装配精度、密封系统优化、润滑系统优化来提高系统的稳定性。

关键词：轧机；支持辊；油膜轴承；润滑油；乳化液

冷轧20辊轧机张力闭环控制技术研究

秦大伟，张 栋，孙瑞琪，张 岩，刘宝权

（鞍钢集团北京研究院有限公司，北京 100010）

摘 要：以冷轧20辊轧机入口张力闭环控制为研究对象，采用响应曲线法建立控制系统模型并进行仿真研究。建立不同卷径状态下的控制对象模型并进行控制器设计，实现了张力自适应控制，达到了超调量小于2%，调节时间小于1s，稳态无静差的设计目标。针对在卷径1米以下薄带钢轧制期间卷取张力波动大的问题，提出卷取偏心补偿控制方法，消除卷取偏心带来的周期性扰动，提出前馈AGC转矩补偿控制方法，消除辊缝调节对张力的扰动。

关键词：冷轧；20辊轧机；直接张力；间接张力；响应曲线法

Research on Tension Closed-loop Control Technology of 20-high Cold Rolling Mill

QIN Dawei, ZHANG Dong, SUN Ruiqi, ZHANG Yan, LIU Baoquan

(Ansteel Group Beijing Research Institute Co., Ltd., Beijing 100010, China)

Abstract: Taking the entry tension closed-loop control of 20-high cold rolling mill as the research object, the control system model was established by response curve method and the simulation research was carried out. The control object model under different coil diameters is established and the controller is designed. The tension adaptive control is realized, and the design goal of overshoot less than 2 %, adjustment time less than 1s and steady state without static error is achieved. Aiming at the problem of large coiling tension deviation during the rolling of thin strip steel with a coil diameter of less than 1 meter, a coiling eccentricity compensation control method is proposed to eliminate the periodic disturbance caused by coiling eccentricity. A feed-forward AGC torque compensation control method is proposed to eliminate the disturbance of roll gap adjustment to tension.

Key words: cold rolling; 20-high rolling mill; direct tension; indirect tension; response curve method

冷连轧带钢变规格控制优化

潘建华，王 弢，柳 军

（鞍钢股份有限公司冷轧厂，辽宁鞍山 114021）

摘 要：动态变规格对于实现全连续轧制方式有着非常重要的意义，它不仅是全连续冷连轧机区别常规冷连轧机最明显的特征，也是全连续冷连轧机生产的核心技术。本文通过对鞍钢冷轧厂某连轧机组动态规格变换策略优化，不但能够保证轧机快速而准确的实现规格变换，有效的提高产品质量和成材率，而且可以通过楔形过渡区域厚度增厚，减少焊缝区域所受到的冲击，降低断带风险。

关键词：冷轧机；带钢变规格；张力控制

Optimization of Dynamic Gauge Changing Control for Tandem Cold Mill

PAN Jianhua, WANG Tao, LIU Jun

(Cold Rolling Mill, Ansteel Company, Ltd., Anshan 114021, China)

Abstract: Dynamic gauge changing is of great significance for realizing the full continuous rolling mode. It is not only the most obvious feature of the full continuous cold rolling mill different from the conventional cold rolling mill, but also the core technology of the full continuous cold rolling mill. By optimizing the dynamic specification transformation strategy of a continuous rolling mill in Angang Cold Rolling Plant, this paper can not only ensure that the rolling mill can realize the specification transformation quickly and accurately, effectively improve the product quality and yield, but also thicken the thickness of the wedge transition area, reduce the impact on the weld area, and reduce the risk of strip break.

Key words: cold rolling mill; dynamic gauge changing; tension control

冷轧厂退火炉辐射管烧嘴失效案例分析

姚 舜

（宝山钢铁股份有限公司冷轧厂，上海 200941）

摘 要： 宝钢某热镀锌机组投产 5 年后发现退火炉的辐射管烧嘴全部开裂，并在烧嘴盘区域引起辐射管大量开裂的现象，对生产顺行造成了很大的影响。冷轧厂与中央研究院、宝钢工程和国内烧嘴供应商一起合作，通过金相分析、建模仿真、试验室燃烧测试等方法，锁定了烧嘴失效的原因，并与外方交涉后形成了后续的解决方案。

关键词： 辐射管；烧嘴；开裂

Failure Case Analysisof Radiant Tube Burner of Annealing Furnace in Cold Rolling Mill

YAO Shun

(Baoshan Iron & Steel Co., Ltd., Cold Rolling Plantt, Shanghai 200941, China)

Abstract: Five years after a hot dip galvanizing line was put into operation in Baosteel, it was found that all the radiant tube burners of the annealing furnace were cracked, and a large number of radiant tube cracks were caused in the area of the burner plate, which had a great impact on the production of the line. In cooperation with Academia Sinica, Baosteel Engineering and domestic burner suppliers, the Cold Rolling Mill identified the cause of burner failure through metallographic analysis, modeling and simulation, and laboratory combustion test, and negotiated with foreign parties to form a follow-upsolution.

Key words: radianttube; burner; crack

650mm 棒材轧机 2 号立式机架地脚螺栓断裂原因分析与修复措施

刘 宏[1]，戴江波[2]

（1. 宝武襄阳重材有限公司，湖北襄阳　441105；
2. 中冶南方武汉钢铁设计研究院，湖北武汉　430223）

摘　要：轧机是轧材厂的主要设备，轧机系统是否稳定直接关系到产品的质量，进而关系到企业的效益。俗话说轧钢就是轧"设备"，轧材厂设备的功能精度是否能保证，直接影响整条轧线生产能否顺行。本文结合轧线 2 号 650mm V 轧机（轧制力矩 500kN·m）地脚螺栓断裂的现场实际，进行事故分析，提出低成本的解决方案并实施，取得了较好的经济效益。

关键词：连铸连轧；轧机；地脚螺栓；断裂；措施

Abstract: Rolling mill is the main equipment of rolling line. Whether the rolling mill system is stable or not is directly related to the quality of products, and then to the benefit of enterprises. As the saying goes, steel rolling is rolling "equipment". Whether the functional accuracy of the equipment in the rolling mill can be guaranteed directly affects the smooth production of the whole rolling line. In this paper, based on the field practice of the anchor bolt fracture of 2#650mmV rolling mill, the accident analysis is carried out, and the low-cost solution is put forward and implemented, which has achieved good economic benefits.

Key words: continuous casting and rolling; rolling mill; foundation bolt; fracture; measure

连续酸洗机组入口段带钢跑偏分析及解决方法

李建文，余卫军，魏　静

（武汉钢铁有限公司硅钢部，湖北武汉　430083）

摘　要：连续酸洗机组入口段一般采用上下双通道设计保证机组连续生产节奏。由于上部通道通钢线较长，设备辊系较多，带钢在甩尾时容易发生跑偏。本文介绍对某发生跑偏的连续酸洗机组设备辊系操控次序和安装精度状态分析，采取调整自动操控次序优化方法，极大的减少带钢甩尾跑偏发生。

关键词：连续机组；入口段；带钢甩尾；跑偏分析

Analysis and Control Method of Strip Deviation in the Entrance Section of Normalized Pickling Unit

LI Jianwen, YU Weijun, WEI Jing

(Silicon Steel Department of Wuhan Iron and Steel Co., Ltd., Wuhan 430083, China)

Abstract: The inlet section of the normalized pickling unit generally adopts the upper and lower double channel design to ensure the high-speed production rhythm of the unit. Due to the long steel wire in the upper channel and the large number of equipment rolls, the strip is prone to deviation when it flicks. This paper introduces the analysis of the control sequence and installation accuracy state of the roll system of a unit that has been deviated, and adopted the method of adjusting the control sequence optimization to greatly improve the occurrence of strip steel drifting.

Key words: high-speed unit; inlet section; strip tail drift; deviation analysis

多区段微尺度辊形电磁调控技术理论研究

杨庭松[1]，袁铁衡[1]，雍晟一[1]，孙文权[1]，荆丰伟[1]，

杜凤山[2]，许志强[2]，何安瑞[1]

（1. 北京科技大学，国家板带生产先进装备工程技术研究中心，北京　100083；
2. 燕山大学，国家冷轧板带装备及工艺工程技术研究中心，河北秦皇岛　066000）

摘　要：作为板带轧机辊缝形状控制的活跃因素，轧辊辊形影响着机组板形控制能力，决定了板形质量水平[1]。为实现冷轧超平板带材的稳顺制备，开发强功能、高功效的新辊形，配合弯辊、窜辊等板形控制手段，已成为应对中边浪、肋浪以及复合浪等多种板形缺陷、实现灵活多变的辊缝形状控制目标的关键方法。然而，机型、辊系的适配性约束与板宽、板厚的适轧性变动，使得辊形的高效性难以覆盖于全机型、全品规，适配性、适轧性与高效性间的权衡是轧辊辊形设计开发的一个共性问题。CVC[2]、Smartcrown[3]、UCM[4]、EVC[5]、ASR[6]等辊形虽然可配合轧辊弯窜策略，实现多样化的板形缺陷治理目标，但其一经磨损则辊形高效性难以保持，在一定程度上约束了此类辊形的使用时效性。相比之下，基于辊形柔性控制理念的 NIPCO 技术、VC 技术[7]、DSR 技术[8]虽然可借助其内源力驱动实现辊形调控，但其亦难长期服役于重载工况，即投入环境受限。由此，本研究提出了多区段微尺度辊形电磁调控技术[9-11]，借助轧辊内源磁致热力驱动的多区段布局来微尺度柔性调节轧辊辊形，进而改善板带板形质量。

多区段、灵活的辊缝控制是辊形电磁调控技术的特点，也是其研究的难点和热点。整理并归纳基本调控组元调控能力的参数化影响关系[12-14]，分析了双棒联控、三棒联控等典型多区联控模式下的辊形叠加关系，揭示了多区段电磁调控辊形的形成机理。提出了辊形电磁调控技术的多区联控方法，探讨了以辊缝形状控制需求为目标辊形的电磁调控辊形预设方法及预设流程。结果表明：无论是双棒联控，还是三棒联控，其多区联控辊形均可视为基本调控组元独立胀形后按空间位置叠加而成的综合辊形。针对辊形调控能力与电磁调控参数间的匹配关系，基于量纲分析理论推导了电磁棒温升功率与电磁调控参数的计算模型。基于自主搭建的板形电磁调控平台，证明了多区段微尺度辊形电磁调控技术及其控制方法的可行性，拓展了辊形电磁调控技术的应用范围，丰富了该技术理论并构建了多区段微尺度辊形电磁调控技术体系雏形。

关键词：辊形电磁调控技术；多区段；微尺度；辊缝形状控制

参 考 文 献

[1] 孔繁甫, 何安瑞, 邵健, 等. 机械工程学报, 2012, 48(22): 87-92.
[2] Ding J, He Y, Song M, et al. The International Journal of Advanced Manufacturing Technology, 2021, 113: 87-97.
[3] Wang Q, Li X, Sun J, et al. Journal of Manufacturing Processes, 2021, 69: 451-472.
[4] 张建雷, 陈卫, 陆佳栋, 等. 钢铁, 2022, 57(8): 123-131.
[5] Cao J, Chai X, Li Y, et al. Journal of Materials Processing Technology, 2017, 252: 432-439.
[6] Cao J, Wei G, Zhang J, et al. Journal of Central South University. 2008, 15(2): 264-270.
[7] 白振华, 王凯, 王亚军, 等. 中国机械工程, 2013, 24(22): 3096-3100.

[8] 刘志亮, 李文强, 王英杰. 中国机械工程, 2011, 22(13): 1624-1628.
[9] Liu W, Feng Y, Yang T, et al. International Journal of Heat and Mass Transfer, 2022, 189: 172-183.
[10] 杜凤山, 刘文文, 冯岩峰, 等. 工程科学学报, 2017, 39(12): 1874-1880.
[11] Feng Y, Liu W, Yang T, et al. Metallurgical Research and Technology, 2019, 116(4): 405.
[12] Yang T, Liu J, Zhou H, et al. The International Journal of Advanced Manufacturing Technology, 2021, 116(1-2): 403-415.
[13] Yang T, Wang Y, Liu J, et al. The International Journal of Advanced Manufacturing Technology, 2021, 114(3-4): 1065-1074.
[14] Yang T, Liu J, Ren X, et al. Metallurgical Research & Technology, 2021, 118(3): 305-313.

轧机齿轮箱输入轴轴承外圈偏心套压溃分析

李建文，熊　勇

（武汉钢铁有限公司硅钢部，湖北武汉　430083）

摘　要：轴承是轧机齿轮箱的主要组成部件，也是主要易损件，轴承质量好坏和使用寿命对轧制成本有着主要影响。本文通过对齿轮箱输入轴轴承外圈偏心套压溃问题机理的分析，从根本上找出偏心套内孔与轴承外圈发生压溃的原因，并指出设计存在的不当。给出适当防控措施，预防齿轮箱输入轴轴承外圈偏心套压溃问题发生，提高轧机齿轮箱运行的可靠性和使用效率。

关键词：齿轮箱；输入轴；偏心套；轴承；压溃分析

Crushing Analysis of Eccentric Outer Ring of Bearing Input Shaft of Rolling Mill Gear Box

LI Jianwen, XIONG Yong

(Silicon Steel Department, Wuhan Iron and Steel Co., Ltd., Wuhan 430083, China)

Abstract: Bearing is the main component of the gear box of rolling mill, and it is also the main wearing part. The quality and service life of bearing have the main influence on the rolling cost. This paper analyzes the mechanism of eccentric sleeve crushing of gear box input shaft bearing outer ring, finds out the reason of the crush between eccentric sleeve hole and bearing outer ring, and points out the improper design. The proper prevention and control measures are given to prevent the eccentric sleeve of gear box input shaft bearing from collapsing, and to improve the reliability and efficiency of gear box operation.

Key words: gear box; input axis; eccentric sleeve; bearings; crushing analysis

转炉支撑机构现状研究与趋势分析

余　杨，王　庆，曾　鸣

（中冶赛迪工程技术股份有限公司炼钢事业部，重庆　401122）

摘　要：转炉运行品质是指有无磨损、冲击、振动和异响等安全性指标，与转炉支撑机构密切相关。本文系统地梳理和分析了从 20 世纪至今各国转炉支撑机构的特点，归纳性描述了转炉支撑机构特点并决定转炉特别是大型转炉运行品质的主要因素。采用综合对比分析方法，首次提出支撑机构影响并决定转炉特别是大型转炉运行品质的主要因素。结果表明，当今各国主要采用六种典型转炉下悬挂支撑机构。影响转炉运行品质的主要因素：（1）吊挂机构部件；（2）吊挂机构数量；（3）吊挂机构布置。在这三方面技术进步以及如何综合运用是提高转炉运行品质的发展趋势和方向。

关键词：转炉；支撑机构；运行品质；振动和异响

Current Research and Trend Analysis of Converter Support Mechanism

YU Yang, WANG Qing, ZENG Ming

(CISDI Engineering Co., Ltd., Steelmaking Business Department, Chongqing 401122, China)

Abstract: The running quality of the converter refers to the safety indexes such as whether there is wear, impact, vibration and abnormal noise, etc., which is closely related to the supporting mechanism of the converter. This paper systematically sorts out and analyzes the characteristics of the supporting mechanism of the converter in various countries from the last century to the present, and summarizes the characteristics of the supporting mechanism of the converter and the main factors that determine the running quality of the converter, especially the large converter. The main factors that influence and determine the running quality of converter, especially large converter, are put forward for the first time by means of comprehensive comparative analysis. The results show that six typical under suspension support mechanisms are mainly used in various countries in the world today; the main factors affecting the running quality of converter are: (1) the components of the hanging mechanism, (2) the number of the hanging mechanism, and (3) the layout of the hanging mechanism. The technical progress in these three aspects and how to use them comprehensively are the development trend and direction of improving the running quality of converter.

Key words: converter; support mechanism; running quality; vibration and abnormal noise

转炉运行品质影响因素分析和研究

余 杨，王 庆，曾 鸣

（中冶赛迪工程技术股份有限公司炼钢事业部，重庆　401122）

摘　要：转炉特别是 100～500t 中大型转炉运行品质指无明显磨损、冲击、振动和异响等安全性指标以及设备运行平稳，具有足够的安全裕度和稳定性等。本文描述了当今世界主要采用转炉下悬挂支撑机构的六种典型结构形式，分别从吊挂机构部件、吊挂机构数量和吊挂机构布置三个方面分析了转炉下悬挂支撑机构的特点；采用有限元分析手段分析了支撑机构应力和变形，并结合大量实际工程验证。研究结果表明，转炉吊挂机构部件的最优选择是采用连杆机构；吊挂机构数量的最优组合是采用四组垂直吊挂和四组水平支撑机构；吊挂机构布置的最优位置是水平支撑机构位于垂直吊挂机构的正上方。如果以上三条都满足，转炉运行品质必定优异，也意味着是今后转炉支撑技术发展的趋势和方向。

关键词：转炉；支撑机构；连杆机构；运行品质

Analysis and Research on Influencing Factors of Running Quality of Converter

YU Yang, WANG Qing, ZENG Ming

(CISDI Engineering Co., Ltd., Steelmaking Business Department, Chongging 401122, China)

Abstract: The running quality of converter especially 100 ~ 500t medium and large converter refers to the safety indicators such as no obvious wear, impact, vibration and abnormal noise, and the stable operation of the equipment with sufficient safety margin and stability, etc. In this paper, six typical under suspension support mechanisms which mainly used in various countries in the world today are described, and the characteristics of the suspension support mechanism under the converter are analyzed from three aspects: components, number of suspension support mechanism and layout of suspension support mechanism. The stress and deformation of the supporting mechanism are analyzed by finite element method and verified by a large number of practical engineering. The research results show that the connecting rod mechanism is the best choice for the hanging mechanism of converter. The optimal combination of the number of hanging mechanisms is to adopt four groups of vertical hanging mechanisms and four groups of horizontal supporting mechanisms; the optimal position of the hanging mechanism is that the horizontal support mechanism is located directly above the vertical hanging mechanism. The running quality of the converter must be excellent if all three of the above are met, which also means the trend and direction of the development of the support technology of the converter in the future.

Key words: converter; support mechanism; link mechanism; running quality

滚筒飞剪传动计算研究

王 蕾，常铁柱，张海东，韩 文，黄晓慧

（中冶京诚工程技术有限公司，北京 100176）

摘 要：介绍滚筒式飞剪工作原理及设备结构，分析滚筒式飞剪剪切功计算，开发出一种滚筒飞剪设计软件。本软件运行后，输入钢板厚度等一系列参数，经程序分析计算，得到合适的飞剪传动电机功率，对飞剪设计工作起到很好的指导作用。

关键词：剪切；剪切力；飞轮矩；设计

Study on Calculation of Roller Flying Shear Transmission

WANG Lei, CHANG Tiezhu, ZHANG Haidong, HAN Wen, HUANG Xiaohui

(MCC Capital Engineering & Research Incorporation Limited, Beijing 100176, China)

Abstract: This paper introduces the working principle and equipment structure of drum flying shear, analyzes the calculation of shearing work of drum flying shear, and develops a kind of software for drum flying shear. After the software runs, a series of parameters, such as steel plate thickness, are inputted, and the appropriate power of flying shear drive motor is obtained through program analysis and calculation, which plays a good guiding role in flying shear design.

Key words: shear; shear force; flywheel moment; design

"一罐到底"铁水罐罐盖开闭盖装置的研制

项克舜

(武汉钢铁有限公司运输部，湖北武汉　430083)

摘　要：文章介绍了"一罐到底"110t铁水罐加盖装置的设计要求及各个组成部分的结构。加盖装置的动力源为车载液压站，地面接电装置和车载受电装置的连结采用自动接电方式。运用表明，铁水罐加盖后，保温效果明显，无可见烟尘排出，在提高铁水温度、保护铁罐耐材等方面均有良好效果，对钢铁厂降本增效发挥了积极作用，自动接电装置解决了铁水车空重车高度变化和停车精度难题。

关键词：一罐到底；铁水罐；开闭盖装置；自动接电装置

The Development of "One Can to the End" Iron Water Can Lid Opening and Closing Device

XIANG Keshun

(Transportation Department of Wuhan Iron and Steel Co., Ltd., Wuhan 430083, China)

Abstract: The article introduces the design requirements and the structure of each component of the "one tank to the end" 110t iron water tank capping device. The power source of the capping device is the truck-mounted hydraulic station, and the connection between the ground receiving device and the truck-mounted receiving device adopts the automatic power connection method. The application shows that, after the lid of the iron tank, the heat preservation effect is obvious, no visible soot discharge, good effect in improving the temperature of the iron and protecting the resistant material of the iron tank, etc., which plays a positive role in reducing the cost and increasing the efficiency of the iron and steel plant, and the automatic power receiving device solves the problem of height change and stopping accuracy of the empty heavy car of the iron tank.

Key words: one can to the end; iron water cans; opening and closing cover device; automatic power connection device

宝钢冷轧激光焊机机械精度智能检测装备的研发

周为民[1]，屈军杰[2]，赵春涛[1]，潘红良[2]

(1. 宝山钢铁股份有限公司冷轧厂，上海　200299；
2. 上海务宝机电科技有限公司智能装备部，上海　200299)

摘　要：针对激光焊接机精度不足而引发的冷轧带钢焊缝不良问题，设计开发了焊机状态智能检测装备代替现场人工测量，提高了测试精准度、降低了操作难度和安全风险。检测装置由滑移平台、传感器固定工装、滑移平台吸盘、2D激光轮廓测量仪、通讯电缆、PC机和校准工装组成，实现了激光焊机双切剪刀片平行度检测、焊接导向轮侧平行度检测、焊接导向轮/预平整轮/焊接平整轮高度的检测。

关键词：激光焊机；状态；智能检测；精度

R&D of Intelligent Inspection Equipment for Welder

ZHOU Weimin[1], QU Junjie[2], ZHAO Chuntao[1], PAN Hongliang[2]

(1. Cold Rolling Mill, Baoshan Steel Co. Ltd., Shanghai 200299, China; 2. Intelligent Equipment Department, Shanghai Wubao Mechatronics Technology Co., Ltd., Shanghai 200299, China)

Abstract: In view of the poor weld of cold rolled strip steel caused by the insufficient accuracy of laser welding machine, intelligent inspection equipment for welder is designed and developed to replace the field manual measurement, which improve the test accuracy, reduce the difficulty of operation and safety risks. The detection device is composed of slip platform, sensor fixed tooling, slip platform sucker, 2D laser profile measuring instrument, communication cable, PC and calibration tooling. The detection device realizes the parallelism detection of the double-cut scissors of the laser welding machine, the parallelism detection of the side of the welding guide wheel, and the height detection of the welding guide wheel/pre-leveling wheel/welding leveling wheel.

Key words: laser welding machine; state; intelligent detection; accuracy

中频感应熔炼炉升级改造研究

徐凤娟，王东海，耿 军，赵福星，杨晓勇，王国良

（通化钢铁股份有限公司，吉林通化 134003）

摘 要：本文介绍了通化钢铁股份有限公司以往使用的中频感应熔炼炉（简称中频炉）中存在的问题，开展相应对策分析，继而按照生产工艺和需求制定了改造设计要求，从炉体、控制系统等两个方向入手，将铝制炉体改造升级为减速机钢壳炉体，选用倍压谐振中频电源，增加了冷却水流量温度保护报警装置，使中频炉投入生产时操作更加方便，运行更加稳定，更加安全可靠。通过改造后一年半的运行数据监测，可以发现，改造后实现了能耗节约5%～8%、电源稳定性提升至98.8%，可控性达99.5%，流量报警5次、超温报警7次，及时有效的防止由于冷却水水流量、水压力和水温度不正常引起的生产事故，为中频炉的生产提供了坚实的保障，有效提高生产效益，企业经济效益得到同步提升，也为同类产线改造提供了参考和借鉴。

关键词：中频炉；改造升级；控制系统；电源

Research on Upgrading and Renovation of Medium Frequency Induction Melting Furnace

XU Fengjuan, WANG Donghai, GENG Jun, ZHAO Fuxing,
YANG Xiaoyong, WANG Guoliang

(Tonghua Steel Co., Ltd., Tonghua 134003, China)

Abstract: This article introduces the problems existing in the medium frequency induction melting furnace (referred to as the medium frequency furnace) used in the past by Tonghua Steel Co., Ltd., Tonghua City, conducts corresponding countermeasures analysis, and then formulates transformation design requirements based on production process and requirements. Starting from the furnace body and control system, the aluminum furnace body is upgraded to a reducer steel shell furnace body, and a double voltage resonant medium frequency power supply is selected. A cooling water flow temperature protection alarm device is added, Make the operation of the intermediate frequency furnace more convenient, stable, and safe when put into production. Through one and a half years of operation data monitoring after the renovation, it can be found that the renovation has achieved energy savings of 5%~8%, power stability improvement to 98.8%, controllability up to 99.5%, flow alarm 5 times, and overtemperature alarm 7 times, timely and effectively preventing production accidents caused by abnormal cooling water flow, water pressure, and water temperature, providing a solid guarantee for the production of intermediate frequency furnaces and effectively improving production efficiency, The economic benefits of enterprises have been synchronously improved, providing reference and reference for the transformation of similar production lines.

Key words: medium frequency furnace; renovation and upgrading; control system; source

绿色数智化流体系统推进酸轧联合机组工艺技术变革再造

毛召芝，沈志前，张海东，常铁柱，

贾大朋，马丽坤，黄晓慧

（中冶京诚工程技术有限公司，北京 100176）

摘 要： 近二十年来，中冶京诚酸轧联合机组技术取得了令人瞩目的进步，伴随着新中国钢铁工业一同成长，紧抓酸轧联合机组发展机遇，发挥多专业技术协同优势，历经众多工程洗礼，与用户一起成长，率先完成全流程酸轧联合机组生产工艺装备自主集成。日趋成熟的绿色钢铁解决方案云集多项自主知识产权，以流体系统装备为重要支撑，深度融合专属中冶京诚的低碳绿色和数智化DNA，以数智化和装备创新为抓手，在碳减排、污染控制、能效提升、循环利用等诸多维度推动酸轧联合机组工艺技术变革再造。

关键词： 酸轧机组；绿色数智化；流体系统；技术及装备创新；工艺技术变革再造

Green Digital Intelligent Fluid System Promotes Process Technology Transformation and Reconstruction of PL-TCM

MAO Zhaozhi, SHEN Zhiqian, ZHANG Haidong, CHANG Tiezhu,

JIA Dapeng, MA Likun, HUANG Xiaohui

(Capital Engineering & Research Incorporation Limited, Beijing 100176, China)

Abstract: In the past two decades, the technology of MCC Jingcheng acid rolling combined unit has made remarkable progress. With the development of new China Steel's iron and steel industry, it has seized the development opportunity of acid rolling combined unit, given play to the advantages of multi professional technology synergy, experienced many engineering baptisms, and grown with users, taking the lead in completing the independent integration of production process equipment of the whole process acid rolling combined unit. The increasingly mature green steel solutions gather multiple independent intellectual property rights, with fluid system equipment as an important support, deeply integrating the exclusive low-carbon green and digital DNA of MCC Jingcheng. With digital and equipment innovation as the starting point, they promote the transformation and reconstruction of acid rolling combined unit process technology in various dimensions such as carbon emission reduction, pollution control, energy efficiency improvement, and recycling.

Key words: cold rolling mill; emulsion system; emulsion injection; low pressure loss; technological and equipment innovation

电工钢常化酸洗机组设备组成及特点

马丽坤，谢天华，耿庆斌，张 华

（中冶京诚工程技术有限公司轧钢工程技术所，北京 100176）

摘 要： 不同牌号的电工钢生产工序不同，本文针对高牌号无取向电工钢和高磁感取向电工钢的常化酸洗机组，主要介绍常化酸洗机组的工艺流程、设备组成及特点。

关键词： 电工钢；常化酸洗；设备组成

Equipment Composition and Characteristic of the Normalizing and Pickling Line for Electrical Steel

MA Likun, XIE Tianhua, GENG Qingbin, ZHANG Hua

(Capital Engineering & Research Incorportation Limited Steel Rolling Division, Beijing 100176, China)

Abstract: Different grades of electrical steel have different production processes. This article mainly introduces the technological process, equipment composition and characteristics of the normalizing and pickling line for high grade non oriented electrical steel and high magnetic induction oriented electrical steel.

Key words: electrical steel; normalizing and pickling; equipment composition

薄带铸轧线分卷飞剪机的设计与应用

常铁柱，王 蕾，张海东，黄晓慧

（中冶京诚工程技术有限公司，北京 100176）

摘 要： 飞剪装备是短流程超薄带铸轧连续生产线上对带材进行分卷的关键设备，通过"滚切"加"斩切"双剪切模式组合实现了厚料和薄料的分别剪切，既有利于降低剪切冲击力，又提高了剪切精度；通过双侧齿轮反向消隙传动结构，以较小的刀轴直径获得更大的抗扭刚度，提高剪刃侧隙的均匀性；通过建立剪切速度的精确控制模型实现了剪切过程的稳定、可靠运行，投产后，运行效果良好。

关键词： 铸轧；飞剪；超薄带

A Type of Flying Shear for Ultra-thin Strip Casting Rolling Line

CHANG Tiezhu, WANG Lei, ZHANG Haidong, HUANG Xiaohui

(MCC Capital Engineering & Research Incorporation Limited, Beijing 100176, China)

Abstract: The flying shear is a key equipment to separate the strip in the continuous production line of short process ultra-thin strip casting and rolling. The combination of "rolling cut" and "chopping cut" dual shear modes is designed to cut the thick and thin materials independently, which is beneficial for reducing shear impact force and improving shear accuracy; By using a dual side gears with reverse clearance elimination, a smaller shaft diameter is obtained with greater torsional stiffness and uniformity of the shear edge backlash; By establishing an accurate control model for shear speed, a stable and reliable operation of the shear process was achieved. After in use, the using effects were good.

Key words: casting rolling line; flying shear; ultra-thin strip

160T 铁水车轮对烧轴故障分析及对策

易文祥

（柳钢铁路运输中心，广西柳州 545000）

摘　要：铁水车作为冶金钢铁企业中重要的铁水运输工具，由于作业环境和产品质量等因素的影响，铁水车轮对轴颈与轴承可能出现异常，甚至发生烧轴现象。为了确保企业铁路运输的安全、稳定、优质和高效发展，有必要规范铁水车轮对轴承的运行管理，并制定可行的有效措施来解决这个问题。

在这种背景下，本文使用了 PDCA 循环法来解决铁水车轮对烧轴问题。具体来说，PDCA 循环法被用于从人、机、料、法、环等多个角度深入分析问题，并提出解决问题的对策和思路。PDCA 循环法的四个步骤包括：计划（Plan）：针对铁水车轮对烧轴问题，可能涉及设定减少烧轴发生率的目标，收集与分析轮对轴颈烧轴的数据，并制定解决问题的计划。执行（Do）：在这个阶段，根据计划开始执行。采取一些针对性的操作，比如对轮对轴颈和轴承进行维护，改善作业环境等。检查（Check）：通过比较实际执行情况与预期目标，收集数据和反馈，来判断是否已经减少了烧轴现象。行动（Act）：如果烧轴现象得到改善，可以将成功的方法固化为标准操作。如果问题没有得到完全解决，可以分析原因，提出改进的措施，然后制定新的计划。通过这个 PDCA 循环的过程，可以不断地优化解决问题的方法，逐步提高铁水车轮对烧轴问题的处理效果。最终，通过持续的改进，为冶金企业的铁水运输工作带来稳定性、安全性和效率的提升，对冶金企业铁水运输工作的日常维护保养及降本工作具有借鉴意义。

关键词：铁水车烧轴；故障分析；思路及对策；检修工艺

精品轨梁轧制工艺、装备与控制关键技术研发及应用

陈莹卷[1]，周　民[2]，樊泽兴[1]，郭显平[3]，谭成楠[1]，马靳江[1]

（1. 中冶赛迪工程技术股份有限公司，重庆 401122；2. 中冶赛迪技术研究中心有限公司，

重庆　401122；3. 中冶赛迪装备有限公司，重庆　401122）

摘　要：轨梁产品是高速铁路、绿色装配式建筑及大型钢结构等国家重大工程建设的核心原材料，在国民经济建设中的作用不可替代。轨梁属于复杂断面型钢，使用工况特殊，对尺寸精度要求极高，产品规格范围广、变形过程中金属流动不对称、截面温度不均匀，生产过程具有多变量、强耦合的特点，尺寸精准控制和高效生产是行业面临的世界性难题。采用国外技术投资成本高、建设周期长，无法适应我国高效高品质轨梁生产需求。

针对上述问题，作者采用有限元模拟、数值分析、实验室实验及工业化验证等方法开展了非往复组合式多机架万能连轧工艺、四机架往复可逆万能轧制工艺研究；构建了 H 型钢、钢轨轧制高精度限制宽展及力能模型，开发了多模式万能轧制及轧后控制冷却技术，并基于小曲率平面弯曲弹复理论提出多辊矫直压下量与曲率关系模型及工艺规程；研究了万能轧机等刚度机架构型及整机耦合刚度分析方法，开发了自动轴向调整及轴承座载荷均布技术，开展了轧机装备系列化参数设计，开发出开坯轧机、万能轧机及矫直机系列化机组；研发出万能轧机升降组合式辊系组、矫直机辊环锁紧机构、万能轧机及矫直机快速换辊装置；构建了机架间微张力与轨梁尺寸精度关系模型，开发出万能轧机工艺控制、液压伺服及智能化适应性控制系统。

研究成果已应用于永洋轻轨、永洋重轨及印度 AARTI 等国内外 10 余条生产线。所生产的轨梁产品尺寸精度高、平直度好、力学性能稳定，广泛应用于高速铁路、各类矿山、码头等，并远销至东南亚、非洲、南美洲等"一带一路"沿线地区。研究成果提升了我国复杂断面型钢高端制造水平，引领了精品轨梁生产技术发展，打破了国外垄断，实现了复杂断面型钢核心技术装备自主可控。

参 考 文 献

[1] 谭成楠, 周民, 马靳江, 等. EH36 球扁钢变形抗力模型研究[J]. 热加工工艺, 2015, 44(3): 178-180.
[2] 龚殿尧, 高志宇, 徐建忠, 等. 小型H型钢超快冷"内并外扩"的有限元模拟[J]. 钢铁, 2022, 57(1): 102-109.
[3] Schleinzer G, Fischer F D. Residual stress formation during the roller straightening of railway rails[J].. International Journal of Mechanical Sciences, 2001, 43(10): 2281-2295.
[4] 殷璟, 赵军, 李跃林, 等. 多辊矫直压下规程的制定[J]. 轧钢, 2013, 30(4): 19-23.

无取向硅钢中超低硫检出下限的延伸及其准确性研究

易赛，王鹏，齐郁，叶静，张兆雄，李三红

（宝钢股份武汉钢铁有限公司质检中心，湖北武汉　430080）

摘　要：本文通过实验研究，以燃烧峰形、分析精度比对为研究主要判定依据，确定合适的研究设备美国 LECO 公司红外碳硫分析仪 LECO CS-844，标准物质 YSBC201117c-2009、YSB14230d-2017 进行线性校准，标准物质 GBW01401a 进行确认，分析方法 5-LOW-CS（集成时间 65s，炉子功率 85%），实现了对钢中超低硫检出下限由 0.001% 延伸至 0.0005% 以下及其准确性的确认，拓展了 CS-844 在超低硫分析中的应用，满足了不断提升的市场化需求。
关键词：红外碳硫分析仪；超低硫；检出下限；准确性

Study on the Extension and Accuracy of the Detection Lower Limit for Ultra Low Sulfur in Non-oriented Silicon Steel

YI Sai, WANG Peng, QI Yu, YE Jing, ZHANG Zhaoxiong, LI Sanhong

(Quality Inspection Center of Baosteel Wuhan Iron and Steel Co., Ltd., Wuhan 430080, China)

Abstract: Through experimental research and based on the peak shape of combustion and the comparison of analysis accuracy., this paper determines the appropriate research equipment the LECO infrared carbon-sulfur analyzer LECO CS-844, linear calibration of standard substances YSBC201117c-2009 and YSB14230d-2017,confirmation of standard substance GBW01401a,the analysis method 5-LOW-CS (integration time of 65 seconds,furnace power of 85%), achieving the confirmation of the detection limit of ultra-low sulfur in steel from 0.001% to below 0.0005% and its accuracy, extending the application of CS-844 in ultra-low sulfur analysis and meeting the constantly improving market demand.

Key words: infrared carbon and sulfur analyzer; ultra low sulfur; detection lower limit; accuracy

智能型多功能油液监测分离的控制技术

沈　强，任佩剑，付乾坤，董耀元，曲晓东，王　健

（鞍钢股份热轧带钢厂，辽宁鞍山　114021）

摘　要： 介绍了鞍钢股份公司热轧带钢厂液压系统的设备组成，由于热轧企业工作环境较为恶劣，液压系统极易出现进水现象并发生油液乳化问题，针对该问题进行了详细分析，并提出了智能型多功能油液监测分离的控制技术，说明了智能型多功能油液监测分离的控制技术的组成，通过关键技术现场创新改造，采用智能化控制，实现油水快速分离、油品污染度实时监测、循环过滤、定时加温等多项功能。创新后消除因系统进水所产生的液压油乳化问题，大大减少液压冲击现象发生，进而减少泄漏事故发生，保证了液压系统稳定正常工作，同时大大减少处理泄漏和更换液压元件所产生的事故时间，节省能源损耗、降低备件成本、延长了液压元件的使用周期，达到降本增效的目的。

关键词： 液压油；乳化；液压系统；泄漏

Control Technology of Intelligent Multifunctional Oil-water Monitoring and Separation

SHEN Qiang, REN Peijian, FU Qiankun, DONG Yaoyuan, QU Xiaodong, WANG Jian

(Angang Steel Company Limited, Hot Rolled Strip Plant, Anshan 114021, China)

Abstract: This article introduces the equipment composition of the hydraulic system in the hot rolling strip steel plant of Ansteel Co., Ltd. Due to the harsh working environment of the hot rolling enterprise, the hydraulic system is prone to water ingress and oil emulsification problems. A detailed analysis is conducted to address this issue, and an intelligent multi-functional oil monitoring and separation control technology is proposed. The composition of the intelligent multi-functional oil monitoring and separation control technology is explained. Through on-site innovation and transformation of key technologies, Adopting intelligent control, it achieves multiple functions such as rapid separation of oil and water, real-time monitoring of oil pollution, cyclic filtration, and timed heating. After innovation, the problem of hydraulic oil emulsification caused by system water ingress is eliminated, greatly reducing the occurrence of hydraulic impact, thereby reducing leakage accidents and ensuring the stable and normal operation of the hydraulic system. At the same time, the accident time caused by handling leaks and replacing hydraulic components is greatly reduced, energy consumption is saved, spare parts cost is reduced, and the service cycle of hydraulic components is extended, achieving the goal of cost reduction and efficiency increase.

Key words: hydraulic oil; emulsification; hydraulic system; leakage

低碳背景下废钢铁及 LIBS 技术在废钢判级的研究进展

刘艳丽，刘 洁，安治国

（河钢材料技术研究院，河北石家庄 050000）

摘 要： 回顾了炼钢技术发展历程及电炉炼钢现状，介绍了河钢集团的低碳发展路径；对钢铁材料快速检测分类大大提高钢铁废弃物的回收利用率；详细分析了废钢铁常用分类方法及标准，并对新旧标准进行了对比解读；依次从废钢外观、机器视觉、锈蚀程度、成分定量分析方面对现场废钢分类现状展开了调研；结合激光诱导击穿光谱（LIBS）技术的优势阐述了其在冶金领域分类技术研究进展，同时总结了手持 LIBS 废钢分类应用研究。对河钢集团基于 LIBS 技术开展废钢分类工作进行了介绍，根据成分判断集成了基于 Stacking 集成学习方法建立多元素定量分析模型，准确率接近 100%；根据光谱直接判断的方式预测准确率大于 98%；最后对 LIBS 技术在废钢分类方面做出展望（引用文献 91 篇）。

关键词： 废钢；钢铁；分类；判级；激光诱导击穿光谱；低碳；电炉

Research Progress of Scrap Steel and LIBS Technology in Scrap Grading under Low Carbon Background

LIU Yanli, LIU Jie, AN Zhiguo

(HBIS Materials Technology Research Institute, Shijiazhuang 050000, China)

Abstract: The development history of steelmaking technology and the current situation of electric furnace steelmaking are reviewed, and the low-carbon development path of Hegang Group is introduced; Rapid detection and classification of steel materials can greatly improve the recycling rate of steel waste. The common classification methods and standards of scrap steel are analyzed in detail, and the new and old standards are compared and interpreted. The current situation of scrap classification on site was investigated from the aspects of scrap appearance, machine vision, corrosion degree and quantitative analysis of components; Combined with the advantages of laser induced breakdown spectroscopy (LIBS) technology, the research progress of LIBS classification technology in metallurgical field is described, and the application research of handheld LIBS scrap classification is summarized. The classification of scrap steel based on LIBS technology in Hegang Group is introduced, According to the component judgment, a multi-element quantitative analysis model is built based on the Stacking integrated learning method, and the accuracy rate is close to 100%; The prediction accuracy of direct judgment based on spectrum is more than 98%; Finally, the prospect of LIBS technology in scrap classification is made. (91

Key words: steel scrap; steel; grading; sentence level; laser induced breakdown spectroscopy; low carbon; electric furnace

新型低重量容积比钢包设计与应用

杨 阳，徐 勇

（北京中冶设备研究设计总院有限公司，北京 100029）

摘 要： 在不降低结构强度的前提下，实现钢包扩容减重是钢铁企业响应国家双碳政策的措施之一。通过分析钢厂工艺现状，本文提出了一种钢包扩容与优化减重的方法。通过该方法设计出的钢包其重量容积比低至 0.16~0.18，较同类型钢包重量减轻 10%~20%。该成果已成功应用于国内某钢厂，设备平稳运行近两年。

关键词： 钢包；重量容积比；有限元分析

Design and Application of the Navel Low Weight-Volume Ratio Ladle

YANG Yang, XU Yong

(Beijing Metallurgical Equipment Research Design Institute Co., Ltd., Beijing 100029, China)

Abstract: Achieving ladle expansion and weight reduction without reducing structural strength is one of the measures for steel enterprises to respond to the national dual-carbon policy. By analyzing the current situation of the steel plant process, this paper proposes a method of ladle capacity expansion and weight reduction optimization. The ladle designed by this method has a weight-volume ratio as low as 0.16-0.18, which is 10%-20% less than the weight of the same type of ladle. The results have been successfully applied to a domestic steel mill, and the equipment has been running smoothly for nearly two years.

Key words: ladle; weight-volume ratio; finite element analysis

中厚板电磁超声探伤设备的开发与应用

徐 科，田士辉，姜家宏，师英杰

（北京科技大学钢铁共性技术协同创新中心，北京 100083）

摘 要： 中厚板作为一种重要的结构材料，在工程建设、交通运输、工业制造等领域发挥着关键的作用。它的性能和质量直接影响到使用场景的安全性和可靠性。随着国民经济的发展，用户对中厚板质量的要求越来越严格。中厚板的内部质量主要通过超声波探伤手段进行检测。国外从20世纪60年代末期开始对超声波探伤技术进行了研发，70年代后期，自动超声波探伤设备在中厚板生产线上得到了应用。但是，传统的超声波探伤设备采用压电探头（Piezoelectric Transducer，简称 PZT），存在适用钢板温度范围窄、需消耗大量耦合水、易受干扰等问题，限制了PZT探伤设备在中厚板生产中的应用。

电磁超声换能器（Electromagnetic Acoustic Transducer，简称 EMAT）是一种新型的超声波激发和接收装置。与 PZT 通过压电晶片的压电效应来发射和接收超声波不同，EMAT 通过电磁效应激发和接收超声波。在非铁磁性材料中，EMAT 通过洛伦兹力激发超声波；在铁磁性材料中，EMAT 通过洛伦兹力和磁致伸缩力共同作用激发超声波。因此，EMAT 不需要耦合剂，并且可以非接触测量，所以非常适用于高温在线检测的应用场合。

本文介绍一套目前已经国产化的中厚板电磁超声探伤设备，应用于国内某钢厂 3800 宽厚板生产线。系统采用非接触式 EMAT 横波探头，每个探头有 8 个激活线圈，产生 8 个独立通道，以与钢板表面垂直方向发射和接收超声波。板体检测系统（BTS）由两排 EMAT 组成，每排 BTS 配置有 30 个 EMAT，共 60 个 EMAT，480 个通道。两排 BTS 探头以交叉形式排布，保证探头有效检测区域达到 105%的覆盖率。纵向边缘检测系统（ETS）检测钢板的左右纵向边缘，左右边缘各采用 1 个 16 通道的探头，以提高小缺陷的检测灵敏度。ETS 可以根据钢板的实际宽度自动调整。系统可检测钢板厚度 5~150mm，检测速度 0~2m/s，检测钢板温度范围–20~350℃。

与 PZT 探伤设备相比，EMAT 探伤设备具有以下优点：（1）不需要耦合介质。相比于 PZT 需要水做耦合介质，EMAT 不需要耦合剂，不仅避免水资源的浪费，而且消除了由于水喷洒在钢板以及其它设备上造成的锈蚀和污染。（2）适用温度范围广。由于 PZT 使用水做耦合剂，检测温度一般不超过 100℃，所以需要对钢板做下线处理，这必然影响生产节奏。而 EMAT 适用温度范围广，可用于冷床后钢板的在线探伤。（3）检测速度快。当钢板通过速度过快时，PZT 与钢板表面之间会形成气泡使耦合失效，而 EMAT 没有这方面的问题，可大大提高检测速度。（4）抗干扰能力强。耦合水的声学噪音会对 PZT 探伤造成一定干扰，产生误判；并且 PZT 与钢板表面不垂直时就会造成声场的失真，产生漏检。而根据电磁效应原理，即使 EMAT 与钢板形成一定倾角的情况下，仍能保证超声波发射和接收的方向垂直于钢板表面，不会造成声场的失真。（5）可用横波检测。PZT 一般使用液态耦合剂，由于横波不能在液体中传播，所以 PZT 很难激发横波，一般用纵波检测。而 EMAT 很容易激发横波。横波的声速仅为纵波的一半左右，所以缺陷检测分辨率更高。

接触式板形仪应用技术综述和发展趋势浅析

俞鸿毅[1]，王 劲[1]，石 燕[2]

（1. 宝山钢铁股份有限公司冷轧厂，上海 201900；
2. 欧冶工业品股份有限公司华东大区，上海 201900）

摘 要：接触式板形仪是高端冷轧带钢必备核心设备之一。本文简要介绍了板形检测和控制系统的基本组成，详细叙述了冷轧产线主要应用的接触式板形辊检测和信号传输技术。内容包括传感器形式和布置方法，信号传输原理和辊面状态等。同时，提出了合理布置传感器、激励和信号无线化、辊面一体无缝等发展趋势的一些认识。

关键词：板形仪；压力传感器；整辊无缝；无线传输

Overview of the Application Technology and Analysis of Development Trends of Contact Shape Meter

YU Hongyi[1], WANG Jin[1], SHI Yan[2]

(1. Baoshan Iron & Steel Co., Ltd., Cold Rolling Plant, Shanghai 201900, China;
2. Obei Co., Ltd., East China Region, Shanghai 201900, China)

Abstract: Stressometer is one important equipment of high quality strip processing line. The paper introduces the basic composition of measuring roll and control system briefly. Detailed description of the main application of contact type strip roller technology and signal transmission in the cold rolling production line, including the type and the arrangement of sensors in the measuring roll, the principle of signal transmit unit, the surface type of measuring roll etc. Meanwhile, The paper proposed some understanding of the development trends such as rational placement of sensors, wireless excitation and signal, and seamless integration of roller surfaces.

Key words: stressometer; press sensor; seamless roll; wireless transmitter

不锈钢复合板在热风炉炉壳工程施工中的应用

何泽凤[1]，李成杰[2]

（1. 中国二十二冶集团有限公司，河北唐山 064000；
2. 金环建设集团有限公司，河北石家庄 050000）

摘 要： 对904L+Q345R不锈钢复合板在冶建热风炉炉壳工程施工中应用进行了研究、介绍了复合板焊接坡口的选取、焊接施工工艺控制及其焊后热处理工艺的制定，得出了工程施工过程中严格控制工艺流程是确保质量精准监控的有效措施等记述方法，为公司积累了施工经验及业绩，同时也获得了社会的认可。

关键词： 不锈钢复合板；剖口形式；焊口工艺检测；焊接工艺；施工应用

Application of Stainless Steel Composite Plate in the Construction of the Hot Blast Furnace Shell

HE Zefeng[1], LI Chengjie[2]

(1. China MCC22 Group Corporation Ltd., Tangshan 064000, China;
2. Jinhuan Construction Group Co., Ltd., Shijiazhuang 050000, China)

Abstract: This paper studies the application of 904L+Q345R stainless steel composite plate in the construction of hot blast furnace shell engineering, introduces the selection of welding groove of the composite plate, the control of welding construction technology and the formulation of post-welding heat treatment technology, and concludes that strict control of process flow is an effective measure to ensure accurate quality monitoring in the construction process. The company has accumulated construction experience and performance, but also gained the social recognition.

Key words: stainless steel composite plate; profile form; weld process testing; welding process; construction applications

浅析多探头中厚板自动探伤仪精度评价方法

王立坚，黄 毅，闫智平

（首钢京唐钢铁联合有限责任公司制造部，河北唐山 063200）

摘 要： 从灵敏度、探测盲区、设备稳定性、缺陷测量精度四个维度入手，探讨影响多探头组合式中厚板自动探伤仪功能精度的因素及检测方法，确保漏报率、误报率、缺陷再现率所代表的整机性能满足使用要求。旨在提升探伤设备管理的规范化、制度化。

关键词：多探头；自动探伤；功能；精度

Analyses on Evaluation Method of More Probe Automatic Ultrasonic Testing Equipment Accuracy for Heavy Plate

WANG Lijian, HUANG Yi, YAN Zhiping

(Manufacturing Department of Shougang Jingtang United Iron & Steel Co., Ltd., Tangshan 063200, China)

Abstract: From sensitivity, detection blind area, equipment stability, defect measuring precision four dimensions, this paper discusses influence factors and testing methods on the function and accuracy of more probe combined automatic ultrasonic testing equipment for heavy plate, to ensure that non-response rates, the rate of false positives, defect reproducibility represented the whole machine performance meets the requirement. Designed to improve ultrasonic testing equipment management standardized and institutionalized.

Key words: more probe; automatic ultrasonic; function; accuracy

大型转炉耳轴轴承损坏原因分析及措施制定

白 华[1]，李 伟[1]，向忠辉[2]，章勤奋[1]

（1. 武汉钢铁有限公司设备管理部，湖北武汉 430083；
2. 武汉钢铁有限公司炼钢厂，湖北武汉 430083）

摘 要：转炉耳轴轴承是固定在转炉耳轴上支撑炉体旋转的关键部件，运行载荷大、经常受冲击，其能否正常运转直接关系到转炉炼钢生产的高效与安全。文章主要对某炼钢厂2号转炉耳轴轴承的基本情况和突发损坏现象，以及耳轴轴承的日常维护进行了详细的介绍，对造成耳轴轴承损坏的原因从系统和局部进行了分析并最终给出结论，最后结合自身维护检修的状况提出有针对性的整改措施。

关键词：转炉；轴承；疲劳失效；分析

Analysis of the Causes of Damage to the Ear Shaft Bearing of a Large Converter and the Formulation of Measures

BAI Hua[1], LI Wei[1], XIANG Zhonghui[2], ZHANG Qinfen[1]

(1. Wuhan Iron and Steel Co., Ltd., Equipment Management Department, Wuhan 430083, China;
2. Wuhan Iron and Steel Co., Ltd., Wuhan 430083, China)

Abstract: The converter trunnion bearing is a key component fixed on the converter trunnion to support the rotation of the converter body. It has large operating load and is often impacted. Its normal operation is directly related to the efficiency and safety of converter steelmaking. The article mainly provides a detailed introduction to the basic situation and sudden damage phenomenon of the 2 # converter trunnion bearing in a certain steelmaking plant, as well as the daily maintenance of the trunnion bearing. The reasons for the damage to the trunnion bearing are analyzed systematically and locally, and a conclusion is finally drawn. Finally, targeted rectification measures are proposed based on the status of self maintenance and repair.

Key words: converter; bearings; fatigue failure; analysis

S214 通廊结构问题分析与加固方案

任 强，廖生伟，谭 磊

（武汉钢铁有限公司设备管理部，湖北武汉 430083）

摘 要：针对钢结构架式通廊下弦杆、腹杆等主要受力构件存在严重锈蚀问题，在不影响皮带正常运行情况下，采用梁式及桁架式结构支顶、原有杆件增大截面等综合加固维修方案，保证皮带通廊的安全运行，取得了良好效果。

关键词：桁架式通廊；锈蚀；加固

Structural Problem Analysis and Reinforcement Scheme of the S214 Corridor

REN Qiang, LIAO Shengwei, TAN Lei

(Equipment Management Department of Wuhan Iron and Steel Co., Ltd., Wuhan 430083, China)

Abstract: In view of the serious corrosion of the main stress components such as the lower chord and the belly bar of the steel structure, the comprehensive reinforcement and maintenance programs such as beam and truss structure support and original bar section are adopted to ensure the safe operation of the belt corridor under the condition of normal operation of the belt.

Key words: trussed corridor; corrosion; reinforcement

浅谈高炉炉役后期炉身故障的处理方法

陈建平[1]，席强华[2]，刘 嘉[2]，陈长鑫[2]

（1. 武汉钢铁有限公司设备管理部，湖北武汉 430083；
2. 武汉钢铁有限公司炼铁厂，湖北武汉 430083）

摘 要：高炉炉役后期难以避免会出现炉身冷却壁漏水损坏、炉皮发红泄漏、冷却壁套管腐蚀泄漏等设备故障，给高炉的生产顺行和安全带来较大影响。本文结合武钢5号高炉炉役后期的日常维护经验，总结炉身故障的处理方法，包含冷却壁穿管、冷却壁漏水快速卡死方法；损坏冷却壁处炉皮温度实时监测，避免炉皮发红的处理手段以及炉皮开裂损坏的处理方法；冷却壁水管损坏后的套管处理方法以及未损坏冷却壁套管的预防性维护方法等。通过这些方法的运用，有效保障了炉役后期的设备状态受控，延长了高炉使用寿命。

关键词：炉役后期；炉身故障；冷却壁；处理方法

Talking about the Handling Methods of Late-stage Furnace Body Faults in Blast Furnaces

CHEN Jianping[1], XI Qianghua[2], LIU Jia[2], CHEN Changxin[2]

(1. Equipment Management Department of Wuhan Iron & Steel Co., Ltd., Wuhan 430083, China;
2. Ironmaking Plant of Wuhan Iron & Steel Co., Ltd., Wuhan 430083, China)

Abstract: In the later stages of blast furnace operation, equipment faults such as furnace body cooling wall water leakage and damage, reddening and leakage of furnace skin, and corrosion and leakage of cooling wall sleeves are difficult to avoid. These faults have a significant impact on the smooth production and safety of blast furnaces. This article combines the daily maintenance experience of Wuhan Iron and Steel's No.5 blast furnace in the later stages of operation to summarize the methods for handling furnace body faults. These methods include quick sealing of cooling wall tube punctures and water leakage, real-time monitoring of furnace skin temperature at damaged cooling wall sections to prevent reddening, as well as methods for handling furnace skin cracking and damage. The article also covers the treatment of damaged cooling wall tubes and preventive maintenance methods for undamaged cooling wall sleeves. By applying these methods, the equipment status in the later stages of blast furnace operation can be effectively controlled, thereby extending the service life of the blast furnace.

Key words: later stages of blast furnace operation; furnace body faults; cooling wall; handling methods

某钢铁基地10kV供电系统继电保护定值整定优化

后程瑞，许子义，成奕佳，李金涛，赵志芸

（武钢集团昆明钢铁股份有限公司能源动力厂供电作业区，云南安宁　650302）

摘　要：本文介绍了某钢铁基地10kV各配电室继电保护定值整定存在的问题，总结了定值误整定现象，并介绍了该基地10kV配电室定值整定的原则。

关键词：继电保护；定值计算

Optimization of Relay Protection Setting for 10kV Power Supply System in a Certain Steel Base

HOU Chengrui, XU Ziyi, CHENG Yijia, LI Jintao, ZHAO Zhiyun

(Power Supply Operation Area of Energy and Power Plant of Kunming Lron and Steel Co., Ltd. of Wuhan Lron and Steel Group, Anning 650302, China)

Abstract: This article introduces the problems in the relay protection setting of 10kV distribution rooms in a certain steel base, summarizes the phenomenon of setting error, and introduces the principles of setting 10kV distribution rooms in the base.

Key words: relay protection; fixed value calculation

考虑温度载荷的桥式起重机疲劳失效原因分析

章勤奋[1]，师骅[1]，孙海峰[2]，邓杨[2]

(1. 武汉钢铁有限公司设备管理部，湖北武汉 430080；
2. 武汉钢铁有限公司热轧厂，湖北武汉 430080)

摘 要： 热轧厂板坯库桥式起重机承担着高温板坯工作，辐射温度高，运行环境恶劣，加之工作繁重，由结构裂纹导致起重机失效是各钢铁企业面临的共性问题。起重机工作环境的温度场对裂纹的产生及拓展有一定影响，基于此，本文对某钢厂热轧板坯库 16+16t 桥式起重机的舱门四周圆角裂纹展开研究，利用有限元软件建立了起重机桥架结构有限元模型，并对桥架结构受温度场的影响进行分析，找出了裂纹产生的原因，为该类起重机的加固和改进提供了理论基础。

关键词： 桥式起重机；温度场；疲劳裂纹；有限元

Analysis of Fatigue Failure Causes of Bridge Crane in Considering with Temperature Load

ZHANG Qinfen[1], SHI Hua[1], SUN Haifeng[2], DENG Yang[2]

(1. Equipment Management Department, Wuhan Iron and Steel Co., Ltd., Wuhan 430080, China;
2. Hot Rolling Equipment Department, Wuhan Iron & Steel Co., Ltd., Wuhan 430080, China)

Abstract: The bridge crane of slab warehouse in hot rolling mill works on the environment of high temperature radiation, beside with the heavy work condition, leading to bad operating environment. The failure of crane caused by structural crack is a common problem faced by all iron and steel enterprises. The temperature field of the working environment of the crane has a certain influence on the generation and development of cracks. Based on this, this paper studies the rounded corner cracks around the hatch door of the 16+16t bridge crane in a hot-rolled slab warehouse of a steel plant. Finite element software is used to establish the finite element model of the crane bridge structure, and the influence of temperature field on the bridge structure is analyzed to find out the causes of cracks. It provides a theoretical basis for the reinforcement and improvement of this kind of crane.

Key words: bridge crane; temperature field; fatigue cracks; CAE simulation

钛、铝和不锈钢连续轧制复合机组装备及工艺的研究

龚辉，刘德华，牛艳，付建辉，胡志忠，
杨可托，李文武，段粉，苗锁周

(中冶陕压重工设备有限公司，陕西西安 710119)

摘 要： 本文详细介绍了超宽幅单机架 1450mm 四辊三层钛、铝和不锈钢连续轧制复合机组的情况，主要包括产品规格、工艺流程、机组装备及主要技术参数、机械设备和电气控制方面的创新点。该机组投产以来，设备运行平稳，产品质量优异，创造了很好的经济效益。

关键词： 单机架 1450mm；四辊；三层轧制复合机组

Abstract: This article introduces in detail the ultra-wide single-frame 1450mm four-roller three-layer titanium, aluminum and stainless steel continuous rolling composite unit, mainly including product specifications, process flow, unit equipment and main technical parameters, mechanical equipment and electrical control innovations. Since the unit was put into operation, the equipment has run smoothly, the product quality is excellent, and good economic benefits have been created.

Key words: single rack 1450mm; four-roll; three-layer rolling compounding mill

板带热轧组合立辊的应用

马忠辉，刘　娜，李艳红

（鞍钢轧辊有限公司，辽宁鞍山　114000）

摘 要： 本文针对热轧立辊在使用存在的质量问题，创新将高铬钢材料以镶套的方式引入热轧立辊制造，提高轧边立辊使用寿命，降低轧辊采购成本。

关键词： 热轧立辊；高铬钢材料；组合立辊

Application of Combined Vertical Roller for Hot Rolling of Plate and Strip

MA Zhonghui, LIU Na, LI Yanhong

(Ansteel Roll Co., Ltd., Anshan 114000, China)

Abstract: This artical focuses on the quality issues existing in the use of hot rolling vertical rollers, and innovatively introduces high chromium steel materials into the manufacturing of hot rolling vertical roller through sleeve insertion, improving the service life of edge rolling vertical roller and reducing the cost of roller procurement.

Key words: hot rolled vertical; high chromium steel materical; combined vertical roller

主从控制在冶金装备中的典型应用及失控分析

程　曦，刘　浩，钟　实，丁继亚，黄亚军

（武钢有限设备管理部，湖北武汉　430083）

摘 要： 变频器主从控制广泛应用于冶金行业中的大型升降、摆动和翻转设备。根据长期跟踪，变频器常用的几种主从控制模式在冶金装备的实际运用中遇到偶发或极端工况时，会处于失控或短时失控状态，对设备和人身的安全带来较大影响,通过对变频器主从控制原理及失控成因的分析,提出了一些变频器主从控制实际应用时的处理措施。有利于减少装备失控后造成的影响,保障设备及人员安全运行。

关键词：冶金装备；主从控制；失控分析

Typical Application and Uncontrolled Analysis of Master Slave Control in Metallurgical Equipment

CHENG Xi, LIU Hao, ZHONG Shi, DING Jiya, HUANG Yajun

(Equipment Management Department of Wuhan Iron and Steel Co., Ltd., Wuhan 430083, China)

Abstract: The master slave control of frequency converters is widely used in large-scale lifting, swinging, and flipping equipment in the metallurgical industry. Based on long-term tracking, several commonly used master-slave control modes of frequency converters can be in a state of loss of control or short-term loss of control when encountering occasional or extreme operating conditions in the practical application of metallurgical equipment, which has a significant impact on the safety of equipment and personnel. By analyzing the principle and causes of master-slave control of frequency converters, some treatment measures for practical application of master-slave control of frequency converters are proposed. It is beneficial to reduce the impact of equipment loss of control and ensure the safe operation of equipment and personnel.

Key words: metallurgical equipment; master slave control; uncontrolled analysis

基于时频包络谱峭度的轴承故障冲击特征提取方法

刘晗，程本俊，邬忠尧

（宝武智能装备科技有限公司，上海　201900）

摘　要：本文结合振动信号的时频分布与包络谱分析,提出了一种基于时频包络谱峭度的轴承冲击故障特征提取方法,利用轴承共振频率处时频包络谱峭度值较大的特点,为共振频带滤波器的设计参数选择提供了参考,并通过仿真信号分析和实际案例验证了所提出方法的可行性与有效性。

关键词：滚动轴承；故障诊断；时频分布；包络谱峭度

Extraction Method of Bearing Fault Impact Characteristics Based on Time-frequency Envelope Spectral Steepness

LIU Han, CHENG Benjun, WU Zhongyao

(Baowu Equipment Intelligence Technology Co., Ltd., Shanghai 201900, China)

Abstract: Combined with the time-frequency distribution and envelope spectrum analysis of vibration signals, this paper proposes a bearing impact fault feature extraction method based on time-frequency envelope spectral steepness, which provides a reference for the design parameter selection of resonance band filters by using the characteristics of large time-frequency envelope spectral slope values at the bearing resonance frequency, and verifies the feasibility and effectiveness of the proposed method through simulation signal analysis and practical cases.

Key words: rolling bearing; fault diagnosis; time-frequency distribution; envelope spectral steepness

智能巡检机器人在 220kV 变电站系统中研究与应用

苟俊涛

（张宜高科科技有限公司，河北宣化　075100）

摘　要：随着科学技术与智能变电站的迅猛发展，为了保证电缆隧道内设备设施的安全稳定运行，隧道智能巡检机器人得到了广泛地应用。通过简要介绍电缆隧道智能巡检机器人在 220kV 智能变电站应用现状，重点对智能巡检机器人的关键技术和巡检要点进行了阐述，对提高隧道巡检工作效率、降低人工成本具有借鉴指导作用。

关键词：智能巡检机器人；220kV 变电站；关键技术和巡检要点；隧道巡检；降低人工成本

Research and Application of Intelligent Inspection Robot in 220kV Substation System

GOU Juntao

(HBZX High Tech Co., Ltd., Xuanhua 075100, China)

Abstract: With the vigorous development of science and technology and intelligent substation, in order to ensure the safe and stable operation of equipment and facilities in cable tunnel, intelligent tunnel inspection robot has been widely used.This paper briefly introduces the application status of intelligent inspection robot for cable tunnel in 220kV substation.The key technologies and inspection points of the intelligent inspection robot are expounded.It can improve the efficiency of tunnel inspection and reduce labor cost.

Key words: intelligent inspection robot; 220kV substation; key technologies and inspection points; tunnel inspection; reduce labor cost

旋风除尘卸灰控制系统改进设计

付守壮

（北京首钢股份有限公司炼铁作业部，河北迁安　064400）

摘　要：本文主要结合 PLC 和触摸屏的综合应用控制理念，对迁钢炼铁 3 号高炉旋风除尘卸灰系统进行改进设计，以实现自动化卸灰功能，既节省人力成本，更符合环保要求。文中简述了设备的硬件选型以及 PLC、触摸屏和计算机三者之间的通讯参数设置，着重论述了 PLC 程序控制和触摸屏画面组态，用以实现对现场开关量和模拟量的综合控制功能，通过触摸屏为系统提了供良好的人机界面，使得操作更加简洁。

关键词：PLC；触摸屏；WinCC flexible；通讯连接；组态画面

Improvement Design of Cyclone Dust Removal and Ash Discharge Control System

FU Shouzhuang

(Ironmaking Operation Department of Beijing Shougang Co., Ltd., Qian'an 064400, China)

Abstract: This article mainly combines the comprehensive application control concept of PLC and touch screen to improve the design of the cyclone dust removal and ash discharge system for the 3# blast furnace of Qiangang Iron and Steel Co., Ltd., in order to achieve automatic ash discharge function, which not only saves labor costs but also meets environmental protection requirements. The article briefly describes the hardware selection of the equipment and the communication parameter settings between PLC, touch screen, and computer. It focuses on PLC program control and touch screen screen screen configuration to achieve comprehensive control of on-site switch and analog quantities. The touch screen provides a good human-machine interface for the system, making the operation more concise.

Key words: PLC; touch screen; WinCC flexible; communication connection; configuration screen

相控阵探伤技术在车轮样轮校准中的问题分析及解决方法

程松林，程旭辉，陶国强，刘海波，张盛华

（中国宝武集团马钢轨交材料科技有限公司，安徽马鞍山 243000）

摘 要：随着中国铁路向高速化、重载化方向不断发展，列车车轮的生产及出厂标准要求越来越高，超声相控阵检测设备因其具有灵敏度高、声束可控性好等优点已逐步推广应用到列车车轮的探伤中。实际应用中，超声相控阵设备在样轮校准过程中存在轮辋内部同深度、同当量、不同位置的人工缺陷孔回波增益差异大的情况，也导致待检车轮在探伤过程中轮辋内部缺陷定量的不准确性，存在缺陷漏检风险。本课题针对上述问题，分析造成这种现象的可能性是探头内部晶片的激发声能存在差异，猜想采用相控阵超声波同底波高度法消除该差异，实现定量的准确性。本课题采用相控阵超声波同底波高度法对轮辋内部 2 个同深度、同当量、不同位置的人工缺陷孔进行了回波增益试验。实验结果显示，使用该方法回波增益由原来的 5.2dB 降低到 0.6dB，下降 88%；继而采用相控阵超声波同底波高度法对轮辋内部 11 个不同深度、同当量、不同位置的人工缺陷孔进行回波增益实验，实验结果显示，使用该方法回波增益极差由原来的 7.3dB 减小到 3.06dB，下降 58%。通过上述试验，验证了相控阵超声波同底波高度法能解决同深度、同当量、不同位置的人工缺陷孔回波增益差异大的问题，同时也能解决不同深度、同当量、不同位置的人工缺陷孔回波增益差异大的问题，有效提高车轮轮辋内部检出缺陷的定量准确性。该方法的应用将极大降低车轮轮辋内部缺陷的漏检风险。

关键词：超声相控阵技术；平底孔；底波

冷轧电工钢薄板月牙剪切过程仿真分析研究

陈 兵，王 洋，徐厚均，何俊烨

（北京科技大学机械工程学院，北京 100083）

摘 要：本文针对某钢企冷轧线电工钢薄板月牙剪切断裂过程及提高月牙断面剪切质量之目的，对月牙剪切工艺参数优化展开技术攻关的实际需求，深入分析了月牙加工过程。首先针对电工钢板材，设计了不同应力状态下的缺口拉伸试样，通过电子万能试验机对缺口拉伸试样进行了拉伸实验研究，获得了板材拉伸过程中的载荷位移曲线，进而通过 Abaqus 有限元软件建立了不同缺口拉伸试样的仿真模型，结合拉伸实验数据反求 Johnson-Cook 本构模型和断裂准则的参数。通过设置典型仿真工况，改变剪切工艺参数，分析剪切工艺参数对带钢缺陷产生的影响，找到带钢月牙剪切区域易发生断裂的原因，为电工钢板材月牙剪剪切参数的选择提供理论指导。

关键词：冷轧；月牙剪；力学行为；有限单元法；仿真

Finite Element Simulation Analysis of Shear Fracture Process of Electrical Steel Thin Plate Crescent

CHEN Bing, WANG Yang, XU Houjun, HE Junye

(School of Mechanical Engineering, University of Science and Technology Beijing, Beijing 100083, China)

Abstract: Given the cutting process of Electrical steel thin plate crescent shear in a cold rolling line of a steel enterprise and the purpose of improving the cutting quality of the crescent section, the actual demand for optimizing the cutting process parameters of the crescent shear is carried out, and the cutting process of Electrical steel crescent shear is deeply analyzed. Firstly, notch tensile specimens under different stress states were designed for oriented Electrical steel sheets. The tensile test of notch tensile specimens was carried out using an electronic universal testing machine, and the load-displacement curve in the tensile process of the sheet was obtained. Then, the simulation models of different notch tensile specimens were established through Abaqus finite element software, and the parameters of the Johnson-Cook constitutive model and fracture criterion were inversely calculated based on the tensile test data. By setting typical simulation conditions, changing shear process parameters, and analyzing the impact of shear process parameters on strip steel damage, finding the reasons for the fracture of the crescent shear area of strip steel, reducing the possibility of band breakage at the weld, and providing theoretical guidance for the selection of shear parameters of electrical steel plate crescent shear.

Key words: cold rolling; crescent scissors; mechanical behavior; finite element method; simulation

锌合金中稀土元素测定方法研究

张 杰，田秀梅，王 伟，李 颖，王一凌

（海洋装备用金属材料及其应用国家重点实验室，鞍钢集团钢铁研究院，辽宁鞍山 114009）

摘　要：采用电感耦合等离子体原子发射光谱法（ICP-AES）定量分析锌合金中的 La、Ce、Pr、Nd。研究了锌合金中各稀土元素谱线干扰的校正，同时用正交设计实验对仪器工作条件进行了优化。在实验波长下，构造了 5 种稀土元素的 K 矩阵模型，利用 K 矩阵校正测定锌合金样品，样品平行测定的相对标准偏差在 0.20 %～0.54%之间，回收率实验结果为 98.33%～110.00%。本实验利用 K 矩阵校正稀土元素谱线干扰，实现了锌合金中稀土元素的快速、准确测定。

关键词：电感耦合等离子体发射光谱；稀土；锌合金；K 矩阵；干扰校正

Study on Determination Method of Rare Earth Elements in Zinc Alloy

ZHANG Jie, TIAN Xiumei, WANG Wei, Li Ying, WANG Yiling

(State Key Laboratory of Metal Material for Marine Equipment and Application, Iron & Steel Research Institute of Ansteel Group, Anshan 114009, China)

Abstract: Rare earth elements La, Ce, Pr, Nd, Sm were quantitatively analyzed by inductively coupled plasma atomic emission spectroscopy(ICP-AES). In the experimental wavelength, the LREEs had measured the standard curve, and the correlation coefficient of the regression equation, the linear range of up to four orders of magnitude. Make use of K matrix correction to measure the rare earth rich residue, rare earth mixture and zinc alloy sample respectively, made sure the content of the LREEs in the above sample. The RSD of the samples parallel measurement is at 0.20%～0.54%, the recovery of the experiment result is 98.33%～110.00%, which express the sophistication and accurate of the experiment is good. This experiment use the K matrix method of chemometrics to handle the spectrum test data of the LREEs in the actual sample. For the mathematics model, which will overcome the mutually interference in the LREEs. Don't need to carry on separation to the LREEs, carried out the fast and accurate measurement of the LREEs.

Key words: inductively coupled plasma atomic emission spectrometry(ICP-AES); rare earth elements(REEs); zinc alloy; K matrix; interference correction

桥式起重机典型故障的研究与解决方案

苏二龙

（承德钒钛新材料有限公司，河北承德　067002）

摘　要：承德钒钛新材料有限公司 1780 生产线投产于 2008 年，共有桥式起重机 41 台，个别起重机因设计缺陷、结构变形、磨损超标、维护不善等原因，频繁发生故障，制约生产顺行。主要表现在：大车运行机构车轮啃轨、主小车行走悬挂减速机力矩臂销轴切断、主起升减速机装配精度差导致各齿轮轴轴向窜动和异响、主起升卷筒与球铰联轴器配合面磨损导致螺栓切断和球铰联轴器碎裂、80t 副钩钩头磨损剧烈导致寿命短等。本文针对以上主要问题，通过采集数据深入分析，制定相应措施并付诸实践，使问题得到彻底解决。

关键词：啃轨；力矩臂；减速机；球铰联轴器；铁包倾翻钩

Research and Solutions for Typical Faults of Overhead Traveling Cranes

SU Erlong

(Chengde Vanadium Titanium New Material Limited, Chengde 067002, China)

Abstract: The 1780 production line of Chengde Vanadium Titanium New Material Limited was put into production in 2008. There are 41 overhead traveling cranes in the company, frequent failures occurred, which restricted the production. Mainly displays in: car trunning mechanism wheel chew rail, the main car walking suspension reducer torque arm pin shaft cutting, main lift reducer assembly accuracy leads to the gear shaft axial channeling and abnormal ring, main lift drum and ball hinge coupling with surface wear bolt cutting and ball hinge coupling fracture, 80t deouty hook hookhead wear severe lead to short life, etc. Inview of the above main problems, this paper collects the data through in-depth analysis, formulates the corresponding measures and puts them into practice, so that the problem is thoroughly solved.

Key words: chew rail; torque arm; reducer; ball hinge coupling; iron bag roll hook

智能化新能源钢卷车在现代钢厂冷轧车间的应用

韦富强[1,2,3]，孟祥军[2,3]，张　建[1,2,3]，李永卿[2,3]

（1. 北京首钢国际工程技术有限公司，北京　100043；
2. 北京市冶金三维仿真设计工程技术研究中心，北京　100043；
3. 北京首钢云翔工业科技有限责任公司，北京　100043）

摘　要： 本文概述了近年来冷轧车间钢卷运输技术的发展，分析了国内冷轧车间建设中钢卷运输技术的应用趋势，介绍了新能源智能化钢卷车在某冷轧车间的工程应用方案及在其他冷轧车间的应用情况，得出了新能源智能化钢卷车在冷轧车间内部钢卷运输领域中具有明显优势等结论。

关键词： 智能化；新能源；钢卷车；钢卷运输；冷轧车间；界面技术

Application of Intelligent New Energy Coil Car in Cold Rolling Workshop of Modern Steel Plant

WEI Fuqiang[1,2,3], MENG Xiangjun[2,3], ZHANG Jian[1,2,3], LI Yongqing[2,3]

(1. Beijing Shougang International Engineering Technology Co., Ltd., Beijing 100043, China; 2. Beijing Metallurgical 3-D Simulation Design Engineering Technology Research Center, Beijing 100043, China; 3. Beijing Shougang Yunxiang Industrial Technology Co., Ltd., Beijing 100043, China)

Abstract: This paper summarizes the development of steel coil transportation technology in cold rolling workshops in recent years, analyzes the application trend of steel coil transportation technology in the construction of domestic cold rolling workshops, introduces the engineering application plan of new energy intelligent steel coil cars in a cold rolling workshop and its application in other cold rolling workshops, and concludes that new energy intelligent steel coil cars have obvious advantages in the field of internal steel coil transportation in cold rolling workshops.

Key words: intelligent; new energy; coil car; steel coil transportation; cold rolling workshop; interface technology

UCMW 轧机万向十字接轴常见损坏现象及应对措施

王承刚，玄利剑

（首钢智新迁安电磁材料有限公司，河北迁安 064400）

摘 要：在 UCMW 轧机的主传动系统中，万向十字接轴是轧制扭矩传递的中心环节。接轴的损坏将直接造成主减速机、轧机出现振动异常并造成轧制产品质量的下降，严重的损坏情况将会直接迫使轧机停机。由于 UCMW 轧机具备工作辊窜辊功能，因此要求其使用的接轴需具备滑动花键结构，该结构的存在使接轴整体的动平衡保持更加困难，更易出现损坏。本文通过对接轴损坏情况进行分析，制定相应接轴结构、管理优化方案，可有效提高接轴使用寿命、降低接轴成本消耗。

关键词：UCMW 轧机；万向十字接轴；寿命提升

Common Damage Phenomena and Countermeasures of Universal Cross Joint in UCMW Rolling Mill

WANG Chenggang, XUAN Lijian

(Shougang Zhixin Qian'an Electromagnetic Materials Co., Ltd., Qian'an 064400, China)

Abstract: In the main transmission system of UCMW rolling mill, the universal cross joint shaft is the central link for rolling torque transmission. The damage of the universal cross joint shaft will directly cause abnormal vibration of the main reducer and rolling mill, leading to a decrease in the quality of the rolled product. Severe damage will directly force the rolling mill to stop. Due to the work roll shifting function of UCMW rolling mill, it is required that the universal cross joint shaft used must have a sliding spline structure. The existence of this structure makes it more difficult to maintain the overall dynamic balance of the universal cross joint shaft and is more prone to damage. This article analyzes the phenomenon damage of the universal joint shaft, and develops corresponding methods. This can effectively improve the service life of the universal cross joint shaft and reduce the cost consumption of the universal cross joint shaft.

Key words: UCMW rolling mill; universal cross joint shaft; increase lifespan

TMEIC 激光焊机光束焦点位置对焊缝质量影响分析

李冠良，唐晓宇，玄利剑，张 静，辛广荣

（首钢智新迁安电磁材料有限公司，河北迁安 064400）

摘 要：激光焊机广泛的应用于钢铁企业板带生产产线，其作用在于将前后卷带钢带头和带尾进行焊接保证产线生产的连续性，因此焊机作为产线咽喉设备其焊接后形成焊缝的焊接质量必须要有良好的保证。不同的激光焦点位置对焊缝影响极大，需要重点关注。本文以 LBW-8S313NP 型 TMEIC 激光焊机为研究对象，探究激光光束焦点位置对焊缝质量的影响，对激光焊接焦点位置的调整具有广泛的借鉴意义。

关键词：激光焦点位置；焊缝横截面成型；焊接质量；激光焊接

Study on the Influence of Beam Focus Position on Weld Quality in TMEIC Laser Welder

LI Guanliang, TANG Xiaoyu, XUAN Lijian, ZHANG Jing, XIN Guangrong

(Shougang Zhixin Qian'an Electromagnetic Materials Co., Ltd,. Qian'an 064400, China)

Abstract: Laser welding machine is widely used in the plate and strip production line of iron and steel enterprises. Its role is to weld the lead and tail of the coil steel before and after to ensure the continuity of the production line. Therefore, the welding machine as the throat equipment of the production line must have a good guarantee of the welding quality of the weld formed after welding. Different laser focus positions have a great impact on the weld and need to be focused on. This paper takes LBW-8S313NP TMEIC laser welding machine as the research object, explores the influence of laser beam focus position on weld quality, and has extensive reference significance for the adjustment of laser welding focus position.

Key words: laser focus position; weld cross section forming; welding quality; laser welding

直接还原铁热装热送新装备技术开发

李少英，潘宏涛，王少臣，李佳辉，习小军

（中冶京诚工程技术有限公司，北京 100176）

摘 要：在"碳达峰、碳中和"的战略背景下，钢铁产业高端化、绿色化、低碳化发展是时代的必然要求。氢基竖炉+电弧炉炼钢工艺作为钢铁工业绿色低碳工艺路线的代表，其节能降耗技术的开发尤为重要。直接还原铁（DRI）输送作为连接氢基竖炉和电弧炉的重要环节，包括冷态和热态两种方式[1]。采用热态 DRI 作为电弧炉炼钢的原料，由于 DRI 热装温度可达 600℃左右，可以给电弧炉带来一部分物理热，因此可以大大降低电弧炉炼钢的电耗及冶炼时间。同时，采用热态 DRI 直接热装电弧炉后，也可减少氢基竖炉冷却热态 DRI 的能源及物料。因此，实现直接还原铁在氢基竖炉工序和电弧炉工序之间的热态储存和转运，对氢基竖炉+电弧炉的低碳炼钢工艺路线的节能降耗具有重要意义。

本文分析了不同 DRI 与废钢比例的条件下，电量消耗、石灰消耗、碳粉消耗、氧气消耗、天然气消耗、铁损、烟尘、渣中铁珠以及冶炼成本随 DRI 入炉温度的变化关系。结果表明，当 DRI 取代废钢进行电弧炉冶炼时，DRI 入炉温度每增高 100℃，吨钢可节约电能 25kW·h，吨钢成本可降低 15 元。目前，DRI 热装热送技术主要包括重力输送、风力输送、链板输送和链斗输送[2]。国内应用尚未见报道，国外虽少量应用，但仍有不足。例如，重力直送系统要求厂房高度高；气力输送系统的动力消耗大；耐热容器罐车输送系统难以实现连续上料；链斗转运热送系统易于卡料。因此，开发 DRI 热装热送新技术十分必要。

针对以上问题，结合热态 DRI 的特点，以"输送能耗最低、输送装置简单"为原则，设计出一种 DRI 转运密闭、热态保温、计量可控的输送设备，同时实现了造渣料与直接还原铁分层协同加料的功能。装备设计如图 1 所示。在竖直方向上，该装备借助倾斜的耐高温溜管将热态直接还原铁储罐的底部与水平输送装置的第一端连通，在实现热态 DRI 重力输送的同时，降低了厂房的基建高度；在水平方向上，该装备通过非谐式振动的密闭水平槽将热态

DRI 输送至电炉，与耐材罐运输和料斗式运输相比，在实现连续上料的同时，避免了卡料。

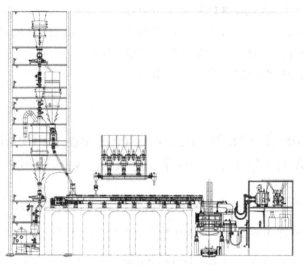

图 1　新型直接还原铁装备设计图

参 考 文 献

[1] 张建良, 刘玉全, 张宗旺, 等. 一种新型优质电炉炉料[J]. 钢铁研究学报, 2001, 13(4): 69-72.
[2] 花皑, 崔于飞, 吴培珍, 等. 直接还原铁的制造工艺及设备[J]. 工业加热, 2011, 40(1):1-3.

轨梁厂 BD 轧辊的堆焊再制造技术

魏占山，孙贻宝，王金生，张金龙

（包钢集团冶金轧辊制造有限公司，内蒙古包头　014010）

摘　要：本文针对轨梁厂生产中所使用的 BD 轧辊的技术要求、工作条件及磨损状况，开发了一种与其母材相匹配的药芯焊丝，通过选择与之相匹配的焊剂，在试板上进行了堆焊试验。通过对堆焊试样的成分检测、硬度测试及组织观察分析，对 BD 轧辊堆焊工艺进行了研究，并对 BD 轧辊进行了表面堆焊再制造。经过上机轧制，过钢量达到了修复技术要求。

关键词：BD 轧辊；堆焊；药芯焊丝；再制造

Remanufacturing Technology of Breaking-down Roll Surface Welding in Rail and Beam Factory

WEI Zhanshan, SUN Yibao, WANG Jinsheng, ZHANG Jinlong

(Baotou Steel Group Metallurgical Roll Manufacturing Co., Ltd., Baotou 014010, China)

Abstract: According to the technical requirements, working conditions and wear condition of BD roll used in rail and beam factory, a kind of flux-cored wire matching its base material is developed in this paper. The surfacing process of BD roll was studied by the composition test, hardness test and microstructure observation and analysis of the surfacing sample, and the surface surfacing of BD roll was remanufactured. After machine rolling, the amount of steel passed has reached the technical requirements of repair.

Key words: breaking-down roll; surface welding;flux core welding wire; remanufactrting

无人自动抓渣天车关键技术研究及应用

刘海峰，薛　鑫，张　灿，石玉龙，琚小然

（河钢集团河钢材料技术研究院中试试验中心，河北石家庄　050000）

摘　要： 在天车行业智能无人天车是未来天车发展的方向，文章探讨了智能无人天车的定义、结构、发展方向，重点介绍了在河钢材料院在石钢对一部轧钢线上的抓渣天车初步无人自动化实践，利用 PLC 控制、变频传动、无线通讯等技术实现无人抓渣工艺。

关键词： 天车；智能；无人化；自动抓渣

Study and Practice of Key Unmanned Slag Grabbing Crown Technologies

LIU Haifeng, XUE Xin, ZHANG Can, SHI Yulong, JU Xiaoran

(HBIS Material Technology Research Institute Pilot Test Center, Shijiazhuang 050000, China)

Abstract: In the crown block industry, intelligent unmanned crown blocks are the future direction of crown block development. This article explores the definition, structure, and development direction of intelligent unmanned crown blocks, and focuses on the preliminary unmanned automation practice of slag grabbing crown blocks on a rolling line at Shigang in HBIS. The unmanned slag grabbing process is achieved by using PLC control, frequency conversion transmission, wireless communication and other technologies.

Key words: crown block; intelligent; unmanned; automatic slag grabbing

热轧板带工艺建模及软件开发

王　涛[1,2,3]，刘元铭[1,2,3]，黄庆学[1,2,3]

（1. 太原理工大学　机械与运载工程学院，山西太原　030024；
2. 太原理工大学　先进金属复合材料成形技术与装备教育部工程研究中心，山西太原　030024；
3. 金属成形技术与重型装备全国重点实验室，山西太原　030024）

摘　要： 板带作为高附加值、高技术含量的钢铁产品，广泛应用于航空航天、海洋工程、核电装备等行业领域。针对目前板带热轧生产线设计领域主要依靠设计人员经验，难以应对客户定制化、个性化产品不断增长需求的问题，构建了基于数据驱动与专家知识的板带热轧生产线智能化设计系统，实现了多种类型板带热轧线总体布局、设备配置等数字化柔性设计。建立了板带生产全流程高精度的温度、变形抗力、轧制力等计算模型，提出了高效高质轧制规程制定方法，开发了轧件塑性变形二/三维有限元法与辊系弹性变形影响函数法耦合的有载辊缝快速计算模型，提高了板凸度和板形计算效率和精度。最终开发出具有完全自主知识产权的板带热轧生产线智能化设计及工艺优化软件，为我国热轧技术高质量发展提供支撑。

关键词：板带；热轧；模型；智能化；软件

Modeling and Software Development for Hot-rolled Production of Plate and Strip

WANG Tao[1,2,3], LIU Yuanming[1,2,3], HUANG Qingxue[1,2,3]

(1. College of Mechanical and Vehicle Engineering, Taiyuan University of Technology, Taiyuan 030024, China;
2. Engineering Research Center of Advanced Metal Composites Forming Technology and Equipment, Ministry of Education, Taiyuan University of Technology, Taiyuan 030024, China;
3. National Key Laboratory of Metal Forming Technology and Heavy Equipment, Taiyuan 030024, China)

Abstract: Hot-rolled plate and strip, as a high-value-added and high-tech steel product, is widely used in industries such as aerospace, marine engineering, and nuclear power equipment. In response to the current challenges where the design of hot-rolled plate and strip production lines heavily relies on the experience of design personnel and struggles to meet the growing demand for customized and personalized products, we have developed an intelligent design system for hot-rolled plate and strip production lines based on data-driven approaches and expert knowledge. This system enables the digital and flexible design of various types of hot-rolled plate and strip production lines, including overall layouts and equipment configurations. We have established highly accurate computational models for temperature, deformation resistance, rolling force, and other parameters throughout the entire process of hot-rolled plate and strip production. We have also proposed an efficient and high-quality rolling process development method. Additionally, we have developed a rapid calculation model for loaded roll gap based on the coupling of the two/three-dimensional finite element method for plastic deformation of rolled pieces and the elastic deformation influence function of the roll system. This has significantly improved the efficiency and accuracy of calculations for plate curvature and shape. Ultimately, we have developed intelligent design and process optimization software for hot-rolled plate and strip production lines with complete intellectual property rights, providing support for the high-quality development of hot-rolling technology in China.

Key words: plate and strip; hot rolling; modeling; intelligent; software

14 冶金自动化与智能化

大会特邀报告
第十三届冶金青年科技奖获奖人特邀报告
分会场特邀报告
矿业工程
焦化及节能环保
炼铁与原料
炼钢与连铸
电冶金与废钢铁
轧制与热处理
表面与涂镀
金属材料深加工
粉末冶金
先进钢铁材料及应用
节能与低碳技术
冶金环保与资源利用
冶金设备与工程技术
★ 冶金自动化与智能化
冶金物流
冶金流程工程学
其他

钢铁企业电力负荷预测方法综述

张效华

（中国钢研科技集团有限公司钢铁绿色化智能化技术中心，北京 100081）

摘　要：电力负荷预测对于钢铁企业电力系统经济稳定运行具有重要意义。钢铁企业电力负荷影响因素多，呈现负荷冲击大、随机性强的特点，因此电力负荷预测一直是钢铁企业具有挑战性的问题。电力负荷预测方法主要包括参数方法、机器智能算法、深度学习方法和组合预测方法等。本文基于近年来电力负荷预测相关的文献，分析了相关算法的理论及其改进算法的应用，对钢铁企业电力负荷预测相关研究做了综述。最后对在钢铁企业进行电力负荷预测时可以考虑的方法做出了讨论。

关键词：电力负荷预测；回归模型；时间序列；机器学习；深度学习

A Review of Load Forecasting in Iron and Steel Plants

ZHANG Xiaohua

(Steel Green Intelligent Technology Center, China Iron & Steel Research
Institute Group, Beijing 100081, China)

Abstract: Power load forecasting is of great significance to the economic and stable operation of the power system in iron and steel enterprises. Many factors affect the power load of iron and steel enterprises, showing the characteristics of significant load impact and strong randomness. Therefore, power load forecasting has always been challenging for iron and steel enterprises. Power load forecasting methods mainly include parameter methods, machine intelligence algorithms, deep learning, and combined forecasting methods. This paper investigates the literature on power load forecasting in recent years, analyzes the theory of related algorithms and the application of improved algorithms, and summarizes the related research on power load forecasting in iron and steel enterprises. Finally, the methods that should be considered in the power load forecasting of iron and steel enterprises are discussed.

Key words: load forecasting; regression model; time sequence; machine learning; deep learning

一种工业机器人视觉定位技术的原理及应用

陈　贝

（宝武集团广东中南钢铁股份有限公司，广东韶关 512123）

摘　要：本文介绍了一种典型的工业机器人视觉定位技术及其系统构成。该系统由工业机器人、PLC 系统、视觉检测系统、视觉定位系统、视觉检测辅助设备构成。该系统适合应用于环境复杂多变的钢铁冶金生产现场。本文通过对该种系统的构成和工作原理进行剖析，并通过举例说明其工作流程，从而阐明了该种工业机器人视觉定位技术的原理及其特点。

关键词：智能制造；钢铁冶金；工业机器人；视觉检测

The Principle and Application of Visual Positioning Technology for Industrial Robot

CHEN Bei

(Baowu Group Guangdong Zhongnan Iron and Steel Co., Ltd., Shaoguan 512123, China)

Abstract: This paper introduces a typical industrial robot visual positioning technology and its system structure. The system consists of industrial robot, PLC system, visual inspection system, visual positioning system and visual inspection auxiliary equipment. The system is suitable for iron and steel metallurgy production site with complex and changeable environment. In this paper, the structure and working principle of the system are analyzed, and its working flow is illustrated by an example, so as to clarify the principle and characteristics of the visual positioning technology of industrial robots.

Key words: intelligent manufacturing; iron and steel metallurgy; industrial robot; visual inspection

连铸大包自动浇钢技术开发与应用

唐伟新，陈 贝

（宝武集团广东中南钢铁股份有限公司，广东韶关 512123）

摘 要： 本文介绍了在中南钢铁炼钢厂应用的连铸大包自动浇钢技术，本技术主要配设3台6轴铸造级工业机器人，自动完成机器人自动拆装滑动水口滑板油缸、机器人自动拔插介质管线、机器人自动套装拆卸大包长水口、机器人自动测温取样、长水口碗部自动清理及自动加大包长水口密封圈、机器人自动分拣及自动加覆盖剂等若干工作。此技术的应用效果良好，具有广阔的应用前景。

关键词： 钢铁冶金；连铸；机器人；浇钢

Development and Application of Automatic Casting Technology for Continuous Casting Ladle

TANG Weixin, CHEN Bei

(Baowu Group Guangdong Zhongnan Iron and Steel Co., Ltd., Shaoguan 512123, China)

Abstract: This paper introduces the continuous casting automatic ladle casting technology applied in Zhongnan Iron and Steel Plant. This technology is mainly equipped with three industrial robots of 6-axis casting class. The robot can automatically disassemble and install the sliding nozzle slide cylinder, automatically pull and insert the medium pipeline, automatically disassemble the large bag long nozzle, automatically measure and sample the temperature of the robot, automatically clean the bowl part of the long nozzle and automatically increase the sealing ring of the long nozzle, automatically sort and automatically add the covering agent. The application effect of this technique is good and it has broad application prospect.

Key words: iron and steel metallurgy; continuous casting; robot; casting steel

基于 5G-TSN 架构的云化 PLC 时延保障方法研究

张 岩[1]，王军生[1]，孙 雷[2]，孙瑞琪[1]，秦大伟[1]

（1. 鞍钢集团北京研究院有限公司，北京 102201；2. 北京科技大学自动化学院，北京 100083）

摘 要：工业控制系统对承载网络性能要求极为严格，不仅需要具备低时延、低抖动和高可靠能力，确定性的时延保证是其系统安全可控的基础，为此提出了一种基于 5G 与时间敏感网络（TSN）的云化 PLC 时延保障方法。在 TSN 中，改进了业务流处理架构，优先处理承载于较差质量无线信道的工业业务流；在 5G 网络中，分析了 5G 系统传输时延影响因素，对无线信道质量、无线资源数量与空口最大传输次数的关系进行建模，基于该模型动态规划承载于不同无线资源上工业控制信号的 5G 系统传输时延预算。仿真结果表明，提出的联合调度时延保障方法能降低承载于较差信道质量的云化 PLC 时间敏感网络域传输时延，并能有效消除 5G 空口随机变化导致的重传对传输时延造成的抖动，实现云化 PLC 端到端确定性时延保障。

关键词：5G 与 TSN；联合调度；云化 PLC；跨网确定性传输；确定性时延

Research on Delay Guarantee Mechanism of Cloud Programmable Logic Controller Based on 5G-TSN Architecture

ZHANG Yan[1], WANG Junsheng[1], SUN Lei[2], SUN Ruiqi[1], QIN Dawei[1]

(1. Beijing Research Institute of Ansteel Co., Ltd., Beijing 102201, China;
2. University of Science and Technology Beijing, Beijing 100083, China)

Abstract: The industrial control system has very strict requirements on the performance of the bear-networks, which requires not only low delay, low shake and high reliability. The deterministic delay guarantee is the basis of system security and control. A delay guarantee method of cloud PLC based on 5G and time sensitive network (TSN) is considered. The architecture of service flow processing is improved to give priority to the industrial service flow carried by the poor quality wireless channel in TSN. In 5G network, factors affecting the transmission dela of 5G are analyzed, and the relationship between the quality of wireless channel, the number of wireless resources and the maximum transmission times of air interface is modeled. Based on this model, the maximum retransmission times meeting the requirements of data transmission reliability can be obtained. The retransmission factor can be set dynamically to dynamically plan the transmission delay budget of 5G system bearing industrial control signals on different wireless resources. Simulation results demonstrate that the proposed joint scheduling mechanism achieves less TSN transmission delay of industrial traffics bearing on worse quality radio resources and can eliminate shaker caused by retransmission due to variation of 5G radio channel, which can guarantee end-to-end deterministic delay for industrial service.

Key words: 5G and TSN; joint scheduling; cloud PLC; deterministic transmission cross networks; deterministic delay

炼钢厂行车实时调度系统研究及应用

徐林伟，卢　义，张　飞，徐超琼，丁昭祥

（中冶赛迪重庆信息技术有限公司，重庆　404100）

摘　要：针对钢铁企业炼钢厂行车调度问题，考虑到炼钢生产过程各个工序较难形成准确的标准工时以及各种突发问题较多，很难根据炉次计划和浇次计划分解出长时间的行车需要执行的任务，而目前论文更多是通过炉次计划和浇次计划分解出行车作业任务再对任务分配进行行车调度的优化，难以实现工程的应用，而且没有考虑某一行车完成一系列作业任务的情况。本文将通过对比精益管理标准化调度规则方法和通过考虑复杂的系列动作的高效的启发式优化算法与仿真的方法实现行车明确任务的实时调度，最后选取效果较好的方法应用于炼钢行车调度并实现工程化应用。

关键词：精益；标准化；天车调度；调度规则；输入输出

Research and Application of Real-time Operation Scheduling System in Steelworks

XU Linwei, LU Yi, ZHANG Fei, XU Chaoqiong, DING Zhaoxiang

(CISDI Information Technology Co., Ltd., Chongqing 404100, China)

Abstract: In view of the problem of traveling car scheduling in steelmaking plants of iron and steel enterprises, it is difficult to decompose the tasks to be performed by traveling car for a long time according to furnace plan and pouring plan. Because it is difficult to form prepared standard working hours for each process of steelmaking production process and there are many kinds of unexpected problems. Many papers currently decompose the traveling car operation tasks by furnace plan and pouring plan and then optimize the traveling car scheduling by task assignment, which is difficult to realize the application of engineering. In this paper, we will compare the rules of forming standardized traveling crane scheduling by summarizing traveling crane scheduling through lean management and the optimization effect by optimizing traveling crane with more definite tasks in real time, and select the method with better effect to apply to steelmaking traveling crane scheduling and realize the application of engineering.

Key words: lean; standardization; overhead crane scheduling; scheduling rules; input and output

基于大数据分析的热镀锌机组关键设备故障诊断技术

王　鲁[1]，吴　军[1]，顾成勇[1]，王学敏[2]

（1. 宝钢日铁汽车板公司，上海　201900；2. 宝钢股份中央研究院，上海　201900）

摘 要：目前热镀锌机组在设备运维方面主要存在以下问题：在线监控比例低，使得异常发现与维修不及时，如部分射线区域，日常设备管理人员不能进入，一些关键设备的齿轮箱未能实现实时设备的状态把控；点检方式原始、低效，数据采集单一，未将各数据系统的数据统一管理和有效利用；高度依赖诊断人员的业务水平；已有的一些在线诊断系统由于数据单一，模型不准，造成过度告警或故障遗漏；全流程（端到端）及支持系统集成度低，本文对热镀锌机组关键设备常见故障机理分析，以尽可能少的安装硬件传感器为宗旨，将大数据与人工智能作为主要技术手段，构建以运行机理为基础、数据驱动为主线、采集现场已有生产数据进行分工况设备状态分析和建模。通过设备自身过程数据"以软代硬"进行健康度预诊。

关键词：大数据；数据清洗；模型；健康度；设备状态

Fault Diagnosis Technology for Critical Equipment of Hot Dip Galvanizing Unit Based on Big Data Analysis

WANG Lu[1], WU Jun[1], GU Chengyong[1], WANG Xuemin[2]

(1. Baosteel-Nippon Steel Automotive Steel Sheets Co., Ltd., Shanghai 201900, China;
2. Baosteel Academia Sinica, Shanghai 201900, China)

Abstract: At present, there are mainly the following problems in equipment operation and maintenance of hot-dip galvanizing units: low proportion of online monitoring, resulting in delayed detection and maintenance of abnormalities, such as some radiation areas where daily equipment management personnel cannot enter, and some key equipment gearboxes fail to achieve real-time equipment status control; The inspection method is primitive and inefficient, with single data collection, and the data of each data system is not uniformly managed and effectively utilized; Highly dependent on the professional level of diagnostic personnel; Some existing online diagnosis systems cause excessive alarm or fault omission due to single data and inaccurate model; The integration of the entire process (end-to-end) and supporting systems is low. This article analyzes the common fault mechanisms of key equipment in hot-dip galvanizing units, with the aim of minimizing the installation of hardware sensors. Big data and artificial intelligence are used as the main technical means, and the construction is based on operating mechanisms, data-driven as the main line, and the collection of existing production data on site for equipment status analysis and modeling under different working conditions. Pre diagnosis of health status is carried out through the device's own process data we called 'replacing hardware with software'.

Key words: big data; data cleaning; model; health level; device status

焦化厂能源管理中心方案研究及应用

孙 玲，刘亚玲，高凤君，李明杰，樊艳伟

（北京首钢自动化信息技术有限公司信息事业部，北京 100041）

摘 要：本文根据焦化厂能源管理特点和生产实际情况研究了焦化厂能源管理中心解决方案并进行了应用总结。将大数据技术等新信息技术同智能化能源管理中心实际需求相结合，提出能源管理中心在焦化行业中应用的新思路。构建全面的能源统一管理体系重点介绍和阐述了焦化厂能源管理中心的架构设计、主要功能、技术架构，通过实施能源管理中心，归纳了能源管理中心在焦化企业中的应用效果及发挥的优势作用，为焦化企业能源管理中心的建设提供借鉴和参考。

关键词：焦化厂；能源管理中心；智能化；信息技术

Research and Application of Energy Management Center Solution in Coking Plant

SUN Ling, LIU Yaling, GAO Fengjun, LI Mingjie, FAN Yanwei

(Information Department, Beijing Shougang Automation & Information Technology Co., Ltd., Beijing 100041, China)

Abstract: According to the characteristics of energy management in coking plant and the actual production situation, this paper studies the solution of energy management center in coking plant and summarizes its application.Combine big data technology and other new information technologies with the actual needs of intelligent energy management center, and propose new ideas for the application of energy management center in coking industry. Building a comprehensive unified energy management system focuses on the introduction and elaboration of the architecture design, main functions and technical framework of the energy management center of the coking plant. Through the implementation of the energy management center, the application effects and advantages of the energy management center in coking enterprises are summarized, providing reference and reference for the construction of the energy management center of coking enterprises.

Key words: coking plant; energy management center; intelligent; information technology

武钢 CSP 转炉在低铁钢比下氮的控制

徐培春，邱 晨，陈尚波，叶 飞

（武钢有限热轧厂，湖北武汉 430080）

摘 要：根据转炉冶炼过程的增氮原理进行分析，确定了在低铁钢比下转炉主要的增氮因素，并通过数据统计分析，进行了验证。最后总结出了转炉在低铁钢比下的控氮措施，取得了较好效果。

关键词：转炉；控氮；低铁钢比

煤气透平机组智能运维解决方案关键技术研究

蔡正国，李征磊，陈红琳，章天一

（宝武装备智能科技有限公司，上海 201900）

摘 要：研究煤气透平机设备的状态异常和维护策略，提出煤气透平机组智能运维的数据采集方案和智能诊断方法，形成了透平机发电机、大型阀门管道系统、液压润滑系统、氮气系统、循环水系统的预警诊断模型和智能运维解决方案。通过对煤气透平机组的在线状态监测和故障诊断，向用户实时推送设备智能运维检修建议，避免因煤气透平机组转子不平衡、安装不对中、油膜轴承故障、喘振和油膜振荡等导致的非计划停机，支撑煤气透平机组的安全运行。

关键词：煤气透平机组；在线监测与故障诊断；远程运维

Research on Key Technology of Smart Operation & Maintenance Solution for Gas Turbine Units

CAI Zhengguo, LI Zhenglei, CHEN Honglin, ZHANG Tianyi

(Baowu Equipment Intelligence Technology Co., Ltd., Shanghai 201900, China)

Abstract: The present situation of Blast Furnace Gas Turbine Units is introduced and solution of smart operation & maintenance is put forward through analyzing the turbine abinomal and maintenance strategy. Real-time alarm & diagnosis criterion together with key technology of smart operation & maintenance are determined for turbine, large valve system, hydraulic and lubrication system, nitrogen system and cycling water system etc. By on-line surveillance and fault diagnosis through the vibration and process parameters of gas turbine, the operation & maintenance suggestion are pushed to handle the shutdown caused by rotor imbalance, misalignment, sliding bearing failure, surge as well as oil whip.

Key words: gas turbine; on-line sueveilaance & fault diagnosis; remote operation & maintenance

大型钢铁企业生产系统存储双活的构建与实践

樊贵先[1]，程志[1]，彭超[2]，彭震[1]，胡炜[1]

（1. 武汉钢铁有限公司，湖北武汉　430000；2. 宝信软件（武汉）有限公司，湖北武汉　430080）

摘　要： 伴随信息技术的飞速发展，存储系统与其自身的高可靠性，在数据中心建设中占有重要地位，但在实际运行过程中经常会出现，因存储单点故障造成系统停机、数据丢失等案例。本文以大型钢铁企业生产系统存储双活的构建与实践，通过 MCC 存储架构，搭建数据存储双活站点，满足钢铁企业生产系统业务运行的连续性和高可用性，实现数据中心容灾，避免因存储单点故障造成系统运行风险。

关键词： 存储；双活；脑裂；构建

Construction and Practice of Production System Storage in Large Steel Enterprises

FAN Guixian[1], CHENG Zhi[1], PENG Chao[2], PENG Zhen[1], HU Wei[1]

(1. Wuhan Iron & Steel Co., Ltd., Wuhan 430000, China;
2. Baoxin Software (Wuhan) Co., Ltd., Wuhan 430080, China)

Abstract: With the rapid development of information technology, the storage system and its own high reliability play an important role in the construction of data center, but often appear in the actual operation process, caused by a single point of storage failure. This paper takes the construction and practice of the production system storage of large steel enterprises, through the MCC storage architecture, build the data storage dual activity site, to meet the continuity and high availability of the production system business operation of steel enterprises, realize the disaster recovery of the data center, and avoid the risk of system operation caused by a single point of storage fault.

Key words: storage; double-live brain; crack; construction

双交叉限幅控制在冷轧连续退火炉的应用

兰晓栋，杨红伟，周惠娟，李文静，李 锐，王承刚

（首钢智新迁安电磁材料有限公司，河北唐山 064400）

摘 要： 本文论述了冷轧连续退火炉基于 S7-400 系列 PLC 为基础下的燃烧控制技术，重点针对燃烧控制系统中采用的双交叉限幅控制进行了分析，针对在实际生产应用中当系统负荷发生变化时不能保证空气和燃气的最佳配比的问题和系统响应速度慢的问题，提出根据系统负荷变化，对双交叉限幅模型进行实时修正的双交叉限幅控制。

关键词： 连续退火炉；燃烧；交叉限幅

The Application of Double Cross Control in Continuous Annealing Furnace

LAN Xiaodong, YANG Hongwei, ZHOU Huijuan,
LI Wenjing, LI Rui, WANG Chenggang

(Shougang Zhixin Qian'an Electromagnetic Materials Co., Ltd., Tangshan 064400, China)

Abstract: This paper discusses the control method of combustion based on S7-400 series PLC, mainly about double cross control. In practical production, when the system load changes, it can not guarantee the best ratio between air and gas and system response delay with the consideration of the varying system load. It presents a real time correction for the double cross control.

Key words: continuous annealing furnace; combustion; double cross control

宝山基地 4 号连铸 L2 过程机改造实践

阎建兵，黄晔华，陈建尧，陆 华，朱海健

（宝山钢铁股份有限公司宝山基地设备部，上海 201900）

摘 要： 本文分析了宝钢股份宝山基地 4 号连铸 L2 过程机存在的问题、改造的必要性，介绍了改造技术路线选择，从系统集成设计、数据库表设计、应用功能设计、画面设计、模型设计等方面介绍了自主集成改造方案、实施技术，从改造效果、改造创新点、改造经验方面对自主集成进行了总结。4 号连铸 L2 自主集成项目不仅实现了 5 个月的敏捷设计、投运，同时为公司节约几百万元软件开发投资费用。

关键词： 连铸；改造；模型；接口

Renovation Practice of 4 # CC L2 System in Baosteel

YAN Jianbing, HUANG Yehua, CHEN Jianyao, LU Hua, ZHU Haijian

(Department of Equipment, Baosteel Co., Ltd., Shanghai 201900, China)

Abstract: This paper analyzes the existing problems in the 4 #CC L2system in Baoshan Base of Baosteel Co., Ltd., and the necessity of revamping, introduces the selection of revamping Technology roadmap, introduces the revamping scheme and implementation technology of from the aspects of system integration design, database table design, application function design, HMI design, model design, etc., and summarizes the revamping from the aspects of revamping effects, innovation points and experiences. The 4 #CC L2system revamping project not only achieved 5 months of agile design and operation, but also saved the company several million yuan in software development investment costs.

Key words: casting; renovation; model; interface

"手"+"眼"结合的新一代智能化装备

邓能辉[1]，王孝敏[2]，吴昆鹏[1]，崔广礼[1]

（1. 北京科技大学国家板带生产先进装备工程技术研究中心，北京　100083；
2. 江阴兴澄特种钢铁有限公司厚板分厂，江苏无锡　214400）

摘　要： 中厚板生产过程存在部分长期依赖人工的重复性操作工序，严重制约智能化生产制造的进一步发展。为此北科工研依靠在视觉检测方面积累的大量经验和在自动化控制方面的深厚底蕴为基础设计出粗轧自动转钢、双边剪自动对中、冷床智能上下料三大智能装备，"手""眼"协动实现产线智能化的大幅提升，节约生产成本，提升生产效益。

粗轧自动转钢采用宽泛的角度识别策略解决遮挡、雾气干扰条件下角度计算问题，实现准确全面的角度度量。设计完整时序的转动角度跟踪模块，实现360°内实际角度跟踪，满足复杂因素影响下的角度跟踪需求。采取安全限位和位置优选策略，后台自主学习人工转钢经验，轮询位置组合，实现快速转钢，提升转钢效率。利用多级转速调控模型动态调整转钢角度，保证钢坯转动稳定，降低"欠转""过转"的概率。

双边剪自动对中采用排布多组面阵相机进行辊道区域数据采集，具备超高分辨率，实现钢板边缘细节化的清晰展示。采用级联的视觉识别算法模型，逐层次的优化边缘细粒度分割，提升边缘检测效果，钢板宽度尺寸测量误差控制在5mm以内。利用标定技术并结合先验知识自动虚拟出双边剪切位置线，实时控制磁头移动，剪切位置量化更加精准。本系统可有效去除生产现场对于激光划线设备的依赖，降低设备的维护和更新成本，摆脱人工观测剪切线的主观判定误差，实现更加精确的剪切余量控制。

冷床智能上下料采取双冷床联动模式，利用辊道拼板策略实现多块钢板同步上料，提升冷床区域布料效率，减少设备性能损失，保证冷床最大化布料能力。利用多进程图像同步采集流处理，完成多路相机实时同步位置分析，结合新型语义分割模型，保证钢板重复测量位置误差在要求范围内，实现钢板的精确位置停放。结合逻辑跟踪与视觉检测交叉验证，实现监控盲区下的钢板位置精准跟踪，满足生产现场对于冷床全域的钢板跟踪需求。以行车识别模型、冷床余量判定模型、温度控制模型为基础，综合辊道钢板信息给出时空条件下最优下料时机，实现智能、高效、安全、稳定的冷床上下料联动控制。

视觉和控制结合的新一代智能化装备打破了传统自动化控制依靠单一逻辑判断不足的弊端，真正实现比拟人工操作的高效操作方式，将在关键工序的效率提升上发挥出重要作用。

冶金机器人的可预测碰撞处理

孙 睿, 窦 刚

（中冶南方工程技术有限公司技术研究院, 湖北 武汉 430223）

摘 要: 利用冶金机器人代替人工进行自动贴标[1]、自动拆捆带[2,3]等重复性劳动已逐渐成为各大钢厂自动化产线建设的重要内容之一。在这些应用场景中，机器人的末端执行器需要和工件紧密接触。当工件的尺寸测量存在误差，或工件本身存在一些缺陷时，机器人的实际压下量可能超出设定范围，从而导致机器人触发运动监测报警，需操作工手动操作机器人恢复功能，极大影响了生产效率，且存在较大的安全隐患。

针对上述可预测碰撞问题，本文开发了一种机器人碰撞恢复程序，该碰撞恢复程序能自动识别机器人运动监测错误，并调整机器人压下量，重复原工作过程，避免再次触发运动监测错误，保障生产工序正常运作。值得注意的是，压下量的调整量需要根据现场情况进行调节，若调整后的压下量仍然过大，机器人可能再次触发运动监测错误，而若调整后的压下量过小，可能无法保证工序的正常运作。因此程序可自由设置压下量大小与可允许连续碰撞的最大次数。此外，配置相应的 IO 信号，通过 PROFINET 与 PLC 通讯，使得错误报警信号能正确传输到 PLC 上，保证错误信息不遗漏，方便调试人员了解机器人工作状态。

为检验碰撞恢复程序的可行性，利用机器人实训平台进行测试实验，其中机器人型号为 ABB IRB1200 5kg/0.9m，机器人第六轴上安装了爪状的末端执行器。测试中，机器人从初始位置缓慢下降，直至机器人末端执行器与工作台发生碰撞，触发运动监测错误并进入碰撞恢复程序，调整压下量之后再次从初始位置再次缓慢下降。在这个过程中，实时记录机器人各轴的角位移、扭矩以及 TCP 坐标位置随时间的变化关系。

各轴扭矩随时间的变化关系如图 1、图 2 所示，可以发现机器人在 30s 附近发生碰撞，并成功运行碰撞恢复程序，随后第二次下降时并未发生二次碰撞，验证了该程序的可行性。与此同时，由图 1、图 2 可知，测试实验中，当机器人的末端执行器与工作台发生碰撞时，各轴的扭矩均会发生一定程度的突变，其中 2 轴和 3 轴的扭矩变化最大，说明 2 轴和 3 轴受到的冲击力更大。此外，与姿态 2 相比，采用姿态 1 发生碰撞时，扭矩的变化更小，这是因为采用姿态 1 下降时，所受反作用力与 6 轴平行，碰撞时的刚性更大。因此在采用姿态 1 的方式进行自动贴标、自动拆捆等工序时需要在末端执行器上增加一定的缓冲结构以减少机器人各轴的刚性冲击。

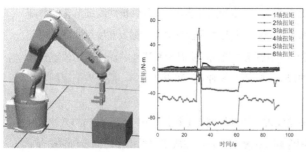
图 1 采用姿态 1 发生碰撞时，各轴扭矩随时间的变化

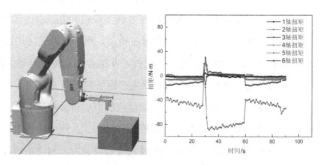
图 2 采用姿态 2 发生碰撞时，各轴扭矩随时间的变化

参 考 文 献

[1] 刘青松, 王磊. 高温板坯智能贴标系统的研究与应用[J]. 冶金自动化, 2022, 46(S1): 446-447.
[2] 李康宇, 李柳, 赵东洋, 等. 基于工业机器人的全自动钢卷拆捆系统研发与应用[J]. 制造业自动化, 2020, 42(8): 43-46.
[3] 王荣浩, 徐斌, 张文博, 等. 钢卷自动拆捆机器人捆带定位系统研究与应用[J]. 制造业自动化, 2022, 44(9): 155-158, 188.

鞍钢1450连退线2号卷取机带尾定位的控制过程

张国利,林 彬,郑大勇

(鞍钢股份有限公司冷轧厂,辽宁鞍山 114000)

摘 要:鞍钢股份冷轧厂1450连退机组为成品直下机组,部分成品产生钢卷带尾边浪、带尾溢出边、钢卷外圈划伤的质量缺陷,采用卷取机带尾定位速度分段控制使带尾快速通过卷取机转换导板,对转换导板进行延缓控制,控制抬起和压下时机,在程序中创造出可以控制的空间和时间,以解决穿带和抛尾带钢无间隙的问题,减少带尾边浪;增加永磁铁和电磁铁产生微小张力,配合 EPC 控制系统消除带尾溢出边;改变卸卷车卸卷自动过程,避免卸卷车与钢卷产生相对滑动摩擦,消除钢卷外圈划伤。

关键词:质量缺陷;卷取机带尾定位速度;卷取机转换导板;EPC 控制;卸卷自动过程

Abstract: In the 1450 continuous annealing unit of Ansteel Cold Rolling Plant, there are quality defects such as wave on the tail edge of the steel coil, overflow edge of the tail, and scratch of the outer ring of the steel coil. The section control of the positioning speed of the recoiling machine tail is adopted to make the tail quickly pass through the recoiling machine to convert the guide plate, delay the control of the conversion guide plate, control the timing of lifting and lowering, and create controllable space and time in the program. In order to solve the problem of no gap between the steel plate of the wearing belt and the throwing tail, reduce the trailing edge wave; Add permanent magnet and electromagnet with EPC control system to eliminate tail overflow edge; Change the automatic unloading process of the uncoil truck to avoid the relative sliding friction between the uncoil truck and the steel coil and eliminate the scratches.

Key words: quality defect; recoiling machine tail positioning speed; recoiling machine conversion guide; EPC control; automatic uncoiling process

面向智能制造的知识图谱驱动设备故障诊断方法研究

刘佳伟[1],王军生[2],金 鹏[1],蔡 畅[3],孙瑞琪[2]

(1. 鞍钢集团自动化有限公司,辽宁鞍山 114000;2. 鞍钢集团北京研究院有限公司,北京 102200;3. 辽宁科技大学电子与信息工程学院,辽宁沈阳 110325)

摘 要:针对旋转机械故障诊断过程中,征兆条件缺失导致传统基于规则的故障诊断推理方法准确率低的问题,本文提出了一种知识图谱驱动设备故障诊断方法。首先,利用本体建模技术,建立了一种基于本体的知识表示方法,用于构建旋转机械故障诊断的知识本体表示模型并在此基础上,构建了旋转机械故障诊断的知识图谱。其次,结合基于关系路径的知识图谱推理方法,提出了基于知识图谱的旋转机械故障诊断方法,利用旋转机械设备结构间关系进行故障原因推理。最后,通过以钢厂电机为例的实验验证,得出基于知识图谱的故障诊断方法在输入征兆缺失的条件下,其准确率达到了 91.1%,明显高于传统基于规则的故障诊断推理方法的准确率,有效解决了征兆缺失导致的诊断准确率低的问题。本文的研究成果表明,使用知识图谱来驱动设备故障诊断是一种新颖的方法,可以应用于旋转机械故障诊断，同时还有望为其他机械设备的智能诊断方法的发展提供有益的经验。

关键词：知识图谱；旋转机械；故障诊断；基于规则的推理方法

Research on Knowledge Graph-driven Equipment Fault Diagnosis Method for Intelligent Manufacturing

LIU Jiawei[1], WANG Junsheng[2], JIN Peng[1], CAI Chang[3], SUN Ruiqi[2]

(1. Ansteel Group Automation Co., Ltd., Anshan 114000, China; 2. Beijing Research Institute of Ansteel Group Co., Ltd., Beijing 102200, China; 3. School of Electronics and Information Engineering, Liaoning University of Technology, Shenyang 110325, China)

Abstract: This article proposes a knowledge graph-driven method for diagnosing faults in rotating machinery, addressing the problem of low accuracy in traditional rule-based diagnostic reasoning methods caused by missing symptom conditions. Firstly, using ontology modeling technology, a knowledge representation method based on ontology was developed to construct a knowledge ontology representation model for diagnosing faults in rotating machinery. Based on this, a knowledge graph for diagnosing faults in rotating machinery was successfully constructed. Secondly, a knowledge graph-based rotating machinery fault diagnosis method is proposed, combining a relation path-based knowledge graph inference method to infer the causes of faults using the relationships between the structures of rotating machinery equipment. Finally, an experimental verification is conducted using a motor as an example, showing that the accuracy of the knowledge graph-based fault diagnosis method reached 91.1% under the condition of missing symptom inputs, which is significantly higher than the accuracy of traditional rule-based diagnostic reasoning methods, effectively solving the problem of low diagnostic accuracy caused by missing symptoms. The research findings of this article indicate that using a knowledge graph to drive equipment fault diagnosis is a novel method that can be applied to the diagnosis of faults in rotating machinery, and is also expected to provide valuable experience for the development of intelligent diagnosis methods for other mechanical equipment.

Key words: knowledge graph; rotating machinery; fault diagnosis; rule-based reasoning method

高可用性缓存管理系统的构建

车 兰，赵 勐，刘 星，吴长河，孙上海，金树来

（鞍钢集团信息产业有限公司，辽宁鞍山 114000）

摘 要：鞍钢股份热轧带钢厂 EPS 生产线 MES 系统基于 keepalived+redis 构建高可用性缓存管理系统，集先进的信息化技术及实用性于一体，具有稳定性高、响应速度快、配置简便、成本低等特点。本文对系统的工作原理、方案思路、keepalived 和 redis 的特点与优势分析、系统参数配置等方面进行了论述。该系统能够完全满足现场生产的要求，适合在其他类似生产线上进行推广应用。

关键词：高可用性；缓存管理；keepalived；redis；配置

Construction for High Availability Cache Management System

CHE Lan, ZHAO Meng, LIU Xing, WU Changhe, SUN Shanghai, JIN Shulai

(Ansteel Group Information Industry Co., Ltd., Anshan 114000, China)

Abstract: The MES system of the EPS production line of Angang Steel Hot Rolling Mill is based on keepalived+redis to build a high availability cache management system, which integrates advanced information technology and practicality. It has the characteristics of high stability, fast response speed, simple configuration, and low cost. This article discusses the working principle, solution ideas, characteristics and advantages analysis of keepalived and redis, and system parameter configuration of the system. This system can fully meet the requirements of on-site production and is suitable for promotion and application on other similar production lines.

Key words: high availability; cache management; keepalived; redis; configuration

鞍钢销售物流管理系统的设计与实现

韩奇颖

（鞍钢集团信息产业有限公司智慧运维事业部，辽宁鞍山　114000）

摘　要： 随着互联网与信息技术在物流环节的广泛应用，钢铁企业也迫切需要借助信息技术手段，构建适合自身物流业务及流程的信息化平台，并将上游及下游环节串联起来，形成完整的供应链。鞍钢是中国的大型钢铁企业，年产能超四千万吨。本论文结合当前鞍钢销售物流的业务需求及目前钢铁企业物流系统的现状分析，介绍了鞍钢销售物流管理系统所涵盖内容及其大致的研究情况，深入地研究鞍钢销售物流管理系统的设计与实现方法。

关键词： 钢铁企业；销售物流管理；信息化

Design and Implementation of Ansteel Sales Logistics Management System

HAN Qiying

(Smart Operation and Maintenance Business Unit, Information Industry Co., Ltd. of Ansteel Group, Anshan 114000, China)

Abstract: With the widespread application of the Internet and information technology in logistics, steel enterprises urgently need to use information technology means to build an information platform suitable for their own logistics business and processes, and connect upstream and downstream links to form a complete supply chain. Ansteel is a large steel enterprise in China with an annual production capacity of over 40 million tons. This paper combines the current business needs of Ansteel's sales logistics and the analysis of the current situation of the logistics system of steel enterprises, introduces the content and general research situation of Ansteel's sales logistics management system, and deeply studies the design and implementation methods of Ansteel's sales logistics management system.

Key words: steel enterprises; sales logistics management; promotion of information technology

基于 BP 神经网络的钢包渣眼演化行为的预测

刘晓航[1,2]，王盱杰[3]，刘　畅[1,2]，贺　铸[1,2]，李光强[1,2]，王　强[1,2]

（1. 武汉科技大学省部共建耐火材料与冶金国家重点实验室，湖北武汉　430081；
2. 武汉科技大学钢铁冶金及资源利用省部共建教育部重点实验室，湖北武汉　430081；
3. 武汉工程大学资源与安全工程学院，湖北武汉　430073）

摘　要：钢包底吹精炼过程中，渣眼行为对钢成分和夹杂物数量的调控有着至关重要的作用。由于生产现场工况复杂、温度较高、成本较大等原因，现场测试存在一定难度和危险，测试结果的精度也无法保证。为准确探究钢包底吹孔位置、单孔底吹气体流量和渣层厚度对渣眼面积的影响，本文采用 1∶5 的相似比，以 150t 实际工业钢包为原型建立水模型，利用 BP 神经网络算法拟合实验数据生成模型，对精炼过程中渣眼的演化行为进行预测。分析表明，当隐藏层神经元数目 $n=16$，迭代次数 Epoch=60000 时，模型损失函数 Error 值达到最小，决定系数 $R2=93.439\%$，模型性能优异，预测精度满足工业需求，可有效地指导工业生产。

关键词：BP 神经网络；150t 工业钢包；渣眼演化；水模型；单孔底吹

Prediction of Evolution Behavior of Slag Eye in Steel Ladle Based on BP Neural Network

LIU Xiaohang[1,2], WANG Xijie[3], LIU Chang[1,2], HE Zhu[1,2],
LI Guangqiang[1,2], WANG Qiang[1,2]

(1. The State Key Laboratory of Refractories and Metallurgy, Wuhan University of Science and Technology, Wuhan 430081, China; 2. Key Laboratory for Ferrous Metallurgy and Resource Utilization of Ministry of Education, Wuhan University of Science and Technology, Wuhan 430081, China;
3. School of Resources and Safety Engineering, Wuhan Institute of Technology, Wuhan 430073, China)

Abstract: The behavior of slag eye during the refining process of bottom blown steel ladle plays a crucial role in regulating the steel composition and the quantity of inclusions. Due to the complex working conditions, high temperatures, and high costs at the production site, on-site testing is challenging and carries certain risks, and the accuracy of the test results cannot be guaranteed. In order to accurately investigate the effects of the position of the bottom blown hole, the gas flow rate of a single hole, and the thickness of the slag layer on the slag eye area in the steel ladle refining process, this study used a 1∶5 similarity ratio and established a water model based on a 150-ton actual industrial ladle prototype. The BP neural network algorithm was employed to fit the experimental data and generate a predictive model for the evolution behavior of slag eye during the refining process. The analysis shows that when the number of neurons in the hidden layer is $n=16$ and the number of iterations Epoch is 60000, the model loss function Error value is minimized, and the determination coefficient R2 is 93.439%. The model demonstrates excellent performance, and its prediction accuracy meets industrial requirements, effectively guiding industrial production.

Key words: BP neural network; 150-ton industrial ladle; slag eye; water model; single hole bottom blowing

面向冶金领域的巡检无人机轨迹跟踪控制系统研究

高 艺[1,2]，刘景亚[1,2]，吴曼玲[1,2]

（1. 中冶赛迪技术研究中心有限公司智能机器人研究所，重庆 401120；
2. 冶金智能装备重庆市重点实验室，重庆 401120）

摘 要：针对冶金现场人工巡检效率低、劳动强度大等问题，利用四旋翼无人机高机动性、灵活控制性和经济性的优势，替代工人执行巡检工作。基于牛顿-欧拉方程建立四旋翼的动力学和运动学模型，由于四旋翼无人机系统具有固有的非线性特性，设计一种非线性模型预测控制器（NMPC），利用非线性动力学的状态相关系数（SDC）将运动方程分解，得到模型的伪线性形式，降低优化问题的求解难度和计算量，克服模型线性化带来的误差，改善无人机的稳态误差、动态跟踪特性和抗干扰能力。最后通过仿真验证了所述方法的有效性和可行性。

关键词：冶金现场；巡检无人机；非线性模型预测控制；状态相关系数；轨迹跟踪

Research on Trajectory Tracking Control of Quad-rotor for Metallurgical Industry

GAO Yi[1,2], LIU Jingya[1,2], WU Manling[1,2]

(1. CISDI Research & Development Co., Ltd., Intelligent Robot Research Division, Chongqing 401120, China; 2. Chongqing Key Laboratory for Metallurgical Intelligent Equipment, Chongqing 401120, China)

Abstract: Aiming at the problems such as low efficiency and high labor intensity of manual inspection in metallurgical field, the quad-rotor is used to replace workers to carry out inspection. The dynamics and kinematics models of quad-rotor are established based on Newton-Euler equation. Due to the inherent nonlinear characteristics of quad-rotor UAV system, a nonlinear model prediction controller (NMPC) is designed. The state correlation coefficient (SDC) of nonlinear dynamics is used to decompose the motion equation and obtain the pseudo-linear form of the model, which reduces the solving difficulty and calculation amount of optimization problems, overcomes the errors caused by model linearization, and improves the steady-state errors, dynamic tracking characteristics and anti-interference ability of the UAV. Finally, the effectiveness and feasibility of the proposed method are verified by simulation.

Key words: metallurgical factories; quad-rotor; nonlinear model prediction; state dependent coefficient; trajectory tracking

鞍钢股份炼钢总厂智能铁水倒罐间场景开发与应用

吴丙恒，吴世龙，颜世蛟

（鞍钢股份有限公司炼钢总厂，辽宁鞍山 114000）

摘 要：本文针对铁水倒罐间场景智能化，介绍了智能化效果及其实施的全过程。制定了人工替代、算法辅助和平台赋能的战略目标，搭建了防碰撞、防溢铁、防脱钩的安全架构，对铁水倒运业务流程进行了数字化转型，开发了辅助泊车、柔性插拔电、自动配铁、自动折铁、自动测温取样、辅助摘挂罐等智能模型，总结阐述了智能场景落地生态的充分与必要条件。智能化技术的开发与应用，助力铁水倒罐间 3D 岗位无人化、安全本质化、生产高效化，展示了一种传统制造向智能制造转型的生态模式。

关键词：智能场景；战略目标；安全架构；业务架构；技术模块；落地生态

Development and Application of Intelligent Molten Iron Pouring Tank Scene of AnSteel General Plant

WU Bingheng, WU Shilong, YAN Shijiao

(General Steelmaking Plant of Angang steel Co., Ltd., Anshan 114000, China)

Abstract: Aiming at the intelligence of molten iron pouring tank scene, this paper introduces the whole process of intelligent effect and its implementation. The strategic goals of manual substitution, algorithm assistance and platform empowerment were formulated, a safety architecture for anti-collision, anti-spillage and anti-decoupling was built, the digital transformation of the molten iron reverse transportation business process was carried out, and intelligent models such as auxiliary parking, flexible plug-in and unplugged, automatic iron distribution, automatic iron folding, automatic temperature measurement and sampling, and auxiliary hanging tanks were developed, and the sufficient and necessary conditions for the landing ecology of intelligent scenes were summarized. The development and application of intelligent technology has helped the 3D post in the molten iron pouring tank room to be unmanned, safe and essential, and production efficient, demonstrating an ecological model of transformation from traditional manufacturing to intelligent manufacturing.

Key words: intelligent scenario; strategic goal; security architecture; business architecture; technology module; landing ecology

鞍钢热轧厂数字化车间建设探索与实践

董 广，郑英杰，张 平，乔树峰

（鞍钢股份有限公司，辽宁鞍山 114021）

摘　要：热轧带钢生产工艺在经历以自动化、信息化为主要特征的工业 3.0 阶段后，依靠传统的工艺装备在提升生产效率、提高控制精度及成本控制方面已进入瓶颈，可提升空间有限。随着物联网、云计算、大数据、人工智能等新技术的飞速发展，通过开展工艺模型、智能算法、大数据、可视化方面的能力建设，利用数据挖掘、机器学习，融合专家知识经验，构建智能产线的数字孪生系统成为打造智能化、高效率、低成本的绿色热轧产线关键。鞍钢股份热轧带钢厂通过开发轧线智能测控技术、高精度数学模型、远程运维技术等，实现轧线极度自动化和生产集约化，通过构建全域大数据平台，深度融合智能算法与工艺技术，以工厂绩效作为引领，实现热轧生产过程的生产、质量、设备、能源等任务进行协同管控，进一步为生产质量、效率、成本的精益化管控，提升产线智能化的内涵赋能。

关键词：热轧带钢生产；智能算法；精益化管控；智能产线

Exploration and Practice on the Construction of Digital Workshop in Angang Hot Rolling Mill

DONG Guang, ZHENG Yingjie, ZHANG Ping, QIAO Shufeng

(Angang Steel Co., Ltd., Anshan 114021, China)

Abstract: After experiencing the industrial 3.0 stage characterized by automation and informationization, the production process of hot rolled strip steel has entered the bottleneck of improving production efficiency, improving control accuracy and cost control by relying on traditional process equipment, and the space for improvement is limited.With the rapid development of new technologies such as the Internet of Things, cloud computing, big data and artificial intelligence, through the capacity building of process models, intelligent algorithms, big data and visualization, the use of data mining, machine learning, and the integration of expert knowledge and experience, the construction of digital twin system of intelligent production line has become the key to creating intelligent, efficient and low-cost green hot rolling production line.By developing intelligent rolling line measurement and control technology, high-precision mathematical model and remote operation and maintenance technology, Ansteel Hot Strip Mill realizes extreme automation of rolling line and production intensification. By building a global big data platform, deeply integrating intelligent algorithm and process technology, and taking factory performance as the guide, it realizes coordinated control of production, quality, equipment and energy in hot rolling production process. Further for the lean control of production quality, efficiency and cost, and enhance the connotation of intelligent production lines.

Key words: hot rolled strip steel production; intelligence algorithms; lean control; intelligent production line

面向质量精细管控的型材数字化工厂关键技术与实施方案

董广，于天琦，王久朋，卜俊男，于萍

（鞍钢股份有限公司，辽宁鞍山 114021）

摘 要： 随着工业互联网技术及"中国制造 2025"的推广及应用，结合大数据技术实现钢铁企业智能制造已成为未来钢铁企业发展方向。针对质量精细化管控需求的型材钢企，实现生产过程数字化从而达到智能化制造仍是其未来发展的重点。鞍钢大型厂轨梁分厂采用数字化技术已实现包括机器人贴标、字符识别、轨型仪及能耗监控与分析、设备状态监控等功能，形成了具有鞍钢特色的数字化轨梁生产车间，对鞍钢大型厂轨梁分厂数字化、智能化应用水平提升作用明显，在提高全线管控水平同时也保证了实物质量稳定性水平，为型材钢企数字化发展提供了很好的范例。

关键词： 型材；数字化；轨梁；互联网；大数据

Key Technologies and Implementation Solution of Digitalized Factory in Profile Steel Enterprises for Quality Lean Control

DONG Guang, YU Tianqi, WANG Jiupeng, BU Junnan, YU Ping

(Angang Steel Co., Ltd., Anshan 114021, China)

Abstract: With the promotion and application of industrial Internet technology and "Made in China 2025", the realization of intelligent manufacturing in the steel industry through the integration of big data technology has become the future direction of steel enterprises. For profiled steel enterprises that require quality control, achieving digitalization of the production process to achieve intelligent manufacturing is still a key focus of their future development. The digital technology adopted by the Ansteel Large-Scale Rail Beam Factory has realized functions such as robot labeling, character recognition, rail profile measurement, energy consumption monitoring and analysis, and equipment status monitoring, forming a digital rail beam production workshop with Ansteel characteristics. It has significantly improved the digital and intelligent application level of the Ansteel Large-Scale Rail Beam Factory, and has not only improved the overall control level but also ensured the stability of physical quality, providing a good example for the digital development of profiled steel enterprises.

Key words: profile steel enterprises; digitization; rail beam; internet; big data

红色工业的探索与发现：图像识别技术在冶金领域的应用与前景

张瀚文

（中钢设备有限公司，北京 100080）

摘　要： 红色工业，即冶金工业，是我国重要的基础产业。随着科技的发展，人工智能（AI）在冶金工业中的应用正成为推动行业变革和发展的关键因素。AI 可以提高生产效率、改进质量控制，通过机器学习和深度学习算法，冶金企业可以利用 AI 分析大量数据，预测产品质量和检测缺陷，提高产品质量的稳定性和可靠性。本文通过实际案例的分析，总结了人工智能中图像识别技术在冶金领域的应用案例以及带来的效益。该技术将在冶金工业流程管控监测、设备维护异常检测、产品质量检测与分类等方面展现出巨大潜力。最终提出了采用深度学习和迁移学习算法、增加领域知识以及开发移动边缘计算设备等措施来推动图像识别技术在冶金领域的应用。

关键词： 冶金行业；人工智能；图像识别技术；冶炼过程监测；产品质量检测；前景展望

Abstract: The red industry, namely the metallurgical industry, is a pillar industry in China. With recent advances in science and technology, the application of artificial intelligence (AI) in this industry is becoming key to transforming and improving it. AI can boost productivity, enhance quality control, and enable metallurgical companies to analyze huge amounts of data using machine learning and deep learning algorithms. This allows them to predict product quality, detect defects, and improve the consistency and reliability of quality.

This article reviews how image recognition technology, a branch of AI, has been applied in the metallurgical industry by analyzing real-world cases. This technology shows great promise for process monitoring, equipment maintenance, defect detection, and product quality inspection and classification in the industry. The article proposes measures like adopting deep learning and transfer learning algorithms, incorporating domain knowledge, and developing mobile edge computing to drive the application of image recognition in metallurgy.

In summary, AI facilitates the intelligent transformation of the metallurgical industry. Image recognition, an important AI technology, provides tools for automating metallurgical processes, inspecting quality, and managing warehouses. It reduces workload, improves detection accuracy, and ensures product quality, thereby enhancing the core competitiveness of companies and enabling high-quality development of the industry.

Key words: metallurgy industry; artificial intelligence; image recognition technology; smelting process monitoring; product quality inspection; prospects

冶金企业人工智能的发展与应用综述

陈 琛，倪书权，郝景章，李加旺，史 光

（五矿集团中冶京诚工程技术有限公司，北京 100176）

摘 要：人工智能在钢铁行业有广泛的应用，能够提高生产效率和质量，降低成本和风险。国际和国内先进钢铁企业的实践表明，人工智能技术对于实现智慧化和高质量发展具有重要作用。同时，冶金人工智能的发展仍面临一些挑战，需要加强合作，加大研发投入，推动冶金人工智能的创新与应用，实现更高质量、更高效率、更可持续的发展。

关键词：冶金行业；人工智能；钢铁企业；工业应用；发展方向

"工业4.0"助推中国钢铁行业迈进数字化智能化新时代

佐祥均，代 虹

（中冶赛迪咨询有限公司，重庆 401147）

摘 要：本文全方位的阐释了"工业4.0"对钢铁行业数字化智能化发展的内涵、作用和影响。对世界钢铁工业两化融合现状进行了概括总结，并剖析了存在的问题。从个性化服务、智能化制造、大数据管理、供应链集成等方面讨论了"工业4.0"对钢铁工业转型升级带来的积极作用。阐述了"钢铁4.0"的层次化战略实施要点，并基于智能装备、智能工厂、智能互联加基础设计，提出了打造中国钢铁行业的"工业4.0"企业系统化的实施建议。

关键词：工业4.0；钢铁行业；数字化；智能化

Industry 4.0 Boosts the Steel Industry into a New Era of Digitalization and Smart Technology

ZUO Xiangjun, DAI Hong

(CISDI Engineering Co., Ltd., Chongqing 401147, China)

Abstract: This paper elaborates the connotation, function and influence of industry 4.0 on the digital and intelligent transformation of iron and steel industry, and summarizes the current integration of digital and intelligent transformation in the iron and steel industry worldwide, as well as analyzes the existing problems. The positive effects of industry 4.0 on transformation and upgrading of iron and steel industry are discussed from the aspects of personalized service, intelligent manufacturing, big data management and supply chain integration. It also expounds the key points of implementing the hierarchical strategy of "iron and steel 4.0", and proposes suggestions for steel producers to implement "industry 4.0" systematically in terms of intelligent equipment, intelligent factory, intelligent interconnection and basic design.

Key words: industry 4.0; iron and steel Industry; digitalization; smart technology

高线卸卷实现自动控制，避免擦划伤问题

杜鹏阳，黄宝新，郭思聪，王　新，梁　琪

（鞍钢股份线材厂，辽宁鞍山　114000）

摘　要： 卸卷站是线材生产线的最后一道工序，卸卷小车将 C 型钩上的成品线材运输至卸卷台架。存储四卷线卷后，再由成品吊车将产品调运至成品库。操作工手动操作卸卷小车时，经常会造成产品间发生碰撞，产生划伤，影响产品质量。本文介绍了关于卸卷小车手动控制改型为自动控制的方法及优化。

关键词： 线材；卸卷站；测距传感器；PLC；操作屏

Automatic Control of High-line Unloading Based on Avoiding Scratch Problems

DU Pengyang, HUANG Baoxin, GUO Sicong, WANG Xin, LIANG Qi

(Angang Steel Wire Factory, Anshan 114000, China)

Abstract: The unloading station is the last process of the wire production line, and the unloading trolley will transport the finished wire on the C-hook to the unloading platform. After several coils are stored, the finished product crane transfers the product to the finished product warehouse. When the operator manually operates the unloading trolley, it often causes collisions between products, resulting in scratches and affecting product quality. This paper introduces the method and optimization of changing the manual control to automatic control.

Key words: wire rod; unloading station; distance sensor; PLC; operation panel

基于 5G 和云化 PLC 的工业智能泵站系统设计

秦大伟[1]，王军生[1]，张　岩[1]，孙瑞琪[1]，刘佳伟[2]，吴晓宁[2]

（1. 鞍钢集团北京研究院有限公司，北京　100010；
2. 鞍钢集团自动化有限公司，辽宁鞍山　114001）

摘　要： 随着工业互联网、云计算等技术的兴起和快速发展，工业泵站逐渐走向无人值守和智能化运管。围绕工业

智能泵站管控一体化平台建设，提出了一种基于 5G 和云化 PLC 和云、边、端协同的工业智能泵站系统构建方法。采用工业互联网和分布式云计算架构，融合了集中式云计算易运维和边缘计算快速响应等优点。采用公网专用与专网专用的混合组网模式部署工业 5G 专网，应用 5G LAN 技术实现终端设备之间二层网络通信，简化了网络结构。针对传统冗余控制设备资金投入大的问题，提出了基于 S2 系统冗余的云化 PLC 冗余控制方案，实现了工业智能泵站冗余控制，降低了使用成本，增强了系统的可靠性。

关键词：工业智能泵站；云化 PLC；边缘计算；5G LAN；冗余控制

Design of Industrial Intelligent Pumping Station System Based on 5G and Cloud PLC

QIN Dawei[1], WANG Junsheng[1], ZHANG Yan[1],
SUN Ruiqi[1], LIU Jiawei[2], WU Xiaoning[2]

(1. Ansteel Group Beijing Research Institute Co., Ltd., Beijing 100010, China;
2. Angang Group Automation Co., Ltd., Anshan 114001, China)

Abstract: With the rise and rapid development of technologies such as industrial internet and cloud computing, industrial pumping stations are gradually moving towards unmanned and intelligent operation and management.Focusing on the construction of integrated management and control platform of industrial intelligent pumping station, a construction method of industrial intelligent pumping station system based on 5G and cloud PLC and cloud, edge and end collaboration is proposed.The industrial Internet and distributed cloud computing architecture are adopted, which combines the advantages of centralized cloud computing and fast response of edge computing.The industrial 5G private network is deployed in a hybrid network mode dedicated to public network and private network. The 5G LAN technology is used to realize the two-layer network communication between terminal devices, which simplifies the network structure. Aiming at the problem of large capital investment in traditional redundant control equipment, a cloud PLC redundant control scheme based on S2 system redundancy is proposed, which realizes the redundant control of industrial intelligent pumping station, reduces the use cost and enhances the reliability of the system.

Key words: industrial intelligent pumping station; cloud-based PLC; edge computing; 5G local area network; redundancy control

宝钢湛钢炼铁厂智慧制造探索与实践

韩仁义，梁利生

（宝钢湛江钢铁有限公司，广东湛江 524072）

摘 要：宝钢湛江钢铁炼铁厂自 2015 年 9 月投产以来，认真贯彻公司"简单、高效、低成本、高质量"的管理理念，在稳定生产、提升指标、降低成本的同时，致力于铁前各工序智慧制造的升级改造，在数字化原料场、智慧烧结、焦炉四大车无人化、高炉炉前作业自动化等方面开展了一系列的探索和实践，取得了很好效果，助力公司早日建成世界最高效率的绿色碳钢制造基地。

关键词：智慧制造；数字化；无人化；自动化；实践

Exploration and Practice of Intelligent Manufacturing in Iron Making Plant of Baosteel Zhanjiang Iron and Steel

HAN Renyi, LIANG Lisheng

(Baosteel Zhanjiang Iron & Steel Co., Ltd., Zhanjiang 524072, China)

Abstract: Since the iron making plant of Zhanjiang Iron and Steel was put into operation in September 2015, the company's management philosophy of "simple, efficient, low cost and high quality" has been earnestly implemented. While stabilizing production, improving indicators and reducing costs, the intelligent manufacturing of ironmaking process has been upgraded and transformed. A series of explorations and practices have been carried out in digital raw material field, intelligent sintering, unmanned of coke oven four vehicles, operation automation of tapping and other aspects, and achieved good results, helping the company to build the world's most efficient green steel manufacturing base as soon as possible.

Key words: intelligent manufacturing; digital; unmanned; automation; practice

高分辨率数字钢卷快速搜索系统及应用

刘安平，彭燕华

（中冶赛迪信息技术（重庆）有限公司，重庆 401122）

摘 要： 热轧厂每天生产数百卷钢，在对应的每个高分辨率数字钢卷中记录有数千个信号，每个信号按长度等份为30000 份，一天产生约 100G 个浮点数，这些毫秒级、厘米级的庞大数据，如果采用传统数据库，在数千万块钢卷中搜索统计某种指标，运算时间将会是小时级，实际工作中难以接受。CFS(Coil Fast Search system) 采用 PDA(Process Data Acquisition)高速数据采集分析系统内核，搜索统计速度比传统数据库快数百倍，实现秒级完成千万块钢卷中搜索出符合要求的 10000 块钢，并统计出这 10000 块钢的某种指标、返回抽取数据和统计结果，让数字钢卷真正发挥效用。

关键词： 高分辨率；数字钢卷；快速；搜索；PDA

High Resolution Digital Steel Coil Fast Search System and Its Application

LIU Anping, PENG Yanhua

(CISDI Information Technology (Chongqing) Co., Ltd., Chongqing 401122, China)

Abstract: The hot rolling plant produces hundreds of coils of steel every day, and thousands of signals are recorded in each corresponding high-resolution digital steel coil, Each signal is divided into 30000 equal parts by length, about 100G Floating-point arithmetic numbers are generated a day, These data of millisecond and centimeter levels is huge, if the traditional database is used to save and search and count certain indicators in tens of millions of steel coils, the calculation time will be hours, which is difficult to accept in actual work. The search and statistics speed of CFS (Coal Fast Search System) is hundreds of times faster than traditional databases by adopting PDA high-speed data acquisition and analysis system core, achieving the ability to search for 10000 steel pieces that meet the requirements in tens of millions of steel coils in seconds, and calculate certain indicators of these 10000 steel pieces, return extracted data, and statistical results, making digital steel coils truly effective.

Key words: high resolution; digital steel coil; fast; search; PDA

武钢 CSP 连铸二级过程控制系统

徐 重[1]，尹海峰[2]，马小雨[2]

（1. 宝钢股份武钢有限设备部，湖北武汉 430080；
2. 西马克技术（北京）有限公司，北京 100102）

摘 要： 武钢有限 CSP 厂（以下简称武钢 CSP 厂）连铸采用西马克公司二级过程控制技术，选用了三台 hpDL5800 G10 高性能服务器，组成一个稳定的三角形节点，通过 VMware esxi 虚拟主机软件，搭建出一套具有高级资源管理，功能高效、灵活的虚拟主机平台。操作系统采用了 Ubuntu 系统。应用平台采用的是前端+后端+数据库三层 web 应用架构。应用结果表明，系统运行稳定可靠，较好地满足了 CSP 厂连铸二级过程控制要求。

关键词： CSP；连铸；二级；过程控制系统

CSP Continuous Casting Process Control System of Wisco

XU Zhong[1], YIN Haifeng[2], MA Xiaoyu[2]

(1. Equipment Department of Wuhan Iron and Steel Co., Ltd., Baosteel, Wuhan 430080, China;
2. SMS Siemag Technology (Beijing) Co., Ltd., Beijing 100102, China)

Abstract: Wuhan Steel Co., Ltd., CSP Plant (hereinafter referred to as WISCO CSP Plant) adopts SMS Siemag's level2 process control technology for continuous casting. Three hpDL5800 G10 high-performance servers are selected to form a stable triangular node. Through VMware esxi virtual host software, a set of advanced resource management is built. Function efficient, flexible virtual host platform. The operating system is Ubuntu. The application platform adopts the front end + back end + database three-layer web application architecture. The application results show that the system runs stably and reliably, and meets the requirements of process control of continuous casting in CSP plant.

Key words: CSP; continuous casting; second level; process control system

电液直驱结晶器振动优化控制研究与实践

蔡春扬[1,2]，方学红[3]，王永猛[3]，刘 玉[1,2]，龙 灏[1,2]

（1. 中冶赛迪技术研究中心有限公司，重庆 401122；2. 冶金智能装备重庆市重点实验室，
重庆 401122；3. 中冶赛迪工程技术股份有限公司，重庆 401122）

摘 要： 针对新一代电液直驱连铸结晶器振动方式，并基于电液直驱伺服技术的系统特性，采用传统 PID 控制算法

的振动效果不佳，导致在实际工程应用中，不能满足工艺要求。本文提出了一种电液直驱连铸结晶器振动优化控制方法，并具体介绍了该方法的实现过程。应用该方法后，有效改善了电液直驱连铸结晶器振动效果，满足了实际工程中高拉速连铸结晶器振动的工艺要求以及提高了振动装置的稳定性和延长了振动装置的使用寿命。

关键词：连铸结晶器振动；电液直驱；补偿抛物线；优化控制

Research and Practice on the Optimal Control about Mould Oscillation by Electro-hydraulic Direct Drive

CAI Chunyang[1,2], FANG Xuehong[3], WANG Yongmeng[3], LIU Yu[1,2], LONG Hao[1,2]

(1. CISDI R&D Co., Ltd., Chongqing 401122, China; 2. Chongqing Key Laboratory for Metallurgical Intelligent Equipment, Chongqing 401122, China; 3. CISDI Engineering Co., Ltd., Chongqing 401122, China)

Abstract: According to the continuous casting mould oscillation mode of the new generation by electro-hydraulic direct drive and based on the system characteristics of electro-hydraulic direct drive servo technology, adapting the traditional PID control algorithm which leads to the oscillation effect unsatisfactory can't meet the process requirements in practical engineering applications. In this paper, an optimal control method about continuous casting mould oscillation by electro-hydraulic direct drive is presented, and the realization process of the method is introduced in detail. The application of this method has effectively improved the effect about continuous casting mould oscillation by electro-hydraulic direct drive, met the requirements of process about high casting speed continuous casting mould oscillation, increased the stability of the oscillation device and prolonged the service life of the oscillation device.

Key words: continuous casting mould oscillation; electro-hydraulic direct drive; compensating parabola; optimal control

无人驾驶混铁车制动系统设计研究

王　康，张文龙，于　磊，葛　光

（宝山钢铁股份有限公司运输部，上海　201900）

摘　要：无人驾驶混铁车作为全球首创的智能铁水运输设备，可智能、绿色、高效地完成从高炉到炼钢的铁水运输作业，开启了"罐空即配，召之即来，满罐即走，到站即用"的智能铁水运输新模式。本文以无人驾驶混铁车为研究对象，对其制动系统进行了设计研究。首先介绍了无人驾驶混铁车结构组成，对其机械制动系统、电气制动系统、制动算法和制动目标进行了设计。然后分析了无人驾驶混铁车机械制动和电气制动实现方法，最后对无人驾驶混铁车制动性能进行了试验，符合设计目标。

关键词：无人驾驶混铁车；制动系统；机械制动；电气制动

Research on Braking System of SmartTPC

WANG Kang, ZHANG Wenlong, YU Lei, GE Guang

(Transportation Department, Baoshan Iron & Steel Co., Ltd., Shanghai 201900, China)

Abstract: As the world's first intelligent molten iron transportation equipment, SmartTPC can intelligently, green and efficiently complete the molten iron transportation operation from blast furnace to steelmaking, and it opens a new mode of intelligent molten iron transportation called " Preparing when the tank is empty, Coming when you call it, leaving when the tank is full ". In this paper, the designing of the braking system of SmartTPC is studied. Firstly, the structural composition of SmartTPC is introduced, and the mechanical braking system, electrical braking system, braking algorithm and braking target are designed. Then, the implementation methods of mechanical braking and electrical braking of SmartTPC are analyzed, and finally the braking performance of SmartTPC is tested, which meets the design goals.

Key words: SmartTPC; braking system; mechanical braking; electrical braking

无人框架车动力系统设计研究

王　康，张文龙，于　磊，张子才

（宝山钢铁股份有限公司运输部，上海　201900）

摘　要：无人框架车具有安全性好，运输效率高，使用成本低等优点，在钢厂开始推广使用。本文以无人框架车为研究对象，对其动力系统进行了设计研究。首先介绍了无人框架车动力系统组成和功率分配规则，对车辆运行过程分解得到基本动作。分别对轮边电机参数、"发电机+发动机"参数进行设计，根据功率分配规则对动力电池产品性能进行论证。运用反向设计方法，重新匹配"发电机+发动机"参数，完成无人框架车动力系统设计。

关键词：无人框架车；动力系统；参数设计；反向设计

Research on Power System Design of Automated Guided Vehicle

WANG Kang, ZHANG Wenlong, YU Lei, ZHANG Zicai

(Transportation Department, Baoshan Iron &Steel Co., Ltd., Shanghai 201900, China)

Abstract: Automated Guided Vehicle has the advantages of good safety, high transportation efficiency, and low use costs. This article uses unmanned frame vehicles as a research object, and has a design and research on its power system. First of all, the composition and power distribution rules of the driverless vehicle power system are introduced, and the basic actions of the vehicle operation process have been decomposed. Design the parameters of the wrapping motor and the "generator+ engine" parameter, respectively, and demonstrate the performance of the power battery product according to the power distribution rules. Use the reverse design method to re-match the "generator+engine" parameter to complete the design of the driverless vehicle power system.

Key words: automated guided vehicle; power system; parameter design; reverse design

利用 RFC 接口技术实现 SAP 与轻迈系统数据同步

王国靖

（中国五矿鲁中矿业有限公司，山东济南　271113）

摘　要： 随着企业管理需求的不断提升，SAP（企业资源管理软件）系统逐渐成为企业业务管理的中心，在日常运用中也经常需要与外围系统进行集成，为实现 SAP 及外围系统的数据共享便要用到接口技术。本文结合鲁中矿业有限公司选矿部轻迈系统建设经验，着重介绍通过调用 RFC 接口，将 SAP 系统中的数据传输到轻迈系统中，实现两个系统之间的数据互通，实践结果表明，该方案能够实现高效、准确的数据同步，提升企业数据管理的效率。

关键词： SAP；轻迈系统；RFC 接口；数据同步

Utilizing RFC Interface Technology to Achieve Data Synchronization between SAP and Qingmai Systems

WANG Guojing

(China Minmetals Corporation Minmetals Luzhong MINING Co., Ltd., Jinan 271113, China)

Abstract: With the continuous improvement of enterprise management requirements, SAP (Enterprise Resource Management Software) system has gradually become the center of enterprise business management. In daily applications, it is often necessary to integrate with peripheral systems. In order to achieve data sharing between SAP and peripheral systems, interface technology is needed. This article combines the construction experience of the Qingmai system in the concentrator of Minmetals Luzhong Mining Co., Ltd., and focuses on introducing the use of RFC interfaces to transfer data from the SAP system to the Qingmai system, achieving data exchange between the two systems. The practical results show that this solution can achieve efficient and accurate data synchronization and improve the efficiency of enterprise data management.

Key words: SAP; qingmai system; RFC interface; data synchronization

钢铁业数字化智能化技术创新实践与思考

张　伟，刘　晔

（中国宝武钢铁集团中央研究院，上海　201999）

摘　要： 信息通讯技术（ICT）快速迭代，大数据、人工智能、数字孪生、元宇宙、大模型等概念不断出现，并试图在钢铁工业找到更多应用场景，但其需求导向下的整体蓝图和技术路径并不明确。另一方面，围绕"生产效率、质量管控、智慧决策"，面向极致效率及危险、繁重、重复作业，数字化的质量管控，生产及供应链的智能决策等，钢铁行业在研发设计、生产制造、质量管控、设备管理、能源环保、采购营销等环节还存在大量痛点、难点问题。

宝武自 2019 年提出"四个一律"以来，经过三跨融合的升级，充分利用自动化、信息化技术，提高操作。控制自动化，推动数字化转型，开展各专业、各领域数据的采存储，建立了多层次的数据中心。接下来的重点是推动智能化升级，瞄准各种场景，例如从工厂的效率、质量、成本到公司专业业务、经营决策，利用大数据、人工智能等技术开发数据的应用，让数据反哺控制精度提高、效益提升，让数据支撑经营决策优化，聚焦探索绿色、高效、柔性的下一代数字化、智能化工厂技术路径，推动数字化到智能化升级。

从数字化工厂到智慧化钢铁，宝武技术创新初步技术路径：基于数据驱动，建立高效的基础平台、基础软件和算法库，将智能方法（统计、优化、机器学习、深度学习等），与装备、模型、材料的研发相结合，提升研发和应用整体水平，并在重点领域产生引领性突破。具体包括三方面：（1）推动数字化转型，加强工厂级数据中心与经营决策平台技术研发，开发数据高阶应用，推动研发数字化转型。（2）让数据说话支撑决策，将数据应用从分析、监测类向预测、决策、控制类升级，推动业务决策模型智能化、控制模型管控一体化，服务经营决策、生产运行、质量提升。（3）让数据反哺提升精度，加快推进一批测控技术、智能装备、特种机器人的开发和部署。

PDA 高速数据采集系统通讯协议开发

刘安平，彭燕华

（中冶赛迪信息技术（重庆）有限公司，重庆　401122）

摘　要：PDA(Process Data Acquisition)高速数据采集分析系统是一种集数据采集、压缩、存贮和分析于一体的工业实时高速数据采集与分析平台，具有在线和离线分析功能，同时它也是一个高性能的通用产品，是工业 4.0 时代大数据基础平台。然而所涉及的几乎都是控制系统和通讯协议的底层技术，这是我国的重大技术短板，工业控制领域高性能控制器基本被国外垄断，现场总线标准、通讯协议几乎由国外公司掌控，对工业数据进行高速采集面临诸多技术壁垒和国外的高强度重重加密，自主地拿到我们自己的机台、自己工厂的高频高密数据是一种奢望，这种局面迫切须要改变。

关键词：PDA；高速；数据；采集；通讯协议；开发

Development of Communication Protocol for PDA High Speed Data Acquisition System

LIU Anping, PENG Yanhua

(CISDI Information Technology (Chongqing) Co., Ltd., Chongqing 401122, China)

Abstract: The PDA (Process Data Acquisition) high-speed data acquisition and analysis system is an industrial real-time high-speed data acquisition and analysis platform that integrates data acquisition, compression, storage and analysis. It has Online and offline analysis functions. It is also a high-performance general product and the basic platform of Big data in the Industry 4.0 era. However, almost all of the underlying technologies involved are control systems and communication protocols, which is a major technological weakness in China. High performance controllers in the industrial control field are basically monopolized by foreign companies, and field bus standards and communication protocols are almost controlled by foreign companies. High speed collection of industrial data faces many technical barriers and high intensity encryption from foreign countries, and we independently obtain our own machines The high-frequency and high-density data of one's own factory is an extravagant expectation, and this situation urgently needs to be changed.

Key words: PDA; high-speed; data; acquisition; communication protocol; development

智能制造系统在某双高棒车间的开发和应用

赵　铭[1]，张芙蓉[2]

（1. 中钢设备有限公司，北京　100080；2. 上海宝信软件股份有限公司，上海　201900）

摘　要：发展智能制造是钢铁行业转型升级的重要方向，是我国钢铁工业由大到强的重要保障，本文介绍了某双高棒车间智能制造系统的开发和应用，其主要内容包括互联网平台、智能提升、集控中心、智慧管控等，智能制造系统可以提高轧线生产效率、产品质量，提高能源利用率，降低成本，优化人员配置，是未来棒材厂的发展趋势，具有广阔的应用前景。

关键词：智能制造；双高棒；集控中心；开发和应用

Development and Application of the Intelligent Manufacturing System in a Double High-speed Bar Line

ZHAO Ming[1], ZHANG Furong[2]

(1. Sinosteel MECC, Beijing 100080, China; 2. Baoxin Software Co., Ltd., Shanghai 201900, China)

Abstract: The development of the intelligent manufacturing system is the important direction and guarantee for the transformation and upgrading of our country's steel industry. This paper introduces the development and application of the intelligent manufacturing system in a double high-speed bar line. The system includes internet platform, intelligent improvement, centralized control center, intelligent management and so on. The intelligent manufacturing system can improve productivity, quality and the energy utilization. It also can reduce the production costs and optimize human resources. It is the development trends and it has broad application prospects in the future.

Key words: intelligent manufacturing system; double high-speed bar mill line; centralized control center; the development and application

浅谈水处理分厂 UPS 的演变

李 璐

（宝山钢铁股份有限公司能源环保部，上海 200948）

摘 要：UPS 作为一种无间断供电的电源，用来确保重要设备的电源供给。该装置将交流电整流为直流电，存入自带的电池中，再通过逆变装置，将直流电变成交流供后续设备使用。因此其具备两大作用。一为稳定电压，原始交流电压存在电压波动，后续仪表设备感觉敏感，电压波动对设备影响较大，因为稳压后的供电对设备起到很好的保护作用。二为临时供电，一旦外部交流电停电后，UPS 自带蓄电池将持续对外供电，确保所供设备继续有电，保证系统运行正常。公司自本世纪初使用 UPS 后，前后经历了几次变革，通过不断完善，确保 UPS 自身稳定，且避免本身成为故障点影响系统。

关键词：UPS；演变；完善；优化布局

二十辊轧机自动板形控制系统研究与优化

罗劲松，钱 华，郑 涛

（宁波宝新不锈钢有限公司，浙江宁波 315807）

摘　要：本文通过对板形测量影响因素的梳理分析，建立相关模型来主动识别真实板形，并将对应的处置方法程序化处理，优化现有的板形控制系统。经过优化后的板形控制系统，板形目标曲线可根据轧机实时轧制情况自动响应，可以规避因板形测量问题、轧制条件不符等问题对自动板形系统的干扰，从而在钢卷轧制中能全面实现自动板形控制。

关键词：自动板形控制系统；板形测量偏差；轧制条件；板形目标曲线

Research and Optimization of Automatic Flatness Control System for 20 High Mill

LUO Jinsong, QIAN Hua, ZHENG Tao

(Ningbo Baoxin Stainless Steel Co., Ltd., Ningbo 315807, China)

Abstract: By combing and analyzing the influencing factors of flatness measurement, this paper establishes relevant models to actively identify the real flatness, and processes the corresponding disposal methods to optimize the existing flatness control system. With the optimized flatness control system, the flatness target curve can automatically respond according to the real-time rolling situation of the rolling mill, and can avoid the interference to the automatic flatness system due to the problems of flatness measurement and inconsistent rolling conditions, so as to fully realize the automatic flatness control in coil rolling.

Key words: automatic flatness control system; flatness measurement deviation; rolling conditions; flatness target curve

煤焦检测智能分析技术应用与系统开发

欧连发

（宝武集团中南钢铁股份有限公司制造管理部，广东广州　510335）

摘　要：煤焦检测智能分析系统解决了困扰公司外购焦煤、喷吹煤、动力煤，高炉磨后喷吹煤、自产焦炭等原燃料关键成分检测分析及结果报出不及时的技术难题。通过分析检测缺陷和流程梳理，设计了设备总成平台、机器人控制技术的应用，开发了系统控制功能、软件管理平台和系统通讯及设备通讯接口，满足了煤焦检测流程全自动化管理，样品及时分析、结果及时发布不需要人工干预的技术要求，解放和优化了岗位繁琐的手工操作，提升了钢铁行业智慧制造技术运用水平和设备管理水平，增强了企业竞争力。

关键词：煤焦检测；智能分析；技术应用；系统开发

基于主变励磁涌流引发电动风机误跳机的分析

陶瑞基，阮文茜

（广东中南钢铁股份有限公司，广东韶关　512122）

摘　要：本文就一起主变在检修后重新送电投入运行时产生了励磁涌流，造成挂在同一个供电电源系统的电动风机零序保护误跳机的事故进行了分析，运行人员应该高度重视主变空投时产生的励磁涌流造成的严重后果，需要预防性地调整供配电系统的运行方式，避免因主变励磁涌流造成对供配电系统的不良影响。本文给同行进行借鉴，避免发生同样的事故而造成经济损失。

关键词：励磁涌流；零序保护；电动风机；运行方式

Analysis of Misoperation of Electric Fan Caused by Inrush Current of Main Transformer

TAO Ruiji, RUAN Wenxi

(Guangdong Zhongnan Iron and Steel Co., Ltd., Shaoguan 512122, China)

Abstract: This paper analyzes the accident that an electric fan hanging in the same power supply system jumps out of zero sequence protection due to the inrush current generated when the main transformer is re-powered after maintenance. The operator should pay great attention to the serious consequences caused by the inrush current generated during the air delivery of the main transformer, and it is necessary to adjust the operation mode of the power supply and distribution system precautionary to avoid the inrush current caused by the main transformer. The adverse effects on power supply and distribution system can be used for reference by peers to avoid the same accidents and economic losses.

Key words: excitation inrush current; zero sequence protection; electric fan; mode of operation

基于精炼炉变压器绕组匝间短路故障的分析

陶瑞基，王志康

（广东中南钢铁股份有限公司，广东韶关 512122）

摘　要：钢铁行业中的钢包精炼炉变压器是电炉变压器的一种，其特点是输出电压较低、电流较大、电弧稳定、电流波动小。为了适应炉况的变化，精炼炉变压器均采用精细有载调压，是钢铁行业一种重要的设备。如果精炼炉变压器出现故障，会影响到钢厂的总体生产，特别是品种钢的生产，给企业造成较大的经济损失。本文结合一炼钢厂的 1 台精炼炉变压器发生的电气故障进行原因分析，并提出预防措施，给同行经验及教训借鉴。

关键词：精炼炉变压器；保护跳闸；匝间短路；调压线圈

Analysis of Short Circuit Fault between Turns of Transformer Winding Based on Refining Furnace

TAO Ruiji, WANG Zhikang

(Guangdong Zhongnan Iron and Steel Co., Ltd., Shaoguan 512122, China)

Abstract: Ladle refining furnace transformer in iron and steel industry is a kind of electric furnace transformer, which is characterized by low output voltage, large current, stable arc and small current fluctuation. In order to adapt to the change of furnace conditions, fine on-load voltage regulation is adopted in the transformers of refining furnace, which is an important equipment in iron and steel industry. If the transformer of refining furnace breaks down, it will affect the overall production of steel mills, especially the production of variety steel, causing great economic losses to enterprises. In this paper, the causes of electrical failure of a refining furnace transformer in a steel plant are analyzed, and preventive measures are put forward, which can be used for reference by peers.

Key words: refining furnace transformer; protection tripping; inter-turn short circuit; voltage regulating coil

智慧高炉集约化管控大数据应用平台研究与开发

费 静[1,2]，车玉满[1,2]，郭天永[1,2]，姜 喆[1,2]，孙 波[3]

(1. 海洋装备用金属材料及其应用国家重点实验室，辽宁鞍山 114009；2. 鞍钢集团钢铁研究院，辽宁鞍山 114009；3. 鞍钢股份有限公司炼铁总厂，辽宁鞍山 114021)

摘 要： 鞍钢高炉已经完成大型化、现代化改造，大量信息技术已在高炉工序得到推广与应用，本文以鞍钢高炉群为背景，以大数据和工业互联网技术为基础，结合高炉生产冶炼工艺机理，利用人工智能、云端分析等技术，通过数据治理平台，有效整合高炉各工序数据，设计数据资产目录，建立高炉集约化管控大数据应用平台，对数据进行多维度深度挖掘，构建高炉智能应用模型，实现高炉安全生产运行的可视化智能监控，指导高炉生产操作，提升高炉生产操作、管理整体信息化、智能化水平。

关键词： 高炉炼铁；大数据；集约化管控；智能应用

Research and Development of Big Data Application Platform for Intelligent Blast Furnace Intensive Management and Control

FEI Jing[1,2], CHE Yuman[1,2], GUO Tianyong[1,2], JIANG Zhe[1,2], SUN Bo[3]

(1. State Key Laboratory of Metal Material for Marine Equipment and Application, Anshan 114009, China; 2. Ansteel Iron & Steel Research Institutes, Anshan 114009, China; 3. General Ironmaking Plant of Angang Steel Co., Ltd., Anshan 114021, China)

Abstract: The blast furnaces of Anshan Iron and Steel have completed large-scale and modernization, and a large amount of information technology has been popularized and applied in the process of blast furnaces. This paper takes Anshan Iron and Steel blast furnace group as the research background. Based on big data and industrial internet technology, combing the smelting process mechanism of blast furnace production and using artificial intelligence, cloud analysis and other technologies, the data management platform was used to effectively integrate the data of each process of blast furnace, design the data asset catalog. The big data application platform for the intensive control of the blast furnace was established. The data was carried out multidimensional in-depth mining, and the intelligent application model of blast furnace was established. The visual intelligent monitoring of the safe production and operation of blast furnace was realized, and the production operation of blast furnace was guided. The overall information and intelligent level of production operation and management of blast furnace has been improved.

Key words: blast furnace iron making; big data; intensive control; intelligent application

转炉全自动出渣系统的开发与应用

秦佳星，江腾飞，陶 金

（北京首钢股份有限公司迁安钢铁公司，河北迁安 064400）

摘　要：介绍了迁钢 5 号转炉自动出渣控制技术的设计理念，包括系统安全控制逻辑的搭建，出渣模式的设计，以及系统硬件设备的独立控制方式和联动控制逻辑等。针对渣罐渣量检测、钢包位置识别，先进的图像分析技术在自动出渣系统中得到了的应用。目前迁钢 5 号转炉自动出渣技术已经应用于实现生产过程中，投入率 100%，成功率在 96%以上，代替了人工按钮+手柄多线操作，降低了岗位的操作强度，保障了出渣系统的安全稳定性。

关键词：转炉；自动出渣；渣罐；图像识别

Abstract: The design concept of automatic slagging control technology for No. 5 converter in Qiangang is introduced. Including the construction of system safety control logic, the design of the slag mode, and the independent control mode and linkage control logic of the system hardware equipment. For slag tank slag volume detection、ladle position recognition, it represents the application of image analysis technology in automatic slag discharge system. At present, Qiangang 5# converter automatic slagging technology has been applied in the production process, the input rate is 100% and the success rate is more than 96%, instead of manual button and handle multi-line operation, the system reduce the operating intensity of the post and ensure the safety and stability of the slag discharge system.

Key words: converter; automatic slag discharge; slag pot; image recognition

集中空压机站远控滞后现象的分析与优化

薛冬晨，鲁绍军，李 杨，刘 强，牛勋铎

（北京首钢股份有限公司，北京 100043）

摘　要：集中空压机站运行一段时间后出现偶发性启、停机，加、卸载反应慢问题，本文针对这一实际问题展开分析，利用现有能源网络，从远控中心的操作员站中选出 3 台上位机直接与现场 PLC 控制系统进行对接，不再经过 GR 服务器转接控制，以确保空压机站实时性和可靠性。改造后没有再出现控制滞后的问题，效果良好，成为各集中空压机站实现无人值守的关键，并对能源数据采集、集中控制系统领域提供实际、有效的借鉴意义。

关键词：空压机；远程控制；滞后

Analysis and Optimization of Remote Control Lag in Centralized Air Compressor Station

XUE Dongchen, LU Shaojun, LI Yang, LIU Qiang, NIU Xunduo

(Beijing Shougang Co., Ltd., Beijing 100043, China)

Abstract: After a period of operation, the centralized air compressor station appears the problem of occasional start-up, shutdown, slow response to loading and unloading (compression and decompress). This paper analyzes this practical problem, and selects three upper computers from the operator station of the remote control center to connect directly with the field PLC control system by using the existing energy network, instead of transferring through GR server. Control to ensure the real-time and reliability of air compressor station. The problem of control lag does not appear again after the transformation, and the effect is good. It becomes the key to realize unattended operation in all centralized air compressor stations, and provides practical and effective reference for the field of energy data acquisition and centralized control system.

Key words: air compressor; remote control; lagging

PROFIBUS-DP 网络质量检测技术在钢厂的运用

丁继亚，钟　实，程　曦，黄亚军，刘　浩

（武钢有限设备管理部，湖北武汉　430083）

摘　要： PROFIBUS-DP 网络是工厂基础自动化应用最为广泛的现场总线，其网络质量的好坏直接影响生产是否正常顺行，一旦故障易造成长时间停机或频繁的重复故障，对生产影响较大。本文介绍了如何利用网络诊断检测工具提前发现隐患并予以消除，减少网络通讯故障的发生。

关键词： PROFIBUS-DP；网络质量检测；故障

数字化应变分析技术在汽车零部件冲压成形性能评价中的应用

郝志强[1,2]，刘文博[1,2]，徐　鑫[1,2]，梁　笑[1,2]，李春林[1,2]

（1. 海洋装备用金属材料及其应用国家重点实验室，辽宁鞍山　114009；
2. 鞍钢集团钢铁研究院，辽宁鞍山　114009）

摘　要： 以 210P1 冲压汽车零部件左/右 C 柱内板为例，利用数字化应变分析技术，结合金属塑性成形体积不变原理与 210P1 的成形极限曲线，获得了重点关注区域内钢板的应变分布云图与最大厚度减薄率值云图。结果表明，重点关注区域内钢板的最大厚度减薄率值为 20.67%，小于 25% 的厚度减薄安全裕度值；最小成形极限值为 21.8%，大于 10% 的成形安全裕度值，钢板成形性能优良，且具有较高的使用安全性。

关键词： 数字化应变分析；成形极限；厚度减薄率；安全裕度

Application of Digital Strain Analysis Technology in Evaluation of Stamping Forming Performance of Automotive Parts

HAO Zhiqiang[1,2], LIU Wenbo[1,2], XU Xin[1,2], LIANG Xiao[1,2], LI Chunlin[1,2]

(1. State Key Laboratory of Metal Material for Marine Equipment and Application, Anshan 114009, China;
2. Ansteel Iron & Steel Research Institutes, Anshan 114009, China)

Abstract: Taking the left/right C-pillar inner plate of 210P1 stamping automotive components as an example, using digital strain analysis technology, combined with the principle of volume invariance in metal plastic forming and the forming limit curve of 210P1. Obtained the strain distribution contour map and maximum thickness reduction rate contour map of the steel plate in the key focus area.. The results show that the maximum thickness reduction rate of steel plates in the key focus area is 20.67%, which is less than the safety margin value of 25% thickness reduction; The minimum forming limit value is 21.8%, which is greater than the forming safety margin value of 10%. The steel plate has excellent forming performance and high usage safety.

Key words: digital strain analysis; forming limit; thickness reduction rate; safety margin

智慧矿山生产管控平台

郭武英，霍成全，徐 涛

（富蕴蒙库铁矿有限责任公司，新疆富蕴 836100）

摘 要： 随着全球经济的快速发展和资源的日益消耗，矿产资源的保护和可持续利用成为当今社会关注的焦点。为了实现绿色低碳、高质量发展，智慧矿山生产管控平台被提出并广泛应用于矿业生产过程中。随着互联网、人工智能、大数据、物联网、云计算等先进技术的飞速发展，也使得采矿业向自动化、信息化、智能化乃至智慧化的方向发展成为一种重要趋势。本文将探讨智慧矿山生产管控平台的概念、作用与应用，并分析其对矿业企业的可持续发展带来的益处。

关键词： 采矿生产；管控系统；三维可视化；数字化

Smart Mine Production Control Platform

GUO Wuying, HUO Chengquan, XU Tao

(Fuyun Mengku Iron Mine Co., Ltd., Fuyun 836100, China)

Abstract: With the rapid development of the global economy and the increasing consumption of resources, the protection and sustainable utilization of mineral resources have become the focus of attention in today's society. In order to achieve green, low-carbon, and high-quality development, mining production control systems have been proposed and widely applied in the mining production process. With the rapid development of the Internet, artificial intelligence, Big data, Internet of Things, cloud computing and other advanced technologies, the mining industry has also become an important trend towards automation, informatization, intelligence and even intelligence. This article will explore the concept, role, and application of mining production control systems, and analyze their benefits for the sustainable development of mining enterprises.

Key words: mining production; control system; 3D visualization; digitization

冷轧清洗段刷辊调节系统设计与优化

黄亚军

（武汉钢铁有限公司设备管理部设备技术室，湖北武汉　430081）

摘　要：当前冷轧产品的表面质量要求不断提高，镀锌、连退等处理线的清洗段的重要性也越显突出。本文介绍了某钢厂镀锌机组清洗段刷辊调节系统原设计、存在的问题，以及通过PID控制原理对调节系统进行优化改型，提高刷洗的控制精度和响应速度。

关键词：清洗段；PID控制；优化改型

Design and Optimization of Brush Roll Adjustment System for Cold Rolling Cleaning Section

HUANG Yajun

(Equipment Management Department of Wuhan Iron & Steel Co., Ltd., Wuhan 430081, China)

Abstract: At present, the surface quality requirements of cold-rolled products are constantly improving, and the importance of cleaning sections of galvanizing, continuous withdrawal and other treatment lines is becoming more and more prominent. This paper introduces the original design and existing problems of the brush roll adjustment system of the cleaning section of a galvanizing unit of a steel mill, and optimizes and modifies the adjustment system through the PID control principle to improve the control accuracy and response speed of brushing.

Key words: cleaning section; PID control; optimization modification

高炉鼓风机和TRT集控改造运用实践

杨　奇，尹丽琼，李　祥，陈红云，陈树海，侯化军

（昆明钢铁股份有限公司设备部，云南昆明　650302）

摘　要：集中监控是智慧制造的集中体现，是跨工序、跨区域、远距离、大规模集控，融合自动化、物联网、大数据、人工智能的规模化、集中化应用，本文介绍了昆钢公司高炉鼓风机和TRT集控改造过程中遇到的问题，从系统配置、技术方案、分析了集中监控的功能以及效果。

关键词：集中控制；网络；励磁调节；电力监控

冶金行业安全生产防控技术及应用

葛雨田，王　敏，邢立东

（北京科技大学，金属冶炼重大事故防控技术支撑基地，北京　100083）

摘　要：为有效预防、遏制各类冶金事故的发生，发改委推出了《冶金行业较大危险因素辨识与防范指导手册》。自2020年起，陆续颁布了《冶金工贸企业安全生产标准化基本规范指导》《工贸企业重大事故隐患判定标准》等安全指导性文件，国家对冶金生产安全愈发重视。

金属冶炼过程，主要涉及危险源包括有毒、高温流程产物，高温压力容器及特种设备，这些危险源普遍危险性高，危害大并存在不确定性，突发性和相关因素众多且相互交叉影响等特点。依靠人工判断与设备安全联锁的方式进行生产会导致事故应急救援不及时，处置措施不完善问题，无法保障冶金行业安全生产。开发"以机器识别代替人眼观察，以算法拟合代替人工判断"的冶金企业安全模型为解决以上问题提供了新思路。目前，已有学者从冶金生产安全入手使用AHP模糊综合评价、BP神经网络、SIC模型、计算机视觉等方式对各类风险源进行风险评估及预测，这些成果可有效增强冶金企业的风险预知与事故应急调度能力，提高冶金企业的本质化安全。

开发涵盖多种冶金安全生产模型的一体式智能安全管控平台是未来冶金安全发展新方向，见图1。平台实现钢铁行业较大风险源的辨识、分级与管控，具体到场所、工艺、装备、部件，形成覆盖全面、脉络清晰、专业可靠的分级风险源从属关系图谱，再从员工、设备、环境与管理四个维度对企业内部危险源实现管控，包含人员安全培训考核、人员装备监控、电子围栏、装备数据整合、冶金安全生产模型、危险预测预警模型、厂区实时监控等多种方式及手段，结合实际情况实现智能的调度，在数字时代背景下构建"工业互联网 + 安全生产"的安全生产体系，促使冶金企业趋于本质型和长久型安全。

基于数据治理的高炉运行动态评价系统开发与应用

白　雪[1]，陈英博[2]，高大鹏[1]，李洪亮[1]，吕秀丽[1]，薛芷蘅[1]

（1. 鞍钢集团自动化有限公司，辽宁鞍山　114000；2. 鞍钢股份炼铁总厂，辽宁鞍山　114000）

摘　要：利用工业互联网平台采集和融合钢铁企业高炉工序各级信息系统的工艺、生产、设备相关数据，通过数据有效治理，实现数据标准规范定、数据质量持续改进、数据资源安全共享。基于治理后的数据开发高炉运行多维动态评价系统，从原料成分、高炉操作、设备状态、工艺参数等维度定义高炉运行评价指标，通过对不同工况指标评价标准的动态调整，实现对高炉生产过程中的运行状态进行科学动态的评价和预警，使高炉的管理人员和操作人员可以及时优化调整生产组织，最终达到高炉生产运行"稳定、高效、低耗"的目的。

关键词：数据治理；数据质量；高炉评价；多维动态

Abstract: Utilize industrial internet platforms to collect and integrate process, production, and equipment related data from various levels of information systems in steel and iron enterprises' blast furnace processes. Through effective data governance, achieve standardized definition of data standards, continuous improvement of data quality, and secure sharing of data resources. Based on the data after governance, a multidimensional dynamic evaluation system for blast furnace operation is developed. Evaluation indicators for blast furnace operation are defined from dimensions such as raw material composition, blast furnace operation, equipment status, and process parameters. By dynamically adjusting evaluation standards for different operating conditions, scientific and dynamic evaluation and early warning of the operating status in the blast furnace production process are achieved, enabling blast furnace managers and operators to optimize and adjust production organization in a timely manner, The ultimate goal of achieving stable, efficient, and low consumption in blast furnace production and operation is achieved.

Key words: multidimensional dynamics of data governance; data quality; blast furnace evaluation; multidimensional dynamics

"安全生产+经济运行"双核驱动下的钢铁企业能源智能化管控系统开发与应用

白 雪[1]，王念军[2]，周文旭[2]，高大鹏[1]

（1. 鞍钢集团自动化有限公司，辽宁鞍山 114009；
2. 鞍钢朝阳钢铁有限公司能源管控中心，辽宁鞍山 114009）

摘 要： 建立钢铁企业能源智能化管控系统，采用"安全生产+经济运行"双核驱动的能源精细化管控模式，构建安全生产和经济运行双指标评价指标体系，采用事前有计划、事中有监控和事后有评价的全流程能源智能管控，实现去冗余、去活套，紧平衡生产，打造全流程极致能耗管控模式。

关键词： 能源；智能化；精细化；安全生产；经济运行；极致能耗

Abstract: The intelligent management and control system of energy in iron and steel enterprises is established, and the model of energy fine management and control driven by dual cores of "Safe production + economic operation" is adopted, and the double-index Evaluation Index System of safe production and economic operation is constructed, the whole process energy intelligent control is planned, monitored and evaluated in advance to realize de-redundancy, de-looper, tight balance production, and create the ultimate energy consumption control mode.

Key words: energy; intelligent; fine; safe production; economic operation; ultimate energy consumption

厚板4300mm轧机机架辊控制技术

张 丽，赵中杨，齐 鑫

（鞍钢集团信息产业有限公司，辽宁鞍山 114000）

摘 要： 鞍钢股份厚板厂4300mm轧机改造中机架辊的特点是两电机轴通过机械系统刚性连接并且在低频段运行，主动和从动轴在加减速、停车等动态过程中会出现速度不一致，从而转矩不一致的现象，造成齿轮间隙增大，对机械轴伤害的后果，更有甚的是齿轮间隙的增大又助长了主动和从动轴速度的不一致，形成了恶性循环，针对机架辊的这种特点，我们屏弃了传统的主从控制方式，大胆采用"轻解耦"主从控制方式，同时就低频且频繁正反转，动态时间相对较长的特点进行了低频转矩补偿的传动控制方式。本文就该控制系统的系统配置、"轻解耦"主从控制、动态转矩补偿等方面进行了分析，并对传统主从控制与"轻解耦"主从控制系统进行对比，进一步确定文中所述主从控制系统的先进性。

关键词：刚性连接；系统配置；轻解耦控制；动态转矩补偿

Control Technology of Stand Rolls for Heavy Plate Mill 4300mm

ZHANG Li, ZHAO Zhongyang, QI Xin

(Ansteel Group Information Industry Co., Ltd., Anshan 114000, China)

Abstract: In the project of modernization of heavy plate rolling mill 4300mm, Ansteel Co. Ltd, the two motor shafts are connected to mill stand rolls rigidly through a mechanical system and motors operate in low frequency bands, so that driving and driven shafts have different speed in dynamic processes such as acceleration and deceleration, and stopping, resulting in different torque, increased gear gap, and mechanical shaft damage. Increased gear gaps also lead to difference of speeds of driving and driven shafts. To solve this problem, the modern "partial decoupling" type of control system is applied instead of traditional master-slave control. At the same time, the drive control method of low-frequency torque compensation is applied for the characteristics of low frequency and frequent forward and reverse rotation and relatively long dynamic time. In this paper, the system configuration, "partial decoupling" master-slave control, and dynamic torque compensation of the control system are discussed. The traditional master-slave control and the "partial decoupling" control system are compared to show the advantage of the "partial decoupling" control system described in this paper.

Key words: rigid connection; system configuration; partial decoupling control; dynamic torque compensation

高炉热风炉在线能效分析系统的建立及应用

孙守斌[1,2]，刘常鹏[1,2]，白 雪[3]，鲁 璐[3]，王东山[1,2]

（1. 海洋装备用金属材料及其应用国家重点实验室，辽宁鞍山 114009；2. 鞍钢集团钢铁研究院环境与资源研究所，辽宁鞍山 114009；3. 鞍钢集团自动化有限公司，辽宁鞍山 114000）

摘 要：本文介绍一种集成数据驱动及工艺驱动的热风炉能效分析模型，通过现场测试、大数据分析、在线监测等技术手段对能效参数进行诊断分析，并通过能效模型对热风炉运行过程的在线指导，达到提高设备能效的目标。

关键词：热风炉；能效；在线分析

Establishment and Application of Online Energy Efficiency Analysis System for Hot Blast Stove

SUN Shoubin[1,2], LIU Changpeng[1,2], BAI Xue[3],
LU Lu[3], WANG Dongshan[1,2]

(1. State Key Laboratory of Metal Material for Marine Equipment and Application, Anshan 114009, China;
2. Enviroment and Resource Institute, Iron and Steel Research Institute of Ansteel Group,
Anshan 114009, China; 3. AI Management Division of Information Industry Co.,
Ltd. of Ansteel Group Corporation, Anshan 114000, China)

Abstract: this paper introduces an integrated data-driven and process-driven energy efficiency analysis model for hot blast stoves. The energy efficiency parameters are diagnosed and analyzed through field test, big data analysis, online monitoring and other technical means, and the energy efficiency model is used to guide the operation process of hot blast stoves online to achieve the goal of improving energy efficiency of equipment.

Key words: hot blast stoves; energy efficiency; online analyse

基于物联网技术的井下铲运机遥控改装及应用

霍成全

（富蕴蒙库铁矿有限责任公司，新疆富蕴 836100）

摘　要： 铲装作业是在频繁爆破作业场所中的离散作业，一直是国内外采矿的难题，也是影响采矿活动各项指标的关键因素之一。本次改装利用 RFID 技术、UWB 技术、无线通讯与传感技术、计算机技术，将传统井下无法管控的作业过程变为"在线"可控，将影响矿石生产品质的铲装点、铲装量由人工计车数管理转化为自动化数字计量，让计划推进可实时跟踪、贫化控制有抓手。通过对铲运机的智能化改装实现铲装作业远程可视化摇控，作业流程定位追踪。

关键词： 遥控铲运；网络化协同；智能化生产

Remote Control Modification and Application of Underground Scraper Based on IoT Technology

HUO Chengquan

(Fuyun Mengku Iron Mine Co., Ltd., Fuyun 836100, China)

Abstract: Shovel loading operation is a discrete operation in frequent blasting workplaces. It has always been a difficult problem in mining at home and abroad, and it is also one of the key factors affecting the indicators of mining activities. This modification uses RFID technology, UWB technology, wireless communication and sensing technology, and computer technology to change the traditional underground operation process that cannot be controlled into " online " controllable, and change the shoveling point and shoveling amount that affect the quality of ore production from manual counting to automatic digital measurement, so that the plan can be promoted in real time. Tracking and dilution control have a starting point. Through the intelligent modification of the scraper, the remote visual control of the shoveling operation and the positioning and tracking of the operation process are realized.

Key words: remote control shoveling; networked cooperative; intelligent production

探索无底柱分段崩落采矿中深孔智能凿岩作业

霍成全，王双虎，袁明超

（富蕴蒙库铁矿有限责任公司，新疆富蕴 836100）

摘　要：随着采矿行业的不断发展，凿岩台车作为采矿过程中的重要设备，其智能化应用已经成为行业发展的趋势。凿岩台车智能化应用的实现，不仅可以提高采矿效率，降低生产成本，还可以提高采矿安全性和环保性，符合国家的产业发展和环保要求。传统的凿岩台车操作需要大量人力，操作难度大，效率低下，存在安全隐患。为了解决这些问题，开展无底柱分段崩落采矿中凿岩台车的智能应用，随着智能化技术的不断发展，智能化应用已经成为行业发展的趋势。

关键词：凿岩台车；智能化；数据采集与分析；人工智能；远程控制

Exploration of Intelligent Rock Drilling Operations in Segmental Collapse Mining of Bottomless Pillars

HUO Chengquan, WANG Shuanghu, YUAN Mingchao

(Fuyun Mengku Iron Mine Co., Ltd., Fuyun 836100, China)

Abstract: With the continuous development of the mining industry, rock drilling jumbos, as important equipment in the mining process, have seen an increasing trend towards intelligent applications. The implementation of intelligent applications in rock drilling jumbos not only improves mining efficiency and reduces production costs but also enhances mining safety and environmental friendliness, in line with national industrial development and environmental requirements. Traditional rock drilling jumbos require a large amount of manual labor, have high operational difficulty, low efficiency, and safety hazards. In order to address these issues, the intelligent application of rock drilling jumbos in segmental collapse mining of bottomless pillars has been explored. With the continuous development of intelligent technology, intelligent applications have become a trend in the industry.

Key words: rock drilling jumbos; intelligent applications; data collection and analysis; artificial intelligence; remote control

高压设备在线监测技术应用及探索

钟　实[1]，周文豪[2]，张丁鹏[3]，洪轶雄[3]

（1. 武汉钢铁有限公司设备管理部，湖北 武汉　430080；2. 上海市避雷装置检测站有限公司检测检验部，上海　200030；3. 上海金艺检测技术有限公司智能电气部，上海　201900）

摘　要：变电站是电力系统的关键组成部分，随着经济和电力行业的快速发展，变电站在输变电过程中发挥着重要作用。部分变电站不具备系统联动性，数据相对分离，造成电力监控系统不够完整，设备运行的电压、电流、有功、无功及功率因数等基础数据利用率不高，甚至仍需要依靠运行人员对数据进行分析，结果误差较大。随着智能化技术的不断发展，对变电站进行智能化改造是一大趋势。变电站智能系统可以实现实时数据采集，并具有事故预警和远程遥控功能，能够第一时间对事故进行预判和处理，取代绝大部分的人工巡视，提高了运维效率和安全性，减少现场人力资源的成本。本文详细叙述了武钢在高压设备温度集中监控系统构建上进行的在线监测技术的探索和尝试，为后续技术发展与应用提供了参考和借鉴。

关键词：电力设备；远程运维；在线监测；监控平台

Application and Exploration of Online Monitoring Technology for High Voltage Equipment

ZHONG Shi[1], ZHOU Wenhao[2], ZHANG Dingpeng[3], HONG Yixiong[3]

(1. Baosteel Co., Ltd., Wuhan 430080, China; 2. Shanghai Lightning Protection Device Testing Station Co., Ltd., Shanghai 200030, China; 3. Shanghai Jinyi Inspection Technology Co., Ltd., Shanghai 201900, China)

Abstract: With the rapid development of economy and power industry, substation is a key component of power system and plays an important role in the process of power transmission and transformation. Some substations do not have system linkage and data are relatively separated, resulting in incomplete power monitoring system, low utilization of basic data such as voltage, current, active power, reactive power and power factor of equipment operation, and even need to rely on operators to analyze the data, resulting in large error. With the continuous development of intelligent technology, intelligent transformation of substation is a major trend. The substation intelligent system can realize real-time data collection, and has the functions of accident early warning and remote control. It can predict and handle the accident at the first time, replacing most of the manual inspection, improving the efficiency and safety of operation and maintenance, and reducing the cost of on-site human resources.This article provides a detailed description of the exploration and attempt of online monitoring technology in the construction of a centralized temperature monitoring system for high-pressure equipment at Wuhan Iron and Steel Corporation, providing reference and reference for the subsequent development and application of technology.

Key words: power equipment; remote operation and maintenance; on-line monitoring; monitoring platform

递归滤波算法在低温卷取控制方面的应用

杨哲懿[1,2]，高磊[1,2]，王尊呈[1,2]，王帅[1,2]，于生龙[3]，杨博[3]

（1. 海洋装备用金属材料及其应用国家重点实验室，辽宁鞍山　114009；2. 鞍钢集团钢铁研究院热轧产品研究所，辽宁鞍山　114009；3. 鞍钢股份有限公司热轧带钢厂，辽宁鞍山　114009）

摘　要：本文所使用的递归式滤波算法为卡尔曼滤波算法，应用该算法对热轧厚规格高强钢在低温卷取过程中的温度控制进行优化。通过提取并分析现场厚规格高强钢低温卷取工艺下的卷取温度历史数据，确定造成数据异常波动的原因及数据噪声类型。基于递归式滤波算法搭建出滤波仿真系统平台，采用离线运行的方式分析出滤波控制参数的影响规律和滤波效果。通过对现有层流冷却控制系统进行二次开发，在控制系统中嵌入滤波功能，实现厚规格高强钢低温卷取工艺下的全自动控制，最终通过实际应用验证了该滤波算法能够有效提升低温卷取的温度控制精度。

关键词：低温卷取；卡尔曼滤波算法；厚规格高强钢

基于 GDIA 的钢铁连铸安全生产管控系统信息需求数据模型

张 充[1]，巴烈电[2]，张 伟[1]，赵挺生[1]

（1. 华中科技大学土木与水利工程学院，湖北武汉 430074；
2. 大冶特殊钢有限公司，湖北黄石 435000）

摘 要：应用信息化技术实施安全生产管控是钢铁行业的研究热点，当前研究侧重以技术为中心提出解决方案，缺少探讨人和技术元素之间有效交互、协同方式。本文基于 GDIA 识别钢铁连铸安全生产管控的主要目标、达成的 4 个决策点和用于制定决策的 3 个层次信息需求；构建 UML 用例数据模型和 4 个序列数据模型以准确地描述达成 4 个决策点的过程中，安全生产管控系统应提供的信息和服务、系统内部的工作流程及与安全生产管理人员之间的交互行为逻辑；最后基于 GDI 结构图和 UML 数据模型开展系统功能设计实现与验证，实现基于安全生产人员心智模型的人机高效交互与协同，提供以技术为基础、以人为中心的信息化、智能化安全生产管控服务，促进钢铁连铸安全生产管控效能提升。

关键词：钢铁连铸；安全生产；管控系统；信息需求；数据模型

Information Requirement Data Model of Steel Continuous Casting Safety Management and Control System Based on GDIA

ZHANG Chong[1], BA Liedian[2], ZHANG Wei[1], ZHAO Tingsheng[1]

(1. School of Civil and Hydraulic Engineering, Huazhong University of Science and Technology, Wuhan 430074, China; 2. Daye Special Steel Co., Ltd., Huangshi 435000, China)

Abstract: The application of information technology to implement production safety control is a research hotspot in the iron and steel industry, and the current research focuses on proposing solutions centered on technology, but lacks the way to explore the effective interaction and synergy between human and technology elements. Based on GDIA, this paper identifies the main goals of production safety control in steel continuous casting, the four decision points to be reached and the three levels of information needs for decision making; constructs UML use case data model and four sequence data models to accurately describe the information and services that the production safety control system should provide in the process of reaching the four decision points, the workflow within the system and the interaction behavior logic with the safety management personnel; finally, based on the GDI structure diagram and UML structure model, the system provides the information and services to be provided by the system, and the interaction behavior logic with the safety management personnel. Finally, based on the GDI structure diagram and UML data model, the system function design is implemented and verified to realize efficient human-computer interaction and collaboration based on the mental model of safety production personnel, to provide technology-based, human-centered informatization and intelligent safety production control services, and to promote the efficiency of steel continuous casting safety production control.

Key words: steel continuous casting; safety production; management and control system; information requirements; data model

激光气体分析仪在转炉煤气回收中的应用

张英利

(河钢宣钢检修公司，河北张家口 075100)

摘　要：本文介绍激光在线气体分析仪在转炉煤气回收系统中的应用，主要对激光气体分析仪的安装、调试、操作和维护等内容作了详细的说明，同时也阐述了激光气体分析仪的测量原理、系统构成和性能特点。实践证明该分析仪安装简便、维护量少、故障率低、标定简单、安全可靠。实践证明激光分析仪能够满足在转炉煤气回收利用中煤气浓度的监测功能要求。

关键词：激光在线气体分析仪；煤气回收系统；应用；测量原理

Application of Laser Gas Analyzer in Gas Recovery of Converter

ZHANG Yingli

(Maintenance Company of HBIS Group Xuansteel Company, Zhangjiakou 075100, China)

Abstract: This paper introduces the application of laser on line gas analyzer in converter gas recovery system, mainly elaborates the installation, commissioning, operation and maintenance of laser gas analyzer, and also expounds the measuring principle, system structure and performance characteristics of laser gas analyzer. Practice shows that the analyzer is easy to install, less maintenance, low failure rate, simple calibration, safe and reliable. Practice shows that laser analyzer can meet the need of converter gas recovery.

Key words: laser on-line gas analyzer; gas recovery system; application; measuring principle

低压电气控制整合与提升

李晟京

(河钢宣钢检修公司，河北张家口 075100)

摘　要：本文主要阐述了为保证宣钢能源公司炼焦车间的生产质量，提高净化车间荒煤气回收质量，对净化车间硫铵、粗苯、蒸氨、制酸等重要工段电气控制、供应等进行整体电气重建。包括对各工段高、低压配电室的新建，高压、低压控制方式，电气控制及远程监控手段的技术提升。整体中包含对粗苯低压配电室的切除及整合，对制酸、蒸氨、硫铵等工段的电气、仪表设备进行整体优化，最后形成一个多工段电气控制配电室。

关键词：电力供应；运行质量；整合升级；电气控制

Integration and Improvement of Low-Voltage Electrical Control

LI Shengjing

(Maintenance Company of HBIS Group Xuansteel Company, Zhangjiakou 075100, China)

Abstract: In order to guarantee the production quality of coking workshop in Xuangang energy company and improve the quality of waste gas recovery in purification workshop, the electrical control and supply of the important sections of the purification workshop, such as ammonium sulfate, crude benzene, ammonia distillation, acid making, etc. , are reconstructed. Including the new high-voltage, low-voltage distribution room of each section, high-voltage, low-voltage control mode, electrical control and remote monitoring means of technological upgrading. The whole includes the removal and integration of the low-voltage distribution room of crude benzene, and the optimization of the electric and instrument equipment in the process of acid making, ammonia evaporation and ammonium sulphate.

Key words: power supply; quality of operation; integration upgrade; electrical control

轧钢加热炉基础自动化系统的优化与应用

孔海超

（河钢宣钢检修公司，河北张家口　075100）

摘　要： 本文主要介绍了宣钢二高线厂加热炉基础自动化系统的工作与控制原理。基础自动化系统优化升级后设置电控和仪控两套 PLC 系统，既考虑到电控和仪控的独立性，又考虑到整个炉区的操作统一性，保证了整个系统良好的控制性能。加热炉的主要功能是通过最佳的方式为轧机提供坯料，达到优化容器内少，降低消耗，提升整体生产率的目的。通过自动化系统实现对板坯温度的合理预测分析，在根本上提升了温度控制均匀性，有效的提升了产品质量。确保了炉区生产的节能、高效、安全和稳定运行。

关键词： 加热炉；自动化系统；控制性能；操作

Optimization and Application of Basic Automation System for Steel Rolling Heating Furnace

KONG Haichao

(HBIS Xuangang Maintenance Company, Zhangjiakou 075100, China)

Abstract: This article mainly introduces the working and control principles of the basic automation system for the heating furnace at the Second High Speed Wire Plant of Xuangang. After the optimization and upgrading of the basic automation system, two sets of PLC systems, electronic control and instrument control, were set up, taking into account the independence of electronic control and instrument control, as well as the uniformity of operation in the entire furnace area, ensuring the good control performance of the entire system. The main function of the heating furnace is to provide billet for the rolling mill in the best way, achieving the goal of optimizing the amount of material in the container, reducing consumption, and improving overall productivity. By implementing reasonable prediction and analysis of slab temperature through automated systems, the uniformity of temperature control has been fundamentally improved, effectively improving product quality. Ensuring energy-saving, efficient, safe, and stable operation of furnace production.

Key words: heating furnace; automation system; control performance; operation

还原竖炉冷却系统复杂控制技术设计与实现

郝广春

(河钢宣钢检修公司,河北张家口 075100)

摘 要: 我国已成为世界上 CO_2 排放量最大的国家,钢铁行业是资源能源和 CO_2 排放密集型产业。为早日实现2030年前碳达峰及2060年前碳中和的目标,河钢集团公司,建设一个55万吨还原铁年产能,世界上第一个基于富氢焦炉煤气的生产工厂,工厂主要包括直接还原竖炉,还原铁冷却器。详细介绍了还原铁冷却系统特点,热的还原铁被装入这个冷却系统。压缩的冷却氮气是从还原铁冷却器圆锥段的低点被喷入的,氮气被喷入用以提升冷却效果。冷却氮气回路包含一个急冷/洗涤系统,用来冷却和洗涤离开反应器的冷却区域的冷却氮气。

关键词: 直接还原竖炉;还原铁冷却系统;复杂控制技术;氢冶金

Design and Implementation of Complex Control Technologies for Cooling System in DR Shaft

HAO Guangchun

(Maintenance Company of HBIS Group Xuansteel Company, Zhangjiakou 075100, China)

Abstract: China has become the country with the largest CO_2 emission in the world, and the iron and steel industry is a resource energy and CO_2 emission intensive industry. In order to achieve the goal of carbon peak by 2030 and carbon neutralization by 2060, HBIS Group Co. Ltd.,install a 0.550 MTPA Direct Reduction ("DR") Plant, This will be the first worldwide production plant based on H_2-enriched COG. The Plant mainly consists of one DR shafts,DRI cooler. The thesis expounds the characteristics of DRI cooling system, Hot DRI from the reactor is feed into this cooling system.The compressed cooling gas is injected at the lower conical section of the DRI cooler. The N_2 is injected for the cooling process.The Cooling Gas circuit includes one quenching/scrubbing system,for cooling and cleaning of the Cooling Gas stream leaving the cooling zone of the Reactor.

Key words: dr shafts; dri cooling system; complex control technologies; hydrogen metallurgy

脱硝氨逃逸在线监测分析系统优化

贾旭鉴

(河钢宣钢检修公司,河北张家口 075100)

摘 要: 针对脱硫脱硝控制系统在使用过程中存在氨逃逸现象,引发设备管路阻塞腐蚀、加药控制系统调节困难、原料浪费严重等诸多问题，脱硝氨逃逸在线监测分析系统优化从控制烟气温度、调节喷氨分布、催化剂利用、氨水雾化效果等多方面着手，通过 PLC 控制系统编程，提升设备控制精细化程度，提高智能化水平，并添加报警、趋势和报表画面，做到对脱硝系统氨逃逸指数在线监测分析，并能及时自动精准管控。

关键词：环保；脱硫脱硝控制；氨逃逸；PLC

Optimization of Online Monitoring and Analysis System for Denitration Ammonia Escape

JIA Xujian

(Maintenance Company of HBIS Group Xuansteel Company, Zhangjiakou 075100, China)

Abstract: In view of the phenomenon of ammonia escaping in the use of desulfurization and denitrification control system, which leads to blockage and corrosion of equipment pipelines, difficult adjustment of dosing control system, serious waste of raw materials and many other problems. The optimization of the denitration ammonia escape online monitoring and analysis system starts from various aspects such as controlling flue gas temperature, adjusting the distribution of ammonia injection, catalyst utilization, and ammonia atomization effect. Through PLC control system programming, the precision of equipment control is improved, the level of intelligence is improved, and alarm, trend, and report screens are added to achieve online monitoring and analysis of the ammonia escape index of the denitration system, and can be timely and accurately controlled.

Key words: environmental protection; desulfurization and denitrification control; ammonia slip; PLC

基于 YOLOv8 改进的钢水罐精准位置信息识别模型

廖文彬，巴烈电，张 伟，赵挺生，张 充

（华中科技大学土木与水利工程学院，湖北武汉 430070）

摘 要：图像识别广泛应用于钢铁行业安全生产管控技术中，但面临作业环境存在密集分布与遮挡、光照条件不良、设备尺度大幅变换等不利因素导致的识别精度低等问题。本文对 YOLOv8 模型进行改进，提出了 YOLO-Ladle 改进模型：通过整理图像数据集和梳理改进方向，从添加注意力机制、替换轻量化骨干网络、修改特征融合层等 3 个方面对基础模型进行改进；基于改进模型展开消融实验，结果表明模型体积减少了 63.98%，计算需求降低 40.6%；识别精度和召回率指标分别提高了 0.95% 与 0.51%，且 map50 值达到 98.6%，能满足钢包实时精准定位需求，可应用于边缘设备的轻量化部署，能为钢包姿态识别、目标追踪、碰撞预警等高级工业安全检测任务提供有效的基础模型。

关键词：计算机视觉；改进 YOLOv8；工业安全；钢包监测；智能管控；图像分割

Improved Semantic Segmentation Model for Ladle Based on YOLOv8

LIAO Wenbin, BA Liedian, ZHANG Wei, ZHAO Tingsheng, ZHANG Chong

(School of Civil & Hydraulic Engineering Huazhong University of Science and Technology, Wuhan 430070, China)

Abstract: In order to meet the demand for safety control intelligence in the iron and steel industry, and to solve the problems of low recognition accuracy caused by the existence of high-energy equipment in the closed operating environment with dense distribution and frequent occlusion, large scale transformation, poor lighting conditions and other unfavorable factors. In this paper, the YOLOv8 model is improved based on deep learning technology, and the YOLO-Ladle improvement model is proposed. By organizing the characteristics of the image dataset and the model, the model is improved by adding the attention mechanism, modifying the convolution module, replacing the lightweight backbone network, and modifying the feature fusion layer. Compared with the original model, the improved model reduces 63.98% of the model size and 40.6% of the computational requirements, and improves 0.95% and 0.51% of the recognition accuracy and recall rate indexes, respectively, which can satisfy the needs of real-time safety control, and be applied to high-performance recognition of the spatial location information of steel ladle and lightweight deployment of the edge equipment. At the same time, it can provide basic image segmentation models for advanced industrial safety detection tasks such as ladle attitude recognition, target tracking, and collision warning.

Key words: industrial safety; improvement of YOLOv8; attention mechanism; ladle monitoring; intelligent control; semantic segmentation

洗精煤采制样设备的建设研究

宫文浩，张 澎，宋立伟，韩 博，付兴海

（鞍钢股份有限公司质检计量中心，辽宁鞍山 114000）

摘 要： 煤是钢厂生产使用的重要原燃料，对购置的煤准确采样、制样、化验，真实反映出煤的品质，直接影响钢厂的采购成本，进而影响产品利润。本文重点介绍鞍钢股份质检计量中心针对水分较大的洗精煤的采制样设备建设的位置、环境、配置以及自动化智能化等方面进行分享探讨。

关键词： 洗精煤；采样；制样

Research on the Construction of Sampling Equipment for Clean Coal Mining

GONG Wenhao, ZHANG Peng, SONG Liwei, HAN Bo, FU Xinghai

(Angang Steel Co., Ltd., Quality Inspection and Metrology Center, Anshan 114000, China)

Abstract: Coal is an important raw fuel used in the production of steel mills. Accurately sampling, preparing, and testing purchased coal can truly reflect the quality of coal, directly affecting the procurement cost of steel mills and ultimately affecting product profits. This article focuses on exploring and sharing the location, environment, and configuration of sampling and preparation equipment for washed coal with high moisture content.

Key words: clean coal; sampling; sample preparation

行车无人化技术在热轧库区的应用

徐明红

（宝山钢铁股份有限公司，上海　201900）

摘　要：本文介绍了宝钢股份热轧生产线入口板坯库和出口钢卷库行车无人化技术的发展和应用情况，解决了行车无人化相关高精度检测技术，包括三维定位技术、激光扫描和防摇摆检测技术等，通过库区无人化行车全自动系统架构，实现了库区内生产作业信息的一体化。库区行车无人化实现了整体热轧工序的高度自动化，有效提高了劳动生产率，已成为宝钢股份智能制造的重要手段之一。

关键词：热轧库区；行车无人化系统；控制技术

Application of Unmanned Automatic Crane System in Hot Rolling Yards

XU Minghong

(Baoshan Iron & Steel Co., Ltd., Shanghai 201900, China)

Abstract: This paper introduces the development and application of unmanned automatic crane system in slab yards and coil yards of Baosteel's hot rolling production line. It solves high-precision detection technologies related to automatic crane, including three dimension positioning technology, laser scanning and anti swing detection technology. Through the fully automatic system architecture of unmanned automatic crane system in the warehouse area, the integration of production and operation information in the warehouse area is achieved. The unmanned automatic crane system in the warehouse area has achieved a high degree of automation in the overall hot rolling process, effectively improving labor productivity, and has become one of the important means of intelligent manufacturing for Baosteel.

Key words: hot rolling yards; unmanned automatic crane system; control technology

论信息化系统提升安全管理水平的探索与实践

刘晓宇，王　帅，张　靓

（内蒙古包钢钢联股份有限公司轨梁轧钢厂，内蒙古包头　014010）

摘　要：安全网格化智慧管控系统是以安全网格化"一图一表一册"作为基础考评依据，系统将全厂设备设施按照专责网格化区域录入，对于厂级查处的隐患及违章问题按照专责，落实到具体网格管理人员。按照积分制原则，各层级人员个人安全绩效分数实时更新，以每月查处的安全隐患、人员违章内容作为客观评价依据，按相关安全生产管理制度客观公正进行考核；年末兑现安全绩效工资。

  首先由安全管理部制定培训、监督、考核联动的安全管控模式，提出探索安全管理水平提升的要求，电气自动化部通过研究探索，捕获切实需求，制定软件设计方案，实施 BS 架构下以 C#结合 web 技术进行系统开发。

关键词：web 技术；智慧管控；网格化；隐患排查

生产协同智能管控中心

郭武英，徐 涛

（富蕴蒙库铁矿有限责任公司，新疆富蕴 836100）

摘 要：随着人工智能和物联网技术的不断发展，智能制造得到了广泛的关注。智能管控中心作为调度、指挥、管控、发布平台，主要承担物流运输、设备抢修、信息发布等业务，在企业中发挥着重要的作用。平台建设应基于能源、运输、制造、设备、环境、视频监控等现有各系统的数据，实现有效联动，充分发挥各子系统的功能，不断提高管控能力，确保生产稳定、物流畅通。

关键词：管控中心；生产效率和质量；生产协同

Production Collaboration Intelligent Control Center

GUO Wuying, XU Tao

(Fuyun Mengku Iron Mine Co., Ltd., Fuyun 836100, China)

Abstract: With the continuous development of artificial intelligence and the Internet of Things technology, intelligent manufacturing has received widespread attention. As a dispatch, command, control, and release platform, the intelligent control center mainly undertakes logistics transportation, equipment repair, information release, and other businesses, playing an important role in enterprises. The platform construction should be based on the data of existing systems such as energy, transportation, manufacturing, equipment, environment, and video surveillance, to achieve effective linkage, fully utilize the functions of each subsystem, continuously improve control capabilities, and ensure stable production and smooth logistics.

Key words: control center; production efficiency and quality; production collaboration

5G 工控网络下云化 PLC 技术的探索与实践

孙瑞琪，王军生，张 岩，秦大伟

（鞍钢集团北京研究院有限公司，北京 102200）

摘 要：在信息化技术革新的影响下，5G 与工业互联网对钢铁行业应用赋能已经从生产制造的辅助环节逐渐向工业控制环节渗透蔓延。本文构建了 5G 与工控深度融合网络，提出了一种部署在私有云服务器的 PLC 系统应用方案，设计了适用于云化控制的物联网协议及安全策略，并成功应用于某钢铁企业炼钢生产环节设备的远程控制，同时对 5G 云化 PLC 系统的实时性、可靠性、稳定性进行测试，结果表明，5G 工控网络稳定工作在<10ms@99.99%可靠性

状态，云化 PLC 系统连续运行 2000h 且控制功能稳定，此结果可为未来真正实现集中控制、生产管理一站式提供研究基础和技术攻关方向。

关键词：5G；工业互联网；云化 PLC；钢铁工业；智能制造；柔性生产

Exploration and Practice of Cloud-based PLC Technology in 5G Industrial Control Network

SUN Ruiqi, WANG Junsheng, ZHANG Yan, QIN Dawei

(Ansteel Group Beijing Research Institute Co., Ltd., Beijing 102200, China)

Abstract: Under the influence of information technology innovation, the empowerment of 5G and industrial internet in the steel industry has gradually expanded from production assistance to industrial control. In this paper, firstly a deep integration network of 5G and industrial control is constructed, and a PLC system solution deployed on private cloud servers is proposed. Secondly, IoT protocols and security policies for cloud control are designed. Finally, the proposed scheme was successfully applied to the remote control of equipment in the steelmaking production process of a certain steel enterprise. At the same time, the real-time performance, reliability and stability of the 5G cloud-based PLC system were tested. The result indicated that the 5G industrial control network can work stably with a delay of less than 10ms and 99.99% reliability. The cloud-based PLC system run continuously for 2000 hours and maintained stable control functions. This result can provide a research foundation and technical direction for achieving centralized control and one-stop production management in the future.

Key words: 5G; industrial internet; cloud-based PLC; iron and steel industy; intelligent manufacturing; flexible production

简述钢铁工业智能制造技术发展应用

闫子政，翟玉龙

（宝钢股份上海梅山钢铁股份有限公司炼铁厂，江苏南京 210039）

摘　要：能制造作为体现国家竞争力的重要内容之一，从发展之初就受到各国重视，制定了一系列战略及政策，为其发展奠定了坚实的基础。本文简述了钢铁工业智能制造技术的发展和应用现状。首先介绍了智能制造背景及其在全球范围内的应用，以及中国钢铁工业面临的市场需求和环境污染等问题。然后探讨了钢铁工业智能制造的实践情况，包括国外和国内的钢铁企业在智能制造方面的应用实例。最后，本文结合国内外智能制造的发展现状，总结了钢铁工业智能制造技术的应用发展趋势，并提出了智能制造技术在钢铁工业中的应用前景和推广方向，以期对钢铁工业的智能化转型提供参考和借鉴。

关键词：智能制造；钢铁工业；数字化转型；制造业

Brief Overview of the Development and Application of Intelligent Manufacturing Technology in the Steel Industry

YAN Zizheng, ZHAI Yulong

(Baosteel Shanghai Meishan Iron & Steel Co., Ltd., Nanjing 210039, China)

Abstract: Intelligent manufacturing, as an important aspect of national competitiveness, has been highly valued by countries worldwide since its inception. A series of strategies and policies have been formulated to provide a solid foundation for its development. This article briefly describes the development and current status of intelligent manufacturing technology in the steel industry. Firstly, the background of intelligent manufacturing and its global applications, as well as the market demand and environmental pollution issues faced by China's steel industry, are introduced. Then, the practical application of intelligent manufacturing in the steel industry is discussed, including examples of application by foreign and domestic steel enterprises. Finally, based on the current development of intelligent manufacturing at home and abroad, the article summarizes the development trend of intelligent manufacturing technology in the steel industry and proposes the prospects and promotion direction of intelligent manufacturing technology in the steel industry, in order to provide reference and guidance for the intelligent transformation of the steel industry.

Key words: intelligent manufacturing; steel industry; digital transformation; manufacturing industry

15 冶金物流

大会特邀报告
第十三届冶金青年科技奖获奖人特邀报告
分会场特邀报告
矿业工程
焦化及节能环保
炼铁与原料
炼钢与连铸
电冶金与废钢铁
轧制与热处理
表面与涂镀
金属材料深加工
粉末冶金
先进钢铁材料及应用
节能与低碳技术
冶金环保与资源利用
冶金设备与工程技术
冶金自动化与智能化
★ 冶金物流
冶金流程工程学
其他

公路高架交通灯在中南股份铁路道口的应用

刘 鹏

(广东中南钢铁股份有限公司物流部设备管理室,广东韶关 512000)

摘 要:铁路道口防护设备在铁路线路与道路(公路、城乡道路)平交的道口上,为防止铁路车辆与道路车辆以及行人等发生冲撞而采取的安全措施。铁路道口防护对提高铁路、道路的通过能力和保证行车安全具有重要的意义。在道口增设防护设备后,虽然机动车辆不断增多,但交通事故反而会大幅度下降。中南股份全自动道口系统更是升级版,已取消电动栏杆,在道口处加装公路高架交通灯是提高道口安全性能的重要举措。

关键词:全自动道口;控制;交通灯

Application of Highway Elevated Traffic Lights at Zhongnan Joint-stock Railway Crossing

LIU Peng

(Guangdong Zhongnan Iron and Steel Logistics Department Equipment Management Room, Shaoguan 512000, China)

Abstract: Railway crossing protection equipment is a safety measure taken to prevent railway vehicles from colliding with road vehicles and pedestrians at the crossing where railway lines and roads (highways, urban and rural roads) intersect. Railway crossing protection is of great significance for improving the capacity of railways and roads and ensuring traffic safety. After adding protective equipment at the crossing, although the number of motor vehicles is increasing, the traffic accidents will be greatly reduced. The automatic crossing system of Zhongnan Co., Ltd. is an upgraded version, and the electric railings have been cancelled. It is an important measure to improve the safety performance of the crossing by installing overhead traffic lights at the crossing.

Key words: fully automatic mouth; control; traffic light

物联网和大数据技术在物流系统中的应用

任立群

(宝信软件(山西)有限公司信息化事业部,山西太原 030000)

摘 要:近年来随着物联网、大数据等技术的进步,许多企业的供应链物流系统正在经历数字化转型,物联网和大数据技术是支持数字化转型的关键技术,这些技术可以将传感器、智能设备和数据分析方法结合在一起,实现物流信息的实时监测和精细化管理。如今企业需要应对的挑战越来越多,只有利用物联网和大数据技术来优化其供应链物流系统,才能更好地应对挑战,提高运作效率和降低成本。为了更好的研究物联网和大数据技术在智慧物流系统中的应用研究,本文将结合太钢集团供应链物流系统对相关工作进行详细的论述。

关键词：供应链管理；大数据；物联网；智慧物流系统

钢铁供应链数字货运平台管理技术的研究

张丽莉，刘振清，吴 浩，梁 君，姜 欢

（德邻陆港供应链服务有限公司，辽宁鞍山 114000）

摘 要：随着我国社会经济脚步的不断加快，国内对于供应链服务的应用需求不断增强，钢铁作为大宗商品，生产与采购量庞大，核心企业无法大而全的适配整条供应链，需要依靠大量的社会资源共同建设，社会上的铁路、港口、仓库等资源方主体又因为缺乏有效组织而无法将效率最大化。以物流为例，需要港口、货运、水运、汽运等服务商整合服务单元，大力发展多式联运、最后一公里物流、港到门等服务。需要统一的平台通过整合运力并与运输需求匹配，有效组织资源围绕核心企业共建配套设施。钢铁供应链企业为了能在行业中保持竞争优势，必须对大宗物流进行有效的控制和管理。本文对新时期下钢铁物流存在的问题进行研究，并提出建设数字货运平台管理技术的解决方案。

关键词：数字货运；多式联运；大宗物流；智慧物流；供应链管理

FAD 型铁路货运站场机车自动驾驶系统研究与应用

郭 兵，田力男，王 晨，李生意，潘广海

（鞍钢股份鲅鱼圈钢铁分公司物流运输部，辽宁营口 115007）

摘 要：随着 5G 通信、自动控制和环境感知等技术的发展，铁路货运站场机车自动驾驶技术逐步进入国内各大冶金运输企业领导的视线。以鞍钢股份有限公司鲅鱼圈钢铁分公司（以下简称鞍钢鲅鱼圈）为例，鞍钢鲅鱼圈铁路货运站场自动驾驶技术是鞍钢鲅鱼圈联合北京全路通信信号研究设计院集团有限公司（以下简称北京通号）共同研发，在鞍钢鲅鱼圈铁路货运站场进行了实践，证实了冶金企业机车自动驾驶技术是可行的。本文通过论述机车自动驾驶系统中重点技术问题及解决方案，为钢铁企业铁路机车自动驾驶提供参考。

关键词：5G 通信；环境感知；自动驾驶

鞍钢鲅鱼圈铁路智能化无人货场研究

郭 兵，杨星亮，王 晨，李生意

（鞍钢股份鲅鱼圈钢铁分公司物流运输部，辽宁营口 115007）

摘　要：工业 4.0 时代,传统的货场管理模式和作业方式效率低下,作业等待工时浪费严重,不符合精益生产的集约化和少人化管理。基于当前鞍钢鲅鱼圈设备现状和实际需求,契合鞍钢集团数字化、智能化、低碳化发展战略,以鞍钢鲅鱼圈智能无人货场新建工程为背景,深入剖析新一代智能货场生产指挥系统的技术要点,并进一步指出该系统对于完善鞍钢鲅鱼圈铁路货场装备、提高公转铁整体联运效率、压缩货物周转时间等方面的重要意义。

关键词：无人货场；智能管理；数字化指挥；无人吊车

鞍钢股份鲅鱼圈分公司新型铁水运输模式探索与分析

王　晨，郭　兵，田力男，孙东生

（鞍钢股份鲅鱼圈钢铁分公司物流运输部，辽宁营口　115007）

摘　要：在国家大力推进供给侧结构性改革、碳达峰、创新驱动发展和智能制造的大环境下，钢铁企业新型的物流运输模式逐步纳入管理者的视线，对提质增效提出了更高的要求。鞍钢股份鲅鱼圈分公司及时响应国家策略，探索与剖析新型铁水运输模式，将机车牵引鱼雷罐转变为直接给鱼雷罐赋予动力实现其自驱动。本文基于无人化技术的鱼雷罐自驱动的新型运输模式，将充分发挥信息集控优势，最大限度的达成"运输工艺最优化""运输流程便捷化"，实现铁水温降大幅降低和运输效率的提升。

关键词：铁水运输；自驱动；无人化技术；提质增效

超偏载检测装置在鞍钢鲅鱼圈的研究与运用

郭　兵，张晓军，李　丹，张思波

（鞍钢股份鲅鱼圈钢铁分公司物流运输部，辽宁营口　115007）

摘　要：铁路货车超偏载检测装置是一种自动对行进中的列车实行不停车、不摘钩、连续动态称量及超载、偏载检测的安全计量设备[1]，担负着铁路运输过程中货物列车安全运行的重要责任。当货运车辆通过该设备时可动态精确测量货车每节车厢的重量、速度、各项偏移量及该节车厢的车型、车号，并提供各测试点通过货车的偏载报警。该装置的有效运用可极大程度提高车站接发列车作业能力，压缩局车在站停留时间，减少铁路车辆延时费。本文对超偏载检测装置在鞍钢鲅鱼圈中的运行现状进行了分析，并结合存在的问题给出了相应的解决方案，提升了铁路货运车辆行车安全系数和运输效率。

关键词：超偏载；检测；运输安全；延时费

基于一贯制管理的鞍钢质量管理系统构建

王 静，闵承鑫

（鞍钢股份有限公司数智发展部，辽宁鞍山 114000）

摘 要：一贯制质量管理是目前国内大型现代化钢铁企业管理体制中的重要一环，其通过质量管理职能的集中和质量控制关键点信息的一贯确保了产品和服务的可靠性，大幅度提高了企业质量管理的效率和效益。推行一贯制质量管理，建立一贯制质量管理的完整体系必须依赖先进的信息化手段，本文介绍了鞍钢股份的一贯制质量管理系统，包含六大模块，覆盖了从接受合同到成品出厂的整个过程，并简要概括了系统建设过程中结合鞍钢特点所开发的一系列特色功能，以期能为各钢厂技术人员在进行相应系统迭代升级过程中提供借鉴和帮助。

关键词：一贯制；质量；质量管理系统；钢铁企业

The Construction of Ansteel Quality Management System Based on the Consistent Management

WANG Jing, MIN Chengxin

(Digital Intelligence Development Department, Ansteel Co., Ltd., Anshan 114000, China)

Abstract: Consistent quality management is an important part of the current management system of large modern steel enterprises in China. Through the centralized quality management function and the consistency of key quality contral points information, it ensures the reliability of products and greatly improving the efficiency and effectiveness of enterprise quality management. The implementation of consistent quality management and the establishment of a complete system of consistent quality management must rely on advanced information means. This article introduces the consistent quality management system of Angang Steel Co., Ltd., which includes six modules, covering the whole process from contract acceptance to finished products delivery. It also briefly summarizes a series of characteristic functions developed based on the characteristics of Angang Steel during the system construction process, with the aim of providing reference and assistance for technical personnel of various steel mills during the corresponding system iteration and upgrading process.

Key words: the consistent management; quality; quality management system; iron and steel enterprise

高炉铁水敞口加盖保温技术研究与运行实践

徐大勇[1,2]，马光宇[1,2]，刘常鹏[1,2]，袁 玲[1,2]，李卫东[1,2]，孙守斌[1,2]

（1. 海洋装备用金属材料及其应用国家重点实验室，辽宁鞍山 114000；
2. 鞍钢集团公司钢铁研究院，辽宁鞍山 114000）

摘 要：当前国内很多钢铁企业正在开展对敞口运输的铁水罐实施加盖保温运行试验，采用的方式各有千秋，技术各有所长。本文针对敞口铁水罐在运输过程温降大的问题，阐述了敞口铁水罐热量损失构成和原因，提出及时对钢厂转炉翻铁后红热的空铁水罐重点实施加盖保温的技术思想。设计开发了随车行走的铁水罐加盖保温装置，有效降低空罐返回期内衬热量损失和运输过程铁水的热量损失，提高铁水温度，大幅降低铁水罐烟尘污染，提高铁钢界面能效。

关键词：铁水温度；铁水罐加盖；界面技术

Thermal Insulation Technology Research and Operation Practice of Capping for Open Molten Iron Tank of Blast Furnace

XU Dayong[1,2], MA Guangyu[1,2], LIU Changpeng[1,2],
YUAN Ling[1,2], LI Weidong[1,2], SUN Shoubin[1,2]

(1. State Key Laboratory of Metal Material for Marine Equipment and Application, Anshan 114000, China; 2. Iron and Steel Research Institute of Ansteel Group, Anshan 114000, China)

Abstract: At present, many domestic iron and steel enterprises are carrying out the operation test of capping and thermal insulation for open-ended transportation of molten iron tanks. The methods adopted have their own advantages and technologies. Aiming at the problem of large temperature drop of open molten iron tank during transportation, the composition and causes of heat loss of open molten iron tank are expounded, and the technical idea of timely covering and heat preservation of red hot empty molten iron tank after turning over is put forward. The thermal insulation device for the molten iron tank covered with the car is designed, which can effectively reduce the heat loss of the inner lining during the return period of the empty tank and the heat loss of the molten iron during transportation, increase the molten iron temperature, greatly reduce the smoke pollution of the molten iron tank, and improve the energy efficiency of the iron-steel interface.

Key words: molten iron temperature; molten iron tank capping; interface technology

鞍钢股份智慧供应链应用探索

侯海云[1]，荣树强[2]，易　茗[3]

（1. 鞍山钢铁集团有限公司机关，辽宁鞍山　114000；2. 鞍钢股份有限公司数智发展部，辽宁鞍山　114000；3. 鞍钢股份有限公司物流管理中心，辽宁鞍山　114000）

摘　要：鞍钢股份有限公司应对我国经济发展转型升级趋势，实施管理与信息化整体提升项目，推进钢铁产业销售、采购、制造、物流等核心价值链业务在组织职能、业务流程、系统架构等方面进行优化整合，实现从销售到采购、从采购到制造、从制造到运输、从运输到加工配送等整体供应链的流程贯通，初步建立鞍钢股份智慧供应链体系，取得较好的应用效果，公司编制《钢铁行业智慧供应链方向智能制造标准应用试点》入选工信部 2022 年度智能制造标准应用试点，后续将围绕钢铁行业智慧供应链方向持续开展应用探索工作。

关键词：智慧供应链；钢铁；销售；采购；制造；物流

Exploration of Application of Intelligent Supply Chains of Angang Steel Company Limited

HOU Haiyun[1], RONG Shuqiang[2], YI Ming[3]

(1. The Organ of Anshan Iron and Steel Group Company Limited, Anshan 114000, China;
2. The Digital Intelligence Development Department of Angang Steel Company Limited, Anshan 114000, China; 3. The Logistics Manage Center of Angang Steel Company Limiter, Anshan 114000, China)

Abstract: Angang Steel Company Limited responded to the transformation and upgrading trend of China's economic development, implement the overall improvement project of management and informatization, promote the optimization and integration of core value chain businesses such as sales, procurement, manufacturing and logistics in the iron and steel industry in terms of organizational functions, business processes, system architecture, etc., and achieve the integration of the overall supply chain processes from sales to procurement, from procurement to manufacturing, from manufacturing to transportation, from transportation to processing and distribution, Angang Steel Company Limited has initially established a intelligent supply chain system, which has achieved good application results. Our company compiled the Pilot Project of Intelligent Manufacturing Standards in the Direction of Intelligent Supply Chains in the Iron and Steel Industry, which has been selected as the pilot project of intelligent manufacturing standards in 2022 by the Ministry of Industry and Information Technology of the People's Republic of China. Later, we will continue to carry out application exploration work around the direction of intelligent supply chains in the iron and steel industry.

Key words: intelligent supply chain; iron and steel; sales; procurement; manufacturing; transportation

低铁耗生产模式下铁水运输作业分析

王 玮，程献权

（广西柳州钢铁集团，广西柳州 545000）

摘 要： 本文通过对柳钢集团本部转炉炼钢厂的低铁耗生产模式为出发点，钢铁生产单位在环保限产以及原材料成本增长的情况下，为应对低铁耗新作业模式，对铁水运输行车组织和作业模式进行分析，从铁水取送计划着手，凭借定量分析和定性分析相结合的探索手段，对铁水运输组织方案进行优化，提高铁水罐周转率，以达到降本增效的目的。

关键词： 低铁耗；铁水运输；周转率

ESG助推中国钢铁企业低碳转型与高质量发展

段希雅

（北京科技大学机械工程学院，北京 100083）

摘　要： 应对气候变化应在可持续发展框架下，充分立足我国国情和能力，在高质量发展的要求下落实推进钢铁企业绿色低碳转型。本文对ESG内涵及其在钢铁行业的作用进行阐释，并结合中国钢铁行业碳排放现状与挑战分析，从绿色供应链管理、社会责任价值链建设、治理机制的优化角度，提出ESG对中国钢铁企业实现低碳转型和高质量发展的价值贡献。

关键词： ESG；钢铁；碳排放；绿色供应量；可持续；高质量发展

ESG Promotes Low-carbon Transformation and High-quality Development of Chinese Steel Enterprises

DUAN Xiya

(School of Mechanical Engineering, University of Science and Technology Beijing, Beijing 100083, China)

Abstract: In response to climate change, sustainable development should be fully considered based on China's national conditions and capabilities, and the green and low-carbon transformation of steel enterprises should be promoted under the requirements of high-quality development. This article explains the connotation of ESG and its role in the steel industry, and analyzes the current situation and challenges of carbon emissions in the Chinese steel industry. From the perspectives of green supply chain management, social responsibility value chain construction, and optimization of governance mechanisms, the value contribution of ESG to the low-carbon transformation and high-quality development of Chinese steel enterprises is proposed.

Key words: ESG; steel; carbon emissions; green supply; sustainable; high-quality development

浅谈人工智能在供应链管理中的应用研究

王　浩

（北京科技大学机械工程学院，北京　100083）

摘　要： 随着人工智能技术的不断发展，其应用领域也更趋于广泛。在人工智能技术的应用范围中，其在供应链管理的应用备受人们关注。由此本文的目的是探讨人工智能技术在供应管理中各项环节的应用，并分析其对供应链管理的影响与挑战。首先文章将介绍人工智能技术在供应链规划、执行与协同中的应用，然后针对人工智能技术在优化供应链管理决策、提高供应链管理效率与精确度等方面的影响进行分析。最后，对人工智能技术在供应链管理中的挑战进行讨论，并提出了未来的研究方向。

关键词： 人工智能技术；供应链管理；供应链规划；供应链执行；供应链协同

Discussion on the Application of Artificial Intelligence in Supply Chain Management

WANG Hao

(School of Mechanical Engineering, University of Science and Technology Beijing, Beijing 100083, China)

Abstract: With the continuous development of artificial intelligence technology, its application fields are also becoming more widespread. In the application scope of artificial intelligence technology, its application in supply chain management has attracted people's attention. The purpose of this article is to explore the application of artificial intelligence technology in various aspects of supply management, and analyze its impact and challenges on supply chain management. Firstly, the article will introduce the application of artificial intelligence technology in supply chain planning, execution, and collaboration, and then analyze the impact of artificial intelligence technology on optimizing supply chain management decisions, improving supply chain management efficiency and accuracy. Finally, the challenges of artificial intelligence technology in supply chain management were discussed, and future research directions were proposed.

Key words: artificial intelligence technology; supply chain management; supply chain planning; supply chain execution; supply chain collaboration

冶金工矿企业铁路小半径曲线轨道结构创新设计

江宏法

(宝武集团马鞍山钢铁股份有限公司运输部,安徽马鞍山 234000)

摘 要: 本文针对解决冶金工矿企业铁路小半径曲线轨道结构病害问题,提出砼枕轨道分开可调式护轨结构创新设计,技术实用、实施简单,结构安全可靠性高,维修量少,寿命周期长,综合经济效益可观,具有普遍的推广使用价值。

关键词: 小半径曲线;护轨结构;铺设效果

Innovative Design on Small-Radius Curved Railway Tracks Used by Metallurgical and Mining Enterprises

JIANG Hongfa

(Railway Transportation Subsidiary of China Baowu Maanshan Iron & Steel Co., Ltd., Maanshan 243000, China)

Abstract: To solve the structural sufferings of small-radius curved railway tracks used by metallurgical and mining enterprises, this paper provides an innovative design of an assemblable and adjustable guard rail structure with concrete ties. This design is technically practical, simplified in implementation, and highly safe and reliable, requiring less maintenance and enjoying a longer life cycle, with considerable comprehensive economic benefits and worthiness of popularization and application.

Key words: small-radius curved; guard rail structure; railway laying effects

车辆全流程管控系统的研究与实现

王宏,钱松

(包钢集团新联信息产业有限公司,内蒙古包头 014010)

摘　要：随着钢铁企业信息化的建设，钢铁企业内部物流的每一个环节产生数据量越来越多，要体现数据的动态价值，实现过程信息可实时监控、变化通过系统作业指令体现、信息动态共享、提高作业效能。因此，搭建智慧物流管控平台，实现数据的高效共享非常有必要。本文以包钢股份车辆全流程管控系统系统建设为例，阐述了数据的高效共享，实现业务流程重组、改变工作方式及习惯、完成车辆运输全程可视化，实现对包钢股份厂内物流业务的管理提升。

关键词：钢铁物流；平台设计；移动互联网；数据共享；流程重组

Research and Implementation of the Whole Process of Vehicle Control System

WANG Hong, QIAN Song

(Baogang Group Xinlian Information Industry Co., Ltd., Baotou 014010, China)

Abstract: With the construction of iron and steel enterprise information, each link of the internal logistics of iron and steel enterprises produces more and more data, to reflect the dynamic value of data, the process information can be monitored in real time, the change is reflected through the system operation instructions, information dynamic sharing, improve the operation efficiency. Therefore, it is very necessary to build an intelligent logistics control platform and realize the efficient sharing of data. This paper takes the construction of the whole process vehicle control system of Baotou Steel as an example, expounds the efficient sharing of data, realizes the realization of business process restructuring, the change of working methods and habits, completes the visualization of the whole process of vehicle transportation, and improves the management of the logistics business in the factory of Baotou Steel.

Key words: steel logistics; platform design; mobile internet; data sharing; process restructuring

铁路车辆运行中常见故障及应急处理措施

李世锋

（张宜科技物流公司，河北宣化　075100）

摘　要：铁路车辆指宣钢厂内运行铁路车辆（包括宣钢自备车和国铁车辆），主要负责宣钢厂区原燃料、成品材、焦炭等物料拉对。日常运输作业中会出现不同类型的故障，影响车辆安全运行及生产运输效率，如何实现铁路车辆运用故障快速恢复，缩短故障处理用时，及时恢复铁路运输正常运行，在此对常见铁路车辆故障的一些快速处理方法进行浅谈。

关键词：车钩；制动系统；应急救援

Common Failures and Emergency Measures in the Operation of Railway Rolling Stock

LI Shifeng

(Zhang xuan Technology Logistics Company, Xuanhua 075100, China)

Abstract: Rolling stock is xuangang company internal operation rolling stock (including Xuangang's own rolling stock and national railway rolling stock), mainly responsible for the transportation of steel and oter materials within Xuangang company. Different types of faults occur in daily transportetion operations, it affects the safe operation of vehicies and the efficiency of production and transportation, how to realize the quick recovery of railway vehicle operation failure, shorten the troubleshooting time, normal railway transport will be resumed in a timely manner, in this paper, some quick treatment methods for common railway vehicle faults are briefly discussed.

Key words: coupler; braking system; emergency rescue

冶金企业内燃机车空调改造的探索与实践

乔玉栋，范秀川，白　帅，罗　健

（物流公司机务作业区，河北张家口　075100）

摘　要：冶金企业的内燃机车空调系统是实现对内燃机车司机室空气进行制冷、加热、换气和空气净化的装置。它可以为乘务人员提供舒适的驾驶环境，降低乘务员的疲劳强度，提高行车安全。但原有机车空调存在：维修成本高、维修拆装不便、耗油量大等诸多缺点。河钢集团张宣科技公司对原装机车空调进行改造更换，在保证空调运行可靠性、实用性的基础上，实现降低机车空调维修成本、减轻维修保养难度、降低内燃机车油耗等效果。本文将对冶金企业内燃机车空调改造的探索与实践进行简单论证。

关键词：机车空调；民用空调；工作环境；油耗；排放

Exploration and Practice of Air Conditioning Transformation of Internal Combustion Locomotive in Metallurgical Enterprises

QIAO Yudong, FAN Xiuchuan, BAI Shuai, LUO Jian

(Logistics Co., Ltd., Zhangjiakou 075100, China)

Abstract: The air conditioning system of diesel locomotives in metallurgical enterprises is a device to refrigerate, heat, exchange and purify the air in the cab of diesel locomotives. It can provide a comfortable driving environment for crew, reduce the fatigue strength of flight attendants, and improve driving safety. However, the original locomotive air conditioning has many shortcomings, such as high maintenance cost, inconvenient maintenance and folding, large fuel consumption and so on. On the basis of ensuring the reliability and practicability of air conditioning, the original locomotive air conditioning has been reformed and replaced by Zhangxuan Science and Technology Company of Hegang Group, which can reduce the maintenance cost of locomotive air conditioning, reduce the difficulty of maintenance and reduce the fuel consumption of diesel locomotives. This paper will briefly demonstrate the exploration and practice of air conditioning transformation of diesel locomotives in metallurgical enterprises.

Key words: locomotive air conditioning; civil air conditioning; working environment; fuel consumption; emissions ~ 0 introduction

冶金铁路小半径曲线钢轨磨耗的原因及防治

王建栋，闫 炜

（河钢宣钢物流公司，河北张家口 075100）

摘 要：本文分析了冶金厂区内部铁路小半径曲线钢轨频繁磨耗的起因及曲线侧磨产生的原因，针对工务系统多年积累的实际维修经验和科学整治作业方法，提出了小半径曲线钢轨磨耗的整治办法，探索企业铁路小半径曲线的养护维修方法与检查，使之减少工务维修量及维修成本，同时保障冶金企业机车车辆运输安全作业。

关键词：冶金铁路；小半径曲线；磨耗；整治

Causes and Prevention of Wear of Small Radius Curved Rail in Metallurgical Railways

Abstract: This article analyzes the causes of frequent wear and tear of small radius curve rails in the metallurgical plant area, as well as the causes of curve side wear. Based on years of accumulated practical maintenance experience and scientific rectification methods in the engineering system, a rectification method for small radius curve rail wear is proposed. The maintenance and inspection methods for small radius curve rails in enterprise railways are explored to reduce the quantity and cost of engineering maintenance, At the same time, ensure the safe operation of locomotive and vehicle transportation in metallurgical enterprises.

Key words: metallurgical railway; small radius curve; wear and tear; rectification

16 冶金流程工程学

大会特邀报告
第十三届冶金青年科技奖获奖人特邀报告
分会场特邀报告
矿业工程
焦化及节能环保
炼铁与原料
炼钢与连铸
电冶金与废钢铁
轧制与热处理
表面与涂镀
金属材料深加工
粉末冶金
先进钢铁材料及应用
节能与低碳技术
冶金环保与资源利用
冶金设备与工程技术
冶金自动化与智能化
冶金物流
★ 冶金流程工程学
其他

ed
SPHC 钢直上浇注的生产实践

李 超，尹宏军，李文博，冉茂铎

（鞍钢股份鲅鱼圈钢铁分公司炼钢部，辽宁营口 115007）

摘 要：本文通过分析现场工况，使用氩站直上工艺试验，选取 SPHC 钢种进行研究，通过锰控制、Als 控制、温度控制等工艺研究，进行了生产试验，根据工艺试验 14 罐后总结，生产节奏正常，铸机浇铸稳定，无因为该工艺出现断浇、漏钢等事故，钢水成分合格率 100%，温度合格率 100%，经过测算试验的罐次约节约成本 15.6 万元。

关键词：氩站直上；SPHC；锰含量；温度

Production Practice of the Straight Casting in SPHC Steel

LI Chao, YIN Hongjun, LI Wenbo, RAN Maoduo

(Steelmaking Department Bayuquan Iron & Steel Subsidiary of
Angang Steel Co., Ltd., Yingkou 115007, China)

Abstract: The paper through the analysis of field condition, using argon station up process test, by selecting SPHC steel, through manganese control, Als control, temperature control process research, the production test was done, according to the process test 14 tank summary, the production rhythm is normal, casting stability, no the process cutting, steel leakage accident, molten steel composition qualified rate is 100%, temperature qualified rate is 100%, after measuring the test of tank saving cost is about 156000 RMB.

Key words: argon station straight up; SPHC; manganese content; temperature

基于冶金流程工程学的连铸-热轧界面动态设计

张 琼，耿明山

（中冶京诚工程技术有限公司技术研究院，北京 100076）

摘 要：冶金流程工程学提出了动态精准设计理念，为钢厂的设计指出了先进的方向。本文基于冶金流程工程学理论，分析了连铸-热轧区间动态运行过程，利用仿真技术实现了连铸-板坯库-热轧区段的动态设计，并通过实例验证了动态设计的优越性，为工厂布局设计及生产组织提供了科学、高效的建议和指导。

关键词：冶金流程工程学；连铸-热轧界面；动态设计；仿真；优化

Dynamic Design of Continuous Casting-hot Rolling Based on Metallurgical Process Engineering

ZHANG Qiong, GENG Mingshan

(Technical Research Institute, MCC Capital Engineering and Research Incorporation Limited, Beijing 100076, China)

Abstract: Metallurgical process engineering puts forward the concept of dynamic precision design and points out the advanced direction for steel mill design. Based on the theory of metallurgical process engineering, this paper analyzes the dynamic operation process of the continuous cast-hot rolling, and realizes the dynamic design of the continuous cast-slab store-hot rolling with simulation technology. The superiority of dynamic design is verified through examples, providing scientific and efficient suggestions and guidance for factory layout design and production organization.

Key words: metallurgical process engineering; continuous casting-hot rolling interface; dynamic design; simulation; optimization

板坯赋号规则的优化实践

董金刚[1]，陈廷星[1]，谢璞石[2]

（1. 宝钢湛江钢铁有限公司制造管理部，广东湛江　524000；
2. 宝钢湛江钢铁有限公司炼钢厂，广东湛江　524000）

摘　要： 板坯赋号是指板坯切割时按照赋号规则给实物板坯赋予一个材料申请时预先设定的板坯号（预定板坯号），实现虚拟板坯和实物板坯的勾连，满足合同和板坯的匹配、板坯的热装轧制，未赋号的板坯变为无委托板坯，需要重新匹配合同才能轧制生产。预定板坯号的赋号率对板坯的热装轧制和合同完成具有重要影响。随着湛江钢铁公司三号高炉投产，连铸工序成为公司的生产瓶颈，调宽板坯数量明显增加，原来大部分调宽坯不赋预定板坯号的规则将产生大量的无委托板坯，同时影响板坯的热装。生产中发现板坯长度、炉重量的管理也存在影响板坯赋号的问题。

湛江钢铁公司板坯赋号中的长度问题包括板坯设计的长度范围太小、板坯赋号的长度公差要求与板坯输出的长度公差要求不一致、头坯因为长度不合而赋号率低，采取的优化措施：炉次宽度设计时，增加板坯长度范围最小值的限制条件；板坯赋号的长度公差值与板坯输出的长度公差值保持一致，实现不同管理的要求相同；优化头板的切割长度和长度赋号要求，实现大部分头坯赋号。板坯宽度问题包括大部分调宽板坯赋特殊预定板坯号、板坯的实际生产宽度与设计宽度不一致，为此综合考虑热轧侧压能力、钢种强度等级和特殊钢种的要求，对板坯不同的调宽量采取不同的优化措施，提高调宽板坯的赋号率；引入"宽度组"概念，其作用体现为：规范和引导连铸工序按宽度组和预定板坯号赋号规则切割，提高板坯与合同的匹配度；某流的多余板坯（指实际生产的板坯块数大于材料申请时的设计块数）可借用另一流多余的预定板坯号，提高板坯赋号率和热装率；生产异常（如单流浇铸、宽度错流生产、不同宽度错位生产等）时不影响板坯赋号，提高生产容错度。炉重量问题是指炼钢生产计划编制时设定的一炉钢的板坯重量与实际的板坯重量不一致，采取的措施是：根据钢种和精炼工艺修正炼钢计划的炉重量设定值，并根据实际重量和设定重量的差异定期修改炉重量设计值，减少因炉重量波动而出现的多余板坯。同时规范了板坯不赋号的代码管理，板坯不赋号的代码只保留长度不合、宽度不合、多余板坯三个代码，取消特殊预定板坯号的使用。

板坯赋号规则优化后，板坯长度不合而未赋号的比例从 2.1%下降为 1.4%，调宽板坯赋号率从 20%提升到 67.4%，总的板坯赋号率从 93%提高到 95.3%，每天可减少 30~35 块无委托板坯。

典型钢铁制造流程极致能效与极限碳排放分析

张 琦[1,2,3]，籍杨梅[1]，王鑫鑫[1]

（1. 东北大学 国家环境保护生态工业重点实验室，辽宁沈阳 110819；
2. 辽宁省流程工业节能与绿色低碳技术工程研究中心，辽宁沈阳 110819；
3. 低碳钢铁前沿技术教育部工程研究中心，辽宁沈阳 110819）

摘 要：高炉-转炉流程是我国钢铁生产的主要制造流程，该流程也是典型的铁-煤化工过程，其物质、能源消耗和 CO_2 排放一直居高不下，在双碳目标下，钢铁生产流程节能与 CO_2 排放面临巨大压力和挑战。在冶金流程工程学和系统节能理论指导下，中国钢铁工业节能减排工作得到跃升发展，钢铁制造流程的系统能效显著提升，钢铁工业绿色化与智能化协同和低碳转型发展指日可待。

钢铁制造流程动态运行过程的物理本质是铁素流在能量流的驱动和作用下，按照设定的"程序"，沿着特定的"流程网络"做动态-有序的运行，实现从原材料使用到生产出最终产品[1]。该流程是多成分、多相态、多层次、开放性、远离平衡、非稳态、动态有序性等多复杂性表征的集合[2]。其能源消耗、CO_2 排放受原燃料条件、冶炼工艺水平和生产结构等影响较大[3]，是钢铁行业碳达峰碳中和目标实现的重点领域。本研究围绕钢铁生产过程反应机理，采用热力学和热化学理论，对相应的生产工艺进行假设，基于实际的原料、燃料参数和反应条件，分析了典型钢铁制造流程的理论极限能耗、系统能效和理论 CO_2 排放量，揭示了钢铁冶金反应器与各工序真实能效水平。结果表明，轧制过程为冷装料情况下，原料为全铁水时高炉-转炉长流程的理论极限能耗和 CO_2 排放为 388.24 kgce/t 和 824.08 $kgCO_2$；15%废钢比的条件下，高炉-转炉长流程的理论极限能耗和 CO_2 排放减少，为 337.99 kgce/t 和 681.30 $kgCO_2/t$。采用电炉短流程炼钢工艺时，钢铁生产过程的能耗进一步降低。如采用 100%废钢时，短流程炼钢工艺的理论极限能耗和 CO_2 排放分别为 72.36 kgce/t 和 328.79 $kgCO_2/t$。研究炼钢过程不同原料结构、冶炼工艺对全流程理论能耗及 CO_2 排放的影响，为极致能效、极限碳排放分析提供理论基础，提出钢铁生产工序的能源管理目标和方向，指导实际生产过程，实现全流程有效节能、降碳、减排，推动钢铁行业碳达峰碳中和目标早日实现。

将复杂的钢铁制造流程抽象为物质流、能量流流动以及物质流-能量流的协同作用过程，也便于发挥现代钢铁联合企业的产品制造功能、能源转换功能及其废弃物的消纳处理等三大功能，有利于进一步挖掘节能、降耗、减排的潜力并寻求新的节能突破口[4]。本研究围绕冶金流程工程学理论、系统节能理论与方法，从多个层次、多个角度分析了钢铁制造流程系统节能的进展和发展方向，并给出在系统节能理论指导下企业能效提升的途径，旨在为新形势下企业开展系统节能、减排工作提供理论依据和分析工具，同时为中国钢铁工业绿色化、智能化发展提供新思路、新方向。

参 考 文 献

[1] 殷瑞钰. 冶金流程学[M]. 北京: 冶金工业出版社, 2023.
[2] 殷瑞钰. 冶金流程工程学[M]. 北京: 冶金工业出版社, 2009.
[3] 陆钟武. 系统节能基础[M]. 沈阳: 东北大学出版社, 2010.
[4] 张琦, 蔡九菊. 钢铁制造流程系统节能与能效提升[J]. 钢铁, 2021, **56(8)**: 32-42.

CBAM 对中国钢铁行业的影响和应对策略

李 涛[1,2]，上官方钦[1,2]，郦秀萍[1,2]，周继程[1,2]，倪 冰[1,2]

（1. 中国钢研科技集团有限公司钢铁绿色化智能化技术中心，北京 100081；
2. 冶金智能制造系统全国重点实验室，北京 100081）

摘 要：为了更好地应对 CBAM（Carbon Border Adjustment Mechanism）对中国钢铁行业带来的冲击和影响。通过分析中国钢铁行业碳排放结构和现状，基于 CBAM 最新实施进展，以 1t 生铁为例，通过设置不同的碳排放强度和碳市场交易情景，分析和预测了 CBAM 法案可能对生铁出口欧盟成本产生的冲击和影响。以 1t 生铁为例，根据我国钢铁长流程生产工艺碳排放强度特点，设置了 3 种不同高炉炼铁工序碳排放水平情景：$1.2tCO_2/t$、$1.4tCO_2/t$ 和 $1.6tCO_2/t$。基于对欧盟和中国碳交易市场的预测，在欧盟碳市场交易价格为 120 欧元/tCO_2 的情况下，分析了中国钢铁行业未纳入碳市场和中国钢铁行业纳入碳市场（碳价为 80 元/t）时碳市场价格波动和工序碳排放强度对生铁出口欧盟成本的影响。研究结果表明：以 1t 生铁为例，将使我国生铁出口欧盟成本增加 95.52~280.32 元/t。假设当铁水碳排放强度为 $1.2tCO_2/t$ 时，我国出口欧盟的生铁不需要额外缴纳碳关税。当生铁碳排放强度从 $1.4 tCO_2/t$ 升高到 $1.6tCO_2/t$ 时（欧盟碳价为 120 欧元/t，中国碳价为 80 元/t），出口欧盟的生铁成本增加了 159.20 元/t。如果钢铁行业未纳入中国碳市场或者中国碳市场未和欧盟碳市场挂钩（中国碳价为 0 元/t）时，出口欧盟生铁的成本将会增加 105.12~280.32 元/t。过渡期 CBAM 对生铁出口欧盟的成本影响相对较小。过渡期过后，随着 CBAM 免费配额削减速度逐年加快，生铁出口欧盟需要缴纳的碳关税逐年加速增加，到 2034 年 1t 生铁需要缴纳的碳关税已经达到了 1114.40 元/t。为了树立中国钢铁行业低碳形象，提高中国钢铁产品的国际竞争力，建议中国钢铁行业应该加强顶层设计，积极应对碳关税；工序碳排放水平和碳市场价格波动是影响钢材出口欧盟成本的重要因素，在中国钢铁行业碳市场建设初期，工序碳排放强度的影响会更大一些，建议中国钢铁行业应该首先降低工序碳排放强度。此外，应加强中国钢铁行业碳市场和数字化碳管理的能力建设，推进"六化一体+多平台"建设和应用;加强国际合作和交流。在提高钢铁行业的国际影响力和话语权的同时，应该激发钢铁企业节能降碳的主观能动性。尽管我国出口欧盟的钢材比例很少，但是随着不同版本碳关税的陆续出台，因此，CBAM 对中国钢铁行业的影响不容小觑。

关键词：碳边境调节机制；钢铁行业；贸易影响；政策建议

Impact of CBAM on China's Steel Industry and Strategies

LI Tao[1,2], SHANGGUAN Fangqin[1,2], LI Xiuping[1,2], ZHOU Jicheng[1,2], NI Bing[1,2]

(1. Steel Industry Green and Intelligent Manufacturing Technology Center, China Iron and Steel Research Institute Group, Beijing 100081, China; 2. National Key Laboratory of Metallurgical Intelligent Manufacturing System, Beijing 100081, China)

Abstract: To better cope with the impact and impact of CBAM (Carbon Border Adjustment Mechanism) on China's steel industry. By analyzing the carbon emission structure and current situation of China's iron and steel industry, based on the latest implementation progress of CBAM and taking 1t pig iron as an example, this paper analyzed and predicted the impact of CBAM legislation on the cost of pig iron export to the EU by setting different carbon emission intensity and carbon market trading scenarios. Taking 1t pig iron as an example, three different scenarios of carbon emission levels of blast furnace ironmaking processes were set up according to the carbon emission intensity characteristics of BF-BOF production process: 1.2tCO$_2$/t, 1.4tCO$_2$/t and 1.6tCO$_2$/t. Based on the prediction of the EU and China carbon trading markets, the influence of the price fluctuation of the carbon market and the carbon emission intensity of the process on the cost of pig iron export to the EU when China's steel industry is not included in the carbon market and when China's steel industry is included in the carbon market (carbon price is 80 yuan/t) was analyzed. The results have shown that: taking 1t pig iron as an example, the cost of China's pig iron export to the EU would increase by 95.52-280.32yuan/t. When the carbon emission intensity of pig iron is 1.2tCO$_2$/t, China's pig iron exported to the EU does not need to pay additional carbon tariffs. When the carbon emission intensity of pig iron increases from 1.4tCO$_2$/t to 1.6tCO$_2$/t (EU carbon price is 120euros/t, China carbon price is 80yuan/t), the cost of pig iron exported to the EU increased by 159.20yuan/t. If the steel industry is not included in the Chinese carbon market or the Chinese carbon market is not linked to the EU carbon market (China's carbon price is 0yuan/t), the cost of exporting pig iron to the EU will increase by 105.12-280.32yuan/t. Transitional CBAM has a relatively small impact on the cost of pig iron exports to the EU. After the transition period, with the rate of CBAM free quota reduction accelerating year by year, the carbon tariff to be paid for pig iron exports to the EU has accelerated year by year, and the carbon tariff to be paid for 1t pig iron has reached 1114.40yuan/t by 2034. In order to establish the low-carbon image of China's steel industry and improve the international competitiveness of China's steel products, it was suggested that China's steel industry should strengthen the top-level design and actively respond to carbon tariffs; process carbon emission level and carbon market price fluctuation are important factors affecting the cost of steel exports to the EU. In the early stage of carbon market construction in China's steel industry, process carbon emission intensity would have a greater impact. It was suggested that China's steel industry should first reduce process carbon emission intensity. In addition, the capacity building of carbon market and digital carbon management in China's steel industry should be strengthened, and the construction and application of "six integration + multiple platforms" should be promoted. Strengthen international cooperation and exchanges. While improving the international influence and discourse power of the steel industry, the subjective initiative of steel enterprises to save energy and reduce carbon should be stimulated. Although the proportion of China's steel exports to the EU is relatively insignificant, given the implementation of diverse iterations of carbon tariffs, the potential ramifications of CBAM on China's steel industry should not be underestimated.

Key words: CBAM; iron and steel industry; trade impact; policy recommendation

订单快速响应的炼钢-连铸-热轧一体化生产计划方法及系统

郑 忠[1]，王永周[1]，马仕成[1]，高小强[2]

(1. 重庆大学材料科学与工程学院，重庆 400044；
2. 重庆大学经济与工商管理学院，重庆 400044)

摘 要：根据冶金流程学[1]理论，生产计划是钢铁制造流程中物质流和能量流的运行程序，对钢铁企业生产的动态有序、协同连续运行，减少碳排、降本增效等有重要作用。炼钢-连铸-热轧区段的生产批量计划主要包括炉次计划、浇次计划[2]和热轧单元计划[3]，对扁平材还涉及合同计划与坯料计划。由于生产环境复杂多变，特别是外部市场的频繁波动，以及内部生产执行的突发状况，常导致预排的生产批量计划需要进行适时调整。

钢铁企业的生产计划系统中，炼钢-连铸-热轧区段的生产批量计划分工序单独编制。随着连铸中间包快换、热送热装等新技术发展应用，各工序间的衔接更加紧密，需要统筹考虑各工序生产批量计划目标和约束等要求进行一体化的生产批量计划编制。并且，由于内外生产环境复杂多变，如外部市场环境和内部生产资源条件等不确定性变化，亟待快速进行生产批量计划的适应性调整。然而，企业生产计划系统基本是依靠计划员进行生产计划编排，面对变化只能"见招拆招"式调整，信息系统仅起到部分数据收集或辅助计算功能，强烈依赖计划员个人的计划编制经验、计算能力以及对全局信息的掌控和反映能力，工作量极大，计划及调整方案难以全局优化。随着钢铁企业装备技术能力提升和智能化绿色化转型的迫切需求，一体化生产批量计划的快速响应问题变得愈发重要。

炼钢-连铸-热轧生产批量计划可以实现物质流在工序间的衔接、匹配和协调，引导物质流在流程网络中做稳定连续的优化运行[1]。针对分工序编制生产计划协同不足和生产批量难以适应内外环境变化的问题，提出一体化的生产批量计划编制新方法，通过构建多工序生产计划规则库，以模块化方式搭建生产批量计划模型，以数据驱动和人机交互方式提升生产批量计划的适应性，全面改善并提升生产计划对订单变化的及时响应和调整能力。某厂的生产实践案例显示，一体化生产批量计划编制相比于分工序生产批量计划编制方法能够降低余材率1.27%，提升热送热装效果，有利于连铸与热轧工序的生产协同；面对外部市场变化引起的合同订单变化，或者生产设备故障等资源条件变化，系统可以迅速响应，根据订单或产能变化，自适应进行变化场景限制下的生产计划调整。

综上，通过提升炼钢-连铸-热轧一体化生产批量计划系统的智能化水平，提高生产计划的可行性和优化性，实现钢铁生产的高效低成本运行。

参 考 文 献

[1] 殷瑞钰. 冶金流程学[M]. 北京：冶金工业出版社，2023.
[2] Tang L, Liu J, Rong A ,et al. A review of planning and scheduling systems and methods for integrated steel production[J]. *European Journal of Operational Research*, 2001, 133(1):1-20.
[3] Atilla Zgür, Uyguna Y, Marc-Thorsten Hütt. A review of planning and scheduling methods for hot rolling mills in steel production[J]. Computers & Industrial Engineering, 2020, 151 (20): 106606.

基于耗散结构优化的钢铁制造流程设计研究

张福明 [1,2]

（1. 首钢集团有限公司总工程师室，北京 100041；
2. 北京冶金三维仿真工程技术研究中心，北京 100043）

摘 要：钢铁制造流程属于开放、远离平衡、不可逆的耗散结构体系。钢铁制造是铁素物质流在能量流的驱动下、在信息流的调控下做动态运行的过程。阐述了钢铁制造过程的物理本质和运行特征，提出了钢铁制造流程静态物理体系的设计和构建理念、原则和方法，研究了流程工程动态精准设计体系在工程设计中的应用，讨论了智能化钢铁制造流程的系统层次和构建理念。着重阐述了钢铁制造流程信息物理系统的内涵和架构，提出了静态物理系统与信息系统集成、耦合、协同的设计理念和方法，简要介绍了钢铁制造流程信息物理系统的应用实践。

关键词：钢铁；制造流程；耗散结构；智能化；信息物理系统；设计；网络

Research and Design on Steel Manufacturing Process Based on Disatiative Structure Optimization

ZHANG Fuming[1,2]

(1. Chief Engineer Office of Shougang Group Co., Ltd., Beijing 100041, China;
2. Beijing Metallurgical 3D Simulation Engineering Technology Research Center, Beijing 100043, China)

Abstract: Iron and steel manufacturing process belongs to open, away from balance and irreversible dissipation structure system. Iron and steel manufacturing is a dynamic process of iron element material flow driven by energy flow and controlled by information flow. The physical essence and operation characteristics of iron and steel manufacturing process are expounded, the design and construction concept, principle and method of static physical system of iron and steel manufacturing process are put forward. The application of process engineering dynamic precision design system in engineering design is studied, the system level and structure concept of intelligent iron and steel manufacturing process are discussed. The connotation and structure of the cyber physics system of iron and steel manufacturing process are emphasized. The design philosophy and method of the integration, coupling and cooperation between the static physical system and the cyber system are proposed, and a briefly introduction application result of the cyber physics system of iron and steel manufacturing process is illustrated.

Key words: iron and steel; manufacturing process; dissipative structure; intelligentization; cyber physical system; design; network

热轧-冷轧界面装备技术的创新与应用

韦富强[1,2,3], 张 建[1,2,3], 郑江涛[2,3], 孟祥军[2,3], 王 超[2,3], 杨建立[2,3]

（1. 北京首钢国际工程技术有限公司，北京 100043；
2. 北京市冶金三维仿真设计工程技术研究中心，北京 100043；
3. 北京首钢云翔工业科技有限责任公司，北京 100043）

摘 要：本文概述了 20 年来热轧-冷轧界面装备的技术进步，分析了热轧-冷轧界面装备系列创新技术，介绍了这些新装备技术的首次工程应用情况及实际应用效果，得出了自行驶钢卷车是可以实现热轧-冷轧之间智能化的运输方式等结论。

关键词：界面技术；智能化；超级电容器；钢卷运输

Innovation and Application of Interface Equipment Technology between Hot Rolling and Cold Rolling

WEI Fuqiang[1,2,3], ZHANG Jian[1,2,3], ZHENG Jiangtao[2,3],
MENG Xiangjun[2,3], WANG Chao[2,3], YANG Jianli[2,3]

(1. Beijing Shougang International Engineering Technology Co., Ltd., Beijing 100043, China;
2. Beijing Metallurgical 3-D Simulation Design Engineering Technology Research Center, Beijing 100043, China; 3. Beijing Shougang Yunxiang Industrial Technology Co., Ltd., Beijing 100043, China)

Abstract: This paper summarizes the progress of hot rolling & cold rolling interface equipment technology in the past 20 years, analyzes the innovative interface equipment technologies, introduces the first engineering application and actual application results of these new technologies, and concludes that self-driving steel coil cars can achieve intelligent transportation between hot rolling and cold rolling etc.

Key words: interface technology; intelligent; super capacitor; steel coil transportation

基于深度学习的铸坯低倍缺陷智能检测模型及系统

杨灏瀛，魏 星，汪红兵，赵文慈，闫 岩，

温 宁，李 航

（北京科技大学，北京 100083）

摘 要：当前，铸坯低倍缺陷检测主要依赖人工经验，存在人工测量误差、评级具有主观随意性、无法积累缺陷样本等缺点。针对铸坯低倍缺陷图像分辨率比较大、缺陷尺寸较小、缺陷特征不明显等特点，通过对原始图像进行切分和扩增构造训练数据集，构建融合小样本学习、双向特征金字塔网络等关键技术的铸坯低倍缺陷检测模型，实现缺陷定位；根据低倍缺陷的特点融合 HOG 和 LBP 传统纹理特征训练 SVM 模型实现对检测出的缺陷进行精确分类；最后，基于比例尺计算缺陷尺寸并按照 YB/T 4003—2016 标准开发低倍缺陷评级算法，实现对各类低倍缺陷的定级。初步使用效果表明：缺陷检测模型平均精度为 82%，仅对尺度较小、成像特征不明显的缺陷存在部分漏检，评级算法准确度约为 93%，低倍缺陷评定标准最终以某类缺陷最高等级为评级对象，在流程角度，低倍检测精度约为 95%以上，并实现了支持获取低倍图像、缺陷检测、尺寸计算、缺陷评级、输出报告的智能化评定系统。

关键词：深度学习；小样本学习；支持向量机；铸坯低倍缺陷；纹理特征

基于钢水温度预定和预测相结合的 LF 炉供电推荐模型

李晓龙，贺东风

（北京科技大学冶金与生态工程学院，北京 100083）

摘 要：在炼钢过程中，每一工序钢水温度的控制是否精确，不仅会影响该工序物理化学反应的进行和能量的消耗，也关系着能否获得合适、稳定的连铸钢水温度，进一步影响着连铸坯质量和生产安全[1]。即在转炉-LF-连铸的生产过程中，控制 LF 炉钢水终点温度，是控制连铸开浇温度的必要步骤。对此，许多学者建立了基于数据驱动的终点温度预测模型，可以简化影响钢水温度的众多因素之间的复杂非线性关系和熔池反应条件，在生产中使用较广。但在实际应用中，模型所预测的是某一时刻的钢水温度，提供的是一种参考而不是措施，仍然需要操作人员根据人工经验对钢水温度进行纠偏，不便于对连铸钢水温度进行控制。

本文提出一种基于钢水温度预定和预测相结合的 LF 炉供电控制模型。首先，建立基于案例推理的 LF 终点温度预定模型，在目标钢种液相线温度、过热度等温度制度的基础上，根据当前生产计划和所用钢包热状态，预定 LF 终点温度，并认为该预定温度为优化后的、合理的目标终点温度。然后，分别建立基于 BP 神经网络、支持向量机回归和随机森林回归的 LF 供电量推荐模型，在 LF 进站后将预定温度和进站温度、成分、重量和相关电气参数等共同作为输入，通过供电量推荐模型计算 LF 总供电量。之后，建立基于优化案例推理的 LF 终点温度预测模型，将进站温度、成分、重量、相关电气参数和所推荐的供电量作为影响因素，预测 LF 终点温度[2]。最后，判断预测终点温度和预定终点温度之间的误差，若误差小于设定值，则确认供电量推荐；若误差超过设定值，则在不改变其

他参数的情况下增加或减少供电量推荐，重新输入 LF 终点温度预测模型。

选取预定终点温度和实际终点温度误差小于 5℃ 的炉次共 2667 炉，计算模型推荐的供电量和实际供电量之间的误差。其中，基于 BP 神经网络的模型推荐供电量的误差在 5%、7%、10% 之内的炉次分别占 87.14%、93.51%、96.88%，相较其余两种算法误差更小。

为减少 LF 工序钢水温度波动对连铸的影响，同时为 LF 操作人员提供稳定钢水温度的解决策略，本文综合生产前预定和生产中预测的角度，在保证终点温度符合要求的前提下，提出了基于钢水温度预定和预测相结合的 LF 炉供电推荐模型，其中，基于 BP 神经网络的模型推荐供电量误差较小，效果更优。

参 考 文 献

[1] 贺东风，何飞，徐安军，等．北京科技大学学报，2014(S1 vo 36): 200-206.
[2] FENG K, HE D, XU A, et al. Steel Research International, 2016, 87(1): 79-86.

数据与机理联合驱动的转炉终点碳含量预测模型

夏一杰，汪红兵，魏　星，石雪卓，温　宁，解　欢

（北京科技大学，北京　100083）

摘　要：转炉炼钢过程中，终点控制以及终点预测对于实现冶炼目标至关重要。当前，主要存在两种终点预测方式：基于数据驱动的终点预测方法以及基于机理计算的终点预测方法。两种方法都存在着优点与不足。为了结合二者的优势，设计并提出了数据与机理联合驱动的转炉终点碳含量预测模型。基于可测数据对机理公式中的不可测参数进行修正，从而针对每一炉次进行定制优化，同时设计损失函数为机理损失与数据损失的加权之和，采取根据修正后的机理公式计算出的终点碳含量与直接采用数据驱动的方式预测的终点碳含量二者进行加权求和的方式得到最终的预测结果。实验结果表明，数据与机理相结合的方式在命中率与训练速度上优于单纯的数据驱动的方式，显著优于单纯的机理计算的方式。数据与机理相结合的方式，在 (−0.009,0.009) 的误差区间内，达到了 79.27% 的命中率，在 (−0.012,0.012) 的误差区间内，达到了 88.60% 的命中率。未来，基于该模型框架，通过顶吹和底吹相关实时数据进一步优化机理公式中的不可测参数，可充分发挥机理和数据联合驱动的优势，进一步提高模型精度和训练速度。
关键词：终点预测；机理模型；深度学习；转炉炼钢；不可测参数

"碳中和"战略下钢铁制造流程的低碳发展与技术创新

张福明[1]，刘清梅[2]

（1. 首钢集团有限公司，北京　100041；2. 首钢集团有限公司技术研究院，北京　100043）

摘　要：加快钢铁工业低碳转型与发展是落实全球应对气候变化目标及"碳中和"战略的重要途径。我国钢铁粗钢产量规模大，CO_2 排放量处于较高的水平。2022 年，我国粗钢产量 10.13 亿吨，钢铁工业 CO_2 排放量约 18.23 亿吨，占我国 CO_2 总排放量的 15% 以上。钢铁行业低碳发展是实现我国"碳中和"战略目标的重要支撑，同时也存在着

艰巨的挑战。

　　钢铁制造流程以高炉—转炉长流程和废钢—电炉短流程为主，长流程吨钢碳排放强度约电炉短流程的 3 倍。在全球范围内，要实现《巴黎协定》的 2.0℃温控目标，要求 2050 年全球钢铁行业的直接碳排放总量比 2019 年减少 55%，碳排放强度下降 60%，吨钢排放从 1.4t 降到 0.6t。由此认为，推动长流程降碳和短流程有序发展是我国钢铁制造流程低碳发展的主要方向。

　　钢铁制造流程是基于钢铁厂三大主体功能协同发展的生产模式的外在体现。钢铁厂的功能主要涵盖"钢铁产品制造功能""能源转换功能"和"废弃物消纳－处理和再资源化"。在"碳中和"战略的指导下，钢铁工业需要进一步审视钢铁制造流程中铁素流、碳素流和能量流的行为规律，以达到挖掘钢铁制造流程减碳潜力的目的。本文统计了欧盟、亚洲、美洲等区域主要钢铁企业实现"碳中和"战略的减排目标和低碳路线，重点分析钢铁领域在低碳发展过程中开展的流程减碳实践性工作，在此基础上，深入探讨钢铁厂提高其自身绿色综合竞争力所需要关注的钢铁制造流程的发展领域。

　　制造流程减碳技术创新是对所研究对象系统的本质（耗散结构的输入－输出特征）以及开放系统下钢铁与非钢铁领域构成的复杂关联关系的不断认识。钢铁厂碳素能量流的动态运行是低碳减排的驱动力，本文以钢铁冶金减碳创新技术为出发点，剖析关键技术在减碳运行和能源效率最佳化方面的创新点。文章还从"多方位服务于绿色环境社会的可能性"视角出发，归纳总结了助力钢铁工业 CO_2 资源化利用研究的技术措施，这将有利于钢铁厂进一步探讨钢铁制造流程相关理论的建立、技术创新的研究和工程实施的策略等，有助于从全产业链视角考虑钢铁制造流程碳减排与再循环产业转型升级的未来发展。

关键词："碳中和"战略；钢铁制造流程；碳减排路线；低碳转型；技术创新

Low-carbon Progress and Technological Innovation of Iron and Steel Manufacturing Process under the Strategy of "Carbon Neutrality"

ZHANG Fuming[1], LIU Qingmei[2]

(1. Shougang Group Co., Ltd., Beijing 100041, China; 2. Research Institute of Technology of Shougang Group, Co., Ltd., Beijing 100043, China)

Abstract: The accelaration of low-carbon transformation and development in the iron and steel industry is an important way to implement the global climate change target and the "carbon neutral" strategy. Due to the large scale of Chinas crude steel production, the carbon dioxide emission is at a high level.In 2022, China's crude steel production is 1.013 billion tons, and the steel industry CO_2 emissions is about 1.823 billion tons, which accounted for 15% total carbon dioxide emissions in China.The low-carbon development of the iron and steel industry plays an important role in fulfilling the strategic goal of "carbon neutral" in China, while it also shows arduous challenges.

　　The iron and steel manufacturing process mainly includes the BF-BOF process and EAF poreces. The BF-BOF process carbon emission intensity of per ton is about 3 times of that of the EAF process.In the world, for achieving the 2.0℃ temperature controlling target of the Paris Agreement, the total direct CO_2 emissions of the global iron and steel industry in 2050 should be dropped by 55% from 2019, the carbon intensity by 60%, and CO_2 emission is reduced from 1.4 tons to 0.6 tons.Therefore, it is indicated that promoting the carbon reduction in the BF-BOF process and improving EAF process orderly are the main dirctions of the low carbon development of iron and steel manufacturing process in China.

　　The iron and steel manufacturing process is the external profile of the production mode based on the coordinated development of the three main functions of iron and steel plants.The functions of iron and steel plants mainly include "product manufacturing" "energy conversion" and "waste consumption-treatment and recycling". Under the guidance of the "carbon neutral" strategy, the behavior rules of iron element flow, carbon flow and energy flow in the iron and steel manufacturing process should be further developed, so as to achieve the purpose of excavating the carbon reduction capacity.

This paper statistics the carbon reduction targets and low carbon routes for fulfilling "carbon neatral" strategy in regions of the European Union, Asia, America and others, and analyzes the experienced work of the iron and steel fields of low carbon development process.On this basis, the development areas of iron and steel manufacturing process, which are able to enhance green comprehensive competitivemenss of iron and steel plants, are investigated deeply.

Furthermore, the technological innovation of carbon reduction in the manufacturing process is a continuous understanding of the essence of the research object system, which includes the input-output characteristics of dissipative structure, and the complex correlation between steel and non-steel fields within the opening system.The dynamic operation of carbon energy flow in iron and steel plant is the driving force of low carbon emission reduction. This paper takes the innovation technology of carbon reduction in iron and steel metallurgy as the starting point, and analyzes the innovation points of key technologies in carbon reduction operation and the optimization of energy efficiency.

From the point of "the possibility of serving the green environment society" perspective, technical measures to promote CO_2 resource utilization in the iron and steel industry is summarized. It benefits to further explore the establishment of steel manufacturing process related theory, technology innovation research and engineering implementation strategy. It also helps to study the carbon emission reduction and recycling of industrial transformation and upgrading development with the perspective of the whole iron and steel industry chain.

Key words: carbon neutral strategic; iron andsteel manufacturing process; emission reduction routes; low-carbon transformation; technological innovation

基于"炉-机对应"的炼钢-连铸生产调度问题灰狼优化模型

陈 博[1]，邵 鑫[1]，刘 青[1]，王忠刚[2]，高志滨[2]

（1. 北京科技大学钢铁冶金新技术国家重点实验室，北京 100083；
2. 莱芜钢铁集团银山型钢有限公司，山东莱芜 271104）

摘 要：炼钢-连铸区段作为钢铁流程的关键区段，由于多设备的存在及早期车间布局不合理，加之钢产品的生产方式由过去的规模化、大批量生产转变为多品种、小批量、多规格、高质量及快节奏的生产方式，给当前的炼钢-连铸区段的生产运行、计划排产、车间调度带来了较多的困难。典型的炼钢-连铸生产流程包括炼钢、炉外精炼与连铸三道主工序，每个工序一般配有多台设备，因此，炼钢-连铸区段的生产调度问题可归纳为多阶段多并行机的混合流水调度问题。

炉-机对应策略指的是三道工序之间不同设备间的对应关系，一般为"定炉对定机"或"一一对应"两种，本文以国内某无精炼跨炼钢厂为例，以浇次计划内所有炉次总等待时间最小为目标建立炼钢-连铸生产调度模型。用改进的灰狼优化算法求解生产计划，运用Matlab2021编写引入"炉-机对应"的灰狼优化算法。灰狼算法参数少易调节，进化公式简单，收敛速度快，每个个体位置都是一个可能解。生成初始种群时引入"炉-机对应"，在设备指派时遵循两条"炉-机对应"策略：（1）转炉周期不大于连铸周期，一台转炉即可提供满足一台连铸机连浇所需钢水量，否则需要其他转炉适时提供钢水以满足连浇；（2）精炼炉周期不大于连铸周期，一台精炼炉即可提供满足一台连铸机连浇所需钢水量，否则需要其他精炼炉适时提供钢水以满足连浇。个体位置更新时，取前50%精英个体与随机种群合并，采用权重系数法，在算法求解过程中设定最大等待时间限定，即当某一炉次在某工序前等待时间超过30min，则该个体再次更新位置。

以该厂主要生产模式下的实际生产计划为仿真算例进行实验，分别用遗传算法和改进灰狼算法求解以总等待时间最短为目标的生产计划。结果表明：相比于遗传算法，改进的灰狼算法缩短了等待时间，提高了炉-机匹配度，以四炉对三机生产模式为例，采用改进灰狼优化算法得到的总等待时间比遗传算法减少 30%，炉次在工序前等待时间超过 30min 的占比降低 3%；从炉-机匹配度方面看，引入"炉-机对应"后，四炉对三机生产模式下四号转炉向四号精炼炉提供钢水比例由 23%提升至 44%，四号精炼炉向四号连铸机提供钢水比例由 32%提升至 50%，如图 1 所示。

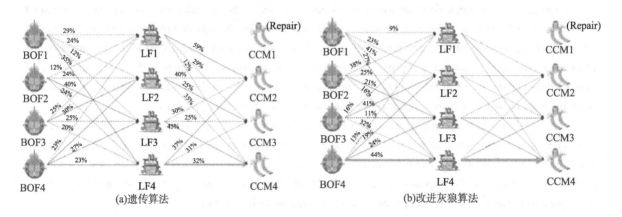

图 1　遗传算法与改进灰狼算法炉-机匹配度对比

钢铁工业极限能耗分析与能效提升措施探讨

上官方钦[1]，周继程[1]，倪　冰[1]，崔志峰[2]

（1. 中国钢研科技集团有限公司钢铁绿色化智能化中心，北京　100081；
2. 北京科技大学冶金与生态工程学院，北京　100083）

摘　要：绿色化、智能化和产品品牌化将引导着中国钢铁工业的转型升级和高质量发展。中国钢铁工业的绿色化发展进程先后经历了节能、减排阶段，目前已经进入脱碳化阶段。钢铁工业是国民经济的重要基础产业，是建设现代化强国的重要支撑，是实现绿色低碳发展的重要领域。虽然在过去三十年间节能减排取得显著成绩，但由于粗钢产量巨大，仍然是 CO_2 排放大户，约占全国碳排放总量的 15%~17%，钢铁工业脱碳化发展任重道远。

本团队构建了双碳分析模型，为中国钢铁行业提出了未来实现双碳战略的三阶段发展目标和路线图设想；同时指出，长远来看，在逐步适应碳达峰、碳中和的过程中，未来钢铁行业的生产制造流程将趋向为三大类，即高炉-转炉长流程、全废钢电炉流程和氢还原-电炉流程，并由此派生出不同情景的钢厂以及不同的降碳路线。此外，通过模型的计算和评估，结果表明能效提升是现阶段钢铁行业落实双碳战略的重点工作。

系统介绍了中国钢铁行业的节能工作历程以及面临的现实挑战、研发挑战、未来发展趋势，指出无论是能量的输入与输出还是能源的供应与需求，钢铁企业二次能源的生产量与使用量、总供应量与总需求量之间始终是动态变化的，不是静止平衡的。因此，必须摆脱传统的"静态""平衡"观点的束缚，用"动态""非平衡""耗散结构"的理论及方法，根据能源介质的数量、品质（热值）和供给需求，合理地规划、设计、调控能源生产运行过程与使用过程，挖掘"动态""非平衡"状态下的节能潜力。未来我国钢铁工业吨钢能耗能否再次出现较大幅度的下降走势，将取决于新一轮节能理论、技术和管理手段的支撑，即钢铁制造流程物质流、能量流和信息流的耦合优化。

对两类理想钢铁制造流程以及烧结、焦化、高炉炼铁的理论极限能耗进行了分析，结果表明：以矿石源头的长流程（理想的一步法）炼钢的理论极限能耗为 299kgce/t 钢；而以废钢为源头的短流程的理论极限能耗为 46kgce/t 钢。

同时，论述了能效与能耗的区别，在此基础上，对烧结、焦化、炼铁、转炉炼钢、电炉炼钢、轧钢、界面技术、自发电等关键环节的能效提升措施进行了详细的分析和探讨。并在此基础上，进一步分析了未来三类钢铁制造流程的降碳路线图，指出长流程的碳排放强度有望从 2.0tCO$_2$/t 钢降低到 0.87tCO$_2$/t 钢，需进一步通过碳汇、碳交易等手段实现碳中和；全废钢电炉短流程的碳排放强度有望从 0.45tCO$_2$/t 钢降低到近零碳排放；氢还原-电炉流程（50%废钢+50%DRI 原料结构）有望从 1.31tCO$_2$/t 钢降低到近零碳排放。

炼钢车间天车调度管理系统设计、开发与应用

冯 凯，徐安军，贺东风，潘 俊，王宇轩，王 磊

（北京科技大学冶金与生态工程学院，北京 100083）

摘 要：随着钢铁生产智能制造的持续发展，钢铁企业炼钢车间智能化升级重点，正在由单体冶炼工序智能管控，转向工序间"亚界面"管控和车间级总体管控。天车调度作为炼钢过程生产组织调度的重要部分，承担炼钢车间工序间物质流位移的重要功能，是促进工序间高效衔接匹配和炼钢过程动态有序运行的主要着力点。目前，由于炼钢车间天车缺少完善的信息化管理、调度过程依赖人工指挥等情况，导致整个炼钢过程数字化管理水平低、物流效率低、钢水温度损失大、温度波动大等问题，已经成为炼钢过程数字化、智能化和安全高效生产的明显短板。

针对该问题，通过完善数据采集和管理、研发天车任务生产模型、天车调度模型等工作，设计、开发炼钢车间一体化的天车智能调度系统，已经成为保障炼钢过程的数字化、智能化、安全有序生产的重要一环。

解决天车调度问题的基础是天车作业过程时间因素的解析。基于冶金流程工程学中时间因素的分析方法，设计了天车任务动作时间解析规范表征方法，进而形成"动作-时间"关系表和时空轨迹图，对实际生产过程进行了时间和空间解析。

解决天车调度问题的关键是制定可行且高效的天车调度方案。通过前期研究和实际生产过程调研，提出基于分层规则的天车调度方法，设计了任务自动生成规则、天车分配规则和任务优先级判定规则，并在应用实际生产中验证了三类规则的合理性和优化性。

在以上研究基础上，设计炼钢车间天车智能调度系统的核心功能，包括实现现场不同系统的数据采集、集成、总表管理，并提供系统状态监测、天车任务指挥管理、异常情况处理等，从而实现天车调度业务流程的数字化管理和智能化驱动，为生产与调度一体化提供保障。同时，面向不同岗位的用户，提供定制化的交互页面，如：面向天车调度管理员，提供天车任务的操作页面；面向天车操作工和其他工序用户，提供主要信息的展示页面；面向系统维护人员，提供后台管理页面；面向管理人员和技术人员，提供系统分析和报表页面。

炼钢车间天车智能调度系统采用数据层、业务层、交互层的 B/S 软件系统架构，模块化的设计开发模式，已经在唐钢新区热轧作业区炼钢车间上线使用。实际生产使用效果表明，天车调度作业组织效率明显提升，工序间非必要等待时间显著下降，天车运行冲突避让情况有所改善。

基于 FlexSim 对炼铁-炼钢界面调度规则的仿真优化

王 磊，徐安军，王宇轩，刘 旋，解 欢

（北京科技大学冶金与生态工程学院，北京 100083）

摘 要：进入 21 世纪，冶金工业的时代命题是绿色化、智能化，其通过原子/分子层次、工序/装置层次、制造流程

三个层次的科学及技术来支撑。现代钢铁企业的生产流程是由炼铁、炼钢、连铸、连轧等主体工序及其之间衔接-匹配、协调-缓冲的"界面技术"构成的复杂系统[1-2]。铁钢界面技术涵盖了界面模式选择、工序设备配置和生产物流组织等多个方面。在实际生产过程中，影响铁钢界面技术的因素较多，如生产调度、天车调度及其之间的互相耦合等，这些复杂因素的影响导致生产计划混乱，无法精准实现铁钢界面生产流程的动态、有序和连续化运行[3-4]。而基于 FlexSim 仿真建立的炼铁-炼钢区段的调度仿真模型作为铁钢界面调度规则优化的重要方法之一，可以用于指导实际生产，故而借助仿真软件来模拟和优化炼铁-炼钢区段的调度规则是有必要的。通过建立与实际生产情况相符的模型，利用 FlexSim 仿真软件进行仿真试验研究，探究炼铁-炼钢区段复杂的生产关系，以实现对铁钢界面调度规则的优化。

在唐钢新区的炼铁-炼钢区段实际生产过程中，炼铁-炼钢区段的调度规则更多地基于现场工作人员的经验进行调整，这容易导致整个钢铁联合生产的节奏出现阶段性的偏差。因此，通过对包括高炉出铁情况、铁水包运输、天车调运、KR 脱硫处理工序以及转炉兑铁等环节的生产统计数据、火车运输情况、各跨区天车任务负荷情况以及转炉兑铁等实际生产问题进行分析，梳理出炼铁-炼钢区段的调度规则，建立基于层流化的运行方案，并对各个工序的功能解析、合理定位以及优化方案进行研究。通过利用 FlexSim 软件对唐钢新区铁钢界面进行仿真试验，实施层流化运行方案，实现对炼铁-炼钢区段调度规则的优化仿真，提高高炉出铁效率，缩短天车运输距离、避免天车工作任务分配不均匀，提高任务的执行效率，更减少任务分工的随机性和人为因素的影响，有效缩短 KR 工序与转路工序之间的传搁时间，以达到提高生产效率，减少物质流和能耗损失的目的。

参 考 文 献

[1] 殷瑞钰. 冶金流程工程学[M]. 2 版. 北京：冶金工业出版社, 2009.
[2] 殷瑞钰. 冶金流程集成理论与方法[M]. 北京：冶金工业出版社, 2013.
[3] 周继程, 郦秀萍, 韩伟刚, 等. 炼铁-炼钢界面模式与铁水罐多功能化技术[J]. 钢铁, 2017, 52(4): 94.
[4] 魏绿坤, 徐安军, 潘俊. 基于 Flexsim 的炼钢-连铸区段流程仿真与优化分析[J]. 中国冶金.

钢铁制造流程能量流网络的有向网分析及建模

杜学强[1,2]，郦秀萍[3]，孙彦广[1,2]，刘 晓[1]

（1. 钢铁研究总院，北京 100081；2. 冶金自动化研究设计院，北京 100071；
3. 中国钢研科技集团有限公司钢铁绿色化智能化技术中心，北京 100081）

摘 要：中国钢铁工业在传统的技术体系下节能难度越来越大，能量流网络化运行与智能化调控是进一步挖掘节能潜力的重要手段。能量流网络建模是能源系统智能调控和集成匹配优化的基础。论文基于冶金流程工程学，从系统的角度解析钢铁制造流程能量流网络作为"有向网"所具有的系统结构及其动态运行特征。通过形式化抽象，提出了基于混杂 Petri 网的能量流网络建模方法。以典型钢厂为例建立了能源转换环节的 Petri 网模型，通过仿真验证了模型的有效性。该方法可为钢铁制造流程能量流网络系统能源利用状况评测和能源调度动态优化提供技术支撑。

关键词：钢铁制造流程；能量流网络；有向网；Petri 网；建模

Analysis and Modeling of Energy Flow Network for Steel Manufacturing Process Based on Directed Network

DU Xueqiang[1,2], LI Xiuping[3], SUN Yanguang[1,2], LIU Xiao[1]

(1. Central Iron and Steel Research Institute, Beijing 100081, China; 2. Automation Research and Design Institute of Metallurgical Industry, Beijing 100071, China; 3. Steel Industry Green and Intelligent Manufacturing Technology Center, China Iron and Steel Research Institute Group, Beijing 100081, China)

Abstract: Energy-saving in steel industry is becoming increasingly difficult under the traditional technological system in China. Networked operation and intelligent dispatching of energy flow is an important way to further tap into energy-saving potential. Energy flow network modeling is the foundation for intelligent dispatching and integrated optimization of energy system. Based on the theory of metallurgical process engineering, the structure and dynamic operating characteristics of the energy flow network for steel manufacturing process as a "directed network" are analyzed from the perspective of "System". A modeling method for energy flow network based on hybrid Petri net is proposed through formal abstraction. As an example, a Petri net model for energy conversion process of a typical steel plant is established, and its effectiveness is verified through simulation. This method can provide technical support for the energy evaluation, dynamic dispatching and optimization of the energy flow network system for steel manufacturing process.

Key words: steel manufacturing process; energy flow network; directed network; Petri net; modeling

转炉出钢过程流体流动特性的研究

袁 飞[1]，刘 旋[2]，徐安军[2]

（1. 北京科技大学金属冶炼重大事故防控技术支撑基地，北京 100083；
2. 北京科技大学冶金与生态工程学院，北京 100083）

摘 要：转炉出钢过程实现了钢渣分离，并促进钢包内夹渣和脱氧产物的上浮，同时，其也对钢水温降和转运时间等炼钢流程工艺参数存在显著影响。在炼钢界面中，转炉出钢是影响钢水质量、生产效率和成本的重要过程因素。本文以某钢厂 200t 转炉为例，采用多相流 VOF 数值模型研究了整个转炉出钢过程中钢水的流动特性，分析了出钢口直径、形状及角度对对出钢过程中出钢口速度、侵蚀程度、出钢时间以及对钢水冲击特性、搅拌效果的影响规律。转炉出钢过程模型的计算域包含转炉熔池、出钢管道、钢包熔池以及出钢口末端与钢包顶面间的空气域共计四个部分，考察了分段变径型（180~160mm）和圆柱形（180mm、160mm）孔径的出钢口，以及出钢口角度 10°和 15°工况下的出钢注流特性。

（1）出钢过程钢水流场的分布规律。

图 1 是在转炉出钢后期，不同工艺参数的出钢口中心截面钢水相的分布云图，五种工况下钢包中钢液卷吸空气的程度不同。圆柱形的出钢口具有卷吸空气量大、气泡小的特点，而分段变径型的出钢口卷吸空气量较小，但是气泡相对比较大。

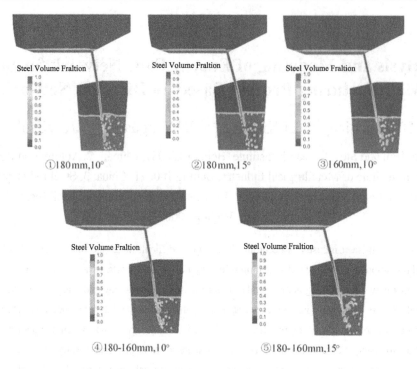

图1 不同工艺参数的出钢口中心截面钢水相分布（出钢后期）

当出钢口工艺参数为①~⑤时，出钢口的平均流速分别为5.61m/s、5.74m/s、5.71m/s、6.34m/s、6.22m/s，出钢时间分别为203.4s、197.4s、250.8s、226.9s、230.3s，出钢口的形状和直径对出钢时间的影响较大，角度的影响较小。当出钢口的工艺参数为180~160mm&10°时，出钢前中后期的出钢口管道内湍动能平均值分别为2.05m^2/s^2、1.99m^2/s^2、1.69m^2/s^2，较低的湍流可以减小钢液对出钢通道壁的磨损，因此选取合理的出钢口工艺参数尤为重要。

（2）出钢过程钢包包底的侵蚀规律。

在转炉出钢过程中，随着钢包内钢水液面的上升，钢液对钢包包底的冲击力逐渐减小。对于①~⑤的钢口工艺参数，钢包包底的平均剪切力分别为1345.58Pa、1418.92Pa、1258.61Pa、1008.38Pa、1110.67Pa。当出钢口形状为分段变径型（180~160mm），出钢口角度为10°时，钢液对钢包包底的冲击剪切力最多可以减少21.9%。

（3）出钢过程钢包的混匀特性规律。

在转炉出钢过程中，转炉中的钢液在重力的作用下经过出钢管道注入钢包。钢液的持续注入会促进钢包熔池的搅拌。将熔池中速度小于0.1m/s的区域定义为死区，不同工艺参数的出钢口死区分布有很大差异。图2展示了钢包中心截面死区百分比统计结果：对于①~⑤的出钢口工艺参数，钢包中心截面的死区百分比分别为26.82%、1.68%、27.04%、1.65%、3.57%，当出钢口工艺参数为180~160mm&10°时，钢液对钢包熔池的搅拌效果最佳。

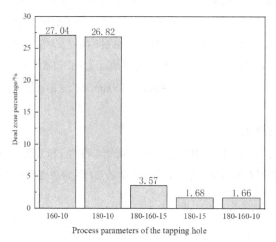

图2 不同工艺参数下出钢口钢包中心段死区百分比（出钢后期）

钢铁制造流程系统实时生产柔性和稳定性度量方法

高小强[1]，白云航[1]，匡静静[1]，丁卓非[1]，郑 忠[2]

（1. 重庆大学经济与工商管理学院，重庆 400044；
2. 重庆大学材料科学与工程学院，重庆 400044）

摘 要：生产的稳定顺行是实现钢铁绿色制造的基本前提，智能制造是实现钢铁绿色制造的重要手段。定量刻画钢铁制造流程系统的实时生产柔性和稳定性对于通过钢铁智能制造实现绿色制造目标具有重要意义。另外，在评价全流程或流程区段的生产计划或生产实绩的优劣时，除了生产效率的考虑外，生产节奏的均匀性也是一个需要考虑的重要因素。寻求能对生产计划或生产实绩的甘特图进行准确分类的方法对于综合多种因素评价生产计划或生产实绩的优劣也有重要价值。

本文从流的视角以钢铁制造流程系统中物质流、能量流和信息流为对象，考虑三流有序流动所需要的各种冶金工艺处理、运输等流动控制资源的可用性，定义了流程系统中实时生产柔性指数来刻画流程系统的生产柔性。在此基础上，采用单位根检验的ADF（Augmented Dickey-Fuller）检验方法对生产柔性指数的时间序列的平稳性进行了检验，以此度量生产系统在一段时间内的稳定性。针对某炼钢厂铁水预处理区段的铁素物质流运行情况，考虑该子系统内的主要流动控制资源，计算该子系统的实时生产柔性指数，并进行了ADF检验。

初步结果表明：生产柔性指数可以实时反映生产系统中由流动控制资源占用情况表征的生产系统柔性；在此基础上对实时生产柔性的时间序列的ADF检验结果可以用于评价生产系统在一段时间内是否稳定；实时生产柔性的时间序列单位根检验的ADF值也是用于对这些序列对应的生产计划或生产实绩的甘特图进行分类的重要特征。

唐钢新区项目设计特点及运行实绩

李建生[1]，郦秀萍[2]，马新光[1]，韩伟刚[3]

（1. 唐山钢铁集团有限责任公司生产制造部，河北唐山 063000；2 中国钢研科技集团有限公司钢铁绿色化智能化中心，北京 100081；3 华北理工大学冶金与能源学院，河北唐山 063210）

摘 要：河钢集团唐钢新区项目全面贯彻冶金流程工程学理论，采用动态精准设计方法，从总图布置设计、车间平面布置和优化、物流运输设计等方面构建优化的流程网络，实现了简捷、紧凑、集约、顺畅的目标，使"流"的行为按照动态有序、协同连续的规范运行，实现运行过程中的耗散"最小化"。投产以来，通过与科研院所合作开展技术攻关，全流程各区段界面运行指标取得显著进步。

关键词：总图布置；流程网络；设计理念；界面技术

Design Characteristics and Operational Performance of Tanggang New Area Project

LI Jiansheng[1], LI Xiuping[2], MA Xinguang[1], HAN Weigang[3]

(1. Manufacturing Department, Tangshan Iron and Steel Group Co., Ltd., Tangshan 063000, China; 2. Steel Industry Green and Intelligent Manufacturing Technology Center, China Iron and Steel ResearchInstitute Group, Beijing 100081, China; 3. School of Metallurgy and Energy, North China University of Science and Technology, Tangshan 063210, China)

Abstract: The Tanggang New Area Project of Hegang Group fully implements the theory of metallurgical process engineering, adopts dynamic and precise design methods, and constructs an optimized process network from the aspects of general layout design, workshop layout and optimization, logistics transportation design, etc., achieving the goals of simplicity, compactness, intensification, and smoothness. This enables the behavior of "flow" to operate in accordance with dynamic, orderly, collaborative and continuous standards, and achieves the "minimization" of dissipation during the operation process. Since its production, significant progress has been made in the operational indicators of various interfaces in the entire process through cooperation with research institutes to carry out technical research.

Key words: general layout; process network; design concept; interface technology

17 其他

- 大会特邀报告
- 第十三届冶金青年科技奖获奖人特邀报告
- 分会场特邀报告
- 矿业工程
- 焦化及节能环保
- 炼铁与原料
- 炼钢与连铸
- 电冶金与废钢铁
- 轧制与热处理
- 表面与涂镀
- 金属材料深加工
- 粉末冶金
- 先进钢铁材料及应用
- 节能与低碳技术
- 冶金环保与资源利用
- 冶金设备与工程技术
- 冶金自动化与智能化
- 冶金物流
- 冶金流程工程学

★ 其他

X射线荧光光谱法测定预熔渣中CaF$_2$含量

牛威斌，禹青霄，葛晶晶

（河北大河材料科技有限公司理化检测中心，河北石家庄 050081）

摘 要：采用熔融制样—X射线荧光光谱法测定预熔渣中CaF$_2$含量，通过实验确定了熔融制样及荧光测定的最佳条件，并建立了一条校准工作曲线。结果表明，CaF$_2$含量在30%~60%时，相关系数为0.9998，样品精密度的相对标准偏差小于0.3%，准确度的相对误差小于0.1%，均符合要求，能够满足实际检测需要。

关键词：熔融制样；X射线荧光光谱；预熔渣；CaF$_2$

Determination of CaF$_2$ Content in Premelted Slag by X-ray Fluorescence Spectrometry

NIU Weibin, YU Qingxiao, GE Jingjing

(Hebei Dahe Material Technology Co., Ltd., Physical and Chemical Testing Center, Shijiazhuang 050081, China)

Abstract: The content of CaF$_2$ in pre-melted slag was determined by fusion sample preparation-X-ray fluorescence spectrometry. The optimal conditions for fusion sample preparation and fluorescence measurement were determined by experiments, and a calibration working curve was established. The results show that when the CaF$_2$ content is between 30% and 60%, the correlation coefficient is 0.9998, the relative standard deviation of sample precision is less than 0.3%, and the relative error of accuracy is less than 0.1%, all of which meet the requirements and can meet the actual detection needs.

Key words: fusion sample preparation; X-ray fluorescence spectroscopy; pre-slag; CaF$_2$

新型管线测绘技术在冶金行业中的应用与探索

刘 光[1]，徐 明[2]，李晓龙[1]，赵逸凡[1]，银振华[1]

（1. 北京首钢国际工程工程技术有限公司北京市冶金三维仿真设计工程技术研究中心，北京 100043；2. 北京首钢建设投资有限公司，北京 100043）

摘 要：准确详实的综合管线资料是冶金行业规划建设、管理技改、运维的最重要的成果。冶金行业管道布设复杂，种类繁多，且最易受环境和使用影响，老化、锈蚀、积灰、积尘、积污、积泥、温度变化、地质地形的变化、生物入侵、临时施工措施等都对其使用构成重要影响。与此同时，由于介质和使用条件的不同，管理也往往存在巨大差异。除此之外复杂的敷设方式，也使管理和维护更为复杂。因此，准确反映管线的现状一直是冶金行业的重点难题。本文结合首钢园区的管道测绘实践，提出地下管线以传统测绘手段+雷达探测、地上以三维激光扫描等新型采集技术，辅以GIS平台综合管理的新型管线测绘技术方法。

关键词：总图管理；综合管线；探地雷达；激光扫描

Application and Exploration of New Pipeline Surveying Technology in the Metallurgical Industry

LIU Guang[1], XU Ming[2], LI Xiaolong[1], ZHAO Yifan[1], YIN Zhenhua[1]

(1. BSIET Beijing Metallurgical 3-D Simulation Design Engineering Technology Research Center, Beijing 100043, China; 2. Beijing Shougang Construction Investment Co., Ltd., Beijng 100043, China)

Abstract: Accurate and detailed comprehensive pipeline information is the most important result for planning and construction, management and renovation, and operation and maintenance in the metallurgical industry. The pipeline layout in the metallurgical industry is complex, with a wide variety of types, and is most easily affected by the environment and usage. Aging, corrosion, dust accumulation, pollution, mud accumulation, temperature changes, geological and topographic variations, biological invasion, and temporary construction measures all have a significant impact on its use. At the same time, due to the different media and usage conditions, even pipelines of the same material or diameter often have significant differences in their use and maintenance methods and management. In addition to complex laying methods, management and maintenance are also complicated. Therefore, accurately reflecting the current state of the pipeline has always been a key issue in the metallurgical industry. This article proposes a new pipeline surveying technology method, combining traditional surveying methods with radar detection for underground pipelines and 3D laser scanning for above-ground pipelines, supplemented with a GIS platform for comprehensive management based on the pipeline surveying experience of Shougang.

Key words: general layout management; integrated pipelines; ground penetrating radar; laser scanning

针状焦预处理工艺对比分析

刘海丰，何 莹，张大奎，穆春丰，刘庆佩

（鞍钢化学科技有限公司，辽宁鞍山 114000）

摘 要： 本文介绍了石油系与煤系针状焦原料的种类及性质，总结了石油系与煤系原料的性质要求；阐述石油系与煤系针状焦的原料预处理技术；提出国内针状焦预处理技术发展方向。

关键词： 针状焦；针状焦原料；预处理；生产

Comparison and Analysis of Needle Coke Pretreatment Process

LIU Haifeng, HE Ying, ZHANG Dakui, MU Chunfeng, LIU Qingpei

(Ansteel Chemical Technology Co., Ltd., Anshan 114000, China)

Abstract: This article introduces the types and properties of petroleum and coal based needle coke raw materials, and summarizes the property requirements of petroleum and coal based raw materials; Elaborate on the raw material pretreatment technology of petroleum and coal based needle coke; Propose the development direction of domestic needle coke pretreatment technology.

Key words: needle coke; needle coke raw material; pretreatment; production

标准曲线法和标准加入法测定萤石中氟化钙

王泽庭，张　健，王炳琨，康宝军，张国龙，单桂艳

(鞍钢股份有限公司质检计量中心中心化验室，辽宁鞍山　114000)

摘　要：氟化钙含量是萤石的重要质量指标。本文以氢氧化钠为熔剂与萤石进行混合，在700℃条件下熔融混合形成熔融物，再将熔融物溶解到水中，用氟离子选择性电极测量溶液的电极电位。根据能斯特方程，采用标准曲线法和标准加入法两种方法对溶液中氟离子浓度进行了计算，进而求出萤石中氟化钙的含量。结果显示，标准曲线法和标准加入法的线性关系均良好，对于氟化钙含量为98.59%和93.28%的萤石，标准曲线法的标准偏差为0.394%和1.24%，标准加入法的标准偏差为0.13%、0.21%。标准加入法可作为标准曲线法的补充，提高检测结果的准确性。

关键词：萤石；氟化钙；离子选择电极；标准曲线法；标准加入法

Determination of Calcium Fluoride in Fluorite by Standard Curve Method and Standard Addition Method

WANG Zeting, ZHANG Jian, WANG Bingkun, KANG Baojun,
ZHANG Guolong, SHAN Guiyan

(Angang Steel Co., Ltd., Quality Neasurement Center Central Laboratory, Anshan 114000, China)

Abstract: The content of calcium fluoride is an important quality index of fluorite. In this paper, sodium hydroxide is mixed with fluorite, which is melted and mixed at 700℃, and then dissolved into water. The electrode potential of the solution is measured by fluoride ion selective electrode. According to Nernst equation, the concentration of fluoride ions in solution is calculated by standard curve method and standard addition method, and the content of calcium fluoride in fluorite is obtained. The results show that the linear relationship between the standard curve method and the standard addition method is good. For the fluorite with 98.59% and 93.28% calcium fluoride content, the standard deviation of the standard curve method is 0.394% and 1.24%, and the standard deviation of the standard addition method is 0.13% and 0.21%. The standard addition method can be used as a supplement to the standard curve method to improve the accuracy of test results.

Key words: fluorite; calcium fluoride; ion-selective electrode; standard curve method; standard addition method

X射线能量色散检测技术在钼铁合金分析中的研究与应用

康宝军[1]，何文英[1]，吴楠楠[1]，唐艳秀[2]，王炳琨[1]，周芊蔚[1]

(1. 鞍钢股份质检计量中心，辽宁鞍山　114000；2. 鞍钢股份制造管理部，辽宁鞍山　114000)

摘　要：采用 X 射线能量色散荧光仪对钼铁合金中钼、铜、磷元素进行检测和研究，通过对制样、分析条件的优化，提高分析的精度和准确度。本课题解决了湿法化学分析钼铁合金周期长，干扰因素多的问题，为钼铁合金快速检测提供全新的分析方法。

关键词：X 射线能量色散光谱仪；钼铁合金；粉末压片

Research and Application of X-ray Energy Dispersion Detection Technology in Ferromolybdenum Alloy Analysis

KANG Baojun[1], HE Wenying[1], WU Nannan[1], TANG Yanxiu[2], WANG Bingkun[1], ZHOU Qianwei[1]

(1. Ansteel Quality Inspection and Metering Center, Anshan 114000, China;
2. Ansteel Manufacturing Management Department, Anshan 114000, China)

Abstract: The detection and research of Mo, Cu, and P in ferromolybdenum alloy were carried out using an X-ray energy dispersive fluorescence analyzer. By optimizing the sample preparation and analysis conditions, the precision and accuracy of the analysis were improved. This subject solved the problem of long cycle and many interference factors of wet chemical analysis of ferromolybdenum alloy, and provided a new analytical method for rapid detection of ferromolybdenum alloy.

Key words: X-ray energy dispersion spectrometer; molybdenum iron alloy; pressed powder

常规仪表故障与维护技术探讨

武弘韬

（鞍山钢铁股份有限公司质检计量中心，辽宁鞍山　114000）

摘　要：在钢铁企业产生中，仪表是必不可少的，对其生产效率与质量产生了重要影响。本文重点对钢铁企业常规仪表故障开展分析，并且针对其具体故障，提出了相应的维护技术与手段。如压力控制仪表系统故障及维护技术、温度控制仪表故障及维护技术、流量控制仪表系统故障及维护技术。同时，还提出了强化常规仪表周期维护、仪表日常维护的要点，以期实现理想的维护效果，为仪表运行创造优越环境，减少故障出现，为钢铁企业稳定生产保驾护航，推动钢铁企业持续进步和发展，为关注此类话题的人们提供参考。

关键词：仪表故障；流量控制仪表；压力控制仪表；流量控制仪表

Discussion on Conventional Instrument Failure and Maintenance Technology

WU Hongtao

(Quality Inspection and Measurement Center of Anshan Iron and Steel Co., Ltd., Anshan 114000, China)

Abstract: In the production of iron and steel enterprises, the instrument is essential, which has an important impact on its production efficiency and quality. This paper focuses on the analysis of conventional instrument faults in iron and steel enterprises, and puts forward the corresponding maintenance technology and means for its specific faults.

Key words: instrument failure; flow control instrument; pressure control instrument; flow control instrument

联系测量技术在陈台沟铁矿超深井及附属工程施工中的研究与应用

刘 骥

(鞍山五矿陈台沟有限公司,辽宁鞍山 114000)

摘 要:为指导井下-780m、-1020m、-1050m 五个水平巷道工程精确施工,开展井上井下联系测量统一坐标系统,分别为陀螺定向统一平面坐标系统,高程联系测量统一高程系统,通过精确的计算和平差,消除误差,达到控制贯通精度的效果。

关键词:统一坐标系统;陀螺定向;高程联系测量;平差;贯通精度

The Research and Application of Contact Survey Technology in the Construction of Super-deep Well and Affiliated Engineering of Chentaigou Iron Mine

LIU Ji

(Anshan Minmetals Chentaigou Co., Ltd., Anshan 114000, China)

Abstract: In order to guide the accurate construction of -780 m, -1020 m and -1050 m underground horizontal laneways, the unified coordinate system of the underground and underground mine contact survey is developed, through accurate calculation and adjustment, the unified elevation system of elevation connection survey can eliminate the error and achieve the effect of controlling the running-through precision.

Key words: unified coordinate system; gyro orientation; elevation measurement; adjustment; through precision

尾矿库挡砂堤稳定性研究

王胜建

(鞍山五矿矿业有限公司,辽宁鞍山 114000)

摘 要:尾矿库作为贮存选矿厂废弃尾矿设施,也是一座人为形成的泥石流危险源。本文以实际案例为基础,运用瑞典圆弧法原理,依靠软件分析挡砂堤稳定性,并在此基础上辨识尾矿库挡砂堤风险源,并在此基础上采相应的防范措施,以提高尾矿库的安全水平,减少尾矿库运行中的危险,降低尾矿库挡砂坝溃坝风险,着重保护库区居民的生命财产安全。

关键词:挡砂堤;稳定性;风险

Study on the Stability of Tailings Dam

WANG Shengjian

(Anshan Minmetals Chentaigou Mining Co., Ltd., Anshan 114000, China)

Abstract: The tailings pond serves as an abandoned tailings facility for storing concentrators and is also a man-made source of debris flow hazards. Based on actual cases, this paper uses the principle of Sweden arc method, relies on software to analyze the stability of the sand retaining embankment, and on this basis, identifies the risk source of the tailings reservoir retaining embankment, and takes corresponding preventive measures on this basis to improve the safety level of the tailing pond, reduce the danger in the operation of the tailing pond, reduce the risk of the dam failure of the sand retaining dam in the tailing pond, and focus on protecting the life and property safety of residents in the reservoir area.

Key words: tailings dam; stability; hazard

高频燃烧红外吸收法测定冶金用二氧化钛中碳

杨晓倩，郭圣洁，谭胜楠，牛威斌

（河钢材料技术研究院理化检测中心，河北石家庄 050000）

摘 要： 利用高频燃烧红外吸收法实现对冶金用二氧化钛中碳含量的测定。实验表明，称取 0.1g 冶金用二氧化钛样品，覆盖 0.2g 纯铁助熔剂和 1.2g 钨助熔剂可使碳完全释放。由于市面上少有冶金用二氧化钛标准样品，不足以覆盖所有冶金用二氧化钛样品中碳含量范围，所以通过选取可以覆盖碳含量的四种钢铁标准物质建立校准曲线。结果表明，碳含量校准曲线线性回归方程相关系数为 $r=0.9998$，检出限为 0.00019%，定量限为 0.00063%。对 2 个牌号的冶金用二氧化钛试样进行加标回收实验，回收率在 95%~105%。

关键词： 冶金用二氧化钛；高频；红外吸收法；碳

Determination of Carbon in Titanium Dioxide for Metallurgy by High Frequency Combustion Infrared Absorption Method

YANG Xiaoqian, GUO Shengjie, TAN Shengnan, NIU Weibin

(HBIS Materials Technology Research Institute Physical and Chemical Testing Center, Shijiazhuang 050000, China)

Abstract: The determination of carbon content in titanium dioxide for metallurgy was achieved by high-frequency combustion infrared absorption. The experiments showed that 0.1g of titanium dioxide for metallurgy was weighed and covered with 0.2g of pure iron flux and 1.2g of tungsten flux to completely release the carbon. Since there are few titanium dioxide for metallurgy standard samples available in the market, it is not enough to cover the carbon content range in all titanium dioxide for metallurgy samples, so the calibration curve was established by selecting four steel standards that can cover the carbon content. The results showed that the linear regression equation correlation coefficient of carbon content calibration curve was $r=0.9998$, the detection limit was 0.00019%, and the quantification limit was 0.00063%. The spiked recovery experiments were carried out for two grades of titanium dioxide specimens for metallurgy, and the recoveries were in the range of 95%~105%.

Key words: titanium dioxide for metallurgy; high frequency;combustion infrared absorption method; carbon

ICP-AES 法测定镀锡板镀层中的铅

戚振南，任玲玲，杨慧贤，谭胜楠，郭圣洁，葛晶晶

（河钢材料技术研究院，河北石家庄 050023）

摘 要： 采用氯化铵溶液溶解，高纯物质进行基体匹配，配置了系列标准溶液，建立了电感耦合等离子体原子发射光谱法（ICP-AES）测定镀锡板镀层中铅元素的方法，铬元素质量分数 0.04~0.3μg/mL 范围内标准曲线呈线性，线性相关系数为 0.9999，检出限为 0.03μg/mL（质量分数）。该方法应用于镀锡板样品测定，结果的相对标准偏差小于 3%，加标回收率在 100%~110%范围，较为满意。

关键词： 镀锡板；镀层；ICP-AES；铅

Determination of Plumbum in Tinplate Layer by ICP-AES

QI Zhennan, REN Lingling, YANG Huixian, TAN Shengnan,
GUO Shengjie, GE Jingjing

(Material Technology Research Institute of HBIS Group, Shijiazhuang 050023, China)

Abstract: The standard solution series and the method for the determination of chromium in tinplate coating by inductively coupled plasma atomic emission spectrometry (ICP-AES) was established by matrix matching method in Ammonium chloride solution. The calibration curve was linear in the mass fraction of chromium in the range of 0.04~0.3μg/mL, the linear correlation coefficient was 0.9999 and the detection limit was 0.03μg/mL. The method was applied to the determination of tinplate samples. The relative standard deviations (RSD, n=6) of the results were less than 3%, and the spiked recoveries were in the range of 100%~110%.

Key words: tinplate; coating; inductively coupled plasma atomic emission spectrometry (ICP-AES); plumbum

钢水成分查询系统的程序设计与实现

易晓珊

（湖南钢铁集团有限公司，湖南湘潭 411100）

摘 要： 宽厚板厂炼钢区域（转炉、精炼及连铸）大屏幕成分查询系统有较大局限性，现场操作人员不能全面掌握钢水成分，及其一些冶金性能数据，不能满足精益生产和精细管理的要求。针对现实情况，我们通过自主设计开发"钢水成分查询系统"，增加多样品成分查询、"碳当量值"和"焊接敏感系数值"等性能指标的功能查询以及残余元素超范围自动报警等功能，极大提高了生产、工艺的过程控制和工作效率。

关键词： 钢水成分查询；数据库；Visual studio C#

集中计量管理系统研究与应用

高启胜，翟英杰

（鞍钢股份有限公司质检计量中心，辽宁鞍山　114002）

摘　要：集中计量管理系统通过调整、优化原有计量业务流程，以采购物流、内转物流和销售物流为主轴，充分整合衡器设备，并结合多种计量防作弊手段，利用自动化控制技术、网络技术、多媒体技术、计算机技术、互联网技术、物联网技术、数据库技术等技术，构建完善的远程可操控集中计量管理系统。

关键词：集中计量管理系统；汽车衡；轨道衡；系统架构；业务流程；数据管理

Research and Application of Centralized Measurement Management System

GAO Qisheng, ZHAI Yingjie

(Quality Inspection and Measurement Center, Angang Steel Company Limited, Anshan 114002, China)

Abstract: The centralized measurement management system build a perfect remote controllable centralized measurement management system by adjusting and optimizing the original measurement business process. It takes procurement logistics, internal logistics and sales logistics as the main axis, fully integrates weighing equipment, combines a variety of measures to prevent cheating, and uses automation control technology, network technology, multimedia technology, computer technology, internet technology, internet of things technology, database technology and other technologies.

Key words: centralized measurement management system; truck scale; rail scale; system architecture; business process; data management

利用不确定度评价熔融炉性能对结果的影响

王丽晖，李　奎，孙　伟，高　巍，郝延楠，潘　忠

（鞍钢股份有限公司鲅鱼圈钢铁分公司，辽宁营口　115007）

摘　要：本文针对 X 射线荧光谱钴内标熔片法测定铁矿石中全铁含量，评定测量过程中的样片制备重复性、光谱仪测量重复性、校准曲线变动性、标准物质定值、称量的不确定度分量来源进行了评定，详细阐述了计算过程及各不确定度的贡献。从不确定度各分量评定的结果得出，熔融炉制备样品的重复性对测量结果的贡献最大。对于铁含量为 61.96% 的样品，采用不同熔融炉制备样片，检测结果的扩展不确定度分别为 0.62%、0.74%、1.30%。由此看出，熔融炉性能对 X 射线荧光谱钴内标熔片法测定铁矿石中全铁含量不确定度的贡献率差别较大，可作为评价熔融炉的性能优劣的判定依据。

关键词：X 射线荧光光谱仪；不确定度；评价；熔融炉；性能

The Uncertainty Was Used to Evaluate the Influence of Melting Furnace Performance on the Results

WANG Lihui, LI Kui, SUN Wei, GAO Wei, HAO Yannan, PAN Zhong

(Angang Steel Co., Ltd., Bayuquan Iron & Steel Subsidiary, Yingkou 115007, China)

Abstract: In this paper, Melt sheet method of X-ray fluorescence is studied .That measures the determination of total iron content in iron ore by cobalt internal standard. The uncertainty components of sample preparation repeat-ability, X-ray fluorescence spectrometer repeat-ability, calibration cure variability, reference material setting and weighing were evaluated. From the evaluation results, the repeat-ability of sample prepared by the melting furnace has the greatest contribution to the measurement result. In this process, the expanded uncertainty of samples prepared by different melting furnaces are 0.62%, 0.74%, 1.30%for sample with iron content of 61.96%. It is concluded that the contribution rate of different melting furnace properties to the uncertainty of analysis results is different. It can be used as the basis for judging the performance index of melting furnace.

Key words: X-ray fluorescence spectrometer; uncertainty; evaluate; melting furnace; performance

储氢容器用钢设计及其氢敏感性机理研究

范秀如，米志杉，苏 航，杨 丽

（中国钢研科技集团有限公司，北京 100081）

摘 要： 氢能是未来的最重要的清洁能源之一，而在大规模储存和运输氢气方面，高压气态储氢具有设备结构简单、能耗低、响应速度快等优点，是目前工业发展中最常用的储氢技术。针对高压气态储氢容器这一重大应用需求，高强度低合金（HSLA）钢兼具强度、韧性、焊接性的优越性，被认为是气态氢储存和运输的候选材料。然而，氢气环境下，一些金属材料的韧性与塑性会降低，通常称为氢脆（HE），并且随着高压储氢容器用钢强度的提升，其氢脆敏感性增加，这会给高压氢储运产业带来巨大的损失与安全隐患。

影响材料氢脆敏感性的因素较多，机理十分复杂，但受限于氢原子尺寸，现有的实验手段无法直接观测到氢与材料的交互作用，因而氢脆的成因目前仍存在争议。因此，通过原子水平的计算模拟对氢与材料的相互作用进行全面研究，有望提高对金属材料氢脆敏感性的理解，为低氢脆敏感性合金的设计提供理论依据。

基于此，本研究采用密度泛函理论（DFT）方法模拟合金元素掺杂对bcc-Fe本体结构和晶界的及其与氢相互作用的影响。为了在原子电子层面上理解氢脆机理，分别通过密度泛函理论方法模拟了掺杂合金元素（Cr、Mn、Ni和Mo元素）在体心立方铁晶格结构和晶界结构对氢结合与氢扩散的影响。

经密度泛函理论计算发现，在体心立方铁晶格结构和晶界结构中掺杂合金元素Cr、Ni和Mo会增加氢结合能并减弱氢损伤的可能性，而Mn掺杂会降低氢结合能并导致更多的氢原子积累。同时，合金元素显著影响铁晶格结构和晶界结构的氢扩散行为。在体心立方铁晶格结构中掺杂Cr可以降低氢扩散系数，而在体心立方铁晶格结构中掺杂Ni可以显著增加氢扩散系数。基于以上结果，可认为合理控制Cr、Ni和Mo的合金含量可一定程度降低氢脆敏感性，而合金元素Mn的提高可能导致氢脆敏感性增加。

在DFT计算结果的基础上，针对两种试验钢种开展了氢脆敏感性测试试验，其中Ni、Cr、Mo含量较高、Mn含量较低的试验钢1号钢在高压氢气中的氢脆敏感性明显优于对照钢2号钢，与计算预测吻合。因此，我们认为密度泛函理论方法不仅可以提高对低合金高强钢的氢脆机理的理解，同时还可以为合金的理性设计提供理论指导。

关键词： 密度泛函理论；氢脆；低合金高强钢；氢脆敏感性；氢扩散系数

液压系统功率损耗原因及对策分析

雷丛卉，宁　博，陈国防，郭　佳，彭立广

（中国重型机械研究院股份公司，陕西西安　710018）

摘　要：21 世纪至今，以工程机械、冶金机械、矿山机械、农业机械、航空航天、智能机床等为代表的装备制造业取得快速发展，我国液压行业进入成熟发展阶段。液压技术存在耗能高、效率低的缺点，液压系统的功率损耗往往会转换为热量，使系统温度升高，本文通过分析液压系统功率损耗的危害及原因，给出了降低液压系统功率损耗的有效途径。

关键词：液压系统；功率损耗；发热；解决途径

Analysis on the Causes and Countermeasures of Power Loss in Hydraulic Systems

LEI Conghui, NING Bo, CHEN Guofang, GUO Jia, PENG Liguang

(China Heavy Machinery Research Institute Co., Ltd., Xi'an 710018, China)

Abstract: Since the 21st century, the equipment manufacturing industry represented by engineering machinery, metallurgical machinery, mining machinery, agricultural machinery, aerospace, intelligent machine tools, etc. has achieved rapid development, and China's hydraulic industry has entered a mature development stage. Hydraulic technology has the drawbacks of high energy consumption and low efficiency. The power loss of hydraulic systems is often converted into heat, causing the system temperature to rise. This article analyzes the hazards and reasons of power loss in hydraulic systems and provides ways to reduce power loss in hydraulic systems.

Key words: hydraulic systems; power loss; heating; solution

基于高频数据可视化分析的设计与实现

李保琴，张　强，俞烨波

（上海宝信软件股份有限公司工业互联网研究院，上海　219000）

摘　要：本文针对钢铁行业生产过程中的时序数据海量高频、数据价值挖掘困难等特点，就如何存储、展示和分析进行探讨。基于实时数据库，提出一种高性能的数据可视化方案。为钢铁行业有效挖掘数据价值提供一种高效工具，针对生产过程中遇到的多种场景形成标准模板，结合数字钢卷助力带钢时序数据建模。

关键词：数字钢铁；高频数据；可视化分析

预处理工艺对 H13 钢碳化物析出行为的影响

樊明强[1,2]，赵英利[1]，张利江[2]，赵峥嵘[1]，王建强[2]

（1. 河钢材料技术研究院特种材料研发中心，河北石家庄 052165；
2. 河钢集团张宣科技特种材料研究制造公司，河北宣化 075100）

摘　要：采用高温固溶和高温固溶+高温回火工艺研究了不同预处理组织对 H13 钢的球化退火行为的影响。结果表明，与传统的高温固溶预处理工艺相比，高温固溶+高温回火预处理可获得回火索氏体组织，未溶碳化物数量偏多，尺寸偏大，球化机理由马氏体球化逐渐发展为珠光体球化，退火组织中碳化物析出特点是数量少、尺寸偏大、圆整度高、总量不变，抗热疲劳性能更优。高温固溶+高温回火的预处理工艺可控制退火过程碳化物析出形态，且工艺可操控性强，是一种实用性强的 H13 钢预处理手段。

关键词：H13 钢；预处理工艺；碳化物；球化退火

Effect of Pretreatment Technology on Carbide Precipitation Behavior of H13 Steel

FAN Mingqiang[1,2], ZHAO Yingli[1], ZHANG Lijiang[2],
ZHAO Zhengrong[1], WANG Jianqiang[2]

(1. HBIS Group Technology Research Institute, Special Materials Research and Development Center, Shijiazhuang 052165, China; 2. HBIS Group Zhangxuan Technology, Special Materials Research and Manufacturing Company, Xuanhua 075100, China)

Abstract: The effect of different pretreatment structures on the spheroidizing annealing behavior of H13 steel was studied by high temperature solution and high temperature solution + high temperature tempering process. The results show that compared with the traditional high temperature solid solution pretreatment process, high temperature solid solution + high temperature tempering pretreatment can obtain tempered sortenite structure, the quantity of undissolved carbide is more, the size is larger, the nodularization of martensite gradually develops into pearlite nodularization, and the carbide precipitation in the annealed structure is characterized by small quantity, large size, high round degree and constant total amount. Better thermal fatigue resistance. The pretreatment process of high temperature solution + high temperature tempering can control the carbide precipitation form in the annealing process, and the process is controllable, which is a practical method for H13 steel pretreatment.

Key words: H13 steel; pretreatment process; carbide; spheroidizing annealing

轻量化低应力机车车轮设计及验证

黄孝卿[1,2]，刘智[1]，陈刚[1]，国新春[1]

（1. 宝武集团马钢轨交材料科技有限公司，安徽马鞍山 243002；
2. 西南交通大学牵引动力国家重点实验室摩擦学研究所，四川成都 610031）

摘　要：为提高机车轮的服役安全性能，优化辐板形状是其重要途径之一。基于国外客户提供的接口尺寸及主要设计参数，对出口机车车轮进行了结构设计，按照 AAR S-660 标准，对新设计车轮进行了制动热力学分析计算。结果表明：（1）重量仅为 380kg，较普通 914 车轮减重 10kg 以上；（2）最大热应力为 533MPa，较北美成熟 914 车轮最大应力降低 84MPa。

关键词：轻量化；低应力；车轮设计；机车车轮

Structure Design and Verification of Lightweight and Low Stress Locomotive Wheels

HUANG Xiaoqing[1,2], LIU Zhi[1], CHEN Gang[1], GUO Xinchun[1]

(1. Baowu Group Masteel Rail Transit Materials Technology Company Limited, Maanshan 243002, China;
2. Tribology Research Institute, State Key Laboratory of Traction Power, Southwest Jiaotong University, Chengdu 610031, China)

Abstract: Optimizing the shape of the wheel plate is one of the important ways to improve the service safety performance of locomotive wheels. Based on the interface dimensions and main design parameters provided by foreign customers, the structural design of exported locomotive wheels was carried out. According to the AAR S-660 standard, braking thermodynamic analysis and calculation were conducted on the newly designed wheels. The results show that: (1) the weight is only 380kg, which is more than 10kg less than the ordinary 914 wheel; (2) The maximum thermal stress is 533MPa, which is 84 MPa lower than the maximum stress of mature 914 wheels in North America.

Key words: lightweight; low stress; structure design; locomotive wheels

高氮钒铁中钒元素化学分析方法的研究与讨论

贺红侠

（河钢张宣科技技术中心，河北张家口 075100）

摘　要：高氮钒铁氮含量较高，内部形成的氮化物比较稳定，采用硫酸亚铁铵滴定法分析时试样不能充分溶解，溶液内有明显残渣，滴定结果偏差很大，本文对硫酸亚铁铵滴定法进行优化，使样品溶解完全的同时操作更加简单快捷，通过大量标准样品分析和实验室比对证明该法分析快速、所得数据准确可靠，满足日常分析需求。

关键词：高氮钒铁；钒元素；氢氟酸；硫酸亚铁铵

Study and Discussion on Chemical Analysis Methods of Vanadium in High Nitrogen Ferric Vanadium

HE Hongxia

(Hegang Zhangxuan Science and Technology Center, Zhangjiakou 075100, China)

Abstract: Fe-vanadium with high nitrogen content has a high nitrogen content, and the nitride formed inside is relatively stable. The sample cannot be fully dissolved when analyzed by ammonium ferrous sulfate titration method, and there are obvious residues in the solution, resulting in a large deviation in the titration results. In this paper, ammonium ferrous sulfate titration method is optimized to make the sample completely dissolved and the operation easier and faster. Through the analysis of a large number of standard samples and laboratory comparison, it is proved that the method is fast, accurate and reliable, and can meet the needs of daily analysis.

Key words: high nitrogen ferric vanadium; vanadium; hydrofluoric acid; ammonium ferrous sulfate

基于 OpenCV 的石墨球化率测定

郑秋薇，邢佳文

（燕山大学机械工程学院，河北秦皇岛　066044）

摘　要：球墨铸铁综合性能优异，价格低廉，重量较轻，自被发现以来就备受人们的关注。随着中国生产技术和生产设备的不断进步，球墨铸铁得到了广泛的应用，保持着良好的发展趋势。球化率是评价墨铸铁质量的重要指标，球墨铸铁的力学性能在很大程度上决定于球化率。本文根据2022年7月1日国家正式施行的球墨铸铁检验标准GB/T 9441—2021《球墨铸铁金相检验》，基于OpenCV库完成了石墨球化率的测定，经检测，程序结果可靠，操作简便。

关键词：球墨铸铁；OpenCV；球化率；GB/T 9441—2021

Determination of Nodularity of Graphite Based on OpenCV

ZHENG Qiuwei, XING Jiawen

(School of Mechanical Engineering, Yanshan University, Qinhuangdao 066044, China)

Abstract: Ductile iron has excellent comprehensive properties, low price and light weight, and has attracted much attention since its discovery. With the continuous improvement of China's production technology and production equipment, ductile iron has been widely used and maintained a good development trend. Nodularity is an important index for evaluating the quality of cast iron, and the mechanical properties of ductile iron are largely determined by the nodularity rate. According to the national nodular cast iron inspection standard GB/T 9441—2021 "Ductile Iron Metallographic Test" officially implemented on July 1, 2022, the determination of graphite nodularity was completed based on the OpenCV library.

Key words: spheroidal graphite cast iron; OpenCV; nodularizability; GB/T 9441—2021

S 钢铁集团加工配送中心加工产线配置策略研究

王 兴，陈利成，刘昊林

（中冶赛迪工程技术股份有限公司咨询事业本部，重庆 401122）

摘 要： 国内钢材加工配送中心的发展存在设备利用率不高、产品结构单一，无法为客户提供个性化服务等问题，本文结合市场数据分析及对不同类型钢材加工配送中心的走访调研，综合考虑不同钢材品种对应的加工产线对 S 钢铁集团加工配送中心的贡献程度，科学合理配置加工产线及规模，选择合适的运营模式，重点解决上述难点、痛点，进而确保本项目配置的深加工产品结构丰富，具备市场竞争力，符合满足钢材加工中心未来发展趋势，契合 S 钢铁集团未来战略转型、多元化发展的需求。

关键词： 钢材深加工；运营模式；钢材品种；加工产线；加工规模

Research on the Configuration Strategy of Processing Production Line of S Steel Group's Processing and Distribution Center

WANG Xing, CHEN Licheng, LIU Haolin

(CISDI Engineering Co., Ltd., Consulting Business Division, Chongqing 401122, China)

Abstract: The development of domestic steel processing and distribution centers faces problems such as low equipment utilization, single product structure, and inability to provide personalized services to customers. This article combines market data analysis and visits to different types of steel processing and distribution centers, comprehensively considering the contribution of processing production lines corresponding to different steel products to S Steel Group's processing and distribution centers, and scientifically and reasonably configuring processing production lines and scales, Select an appropriate operating mode, focus on solving the above difficulties and pain points, and ensure that the deep processing products configured for this project have a rich structure, market competitiveness, and meet the future development trend of steel processing centers, in line with the needs of S Steel Group's future strategic transformation and diversified development.

Key words: steel further processing; operation mode; steel grade; processing production line; processing scale

钢铁企业系统诊断与竞争力分析

陈利成，陈琼妮，吴 波，王 兴

（中冶赛迪工程技术股份有限公司咨询事业本部，重庆 400013）

摘 要： 目前国内外发布的钢铁企业竞争力评级研究，主要用于钢铁企业竞争力评价。而对钢铁企业来说，从竞争力分析角度系统性地进行剖析诊断、对比差距、发现问题、从而有针对性地提出持续改善其竞争力的解决方案尤为重要。本文根据多年来服务于钢铁企业的经验，以财务分析的视角，尝试从企业的盈利能力、生存能力和可持续发展能力维度出发构建竞争力分析框架，为实现企业竞争力持续改善和提升的高质量发展提供相应的分析方法和思路参考。

关键词：系统诊断；竞争力分析；盈利能力；生存能力；可持续发展能力

System Diagnosis and Competitiveness Analysis of Steel Enterprises

CHEN Licheng, CHEN Qiongni, WU Bo, WANG Xing

(CISDI Engineering Co., Ltd., Chongqing 400013, China)

Abstract: At present, research on the competitiveness rating of steel enterprises released domestically and internationally is mainly used for evaluating the competitiveness of steel enterprises. For steel enterprises, it is particularly important to systematically analyze and diagnose, compare gaps, identify problems, and propose targeted solutions to continuously improve their competitiveness from the perspective of competitiveness analysis. Based on years of experience in serving steel enterprises, this article attempts to construct a competitiveness analysis framework from the perspective of financial analysis, starting from the dimensions of profitability, survival ability, and sustainable development ability of enterprises. It provides corresponding analytical methods and ideas for achieving high-quality development of continuous improvement and enhancement of enterprise competitiveness.

Key words: system diagnosis; competitiveness analysis; profitability; survival ability; sustainable development ability

Nb 元素对高强度管线钢抗 H_2S 性能的影响研究

樊艳秋 [1,2]，马长文 [1]，李少坡 [2]，丁文华 [1]，李战军 [1,2]

（1. 首钢集团有限公司技术研究院，北京 100043；
2. 北京市能源用钢工程技术研究中心，北京 100043）

摘　要：随着国民经济的发展，石油、天然气等能源消费总量持续增长，这就对输送用管道用钢的提出了更高的要求。H_2S 是石油和天然气中最具有腐蚀作用的有害介质之一，虽然目前油田已对油气进行净化处理，但是当净化处理不善时，输送介质中仍含有较高含量 H_2S，管道用钢在湿 H_2S 的服役环境中容易产生氢致裂纹，一旦发生氢致断裂，会对周围环境产生较大的负面影响和安全隐患，严重影响着油气输送管线的使用寿命。因此，在如何改善油气管线的服役性能，尤其是抗氢致开裂性能，是影响管道服役期限和安全的重要因素之一。在高强度级别管线用钢中，通常需要添加 Nb、V、Ti 等微合金元素以保证钢板的强韧性，尤其是 Nb 元素是可有效提高钢板的性能。虽然管线钢在 H_2S 环境下的氢致开裂行为已有较多研究，但是 Nb 元素对高强度管线钢抗 H_2S 性能的影响程度和作用规律还有待研究。

因此本文研究了不同 Nb 含量对高强度管线钢抗 H_2S 性能的影响，即以 X80 为研究对象，分别添加不同含量的 Nb 元素，通过抗酸试验、氢渗透试验等分析了组织的氢渗透能力以及氢陷阱的氢捕捉效率，并通过 EBSD 等试验手段观察了不同类型氢陷阱的分布情况。研究发现：当 Nb 元素添加量为 0.05%时，钢板内部未发现氢致裂纹，抗 H_2S 性能优异。其晶粒尺寸得到了明显细化，平均晶粒尺寸达到 13μm；且有效晶粒尺寸细小，为 4.56μm。此外，组织具备较强的氢渗透能力，其氢渗透通量和氢扩散系数分别达到 $4.72×10^{-10}$mol/(m²·s)和 $2.49×10^{-6}$cm²/s。同时，均匀弥散分布的大角度晶界使氢分布更加均匀，有效减缓了局部氢压，对抗 H2S 性能起到了明显的贡献作用。

参 考 文 献

[1] 李晶. 抗硫化氢应力腐蚀管线钢组织性能控制与腐蚀行为研究[D]. 沈阳: 东北大学, 2018.
[2] 孙宪进. 第十三届中国钢铁年会论文集[M]. 北京: 冶金工业出版社, 2022: 66.
[3] Li L F, Song B, Cheng J, et al. Effects of vanadium precipitates on hydrogen trapping efficiency and hydrogen induced cracking resistance in X80 pipeline steel[J]. Int. J. Hydrogen Energy, 2018, 43: 17353-17363.
[4] Li X, Song J, Gan K, et al. Identifying sulfide stress cracking stages on a HSLA pipeline steel in H_2S environment by electrochemical noise[J]. Journal of Electroanalytical Chemistry, 2020, 114480-114494.